Thiophenes

BEST SYNTHETIC METHODS

Gronowitz and Hörnfeldt: Thiophenes, 2004

Other Volumes in the Series

Brandsma: Synthesis of Acetylenes, Allenes and Cumulenes: Methods and Techniques, 2004
Osborn: Carbohydrates, 2003
Jones: Quaternary Ammonium Salts: Their Use in Phase-Transfer Catalysed Reactions, 2001
Varvoglis: Hypervalent Iodine in Organic Synthesis, 1997
Grimmett: Imidazole and Benzimidazole Synthesis, 1997
Wakefield: Organomagnesium Methods in Organic Synthesis, 1995
Metzner: Sulfur Reagents in Organic Synthesis, 1994
Pearson: Iron Compounds in Organic Synthesis, 1994
Petragnani: Tellurium in Organic Synthesis, 1994
Motherwell: Free Radical Chain Reactions in Organic Synthesis, 1991

Related Titles of Interest

Books
Cordell: The Alkaloids
Levy & Tang: Chemistry of C-Glycosides
Pietra: Biodiversity and Natural Product Diversity
Rahman: Studies in Natural Product Chemistry
Sturgeon: Advances in Macromolecular Carbohydrate Research

Major Reference Works
Barton, Nakanishi & Meth-Cohn: Comprehensive Natural Products Chemistry
Katritzky, Meth-Cohn & Rees: Comprehensive Organic Functional Group Transformations
Katritzky, Rees & Scriven: Comprehensive Heterocyclic Chemistry II
Sainsbury: Rodd's Chemistry of Carbon Compounds

Journals
Bioorganic and Medicinal Chemistry
Bioorganic and Medicinal Chemistry Letters
Carbohydrate Research
Heterocycles (distributed by Elsevier)
Journal of Supramolecular Chemistry
Phytochemistry
Tetrahedron
Tetrahedron: Asymmetry
Tetrahedron Letters

Thiophenes

Salo Gronowitz and Anna-Britta Hörnfeldt

Department of Organic Chemistry 1
Lund University, Lund
Sweden

ELSEVIER
ACADEMIC
PRESS

Amsterdam · Boston · Heidelberg · London · New York · Oxford · Paris
San Diego · San Francisco · Singapore · Sydney · Tokyo

ELSEVIER Ltd
The Boulevard, Langford Lane
Kidlington, Oxford OX5 1GB, UK

First edition 2004

ISBN: 0-12-303953-3
ISSN: 1478-9914

∞ The paper used in this publication meets the requirements of ANSI/NISO Z39.48-1992 (Permanence of Paper).
Printed and bound in Great Britain.

Preface

The most recent comprehensive reviews treating methods for the preparation of thiophenes are given in the five parts of "Thiophene and its Derivatives", edited by Gronowitz [1]. These volumes were published between 1985 and 1992.

A more recent review published in 1994 and exclusively treating the synthesis of thiophenes is given in a chapter by W. D. Rudorf in Houben-Weyl [2]. In Houben-Weyl the chapters are based on the starting materials and reactions used to obtain the thiophenic products.

In the present handbook a completely different approach is used, focusing on the structures of the products, which are treated in a systematic way following the periodic table. In Chapter 1 the preparation of thiophenes carrying group I substituents, such as lithium, sodium etc. are given, followed in Chapter 2 by treatment of compounds having group II substituents, such as magnesium, and in Chapter 3 methods for the preparation of boron and aluminium substituted thiophenes are presented. In Chapter 4A to 4F compounds containing carbon-bound substituents are discussed, starting with alkyl and functionalized alkyl groups in Chapter 4A, vinyl-substituted derivatives in Chapter 4B, and in Chapter 4C methods for the preparation of compounds with carbon–carbon triple bonds bound to the thiophene ring are given. Aryl-substituted thiophenes are treated in Chapter 4D. Then methods for acyl-substituted thiophenes are given in Chapter 4E and thiophenecarboxylic acids, carboxylic acid and cyano-substituted thiophenes are dealt with in Chapter 4F. In Chapter 4G thiophenes containing other group IV elements such as silicon, germanium and tin are presented. Then in Chapter 5 the preparation of thiophenes containing group V elements directly bound to the ring, such as amino- and nitrothiophenes as well as thiophene phosphorus derivatives, are treated. Chapter 6 describes synthetic methods for thiophenes containing group VI elements, such as oxygen, sulfur and selenium bound to the thiophene ring. Chapter 7 gives synthetic methods for the halothiophenes.

The preparation of a specifically substituted thiophene will be given in the chapter treating compounds with the substituent with the highest number. Thus methods for the preparation of bromothiophene aldehydes are treated in Chapter 7 and not in Chapter 4E. Methods for the syntheses of formylthiophenecarboxylic acids are found in Chapter 4F and not in Chapter 4E. The very detailed list of contents makes it easy to find the section in which a specific substituted thiophene can be found.

Finally a special chapter 8 is devoted to the preparation of compounds containing several thiophene rings bound to each other, such as the bithienyls, terthienyls and quaterthienyls etc.

Each chapter will be richly illustrated by experimental procedures, mostly published after 1992 and thus not treated at all in the reviews mentioned above.

REFERENCES

1. S. Gronowitz, "Thiophene and its Derivatives" in *The Chemistry of Heterocyclic Compounds*, A. Weissberger and E. C. Taylor. eds., Vol. 44, Part 1–5, 1985–1992, John Wiley & Sons.
2. W. D. Rudorf, Houben-Weyl, *Methoden der Organischen Chemie*, Vierte Auflage, Band E6a Hetarene I, Teil 1 Thiophene, pp 186-556, Editor: R. Kreher. Georg Thieme Verlag, Stuttgart (New York) 1994.

Contents

Detailed Contents

4C Thienylacetylenes

4D Arylthiophenes

4E Acylthiophenes

4F Thiophenecarboxylic Acids and their Derivatives

– 1 –

Syntheses of Thiophenes with Group I Substituents

1.1 LITHIUM DERIVATIVES

1.1.1 General aspects

Thienyllithium derivatives are of much greater importance in syntheses than phenyllithium derivatives. This is due to the easy and convenient preparation of 2-thienyllithium derivatives by metalation of thiophene with butyllithium in ether at reflux temperature or lower [1,2]. Other metalating reagents such as lithium diisopropylamide, which is especially useful if the thiophene ring also carries iodine-, bromine- or carbonyl-containing substituents, can also be used. Secondly, 3-thienyllithium derivatives can be prepared by halogen–metal exchange with butyllithium at −70 °C [3–6]. Thirdly, *ortho*-halothienyllithium derivatives, prepared by halogen–metal exchange of *o*-dibromothiophenes at −70 °C, are reasonably stable at this temperature and do not split off lithium bromide (as does *ortho*-bromophenyllithium) to give dehydrothiophenes. This has opened the possibility to prepare a great variety of disubstituted thiophenes, some of them in one-pot procedure [7–11].

4-(1-Hydroxycyclohexyl)-α-phenyl-3-thiophenemethanol [9]

To a solution of 3,4-dibromothiophene (4.0 g, 16.5 mmol) in anhydrous diethyl ether (15 ml) under nitrogen at −80 °C is added 0.73 *M* butyllithium in ether (22.6 ml, 15.5 mmol). After stirring at this temperature for 10 min, a solution of cyclohexanone (1.6 g, 16.5 mmol) in anhydrous ether (15 ml) is added and the reaction mixture is stirred at −80 °C for 30 min. Then it is diluted with anhydrous tetrahydrofuran (100 ml) and one more batch of the butyllithium solution (22.6 ml, 16.5 mmol) is added followed by benzaldehyde (1.75 g, 16.5 mmol) in anhydrous tetrahydrofuran (20 ml). The reaction mixture is allowed to warm up to room temperature slowly. After removing most of the solvent *in vacuo* the residue is poured into water and the product is extracted with ether. The combined ether extracts are dried over magnesium sulfate and evaporated, the residue recrystallized from cyclohexane giving 2.0 g (42%) of the title compound as colorless crystals mp 82–96 °C.

However, one should always be aware of the possibility of rearrangement reactions, when lithium does not substitute the most acidic hydrogen. Recently it has been shown that 3-thienyllithium prepared by halogen–metal exchange in non-ethereal solvents is stable even at room temperature [12].

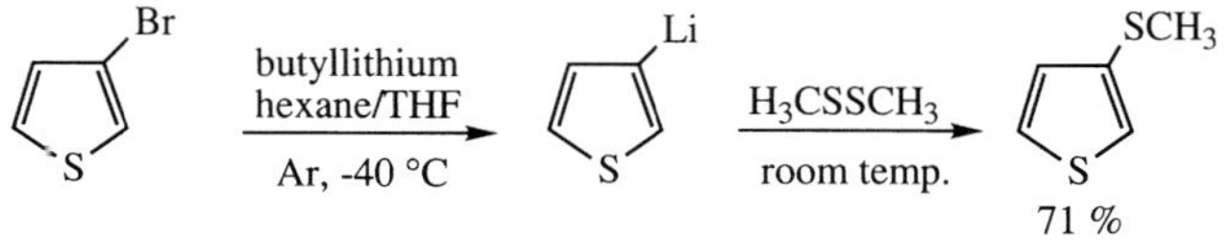

In the present chapter the methods to obtain the various thienyllithium derivatives will be discussed. The different functionalized thiophenes obtained through reactions with a manifold of electrophiles are treated in the appropriate chapters.

1.1.2 Metalation of thiophenes

2-Thienyllithium has been obtained by metalation with butyllithium and ethyllithium in ether. The less reactive phenyllithium has also been used, and this seems to give metalation also in the presence of halogens in β-position. This is also true for lithium diisopropylamide, which is particularly useful for the metalation of thiophenes carrying carbonyl-containing substituent(s) having azomethine bonds, which are incompatible with alkyl or aryllithium [3,13–22]. Polystyryllithium has also been used to metalate thiophene [23].

The convenient metalation of thiophene with butyllithium has led to the situation where only few attempts have been made to prepare 2-thienyllithium directly from lithium metal and a halothiophene. However, it has been obtained from 2-iodothiophene using the following methods [24]. Also direct treatment of thiophene with lithium metal in the presence of an electron sink has not been a widely used method, although treatment of thiophene with lithium metal in the presence of diphenylethylene followed by carbon dioxide gave 2-thiophenecarboxylic acid [25].

Metalation of thiophene with lithium dihydronaphthylide in the presence of 1,1-diphenyl-ethylene [25]

To 1.0 *M* lithium dihydronaphthylide (20 ml) is added 1,1-diphenylethylene (4.0 ml) and thiophene (5.0 ml). An exothermic reaction takes place and the mixture turns red-brown instantly. After 2 h at room temperature the mixture is carbonated and after workup 2-thiophenecarboxylic acid is obtained in 92% yield.

Coordinating solvents and/or additives with Lewis base character are often beneficial for the metalation reaction and this is the reason why not only diethyl ether and tetrahydrofuran, but also dioxane, methylal and the glymes, are used as solvents. Enhanced reactivity of the metalating agent has also been achieved by the addition of *N*,*N*,*N*′,*N*′-tetramethylethylenediamine (TMEDA), 1,4-diazabicyclo[2.2.2]octane (DABCO) or other similar tertiary amines [26–28].

The effect of the solvent system is demonstrated by the fact that thiophene is not metalated at all with butyllithium in hexane, as judged by the yield of 2-thiophencarboxylic acid obtained upon reaction with carbon dioxide. However, upon addition of triethylamine a quantitative yield of 2-thiophene-carboxylic acid was obtained [29]. The best method for the di-metalation of thiophene appears to be reflux under argon with butyllithium in hexane and tetramethylethylenediamine.

3-Substituted thiophenes have, of course, two non-equivalent α-positions and the ratio of isomeric lithium derivatives obtained depends primarily on the nature of the substituent, and of the metalating agent. Substituents carrying electronegative elements such as the halogens, oxygen or nitrogen, which might coordinate the lithiating reagent, react selectively with the 2-position. Thus 3-methoxythiophene [30] and 3-phenoxythiophene [31] yield predominantly 3-methoxy- and 3-phenoxy-2-thienyllithium, which might be stabilized by intra- or intermolecular complexation. Also 3-dimethylaminomethyl thiophene [32,33], 3-*N*-pivaloylamino [34], *N*-methyl [32,33] as well as the *N*,*N*-diethylamide of 3-thiophenecarboxylic acid give selective metalation in the 2-position [35,36]. This is also the case with the dietyl acetal of 3-thiophene aldehyde, 3-ethylthiothiophene [37], 3-phenylthiothiophene [31], 3,3-dithienyl sulfide and 3,3′-dithienyl sulfone [38], di-(3-thienyl) phenyl phosphine oxide [39] and 3-nitrothiophene [14].

Lithiation of 3-*N*,*N*-dimethylcarboxamide also occurs in the 2-position and has been used for an elegant synthesis of a dithienoquinone [33,40]. 3-Fluoro- [41], 3-chloro- [42], 3-bromo- [15] and 3-iodothiophene [43] are metalated selectively in the 2-position. In the two latter cases lithium diisopropylamide is used as the metalating agent [9]. 3-Bromothiophene can also be used with phenyllithium for metalation [9,44]. Metalation of the most acidic α-position is also the case with 3-cyanothiophene, which upon reaction with carbon dioxide yield 3-cyano-2-thiophenecarboxylic acid together with minor amounts of a product due to addition of the lithiating reagent to the cyano group [45]. 3-(4,4-Dimethyl-2-isoxazolinyl) thiophene also gives selectively the 2-lithium derivative [46,47]. A very interesting case of *ortho*-directivity occurs in the metalation of 3-hydroxymethyl-2-*t*-butyldimethylsilylthiophene with two equivalents of butyllithium. The metalation takes place in the 4-position and not in the free 5-position as in the reaction with various electrophiles. Desilylation gave 3,4-disubstituted thiophenes [48].

Metalation of 3-alkylthiophenes is not selective and the ratio of 2-lithium to 5-lithium derivatives is dependent on the size of the alkyl group and on the metalating agents. Thus with 3-methylthiophene the most favorable 5-lithium to 2-lithium derivative ratio (93:7) was obtained using butyllithium/*N*,*N*, *N'*,*N'*-tetramethylethylenediamine, while the worst (74:26) was obtained with lithium diisopropylamide. Using butyllithium/*N*,*N*,*N'*,*N'*-tetramethylethylenediamine almost complete selectivity for the 5-position was obtained with 3-ethyl- (97%), 3-isopropyl- (99%) and 3-tert-butylthiophene (100%) [28,49].

Metalation of 3-methylthiophene with butyllithium-/N,N,N',N'-tetramethylethylenediamine complex in ether; analysis via methyl derivatives [28].

N,*N*,*N'*,*N'*-tetramethylethylenediamine (6.4 g, 0.055 mol), distilled from calcium hydride, is mixed with 1.03 *M* butyllithium in diethyl ether in a nitrogen-swept funnel. Under nitrogen with stirring this mixture is added dropwise at room temperature to 3-methylthiophene (4.9 g, 0.05 mol) in anhydrous diethyl ether (50 ml). When the addition is completed, the reaction mixture is refluxed for 15 min. The heating is interrupted and dimethyl sulfate (87.0 g, 0.056 mol) in anhydrous ether (50 ml) is slowly added, after which the reaction mixture is refluxed for another 15 min. After removing the heating mantle the reaction mixture is, under vigorous stirring, treated with concentrated ammonium hydroxide solution (25 ml) for 1 h. The phases are separated and the water phase is extracted twice with ether. The combined ether phases are washed with 2 *M* hydrochloric acid and water, dried over calcium chloride and distilled under nitrogen giving 0.5 g (10%) of 3-methylthiophene and 4.4 g (79%) of 2,3- and 2,4-dimethylthiophene (7:93) bp 138–140 °C.

1.1.3 *β*-Metalation of thiophenes

In a few cases, when strongly coordinating groups are present in the 2-position, metalation occurs in the 3-position, even in the presence of a free 5-position. In such cases the isomer distribution is dependent on metalating agent and solvent. Lithium 2-thiophenecarboxylate reacts with butyllithium in tetrahydrofuran at −78 °C to give, selectively, the 3-lithium derivative, which upon reaction with a number of electrophiles gives in good yields 3-substituted 2-thiophenecarboxylic acids [50]. On the other hand metalation of 2-thiophenecarboxylic acid with two equivalents of lithium diisopropylamide led selectively to the 5-lithium isomer [51,52]. 2-*N*- (tert-butyl carboxamido)thiophene is selectively metalated in the 3-position with butyllithium in dimethoxyethane and various 2,3-disubstituted thiophenes are obtained upon reaction with different electrophiles [53]. Also 2-(4,4′-dimethyl-2-isoxazolinyl)thiophene gives selectively the 3-lithium derivative upon reaction with butyllithium in hexane at −78 °C. Butyllithium/*N*,*N*,*N*′,*N*′-tetramethylethylenediamine at −78 °C leads to a mixture of 3-lithium and 5-lithium isomers in the proportion of 1:2, while lithium diisopropylamide in tetrahydrofuran at −78 °C selectively gives the 5-lithium derivative [54]. The metalation with butyllithium was thus shown to be dependent on solvent, temperature and starting material concentration. Under kinetic control the *ortho*-disubstituted thiophene is preferentially obtained, while the 2,5-disubstituted product is thermodynamically controlled and is formed by transmetalation of the 3-lithio derivative [55]. However, in order to obtain good yields of 2,3-disubstituted thiophenes upon reaction with electrophiles, butyllithium in diethyl ether at −78 °C has to be used. Lithiation of 2-(2-thienyl)imidazoline with butyllithium occurs very selectively in the 3-position at −70 °C and no product derived from the 5-lithium derivative is observed. Metalation with lithium diisopropylamide in the absence of *N*,*N*,*N*′,*N*′-tetramethylethylenediamine, however, only yields the 5-lithium derivative [56]. Metalation of *N*-tert-butylthiophene-2-sulfonamide with butyllithium occurs competively at the 3- and 5-positions. Equilibration of the mixture of lithium derivatives obtained upon metalation with lithium disopropylamide allows selective formation of the *N*-5-dilithiothiophene sulfonamide [57]. 5-(2-Pyridinyl)thiophene is predominantly metalated in the 3-position with ethereal butyllithium at 0 °C (the ratio 3 to 5 is 83:17), while butyllithium in tetrahydrofuran at the same temperature yields predominantly the 5-isomer (3 to 5 ratio is 5:95). The ratios of lithium derivatives were determined by reaction with trimethylsilyl chloride [58–60].

The *ortho*-directing ability of α-aminoalkoxides obtained by reaction of aldehydes with lithium *N*,*N*′,*N*′-trimethylethylene diamine (LTMDA) in metalation is synthetically very useful. However, with 2-thiophene aldehyde followed by methyl iodide, a mixture of 67% of 3-methyl-2-thiophene aldehyde and 33% of 5-methyl-2-thiophene aldehyde was obtained [61]. Blocking of the aldehyde function with lithium *N*-methyl piperazide followed by metalation with butyllithium gave only the 5-lithium derivative, which was used for the preparation of trimethylsilyl-, trimethyltin- and trimethylgermanium derivatives [62]. If both α-positions are blocked, metalation occurs in the β-position in yields which are preparatively useful. Both 2-methoxy-5-methylthiophene [63], 5-methyl-2-methylthiothiophene [64] and 5-methyl-2-dimethyl aminomethylthiophene [33] are metalated in the 3-position. 2,5-Dibromothiophene is an especially interesting case. With butyllithium in ether halogen–metal exchange is obtained. Upon reaction with lithium diisopropylamide, it has recently been shown that after quenching with trimethylsilyl chloride 3,5-dibromo-2-trimethylsilylthiophene is formed [65] and not 2,5-dibromo-3-trimethylsilylthiophene, as previously claimed [66,67]. Therefore, the assignment of the structures of the reaction products obtained with trimethyltin chloride and phenylselenyl chloride are probably also erroneous [66]. On the other hand, the products obtained with other electrophiles, such as water, dimethyl disulfide and carbon electrophiles, were correctly described as 2,4-dibromo-5-substituted thiophenes [67]. Thus none of the electrophiles appear to be reactive enough to trap the 2,5-dibromo-3-thienyllithium before it rearranges to the thermodynamically more stable 2,4-dibromo-5-thienyllithium. It should be noted that 2,5-dimethoxythiophene is not metalated in the 3-position, instead the butyllithium attacks the thiophenic sulfur giving ring-opened products [68,69]. 2,5-Dimethylthiophene is metalated by ethyllithium in the 3-position and on the methyl group, as a mixture of 2,5-dimethyl-3-thiophenecarboxylic acid and 3-carboxy-5-methyl-2-thiophenecarboxylic acid was obtained in low total yield, upon reaction with carbon dioxide [70].

1.1.4 Halogen–metal exchange

Halogen–metal exchange between bromo- or iodothiophenes and organolithium reagents are of immense synthetic importance in the thiophene series. The reaction is very fast, even at −70 °C, and is usually carried out at this temperature. The reaction is selective and in dibromo- or diiodothiophenes the α-halogens are exchanged selectively before β-halogen. In general, iodine exchanges faster than bromine. Chlorine undergoes halogen–metal exchange only in special cases, when metalation of an α-position is not available, while fluorothiophenes do not undergo halogen–metal exchange at all. Halogen–metal exchange of bromo- and iodothiophenes is in most cases much faster than metalation with alkyllithium reagents, although exceptions have been found. Two complications should be noted (1) rearrangements occur easily if the lithium is not substituting the most acidic hydrogen. Temperatures have to be carefully controlled and in some cases reverse addition (adding the thiophene derivative to the alkyllithium) is necessary. (2) 3-Thienyllithium derivatives may ring open [71,72], when the temperature is raised. However, it has recently been found that 3-thienyllithium can be obtained at −40 °C by halogen–metal exchange with butyllithium in hexane, in the presence of small amounts of tetrahydrofuran. 3-Thienyllithium prepared in this way was stable at room temperature and reacted readily with electrophiles [12].

An α-positioned chlorine is less prone to halogen–metal exchange than a β-bromine and 3-bromo-2-chlorothiophene gives selectively 2-chloro-3-thienyllithium upon reaction with ethyllithium at −70 °C [73]. Also 2,5-dichloro-3-thienyllithium prepared by halogen–metal exchange of 2,5-dichloro-3-iodothiophene was stable and did not rearrange [74]. The complication due to rearrangements, caused by halogen–metal exchange and metalation (halogen-dance) is especially noticeable with 2-bromo-3-iodothiophene. At −100 °C halogen–metal exchange with ethyllithium of the β-iodine was fastest, and upon hydrolysis 2-bromothiophene was obtained and reaction with carbon dioxide gave 84% yield of 2-bromo-3-thiophenecarboxylic acid. Carrying out the halogen–metal exchange at −60 °C led to halogen-dance and the main product was derived from 3-bromo-2-thienyllithium, and upon reaction with carbon dioxide 71% of 3-bromo-2-thiophenecarboxylic acid was obtained [75]. Great care has especially to be excercised in the halogen–metal exchange of 3,4-dibromothiophene as 4-bromo-3-thienyllithium easily rearranges to 3-bromo-2-thienyllithium and finally to the most stable 3,5-dibromo-3-thienyllithium, but 4-bromo-2-thienyllithium obtained from 2,4-dibromo thiophene also rearranges through halogen-dance upon standing at higher temperatures [76].

The selectivity in the halogen–metal exchange of 3-alkyl-2,5-dibromothiophenes with various organic lithium reagents has been studied. The selectivity was in the same direction as in metalation but less pronounced. Especially noteworthy is, that probably due to release of steric strain, 3-tert-butyl-2,5-dibromothiophene with uncomplexed butyllithium reagent shows a preference for the 2-position (57:43), while butyllithium/*N*,*N*,*N*′,*N*′-tetramethylethylenediamine gave a ratio of 2-lithium to 5-lithium reagents of 13:87 [28].

Chlorothiophenes undergo halogen–metal exchange when no α-hydrogens, which can be metalated, are present. Thus tetrachlorothiophene yields 3,4-dichloro-2,5-dilithiothiophene upon treatment with two equivalents of butyllithium [70]. Halogen–metal exchange of 3,4,5-trichloro-2-pentachlorophenylthiophene, with two equivalents of butyllithium in tetrahydrofuran at −78 °C, gives a mixture of α-lithium and *para*-dilithium derivative. If three equivalents of butyllithium were used the α, *para* dilithium derivative was obtained in 67% yield [77]. 2,5-Dimethyl-3,4-dichlorothiophene upon treatment with ethyllithium/*N*,*N*,*N*′,*N*′-tetramethylethylenediamine gives halogen–metal exchange in the 3-position accompanied by metalation in the 2-position in low total yields [70]. 3,4-Dichloro-2,5-dimethoxythiophene does not give halogen–metal exchange, but is attacked by butyllithium on the sulfur, giving ring-opened products, while 2,5-dimethoxy -3-chlorothiophene is metalated in the free β-position [78].

It has been shown that halogen–metal exhange can be carried out selectively at −78 °C with tert-butyllithium in tetrahydrofuran with the sterically hindered ester, 2-bromo-3-(*para*-fluorophenyl)-4-carboethoxy-5-isopropylthiophene, without interference from addition to the ester function. Upon reaction of the lithium intermediate with *N*,*N*-dimethylformamide a 72% yield of 2-formyl-3-(*para*-fluorophenyl)-4-carboethoxy-5-isopropylthiophene was obtained. Direct metalation of 3-(*para*-fluorophenyl)-4-carbethoxy-2-isopropylthiophene failed [79].

1.1.5 Di- and polylithium derivatives

2,5-Dilithiothiophene and 3-methyl-2,5-dilithiothiophene have been prepared by metalation with butyllithium/*N*,*N*,*N*′,*N*′-tetramethylethylenediamine [80,81].

3,4-Dimethyl-2,5-dilithiothiophene could be obtained by metalation with butyllithium in refluxing hexane/ether [82]. Selective 3,5-dilithiation in good yield has been achieved with 2-diethylcarboxamidothiophene [83] and 2-tert-butylsulfonylthiophene [84] using *sec*-butyllithium/*N*,*N*,*N*′,*N*′-tetramethylethylenediamine or butyllithium as metalating agents. However, for preparative purposes, halogen–metal exchange is often more useful. Thus 2,5-dibromothiophene upon treatment with ethyllithium at −70 °C yields 2,5-dilithiothiophene, which upon reaction with perchloryl fluoride gives 2,5-difluorothiophene [85].

2,3,5-Tribromothiophene gives, with two equivalents of butyllithium, 3-bromo-2,5-dilithiothiophene, the hydrolysis of which was once the best method for the preparation of 3-bromothiophene and which upon reaction with perchloryl fluoride yielded 3-bromo-2,5-difluorothiophene [85]. Even 3,4-dichloro-2,5-dilithiothiophene was obtained by halogen–metal exchange with two equivalents of butyllithium of tetrachlorothiophene and could be used for the preparation of a number of 3,4-dichloro-2,5-disubstituted thiophenes [86]. 3,4-Diiodothiophene has been used as starting material for 3,4-dilithiothiophene through halogen–metal exchange at −60 °C, and upon reaction with *N*,*N*-dimethylformamide, 3,4-diformylthiophene was obtained [87].

It has been reported that 3,4-dibromo- and 2,3-dibromothiophene also give the 3,4- and 2,3-dilithiothiophenes upon treatment with 2.5 equiv. of butyllithium, in high yields as judged by the yields of 3,4-dimethyl- and 2,3-dimethylthiophene obtained upon reaction with dimethyl sulfate [88]. However, it could be possible that in these cases there was stepwise halogen–metal exchange, followed by reaction with dimethyl sulfate, renewed halogen–metal exchange and reaction with dimethyl sulfate, as it has been shown that dimethyl sulfate might exist together with butyllithium at low temperature [89]. Using large excess of butyllithium, tri- and tetralithiation has been achieved with 2,3,4-, 2,3,5-tribromothiophene and tetrabromothiophene [88].

1.2 SODIUM DERIVATIVES

1.2.1 Metalation of thiophenes with alkyl or aryl sodium derivatives prepared *in situ*

Before the development of the thienyllithium derivatives, as versatile reagents for the preparation of various thiophene derivatives, Schick and Hartough at Socony Oil Co. found two ways to prepare 2-thienylsodium derivatives. They could easily be prepared by a *trans*-metalation procedure, adding ethyl chloride or aryl bromide to a mixture of sodium amalgam and thiophene or an alkylthiophene in diethyl ether, which yielded the 2-thienylsodium derivatives in 60–68% yield [90,91]. 3-Methylthiophene, upon reaction with sodium amalgam and bromobenzene followed by carbon dioxide, gave 4-methyl-2-thiophenecarboxylic acid [91].

It is essential that an alkyl chloride is used, as with ethyl bromide or butyl bromide only about 20% yield of the acid was obtained, due to competing Wurtz couplings [92]. In a later work metalation of 3-methylthiophene with phenylsodium gave, upon carbonation, 4-methyl-2-thiophenecarboxylic acid (58%) and 3-methylthiophene-2,5-dicarboxylic acid (11%) [93].

Metalation of thiophene with amylsodium in the presence of sodium tert-amylate gave 2,5-thiophenedicarboxylic acid in 50% yield [94]. Thiophene can also be metalated with a benzene soluble ethylsodium-diethyl zinc complex [95]. Butylcesium and butylpotassium have also been used for the metalation of thiophene at low temperatures [96].

1.2.2 Reaction of 2-halothiophenes with sodium amalgam

Alternatively, the direct reaction of 2-chlorothiophene with sodium or 20–50% sodium amalgam in inert solvents, such as benzene (but not diethyl ether), at temperatures above 60 °C, and after reaction with carbon dioxide gave 2-thiophenecarboxylic [90]. On the other hand the reaction of 2-chlorothiophene with sodium amalgam in refluxing diethyl ether gave, upon carbonation, 5-chloro-2-thiophenecarboxylic acid in 92% yield. Also 2-bromo- and 2-iodothiophene are metalated in the same way albeit in lower yields. The thienylsodiums are ether-insoluble, brown to black solids. They are rapidly oxidized by air.

1.2.3 Reaction of 3-halothiophenes with sodium sand

Wurtz–Fittig reaction of 3-bromothiophene with sodium sand and trimethylsillyl chloride in dioxane at 60 °C gives isomer-free trimethyl-3-thienylsilane in

80% yield. In contrast to the behavior of 3-thienyllithium derivatives no tendency towards rearrangement to the 2-substituted derivatives was observed [97].

1.2.4 Reactions of thienylsodium derivatives

The thienylsodium derivatives have mostly been characterized through the reaction with carbon dioxide. However, the reaction of thienylsodium derivatives with various epoxides, styrene oxide and *epi*-chlorohydrin have been used for the preparation of 1-(2-thienyl)ethanol, 1-(2-thienyl)-2-propanol, 1-(2-thienyl)-2-phenyl-2-ethanol and 1-(2-thienyl)-3-chloro-2-propanol in reasonable yields [98,99]. 2-Thienylsodium in contrast to 2-thiophenemagnesium bromide gave 1,2-addition with butadienemono oxide [100]. A comparison of the reaction of 2-thienylsodium, 2-thienyllithium and 2-thiophenemagnesium bromide with aliphatic aldehydes indicated that in many cases the sodium derivative gave highest yields and their use certainly constitute the most economical process for the preparation of 2-thienylcarbinols. Thus di-(2-thienyl)carbinol was obtained upon reaction with 2-thiophene aldehyde [101].

1.3 COPPER DERIVATIVES

1.3.1 General aspects

Thienylcopper derivatives, as defined starting materials or as suspected intermediates, play an important role in thiophene chemistry. In this section the preparation of thienyl copper derivatives and 2-thienyl cuprates and their use in synthesis will be treated first. Then the synthetic use of the reaction of thienyllithium derivatives with cupric chloride yielding bithienyls will be discussed. Thirdly the Ullman reaction, probably proceeding *via* copper intermediates will be dealt with and finally some copper-promoted nucleophilic aromatic substitutions will be described.

1.3.2 Thienyl copper derivatives

2-Thienylcopper was first prepared by the reaction of 2-thienylmagnesium iodide with copper(I) iodide in diethyl ether [102–104].

2-Thienylferrocene [104]

2-Thienylmagnesium iodide, prepared from 2-iodothiophene (1.75 g, 8.3 mmol) and magnesium (0.4 g) in anhydrous diethyl ether, is added to a mixture of bromoferrocene (0.5 g), copper bromide (0.1 g) and molecular copper (0.05 g). The air is displaced with nitrogen, the ether is distilled off and the reaction mixture is heated on a silicone bath at 130 °C for 40 min. After cooling ether is added followed by water and dilute hydrochloric acid, dropwise. The phases are separated, the ether phase washed with water, dried over magnesium sulfate and evaporated. The residue is chromatographed on aluminium oxide using heptane as eluent giving 0.44 g (87%) after recrystallization from ethanol mp 116.5–117.7 °C.

Perhalogenated unsymmetrical biaryls were prepared from perhalo-2-thienyl copper, obtained from the corresponding Grignard reagents and perhalo iodoarenes in refluxing dioxane [105]. 3,4-Dichlorothienyl-2,5-dicopper was prepared from 3,4-dichloro-2,5-dilithiothiophene and cuprous chloride or iodide and characterized through reactions with acetyl chloride, iodine and allyl bromide [86].

3,4-Dichlorothienyl-2,5-dicopper [86]

3,4-Dichloro-2,5-dilithiothiophene (0.05 mol) is prepared in diethyl ether and then cooled to −70 °C, after which tetrahydrofuran (150–200 ml) is slowly added and immediately followed by cuprous halide (CuCl or CuI, 0.10 mol). The stirring is continued for 3–5 h, while the reaction mixture is warmed to 0 °C. As indicated by the color of the reaction mixture little or no reaction takes place below −20 °C. When the formation of the copper compound is completed, the reaction mixture consists of a dark-brown oil suspended in a light-brown solution. The compound is then derivatized.

The synthetic breakthrough came, when Nilsson *et al.* found that 2-thienylcopper derivatives prepared by transmetalation of 2-thienyllithium with cuprous bromide or iodide in diethyl ether, followed by exchange of the solvent for pyridine or quinoline upon reaction with aryl halides led to a convenient method for the synthesis of 2-arylthiophenes [106–108].

This method has also been used by other groups for the preparation of aryl and hetaryl thiophenes [103,109–111]. In the preparation of hetarylthiophenes the halogen and copper compounds can be interchanged. Better yields of 2-(2-thienyl)-1-methylpyrrole were obtained when *N*-methyl-2-pyrrolylcopper was reacted with 2-iodothiophene than when 2-thienylcopper was reacted with *N*-methyl 2-iodopyrrole, in part due to the instability of the iodopyrrole [112].

4-Bromo-2′,3-bithienyl [110]

To a solution of 2-thienylcopper, prepared from thiophene (10.0 g, 0.12 mol), butyllithium (0.10 mol) and cupric bromide (14.3 g, 0.10 mol), in anhydrous pyridine (80 ml) is added a solution of 3-bromo-4-iodothiophene (30.0 g, 0.10 mol) in anhydrous *N,N,N′,N′*-tetramethylethylenediamine. The reaction mixture is refluxed and the progress of the reaction is followed by GLC. When the ratio of the dihalothiophene and the title compound is constant (3–4 h), the reaction mixture is cooled and cold water (200 ml) is added. The solid copper iodide is filtered off and the filtrate extracted three times with ether. The combined organic phases are washed with cold 0.5 *M* hydrochloric acid, 10% sodium carbonate solution and water and dried over sodium sulfate. After evaporation the crude product is chromatographed on silica gel using petroleum ether (bp 30–60 °C) as eluent giving 13.5 g (55%) of the title compound as a colorless oil bp 98–100 °C/0.5 mm Hg.

Especially, 3,4,5-trichloro-2-thienylcopper was quite stable and was formed by heating cuprous 3,4,5-trichloro-2-thiophenecarboxylate [108].

3,4,5-Trichloro-2-thienylcopper [108]

Butyllithium (15 ml, 20–25% in hexane) diluted with anhydrous diethyl ether (50 ml) is added dropwise over a period of 33 min to tetrachlorothiophene (8.88 g, 40 mmol) in anhydrous diethyl ether (50 ml) at −30 °C. After additional stirring for 2.5 h the yellow trichloro-2-thienyllithium solution is transferred to a dropping funnel. Anhydrous cuprous bromide (25 mmol) is suspended in anhydrous diethyl ether (50 ml) in a reaction flask cooled to −30 °C, the trichloro-2-thienyllithium solution is added over a period of 40 min with magnetic stirring. During the addition the color of the suspension changed from pale-yellow to bright-yellow to brown and the cuprous bromide dissolved slowly. The brownish solution was stirred for an additional hour and then kept in a refrigerator overnight.

Meisenheimer compounds and aryl 2,4,6-trinitrobenzenes are obtained from 2-thienylcopper and 1,3,5-trinitrobenzene [113]. The reaction of 2-thienylcopper with 2-(bromomethyl)hexene and a iodomethylzinc derivative gave an excellent yield of the alkylated product shown below [114].

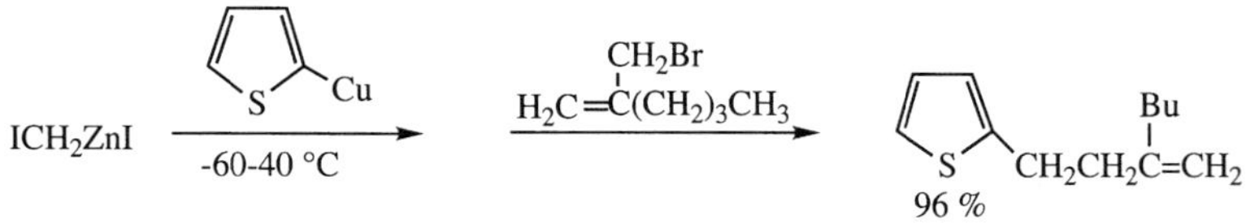

1.3.3 Lithium organothienyl cuprates (LiRThCu)

Thienylcopper gives mixed lithium organothienyl cuprates (LiRThCu) upon reaction with organolithium compounds. These cuprates react with enones or enolates adding the R group 1,4- to the substrate, 2-thienylcopper being regenerated [115]. It was found that trimethylchlorosilane had a favorable influence on the conjugate addition of lithium methyl(2-thienyl)cuprate to methyl cinnamate [116].

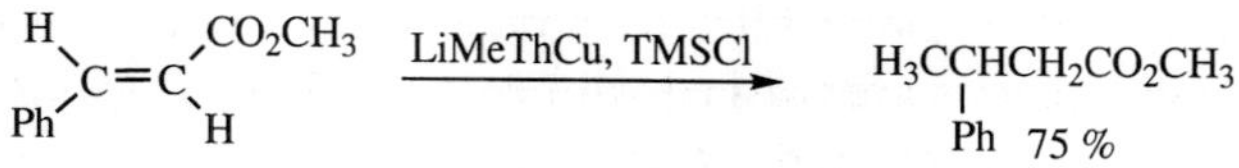

Higher order cuprates such as Me(2-Th)$CuCNLi_2$, prepared from commercially available 2-Th(CuCN)Li_2 has been *trans*-metalated with vinylstannanes and used for 1,4-addition [117–120].

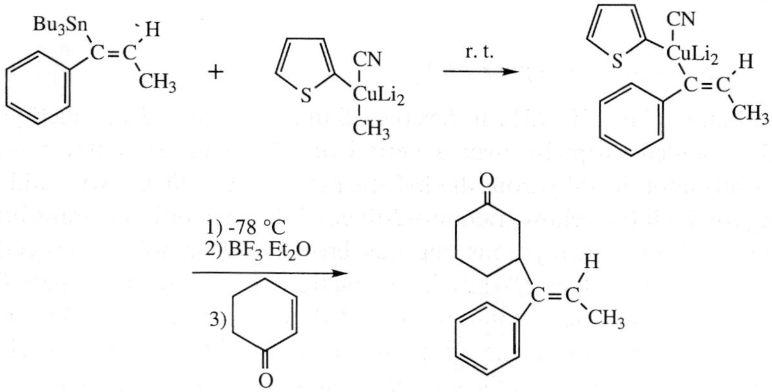

Highly reactive zerovalent copper complexes have been prepared by the direct reduction of lithium (2-thienylcyano)cuprate with preformed lithium naphthalenide [121,122].

1.3.4 The reaction of thienyllithia with cupric chloride

The reaction of thienyllithium derivatives with cupric chloride at low temperatures is one of the best methods for the preparation of symmetrical bithienyls and was first applied for the preparation of 2,2′- and 3,3′-bithienyl [123]. The detailed mechanism is not known, but it is likely that the reaction proceeds *via* copper intermediates. Numerous substituted bithienyls have been prepared in this way [124,125], especially in connection with investigations on atropisomerism in 3,3′-bithienyls [126–135].

This reaction is also useful for intramolecular carbon–carbon bond formation. Thus dithieno analogs of fluorene have been prepared through halogen–lithium exchange of di(*ortho*-iodothienyl)methanes, followed by reaction with cupric chloride [136,137]. Dimetalation of di(3-thienyl)sulfide with butyllithium followed by reaction with cupric chloride gave dithienothiophene in 50% yield [138].

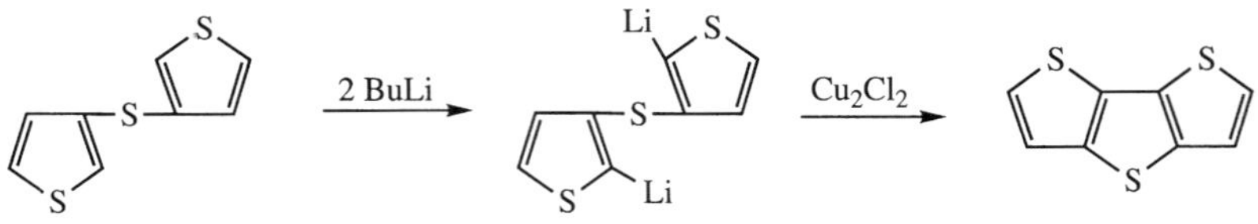

Similarly the reaction of di(3-thienyl)phenylphosphine oxide with butyllithium followed by cupric chloride gave the tricyclic compound, shown below [139].

All isomeric dithienobenzenes could be prepared by the reaction of *cis*-di-(*ortho*-bromothienyl)ethenes with butyllithium followed by cupric chloride [140]. Halogen–lithium exchange of 1,3-di(3-bromo-2-thienyl)propane followed by reaction with cupric chloride was used for the preparation of 7,8,9-trihydrocyclohepta[1,2-*b*;4,3-*b*′]dithiophene in low yield [141].

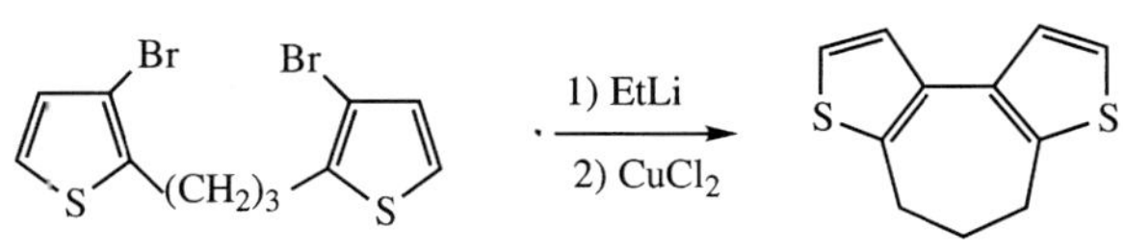

1.3.5 Ullmann reactions

The classical Ullmann coupling, which consists of heating with copper powder without any solvent, gives mostly low yields when applied to thiophenes [142,143]. By using nitroiodothiophene in xylene, good yields of dinitrobithienyls were obtained [144]. Wynberg found that nonactivated derivatives, such as 2-iodothiophene, also underwent coupling in *N*,*N*-dimethylformamide and that reduction was supressed in this solvent [145]. In connection with work on optically active bithienyls, extensive use of the Ullmann reaction has been made. Thus from 3-bromo-4-nitro-2-carbomethoxythiophene in *N*,*N*-dimethylformamide a 73% yield of 2,2′-dicarbomethoxy-4,4′-dinitro-3,3′-bithienyl was obtained, when using electrolytically prepared copper powder, which supressed dehalogenation [146]. In the same way 2,2′,4,4′-tetracarbomethoxy-3,3-bithienyl was obtained from 2,4-dicarbomethoxy-3-bromothiophene [147] and 4,4′-di-bromo-5,5′-dicarbomethoxy-3,3′-bithienyl from methyl 3,4-dibromo-2-thiophenecarboxylate, albeit in low yield [130]. The yields in these couplings are thus sensitive to the quality of the copper powder used and attempts to couple 3-bromo-2-formyl-3-nitrothiophene failed [146]. The advantage of the Ullmann reaction is of course that functional groups are tolerated, in contrast to the route *via* coupling of thienyllithium derivatives with cupric chloride. The dithiophene analogs of fluorenone have been obtained *via* intramolecular Ullmann reaction of di(*ortho*-iodothienyl)ketone [148].

Mixed Ullmann couplings have been carried out between methyl *ortho*-iodobenzoate and 2-iodothiophene [149,150] and 2-iodo-3-phenylthiophene [143]. The unsymmetrical compound from *ortho*-bromonitrobenzene and iodothiophene has also been prepared [151]. The reactions were carried out without solvents at about 210 °C. Due to the formation of symmetrical byproducts, separation can be tedious and yields mediocre.

1.3.6 Other copper(0)-promoted reactions

The reaction of 3-iodothiophene with iodotrifluoromethane with copper in *N*,*N*-dimethylformamide at 130 °C gave a low yield of 3-trifluoromethylthiophene [152]. The reaction of 3-bromothiophene with copper and perfluoroalkyl iodides gave 3-perfluoroalkylthiophenes containing minor amounts of the 2-isomer [152]. The reaction of 2-iodothiophene with perfluoro-1,3-diiodopropane with copper powder in pyridine gave 1,3-di(2-thienyl)perfluoropropane [153].

1.3.7 Copper-promoted nucleophilic aromatic substitution

The best method for the preparation of 2- [63] and 3-methoxythiophene [30] still is the reaction of the isomeric bromothiophenes with sodium methylate in methanol in the presence of potassium iodide and cupric oxide. 2-Methoxy-5-metylthiophene was prepared similarly from 2-iodo-5-methylthiophene [74]. The copper-promoted reactions have mostly been used with sulfur nucleophiles. Through the reactions of the isomeric bromothiophenes with the isomer thiophenethiols with cuprous oxide in *N*,*N*-dimethylformamide all four isomeric di(thienyl) sulfides were obtained in good yields [154]. The reaction of cuprous trifluoromethylthiolate with 2,4-diiodothiophene was used for the preparation of di(2,4-trifluoromethylthio)thiophene [15].

1.3.8 Various copper-promoted reactions

Decarboxylation of cuprous salts of thiophenecarboxylic acids, which most probably proceeds *via* thienylcopper intermediates, has been achieved by heating in quinoline [108,156]. Dehalogenation of thiophenes can be carried out with copper in quinoline [157] or propionic acid [158].

REFERENCES

1. R. R. Benkeser and R. B. Currie, *J. Am. Chem. Soc.* **70**, 1780 (1948).
2. H. Gilman and D. Shirley, *J. Am. Chem. Soc.* **71**, 1870 (1949).
3. S. Gronowitz, *Arkiv Kemi* **7**, 361 (1954).
4. S.-O. Lawesson, *Arkiv Kemi* **11**, 317 (1957).
5. S. O. Lawesson, *Arkiv Kemi* **11**, 325 (1957).
6. P. Moses and S. Gronowitz, *Arkiv Kemi* **18**, 119 (1961).
7. U. Michael and S. Gronowitz, *Acta Chem. Scand.* **22**, 1353 (1968).
8. S. Gronowitz, T. Dahlgren, J. Namtvedt, C. Roos, G. Rosén, B. Sjöberg and U. Forsgren, *Acta Pharm. Suecica* **8**, 623 (1971).
9. F. Sauter, P. Stanetti, H. Fröhlich and W. Ramer, *Heterocycles* **26**, 2639 (1987).
10. F. Sauter, P. Stanetti, and H. Fröhlich, *Heterocycles* **26**, 2657 (1987).
11. A. Schöning and W. Friedrichsen, *Tetrahedron Lett.* **29**, 1137 (1988).
12. X. Wu, T. -A. Chen, L. Zhu and R. D. Rieke, *Tetrahedron Lett.* **35**, 3673 (1994).
13. N. Gjös and S Gronowitz, *Acta Chem. Scand.* **25**, 2596 (1971).
14. G. M. Davies and P. S. Davies, *Tetrahedron Lett.* 3507 (1972).
15. S. Gronowitz, T. Frejd, O. Karlsson, K. Lawitz, P. Pedaja and K. Pettersson, *Chem. Scripta* **18**, 192 (1981).
16. S. Gronowitz, L. Svensson, M. Herslöf, A, Tjörnebo, N. Stjernström and S. O. Ögren, *Acta Pharm. Suecica* **16**, 376 (1979).
17. S. Gronowitz and T. Frejd, *Acta Chem. Scand.* **B30**, 485 (1976).
18. A. Hallberg and S. Gronowitz, *Chem. Scripta* **16**, 42 (1980).
19. J. O. Karlsson, S. Gronowitz and T. Frejd, *J. Org. Chem.* **47**, 374 (1982).
20. E. C. Taylor and D. E. Vogel, *J. Org. Chem.* **50**, 1002 (1985).
21. S. Gronowitz and C. Glennow, *Chem. Scripta* **11**, 76 (1977).
22. S. Gronowitz, K. Stenhammar and L. Svensson, *Heterocycles* **15**, 947 (1981).
23. D. Braun and E. Seelig, *Chem. Ber.* **97**, 3098 (1964).
24. D. Bryce-Smith and A. C. Skinner, *J. Chem. Soc.* 577 (1963).
25. C. G. Screttas, *J. Chem. Soc. Perkin II* 745 (1974).
26. D. E. Seitz, S. -H. Lee, R. N. Hanson and J. C. Bottaro, *Synth. Comm.* **13**, 121 (1983).
27. B. Abarca, R. Ballesteros and G. Jones, *Ann. Quim.* **79**, 23 (1983).
28. S. Gronowitz, B. Cederlund and A.-B. Hörnfeldt, *Chem. Scripta* **5**, 217 (1974).
29. C. G. Screttas and J. F. Eastham, *J. Am. Chem. Soc.* **87**, 3276 (1965).
30. S. Gronowitz, *Arkiv Kemi* **12**, 239 (1958).
31. W. H. Watthey and M. Desai, *J. Org. Chem.* **47**, 1755 (1982).
32. D. W. Slocum and P. L. Gierer, *J. Chem. Soc. D* 305 (1971).
33. D. W. Slocum and P. L. Gierer *J. Org. Chem.* **42**, 3668 (1976).
34. M. Prats, C. Galvez, Y. Gasanz and A. Rodriguez, *J. Org. Chem.* **57**, 2184 (1992).
35. M. Iwao and T. Kuraishi, *Tetrahedron Lett.* **26**, 6213 (1985).
36. M. Watanabe and V. Snieckus, *J. Am. Chem. Soc.* **102**, 1457 (1980).

37. Ya. L. Gol'dfarb, M. B. Ibrazimowa and O. A. Kalinovskii, *Bull. Acad,. Sci. USSR* 1029 (1962).
38. F. M. Stoyanovich and B. P. Fedorov, *J. Org. Chem. USSR* **1**, 1269 (1965).
39. J.-P. Lampin and F. Mathey, *J. Organometal. Chem.* **71**, 239 (1974).
40. G. Kossmehl, P. Beimling and G. Manecke, *Makromol Chem.* **184**, 27 (1983).
41. S. Gronowitz and U. Rosén, *Chem. Scripta* **1**, 33 (1971).
42. ICI Fr Patent 2,140,127 (1973); *Chem. Abstr.* 79, 42677n.
43. Y. Yang, A.-B. Hörnfeldt and S. Gronowitz, *Chem. Scripta* **28**, 275 (1988).
44. D. C. Harrowven, *Tetrahedron Lett.* **34**, 5653 (1993).
45. S. Gronowitz and B. Eriksson, *Arkiv Kemi* **21**, 335 (1963).
46. M. Iwao, M. L. Lee and R. N. Castle, *J. Heterocycl. Chem.* **17**, 1259 (1980).
47. R. Pratap, M. L. Lee and R. N. Castle, *J. Heterocycl. Chem.* **18**, 1457 (1981).
48. P. G. Spinazzé and B. A. Keay, *Tetrahedron Lett.* **30**, 1765 (1989).
49. A. J. Clarke, S. M. McNamara and O. Meth-Cohn, *Tetrahedron Lett.* **27**, 2373 (1974).
50. A. J. Carpenter and D. J. Chadwick, *Tetrahedron Lett.* **26**, 1777 (1985).
51. D. W. Knight and A. P. Nott, *Tetrahedron Lett.* **21**, 5051 (1981).
52. D. W. Knight and A. P. Nott *J. Chem. Soc., Perkin I* 791 (1983).
53. A. J. Carpenter and D. J. Chadwick, *J. Org. Chem.* **50**, 4362 (1985).
54. A. J. Carpenter and D. J. Chadwick, *J. Chem. Soc., Perkin I* 173 (1985).
55. P. Riberaux and G. Quéguiner, *Tetrahedron* **40**, 2107 (1984).
56. D. J. Chadwick and D. S. Ennis, *Tetrahedron* **47**, 9901 (1991).
57. S. A. Graham, and Th. H. Scholz, *J. Org. Chem.* **56**, 4260 (1991).
58. Th. Kauffmann, A. Mitschker and A. Woltermann, *Chem. Ber.* **116**, 992 (1983).
59. T. Kauffmann, R. Otter, B. Greving, J. König, A. Mitschker and E. Weinhöfer, *Chem. Ber.* **116**, 479 (1983).
60. P. Ribenaux and G. Quequiner, *Tetrahedron* **39**, 3593 (1983).
61. D. L. Comins and M. O. Killpack, *J. Org. Chem.* **52**, 104 (1987).
62. F. Denath, H. Gaspard-Iloughmane and J. Dubac, *Synthesis* 954, (1992).
63. J. Sicé, *J. Am. Chem. Soc.* **75**, 3697 (1953).
64. L. Gol'dfarb, M. A. Kalik amd M. L. Kirmalova , *J. Gen. Chem. USSR* **29**, 3952 (1959).
65. H. Fröhlich and W. Kalt, *J. Org. Chem.* **55**, 2993 (1990).
66. C. V. Pham, R. S. Macomber, H. B. Mark and H. Zimmer, *J. Org. Chem.* **49**, 5250 (1984).
67. S. Kano, Y. Yuasa, T. Yokomatsu and S. Shibuya, *Heterocycles* **20**, 2035 (1983).
68. J. M. Barker, P. R. Huddleston and and S. W. Shutler, *J. Chem. Soc., Perkin I* 2483 (1975).
69. A. Hallberg, T. Frejd and S. Gronowitz, *J. Chem. Soc., Perkin I* 1390 (1980).
70. S. Gronowitz and T. Frejd, *Acta Chem. Scand.* **B29**, 818 (1975).
71. S. Gronowitz and T. Frejd, *Chem. Heterocycl. Comp. USSR* **14**, 353 (1978).
72. B. Iddon, *Heterocycles* **20**, 1127 (1983).
73. B. Yom-Tov and S. Gronowitz, *Chem. Scripta* **3**, 37 (1973).
74. S. Gronowitz and T. Frejd, *Acta Chem. Scand.* **B30**, 439 (1976).
75. S. Gronowitz and B. Holm, *Acta Chem. Scand.* **23**, 2207 (1969).
76. P. Moses and S. Gronowitz, *Arkiv Kemi* **18**, 119 (1961).
77. M. T. Rahman and H. Gilman, *J. Ind. Chem. Soc.* **56**, 346 (1979).
78. A. Hallberg, T. Frejd and S. Gronowitz, *Chem. Scripta* **13**, 186 (1978).
79. G. M. Coppola, R. E. Damon and H. Yu, *J. Heterocycl. Chem.* **33**, 687 (1996).
80. D. J. Chadwick and C. Willbe, *J. Chem. Soc., Perkin I* 887 (1977).
81. B. Feringa, R. Hulst, R. Rikers and L. Brandsma, *Synthesis* 316 (1988).
82. C. Hoogzand, J. Nielsen and E. H. Braye, *J. Chem. Soc., Chem. Commun.* 1520 (1971).
83. E. G. Doadt and V. Snieckus, *Tetrahedron Lett.* **26**, 1149 (1985).
84. F. M. Stoyanovich and B. P. Fedorov, *Chem. Heterocycl. Comp.* **3**, 823 (1967).
85. H. Christiansen, S. Gronowitz, B. Rodmar, S. Rodmar, U. Rosén and M. K. Sharma, *Arkiv Kemi* **30**, 561 (1969).

86. M. R. Smith and H. Gilman, *J. Organometal. Chem.* **42**, 1 (1972).
87. M. Robba, B. Roques and M. Bonhomme, *Bull. Soc. Chim. Fr.* **7**, 2495 (1967).
88. M. Janda, J. Strogl, J. Stibor, M. Nemec and P. Vopatrana, *Synthesis* 545 (1972).
89. G. C. Nwokogu and H. Hart, *Tetrahedron Lett.* **24**, 5725 (1983).
90. J. W. Schick and H. D. Hartough, *J. Am. Chem. Soc.* **70**, 286 (1948).
91. J. W. Schick and H. D. Hartough, *J. Am. Chem. Soc.* **70**, 1645 (1948).
92. J. A. Blanchette and E. V. Brown, *J. Am. Chem. Soc.* **74**, 1848 (1952).
93. V. Ramanathan and R. Levine, *J. Org. Chem.* **27**, 1667 (1962).
94. A. A. Morton and C. E. Claff Jr, *J. Am. Chem. Soc.* **76**, 4935 (1954).
95. J. H. Ludwig and H. Schulze *J. Org. Chem.* **24**, 1573 (1959).
96. P. Benoit and N. Collington, *Bull. Soc. Chim. Fr.* 1302 (1975).
97. F. Effenberger and D. Häbich, *Liebigs Ann. Chem.* 842 (1979).
98. J. W. Schick and H. D. Hartough, *J. Am. Chem. Soc.* **70**, 1646 (1948).
99. G. van Zyl, J. F. Zack, Jr, E. S. Huyser and P. L. Cook, *J. Am. Chem. Soc.* **76**, 707 (1954).
100. G. T. Gmitter and F. L. Benton, *J. Am. Chem. Soc.* **72**, 4586 (1950).
101. G. van Zyl, R. J. Langenburg, H. H. Tam and R. N. Schut, *J. Am. Chem. Soc.* **78**, 1955 (1956).
102. H. Gilman and M. Straley, *Rec.* **55**, 821 (1936).
103. M. T. Rahman, *Monatsh. Chemie* **113**, 91 (1982).
104. A. N. Nesmeyanov, V. A. Sazonova and V. N. Drozd, *Proc. Acad. Sci. USSR* **165**, 1121 (1965).
105. M. T. Rahman and H. Gilman, *J. Indian Chem. Soc.* **56**, 299 (1979).
106. M. Nilsson, *Tetrahedron Lett.* **7**, 679 (1966).
107. M. Nilsson and C. Ullenius, *Acta Chem. Scand.* **24**, 2379 (1970).
108. M. Nilsson and C. Ullenius, *Acta Chem. Scand.* **25**, 2428 (1971).
109. S. Liljefors and S. Gronowitz, *Chem. Scripta* **15**, 102 (1980).
110. P. Spagnolo, P. Zanirato and S. Gronowitz, *J. Org. Chem.* **47**, 3177 (1982).
111. L. J. Baldwin, S. Pakray, R. N. Castle and M. L. Lee, *J. Heterocycl. Chem.* **22**, 1667 (1985).
112. N. Gjös and S. Gronowitz, *Acta Chem. Scand.* **25**, 2596, (1971).
113. M. Nilsson, C. Ullenius and O. Wennerström, *Tetrahedron Lett.* **29**, 2713 (1971).
114. P. Knochel, N. Jeong, M. J. Rozema and M. C. P. Yeh, *J. Am. Chem. Soc.* **111**, 6474 (1989).
115. E. L. Lindstedt and M. Nilsson, *Acta Chem. Scand.* **B40**, 466 (1986).
116. E.-L. Lindstedt, M. Nilsson and T. Olsson, *J. Organomet. Chem.* **334**, 255 (1987).
117. E. Pierce and R. D. Tillyer, *J. Org. Chem.* **53**, 5366 (1988).
118. B. H. Lipshutz, E. L. Elsworth, J. E. Behling and A. L. Campbell, *Tetrahedron Lett.* **29**, 893 (1988).
119. J. R. Behling, K. A. Babiak, J. S. Ng, A. L. Campbell, R. Moretti, M. Koerner and B. H. Lipshutz, *J. Am. Chem. Soc.* **110**, 2641 (1988).
120. J. R. Behling, J. S. Ng, K. A. Babiak, A. L. Campbell, E. Elsworth and H. Lipshutz, *Tetrahedron Lett.* **30**, 27 (1989).
121. W. R. Klein and R. D. Rieke, *Synth. Comm.* **22**, 2635 (1992).
122. R. D. Rieke, W. R. Klein and T.-C Wu, *J. Org. Chem.* **58**, 2492 (1993).
123. S. Gronowitz and H.-O. Karlsson, *Arkiv Kemi* **17**, 89 (1960).
124. S. Gronowitz and V. Vilks, *Arkiv Kemi* **21**, 191 (1963).
125. A.-B. Hörnfeldt, *Acta Chem. Scand.* **21**, 1952 (1967).
126. S. Gronowitz and R. Beselin, *Arkiv Kemi* **21**, 335 (1963).
127. S. Gronowitz and P. Gustafson, *Arkiv Kemi* **20**, 289 (1963).
128. S. Gronowitz and H. Frostling, *Tetrahedron Lett.* 604 (1961).
129. S. Gronowitz, *Acta Chem. Scand.* **15**, 1393 (1961).
130. R. Håkansson and E. Wiklund, *Arkiv Kemi* **31**, 101 (1969).
131. R. Håkansson, A. Ask and A. Almquist, *Chem. Scripta* **2**, 72 (1972).
132. E. Wiklund and R. Håkansson, *Chem. Scripta* **3**, 220 (1973).

133. E. Wiklund and R. Håkansson, *Chem. Scripta* **6**, 174 (1974).
134. A. Almquist and R. Håkansson, *Chem. Scripta* **11**, 57 (1977).
135. S. Gronowitz and S. Hagen, *Arkiv Kemi* **27**, 153 (1967).
136. H. Wynberg and A. Kraak, *J. Org. Chem.* **29**, 2455 (1964).
137. A. Wiersema and S. Gronowitz, *Acta Chem. Scand.* **24**, 2593 (1970).
138. F. M. Stoyanovich and B. P. Fedorov, *J. Org. Chem. USSR* **1**, 1296 (1965).
139. J.-P. Lampin and F. Mathey, *J. Organometal. Chem.* **71**, 239 (1974).
140. S. Gronowitz and T. Dahlgren, *Chem. Scripta* **12**, 57 (1977).
141. S. Gronowitz and P. Pedaja, *Chem. Scripta* **15**, 187 (1980).
142. G. N. Jean and F. F. Nord, *J. Org. Chem.* **20**, 1363 (1955).
143. A. I. Johnson, *J. Org. Chem.* **41**, 1320 (1976).
144. C. Carpanelli and G. Leandri, *Ann. Chim. (Rome)* **51**, 181 (1961).
145. H. Wynberg and A. Logothetis, *J. Am. Chem. Soc.* **78**, 1958 (1956).
146. S. Gronowitz and K. Dahlgren, *Arkiv Kemi* **21**, 201 (1963).
147. R. Håkansson and A. Svensson, *Chemica Scripta* **7**, 186 (1975).
148. P. Jordens, G. Rawson and H. Wynberg, *J. Chem. Soc.C* 273 (1970).
149. A. W. Chow, N. M. Hall, J. R. E. Hoover, M. M. Dolan and R. J. Ferlauto, *J. Med. Chem.* **9**, 551 (1966).
150. D. W. H. MacDowell and A. T. Jeffries, *J. Org. Chem.* **35**, 871 (1970).
151. J. W. Barton, D. J. Lapham and D. J. Rowe, *J. Chem. Soc., Perkin I* 131 (1985).
152. J. Leroy, M. Rubinstein and C. Wakselman, *J. Fluorine Chem.* **27**, 91 (1985).
153. V. C. R. McLoughlin and J. Thrower, *Tetrahedron* **25**, 5921 (1969).
154. E. Jones and I. M. Moodie, *Tetrahedron* **21**, 2413 (1965).
155. P. Hurtel, B. Decroix, J. Morel and F. Terrier, *J. Chem. Res. (S)*, 58 (1983).
156. S. Gronowitz and G. Borgen, *Acta Chem. Scand.* **19**, 1180 (1965)
157. R. Motayama, S. Nishimura and J. Ogawa, *Nippon Kagaku Zasshi* **78**, 950 (1957).
158. J. Skramstad, *Acta Chem. Scand.* **23**, 703 (1969).

– 2 –

Syntheses of Thiophenes with Group II Substituents

2.1 MAGNESIUM DERIVATIVES

2.1.1 Introduction

Grignard reagents play an important role in the syntheses of thiophene derivatives, although they have been superseded by thienyllithium derivatives to a great extent. They are today especially of interest when, in their reactions with electrophiles, they give different products than the corresponding lithium derivatives. In this connection they are sometimes prepared from the corresponding lithium derivatives through reaction with anhydrous magnesium bromide.

2.1.2 Synthesis from halothiophenes and magnesium

2-Bromo- and 2-iodothiophene easily give the Grignard reagents in the same way as bromo- or iodobenzene in anhydrous diethyl ether as solvent; tetrahydrofuran has also been used in some cases [1–4]. 3-Thiophenemagnesium iodide was recently prepared from 3-iodothiophene by use of the active Riekes metal [5]. The 2-thiophenemagnesium halide has been reacted with a variety of electrophiles to produce 2-substituted thiophenes, for which there will be examples in the appropriate chapters. Only in the reaction of 2-thiophenemagnesium bromide and also of the 3-isomer with alkyl *para*-toluene sulfonates was an isomer mixture of 2- and 3-alkyl derivatives obtained [6]. 2-Chlorothiophene reacts very sluggishly, but its use has been claimed in patents. From 2,5-dibromothiophene in diethyl ether, 5-bromo-2-thiophenemagnesium bromide is obtained, also when excess magnesium is used [7,8]. However, in tetrahydrofuran the 2,5-thiophenedimagnesium bromide has been obtained [3]. Upon reaction with magnesium in diethyl ether, 2-bromo-5-chlorothiophene gives 5-chloro-2-thiophenemagnesium bromide [9]. The selectivity in the reaction of 3-alkyl-2,5-dibromothiophene with magnesium in ether is somewhat susceptible to steric hindrance, as the proportions of 3-alkyl-5-thiophenemagnesium bromide to 3-alkyl-2-thiophenemagnesium bromide change from 85:15 to 87:13 to 89:11 and to 94:6, when the alkyl group is changed from methyl to ethyl to isopropyl and to *tert*-butyl, respectively [10].

2.1.3 Formation of Grignard reagents through the entrainment procedure

In contrast to 2-bromo- and 2-iodothiophene, 3-bromo- and 3-iodothiophene do not form Grignard reagents in the usual way. Grignard [11] and Steinkopf *et al.* [12]. found that Grignard reagents can be formed from inert halides forming insoluble organomagnesium compounds, such as 3-bromothiophene, when they were reacted together with reactive halides such as ethyl bromide and excess magnesium. This procedure, called the entrainment method depends, according to Grignard, upon the formation of an ether-soluble complex and the function of the auxiliary halide is to keep the magnesium surfaces clean and reactive. It was demonstrated that the mechanism of the entrainment method is not a halogen–magnesium exchange between the halide and ethylmagnesium [13] as 3-bromothiophene did not give halogen–magnesium interconversion with ethylmagnesium bromide [14]. Reagents have been prepared from 3-bromothiophene [15], 3-bromo-2,5-dialkylthiophenes [16] and 2,5-dialkyl-3-iodothiophenes [17] by the entrainment methodology. 3-Thiophenemagnesium bromide has been prepared by entrainment with methyl iodide [18] and has been used for 1,4-additions where the lithium derivative does not work [19]. The entrainment reaction was previously applied to dibromo- and polybromothiophenes in connection with preparation of 3-bromothiophene from 2,3,5-tribromothiophene and mixtures of 2,3- and 2,4-dibromothiophene [14] and 2,3,4-tribromothiophene from tetrabromothiophene [20] through hydrolysis of the magnesium derivatives. From 2,4-dibromothiophene, 4-bromo-2-thiophenecarboxylic acid was obtained upon reaction with carbon dioxide [21]. The entrainment method has been applied to tetrachlorothiophene in connection with preparation of a number of 2-substituted 3,4,5-trichlorothiophenes [9,22].

The drawback of the entrainment method is that excess of sometimes, expensive electrophiles have to be used for the reaction with ethylmagnesium bromide, and in addition separation problems could occur. This can be circumvented by using 1,2-dibromoethane as the auxiliary halide, which gives magnesium bromide upon reaction with magnesium. In this way

2-chloro-3-formylthiophene and 2-bromo-3-formylthiophene were obtained from 2-chloro-3-bromo- and 2-bromo-3-iodothiophene [23].

2.1.4 Grignard reagents through metalation and halogen–magnesium exchange

2,3,4-Trichlorothiophene has been metalated with ethylmagnesium bromide [24]. In contrast to 3-bromothiophene, α-halogenated thiophenes show magnesium exchange with ethylmagnesium bromide [10,13]. Halogen–magnesium exchange of 2,3-diiodo- and 3,4-diiodothiophene with ethylmagnesium bromide was used in the synthesis of 2-*tert*-butoxy- and 4-*tert*-butoxy-3-iodothiophene [25]. 3,4-Diiodothiophene was prepared from tetraiodothiophene in tetrahydrofuran upon reaction with ethylmagnesium bromide followed by hydrolysis [26]. As in the reaction of 3-alkyl-2,5-dibromothiophene with butyllithium, in the halogen–metal exchange with ethylmagnesium bromide also the release of steric strain is important. With 3-methyl-2,5-dibromothiophene 83% of the 5-magnesium and 17% of the 2-magnesium derivatives are formed, while the proportions were changed to 36% of the 5-isomer and 64% of the 2-isomer for 2,5-dibromo-3-*tert*-butylthiophene [10].

2.1.5 Magnesium derivatives from organolithium derivatives and magnesium bromide

In some cases it is advantageous to use Grignard reagents instead of lithium compounds, even if the lithium derivatives are more easily available than the corresponding magnesium compounds, as the latter give higher yields with some electrophiles. Grignard reagents give much better yields of *tert*-butoxy derivatives in the reaction with *tert*-butyl perbenzoate. Therefore a number of tert-butoxythiophenes have been prepared from magnesium derivatives obtained from the corresponding lithium derivatives through reaction with magnesium bromide, prepared from magnesium and bromine in anhydrous diethyl ether [27–29].

In some cases Grignard and lithium derivatives react differently, magnesium derivatives giving 1,4-addition instead of 1,2-addition. Furthermore 2,5-dialkyl-3-thiophenemagnesium derivatives are stable [30], while the corresponding lithium derivatives undergo ring-opening reactions at room temperature [31]. 3-Thiophenemagnesium bromide prepared from 3-thienyllithium gives upon reaction with acetic anhydride at −70 °C, 3-acetylthiophene in 68% yield [32], and has also been reacted with methyl 3-thienylglyoxalate to give methyl 3,3′-dithienylglycolate [33].

2.2 CALCIUM AND STRONTIUM DERIVATIVES

2.2.1 From halothiophenes and metal

2-Thienylcalcium bromide was obtained from calcium amalgam and 2-bromothiophene and 2-thienylstrontium iodide from 2-iodothiophene and metallic strontium. The rate of formation of α,β-difluoro-β-chlorovinylthiophene from 2-thienyl calcium bromide and 2-thienyl strontium iodide reagents was compared with that of 2-thienyllithium and found to be Li > Ca > Sr [34]. Reaction of 2-iodothiophene with calcium followed by carboxylation anomalously yielded a mixture of 2,5-thiophenedicarboxylic acid and 5-iodo-2-thiophenecarboxylic acid, as probably both metalation and metal–halogen exchange occurred [35]. In their reactivity organocalcium compounds are more similar to organolithium reagents than Grignard reagents [35].

2.2.2 Metalation

Metalation of thiophene with methylcalcium iodide [35,36] and with phenylcalcium iodide-1,4-diazabicyclo[2.2.2]octane (DABCO) [37] gives 2-thienylcalcium iodide characterized in low yields as the carboxylic acid, upon reaction with carbon dioxide. Thiophene and calcium vapor react at 900 °C and 10^{-3} torr to give 2-thienylcalcium hydride, which upon reaction with trimethylchlorosilane gives 2-trimetylsilylthiophene [38].

2.3 ZINC DERIVATIVES

2.3.1 Introduction

Organozinc compounds are definitely superior to magnesium and lithium derivatives for the preparation of ketones from acid chlorides. During recent years increased use of organozinc derivatives has occurred due to the discovery of nickel and palladium-catalyzed couplings to give new carbon–carbon bonds. Most organozinc derivatives were first prepared from Grignard reagents, later from organolithium derivatives and anhydrous zinc chloride [39].

2.3.2 From organolithium derivatives and zinc chloride

The first thiophene derivative was prepared from 3-thienyllithium and zinc chloride. However, its use for the synthesis of 3-acetylthiophene was unsuccessful, as a rearrangement occurred giving a mixture of 2- and

3-acetylthiophene [32]. 2-Thienylzinc chloride was later prepared from 2-thienyllithium and anhydrous zinc chloride [40–42] and used to prepare 2-phenylthiophene, through palladium(0)-catalyzed coupling with iodobenzene and ethyl 2-thiophene acetate and 2-thenyl cyanide by Ni(II)-acetylacetonate catalyzed couplings with the appropriate reagents. 2-Thienylzinc bromide [43] and 3-bromo-5-(1-propynyl)-2-thienylzinc chloride [44] were used in palladium(0)-catalyzed Heck couplings.

2.3.3 From iodothiophenes and zinc metal

Highly activated zinc dust reacts at room temperature in *N,N*-dimethylacetamide with 2-iodothiophene and 2-benzoyl-5-iodothiophene to give the corresponding thienylzinc derivatives in over 60% yield [45]. After transmetalation to the corresponding copper derivatives, they were reacted with various electrophiles. Recently Rieke using finely divided zinc, generated from the reduction of zinc salt with lithium using naphthalene or biphenyl as electron carrier in tetrahydrofuran, for the preparation of 3-thienylzinc iodide from 3-iodothiophene in 97–100% yield [5]. Such Rieke zinc is used for the preparation of 4-methylthio-2-thienylzinc bromide from the bromo derivative [46].

2.4 CADMIUM DERIVATIVES

2.4.1 Introduction

Thienylcadmium derivatives have similar reactivities as the corresponding zinc derivatives, but have been used to a lesser extent due to their toxicity. They have mostly been used for the preparation of acetyl derivatives through the reaction with acetyl chloride.

2.4.2 From Grignard reagents and cadmium chloride

Organocadmium derivatives were first prepared from Grignard reagents and cadmium chloride by Gilman and Nelson [47]. From 2-thiophenemagnesium bromide, 2-thienylcadmium chloride was prepared, giving 57% of 2-acetylthiophene upon reaction with acetyl chloride [32]. Similarly, 2-acetyl-3,4,5-trichlorothiophene was prepared from trichloro-2-thienylcadmium chloride [22]. 2-Thienylcadmium iodide prepared from 2-thiophenemagnesium iodide was reacted with β-propiolactone to give β-(2-thienyl)propionic acid [48].

2.4.3 From thienyllithium derivatives and cadmium chloride

It was claimed that organolithium derivatives easily give organocadmium derivatives with cadmium chloride [47]. However, it was found that the reaction of 2-thienyllithium was much slower with cadmium chloride than with the corresponding Grignard reagent. It took twelve hours of reflux before Michler's ketone test resulted negative, while with 2-thiophenemagnesium bromide the test was negative after only 20 min of reflux [32]. Reaction with acetyl chloride gave 55% of 2-acetylthiophene. Converting 3-thienyllithium to the cadmium derivative gave extensive decomposition and only 15% of impure 3-acetylthiophene was obtained [32]. These results were later confirmed in attempts to prepare 3,4-dichloro-2,5-thiophenecadmium chloride from 3,4-dichloro-2,5-dilithiothiophene, when only low yields of the diacetyl derivative were obtained [49].

2.5 MERCURY DERIVATIVES

2.5.1 Introduction

Mercury derivatives have played an important role in the development of thiophene chemistry. Historically they were of utmost importance for separation, purification and identification of thiophene derivatives, as they are stable and can also be recrystallized from hydroxylic solvents. Steam distillation of the thienylmercuric chlorides in the presence of hydrochloric acid regenerates the original thiophene derivative in quantitative yields. The thienylmercury chlorides have also been used in the synthesis of bromo-, iodo-, thiocyano- and acyl derivatives. During recent years new uses of thienylmercury derivatives have been described. They have been used for palladium(0)-catalyzed cross-coupling reactions [50] and in palladium(0)-catalyzed acyl demetallation, providing a general method for the synthesis of unsymmetrical heterocyclic ketones [51].

2.5.2 Thienylmercury halides

2.5.2.1 Electrophilic mercuration of thiophenes

Volhard prepared 2-thienylmercury chloride in almost quantitative yield by reacting thiophene in dilute alcoholic solution with a saturated aqueous solution of mercuric chloride and sodium acetate at room temperature. Upon reflux 2,5-dithiophenemercuric chloride was obtained [52].

Tri- and tetramercury halides cannot be obtained by direct mercuration. However, the method has been used for the preparation of numerous alkyl-substituted thienylmercuric chlorides, as well as aryl-substituted thienyl mercuric chlorides. A large number of halothiophenes have been characterized and purified *via* their chloromercury derivatives [53–57]. When the 2- and 5-positions are blocked with electron-donating groups such as alkyl, alkoxy or alkylthio groups, chloromercuration occurs in the 3-position [58,59]. The mercuration of 2,5-and 2,4-dimethylthiophene proceeds *via* an addition elimination mechanism [53,54,60].

2.5.2.2 From dithienylmercury derivatives

Thienylmercury bromides have mostly been prepared through the reaction of di(thienyl)mercury derivatives with mercuric bromide [54,55,61,62]. Thienylmercury iodides are obtained similarly from di(2-thienyl)mercury and mercuric iodide in boiling acetone [53,54].

2.5.2.3 Exchange of halogen

Thienylmercury iodides were also prepared through the reaction of the chloride with more than two equivalents of sodium iodide in acetone [54,61,63]. In refluxing acetone di(2-thienyl)mercury is formed [53]. The reaction between thienylmercury chlorides and sodium thiocyanate is used for the preparation of the more soluble thienylmercury thiocyanates [54,64,65]. Thienylmercury cyanates have been prepared from thiophenes and mercuric cyanate [66].

2.5.2.4 Various methods

2-Thienylmercury chloride is immediately precipitated upon treatment of an aqueous solution of 2-thiopheneboronic acid with a solution of mercuric chloride [67,68].

2.5.3 Thienylmercury acetates

2.5.3.1 Electrophilic mercuration

Mercuric acetate reacts rapidly with thiophene and its derivatives to form poly mercury acetates. Thus treatment of thiophene with mercuric oxide in glacial acetic acid at 80–90 °C gives a quantitative yield of tetraacetoxymercurythiophene [69]. If 50% acetic acid is used as a solvent at 50 °C only the α-positions are mercurated. If the α-positions are blocked, the 3,4-dimercuryacetate is formed in 50% acetic acid. In general, in the method employing glacial acetic acid all unsubstituted nuclear hydrogens are replaced. In contrast to mercuric chloride, mercuric acetate also reacts with strongly deactivated thiophenes.

From 2-nitrothiophene the 4,5-dimercury acetate was obtained, while 3-nitrothiophene yields 3-nitro-3,4,5-thiophenetrimercury acetate [70]. Also 2-benzoylthiophene has been converted to the 4,5-dimercuryacetate, which with bromine and potassium bromide in water was converted to 2-benzoyl-4,5-dibromothiophene in quantitative yield [71]. 2-Phenylthiophene [72] and 2,4-dichlorothiophene [73] could be acetoxymercurated in the remaining α-position, while 3,4-dimethoxythiophene yielded 2,5-diacetoxymercury-3,4-dimethoxythiophene [74].

2.5.3.2 Decarboxylative acetoxymercuration

When mercuric acetate in glacial acetic acid is heated with thiophenecarboxylic acid, mercuration of the unsubstituted positions and decarboxylative mercuration of the carboxylic group occurs. Thus 2-thiophenecarboxylic acid yields tetra(acetoxymercury)thiophene [75,76]. 2,5-Diiodo-3,4-dimethoxythiophene was prepared from the 2,5-diacetoxymercury-3,4-dimethoxythiophene, prepared in low yield from 3,4-dimethoxy-2,5-thiophenedicarboxylic acid [77]. This method has especially been useful, when applied to halogenated thiophenecarboxylic acids in connection with syntheses of 2,3-dibromo- and 2,3-dichlorothiophene [78]. The mercury acetates are first converted to the mercury chlorides by treatment with sodium chloride solution before steam distillation in the presence of hydrochloric acid.

2.5.4 Dithienylmercury derivatives

2.5.4.1 From thienylmercury halides

Thienylmercuric halides upon refluxing with sodium halides in acetone give dithienylmercury derivatives [53–55,64,65,79,80]. They have also been obtained through the reaction of 2-thienylmercury chloride with sodium in xylene [53] and with hydrazine [81]. Reaction of thiophenehydrazine derivatives such as 3,5-dinitro-2-thienylhydrazine with mercuric oxide in ethanol leads to di(3,5-dinitro-2-thienyl)mercury [82].

2.5.4.2 From thienyllithium derivatives and mercuric chloride

In certain cases when direct mercuration is not possible, dithienyl mercuric derivatives have been prepared from lithium derivatives obtained through halogen–metal exchange, as in the preparation of di(4-iodo-3-thienyl)mercury from 3,4-diodothiophene [83] and di(trichloro-2-thienyl)mercury from tetrachlorothiophene [84].

REFERENCES

1. J. Meijer, P. Vermeer and L. Brandsma, *Recueil* **92**, 601 (1973).
2. L. M. Weinstock, R. B. Currie and A. V. Lovell, *Synth. Commun.* **11**, 943 (1981).
3. I. G. C. Coutts, H. R. Goldschmid and O. C. Musgrave, *J. Chem. Soc. C* 488 (1970).
4. C. Feugeas, *Bull. Soc. Fr.* 2579 (1963).
5. X. Wu and R. D. Rieke, *J. Org Chem.* **60**, 6658 (1995).
6. A.-B. Hörnfeldt and S. Gronowitz, *Acta Chem. Scand.* **17**, 1163 (1963).
7. C. D. Hurd and R. P. Holysz, *J. Am. Chem. Soc.* **72**, 1732 (1950).
8. R. A. Benkeser and A. Torkelson, *J. Am. Chem. Soc.* **76**, 1252 (1954).
9. G. Bachman and L. V. Heisey, *J. Am. Chem. Soc.* **70**, 2378 (1948).
10. S. Gronowitz, B. Cederlund and A.-B. Hörnfeldt, *Chem. Scripta* **5**, 217 (1974).
11. V. Grignard, *Compt. Rend.* **198**, 625 (1934).
12. W. Steinkopf, H. Jacob and H. Penz, *Liebigs Ann.* **512**, 136 (1934).
13. S. Gronowitz, *Arkiv Kemi* **12**, 115 (1958).
14. S. Gronowitz, *Arkiv Kemi* **7**, 267 (1954).
15. F. B. Deans and C. Eaborn, *J. Chem. Soc.* 2303 (1959).
16. Ya. L. Gol'dfarb and P. A. Konstantinov, *Bull Acad. Sci. USSR* 113 (1957).
17. D. Cagniant and P. Cagniant, *Bull. Soc. Chim. Fr.* 713 (1953).
18. S.-O. Lawesson, *Arkiv Kemi* **11**, 387 (1957).
19. R. L Clarke, M. L. Heckeler, A. J. Gambino and S. J. Daum, *J. Med. Chem.* **21**, 1243 (1978).
20. S.-O. Lawesson, *Arkiv Kemi* **11**, 325 (1957).
21. S.-O. Lawesson, *Arkiv Kemi* **11**, 317 (1957).
22. M. T. Rahman, M. R. Smith Jr, A. F. Webb and H. Gilman, *Organometh. Chem. Synth.* **1**, 105 (1970/71).
23. S. Gronowitz and K. Pettersson, *J. Heterocycl. Chem.* **13**, 1099 (1976).
24. M. T. Rahman, *J. Indian Chem. Soc.* **58**, 21 (1981).
25. J.-M. Meunier and P. Fournari, *Bull. Soc. Chim. Fr.* 3343 (1971).
26. M. Hori, T. Kataoka, H. Shimizu, and S. Yoshimura, *Yakugaku Zasshi* **94**, 1429 (1974).
27. S. Gronowitz, *Arkiv Kemi* **16**, 363 (1960).
28. J. Z. Mortensen and S.-O. Lawesson, *Acta Chem. Scand.* **22**, 1056 (1968).
29. H. J. Jacobsen and S.-O. Lawesson, *Tetrahedron* **21**, 3331 (1965).
30. W. Steinkopf, I. Poulsen and O. Herday, *Liebigs Ann.* **536**, 128 (1938).
31. S. Gronowitz and T. Frejd, *Acta Chem. Scand.* **24**, 2656 (1970).
32. S. Gronowitz, *Arkiv Kemi* **12**, 533 (1958).
33. K. Nyberg, B. Östman and G. Wallerberg, *Acta Chem. Scand.* **24**, 1590 (1970).
34. T. A. Storostina, I. E. Paleeva, L. F. Kozhemiakyna, L. F. Rybakova, R. Shifrina, V. A. Chernoplekova, and K. A. Kocheskov *J. Org. Chem. USSR* **14**, 2600 (1978).
35. D. F. Bryce-Smith and A. C. Skinner, *J. Chem. Soc.* 577, (1963).
36. M. Chastrette and R. Gauthier, *Compt. rend. Acad. Sci. (Paris)* **277**, 805 (1973).
37. M. A. Zemlyianichenko, N. I. Sheverdina and K. A. Kocheshkov. *Proc. Acad. Sci. USSR* **202**, 70 (1972).
38. K. Mochida, K. Kojima and Y. Yoshida, *Bull. Chem. Soc. Japan* **60**, 2255 (1987).
39. D. A. Shirley, *Organic reactions* **VIII**, 28 (1954).
40. E. Negishi, F. T. Luo, R. Frisbee and H. Matahita, *Heterocycles* **18**, 117 (1982).
41. T. Klingstedt and T. Frejd, *Organometallics* **2**, 598 (1983).
42. T. Klingstedt and T. Frejd, *Synthesis* 40 (1987).
43. T. L. Gilcrist and R. J. Summersell, *Tetrahedron Lett.* **28**, 1469 (1987).
44. J. O. Karlsson, S. Gronowitz and T. Frejd, *J. Org. Chem.* **47**, 374 (1982).
45. T. N. Majid and P. Knochel, *Tetrahedron Lett.* **31**, 4413 (1990).

46. U. Folli, D. Iarossi, M. Montorsi, A. Mucci and L. Schenetti, *J. Chem. Soc., Perkin Trans. 1* 537 (1995).
47. H. Gilman and J. F. Nelson, *Rec. trav. chim.* **55**, 518 (1936).
48. C. G. Stuckwisch and J. V. Bailey, *J. Org. Chem.* **28**, 2362 (1963).
49. M. R. Smith and H. Gilman, *J. Organomet. Chem.* **42**, 1 (1972).
50. N. A. Bumagin, P. G. More and I. P. Beletskaya, *J. Organomet. Chem.* **364**, 231 (1989).
51. N. A. Bumagin, P. G. More and I. P. Beletskaya, *J. Organomet. Chem.* **365**, 379 (1989).
52. J. Volhard, *Ann.* **267**, 172 (1892).
53. W. Steinkopf and M. Bauermeister, *Ann.* **403**, 50 (1914).
54. W. Steinkopf, *Ann.* **424**, 23 (1920).
55. W. Steinkopf, *Ann.* **413**, 310 (1916).
56. W, Steinkopf, H. Frömmel and J. Leo, *Ann.* **546**, 199 (1941).
57. W. Steinkopf, H. F. Schmitt and H. Fiedler, *Ann.* **527**, 237 (1937).
58. W. Steinkopf and P. Leonhardt, *Ann.* **495**, 166 (1923).
59. M. P. Cagniant and P. Cagniant, *Bull. Soc. Chim. Fr.* 713 (1953).
60. W. Steinkopf, *Ann.* **430**, 78 (1923).
61. W. Steinkopf, A. Meckoll and H. Strauch, *Ann.* **545**, 45 (1940).
62. M. Temciuc, A.-B. Hörnfeldt and S. Gronowitz, *J. Heterocycl. Chem.* **32**, 791 (1995).
63. E. Cherbuliez and C. Giddy, *Helv. Chim. Acta* **35**, 160 (1952).
64. W. Steinkopf, *Ann.* **428**, 123 (1921).
65. A. F. Shepard A. L. Henne and T. Midgeley Jr, *J. Am. Chem. Soc.* **56**, 1355 (1934).
66. E. Söderbäck, *Acta Chem. Scand.* **13**, 1221 (1959).
67. E. Krause and G. Renwanz, *Ber.* **65**, 777 (1932).
68. J. R. Johnson, M. G. Van Campen and O. Grummitt, *J. Am. Chem. Soc.* **60**, 111 (1938).
69. Paolini and Silbermann, *Gazzetta Chimica Italia* **45 II** 388 (1915).
70. W. Steinkopf, *Ann.* **545**, 38 (1940).
71. A. W. Weitkamp and C. S. Hamilton, *J. Am. Chem. Soc.* **59**, 2699 (1937).
72. A. W. Horton *J. Org. Chem.* **14**, 761 (1949).
73. E. Profft and A. Kubat, W*iss. Z. Tech. Hoch. Chem. Leuna-Merseburg*, **2**, 243 (1959/1960); *Chem. Abstr.* **55**, 1573 (1961).
74. G. Henrio and J. Morel, *Tetrahedron Lett.* **25**, 2167 (1974).
75. N. P. Buu-Hoi, *Bull. Soc. Chim. Fr.* 1407 (1958).
76. J.-L. Chanal, M.-T. Calmette, B. Bonnaud and H. Cousse, *Eur. J. Med. Chem.* **9**, 641 (1974).
77. M. Morel and P. Pastour, *Bull. Soc. Chim. Fr.* 737 (1968).
78. W. Steinkopf and W. Köhler, *Ann.* **532**, 281 (1937).
79. W. Steinkopf, H. Jacob and Penz, *Ann.* **512**, 136 (1934).
80. W. Steinkopf, *Ann.* **513**, 281 (1934).
81. D. Spinelli and A. Salvenini, *Ann. Chim. (Rome)* 1423 (1960).
82. H. Beyer and S. Melde, *J. Prakt. Chem.* **24**, 91 (1964).
83. G. Wittig and M. Rings, *Liebigs Ann. Chem.* **719,** 127 (1968).
84. M. D. Rausch T. R. Criswell and A. K. Ignatowicz, *J. Organomet. Chem.* **13**, 419 (1968).

– 3 –

Syntheses of Thiophenes with Group III Substituents

3.1 BORON DERIVATIVES

3.1.1 Introduction

Thiopheneboron derivatives, especially thiopheneboronic acids, are of great synthetic importance for the preparation of various types of thiophene derivatives. Oxidation of thiopheneboronic acids with hydrogen peroxide was extensively used for the preparation of hydroxythiophene systems (for review cf. ref. [1]). The reaction of *ortho*-carbonyl-substituted thiopheneboronic acids with hydrazine derivatives opened the route to heteroaromatic boron compounds (for review cf. ref. [2]). During recent years the palladium(0)-catalyzed Suzuki coupling between thiopheneboronic acids and aromatic and heterocyclic halides has been used for the preparation of aryl-, hetarylthiophenes and unsymmetrical bithienyls, as well as dithienopyridines and thieno-fused naphthyridines (for review cf. ref. [3]).

3.1.2 Thiopheneboronic acids from Grignard reagents

2-Thiopheneboronic acid was first prepared through the reaction of 2-thiophenemagnesium bromide and boron trifluoride in ether, followed by reaction with water [4] and later from thiophenemagnesium iodide and methyl borate [5]. Also, the reaction of borane complexes with 2-thiophenemagnesium bromide followed by hydrolysis has been used for the preparation of 2-thiopheneboronic acid [6].

General procedure for preparation of arylboronic acids [6]

Magnesium turnings (12 mmol) are placed in a round-bottomed flask and then flame-dried under nitrogen atmosphere. After cooling to 0 °C borane (20.0 ml of a 2.0 *M* solution in tetrahydrofuran, 40 mmol) is added to the flask followed by the slow addition of aryl bromide (5 ml of a 2.0 *M* solution in tetrahydrofuran, 10 mmol). The reaction mixture is stirred at room temperature until most of the magnesium has disappeared (normally 12–16 h). The mixture is then poured into ice-water and acidified with 10% hydrochloric acid (5 ml). The product is extracted into ether (3 × 75 ml). The combined etheral phases are dried over sodium sulfate and evaporated. The product is precipitated by trituration with petroleum ether and recrystallized from water.

2-Thiopheneboronic acid, 68% yield, mp 132–133 °C was prepared in this way. 3-Thiophenemagnesium iodide, prepared by the entrainment method has been used for the preparation of 3-thiopheneboronic acid in low yield [7]. From 2-thiophenemagnesium iodide and triethyl borate, 2-thiopheneboronic acid was obtained in 61% yield [8]. 2,5-Thiophenediboronic acid was prepared from the Grignard reagent derived from 2,5-dibromothiophene in 62% yield [9].

Using two equivalents of the Grignard reagent and one equivalent of tri(isobutyl)borate led to di(2-thienyl)borinic acid in 57% yield [8]. This compound has also been prepared from 2-thiophenemagnesium bromide and was characterized as the 2-aminoethyl ester through reaction with ethanolamine [10].

2-Aminoethyl di(2-thienyl)borinate [10]

The Grignard reagent prepared from 2-bromothiophene (41 g, 0.25 mol), magnesium (7 g) and anhydrous diethyl ether (60 ml) is added with stirring to a cooled solution of freshly distilled tributyl borate (28 g) in anhydrous ether (150 ml) at such a rate that the temperature does not rise above −70 °C. The reaction mixture is allowed to warm to room temperature overnight and kept for three days, after which it is acidified with 1 *M* hydrochloric acid. The phases are separated, the aqueous phase extracted with ether and the combined organic phases evaporated to *ca.* 250 ml. The solution of the crude borinic acid is added to a mixture of 2-aminoethanol (20 ml), methanol (50 ml) and water (150 ml) at 0 °C. The resulting precipitate is filtered off, washed with water and dried. Recrystallization from methanol/chloroform gives 20 g (70%) of the title compound as needles mp 200–202 °C.

Mixed aryl thienylborinic acids are also of synthetic use for the preparation of aryl thiophenes [11,12].

3.1.3 Thiopheneboronic acids from thienyllithium derivatives

The most common method for the preparation of 2-thiopheneboronic acids is the metalation of thiophenes with alkyllithium derivatives followed by the reaction with alkyl borates at −70 °C and hydrolysis [13–18]. In a recent case, 3-dodecylthiophene was lithiated in the 5-position with lithium diisopropyl amide and reacted with trimethyl borate. After hydrolysis of the borate followed by addition of sodium hydroxide sodium 4-dodecyl-2-thienylboronate precipitated and was used directly in the Suzuki reaction [19].

Sodium 4-dodecyl-2-thienylboronate [19]

To a stirred solution of 3-dodecylthiophene (32.0 g, 126 mmol) in anhydrous tetrahydrofuran (200 ml) at −60 °C 2 *M* lithium diisopropylamide (63 ml, 126 mmol) is added dropwise. The resulting mixture is stirred for 1 h and then allowed to warm to room temperature over a 2 h period. After cooling the mixture to −60 °C, trimethyl borate (30 ml, 240 mmol) was slowly added. The reaction mixture is allowed to warm to room temperature overnight and then hydrolyzed with aqueous hydrochloric acid. The phases are separated and the aqueous phase extracted with diethyl ether (150 ml). The combined organic phases are dried over sodium sulfate and then stirred with sodium hydroxide (20 g). The resulting precipitate is filtered off and washed with cold diethyl ether giving 15.0 g (36%) of the title compound as a colorless solid, which is used in next step without further purification.

In some cases halogen–metal exhange is used, as in the preparation of 4-bromo-3,5-dimethyl-2-thiopheneboronic acid from 2,4-dibromo-3,5-dimethylthiophene, butyllithium and tributyl borate [20,21].

In the same way, the reaction between alkyl borate and 3-thienyllithium derivatives, prepared by halogen–metal exchange of 3-halothiophenes has been used for the preparation of 3-thiopheneboronic acids [7,15–17]. Of great synthetic value is the preparation of 3-formyl-2-thiopheneboronic acid and 2-formyl-3-thiopheneboronic acid, both of which can be prepared in a one-pot procedure from 2,3-dibromothiophene by two consecutive halogen–metal exchange reactions, interfaced with reactions with *N*,*N*-dimethylformamide and ethyl borate [16,22].

In many cases the thiopheneboronic acids or their esters are not isolated, but directly transformed to the desired products, as in the preparation of the hydroxythiophene systems through oxidation with hydrogen peroxide [23–27].

3.1.4 Other methods for preparation of thiopheneboronic acids

2-Thiopheneboronic acid has also been obtained by the reaction of 2-thiophenemercuric chloride or di(2-thienyl)mercury with borane followed by hydrolysis [28]. The carbon–boron bond in thiophenes is in some cases stable enough to survive electrophilic aromatic substitution such as halogenation and nitration. Compounds prepared in this way are given in the appropriate chapters.

Oxidation of 2-formyl-3-thiopheneboronic acid with alkaline silver oxide gave 2-carboxy-3-thiopheneboronic acid in low yield [29].

3.1.5 Other derivatives containing thiophene boron bonds

3.1.5.1 Derivatives of thiopheneboronic acids

Many esters of thiopheneboronic acids have been prepared from thienyllithium derivatives and alkyl borates. However, in most cases they were not characterized, but immediately used for synthetic purposes [26]. Treatment of the Grignard reagent derived from 2-bromo-3-octylthiophene with trimethyl borate followed by hydrolysis and treatment with 2,2-dimethyl-1,3-propandiol is used for the preparation of [1′,3′-(2′,2′-dimethylenepropylene)]-3-octyl-2-thienylboronate [30].

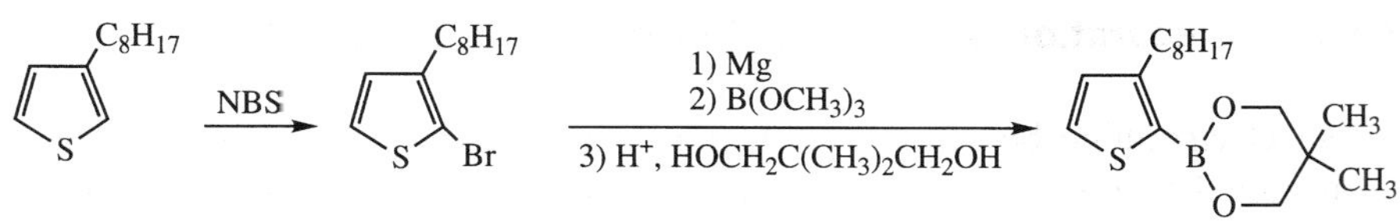

[1′,3′-(2′,2′-Dimethylenepropylene)]-3-octyl-2-thienylboronate [30]

To magnesium (1.34 g, 55 mmol) in anhydrous tetrahydrofuran (50 ml) heated to maintain a mild reflux 2-bromo-3-octylthiophene (10.00 g, 36.3 mmol) is added dropwise. When the addition is completed the reaction mixture is refluxed for 1 h and then transferred *via* cannula to a solution of trimethyl borate (16.0 ml, 143 mmol) in anhydrous tetrahydrofuran (50 ml) at −78 °C. The mixture is allowed to warm to room temperature and stirred for 30 min before being poured into 10% hydrochloric acid (50 ml). The phases are separated and the aqueous phase extracted with diethyl ether. The combined organic phases are successively dried over sodium sulfate and molecular sieves in the presence of 2,2-dimethyl-1,3-propandiol (4.78 g, 36.3 mmol). Evaporation provides 10.86 g of a slightly yellow liquid contaminated by white crystals. This crude product is dissolved in hexane, filtered on Celite and heated for one day at 100 °C under vacuum giving 8.32 g (74%) of the title compound as a thick slightly yellow liquid.

The reaction of 2-thiophenemagnesium iodide with tri(isopropyl)borate was used for the preparation of di(isopropyl) 2-thiopheneborate [8]. Halogen–metal exchange of *cis* 1,2-di-(3-bromo-2-thienyl)ethene and ethyllithium followed by ethyl borate led to the ether of a dithienoborepin derivative [31]. The reaction of the lithium derivative with dietylaminoboron dichloride gave the 2-aminoethoxy borepin derivative in 85% yield, which through reaction with methylmagnesium iodide was transformed to 4-methyl-4*H*-borepino[3,2-b;6,7-*b*′]dithiophene [32].

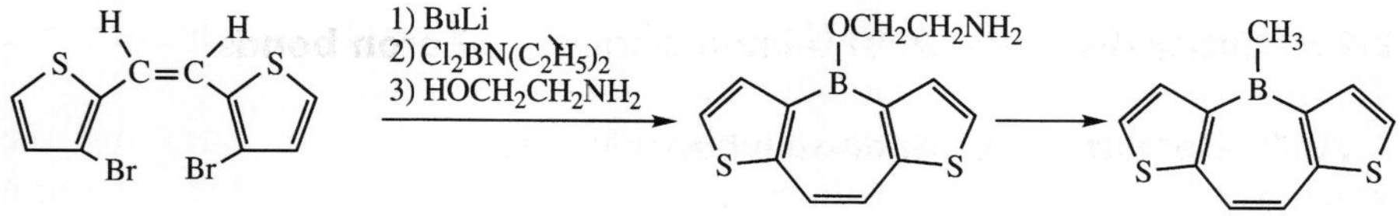

The palladium(0)-catalyzed reaction of 2-iodothiophene with pinacolborane is a good method for the preparation of 2-thienylpinacol boronate [33].

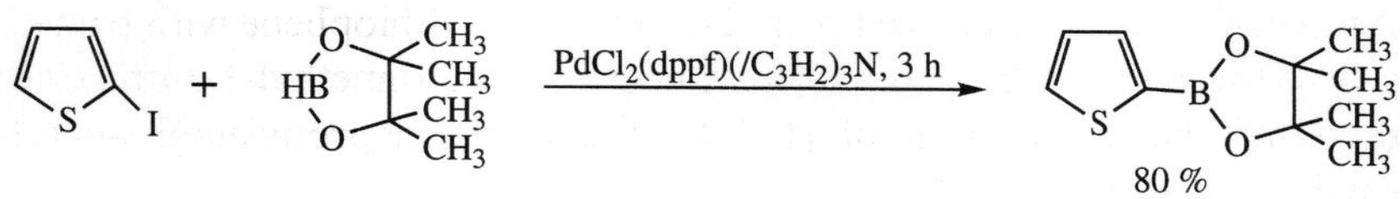

3.1.5.2 Preparation of thiopheneboron halides and amides

Direct electrophilic boronation with boron trichloride or alkyl or arylboron dichloride has not been used extensively due to polymerization reactions, except for the reaction of *ortho*-aminobithienyls with aryl and hetarylboron dichloride in connection with the preparation of borazarobenzodithiophenes [34,35].

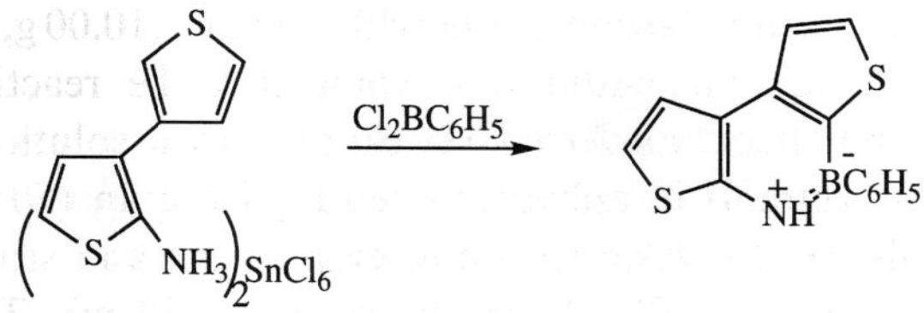

Therefore 2-thiopheneboron dichlorides are best obtained through the reaction of tetra-(2-thienyl)stannanes with boron trichloride in benzene [36,37]. 3-Alkyl-4-trimethylsilylthiophenes, upon treatment with boron trichloride in methylene chloride at −78 to 0°C, give the corresponding boron dichlorides, which were converted to the boroxines upon treatment with 0.5 *M* sodium carbonate [38].

Tris(heptylthien-3-yl)boroxine [38]

To a solution of 3-(trimethylsilyl)-4-heptylthiophene (50.8 mg, 0.2 mmol) in dichloromethane (8 ml) a 1.0 *M* solution of boron trichloride in dichloromethane (0.3 ml) is added under nitrogen at −78 °C. The reaction mixture is stirred for 2 h and then allowed to slowly warm to 0 °C for 8 h. The reaction is quenched with 0.5 *M* sodium carbonate solution (10 ml) followed by extraction with diethyl ether (3 × 15 ml). The combined organic phases are dried over magnesium sulfate and evaporated. The crude product is chromatographed on silica gel (10 g) using hexane/ether (1:1) as eluent giving 26.5 mg (64%) of the title compound as a colorless oil.

Another method to achieve thiophene–boron bonds is to carry out the reaction of iodothiophene with boron triiodide in carbon disulfide at room temperature. With 2-iodo-5-methylthiophene, 5-methyl-2-thiopheneboron diiodide was obtained together with iodine. 2,5-Diiodothiophene gave 2,5-dithiopheneboron diiodide [39], while the product from 2-iodothiophene polymerized.

5-Methyl-2-thiopheneboron diiodide [39]

To a solution of boron triiodide (3.91 g, 10.0 mmol) in carbon disulfide (10 ml) 2-iodo-5-methylthiophene (2.24 g, 10.0 mmol) in carbon disulfide (10 ml) is added dropwise with stirring at room temperature leading to an exothermic reaction. After 2 h iodine is precipitated with mercury; and the product of the reaction is distilled giving 1.5 g (41%) of the title compound bp 90–92 °C/ 0.1 mm Hg, mp 29–30 °C.

This iodide can be transformed to the more stable bismethylmercapto derivatives through reaction with dimethyl disulfide [39]. The boron diiodides are easily dimerized to dithieno-1,4-diborins. Thus 3,4-diiodo-2,5-dimethylthiophene gives, with equivalent amounts of boron triiodide, 4,8-diiodo-1,3,5,7-tetramethyl-4*H*,8*H*-dithieno[3,4-*b*;3′,4′-*e*]1,4-diborin in quantitative yield [40]. This compound reacts with triiodoboron thiine to give 1,3-diiodo-4,6-dimethyl-1*H*,3*H*-thieno[3,4-*c*]thiadiborole, which is also obtained in a one-pot procedure from 3,4-diiodo-2,5-dimethylthiophene boron triiodide and sulfur [40]

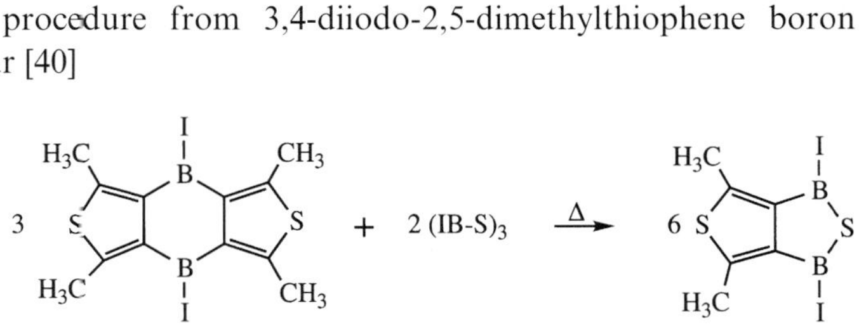

This compound was useful for the preparation amino boron derivatives through the reaction with methylamine and dimethyl amine [40]. Reaction of 2-thienyllithium with dimethylaminoboron dichloride or diisopropylaminoboron dichloride gives di(2-thienyl)dimethylaminoborane and di(2-thienyl)-diisopropylaminoborane, respectively [41]. Dilithiation of this compound with lithium 2,2,6,6-tetramethylpiperidine followed by reaction with diisopropylaminoboron dichloride gave a boronbridged tetrathiaporphyrinogene.

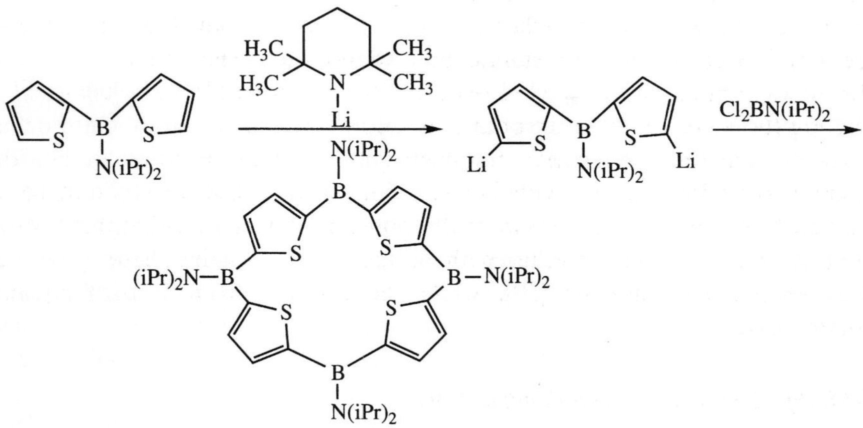

Di(2-thienyl)diisopropylaminoborane [41]

Thiophene (10.5 g, 125 mmol) is added dropwise from a syringe to a solution of butyllithium (125 mmol) and *N,N,N′,N′*-tetramethylethylenediamine (20 ml, 133 mmol) in hexane (100 ml) at 0 °C. After the reaction mixture has been allowed to warm to room temperature, it is stirred for 1 h, then cooled to −60 °C and treated drop-by-drop with diisopropylaminoboron dichloride (11.5 g, 63 mmol) in hexane (50 ml). The stirred reaction mixture is allowed to warm to room temperature overnight. The precipitated lithium chloride is removed by filtration through a glass frit (porosity 3). The solvent and *N,N,N′,N′*-tetramethylethylenediamine are removed under vacuum. Distillation of the residue gives 13.9 g (80%) of the title compound bp 90 °C/ 7.5×10^{-3}mm Hg, mp 37 °C.

Boronbridged tetrathiaporphyrinogene [41]

A solution of di(2-thienyl)diisopropylaminoborane (2.5 g, 9 mmol) in hexane (10 ml) is added dropwise to a solution of lithium 2,2,6,6-tetramethylpiperidine, freshly prepared by treating 2,2,6,6-tetramethylpiperidine (2.54 g, 18 mmol) with butyllithium in hexane (18 mmol). After a few seconds a dispersion of the dilitiated product is formed. The reaction mixture is stirred for 1 h at room temperature and then a solution of diisopropylaminoboron dichloride (0.82 g, 4.5 mmol) in hexane (10 ml) is added dropwise at room temperature. The reaction mixture is stirred at 50 °C for 3 h. Volatile material is pumped off and the crude product is taken up in toluene (10 ml). The resulting solution is filtered through a glass frit (porosity 3), the filtrate evaporated and the residue reystallized from dichloromethane/hexane (2:1) giving 62% of the title compound mp 182 °C (decomp).

3.1.5.3 Preparation of thienyl alkyl and aryl boranes

The reaction of thienyllithium derivatives with dialkyl boron chloride at −70 °C offers a convenient route to dialkyl thienylboranes and alkyl dithienylboranes. Using dimethylaminoboron dichloride gave dimethyl-amino-di(2-thienyl)borane. However, yields are only between 10 and 50%. Working at high concentrations often gives diborylated compounds as byproducts. A second general method consists in the reaction of tetra(2-thienyl)stannane with dialkylboron chloride or alkylboron dichloride, which has been used for the preparation of dimethyl- and diethyl 2-thienylborane in about 50% yield and methyl and ethyl di(2-thienyl)borane, respectively. It is best to use the borane in excess in order to facilitate separation of the product from the less volatile di(2-thienyl)stannane dichlorides. 1-(2-Thienyl)- and 1-(3-thienyl)-borolane are synthesized from 2- and 3-thienyllithium and 1-chloro-borolane. Tri(2-thienyl)borane was prepared through the reaction of di(2-thienyl)boron chloride and trimethyl (2-thienyl)stannane [36].

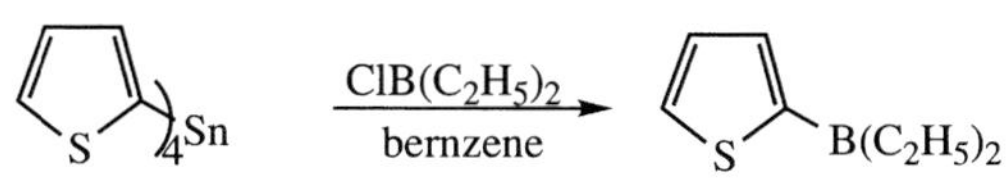

Diethyl(2-thienyl)borane [36]

To a suspension of tetra(2-thienyl)stannane (3.4 g, 7.5 mmol), diethylboron chloride (1.6 g, 15.3 mmol) in benzene (10 ml) is added dropwise under stirring. After 3 h of reflux, the diethylboron chloride is consumed according to ^{11}B NMR. Distillation gives 1.15 g (51%) of the title compound as a colorless oxidation-sensitive liquid bp 44–46 °C/1 mm Hg.

Tri(2-thienyl)borane [36]

Trimethyl(2-thienyl)stannane (4.3 g, 17 mmol) in hexane (10 ml) is cooled to −78 °C and under stirring chlorodi(2-thienyl)borane (3.7 g, 17 mmol) in hexane (10 ml) is added. The reaction mixture is allowed to warm to room temperature, evaporated and sublimed (40 °C/1 mm Hg) in order to separate trimethylchlorostannane. The residue is crystallized from hexane giving 2.8 g (65%) of the title compound as colorless crystals mp 98–104 °C.

1-(2-Thienyl)borolane [36]

Thiophene (1.4 g, 20 mmol) in anhydrous diethyl ether (10 ml) is refluxed for 3 h with an equivalent amount of butyllithium. To the 2-thienyllithium so obtained as a suspension, 1-chloroborolane (2.0 g, 20 mmol) in hexane is added dropwise under stirring at room temperature. This reaction is in general exothermic. After evaporation the residue is distilled giving 1.1 g (35%) of the title compound as a colorless liquid bp 39–41 °C/1 mm Hg.

Reaction of 2-iodo-5-methylthiophene with one-third of an equivalent of boron triiodide can be used for the preparation of tris(5-methyl-2-thienyl)-borane in 60% yield. The reaction of 2-iodo-5-methylthiophene with dimethylboron iodide and methylboron diiodide was similarly used for the preparation of dimethyl (5-methyl-2-thienyl)borane and methyl di(5-methyl-2-thienyl)-borane [39].

The reaction of dimesitylboron fluoride with thienyllithium derivatives is used for the preparation of dimesityl bithienylboranes [42], and was recently also used in the reaction of dimesitylboron fluoride with lithiated oligothiophenes [43].

The reaction of diethyl (5-methyl-2-thienyl)borane with pentacarbonylchromium tetrahydrofuran gives the stable pentacarbonyldiethyl(5-methyl-2-thienyl)boranechromium(0) [44].

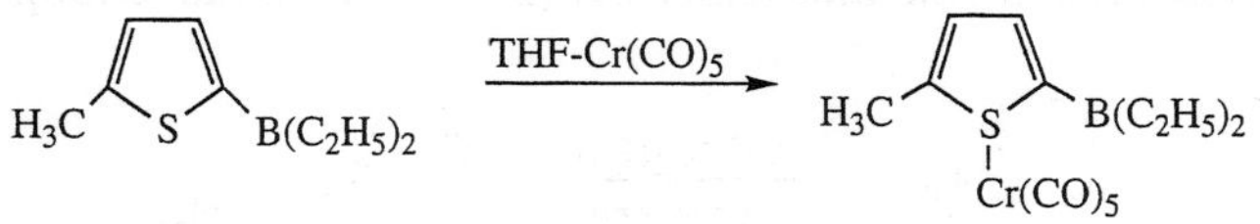

3.1.5.4 Tetrathienylborates

Potassium and cesium tetra(thienyl)borates have been prepared through the reaction of 2-thiophenemagnesium halides with potassium fluoroborate [45], thienyllithium derivatives with boron trifluoride etherate at −70 °C and halothiophenes with magnesium and sodium boron tetrafluoride in tetrahydrofuran [46,47].

Cesium tetrakis(2-thienyl)borate [46]

Anhydrous diethyl ether (200 ml) is cooled to −70 °C and 2-bromothiophene (25.0 g, 0.153 mol) is added. While the mixture is vigorously stirred, 1.6 *M* butyllithium in hexane (96 ml, 0.154 mol) is added over a 5 min period. The reaction is instantaneous and then boron trifluoride etherate (5.4 g, 0.038 mol) is added. The reaction is allowed to proceed at −70 °C for 1 h and then allowed to come to room temperature, which requires about 1 h. Lithium fluoride starts to appear before the reaction mixture reaches room temperature, when it is poured into ice water (200 ml) and stirred for 10 min. The phases are separated, the aqueous phase saturated with sodium chloride and extracted with diethyl ether (3 × 75 ml). The combined ether phases are poured into deionized water, the ether is evaporated and the ether-free solution filtered through a Celite pad. The filtrate is diluted with deionized water (300 ml) and the borate is precipitated as the cesium salt. Recrystallization of the cesium salt is accomplished by dissolving in acetone, filtering the solution and adding water. The precipitate is filtered off and dried under vacuum over calcium chloride at 80 °C giving 12.1 g of the title compound.

Lithium tributyl-2-thienylborate was prepared through the reaction of thiophene with *tert*-butyllithium followed by tributylborane [48]. The borate complexes prepared from trialkylboranes and 2-thienyllithium are useful for the preparation of 2-alkylthiophenes through the reaction with iodine [49].

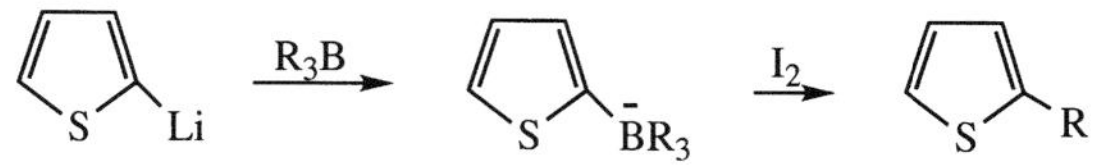

3.2 ALUMINIUM DERIVATIVES

Tri(2-thienyl)aluminium etherate was prepared in 85% yield from 2-thiophenemagnesium bromide and aluminium trichloride [50,51].

Tri(2-thienyl)aluminium etherate [50]

A solution of aluminium trichloride in diethyl ether is prepared by placing aluminium trichloride (32.71 g, 0.245 mol) in a 250 ml flask and cooling to −78 °C. Diethyl ether (100 ml) is added slowly with stirring to prevent overheating from the extremely exothermic formation of the tri(chloro)aluminium etherate complex. When the addition is completed the solution is brought to room temperature, transferred to a dropping funnel and added dropwise to the Grignard reagent of 2-bromothiophene (0.736 mol) in diethyl ether. The reaction mixture is refluxed for 4 h and the diethyl ether evaporated. The solid residue is extracted with diethyl ether/pentane (1:3) (200 ml). The supernatant liquid is decanted and when it is cooled to −20 °C it gives the title compound as a white crystalline solid. The ether is easily removed and tri(2-thienyl)aluminium is obtained in a yield of 85% mp 82–83 °C.

3.3 GALLIUM DERIVATIVES

Tri(2-thienyl)gallium has been prepared through the reaction of three equivalents of 2-thiophenemagnesium bromide with gallium trichloride [52].

It was purified as a dioxane complex, which was heated *in vacuo*. The reaction of the dioxane complex of 2-thienylgallium dichloride with phenylmagnesium bromide gave the dioxane complex of diphenyl 2-thienylgallium. Similarly, phenyl di(2-thienyl)gallium was obtained from 2-thiophenemagnesium bromide and phenylgallium dichloride [53].

3.4 INDIUM DERIVATIVES

The dioxane complex of tri(2-thienyl)indium has been prepared from indium trichloride and 2-thiophenemagnesium bromide. Upon heating *in vacuo* it loses dioxane giving pure tri(2-thienyl)indium [50,54].

3.5 THALLIUM DERIVATIVES

2-Thienylthallium difluoroacetate is most conveniently prepared in 82% yield through aromatic substitution using thallium(III) trifluoroacetate as the electrophile [55].

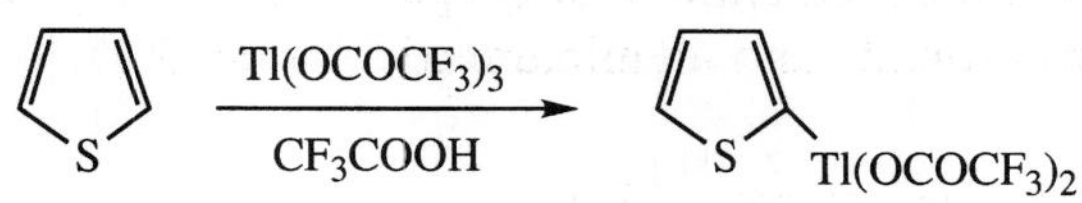

Electrophilic thallation of thiophenes with phenylthallium(III) 18-crown-6 diperchlorate affords the corresponding (18-crown-6)phenyl thienylthallium(III) perchlorates in high yields [56].

3.6 CERIUM DERIVATIVES

2-Thienylcerium dichloride is prepared from 2-thienyllithium and cerium trichloride [57]. In contrast to 2-thienyllithium it gives a diol upon reaction with the spiroketone [57].

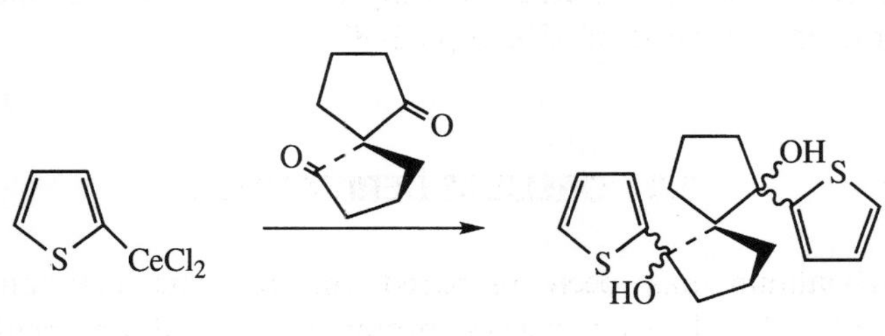

REFERENCES

1. S. Gronowitz and A.-B. Hörnfeldt, *The Chemistry of Heterocyclic Compounds* Vol. 44, Part 3, 24–25, (1986)
2. S. Gronowitz, *J. Heterocycl. Chem.* **12S**, 17 (1975).
3. S. Gronowitz, *J. Heterocycl. Chem.* **31**, 641 (1994).
4. E. Krause and G. Renwanz, *Ber.* **65**, 777 (1932).
5. J. R. Johnson, M. G. Van Campen Jr and O. Grummitt, *J. Am. Chem. Soc.* **60**, 111 (1938).
6. G. W. Kabalka, U. Sastry, K. A. R. Sastry, F. F. Knapp Jr and P. C. Srivastava, *J. Organomet. Chem.* **259**, 269 (1983).
7. S.-O. Lawesson, *Arkiv kemi* **11**, 387 (1957).
8. H. J. Roth and B. Miller, *Arch. Pharm.* **297**, 513 (1964).
9. I. G. C. Coutts, H. R. Goldschmid and O. C. Musgrave, *J. Chem. Soc. C* **488** (1970).
10. I. G. C. Coutts and O. C. Musgrave, *J. Chem. Soc. C* **2225** (1970).
11. G. M. Davies, P. S. Davies, W. E. Paget and J. M. Wardleworth, *Tetrahedron Lett.* **1976**, 795.
12. A. Pelter, H. Williamson and G. M. Davies *Tetrahedron Lett.* **25**, 453 (1984).
13. J. Namtvedt, *Acta Chem. Scand.* **22**, 1611 (1968).
14. A.-B. Hörnfeldt and S. Gronowitz, *Arkiv Kemi* **21**, 239 (1963).
15. S. Gronowitz T, Dahlgren, J. Namtvedt, C. Roos, G. Rosén , B. Sjöberg and U. Forsgren, *Acta Pharm. Suecica* **8**, 623 (1971).
16. U. Michael and S. Gronowitz, *Acta Chem. Scand.* **22**, 1353 (1968).
17. S. Gronowitz and A. Bugge, *Acta Chem. Scand.* **19**, 1271 (1965).
18. S. Gronowitz and C. Glennow, *Chem. Scripta* **11**, 76 (1977).
19. T. Kirschbaum, R. Azumi, E. Mena-Ostreritz and P. Bäuerle, *New J. Chem.* 241 (1999).
20. M. Takeshita and M. Irie, *Tetrahedron Lett.* **39**, 613 (1998).
21. M. Takeshita and M. Irie, *J. Org. Chem.* **63**, 6643 (1998).
22. S. Gronowitz and A. Maltesson, *Acta Chem. Scand.* **25**, 2435 (1971).
23. A.-B. Hörnfeldt, *Arkiv Kemi* **22**, 211 (1964).
24. R. Kreiswetter and P. Margaretha, *Helv. Chim. Acta* **68**, 2350 (1985).
25. A.-B. Hörnfeldt, *Acta Chem. Scand.* **21**,1952 (1967).
26. R. Lantz and A.-B. Hörnfeldt, *Chem. Scripta* **2**, 9 (1972).
27. A.-B. Hörnfeldt and P. O. Sundberg, *Acta Chem. Scand.* **26**, 31 (1972).
28. S. W. Breuer, F. G. Thorpe and J. C. Podesta, *Tetrahedron Lett.* **3719** (1974).
29. S. Gronowitz and A. Bugge, *Acta Chem. Scand.* **20**, 261 (1966).
30. G. Bidan, A. De Nicola, V. Enée and S. Guillerez, *Chem. Mater.* **10**, 1052 (1998).
31. S. Gronowitz, P. Gassne and B. Yom-Tov, *Acta Chem. Scand.* **23**, 2927 (1969).
32. A. T. Jeffries III and S. Gronowitz, *Chem. Scripta* **4**, 183 (1973).
33. M. Murata, S. Watanabe and Y. Masuda, *J. Org. Chem.* **62**, 6458 (1997).
34. S. Gronowitz and I. Ander, *Chem. Scripta* **15**, 23 (1980).
35. S. Gronowitz, I. Ander and P. Zanirato, *Chem. Scripta* **22**, 55 (1983).
36. B. Wrackmeyer and H. Nöth, *Chem. Ber.* **109**, 1075 (1976).
37. S. Gronowitz and I. Ander, *Chem. Scripta* **15**, 135 (1980).
38. X.-S. Ye and H. N. C. Wong, *J. Org. Chem.* **62**, 1940 (1997).
39. W. Siebert, *Chem. Ber.* **103**, 2308 (1970).
40. B. Asgouraladi, R. Full, K.-J. Schaper and W. Siebert, *Chem. Ber.* **107**, 34 (1974).
41. F. H. Carré, R. J.-P. Corriu, T. Deforth, W. E. Douglas, W. Siebert and W. Weinmann, *Angew. Chem. Int. Ed.* **37**, 652 (1998).
42. C. Branger, M. Lequan, R. M. Lequan, M. Barzoukas and A. Fort, *J. Mater. Chem.* **6**, 555 (1996).
43. T. Noda and Y. Shirota, *J. Am. Chem . Soc.* **120**, 9714 (1998).
44. H. Nöth and U. Schuchardt, *J. Organomet. Chem.* **125,** 155 (1977).

45. V. A. Sozonova and E. P. Serebryakov, *Doklady Akad. Nauk SSSR* **113**, 1295 (1957).
46. G. E. Pacey and C. E. Moore, *Anal. Chim. Acta* **105,** 353 (1979).
47. G. E. Pacey and C. E. Moore, *Tetrahedron* **27**, 1013 (1980).
48. M. Ishikura, M. Kamada, I. Oda, T. Ohta and M. Terashima, *J. Heterocycl. Chem.* **24**, 377 (1987).
49. T. Sotoyama, S. Hara and A. Suzuki, *Bull. Chem. Soc. Japan* **52**, 1865 (1979).
50. H. Rahrbarnoori, R. Kumar, M. J. Heeg and J. P. Oliver, *Organometallics* **13**, 3300 (1994).
51. R. Kumar, H. Rahbarnoohi, M. J. Heeg, D. G. Glick and J. P. Oliver, *Inorg. Chem.* **33**, 1103 (1994).
52. I. M. Viktorova, N. I. Sheverdina and K. A. Kocheshkov, *J. General Chemistry USSR (Engl. Transl.)* **37**, 2547 (1967).
53. I. M. Viktorova, N. I. Sheverdina, A. N. Rodionov and K. A. Kocheshkov, *Proc. Acad. Sci. USSR (Engl. Transl.)* **965** (1969).
54. I. M. Viktorova, N. I. Sheverdina, Yu. P. Endovin and K. A. Kocheshkov, *Bull. Acad. Sci. USSR Ser. Khim. (Engl. Transl.)* **2288** (1968).
55. A. McKillop, J. S. Fowler, M. J. Zelesko, J. D. Hunt, E. C. Taylor and G. McGillivray, *Tetrahedron Lett.* **29**, 2423 (1969).
56. F. Kakiuchi, S. Murai and Y. Kawasaki, *Organometallics* **11**, 4352 (1992).
57. J. Nakayama and T. Fujimori, *J. Chem. Soc. Chem. Commun.* **1991**, 1614.

— 4 —

Syntheses of Thiophenes with Group IV Substituents

4A. ALKYL- AND FUNCTIONALIZED ALKYLTHIOPHENES
4B. VINYLTHIOPHENES
4C. THIENYLACETYLENES
4D. ARYLTHIOPHENES
4E. ACYLTHIOPHENES
4F. THIOPHENECARBOXYLIC ACIDS AND THEIR DERIVATIVES
4G. THIOPHENE DERIVATIVES CONTAINING SILICON, GERMANIUM, TIN AND LEAD

– 4A –

Alkyl- and Functionalized Alkylthiophenes

4A.1 ALKYLTHIOPHENES AND SUBSTITUTED ALKYLTHIOPHENES

4A.1.1 Cyclization of acyclic compounds

Alkylthiophenes have been prepared from alkanes, alkenes and alkadienes, and sulfur, hydrogen sulfide, carbon disulfide and other sulfur-containing reagents, using various types of catalysts and high temperatures. These methods are mostly of technical interest. The reaction of 2-hexanol with carbon disulfide over a chromium–aluminium oxide catalyst at about 55 °C gave 2,5-dimethylthiophene in 68% yield, while 2-heptanol gave 2-ethyl-5-methylthiophene in 84% yield [1,2].

Recently 3-methylthiophene was synthesized in over 95% yield from the reaction of 2-methylbutanol with carbon disulfide over magnesium oxide supported potassium-promoted $Fe_{0.95}Cr_{0.05}OOK$ catalyst [3].

Alkylthiophenes have been prepared by dehydrocyclization of dialkyl sulfides [4,5]. A classical method, still very useful for the preparation of alkylthiophenes, is the action of hydrogen sulfide on 1,4-difunctionalized acyclic compounds. Treatment of acetylenic epoxides with hydrogen sulfide in barium hydroxide [6–8] has been used for the preparation of 2,3,5-trimethylthiophene (73%). Another method for cyclizing 1,4-diketones is to use the classical Paal–Knorr reaction with phosphorus pentasulfide [9–12]. This reaction has been used for the preparation of (2,5)heterocyclophanes and (2,4)heterophanes [6,8,13–15].

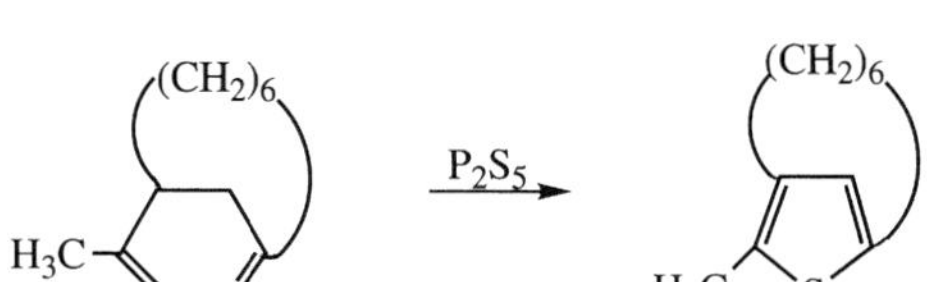

8-Methyl[6](2,4)thiophenophane [15]

Treatment of 3-acetylcyclononane (50 mg, 0.275 mmol) with phosphorus pentasulfide (267 mg) in toluene (4 ml) gives 30 mg (61%) of the title compound after purification by preparative thin layer chromatography using hexane as eluent.

Strained systems, such as cyclobutanothiophenes, have also been prepared by the Paal–Knorr method albeit in low yields [16]. Later Lawessons reagent, sometimes together with phosphorus pentasulfide and sodium bicarbonate in ether, was used in the reaction with 1,4-diketones [17–19].

A more convenient method than the Paal–Knorr reaction is in many cases the reaction of 1,4-dicarbonyl compounds with hydrogen sulfide and hydrogen chloride. It has been used for the preparation of various alkylthiophenes such as 2-*tert*-butyl-5-methylthiophene (85%) and 2,5-dimethylthiophene (49%) [20–22]. From the aldehyde acetal optically active 3-alkylthiophenes have been prepared [23].

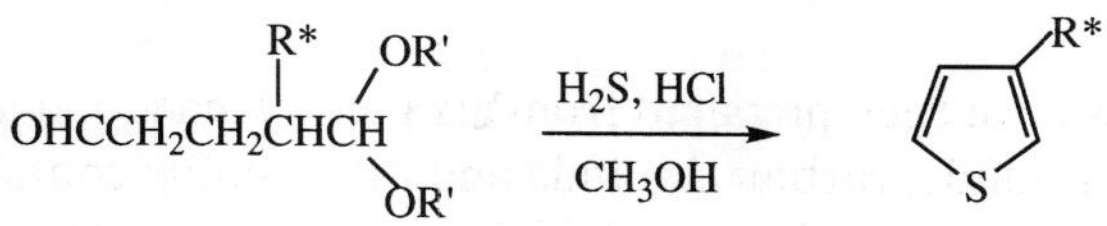

2,5-Dimethylthiophene and 2-methyl-5-propylthiophene were prepared from 5-oxo derivatives of 1-alkynes using hydrogen sulfide and hydrogen chloride at −20 °C [24]. Various 2,5-dialkylthiophenes have been prepared by saturating an alkaline alcoholic solution of 1,4-dialkyldiacetylenes with hydrogen sulfide [25–27].

2,4-Dialkylthiophenes such as 2-cyclohexyl- and 2-cyclopentyl-4-methylthiophene have been prepared in high yields through the reaction of β,γ-dichloroketones, easily prepared by electrophilic addition of acid chlorides to allyl or methallyl chlorides in the presence of aluminium trichloride and phosphorus pentasulfide [28].

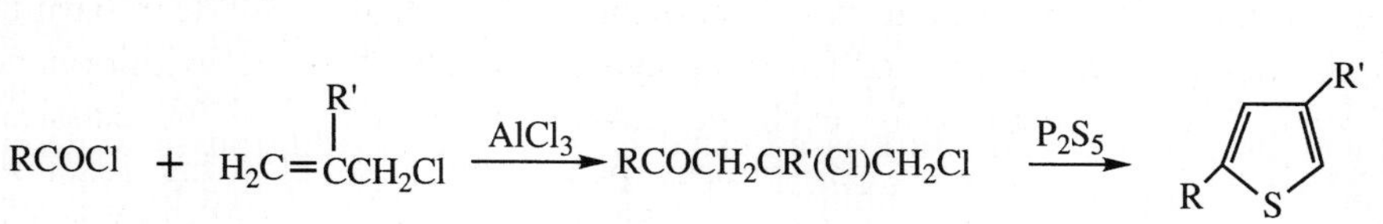

A modification of Victor Meyer's classical synthesis of thiophene from sodium succinate and P_4S_7 has been published [29]. A detailed procedure for the synthesis of 3-methylthiophene in 60% yield from disodium methylsuccinate, using mineral oil as diluent, is available [30]. The reaction of optically active sec. butylsuccinate gave a 40% yield of (+)-*S*-3-*sec*-butylthiophene with an optical purity of 43% [23].

A new ring-closure reaction to alkyl- and also 2-benzylthiophenes consists in the palladium-catalyzed cycloisomerization of (*Z*)-2-en-4-yne-1-thiols [31]. A method for the preparation of 2-alkyl- and 2,3-dialkylthiophenes from ketones as C_2-fragments and an S-protected carbonodithionic acid as C_2S-fragment, *O*-ethyl[(2-oxoethyl-thio]thioformate, has been used for the preparation of 3-ethyl-2-propylthiophene in 67% yield and 4,5,6,7-tetrahydrobenzo(*b*)thiophene in 70% yield [32].

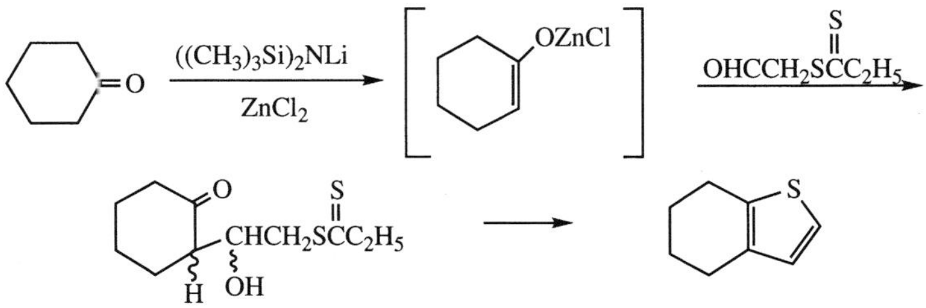

Highly congested thiophenes such as 3,4-di-*tert*-butyl-, 3,4-di-(1-adamantyl), 3,4-di-neopentyl-5 and 3-(1-adamantyl)-4-*tert*-butylthiophene are conveniently prepared by intramolecular reductive coupling of 3-thiapentane-1,5-diones with low-valent titanium, followed by acid-catalyzed dehydration of the resulting thiolane-3,4-diols [33].

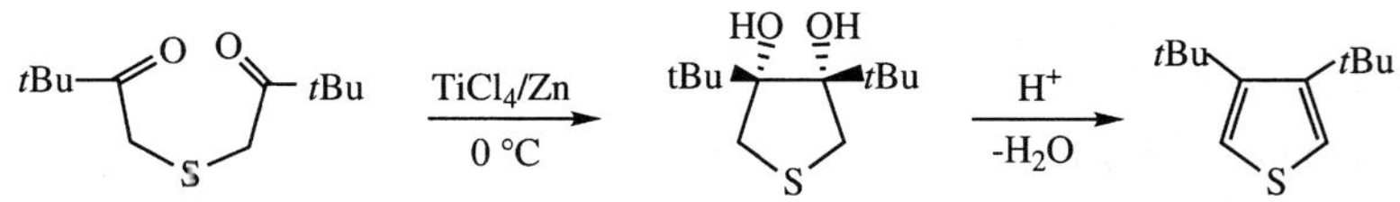

3,4-Di-tert-butyl-cis-thiolane-3,4-diol [33]

Titanium(IV) chloride (49.3 ml, 0.45 mol) is added dropwise over a period of 3–4 h to a stirred mixture of 1,5-di-*tert*-butyl-3-thiapentane-1,5-dione (34.5 g, 0.15 mol) and zinc powder (59 g, 0.9 mol) in tetrahydrofuran (800 ml) at −18 °C under argon. When the addition is completed the reaction mixture is stirred at −10 °C for 5 h. The reaction is quenched by addition of crushed ice (500 g) and then the pH of the mixture is adjusted to about 8 by addition of aqueous sodium carbonate. After addition of hexane the mixture is stirred for 1 h and then filtered by use of a Büchner funnel (diameter about 25 cm) on which a 3 cm thick pad of Celite is placed. The Celite and the solid material on the Büchner funnel are washed with hexane (500 ml). The organic layer of the filtrate is concentrated to about 500 ml, washed with water (10 × 100 ml), dried over magnesium sulfate, and evaporated. Recrystallization of the residue gives 21.0–27.8 g (60–80%) of the title compound as colorless needles mp 108.5–110 °C.

3,4-Di-tert-butylthiophene [88]

A mixture of 3,4-di-*tert*-butyl-*cis*-thiolane-3,4-diol (1.07 g, 4.6 mmol) and *para*-toluenesulfonic acid (100 mg) in benzene (60 ml) is heated under reflux for 1 h. The reaction mixture is washed with aqueous sodium bicarbonate, dried over sodium sulfate and evaporated. The residue is chromatographed on silica gel using hexane as eluent giving 803 mg (89%) of the title compound bp 75 °C/ 4 mm Hg (bulb-to-bulb distillation) mp 43.0–43.5 °C.

Tetra-*tert*-butylthiophene has been prepared in 53% yield from a seven-membered cycloalkyne and a reductive desulfurization in the last step [34].

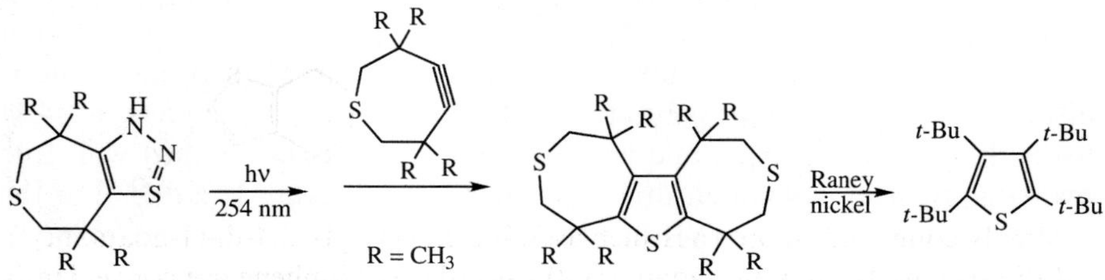

4A.1.2 Alkylation of thiophene and alkylthiophenes

Many different catalysts can be used for the preparation of alkylthiophenes from thiophene and alkenes, such as zeolites, zinc dichloride, aluminium trichloride, sulfuric acid, phosphoric acid, and boron trifluoride etherate.

In the laboratory the reaction of thiophene with isobutylene is a convenient method for the preparation of a mixture of 2- and 3-*tert*-butylthiophene and 2,4- and 2,5-di-*tert*-butylthiophene [35]. The choice of catalyst and proportions of reagents depend upon whether a higher or lower proportion of mono- to dialkylthiophenes is desired. This procedure can be used for the preparation of amyl-, *tert*-octyl and hexadecylthiophenes.

Both 2,5- and 2,4-di-tert-butylthiophene can easily be separated by a chemical procedure. Metalation of the mixture with the amount of butyllithium corresponding to the amount of 2,4-di-tert-butylthiophene followed by reaction with carbon dioxide gives pure 2,5-di-tert-butylthiophene and 3,5-di-tert-butylthiophenecarboxylic acid, which are easily separated. The latter upon decarboxylation gives pure 2,4-di-tert-butylthiophene [36].

2,4-Di-tert-butyl-5-thiophenecarboxylic acid and 2,5-di-tert-butylthiophene [36]

A mixture of 2,5-di-*tert*-butylthiophene (75%) and 2,4-di-*tert*-butylthiophene (25%) (627 g, 3.20 mol) is dissolved in anhydrous ether (300 ml) and added under nitrogen with stirring to 0.9 *M* butyllithium (1800 ml). The reaction mixture is refluxed for 4 h, cooled and poured onto solid carbon dioxide. At −10 °C the mixture is hydrolyzed with 2 *M* hydrochloric acid in excess. The phases are separated and the aqueous phase extracted with ether. The combined ether phases are extracted several times with 2 *M* sodium hydroxide solution. Acidification yields 111 g (58%) of 2,4-di-*tert*-butyl-5-thiophenecarboxylic acid mp 197–198 °C from ethanol. The ether phase is dried and fractionated giving 455 g of 2,5-di-tert-butylthiophene.

2,4-Di-tert-butylthiophene [36]

2,4-Di-*tert*-butyl-5-thiophenecarboxylic acid (20 g, 83 mmol), quinoline (50 ml), and copper powder (4 g) are carefully heated under nitrogen until the reaction starts and then the reaction mixture is refluxed for 2 h. Most of the liquid is distilled from the copper and the distillate treated with excess of 4 *M* hydrochloric acid. The organic layer is taken up in ether, washed with 2 *M* hydrochloric acid and water, dried and fractionated giving 15 g (92%) of the title compound bp 88–89 °C/10 mm Hg.

4A.1.3 From metalorganic reagents

The reaction of the appropriate thienyllithium derivatives with dimethyl sulfate has been used for the preparation of 2-methylthiophene [37], 2,3-dimethylthiophene [38], 3-methylthiophene, 2,4-dimethylthiophene and 2,3,5-trimethylthiophene [39], as well as 2,3,4-trimethylthiophene [40].

2,3,4-Trimethylthiophene [40]

A solution of 2,3,4-tribromothiophene (16.0 g, 0.05 mol) in diethyl ether (100 ml) placed in a flame-dried flask previously flushed with anhydrous nitrogen under nitrogen at −70 °C is treated with a precooled (−70 °C) ethereal solution of butyllithium (1.04 mol). After 3 h at −70 °C a precooled (−70 °C) ethereal solution of dimethyl sulfate is added and the reaction mixture is stirred for another 45 min. The cooling bath is removed and the flow of nitrogen is stopped, after which a 4 *M* solution of hydroxide (1.2 equiv. based on dimethyl sulfate) is added and the mixture vigorously stirred for 2 h. The phases are separated and the organic phase washed twice with water, dried over magnesium sulfate, evaporated, and distilled giving 5.8 g (92%) of the title compound bp 75 °C/30 mm Hg.

Care has to be taken not to work on a large scale and high concentrations as the reaction can go out of hands. It is better to use dimethyl sulfate than methyl tosylate as methylating agent as the latter reacts more slowly, giving instable lithium derivatives an opportunity to rearrange. Yields are often low and sulfones are formed as by-products [41]. Alternatively alkylthiophenes have been obtained by the reaction of thienyllithium derivatives with trialkyl boranes [42,43].

4A.1.4 From functionalized thiophenes

Clemmensen [44–46] and especially Wolff–Kishner [47–50] reduction of formyl- and acylthiophenes have extensively been used for the preparation of alkyl thiophenes in high yields. Also semicarbazones have sucessfully been reduced by the Wolff–Kishner procedure [51–55]. Sterically hindered ketones which do not form hydrazones cannot be reduced through Wolff–Kishner reduction [56]. In such cases and in cases where easily reducible halogens are present, the Clemmensen reduction is to be preferred [56,57].

4A.1.5 Cross-coupling of halothiophenes with metalorganic reagents

The reaction of butylmagnesium bromide with 3,4-dibromothiophene using [1,3-bis(diphenylphosphino)propane]nickel(II) chloride is an excellent method for the preparation of 3,4-dibutylthiophene [58,59].

3,4-Dibutylthiophene [59]

A solution of 2 *M* butylmagnesium chloride in diethyl ether (165 ml, 0.33 mol) is added dropwise to a solution of 3,4-dibromothiophene (34 g, 0.14 mol) and [1,3-bis(diphenylphosphino)propane]nickel(II) chloride (0.5 g) in anhydrous diethyl ether (400 ml) at 0 °C. The solution is warmed slowly to 50 °C and held at this temperature for 16 h. Water (50 ml), containing a few drops of concentrated hydrochloric acid, is then added dropwise to the reaction mixture with ice-bath cooling. After filtration to remove solid impurities the filtrate is extracted with ether (3 × 100 ml). The combined ether phases are washed with sodium chloride solution and dried over magnesium sulfate. Upon distillation 27.7 g (92%) of the title compound is obtained, bp 92–94 °C/1 mm Hg.

Hexylmagnesium bromide and the Grignard reagent from 1-bromo-undec-10-ene give upon [1,3-bis(diphenylphosphino)propane]dichloro-nickel(II) catalyzed coupling with 3-bromothiophene, 3-hexylthiophene, and 3-(undec-10-enyl)thiophene in good yields [60].

3-(Undec-10-enyl)thiophene [60]

A solution of 1-bromoundec-10-ene (11.1 g, 61 mmol) in anhydrous diethyl ether (15 ml) is added dropwise to a stirred suspension of magnesium turnings

(1.49 g, 61 mmol) in anhydrous diethyl ether (5 ml) and stirred for 1 h at room temperature. The gray solution is transferred *via* a cannula to a dropping funnel and is added dropwise to a suspension of [1,3-bis(di-phenylphosphino)-propane]dichloronickel(II) (0.11 g, 0.45 mol%) and 3-bromothiophene (7.3 g, 45 mmol) in anhydrous diethyl ether (15 ml) at such a rate that gentle reflux is maintained. The reaction mixture is then stirred for three days at room temperature and carefully quenched with 1 *M* hydrochloric acid (20 ml). The orange organic phase is washed with water (twice) and sodium chloride solution, dried and evaporated. The orange liquid residue is distilled (Kugelrohr, 100–130 °C /0.1 mm Hg) giving 5.81 (55%) of a mixture of the title compound (96 mol%) and 3-(undec-9-enyl)thiophene (4 mol%) as a colorless liquid.

4A.2 FLUOROALKYLTHIOPHENES

4A.2.1 Trifluoromethylthiophenes

2-Trifluoromethylthiophene is prepared in 72% yield by the reaction of thiophene with bis(trifluoroacetyl) peroxide in Freon 113 at 60 °C [61]. Photochemical reaction of thiophene with trifluoromethyl iodide, ditrifluoromethyl telluride or di(trifluoromethyl)mercuride also yields predominantly 2-trifluoromethylthiophene together with minor amounts of the 3-isomer [62]. The 2-isomer can also be prepared in two steps through reaction of 2,5-thiophenedicarboxylic acid with sulfur tetrafluoride and anhydrous hydrogen fluoride followed by decarboxylation of the 5-trifluoromethyl-2-thiophene-carboxylic acid with copper in quinoline [63]. This approach is also used for the preparation of 2,5-bis(trifluoromethyl)thiophene [63].

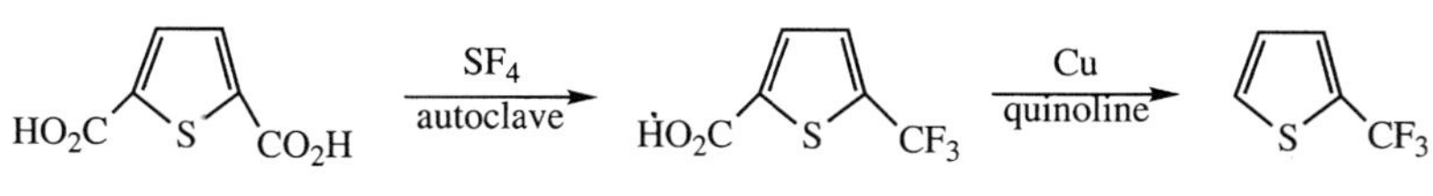

The reaction of trifluoromethylcopper with tetraiodothiophene is used for the preparation of tetrakis(trifluoromethyl)thiophene [64].

4A.2.2 Other perfluoroalkylthiophenes

Thiophene can be fluoroalkylated in moderate yield with polyfluoroalkyl iodides in the presence of catalytic amounts of tetrakis(triphenylphosphine)-nickel [65]. The reaction of various bromothiophenes with perfluorohexyl iodide and copper in dimethylsulfoxide provides a convenient method for the preparation of perfluoroalkylthiophenes [66].

4A.3 ARALKYLTHIOPHENES

4A.3.1 Reduction of diaryl ketones

An often used method for the preparation of diarylmethane derivatives is the reduction of thiophene ketones, easily prepared by electrophilic acylation, Thus 2-benzyl-5-alkylthiophenes were prepared by benzoylation of 2-alkylthiophenes followed by Wolff–Kishner reduction. However, the yields are rather low. A much better method is reduction of the ketone with aluminium chloride/lithium aluminium hydride, which in most cases gives very good yields [38,67–69].

1,3,4,6-Tetraethyl-7H-cyclopenta[1,2-c:3,4-c′]dithiophene [67]

Lithium aluminium hydride (0.87 g, 23 mmol) is dissolved in a solution of aluminium chloride (6.15 g, 46 mmol) in anhydrous diethyl ether (50 ml). After the vigorous reaction has subsided, 1,3,4,6-tetraethyl-7*H*-cyclopenta[1,2-*c*:3,4-*c*′]dithiophene-7-one (4.0 g, 13 mmol) dissolved in anhydrous diethyl ether (10 ml) is added during 10 min, after which the reaction mixture is refluxed for 30 min. Excess lithium aluminium hydride is destroyed by addition of ethyl acetate and water. More ether is added, the phases are separated, and the organic phase washed with 2 *M* hydrochloric acid and water, dried, and evaporated giving 3.4 g (90%) of the title compound mp 63.0–63.5 °C after sublimation at 0.02 mm Hg and 128 °C.

4A.3.2 Reduction of diaryl carbinols

Another good and more general method for the preparation of aralkylthiophenes is the reduction of diarylcarbinols which are easily available through the reaction of thienyllithium derivatives with aryl aldehydes, ketones, or ester, or alternatively, through the reaction of thienylcarbonyl derivatives with various aromatic lithium compounds. Also, in this case, the best method consists in reduction with aluminium chloride/lithium alumnium hydride [38,70].

Alternatively a one-pot procedure, consisting in the reaction of a number of thiophene aldehydes and ketones with phenyl-, 4-methylphenyl-, and 4-methoxyphenylphenyllithium followed by lithium–ammonia/ammonium chloride reduction, gave the corresponding benzyl thiophenes mostly in very good yields [71].

2-(Phenylmethyl)thiophene [71]

To a vigorously stirred mixture of lithium–sodium alloy (67 mg, 9.7 mmol) in anhydrous ether (10 ml), under argon, a solution of bromobenzene (786 mg, 5.00 mmol) is slowly added (25 min). The mixture turn, cloudy gray and within 45 min after the addition is complete the alloy slivers are consumed and the mixture turns dark grey. After cooling the reaction mixture to −78 °C a solution of thiophene aldehyde (343 mg, 3.06 mmol) in anhydrous ether (10 ml) is slowly added (5 min). After 10 min the cooling bath is removed and the stirred mixture is allowed to return to room temperature (45 min). After adding an additional 25 ml of anhydrous ether and distilling, a 100 ml of anhydrous ammonia is quickly added into the mixture lithium wire (63 mg, 9.1 mmol, 4 pieces). After 2 min the dark blue–black color of the mixture is discharged by the addition (1 min) of ammonium chloride (1.2 g) and the ammonia is then allowed to evaporate. The residue is treated with ether (50 ml) and water (50 ml), the phases separated and the aqueous phase is extracted with ether (4 × 30 ml). The combined ether phases are dried over magnesium sulfate and evaporated giving 530 mg of a yellow oil, which is flash chromatographed and 392 mg (74%) of the title compound is then obtained as a colorless oil.

A third method consists in the reduction of thienyl arylcarbinols with sodium borohydride/trifluoracetic acid, giving in some cases good yields [72]. In this way 2-benzyl- and 3-benzylthiophene were obtained in 85 and 95% yield, respectively [72].

2-Benzo[b]thienyl-3-thienylmethane [72]

To a magnetically stirred mixture of 2-benzo[*b*]thienyl-3-thienylmethanol (0.34 g, 1.4 mmol) and sodium borohydride powder (0.26 g, 6.9 mmol) in anhydrous diethyl ether (25 ml) at 25 °C, trifluoroacetic acid (2 ml) is added over 10 min. The reaction mixture is stirred at 25 °C for 1 h and then poured into 10% aqueous sodium hydroxide solution (25 ml). After 20 min the phases are separated and the aqueous phase extracted with ether (2 × 50 ml). The combined ether phases are dried over sodium sulfate. Evaporation affords a pale, yellow solid, which is flash chromatographed using hexane as eluent giving 0.25 g (78%) of the title compound as a white solid mp 51–53 °C.

A method developed recently allows a large scale preparation of biarylmethanes *via* crude biarylmethanols, formed through the reaction of aryl Grignard reagents with carbonyl compounds, by reduction with iodotrimethylsilane. 2-Thiophenemagnesium bromide gave excellent yields of 2-(4-fluorobenzyl)-, 2-(4-nitrobenzyl)-, 2-(3-nitrobenzyl)thiophene, and 2-benzylthiophene [73].

2-(4′-Fluorobenzyl)thiophene [73]

A 50-l three-necked flask equipped with a mechanical stirrer, a thermometer probe, and a nitrogen inlet adapter is charged with sodium iodide (4189 g, 4.36 mol), acetonitrile (1500 ml), and chlorotrimethylsilane (3554 ml, 28 mol). The reaction mixture is stirred at room temperature for 15 min prior to cooling to 0 °C with an ice-water bath. A solution of 2-(4′-fluorophenyl)thienylcarbinol (1470 g, 91% pure, 6.42 mol) in acetonitrile (1500 ml) is added slowly *via* pressure-equalizing addition funnel over 3.5 h to maintain the reaction temperature below 10 °C. The reaction mixture is allowed to warm to room temperature overnight (12 h) and is recooled to 5 °C prior to work-up. A solution of sodium hydroxide (760 g in 5 l of distilled water) is added over 10 min, the internal temperature of the reaction is risen to 40 °C before recooling to room temperature. The stirring is continued for an additional 45 min to ensure complete neutralization, pH 7 in the aqueous layer, after which ethyl acetate (3500 ml) and a solution of sodium thiosulfate pentahydrate (3598 g, 14.50 mol) in distilled water (3500 ml) is added. The resulting pale-brown mixture is stirred for 1 h. The layers are separated and the organic layer is washed with a solution of sodium hydroxide (140 g)/sodium thiosulfate pentahydrate (553 g) in distilled water (4000 ml) followed by stirring with distilled water (3000 ml) containing triethylamine (6.5 ml) in order to facilitate removal of 4-fluorobenzyl iodide for 1 h prior to addition of sodium chloride (500 g). After a short mixing period the layers are separated and the organic layer dried over magnesium sulfate. Evaporation affords 1319 g (97%, 93% pure) of the title compound as a pale-brown oil bp 110–115 °C/2 mm Hg.

3,3′-Dibromo-2,2′-dithienylmethane, prepared by reduction of the ketone [74,75] gives, upon halogen–metal exchange with butyllithium followed by reaction with cuprous cyanide, a Gilman cuprate, which upon oxidation with molecular oxygen gives the 10-membered ring cyclophane [76].

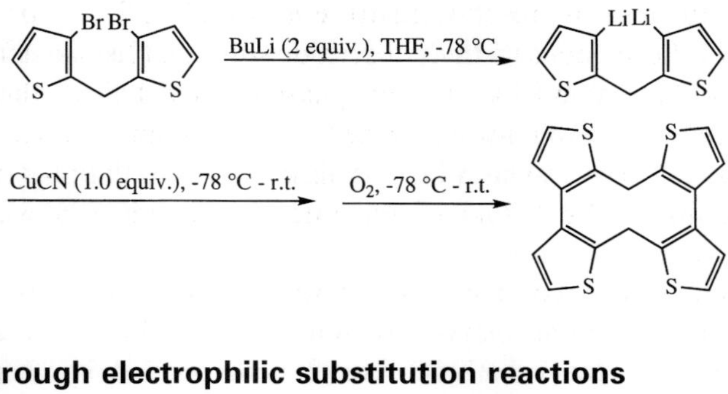

4A.3.3 Through electrophilic substitution reactions

The condensation of thiophene with ketones under the influence of 70% sulfuric acid leads to dithienylmethane derivatives. With acetone,

2,2-di-(2-thienyl)propane is obtained [77] and from 2-butanone 2,2-di(2-thienyl)butane was prepared [78].

Condensation of 2- and 3-thiophenealdehyde with azulenes in acetic acid gives bis(3-methyl-1-azulenyl)thienylmethanes, which upon treatment with 2,3-dichloro-5,6-dicyano-1,4-benzoquinone and hexafluorophosphoric acid give stable monocations [79].

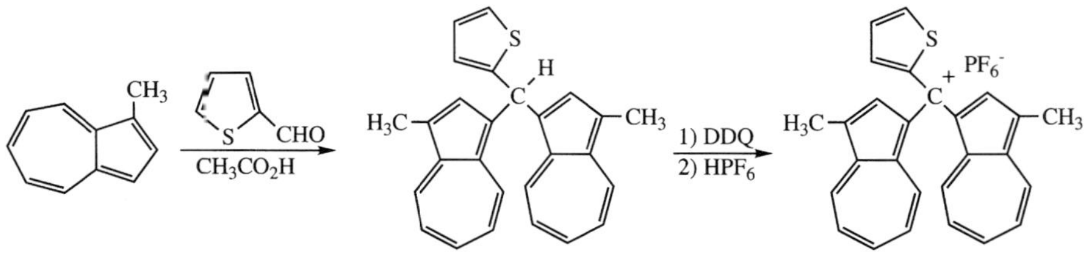

An excellent method for the preparation of 1,1-di(2-thienyl)ethane is the reaction of a mixture of thiophene and hydrochloric acid with acetetaldehyde acetal [80].

1,1-Di(2-thienyl)ethane [80]

A mixture of thiophene (900 ml, 11.3 mol) and concentrated hydrochloric acid is cooled in ice and stirred vigorously while acetaldehyde acetal (202 g, 1.7 mol) is added dropwise over a period of 1 h. The cooling bath is removed and the reaction mixture stirred at room temperature for 65 h, after which water (400 ml) is added. The phases are separated and the organic phase washed with water until the washings are neutral, followed by sodium chloride solution and dried. Evaporation and distillation gives 173.4 g (52%) of the title compound as a pale-green oil bp 101–110 °C/12 mm Hg.

The condensation of 2-methylthiophene with various aldehydes using a macroporous ion-exchange resin as catalyst gives excellent yields of alkyl- and aryl-substituted dithienylmethanes [81]. Tris(2-thienyl)methane can be prepared through the condensation of thiophene and 2-thiophene aldehyde in the proportions of 1:2, using a catalytic amount of 70% sulfuric acid. Alternatively catalytic amounts of sulfuric acid absorbed on silica gel have been used [82]. Phosphorus pentoxide in benzene can also be used for this preparation [83], but was inferior to the first-mentioned conditions in the preparation of tris(5-methyl-2-thienyl)methane and tris(2,5-dimethyl-3-thienyl)methane.

1,1-Bis(5-methyl-2-thienyl)alkanes can be prepared in excellent yield from the reaction of *N*-α-benzotriazolylalkyl)carbamate with an excess of 2-methylthiophene in methylene chloride in the presence of zinc chloride [84].

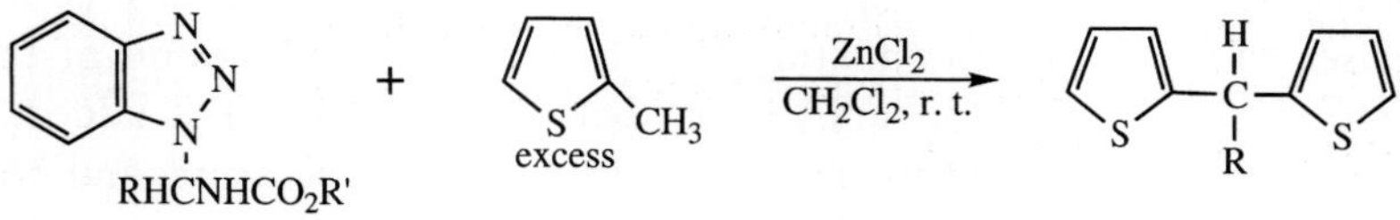

1,1-Bis(5-methyl-2-thienyl)-2-methylpropane [84]

A mixture of methyl *N*-(α-benzotriazolylisopropyl)carbamate (2.48 g, 10 mmol), 2-methylthiophene (2.0 g, 22 mmol), and zinc chloride (2.73 g, 20 mmol) in anhydrous methylene chloride (50 ml) is stirred overnight and poured into ice-water (50 ml). The phases are separated and the water phase extracted with chloroform (2 × 20 ml). The combined organic phases are washed with 2% sodium hydroxide solution (30 ml) and water (30 ml) and dried over magnesium sulfate (10 mg). After evaporation the residue is chromatographed on silica gel using methylene chloride as eluent giving 2.25 g (90%) of the title compound mp 45–46 °C.

Chloromethylation of thiophene in the presence of zinc chloride gives di(2-thienyl)methane as the main product [68,85]. Reaction of 2,5-dichloro-3-chloromethylthiophene with benzene and some alkylbenzenes in the presence of aluminium chloride has been used for an unexpected preparation of 4-aryl-3-arylmethyl-2-chlorothiophenes [86].

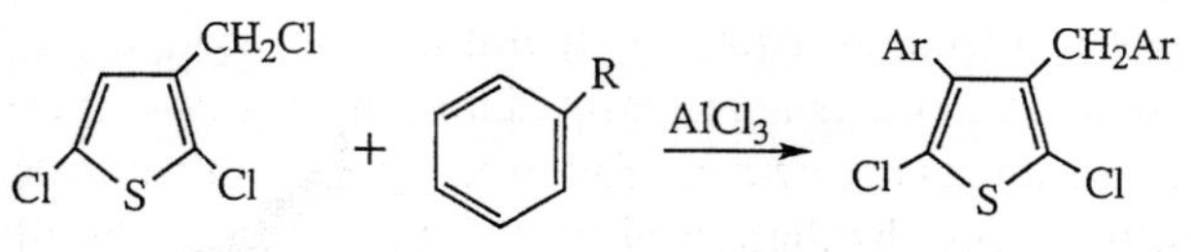

Cupric trifluoromethane sulfonate catalyzed reaction of thiophene with methyl vinylketone at room temperature gives the compound shown below [87].

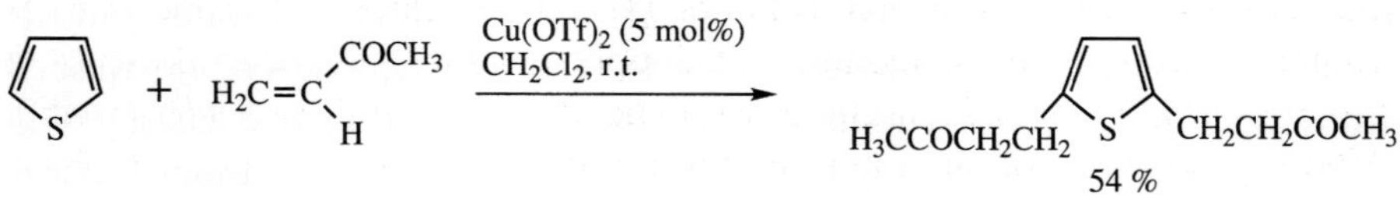

4A.3.4 From thienyllithium derivatives

Reaction of 2-thienyllithium and benzyl bromide is a good method for the preparation of 2-benzylthiophene [88].

2-Benzylthiophene [88]

To a solution of butyllithium in diethyl ether (0.3 mol) at 0 °C a solution of thiophene (25.2 g, 0.3 mol) in anhydrous diethyl ether (25 ml) is added dropwise. After cooling the mixture to −15 to −20 °C a solution of benzyl bromide (51.3 g, 0.3 mol) in anhydrous diethyl ether (50 ml) is added. The reaction mixture is refluxed for 18 h, cooled to room temperature, and poured into crushed ice. The phases are separated and the organic phase washed with water, dried, and evaporated giving 32.4 g (63%) of the title compound mp 132–134 °C.

4A.3.5 From bromothiophenes

[1,3-Bis(diphenylphosphino)propane]dichloronickel(II)-catalyzed coupling of bromothiophenes with benzylmagnesium chloride gives almost quantitative yields of benzylthiophenes [89].

4A.4 AMINOALKYLTHIOPHENES

4A.4.1 From halomethylthiophenes

The reaction of 2-thenyl chloride or bromide with hexamethylene tetramine followed by degradation with hydrogen chloride is an old but useful method for the preparation of 2-thenyl amine in high yield [90,91]. In order to obtain primary amines the route *via* the azide has been used as in the preparation of 1-(3-thienyl)ethyl amine from 1-(3-thienyl)chloroethane [92].

1-(3-Thienyl)ethyl amine [92]

A mixture of 1-(3-thienyl)chloroethane (5.0 g, 34 mmol) and sodium azide (2.65 g, 41 mmol) in *N,N*-dimethylformamide is stirred at room temperature for three days, after which the reaction mixture is poured into water and the product extracted with methylene chloride. The organic extracts are washed with water, dried over magnesium sulfate, and evaporated to give 5.02 g (96%) of 1-(3-thienyl)ethylazide as a brown oil which is used without further purification. A solution of this compound (5.0 g, 32.6 mmol) in anhydrous diethyl ether (80 ml) is added to a suspension of lithium aluminium hydride (1.7 g, 44.8 mmol) under nitrogen at such a rate that gentle reflux is maintained. The reaction mixture is heated to reflux for an additional hour, water (2 ml), 15% sodium hydroxide solution (2 ml), and water (6 ml) are used in a standard work-up to give 3.7 g (90%) of the title compound as an amber oil.

Secondary amines can be prepared by the reaction of the thienyl halides with alkylamines such as methyl amine [90,93], 2-thenyl amine [94], and benzyl amine [95]. 2-Dimethylaminomethylthiophene can conveniently be prepared from 2-chloromethylthiophene and dimethylamine [96]. In connection with preparation of compounds of medicinal interest, halomethylthiophenes have been reacted with other secondary amines such as the sodium salt of 5-oxoproline esters [97], L-methyl prolinate [98], 2-piperidinemethanol [99] (cf. below), and ethyl nipecotate [100] with potassium carbonate in acetonitrile.

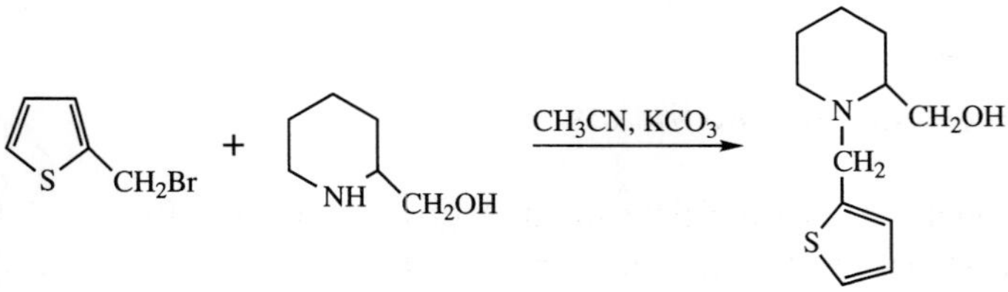

Potassium salts of pyrroles have been reacted with halomethylthiophenes to give 1-(thienylmethyl)pyrroles [101]. Quaternary ammonium salts have been obtained from the reaction of halomethylthiophenes with tertiary amines [95,102]. The products shown below can be prepared [103].

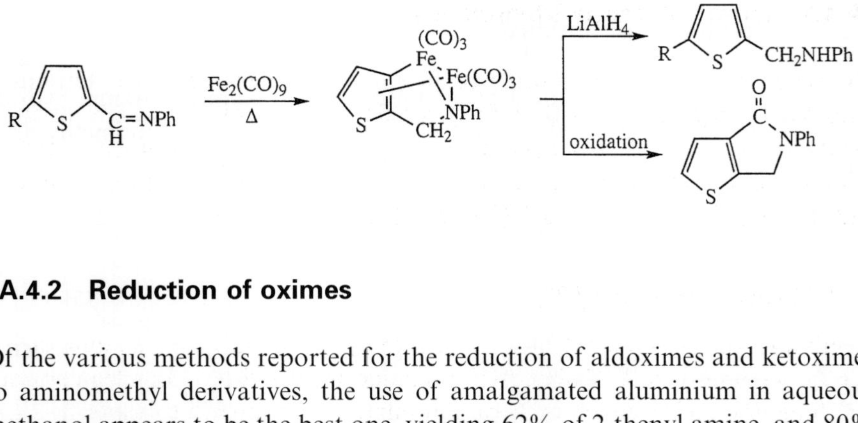

4A.4.2 Reduction of oximes

Of the various methods reported for the reduction of aldoximes and ketoximes to aminomethyl derivatives, the use of amalgamated aluminium in aqueous methanol appears to be the best one, yielding 62% of 2-thenyl amine, and 80% of 5-*tert*-butyl-2-thenylamine respectively [104]. Zinc dust in aqueous ammonia and ammonium acetate in alcohol has been used for the reduction of the oxime from 3-phenyl-2-thiophene aldehyde to 2-(aminomethyl)-3-phenylthiophene [105].

2-(Aminomethyl)-3-phenylthiophene [105]

3-Phenyl-2-thiophene aldehyde oxime (3.30 g, 16.2 mmol), zinc dust (8.49 g, 0.13 mmol), ammonium acetate (1.00 g, 13 mmol), aqueous ammonia (*d* 0.88;

97 ml), and ethanol (50 ml) are refluxed for 18 h, after which the reaction mixture is evaporated and the residue stirred with aqueous potassium hydroxide (33% w/v; 100 ml). Diethyl ether (30 ml) is added to the mixture, which is then filtered through a pad of Celite. The phases of the filtrate are separated, the organic phase dried, and evaporated giving 2.89 g (94%) of the title compound as a brown oil.

A highly 2-substituted 3-thiophenealdoxime was reduced in high yield to the aminomethyl derivative with lithium aluminium hydride in connection with syntheses of thiophene analogs of isoquinoline [106].

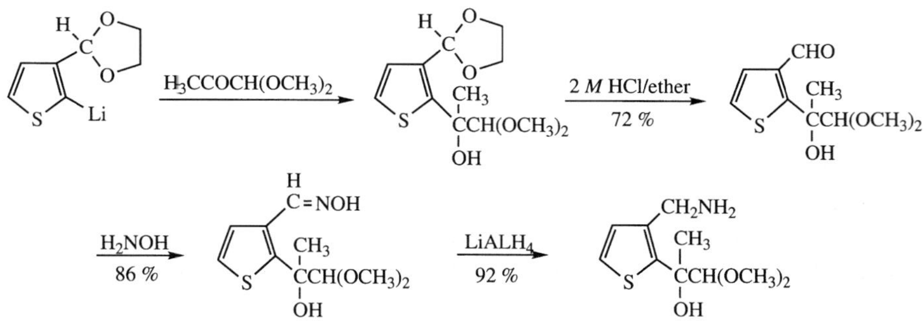

4A.4.3 Reduction of amides

Reduction of thiophenecarboxylic acid amides with lithium aluminium hydride is an excellent method for the preparation of *N,N*-dimethylaminomethylthiophenes. Amides prepared from thiophene acid chlorides and *N*-methyl(trimethylsilyl)methylamine are reduced by lithium aluminium hydride to *N,N*-dimethyl-*N*-(trimethylsilyl)methylthenyl)ammonium iodide [107].

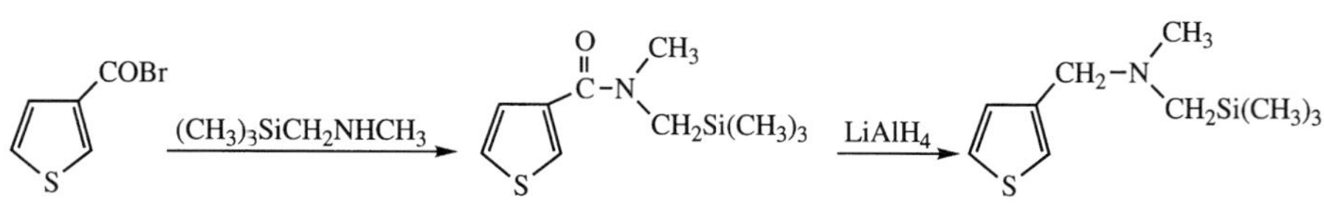

4A.4.4 From aldehydes and ketones

2-Thiophene aldehydes undergo the Leuckart reaction in the presence of formic acid with *N,N*-dimethylformamide [108], secondary aliphatic amines [109,110], or formamide [104] to give 2-dialkylaminomethylthiophene and 2-aminomethylthiophene, respectively. Heating a mixture of 2-thiophene aldehyde and piperidine with anhydrous formic acid yields 1-(2-thenyl)piperidine [111].

1-(2-Thenyl)piperidine [111]

A mixture prepared under cooling from 2-thiophene aldehyde (35 g, 0.31 mol), piperidine (54 g, 0.63 mol), and anhydrous formic acid (35 ml) is heated at 110 °C for 4 h. After cooling hydrochloric acid is added, the reaction mixture filtered, the phases are separated and the aqueous phase washed two times with ether It is then made alkaline with ammonia and the precipitate formed taken up in ether. The organic phase is washed with water, dried over sodium sulfate, and evaporated giving 46.5 g (83%) of the title compound as an oil.

2-Acetylthiophene gave with ammonium formate at 180–190 °C followed by treatment with 30% sodium hydroxide at 130 °C 1-(2-thienyl)ethylamine in 50–60% yield [90]. α-Amino-3-thienylacetic acid is best synthesized through the Strecker reaction of 3-thiophene aldehyde and has been resolved into its enantiomers through crystallization of the salts with *d*-camphor-10-sulfonic acid and α-bromo-*d*-camphor-π–sulfonic acid [112]. The reaction of 2- thiophene aldehyde and 3-thiophene aldehyde with malonic acid and 5% ammonia in ethanol or with ammonium acetate are good methods for the preparation of 3-amino-3-(2-thienyl)- and 3-amino-(3-thienyl)propionic acid giving yields of about 50% [113–115]. About 10% of the corresponding (thienyl)acrylic acids are obtained as byproducts under these conditions. These amino acids can be cyclized to 4-amino-4,5-dihydrocyclopenta[*b*]thiophene-6-one derivatives [115–119], 6-aminocyclopenta[*b*]thiophene derivatives [120] and 6-aminocyclopenta[*c*]thiophene-4-ones [121].

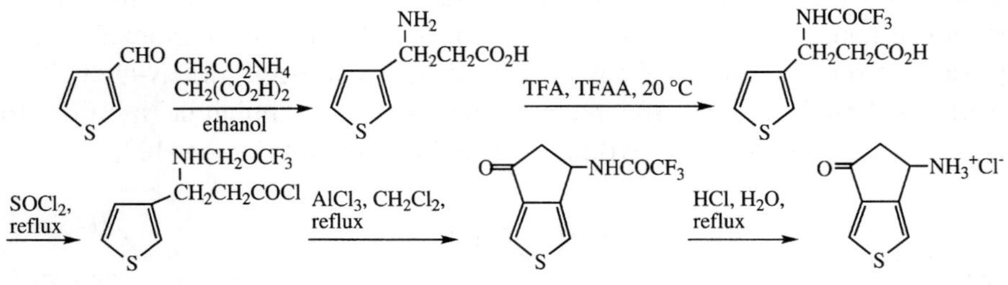

The condensation of 3-thiophenealdehyde with aminoacetaldehyde dimethyl acetal followed by hydrogenation over palladium on charcoal [122] or sodium borohydride [123] gives *N*-(2,2-dimethoxyethyl)-3-thenylamine. The reaction of the Schiff base from 2-thiophene aldehyde and aniline with ketene silylacetals, catalyzed by phosphonium salts gave the α-aminoester in 71% yield at −78 °C [124].

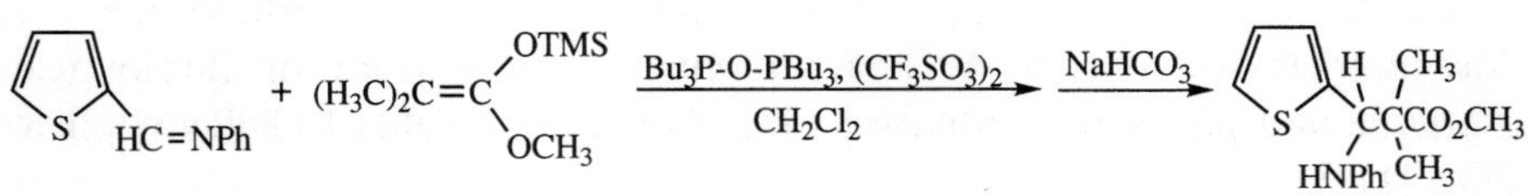

The azomethine derivative prepared through reaction of 2-thiophene aldehyde with an excess of 25% ammonium hydroxide gave di(2-thenyl)amine upon reduction with sodium borohydride in alcohol [125].

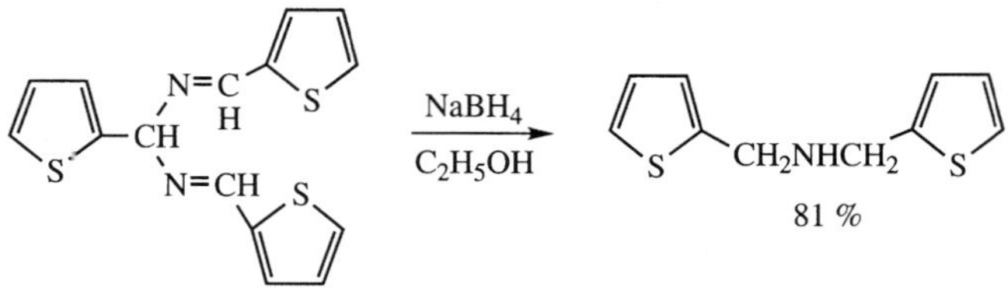

Also other Schiff bases between thiophene aldehydes and substitute anilines such as 4-ethoxyaniline have been reduced to thenylamines with sodium borohydride [126]. The reaction of the Schiff base between 2-thiophenealdehyde and 2-aminobenzonitrile gave 2-[di(2-thienyl)methylamino]benzonitrile upon reaction with 2-thienyllithium [127].

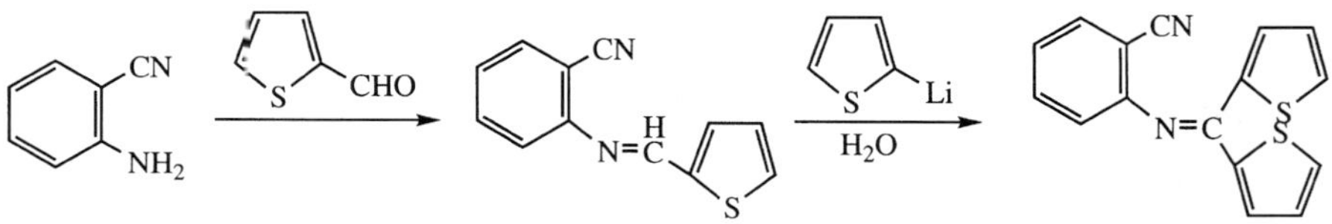

Cross cyclotetramerization of 4-tert-butylphtalonitrile and 3,4-dicyanothiophene in refluxing 2-(dimethylamino)ethanol gives the porphyrazine [128].

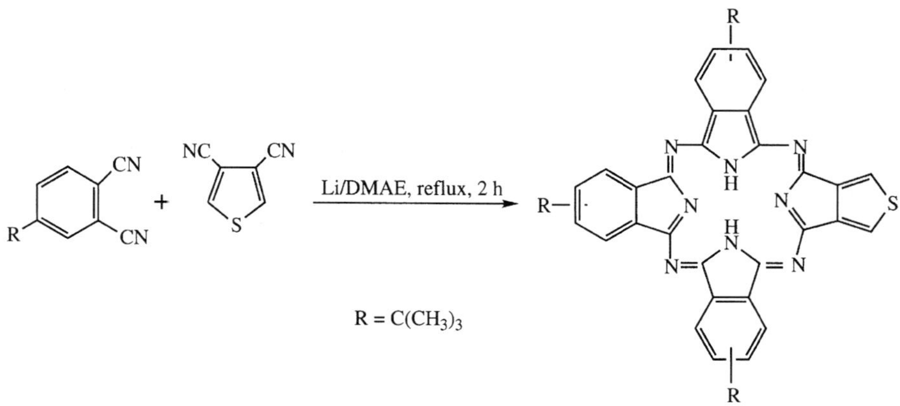

4A.4.5 Aminomethylation of thiophenes

The reaction of thiophene, 2-alkylthiophenes, and 2-halothiophenes with ammonium chloride and formaldehyde developed by Hartough and coworkers [93,129,130] in the 1940s, is an inexpensive method for the preparation of aminomethylthiophenes, although yields are not very high. The reaction mechanism has not been elucidated in detail. It is assumed that the Schiff base *N*-(2-thenylformaldimine), existing as a cyclic trimer [131], which upon protonation gives the electrophilic aminomethylating reagent attacking the thiophenes. Hartough has described a detailed procedure for the preparation of 2-aminomethylthiophene on a large scale in 44–59% yield [132]. This method has not been used very much during recent years, but certainly is one of the best and most inexpensive methods for the preparation of 2-thenyl amines. The Mannich reaction with primary amines is a convenient method for the preparation of 3-acylaminomethyl derivatives of 2-acylaminothiophenes [133].

4A.4.6 Aminomethylation of thienylmetal derivatives

A new method for the introduction of a primary aminomethyl group consists in lithiation of a thiophene, transmetalation to an organozinc reagent, and treatment with 1-(triphenylphosphoroylideneaminomethyl)benzotriazole [134].

2-Aminomethyl-5-methylthiophene [134]

To a solution of 2-methylthiophene (490 mg, 5 mmol) and *N,N,N′,N′*-tetramethylethylenediamine (0.83 ml, 5.5 mmol) in hexane (5 ml) is added 2 *M* butyllithium (2.75 ml, 5.5 mmol) at −78 °C. After stirring for 10 min, the reaction mixture is refluxed for 0.5 h. The solution is cooled to room temperature, anhydrous zinc bromide (0.73 g, 5.5 mmol) is added, and the mixture stirred at room temperature for 1 h. A solution of 1-(triphenylphosphoroylideneaminomethyl)benzotriazole 2.45 g, 6 mmol) in tetrahydrofuran (30 ml) is added and the reaction mixture stirred at 20 °C for 48 h. The mixture is worked up with aqueous ammonium solution, extracted with diethyl ether (3 × 50 ml), washed with water (3 × 30 ml), dried over magnesium sulfate, and the solvent evaporated. The amine is purified by column chromatography on silica gel using ethyl acetate as eluent giving 241 mg (38%) of the title compound as an oil.

The reaction of 2-thiophenemagnesium bromide with *N,N*-bis(trimethylsilyl)methylthiomethylamine is an excellent method for the preparation of *N,N*-di-(trimethylsilyl)-2-aminomethylthiophene in 95% yield, with organolithium reagents the aminomethylation reaction is unsuccessful [135]. 3-Thienyl trimethylstannane upon reaction with Eschenmosers salts give the *N,N*-dialkylaminomethyl derivatives in reasonable yields [136].

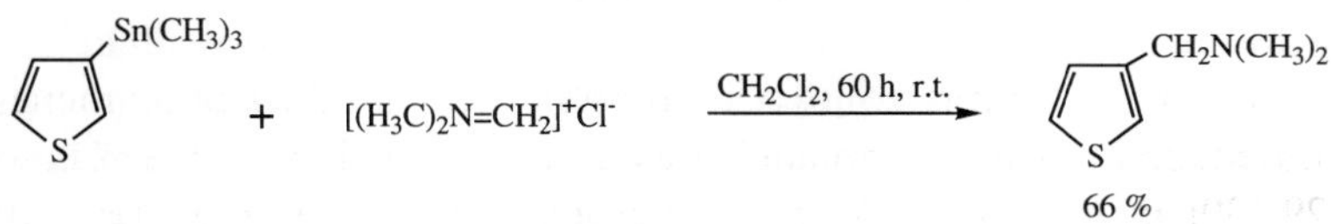

The reaction of activated arylsulfonylaziridines gives sulfonamides, which upon reductive cleavage with sodium bis(2-methoxyethoxy)aluminium hydride yields 2-(2-aminoethyl)thiophene [137].

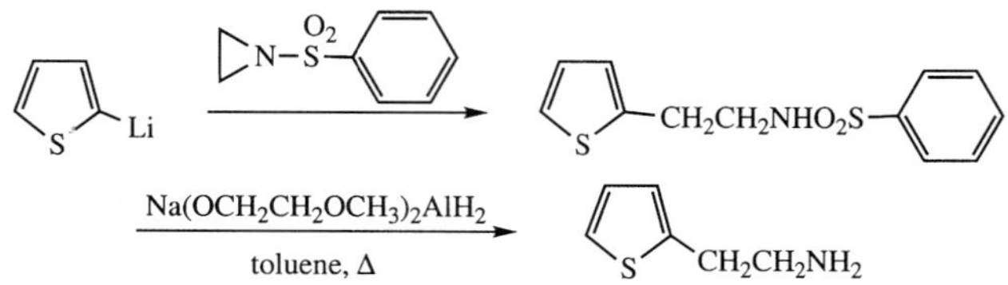

2-(2-Aminoethyl)thiophene [137]

To a solution of *N*-[2-(2-thienyl)ethyl]benzsulfonamide (10.6 g, 0.04 mol) in toluene (50 ml) sodium bis(2-methoxyethoxy)aluminiumdihydride (70% in toluene) (46.2 g, 0.16 mol) is slowly added dropwise. The reaction mixture is refluxed for 72 h and then neutralized with 10% aqueous sodium hydroxide solution (100 ml). The phases are separated and the water phase extracted three times with diethyl ether. The combined organic phases are washed with sodium chloride solution and then treated with an oxalic acid solution (125 ml). The phases are separated, the aqueous phase washed with diethyl ether and then neutralized with concentrated sodium hydroxide solution. The solid material is filtered off and the filtrate extracted three times with 50 ml portions of diethyl ether. The combined organic phases are dried over sodium sulfate, evaporated at 30 °C/14 mm Hg and distilled giving 1.4 g (25%) of the title compound bp 24–27 °C/0.001 mm Hg.

4A.4.7 Ring-closure reactions

The reaction of diacetylenic amines with sodium hydrogen sulfides is used for the preparation of 2,5-bis(dimethylaminomethylene)thiophene [138].

2,5-Bis(dimethylaminomethylene)thiophene [138]

To a solution of the diacetylene (4.15 g, 25.3 mmol) in ethanol/water (1:1) 100 ml sodium hydrosulfide (2.9 g) is added. The reaction mixture is stirred at 100 °C for 15 h, after which another portion of sodium hydrosulfide (1.4 g) is added and the heating is continued for 14 h. The product is extracted with diethyl ether, the combined organic phases are dried over potassium carbonate and distilled giving 2.00 g (44%) of the title compound bp 96–97 °C/3 mm Hg.

4A.5 AZIDOMETHYLTHIOPHENES

α-Azido-3-thienylacetic acid can be prepared in 74% yield through the reaction of α-bromo-3-thienylacetic acid with sodium azide in acetone [112]. The reaction fails with the 2-isomer.

4A.5.1 Other thenyl-nitrogen derivatives

Thenyl *O*-alkyl hydroxylamines can be prepared by the addition of thienyllithium derivatives to *O*-alkylaldoximes in the presence of boron trifluoride etherate [139].

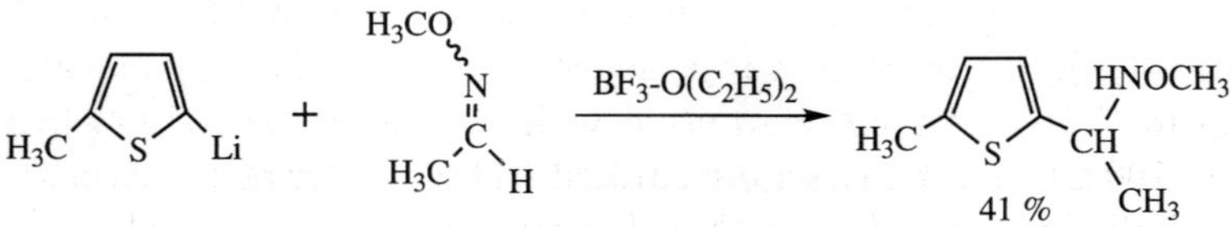

2-Thenylisocyanate is prepared by the reaction of 2-thiopheneacetyl chloride with trimethylsilyl azide. Alternatively, as in the preparation of 3-thenylisocyanate, 3-thenylacetic acid in acetonitrile is reacted with triethylamine followed by diphenylphosphoryl azide [140].

3-Thienylmethyl isocyanate [140]

To a solution of 3-thienylacetic acid (2.0 g, 14.1 mmol) in anhydrous acetonitrile (15 ml) triethylamine (2.1 ml, 15.0 mmol) is slowly added, followed by slow addition of diphenylphosphoryl azide 3.24 ml, 15.0 mmol). The deep-green reaction mixture is heated at 55 °C for 2 h and evaporated. The residue is dissolved in dichloromethane and washed two times with hexane, dried *in vacuo* at 35 °C giving 1.43 g (73%) of the title compound bp 67–88.5 °C/0.04 mm Hg.

4A.6 FLUOROMETHYLTHIOPHENES

Tetra(fluoromethyl)thiophene is prepared from the diene either with potassium sulfide or thiourea [141].

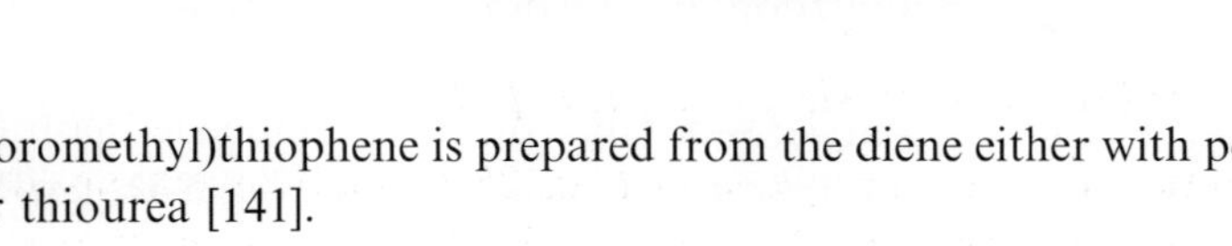

Tetra(fluoromethyl)thiophene [141]

A mixture of the diene (2.0 g, 5.5 mmol), thiourea (0.5 g, 6.6 mmol), and acetonitrile (20 ml) is heated under reflux for 16 h. Volatile material is transferred *in vacuo* to a cold trap and water (30 ml) is added to it. The lower phase is separated and purified by distillation giving 1.0 g (51%) of the title compound bp 134–135 °C.

2-(1,1-Difluoroalkyl)thiophenes are best prepared in high yields by fluorodesulfurization of the corresponding 1,3-dithiolanes using nitrosyl tetrafluoroborate/pyridinium polyhydrogen fluoride [142].

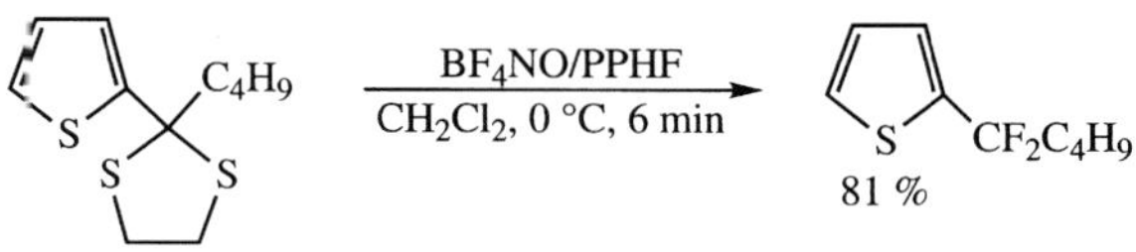

4A.7 CHLOROMETHYLTHIOPHENES

4A.7.1 General

Haloalkylthiophenes especially the chloromethyl and bromomethyl derivatives are very useful for the preparation of numerous thiophenes through nucleophilic substitution. However, most of them are highly lachrimatory and quite unstable. They decompose to polymers with evolution of hydrogen chloride or hydrogen bromide and should not be stored in closed flasks. They are best used immediately for the next step in one-pot procedures and without purification.

4A.7.2 Electrophilic halomethylation

Until recently direct chloromethylation, using formalin and hydrogen chloride [143–146] or chloromethyl ether in acetic acid [147,148] has been used for the chloromethylation of thiophene and 2-alkyl and 2-halothiophenes. A detailed study of the chloromethylation of 2,5-dimethylthiophene and 2,5-di-*tert*-butylthiophene with formalin and hydrochloric acid was carried out already in 1956 [149]. 2,5-Di-*tert*-butyl-3,4-bis(chloromethyl)thiophene has also been obtained by reduction and successive chlorination of 2,5-Di-*tert*-butyl-3,4-bis(methoxycarbonyl)thiophene [150].

Bis(chloro)methylation can also be carried out. Thus 2,5-dichloromethylthiophene and 3,4-dichloromethyl-2,5-dimethylthiophene were prepared in 59 and 52% yield respectively, from thiophene and 2,5-dimethylthiophene and excess chloromethyl methyl ether in phosphoric acid/acetic acid [151].

4A.7.3 From hydroxymethyl derivatives

The carcinogenic properties of the chloromethylation reagents have, in recent years, led to alternative methods for the preparation of chloroalkyl derivatives. Thus treatment of thenyl alcohols obtained upon reduction of thiophene aldehydes with sodium borohydride, with thionyl chloride in the presence of a base such as triethyl amine [152–154] or pyridine has been used for the preparation of such compounds [155]. 1-(3-Thienyl)ethanol was converted to 1-(3-thienyl)chloroethane upon treatment with thionyl chloride [92]. Alternatively phosphorus trichloride [156], phosphorus pentachloride, and zinc chloride [157] have been used as chlorinating agents.

1-(3-Thienyl)chloroethane [92]

Thionyl chloride (95 ml, 0.819 mol) is added to 1-(3-thienyl)ethanol (100 g, 0.780 mol) at 0 °C over 1 h period and the mixture is stirred at room temperature for 2.5 h and ice-water is added to the mixture. Extracting with diethyl ether and subsequent treatment of the combined organic phases with magnesium sulfate and charcoal and removal of the solvent *in vacuo* at 40 °C gives 99.6 g (87%) of the title compound as an amber oil.

4A.7.4 From tetramethylthiophene

Reaction of tetramethylthiophene with one equivalent of *N*-chlorosuccinimide in methylene chloride/acetic acid at −18 °C yields selectively 2-chloromethyl-3,4,5-trimethyl thiophene, while the reaction with two equivalents of sulfuryl chloride in methylene chloride at −18 °C gives 61% yield of 2,5-dichloromethyl-3,4-dimethylthiophene [158].

4A.8 BROMOMETHYLTHIOPHENES

4A.8.1 From alkylthiophenes

Side chain bromination with *N*-bromosuccinimide of especially 3-methylthiophene is synthetically very useful. Such reactions are most probably radicaloid and are favored by the presence of radical initiators such as dibenzoyl peroxide and azobisisobutyronitrile and disfavored in their absence relative to ring substitution. However, cases have been found, where ring substitution predominates also in the presence of radical initiators [50,159] and where side chain bromination occurs in their absence [160–162]. This discrepancy, which often resulted in a lack of reproducibility [161,162] is supposed to arise from adventitious catalysts for either reaction present as impurities in the *N*-bromosuccinimide or the solvent [163,164]. Side chain bromination is favored by using freshly prepared *N*-bromosuccinimide [165,166] and non-polar solvents such as carbon tetrachloride or benzene.

2-Bromomethylthiophenes have been prepared by the reaction of 2-methylthiophenes with *N*-bromosuccinimide in the presence of benzoyl peroxide. Best yields and less by-products, due to nuclear bromination, are obtained with 2-methylthiophenes containing electron-withdrawing groups [168,169]. The side chain bromination of 3-methylthiophene to 3-thenyl bromide using either benzoyl peroxide or azobisisobutyronitrile as initiator is one of the main synthetic routes to 3-substituted thiophenes, but special precautions have to be observed in order to avoid bromination in the 2-position [166,170–175].

Side chain bromination with *N*-bromosuccinimide can also be obtained in the presence of double bonds, as diethyl [(3-methyl-2-thienyl)methylene]propanedioate gives the bromomethyl derivative [176].

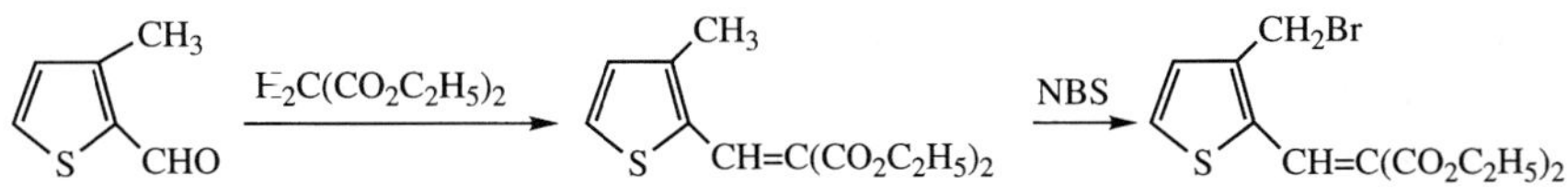

Diethyl [[3-(bromomethyl)-2-thienyl]methylene]propanedioate [176]

To a solution of diethyl [[(3-methyl)-2-thienyl]methylene]propanedioate (2.68 g, 10 mmol), in carbon tetrachloride (100 ml) *N*-bromosuccinimide (2.12 g, 12 mmol) and dibenzoyl peroxide (0.05 g) are added. The reaction mixture is, under stirring, refluxed for 5 h, after which it is cooled in an ice bath and filtrated. The solid is washed with carbon tetrachloride and the combined filtrates evaporated. The residue is purified by flash chromatography on silica gel using hexane/ethyl acetate (6:1) giving 2.96 g (85%) of the title compound as white crystals mp 58–59 °C.

Bromomethylthiophenes have also been obtained by brominating alkylthiophenes in the presence of azobisisobutyronitrile and light [177]. Reaction of tetramethylthiophene with one equivalent of bromine in methylene chloride at −18 °C gives 2-bromomethyl-3,4,5-trimethylthiophene in 76% yield and with two equivalents of bromine a quantitative yield of 2,5-dibromomethyl-3,4-dimethylthiophene was obtained [158].

4A.8.2 From hydroxymethyl derivatives

Hydroxymethylthiophenes have been converted in almost quantitative yields into the corresponding bromomethyl derivatives by reaction with phosphorus tribromide in ether [95,178] or through the reaction with triphenyl phosphine and carbon tetrabromide in tetrahydrofuran [179]. This approach was used for the preparation of 2,3-di(bromomethyl)thiophene from 2,3-dihydroxymethylthiophene obtained from 2,3-thiophenedicarboxylic acid through reduction with lithium aluminium hydride [179,180] and for 2,4-di(bromomethyl)thiophene from 2,4-dihydroxymethylthiophene obtained from 2,4-diformylthiophene through reduction with sodium borohydride [181].

2,3-Di(bromomethyl)thiophene [179]

To a solution of 2,3-di(hydroxymethyl)thiophene (0.60 g, 4.16 mmol) and triphenylphosphine (3.3 g, 12.5 mmol) in tetrahydrofuran (12 ml) at 5 °C under argon is added carbon tetrabromide (3.0 g, 9.15 mmol). The reaction mixture is allowed to warm to 25 °C where it is stirred for 1 h, after which it is diluted with diethyl ether (24 ml) to precipitate triphenylphosphine oxide, which is removed by filtration. The filtrate is concentrated under reduced pressure to afford a crude solid which is purified by column chromatography to provide 0.97 g (87%) of the title compound mp 47.5–49.0 °C.

The reaction of hydroxymethylthiophenes with hydrobromic acid has also been used for the preparation of bromomethylthiophenes [182]. Thus α-bromo-3-thienylacetic acid is best obtained through the reaction of 3-thienylglycolic acid with 30% hydrobromic acid in glacial acetic acid [112].

α-Bromo-3-thienylacetic acid [112]

To a solution of 3-thienylglycolic acid (13.5 g, 0.085 mol) in acetic acid (50 ml) is added 30% hydrobromic acid in acetic acid (150 ml). The reaction mixture is allowed to stand at room temperature for 24 h and is then poured into ice-water and extracted with ether. The combined ether phases are treated with magnesium sulfate and charcoal, after which the ether and acetic acid is removed *in vacuo*, giving 14.1 g (75%) of the title compound after recrystallization from ligroin mp 91.5–92.0 °C.

A convenient one-pot method for the preparation of aralkyl bromides, consisting in quenching of the complex formed upon reduction of aromatic aldehydes, ketones, and carboxylic acids with lithium aluminium hydride with hydrogen bromide, has been used for the preparation of 2,5-bis(bromomethyl)thiophene from 2,5-diformylthiophene in almost quantitative yield. However, this compound is unstable and has to be used within two days [183].

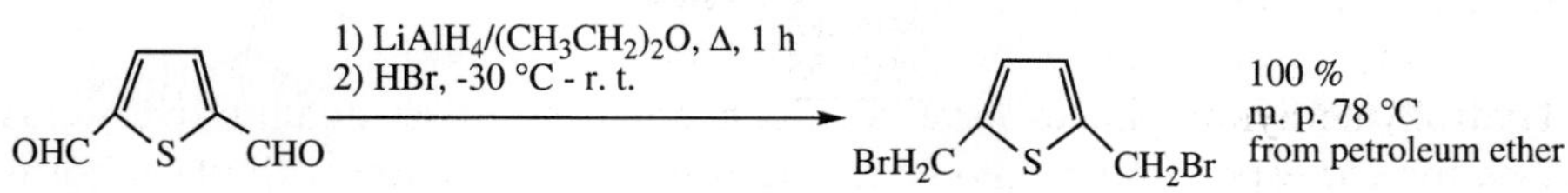

4A.9 IODOMETHYLTHIOPHENES

Due to their sensitivity to air and light very little is known about this type of compounds. 2-Thenyl iodide [108], 4,5-dichloro-2-thenyl iodide [184], and 3,4-bis(iodomethyl)-2,5-dichlorothiophene [155] have been prepared from the corresponding thenyl chloride and sodium iodide in anhydrous acetone.

4A.10 CYANOMETHYLTHIOPHENES

4A.10.1 From halomethylthiophenes

The most commonly used approach for the preparation of thiopheneacetonitriles is the reaction of chloromethyl- or bromomethylthiophenes with sodium or potassium cyanide under various reaction conditions [143,147,153, 185–189]. 2,3,5-Trimethyl-4-(cyanomethyl)thiophene can be prepared from the chloromethyl derivative and sodium cyanide under phase transfer conditions [190].

2,3,5-Trimethyl-4-cyanomethylthiophene [190]

Sodium cyanide (28 g, 0.48 mol) and triethylbenzylammonium chloride (1.0 g, 4.8 mmol) are dissolved in water (30 ml) at 30 °C. To this solution 2,3,5-trimethyl-4-(chloromethyl)thiophene (57 g, 0.32 mol) is added and the reaction mixture is refluxed for 1 h. After being cooled to room temperature the reaction mixture is poured into water and the product is then extracted with ether. Removal of the ether and distillation of the oily residue gives 45 g (85%) of the title compound bp 78 °C/2 mm Hg.

4A.10.2 Various methods

In some cases reaction between thenyl halides and cyanide fails, as in an attempted synthesis of 5-nitro-2-thenylcyanide. 2-Thenyl cyanide has been prepared through the nickel-catalyzed coupling of 2-thienylzinc chloride with bromoacetonitrile [191].

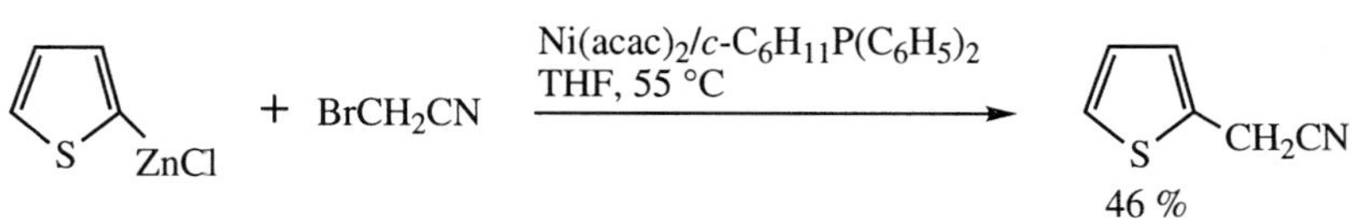

4A.11 HYDROXYMETHYLTHIOPHENES

4A.11.1 Reaction of halomethylthiophenes with oxygen nucleophiles

Resinification often occurs upon hydrolysis of thenyl chlorides with sodium hydroxide or acetates. Therefore thenyl alcohols are best prepared through the reaction of halomethyl thiophenes with sodium or potassium acetate followed by alkaline [182,192] or acidic hydrolysis [157] of the acetates. It has been suggested that hydroxymethylthiophenes are best prepared by reaction of chloromethylthiophenes with sodium formate followed by hydrolysis with hydrochloric acid [193].

4A.11.2 Reduction of thiophene aldehydes and ketones

A very useful method for the preparation of thenyl alcohols is the reduction of the corresponding aldehydes with sodium borohydride [37,152,153, 181,194].

2-Thiophenemethanol [153]

Sodium borohydride (20 g, 0.53 mol) is dissolved in ice-water (600 ml). With vigorous stirring 2-thiophene aldehyde (112.2 g, 1.00 mol) is added dropwise at such a rate that the temperature is kept between 15 and 20 °C by using an external ice-water bath. When the addition is completed the resulting emulsion is stirred at room temperature for 90 min. Excess sodium borohydride is destroyed by adding acetic acid until pH *ca.* 5, after stirring for 5 min the reaction mixture is neutralized with sodium hydrogen carbonate followed by extraction with ether (3 × 150 ml). The combined ether phases are washed with water, dried over magnesium sulfate, and evaporated. The residue consists of 110.2 g (96%) of the title compound with the same IR and GLC data as a distilled sample 89–99 °C/14 mm Hg.

Also lithium aluminium hydride [178,195], dibutyl tin dihydride [196], thiourea dioxide [197], and iron and nickel chloride in acetic acid [192] can be used for the reduction. However, the Cannizzaro reaction of 2-thiophene aldehyde using sodium hydroxide in methanol in the presence of formaldehyde at 65 °C, is still useful and yields 2-hydroxymethylthiophene in 59% yield [198]. Refluxing 2- and 3-thienylglyoxal hydrate with 1 *M* sodium hydroxide solution for 10 min is the best method for the preparation of 2- and 3-thienylglycolic acids [199]. These acids have been resolved into their optical enantiomers and their steric relation to mandelic acid was determined [200,201].

2-Thienylglycolic acid [199]

2-Thienylglyoxal hydrate (48 g, 0.3 mol) is dissolved in 2 *M* sodium hydroxide solution (1000 ml) and the reaction mixture is then refluxed for 10 min. After cooling, the alkaline solution is extracted twice in order to remove possible nonacidic components followed by acidification in the cold with 6 *M* hydrochloric acid saturated with sodium chloride and then extracted with ether (10 × 150 ml). The combined ether phases are treated with charcoal, dried over magnesium sulfate and evaporated. The crystalline residue is recrystallized from benzene giving 44 g (92%) of the title compound as colorless prisms mp 81–82 °C.

Also thienyl β-ketosulfoxides can be reduced to the β-hydroxysulfoxides with sodium borohydride in methanol [202]. The Meerwein–Verley–Ponndorf reaction of alkyl or aryl thienyl ketones is still a good method for the preparation of alkyl thienyl carbinols [203–205] and aryl thienyl carbinols [206]. Ruthenium diamine complexes shown below catalyze the asymmetric hydrogenation of a wide range of aromatic and heterocyclic ketones among them are 2- and 3-acetylthiophene with very high yields [207,208].

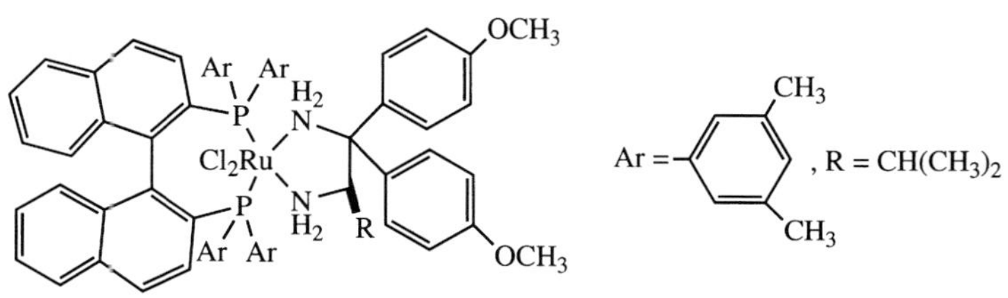

4A.11.3 Reduction of thiophenecarboxylic acids and esters

2,3-Di(hydroxymethyl)thiophene is prepared in good yield by reduction of 2,3-thiophenedicarboxylic acid with lithium aluminiumhydride in tetrahydrofuran [179,180].

2,3-Di(hydroxymethyl)thiophene [179]

To a slurry of lithium aluminiumhydride (1.1 g, 29 mmol) in tetrahydrofuran (30 ml) under argon at 5 °C 2,3-thiophenedicarboxylic acid (1.0 g, 5.5 mmol) is added. The reaction mixture is allowed to warm to 25 °C and stirred for 6 h, after which it is cooled to 5 °C, quenched with saturated ammonium chloride solution (30 ml), acidified with 2 *M* hydrochloric acid (40 ml) to dissolve the aluminium salts and extracted with dichloromethne (3 × 50 ml). The combined organic phases are dried over sodium sulfate and evaporated. The residual oil is purified by column chromatography using dichloromethane/methanol (98:2–95:5) affording 0.66 g (79%) of the title compound as an oil.

4A.11.4 Reaction of thiophene aldehydes and ketones with metalorganic reagents

The addition of aliphatic or aromatic Grignard reagents or lithium derivatives to thiophene aldehydes or ketones is in most cases a very useful method for the preparation of thienyl carbinols [209–221].

1-Methyl-2-[(1-hydroxy-1-thien-3-yl)methyl]-3-[(benzotriazol-1-yl)-methyl]indole [221]

To a solution of 1-methyl-2-bromo-3-[(benzotriazol-1-yl)methyl]indole (1.2 g, 3.5 mmol) in tetrahydrofuran (40 ml) at −78 °C under argon is added 1.7 *M* *tert*-butyllithium in pentane (4.2 mmol). After 5 min, 3-thiophene aldehyde (0.47 g, 4.2 mmol) in tetrahydrofuran (8 ml) is added dropwise. The reaction mixture is stirred at −78 °C for 2 h, after which the reaction is quenched with water. The phases are separated and the organic phase washed with water, dried over magnesium sulfate and evaporated. The residue is subjected to column chromatography using hexane/ethyl acetate (3:1) as eluent giving 1.2 g (91%) of the title compound mp 78–79 °C.

In this way 4-pyrazolylthienyl carbinols have been prepared [222]. Thienylcarbinols have been resolved into optically active forms by kinetic resolution using the Sharpless reagent [221].

(R)-1-(2-Thienyl)pentan-1-ol [221]

To a solution of titanium tetraisopropoxide (1.87 g, 6.57 mmol) in dichloromethane (5 ml) L-(+)-DIPT (1.85 g, 7.88 mmol) is added at −20 °C and the mixture is stirred for 10 min, after which 1-(2-thienyl)pentan-1-ol (1.12 g, 6.57 mmol) in dichloromethane (2 ml) is added and the solution so obtained is stirred for 20 min at −20 °C. After addition of 3.24 *M* *tert*-butyl hydrogen peroxide in dichloromethane (6.08 ml, 19.7 mmol) at 0 °C the stirring is continued at this temperature for 18 h. Dimethyl sulfide (5 ml) is added at −20 °C and the reaction mixture is stirred at this temperature for 30 min and poured into a mixture of 10% tartaric acid solution (0.5 ml), diethyl ether (20 ml), sodium fluoride (1.5 g), and Celite (1.5 g). After stirring vigorously for 1 h at room temperature the precipitate formed is filtered off through a pad of Celite. The filtrate is concentrated giving an oil, which is dissolved in diethyl ether (50 ml) and treated with 1 *M* sodium hydroxide solution (25 ml) for 15 min at 0 °C under vigorous stirring. The phases are separated and the organic phase is washed with sodium chloride solution, dried over sodium sulfate and evaporated. The residue is chromatographed on silica gel using diethyl ether/hexane as eluent giving 436 mg (78%) of the title compound.

The Reformatskii reaction of thiophene aldehydes with α-bromoesters is a good method for the preparation of β-hydroxyesters [223–225].

4A.11.5 Reaction of thienylmetal derivatives with aldehydes, ketones, and esters

Before the development of organolithium derivatives, the reaction of thienylmagnesium derivatives with aldehydes and ketones were used for the preparation of thienyl-, alkyl or arylcarbinols. Thus 2-thenylalcohol was prepared in 65% yield through the reaction of 2-thiophenemagnesium iodide with anhydrous formaldehyde [226,227]. However, as late as 1995, the reaction between 2-thienylmagnesium bromide and 4-fluorobenzaldehyde was used for the preparation of 5-fluorophenyl-2-thienyl carbinol in quantitative yield on a very large scale [73].

α-(4′-Fluorophenyl)-2-thiophenemethanol [73]

A 22-l, four-necked, round-bottomed flask equipped with a mechanical stirrer, a 2-l pressure equalizing addition funnel, two 600-mm coiled condensers, and a nitrogen inlet adapter are charged with magnesium metal turnings (332 g, 13.70 mol), tetrahydrofuran (5.2 l), and iodine (0.63 g, 2 mmol). The reaction mixture is heated at reflux until the purple iodine color disappeared (30 min). To this is added 25 ml of a solution of 2-bromothiophene (2184 g, 13.40 mol) in tetrahydrofuran (2.6 l). The reaction mixture is heated at reflux until a cloudy gray color is formed (30 min). The heating mantle is turned off and the remaining solution of 2-bromothiophene in tetrahydrofuran is added at such a rate that the refluxing is continued (about 2 h). The reaction mixture is then heated to reflux for an additional 3 h prior to cooling to 10 °C with an ice-water bath, after which a solution of 4-fluorobenzaldehyde (1580 g, 12.73 mol) in tetrahydrofuran (1.4 l) is added dropwise at such a rate that the internal temperature is kept below 20 °C (4 h). The resulting mixture is allowed to warm to room temperature (22 °C) and the stirring is continued for 3 h. The reaction mixture is quenched by addition of a solution of ammonium chloride (3010 g, 56.3 mol) in distilled water (5 l) while keeping the internal temperature below 40 °C (90 min). After stirring for 2 h, the layers are separated and the aqueous layer is extracted with ethyl acetate (3 l). The combined organic layers are washed with water (5 l), sodium chloride solution (5 l), and evaporated, followed by azeotrope distillation with acetonitrile (1 l) to afford

2652 g (100%) of a dark viscous oil, which solidifies upon standing containing 91% of the title compound, which can be recrystallized from diethyl ether/hexane mp 44–46 °C.

Also sodium compounds have previously been used for the preparation of di(2-thienyl)carbinol [228].

Di(2-thienyl)carbinol [228]

A sodium amalgam sand, containing sodium (25 g, 1.1 mol), and mercury (10 g, 0.05 mol), is covered with anhydrous diethyl ether (200 ml) and thiophene (63 g, 0.75 mol) is added. After cooling this mixture to 5 °C bromobenzene (78 g, 0.5 mol) is added over a period of 1 h. When the addition is complete the stirring is continued at this temperature for 1 h and then the reaction mixture is refluxed for 2 h, after which it is cooled and 2-thiophene aldehyde (77.6 g, 0.4 mol) is slowly added under cooling. When the addition is complete the stirring is continued for 1 h and the reaction mixture is then hydrolyzed with a saturated ammonium chloride solution. The quite clear solution is filtered, the ether phase separated, dried over magnesium sulfate and evaporated. The residue is poured into petroleum ether with vigorous agitation, the tan solid is collected and dried giving 84.5 g (76%) of the title compound mp 74–76 °C.

However, in most cases the reaction of various thienyllithium derivatives, especially 3-thienyllithia with aliphatic, aromatic, and heterocyclic aldehydes constitute the best methods for the preparation of thienylmethanol and thienyl alkyl and aryl carbinols [38,229–237]. The reaction of thienyllithium derivatives with ethyl chloroformate is an often used route to tris(thienyl) derivatives as illustrated in the preparation of tris(3-methyl-2-thienyl)-carbinol [238].

Tris(3-methyl-2-thienyl)carbinol [238]

To a solution of 3-methyl-2-thienyllithium (37.6 mmol) in anhydrous ether an ethereal solution of ethyl chloroformate (12.5 mmol) is added dropwise with stirring at −70 °C under argon. The stirring is continued at this temperature for 2 h, after which the reaction mixture is allowed to warm to room temperature and the stirring continued for 24 h. After hydrolysis with a saturated ammonium chloride solution, the phases are separated and the aqueous phase extracted with ether. The combined ether phases are dried over sodium sulfate and evaporated at room temperature. The residue is recrystallized from hexane giving 31% of the title compound mp 92 °C.

In order to obtain mixed tris(thienyl)carbinols, thienyllithium derivatives are reacted with thiophenic esters as in the reaction of 2-thienyllithium with 2-carbethoxy-5-methylthiophene, which gives (5-methyl-2-thienyl)-bis(2-thienyl)carbinol in 75% yield [238].

2-Thienyllithium adds diastereoselectively to optically active *N*,*N*-dibenzyl-amino aldehydes constituting a good method for the preparation of β-amino-substituted 2-thenyl alcohols [239]. The reaction of 2,5-dilithiothiophene with substituted benzaldehydes is a good method for the preparation of 2,5-bis-(phenylhydroxymethyl)thiophenes [240]. The reaction of 2- and 3-thienyllithium with chloral are very convenient methods for the preparation of trichloromethyl 2- and 3-thienylcarbinol [241]. The reaction of thienyl-lithium derivatives with various phenyl- or alkylglyoxylic acids [242] or ethyl phenylglyoxylate [243,244] are good methods for the preparation of α-(3-thienyl)glycolic acids.

3-Thienylmandelic acid [243]

3-Bromothiophene (12.0 g, 0.0736 mol) in anhydrous diethyl ether (120 ml) is added to 1.32 *M* butyllithium (60 ml, 0.079 mol) at −70 °C. After stirring for 5 min at −70 °C, this solution is added to ethyl phenylglyoxalate (12.36 g, 0.0695 mol) in anhydrous diethyl ether (60 ml) at −70 °C over a period of 27 min. When the addition is complete the cooling bath is removed, the reaction mixture is allowed to warm to room temperature and poured into a slurry of ice and ammonium chloride. The phases are separated and the aqueous phase extracted with two portions of ether. The combined organic phases are washed twice with water and twice with sodium chloride solution, dried over magnesium sulfate and evaporated. The residue is hydrolyzed by refluxing for 5 h in 10% ethanolic potassium hydroxide solution (70 ml). The ethanol is evaporated and the residue dissolved in water (300 ml). This solution is extracted four times with ether, cooled and acidified with 1 *M* hydrochloric acid (130 ml). The aqueous solution so obtained is extracted with three portions of ether and the combined organic phases are washed with water and sodium chloride solution, dried over magnesium sulfate and evaporated giving 10.9 g (67%) of the title compound mp 129–130 °C after recrystallization from benzene/hexane (1:1).

4A.11.6 Various methods for the preparation of hydroxymethyl derivatives

A recent reinvestigation of the benzoin condensation has shown that 3-thiophene aldehyde, selectively gives 3,3′-thenoin in 64% yield [170], while 2-thiophene aldehyde only gives low yields. Using a thiazolium salt as catalyst, only 20% of the 2,2-thenoin together with 55% of the 2,2′-thenil were obtained [245]. The uncatalyzed ring-opening of the 2-thienylethylene oxide with benzylamine yields selectively the β-amino-α-hydroxy derivative, while trimethylsilyl azide gives the opposite regioisomer [246].

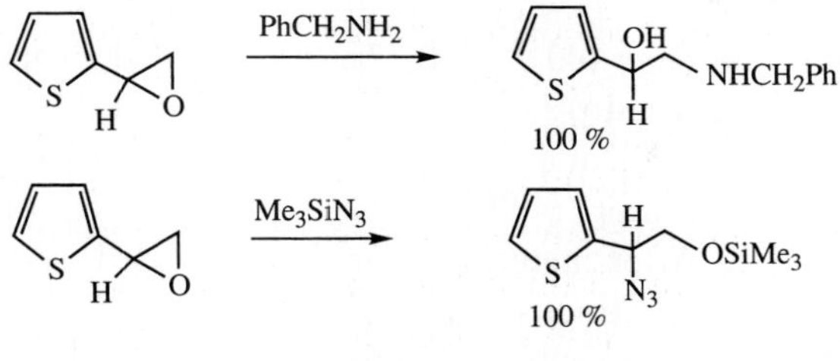

Substituted hydroxy alkyl thiophenes are prepared by the reaction of 1-alkynyl-2,3-epithio- alcohols with a catalytic amount of mercury(II) prepared from mercuric oxide and dilute sulfuric acid [247].

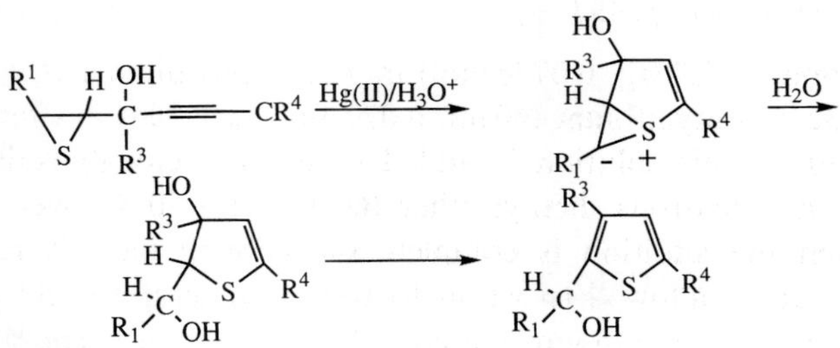

The reaction of 1,6-diethoxy-2,4-hexadiyne and sodium ethoxide in ethanol with hydrogen sulfide is used for the preparation of 2,5-bis(ethoxymethylene)-thiophene [248].

2,5-Bis(ethoxymethylene)thiophene [248]

Hydrogen sulfide is passed into a solution of 1,6-diethoxy-2,4-hexadiyne (4.68 g, 28 mmol) and sodium ethoxide is obtained from sodium (1.38 g) in absolute ethanol (120 ml). The reaction mixture is heated at 80 °C for 20 min. After addition of water the product is extracted with benzene. The combined organic phases are dried over calcium chloride, evaporated, and distilled giving 2.79 g (53%) of the title compound bp 106–107 °C/3 mm Hg.

According to Hartough [249], it is not possible to prepare 2-thenylalcohol through electrophilic substitution of thiophene with formaldehyde in the presence of a Lewis acid catalyst. This is probably due to the much faster condensation of 2-thenyl alcohol with itself or with thiophene than the primary condensation of thiophene with formaldehyde. However, if the 5-position is blocked, as in 2-(1-piperidinomethyl)thiophene, hydroxymethylation can be achieved, as reaction with a 40% formalin solution, saturated with hydrogen chloride, gives 5-(1-piperidinomethyl)-2-thiophenemethanol [111].

5-(1-Piperidinomethyl)-2-thiophenemethanol [111]

A mixture of concentrated hydrochloric acid (21 ml) and 40% formalin solution (17.5 ml) saturated with anhydrous hydrochloric acid gas 2-(1-piperidinomethyl)thiophene (40.65 g, 0.22 mol) is added dropwise at 0–5 °C. The reaction mixture is stirred under cooling and a flow of anhydrous hydrochloric acid gas for 1 h and then heated at 60 °C for 15 min. After cooling through addition of water (100 ml) the stirring is continued at room temperature for 3 h. After neutralization with diluted sodium hydroxide solution, the product is extracted with diethyl ether. The combined organic phases are dried over sodium sulfate, evaporated, and distilled bp (130–140 °C/0.2 mm Hg). Under cooling the title compound crystallizes.

Reaction of 5-methyl-2-thiophene aldehyde with 2-trimethylsilylkoxyfuran in the presence of boron trifluoride-etherate in methylene chloride is used for the preparation of the alcohols shown below, which can undergo β-elimination [250].

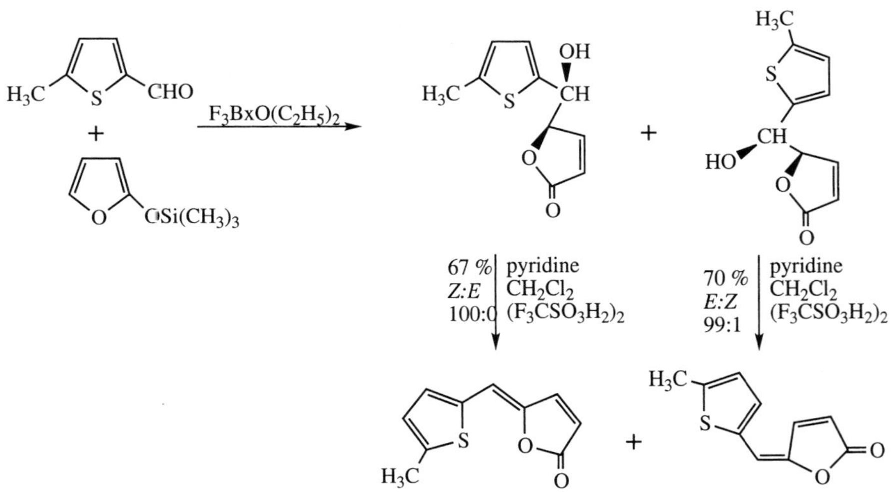

4A.12 THENYL ETHERS

4A.12.1 From halomethylthiophenes

Thenyl ethers are conveniently prepared from the reaction of halomethylthiophenes with methoxide [96,178,192,251–254], ethoxide [253,254], other alkoxides [253,254], and phenoxide ions [252,255,256].

5-Phenoxymethyl-3-thiophenecarboxylic acid [252]

A solution of ethyl 5-chloromethyl-3-thiophenecarboxylate (4.0 g, 20 mmol) and sodium phenoxide (2.3 g, 20 mmol) in methanol (50 ml) is refluxed for 1 h. After evaporation of the methanol the residue is poured into water and the product is extracted with diethyl ether. The combined organic phases are washed with 5% sodium hydroxide solution and water and concentrated. The residue is refluxed with 10% sodium hydroxide solution and this solution is acidified with concentrated hydrochloric acid giving the title compound mp 124 °C after crystallization from heptane.

4A.12.2 From hydroxymethylthiophenes and alkyl halides

Substituted geminal dithienyltetrahydrofurans have been prepared *via* an intramolecular Williamson reaction [257]. 2-Trialkylsilyl-3-hydroxymethylthiophenes undergo a 1,4-carbon-oxygen-silyl-migration, when treated with base containing sodium or potassium counterions to produce 3-[(trialkylsilyl)-oxymethyl]thiophenes in excellent yields [258].

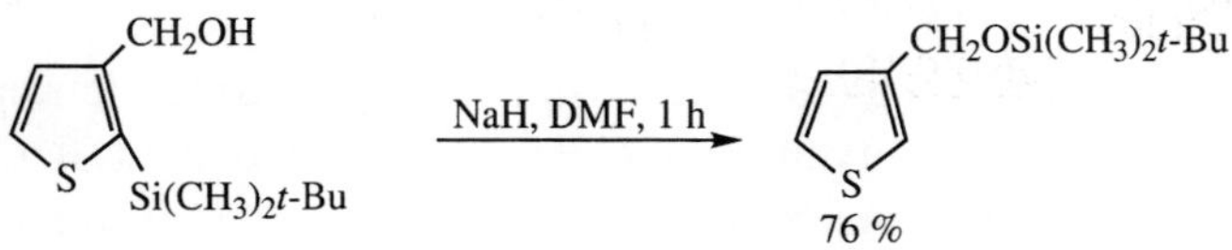

4A.12.3 Various methods

The best method for the preparation of the α-methoxythienylacetic acids is the reaction of trichloromethyl thienyl carbinols with methanolic potassium hydroxide [241].

α-Methoxy-2-thienylacetic acid [241]

To a solution of trichloromethyl 2-thienylcarbinol (50 g, 216 mol) in anhydrous methanol (80 ml), 3.7 *M* methanolic potassium hydroxide solution (238 ml) is added dropwise at 45 °C. The reaction mixture is then refluxed for 1 h, cooled and the potassium chloride is filtered off. The clear filtrate is evaporated and the residue dissolved in hot water (127 ml) and sodium chloride is added. Upon cooling, the sodium salt (41.3 g) precipitates. It is filtered off and carefully washed several times with acetone in order to remove oily impurities. Acidification of a water solution of the salt, extraction with ether and evaporation gives 18.5 g (50%) of the title compound, which after recrystallization yields 17.3 g (47%) mp 59–70 °C.

However, also methylation of the thienylglycolic acids with dimethyl sulphate or with diazomethane, using aluminium *tert*-butoxide as catalyst, as well as the base-catalyzed condensation of thiophene aldehydes with chloroform and methanol can be used [241].

4A.13 THIOPHENEMETHANETHIOLS

4A.13.1 From thenyl halides

Thiophenemethanethiols, which are not very stable compounds can be prepared by reacting thenylisothiuronium halides obtained from thenyl halides and thiourea, with sodium hydroxide [182,252,259–261].

Metalation of thiophenes with butyllithium/magnesium bromide followed by reaction with *O*-propyl chlorothioformate and reduction with lithium aluminium hydride or sodium borohydride is another method for the preparation of 2-mercaptomethylthiophenes [262].

2-Mercaptomethylthiophene [262]

To a solution of thiophene (252 mg, 3.0 mmol) in anhydrous tetrahydrofuran (4 ml) under nitrogen, butyllithium (3.3 mmol) is added at 0 °C followed by addition of magnesium bromide diethyl etherate (6.0 mmol) at room temperature. To this solution, cooled to −78 °C, *O*-propyl chlorothioformate in tetrahydrofuran (3.0 mmol) is added in one portion and the stirring is continued at this temperature for about 1 h, after which anhydrous hexane (10 ml) is added and the yellow mixture is allowed to warm to room temperature. The solution is evaporated and the residue filtered through a thin layer of silica gel using hexane as eluent. To the yellow filtrate a solution of lithium aluminium hydride (6.0 mmol) is added in one portion under nitrogen at 0 °C, the stirring is continued for 1 h and under this period the yellow color disappears. Crushed ice is carefully added under nitrogen followed by acidification with aqueous hydrochloric acid at 0 °C. The product is extracted with diethyl ether and the combined organic phases dried over sodium sulfate and evaporated giving 172 mg (44%) of the title compound as a pale yellow oil after chromatography on silica gel using hexane as eluent.

4A.14 THENYL PHOSPHOROUS DERIVATIVES

Interest in the preparation of thenyl phosphorus derivatives is connected with the development of the Wittig reaction and related reactions.

4A.14.1 Thenyl triaryl phosphonium halides

Thenyl triaryl phosphonium halides are best prepared in almost quantitative yield by the reaction of thenyl halides with triaryl phosphines in benzene [263]. A large number of halo-substituted thenyl triphenyl phosphonium halides have been prepared in this way [264].

Heating 2-(chloromethyl)-5-(prop-1-ynyl)thiophene with triphenylphosphine for 15 min at 110 °C gave (5-propynyl-2-thienyl)triphenylphosphonium chloride in 82% yield [265]. Bifunctional phosphonium salts like 2,5-dimethylthiophene bis(3,4-methylenetriphenyl)phosphonium chloride [266] and 2,5-bis(thiophenemethylene)triphenylphosphonium chloride [267] can also be prepared and was used in connection with preparation of macroyclic compounds and polymers by the Wittig reaction [268].

4A.14.2 Thenyl diethyl phosphonates

Thenyl diethyl phosphonates, useful for the preparation of ethylenes according to the Horner–Wadworth–Emmons' reaction, are conveniently prepared by the Arbusov reaction from halomethylthiophenes with triethyl phosphite [173,269,270]. 2-Thenylphosphonic acid can be prepared by hydrolyzing the dibutyl ester with hydrochloric acid [271].

4A.14.3 Various thenylphosphorus derivatives

2-Thienylphosphonyl carbinols are easily prepared in nearly quantitative yields by reaction of 2-thiophene aldehydes with diethyl or dimethyl phosphite in the presence of triethyl amine [272].

Dimethyl 2-thienylhydroxymethylphosphonate [272]

A mixture of 2-thiophene aldehyde (11.2 g, 0.1 mol), dimethyl phosphite (11.0 g, 0.1 mol), and triethylamine (10.1 g, 0.1 mol) is stirred at 25 °C for 24 h. After removal of the amine at reduced pressure the crude product is chromatographed on silica gel using ethyl acetate as eluent giving 18.1 g (88%) of the title compound as a dense oil.

α-Aminothenylphosphonic acids can be prepared in high yields by the thermal addition of a dialkyl phosphonate to a Schiff's base at 60 °C, which often requires heating for 72 h or more followed by ester cleavage. Through sonochemical activation the rate of formation is increased and a reaction time of 5–120 min can be used [273].

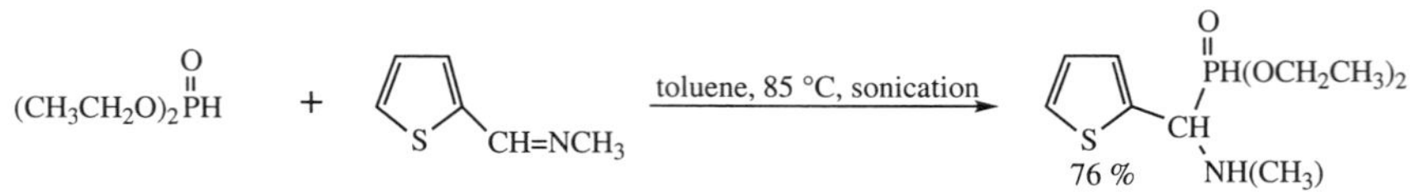

Ring-closure reactions have also been used for the preparation of phosphonic acid derivatives. Thus the reaction of 3-methyl-1,2,4-pentatrienyl-1-phosphonic dichloride with methylsulfenyl chloride in carbon tetrachloride at −12 to −10 °C is a convenient method for the preparation of 3-methyl-2-thenylphosphonic dichloride [274].

4A.15 THENYL SILYL DERIVATIVES

2-Trialkylsilyl-3-hydroxymethylthiophenes undergo a 1,4-carbon-oxygen silyl-migration when treated with bases containing either potassium or sodium counterions to produce 3-[(trialkylsilyl)oxymethyl]thiophenes in excellent yield [258].

The bis(diphenylphosphino)propane nickel chloride-catalyzed coupling of various bromothiophenes with trimethylsilylmethylmagnesium chloride has been exploited for the preparation of various trimethylsilylmethylthiophenes [89,102].

2-(Trimethylsilylmethyl)thiophene [89]

To a mixture of 2-bromothiophene (8.2 g, 50 mmol), $NiCl_2$(dppp) (0.15 g, 0.3 mmol) and anhydrous diethyl ether (100 ml) 1 *M* trimethylsilylmethylmagnesium chloride in ether (70 mmol) is added at 0 °C. The reaction mixture is refluxed for 6 h. After usual work-up distillation gives 5.68 g (67%) of the title compound bp 72 °C/13 mm Hg.

4A.16 THENYL SULFIDES

4A.16.1 From thenyl halides

Nucleophilic substitution of thenyl halides with alkyl [252,275,276] or aryl thiolates [158,256,277], including thienylthiolates [178] often constitutes the best method for the preparation of thenyl sulfides. The reaction of 3,4-bis(chloromethyl)thiophenes with sodium sulfide, using phase transfer technique has been used in the synthesis of 4,6-di-tert-butyl-1*H*,3*H*-thieno[3,4-*c*]thiophene [150].

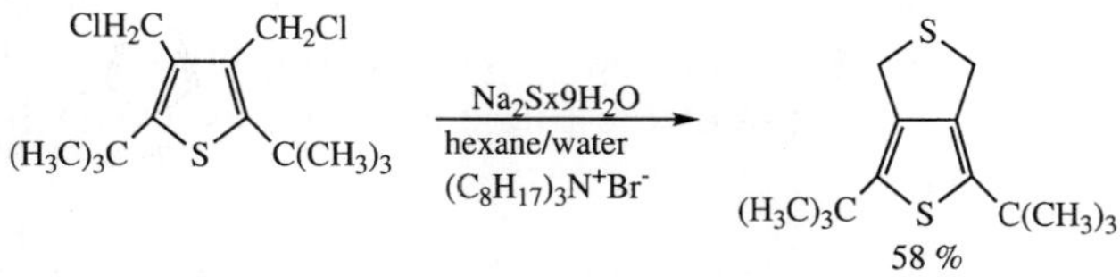

4,6-Di-tert-butyl-1H,3H-thieno[3,4-c]thiophene [150]

To a solution of 2,5-di-tert-butyl-3,4-bis(chloromethyl)thiophene (1.01 g, 3.44 mmol) in hexane (50 ml) a solution of sodium sulfite nonahydrate (1.83 g, 7.62 mmol) in water (20 ml) and a small amount of trioctanemethylammonium chloride are added. The mixture is stirred at room temperature for 5 h, after which water and diethyl ether are added. The phases are separated and the organic phase is washed with water, dried over magnesium sulfate and evaporated. The residue is purified by chromatography on silica gel using carbon tetrachloride as eluent giving 512 mg (58%) of the title compound as colorless crystals mp 117–118 °C.

The reaction of 2,5-bis(chloromethyl)-3,4-dibromothiophene with 1,4-bis(methylmercapto)-2,5-dimethoxybenzene can be used for the preparation of a paracyclo[3]-(2,5)-thiophenophane [279]. The reaction of 2,5-dichloromethylthiophene with the dianion of 1,4,7,10-tetrathiadecane gave 2,5,7,10-tetrathia[12]-(2,5-thiophenophane [280].

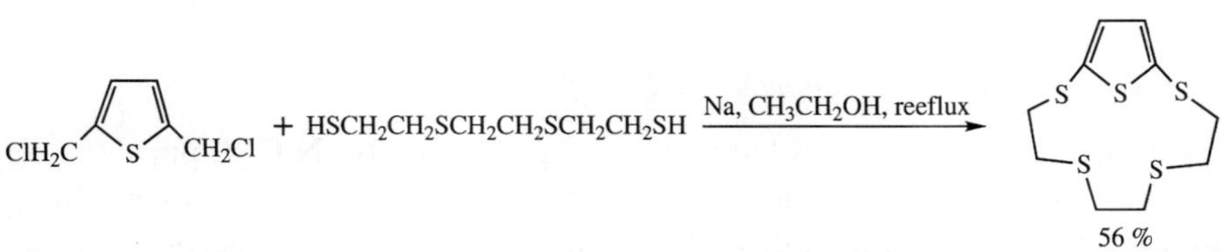

4A.16.2 From thiophenemethanethiols

The reaction of sodium salts of thenylmercaptans with aliphatic bromo esters is a convenient method for the preparation of thenylthio-substituted acids [259]. The isomeric dithenyl sulfides have been prepared by the reaction of 2- and 3-thenyl mercaptan with 2- or 3-thenyl halides in the presence of potassium hydroxide [281].

2,3′-Dithenyl sulfide [281]

3-Thenylmercaptan (14.3 g, 0.11 mol) is added slowly to potassium hydroxide (5.6 g, 0.1 mol) in ethanol (100 ml) and the temperature is maintained at 25 °C.

2-Thenyl chloride (13.3 g, 0.1 mol) is then added and the reaction mixture is refluxed for 3 h, cooled and poured into ice-water. The phases are separated and the aqueous phase extracted with ether. The combined organic phases are washed with water, dried over sodium sulfate and evaporated. Distillation of the residue gives 12.7 g (80%) of the title compound as a pale yellow oil bp 125–130 °C/0.02 mm Hg.

Dithia[3]metacyclo[3]-(2,3), -(2,4)-, (2,5)- and -(3,4)-thiophenophanes can be prepared by dithiol bisalkylations [151]. The reaction of 2,5-bis(methylmercapto)-3,4-dibromothiophene or 2,4-bis(methylmercapto)-3-bromo-5-methylthiophene with 1,4-bis(chloromethyl)benzene yields [2]paracyclo[2](2,5)- and (2,4)thiophenophanes in good yields [279]. Similarly the reaction of bis-2,4-(methylmercapto)thiophenes with 1,3-bis(bromomethyl)benzene constitutes an excellent method for the preparation of metacyclo[3](2,4)thiophenophanes [261].

The reaction of 2,4-bis(mercaptomethyl)thiophene with 2,4-bis-(bromomethyl)thiophene gave a 1:1 mixture of two isomeric 2,10-dithia[3.3](2,4)thiophenophanes, which could be separated by fractional crystallization from chloroform [181]. The best method for the preparation of thiophenophane shown below is the high dilution base coupling of 2,5-bis(bromomethyl)thiophene with 2,6-bis(methylsulfanyl)pyridine [282].

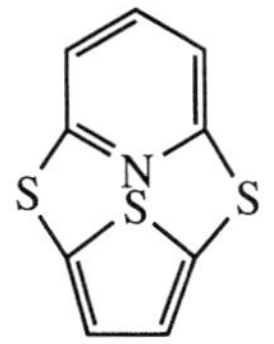

4A.16.3 Various methods

3-(Dimethylaminomethyl)-2-thienylmethylthioethamine was prepared by the reaction of cysteamine hydrochloride with 3-dimethylaminomethyl-2-hydroxymethylthiophene in concentrated hydrochloric acid [235].

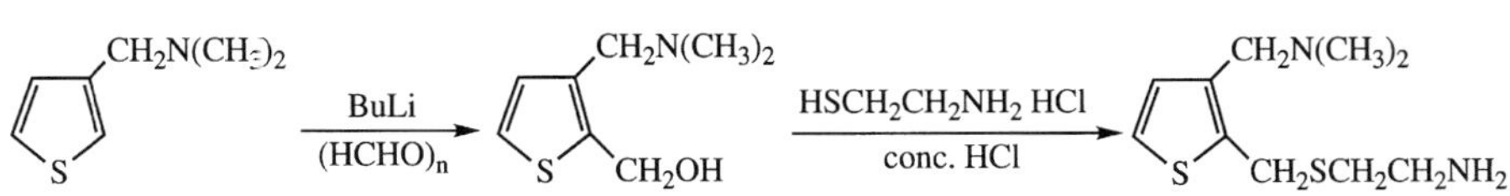

Electrophilic substitution of thiophene with 2,2,2-trifluoro-1-phenylsulfenylethyl carbocation derived through Lewis acid catalyzed dechlorination of

1-chloro-2,2,2-trifluorethyl phenyl sulfide is a good method for the preparation of the compound shown below [283].

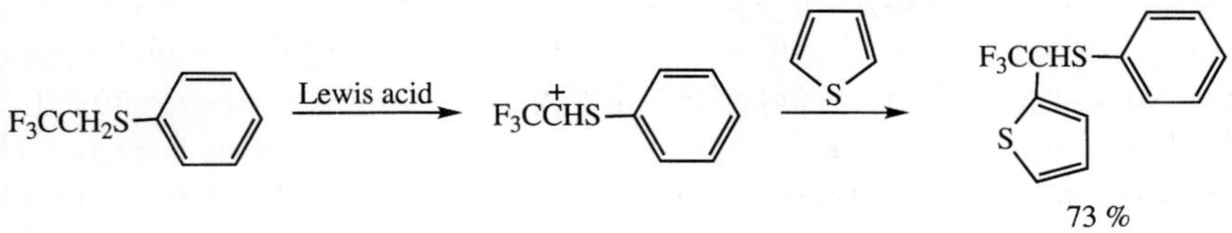

Sommelet–Hauser rearrangement of 3-thenyl dimethyl sulfonium bromide, obtained from the reaction of 3-thenyl bromide with dimethyl sulfide, by treatment with sodium hydride, gives 3-methyl-2-(methylthiomethyl)thiophene [284].

3-Methyl-2-(methylthiomethyl)thiophene [284]

3-Bromomethylthiophene (1.04 g, 5.9 mmol) is mixed with dimethyl sulfide (733 mg, 11.8 mmol) and the mixture is kept at room temperature overnight. The resulting brown paste is washed several times with ether and dried giving 959 mg of the crude sulfonium bromide, which is solved in tetrahydrofuran (25 ml) and treated with sodium hydride (106 mg, 4.4 mmol) with stirring at room temperature for 22 h. The reaction mixture is quenched with water, extracted with ether and dried. After evaporation the residue is column chromatographed on silica gel using hexane as eluent giving 628 mg (67%) of the title compound as a colorless oil.

Michael addition of 2-aminobenzenethiolate to 2-thienylmethylidene-malonate gives the compound below [285].

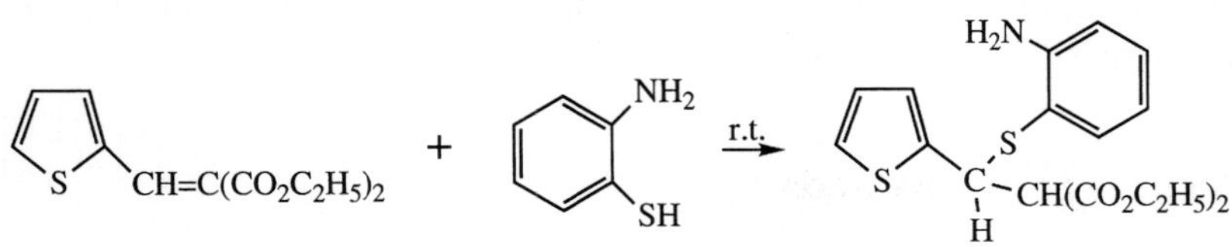

4A.17 THENYL SULFOXIDES

Oxidation of the cyclic sulfide with sodium periodate in aqueous methanol or in benzene/water in the presence of phase transfer agents gives the sulfoxide in excellent yield [286].

4,6-Di-tert-butyl-1H,3H-thieno[3,4-c]thiophene 2-oxide [150]

To a solution of 4,6-di-*tert*-butyl-1*H*,3*H*-thieno[3,4-*c*]thiophene (103 mg, 0.40 mmol) in benzene (20 ml) a solution of sodium periodate (112 mg, 0.52 mmol) in water (5 ml) is added trioctylmethylammonium chloride (a small amount), and methanol (15 ml). The reaction mixture is refluxed for 17 h, allowed to cool to room temperature, and extracted with benzene. The combined extracts are washed with water, dried over magnesium sulfate, and evaporated. The residue is column chromatographed on silica gel using dichloromethane/diethyl ether (9:1) to provide 82 mg (75%) of the title compound as colorless crystals mp 170–171 °C.

4A.18 THENYL SULFONES

4A.18.1 Oxidation of thenyl sulfides

The best and most commonly used method for the preparation of thenyl sulfones is the oxidation of thenyl sulfides with *meta*-chloroperbenzoic acid in methylene chloride, the thiophenic sulfur is not oxidized under these conditions [151,261,279]. The three isomeric dithenyl sulfides have been oxidized to the sulfones with potassium permanganate in acetic acid [281].

4A.18.2 Various methods

Phenyliodonium bis(phenylsulfonyl)methylide reacts with thiophenes to give 2-bis(phenylsulfonyl)methylthiophene [287].

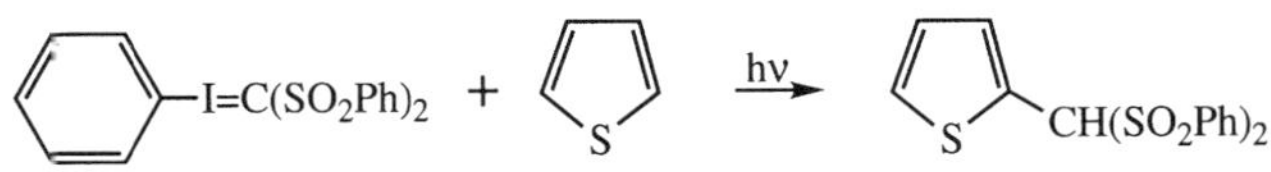

1-(2-Thienyl)ethanol yields upon reaction with sodium *para*-toluenesulfinate and 85% formic acid 1-(2-thienyl)ethyl *para*-tolylsulfone [210].

1-(2-Thienyl)ethyl para-tolylsulfone [210]

A solution of freshly distilled 2-thiophene aldehyde (300 mg, 2.67 mmol) in anhydrous tetrahydrofuran (3 ml) is cooled to −80 °C and under argon 1.5 *M* methyllithium in hexane (1.93 ml, 3.55 mmol) is added. After stirring for 2 h, methanol (2 ml) is added to destroy excess methyllithium followed by removal of most of the methanol. To the resulting residue 85% formic acid (2 ml) and sodium *para*-toluenesulfinate (2.00 g, 9.34 mmol) are added. The reaction mixture is stirred at room temperature for 2 h, after which it is washed with water. The water phases are extracted with dichloromethane and the combined organic phases are dried over sodium sulfate and concentrated. Upon purification by flash chromatography on silica gel using dichloromethane/hexane (9:1) as eluent 216 mg (31%) of the title compound is obtained mp 91–93 °C from ethyl acetate/hexane.

4A.19 THENYL THIOCYANATES

4A.19.1 From thenyl halides and alkali thiocyanates

Thenyl thiocyanates are best prepared through the reaction of thenyl halides with sodium or potassium thiocyanate [253,288].

4A.20 THIOPHENEACETIC ACIDS AND THEIR DERIVATIVES

4A.20.1 Thiopheneacetic acids and esters directly from thiophenes

Homolytic substitution of thiophene using ethyl iodoacetate as radical source promoted by hydrogen peroxide and ferrous salts in dimethylsulfoxide gives ethyl 2-thiopheneacetate 62% yield [289]. Alternatively the radical has been generated from ethyl iodoacetate and ethyl α-bromopropionate through autooxidation of triethylborane in dimethylsulfoxide [290]. The reaction of thiophene with cerium (IV) sulfate and an excess of dimethyl malonate constitutes an excellent method for the direct preparation of dimethyl (2-thienyl) malonate in high yield [291]. 2-Methylthiophene reacts in the same way and various monosubstituted malonate can also be used in this reaction.

Dimethyl (2-thienyl) propanedioate [291]

A solution of thiophene (2 ml, 25 mmol) and dimethyl malonate (5 ml, 45 mmol) in 90% methanol/water (50 ml) is treated with ceric sulfate (4.04 g, 10 mmol). The heterogenous reaction mixture is stirred at ambient temperature under nitrogen until a negative starch–iodine test is obtained after about 4 h. The solids are then filtered and rinsed with methanol (25 ml). The combined filtrate is concentrated to 10 ml at reduced pressure and then partitioned between dichloromethane (50 ml) and water (50 ml). The phases are separated and the organic phase washed with water (50 ml) and saturated sodium chloride solution (50 ml). After drying over sodium sulfate, the solvent is evaporated and the excess of dimethyl malonate is distilled off at 60 °C/ 0.5 mm Hg. The residue is column chromatographed on silica gel using dichloromethane as eluent giving 1.82 g (85%) of the title compound as a colorless oil.

From thiophene and also 2-acetylthiophene reaction with triethyl methanetricarboxylate in the presence of manganese(III) acetate at 60–65 °C in acetic acid solution gives the thienylmethane tricarboxylate in fair yields [292]. Another good method for the preparation of ethyl 2-thiopheneacetate is the bis(acetylacetonato)nickel(II)phosphine-catalyzed coupling of 2-thienylzinc chloride with ethyl bromoacetate [293].

4A.20.2 Thiopheneacetic acids and esters from thenyl cyanides

Thiopheneacetic acids are conveniently obtained through hydrolysis of thenyl cyanides with potassium hydroxide in aqueous or alcoholic media [147, 170,294]. In some cases also acidic hydrolysis has been used successfully [295].

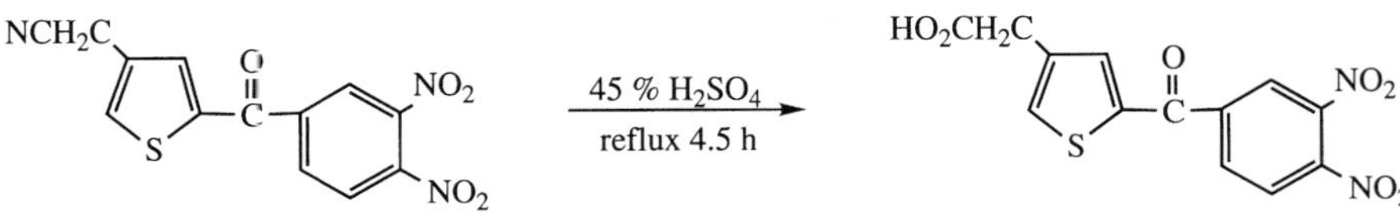

If esters are desired the thenyl cyanides have been reacted with methyl or ethyl alcohol in the presence of hydrogen chloride or sulfuric acid [153,294,296,297]. These esters can then conveniently be hydrolyzed to the acids using alkaline conditions.

4A.20.3 Thiopheneacetic acids from acetylthiophenes

Reaction of acetylthiophenes with sulfur and morpholine followed by alkaline hydrolysis of the intermediate amide is in many case an alternative method for the synthesis of thiopheneacetic acids [156,217,296,298,299].

4A.20.4 Thiopheneacetic acids from thiophene aldehydes

A novel synthetic method for thiopheneacetic acids consists in the reaction of 2-thiophene aldehyde with formaldehyde dimethylmercaptal-S-oxide and reaction of the resulting condensation product with ethanolic hydrogen chloride gave ethyl 2-thiophene acetate [300,301]. However, with strongly electron-donating substituents such as methoxy and also phenyl and *para*-methoxphenyl groups in the 5-position the main products are the (α-methylthio)thiophene acetates [301].

Ethyl 2-thiophene acetate [300]

To a solution of 2-thiophenealdehyde (10.32 g, 92 mmol) in tetrahydrofuran (50 ml) 40% methanolic solution of benzyltrimethylammonium hydroxide (Triton B) (3 ml) is added and the reaction mixture is refluxed for 6 h. After the addition of dichloromethane (100 ml) the mixture is washed with 0.5 *M* sulfuric acid, dried over sodium sulfate, and evaporated. The residue is distilled under reduced pressure giving 17.31 g (86%) of 1-(methylsulfinyl)-1-(methylthio)-2-(2-thienyl)ethylene as a pale yellow oil bp 147–157 °C/011–013 mm Hg. This compound is transformed to the title compound by dissolving 1.76 g of it in ethanol (30 ml) and then adding a saturated ethanolic solution of hydrogen chloride (3 ml). The reaction mixture so obtained is refluxed for 22.5 h. After evaporation, the residue is chromatographed on silica gel using benzene/hexane (1:1) as eluent, giving 1.04 g (79%) of the title compound.

4A.20.5 Various methods for thiopheneacetic acids

Methyl 3-thiopheneacetate has been prepared by a direct ring-closure reaction from methyl 5-chloro-3-chloromethylpenta-2,4-dienoate and sodium sulfide [302]. Thiophenobis(alkoxycarbonyl)methanides upon prolonged heating undergo intramolecular rearrangement to give thiophene-2-malonates in good yields [303]. Ethyl 3-thienylacetate can be used as starting material for α-alkylated thienylacetic acids by deprotonation with lithium isopropylhexyl amide at −78 °C and alkylation of the resulting enolate with iodomethane or 1-iodopropane followed by alkaline hydrolysis [304].

Ethyl 2-(3-Thienyl)propanoate [304]

To a solution of *N*-isopropylcyclohexylamine (968 mg, 6.86 mmol) in anhydrous tetrahydrofuran (30 ml) under nitrogen at −78 °C 1.5 *M* butyllithium in hexane (4.6 ml) is added dropwise. The mixture is allowed to warm to 0 °C and stirred at this temperature for 5 min and then recooled to −78 °C. Ethyl 3-thienylacetate 1.06 g, 6.23 mmol) in anhydrous tetrahydrofuran (15 ml) is added dropwise. This mixture is allowed to warm to room temperature and then added dropwise to a solution of methyl iodide (2.65 g, 18.7 mmol) in anhydrous dimethyl sulfoxide (3 ml) under nitrogen. The reaction mixture is stirred at room temperature overnight, after which water (100 ml) is added and the product extracted with diethyl ether. The combined organic phases are washed with water and sodium chloride solution, dried over magnesium sulfate and evaporated. The residue is purified by chromatography using diethyl ether/light petroleum (1:4) as eluent giving 963 mg (84%) of the title compound as a colorless oil.

The Suzuki coupling of arylboronic acids or esters with α-bromoacetic acid derivatives using tri(1-naphtyl)phosphine as ligand has also been applied to 2-thiopheneboronic acid and ethyl α-bromoacetate. However, with the 2-thiopheneboronic acid only a 33% yield was obtained [305].

4A.21 THIENYLGLYCOLIC ACIDS

Both 2- and 3-thienylglycolic acids are still best prepared by treatment of 2- and 3-thienylglyoxal hydrate with dilute sodium hydroxide solution (cf. [199]).

REFERENCES

1. F. Azizian and J. S. Pizey, *J. Chem. Tech Biotechnol.* **30**, 429 (1980).
2. F. Azizian and J. S. Pizey, *J. Chem. Tech. Biotechnol.* **30**, 648 (1980).
3. B. W. L. Southward, L. S. Fuller, G. J. Hutchings, R. W. Joyner and R. A. Stewart, *Chem. Commun.* 369 (1999).
4. D. Schuyl-Laros, P. J. W. Schuyl and L. Brandsma, *Rec. Trav. Chim.* **91**, 785 (1972).
5. C. A. H. Rasmussen and A. de Groot, *Synthesis* **7**, 575 (1983).
6. F. Ya. Perveev and N. Y. Kudryashova, *Zh. Obshchh. Khim.* **23**, 976 (1953).
7. F. Ya. Perveev and T. N. Kuren'gina, *Zh. Obshch. Khim.* **25**, 1619 (1955).
8. F. Ya Perveev and N. L. Kudryaskova, *Zh. Obshch. Khim.* **24**, 1019 (1954).
9. M. W. Farrer and R. Levine, *J. Am. Chem. Soc.* **72**, 4433 (1950).
10. E. Campaigne and W. O. Foye, *J. Org. Chem.* **17**, 1405 (1952).
11. N. Messina and E. V. Brown, *J. Am. Chem. Soc.* **74**, 920 (1952).
12. R. Ramasseul and A. Rassat, *Bull. Soc Chim. Fr.* 3136 (1965).
13. H. Nozaki, T. Koyama and T. Mori, *Tetrahedron* **25**, 5357 (1969).
14. S. Fujita, T. Kawaguti and H. Nozaki, *Tetrahedron Lett.* 1119 (1971).
15. S. Hirano, T. Hiyama, S. Fujita, T. Kawaguti, Y. Hayashi and H. Nozaki, *Tetrahedron* **30**, 2633 (1974).
16. P. J. Garrat and S. B. Neoh, *J. Org. Chem.* **40**, 970 (1975).
17. H. Wynberg and J. Metselaar, *Synth. Comm.* **1**, 1 (1984).
18. T. Thomsen, V. Pedersen, P. B. Rasmussen, B. Yde, T. P. Andersen and S. O. Lawesson, *Chem. Lett.* 809 (1983).
19. D. R. Shridar, M. Joghibukta, P. Shantan Rao and V. K. Handa, *Synthesis* 1061 (1982).
20. F. Duus, *Acta Chem. Scand.* **27**, 466 (1973).
21. F. Duus, *Tetrahedron* **32**, 2817 (1976).
22. H. Wynberg and V. E. Wiersum, *J. Org. Chem.* **30**, 1058 (1965).
23. C. Botteghi, L. Lardicci and R. Meniacagli, *J. Org. Chem.* **38**, 2361 (1973).
24. K. E. Schulte, J. Reisch and D. Bergenthal, *Chem. Ber.* **101**, 1540 (1968).

25. K. E. Schulte, J. Reisch and L. Hörner, *Angew. Chem.* **72**, 920 (1960).
26. K. Schulte, J. Reisch and L. Hörner, *Chem. Ber.* **95**, 1943 (1962).
27. K. E. Schulte, J. Reisch, W. Herrmann and G. Bohn, *Arch. Pharm.* **296**, 456 (1963).
28. A. G. Ismailov, E. I. Mamedov and V. G. Ibragimov, *J. Org. Chem. USSR* **13**, 2424 (1977).
29. I. Hirao, *J. Pharm. Soc. Jpn* **73**, 1023 (1953).
30. R. F. Feldkamp and B. F. Tullar, *Org. Synth.* **34**, 73 (1954).
31. B. Gabriele, G. Salerno and A. Fazio, *Organic Lett.* **2**, 351 (2000).
32. E. Waldvogel, *Helv. Chim. Acta* **75**, 907 (1992).
33. J. Nakayama, R. Hasemi, K. Yoshimura, Y. Sugihara S. Yamaoka and N. Nakamura, *J. Org. Chem.* **63**, 4912 (1998).
34. A. W. Krebs, E. Franken, M. Müller, H. Collberg, W. Cholcha, J. Wilken and J. Ohrenberg, *Tetrahedron Lett.* **33**, 5947 (1992).
35. P. D. Caesar, *J. Am. Chem. Soc.* **70**, 3623 (1948).
36. S. Gronowitz and G. Borgen, *Acta Chem. Scand.* **19**, 1180 (1965).
37. J. Sicé, *J. Am. Chem. Soc.* **75**, 3697 (1953).
38. A. Wiersema and S. Gronowitz, *Acta Chem. Scand.* **24**, 2593 (1970).
39. E. Wiklund and R. Håkansson, *Chem. Scr.* **3**, 220 (1973).
40. M. Janda, J. Strogl, J. Stibor, M. Nemec and P. Vopatrana, *Synthesis* 545 (1972).
41. S. Gronowitz and P. Pedaja, *Chem. Scr.* **15**, 187 (1980).
42. T. Satayama, S. Hara and A. Suzuki, *Bull. Chem. Soc. Jpn* **52**, 1865 (1979).
43. I. Akomoto, M. Sano and A. Suzuki, *Bull. Soc. Chem. Jpn.* **54**, 1587 (1981).
44. E. Profft, *Chem. Z.* **82**, 295 (1958).
45. A. Logothetis and H. Wynberg, *J. Am. Chem. Soc.* **78**, 1958 (1956).
46. Y. L. Gol'dfarb, S. Z. Taits and L. I. Belen'kii, *Tetrahedron* **19**, 1851 (1963).
47. W. J. King and F. F. Nord, *J. Org. Chem.* **14**, 638 (1949).
48. M. Sy, N. P. Buu-Hoi, and N. D. Xuong, *J. Chem. Soc.* 1975 (1954).
49. J. Lamy, D. Lavit and N. P. Buu-Hoi, *J. Chem. Soc.* 4202 (1958).
50. P. Cagniant and D. Cagniant, *Bull. Soc. Chim. Fr.* 713 (1953).
51. Y. Poirier, L. Legrand and N. Lozac'h, *Bull. Soc. Chim. Fr.* 1054 (1966).
52. H. D. Hartough, *J. Am. Chem. Soc.* **73**, 4033 (1951).
53. P. Cagniant and D. Cagniant, *Bull. Soc. Chim. Fr.* 359 (1955).
54. S. Gronowitz and H. Frostling, *Tetrahedron Lett.* 604 (1961).
55. S. Gronowitz and H. Frostling, *Acta Chem. Scand.* **16**, 1127 (1962).
56. N. P. Buu-hoi, N. D. Xuong and N. Hoan, *Rec. Trav. Chim.* **71**, 285 (1952).
57. S. Z. Taits and Ya. L. Gol'dfarb, *Bull Acad. Sci USSR, Div. Chem Sci. (Engl transl.)* 1574 (1960).
58. J. M. Tour and R. Wu, *Macromol.* **25**, 1901 (1992).
59. C. W. Spangler and M. He, *J. Chem. Soc. Perkin Trans. 1* 715 (1995).
60. K. A. Murray, A. B. Holmes, S. C. Moratti and G. Rumbles, *J. Mater. Chem.* **9**, 2109 (1999).
61. H. Sawada, M. Nakayama, M Yoshida, T. Yoshida and N. Kamigata, *J. Fluorine Chem.* **46**, 423 (1990).
62. D. Naumann and J. Kischkewitz, *J. Fluorine Chem.* **46**, 265 (1990).
63. M. Nishida, S. Fujii, T. Aoki Y. Hayakawa, H. Muramatsu and T. Morita, *J. Fluorine Chem.* **46**, 445 (1990).
64. G. J. Chen and C. Tamborski, *J. Fluorine Chem.* **43**, 207 (1989).
65. Q.-L. Zhou and Y.-Z. Huang, *J. Fluorine Chem.* **43**, 385 (1989).
66. G. J. Chen and C. Tamborski, *J. Fluorine Chem.* **46**, 137 (1990).
67. S. Gronowitz, J. E. Skramstad and B. Eriksson, *Arkiv Kemi* **28**, 99 (1967).
68. M. Ahmed, J. Ashby, M. Ayad and O. Meth-Cohn, *J. Chem. Soc., Perkin Trans. 1* 1099 (1973).
69. M. Ahmed and O. Meth-Cohn, *J. Chem. Soc.* (C) 2104 (1971).
70. J. Skramstad and O. Eriksen, *Acta Chem. Scand.* **45**, 919 (1991).
71. S. S. Hall and S. E. Farahat, *J. Heterocycl. Chem.* **24**, 1205 (1987).

72. C. F. Nutaitis, R. Patragoni, G. Goodkin, B. Neighbour and J. Obaza-Nutaitis, *Organic Preparations and Procedures* **23**, 403 (1991).
73. E. J. Stoner, D. A. Cothron, M. K. Balmer and B. A. Roden, *Tetrahedron* **51**, 11043 (1995).
74. H. Wynberg and A. Kraak, *J. Org. Chem.* **29**, 2455 (1964).
75. A. Kraak, A. K. Wiersema, P. Jordens and H. Wynberg, *Tetrahedron* **24**, 3381 (1968).
76. S. M. H. Kabir and M. Iyoda, *Chem. Commun.* 2329 (2000).
77. J. W. Schick and D. J. Crowley, *J. Am. Chem. Soc.* **73**, 1377 (1951).
78. T. Sone, Y. Ohba and R. Watanabe, *Bull. Chem. Soc. Jpn.* **62**, 1346 (1989).
79. S. Ito, S. Kikuchi, T. Okujima, N. Morita and T. Asao, *J. Org. Chem.* **66**, 2470 (2001).
80. J. M. Barker, P. R. Huddleston and R. Smith, *J. Chem. Res.* (M) **1994**, 0236.
81. A. Riad, Z. Mouloungui, M. Delmas and A. Gaset, *Synth. Comm.* **19**, 3169 (1989).
82. M. Temciuc A.-B. Hörnfeldt and S. Gronowitz, *Heterocycl. Comm.* **1**, 411 (1995).
83. J. Nakayama, M. Sugino, A. Ishii and M. Hoshino, *Chem. Lett.* **1992**, 703.
84. A. R. Katritzky, L. Xie and W.-Q Fan, *J. Org. Chem.* **58**, 4376 (1993).
85. Ya. L. Gol'dfarb and Y. L. Danyushevskii, *Bull. Acad. Sci. USSR., Div Chem. SSSR (Eng. transl.)* 1395 (1956).
86. T. Sone, H. Kawasaki, S. Nagasawa, K. Tate and K. Sato, *Chem. Lett.* 399 (1981).
87. V. J. Bulbule, V. H. Deshpande and A. V. Bedekar, *J. Chem. Res.* (S) 220 (2000).
88. V. Ramanathan and R. Levine, *J. Org. Chem.* **27**, 1667 (1962).
89. K. Tamao, S. Kopdama, I. Nakajima, M. Kumada, A. Minato and K. Suzuki, *Tetrahedron* **38**, 3347 (1982).
90. F. F. Blicke and J. H. Burckhalter, *J. Am. Chem. Soc.* **64**, 478 (1942).
91. N. K. Kochetkov and N.V. Dudikyna, *Zh. Obshchei Khim.* **27**, 1399 (1957).
92. J. B. Press and J. J. McNally, *J. Heterocycl. Chem.* **25**, 1571 (1988).
93. H. D. Hartough, S. L. Meisel, E. Koft and J. W. Schick, *J. Am. Chem. Soc.* **70**, 413 (1948).
94. E. H. Lincoln, R. W. Heinzelmann and J. H. Hunter, *J. Am. Chem Soc.* **71**, 2902 (1949).
95. J. V. Braun R. Fussgänger and M. Kühn, *Liebigs Ann.* **445**, 201 (1925).
96. D. W. Slocum and P. L. Gierer, *J. Org. Chem.* **41**, 3668 (1976).
97. S. Marchalin, B. Decroix and J. Morel, *Acta Chem. Scand.* **47**, 287 (1993).
98. D. Lebosquain and B. Decroix, *Heterocycles* **36**, 2303 (1993).
99. D. Berkes and B. Decroix, *Bull. Soc. Chim. Fr* **131**, 986 (1994).
100. D. Berkes, N. Bar and B. Decroix, *J. Heterocycl. Chem.* **32**, 403 (1995).
101. B. Decroix and J. Morel, *J. Heterocycl. Chem.* **28**, 81 (1991).
102. A. M. van Leusen and K. J. van den Berg, *Tetrahedron Lett.* **29**, 2689 (1988).
103. W. S. Hwang, D.-L. Wang and M. Y. Chiang, *J. Organomet. Chem.* **623**, 231 (2000).
104. Ya. L. Gol'dfarb,, E. A. Krasnyanskaya and B. P. Fabrichnyi, *Bull. Acad. Sci. USSR Div. Chem. Sci. (Engl. transl.)* 1731 (1962).
105. H. Finch, D. H. Reece and J. T. Sharp, *J. Chem. Soc. Perkin Trans. 1* 1193 (1994).
106. E. Sandberg, *Chemica Scr.* **2**, 241 (1972).
107. T. Usami, N. Shirai and Y. Sato, *J. Org. Chem.* **57**, 5419 (1992).
108. A. G. Giumanini, C. Trombini, G. Lercker and A. R. Lepley, *J. Org. Chem.* **41**, 2187 (1976).
109. Y. L. Gol'dfarb and M. B. Ibragimova, *Dokl, Akad. Nauk SSSR (Engl. transl.)* **106**, 71 (1956).
110. Y. L. Gol'dfarb and M. B. Ibragimova, *Dokl. Akad. Nauk SSSR (Engl. transl.)* **113**, 261 (1957).
111. R. Mohr, A. Buschauer and W. Schunack, *Arch. Pharm. (Weinheim)* **321**, 221 (1988).
112. S. Gronowitz, I. Sjögren, L. Wernstedt and B. Sjöberg, *Arkiv Kemi* **23**, 129 (1964).
113. V. P. Mamaev, N. N Suworov and E.M. Rokhlin, *Doklady Akad. Nauk SSSR* **101**, 269 (1955).
114. V. P. Mamaev and T. D. Rubina, *J. Gen. Chem USSR* (*Engl. Transl.*) **27**, 525 (1957).
115. P. Dallemagne, S. Rault, M. Cugnon de Sévricourt, K. M. Hassan and M. Robba, *Tetrahedron Lett.* **27**, 2607 (1986).

116. P. Dallemagne, S. Rault, D. Maume and M. Robba, *Heterocycles* **26**, 1449 (1987).
117. P. Dallemagne, S. Rault and M. Robba, *Bull. Soc. Chim. Fr.* **128**, 260 (1991).
118. P. Dallemagne, J. V. Pilo, S. Rault, M. Robba, M. Saux and A. Carpy, *Heterocycles* **36**, 63 (1992).
119. P. Dallemagne, M. Boulouard, S. Rault and M. Robba, *J. Heterocycl. Chem.* **30**, 799 (1993).
120. P. Dallemagne, S. Rault, M. Gordaliza and M. Robba, *Heterocycles* **26**, 3233 (1987).
121. P. Dallemagne, A. Alsaidi, M. Boulouard, S. Rault and M. Robba, *Heterocycles* **36**, 287 (1993).
122. J. H. Wikel, M. L. Denney and R. T. Vasileff, *J. Heterocycl. Chem.* **30**, 289 (1993).
123. J. R. McCarthy, D. P. Matthews, R. J. Broersma, R. D. McDermott, P. R. Castner, J.-M. Hornsperger, D. A. Demeter, H. J. Weintraub and J. P. Whitten, *J. Med. Chem.* **33**, 1866 (1973).
124. T. Mukaiyama, K. Kashiwagi and S. Matsui, *Chem. Lett.* 1397 (1989).
125. Ya. L. Gol'dfarb, E. G. Ostapenko, V. G. Vinogradova, A. N. Zverev, A. V. Polyakov, A. I. Yanovskii, D. S. Yufit and Yu. T. Struchkov, *Chem. Heterocycl. Comp. USSR* 740 (1987).
126. D. Prajapati and J. S. Sandhu, *Synthesis* 342 (1988).
127. L. Strekowski, M. T. Segla, S. Bin Kong and D. B. Harden, *J. Heterocycl. Chem.* **26**, 923 (1989).
128. V. N. Nemykin, A. E. Polshina and N. Kobayashi, *Chem. Lett.* 1236 (2000)
129. H. D. Hartough, S. J. Lukasiewicz and E. H. Murray, Jr, *J. Amer. Chem. Soc.* **68**, 1389 (1946).
130. H. D. Hartough and S. L. Meisel, *J. Amer. Chem. Soc.* **70**, 4018 (1948).
131. S. J. Angyal, D. R. Penman and G. P. Warwick, *J. Chem. Soc.* 1742 (1953).
132. H. D. Hartough The Chemistry of Heterocyclic Compunds, Vol. 3, (Ed. A. Weissberger), *Thiophene and Its Derivatives.* Interscience Publishers, New York, 1952. p. 510.
133. V. I. Shvedov, I. A. Kharizomenova, N. V. Medvedeva and A. N. Grinev, *Chem. Heterocycl. Comp. USSR* 805 (1975).
134. A. R. Katritzky, J. Wang and B. Yang, *Synth. Comm.* **25**, 2631 (1995).
135. L. Fiocca, M. Fiorenza, G. Reginato, A. Ricci, P. Dembach and G. Seconi, *J. Organomet Chem.* 341, C23 (1988).
136. M. S. Cooper, R. A. Fairhurst, H. Heaney, G. Papageorgiou and R. F. Wilkins, *Tetrahedron* **45**, 1155 (1989).
137. R. Fikentscher, R. Brückmann and R. Betz, *Liebigs Ann. Chem.* 113 (1990).
138. A. S. Zanina, G. N. Khabibulina, V. V. Legkoderya and I. L. Kotlyarevskii, *J. Org. Khim. USSR (Engl transl.)* **8**, 1556 (1972).
139. K. E. Rodriques, A. Basha, J. B. Summers and D. W. Brooks, *Tetrahedron Lett.* **29**, 3455 (1988).
140. J. Heerklotz, A. Linden and M. Hesse, *Tetrahedron* **56**, 7205 (2000).
141. M. W. Briscoe, R. D. Chambers, S. J. Mullins, T. Nakamura and J. F. S. Vaughan, *J. Chem. Soc. Perkin Trans. 1* **3119** (1994).
142. A. A. Kiryanov, A. J. Seed, P. Sampson, *Tetrahedron* **57**, 5757(2001).
143. F. F. Blicke and F. Leonard, *J. Am. Chem. Soc.* **68**, 1934 (1946).
144. A. Giumanini and G. Lercker, *Gazz. Chim. Ital.* **104**, 415 (1974).
145. J. H. Ford G. C. Prescott, and D. L. Coolingworth, *J. Am. Chem. Soc.* **72**, 2109 (1950).
146. R. C. Clapp, J. H. Clarc, J. R. Vaughan, J. P. English, and G. W. Anderson, *J. Am. Chem. Soc.* **69**, 1549 (1947).
147. P. Cagniant, *Bull. Soc. Chim. Fr.* 847 (1949).
148. P. Cagniant and D. Cagniant, *Bull. Soc. Chim. Fr.* 713 (1952).
149. Y. L. Gol'dfarb and M.S. Kondakova, *Izv. Akad. Nauk SSSR, Otdel. Khim Nauk.*, 495 (1953).
150. A. Ishhii, Y. Ida, J. Nakayama and M. Hoshino, *Bull. Chem. Soc. Jpn.* **65**, 2821 (1992).
151. M. Takeshita and M. Tashiro, *J. Org. Chem.* **56**, 2837 (1991).
152. S. Gronowitz and T. Dahlgren, *Chem. Scr.* **12**, 57 (1977).

153. S. Gronowitz and S. Liljefors, *Chem. Scr.* **13**, 39 (1979).
154. H. E. Winberg, F. S. Fawcett, W. E. Mochel and C. W. Theobald, *J. Am. Chem. Soc.* **82**, 1428 (1960).
155. D. J. Zwanenburg and H. Wynberg, *J. Org. Chem.* **34**, 333 (1969).
156. R. Gaertner, *J. Am. Chem. Soc.* **73**, 3934 (1951).
157. T. Sone and Y. Matsuki, *Bull. Soc. Chem. Jpn.* **41**, 1423 (1968).
158. J. Nakayama, T. Kawamura, K. Kuroda and A. Fujita, *Tetrahedron Lett.* **34**, 5725 (1993).
159. T. W. Campbell and W.W. Keading, *J. Am. Chem. Soc.* **73**, 4018 (1951).
160. M. G. Reinecke, H. W. Adickes and C. Pyun, *J. Org. Chem.* **36**, 3820 (1971).
161. K. Dittmer, R. P. Martin, W. Herz and S. J. Cristol, *J. Am. Chem. Soc.* **71**, 1201 (1949).
162. N. P. Buu-Hoi and J. Lecocq, *Compt. Rend.* **222**, 1441 (1946).
163. N. B. Chapman and J. F. A. Williams, *J. Chem. Soc.* 5014 (1952).
164. H. J. Dauben, Jr and L.L. McCoy, *J. Am. Chem. Soc.* **81**, 4863 (1959).
165. E. Campaigne and B. F. Tullar, *Org. Synth.* **33**, 96 (1953).
166. S. Gronowitz, *Arkiv Kemi* **6**, 283 (1953).
167. J. H. Clarc, R. C. Clapp, J. R. Vaughan, Jr, L. H. Sutherland, R. Winterbottom, G. W. Anderson, J. D. Forsythe, J. Blodinger, S. L. Eberlin and J. P. English, *J. Org. Chem.* **14**, 216 (1949).
168. H. R. Snyder, L. A. Carpino, J. F. Zack, Jr, and J. F. Mills, *J. Am. Chem. Soc.* **79**, 2556 (1957).
169. V. N. Gogte, B. D. Tilak, K. N. Gadekar and M. B. Sahasrabude, *Tetrahedron* **3**, 2443 (1967).
170. E. Campaigne and W. N. LeSuer, *J. Am. Chem. Soc.* **70**, 1555 (1948).
171. S. Gronowitz and T. Frejd, *Synth. Commun.* **6**, 475 (1976).
172. E. Campaigne and W. N. LeSuer, *J. Am. Chem. Soc.* **71**, 333 (1949).
173. B. Yom-Tov and S. Gronowitz, *Chem. Scr.* **3**, 37 (1973).
174. H. Wynberg and D. Zwanenburg, *J. Org. Chem.* **29**, 1919 (1964).
175. D. J. Zwanenburg and H. Wynberg, *Rec. Trav. Chim. Pays-Bas* **88**, 321 (1969).
176. C.-K. Sha and C.-P Tsou, J. *Org Chem.* **55**, 2446 (1990).
177. J. A. Clarke and O. Meth-Cohn *Tetrahedron Lett.* 4705 (1975).
178. D. W. H. MacDowell and T. B. Patrick, *J. Org. Chem.* **31**, 3592 (1966).
179. R. J. Heffner and M. M. Joullié, *Synth. Comm.* **21**, 1055 (1991).
180. D. J. Chadwick and A. Plant, *Tetrahedron Lett.* **28**, 6085 (1987).
181. K. Watanabe, Y. Aso, T. Otsubo and F. Ogura, *Chemistry Lett.* 1233 (1992).
182. W. J. Raich and C. S. Hamilton, *J. Am. Chem. Soc.* **79**, 3800 (1957).
183. C. Bilger, R. Royer and P. Demerseman, *Synthesis* 902 (1988).
184. E. Profft and P. Lux, *J. prakt. Chem.* **16**, 18 (1962).
185. S. Gronowitz and P. Pedaja, *Tetrahedron* **34**, 587 (1978).
186. F. F. Blicke and M. F. Zienty, *J. Am. Chem. Soc.* **63**, 2945 (1941).
187. K. Petterson, *Acta Chem. Scand.* **4**, 395 (1950).
188. W. Herz, *J Am. Chem. Soc.* **73**, 351 (1951).
189. M. Winn and G. Bordwell, *J. Org. Chem.* **32**, 1610 (1967).
190. M. Irie and M. Mohri, *J. Org. Chem.* **53**, 803 (1988).
191. T. Frejd and T. Klingstedt, *Synthesis* 40 (1987).
192. W. S. Emerson and T. M. Patrick, Jr., *J. Org. Chem.* **14**, 790, (1949).
193. M. Bercot-Vatterroni, R. C. Moreau and P. Raynaud, *Compt. Rend. Acad. Sci.* **252**, 2419 (1961).
194. Y. Amemiya, A. Terada, K. Wachi, H. Miyazawa, N. Hatakayama, K. Matsuda and T. Oshima, *J. Med. Chem.* **32**, 1265 (1989).
195. M. Janda, J. Srogl and M. Synackova, *Collect. Czech. Chem. Commun.* **37**, 2584 (1972).
196. R. Knocke and W. P. Neumann, *Liebigs Ann. Chem.* 1486 (1974).
197. S. L. Huang and T.-Y. Chen., *J. Chin. Chem. Soc. (Taipei)* **21**, 235 (1974).

198. F. W. Dunn and K. Dittmer, *J. Am. Chem. Soc.* **68**, 2561 (1946).
199. S. Gronowitz, *Arkiv Kemi* **11**, 519 (1957).
200. S. Gronowitz, *Arkiv Kemi* **13**, 87 (1958).
201. S. Gronowitz, *Arkiv Kemi*, **13**, 231 (1958).
202. F. Alcudia, I. Fernandez, J. M. Llera and F. Zorrilla, *Anales de Quimica* (1988).
203. R. Kuhn and O. Dann, *Ann.* **547**, 293 (1941).
204. D. T. Mowry, M. Renoll and W. F. Huber, *J. Am. Chem. Soc.* **68**, 1105 (1946).
205. E. Campaigne and J. L. Diedrich, *J. Am. Chem. Soc.* **70**, 391 (1948).
206. R. Dahlbom, *Acta Chem. Scand.* **3**, 93 (1949).
207. T. Ohkuma, M. Koizumi, M. Yoshida and R. Noyori, *Org. Lett.* **2**, 1749 (2000).
208. M. J. Burk, W. Hems, D. Herzberg, C. Malan and A. Zanotti-Gerosa, *Org. Lett.* **2**, 4173 (2000).
209. J. Teste and N. Lozac'h, *Bull. Soc. Chim. Fr.* 492 (1954).
210. L. Castedo, J. Delamo, C. López, M. L. López and G. Tojo, *Heterocycles* **38**, 495 (1994).
211. Y. Kitano, M. Kusakabe, Y. Kobayashi and F. Sato, *J. Org. Chem.* **54**, 994 (1989).
212. S. J. Phythian, R. J. K. Taylor and J. Bantick, *J. Chem. Soc., Perkin Trans. 1*, 194 (1990).
213. N. P. Buu-Hoi, N. D. Xuong, and B. K. Diep, *J. Org. Chem.* **26**, 1673 (1961).
214. K.-H. Pfoertner, R. Zell, *Helv. Chimica Acta* **63**, 645 (1980).
215. J. Farkas and J. J. K. Novák, *Coll. Czech Chem. Commun.* **25**, 1815 (1960).
216. K. E. Hamlin, A. W. Weston, F. E. Fischer and R, J. Michael, Jr, *J. Am. Chem. Soc.* **71**, 2731 (1949).
217. J. A. Blanchette and E. V. Brown, *J. Am. Chem. Soc.* **73**, 2779 (1951).
218. M. Goemen, G. Soussan, and P. Freon, *Bull. Soc. Chim. Fr.* 1310 (1973).
219. H. J. J. Loozen and E. F. Godefroi, *J. Org. Chem.* **38**, 1053 (1973).
220. S. Gronowitz and B. Eriksson, *Arkiv Kemi* **21**, 335 (1963).
221. A. R. Katritzky and L. Xie, *J. Org. Chem.* **60**, 3707 (1995).
222. M. Hahn, G. Heinisch W. Holzer and H. A. Schwarz, *J. Heterocycl. Chem.* **28**, 1189 (1991).
223. M. Janda and J. Radouch, *Collect. Czech Chem. Commun.* **32**, 2672 (1967).
224. R. D. Schuetz and W. H. Houff, *J. Am. Chem. Soc.* **77**, 1836 (1955).
225. R. E. Miller and F. F. Nord, *J. Org. Chem.* **15**, 89 (1950).
226. W. Steinkopf, *Ann.* **540**, 14 (1940).
227. Blicke and Burckhalter, *J. Am. Chem. Soc.* **64**, 477 (1942).
228. G. van Zyl, R. J. Langenburg, H. H. Tam and R. N. Schut, *J. Am. Chem. Soc.* **78**, 1955 (1956).
229. S. Gronowitz, *Arkiv Kemi* **12**, 533 (1958).
230. J. Skramstad and P. Fröyen, *Acta Chem. Scand.* **47**, 131 (1993).
231. S. H. Lee, R. N. Hansson and D. E. Seitz, *Tetrahedron Lett.* **25**, 1751 (1984).
232. F. Sauter, P. Stannetty, H. Fröhlich and W. Ramer, *Heterocycles* **26**, 2639 (1987).
233. F. Sauter, P. Stanetty and H. Fröhlich, *Heterocycles* **26**, 2657 (1987).
234. T. Kuroda, M. Takahashi, T. Ogiku, H. Ohmizu, T. Nishitani K. Kondo and T. Iwasaki, *J. Org. Chem.* **59**, 7353 (1994).
235. T. H. Brown, M. A. Armitage, R. C. Blakemore, P. Blurton, G. J. Durant, C. R. Ganellin, R. J. Ife, M. E. Parsons, D. A. Rawlings and B. P. Slingsby, *Eur. J. Med. Chem.* **25**, 217 (1990).
236. A. Schöning and W. Friedrichsen, *Tetrahedron Lett.*, **29**, 1137 (1988).
237. S. Gronowitz and L. Svensson, *Chem. Scripta* **15**, 169 (1981).
238. B. Abarca, G. Asensio, R. Ballesteros and T. Varea, *J. Org Chem.* **56**, 3224 (1991).
239. M. T. Reetz, W. Reif and X. Holdgrün, *Heterocycles* **28**, 707 (1989).
240. A. Ullmann and J. Manassen, *J. Chem. Soc. Perkin Trans. 1*, 1066 (1979).
241. S. Gronowitz and T. Raznikiewicz, *Arkiv Kemi*, **17**, 561 (1961).
242. M. Robba and Y. LeGuen, *Chimie Therap.* **1967**, 120.
243. D. W. H. MacDowell and A. T. Jeffries, *J. Org. Chem.* **36**, 1053 (1971).
244. S. Raucher, A. S.-T. Lui and J. E. Macdonald, *J. Org. Chem.* **44**, 1885 (1979).

245. M. S. Kim, J. S. Gong and I.-S. H. Lee, *J. Heterocycl. Chem.* **29**, 149 (1992).
246. B. Alcaide, C. Biurrun, J. Plumet and E. Borredon, *Tetrahedron Lett.* **33**, 7413, (1992).
247. C. M. Marson and J. Campbell, *Tetrahedron Lett.* **38**, 7785 (1997).
248. A. S. Zanina, G. N. Khabibulina, V. V. Legkoderya and I. L. Kotlyarevskii, *J. Org. Khim. USSR* (*Engl transl.*) **8**, 1556 (1972).
249. H. D. Hartough in *Thiophene and its Derivatives*, (Ed. A. Weissberger), Interscience, New York 1952 p 296.
250. R. Brückner *Chem. Commun.* 141 (2001).
251. D. P. Sharp and T. M. Patrick, Jr, *J. Org. Chem.* **26**, 1389 (1961).
252. M. Janda, J. Srogl, M. Nemec and K. Kalfus, *Collect. Czech. Chem. Commun.* **41**, 1541 (1976).
253. T. L. Cairns and B. C. McKusick, *J. Org. Chem.* **15**, 790 (1950).
254. M. Lemaire, R. Garreau, J. Roncali, D. Delabouglise, H. K. Youssoufi and F. Garnier, *New J. Chem.* **13**, 863 (1989).
255. D. E. Aultz, A. R. McFadden and H. B. Lassman, *J. Med. Chem.* **20**, 456 (1977).
256. A. Jilale and B. Decroix, *Chem. Scr.* **27**, 411 (1987).
257. U. Sonnewald, *Acta Chem. Scand.* B **42**, 567 (1988).
258. P. G. Spinazzé and B. A. Keay, *Tetrahedron Lett.*, **30**, 1765 (1989).
259. P. Cagniant, and D. Cagniant, *Bull. Soc. Chim. Fr.* 2597 (1967).
260. V. N. Gogte, L. H. Shah, B. D. Tilak, K. N. Kadegar and M. B. Sahasrabudhe, *Tetrahedron* **23**, 2437 (1967).
261. M. Takeshita and M. Tashiro, *J. Org. Chem.* **57**, 746 (1992).
262. T. Nakamura and M. Matsumoto, *Synth. Commun.* **29**, 201 (1999).
263. K. Schlögl and H. Egger, *Liebigs Ann.*, **676**, 76 (1964).
264. B. Yom-Tov and S. Gronowitz, *Chem. Scripta* **3**, 165 (1973).
265. I. W. J. Still, N. S. Benait and D. V. Frazer, *Synth. Comm.* **18**, 1461 (1988).
266. W. Carruthers and M. G. Pellatt, *J, Chem. Soc., Perkin Trans. 1* 1136 (1973).
267. T. M. Cresp and M. V. Sargent, *J. Chem. Soc., Perkin Trans. 1* 1786 (1973).
268. G. Kossmehl, M. Härtel and G. Manecke, *Makromol. Chem.* **131**, 15 (1970).
269. J. M. Bastian, A. Ebnöther, D. Jucker, E. Rissi and A. P. Stoll, *Helv. Chim. Acta* **54**, 277 (1971).
270. B. Yom-Tov, S. Gronowitz, S. B. Ross and N. E. Stjernström, *Acta Pharm. Suec.* **11**, 149 (1974).
271. G. M. Kosapoloff, *J. Am. Chem. Soc.* **69**, 2248 (1947).
272. M. D'Auria, F. D'Onofrio and F. Sciarroni, *Synth. Comm.* **22**, 698 (1992).
273. C. Hubert, B.Oussaid, G. Etemad-Moghadam, M. Koenig and B. Garrigues, *Synthesis* 1994, 51.
274. C. M. Angelov, D. D. Enchev and M. Kirilov, *Phosphorus and Sulfur*, **35**, 35 (1988).
275. D. J. Zwanenburg, H. de Haan and H. H. Wynberg, *J. Org. Chem.* **31**, 3363 (1966).
276. I. Lapkin and N. V. Bogoslavskii *Khim. Geterosicl. Soedin.*, 53 (1968).
277. K. Sindelar, J. Holubek, O. Matousová E. Svátek, M. Valchár, A. Dlabac, N. Dlohozková, M. Hrubantová and M. Protiva, *Coll. Czech. Chem. Commun.* **53**, 340 (1987).
278. G. Marchand, B. Decroix and J. Morel, *Chem. Scripta* **23**, 80 (1984).
279. M. Takeshita, M. Tashiro and A. Tsuge, *Chem. Ber.* **124**, 1403 (1991).
280. C. R. Lucas, S. Liu, M. J. Newlands, J.-P. Charland and E. J. Gabe, *Can. J. Chem.* **68**, 644 (1990).
281. E. Jones and I. M. Moodie, *J. Chem. Soc. (C)* 1443 (1967).
282. L. R. Hanton, C. Richardson, W. T. Robinson and J. M. Turnbull, *Chem. Commun.* 2465 (2000).
283. K. Uneyama and M. Momota, *Tetrahedron Lett.* **30**, 2265 (1989).
284. M. Yamamoto, M. Kakinuma, S. Kohmoto and K. Yamada, *Bull. Chem. Soc. Jpn* **62**, 9958 (1989).

285. B. Letois, J. C. Lancelot, C. Saturnino, M. Robba and P. De Caprariis, *J. Heterocycl. Chem.*, **29**, 1769 (1992).
286. J. Nakayama, A. Ishii, Y. Kobayashi and M. Hoshino, *J. Chem. Soc., Chem. Commun.* 959 (1988).
287. L. Hatjiarapoglou and A. Varvoglis, *J. Heterocycl. Chem.* **25**, 1599 (1988).
288. C. Sone and Y. Matsuki, *Bull. Chem. Soc. Jap.* **36**, 618 (1963).
289. E. Baciocchi, E. Muraglia and G. Seiter, *J. Org. Chem.* **57**, 6817 (1992).
290. E. Baciocchi and E. Muraglia, *Tetrahedron Lett.* **34**, 5015 (1991).
291. L. M. Weinstock, E. Corley, N. L. Abramson, A. O. King and S. Karady, *Heterocycles* **27**, 2627 (1988).
292. I.-S. Cho and J. M. Muchowski, *Synthesis* 567 (1991).
293. T. Klingstedt and T. Frejd, *Organometallics* **2**, 598 (1983).
294. B. F. Crowe and F. F. Nord, *J. Org. Chem.* **15**, 81 (1950).
295. L. I. Kruse, D. L. Ladd, P. B. Harsch. F. L. McCabe, S.-M. Mong L. Faucette and R. Johnson, *J. Med. Chem.* **32**, 409 (1989).
296. O. Dann and H. Distler, *Ber.* **84**, 423 (1951).
297. F. Leonard, J. *Am. Chem. Soc.* **74**, 2915 (1952).
298. J. A. Blanchette and E. V. Brown, *J. Am. Chem Soc.* **74**, 1066 (1952).
299. E. V. Brown and J. A. Blanchette, *J. Am. Chem. Soc.* **72**, 3414 (1950).
300. K. Ogura, Y. Ito and G. Tschushihashi, *Bull. Chem. Soc. Jpn* **52**, 2013 (1979).
301. S. Gronowitz and M. Herslöf, *Chem. Scripta* **10**, 90 (1976).
302. J. P. Clayton, A. W. Guest, A. W. Taylor and R. Ramage, *J. Chem. Soc., Chem. Commun.* 500 (1979).
303. T. Bowles, R. J. Gillespie, A. E. A. Porter, J. A. Rechka and H. S. Rzepa, *J. Chem. Soc., Perkin Trans. 1* 803 (1988).
304. P. M. Jackson, C. J. Moody and P. Shah, *J. Chem. Soc., Perkin Trans. 1*, 2909 (1990).
305. L. J. Goossen, *Chem. Commun.* 669 (2001).

– 4B –

Vinylthiophenes

4B.1 PREPARATION OF THCR=C R′R″, R,R′R″=H, ALKYL, HALOGEN

4B.1.1 Dehydration of 1-(thienyl)carbinols *via* reduction of thiophene ketones

The easily polymerized 2-vinylthiophene can be obtained by reaction of 2-acetylthiophene with lithium aluminium chloride followed by dehydration of the intermediate carbinol by heating at 180–190 °C/100 mm Hg, in the presence of 1.2% of hydroquinone [1].

In a similar way 3-vinylthiophene can be prepared [2]. In other cases 5-alkyl-2-acetylthiophenes are reduced with sodium borohydride followed by dehydration with phosphorus pentoxide in refluxing benzene [3]. 3-Methyl-2-vinylthiophene is prepared by treating the carbinol with fused potassium hydroxide [4]. Carbethoxylation of 4-oxo-4,5,6,7-tetrahydrobenzo[*b*]thiophene and 7-oxo-4,5,6,7-tetrahydrobenzo[*b*]thiophene with dimethyl carbonate in the presence of sodium hydride, followed by sodium borohydride reduction and acidification is a good method for the preparation of methyl 6,7-dihydrobenzo[*b*]thiophene-5-carboxylate and methyl 4,5-dihydrobenzo[*b*]thiophene-6-carboxylate, respectively [5].

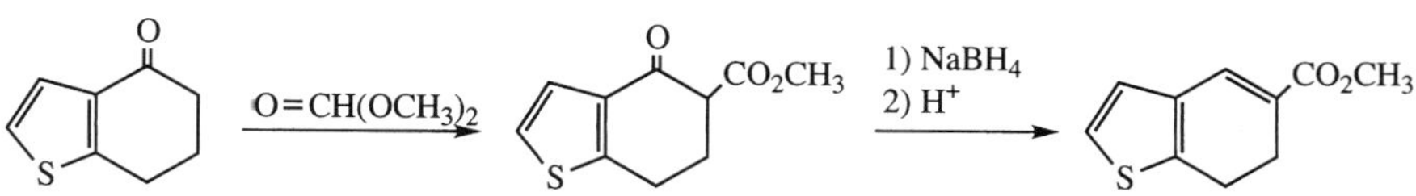

Methyl 6,7-dihydrobenzo[b]thiophene-5-carboxylate [5]

To a stirred suspension of a 55% dispersion of sodium hydride in mineral oil (6.31 g, 0.115 mol) in anhydrous *N*,*N*-dimethylformamide (100 ml), a solution of 4-oxo-4,5,6,7-tetrahydrobenzo[*b*]thiophene (20 g, 0.131 mol) in anhydrous *N*,*N*-dimethylformamide (50 ml) is added dropwise, at room temperature over 30 min under nitrogen. The stirring is continued for 10 min, after which dimethyl carbonate (33 ml) is added over a period of 20 min at 5 °C. After stirring the reaction mixture at room temperature for 90 min, it is poured into water. Extraction with ethyl acetate and evaporation gives methyl 4-oxo-4,5,6,7-tetrahydrobenzo[*b*]thiophene-5-carboxylate as a solid. The solid thus obtained is dissolved in methanol/tetrahydrofuran (1:1) (200 ml). The solution is cooled to −15 °C and treated with sodium borohydride (4.98 g, 0.132 mol) over a period of 1 h. The stirring is continued at this temperature for 30 min after which an excess of crushed solid carbon dioxide is added. The carbon dioxide is allowed to evaporate and the reaction mixture is evaporated. The residue is dissolved in ethyl acetate and this solution is washed with sodium chloride solution, dried over sodium sulfate, and evaporated. The oil thus obtained is dissolved in toluene and refluxed with *para*-toluenesulfonic acid azeotropically for 30 min. After that the reaction mixture is poured into water and the product is extracted with ethyl acetate. The combined organic phases are washed successively with sodium bicarbonate solution and sodium chloride solution, dried over sodium sulfate, and evaporated. The residue is separated chromatographically on silica gel using hexane/ethyl acetate (1:1) as eluent, giving 20.1 g (79%) of the title compound as an oil.

Propenylthiophene is prepared by the Tschugaeff method by preparing the xanthogenate from 1-(2-thienyl)propanol followed by decomposition at 150–200 °C [6].

4B.1.2 *Via* dehydration of alcohols obtained from thienylmetal derivatives and carbonyl compounds

3-Vinylthiophene has been obtained by dehydration of the product from 3-thiophenemagnesium bromide and acetaldehyde with potassium hydrogen sulfate [7] and later from the reaction of 3-thienyllithium with acetaldehyde, followed by thermal dehydration on an alumina column at 250°C/50 mm Hg [8]. 4-Bromo-1,1-bis(3-methyl-2-thienyl)-1-butene, which is an intermediate in the synthesis of a γ-aminobutyric acid uptake inhibitor is prepared through the reaction of 3-methyl-2-thienyllithium with ethyl 4-bromobutyrate followed by acid treatment [9]. 2-(Thienyl)propenes are best prepared through the reaction of 2- and 3-thienyllithium with acetone followed by dehydration with oxalic acid [10].

3-Propylidenethiophene [10]

Butyllithium (1.61 *M*) in hexane (690 ml) is added dropwise to a solution of 3-bromothiophene (151 g, 0.93 mol) in anhydrous diethyl ether (1000 ml) at

−70 °C. When the addition is completed the stirring is continued for 20 min after which acetone (53.7 g, 0.93 mol) in anhydrous diethyl ether (100 ml) is added. The reaction mixture is stirred at low temperature for 4 h, allowed to warm to room temperature, and then poured into 2 *M* hydrochloric acid (650 ml) and ice. The phases are separated and the aqueous phase extracted with ether. The combined organic phases are washed with water, dried over magnesium sulfate, and evaporated. The residue (199 g) and oxalic acid (10.0 g) are heated to 70 °C at 60–70 mm Hg. After 5 h most of the carbinol is consumed and the reaction mixture is cooled and treated with ether. The ether phase is washed with sodium bicarbonate solution, dried over magnesium sulfate, evaporated and distilled under nitrogen giving 62.3 g (54%) bp 56–60 °C/12 mm Hg. This fraction is further purified by chromatography on silica gel using hexane as eluent and the title compound is obtained pure bp 58.5–59.0 °C/12 mm Hg. During the distillation process the flask as well as the receiver are treated with a small amount of hydroquinone in order to prevent polymerization.

For the preparation of 3-(α-methylstyryl)thiophene and 1-phenyl-2-(3-thienyl)propene, 3-thienyllithium was reacted with 2-phenylpropenal and phenylacetone respectively followed by dehydration with refluxing concentrated hydrochloric acid [11]. Cyclohexenylthiophenes, previously important starting materials for phenylthiophenes, are easily prepared by the reaction of various thienyllithium or magnesium derivatives with cyclohexanones [12–15]. The intermediate carbinols are seldom isolated since acidic work-up directly led to the cyclohexenyl derivatives, except for sterically crowded carbinols. In this case refluxing with *para*-toluenesulfonic acid in benzene with water separation was necessary in order to achieve dehydration [12].

Dehydration to the thienylcycloalkenes was achieved with 10% sulfuric acid or *para*-toluenesulfonic acid [16]. Pharmacologically active compounds can be prepared from 2-thiophenemagnesium bromide and thienyl aminoalkyl ketones followed by dehydration [17].

4B.1.3 Dehydration of alcohols obtained from thiophenecarbonyl derivatives and organometallic reagents

Substituted vinylthiophenes have been obtained through the reaction of 2-thiophene aldehydes and 2-acetylthiophene with aryl and alkyl Grignard reagents followed by dehydration [18–20]. A large number of thienylethene derivatives can easily be obtained from thienyl aryl ketones and benzylmagnesium halides. Dehydration can usually be achieved by refluxing the crude carbinol for 5 min with 98% formic acid [19,21–24]. Another route to 4-bromo-1,1-bis(3-methyl-2-thienyl)-1-butene consists in the reaction of cyclopropylmagnesium bromide with bis(3-methyl-2-thienyl)ketone and the intermediate cyclopropylcarbinol opened to the desired product upon treatment with hydrobromic acid in acetic acid [9].

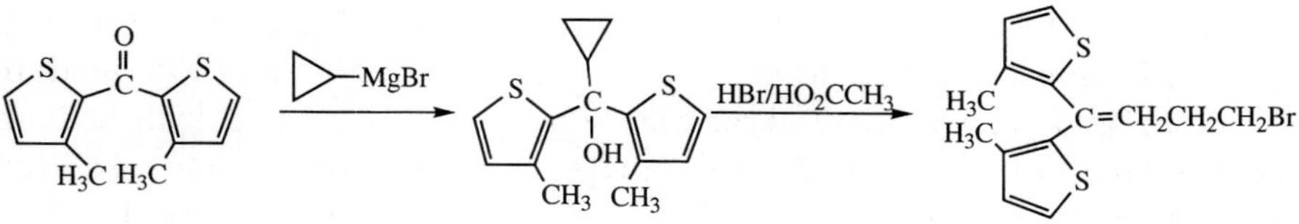

4-Bromo-1,1-bis(3-methyl-2-thienyl)-1-butene [9]

Magnesium turnings (10.7 g, 0.44 mol) in anhydrous tetrahydrofuran (40 ml) are treated with cyclopropyl bromide (53.4 g, 0.44 mol) in anhydrous tetrahydrofuran (60 ml) under nitrogen. After the initial exothermic reaction has subsided, the reaction mixture is heated at reflux for 0.5 h and anhydrous tetrahydrofuran (50 ml) is introduced, after which the reaction mixture is allowed to cool to room temperature. A solution of bis(3-methyl-2-thienyl)methanone (50.6 g, 0.25 mol) in anhydrous tetrahydrofuran (50 ml) is added dropwise. The reaction mixture so obtained is refluxed for 1.5 h and allowed to cool to room temperature and water (200 ml) is carefully introduced. The pH of the aqueous phase is adjusted to 3 with 4 *M* aqueous hydrochloric acid and the phases are separated. The aqueous phase is extracted with tetrahydrofuran (50 ml) and the combined organic phases are dried over sodium sulfate and evaporated giving 63.2 g (98%) of cyclopropyl bis(3-methyl-2-thienyl)methanol as an oil. This carbinol is dissolved directly in acetic acid (300 ml) and 48% aqueous hydrobromic acid is added dropwise at 10 °C. The stirring is continued for 1.5 h and water (1000 ml) is added. The solution so obtained is extracted with diethyl ether (500 ml). The combined organic phases are washed with 10% aqueous potassium carbonate until the washings measured pH 11, dried over sodium sulfate, and evaporated. The residue is subjected to flash chromatography on silica gel using heptane/tetrahydrofuran (9:1) providing 36.5 g (46%) of the title compound as an oil.

Another method to achieve this synthetic goal consists in the reaction of 3-methyl-2-thiophenemagnesium bromide with 4-chloro-1-(2-thienyl)butanone followed by ring opening of the intermediate 2-(3-methyl-2-thienyl)tetrahydrofuran to 4-(3-methyl-2-thienyl)-3-butenol, which then *via* the tosylate is transformed to the chloro derivative [9].

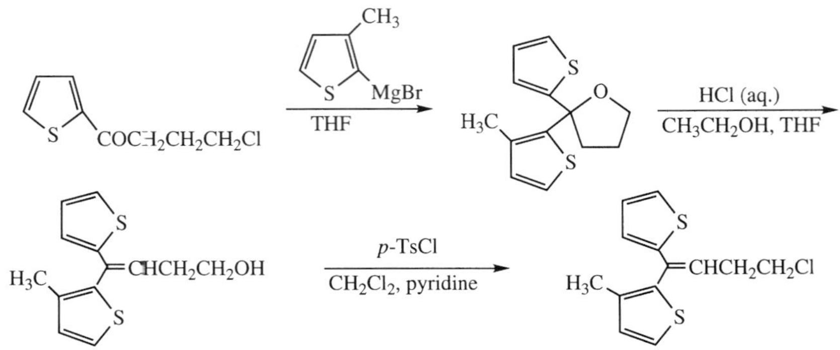

2-(3-Methyl-2-thienyl)-(2-thienyl)tetrahydrofuran [9]

Magnesium turnings (4.7 g, 0.19 mol) in anhydrous tetrahydrofuran (75 ml) are treated with 2-bromo-3-methylthiophene (33.6 g, 0.19 mol) in anhydrous tetrahydrofuran (125 ml) under nitrogen. After the initial exothermic reaction subsides, the reaction mixture is refluxed for 1 h and then allowed to cool to room temperature. A solution of 4-chloro-1-(2-thienyl)butanone (27.8 g, 0.148 mol) in anhydrous tetrahydrofuran (75 ml) is added dropwise. The reaction mixture is refluxed for 0.5 h and then allowed to cool. Concentrated aqueous ammonium chloride (175 ml) is carefully introduced before extraction with ethyl acetate (3 × 200 ml). The combined organic phases are dried over magnesium sulfate and evaporated giving an oil, which is flash chromatographed on silica gel using heptane/ethyl acetate (40:1) as eluent providing 25.6 g (69%) of the title compound as an oil.

4-(3-Methyl-2-thienyl)-4-(2-thienyl)-3-butenol [9]

2-(3-Methyl-2-thienyl)-(2-thienyl)tetrahydrofuran (3.1 g, 12.4 mmol) is dissolved in a mixture of tetrahydrofuran (20 ml) and ethanol (20 ml), after which 2 *M* aqueous hydrochloric acid is introduced. The reaction mixture so obtained is stirred at 80 °C for 1.75 h, evaporated and the residue is purified by flash chromatography using cyclohexane/ethyl acetate (9:1–5:1) as eluent, providing 2.56 g (83%) of the title compound as an oil of *E*- and *Z*-isomers (85:15).

In order to prepare thienylethene derivatives with secondary or tertiary amino groups in the side chain 2-acetyl and 2-propionylthiophene were converted to Mannich bases, that were reacted with benzyl or arylmagnesium halides and the intermediate carbinols dehydrated with 85% sulfuric acid [25,26].

2-Palmityl-5-(5,9-dimethylundecanoyl)thiophene gives 2-palmityl-5-(1,5,9-trimethyl-1-undecenyl)thiophene upon reaction with methylmagnesium iodide followed by dehydration [27]. α-Cyclopentylidene-2-thienylacetic acid is prepared by the reaction of ethyl 2-thienylglyoxylate with cyclopentylmagnesium bromide followed by dehydration [28]. 2-Acetyl- or 2-propionylthiophene can easily be converted to Mannich bases and then reacted with benzylmagnesium halides, arylmagnesium halides, 2-picolyllithium or aryllithium derivatives followed by dehydration giving compounds such as the one below [29–31].

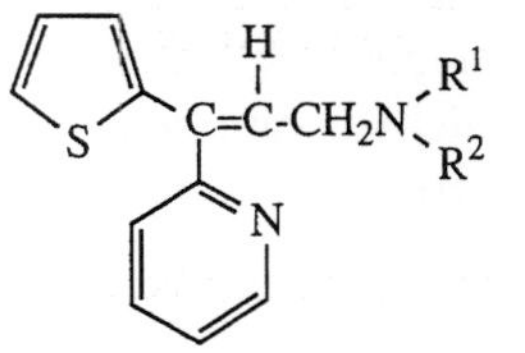

6,7-Dihydrobenzo[*b*]thiophene-5-carbonitrile can conveniently be prepared by sodium borohydride reduction of 5-cyano-6,7-dihydrobenzo[*b*]thiophene-4(5*H*)-one, followed by dehydration through treatment with *para*-toluenesulfonic acid and water separation [32].

4-Hydroxy-6,7-dihydrobenzo[b]thiophene-5-carbonitrile [32]

5-Cyano-6,7-dihydrobenzo[*b*]thiophene-4(5*H*)-one (4.53 g, 25 mmol) is dissolved in twice-distilled methanol (100 ml) with stirring and cooling. A powdered sodium borohydride (1.02 g, 25 mmol) is added in small portions to the cooled mixture, while the temperature is maintained at 0–5 °C. The reaction mixture is kept at room temperature for 16 h after which crushed ice is added followed by careful acidification with hydrochloric acid. The mixture is extracted with diethyl ether, the combined organic phases washed with saturated sodium chloride solution, dried over sodium sulfate and evaporated to provide 4.39 g (96%) of the crude reaction product.

6,7-Dihydrobenzo[b]thiophene-5-carbonitrile [32]

A catalytic amount of *para*-toluenesulfonic acid (200 mg) is added to the crude 4-hydroxy-6,7-dihydrobenzo[*b*]thiophene-5-carbonitrile (4.36 g, 24 mmol) dissolved in anhydrous benzene (125 ml) in a flask fitted with a condenser and a Dean-Stark separator. The reaction mixture is refluxed under nitrogen for 8 h, allowed to cool to room temperature, washed with 5% sodium hydrogen carbonate and saturated sodium chloride solution, dried over sodium sulfate, and evaporated. The residue is purified by chromatography on silica gel using diethyl ether/light petroleum (1:9) as eluent. After recrystallization from diethyl ether/light petroleum 3.33 g (85%) of the title compound is obtained mp 56–57 °C.

4B.1.4 Dehydration of 2-(thienyl)ethanols

2-Vinylthiophene can also be prepared by the reaction of 2-thienylsodium [33] or 2-thiophenemagnesium iodide [34,35] with ethylene oxide followed by dehydration of the 2-(2-thienyl)ethanol with aqueous hydrochloric acid [33,34] or distilling from copper and powdered potassium hydroxide [35]. The Grignard route usually gives higher yields, as found in studies using propylene oxide and styrene oxide and dehydration was achieved by distillation from an equal weight of potassium hydrogen sulfate [36].

4B.1.5 Dehydration of 1-(thienyl)alkyl halides

The best method for the preparation of 2-vinylthiophene still appears to be the chloroethylation of thiophene with paraldehyde and hydrochloric acid to 1-(2-thienyl)ethyl chloride, which is dehydrochlorinated without isolation *via* quaternization with pyridine [37]. According to a detailed procedure published in *Organic Synthesis*, the yield is 50–55% [38]. Side-chain bromination followed by dehydrobromination can also be used for the preparation of thienyl ethene derivatives [39]. Recently it was found that the best method for the preparation of 3-vinylthiophene is a two-step procedure from 2-(3-thienyl)ethanol, which was converted to 3-(2-bromoethyl)thiophene through the reaction with triphenylphosphine dibromide and upon treatment with tetraglyme as solvent and 1,8-diazabicyclo[5.4.0]undec-7-ene (DBU) as base gave the desired compound [40].

3-Vinylthiophene [40]

2-(3-thienyl)ethanol (1.0 g, 7.8 mmol) is added at 0 °C to an anhydrous acetonitrile solution (15 ml) of triphenylphosphine dibromide, prepared *in situ* from triphenylphosphine (2.05 g, 7.8 mmol) and bromine (0.4 ml, 7.8 mmol). The reaction mixture is refluxed for 2–3 h and then cooled to room temperature, diluted with diethyl ether and washed three times with water. The organic phase is dried over magnesium sulfate and evaporated. The residue is chromatographed on silica gel using petroleum ether as eluent, giving 1.41 g (95%) of the halogenated derivative as a clear yellow liquid. This compound (0.95 g, 5.0 mmol) is dissolved in freshly distilled tetraglym (10 ml) and 1,8-diazabicyclo[5.4.0]undec-7-ene (0.90 ml, 6.0 mmol) is added dropwise at room temperature. The reaction mixture is stirred overnight, worked up and distilled at reduced pressure giving 74% of the title compound.

4B.1.6 From thiophene aldehydes and triphenylalkylphosphonium salts by the Wittig reaction

2-Vinyl-3-thiophenaldehyde ethylene acetal was prepared through the reaction of the methylene ylide prepared in tetrahydrofuran with butyllithium and 2-formyl-3-thiophene aldehyde ethylene acetal [41].

4B.1.7 From thienyllithium derivatives and fluoroalkenes

An excellent yield of 1,1-dichloro-2-fluoro-2-thienylethene is obtained in the reaction of 2-thienyllithium with 1,1-dichloro-2,2-difluoroethene [42]. Palladium-catalyzed coupling of 2-thienylzinc bromide with difluoro-1,2-triethylsilyl-1-iodoethylene (*E*) offers a route to difluoro-1,2-triethylsilyl-1-(2-thienyl)ethene [43].

Difluoro-1,2-triethylsilyl-1-(2-thienyl)ethene [43]

Thienylzinc bromide (30 mmol) is added to tetrakis(triphenylphosphine)palladium(0) (0.7 g, *ca* 3%) and (*Z*) 1,2-difluoro-2-triethylsilyl-1-iodoethylene (20 mmol) in tetrahydrofuran (30 ml). The stirring is continued for 4 h, after which the reaction mixture is hydrolyzed with diluted sulfuric acid and extracted with diethyl ether. The combined organic phases are washed with sodium bicarbonate and water, dried over magnesium sulfate, and evaporated. The residue is treated with pentane (100 ml) in order to precipitate the palladium salts and the solution so obtained is filtered through a pad of silica gel, which is then washed with pentane. The filtrate is evaporated and distilled giving 41% of the title compound, 100% *Z*, bp 74–75 °C/0.1 mm Hg.

The analogous chlorofluoro-2,5-disubstituted thiophene is prepared from 2,5-dilithiothiophene and 1,1-dichloro-2,2-difluoroethene [44]. The elimination reaction can also be performed with zinc and zinc chloride [45].

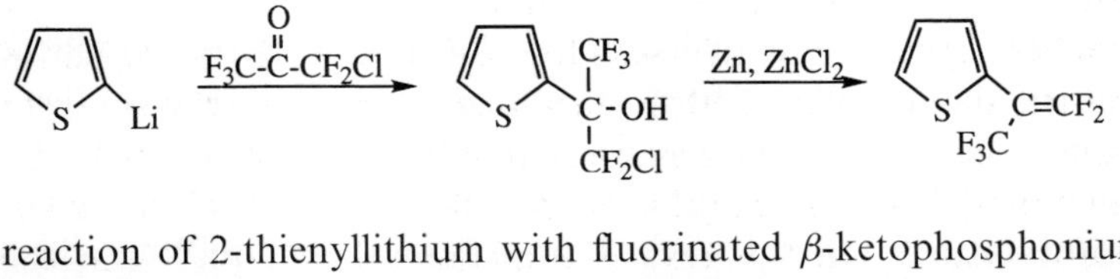

The reaction of 2-thienyllithium with fluorinated *β*-ketophosphonium salts is a convenient method for the introduction of fluoroalkylvinyl group in a one-pot procedure [46].

General procedure for preparation of fluoroalkylvinyl thiophenes [46]

The phosphorane is generated from the corresponding phosphonium salt (3 mmol) and butyllithium (3 mmol) in anhydrous tetrahydrofuran (30 ml) at 0 °C under nitrogen. The reaction mixture is cooled to −78 °C and perfluoroalkanoic anhydride (2.6 mmol) is slowly added until the disappearance of the ylidic color. After stirring at −78 °C for 5 min, thienyllithium (3 mmol) is added. The reaction mixture is allowed to warm to room temperature, stirred for a further 2 h, diluted with petroleum ether (bp 30–60 °C, 100 ml) and filtered. The filtrate is evaporated and the residue purified by chromatography on silica gel using petroleum ether (bp 30–60 °C) as eluent giving the compound shown below.

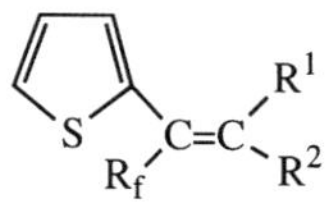

R^1	R^2	R_f	bp °C/mm Hg	Yield (%)	*Z*:*E*
-$(CH_2)_4$-		CF_3	60–61/1.0	72	
-$(CH_2)_4$-		C_2F_5	52–54/0.2	70	
CH_3	CH_3	CF_3	38–40/0.5	56	
CH_3	CH_3	C_2F_5	39–41/0.5	67	
CH_3	Ph	CF_3	48–49 (mp)	82	100:0
CH_3	CH_2Ph	CF_3	82–84/0.3	60	71:29
CH_3	C_3H_7	CF_3	45–47/0.2	69	73:27
CH_3	C_4H_9	CF_3	50–52/0.7	80	67:33

A number of 1,4-bis(thienyl)-1,2,3,4-tetrafluoro-1,3-butadienes have recently been obtained through the reaction of, for instance, 2- or 3-thienyllithium with hexafluoro-1,3-butadiene. The reaction of 2,5-dilithiothiophene with 1-(2-thienyl)-F-1,3-butadiene can be used for the preparation of 2,5-bis[4-(2-thienyl)-1,2,3,4-tetrafluoro-1,3-butadienyl]thiophene [47].

(E,Z)-1-(2-Thienyl)-1,2,3,4,4-pentafluoro-1,3-butadiene [47]

Butyllithium (1.0 *M*) is added dropwise to a solution of thiophene (6.3 g, 75 mmol) in diethyl ether (10 ml) during 30 min, after which the reaction mixture is stirred at 40 °C for 3 h. The solution of 2-thienyllithium is precooled and added to a solution of F-1,3-butadiene (12.5 g, 77 mmol) in anhydrous tetrahydrofuran (40 ml) by transfer tube at −78 °C for 1 h. When the addition is complete the reaction mixture is stirred at −78 °C for 2.5 h, hydrolyzed and extracted with ether. The combined organic phases are dried over magnesium sulfate, evaporated and distilled, giving 5.9 g (35%) of the title compound mainly as *E*-isomer bp 57 °C/8 mm Hg.

2,5-Bis[4-(2-thienyl)-1,2,3,4-tetrafluoro-1,3-butadienyl]thiophene [47]

A solution of 2,5-dilithiothiophene, prepared from thiophene (0.1 g) and butyllithium/tetramethylethylenediamine in hexane (2.4 mmol), is cooled to −70 °C and 1-(2-thienyl)-1,2,3,4,4-pentafluoro-1,3-butadiene(0.55 g) is added. After work-up the crude product is recrystallized from a benzene/hexane mixture and purified by column chromatography on silica gel, using hexane/benzene (10:1) as eluent.

4B.1.8 From thiophenecarbonyl compounds by various other methods

2-Vinylthiophene can also be prepared in 56% yield by the reaction of 2-acetylthiophene with diethyl phosphite and 3–10 mol% of sodium amide followed by heating of the intermediate hydroxyphosphonate at 140 °C [48]. The reaction of 2-acetylthiophene with phosphorus pentachloride followed by dehydrochlorination gives the *E*- and *Z*-isomers of 1,2-dichloro-1-(2-thienyl)-ethene [49]. 1,1-Dibromo-2-(2-thienyl)ethenes are prepared in quantitative yields by the reaction of thiophene aldehydes with carbon tetrabromide and triphenyl phosphine [50–52]. The following reaction can also be performed [53].

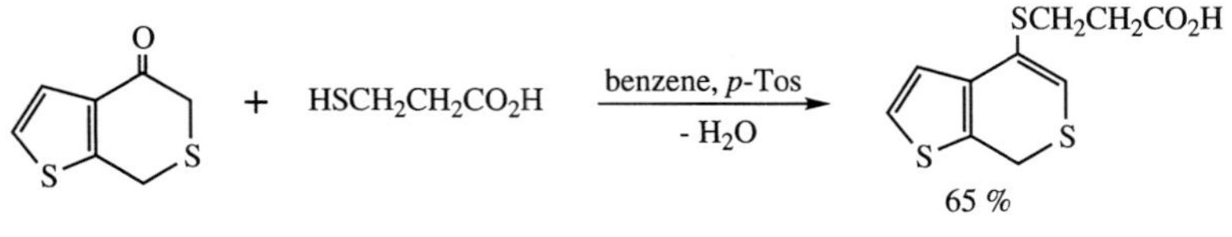

1,1-Dichloro-2,2-di(2-thienyl)ethene and 1,1-dichloro 2-phenyl-2-(2-thienyl)-ethene can be prepared by the reaction of di(2-thienyl)diazomethane or phenyl-(2-thienyl)diazomethane obtained by oxidation of their hydrazones with dichlorocarbene derived from chloroform [54]. The two tautomeric thiophene analogs of methylindene are obtained in the following way [55].

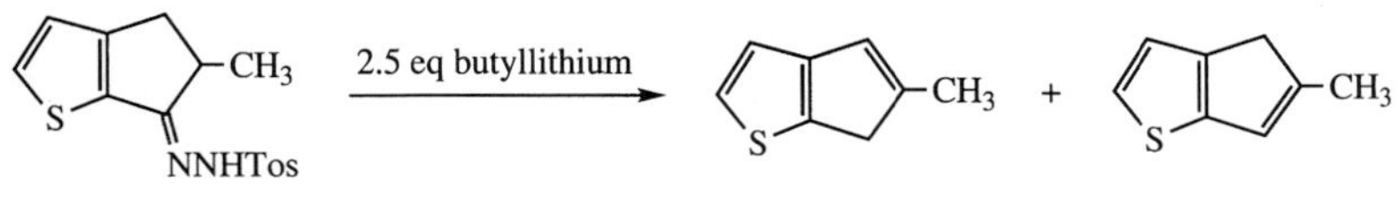

Reaction of 2-acetylthiophene with *N*,*N*-dimethylformamide dimethylacetal, gives *E*-3-dimethylamino-1-(2-thienyl)propenone, which upon reaction in methylene chloride with phosphorus oxychloride, followed by methanol and sodium hexafluorophosphate yields 3-chloro-3-(2-thienyl)prop-2-en-1-ylidene dimethylimminium hexafluorophosphate [56].

E-3-Dimethylamino-1-(2-thienyl)propenone [56]

A mixture of 2-acetylthiophene (10.0 g, 0.079 mol) and *N,N*-dimethylformamide dimethyl acetal (37.8 g, 0.317 mol) is placed in a round-bottomed flask equipped with a reflux condenser and a magnetic stirrer and is heated at reflux overnight. After cooling to room temperature and evaporation the resulting red–orange solid is dried *in vacuo* to yield 14.3 g (95%) of material, which is sufficiently clean for the following step. An analytical sample can be prepared by radial chromatography of a 0.5 g sample product on a 2 mm thick plate of silica gel on a Harrison Chromatotron using hexane/ethyl acetate (60:40) as eluent. The main fraction contains the title compound mp 121–123 °C.

3-Chloro-3-(2-thienyl)prop-2-en-1-yliden-dimethylimminium hexafluorophosphate [56]

Anhydrous dichloromethane (100 ml) and *E*-3-dimethylamino-1-(2-thienyl)-propenone (14.0 g, 0.077 mol) are placed into a dry 250 ml three-necked flask. To this solution phosphorus oxychloride (11.8 g, 0.077 mol) is added dropwise with stirring, after which the stirring is continued for 30 min. After evaporation a cold solution of sodium hexafluorophosphate (26 g, 0.154 mol) in methanol is carefully added to the residue. The yellow solid is isolated by filtration giving 23.7 g (89%) of the title compound, which can be recrystallized from methanol mp 177–178 °C.

Compounds of the type shown below are prepared by the addition of allenic amines to thiophene aldehyde and lithium bromide in tetrahydrofuran followed by treatment of the reaction mixture with a small amount of acetic acid at elevated temperatures. In this reaction morpholine has also been used [57].

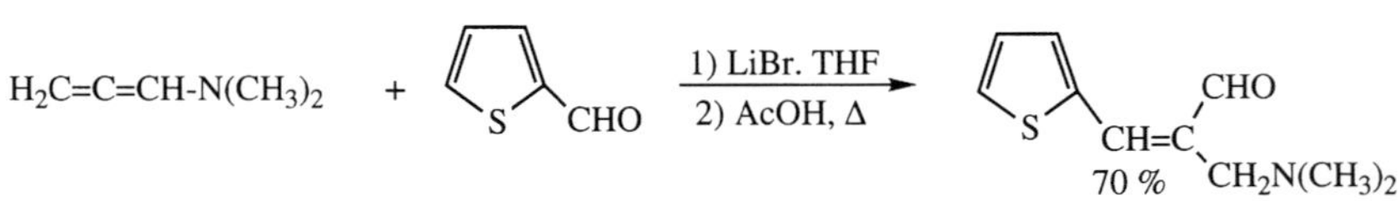

The reaction of 2-thiophenecarbonyl chloride with tetrafluoro-2-cyclobutenyl copper, prepared from 1-chloro-2-iodotetrafluorocyclobutene *via* the zinc derivative, is a method for the preparation of 1-chloro-2-(2-thenoyl)tetrafluorocyclobutene [58].

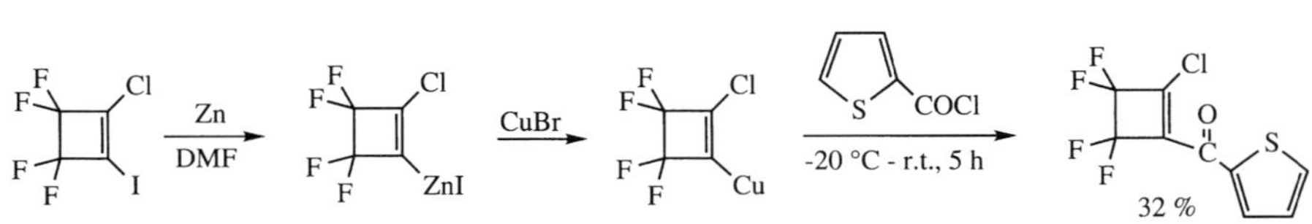

4B.1.9 From vinylmetallic derivatives and halothiophenes

Palladium-catalyzed coupling of 1-trimethylsilylmethylvinylzinc chloride with 2-bromothiophene can be used for the preparation of 1-trimethyl(silylmethyl)-vinylthiophene [59]. The palladium-catalyzed reaction of 2-bromothiophene with the zinc derivative derived from (1-bromoethenyl)trimethylsilane leads to 1-(2-thienyl)ethenyltrimethylsilane [60].

1-(2-Thienyl)ethenyltrimethylsilane [60]

1-(Bromoethenyl)trimethylsilane (1.50 g, 8.37 mmol) in anhydrous tetrahydrofuran (30 ml) is reacted with butyllithium (9.21 mmol) at −78 °C for 1 h. Zinc bromide (12.5 mmol) in anhydrous tetrahydrofuran (30 ml) is added and the reaction mixture stirred at −20 °C for 1.5 h. A solution containing tetrakis-(triphenylphosphine)palladium(0) (0.39 g, 4 mol %) and 2-bromothiophene (0.81 ml, 8.37 mmol) in anhydrous tetrahydrofuran (25 ml) is added. The reaction mixture is allowed to warm to room temperature and then refluxed for 1 h. The reaction is quenched with aqueous ammonium chloride, the phases are separated, the aqueous phase extracted with ether, the combined phases are dried and evaporated. The residue is flash chromatographed and distilled in a kugelrohr apparatus giving 1.34 g (88%) of the title compound bp 105 °C (bath) and 0.2 mm Hg.

The reaction of 1-chloro-trimethylsilylvinylmagnesium bromide or zinc chloride with 3-bromothiophene is used for the preparation of 3-(1-trimethylsilylvinyl)thiophene [61]. Stereodefined 2-arylethenyldimethylphenylsilanes react with 2-iodothiophene under palladium catalysis in the presence of fluoride anions, such as tetrabutylammonium fluoride (TBAF), to give 1-aryl-2-(2-thienyl)ethenes [62].

(E)-1-Phenyl-2-(2-thienyl)ethene [62]

To a solution of tetrakis(triphenylphosphine)palladium(0) (0.47 g, 0.45 mmol) and 2-iodothiophene (5.16 g, 24.6 mmol) in tetrahydrofuran (100 ml) dimethylphenyl-[(*Z*)-2-phenylethenyl]silane (3.74 g, 15.7 mmol), a 1 *M* solution of tetrabutylammonium fluoride (23.5 ml, 23.5 mmol), which was dried over 4 Å molecular sieves for 16 h at room temperature, and triethylamine (2.2 ml, 15.7 mmol) are sequentially added. The reaction mixture is stirred at 65 °C for 112 h under argon, after which it is cooled to room temperature and poured into a large excess of saturated ammonium chloride solution. After extraction with diethyl ether the combined organic phases are washed with water to neutrality, dried, and evaporated. The residue is purified by medium pressure liquid chromatography on silica gel using hexane as eluent, giving 0.9 g (31%) of the title compound mp 108–109 °C from methanol 100% (*E*)-isomer.

The palladium(0)-catalyzed reaction of farnesylzinc iodide with iodothiophenes has been used for the preparation of conformational probes for an isopropenyl subunit [63]. Fluorinated vinylzinc derivatives have been coupled with 2-iodothiophene to give compounds such as 1,1-difluoro-2-fluoro-2-(2-thienyl)ethene and 1-*sec*-butyl-1-fluoro-2-fluoro-2-(2-thienyl)ethene [64,65].

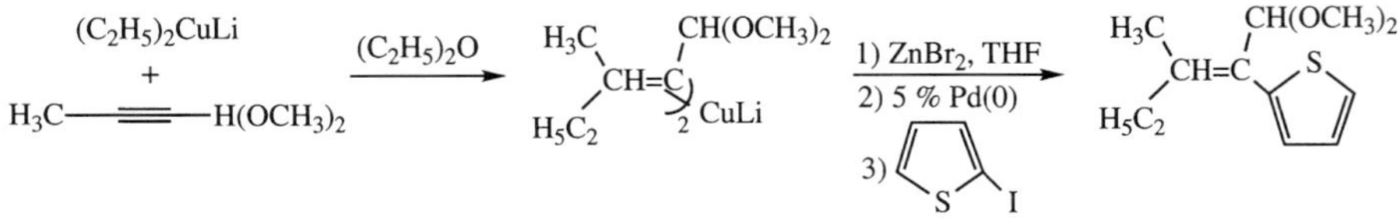

Reactions of the type shown above occur almost exclusively with *cis* stereochemistry [66]. A good method for the synthesis of 2-[(*Z*)-1-decenyl]-thiophene is the reaction, shown below, using tris(diethylamino)sulfonium difluorotrimethylsilicate (TASF) [67].

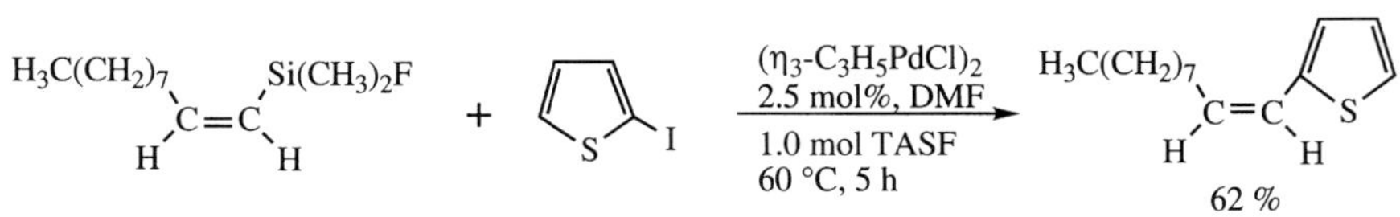

3-(1-Hexenyl)thiophene is conveniently prepared by Suzuki coupling of 3-bromothiophene with 1-hexenylboronic acid [68].

3-(1-Hexenyl)thiophene [68]

A 500 ml three-necked flask is charged with 3-bromothiophene (11.4 g, 70 mmol) and tetrakis(triphenylphosphine)palladium(0) (4.0 g, 5 mol%). After stirring this mixture for 15 min 1-hexenylboronic acid (12.9 g, 98 mmol) in tetrahydrofuran (200 ml) is added together with 2 *M* aqueous sodium hydroxide (80 ml). The flask is fitted with a reflux condenser and the reaction mixture refluxed for 3.5 h, cooled, and diluted with pentane. The phases are separated and the organic phase washed with 2 *M* sodium hydroxide solution (3 × 35 ml) and water, dried over magnesium sulfate, and evaporated. The crude product is flask chromatographed using pentane as eluent giving 55% of the title compound. The yield can be increased by rechromatographing contaminated fractions.

Coupling of 5-methyl-2-thienylmercury chloride with vinyl bromide using Wilkinson's catalyst, chlorotris(triphenylphosphine)rhodium, can be used for the preparation of 5-methyl-2-vinylthiophene [69].

4B.1.10 From thienylmetal derivatives and vinyl halides by transition metal-catalyzed reactions

Palladium-catalyzed reaction of 2-thienylzinc chloride with 2-bromoethenyltrimethylsilane is a good method for the preparation of [2-(2-thienyl)ethenyl]-trimethylsilane in 89% yield [60].

4B.1.11 From vinylboronic acid derivatives and thienyllithia

An early approach to the stereo- and regiospecific preparation of trisubstituted alkenes such as 1-(2-thienyl)-1-phenyl-2-alkylethenes consist in the reaction of 1-bromo-1-hexyne with dibromoborane dimethyl sulfide, followed by treatment with potassium isopropoxide which gives the vinylic boronic acid derivative, which upon reaction with 2-thienyllithium and base yields the boron-containing thienyl derivative. The Suzuki coupling of this compound with iodobenzene gives (*Z*)-1-phenyl-1-(2-thienyl)-2-butylethene. The other isomer is obtained if phenyllithium and 2-iodothiophene are used instead [70].

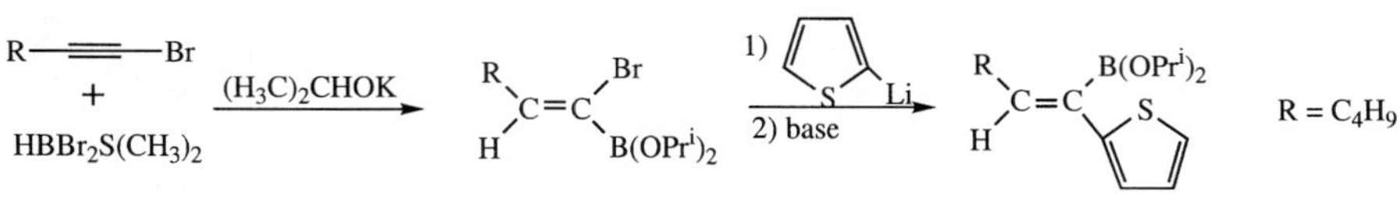

Another excellent approach to the preparation of stereochemical pure unsymmetrically trisubstituted olefins, such as 5-(2-thienyl)-(*E*)-5-decene, in high stereochemical purity, consists in the reaction of (*E*)-2-(1-butyl-1-hexenyl)-1,3,2-dioxaborinane with 2-thienyllthium at −70 °C, followed by reaction with iodine in methanol. Starting from (*Z*)-2-(1-butyl-1-hexenyl)-1,3,2-dioxyborinane the (*Z*) isomer is obtained [71].

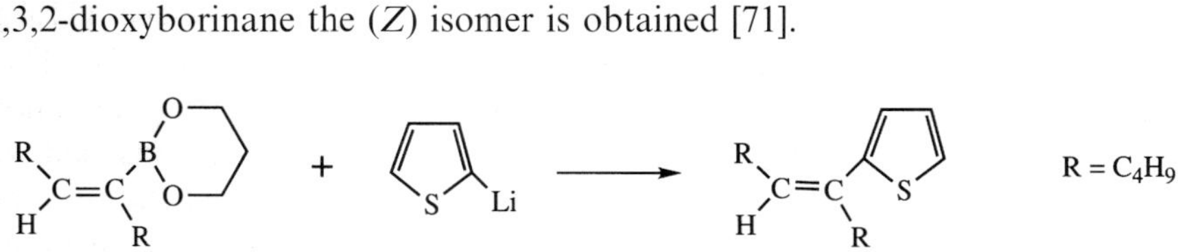

5-(2-Thienyl)-(E)-5-decene [71]

(*E*)-2-(1-butyl-1-hexenyl)-1,3,2-dioxaborinane (2.43 ml, 10 mmol) and anhydrous diethyl ether (20 ml) are placed in a dry 100 ml flask equipped with a magnetic stirring bar and septum inlet. After cooling to −78 °C thienyllithium (10 mmol) is added dropwise. The reaction mixture is stirred at −78 °C for 0.5 h and at 0 °C for 1 h. The solvents are then pumped off and methanol (10 ml) is added at 0 °C. Iodine (2.54 g, 10 mmol) in methanol (40 ml) is then added slowly at −78 °C. After stirring the reaction mixture at −78 °C for 3 h it is allowed to come to room temperature. 3 *M* Sodium hydroxide solution (10 ml) is added, stirring is continued for 15 min, water is added and the solution so obtained is extracted with pentane (3 × 25 ml). The combined organic phases are washed with a 1 *M* aqueous sodium thiosulfate solution (25 ml) and water (2 × 50 ml), dried over potassium carbonate, evaporated and distilled giving 81% of the title compound bp 76–78 °C/0.01 mm Hg.

4B.1.12 Vinylthiophenes *via* side chain modification of vinyl derivatives

Simpler vinylthiophenes can be obtained by decarboxylation of *β*-(thienyl)-acrylic acids with copper and quinoline [72–74]. By oxidation with 2,3-dichloro-5,6-dicyano-1,4-benzoquinone (DDQ) the following product can be obtained [75].

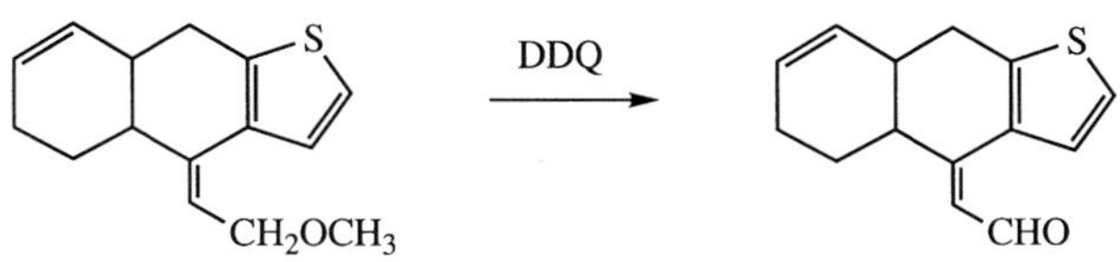

4B.1.13 From thienylacetylene derivatives

Vinylthiophenes are many times conveniently prepared by catalytic hydrogenation of thienylacetylenes using Lindlar's catalyst or other palladium catalysts [76–79]. The reaction of 2- or 3-trimethylsilylethinyl derivatives with diisobutylaluminum hydride (DIBALH) is an excellent method for the preparation of (*Z*)-(*β*-bromo-*β*-trimethylsilylvinyl)thiophenes [80]. The palladium-catalyzed reaction of 2-thienylacetylene with trimethylsilyl cyanide is a convenient route to 2-(2-thienyl-3-trimethylsilyl)-*Z*-prop-2-enenitrile [81].

2-(2-Thienyl-3-trimethylsilyl)-Z-prop-2-enenitrile [81]

Palladium dichloride (36 mg, 0.2 mmol) and pyridine (32 μl, 0.4 mmol) are added to a solution containing thienylacetylene (0.54 g, 4.8 mmol) and trimethylsilyl cyanide (1.34 g, 10 mmol) in toluene (10 ml). The reaction mixture is refluxed with stirring under nitrogen for 20 h. After cooling and evaporation the residue is chromatographed on silica gel using hexane/ethyl acetate (9:1) as eluent followed by bulb-to-bulb distillation (80–90 °C/0.4 mm Hg), giving 39% of the title compound.

Reaction of dimethylphenyl-[(2-thienyl)ethynyl]silane with diisobutylaluminum hydride in hexane is a good method for the preparation of the (*E*)-ethenyl derivative [62]. Another alternative is to react, for instance, 2-ethynyl-5-methylthiophene at −50 °C with the cuprate obtained from dimethylphenylsilyl lithium and copper cyanide which gives the desired dimethyl-[(*E*)- (5-methyl-2-thienyl)ethenyl]phenylsilane in 57% yield [62].

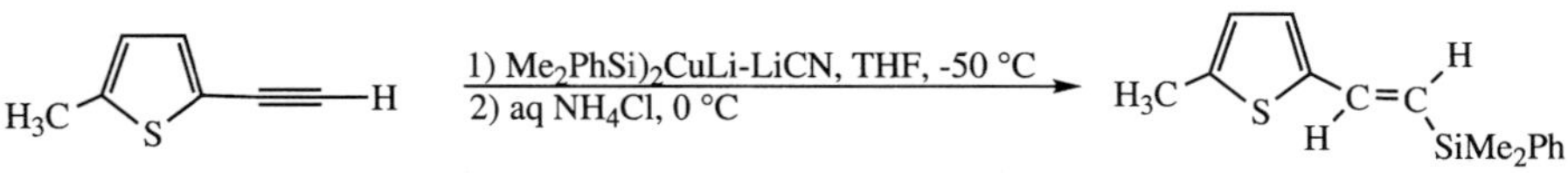

4B.1.14 Vinylthiophenes *via* ring-closure reactions

Ring-closure reactions are especially of importance when other functionalities besides the vinyl group are introduced on the thiophene ring. In the present section only reactions leading to vinyl and alkylsubstituted thiophenes will be mentioned.

4-Methyl-2-vinylthiophene and 4,5-dimethyl-2-vinylthiophene are prepared by the reaction of vinylacetylenic epoxides with hydrogen sulfide and boric acid at 50 °C [82]. The same reaction with 2-methyl-4-(1-hydroxycyclohexyl)-1,2-epoxy-3-butyne gave 4-methyl-2-(1-cyclohexenyl)thiophene after dehydration with 5% sulfuric acid [83]. Reactions of vinyldiacetylenic compounds with hydrogen sulfide is a method for the preparation of the compound shown below [84].

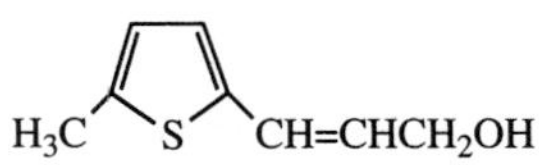

2-Methyl-, 3-ethyl-, and 2-ethyl-4-vinylthiophene are prepared by the reaction of 1-butadienyl triphenylphosphonium salt in pyridine with α-mercaptodiethyl ketone and α-mercapto butyraldehyde, respectively [85]. 3-Alkyl-4-(2-propenyl)thiophenes are conveniently obtained by heating allenyl ethynyl sulfides in isopropanol [86].

4B.1.15 Vinylthiophenes *via* isomerization and rearrangements

The following reaction has been performed [87].

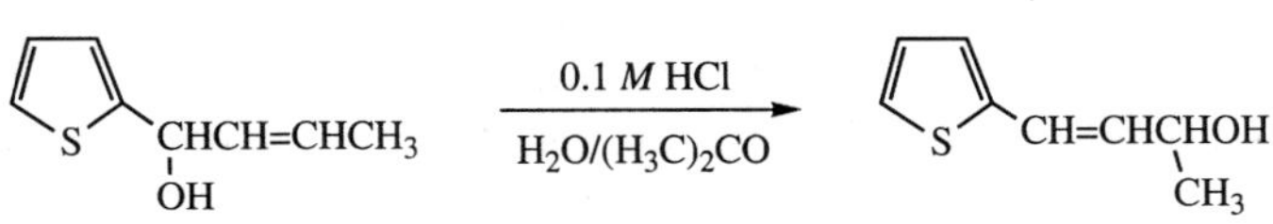

The Meyer–Schuster rearrangenment by refluxing in aqueous formic acid has also been used, giving yields of 45–79% [88,89].

4B.1.16 Ring substitution of vinylthiophenes

The double bond in vinylthiophenes does not interfere in metalation [90–94] or halogen–metal exchange [95–97] of the thiophene ring, making possible the synthesis of a great variety of substituted vinylthiophenes, which will be discussed in other chapters.

In some cases reaction of vinylthiophenes with electrophilic reagents also gives selective ring substitution. In other cases substitution at the vinylic bond or addition to the double bond might occur, depending upon the substituents already present in the thiophene ring or at the double bond. 5-Formyl- and 5-acetyl-2-styrylthiophene can be prepared from 2-styrylthiophene by Vilsmeyer formylation and Friedel-Crafts acetylation [98,99]. Triarylethenes with bromine at the vinylic bond are obtained upon bromination of triarylethenes with bromine or *N*-bromosuccinimide. Dibromination then occurs on the thiophene ring [99–102].

4B.2 PREPARATION OF THCH=CRR′, R=OR, SR, NO_2, CN, CH_2OH

4B.2.1 Condensation of thiophene aldehydes with nitro derivatives

1-(Thienyl)-2-nitroalkenes can be prepared by the condensation of thiophene aldehydes with nitroalkanes such as nitromethane, nitroethane, and phenyl nitromethane [103].

2-(2-Nitropropene-1-yl)thiophene [103]

A mixture of 2-thiophene aldehyde (10 g, 89 mmol), nitroethane (10 g, 134 mmol), and ammonium acetate (4 g, 52 mmol) in acetic acid (40 ml) is refluxed for 2 h. After cooling the reaction mixture to room temperature it is poured into ice-water. The precipitate is filtered off and recrystallized from methanol giving 7.6 g (51%) of the title compound as an orange solid mp 68–70 °C.

The best and most frequently used method in the reaction with nitromethane is an adaption of the preparation of β-nitrostyrene described in *Organic Synthesis* [104], in which an approximately 50% aqueous solution of sodium hydroxide is added to a methanolic solution of the aldehyde and nitromethane and elimination is achieved by adding the reaction product to hydrochloric acid. However, the application of these conditions to thiophene aldehydes gave very varying yields (14–83%) [105,106]. Another modification giving yields of about 50% uses methylamine hydrochloride and sodium carbonate as catalyst in ethanol [107]. A third modification uses primary amines such as butylamine or amylamine and reaction times of 3–4 days, yielding 50–58% of the 1-(thienyl)-2-nitroalkene [105,108] (triethylamine can also be used [109]).

2-(2-Thienyl)-1-nitroethene [109]

A mixture of 2-thiophene aldehyde (11.2 g, 0.10 mol), nitromethane (9.2 g, 0.15 mol), and triethylamine (2.0 g, 0.02 mol) is stirred at room temperature for 17 h and then quenched by addition of 2 *M* hydrochloric acid (50 ml). The resulting solution is washed with water (2 × 50 ml) and sodium chloride solution (50 ml) and dried over sodium sulfate. After evaporation the residue is subjected to chromatography on silica gel using benzene as eluent. The alcohol so obtained is added to a mixture of acetic anhydride (16 g) and pyridine (18.4 g) and the reaction mixture stirred for 2 h, after which it is poured into 2 *M* hydrochloric acid (50 ml). The phases are separated and the aqueous phase extracted with diethyl ether (3 × 30 ml). The combined organic phases are washed with water (3 × 30 ml) and sodium chloride solution (1 × 30 ml) and then dried over sodium sulfate. After evaporation the acetate so obtained is subjected to chromatography on silica gel using hexane/benzene (1:1) as eluent, which is then dissolved in benzene (100 ml) and the resulting solution is refluxed with sodium carbonate for 5 h. The reaction mixture is filtered and the filtrate is washed with water (2 × 50 ml) and sodium chloride solution (1 × 50 ml), dried and evaporated giving 10.2 g (66%) of the title compound mp 79–80 °C.

Another route to 2-(2-thienyl)-1-nitroethene consists in the reaction of 2-vinylthiophene or 2-(2-thienyl)acrylic acid with nitrogen tetroxide [110]. Finally in a fourth development a Schiff's base of the aldehyde prepared first with propyl-, butyl-, or benzylamine and then reacted with glacial acetic acid appears to be the best method of preparation, especially useful with nitromethane and phenyl nitromethane and giving yields higher than 80% [111,112].

3-(2-Thienyl)allyl alcohols are prepared by reduction of the corresponding 3-(2-thienyl)acrolein with sodium borohydride in methanol and can be transformed to the acetate by reaction with acetic anhydride in

4-(*N*,*N*-dimethylamino)pyridine (DMAP) [113,114]. 1-Tosyl-(1-methoxymethyl-2-pyrollyl)-2-(2-thienyl)ethene can be obtained as shown below [115].

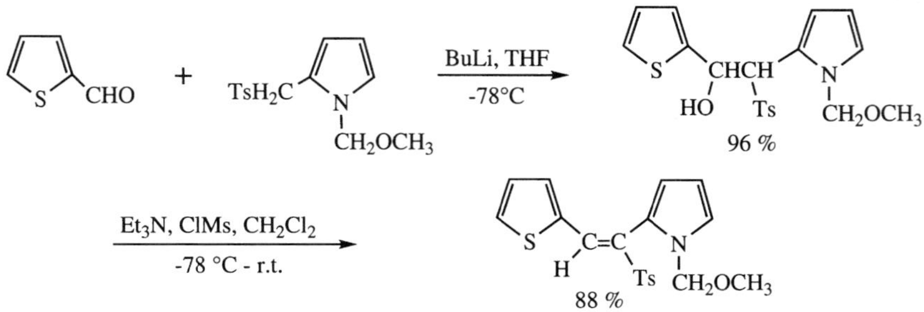

In connection with work on conducting materials and compounds with nonlinear optical properties, compounds of type $ThCH{=}C(SR)_2$ have become of great interest. They are conveniently prepared through the reaction of thiophenealdehydes with Wittig or phosphonate reagents such as(1,3-dithian-2-yl)phosphonium chloride and the corresponding diethylphosphonates of dithiolyl and 1,3-benzodithiolyl derivatives [116–120]. In this connection 1,4-di(1,3-dithiol-2-ylidene)-1,4-di(2-thienyl)but-2-ene has been prepared and found to be a good electron donor [121].

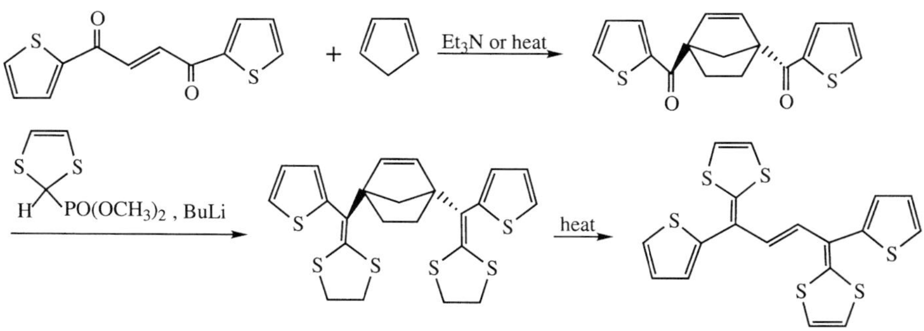

A rigidified tetrathiafulvalene-bithiophene hybride systems can be prepared starting from the dialdehyde derived from a dithiophene analog of fluorenone [122]. In a similar way a number of derivatives have also been prepared from 4H-cyclopenta[2,1-*b*;3,4-*b*′]dithiophene and converted to polymers by electrolytic oxidation [123].

4B.3 PREPARATION OF THC(SR)=C(SR)

Compounds of this type are of interest in connection with preparation of thiophene functionalized tetrathiafulvalene π-electron donors. They can be prepared starting from 2,2′-thenoin. 4,5-Bis(2-Thienyl)-1,3-dithiol-2-one was prepared, which in the usual way can be converted to the tetrathiafulvalene [124].

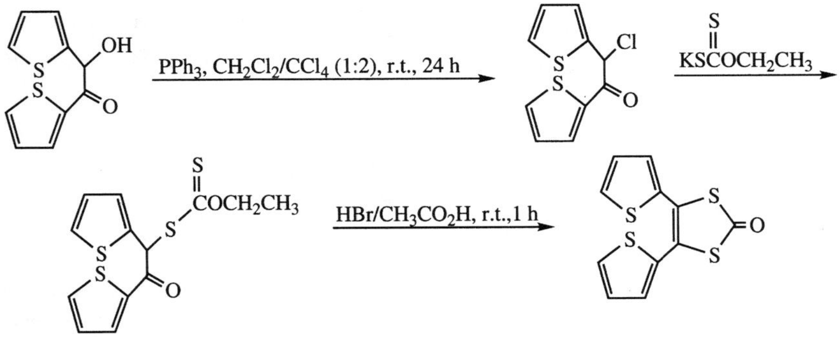

4,5-bis(2-Thienyl)-1,3-dithiol-2-one [124]

2,2′-Thenoin (5.43 g, 24.24 mmol) is dissolved in a mixture of carbon tetrachloride (50 ml) and dichloromethane (25 ml), after which triphenylphosphine (12.7 g, 48.4 mmol) is added and the mixture stirred overnight at room temperature. The solution is diluted with diethyl ether (100 ml) and filtered through a silica plug. The remaining solid residue is dissolved in dichloromethane (20 ml) and precipitated again by addition of diethyl ether (100 ml) and petrol (50 ml). This solution is also passed through the silica pad and after adsorption onto silica the combined filtrates are further purified by flash chromatography using ether/petrol (30:70) as eluent. Evaporation gives the chloro derivative as a yellow oil which is immediately used in the next step. This oil (4.60 g) is dissolved in acetone (20 ml) and potassium *ortho*-ethyl xanthate (3.35 g, 20.9 mmol) is added. After stirring the mixture at room temperature for 15 min, diethyl ether (150 ml) is added and the solution filtered through a silica plug. The filtrate is adsorbed onto silica and is further purified by flash chromatography using ether/petrol (30:70) as eluent. Removal of the solvent gives the xanthate as a yellow oil. This oil (4.20 g) is dissolved in acetic acid (10 ml), hydrogen bromide (48%) solution in acetic acid, 20 ml) is added and the reaction mixture stirred vigorously at room temperature for 1 h. Dichloromethane (50 ml) and water (200 ml) are added and the phases are separated. The aqueous phase is extracted with dichloromethane (3 × 25 ml) and the combined organic phases are washed with saturated sodium bicarbonate (2 × 25 ml) and sodium chloride (2 × 20 ml) solutions. After drying, the solution is adsorbed onto silica and the crude product is flash chromatographed using 3% ether in petrol as eluent. The residue obtained upon evaporation is recrystallized from petrol/ether giving 2.00 g (30%) of the title compound mp 93–95 °C.

4B.4 PREPARATION OF THCH=CHPXY AND THCH=CHPOXY

[2-(2-Thienyl)vinyl]phosphonous dichloride can be prepared through the reaction of 2-vinylthiophene with phosphorus pentachloride in phosphorus trichloride and can be transformed to other phosphorus derivatives [125].

[2-(2-Thienyl)vinyl]phosphonous dichloride [125]

In an atmosphere of argon, 2-vinylthiophene (90 g, 0.82 mol) is added dropwise with stirring to a boiling solution of phosphorus pentachloride (170.1 g, 0.82 mol) in phosphorus trichloride (700 ml). The color of the reaction mixture changes from greenish- to bright-yellow and crystals of the complex are precipitated. The suspension is boiled while methyl phosphorodichloridite (108.6 g) is added dropwise and the clear solution formed is boiled for an additional 30 min. After cooling the reaction mixture is transferred in a stream of argon into an Arbuzov flask, phosphorus trichloride is driven off and the residual volatile products are removed in vacuum. By two successive distillations of the residue in a stream of argon, 88.5 g (51%) of the title compound is obtained as a colorless liquid bp 82 °C/0.02 mm Hg, which is crystallized mp 36–37 °C.

Reaction of 2-thiophenealdehyde and 2-acetylthiophene with the carbanion derived from the *C*-silylated phosphoramine followed by quenching with trimethyl silyl chloride gives the vinylthiophene [126].

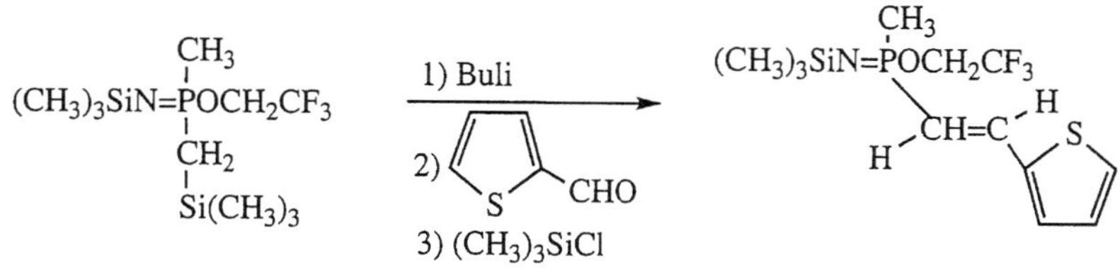

2-Thienyl-β-vinylether is prepared in 78% yield by the reaction of thienylacetaldehyde with butyllithium and water [127]. A very rarely used approach to cyanomethylthiophenes was recently applied to the preparation of 3-methyl-2-cyanomethylthiophene from 3-methyl-2-thiophene aldehyde, which was condensed with rhodanine, saponified to the 3-(3-methyl-2-thienyl)-2-thioketopropionic acid, which was converted to the oxime and dehydrated with acetic anhydride to the desired acetonitrile [128].

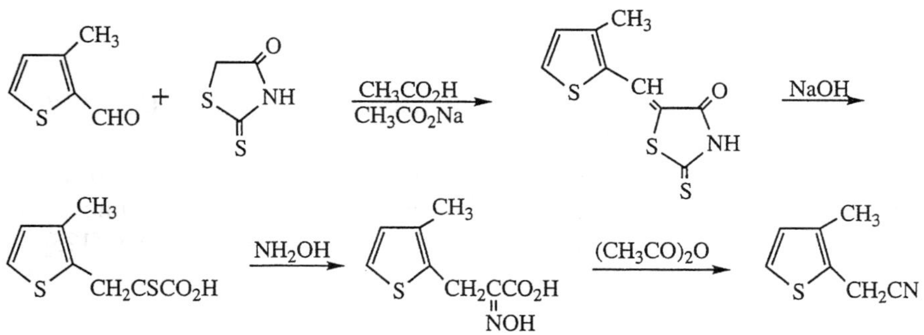

4B.5 PREPARATION OF THCH=CHAr, Ar=ARYL, HETARYL

4B.5.1 Dehydration of alcohols obtained from thiophenecarbonyl compounds and organometallic reagents

Reduction of 2-(1-pyridinium-1-(thienyl)ethanone bromide with sodium borohydride in water, followed by treatment of the alcohol with benzoylchloride under heating gave (*E*)-1-[2-(thienyl)vinyl]pyridinium bromide [129].

A convenient method for the preparation of 1-(phenyl)-2-(2-thienyl)ethene consists in the reaction of 2-thiophenemagnesium bromide with phenylacetaldehyde, followed by dehydration with potassium bisulfate [36]. Thiophene analogs of diethylstilbestrol are prepared through the following reaction [130].

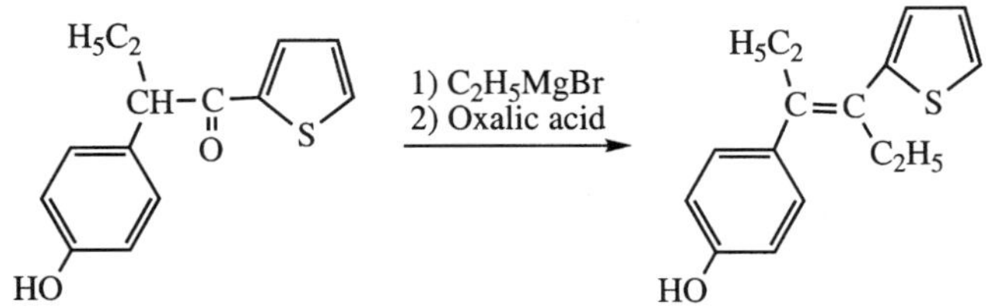

4B.5.2 Condensation of thiophene aldehydes with acidic methyl groups bound to rings

In this type of reaction thiophene aldehydes react similarly to benzaldehyde. From 2,4,6-trinitrotoluene and 2-thiophene aldehyde 5-methyl-2-thiophene aldehyde, and 2,3-dimethyl-3-thiophene aldehyde, the 1-(thienyl)-2-(2,4,6-trinitrophenyl)ethenes were obtained using piperidine in xylene as catalyst and a Dean–Stark trap for removal of the water [21,131]. *E*-1-(3-Methyl-4-nitroisoxazol-5-yl)-2-(2-thienyl)ethene is prepared from 2-thiophene aldehyde and 3,5- dimethyl-4-nitroisoxazole in ethanol with piperidine as catalyst [132].

E-1-(3-Methyl-4-nitroisoxazol-5-yl)-2-(2-thienyl)ethene [132]

3,5-Dimethyl-4-nitroisoxazole (2 g, 14.1 mmol) and 2-thiophene aldehyde (1.58 g, 14.1 mmol) are dissolved in ethanol (15 ml) and refluxed under stirring for 30 min in the presence of piperidine (0.5 ml). The mixture is then cooled and the yellow precipitate is filtered off giving 2.53 g (75%) of the title compound after drying mp 141–142 °C from ethanol.

The active methyl group in ethyl *ortho*-methyl benzoate and *ortho*-methylbenzonitrile have similarly been condensed to the 1,2-diaryl-substituted ethenes. 3-Cyano-4-methylpyridine upon condensation with 2- and 3-thiophene aldehyde using sodium methoxide in methanol lead to simultaneous hydrolysis of the nitrile function, giving the *trans* amides below [133].

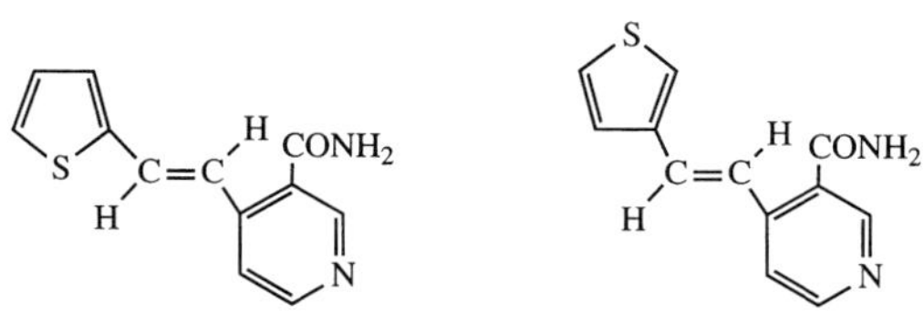

Compounds such as

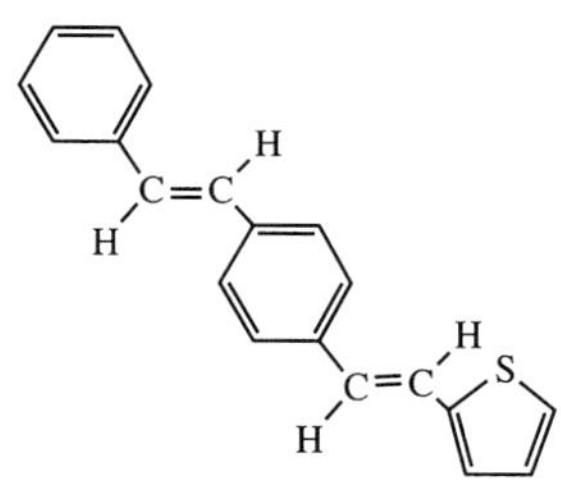

can be obtained if the anil of 2-thiophene aldehyde is condensed with various methyl-substituted carbocycles, and the reaction carried out in *N*,*N*-dimethylformamide in the presence of potassium hydroxide or potassium tert-butoxide [134]. Extensive work on this type of condensation has been carried out in connection with the synthesis of the anthelmintic agent pyrantel and related compounds [135,136].

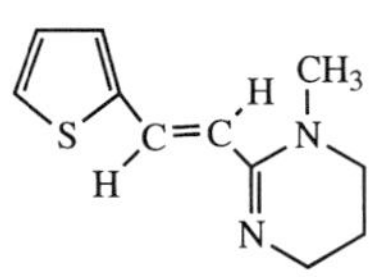

5,6-Dihydro-2-[2-(3-methyl-2-thienyl)ethyl]-4H-1,3-thiazine hydrochloride [136]

A stirred mixture of 3-methyl-2-thiophenepropionic acid (23.5 g, 0.138 mol) and 3-amino-1-propanol is slowly heated to 200 °C. Toward the end of the heating period, a slow stream of anhydrous nitrogen is blown over the melt to remove traces of water. The mixture is allowed to cool to 150 °C and phosphor pentasulfide (6.7 g) is added portionwise. The flask is fitted with vacuum distillation equipment. The more volatile components are removed at 15–20 mm Hg and 5,6-dihydro-2-[2-(3-methyl-2-thienyl)ethyl]-4H-1,3-thiazine is then distilled giving 15.3 g (49%) bp 150–160 °C/0.1 mm Hg. A solution of the free base so obtained and 5 *M* anhydrous hydrogen chloride in methanol (100 ml) is evaporated. The residue is recrystallized from isopropanol/diisopropanol ether mp 137–138 °C.

4B.5.3 From 2-(benzotriazol-1-ylmethyl)thiophenes

2-Alkenylthiophenes are conveniently prepared *via* side-chain elaboration through benzylic metalation of 2-(benzotriazol-1-ylmethyl)thiophenes, which are readily available from the condensation of 1-(hydroxymethyl)benzotriazoles with thiophenes [137].

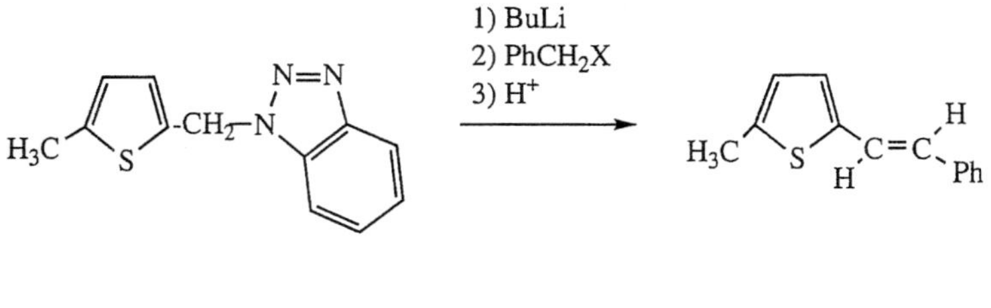

trans-1-(5-Methylthien-2-yl)-2-phenylethene [137]

Butyllithium (2.0 *M*, 2.5 ml) is added to a stirred solution of 2-(benzotriazol-1-ylmethyl)-5-methylthiophene (1.2 g, 5 mmol) in anhydrous tetrahydrofuran (45 ml) under argon at −78 °C. After 1 h benzyl bromide (99 mg, 5.5 mmol) in anhydrous tetrahydrofuran is added. The stirring is continued for 3 h, after which the reaction mixture is allowed to warm to room temperature overnight. Amberlyst-15 resin (10 g) is added and the mixture refluxed under argon for 5 h. Upon cooling the resin is filtered off and the solvent evaporated. The residue is treated with dichloromethane (50 ml) and the solution so obtained is washed with 2 *M* sodium hydroxide solution (30 ml) and water (30 ml), dried over sodium sulfate, and evaporated. The residue is purified by chromatography using ethyl acetate/hexane (1:4) as eluent giving 0.91 g (92%) of the title compound as a colorless solid mp 79–80 °C.

4B.5.4 Low-valent titanium-mediated dimerizations

During recent years this method has become one of the best methods for the preparation of 1,2-diarylethenes and also tetraaryl ethenes. Treatment of 2-thiophene aldehyde with titanium tetrachloride and zinc in refluxing tetrahydrofuran (McMurry coupling) gives 1,2-(2-thienyl)ethene in a 95:5% *E:Z* mixture in 85% yield [50,138–140].

E-1,2-Di-2-thienylethylene [139]

Titanium(IV) chloride (13 ml, 0.12 mol) is added to a stirred solution of 2-thiophene aldehyde (11.2 g, 0.1 mol) in tetrahydrofuran (200 ml) over a period of 0.5 h at 18 °C. The stirring is continued at this temperature for 30 min, after which zinc powder (15.7 g, 0.24 mol) is added in small portions over a period of 30 min. The reaction mixture is stirred at −18 °C for 30 min, warmed to room temperature, and refluxed for 3.5 h. The reaction is quenched by addition of ice-water (150 ml) and the resulting solid is collected by filtration and dried. The solid is dissolved in dichloromethane (150 ml) and the insoluble inorganic material removed by filtration. The filtrate is evaporated and the residue recrystallized from cyclohexane giving 9.4 g (85%) of the title compound mp 133–134 °C.

McMurry reaction with 2,5-dimethyl-3-acetylthiophene gave the *cis* form of 2,3-bis(2,5-dimethyl-3-thienyl)-2-butene in low yield [141]. Tetrathia [22] annulene [2,1,2,1], a new thiophene-derived aromatic macrocycle, is conveniently prepared in 75% yield by titanium-mediated dimerization of 5,5′-diformyl-2,2′-dithienylmethane [142].

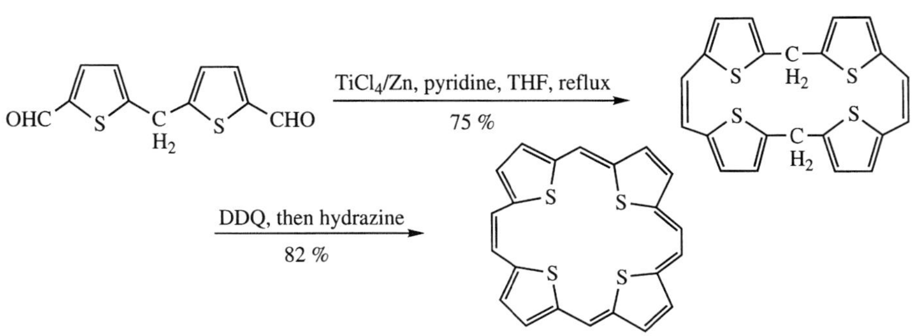

Dioxodithiaporphycenes are prepared by McMurry reaction of 5,5′-diformyl-2,2′-furylthiophene. The benzene-fused dioxodithiaporphycyne derivative was prepared by McMurry reaction of the aldehyde obtained by Suzuki coupling of 5-bromo-2-thiophene aldehyde with the boronic acid derivative shown below [143].

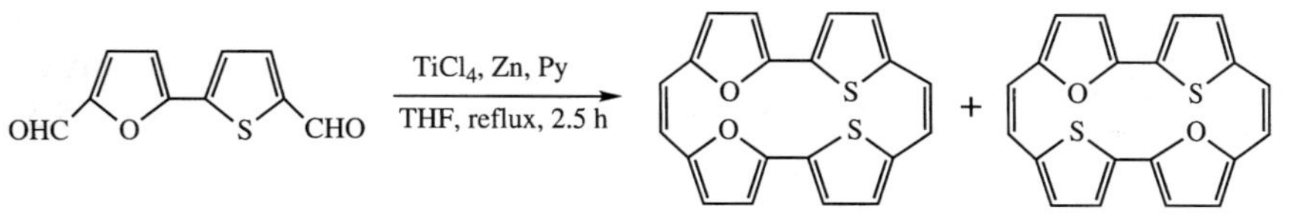

7.2 % 2.6 %

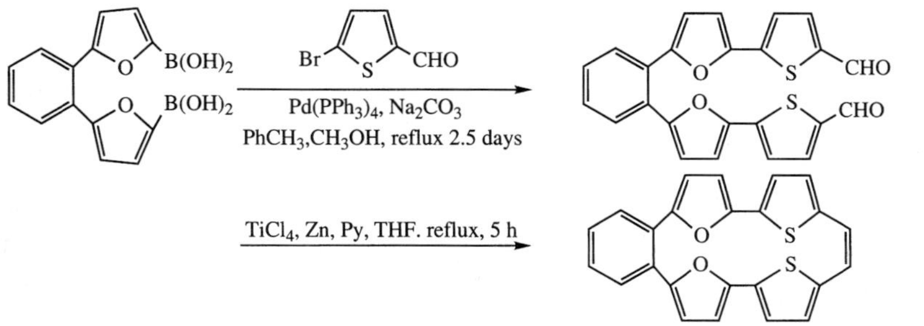

By an intramolecular McMurry reaction the following transformation takes place and the 3,4-di(2-thienyl)-2,5-dihydrothiophene obtained can be oxidized to the sulfone with *meta*- chloroperbenzoic acid [144].

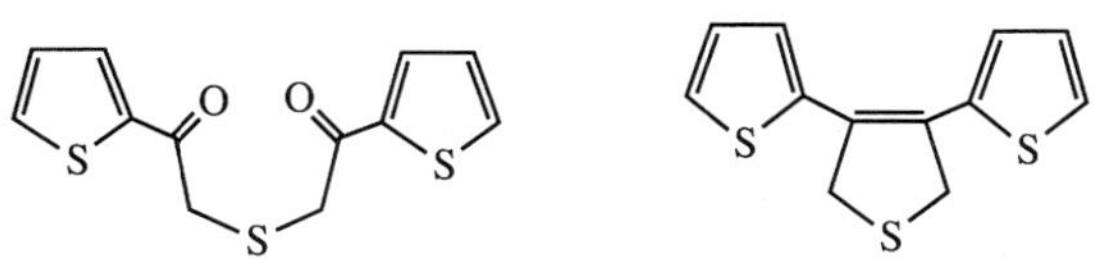

Photochromic 1,2-dithienyl-substituted cycloalkenes and 2,5-dihydrothiophenes have been prepared by intramolecular reductive coupling of diketones and dioxo sulfides, using low-valent titanium derivatives [145].

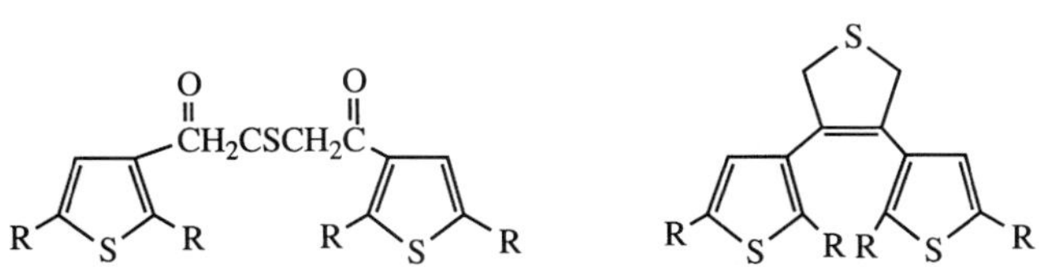

$R = CH_3$

3,4-Di(2,5-dimethyl-3-thienyl)-2,5-dihydrothiophene [145]

Titanium(IV) chloride (1.5 ml, 10 mmol) is added dropwise using a syringe to a stirred suspension of zinc powder (1.3 g, 20 mmol) in freshly distilled anhydrous tetrahydrofuran (100 ml) at −10 °C under argon. After completion of the addition, the mixture is heated to reflux for 1 h. The suspension is cooled to 20 °C, the pyridine (0.5 ml) and the dioxo sulfide (1.7 g, 5 mmol) are added carefully. The reaction mixture is refluxed for 6 h under argon, after which most of the solvent is removed under vacuum. The residue is poured into 10% aqueous potassium carbonate solution (100 ml) and this solution is extracted with diethyl ether (3 × 50 ml). The combined organic phases are dried and evaporated giving an oil. The pure title compound in a yield of 11% is isolated by flash chromatography on silica gel using ethyl acetate/petroleum ether as eluent mp 89–91 °C.

A number of tetrathienylethylenes are best prepared by the titanium-mediated coupling of 2,2′-dithienyl ketones in very good yields [146,147].

Tetra(2-thienyl)ethylene [146]

Pyridine (2.5 ml, 33.2 mmol) is added dropwise followed by a solution of bis(2-thienyl)ketone (5.76 g, 29.6 mmol) in anhydrous tetrahydrofuran (15 ml) to a suspension of low-valent titanium prepared from titanium tetrachloride (5.88 g, 31.0 mmol) and zinc (4.37 g, 66.8 mmol) in anhydrous tetrahydrofuran (100 ml) under nitrogen. The reaction mixture is refluxed for 19 h and poured into 10% potassium carbonate. The product is extracted with dichloromethane and the combined organic phases are washed with sodium chloride, dried over sodium sulfate, and evaporated. The residue is treated with benzene giving pale yellow crystals, which are purified by sublimation at 185 °C/4 × 10^{-2} mm Hg giving 2.16 g (41%) of the title compound.

Diformylation of 1,2-(2-thienyl)ethene followed by renewed titanium-mediated dimerization can be used for the preparation of oligo(thiophene-2,5-diyl)vinylenes [139].

4B.5.5 Other dimerizations

Pyrolysis of the azines derived from 2-thiophene aldehyde and 5-methyl-2-thiophene aldehyde can be used for the preparation of 1,2-(2-thienyl)ethene and the corresponding 5-methyl derivative [148,149]. The reaction shown below has been performed [150].

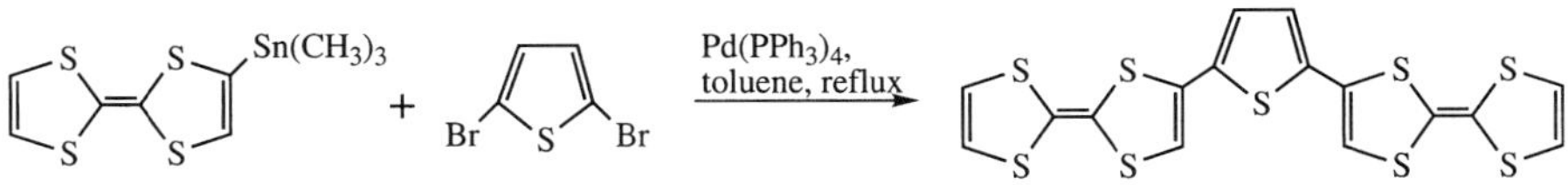

4B.5.6 From thienyllithium derivatives and fluoroalkenes

2-Difluoro-1,2-di-(2-thienyl)ethene and 1-chloro-1-fluoro-1,2-di(2-thienyl)-ethene were obtained in 30 and 55% yield, respectively, upon reaction of 2-thienyllithium with tetrafluoroethene and trifluorochloroethene, respectively [151].

During recent years, there has been great interest in the preparation and light-induced reversible isomerization between two forms of 1,2-bis(2,5-dimethyl-3-thienyl)perfluorocyclopentene derivatives, having different absorption spectra. The radiation-induced coloration of these photochromic compounds in polymer matrices as well as their fatigue mechanism is studied [152,153]. The synthesis of such compounds, because of their potential ability for photonic devices such as optical memory media and photooptical switches, has been extensively studied during recent years.

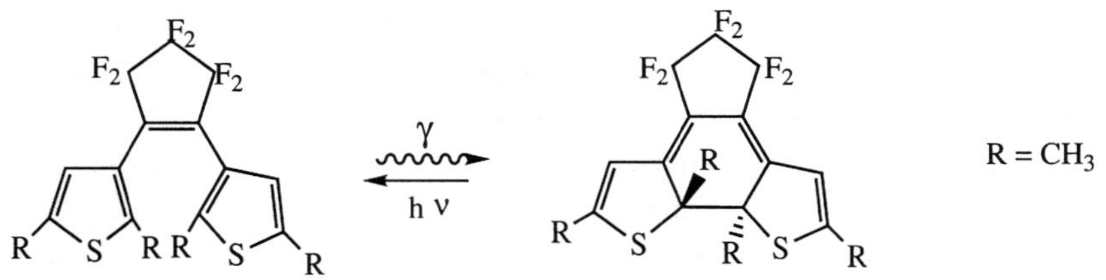

Also diarylperfluorocyclopentenes having 2-isopropyl-5-phenylthiophene and 2-isopropyl-1-benzothiophenearyl groups have been prepared and undergo thermally reversible photochromism in solution. The photogenerated colored closed ring isomers returned to the initial colorlesss open ring form about 60 °C [154].

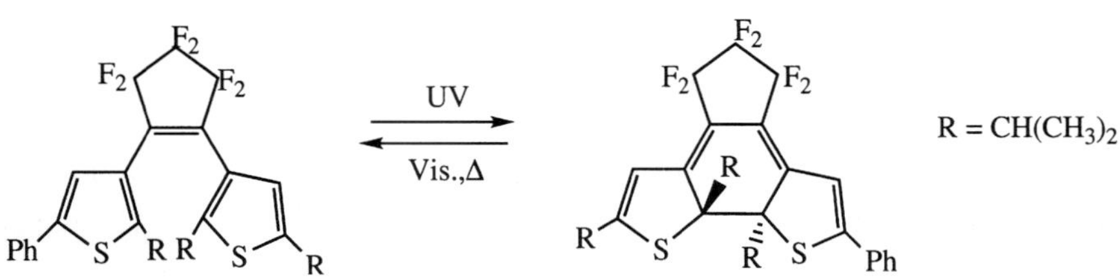

Photochromism in a single crystal of 1,2-bis(2,5-dimethyl-3-thienyl)perfluorocyclopentene [155], as well as 1,2-bis(2-methyl-5-phenyl-3-thienyl)perfluorocyclopentene [156] and 1,2-bis(2-ethyl-5-phenyl-3-thienyl)perfluorocyclopentene, prepared in the usual way from 2-ethyl-5-phenyl-3-thienyllithium and perfluorocyclopentene, were studied by X-ray crystallography [157].

1,2-Bis(2-ethyl-3-thienyl)perfluorocyclopentene [157]

To a solution of 3-bromo-2-ethyl-5-phenylthiophene (9.0 g, 0.034 mmol) in anhydrous tetrahydrofuran (80 ml), 15% butyllithium in hexane (24 ml) is added at −78 °C under nitrogen. The stirring is continued at this temperature for 2 h, after which perfluorocyclopentene (2.3 ml, 0.017 mmol) is slowly added and the stirring continued for 3 h at the same temperature. The reaction is stopped by addition of methanol and the product extracted with diethyl ether. The combined organic phases are dried over magnesium sulfate and evaporated. The residue is purified by chromatography on silica gel using hexane as eluent, giving 6.3 g (68%) of the title compound as colorless crystals mp 164 °C after recrystallization from hexane.

A number 3-bromo-2-methyl-5-arylthiophenes with aryl=phenyl, *para*-tolyl, and tert-butyl give upon reaction with butyllithium followed by perfluorocyclopentene the photochromic systems [156]. Also halogen–metal exchange of 3-bromo-2-methyl-5-(4-methoxyphenyl)thiophene followed by reaction with perfluorocyclopentene gave the tricyclic compound, which upon addition to 5-(4-pyridyl)-2-methyl-3-thienyllithium was used for the preparation of the pentacyclic derivative [158].

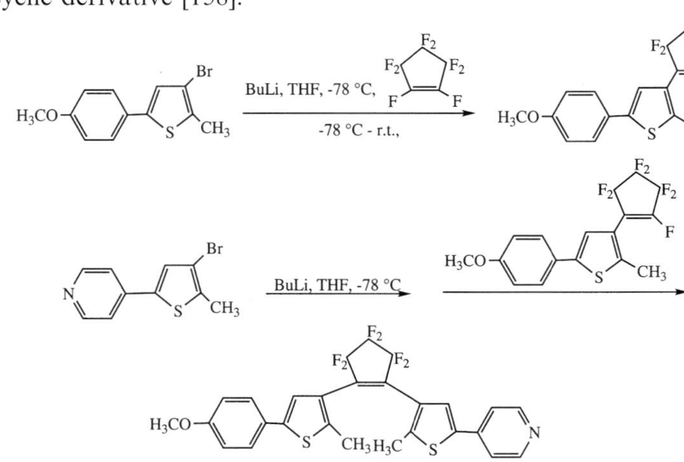

1-[50′-(4″-Methoxyphenyl)-2′-methylthien-3′-yl]perfluorocyclopentene [158]

To a solution of 3-bromo-2-methyl-5-(4-methoxyphenyl)thiophene (767 mg 2.7 mmol) in anhydrous tetrahydrofuran (30 ml) at −78 °C under nitrogen, 1.6 *M* butyllithium in hexane (1.8 ml) is slowly added. After 10 min perfluorocyclopentene is added and the stirring is continued for 1 h after which the reaction mixture is allowed to warm to ambient temperature. The solvent is removed *in vacuo*, the product extracted with dichloromethane and the combined organic phases washed with aqueous sodium hydrogen carbonate, dried and evaporated. The residue is purified by chromatography on silica gel using hexane/dichloromethane (4:1) as eluent, giving 408 mg (38%) of the title compound as a colorless solid mp 65–66 °C.

Novel photochromic amorphous molecular materials such as those shown below were prepared by halogen–metal exchange of 5-[4-[bis(4-methylphenyl)-amino]phenyl]-3-bromo-2-methylthiophene followed by reaction with 1-(2-methylbenzo[*b*]thiophen-3-yl)-2,3,3,4,4,5,5-heptafluorocyclopentene, prepared by the reaction of 3-lithio-2-methylbenzo[*b*]thiophene with 1,2,3,3,4,4,5,5-octafluorocyclopentene, and the second compound was similarly obtained in 45% yield from 5-[4-[bis(4-methylphenyl)amino]phenyl]-3-bromo-2-methyl-thiophene upon reaction with 1,2,3,3,4,4,5,5-octafluorocyclopentene [159].

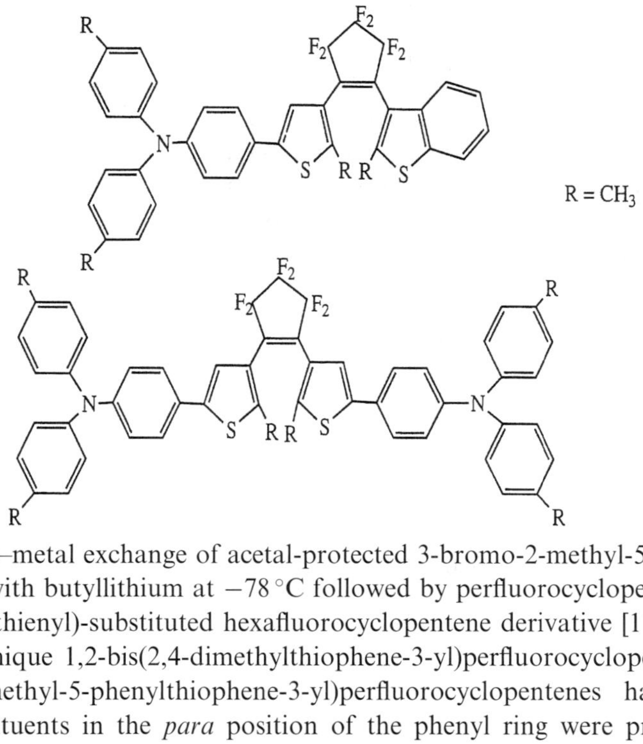

Halogen–metal exchange of acetal-protected 3-bromo-2-methyl-5-thiophene aldehyde with butyllithium at −78 °C followed by perfluorocyclopentene gave the 1,2-(dithienyl)-substituted hexafluorocyclopentene derivative [119]. By the same technique 1,2-bis(2,4-dimethylthiophene-3-yl)perfluorocyclopentene and bis(2,4-dimethyl-5-phenylthiophene-3-yl)perfluorocyclopentenes having various substituents in the *para* position of the phenyl ring were prepared, in connection with investigation of photochromism of dithienylethenes with electron-donating substituents [160].

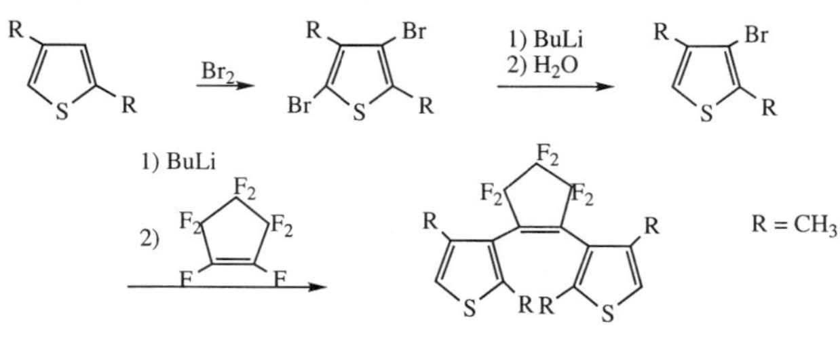

Bis(2,4-dimethyl-5-phenylthiophene-3-yl)perfluorocyclopentene [160]

Butyllithium (1.6 *M*) in hexane (4.3 ml, 6.88 mmol) is added at −60 °C to a solution of 2,4-dimethyl-3-bromothiophene (1.09 g, 5.70 mmol) in anhydrous tetrahydrofuran (50 ml) under nitrogen during 2 min. The stirring is continued at this temperature for 15 min, after which perfluorocyclopentene (0.35 ml, 2.61 mmol) is added and the reaction mixture is stirred at −60 °C for 3 h. The reaction is stopped by the addition of dilute aqueous hydrochloric acid and the product extracted with diethyl ether. The combined organic phases are washed with water, dried, and evaporated. The residue is purified by column chromatography on silica gel using hexane as eluent, giving 580 mg (56%) of the title compound as a white solid mp 133–134 °C.

Other recent examples of the syntheses of photochromic dithienylethenes are the preparation of 1, 2-(2-methyl-5-hydroxymethyl)hexafluoro-1,2-cyclopentene through the reaction of 2-methyl-5-silyl-protected hydroxymethyl-3-thienyllithium with octafluorocyclopentene and the preparation of photochromic bis(monoaza-crown) ethers, through the reaction of 2-methyl-5-aryl-3-thienyllithium with octafluorocyclopentene [161,162], as well as the reaction of 2-aryl-4-methylthiophene with butyllithium followed by octafluorocyclopentene [163]. Mixed photochromic hexafluorocyclopentene derivatives have been prepared by the reaction of 2,4-dimethyl-5-aryl-3-thienyllithium with 1-(2-methylbenzothienyl)perfluorocyclopentene [164].

A novel photochromic molecule containing two photochromic dithienyl ethene moieties linked to a fluorescent bis(phenylethynyl)anthracene unit has recently been prepared [165].

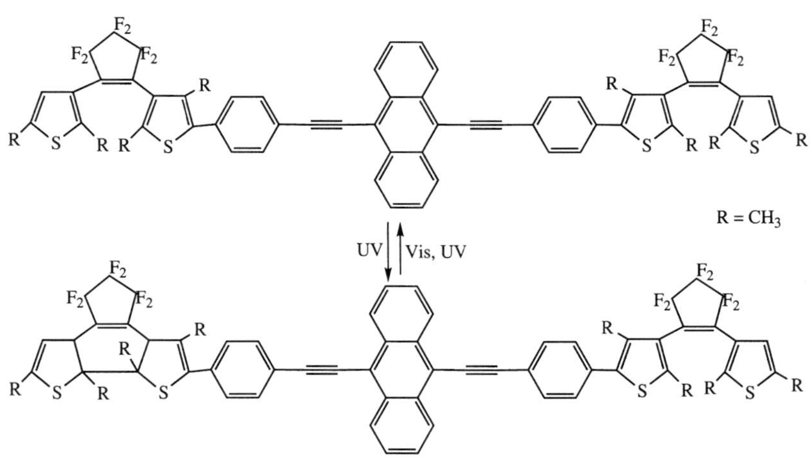

Also more elaborate systems such as dithienylethene-bridged diporphyrins have been prepared recently, using the reaction of thienyllithium derivatives with perfluorocyclopentene derivatives as the key step. 1,2-bis(5-(4-Trimethylsilylethenylphenyl)-2-methylthiophene-3-yl)perfluorocyclopentene was obtained from 3-bromo-2-methyl-5-(4-trimethylsilylethynylphenyl)thiophene by halogen–metal exchange followed by perfluorocyclopentene and 1,2-bis[5-(5,5-dimethyl-1,3-dioxacyclohex-2-yl)-2-methylthiophene-3-yl]perfluorocyclopentene from the protected aldehyde [166].

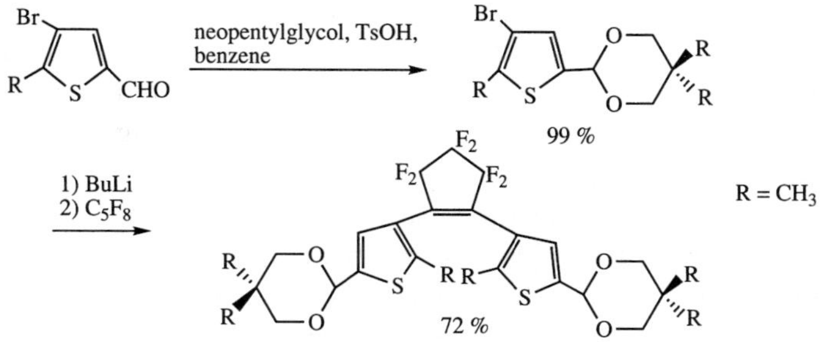

Another recent example is the preparation of 1,2-bis[2,4-dimethyl-5-[4-*N*,*N*-bis(4-methylphenylamino]phenyl-3-thienyl]perfluorocyclopentene shown below [167].

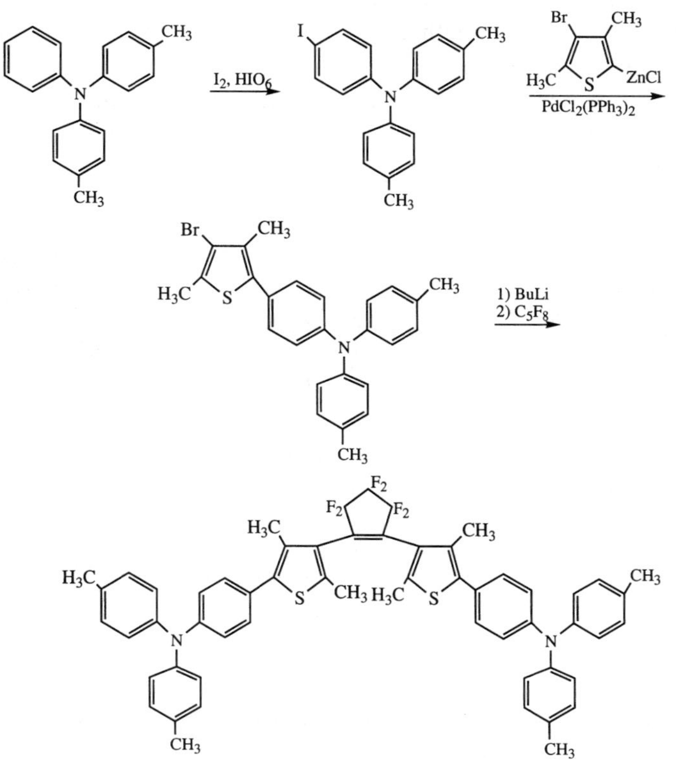

The syntheses of fluorescent diarylethenes having a 2,4,5-triphenylimidazole have recently been described [168].

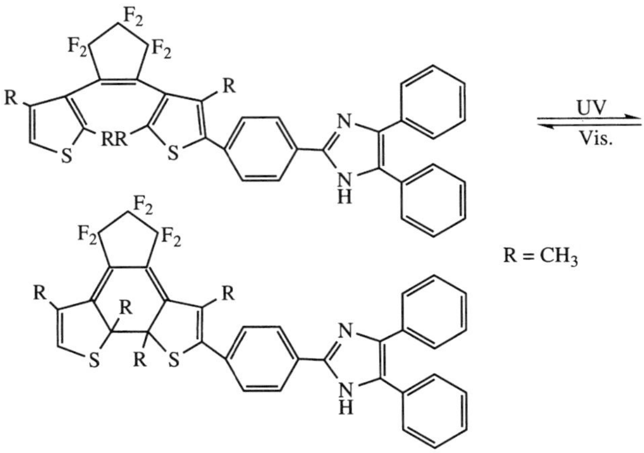

1-(2,4-Dimethyl-3-thienyl)-2-[2,4-dimethyl-5-(2″,4″,5″-triphenylimidazol-4′-yl)-3-thienyl]- perfluorocyclopentene [168]

Butyllithium (1.6 *M*) in hexane (0.72 ml, 1.15 mmol) is added dropwise at −78 °C to a solution of bis(2,4-dimethyl-3-thienyl)perfluorocyclopentene (200 mg, 0.51 mol) and *N*,*N*,*N*′,*N*′-tetramethylethylenediamine (0.17 ml, 1.15 mmol) in anhydrous diethyl ether (3 ml) under argon. The reaction mixture is stirred for 20 min at room temperature, and then tributyl borate (0.38 ml, 1.38 mmol) is added in one portion. The red-brown solution is stirred for 1 h and then diluted with anhydrous tetrahydrofuran (3 ml). Ethylene glycol (0.3 ml), tetrakis(triphenylphosphine)palladium(0) (20 mg), aqueous sodium carbonate (20%; 3 ml) and anhydrous tetrahydrofuran (50 ml) is added to the solution. The solution is refluxed for 12 h at 70 °C, 2 *M* hydrochloric acid is poured into the reaction mixture and then the product is extracted with chloroform, dried over magnesium sulfate, and concentrated. Purification is performed by column chromatography, yielding 66 mg (19%) of the title compound.

In connection with a study of photoswitching of an intramolecular magnetic interaction diarylethenes with two nitronyl nitroxides have been prepared [169,170]. 2,3-Bis(3-thienyl)quinone has been prepared and its novel acid-mediated ring-closing–photochemical ring-opening studied [171].

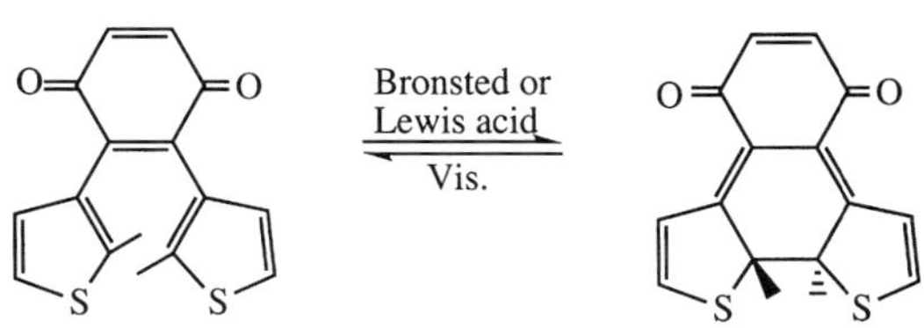

4B.5.7 Wittig reaction

4B.5.7.1 From triphenylbenzylphosphonium halides and aldehydes

Wittig and related reactions are among the most important for the preparation of 1-thienyl-2-arylethenes (thiophene analogs of stilbene) [95,112,172–180]. The preparation of the necessary thenyl halides and triphenylphosphonium halides is described in appropriate sections. Various bases and solvents have been used in this reaction, for instance potassium tert-butoxide [174], sodium methoxide [177], and sodium or lithium ethoxide in the corresponding alcohols [173,179]. In particular the use of sodium methoxide in *N*,*N*-dimethylformamide seems to be the best condition [95,176,178]. Butyllithium in hexane can also be used.

As is well known, the *cis*:*trans* ratio of the olefins obtained in the Wittig reaction depends on the reaction conditions. Normally selectivity is very low, as is also the case in the reaction between triphenyl thenylidene phosphorane and 2-thiophene aldehyde. It should, however, be mentioned that unusually high amounts of *cis* isomers are obtained in the reaction of *ortho*-halo substituted thiophenealdehydes with thenylidene ylides [95]. The pure *trans* isomers are easily obtained pure by recrystallization because of their greater insolubility. It is also possible to transform the *cis* isomer to the *trans* isomer by iodine catalysis [18,182].

1-(4-Nitrophenyl)-2-(3-thienyl)ethene was recently prepared from 4-nitrobenzaldehyde and 3-thenyltriphenylphosphonium bromide using sodium ethoxide in ethanol [183].

1-(4-Nitrophenyl)-2-(3-thienyl)ethene [183]

4-Nitrobenzaldehyde (3.4 g, 22.5 mmol) and 3-thenyltriphenylphosphonium bromide (9.9 g, 22.5 mmol) are dissolved in absolute ethanol (63 ml). 0.4 *M* Sodium ethoxide in ethanol (200 ml) is added and precipitation of a yellow product occurs immediately. The reaction mixture is heated up to reflux for 2 h and then cooled to ambient temperature. One-third of the ethanol is evaporated and water is added instead. The precipitate is filtered off and purified by flash chromatography using dichloromethane as eluent, giving 3.015 g (58%) of the title compound as a *cis*/*trans* mixture mp 166 °C.

The reaction of *ortho*-carbomethoxybenzyltriphenylphosphonium bromide with 3-thiophene aldehyde gave a mixture of *cis* and *trans* 3-(*o*-carboxymethylstyryl)thiophene in the proportions of 7:3 [94]. The synthesis of 4-(5-alkyl-2-thienylvinyl)benzoic acid can be achieved in high yield from 4-methoxycarbonylbenzyltriphenylphosphonium bromide and 5-substituted 2-thiophenealdehydes [182].

Through the Wittig reaction between *o*-xylylenebistriphenylphosphonium bromide with 2-thiophene aldehyde in the presence of base a mixture of *cis* and *trans* 1-(2-thienyl)-2-(*ortho*-vinylphenyl)ethene was obtained in 76% yield [184]. A *cis–trans* mixture of 2-(2-thienylvinyl)-1-benzylpyrrole is prepared from 2-thiophene aldehyde and 2-pyrrolylmethylphosphonium iodide [185]. The Wittig reaction has also been applied to the synthesis of numerous unsymmetrically substitituted di(2-thienyl)ethenes, containing halogen, nitro and methoxy groups in the 5-positions, also formally this should be treated in other chapters [186]. This synthetic approach has recently been of great importance for the preparation of push–pull thiophenes for second order non-linear optical applications [117,187–189], as exemplified in the synthesis of 1-(5-piperidino-2-thienyl)-2-(5-nitro-2-thienyl)ethene from 5-piperidino-2-thiophene aldehyde and the ylide derived from 5-nitro-2-thenyltriphenylphosphonium chloride [187].

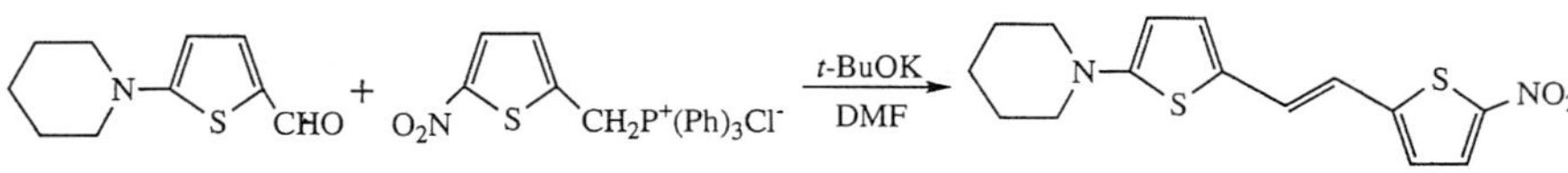

Bridged analogs of open chain dithienylethylene spacers in push–pull NLO chromophosres were prepared by Wittig reaction according to the following scheme [190].

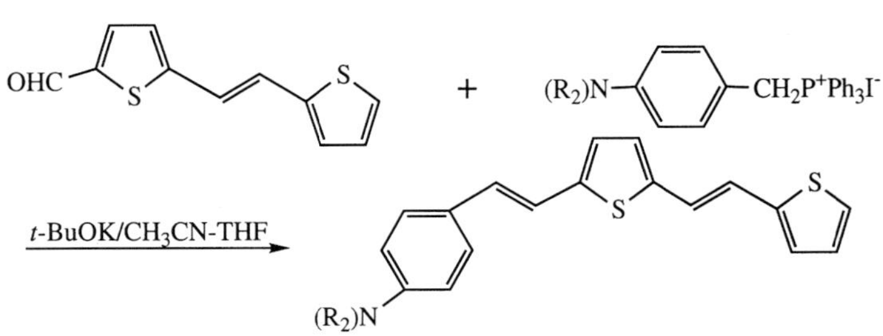

4B.5.7.2 The phosphonate method

The phosphonate method (Wadworth–Emmons–Horner reaction) is often a useful alternative to the classical Wittig reaction, if the *trans*-isomers of 1,2-diarylethenes are desired. The necessary diethyl thenylphosphonates are as mentioned previously, prepared by heating thenyl halides with triethyl phosphites to 100–160 °C for a few hours. The condensation of the thenyl phosphonates with aldehydes and ketones is in most cases carried out using a 50% excess of sodium ethylate in *N*,*N*-dimethylformamide at 25–40 °C [96,175,178,191–194].

trans-1,2-Di(2-thienyl)ethene [195]

A 100-ml flask is charged with 2-thiophene aldehyde (5.6 g, 0.05 mol), diethyl 2-thenylphosphonate (11.7 g, 0.05 mol), and sodium-dried dimethoxyethane (50 ml). Sodium hydride (2.4 g, 0.05 mol) as a 50% suspension in mineral oil is added and the reaction mixture heated to 60 °C, when evolution of hydrogen immediately begins. The reaction is completed by reflux for 0.5 h. After cooling, the reaction mixture is poured into water (500 ml). The precipitate formed is filtered off giving the title compound mp 133–134 °C.

Various aromatic aldehydes have successfully been used such as benzaldehydes [11,196,197], thiophene aldehydes [96,175,191,194,195], furan aldehydes [178], and pyrrole aldehydes [180,198].

(E)-1-(N-Methylpyrrol-2-yl)-2-(2-thienyl)ethylene [198]

A suspension of sodium hydride in oil (50% by weight; 1.08 g, 22. 5 mmol) is repeatedly washed with anhydrous tetrahydrofuran and finally suspended in tetrahydrofuran (40 ml). A solution of diethyl 2-thenylphosphonate (4.8 g, 20.5 mmol) tetrahydrofuran (5 ml) is first added to this suspension, kept under nitrogen and then a solution of *N*-methyl-2-pyrrole aldehyde (1.86 g, 17.1 mmol) in tetrahydrofuran (5 ml) is added. The reaction mixture is cautiously heated on an oil bath at 50 °C until the evolution of hydrogen had ceased and then refluxed for 1 h. The reaction is monitored by TLC using hexane/ethyl acetate (9:1) as eluent. The mixture is then poured onto ice (200 g) followed by extraction with diethyl ether (350 ml). The combined organic phases are washed, dried, and evaporated to leave a solid which, after chromatography on silica gel (350 g) using hexane/ethyl acetate (9:1) as eluent, gives 2.45 g (75%) of the title compound mp 65 °C after sublimation 55 °C/0.5 mm Hg.

The phosphonate method also works well with 1-naphthaldehyde [199], benzo[*b*]thiophene aldehydes, and dibenzothiophene aldehydes [200].

1-(3-Benzo[b]thienyl)-2-(2-thienyl)ethene [200]

Sodium hydride (50%, 1.5 g, 58 mmol) washed twice with hexane (40 ml) is placed in anhydrous 1,2-dimethoxyethane (100 ml). The slurry is cooled to 20 °C and diethyl 3-benzo[*b*]thenylphosphonate (7.1 g, 25 mmol) is added dropwise with stirring. When the addition is completed the stirring is continued for 20 min. 2-Thiophene aldehyde is added dropwise to the pale yellow solution, maintained below 25 °C. The reaction mixture is stirred at room temperature for 3.5 h. A large amount of water is added and the resulting precipitate is collected by filtration giving 4.8 g (80%) of the title compound as pale-yellow prisms mp 89–90 °C from methanol.

Almost exclusively the *trans*-isomers are formed, however, some exceptions have been observed as in the reaction of 2,5-dichloro-3-thenylphosphonate with 4-bromo-3-thiophene aldehyde, which gives equal amounts of *cis* and *trans* isomers [191]. 2-Acetylthiophene [195] and 1-acetylnaphthalene [149] have also been used in the reaction with thenylphosphonates. The reaction of 4-butoxybenzylphosphonate with 2-thiophene aldehyde in tetrahydrofuran with sodium hydride as base, gave 2-(4-butoxystyryl)thiophene, which without isolation was transformed to 5-(4-butoxystyryl)-2-thiophene aldehyde in a total yield of 91%, by metalation with butyllithium and reaction with *N,N*-dimethylformamide.

The phosphonate method has also been extensively used in connection with the syntheses of unsymmetrical *trans*-1,2-hetaryl-substituted ethylenes as potential second order nonlinear optical materials [187,189,201]. In this way 1-(5-diethylamino-2-thienyl)-2-(5-methylsulfonyl-2-thienyl)ethene and 1-(5-phenylthio-2-thienyl)-2-(5-methylsulfonyl-2-thienyl)ethene as well as 1-(4-dimethylaminophenyl)-2-(5-methylsulfonyl-2-thienyl)ethene and some related compounds have recently been prepared in excellent yield using 5-methylsulfonyl-2-thenylphosphonate and the appropriate substituted 2-thiophene aldehydes and benzaldehydes [189].

The phosphonate method using various phosphonates and thiophene carbonyl compounds has also been extensively applied to diethyl benzylphosphonate and 2-thiophene aldehyde [202,203], 4-methyl-2-thiophene aldehyde [204], 5-methyl-2-thiophene aldehyde [196], and also 3-thiophene aldehyde, 3-acetylthiophene [11], and 5-styryl-2-thiophene aldehyde [293], other phosphonates which have been reacted with various thiophene aldehydes are diethyl *meta*- and *para*-methylbenzyl phosphonate [196], ethyl 2,4- and 3,4-dichlorobenzylphosphonate [11,205], diethyl(2-carboethoxy-3-furylmethyl)phosphonate [178], and diethyl 1-naphthylmethyl phosphonate [204]. 1-(4-Pyridyl)-2-(2-thienyl)ethene is prepared by applying the phosphonate method to 4-pyridine aldehyde and the thenylphosphonate. Metalation with lithium diisopropylamide and reaction with *N,N*-dimethylformamide gave the 5-formyl-2-thienyl derivative, which upon renewed condensation with 5-bromo-2-thenyl phosphonate can be used for the preparation of 1-[5-[-(4-pyridyl)thien-2-yl]-2-(5-bromothien-2yl)ethene [206].

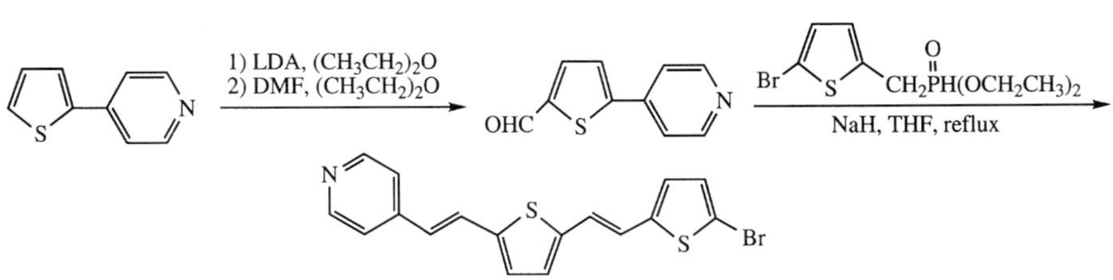

4B.5.8 Reduction of diarylacetylenes

The reduction of 1,2-di(thienyl)acetylenes with diisobutylaluminium hydride (DIBAL-H) is a good method for the stereoselective preparation of *cis* 1,2-di(thienyl)ethenes [207].

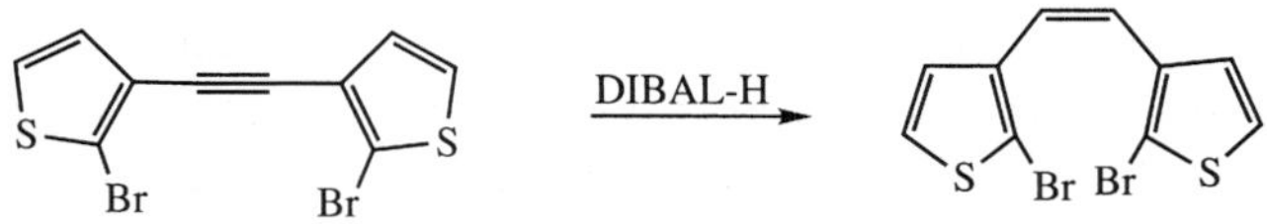

4B.5.9 From thiophenes and olefins

A mixture of *trans*-2-styrylthiophene and *trans,trans*-2,5-distyrylthiophene was obtained upon reaction of thiophene with stoichiometric amounts of palladium(II) chloride and styrene [208]. This field has been the most recently developed for the preparation of vinylthiophenes and related compounds. One of the first examples is the preparation of *trans*-2-styrylthiophene from 2-thienylmagnesium bromide and *trans*-styrylbromide using nickel(II) acetylacetonate as catalyst [209].

During recent years palladium-based catalysts have become more popular than nickel catalysts and 1,4-bis(diphenylphosphino)butanedichloropalladium(II) has been used in the preparation of 2-1-(trimethylsilyl)vinyl)thiophene and 1-(trimethylsilylmethyl)vinylthiophene from thiophenemagnesium bromide and the corresponding silylsubstituted vinyl bromide [59]. Other metalorganic reagents than magnesium derivatives can also be used. Thus 2-thienylzinc bromide gives upon coupling with vinyl bromide a 66% yield of 2-vinylthiophene [210].

Polythiophenediylvinylenes can be prepared by cross-coupling of 2,5-dibromo- or diiodothiophene with 1,2-bistributylstannylethylene using palladium phosphine complexes as catalysts [211]. 2,5-Thiophenediyl bis-tetrathiafulvalene is obtained through palladium-catalyzed cross-coupling of 2,5-dibromothiophene with trialkylstannyl tetrathiafulvalene [212].

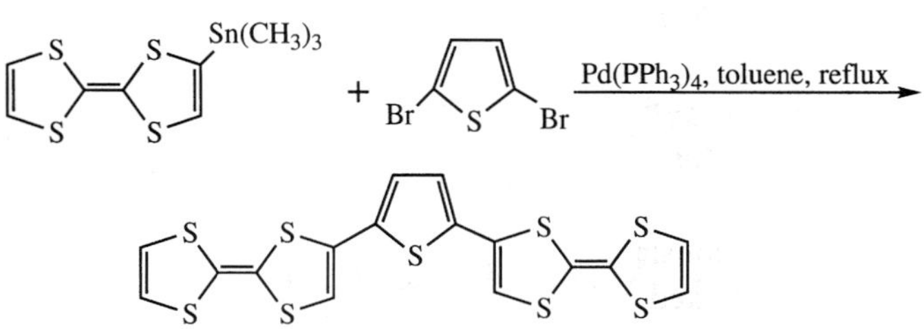

Palladium(0)-catalyzed coupling between 2-iodothiophene and stereodefined 2-arylethenyl dimethyl phenylsilanes can be used for the preparation of (*E*)-1-phenyl-2-(2-thienyl)ethene [62].

4B.5.10 Heck reaction

The palladium(0)-catalyzed coupling between halothiophenes and various vinyl derivatives using triethylamine as base is an excellent method for the preparation of numerous vinyl-substituted thiophenes. Thus (*E*)-2-(phenylvinyl)thiophene was prepared from 3-iodothiophene and styrene [213], and from 2-bromothiophene and 4-vinylpyridine, (*E*)-2-(4′pyridylvinyl)thiophene was obtained [214].

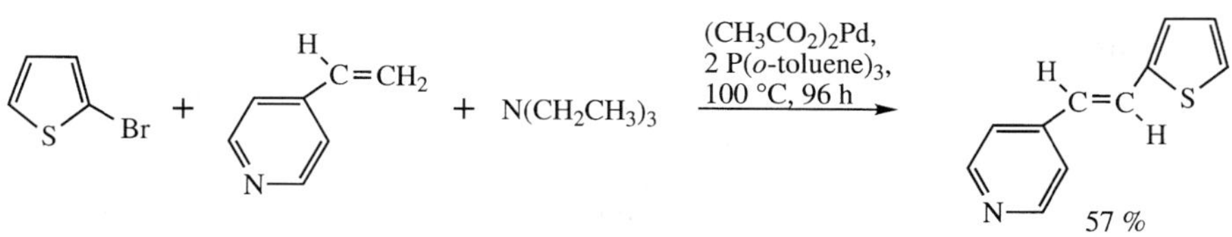

4B.5.11 From thiophenes and acetylenes

The addition of diphenylacetylene to thiophene in the presence of catalytic amounts of $Rh_4(CO)_{12}$ gives 1-(2-thienyl)-1,2-diphenyl)ethene in 48% yield [215]. The following product has also been prepared [216].

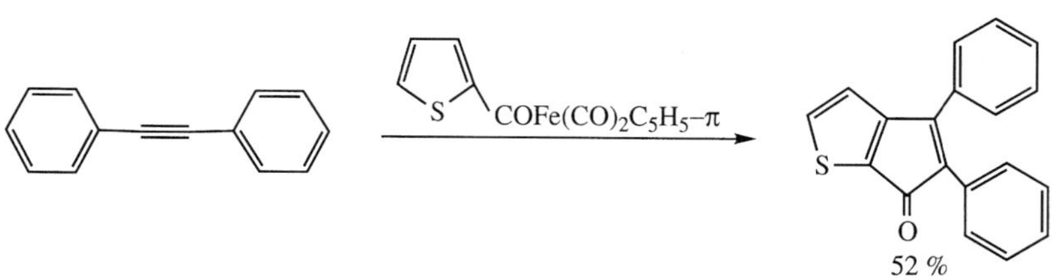

4B.6 PREPARATION OF THC(R′) = C(R″)COR, R = H, ALKYL, ARYL, ALKOXY, R′ = H, ALKYL. ARYL, R″ = H, N_3, COR, CH_2COOH, CH_2CH_2COOH, CN

4B.6.1 Aldol condensation of thiophene aldehydes with aldehydes and ketones

The classical method consists in the reaction of thiophene aldehydes in aldol type condensations. Mixed aldol condensation with acetaldehyde using 10%

sodium hydroxide in aqueous ethanol yields β-[thienyl]acroleins in 30–50% yield [217,218]. The best method for the vinologation of thiophene aldehydes appears to be their reaction with (*Z*)-2-(trimethylsilyl)oxyvinyllithium at −70 °C giving (*E*)-3-(thienyl)propenal in quantitative yield [219]. Chloroacetaldehyde [220] and propionaldehyde [221] are among other aldehydes which can be mentioned to have successfully been condensed with thiophene. The mixed aldol condensation between thiophene aldehydes and aliphatic ketones such as acetone gives 4-(thienyl)-3-butene-2-ones.

4-(Thienyl)-3-butene-2-one [224]

A 33% sodium hydroxide solution (3.8 ml) is added to a mixture of 2-thiophene aldehyde (22.4 g, 200 mmol), acetone (26.3 g, 453 mmol), and water (150 ml). The reaction mixture is stirred at room temperature for 4 h and then neutralized with 10% sulfuric acid. The organic phase is separated, dried over magnesium sulfate, and evaporated. The resulting liquid is distilled giving 24.1 g (79%) of the title compound bp 103–104 °C/6 mm Hg.

Cyclic ketones such as cyclopentanone [225], cyclohexanone [226], cycloheptanone, and cyclo-octanone [227] are also reacted with thiophene aldehydes. Cu(II)-promoted aldol condensation between 2-thiophenealdehyde and butanone occurs selectively at the methylene group [228].

4B.6.2 Claisen–Schmidt condensation between thiophene aldehydes and aryl methyl ketones

Numerous aryl methylketones such as acetophenones [220,225,226,229,230] and various acetyl-substituted heterocyclics, such as acetylthiophenes [229,231–233] and acetylfurans [229,234] and others [229,231,234] can be condensed with thiophene aldehydes to the corresponding chalcone analogs in excellent yields. In most cases aqueous and ethanolic sodium or potassium hydroxide has been used for the condensation.

1-(2-Thienyl)-3-(3-thienyl)-2-propen-1-one [233]

This enone is made from 3-thiophene aldehyde and 2-acetylthiophene in 94% yield following the standard chalcone synthesis [235]. Recrystallization from isopropylalcohol gives 100% pure off-white crystals, by HPLC mp 77–78 °C.

4B.6.3 Mixed Claisen condensation

3-(Thienyl)acrylates can also be prepared through a mixed Claisen condensation of thiophene aldehydes with methyl or ethyl acetate, utilizing sodium ethoxide in xylene. Yields of up to 50% were obtained carrying out the reaction at −10–15 °C [236,237]. The reaction of 2-thiophene aldehyde with ethoxyacetylene in the presence of one equivalent of BF_3-etherate gives ethyl 2-(2-thienyl)acrylate in a 53% yield [238].

4B.6.4 Dehydration of alcohols obtained through the Reformatsky reaction

The reaction of thiophene aldehydes or ketones with α-bromo esters and zinc, followed by dehydration with 6% aqueous oxalic acid, is a useful approach to β-thienylacrylates [239,240]. The Reformatsky reaction followed by dehydration with oxalic acid can be used with phenyl 2-thienyl ketone and dithienyl ketone [241]. Ethyl α-isopropyl-β-methyl-β-(2-thienyl)acrylate and ethyl α-carboethoxy-β-methyl-β-(2-thienyl)acrylate are prepared by the condensation of ethyl α-bromoisovalerate and ethyl bromomalonate with 2-acetylthiophene using magnesium and mercuric chloride [242].

4B.6.5 Condensation of thiophene aldehydes with cyclic active methylene derivatives: Erlenmeyer azlactone synthesis

4-(Thenal)-5-oxazolines are best prepared through the reaction of thiophene aldehydes with fused sodium acetate, acetic anhydride, and hippuric acid [243–246]. Treatment with sodium methoxide in methanol then gives methyl (*Z*)-2-benzamido-3-(thienyl) propenoate [246].

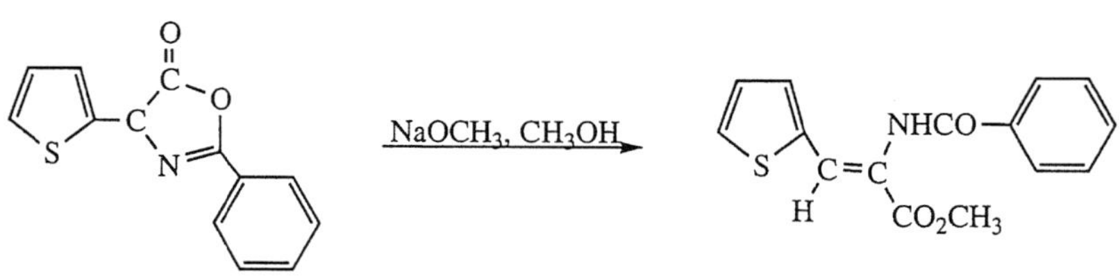

Acetylglycine can also be used in the Erlenmeyer synthesis [242,248]. Under similar reaction conditions the reaction of thiophene aldehydes with rhodanine gives thenal rhodanines in almost quantitative yields [131,244]. The thenal rhodanines are useful intermediates for the synthesis of α-amino acids [249] and β-thienyl-α-mercaptoacrylic acids, obtained through alkaline hydrolysis [250]. Oxidation with iodine in ethanol or benzoyl peroxide leads to the corresponding disulfide. 4-(2-Thenylidene)-1,2-diphenyl-3,5-pyrazolidinedione is prepared by condensation of 2-thiophene aldehyde with 1,2-diphenyl-3,5-pyrazolidinedione in benzene and water separation, using a drop of piperidine as catalyst [251].

4B.6.6 Condensation of thiophene aldehydes or ketones with active methylene derivatives such as malonic esters, malonitriles and related compounds

4B.6.6.1 Knoevenagel reaction

3-(Thienyl)acrylic acids have been prepared in high yields by the Doebner modification using malonic acid, pyridine and piperidine, and 2- and 3-thiophene aldehydes [218,223,226,236,237,252–267].

3-(2-Bromo-5-phenyl-3-thienyl)acrylic acid [266]

A mixture of 2-bromo-5-phenyl-3-thiophene aldehyde (2.66 g, 10 mmol), malonic acid (3.12 g, 30 mmol), pyridine (50 ml), and piperidine (4 ml) is refluxed for 4 h. After cooling, the solution is poured into cold 10% hydrochloric acid (400 ml). The precipitate is collected, washed, dried, and recrystallized twice from benzene giving 2.03 g (66%) of the title compound as pale yellow needles mp 203–204 °C.

2-Thiopheneacetonitrile in ethanol is condensed with 2-thiophene aldehyde 3-methyl-2-thiophene aldehyde and furfural in the presence of potassium tert-butoxide to give *E*-α-[(2-thienyl)methylene]-2-thiopheneacetonitrile, *E*-α-[(3-methyl-2-thienyl)methylene]-2-thiopheneacetonitrile and *E*-α-[(2-furanyl)-methylene]thiopheneacetonitrile. These compounds were electrochemically polymerized to poly(cyanosubstituted diheteroareneethylene) which might be active electrode material for electrochemical super capacitors [128,268]. The Knoevenagel condensation of 2-thiopheneacetonitrile with 3,4-dimethoxybenzaldehyde was used for the preparation of (2*E*)-3-(3,4-dimethoxyphenyl)-2-(2-thienyl)prop-2-enenitrile [268].

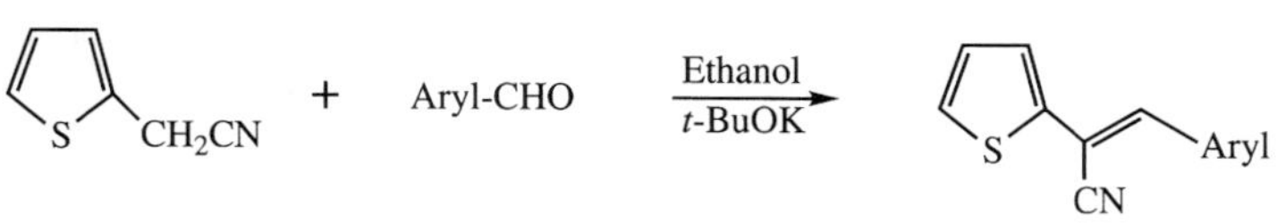

In many cases also halo and otherwise substituted thiophene aldehydes have been used, which are mentioned in other chapters according to the systematics of this book. If diethyl 2-malonates are desired the Knoevenagel condensation is carried out with diethyl malonate using piperidine and benzoic acid as catalyst and benzene as solvent, using a Dean–Stark trap [259–261,269–275].

Diethyl [(3-Methyl-2-thienyl)methylene]propanedioate [269]

To a solution of 3-methyl-2-thiophene aldehyde (3.78 g, 30 mmol) in benzene (120 ml) is added diethyl malonate (7.20 g, 45 mmol), piperidine (0.3 ml), and acetic acid (0.2 ml). The reaction mixture is refluxed for 20 h with a Dean–Stark water separator attached and after cooling it is washed with water, sodium carbonate solution, and sodium chloride solution followed by evaporation. The residue is recrystallized from 5% ethyl acetate in hexane giving 7.40 g (92%) of the title compound as white crystals mp 67–68 °C.

4B.6.6.2 Perkin reaction

The Perkin reaction of 2-thiophenealdehyde using acetic anhydride and sodium acetate does not work well for the preparation of 3-(2-thienyl)acrylic acid [236,256,257]. However, the reaction appears to work satisfactorily with 2,3-dichloro-5-thiophene aldehyde [260] and 5-nitro-2-thiophene aldehyde [276]. The Perkin reaction has also been carried out between 2-thiophene aldehyde and propionic acid [275].

Methyl 2-methyl-3-(2-thienyl)propionate [275]

A mixture of 2-thiophene aldehyde (195 g, 1 mol) and propionic anhydride (195 g, 1.5 mol), and potassium propionate (120 g, 1.25 mol) is heated at 160 °C for 7 h. The reaction mixture is poured into crushed ice (2 l), the crude product is filtered off and esterified with methanol in the presence of sulfuric acid giving 105 g (58%) of the title compound.

4B.6.6.3 Condensation with arylacetic acids

Compounds of type ThCH=C(Ar)COOH are obtained as *cis–trans* mixtures through the reaction of thiophene aldehydes with arylacetic acids using triethyl amine and acetic anhydride [277–282].

2-(1-Naphthyl)-3-(2-thienyl)propanoic acid [282]

A mixture of 1-naphthylethanoic acid (18.6 g, 100 mmol), 2-thiophene aldehyde (17.0 g, 150 mmol), acetic anhydride (40 ml), and triethylamine is refluxed for 6 h. The reaction mixture is then poured into 10% hydrochloric acid and stirred at room temperature for 1 h, after which the product is extracted with benzene (300 ml). The combined benzene phases are extracted with 5% aqueous sodium hydroxide (200 ml). The alkaline solution is acidified with 10% hydrochloric acid and stirring is continued for 1 h. The crystals are collected by filtration, dried in air, and recrystallized from benzene giving 17.4 g (62%) of the title compound as colorless prisms mp 220–222 °C.

4B.6.6.4 Condensation with ethyl azidoacetate

Compounds of type ThCH=C(N_3)COOR are prepared through the condensation of thiophene aldehydes with ethyl azidoacetate using sodium ethoxide in ethanol [273,283–286]. From the isomeric dialdehydes both mono and dicondensation products can be prepared [284,287].

Ethyl 2-azido-3-(2-thienyl)acrylate [286]

A solution of sodium (3.68 g, 0.16 mol) in absolute ethanol (150 ml) is cooled in an ice bath and a mixture of 2-thiophene aldehyde (4.48 g, 0.04 mol) and ethyl azidoacetate (20.64 g, 0.16 mol) is added during 30 min keeping the temperature at 5–10 °C. The stirring is continued for 30 min, after which a cold solution of ammonium chloride is added. The resulting solution is extracted with diethyl ether. The combined organic phases are washed dried and evaporated. The crude product is purified by chromatography on silica gel using benzene/hexane (2:1) as eluent giving 4.73 g (55%) of the title compound as a white solid mp 40–41 °C.

4B.6.6.5 Condensation with ethyl cyanoacetate

Compounds of type ThCH=C(CN)COOR are conveniently prepared through the condensation of thiophene aldehydes with ethyl cyanoacetate in the absence of solvents using piperidine as catalyst [261,288]. Alternatively, sodium ethoxide in ethanol [289] and piperidine in ethanol [290] can be used. This method can also be used for preparing ThC(Ph)=C(CN)COOR from phenyl 2-thienyl ketone using ammonium acetate and acetic acid in toluene as condensating agent [291].

Methyl 2-cyano-3-(2-thienyl)acrylate [288]

Pyridine (0.1 ml) is added to a mixture of 2-thiophene aldehyde (6.0 g, 0.054 mol) and methyl cyanoacetate (6.4 g, 0.065 mol). The reaction mixture is allowed to stand at room temperature for 4 h, after which the separated solid is filtered off, washed with dilute methanol, and dried giving 8.4 g (96%) of the title compound as faintly yellow needles mp 111–112 °C after recrystallization from methanol.

4B.6.6.6 Condensation with β-diketones

2-Nitro-3-thienylmethyleneacetylacetone is prepared by adding a few drops of piperidine to a mixture of 2-nitro-3-thiophene aldehyde and acetylacetone [292].

2-Nitro-3-thienylmethyleneacetylacetone [292]

A mixture of 2-nitro-3-thiophene aldehyde (5.13 g, 32.7 mmol) and acetylacetone (3.27 g, 32.7 mmol) containing piperidine (10 drops) is allowed to stand for two days. The resulting solid is recrystallized from ethanol giving 6.99 g (52%) of the title compound as yellow prisms mp 90–91 °C.

4B.6.6.7 Condensation with β-ketoesters

Compounds of type ThCH=C(COCH$_3$)COOR can be prepared through the condensation of thiophene aldehydes with ethyl acetoacetate using piperidine in ethanol as catalyst [261,293,294]. Another example is the following preparation [295].

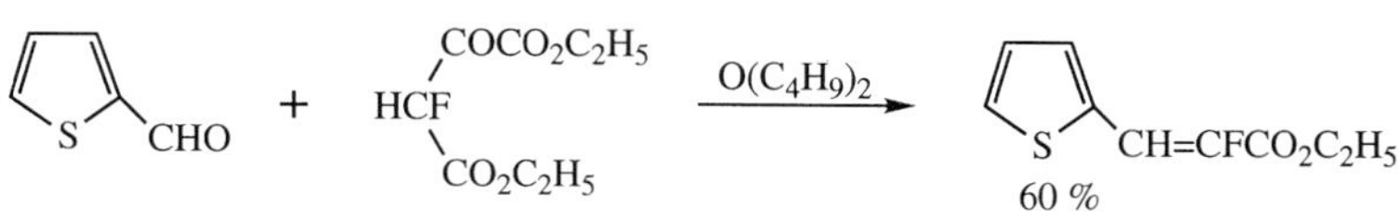

A new methodology for the condensation of thiophene aldehydes with many active methylene compounds is the reaction in the presence of titanium tetrachloride and an organic base in tetrahydrofuran. Ethyl 2-acetyl-3-(2-thienyl)acrylate is obtained in the reaction between triethyl phosphonoacetic acid and 2-thiophene aldehyde [296].

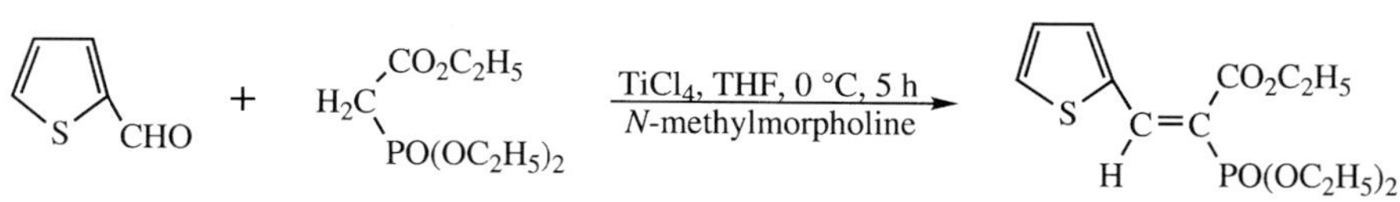

Ethyl 2-acetyl-3-(2-thienyl)acrylate [296]

To anhydrous tetrahydrofuran (200 ml) at 0 °C, under stirring, titanium tetrachloride (11 ml, 0.1 mol) in anhydrous carbon tetrachloride (25 ml) is added and a yellow precipitate is formed. 2-Thiophene aldehyde (5.6 g, 0.05 mol) in anhydrous tetrahydrofuran (25 ml) and ethyl acetoacetate (6.4 g, 0.05 mol) in anhydrous tetrahydrofuran (25 ml) are added and the stirring is continued for 1–2 h at 0 °C and pyridine (16 ml, 0.2 mol) is slowly added dropwise. The stirring is continued at 0 °C for 17 h, after which water and diethyl ether are added. The phases are separated and the aqueous phase is extracted with diethyl ether (2 × 50 ml). The combined organic phases are washed with sodium chloride solution, dried over magnesium sulfate, and evaporated at

30 °C. The residue is recrystallized from ethanol giving 10.2 g (92%) of the title compound mp 78–79 °C.

This methodology has also been applied for the preparation of ethyl 2-acetyl- and ethyl 2-nitro-3-(2-thienyl)acrylate from ethyl acetoacetate and ethyl nitroacetate at temperatures below room temperature [297].

4B.6.6.8 Condensation with malonitrile and other activated nitriles

Compounds of the type ThC(CN)=CHTh, such as 1,2-di(thienyl)acrylonitrile, are very conveniently prepared by alkaline condensation of thiophene aldehydes with thenylcyanides and can be electropolymerized [298].

Compounds of type ThC(R)=C(CN)CN have attracted great interest during recent years in connection with conducting charge transfer complexes. They are easily prepared through the condensation of thiophene aldehydes and ketones with malonitrile using a few drops of pyridine or triethylamine [117,259,271, 299,300] sodium ethoxide in ethanol [289] as solvent, or when applied to alkyl 2-thienyl ketones ammonium acetate in glacial acetic acid in toluene is used and water separated with a Dean–Stark trap [301,302]. In this way 6,7-dihydrobenzo[*b*]thiophen-4(5*H*)-ylidenpropandinitrile was obtained [302].

6,7-Dihydrobenzo[b]thiophen-4(5H)-ylidenpropandinitrile [302]

A mixture of 4-keto-4,5,6,7-tetrahydrothianaphthene (146.3 g, 0.96 mol), malonitrile (63.5 g, 0.96 mol), ammonium acetate (14.8 g, 0.19 mol), acetic acid (46 ml), and benzene (400 ml) is refluxed for 15 h using a Dean–Stark trap. After cooling the reaction mixture is diluted with benzene (200 ml) and washed with water, dried over sodium sulfate, and evaporated, giving 180 g (94%) of the title compound as a yellow powder mp 131–133 °C.

Extended μ-vinylidenediiron donor-based organometallic merocyanines have been prepared by conventional Knoevenagel reaction with malonitrile according to the scheme shown below [303].

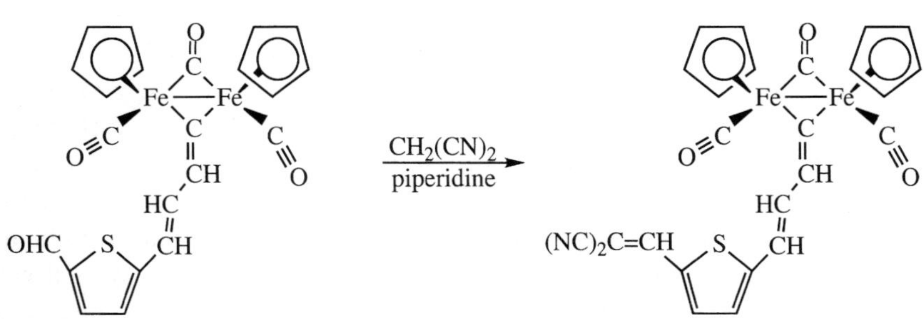

Through titanium tetrachloride-mediated condensation with malonitrile in pyridine/chloroform the following compounds are prepared from the diones [304–306].

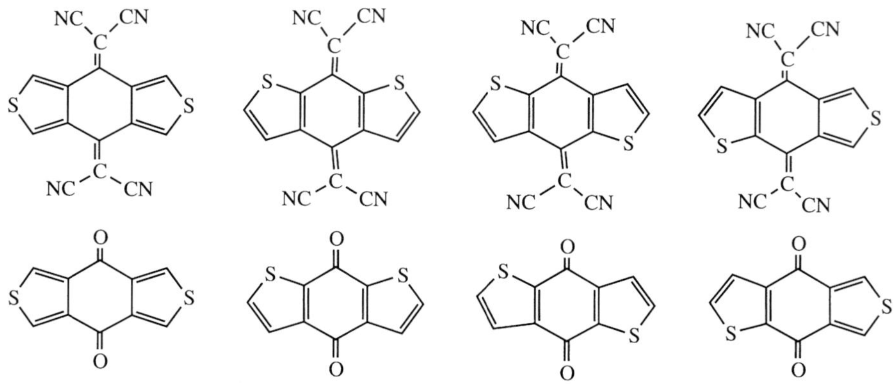

4,8-Bis(dicyanomethylene)-4,8-dihydrobenzo[1,2-b:4,5-b′]dithiophene [306]

Titanium tetrachloride (0.8 ml, 5.8 mmol) is added under nitrogen to a magnetically stirred solution of 4,8-dihydrobenzo[1,2-*b*:4,5-*b*′]dithiophene-4,8-dione (0.635 g, 2.88 mmol) and malonitrile (3.80 g, 57.5 mmol) in a mixture of anhydrous chloroform (200 ml) and pyridine (10 ml). The resulting orange suspension is stirred for 5 h at reflux. After cooling and addition of water (400 ml) the insoluble solid is filtered off by using an aspirator. The phases are separated and the aqueous phase extracted with chloroform (2 × 200 ml). The combined organic phases are dried over magnesium sulfate and evaporated and the residue chromatographed on silica gel using chloroform as eluent. The first fractions contains the mono dicyanomethylene compound and the following fraction gives 0.336 g (37%) of the title compound as red prisms mp 332 °C.

2-Cyano-3-methylthio-3-(2-thienyl)acrylonitrile is prepared through the reaction of methyl thiophene-2-dithiocarboxylate with malonitrile in the presence of sodium hydride followed by methylation with dimethyl sulfate [307].

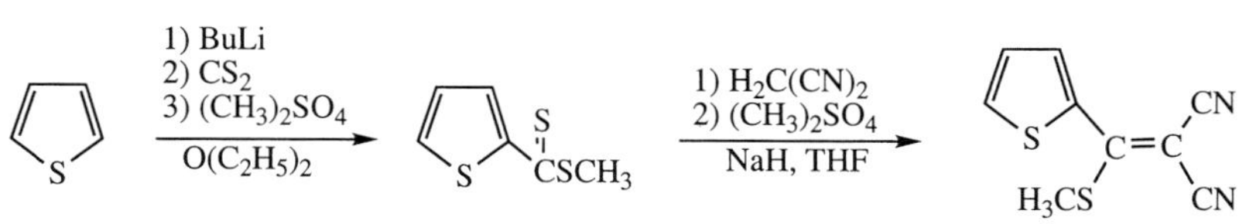

A series of thiophenes and bithienyls containing dimethylamino- or dimethylhydrazono groups as π-donors and dicyanovinyl groups as π-acceptors in the 2- and 5-positions for study of their quadratic nonlinear optical properties have been prepared [308]. Another example is the preparation of a conformation-locked *trans*-polyene shown below, prepared by condensation of the 5-alkenyl substituted 2-thiophene aldehyde and tetracyanoethylene [309].

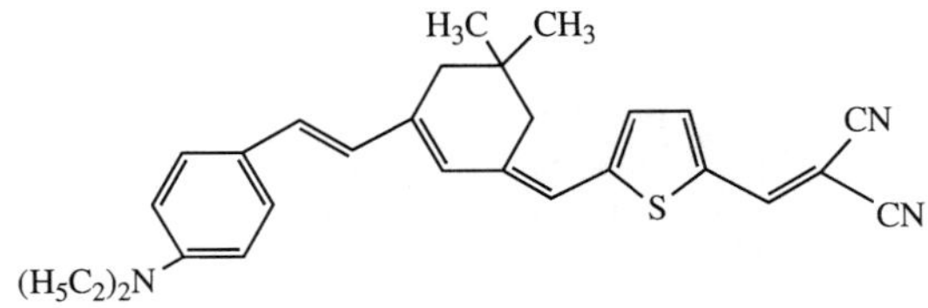

An alternative route to this type of compounds consists in the reaction of a tricyanovinyl-substituted compound with phenyllithium at 0°C, leading to the substitution of the cyano group for a phenyl group [310].

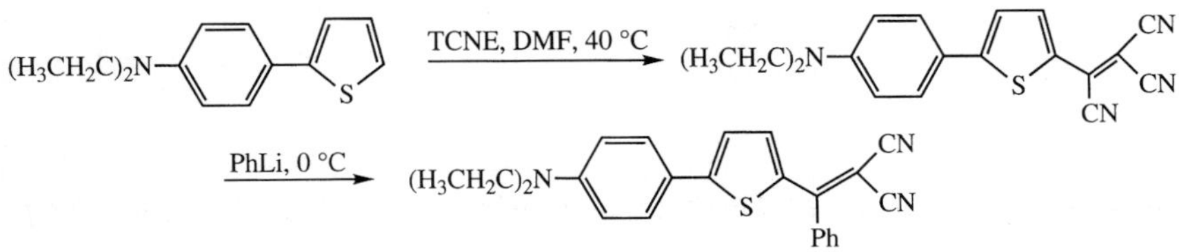

Tricyanovinyl-substituted thiophenes are prepared in a one-step procedure through the reaction of donor-substituted thiophenes with tetracyanoethylene [116,117,188,311].

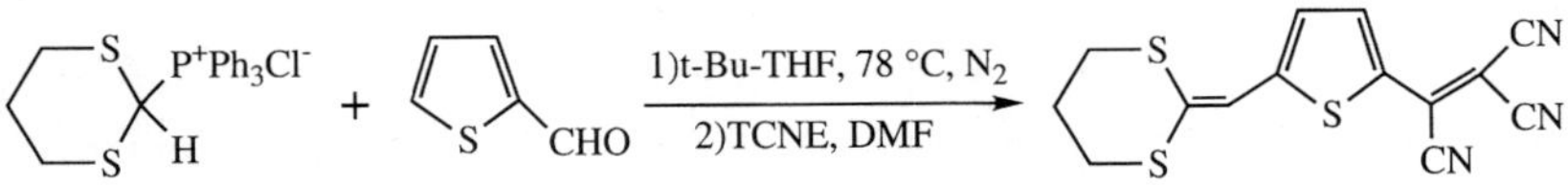

Radical ions can be obtained from 2,5-bis(tricyanovinyl)thiophene [312]. Thiophenes, carrying a tricyanoquinodimethane acceptor group can be prepared to show extremely large second-order optical nonlinearities [313].

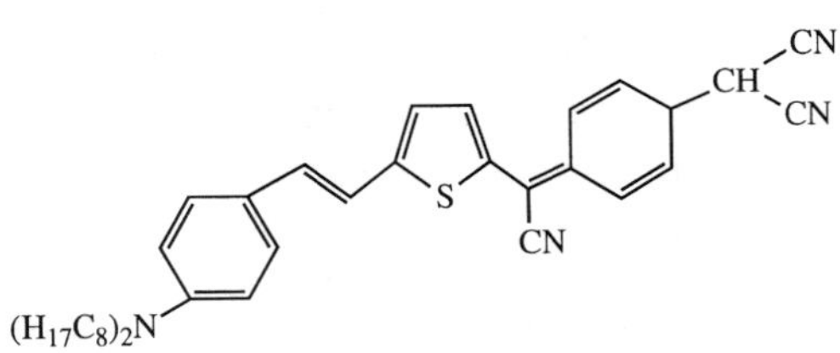

3-(2-Thienyl) and 3-(3-thienyl)acrylonitrile are prepared as *cis–trans* mixtures through the reaction of thiophene aldehydes with cyanoacetic acid in pyridine/toluene using a Dean–Stark trap followed by decarboxylation [265, 314]. 2-Carbamido-3-(thienyl)acrylonitriles are prepared through condensation of thiophene aldehydes and acetylthiophenes with cyanoacetamide and then with ammonium acetate and acetic acid in ethanol [301]. Numerous 2-aryl-3-(thienyl)acrylonitriles have been prepared by condensation of thiophene aldehydes with arylacetonitriles using ethoxide, hydroxide, or piperidine in ethanol [226,244,261,315–323]. 2-Benzoyl-3-(2-thienyl) is prepared by the reaction of 2-thiophene aldehyde with benzoyl acetonitrile in acetic acid and triethylamine [324]. The condensation of 2-thiophene aldehyde with methylene aminoacetonitrile in the presence of sodium ethoxide or Triton B leads to the compound shown below [325].

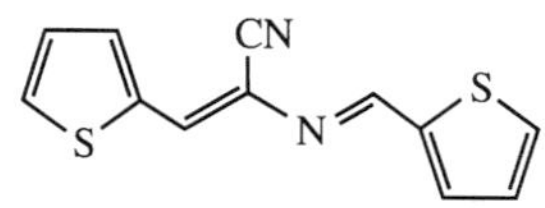

4B.6.6.9 Stobbe condensation

3-(Thienyl)acrylic acids of type ThC(R)=C(COOH)CH_2COOH, useful for the preparation of benzo[*b*]thiophene derivatives, have been prepared by Stobbe condensation between thiophene aldehydes and thienyl ketones with succinates [326–332]. A detailed description of the separation of the *E* and *Z* isomers of 3-carbethoxy-4-(2-thienyl)-3-pentenoic acid obtained from the Stobbe condensation of 2-acetylthiophene and ethyl succinate has been published [333].

(E,Z) 3-Carbethoxy-4-(2-thienyl)-3-pentenoic acid [333]

A 500 ml round bottom flask, fitted with condenser, drying tube, addition funnel, and magnetic bar, is charged with a suspension of sodium hydride, as 50% suspension in oil (4.8 g, 0.2 mol) in anhydrous benzene (120 ml). A mixture of 2-acetylthiophene (12.62 g, 0.1 mol), diethyl succinate (34.8 g, 0.2 mol) and anhydrous benzene (120 ml) is added dropwise. The reaction mixture is stirred at room temperature for 3 h, cooled in an ice-water bath and methanol (2 ml) is added to destroy the sodium hydride. After acidification with hydrochloric acid, the solvent is removed by evaporation and water (200 ml) is added. The aqueous solution is extracted by diethyl ether (3 × 150 ml) and the combined organic phases are extracted with a 5% solution of potassium hydroxide (3 × 60 ml). The combined alkaline solutions are acidified and extracted with diethyl ether. The organic phases are dried over sodium sulfate and evaporated giving 19.0 g (75%) of an oily product.

Separation of the E and Z isomers

The oily product described above is crystallized partially on standing overnight. The solid is collected, washed with hexane and chromatographed on silica gel using ethyl acetate/hexane (8:2) as eluent. The solid obtained after evaporation of the eluent is recrystallized from ethyl acetate giving 6.96 g (27%) of the *Z*-isomer of the title compound mp 93–95 °C. The filtrate constituted mainly of the *E* isomer, which is purified by chromatography on silica gel using ethyl acetate/hexane (8:2) as eluent followed by preparative thin layer chromatography giving 6.52 g (26%) of the pure *E*-isomer as an oil.

Compounds of type $ThCH{=}C(COOH)CH_2CH_2COOCH_3$ have been prepared through the condensation of thiophene aldehydes with dimethyl glutarate in the presence of sodium hydride, predominantly the *E*-half esters were obtained [334].

Compounds of type ArCH=C(Th)COOH are easily prepared by the condensation of thiopheneacetic acids with aromatic aldehydes in a modified Perkin reaction using acetic anhydride and triethylamine, giving predominantly the isomer in which the aryl groups are *cis*-related [335–339]. 3-(2-Thienyl)- and 3-(3-thienyl)coumarin are directly formed in the condensation of 2- and 3-thiopheneacetic acid with *ortho*-fluorobenzaldehyde [340].

3-(2-Thienyl)coumarin [340]

A mixture of *ortho*-fluorobenzaldehyde (124 g, 1.0 mol), 2-thiopheneacetic acid (142 g, 1.0 mol), acetic anhydride (3.2 mol), and freshly distilled triethylamine (1.6 ml) are refluxed for 8 h. The reaction mixture is allowed to cool, poured into water, and extracted with diethyl ether. The combined organic phases are washed with water, 5% aqueous sodium hydroxide, again with water until neutral pH, dried over sodium sulfate and evaporated giving the title compound mp 167–168 °C after recrystallization from ethanol.

Compounds of type ArCH=C(Th)CN can be prepared from various thenyl cyanides and aromatic aldehydes by using a few drops of 50% aqueous potassium hydroxide as solvent [21,22,323,341–344] or sodium ethoxide or potassium *tert*-butoxide in the corresponding alcohol [244,336,345.346]. Recently bis(2-cyano-2-α-thienylethenyl)arylenes have been prepared from 2-thenyl cyanide and aromatic dialdehydes such as isophthalaldehyde and terephthalaldehyde, for study of their electropolymerization [347].

Bis(2-cyano-2-α-thienylethenyl)benzene [347]

A solution of isophthaldehyde (1.5 g, 0.011 mol) and 2-(cyanomethyl)thiophene (2.8 g, 0.023 mol) in ethanol (10 ml) is added to a suspension of potassium tert-butoxide (2.8 g, 0.023 mol) in ethanol (50 ml) under nitrogen. The reaction mixture is stirred at room temperature for 3 h. The reaction is quenched by evaporation of the ethanol followed by addition of water. The product is extracted with dichloromethane, dried over magnesium sulfate, and purified by chromatography on silica using dichloromethane/hexane (1:1) as eluent to obtain 3.1 g (82%) as a yellow solid. Recrystallization from dimethylsulfoxide/ethyl acetate (2:1) gives the title compound as yellowish crystals mp 126–128 °C.

By treatment of 2,3,5-trimethyl-4-thenylcyanide with aqueous sodium hydroxide containing triethyl benzylammonium chloride and carbon tetrachloride a mixture of *cis* and *trans* 1,2-dicyano-1,2-bis(2,3,5-trimethyl-4-thienyl) ethene was obtained in 47% yield, from which the pure *cis* form could be obtained by photoisomerization followed by recrystallization [141].

1,2-Dicyano-1,2-bis(2,3,5-trimethyl-4-thienyl)ethene [141]

A 50% aqueous solution of sodium hydroxide containing triethyl benzylammonium chloride (0.21 g 1.0 mmol) (20 ml) at 40 °C is added to a solution of 2,3,5-trimethyl-4-thenylcyanide (16 g, 0.10 mol) and tetrachloromethane (15 g, 0.10 mol). The reaction mixture is stirred for 1.5 h at 45 °C and then poured into water. The product is extracted with diethyl ether and chloroform, the solvent evaporated and the residue chromatographed on silica gel giving 47% of the title compound as a *cis*/*trans* mixture. This mixture is dissolved in acetonitrile and exposed to ultraviolet light (λ > 350 nm). The precipitated *cis*-form is filtered off and purified by recrystallization from hexane/diethyl ether mp 164 °C.

4B.6.6.10 Other condensation reactions

3-(3-Azido-2-thienyl)-2-alkylthioacrylic esters are conveniently prepared by condensation of 3-azido-2-thiophene aldehyde with ethyl phenylthioacetate and ethyl ethylthioacetone using sodium ethoxide in ethanol [293,348]

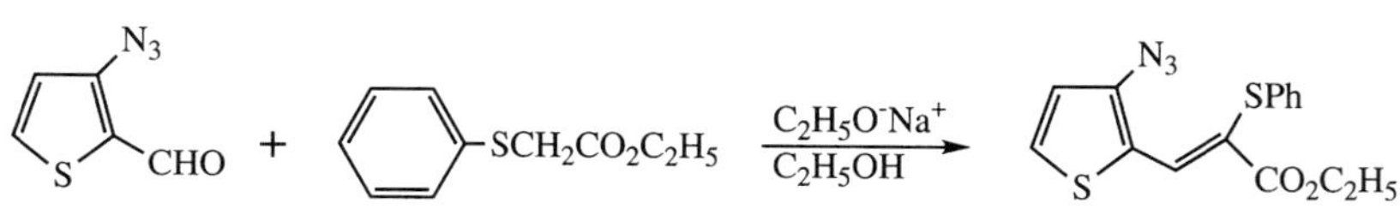

3-Azido-2-thiophene aldehyde has also been condensed with ethyl phenylsulfinyl- and ethyl phenylsulfonylacetate using piperidinium acetate in ethanol to give the corresponding 3-(3-azido-2-thienyl)acrylic esters [393]. Condensation of 2-thiophene aldehyde with (*ortho*-nitrophenylthio)acetic acid was carried out using ammonium acetate and pyridine in acetic acid [349]. Condensation of methyl methyl methylthiomethyl sulfoxide with thiophenealdehydes using triton B in methanol is an excellent method for the preparation of 1-methylsulfinyl-1-methylthio-2-thienylethylenes [350,351]. Thienylsubstituted vinyl sulfones are prepared through the reaction of thiophenealdehydes with sulfonylacetic acid in glacial acetic acid and benzylamine [352]. Cyclopenta[2,1-*b*;4,3-*b*′]dithiophene-4-one is condensed with bis(phenylsulfonyl)methane nonafluorobutyl-sulfonylacetonitrile or hepta-decafluorooctylsulfonylacetonitrile, using a combination of a Grignard reagent of the methylene compound, titanium tetrachloride activation of the ketone and *N*-methylmorpholine as base for the final deprotonation-elimination step [353].

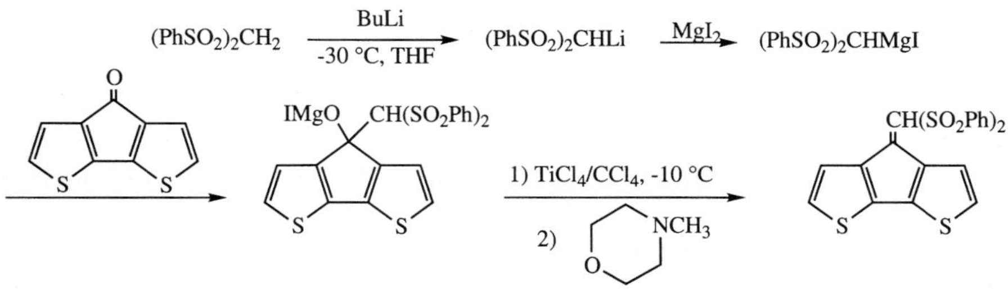

Methyl α-[(amino)methylene]acrylates are prepared by the reaction of 2-thiophene aldehyde with methyl isocyanoacetate in the presence of secondary amines. If the reaction is instead carried out in the presence of sodium hydride as base, α-formylaminoacrylates were obtained [354]. 4H-5-amino-6-cyano-cyclohepta[2,1-*b* 3,4-*b*′]dithiophene is prepared by the reaction of 3,3′-biscyanomethyl-2,2′-bithienyl with sodium ethoxide in ethanol [355].

Thienotropone derivatives were obtained in excellent yield through the reaction of 2,5-dimethyl-3,4-thiophene dialdehyde with 1,3-bis-methylthioacetone in methanol using triethylamine as base [356].

4B.6.6.11 Wittig reaction of functionalized ylides with thiophene aldehydes

The reaction of the ylide derived from with 2- and 3-thiophenealdehyde gives 2- and 3-thienylacrolein in 70–80% yield [357].

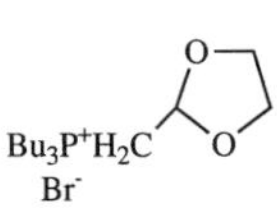

3-(2′-Thienyl)acrolein [357]

A 1 *M* solution of sodium ethoxide in absolute ethanol (150 ml) is added dropwise to a solution of 2-thiophene aldehyde (11.2 g, 0.1 mol) and 1,3-dioxane-2-ylmethyltributylphosphonium bromide (40.6 g, 0.11 mol) in anhydrous *N*,*N*-dimethylformamide (300 ml) at 90 °C. The reaction mixture is then stirred for an additional 16–20 h, after which it is poured into water (1000 ml). The product plus tributylphosphine oxide are extracted with diethyl ether (3 × 300 ml), washed with saturated aqueous sodium chloride solution (2 × 200 ml), dried over magnesium sulfate, filtered, and evaporated. The residue is dissolved in tetrahydrofuran (250 ml), a 10% aqueous solution of hydrochloric acid is added rapidly, the resulting mixture stirred at room temperature for 2 h, after which it is poured into water (1000 ml). The product is extracted with diethyl ether, washed with saturated aqueous sodium chloride solution, dried over magnesium sulfate, filtered, and evaporated. The residue is dissolved in a minimal quantity of ethyl acetate and chromatographed on silica gel using ethyl acetate/hexane (1:4) as eluent giving 11.3 g (82%) of the title compound as *E*-isomer.

Special ylides are used for the preparation of 5-formyl-2-thienylacrolein [358],

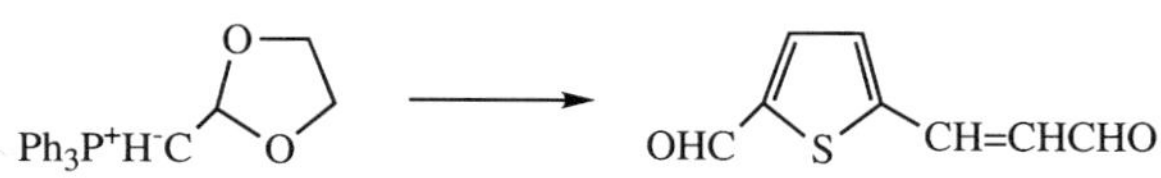

and in the following reaction [359].

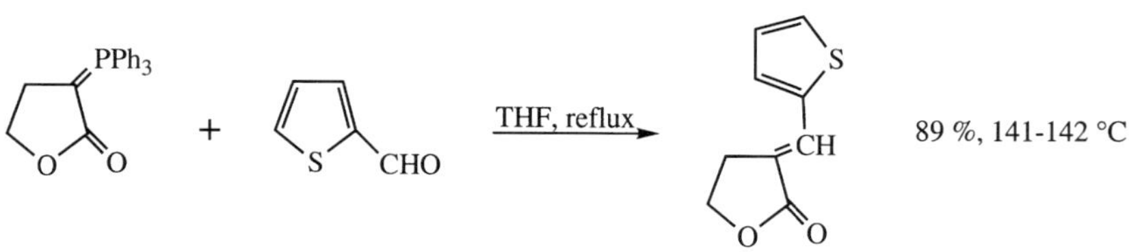

The reaction of the anion from ethyl triphenylphosphinoacetate with various thiophene aldehydes is a good method for the preparation of ethyl thienylacrylates [360]. The following reaction has been performed in 70–80% yield [361].

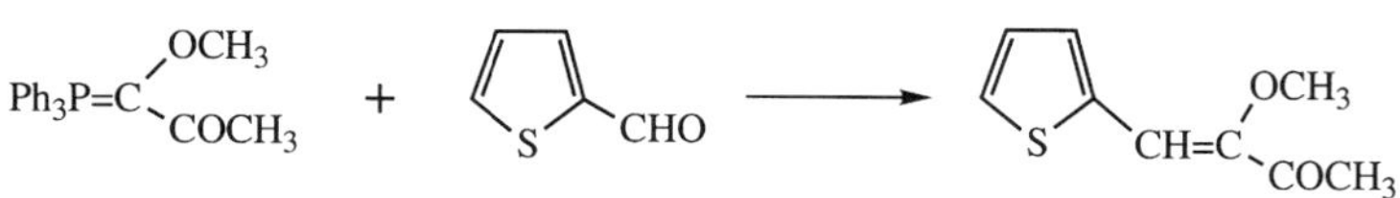

Compounds of type

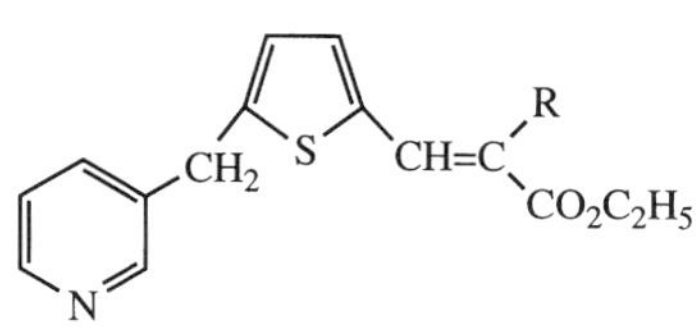

are obtained through the Wittig reaction between 5-(3-pyridylmethyl)-2-thiophene aldehyde and the ylide shown below [362].

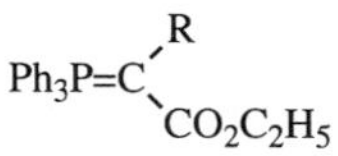

A good method for the preparation of thienylvinyl phosphonates is the reaction of the Wittig reagent from diphenyl triphenylphosphoranylidene-methylphosphonate with thiophene aldehydes [363].

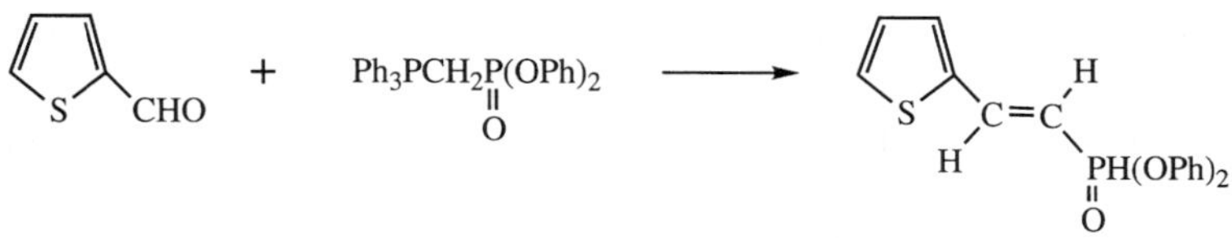

Diphenyl [2-(2-thienyl)vinyl]phosphonate [363]

A mixture of diphenyl triphenylphosphoranylidenemethylphosphonate (2.5 mmol) and 2-thiophene aldehyde (280 mg, 2.5 mmol) in anhydrous benzene (30 ml) is heated at 70 °C for 36 h. The solvent is evaporated and the residue purified by chromatography on silica gel using benzene/diethyl ether (3:1) as eluent giving 770 mg (77%) of the title compound mp 118–119 °C.

4B.6.7 Vinylthiophenes through transition metal catalyzed couplings

4B.6.7.1 From thiophenes and olefins

The reaction of 2-thiophenealdehyde with methyl acrylate with stoichiometric amounts of palladium acetate gave methyl 2-formyl-5-thienylacrylate [364].

(E)-Methyl 2-(5-formylthienyl)acrylate [364]

A solution of 2-thiophene aldehyde (112 mg, 1 mmol), methyl acrylate (258 mg, 3 mmol) and palladium acetate (225 mg, 1 mmol) in acetic acid (40 ml) is heated at reflux temperature in air for 7 h. After evaporation the residue is purified by TLC over silica gel (chloroform) giving 67 mg (34%) of the title compound mp 68–69 °C.

However, this methodology became first of practical use when the reaction could be made catalytic in palladium, by the use of a palladium(II) acetate–cupric acetate catalytic system. With acrylonitrile and methyl acrylate a mixture of mono and divinylated products was obtained [365].

4B.6.7.2 Reaction of thienylzinc reagents with olefins

Ethyl (*E*)-3-phenyl-2-(2-thienyl)-2-propenoate can be obtained through the palladium-catalyzed reaction of 2-thienylzinc bromide with ethyl (*E*)-2-iodo-3-phenyl-2-propenoate, prepared from ethyl (*E*)-2-tributylstannyl-2-propenoate and iodine [366].

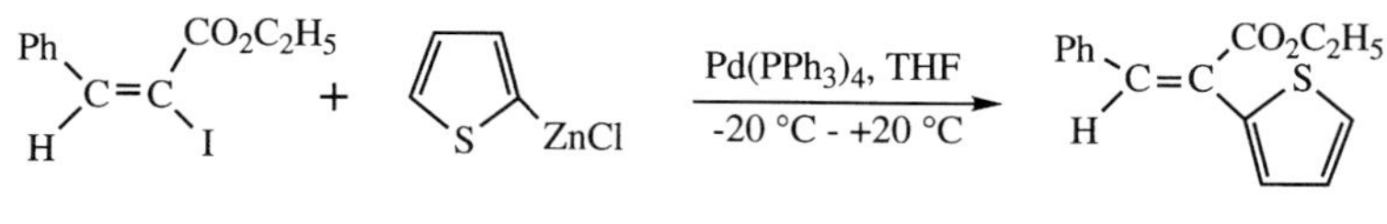

Ethyl (E)-3-phenyl-2-(2-thienyl)-2-propenoate [366]

A 1.17 *M* solution of 2-thienylmagnesium bromide in tetrahydrofuran (33 ml, 38.6 mmol) is added to a stirred solution of zinc chloride (5.80 g, 42.57 mmol) in tetrahydrofuran (50 ml) maintained at −25 °C. After stirring for 0.5 h, a solution of (*E*)-2-iodo-3-phenyl-2-propenoate (6.50 g, 21.5 mmol) and tetrakis(triphenylphosphine)palladium(0) (2.08 g, 1.80 mmol) in tetrahydrofuran (110 ml) is added and the reaction mixture stirred at −25 °C for 3 h and then at room temperature for 3.5 h, after which it is poured into a large excess of a saturated aqueous ammonium chloride solution. The phases are separated and the aqueous phase extracted with ether. The combined organic phases are washed with water, dried and evaporated. The residue is purified by medium pressure liquid chromatography on silica gel using hexane/benzene (55:45) as eluent giving 4.06 g (73%) of the title compound. GLC showed a stereoisomeric purity higher than 98%.

The reaction of 2-thienylzinc bromide with diketene constitutes a good method for the preparation of 3-(2-thienyl)-3-butenoic acid [367].

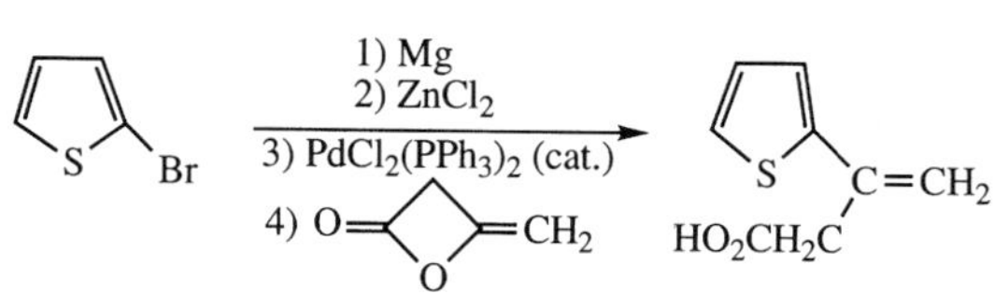

3-(2-Thienyl)-3-butenoic acid [367]

The Grignard reagent prepared from magnesium turnings (7.3 g, 0.3 mol) and 2-bromothiophene (46.7 g, 0.285 mol) and diethyl ether (200 ml) is diluted with additional diethyl ether (100 ml). Anhydrous zinc chloride (35 g, 0.257 mol) is gradually added and the solution is stirred for 45 min giving a white solid in a brown solution. Then dichlorobis(triphenylphosphine)palladium(II) (0.762 g, 0.001 mol) is added followed by a solution of the diketene (16 ml, 0.207 mol) in diethyl ether, which is added dropwise. The reaction mixture is stirred for

30 min, after which it is poured into cold 2 *M* hydrochloric acid and the product is extracted with diethyl ether. The combined organic phases are extracted with 3 *M* sodium hydroxide solution and the aqueous phases acidified with 6 *M* hydrochloric acid. The acid is taken up in diethyl ether and the combined ether phases are dried over magnesium sulfate and evaporated giving 2.52 g (53%) of the title compound as a pale yellow solid mp 65–66 °C.

4B.6.7.3 From vinylmetallic derivatives and halothiophenes

(*Z*)-1-Ethoxy-(2-tributylstannyl)ethene is a convenient reagent for the palladium-catalyzed conversion of 2-bromothiophene to (*Z*)-1-ethoxy-2-(2-thienyl)ethene [368]. A mixture of the *E* and *Z*-isomers of methyl 3-methoxy-2-(2-thienyl)propenoate are obtained through the copper promoted palladium-catalyzed reaction of methyl (*Z*)-2-tributylstannyl-3-methoxypropenoate with 2-iodothiophene [369].

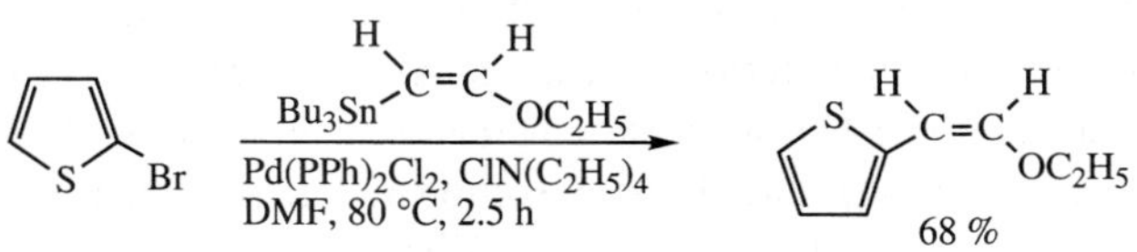

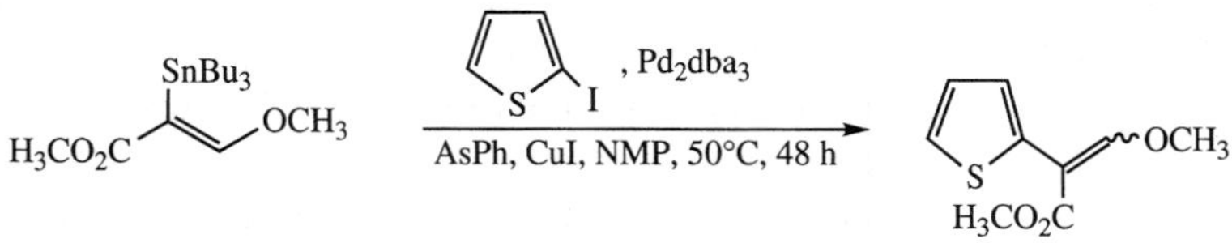

A stereo- and regiospecific synthesis of trisubstituted alkenes can elegantly be achieved *via* palladium-catalyzed cross-coupling reactions of diisopropyl (*E*)-1-alkenylboronates, obtained from (*Z*)-1-bromo-1-alkenylboronates with organic halides. Reaction with phenyllithium followed by 2-iodothiophene gives 1-phenyl-1-(2-thienyl)-1-hexene, while reaction with 2-thienyllithium followed by iodobenzene yields the isomeric 1-hexene in almost quantitative yields [370].

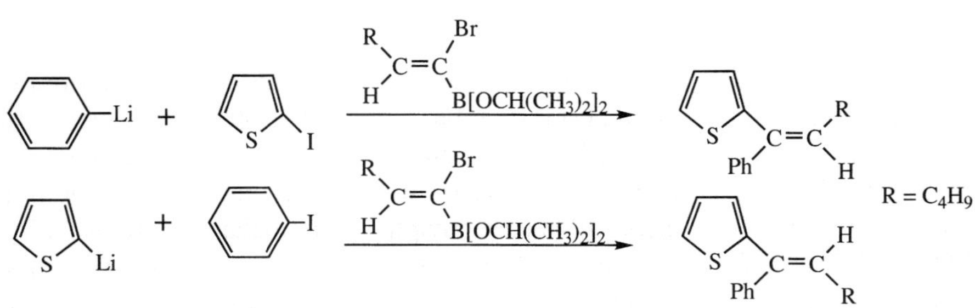

β-Mono- and *β*,*β*-disubstituted *α*,*β*-unsaturated ketones, such as the 2-thienyl derivatives can be prepared stereoselectively by the stepwise alkylation and alkoxycarbonylation of 2-bromo-1-alkenylboronates with 2-thienylzinc bromide under palladium catalysis [371].

4B.6.7.4 Heck reaction of halothiophenes with vinyl derivatives

The Heck reaction of 2,5-dibromothiophene with ethyl acrylate gives 2,5-thiophenedi(2-ethenyl)carboxylic acid after alkaline hydrolysis [372,373].

(E,E)-2,5-Thiophene-di-(2-ethenyl)carboxylic acid [373]

2,5-Dibromothiophene (4.5 g, 0.018 mol) and ethyl acrylate (9.3 g, 0.093 mol) is added to a solution of palladium(II) acetate (0.12 g, 0.52 mmol) and triphenylphosphine (0.58 g, 0.46 mmol) in acetonitrile (60 ml) and triethylamine (40 ml). The reaction mixture is sealed in a tube and heated at 100 °C for 20 h. The mixture is cooled, the content evaporated, the residue is dissolved in methanol (50 ml) and the solution so obtained filtered. The filtrate is added to a solution of sodium hydroxide (7 g) in water (200 ml) and the mixture refluxed for 1 h, after which charcoal is added and the reaction mixture filtered. After cooling, the filtrate is acidified with 20% hydrochloric acid and the precipitate filtered off giving 3.3 g (83%) of the title compound as yellow crystals mp > 310 °C.

An excellent method for the preparation of (*E*)-trimethyl(thienylvinyl)silanes is the palladium-catalyzed reaction of 2- and 3-iodothiophene with vinyl trimethylsilane in the presence of silver nitrate, which enhances the rate of the reaction and completely suppressed desilylation [374].

(E)-Trimethyl(2-thien-2-ylethenyl)silane [374]

Each of the reactants is dissolved or dispersed in acetonitrile (totaling 150 ml) and added to a 250-ml round-bottomed flask in the following order: palladium acetate (67 mg, 0.3 mmol), triphenylphosphine (157 mg, 0.6 mmol), silver nitrate (1.70 g, 10 mmol), 2-iodothiophene (2.10 g, 10 mmol) triethylamine (1.21 g, 12 mmol), and vinyltrimethylsilane (2.00 g, 20 mmol). The flask is closed and the content magnetically stirred at 50 °C for 5 h. After cooling the content is filtered and poured into water (100 ml). The product is extracted with diethyl ether (4 × 50 ml) and the combined organic phases are washed with water, dried over magnesium sulfate and evaporated. The residue is purified by chromatography giving 1.35 g (74%) of the title compound as a yellow oil.

A recent example is the preparation of 3-phenyl-4-[*trans*-(2′-ethylcarbonyl)-vinyl]thiophene and 3-phenyl-4-[*trans*-(methoxycarbonyl)vinyl]thiophene from

4-iodo-3-phenylthiophene and ethyl vinyl ketone and methyl acrylate, respectively. A modification was worked out in which the reaction is carried out with palladium(II) acetate under phase transfer conditions in *N,N*-dimethylformamide, using potassium carbonate and tetrabutyl ammonium iodide [375].

3-Phenyl-4-[trans(methoxycarbonyl)vinyl]thiophene [375]

A mixture of 4-iodo-3-phenylthiophene (28.6 mg, 0.1 mmol), ethyl vinyl ketone (33.6 mg, 0.4 mmol), palladium(II) acetate (0.5 g, 0.002 mmol), potassium carbonate (34.5 mg, 0.25 mmol) and tetrabutylammonium iodide (36.9 mg, 0.1 mmol) in *N,N*-dimethylformamide (3 ml) is stirred at 90 °C for 8 h under nitrogen. The reaction mixture is diluted with diethyl ether (10 ml) and water (10 ml). The phases are separated and the aqueous phase extracted with diethyl ether (2 × 10ml). The combined organic phases are dried over magnesium sulfate and evaporated. The residue is purified by column chromatography on silica gel (10 g) using hexane/ethyl acetate (4:1) as eluent, giving 12.1 mg (50%) of the title compound as a solid mp 80–82 °C.

The palladium-catalyzed Heck reaction of 2-amidoacrylates with 5-methyl-2-iodothiophene under phase transfer conditions is a good method for the preparation of the corresponding didehydroamino acid derivatives which are excellent starting materials for various protected aromatic amino acids [376].

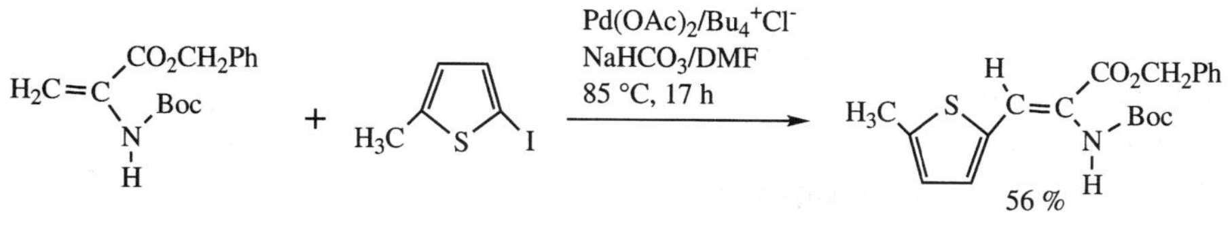

4B.6.8 From thiophenes and alkenes by electrophilic reactions

Electrophilic condensation of thiophene with 2-acetylvinyl chloride using tin tetrachloride as catalyst gives 2-acetylvinylthiophene [377]. Thiophene analogs of chalcones, such as 1-(2-thienyl)-3-(*p*-nitrophenyl)prop-1-en-3-one, are prepared in 53–78% yield through tin tetrachloride catalyzed substitution of thiophene with *β*-chlorovinyl-*para*-nitrophenylketone [378].

1-(2-Thienyl)-3-(p-nitrophenyl)prop-1-en-3-one [378]

A three-necked flask fitted with stirrer, reflux condenser, and dropping funnel is charged with *β*-chlorovinyl-*para*-nitrophenylketone (4.0 g, 0.019 mol) and thiophene (1.54 g, 0.019 mol) in anhydrous benzene (50 ml). The flask is cooled with ice and sodium chloride to −10 °C. The content is stirred vigorously and tin tetrachloride (4.8 g, 0.019 mol) is added over a period of 30 min from the dropping funnel. The reaction mixture is stirred for an additional 1 h, after which diethyl ether (50 ml) and water (50 ml) are added. The phases are separated and the aqueous phase extracted with benzene. The combined organic phases are washed with 5% sodium carbonate until tin is completely removed, dried over calcium chloride and evaporated. Upon cooling the residue 3.60 g (78%) of the title compound is crystallized as yellow needles, mp 153 °C after recrystallization from ethanol.

The condensation of thiophene and 3-methylthiophene with 2,2,6,6-tetramethylpiperid-4-one at 60–100 °C in the presence of 72% sulfuric acid gives the 2,5-bis(2,2,6,6-tetramethyl-1,2,5,6-tetrahydropyrid-4-yl)thiophene [379].

4B.6.9 From thiophenes and acetylenes

A mixture of stereoisomers is prepared by the reaction of a 2-thienylmethoxychromium carbene complex with ethyl propiolate in the presence of alcohol [380].

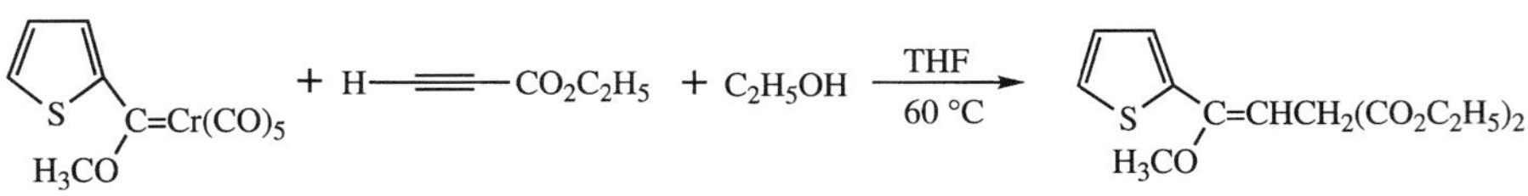

4B.6.10 Synthesis of thienyl-substituted fulvalenes

This type of compound is conveniently prepared by the condensation of thiophene ketones with cyclopentadiene in tetrahydrofuran in the presence of potassium tert-butoxide. Thus 1,4-di(2-thenoyl)benzene gives 1,4-bis([6-(2-thienyl)fulven-6-yl]benzene [121,381].

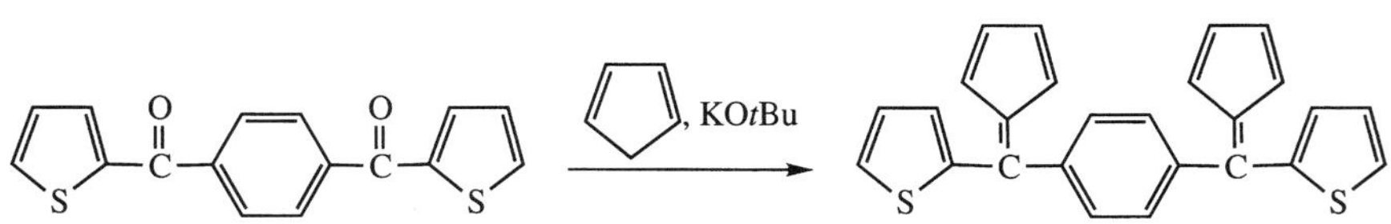

1,4-Bis[6-(2-thienyl)fulven-6-yl]benzene [381]

Cyclopentadiene (0.3 ml, 3.6 mmol) and potassium tert-butoxide (75 mg, 0.67 mmol) are added to a suspension of 1,4-di(2-thenoyl)benzene (100 mg, 0.34 mmol) in tetrahydrofuran (10 ml). The reaction mixture is stirred at room temperature for 23 h under nitrogen, after which it is poured into water and extracted with dichloromethane. The extract is washed with water and dried over sodium sulfate and evaporated. The residue is recrystallized from dichloromethane giving 57 mg (43%) of the title compound as orange needles mp 193–195 °C (decomp.).

Starting from 2,5-thiophenedicarboxylic acid dimethylamide and 2-thienyllithium, rather unstable 2,5-di(2-thenoyl)thiophene was obtained, which without isolation was reacted with cyclopentadiene to give the dimer [382].

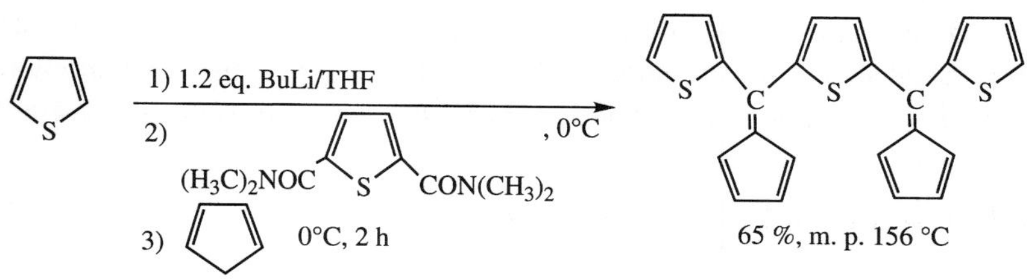

An alternative route starts with dilithiation of 6-(2-thienyl)fulvalene and reaction with dimethylamino chloroformate and 5-thiophene dicarboxylic amide followed by cyclopentadiene to give the trimer and tetramer, respectively [382].

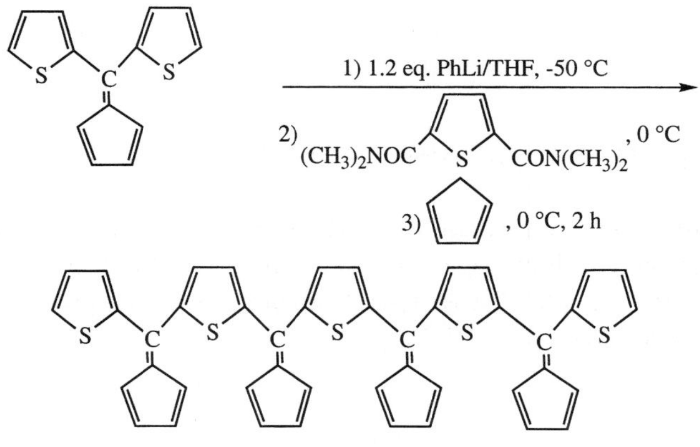

4B.6.11 Modification in the side chains of vinylthiophenes

Many esters and amide derivatives of 2-(thienyl)acrylic acids are prepared by conventional methods in connection with the preparation of compounds of potential medicinal interest [255,260,383–386]. Many thiophene analogs of chalcones are reduced to the corresponding alcohols with sodium borohydride [387,388]. Many other carbonyl compounds, containing the vinylthiophene unit are reduced to the alcohol stage [89,217,389] or to a methylene group

[93,95,96,360] without reduction of the double bond. In the latter case lithium aluminium hydride or lithium aluminium hydride/aluminium chloride is used. Allylic alcohols are obtained upon addition of arylmagnesium derivatives to thenalacetone [391]. Carbonyl derivatives from thiophene analogs of chalcones and *β*-2-thienylacrolein, such as 2,4-dinitrophenylhydrazones, semicarbazones, thiosemicarbazones, and anils are prepared by standard methods [392–395].

Mannich condensation products can be prepared from thienyl-3-butene-2-ones [396,397] and *α*-cyclopentylidene-2-thienylacetonitrile [346]. 2-(5-Methyl-2-thienyl)acrolein gives 1-dimethylaminomethyl-2-(5-methyl-2-thienyl)ethene upon reaction with formic acid and *N,N*-dimethylformamide, which is quaternized by methyl iodide in ether to the trimethyl ammonium iodide, which upon Hoffmann degradation in the presence of silver oxide yields (*E,E*)-[6,2)- (2,5)-thiophenophane-1,5-diene [398].

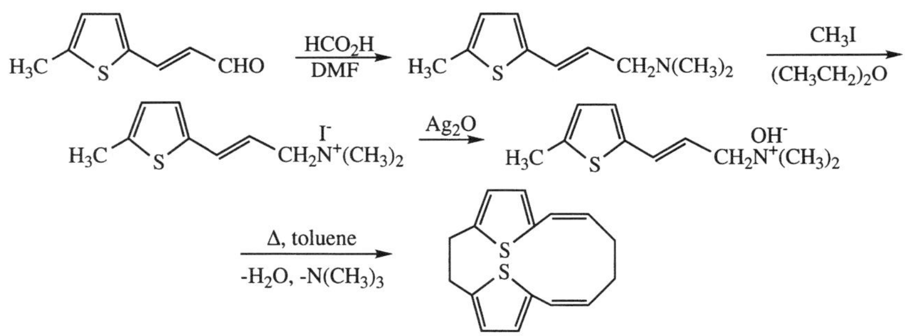

2-Cyano-3-(2-thienyl)thioacrylamide is prepared by the reaction of (2-thienyl)methylene malonitrile with hydrogen sulfide in benzene containing catalytic amounts of triethyl amine [399].

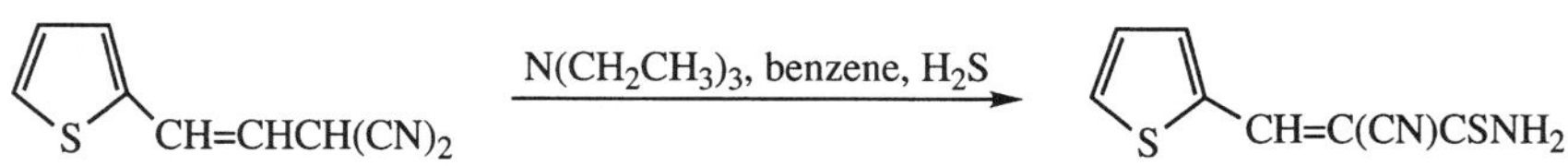

m. p. 169-171 °C

α-Methylsulfinyl- and *α*-methylsulfonyl-*β*-(2-thienyl)acrylic acid are obtained by oxidation of the methylthio substituted acid by hydrogen peroxide in acetic acid [400]. 1,1-Diformyl-2-(thienyl)ethenes are best prepared in excellent yields by the condensation of 2- and 3-thiophene aldehyde with the malonaldehyde equivalent shown below [401].

$$(H_3C)_2NCH=CH=N^+(CH_3)_2ClO_4^-$$

Treatment of trichloroethylene with three equivalents of butyllithium followed by addition of one equivalent of 3-acetyl-2,5-dimethylthiophene gave the acetylendiol, which by palladium-catalyzed carbonylation gave the anhydride derivative [402].

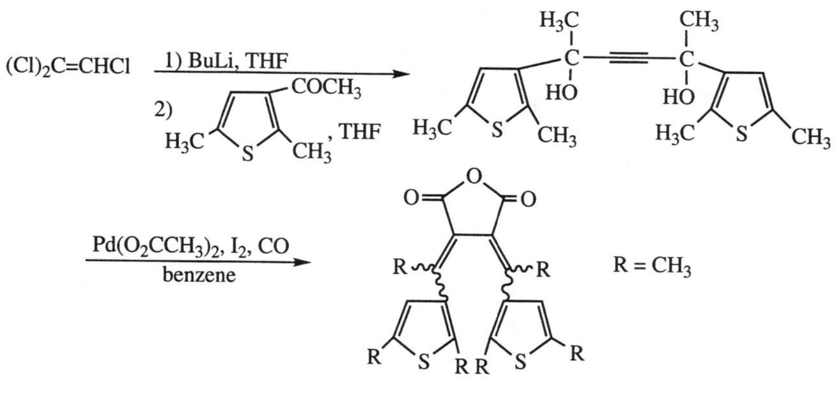

4B.6.12 *Via* ring-closure reactions

Double radical cyclization followed by aromatization is used for the preparation of the thienyl derivative shown below [403].

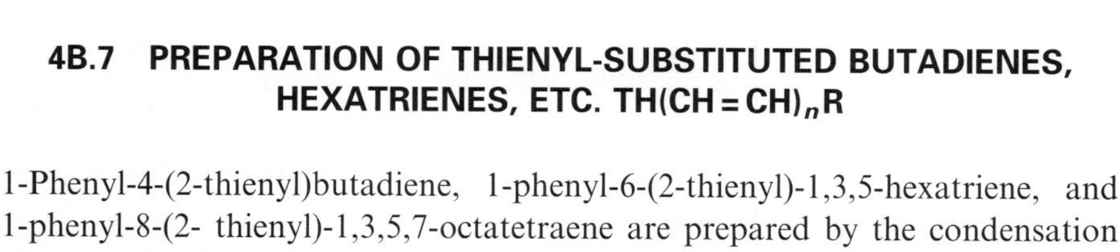

4B.7 PREPARATION OF THIENYL-SUBSTITUTED BUTADIENES, HEXATRIENES, ETC. TH(CH = CH)$_n$R

1-Phenyl-4-(2-thienyl)butadiene, 1-phenyl-6-(2-thienyl)-1,3,5-hexatriene, and 1-phenyl-8-(2- thienyl)-1,3,5,7-octatetraene are prepared by the condensation of β-2-thienylacrolein and phenylacetic acid with β-benzalpropionic acid, β-cinnamylpropionic acid, and acetic anhydride, respectively [404]. Compounds such as ThCH=CH-CH=CHCOR are prepared through the condensation of thienylacryl aldehydes with ketones [405,406]. The reaction of thiophene aldehydes with the ylide from allyl triphenylphosphonium bromide leads to thienylbutadiene [407,408]. 2-(2-Thienyl)butadiene is best prepared by the reaction of 2-thiophene magnesium bromide with methyl vinyl ketone followed by dehydration or by the reduction of 1-methyl-1-(2-thienyl)propargyl alcohol with liquid ammonia and dehydration [410]. An elegant method for the preparation of 2,3-di(2-thienyl)butadiene consists in heating to 170 °C at 0.3 mm Hg of the sulfone [144].

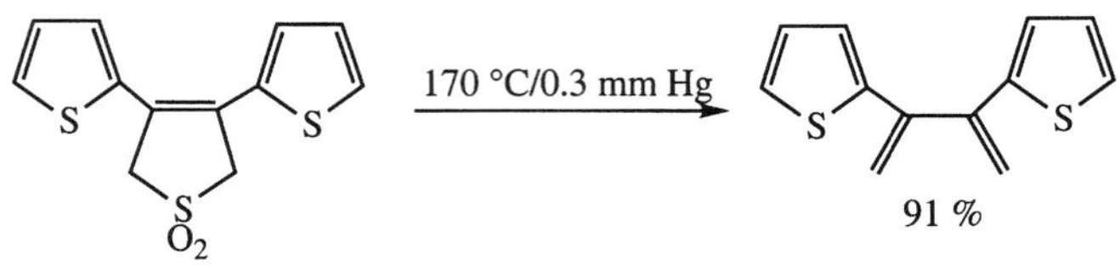

2-Thienylacrolein and analogous 2,2′-bithienyl derivatives give upon condensation with malonitrile in chloroform and using triethylamine as catalyst, 1,1-dicyano-4-thienylbutadienes. From the thienylacroleins condensation with 3-phenyl-5-isoxazolone or *N*,*N*-diethylthiobarbituric acid gives other types of substituted thienylbutadienes [410].

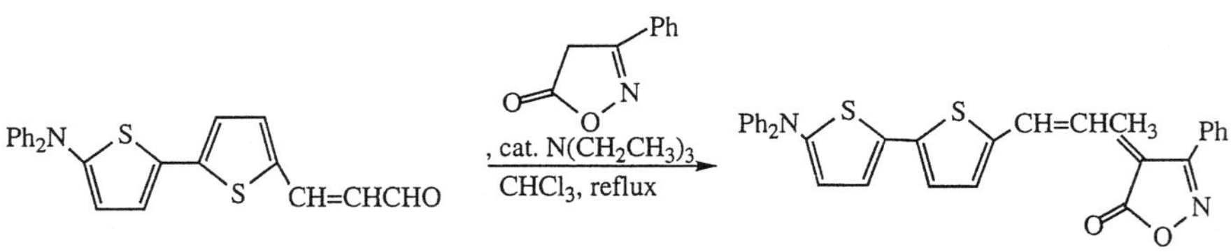

3-(2-Thienyl)acrolein gives 5[3-(2-thienyl)acrylidene]rhodanine upon reaction with rhodanine and anhydrous sodium sulfate in glacial acetic acid, which upon treatment with sodium hydroxide solution yields 2-mercapto-5-(2-thienyl)-2,4-pentadienoic acid in excellent yield [406].

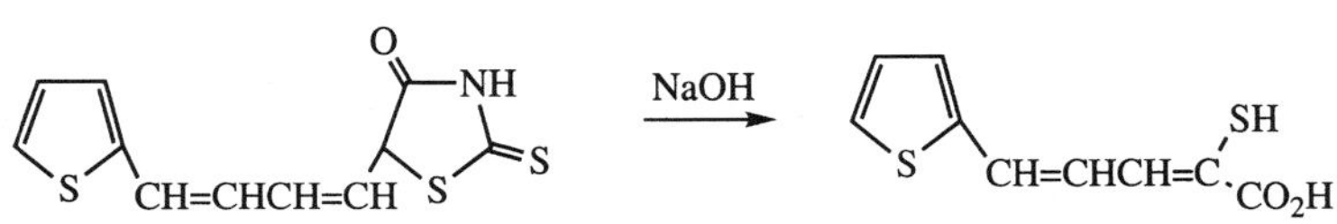

A fluoride ion promoted palladium catalyzed cross coupling is used in the preparation of a 3-thienylbutadiene derivative [411].

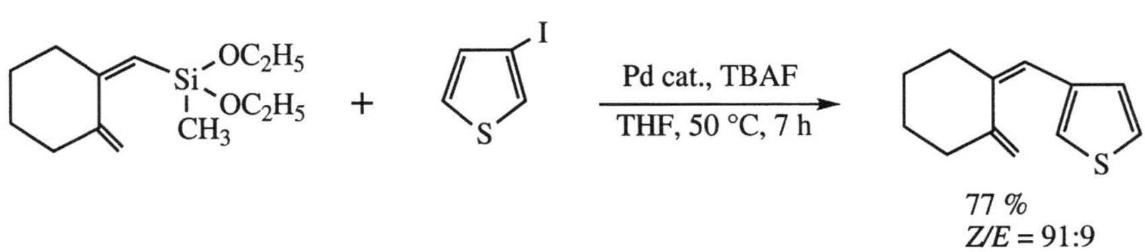

The Wittig reaction between 2,5-thiophene dialdehyde and allyl- or 2,4-pentadienylphosphonium bromide can be used for the preparation of

2,5-polyene disubstituted thiophenes [412]. The following transformation can be performed in 88% yield [413].

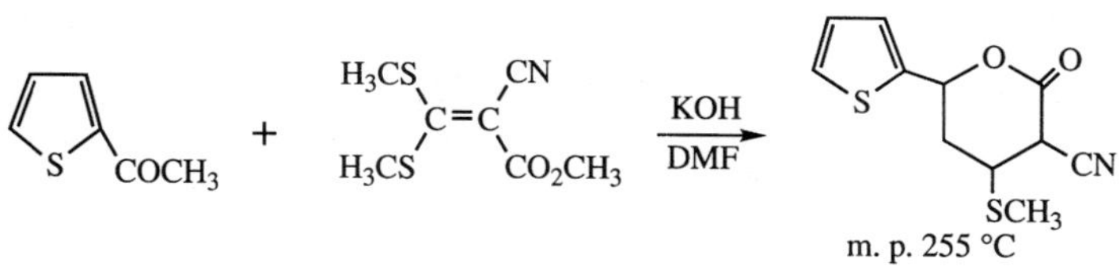

The condensation below is performed by using piperidine and acetic acid in benzene with water separation [414].

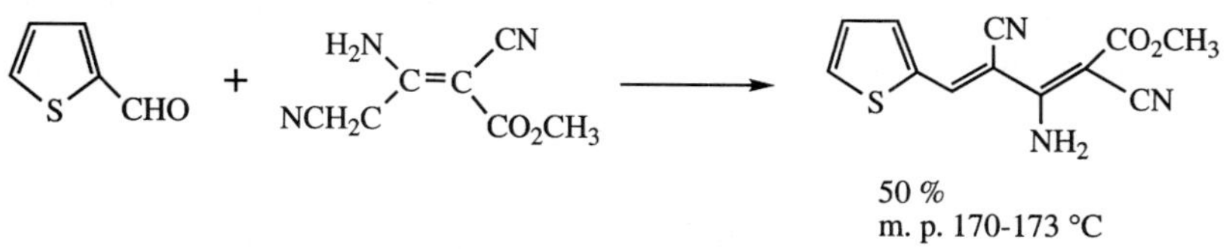

2-Bromo-1-vinyl-3,4-dihydronaphthalene was reacted with 2-thienylzinc chloride under palladium catalysis to give 2-(2-thienyl)-1-vinyl-3,4-dihydronaphthalene in good yield [415].

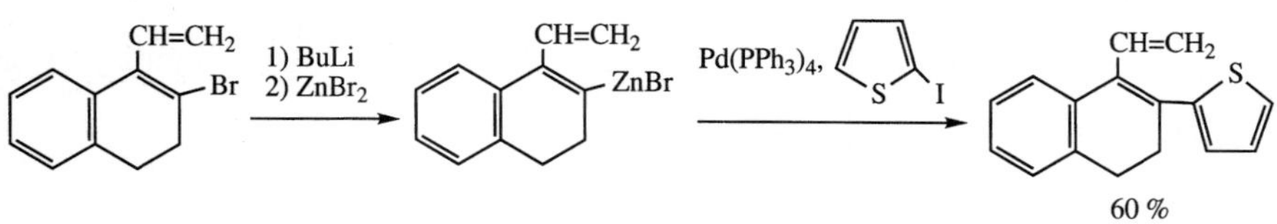

3,4-Dihydro-1-(2-thienyl)naphthalene-2-carboxaldehyde can be prepared through the palladium catalyzed reaction of 2-thienylzinc bromide with 1-bromo-3,4-dihydronaphthalene-2-carbaldehyde [416].

3,4-Dihydro-1-(2-thienyl)naphthalene-2-carboxaldehyde [416]

Thiophene (0.40 ml, 5.0 mmol) is treated with 1.35 *M* butyllithium (4.10 ml, 5.54 mmol) and *N,N,N′,N′*-tetramethylethylenediamine (0.75 ml, 4.97 mmol) in anhydrous tetrahydrofuran 15 ml) at 20 °C for 0.5 h. A solution of zinc chloride (8.40 mmol) in tetrahydrofuran is then added to the organolithium intermediate at −20 °C. After stirring the mixture at −20 °C for 1 h, a solution of tetrakis(triphenylphosphine)palladium(0) (0.23 g, 4 mol%) and 1-bromo-3,4-dihydronaphthalene-2-carbaldehyde (1.18 g, 4.98 mmol) in tetrahydrofuran (25 ml) is added. The reaction mixture is allowed to warm to room temperature, after which it is heated under reflux for 3 h. Ammonium chloride is then added and the mixture is extracted with ethyl acetate, dried over magnesium sulfate and evaporated to afford the crude product as a yellow oil. Purification by flash chromatography using diethyl ether/light petroleum (1:4) as eluent gives 0.80 g (67%) of the title compound as a yellow solid mp 90–91°C after recrystallization from dichloromethane/light petroleum.

The reaction of 2-thienylzinc iodide with cuprous cyanide and 3-iodocyclohexenone affords 3-(2-thienyl)cyclohexenone in 78% yield [417]. By the condensation of both methyl groups of 2-acetyl-5-methylthiophene the following transformation can be performed [418].

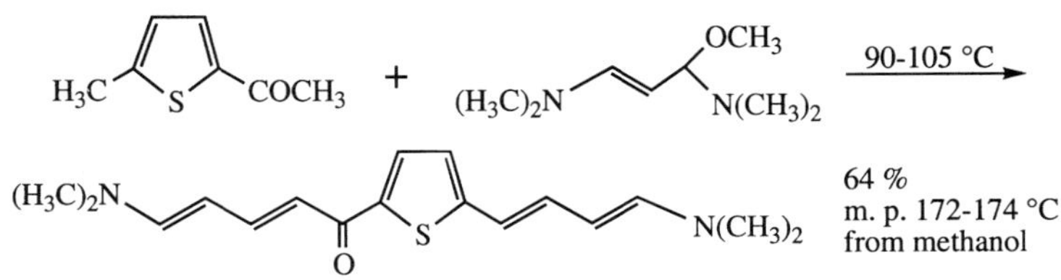

α,ω-Dithienyl polyenes are prepared from appropriately substituted 2- or 3-thiophene aldehydes or propenals by condensation with either bis-Wittig reagents or bisphosphonate esters containing one or two double bonds. In this way polyenes containing either 3, 4, 5, or 6 conjugated double bonds can be prepared [419].

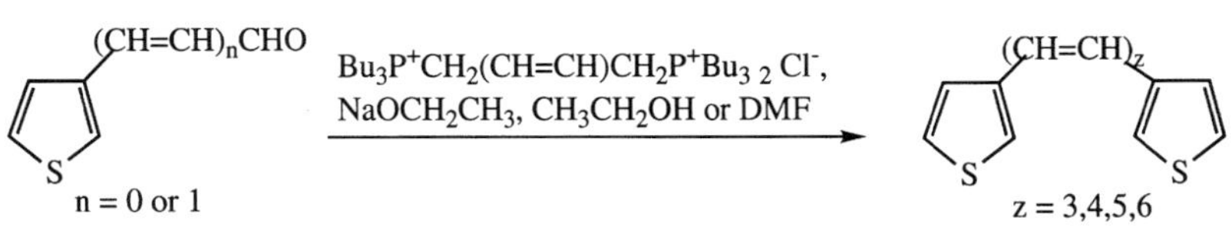

1,6-Bis(2-thienyl)hexa-1,3,5-triene [419]

A solution of potassium tert-butoxide (6 g) in 1,2-dimethoxyethane (200 ml) is added dropwise to a solution of (*E*) tetraethylbut-2-ene-1,4-diyldiphosphonate (6.56 g, 0.02 mol) and 2-thiophene aldehyde (4.48 g, 0.04 mol) in 1,2-dimethoxyethane (100 ml) at room temperature. The resulting mixture is stirred for 16 h, heated at 60–70 °C for 2 h, and then poured into cold water (250 ml). The crude product is isolated by vacuum filtration and upon crystallization from toluene/*N*,*N*-dimethylformamide, 2.63 g (54%) of the title compound is obtained as yellow crystals mp 212–213 °C.

In order to obtain more stable polyenes, these syntheses were carried out with 3,4-dibutyl-2-thiophene aldehyde or propenals [420]. In this way α,ω-(3,4-dibutyl-2-thienyl) polyenes with up to 10 double bonds have recently been prepared [420].

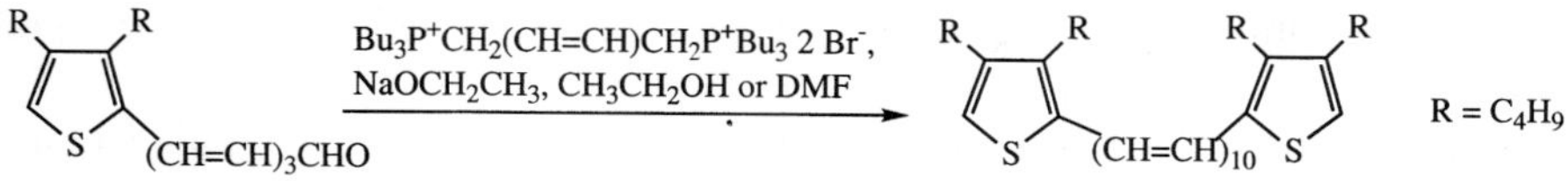

(*E*,*E*,*E*)-1,6-Bis(2′-thienyl-3′,4′-dioctyl)hexa-1,3,5-triene is prepared by the reaction of two equivalents of 3,4-dioctyl-2-thiophene aldehyde with tetraethyl *E*-but-2-ene-1,4-diyldiphosphonate [421]. Metalation of the triene followed by tributyltin chloride gave the 5′-tributylstannyl derivative [421].

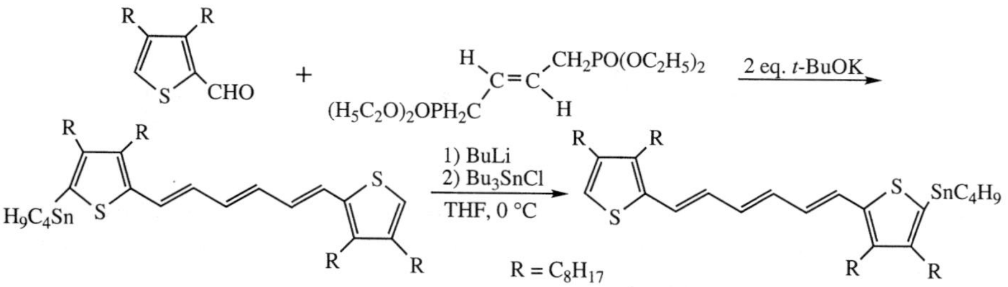

3,4-Dioctyl-2-thiophene aldehyde [421]

In a Schlenk tube under nitrogen atmosphere 1,2-dichloroethane (40 ml), 3,4-dioctylthiophene (7.07 g, 23 mmol), and *N*,*N*-dimethylformamide (2.1 g, 28.7 mmol) are successively introduced. The mixture is cooled to 0 °C and phosphorus oxychloride (4.4 g, 28.7 mmol) is added dropwise, after which the reaction mixture is refluxed for 3 h, cooled to room temperature and poured into 10% hydrochloric acid. This solution is stirred for 1 h and the product is extracted with dichloromethane. The combined organic phases are washed several times with aqueous sodium bicarbonate solution, dried over magnesium sulfate, evaporated and distilled giving 7.50 g (97%) of the title compound bp 165 °C/0.3 mm Hg.

(E,E,E)-1,6-Bis(2′-thienyl-3′,4′-dioctyl)hexa-1,3,5-triene [421]

A three-necked flask equipped with a condenser and a dropping funnel under nitrogen is charged with tetraethyl *E*-but-2-ene-1,4-diyldiphosphonate (3.34 g, 8.66 mmol) and 3,4-dioctyl-2-thiophene aldehyde (6.45 g, 19.08 mmol) in anhydrous tetrahydrofuran (100 ml). Potassium *tert*-butoxide in anhydrous tetrahydrofuran is added dropwise. The dark reaction mixture is stirred at room temperature for 12 h and poured into distilled water at 0 °C, after which the stirring is continued for 1 h. The solid phase formed is filtered off and dissolved in dichloromethane. This solution is dried over magnesium sulfate and evaporated. The residue is purified by chromatography on silica gel using hexane/ethyl acetate (95:5) as eluent giving 5.09 g (85%) of the title compound mp 47 °C.

Reduction of ThC=C-CH=CHSC$_2$H$_5$ with activated zinc in refluxing ethanol gives *Z*-Th(CH=CH)$_2$SC$_2$H$_5$ in good yield. Upon treatment with excess sodium amide in liquid ammonia the ethylmercapto group is removed [422].

4B.8 THIOPHENES WITH CUMULATIVE DOUBLE BONDS

4B.8.1 With two cumulative bonds

The following reaction is an example for preparation of a thiophene with two cumulative bonds [423].

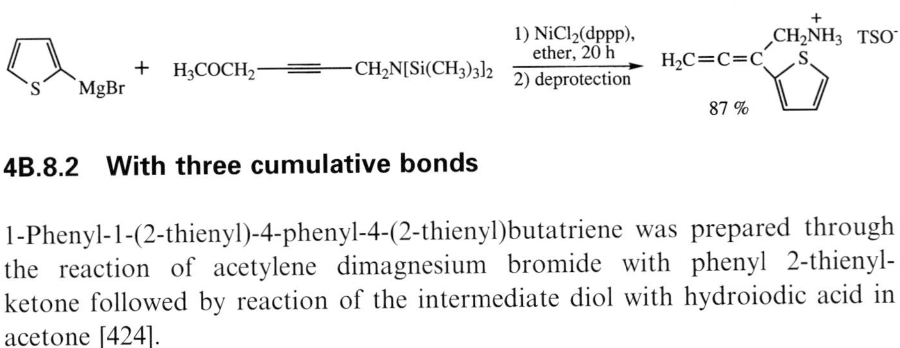

4B.8.2 With three cumulative bonds

1-Phenyl-1-(2-thienyl)-4-phenyl-4-(2-thienyl)butatriene was prepared through the reaction of acetylene dimagnesium bromide with phenyl 2-thienyl-ketone followed by reaction of the intermediate diol with hydroiodic acid in acetone [424].

Tetra(2-thienyl)butatriene is best prepared by treating 1,1-dichloro-2,2′-di-(2-thienyl)ethene with one equivalent of butyllithium in tetrahydrofuran at −90 °C, followed by 0.5 equivalents of cuprous cyanide and warming.

The tetra(2-thienyl)butatriene can be functionalized in the four 5-positions of the thiophene rings by treatment with excess lithium diisopropylamide at −50 °C followed by reactions with electrophiles such as dimethyl disulfide or trimethylsilyl chloride [425].

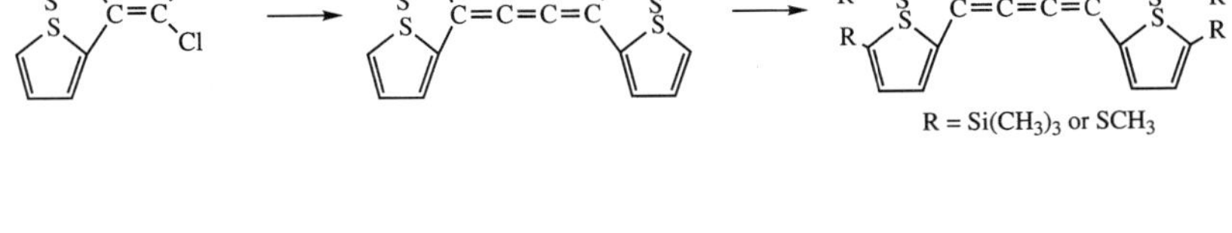

4B.8.3 With four or more cumulative bonds

The reaction of phenyl-2-thienyl ketone with the dimagnesium salt of diacetylene gave 1-phenyl-1(2-thienyl)-6-phenyl-6-(2-thienyl)hexatetraene, albeit in low yield [426].

REFERENCES

1. F. Clesse and H. Quiniou, *Bull. Soc. Chim. Fr.* 1940 (1969).
2. Y. Tominaga, M. L. Lee and R. N. Castle, *J. Heterocycl. Chem.* **18**, 967 (1981).
3. S. Rao, T. Balerao and B. D. Tilak, *Indian J. Chem.* 26B, 208 (1987).
4. N. N. Volkov and O. A. Kazakova, *Chem. Heterocycl. Compds (Engl. Transl.)* 758 (1981).
5. Y. Amemiya, A. Terada, K. Wachi, H. Miyazawa, N. Hatakayama, K. Matsuda and T. Oshima, *J. Med. Chem.* **32**, 1265 (1989).
6. M. Janda and F. Dvorak, *Coll. Czech. Chem. Commun.* **27**, 372 (1962).
7. C. Troyanowsky, *Bull. Soc. Chim Fr.* 424 (1955).
8. B. Abarca, R. Ballesteros, E. Enriquez and G. Jones, *Tetrahedron* **43**, 269 (1987).
9. K. E. Andersen, C. Braestrup, F. C. Grønwald, A. S. Jørgensen, E. B. Nielsen, U. Sonnewald, P. O. Sørensen, P. D. Suzdak and L. S. J. Knutsen, *J. Med.Chem.* **36**, 1716 (1993).
10. S. Gronowitz, B. Cederlund and A.-B. Hörnfeldt, *Chem. Scripta* **5**, 217 (1974).
11. M. L. Tedjamulia, J. G. Stuart, Y. Tominaga, R. N. Castle and M. L. Lee, *J. Heterocycl. Chem.* **21**, 1215 (1984).
12. S. Gronowitz, J. Rehnö, K. Titlestad, M. Vadzis, B. Sjöberg, P. Bamberg, B. Ekström and U. Forsgren, *Acta Pharm Suec.* **9**, 381 (1972).
13. A. Svensson and R. Håkansson, *Chem. Scripta.* **18**, 202 (1981).
14. A. Svensson and R. Håkansson, *Chem. Scripta.* **20**, 188 (1982).
15. H. C. Brown, A. K. Gupta and J. V. N. Vara Prasad, *Bull. Chem. Soc. Jpn.* **61**, 93 (1988).
16. J. Szmuszkovicz and E. J. Modest, *J. Am. Chem. Soc.* **72**, 571 (1950)
17. K. Thiele, K. Posselt, A. Gross and A. W. Schuler, *Chim. Therap.* 228 (1969).
18. F. Bohlmann and P. Herbst, *Chem. Ber.* **95**, 1733 (1962).
19. W. Caruthers and H. N. M. Stewart, *J. Chem. Soc.* 6221 (1965).
20. D. Mukherjee, I. C. Dunn and K. N. Houk, *J. Am. Chem. Soc.* **101**, 251 (1979).
21. Ng. Ph. Buu-Hoi, Ng Hoan and D. Lavit, *J. Chem. Soc.* 2130 (1950).
22. Ng. Ph. Buu-Hoi and Ng. Hoan, *J. Org. Chem.* **17**, 350 (1952).
23. Ng. Ph. Buu-Hoi, E. Lescot, Jr and Ng. D. Xuong, *J. Org. Chem.* **22**, 1057 (1957).
24. Ng. Ph. Buu-Hoi, E. Lescot, Jr and Ng. D Xuong, *J. Org. Chem.* **23**, 1261 (1958).
25. D. W. Adamson, P. A. Barrett, J. W. Billinghurst and T. S. G. Jones, *J. Chem. Soc.* 23415 (1957).
26. A. F. Casey and A. P. Parulkar, *Can. J. Chem.* **47**, 423 (1969).
27. M. M. Martin and J. G. MacConnell, *Tetrahedron* **26**, 307 (1970).
28. K. Pelz and M. Protiva, *Coll. Czech. Commun.* **32**, 2840 (1967).
29. D. W. Adamson, and J. W. Billinghurst, *J. Chem. Soc.* 1039 (1950).
30. A. S. F. Casy and J. L. Meyers, *J. Chem. Soc.* 4092 (1965).
31. P. A. Barrett, *J. Chem. Soc.* 326 (1958).
32. S. Datta and A. De, *J. Chem. Soc., Perkin Trans. 1*, 603 (1989).
33. J. W. Schick and H. D. Hartough, *J. Am. Chem. Soc.* **70**, 1646 (1948).
34. J. F. Scully and E. V. Brown, *J. Am. Chem. Soc.* **75**, 6329 (1953).
35. I. V. Andreeva and M. M. Koton, J. *Gen. Chem. USSR (Engl Transl.)* **27**, 1079 (1957).
36. G. Van Zyl, J. F. Zack Jr, E. S. Huyser and P. L. Cook, *J. Am. Chem. Soc.* **76**, 707 (1954).

37. W. S. Emerson and T. M. Patrick, Jr, *J. Org. Chem.* **13**, 729 (1948).
38. W. S. Emerson and T. M. Patrick, Jr, *Organic Synth.* **38**, 86 (1958).
39. S. Gronowitz and I. Ander, *Acta Chem. Scand.* B**29**, 513 (1975).
40. M. Nicolas, B. Fabre, J. F. Pilard and J. Simonet, *J. Heterocycl. Chem.* **36**, 1105 (1999).
41. S. Hibino, S. Kano, M. Mochizuki and E. Sugino, *J. Org. Chem.* **49**, 5006 (1984)
42. K. Okuhara, *J. Org. Chem.* **41**, 1487 (1976).
43. P. Martinet, R. Sauvetre and J.-F. Normant, *J. Organomet. Chem.* **367**, 1 (1989).
44. K. Okuhara, *Bull. Chem. Soc. Jpn.* **54**, 2045 (1981).
45. W. J. Middleton, D. Metzger and J. A. Snyder, *J. Med. Chem.* **14**, 1193 (1971).
46. Y. Shen and Q. Liao, *J. Fluorine Chem.* **47**, 137 (1990).
47. A. B. Shtarev and Z. Chvátal, *J. Org. Chem.* **62**, 6508 (1997).
48. U. Hasselrodt and F. Korte, *Angew. Chem.* **75**, 138 (1963).
49. J. Kagan, S. K. Aurora, M. Bryzgis, S. N. Dhawan, K. Reid, S. P. Sing and L. Tom *J. Org. Chem.* **48**, 703 (1983).
50. J. Kagan and S. K. Aurora, *J. Org. Chem.* **48**, 4317 (1983).
51. J. P. Beny, S. N. Dhawan, J. Kagan and S. Sundlass, *J. Org. Chem.* **47**, 2201 (1982).
52. D. M. Perrine and J. Kagan, *Heterocycles* **24**, 365 (1986)
53. S. R. Ramadas and N. S. Chandrakumar, *Phosphorus, Sulfur Relat. Elem.* **13**, 79 (1982).
54. H. L. Reimlinger, F. Billiau and M. Peiren, *Chem. Ber.* **97**, 3503 (1964).
55. H. Volz and H. Kowarsch, *Tetrahedron Lett.* 4375 (1976).
56. J. T. Gupton, F. A. Hicks, D. B. Wilkinson and S. A. Petrich, *Heterocycles* **37**, 487 (1994).
57. R. Nijs, H. D. Verkkruijsse, S. Harder, A. C. H. T. M van der Kerk and L. Brandsma, *Synth. Comm.* **21**, 653 (1991).
58. S.-K. Shin and S.-K. Choi, *J. Fluorine Chem.* **43**, 439 (1989).
59. Z. V. Todres, N. G. Furmanova, S. P. Avagyan, Y. T. Struchkov and D. N. Kursanov, *Phosphor, Sulfur Relat. Elem.* **5**, 309 (1979).
60. D. S. Ennis and T. L. Gilchrist, *Tetrahedron* **46**, 2623 (1990).
61. A. Minato, K. Suzuki, K. Tamao and M. Kumada, *Tetrahedron Lett.* **25**, 83 (1984).
62. B. Rossi, A. Carpita and T. Messeri, *Gazz. Chim. Ital.* **122**, 65 (1992).
63. S. A. Biller, J. W. Abt, A. T. Pudzianowski, L. C. Rich, D. A. Slusarchyk and C. P. Ciosek, Jr, *Biorg. Med. Chem Lett.* **3**, 595 (1993).
64. F. Tellier, R. Sauvêtre, J. F. Normant, Y. Dromzee and Y. Jeannin, *J. Organomet. Chem.* **331**, 281 (1987).
65. J. P. Gillet, R. Sauvêtre and J. F. Normant, *Synthesis* 538 (1986).
66. N. Jabri, A. Alexakis and J. F. Normant, *Tetrahedron Lett.* **22**, 3851 (1981).
67. Y. Hatanaka, S. Fukushima and T. Hiyama, *Heterocycles* **30**, 303 (1990).
68. U. Annby, S. Gronowitz and A. Hallberg, *J. Organomet. Chem.* **365**, 233 (1989).
69. R. C. Larock, K. Narayanan and S. S. Hershberger, *J. Org. Chem.* **48**, 4377 (1983).
70. M. Satoh, N. Miyaura and A. Suzuki, *Chem. Lett.* 1329 (1986).
71. H. C. Brown and N. G. Bhat, *J. Org. Chem.* **53**, 6009 (1988).
72. G. Karminskizamola and M. Bajic, *Heterocyles* **22**, 1497 (1985).
73. G. M. Badger, J. H. Bowie, J. A. Elix, G. E. Lewis and U. P. Singh, *Aust. J. Chem.* **20**, 2669 (1967).
74. F. Bottari, G. Lippi and B. Macchia, *Gazz. Chim. Ital.* **99**, 762 (1969),
75. E. Lee-Ruff and F. J. Ablenas, *Can. J. Chem.* **67**, 699 (1989).
76. T. V. Brown, W. Caruthers and M. G. Pellatt, *J. Chem. Soc., Perkin Trans. 1* 483 (1982).
77. W. P. Trompen and H. O. Huisman, *Rec. trav. Chim.* **85**, 175 (1966).
78. R. E. Atkinson, R. F. Curtis and G. T. Phillips, *J. Chem. Soc.* 7109 (1965).
79. F. Bohlmann, H. Bornowski and H. Schoenowsky, *Chem. Ber.* **95**, 1733 (1962).
80. S. Yasuike, J. Kurita and T. Tsuchiya, *Heterocycles* **45**, 1891 (1997).
81. N. Chatani, T. Takeyasu, N. Horiuchi and T. Hanafusa, J. *Org. Chem.* **53**, 3539 (1988).
82. F. Ya. Perveev and N. Y. Kudryashova, *Doklady Akad. Nauk. SSSR* **98**, 975 (1954).

83. F. Ya. Perveev and T. N. Kuren'gina, *J. Gen. Chem. USSR*, **25**, 1579 (1955).
84. F. Bohlmann, P. Herbst and I. Dohrmann, *Chem. Ber.* **96**, 226 (1963).
85. J. M. Mcintosh and F. P. Seguin, *Can. J. Chem.* **53**, 3526 (1975).
86. S. Braverman, Y. Duar and M. Freund, *Isr. J. Chem.* **26**, 108 (1985).
87. E. A. Braude, J. S. Fawcett and D. D. E. Newman, *J. Chem. Soc.* 4155 (1952).
88. J. Jeganathan and M. Srinivasan, *Indian J. Sect., B.* **19**, 312 (1980).
89. J. A. Gauthier and C. Combet-Farnoux, *Bull. Soc. Chim. Fr.* 2145 (1964).
90. Ya. L. Gol'dfarb, B. P. Fabrichnyi and V. I. Rogovik, *Bull. Acad. Sci. SSSR*, 492 (1965).
91. J. Skramstad, *Chem. Scripta* **4**, 81 (1973).
92. R. Grafing and L. Brandsma, *Rec. Trav. Chim.* **99**, 23 (1980).
93. M. W. Majchrzak, *J. Heterocycl. Chem.* **22**, 1205 (1985).
94. A. Hallberg and P. Pedaja, *Tetrahedron* **39**, 819 (1983).
95. B. Yom-Tov and S. Gronowitz, *Chem. Scripta* **3**, 169 (1973).
96. B. Yom-Tov and S. Gronowitz, *J. Heterocycl. Chem.* **15**, 285 (1978).
97. A. T. Jeffries III and S. Gronowitz, *Chem. Scripta* **4**, 183 (1973).
98. M. Muradinszweykowska, A. J. Peters and J. Lugtenburg, *Rec. Trav. Chim.* **103**, 105 (1984).
99. Ng. H. Nam, Ng. Ph. Buu-Hoi and Ng. D. Xuong, *J. Chem. Soc.* 1690 (1954).
100. Ng. Ph. Buu-Hoi and Ng. Hoan, *Rec. trav. Chim.* **69**, 1455 (1950).
101. M. Sy, Ng. Ph. Buu-Hoi and N. D. Xuong, *J. Chem. Soc.* 1975 (1954).
102. Ng. Ph. Buu-Hoi and N. Hoan, *Rec. trav. Chim.* **68**, 441 (1949).
103. M. M. Goodman, G. W. Kabalka, R. C. Marks, F. F. Knapp, Jr, J. Lee and Y. Liang, *J. Med. Chem.* **35**, 280 (1992).
104. D. E. Worrall, *Org. Synth. Coll.*, **1**, 413 (1941).
105. M. L. Dressler and M. M. Joullié, *J. Heterocycl. Chem.* **7**, 1257 (1970)
106. S. Gronowitz and E. Sandberg, *Arkiv Kemi.* **32**, 217 (1970).
107. Y. L. Gol'dfarb, M. A. Kalik, V. K. Zavyalova, E. F. Lavretskaya and S. A. Sukhanova, *Chem. Heterocycl. Compds. (Engl. Transl.)* 962 (1984).
108. S. Conde, R. Madronero, M. P. Fernandez-Tomé and J. Delrio, *J. Med. Chem.* **21**, 978 (1978).
109. M. Fuji, *Bull. Chem. Soc. Jpn.* **61**, 4029 (1988).
110. A. I. Sitkin and V. I. Klimenko, *Chem. Heterocycl. Compds, (Eng. Transl.)*, 34 (1979).
111. D. N. Robertson, *J. Org. Chem.* **25**, 47 (1960).
112. A. Arcoria, E. Maccarone and G. A. Tomaselli, *J. Heterocycl. Chem.* **10**, 191 (1973).
113. M. Iwasaki, J. Li, Y. Kobayashi, H. Matsuzaka, Y. Ishii and M. Hidai, *Tetrahedron Lett.* **30**, 95 (1988).
114. M. Iwasaki, Y. Kobyashi, J.-P. Li, H. Matsuzaka, Y. Ishii and M. Hidai, *J. Org. Chem.* **56**, 1922 (1991).
115. B. Antelo, L. Castedo, J. Delamano, A. Gomez, C. Lopez and G. Tojo, *J. Org. Chem.* **61**, 1188 (1996).
116. A. K.-Y. Jen, V. P. Rao, K. J. Drost, K. Y. Wong and M. P. Cava, *J. Chem. Soc. Chem. Commun.* 2057 (1994).
117. A. I. de Lucas, N. Martin, L. Sanchez, C. Santos, J. Garin, J. Orduna, R. Alcala and B. Villacampa, *Tetrahedron Lett.* **38**, 6107 (1997).
118. T. K. Hansen, M. V. Lakshmikantham, M. P. Cava, R. E. Niziurski-Mann, F. Jensen and J. Becher, *J. Am. Chem. Soc.* **114**, 5035 (1992).
119. S. L. Gilat, S. H. Kawai and J.-M. Lehn, *Chem. Commun.* 1439 (1993).
120. A. S. Benahmed-Gasmi, P. Frere, B. Garrigues, A. Gorgues, M. Jubeault, R. Carlier and F. Texier, *Tetrahedron Lett.* **33**, 6457 (1992).
121. A. Ohta, T. Kobayashi and H. Kato, *J. Chem. Soc. Perkin Trans. 1* 905 (1993).
122. H. Brisset, C. Thobie-Gautier, M. Jubeault, A. Gorgues and J. Roncali, *J. Chem. Soc., Chem. Commun.* 1765 (1994).
123. M. Kozaki, S. Tanaka and Y. Yamashita, *J. Chem. Soc. Chem. Commun.* 1137 (1992).

124. A. Charlton, A. E. Underhill, G. Williams, M. Kalaji, P. J. Murphy, K. M. A. Malik and M. B. Hursthouse, *J. Org. Chem.* **62**, 3098 (1997).
125. V. K. Khairullin, R. R. Shagidullin, M. A. Vasyani, I. K. Pokrovskaya and A. N. Chernov, *J. General Chem. USSR (Engl Transl.)* **54**, 91 (1983).
126. G. M. Scheide and R. H. Neilson, *Phosphorus, Sulfur and Silicon* **46**, 139 (1989).
127. P. L. Kelly, S. F. Thames and J. E., McCleskey, *J. Heterocycl. Chem.* **9**, 141 (1972).
128. F. Fusalba, H. A. Ho, L. Breau and D. Béranger, *Chem. Mater.* **12**, 2581 (2000).
129. K. Sato, S. Arai and T. Yamagishi, *J. Heterocycl. Chem.* **33**, 57 (1996).
130. W. R. Biggerstaff and O. L. Stafford, *J. Am. Chem. Soc.* **74**, 419 (1952).
131. W. S. Emerson and T. M. Patrick, Jr, *J. Org. Chem.* **14**, 790 (1949).
132. A. Baracchi, S. Chimici, F. De Sio, C. Polo, P. Santi-Fantoni and T. Torroba, *Heterocycles* **29**, 2023 (1989).
133. F. J. Villani, E. A. Wefer, T. A. Mann, J. Mayer, L. Peer and A. S. Levy, *J. Heterocycl. Chem.* **9**, 1203 (1972).
134. A. E. Sigrist, P. Liechti, H. R. Meyer and K. Weber, *Helv. Chim. Acta* **52**, 2521 (1969).
135. J. W. McFarland, L. H. Conover, H. Howes, W. C. Austin, R. L Cornwell and J. Danilewicz, *J. Med. Chem.* **12**, 1066 (1969).
136. J. W. McFarland, H. L. Howes, Jr, L. H. Conover, J. E. Lynch, W. C. Austin and D. H. Morgan, *J. Med. Chem.* **13**, 113 (1970).
137. A. R. Katritzky, L. Serdyuk, L. Xie and I. Ghiviriga, *J. Org. Chem.* **62**, 6215 (1997).
138. L. Castedo, M. M. Cid, R, Dominguez, J. A. Seijas and M. C. Villaverda, *Heterocycles* **31**, 1271 (1990).
139. J. Nakayama and T. Fujimoro, *Heterocycles* **32**, 991 (1991).
140. J. Nakayama, S. Murabayashi and M. Hoshino, *Heterocycles* **24**, 2639 (1986).
141. M. Irie and M. Mohri, *J. Org. Chem.* **53**, 803 (1988).
142. Z. Hu and M. P. Cava, *Tetrahedron Lett.* **35**, 3493 (1994).
143. W.-M. Dai and W. L. Mark, *Tetrahedron Lett.* **41**, 10277 (2000).
144. J. Nakayama, H. Machida, R. Saito, K. Akimoto and M. Hoshino, *Chem. Lett.* 1173 (1985).
145. Z.-N. Huang, B.-A. Xu, S. Jin and M.-G. Fan, *Synthesis* 1092 (1998).
146. T. Suzuki, H. Shiohara, M. Monobe, T. Sakimura, S. Tanaka, Y. Yamashita and T. Miyashi, *Angew. Chem. Int. Ed. Engl.* **31**, 455 (1992).
147. E. Fischer, J. Larsen, J. B. Christensen, M. Fourmigué, H. G. Madsen and N. Harrit, *J. Org. Chem.* **61**, 6997 (1996).
148. Ng. Ph. Buu-Hoi and G. Saint-Ruf, *Bull. Soc. Chim. Fr.* 2489 (1968).
149. R. Pratap, Y. Tominaga, M. L. Lee and R. N. Castle, *J. Heterocycl. Chem.* **18**, 973 (1981).
150. T. M. Brown and W. Carruthers, *J. Chem. Soc., Perkin Trans. 1* 2904 (1981).
151. S. Dixon, *J. Org. Chem.* **21**, 400 (1956).
152. S. Irie and M. Irie, *Bull. Chem. Soc. Jpn.* **73**, 2385 (2000).
153. K. Higashiguchi, K. Matsuda, S. Kobatake, T. Yamada, K. Kawai and M. Irie, *Bull. Chem. Soc. Jpn.* **73**, 2389 (2000).
154. S. Kobatake, K. Uchida, E. Tsuchida and M. Irie, *Chem. Lett.* 1340 (2000).
155. T. Yamada, S. Kobatake and M. Irie, *Bull. Chem. Soc. Jpn.* **73**, 2179 (2000).
156. M. Irie, T. Lifca, S. Kobatake and N. Kato, *J. Am. Chem. Soc.* **122**, 4871 (2000).
157. S. Kobatake, K. Shibata, K. Uchida and M. Irie, *J. Am. Chem. Soc.* **122**, 12135 (2000).
158. A. Fernández-Acebes and J. M. Lehn, *Chem. Eur. J.* **5**, 3285 (1999).
159. H. Utsumi, D. Nagahama, H. Nakano and Y. Shirota, *J. Mater. Chem.* **10**, 2436 (2000).
160. M. Irie, K. Sakemura, M. Okinaka and K. Uchida, *J. Org. Chem.* **60**, 8305 (1995).
161. Y. Yokoyama, N. Hosoda, Y. T. Osano and C. Sasaki, *Chem. Lett.* 1093 (1998).
162. S. H. Kawal, *Tetrahedron Lett.* **39**, 4445 (1998).
163. K. Uchida, T. Matsuoka, S. Kobatake, T. Yamaguchi and M. Irie, *Tetrahedron* **57**, 4559 (2001).
164. M. Takeshita and M. Irie, *Chem. Lett.* 1123 (1998).

165. T. Kawai, T. Sasaki and M. Irie, *Chem. Commun.* 711 (2001).
166. A. Osuka, D. Fujikane, H. Shinmori, S. Kobatake and M. Irie, *J. Org. Chem.* **66**, 3913 (2001).
167. M. Fukudomo, K. Kamiyama, T. Kawai and M. Irie, *Chem. Lett.* 70 (2001).
168. K. Yagi, C. F. Soong and M. Irie, *J. Org. Chem.* **66**, 5419 (2001).
169. K. Matsuda, M. Matsuo and M. Irie, *Chem. Lett.* 436 (2001).
170. K. Matsuda and M. Irie, *Chem. Eur. J.* **7**, 3466 (2001).
171. X. Deng and L. S. Liebeskind, *J. Am. Chem. Soc.* **123**, 7703 (2001).
172. A. Arcoria, S. Fisichella, G. Scarlata and M. Torre, *Ann. Chim. Rome* **62**, 723 (1973).
173. D. P. Munro and J. T. Sharp, *J. Chem. Soc. Perkin Trans. 1*, 1718 (1980).
174. C. E. Loader and C. J. Timmons, *J. Chem. Soc. C.* 1677 (1967).
175. A. -B. Hörnfeldt, J. S. Gronowitz and S. Gronowitz, *Acta Chem. Scand.* **22**, 2725 (1968).
176. S. Gronowitz and B. Yom-Tov, *Z. Chem.* **10**, 389 (1970).
177. M. B. Groen, H. Schadenberg and H. Wynberg, *J. Org. Chem.* **36**, 2797 (1971).
178. U. Michael and S. Gronowitz, *Chem. Scripta* **4**, 126 (1973).
179. F. A. Bottino, G. C. Pappalardo, G. Scarlata, D. Sciotto and M. Torre, *Can. J. Chem.* **56**, 2755 (1978).
180. W. Hinz, R. Jones and T. Anderson, *Synthesis* 620 (1986).
181. S. Gronowitz and T. Dahlgren, *Chem. Scripta* **12**, 57 (1977).
182. G. Kossmehl and C. Pithart, *Z. Naturforsch.* 47b, 567 (1992).
183. Y. Greenwald, G. Cohen, J. Poplawskki, M. Ehrenfreund, S. Speiser and D. Davidov, *J. Am. Chem. Soc.* **118**, 2980 (1996).
184. M. Sindler-Kulik, Z Stiplosek, D. Vojvonic, B. Matalko and Z. Marinic, *Heterocycles* **32**, 2357 (1991).
185. V. H. Rawal, R. J. Jones and M. P. Cava, *J. Org. Chem.* **52**, 19 (1987).
186. G. Manecke and M. Härtel, *Chem. Ber.* **106**, 655 (1973).
187. V. P. Rao, A. K.-Y. Jen, K. Y. Wong and K. J. Drost, *Tetrahedron Lett.* **34**, 1747 (1993).
188. V. P. Rao, A. K.-Y. Jen, K. Y. Wong and K. J. Drost, *J. Chem. Soc., Chem. Commun.* 1118 (1993).
189. S.-S. Chou, D.-J. Sung, J.-Y. Huang, P.-K. Yang and H.-C. Lin, *Tetrahedron Lett.* **37**, 7279 (1996).
190. J.-M. Raimundo, P. Blanchard, I. Ledoux-Rak, R. Hierle, L. Michaux and J. Roncali, *Chem. Commun.* 1597 (2000).
191. B. Yom-Tov and S. Gronowitz, *Chem. Scripta* **3**, 37 (1973).
192. S. Gronowitz and L. Svensson, *Chem. Scripta* **15**, 169 (1980).
193. B. Yom-Tov, S. Gronowitz, S. B. Ross and N. E. Stjernström, *Acta Pharm. Suec.* **11**, 149 (1974).
194. M. Farnier, M. Brost, B. Hanquet and R. Guilard, *J. Heterocyl. Chem.* **23**, 517 (1986).
195. R. M. Kellogg, M. B. Groen and H. Wynberg, *J. Org. Chem.* **32**, 3093 (1967).
196. Y. Tominaga, M. L. Tedjamulia, R. N. Castle and M. L. Lee, *J. Heterocycl. Chem.* **20**, 487 (1983).
197. M. L. Tedjamulia, Y. Tominaga, R. N. Castle and M. L. Lee, *J. Hetrocycl. Chem.* **20**, 1143 (1983).
198. A. Berlin, S. Bradamante, R. Ferracioli, G. A. Pagani and F. Sannicolo, *J. Chem. Soc., Perkin Trans. 1* 2631 (1987).
199. M. Iwao, M. L. Lee and R. N. Castle, *J. Heterocycl. Chem.* **17**, 1259 (1980).
200. H. Kudo, M. L. Tedjamulia, R. N. Castle and M. L. Lee, *J. Heterocycl. Chem.* **21**, 185 (1984).
201. A. K.-Y. Jen, V. P. Rao, K. Y. Wong and K. J. Drost, *J. Chem. Soc., Chem. Commun.* 90 (1993).
202. E. J. Seus and C. V. Wilson, *J. Org. Chem.* **26**, 5243 (1961).
203. E. J. Seus, *J. Heterocycl. Chem.* **2**, 318 (1965).
204. Y. Tominaga, M. L. Lee and R. N. Castle, *J. Heterocycl. Chem.* **18**, 977 (1981).
205. R. S. Green and L. Heller, *J. Org. Chem.* **39**, 196 (1974).

206. A. Abbotto, S. Bradamante, A. Facchetti and G. A. Pagani, *J. Org. Chem.* **62**, 5755 (1997).
207. S. Yasuike, F. Nakashima, J. Kurita and T. Tsuchiya, *Heterocycles* **45**, 1899 (1997).
208. R. Asano, I. Moritani, Y. Fujiwara and S. Teranishi, *Bull. Chem. Soc. Jpn.* **46**, 663 (1973).
209. R. J. P. Corriu and J. P. Masse, *J. Chem. Soc., Chem. Commun.*144 (1972).
210. E. Negishi, F. T. Luo, R. Frisbee and H. Matsushita, *Heterocycles* **18**, 117 (1982).
211. R. Galerini, A. Musco, R. Pontellini, A. Bolognesi, S. Destri, M. Catellani, M. Mascherpa and G. Zhuo, *J. Chem. Soc., Chem. Commun* 364 (1991).
212. M. Iyoda, Y. Kuwatani, N. Ueno and M. Oda, *J. Chem. Soc., Chem. Commun.* 158 (1992).
213. T. N. Sidorenko, G. A. Terenteva, O. S. Andrienko, Y. V. Savinykh and V. S. Aksenov, *Chem. Heterocycl. Compd. USSR (Engl. Transl.)* 192 (1983).
214. W. C. Frank, Y. C. Kim and R. F. Heck, *J. Org. Chem.* **43**, 2947 (1978).
215. P. Hong, B. R. Cho and H. Yamazaki, *Chem. Lett.* 507, (1980).
216. N. K. DasGupta and F. W. Birss, *Tetrahedron* **36**, 2711 (1980).
217. C. P. Dell, E. H. Smith and D. Warburton, *J. Chem. Soc., Perkin Trans. 1* 787 (1985).
218. R. M. Kellog and J. Buter, *J. Org. Chem.* **36**, 2236 (1971).
219. L. Duhamel, G. Ple and B. Contreras, *Org. Prep. Proc. Int.* **18**, 219 (1986).
220. R. E. Miller and F. F. Nord, *J. Org. Chem.* **16**, 1720 (1951).
221. G. Jones and M. J. Robinson, *J. Chem. Soc., Perkin Trans. 1* 505 (1977).
222. H. Keskin, R. E. Miller and F. F. Nord, *J. Org. Chem.* **16**, 199 (1951).
223. C. S. Marvel, J. M. Quinn and J. S. Showell, *J. Org. Chem.* **18**, 1730 (1953).
224. Y. Nishima, Y. Makino, S. Hamanaka, A. Ogawa and N. Sonoda, *Bull. Chem. Soc. Jpn.* **62**, 1682 (1989).
225. R. Pallaud and F. Delaveau, *Bull. Soc. Chim. Fr.* 1220 (1955).
226. M. L. Mihailovic and M. Toth, *J. Org. Chem.* **22**, 652 (1957).
227. R. Guilard, P. Fournari and M. Fontesse, *Bull. Soc. Chim. Fr.* 4349 (1972).
228. M. Iwata and S. Emoto, *Bull. Chem. Soc. Jpn.* **49**, 1369 (1976).
229. J. M. Meunier and P. Fournari, *Bull. Soc. Chim. Fr.* 3343 (1971).
230. M. Varache-Béranger, A. Nurich and G. Devaux, *Eur. J. Med. Chem.* **23**, 501 (1988).
231. P. Fournari, R. Guilard and M. Person, *Bull. Soc Chim. Fr.* 4115 (1967).
232. P. Pedaja and S. Gronowitz, *Chem. Scripta* **20**, 53 (1982).
233. D. M. Perrine, D. M. Bush, E. P. Kornak, M. Zhang, Y. H. Cjo and J. Kagan, *J. Org Chem.* **56**, 5095 (1991).
234. S. V. Tsukerman, L. N. Thiem, V. N. Nikitchenko and V. F. Lavrushin, *Chem. Heterocycl. Compds. (Engl. Transl.)* 793 (1967).
235. E. P. Kohler and H. M. Chadwell, *Organic Syntheses;* Wiley: New York, 1941; Collect. Vol. I. pp 78–80.
236. W. J. King and F. F. Nord, *J. Org. Chem.* **14**, 405 (1949).
237. H. Satonaka, *Bull. Chem. Soc. Jpn.* **57**, 473 (1984).
238. H. Vieregge, H. M. Schmidt, J. Renema, H. J. T. Bos, and J. F. Arens, *Rec. trav. Chim. Pays-Bas* **85**, 929 (1966).
239. R. E. Miller and F. F. Nord, *J. Org, Chem.* **15**, 89 (1950).
240. R. D. Schuetz and W. H. Houff, *J. Am. Chem. Soc.* **77**, 1836 (1955).
241. R. M. Acheson, K. E. MacPhee, P. G. Philpoptt and J. A. Barltrop, *J. Chem. Soc.* 698 (1956).
242. R. E. Miller and F. F. Nord, *J. Org. Chem.* **16**, 728 (1951).
243. R. Filler and H. A. Leopold, *J. Org. Chem.* **27**, 4440 (1962).
244. B. F. Crowe and F. F. Nord, *J. Org. Chem.* **15**, 1177 (1950).
245. C. Cativiela, J. A. Mayoral, E. Melendez, L. A. Oro, M. T. Pinillos and R. Uson, *J. Org. Chem.* **49**, 2502 (1984).
246. C. Cativiela, M. D. Diaz de Villegas, J. A. Mayoral, A. Avenoza, and M. A. Roy, *J. Heterocycl. Chem.* **25**, 851 (1988).
247. H. R. Snyder, L. A. Carpino,. J. F. Zack Jr., and J. F. Mills, *J. Am. Chem. Soc.* **79**, 2556 (1957).

248. J. Meiwes, M. Schudok and G. Kretzschmar, *Tetrahedron Assymetri* **8**, 527 (1997).
249. B. F. Crowe and F. F. Nord, *J. Org Chem.* **15**, 81 (1950).
250. E. E. Campaigne and R. E. Cline, *J. Org. Chem.* **21**, 32 (1956).
251. A. M. Islam, K. A. M. El-Bayouki and H. H. Moharam, *Pharmazie* **39**, 382 (1984).
252. E. Campaigne, L. Fedor and R. E. Johnson, *Heterocycl. Chem.* **1**, 242 (1964).
253. K. E. Schulte, J. Reisch and L. Horner, *Ber.* **95**, 1943 (1962).
254. F. Eloy and A. Deryckere, *Bull. Soc. Chim. Belg.* **79**, 301 (1970).
255. R. Kimura, T. Yabuuchi and M. Hisaki, *Chem. Pharm. Bull.* **10**, 1232 (1962).
256. W. Freund, *J. Chem. Soc.* 2889 (1953).
257. W. Freund, *J. Chem. Soc.* 3073 (1952).
258. E. Wiklund and R. Håkansson, *Chem. Scripta* **6**, 226 (1974).
259. A. S. Yoshimura, S. Takahashi, A. Kawamata, K. Kikugawa, H. Suchiro and A. Aoki, *Chem. Pharm. Bull.* **26**, 685 (1978).
260. E. Profft and H. Mitternacht, *J. Prakt. Chem.* **16**, 13 (1962).
261. E. Profft and D. Gerber, *J. Prakt. Chem.* **16**, 8 (1962).
262. A. I. Johnson, *J. Org. Chem.* **41**, 1320 (1976).
263. W. B. Wright, *J. Heterocycl. Chem.* **9**, 879 (1972).
264. W. J. Raich and C. S. Hamilton, *J. Am. Chem. Soc.* **79**, 3800 (1957).
265. S. Gronowitz and I. Ander, *Chem. Scripta* **15**, 145 (1980).
266. E. Campaigne and R. L. White, Jr, *J. Heterocycl. Chem.* **25**, 167 (1988).
267. S. L. Castle, J.-K. Luo, H. Kodo, R. N. Castle and M. L. Lee, *J. Heterocycl. Chem.* **25**, 1363 (1988).
268. P. Soudan, P. Lucas, H. Ang Ho, D. Jobin, L. Breau and D. Bélanger, *J. Mater. Chem.* **11**, 773 (2001).
269. C.-K. Sha and C.-P. Tsou, *J. Org. Chem.* **55**, 2446 (1990).
270. D. Muller, J. F. Muller and D. Cagniant, *J. Chem. Res.* 3673 (1977).
271. C. Sone, *Bull. Chem, Soc. Jpn.* **37**, 1197 (1964).
272. K. Pettersson, *Arkiv Kemi* **7**, 39 (1954).
273. C. J. Moody, C. W. Rees and S. C. Tsoi, *J. Chem. Soc. Perkin Trans. 1* 915 (1984).
274. C.-K. Sha, C.-P. Tsou and S.-L. Wang, *J. Chem. Soc., Chem. Commun.* 320 (1988).
275. V. Zvak, J. Kovac, M. Dandárová, T. Grasza and M. Kriz, *Coll. Czech. Commun.* **49**, 1764 (1984).
276. G. Carrara, R. Ettore, F. Fava. G. Rolland, E. Testa and A. Vecchi, *J. Am. Chem. Soc.* **76**, 4391 (1954).
277. F. Freeman and L. Y. Chang, *J. Am. Chem. Soc.* **108**, 4504 (1986).
278. S. Fisichella, G. Scarlata and D. Sciotto, *Ann. Chim (Rome)* **63**, 55 (1973).
279. B. P. Das, R. T. Cunningham and D. W. Boykin, *J. Med. Chem.* **16**, 1361 (1973).
280. E. Maccarone, A. Mamo, G. Perrini and M. Torre, *J. Heterocycl. Chem.* **18**, 395 (1981).
281. P. S. Bailey and H. H. Hwang, *J. Org. Chem.* **50**, 1778 (1985).
282. Y. Tominaga, L. W. Castle and R. N. Castle, *J. Heterocycl.Chem.* **33**, 1017 (1996).
283. H. Emetsberger and D. Knittel, *Monatsh. Chem.* **103**, 194 (1972).
284. M. Farnier, S. Soth and P. Fournari, *Can. J. Chem.* **54**, 1074 (1976).
285. S. Soth, M. Farnier and P. Fournari, *Bull. Soc. Chim. Fr.* 2511 (1975).
286. J. Eras, C. Galvez and F. Garcia, *J. Heterocycl. Chem.* **21**, 215 (1984).
287. M. Farnier, S. Soth and P. Fournari, *Can. J. Chem.* **54**, 1066 (1986).
288. H. Yasuda and H. Midorikawa, *J. Org. Chem.* **36**, 2196 (1971).
289. R. Kada, V. Knoppova, J. Kovac and M. Balag, *Coll. Czech. Chem. Commun.* **45**, 2360 (1980).
290. S.-O. Lawesson, E. H. Larsen and H. J. Jakobsen, *Arkiv Kemi* **23**, 453 (1965).
291. E. J. Cragoe Jr, C. M. Robb and J. M. Sprague, *J. Org. Chem.* **15**, 381 (1950).
292. T. Kurihara, J. Sasaki, K. Santo Y. Nakamura and S. Harusawa, *Heterocycles* **29**, 2007 (1989).

293. R. S. Gairns, C. J. Moody and C. W. Rees, *J. Chem. Soc., Perkin Trans. 1* 501 (1986).
294. S.-O. Lawesson, E. H. Larsen, G. Sundström and H. J. Jakobsen, *Acta Chem. Scand.* **17**, 2216 (1963).
295. E. D. Bergmann, I. Shahak, E. Salí and Z. Aizenshtat, *J. Chem. Soc. C* 1232 (1968).
296. W. Lehnert, *Tetrahedron* **30**, 301 (1974).
297. W. Lehnert, *Tetrahedron* **28**, 663 (1972).
298. H. Anh Ho, H. Brisset, P. Frere and J. Roncali, *J. Chem. Soc., Chem. Commun.* 2309 (1995).
299. H. G. Sturz and C. R. Noller, *J. Am. Chem. Soc.* **71**, 2949 (1949).
300. A. G. A. Elagamay, M. A. Sofan, Z. E. Kandeel and M. H. Elnagdi, *Coll. Czech. Chem. Commun.* **52**, 1561 (1987).
301. S. W. Schneller and D. R. Moore, *J. Org. Chem.* **40**, 1840 (1975).
302. P. Stanetty, H. Fröhlich and F. Sauter, *Arch. Pharm.* **317**, 168 (1984).
303. T. Farrell, T. Meyer-Friedrichsen, M. Malessa, C. Wittenberg, J. Heck and A. R. Manning, *J. Organomet. Chem.* **625**, 32 (2001).
304. V. G. K. Das, L. K. Mun, C. Wei and C. V. Mak, *Organomet.* **6**, 10 (1987).
305. K. Kobayashi and C. L. Gajurel, J. *Chem. Soc., Chem. Commun.* 1779 (1986).
306. K. Kobayashi, C. L. Gajurel, K. Umemoto and Y. Mazaki, *Bull. Chem. Soc. Jpn.* **65**, 2168 (1992).
307. S. Fukuda, Y. Tominaga, Y. Matsuda and G. Kobayashi, *Heterocycles* **20**, 1793 (1983).
308. M. G. Hutchings, I. Ferguson, D. J. McGeein, J. O. Morley, J. Zyss and I. Ledoux, *J. Chem. Soc., Perkin Trans. 2* 171 (1995).
309. C. F. Shu, W. J. Tsai, J.-Y. Chen, A. K.-Y. Jen, Y. Zhang and T.-A. Chen, *Chem. Commun.* 2279 (1996).
310. V. P. Rao, A. K.-Y. Jen and Y. Cai, *Chem. Commun.* 1237 (1996).
311. K. J. Drost, V. P. Rao and A. K.-Y. Jen, *J. Chem. Soc., Chem. Commun.* 3639 (1994).
312. M. Scholz, G. Gescheidt, U. Schöberl and J. Daub, J. *Chem. Soc., Perkin Trans. 1* 209 (1995).
313. P, Boldt, G. Bourhill, C. Braüchle, Y. Jim. R. Kammler C. Müller, J. Rase and J. Wichern, *Chem. Commun.* 793 (1996).
314. J. Benko, *Acta Chim. Acad. Sci.* **34**, 217 (1962).
315. E. R. Lavagnino and E. R. Shepard, *J. Org. Chem.* **22**, 457 (1957).
316. P. Cagniant and D. Cagniant. *Bull. Soc. Chim. Fr.* 713 (1952).
317. Ng. Ph. Buu-Hoi and M. Sy, *J. Org. Chem.* **23**, 97 (1958).
318. Ng. Ph. Buu-Hoi, V. Q. Yen and N. D. Xuong, *J. Org. Chem.* **23**, 189 (1958).
319. Ng. Ph. Buu-Hoi and D. Lavit, *J. Chem. Soc.* 1721 (1958).
320. N. Messina and E. V. Brown, *J. Am. Chem. Soc.* **74**, 920 (1952).
321. G. Alberghina, M. E. Amato, A. Corsaro, S. Fisichella and G. Scarlata, *J. Chem. Soc., Perkin Trans. 2* 353 (1985).
322. P. Demerseman, Ng. Ph. Buu-Hoi and R. Royer, *J. Chem. Soc.* 4193 (1954).
323. Ng. Ph. Buu-Hoi and Ng. Hoan, *J. Chem. Soc.* 251 (1951).
324. S. A. M. Osman, G. E. H. Elgemeie, G. A. M. Nawar and M. H. Elnagdi, *Monatsh. Chem.* **117**, 105 (1986).
325. J. R. Merchant, H. C. Kaushik and D. D. Vaghani, *Indian J. Chem.* **4**, 27 (1966).
326. S.-M. Abdel-Wahab, and N. R. El-Reyyes, *J. Prakt. Chem.* **313**, 247 (1971).
327. H. H. Moussa, *Indian J. Chem. Section B* **14**, 707 (1976).
328. N. R. El-Rayyes, *J. Prakt. Chem.* **315**, 300 (1973).
329. N. R. Elrayyes and N. A. Alsalman, *J Heterocycl. Chem.* **13**, 285 (1976).
330. H. H. Moussa and D. Zaki, *Indian J. Chem. Sect. B* **15**, 555 (1977).
331. N. R. El-Rayyes and N. A. Alsalman, *J. Prakt. Chem.* **317**, 552 (1973).
332. H. H. Moussa and S. Abdelmeguiv, *J. Heterocycl. Chem.* **18**, 1519 (1981).
333. R. Cruz-Almanza, F. Garcia, B. Ramirez and M. Ordonez, *Organic. Prep. and Procedures* **20**, 265 (1988).
334. C. W. Bird, C. K. Wong and D. Y. Wong, *J. Prakt. Chem.* **314**, 915 (1972).

335. B. P. Das, M. E. Nuss and D. W. Boykin, *J. Med. Chem.* **17**, 516 (1974).
336. B. P. Das, J. A. Campbell, F. B. Samples, R. A. Wallace, L. K. Whisenant and R. W. Woodard, *J. Med. Chem.* **15**, 370 (1972).
337. S. Kusuma, W. D. Wilson and D. W. Boykin, *J. Heterocycl. Chem.* **22**, 1229 (1985).
338. B. P. Das and D. W. Boykin, *J. Med. Chem.* **16**, 413 (1973).
339. Y. Ming and D. W. Boykin, *J. Heterocycl. Chem.* **25**, 1729 (1988).
340. Y. Ming and D. W. Boykin, *Heterocycles* **26**, 3229 (1987).
341. D. Cagniant, *Bull. Soc. Chim. Fr.* 847 (1949).
342. Ng. Hoan, *Bull. Soc. Chim. Fr.* 309 (1953).
343. Ng. Ph. Buu-Hoi, N. G. Hoan and D. Lavit, *J. Chem. Soc.* 4590 (1952).
344. Ng. Ph. Buu-Hoi, Ng. Hoan and D. Lavit, *J. Chem. Soc.* 485 (1953).
345. A. S. Zektzer, J. G. Stuart, G. E. Martin and R. N. Castle, *J. Heterocycl. Chem.* **23**, 1587 (1986).
346. M. Jackman, C. Bolen, F. C. Nachod, B. F. Tullar and S. Archer, *J. Am. Chem. Soc.* **71**, 2301 (1949).
347. S.-C. Lin, J. A. Chen, M.-H. Liu, Y. O. Su and M. Leung, *J. Org. Chem.* **63**, 5059 (1998).
348. C. J. Moody, C. W. Rees, S. C. Tsoi and D. J. Williams, *J. Chem. Soc., Chem. Commun.* 927 (1981).
349. V. Baliah and T. Rangarajan, *J. Chem. Soc.* 4703 (1960).
350. S. Gronowitz and M. Herslöf, *Chem. Scripta* **10**, 90 (1976).
351. K. Ogura, Y. Ito and G. Tsuchihashi, *Synthesis* 736 (1980).
352. V. Baliah and M. Seshapathirao, *J. Org. Chem.* **24**, 867 (1959).
353. D. A. Torres and J. P. Ferraris, *Tetrahedron Lett.* **35**, 7589 (1994).
354. M. Suzuki, K. Nunami, T. Moryia, K. Matsumoto and N. Yoneda. *J. Org. Chem.* **43**, 4933 (1978).
355. S. Gronowitz and P. Pedaja, *Tetrahedron* **34**, 587 (1978).
356. G. Seitz and H. S. The, *Arch. Pharm.* 316 (1983).
357. C. W. Spangler and R. K. McCoy, *Synth. Commun.* **18**, 51 (1988).
358. T. M. Cresp, M. V. Sargent and P. Vogel, *J. Chem. Soc. Perkin Trans. 1* 37, (1974).
359. D. C. Lankin, M. R. Scalise, J. C. Schmidt and H. Zimmer, *J. Heterocycl. Chem.* **11**, 631 (1974).
360. P. Berthelot, C. Vaccher, N. Flouquet, M. Debaert, M. Luycks and C. Brunet, *J. Med Chem.* **34**, 2557 (1991).
361. E. Zbiral, *Tetrahedron Lett.* 1483 (1965).
362. T. Tanouchi, M. Kawamura, I. Ohyama, I. Kajiwara, Y. Iguchi, T. Okada, T. Miyamoto, K. Taniguchi, M. Hayashi, K. Iizuka and M. Nakazawa, *J. Med. Chem.* **24**, 1149 (1981).
363. E. Castagnino, S. Corsano and B. Serena, *Gazz. Chim. Ital.* **113**, 97 (1983).
364. T. Itahara and F. Ouseto, *Synthesis* 488 (1984).
365. Y. Fujiwara, D. Maruyama, M. Yoshidomi and H. Taniguchi, *J. Org. Chem.* **46**, 851 (1981).
366. R. Rossi, A. Carpita and P. Cossi, *Tetrahedron* **48**, 8801 (1992).
367. K. Itoh, T. Harada and H. Nagashima, *Bull. Chem. Soc. Jpn.* **64**, 3746 (1991).
368. T. Sakamoto, Y. Kondo, A. Yasuhara and H. Yamanaka, *Heterocycles* **31**, 219 (1990).
369. D. M. Hodgson, J. Witherington, B. A. Moloney I. C. Richards and J.-L Brayer, *Synlett* 32 (1995).
370. M. Satoh, N. Miyaura and A. Suzuki, *Chem. Lett.* 1329 (1986).
371. N. Yamashina, S. Hyuaga, S. Hara and A. Suzuki, *Tetrahedron Lett.* **30**, 6555 (1989).
372. G. Karminski-Zamola, J. Dogan, M. Bajic, J. Blazevic and M. Malesevic, *Heterocycles* **38**, 759 (1994).
373. M. Malesevic, G. Karminski-Zamola, M. Bajic and D. W. Boykin, *Heterocycles* **41**, 2691 (1995).
374. K. Karabelas and A. Hallberg, *J. Org. Chem.* **51**, 5286 (1986).
375. X. S. Ye and H. N. C. Wong, *J. Org. Chem.* **62**, 1940 (1997).

376. A.-S. Carlström and T. Frejd, *Synthesis* 414 (1989).
377. A. N. Nesmeyanov, N. K. Kochetkov and L. A. Matov, *Dokl. Akad. Nauk. SSSR*, **92**, 85 (1953).
378. V. F. Belyaev and A. I. Abrazhevich, *Chem. Heterocycl. Compds, USSR (Engl. Transl.)* **3**, 170 (1967).
379. L. A. Myshjkina, F. M. Stoyanovbich and Y. L. Gol'dfarb, *Bull. Acad. Sci USSR* 357 (1982).
380. A. Yamashita and T. A. Scahill, *Tetrahedron Lett.* **23**, 3765 (1982).
381. A. Ohta, T. Kobayashi and H. Kato, *J. Chem. Soc., Chem. Commun.* 431 (1993).
382. T. Kawase, H. Kurata, T. Murikawa and M. Oda, *Tetrahedron Lett.* **34**, 3449 (1993).
383. D. W. Henry, V. H. Brown, M. Cory, J. G. Johansson and E. Bueding, *J. Med. Chem.* **16**, 1287 (1973).
384. H. C. Caldwell, J. A. Finkelstein, P. P. Goldman, A. J. Sivak, J. Schlosser C. Pelikan and W. G. Groves, *J. Med. Chem.* **13**, 1076 (1970).
385. M. Lipp, F. Dallacker and G. Koenen, *Chem. Ber.* **91**, 1660 (1958).
386. G. Alberghina, M. E. Amato, S. Fisichella and S. Occhipinti, *Gazz. Chim. Ital.* **111**, 231 (1981).
387. R. I. Pogonina, N. S. Pivnenko, V. P. Izvekov and V. F. Lavrushin, *Chem. Heterocycl. Compds. USSR*, 24 (1971).
388. V. F. Kavrushin, R. I. Pogonina, N. S. Pivnenko and V. P. Izvekov, *Chem. Heterocycl. Compds. USSR*, 1498 (1970).
389. J. B. Carr, P. Kirby, M. H. Hoodrow, H. G. Durham, D. K. Hass and J. J. Boudreau, *J. Med. Chem.* **15**, 1231 (1972).
390. P. Pedaja and S. Gronowitz, *Chem. Scripta* **22**, 53 (1983).
391. Yu. D. Churkin and N. I. Putokhin, *Zh. Org. Khim.* **1**, 596 (1965).
392. V. F. Lavrushin, S. V. Tsukerman and V. M. Nikitchenko, *Zh. Obshch. Khim.* **31**, 2845 (1961).
393. G. Combes, M. Hebbelynck and J. Ledrut, *Bull. Soc. Chim. Fr.* 315 (1953).
394. H. C. Caldwell and W. L. Nobles, *J. Am. Chem. Soc.* **76**, 1159 (1954).
395. Z. M. Garashchenko, G. G. Skovortsova and N. S. Krishtal, *Khim. Geterot. Soedin.* **7**, 1626 (1971).
396. G. Pappalardo, *Gazz. Chim. Ital.* **89**, 1736 (1959).
397. S. Britton and W. L. Nobles, *J. Am. Pharm. Assoc.* **44**, 717 (1955).
398. D. T. Glatzhofer, D. J. Guerrero and M. A. Khan, *Tetrahedron Lett.* **33**, 2633 (1992).
399. M. A. McCall, *J. Org. Chem.* **27**, 2433 (1962).
400. M. Nishio and T. Ito, *Chem. Pharm. Bull. (Tokyo)*, **13**, 1392 (1965).
401. Z. Arnold, D. Dvorak and V. Kral, *Coll. Czech. Chem. Commun.* **49**, 2613 (1984).
402. Y. Yokoyama, T. Sagisaka, Y. Yamaguchi, Y. Yokoyama, J. Kiji, T. Okano, A. Takemoto and S. Mio, *Chem. Lett.* 220 (2000).
403. M. Journet, A. Rouillard, D. Cai and R. D. Larsen, *J. Org. Chem.* **62**, 8630 (1997).
404. R. E. Miller and F. F. Nord, *J. Org. Chem.* **16**, 1380 (1951).
405. J. P. Verge and P. Roffey, *J. Med. Chem.* **18**, 794 (1975)
406. E. Campaigne and R. L. White, Jr, *J. Heterocycl. Chem.* **25**, 367 (1988).
407. L. Skatteböl, *Acta Chem. Scand.* **15**, 2047 (1961).
408. M. D'Auria, A. DeMico, F. D'Onofrio and G. Piancatteli, *J. Org. Chem.* **52**, 5243 (1987).
409. G. T. Gmitter and T. L. Benton *J. Am. Chem. Soc.* **72**, 4586 (1950).
410. P. V. Bedworth, Y. Cai, A. Jen and S. R. Marder, *J. Org. Chem.* **61**, 2242 (1996).
411. K. Tamao, K. Kobayashi and Y. Itoh, *Tetrahedron Lett.* **30**, 6051 (1989).
412. J. W. van Reijendam, G. J. Heeres and M. J. Janssen, *Tetrahedron* **26**, 1291 (1970).
413. Y. Tominaga, Y. Matsuda and G. Kobayashi, *Heterocycles* **4**, 1493 (1976).
414. H. Junek, B. Thierrichter, and P. Widmer, *Monatsh. Chem.* **110**, 483 (1979).
415. T. L. Gilchrist and F. J. Summersell, *J. Chem. Soc., Perkin Trans. 1* 2595 (1988).
416. T. L. Gilchrist and M. A. M. Healy, *Tetrahedron* **49**, 2543 (1993).

417. T. H. Majid and P. Knochel, *Tetrahedron Lett.* **31**, 4413 (1990).
418. Zh. Krasnaya, T. S. Stytsenko, E. P. Prokofé, V. F. Kucherov, *Izv. Akad. Nauk. SSSR, Ser. Khim 1* 362 (1980).
419. C. W. Spangler, P.-K. Liu A. A. Dembek and K. O. Havalka, *J. Chem. Soc., Perkin Trans. 1* 799 (1991).
420. C. W. Spangler and M. He, *J. Chem. Soc. Perkin Trans. 1* 715 (1995).
421. F. Embert, J.-P- Lère-Porte, J. J. E. Moreau, F. Serain-Spirau, A. Righi and J.-L. Sauvajol, *J. Mater. Chem.* **11**, 718 (2001).
422. M. Fosatelli, A. C. M. T. van der Kerk, S. F. Vasilevsky and L. Brandsma, *Tetrahedron Lett.* **33**, 4229 (1992).
423. J. R. McCarthy, C. Barney, D.P. Mathews and T. M. Barger, *Tetrahedron Lett.* **28**, 2207 (1987).
424. V. R. Kuhn and J. Jahn, *Chem. Ber.* **86**, 759 (1953).
425. T. Kawase, S. Muro, H. Kurata and M. Oda, *Chem. Commun.* 778 (1992).
426. P. Cadiot and A. Willemart, *Bull. Soc. Chim. Fr.* 100 (1951).

– 4C –
Thienylacetylenes

4C.1 COMPOUNDS ThC≡CH

4C.1.1 From thiophenecarbonyl compounds

Recent reinvestigation of a classical method for the preparation of 2-thienylacetylene through the reaction of 2-acetylthiophenes with phosphorus pentachloride, followed by dehydrochlorination using sodium amide in liquid ammonia [1–3], indicates that only small amounts of the thienylacetylenes are formed and that the major products are the *E* and *Z*-isomers of 1,2-dichloro-1-(2-thienyl)ethene [4]. These authors claim that the best method for converting 2-acetylthiophene is through the reaction of 2-(bromoacetyl)thiophene with triethyl phosphite followed by reaction with sodium amide [4], which was previously used for the preparation of the 5-phenyl substituted derivative [5].

2-Thienylacetylene [5]

A mixture of 2-(bromoacetyl)thiophene (6.1 g, 0.03 mol) and triethyl phosphite (15 ml, 0.09 mol) under nitrogen is stirred at 90 °C for 1.5 h. The excess of triethyl phosphite is evaporated and the residue is dissolved in diethyl ether (15 ml). This solution is added over a period of 30 min to a suspension of sodium amide prepared from sodium (3.35 g) and anhydrous liquid ammonia at −78 °C. The stirring is continued at this temperature for 1 h, ammonium chloride is added and the ammonia is evaporated in a stream of nitrogen. The residue is poured into water and the product extracted with diethyl ether. The combined organic phases are dried over magnesium sulfate, evaporated and distilled, giving 2.44 g (76%) of the title compound bp 46–46.5 °C/15 mm Hg.

The best method for the preparation of thiophenacetylenes from aldehydes, introduced by Corey, is the reaction with zinc carbon tetrabromide and triphenylphosphine, which with thiophene aldehydes yields the 1,1-dibromo-2-(thienyl)ethenes, which upon treatment with butyllithium or methyllithium at −70 °C gives the desired thienylacetylenes in excellent yields [6–12].

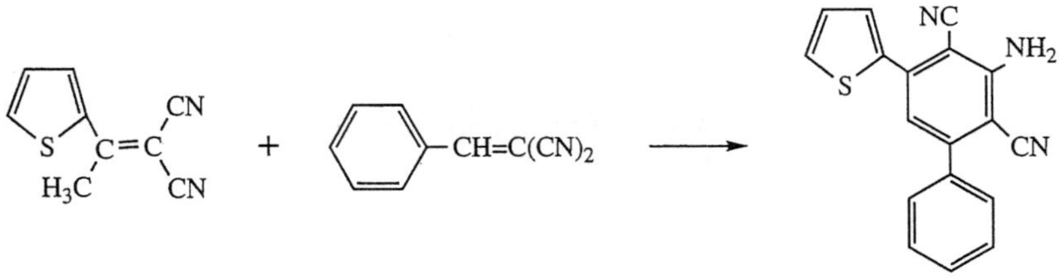

Excess butyllithium and high concentrations must be avoided, since this leads to polymerization and alkylation of the free α-position of the thiophene ring [13]. This method tolerates acetylenic functions very well, as demonstated in the following transformation [10].

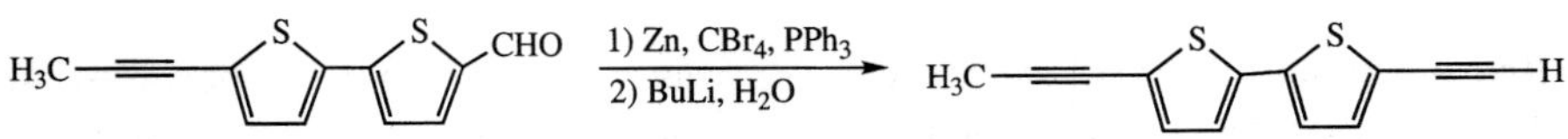

4C.1.2 From vinylthiophenes

Thienyl acetylenes are prepared by the addition of bromine to various vinylthiophenes such as 3-vinylthiophene [14], followed by treatment of the vicinal dibromo derivative with a strong base. 2-Thienylacetylene is prepared by the reaction of 2-thienyllithium with 1,1-dichloro-2,2-difluoroethene followed by treatment of the intermediate 1,1-dichloro-2-fluoro-2-(2-thienyl)-ethene with excess butyllithium [15]. 2,5-Diethynylthiophene is prepared in a similar way [16].

2,5-Diethynylthiophene [16]

Butyllithium in diethyl ether (20 mmol) is added below −58 °C over a period of 10 min to a cooled stirred slurry of 1,1-dichloro-2-fluoro-2-(2-thienyl)ethene (1.00 g, 3.2 mmol) in anhydrous diethyl ether (50 ml). The mixture is allowed to warm to −10 °C and poured onto a mixture of concentrated hydrochloric acid (4 ml) and crushed ice. After shaking the mixture, the phases are separated and the organic phase washed with water and sodium bicarbonate solution, dried over sodium sulfate and briefly evaporated. The oil obtained is dissolved in ethanol (20 ml) and a solution of silver nitrate (2.00 g) in water (2 ml) is added, giving a bright-yellow precipitate, which is filtered off and washed with ethanol and diethyl ether and then stirred in diethyl ether (100 ml). An aqueous solution (15 ml) containing concentrated hydrochloric acid (3 ml) is added and after complete conversion of the precipitate the organic phase is separated and washed with dilute hydrochloric acid, water and sodium bicarbonate solution, dried and evaporated up to 45 °C of the bath temperature giving 0.37 g (87%) of the title compound.

4C.1.3 Palladium(0)-catalyzed coupling reactions of halothiophenes with acetylenic derivatives

Reaction of 3-iodothiophene and trimetylsilylacetylene using palladium(0) catalyst as well as copper iodide in diisopropylamine is an excellent method for the preparation of 3-[(trimethylsilyl)ethynyl]thiophene in quantitative yields. Upon treatment with potassium fluoride in aqueous methanol, 3-ethynylthiophene is obtained in almost quantitative yields [17].

3-[(Trimethylsilyl)ethynyl]thiophene [17]

To a mixture of 3-iodothiophene (13.44 g, 0.064 mol), dichlorobis(benzonitrile)-palladium(II) (0.49 g, 0.0013 mol), triphenylphosphine (0.68 g, 0.0026 mol), cuprous iodide (0.26 g, 0.0013 mol) and diisopropylamine (250 ml) is added trimethylsilyl)acetylene (10.00 ml, 0.071 mol) *via* syringe. The solution is stirred for 18 h at room temperature and filtered. The solvent is removed at reduced pressure and the residue chromatographed on silica gel using hexane as eluent, giving 11.31 g (98%) of the title compound. This compound can also be prepared from 3-bromothiophene at reflux for 18 h. The yield in this reaction is also 98%.

3-Ethynylthiophene [17]

To a solution of 3-[(trimethylsilyl)ethynyl]thiophene (11.50 g, 0.064 mol) in methanol (100 ml) is added potassium fluoride (3.71 g, 0.064 mol) and water (5 ml), after which it is stirred for 4 h and diluted with water. The product is extracted with dichloromethane (3 × 50 ml). The combined organic phases are dried over magnesium sulfate and the solvent removed at reduced pressure. The residue is distilled (trap-to-trap under vacuum) giving 6.77 g (98%) of the title compound as a colorless liquid.

Alternatively diethylamine was used as in the preparation of 3-(trimethylsilyl)ethynylthiophene, 2-bromo-3-trimethylsilylethynylthiophene, (from 3-iodo-2-bromothiophene), 4-bromo-3-trimethylsilylethynylthiophene and 3-bromo-2-trimethylsilylethynylthiophene [18]. Using piperidine as the base 3,4-di(trimethylsilylethynyl)thiophene is obtained upon reflux of 3,4-dibromothiophene and trimethylsilylacetylene and the usual palladium/cuprous iodide catalyst [19].

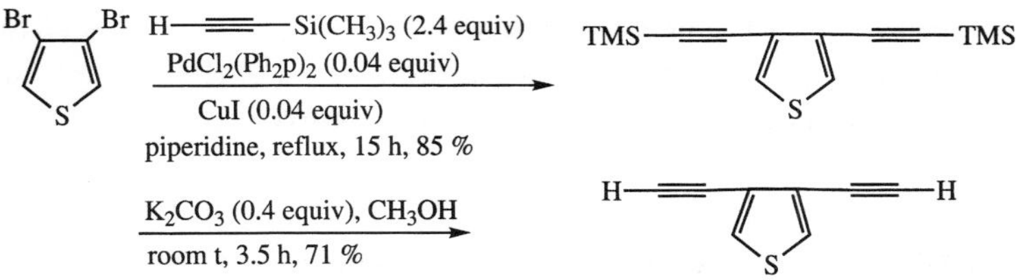

A water-soluble palladium(0) catalyst, prepared from palladium(II) acetate/trisodium 3,3′,3″-phosphinetriyltribenzenesulphonates has been used in the reaction of 2-iodothiophene with trimethylsilylacetylene and propargyl alcohol at room temperature to give 2-(trimethylsilylethynyl)thiophene and 3-(2-thienyl)propargyl alcohol in almost quantitative yield [20].

Stille coupling of tributylstannylacetylene with 4-trimetylsilyl-3-iodothiophene in dioxane under reflux using palladium(0) as catalyst is an excellent method for the preparation of 3- ethynyl-4-(trimethylsilyl)thiophene [21].

3-Ethynyl-4-(trimethylsilyl)thiophene [21]

To a mixture of 4-trimethylsilyl-3-iodothiophene (705 mg, 2.5 mmol), ethynyltributyltin (788 mg, 2.5 mmol) in dioxane (30 ml), tetrakis(triphenylphosphine)palladium(0) (290 mg, 0.25 mmol) is added. The reaction mixture is heated at 90 °C for 6 h under nitrogen, after cooling it is diluted with diethyl ether (150 ml), washed with water (2 × 25 ml), dried over magnesium sulfate and evaporated. The residue is purified by column chromatography on silica gel (150 g) using pentane as eluent, giving 406 mg (90%) of the title compound as a colorless oil.

Stille coupling of two equivalents of 2-iodo-5-(trimethylsilylethinyl)thiophene with one equivalent of bis(tributylstannyl)acetylene is a better method for the preparation of the compound shown below than the coupling of one equivalent of 2-iodo-5-(trimethylsilylethynyl)thiophene with one equivalent of 2-[(trimethylsilyl)ethynyl]-5-(trimethylstannyl)ethynyl]thiophene [22].

1,2-Di(5,5′-trimethylsilylethynyl-2-thienyl)ethyne[22]

To a solution of 2-iodo-5-(trimethylsilylethinyl)thiophene (0.943 g, 3.079 mmol) and bis(tributylstannyl)acetylene (1.023 g, 1.693 mmol) in *N*,*N*-dimethylformamide (15 ml), bis(acetonitrile)dichloropalladium (II) (0.035 g, 0.135 mmol) is added. The reaction mixture is stirred at room temperature for 3 h and then transferred to a separatory funnel with the aid of dichloromethane (50 ml). This solution is washed with water (3 × 50 ml), the combined aqueous phases are extracted with dichloromethane (2 × 50 ml). The combined organic phases are washed with water, dried over sodium sulfate and filtrated. Celite (5 g) is added to the filtrate followed by evaporation. The coated mixture is purified by column chromatography using hexane/tetrahydrofuran (9:1) as eluent, giving 1.13 g (96%) pure product as a yellow solid.

The reaction of 3,4-diiodothiophene with excess acetylene gas in the presence of a palladium catalyst and cuprous iodide in morpholine or trietylamine/*N*,*N*-dimethylformamide gave the trithienocycline in 37% yield [23].

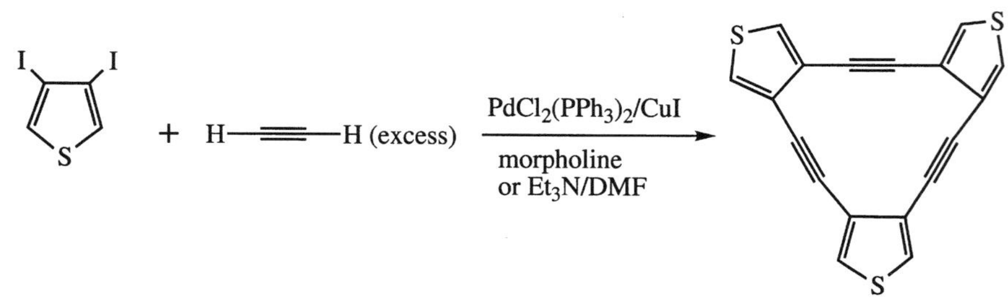

4C.1.4 From thienylacetylene derivatives

The trimethylsilyl protecting groups on acetylenic bonds are conveniently removed by 0.1 to 1 *M* aqueous potassium hydroxide in methanol [24–26]. 2-Thienylacetylene is also obtained by treatment of 2-(thienylethinyl)dimethyl carbinol with potassium hydroxide in benzene [27]. Recently, small amounts of potassium carbonate in anhydrous methanol are used for the preparation of 3-ethynyl-2-thiophene aldehyde from the trimethylsilyl protected derivative [28]. 2- and 3-Thienylacetylene are prepared by decarbonylation of the corresponding thienylpropargyl aldehydes with 4 *M* sodium hydroxide [29,30]. Thienylacetylenes are also prepared by decarboxylation of thiophenepropiolic acids under cupric ion catalysis [31].

4C.1.5 Various methods

2-Thienylacetylene was prepared in 79% yield upon treatment of 5-(2-thienyl)-3-nitroso-2-oxazolone, prepared from 3-hydroxy-3-(2-thienyl)propionic acid hydrazide *via* 5-(2-thienyl)-2-oxazolidone, with sodium methoxide in methanol [32].

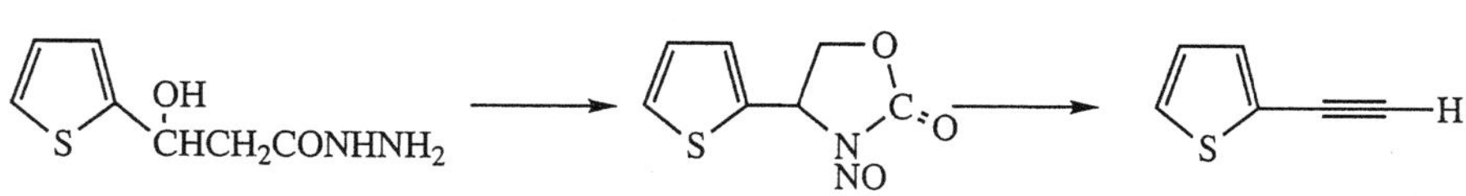

Flash photolysis of thienyl substituted 4-methylene-5(4*H*)-isoxazolone derivatives gives 2- and 3-thienylacetylene, respectively [33].

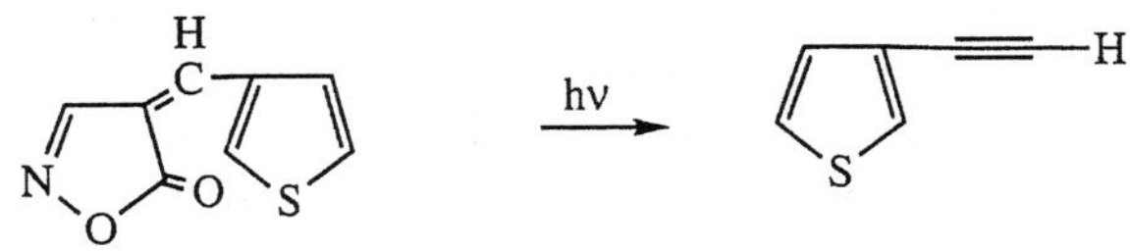

4C.2 COMPOUNDS ThC≡CR, R=ALKYL, CARBOXYL AND SUBSTITUTED ALKYL

4C.2.1 From thienylacetylenes

If thienylacetylenes are available, the alkylation of their lithium or magnesium salts is a convenient method for the preparation of thienylalkylacetylenes [5,34]. Thienylpropiolic acids are obtained upon the reaction of these salts with carbon dioxide [15,16,32]. Sodium acetylides did not react directly with carbon dioxide to give propiolic acids. However, upon reaction with ethyl chloroformate, ethyl thienylpropiolates are obtained [3]. The lithium and magnesium derivatives give acetylenic carbinols upon reaction with aldehydes [35,36].

Methyl (E)-5-(2-thienyl)-2-penten-4-ynoate [37]

A deaerated solution of 2-ethynylthiophene (4.21 g, 39 mmol) in benzene (20 ml) is added to a stirred mixture of tetrakis(triphenylphosphine)palladium(0) (0.68 g, 3.6 mmol), triethylamine (13.9 ml, 100 mmol) and methyl (*E*)-3-iodoacrylate (6.15 g, 19 mmol) in benzene (60 ml) which is maintained under argon. The reaction mixture is stirred for 6 h at room temperature, after which it is diluted with diethyl ether and washed several times with a saturated ammonium chloride solution, with water, dried and evaporated. The residue is purified by chromatography on silica gel using benzene/hexane (1:1) as eluent, giving 4.77 g (86%) of the title compound mp 70 °C.

1-(Dimethylaminomethyl)-2-(2-thienyl)acetylene is prepared in excellent yield by the Mannich reaction of 2-thienylacetylene with paraformaldehyde and dimethylamine [38]. The reactions between the lithium salts of 2-thienylacetylenes and 1,1-dichloro-2,2′-difluoroethene have been performed [16]. Lithium thienyl acetylides give protected derivatives upon reaction with trimethylsilyl chloride [15,16].

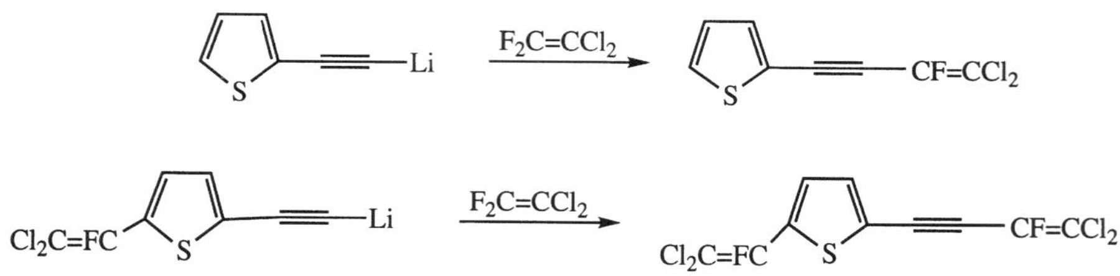

Both 2- and 3-thienylpropargyl alcohols are oxidized by manganese dioxide or nickel oxide to the propargylic aldehydes [29–32], which can be reduced to the thienylpropargyl alcohols with lithium aluminium hydride or sodium borohydride [40]. 1-Methyl-2-(thienyl)acetylene can also be prepared by reduction of the tosylate from thienylpropargyl alcohols with lithium aluminium hydride [31].

Compounds carrying halogen at the acetylenic bonds, such as 1-chloro-2-(2-thienyl)acetylene and 1-iodo-2-(thienyl)acetylene are prepared from 2-thienylacetylene and hypochlorite and from 2-thienylmagnesium iodide and iodine, respectively [41]. 1-Bromo-2-(2-thienyl)actylene is prepared from the acetylene and bromine in 10 *M* sodium hydroxide [25]. The synthetically very useful 1-(trimethylstannyl)-2-(2-thienyl)acetylene is prepared through the reaction of 2-thienylacetylene with di(methylamino)trimethylstannane [42], which on palladium(0)-catalyzed coupling with the protected diene gave the thiophene derivative, which was deprotected to a natural occurring insecticide [42].

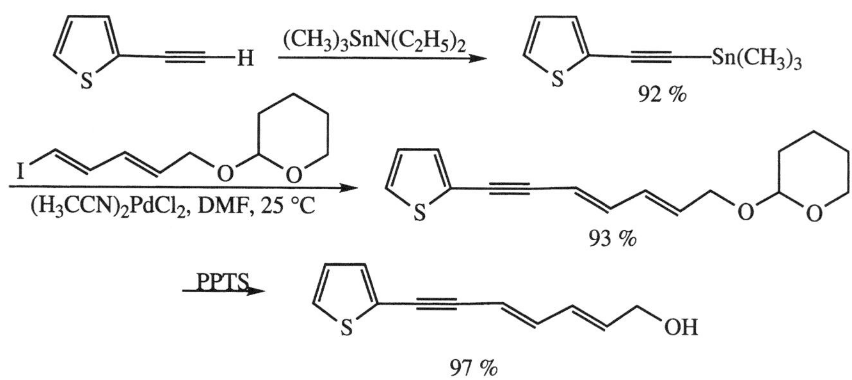

The following compound can be prepared by a normal Claisen condensation of ethyl 2- thienylpropiolate with dimethyl lithiomethanephosphonate [43].

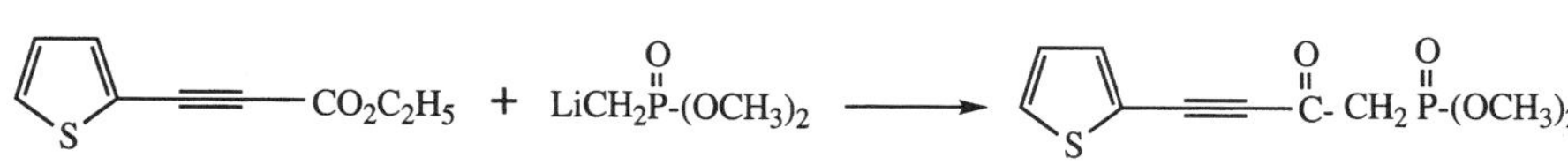

4C.2.2 From vinylthiophenes

The reaction of 2-propenylthiophene [44] and ethyl 3-(thienyl)acrylates [45] with bromine followed by dehydrobromination with strong base is used for the preparation of 2-(1-propynyl)thiophene and 3-(2-thienyl)propiolic acid [46]. This acid is also obtained in 51% yield in the following reaction [47].

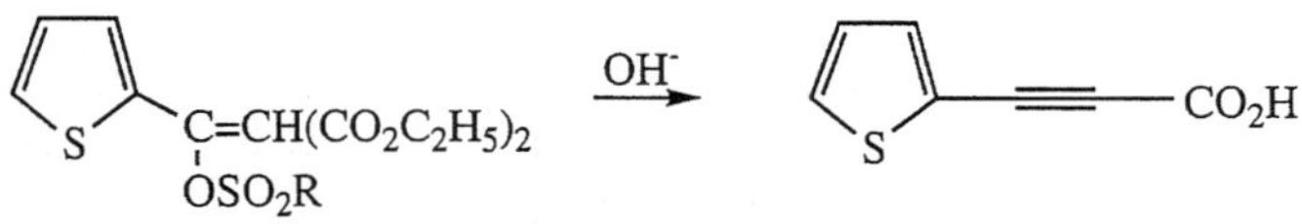

The following reaction giving 1-(thienyl)-2-cyanoacetylene has been performed [48].

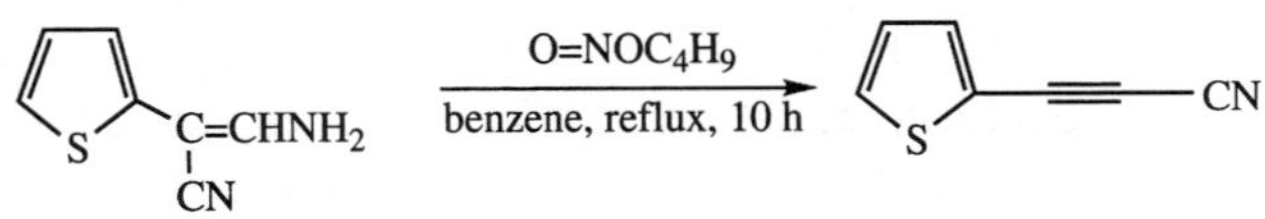

Pyrolysis of compounds containing electron-withdrawing perfluorinated alkyl groups is a very useful method for the preparation of 1-(2-thienyl)-2-perfluoroalkyl acetylenes [49].

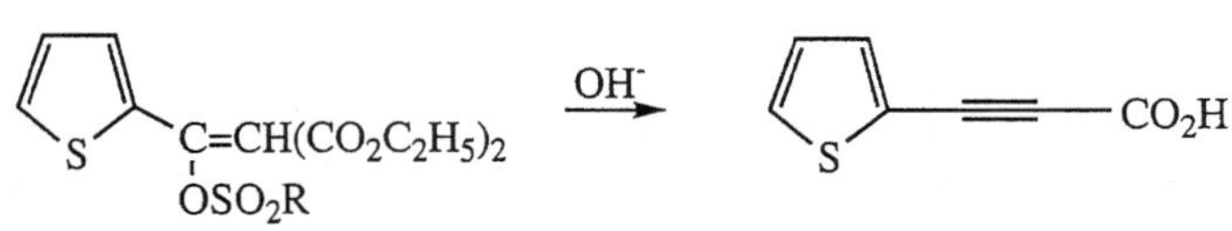

The reaction of 4-(2-thienyl) (thia-2,3-diazole) with strong bases results in the cleavage of the thiadiazole ring with evolution of nitrogen and formation of alkali methyl alkynethiolates, which upon reaction with alkyl halides or aryl halides give 1-alkylthio- or 1-arylthio(2-thienyl)acetylenes [50,51].

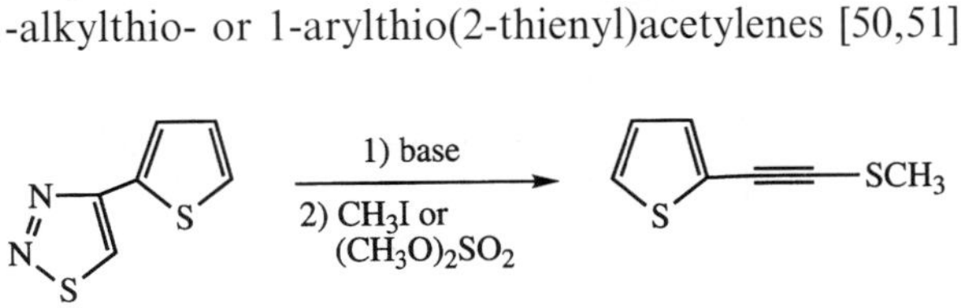

1-Methylthio-2-(2′-thienyl)ethyne [50]

A 2 *M* solution of butyllithium in hexane (260 ml) is added to a stirred suspension of 4-(2′-thienyl)-1,2,3,-thiadiazole (87.1 g, 0.52 mol) in anhydrous tetrahydrofuran under nitrogen, and cooled in an dry ice–acetone bath. The addition is at such a rate that the temperature does not exceed −60 °C. After completed addition the stirring is continued at −60 °C for 10 min after which methyl iodide (110.8 g, 0.78 mol) is added in one portion. The reaction mixture is allowed to warm to 0 °C and added to ice-water. The phases are separated and the aqueous phase extracted twice with diethyl ether. The combined organic phases are dried and evaporated giving 68.7 g (86%) as a yellow liquid after distillation bp 72–74 °C/2 mm Hg.

Ethyl 2-thienylpropiolate can be prepared by the reactions shown below [52].

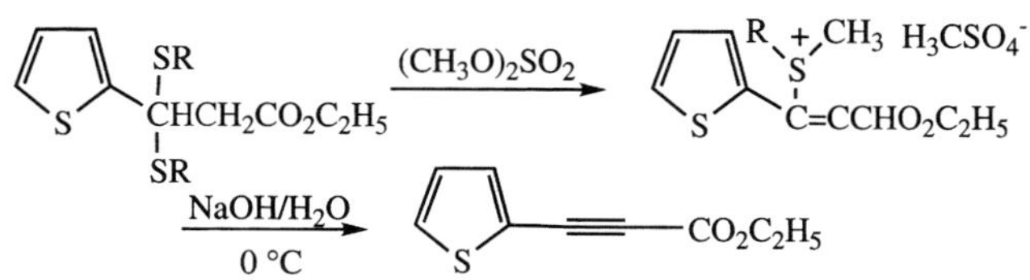

4C.2.3 From iodothiophenes and cuprous acetylides

Many thiophene acetylene derivatives are prepared through the reaction of iodothiophenes and copper acetylides especially in connection with the synthesis of naturally occurring derivatives [29–31,39,53,62]. The cuprous acetylides were generally reacted with the iodothiophenes in refluxing pyridine for 2–6 h. The cuprous acetylides were prepared by quickly adding a large excess of cuprous chloride or cuprous iodide, freshly dissolved in 12 *M* aqueous ammonia to the acetylene in ethanol. All cuprous acetylides are explosive when dry and vary in ease of detonation. In general the acetylides were freshly prepared and stored under water and then dried before use. No marked deterioration in yields was observed when slightly damp cuprous acetylides were used [62].

2-Phenyl-5-(1-propynyl)thiophene [62]

A deaerated solution of 5-phenyl-2-iodothiophene (7 g, 24.5 mmol) in benzene (40 ml) is added to a mixture of benzyltriethylammonium chloride (0.18 g, 0.86 mmol), cuprous iodide (0.25 g, 1.22 mmol) and tetrakis triphenylphosphinepalladium(0) (0.994 g, 0.86 mmol). The mixture is saturated at 0 °C with propyne, after which deaerated 2.5 *M* sodium hydroxide solution (44 ml) is added. The reaction mixture is stirred under propyne atmosphere at room temperature until 5-phenyl-2-iodothiophene is consumed, about 7 h. Saturated ammonium chloride is added and the stirring is continued for 0.5 h, the product is extracted with hexane and evaporated. The residue is purified by chromatography on silica gel using hexane as eluent, giving 3.88 g (80%) of the title compound mp 44–45 °C.

2-(1-Propynyl)thiophene is prepared in the same way in a yield of 93% from 2-iodothiophene. Acetylenic thiophenes can also be prepared by copper- or cobalt-catalyzed coupling of iodothiophene and magnesium acetylide [63].

4C.2.4 Palladium-catalyzed preparation from halothiophenes and acetylides

Thienylacetylenes are prepared by the reaction of bromo- and preferentially iodothiophenes with acetylenic zinc derivatives [64], magnesium derivatives especially trimethylsilyl acetylenemagnesium [24], or tin derivatives in the presence of catalytic amounts of palladium(0) complexes [21,42].

The palladium-catalyzed coupling in the presence of cuprous iodide of 3-ethyl-, 3-butyl- and 3-hydroxyethyl-2-iodothiophene with trimethylsilyl-acetylene is a good method for the preparation of the corresponding 3-alkyl-2-trimethylsilylethynylthiophenes [65].

3-Butyl-2-[(trimethylsilyl)ethynyl]thiophene [65]

To a 100 ml flame-dried vessel containing 3-butyl-2-iodothiophene (20.93 g, 78.7 mmol), bis(triphenylphosphinepalladium(II) chloride (1.1 g, 1.57 mmol) cuprous iodide (0.44 g, 2.3 mmol) and anhydrous tetrahydrofuran (150 ml), diisopropylamine (11.9 ml, 85 mmol) is added at room temperature. The resulting clear brown solution is stirred for 5 min before trimethylsilylacetylene (12.0 ml, 85 mmol) is added. The nitrogen outlet is removed and the septum capped. The reaction mixture is stirred overnight at room temperature, after which it is poured into water. The phases are separated and the aqueous phase extracted with diethyl ether (3 × 20 ml). The combined organic phases are dried over magnesium sulfate, evaporated and distilled, giving 15.0 g (81%) of the title compound as a clear yellow liquid bp 100 °C/1 mm Hg.

In another modification of the palladium(0)-catalyzed reaction (Heck-Sonogashira reaction), the halothiophenes are coupled with the acetylene with cuprous iodide in benzene under phase transfer conditions using benzyltri-methylammonium chloride and sodium hydroxide [6,7,10,62]. Alternatively isopropylamine or triethylamine and cuprous iodide can be used [62], as in the preparation of methyl 5-hydroxy-7-(2-thienyl)hept-6-ynoate and methyl 5-hydroxyhept-6-ynoate [66].

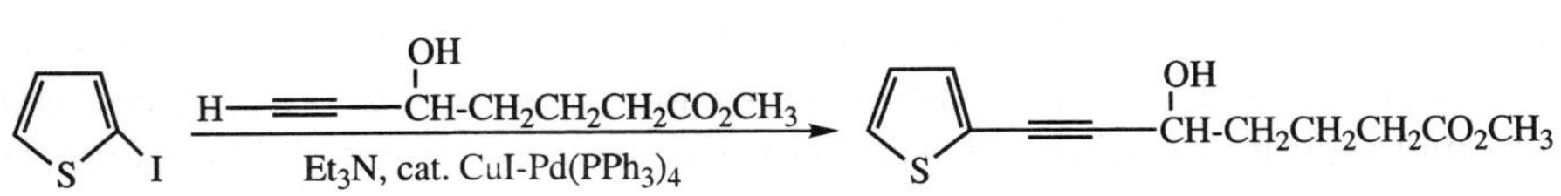

Methyl 5-hydroxy-7-(2-thienyl)hept-6-ynoate [66]

To a mixture of 2-iodothiophene (4.44 g, 21.14 mmol) and methyl 5-hydroxyhept-6-ynoate (1.65 g, 10.58 mmol) in anhydrous triethylamine (165 ml), tetrakistriphenylphosphinepalladium(0) (0.24 g, 0.21 mmol) and cuprous iodide (0.02 g, 0.105 mmol) are added. The reaction mixture is stirred at room temperature under nitrogen for 93 h. The catalyst is filtered off and the filtrate evaporated. The residue is purified by chromatography on silica gel using light petroleum/ethyl acetate (2:1) as eluent, giving 2.30 g (91%) of the title compound as a yellow oil bp (Kugelrohr) 210 °C/0.5 mm Hg.

Very often reflux in triethylamine/acetonitrile is used. 3-Iodo-4-(trimethylsilyl)thiophene has been reacted with a large number of terminal acetylenes [21].

General procedure [21]

A mixture of 3-iodo-4-(trimethylsilyl)thiophene (141 mg, 0.5 mmol), the acetylene (0.75 mmol), tetrakis(triphenylphosphine)palladium(0) (58 mg, 0.05 mmol), cuprous iodide (19 mg, 0.1 mmol), anhydrous acetonitrile (2 ml) and anhydrous triethylamine (5 ml) is stirred at reflux temperature under nitrogen for 18 h. After evaporation of the solvent the residue is purified by chromatography on silica gel.

Acetylene substituent	Yield (%)	Eluent	
C_5H_{11}-	88	hexanes	oil
C_7H_{15}-	89	hexanes	oil
Ph-	85	hexanes	oil
cyclohexene-	88	hexanes	oil
1-hydroxycyclopentyl-	93	hexanes/EtOAc (6:1)	oil
$HOCH_2CH_2$-	92	hexanes/EtOAc (3.5:1)	oil
$HO(CH_2)_4$-	91	hexanes/EtOAc (4:1)	oil
$PhCH_2N(CH_3)CH_2$-	91	hexanes/EtOAc (6:1)	oil
$HO(CH_2)_4$– $PhCH_2N(CH_3)CH_2$–	74	hexanes/EtOAc (5:1)	oil
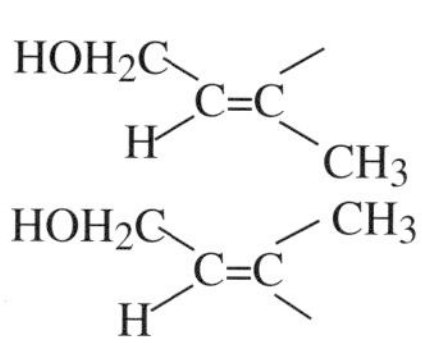	68	hexanes/EtOAc (5:1)	oil

3-(3,3-Dimethyl-1-butynyl)-2-thiophene aldehyde is prepared from 3-bromo-2-thiophene aldehyde and *tert*-butylacetylene in triethylamine in the presence of catalytic amounts of cuprous iodide and a palladium catalyst [28,67].

3-(3,3-Dimethyl-1-butynyl)-2-thiophene aldehyde [28]

To a suspension of 3-bromo-2-thiophene aldehyde (5.00 g, 26.2 mmol), bis(triphenylphosphinepalladium(II) chloride (355 mg) and cuprous iodide in triethylamine (100 ml) *tert*-butylacetylene (2.56 g, 31.2 mmol) is added. The reaction mixture is stirred at room temperature for 12 h. After filtration and evaporation the residue is purified by flash chromatography on silica gel using dichloromethane/petroleum ether (1:1) as eluent giving 4.80 g (96%) of the title compound as a yellow-brown oil.

Similarly, 3-trimethylsilylethinyl-2-thiophene aldehyde was obtained using trimethylsilylacetylene [68]. Under the reaction conditions described in the earlier paragraph, tetraiodothiophene gave, with two equivalents of trimethylsilylacetylene, 2,5-bis[(trimethylsilyl)ethynyl]-3,4-diiodothiophene in 84% yield, which was then coupled with two equivalents of the same reagent to give the tetraacetylene-substituted thiophene tetrakis[(trimethylsilyl)ethynyl]thiophene in 45% yield. If 2,5-bis[(trimethylsilyl)ethynyl]-3,4-diiodothiophene was instead coupled with 2-methyl-3-butyne-2-ol-2,5-bis[(trimethylsilyl)- ethynyl]-3,4-bis(3-hydroxy-3-methyl-1-butynyl]thiophene was obtained. Upon desilylation of tetrakis[(trimethylsilyl)ethynyl]thiophene and 2,5-bis[(trimethylsilyl)ethynyl]-3,4-bis(3-hydroxy-3-methyl-1-butynyl]thiophene by treatment with potassium hydroxide in methanol, highly explosive acetylenes were obtained in high yield [26].

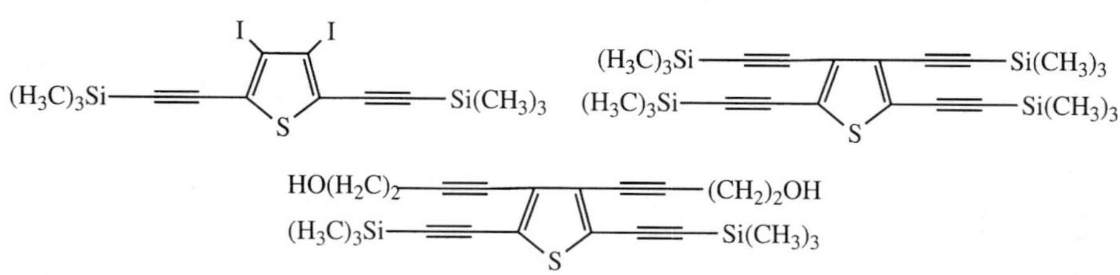

2,5-Bis[(Trimethylsilyl)ethynyl]-3,4-bis(3-hydroxy-3-methyl-1-butynyl] thiophene [26]

To a solution of 2,5-bis[(trimethylsilyl)ethynyl]-3,4-diiodothiophene (5 g, 9.4 mmol) in diisopropylamine (150 ml), dichlorobis(benzonitrile)palladium (36 mg, 0.09 mmol), triphenylphosphine (47 mg, 0.18 mmol) and cuprous iodide (18 mg, 0.09 mmol) are added. The solution is degassed with a stream of argon, after which 2-methyl-3-butyne-2-ol (2.3 g, 28 mmol) is added dropwise at room temperature. The reaction mixture is refluxed for 4 h, cooled and filtered to remove salts. The filtrate is evaporated and the residue dissolved in a minimum amount of tetrahydrofuran/dichloromethane (1:9) and purified by chromatography on neutral alumina first using dichloromethane as eluent and then tetrahydrofuran/dichloromethane (1:9) giving 3.59 g (86%) of the title compound as colorless cubic crystals, which darken upon standing, mp 153–155 °C after recrystallization from methanol/benzene (1:1).

2-(Methoxycarbonyl)-(4-phenyl-1-butynyl)thiophene was prepared by standard Sonogashira reaction between 3-bromo-2-(methoxycarbonyl)thiophene and 4-phenyl-1-butyne and its ester function transformed to a diazoketone function [69].

2-Bromothiophene as well as 3-bromothiophene undergoes the following coupling reaction in good yields [70].

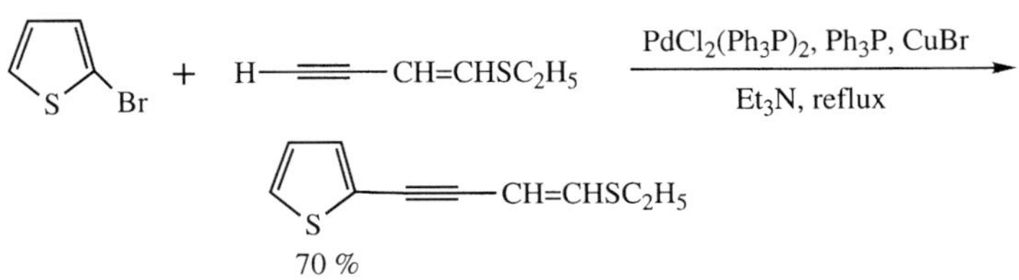

Sonogashira conditions are used in the following three step synthesis [71].

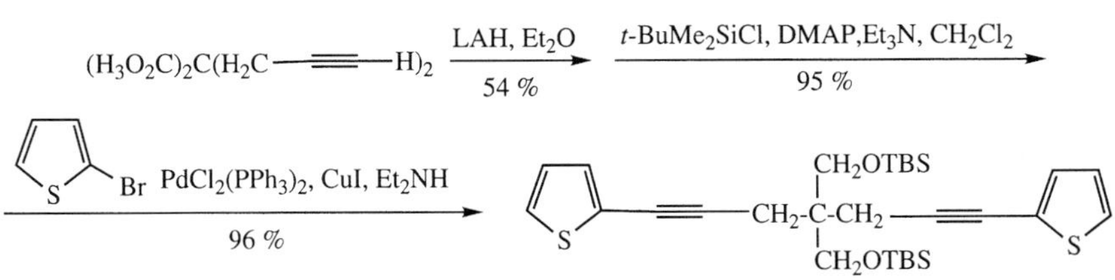

The following palladium-catalyzed one-pot sequential reaction gives the conjugated 2-thienyldienyne with high stereospecificity [72].

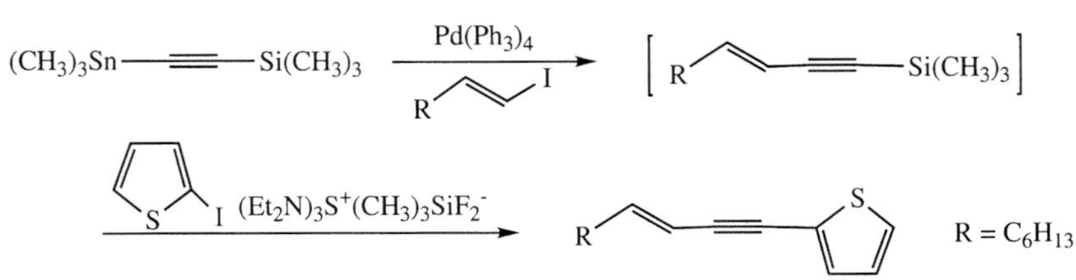

Tetraethynylethenes functionalized with 2-thienyl and 5-nitro-2-thienyl groups are prepared by protodesilylation followed by Sonogashira palladium-catalyzed coupling with 2-iodothiophene or 5-nitro-2-iodothiophene [73].

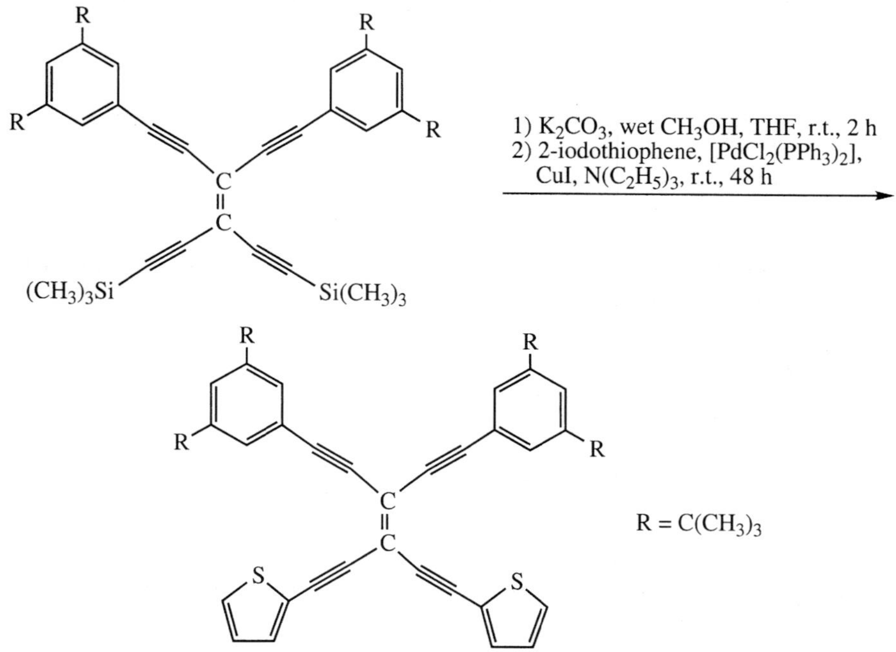

1-[3,5-Di(tert-butyl)phenyl]-3-{2-[3,5-di(tert-butyl)phenyl]ethynyl}-4-[2-(2-thienyl)ethynyl]-6-(2-thienyl)-hex-3-ene-1,5-diyne [73]

A mixture of the silylated derivative (0.20 g, 0.31 mmol) and potassium carbonate (10 mg, 0.07 mmol) in wet methanol (20 ml) and tetrahydrofuran (5 ml) is stirred at room temperature for 2 h. Diethyl ether and water are added, the phases are separated and the organic phase is dried, concentrated to 5 ml and added to triethylamine (30 ml). This solution is degassed, after which 2-iodothiophene (0.143 g, 0.68 mmol), bis(triphenylphosphine)palladium(II) chloride (20 mg, 0.03 mmol) and cuprous iodide (10 mg, 0.06 mmol) are added and the mixture is stirred at room temperature for 48 h. Workup and chromatography on silica gel using hexane/dichloromethane (4:1) as eluent gives 31 mg (15%) of the title compound as an orange solid mp 178–180 °C (dec.).

A protocol for efficient coupling of alkynes with a palladium on carbon–cuprous iodide–triphenylphosphine catalyst system in aqueous media has been applied to 2-iodothiophene and phenyl propargylacetylene [74].

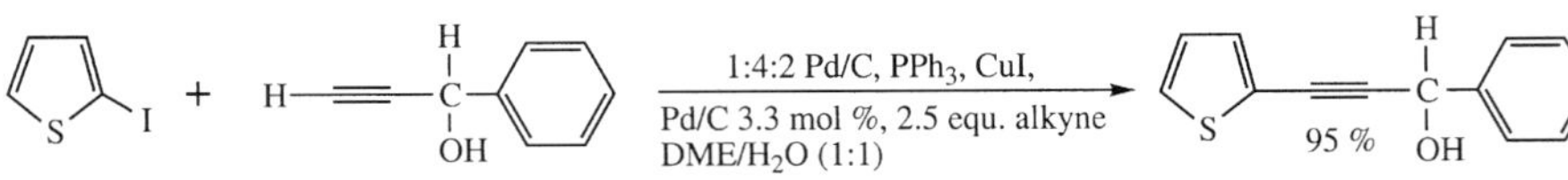

4C.2.5 By ring-closure reactions

The triacetylenes shown below, yield upon reaction with hydrogen sulfide under weakly alkaline conditions, only the thiophene derivatives [75].

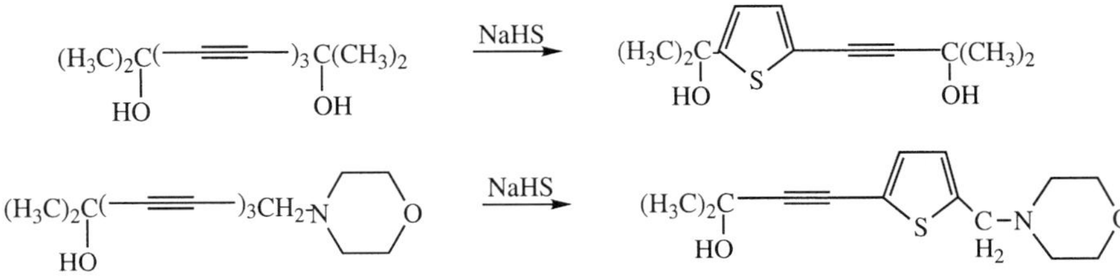

4C.2.6 Other methods

Methyl 2-thienylpropiolate is prepared in excellent yield by the addition of thiophene to trichlorocyclopropenium ion followed by reaction of the intermediate 2-thienyl dichlorocyclopropenium ion with methanol [76].

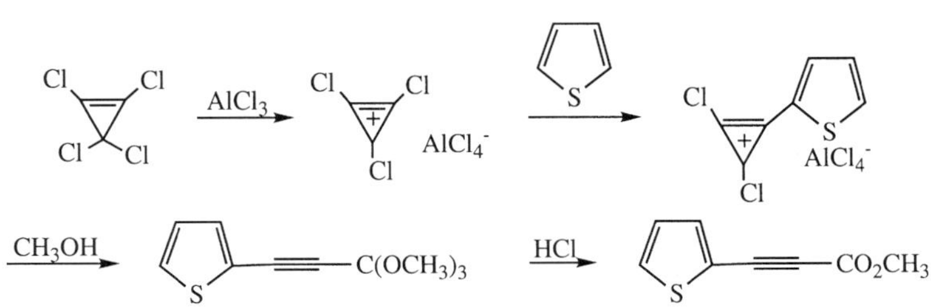

Methyl 2-thienylpropiolate [76]

To anhydrous aluminum chloride (1.4 g, 10 mmol) in dichloromethane (30 ml), tetrachlorocyclopropene (1.8 g, 10 mmol) is added. The resulting suspension is cooled to −30 °C and thiophene (84 mg, 10 mmol) in dichloromethane (10 ml) is added dropwise. The reaction mixture is allowed to warm to room temperature and then briefly heated to 50 °C in order to remove hydrogen chloride, after which it is cooled to 0 °C and vigorously shaken with ice-water. The phases are separated, the organic phase dried over magnesium sulfate and then treated with solid sodium bicarbonate (10 g) and methanol (10 ml) under stirring for 0.5 h. The reaction mixture is filtered, washed with 1 *M* hydrochloric acid and sodium bicarbonate solution and evaporated giving 1.10 g (95%) of the title compound mp 54–55 °C after recrystallization from methanol.

Halogen–metal exchange of 3′-iodo-2,2′:5′,2″-terthienyl with phenyllithium, which leads to ring opening, followed by alkylation with methyl iodide gives (*E*,*Z*)-1-methylthio-1,4-di(2-thienyl)-1-buten-3-yne [77].

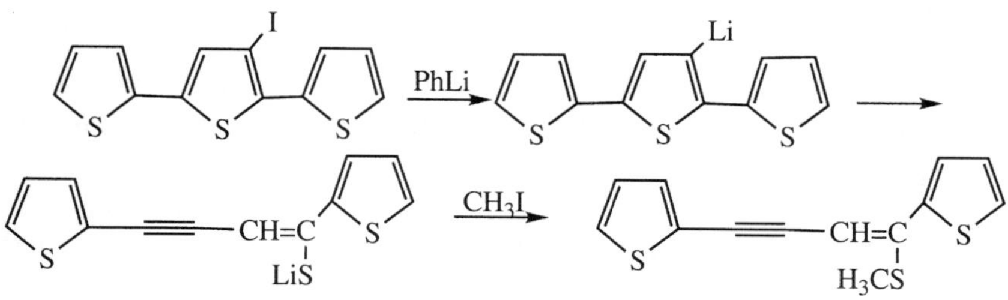

Electrophilic substitution on the thiophene ring of thienylacetylenes has not been very successful. However, functionalization can be achieved through metalation of the 5-position with butyllithium followed by reaction with electrophiles as in the synthesis of junipal by metalation of 2-propynylthiophene followed by reaction with *N*,*N*-dimethylformamide [34,78].

1-Methylthio-2-(5-hydroxymethyl)-2-thienyl)acetylene is prepared by metalation of 1-methylthio-2-(2-thienyl)acetylene followed by reaction with formaldehyde [50].

4C.3 COMPOUNDS ThC≡CAr, Ar=ARYL AND HETARYL

4C.3.1 By bromination and hydro-debromination and other methods

1-Phenyl-2-(2-thienyl)acetylene and 1,2-di-(2-thienyl)acetylene are prepared by the bromination of 1-phenyl-(2-thienyl)ethene and 1,2-di(2-thienyl)ethene followed by treatment with a strong base [35]. 2-Propinyl-5-(2-thienylethenyl)thiophene is prepared through the reaction of 5-(2-thienylethynyl)-2-thiophene aldehyde, zinc, carbon tetrabromide and triphenylphosphine followed by treatment with excess butyllithium and alkylation with methyl iodide [79].

2-Propinyl-5-(2-thienyl)ethenyl)thiophene [79]

Zinc (272 mg), carbon tetrabromide (1.38 g) and triphenylphosphine (1.1 g) are stirred in anhydrous dichloromethane (20 ml) for 48 h at room temperature. Then 5-(2-thienylethynyl)-2-thiophene aldehyde (456 mg, 2.1 mmol) is added. After 1 h the usual work-up gives a crude product that is dissolved in anhydrous diethyl ether (20 ml) and treated at −78 °C with 1.3 *M* butyllithium (4 ml). The reaction mixture is stirred at −78 °C for 1 h and then at 25 °C for 1 h. After addition of a saturated solution of ammonium chloride the mixture is extracted with diethyl ether. The neutral extracts are dried over sodium sulfate and evaporated, giving 358 mg of crude 2-ethinyl-5-(2-thienylethenyl)thiophene. Without further purification it is dissolved in anhydrous tetrahydrofuran (10 ml) and treated at 0 °C with 1.3 *M* butyllithium (1.3 ml) under nitrogen. In another flask magnesium (100 mg) is suspended in anhydrous diethyl ether and treated with methyl iodide (568 mg) dissolved in anhydrous diethyl ether. The mixture is stirred at 25 °C for 1 h. The boron trifluoride etherate (567 mg) is added dropwise to the Grignard reagent. The resulting gaseous trimethylborane is introduced to the reaction mixture containing the lithium acetylide. The reaction mixture is cooled to −78 °C and a solution of iodine (525 mg) in anhydrous diethyl ether (10 ml) is added during 0.5 h. After an additional 45 min at −78 °C, the reaction mixture is allowed to warm to room temperature. The solution is then washed twice with 3 *M* sodium hydroxide solution containing sodium thiosulfate to reduce unreacted iodine. The phases are separated and the aqueous phase extracted with diethyl ether. The combined organic phases are treated with 3 *M* sodium hydroxide solution (2.1 ml) followed by addition of 30% hydrogen peroxide (0.7 ml) and then saturated with potassium carbonate, separated, dried over potassium carbonate and evaporated. The crude product is purified by chromatography on silica gel using hexane as eluent giving 324 mg (85%) of the title compound as an oil.

4C.3.2 Ring-closure reactions

2-Phenylethinyl-5-phenylthiophene is obtained through the reaction of 1,6-diphenyltriacetylene with hydrogen sulfide under weakly alkaline conditions [80]. 1-(5-Vinyl-2-thienyl)-2-(2- thienyl)acetylene is prepared upon ring-closure of 1-(2-thienyl) octatriyn(1,3,5,)-7-ol with hydrogen sulfide [35].

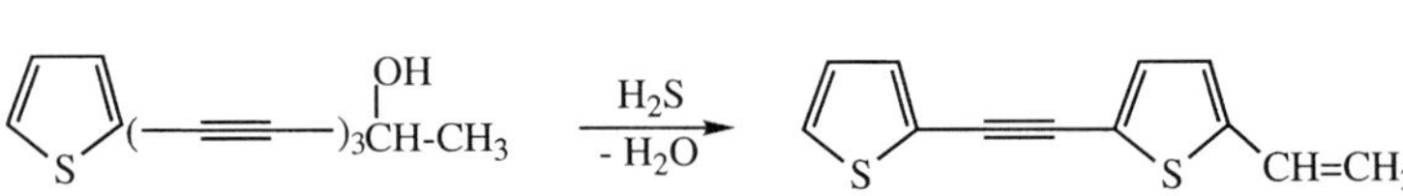

4C.3.3 Photochemical methods

2-Acetyl-5-(5′trimethylsilyl-2′-thienylethinyl)thiophene and 2-hydroxyacetyl-5-(5′-trimethylsilyl-2′-thienylethinyl) thiophene are obtained in 81% yield upon irradiation of 2-acetyl-5-iodothiophene and 2-hydroxyacetyl-5-iodothiophene, respectively in the presence of 2-ethynyl-5-trimethylsilylthiophene [81]. These compounds can be desilylated upon treatment with acid and the side chain of the latter product could be reduced to a glycol group [81].

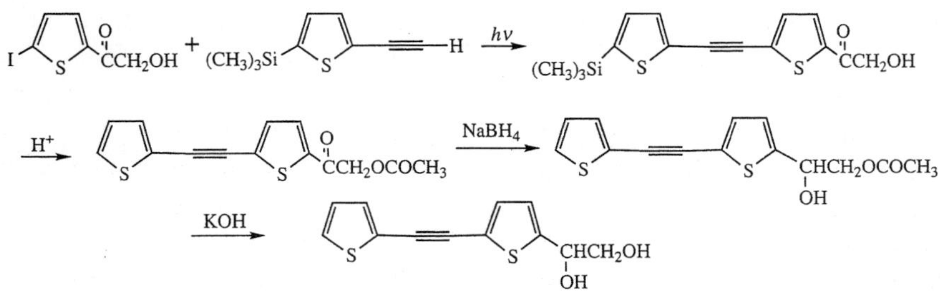

4C.3.4 Through palladium(0)-catalyzed reactions of iodothiophenes with terminal acetylenes (Sonogashira reaction)

1,2-Di(2-thienyl)acetylene is prepared in 68% yield by palladium(0)-catalyzed coupling of 2-iodothiophene with 1-(2-thienyl)-2-trimethylsilylacetylene [24]. Treatment of 2-iodothiophene with trimethylsilylacetylene in the presence of both tetrakis(triphenylphosphine)palladium(0) and cuprous iodide under phase transfer conditions at 40 °C, directly gives 1,2-di(2-thienyl)acetylene in 91% yield [82]. This method can also be used for the preparation of 1,2-di(2-bromo-3-thienyl)acetylene and 1,2-di(4-bromo-3-thienyl)acetylene [83]. Recently 2-trimethylsilylethenyl-5-dodecylthiophene was prepared from 2-iodo-5-dodecylthiophene and trimethylsilylacetylene by the Sonogashira reaction, the product was desilylated to 2-ethynyl-5-dodecylthiophene [84].

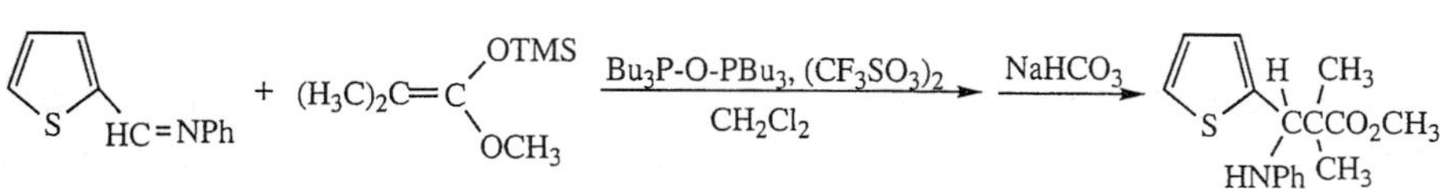

2-Trimethylsilylethenyl-5-dodecylthiophene [84]

A mixture of 2-iodo-5-dodecylthiophene (3.00 g, 7.94 mmol), bis(triphenylphosphinepalladium(II) chloride (147 mg, 0.209 mmol) and cuprous iodide

(79 mg, 0.41 mmol) is degassed, after which tetrahydrofuran (20.0 ml) and triethylamine (5.0 ml) are added. The mixture is stirred for 5 min and trimethylsilylacetylene (1.4 ml, 9.83 mmol) is added dropwise. The reaction mixture is stirred at room temperature overnight and poured into petroleum ether (200 ml). The precipitate formed is removed by filtration and the eluate distilled at reduced pressure. The residue is purified by chromatography on silica gel using petroleum ether as eluent giving 2.61 g (95%) of the title compound as a colorless oil.

Alternatively 5-trimethylsilyl-2-iodothiophene can be reacted with trimethysilylacetylene to give the 1,2-di-(5′-trimethylsilyl-2′-thienyl)acetylene, which upon treatment with potassium hydroxide is desilylated in 92% yield. Similarly 1,2-di-(5′-ethyl-2′-thienyl)acetylene was prepared in 75% yield [82]. This reaction does not work with bromothiophenes. Heck coupling of 3-iodo-4-trimethylsilylthiophene with di(tributylstannyl)acetylene is a good method for the preparation of bis[4-(trimethylsilyl)-3-thienyl]acetylene [21].

Bis[4-(trimethylsilyl)-3-thienyl]acetylene [21]

A mixture of 4-iodo-3-(trimethylsilyl)thiophene (70.5 mg, 0.25 mmol), bis(tributylstannyl)acetylene (75.5 mg, 0.125 mmol) and tetrakis(triphenylphosphine)palladium(0) (29 mg, 0.025 mmol) in dioxane (3 ml) and triethylamine (1 ml) is heated at 90 °C for 8 h under nitrogen. The reaction mixture is cooled, diluted with diethyl ether (15 ml), washed with water (2 × 2.5 ml), dried over magnesium sulfate and evaporated. The residue is purified by chromatography on silica gel (20 g) using hexane containing 2% triethylamine as eluent, yielding 23.5 mg (56%) of the title compound as pale yellow crystals mp 69–71 °C.

Excellent yields and reaction times of 5–25 min have beeen achieved in the Sonogashira reaction of 2- and 3-iodothiophene and 3-bromothiophene with trimethylsilylacetylene by using controlled microwave heating [85].

2-Trimethylsilylethynylthiophene [85]

2-Iodothiophene (189 mg, 0.90 mmol), bis(triphenylphosphine)palladium(II) chloride (12.8 mg, 0.02 mmol), cuprous iodide (6.9 mg, 0.04 mmol), trimethylsilylacetylene (0.14 ml, 1.00 mmol), diethylamine (1.5 ml, 13.60 mmol) and *N*,*N*-dimethylformamide (0.5 ml) are stirred under nitrogen with a magnetic bar in a heavy-walled Smith process vial at 120 °C for 5 min in a microwave cavity. The microwave heating is performed in a Smith Synthesizer singlemode cavity producing continuous irradiation at 2450 MHz (Personal Chemistry AB,

Uppsala, Sweden). The temperature, pressure and irradiation power are monitored during the course of the reaction. The average pressure during the reaction is 3–4 bar. After completed irradiation the reaction tube is cooled with high pressure air until the temperature has fallen below 39 °C. The mixture is then poured into 0.1 *M* hydrochloric acid (5–10 ml) and the product extracted with diethyl ether (3 × 5–10 ml). The combined organic phases are washed with saturated sodium carbonate solution and water, and evaporated. The residue is taken up in pentane and filtered through Celite and the filtrate evaporated, giving 139 mg (86%) of the title compound.

The Sonogashira reaction of two equivalents of 3-bromo-2-iodothiophene with 1,4-diethynylbenzene and 1,4-diethenyl-2,3,5,6-tetrafluorobenzene gives 1,4-di(3-bromo-2-thienylethynyl)benzene and 1,4-di(3-bromo-2-thienylethynyl)- 2,3,5,6-tetrafluorobenzene which by halogen–metal exchange and reaction with iodine gives 1,4-di(3-iodo-2-thienylethynyl)benzene. Renewed Sonogashira reaction with trimethylsilylacetylene followed by removal of the trimethylsilyl group gave the tetraacetylenic compound [86].

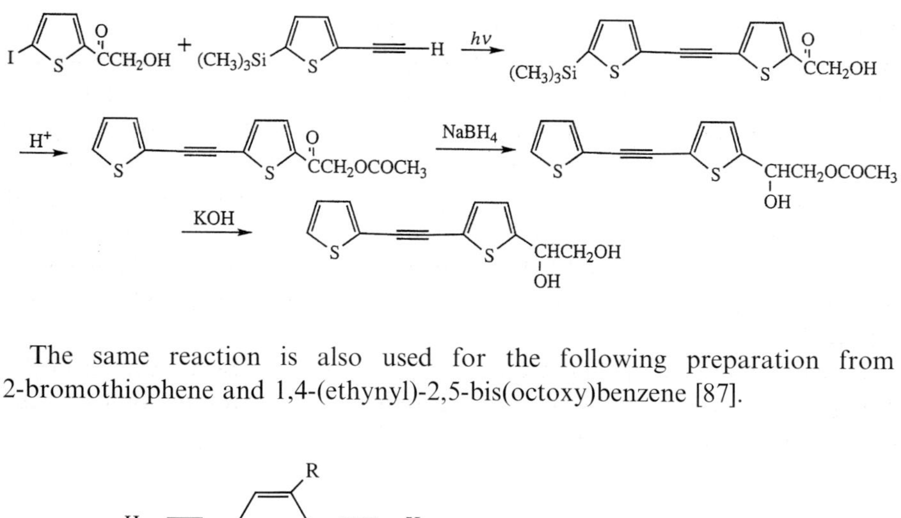

The same reaction is also used for the following preparation from 2-bromothiophene and 1,4-(ethynyl)-2,5-bis(octoxy)benzene [87].

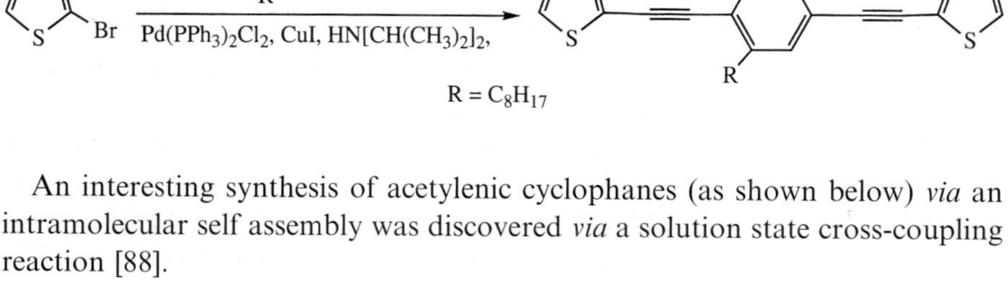

An interesting synthesis of acetylenic cyclophanes (as shown below) *via* an intramolecular self assembly was discovered *via* a solution state cross-coupling reaction [88].

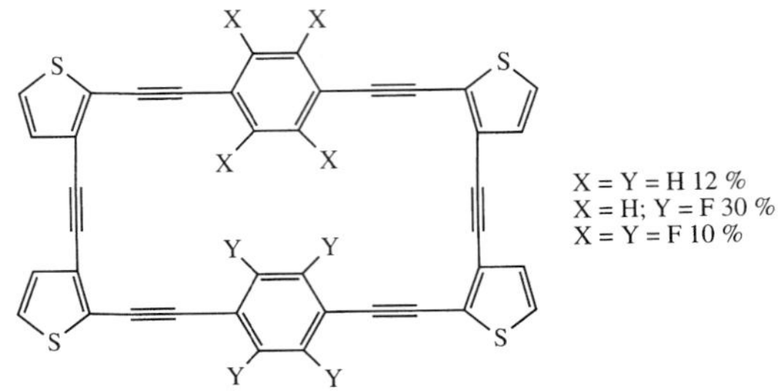

3-Alkynyl-4-phenylthiophenes are readily obtained in high yields by the Sonogashira coupling reaction between 3-iodo-4-phenylthiophene and terminal acetylenes [21].

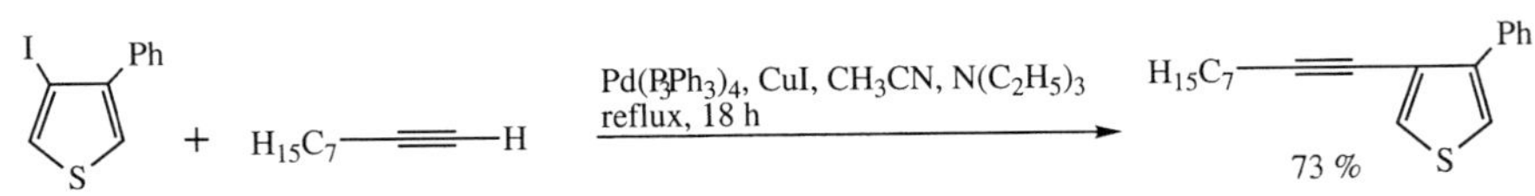

Iododesilylation of 3-(phenylethynyl)-4-trimethylsilylthiophene with iodine and silver triflate in tetrahydrofuran/methanol occurs without serious interference from the triple bond yielding 3-iodo-4-phenylethynylthiophene, which then by Sonogashira coupling with 1-nonyne and 3-butyn-1-ol gives (3-nonyn-1-yl)-4-(phenylethynyl)thiophene and 3-(4′-hydroxybutyn-1′-yl)-4-(phenylethynyl)thiophene, respectively, in excellent yields [21].

This coupling has also been used between silyl-protected 4,5-diiodocatechol and 3-ethynylthiophene for the preparation of 1,2-bis(*tert*-butyldimethylsilyloxy)-4,5-bis(3-ethynylthiophene-yl)benzene, in connection with synthesis of platinum catecholate complexes containing two 3-ethynylthiophene groups [89].

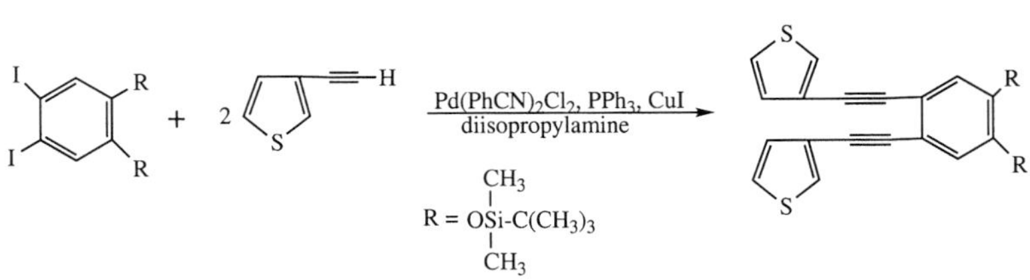

Bis(tert-butyldimethylsilyloxy)-4,5-bis(3-ethynylthiophene-yl)benzene [89]

Diisopropylamine (150 ml) is added to dichlorobis(benzonitrile)palladium(II) (0.4547, 1.2 mmol), triphenylphosphine (0.624, 2.4 mmol), cuprous iodide (0.1142, 0.6 mmol) and 1,2-bis(*tert*-butyldimethylsilyloxy)-4,5-diiodobenzene (7.0 g, 12 mmol). To this solution 3-ethynylthiophene is added slowly using a syringe. The reaction mixture is heated to 60 °C for 1 h and then cooled to room temperature and stirred for 11 h. The mixture is filtered and the solvent evaporated. Chromatography on silica gel using dichloromethane/hexane (1:9) gives 4.046 g (62%) of the title compound as a yellow powder.

In connection with work on the synthesis of photochromic 3*H*-naphtho[2,1-*b*]pyrans, the Sonogashira reaction was used under phase transfer conditions between 2-iodothiophene and the protected terminal acetylene 8-(3-hydroxy-3-methylbut-1-ynyl)-3,3-diphenyl-3*H*-naphtho[2,1-*b*]pyran in the presence of benzyltriethylammonium chloride, 5.5 *M* sodium hydroxide as base, benzene as solvent and a mixture of tetrakis(triphenylphosphine)palladium(0) as catalyst. The same method is used for the following preparation from 2,5-diiodothiophene [90].

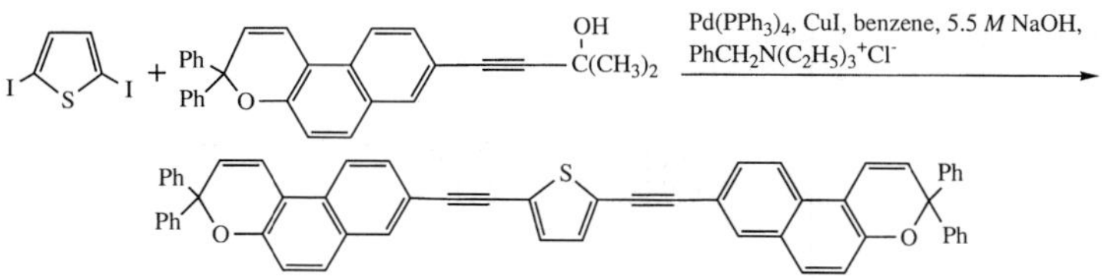

In connection with work on photoswitchable tetraethynylethene-dihydroazulene chromophores the Sonogashira reaction was used for the reaction route using 2-(4-iodophenyl)-8a*H*-azulene-1,1-dicarbonitrile [91].

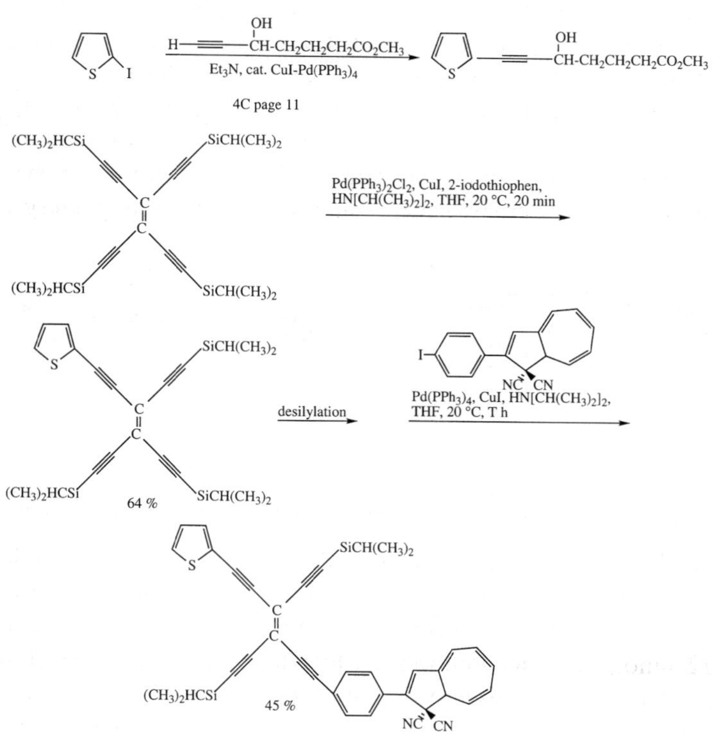

In connection with the study on the synthesis of oligo(2,5-thiophene-ethynylene)s the Sonogashira coupling of 5-iodo-3-ethyl 2-thienylacetylene is used for the preparation of the dimer, which by iodination and renewed Sonogashira coupling gives the tetramer shown below [65].

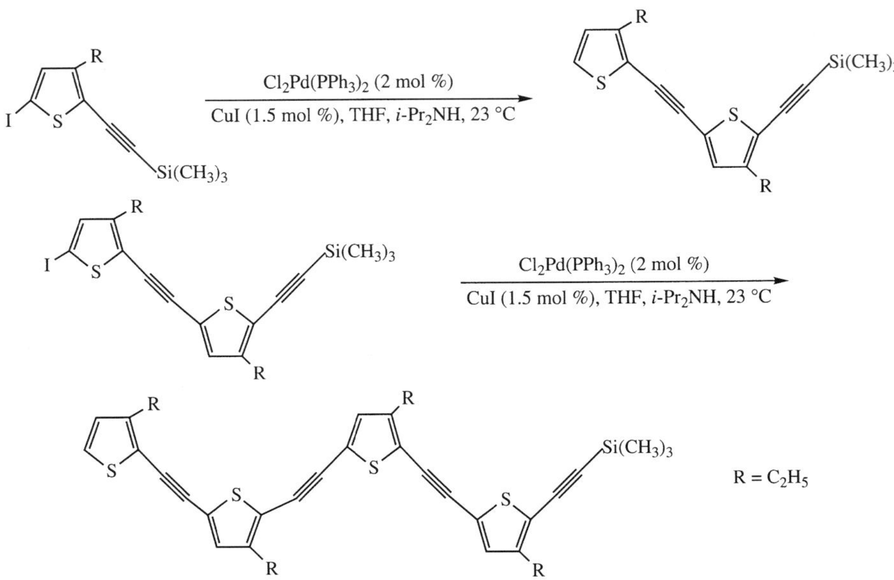

The ligand-free palladium acetate-catalyzed coupling of phenylacetylene with 2-(thienyl)phenyliodonium tosylate, in the presence of sodium bicarbonate in aqueous media, (acetonitrile/water, 4:1) yields only 1-phenyl-2-(2-thienyl)acetylene in 89% yield. At room temperature also other terminal alkynes have sucessfully been used with the same excellent yields [93].

Palladium-catalyzed coupling reactions of 3-hexyl-2,5-diiodothiophene with 1,4-diethynylbenzene or 2,5-diethynylpyridine can be used for the preparation of π-conjugated soluble poly(arylenethienyl) type polymers shown below [93].

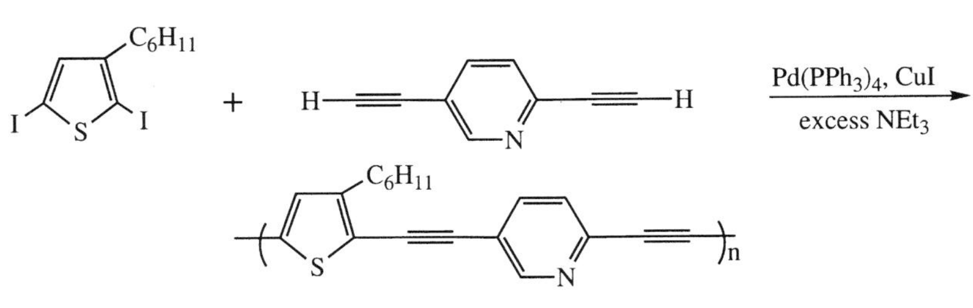

4C.3.5 Electrophilic substitution, metalation and halogen–metal exchange on thienyl acetylenes

The reaction of 3-ethynylthiophene with butyllithium and potassium *tert*-butoxide followed by magnesium bromide in ether and iodine and in tetrahydrofuran gives 3-ethynyl-2-iodothiophene and 3-ethynyl-5-iodothiophene in a 3:1 ratio, which can be chromatographically separated [17]. Reflux of the copper(I) salt of 3-ethynyl-2-iodothiophene in pyridine yields trithienocyclotriyne shown below [17].

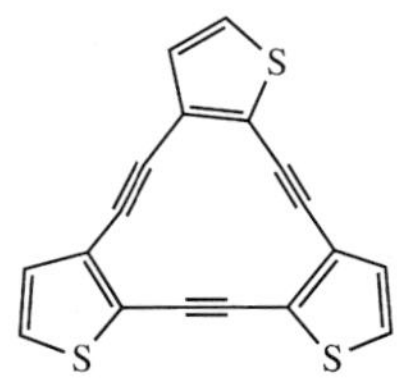

The naturally occurring *O*-benzyl derivative of the naturally occurring thiophene derivative, ineupatoriol, is prepared starting from junipal, by modifying the aldehyde group [94].

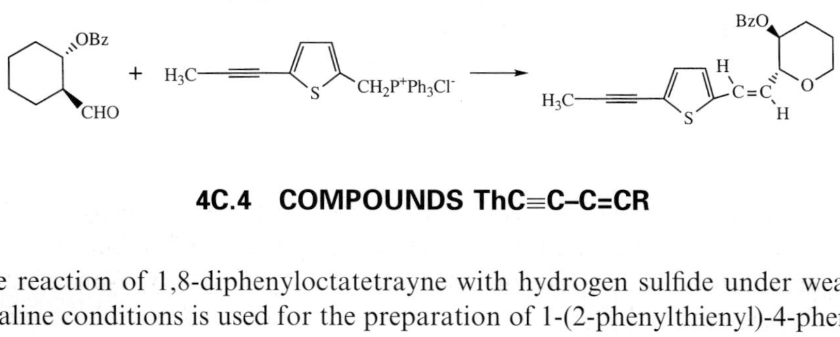

4C.4 COMPOUNDS ThC≡C–C=CR

The reaction of 1,8-diphenyloctatetrayne with hydrogen sulfide under weakly alkaline conditions is used for the preparation of 1-(2-phenylthienyl)-4-phenyl-1,3-butadiyne [80].

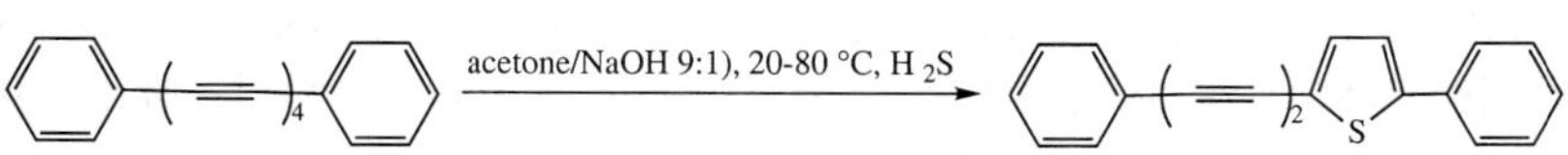

The following reaction sequences have been performed [16].

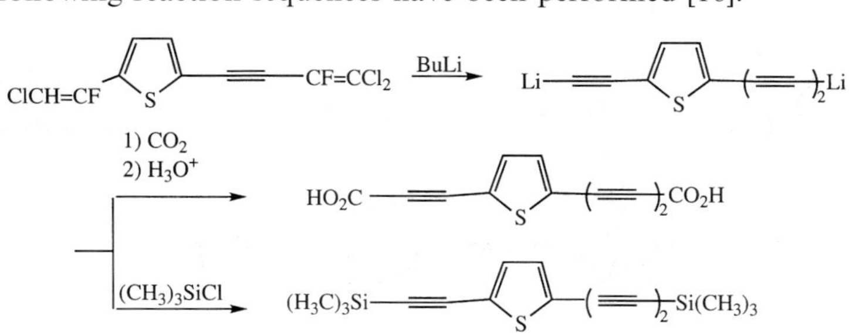

The Cadot–Chodkiewicz reaction, consisting in the coupling of a thienylacetylene with a thienyl bromoacetylene using cuprous chloride, hydroxylamine hydrochloride and ethylamine in methanol is an excellent method for the preparation of 1,4-di(thienyl)diacetylenes. Thus 1,4-di(2-thienyl)diacetylene [25], 1-(2-thienyl)-4-(3-thienyl)diacetylene [12] and 1-(2-thienyl)-4-(2,2′-bithienyl-5-)diacetylene [9] are prepared by the reaction of 1-(bromo)-2-(2-thienyl)acetylene with 2- and 3-thienylacetylene and 5-ethinyl-2,2′-bithienyl, respectively, and other derivatives [27].

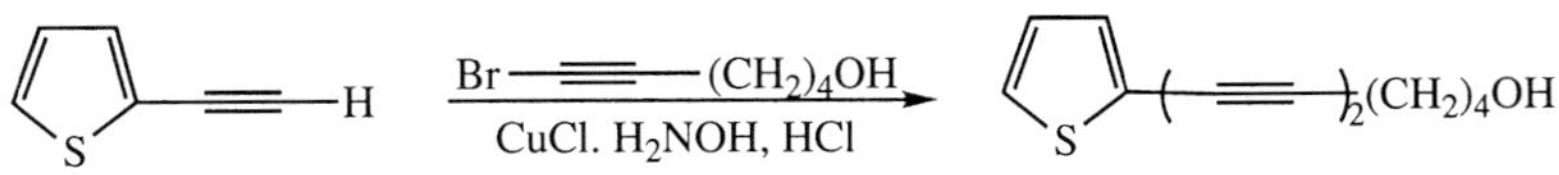

Other bromo acetylenes can also be used for the preparation of thienyltriacetylenes [35].

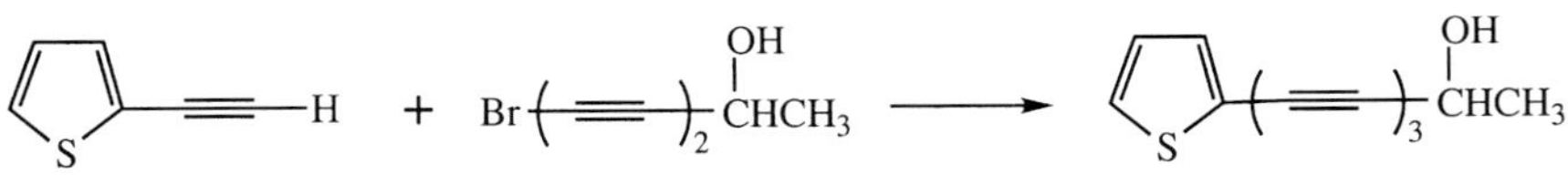

Another often used method for the preparation of 1,4-di(thienyl)diacetylene is the reaction of thienylacetylenes with potassium ferricyanide [95] or cupric chloride [96] with oxygen in the presence of copper (Glaser coupling) [8,9,13,16,26,95–97]. Most of the 1,4-di(thienyl)diacetylenes are used for the preparation of terthienyls through their reaction with hydrogen sulfide under weakly alkaline conditions [8,9,13,25,98]. The Glaser coupling can also be used for unsymmetrical coupling [26]. The reaction of 2-thienylacetylene with palladium(0), cuprous iodide, triethylamine and chloroacetone in benzene gives 1,4-di(2-thienyl)butadiyne [25].

1,4-Bis(2-thienyl)butadiyne [25]

Triethylamine (10.12 g, 0.1 mol) is added under nitrogen to a stirred mixture of tetrakis(triphenylphosphine)palladium(0) (1.17 g, 1.01 mmol) and cuprous iodide (0.72 g, 3.79 mmol) in anhydrous benzene (20 ml). This mixture is sequentially treated with chloroacetone (4.63 g, 0.050 mol) and ethynylthiophene (5.4 g, 50 mmol). After stirring at room temperature for 16 h, an aqueous saturated solution of ammonium chloride is added. The phases are separated and the aqueous phase extracted with benzene. The combined organic phases are filtered and concentrated. The residue is purified by chromatography on silica gel using hexene as eluent giving 87% of the title compound mp 92.5–93 °C.

Another approach for the preparation of 1,4-di(thienyl)diacetylene using an organoborane-mediated approach is useful for the preparation of unsymmetrically substituted diacetylenes. The lithium salt of each acetylenic reagent are introduced successively onto a boron compound to form a complex, which is coupled to the diacetylene by iodine oxidation. In this way 1-(2-thienyl)-4-(2,2′-bithienyl-5)diacetylene is prepared [9].

An elegant synthesis of dehydrothieno[18]annulenes has been achieved by extensive use of palladium/copper alkyne cross-coupling stategy. The compound below is prepared from 3-iodo-2-bromothiophene by palladium/copper cross-coupling first with 1-(trimethylsilyl)buta-1,3-diyne followed by coupling with tri(isopropylsilyl)ethyne.

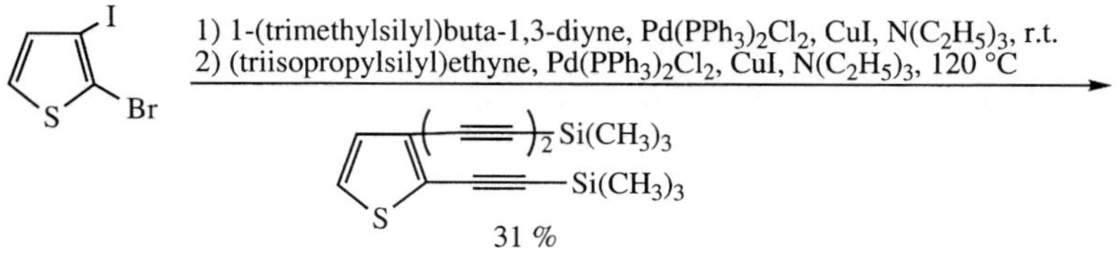

Coupling with 3-bromo-2-iodothiophene yields the α,ω-bisprotected polyyne. Removal of the protecting groups with and subsequent cyclization using cuprous chloride/cupric acetate in pyridine gives the cyclic compound. Analogs of the cyclic compound have been obtained by the same strategy [99].

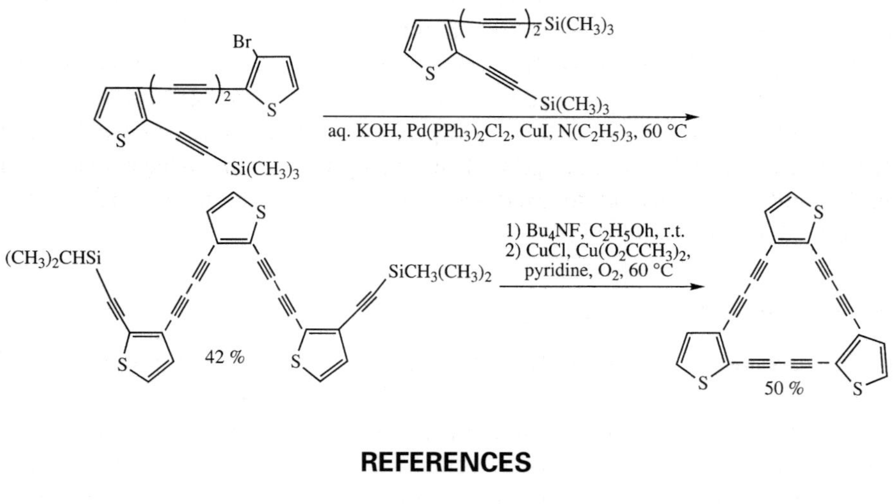

REFERENCES

1. A. Vaitiekunas and F. F. Nord, *J. Org. Chem.* **19**, 902 (1954).
2. K. E. Schulte and G. Bohn, *Arch. Pharm.* **297**, 179 (1964).
3. A. J. Osbahr, A. Vaitiekunas and F. F. Nord, *J. Am. Chem. Soc.* **77**, 1911 (1955).
4. J. Kagan, S. K. Arora, M. Bryzgis, S. N. Dhawan, K. Reid, S. P. Singh and L. Tow, *J. Org. Chem.* **48**, 703 (1983).
5. J. C. Craig and M. Moyle, *J. Chem. Soc.* 1907 (1963).

6. M. D'Auria, A. De Mico, F. D'Onifri and G. Piancatelli, *Gazz. Chim. Ital.* **116**, 747 (1986).
7. M. D'Auria, A. De Nico, F. D, Onofrio and G. Piancatelli, *J. Org. Chem.* **52**, 5243 (1987).
8. J. P. Beny, S. N. Dhawan, J. Kagan and S. Sundlass, *J. Org. Chem.* **47**, 2201 (1982).
9. J. Kagan and S. K. Aurora, *J. Org. Chem.* **48**, 4317 (1983).
10. M. D'Auria, A. Demico, F. D'Onofrio and G. Piancatelli, *Synth. Commun.* **17**, 491 (1987).
11. M. D'Auria, A. De Mico, F. D'Onofrio and G. Piancatelli, *J. Chem. Soc., Perkin Trans. 1* 1777 (1987).
12. J. Kagan, S. K. Aurora, I. Prakash and A. Ustunol, *Heterocycles* **20**, 1341 (1983).
13. D. M. Perrine and J. Kagan, *Heterocycles*. **24**, 365 (1986).
14. C. Troyanowsky, *Bull. Soc. Chim Fr.* 424 (1955).
15. K. Okuhara, *J. Org. Chem.* **41**, 1487 (1976).
16. K. Okuhara, *Bull. Chem. Soc. Jpn.* **54**, 2045 (1981).
17. D. Sooloki, J. D. Bradshaw, C. A. Tessier and W. J. Youngs, *Organometallics* **13**, 451 (1994).
18. S. Yasuike, J. Kurita and T. Tsuchiya, *Heterocycles* **45**, 1991 (1997).
19. Y. Sugihara, T. Yagi and I. Murata, *J. Am. Chem. Soc.* **114**, 1479 (1992).
20. J. P. Genet, E. Blart and M. Savignac, *Synlett.* 715 (1992).
21. X.-S. Ye and H. N. C. Wong, *J. Org. Chem.* **62**, 1940 (1997).
22. A. Buttinellio, E. Viola, E. Antonelli and C. Lo Sterzo, *Organometallics* **17**, 2574 (1998).
23. M. Iyoda, A. Vorasingha, Y. Kuwatami and M. Yoshida, *Tetrahedron Lett.* **39**, 4701 (1998).
24. R. Rossi, A. Carpita and A. Lezzi, *Tetrahedron* **40**, 2773 (1984).
25. A. Carpita, R. Rossi and C. A. Veracini, *Tetrahedron* **41**, 1919 (1985).
26. T. X. Neenan and G. M. Whitesides, *J. Org. Chem.* **53**, 2489 (1988).
27. A. Sarkar, S. Okada, H. Matsuda and H. Nakanishi, *Chem. Lett.* 1073 (1998).
28. J. Bussenius, N. Laber, T. Müller and W. Eberbach, *Chem. Ber.* **127**, 247 (1994).
29. R. E. Atkinson, R. F. Curtis and D. M. Jones, *Chem. Commun.* 718 (1967).
30. R. E. Atkinson, R. F. Curtis D. M. Jones and J. A. Taylor, *J. Chem. Soc. C* 2173 (1969).
31. F. Bohlmann, F. Zdero and H. Kapteyn, *Chem. Ber.* **106**, 2755 (1973).
32. T. B. Patrick, J. M. Disher and W. J. Probst, *J. Org. Chem.* **37**, 4467 (1972).
33. C. Wentrup and H. W. Winter, *Angew. Chem.* **90**, 643 (1978).
34. L. Skattebøl, *Acta Chem. Scand.* **13**, 1460 (1959).
35. F. Bohlmann and P. Herbst, *Chem. Ber.* **95**, 2945 (1962).
36. F. Bohlmann, K. M. Kleine and C. Arndt, *Chem. Ber.* **97**, 2125 (1964).
37. A. Carpita, D. Neri and R. Rossi, *Gazz. Chim. Ital.* **117**, 481 (1987).
38. K. Schlögl and H. Pelousek, *Monatsh. Chem.* **92**, 51 (1961).
39. F. Bohlmann and C. Huhn, *Chem. Ber.* **110**, 1183 (1977).
40. F. Bohlmann and J. Kocur, *Chem. Ber.* **107**, 2115 (1974).
41. M. C. Verploegh, L. Donk, H. J. T. Bos and W. Drenth, *Rec. Trav. Chim. Pays-Bas*, **90**, 765 (1971).
42. J. K. Stille and J. H. Simpson, *J. Am. Chem. Soc.* **109**, 2138 (1987).
43. D. G. Fletcher, K. H. Gibson, H. R. Moss, D. R. Sheldon and E. R. H. Walker, *Prostaglandins* **12**, 493 (1976).
44. M. Janda and F. Dvorak, *Coll. Czech. Chem. Commun.* **27**, 372 (1962).
45. H. Keskin, R. E. Miller and F. F. Nord, *J. Org. Chem*, **16**, 199 (1951).
46. D. Vegh, J. Kovac and M. Dandarova, *Tetrahedron Lett.* **21**, 969 (1980).
47. E. J. D. Brown and J. Harley-Mason, *J. Chem. Soc. C* 1390 (1966).
48. M. Cariou, *Bull. Soc. Chim. Fr.* 651 (1979).
49. Y. Xin, X. Wu and Y. Shen, *J. Fluorine Chem.* **40**,15 (1988).
50. R. Raap, *Can. J. Chem.* **49**, 2155 (1971).
51. R. Raap and R. G. Micetich, *Can. J. Chem.* **46**, 1057 (1968).
52. J. Goselck, L. Beress, H. Schenk and G. Schmidt, *Angew. Chem., Int. Ed.* **4**, 1080 (1965).
53. R. E. Atkinson, R. F. Curtis and G. T. Phillips, *Tetrahedron Lett.* **5**, 3159 (1964).
54. R. F. Curtis and J. A. Taylor, *J. Chem. Soc. C* 1813 (1969).

55. T. V. Brown, W. Carruthers and M. G. Oellatt, *J. Chem. Soc., Perkin Trans. 1* 483 (1982).
56. F. Bohlmann, M. Wotschokowsky, U. Hinz and W. Lucas, *Chem. Ber.* **99**, 984 (1966).
57. R. E. Atkinson and R. F. Curtis, *Tetrahedron Lett.* **6**, 297 (1965).
58. M. D. Rausch, A. Siegel and L. P. Klemann, *J. Org. Chem.* **31**, 2703 (1966).
59. R. E. Atkinson, R. F. Curtis and J. A. Taylor, *J. Chem. Soc. C* 578 (1967).
60. F. Bohlmann and W. Skuballa, *Chem. Ber.* **106**, 497 (1973).
61. F. Bohlmann and P. D. Hopf, *Chem. Ber.* **106**, 3621 (1973).
62. A. Carpita, A. Lezzi, R. Rossi, F. Marchetti and S. Merlino, *Tetrahedron* **41**, 621 (1985).
63. D. Brown, J. Craig, N. H. Dyson, and J. W. Westley, *J. Chem. Soc.* 89 (1966).
64. J. O. Karlsson, A. Svensson and S. Gronowitz, *J. Org. Chem.* **49**, 2018 (1984).
65. D. L. Pearson and J. M. Tour, *J. Org. Chem.* **62**, 1376 (1997).
66. S. J. Phytian, R. J. K. Taylor and J. R. Bantick, *J. Chem. Soc., Perkin Trans. 1* 194 (1990).
67. W. Eberbach and N. Laber, *Tetrahedron Lett.* **33**, 61 (1992).
68. W. Eberbach, N. Laber, J. Bussenius, H. Fritz and G. Rihs, *Chem. Ber.* **126**, 975 (1993).
69. A. Padwa, U. Chiacchio, D. J. Fairfax, J. M. Kassir, A. Litrico, M. A. Semones and S. L. Xu, *J. Org. Chem.* **58**, 6429 (1993).
70. M. Fosatelli, A. C. T. M. van der Kerk, S. F. Vasilevsky and L. Brandsma, *Tetrahedron Lett.* **33**, 4229 (1992).
71. K. Tamao, S. Yamaguchi, M, Shiozaki, Y. Nakagawa and Y. Ito, J. *Am. Chem. Soc.* **114**, 5867 (1992).
72. Y. Hatanaka, K. Matsui and T. Hiyama, *Tetrahedron Lett.* **30**, 2403 (1989).
73. R. R. Tykwinski, A. Hilger, F. Diederich, H. P. Lüthi, P. Seiler, V. Gramlich, J.-P. Gisselbrecht, C. Boudon and M. Gross, *Helv. Chim. Acta* **83**, 1484 (2000).
74. L. Bleicher and N. D. P. Cosford, *Synlett.* 1115 (1995).
75. A. S. Zanina, G. N. Khabibulina, V.V. Legkoderia and I. I. Kotlyarevskii, *Zh. Org. Khim.* **8**, 1527 (1972).
76. D. H. Wadworth, S. M. Geer and M. R. Detty, *J. Org. Chem.* **52**, 3662 (1987).
77. S. Gronowitz and A. Svensson, *Isr. J. Chem.* **27**, 25 (1986).
78. P. J. Kocienski, J, M. Ansell and B. E. Cross, *J. Org. Chem.* **41**, 3650 (1976).
79. M. D'Auria, A. De Mico, F. D'Onofrio and G. Piancatelli, *Gazz. Chim. Ital.* **119**, 201 (1989).
80. K. E. Schulte, J. Reisch and L. Hoerner, *Chem. Ber.* **95**, 1943 (1962).
81. M. D'Auria and D. Tofani, *Tetrahedron* **42**, 9315 (1992).
82. M. D'Auria, *Synth. Commun.* **22**, 2393 (1992).
83. S. Yasuike, F. Nakashima, J. Kurita and T. Tsuchiya, *Heterocycles* **45**, 1899 (1997).
84. Y. Geng, A. Fechtenkötter and K. Müllen, *J. Mater. Chem.* **11**, 1634 (2001).
85. M. Erdélyi and A. Gogoll, *J. Org. Chem.* **66**, 4165 (2001).
86. M. J. Marsella, Z.-Q. Wang, R. J. Reid and K. Yoon, *Org. Lett.* **3**, 885 (2001).
87. J.-P. Lère-Porte, J. J. E. Moreau, F. Serein-Spireau and S. Wakim, *Tetrahedron Lett.* **42**, 3073 (2001).
88. M. J. Marsella, Z.-Q. Wang, R. J. Reid and K. Yoon, *Org. Lett.* **3**, 885 (2001).
89. J. D. Kinder and W. J. Youngs, *Organometallics* **15**, 460 (1996).
90. M. Frigoli, C. Moustrou, A. Samat and R. Guglielmetti, *Helv. Chim. Acta* **83**, 3043 (2000).
91. L. Gobbi, P. Seiler, F. Diederich, V. Gramlich, C. Boudon, J.-P. Gisselbrecht and M. Gross, *Helv. Chim. Acta* **84**, 743 (2001).
92. S.-K. Kang, H.-W. Lee, S.-B Jang and P.-S. Ho, *Chem. Commun.* 835 (1996).
93. M. Takagi, K. Kizu, T. Maruyama, K. Kubota and T. Yamamoto, *Chem. Lett.* 913 (1993).
94. I. W. J. Still, N. S. Banait and D. V. Frazer, *Synth. Commun.* **18**, 1461 (1988).
95. W. T. Smith, Jr and R. W. Shelton, *J. Am. Chem. Soc.* **76**, 2731 (1954).
96. P. J. Elving and C. M. Callahan, *J. Am. Chem. Soc.* **77**, 2077 (1955).
97. F. Bohlmann and A. Suwita, *Chem. Ber.* **108**, 515 (1975).
98. J. Kagan, S. K. Aurora, I. Prakash and A. Ustunol, *Heterocycles* **20**, 1341 (1988).
99. A. Sarkar and M. M. Haley, *Chem. Commun.* 1733 (2000).

– 4D –
Arylthiophenes

4D.1 INTRODUCTION

In this chapter, the preparation of simple arylthiophenes, functionalized in the aryl group and containing alkyl, vinyl and acetylenic groups in the thiophene ring will be discussed.

Arylthiophenes were previously mainly prepared:

(1) By aromatization of cyclohexeneylthiophenes, obtained by reaction of thienyllithium or thienylmagnesium derivatives with cyclohexanones, followed by aromatization or by various ring-closure reactions, which is especially useful for preparation of arylthiophenes functionalized in the thiophene ring.
(2) Ring-closure reactions forming the thiophene ring. Through this method arylthiophenes, highly functionalized in the thiophene ring, are obtained.
(3) During the last 20 years, most of these methods have been superseded by transition metal-catalyzed cross-coupling reactions.

4D.2 REACTION OF THIOPHENEMAGNESIUM HALIDES OR THIENYLLITHIUM DERIVATIVES WITH CYCLOHEXANONES FOLLOWED BY AROMATIZATION

Both 2- and 3-phenylthiophene have been prepared by the reaction of 2- and 3-thienyllithium with cyclohexanone followed by aromatization of the intermediate cyclohexenylthiophenes with chloranil [1,2]. Another way of forming a highly substituted aryl ring of an arylthiophene is by the refluxing of 2-cyano-3-(2-thienyl)crotonitrile with benzaldehydes and malonitrile in ethanol in the presence of catalytic amounts of piperidine, leading to 3-amino-2,4-dicyano-5-thienylbiaryls [3].

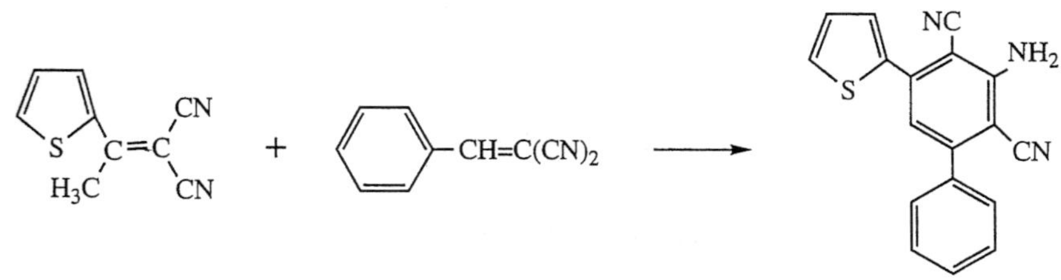

1-Phenyl-3-amino-2,4-dicyano-5-(2-thienyl)benzene [3]

A mixture of 2-cyano-3-(2-thienyl)crotonitrile (1.74 g, 0.01 *M*), benzaldehyde (1.06 g, 0.01 *M*) and malonitrile (0.68 g, 0.01 *M*) is refluxed in ethanol (30 ml) containing piperidine (0.1 ml) for 2 h. The solid formed is collected by filtration and recrystallized from *N,N*-dimethylformamide giving 2.56 g (85%) mp 235°C.

Another example of the synthesis of highly substituted artylthiophenes starting from thiophene derivatives is the preparation of ethyl 4-(2-thienyl)-2-hydroxy-6-methylthiobenzoate by condensation of the 2-thienyl α-oxoketenedithioacetal with excess of Reformatsky reagent derived from ethyl bromoacetate [4].

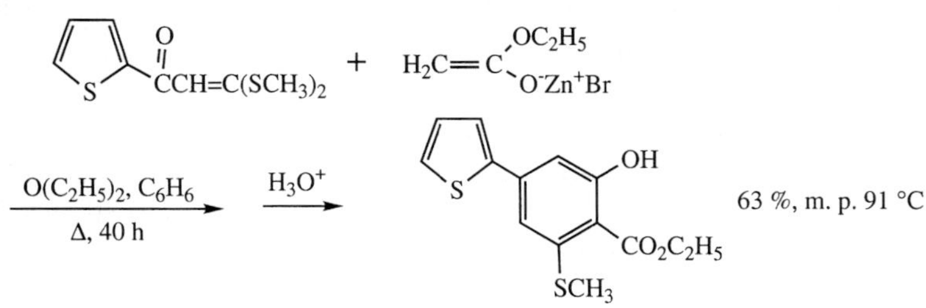

Upon reaction with 3-methyl-5-lithiomethylisoxazole the same α-oxoketenedithioacetal gives 3-methyl-5-(2-thienyl)-4-methylthiobenzisoxazole [5].

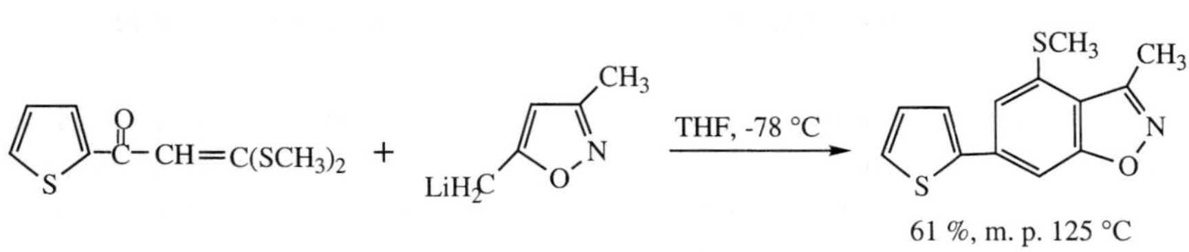

4D.3 RING-CLOSURE REACTIONS LEADING TO ARYLTHIOPHENES

2-Phenylthiophene can be prepared in two steps through the reaction of acetophenone with lithium hexadimethylsilazane, zinc dichloride and

O-ethyl-[(2-oxoethyl)thio]thioformate followed by splitting of the ethoxythiocarbonyl protecting group by treatment with 1-methyl piperazine, followed by ring closure with hydrochloric acid [6].

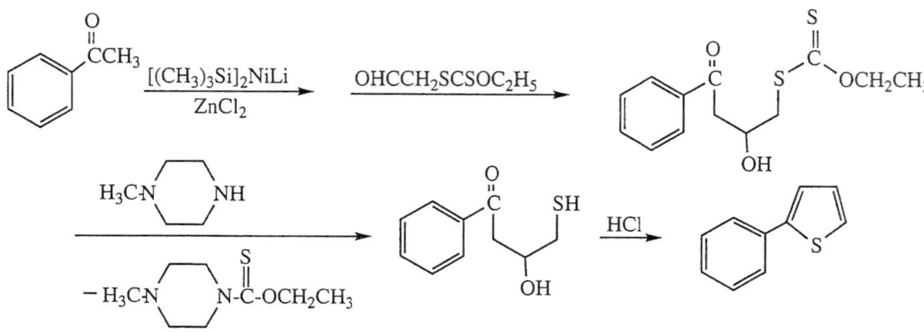

A modification of the Knorr–Paal reaction, consisting in phase transfer catalyzed cyclization of disodium 2-arylsuccinates in *ortho*-dichlorobenzene, using a mixture of diphosphorus pentasulfide and red phosphor, is a highly yielding method (over 80%) for the large scale preparation of 3-arylthiophenes [7]. Later, quantitative yields were obtained using crown ether together with onium salts as phase transfer cocatalysts [8].

3-(4-Methylphenyl)thiophene [8]

A suspension of disodium 2-(4-methylphenyl)succinate (5 g, 0.02 *M*), diphosphorus pentasulfide (8 g, 0.04 *M*), red phosphorus (2.5 g, 0.08 *M*), triethylbenzylammonium chloride (0.273 g, 6 mol.%) and 18-crown-6 (0.32 g, 6 mol.%) in *ortho*-dichlorobenzene (50 ml) and water (2 ml) is slowly heated to 130–135°C with constant stirring for 12 h. *ortho*-Dichlorobenzene is removed under reduced pressure, toluene (50 ml) is added followed by the addition of water (50 ml) and 10% aqueous sodium hydroxide solution to pH = 7. The reaction mixture is kept at 70°C for 1 h, cooled and the phases are separated. The aqueous phase is extracted with toluene (2 × 15 ml). The combined organic phases are washed with water, dried and evaporated. The residue is filtered through a pad of silica gel using hexane as eluent giving 3.4 g (98%) of the title compound mp 112–113°C.

Reacting 1,4-dicarbonyl compounds such as 1,2-dibenzolethane with 2,4-bis(4-methoxyhenyl)-1,3,2,4-dithiadiphosphoethane-2,4-disulfide (Lawesson's reagent) in toluene can be used for the preparation of 2,5-diphenylthiophene [9,10].

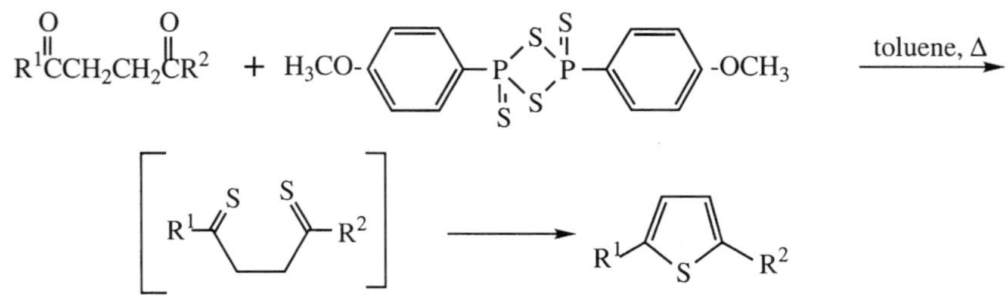

2,5-Diarylthiophenes with various substituents in the aryl groups such as 2,5-bis(*ortho*-carboethoxyphenyl) and 2,5-bis(*ortho*-bromophenyl)thiophene are prepared through the reaction of 1,4-diarylbutadiynes with sodium sulfide in ethanol [11]. The latter compound is converted to the *bis*(*ortho*-formyl-phenyl)thiophene by halogen–metal exchange and reaction with *N*,*N*-dimethyl-formamide [11]. Through Wittig reaction of the dialdehyde with methyl triphenyl phosphonium bromide, 2,5-*bis*(2-vinylphenyl)thiophene was obtained in mediocre yields [12].

Another way of preparing 2,5-diarylthiophenes, a one-pot procedure, consists in the reaction of substituted acetophenones with bromine in methanol followed by reaction with Lawesson's reagent giving the desired compounds in mostly mediocre yields [13].

A much better method for preparing unsymmetrical 2,5-diarylthiophenes consists in the reaction of acetophenones with Vielsmeyer reagent to β-chloro-acroleins followed by their condensation with sodium sulfide and various benzyl bromides [14].

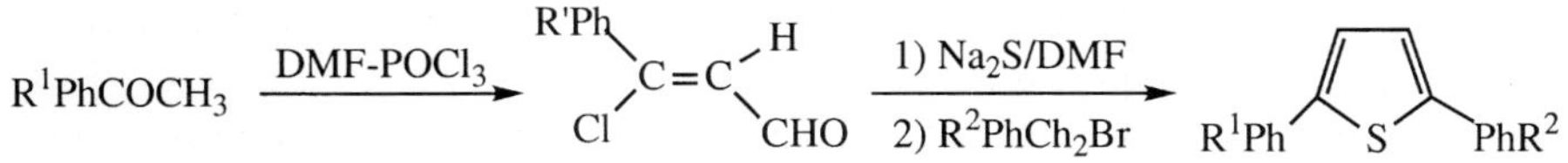

R^1 = H, CH_3, OCH_3, F, Cl, Br, $N(CH_3)_2$ R^2 = CO_2CH_3, SO_2CH_3, CN, NO_2

β-Chloroacroleins [14]

N,*N*-Dimethylformamide (11 ml) is added to an ice cooled solution of phosphoryl chloride (11 ml) and the mixture is stirred for 10 min. To the Vilsmer–Haack reagent the acetophenone (0.1 mol) in *N*,*N*-dimethylformamide (30 ml) is added dropwise and the mixture is stirred at 60°C for 3 h. The reaction mixture is poured into an aqueous sodium acetate solution. The title compounds are either filtered off or extracted with diethyl ether and purified by recrystallization or distillation. Yields 70–90%.

2,5-Diphenylthiophene [14]

The β-chloroacrolein (0.1 mol) in *N,N*-dimethylformamide (50 ml) is added dropwise to a suspension of sodium sulfide nonahydrate (0.1 mol) in *N,N*-dimethylformamide (50 ml). After 2 h the benzyl bromide (0.1 mol) in *N,N*-dimethylformamide is added and the mixture is heated at 50°C for 3 h. Sodium methoxide (0.1 mol) in methanol is added and after 10 min the mixture is poured into cold water and acidified. The crude product is isolated by filtration and recrystallized. Yields 34–55%.

2,3-Diarylthiophenes, such as 2-(4-fluorophenyl)-3-(methylthiophenyl)thiophene, are prepared by reacting 2-[(4-methylthio)phenyl]-1-(4-fluorophenyl)-ethanone with Vilsmeyer reagent followed by reaction of the thus formed 3-chloro-3-(4-fluorophenyl)-2-[4-methylthio)phenyl]propenal with thioglycolic acid and triethylamine, followed by decarboxylation [15].

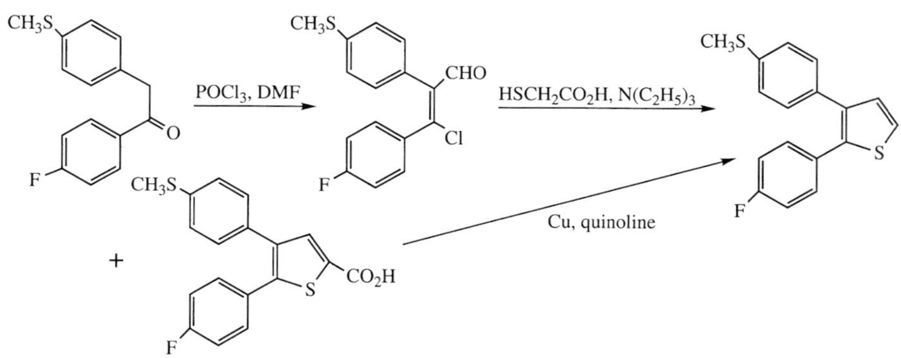

3-Chloro-3-(4-fluorophenyl)-2-[4-(methylthio)phenyl] propenal [15]

A mixture of phosphoryl chloride (20.6 ml, 221 m*M*) and *N,N*-dimethylformamide (22.8 ml, 295 m*M*) in 1,2-dichloroethane (74 ml) is stirred at room temperature for 30 min. To this mixture, 2-[(4-methylthio)phenyl]-1-(4-fluorophenyl)ethanone (38.3 g, 148 mmol) in 1,2-dichloroethane (139 ml) is added. This reaction mixture is refluxed for 10 h, cooled, washed twice with water, dried and evaporated. The residue is washed with toluene giving 28.8 g (64%) of the title compound as a pale brown solid, which is used in the next step without further purification.

2-(4-Fluorophenyl)-3-[4-(methylthio)phenyl]thiophene [15]

A mixture of 3-chloro-3-(4-fluorophenyl)-2-[4-(methylthio)phenyl]propenal (16 g, 55.2 m*M*), thioglycolic acid (3.81 ml, 54.7 m*M*) and triethylamine (16 ml, 115 tm*M*) in pyridine is stirred at 70°C for 1 h and then refluxed for 3 h.

The solvent is evaporated and the residue dissolved in a mixture of ethyl acetate and water. The phases are separated and the organic phase washed with dilute hydrochloric acid, dried and evaporated. The oily residue (20 g) is chromatographed on silica gel using first hexane/toluene (10:1) followed by toluene, chloroform, and finally chloroform/methanol (5:1) as eluent giving 5.8 g (37%) of the title compound as an off-white solid mp 81–83°C and 10.8 g (60%) of the 5-carboxylic acid derivative, which without further purification is used in the next step.

A mixture of the acid (17.7 g, 51.3 m*M*) and copper powder (3.6 g, 56.7 m*M*) in quinoline (31 ml) is refluxed for 5 h. Ethyl acetate is added and the so obtained solution filtered. The filtrate is washed successively with water, dilute hydrochloric acid and water, dried and evaporated. The residue is recrystallized from ethanol giving 13.3 g (86%) of the title compound as pale brown crystals identical with the first crop.

2,5-Diaryl-3-(phenylmethyl)thiophenes are conveniently prepared through the reaction of benzylthiols with butadiynes in potassium hydroxide/dimethyl sulfoxide [16].

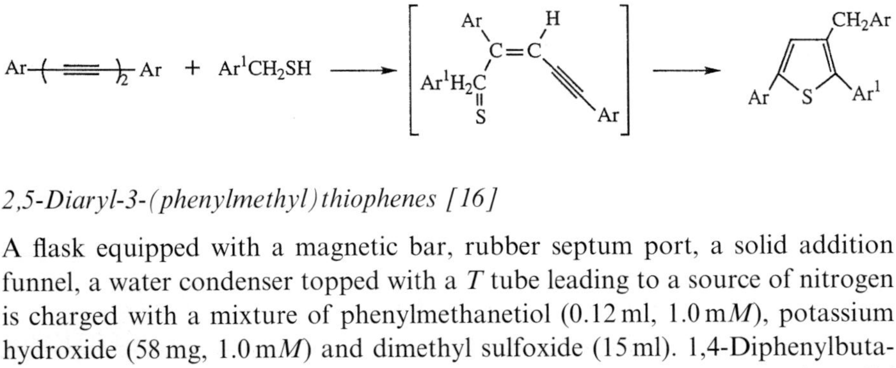

2,5-Diaryl-3-(phenylmethyl)thiophenes [16]

A flask equipped with a magnetic bar, rubber septum port, a solid addition funnel, a water condenser topped with a *T* tube leading to a source of nitrogen is charged with a mixture of phenylmethanetiol (0.12 ml, 1.0 m*M*), potassium hydroxide (58 mg, 1.0 m*M*) and dimethyl sulfoxide (15 ml). 1,4-Diphenylbutadiyne (200 mg, 1.0 mmol) is added slowly in small portions *via* the solid addition funnel, after which the reaction mixture is stirred at 23°C for 1 h. Diethyl ether (50 ml) is added and the so obtained solution is poured into a mixture of ice and 10% ammonium chloride solution (15 ml). The phases are separated and the organic phase washed with saturated sodium chloride solution (25 ml) and dried over 4 Å molecular sieves 10 h. The volatile materials are removed by evaporation and the residue is chromatographed on silica gel (40 g, 225–400 mesh) using hexane as eluent. After recrystallization from methanol the title compounds are obtained in yields of 51–66%.

A recent synthesis of 2-aryl-5-acetylmethylthiophene has been achieved by reaction of 1,6-dioxo-2,4-dienes with Lawesson's reagent in the presence of boron trifluoride-etherate [17]. Unfortunately with other aryl groups, mixtures with 5-methyl-2-aroylmethylthiophenes are obtained.

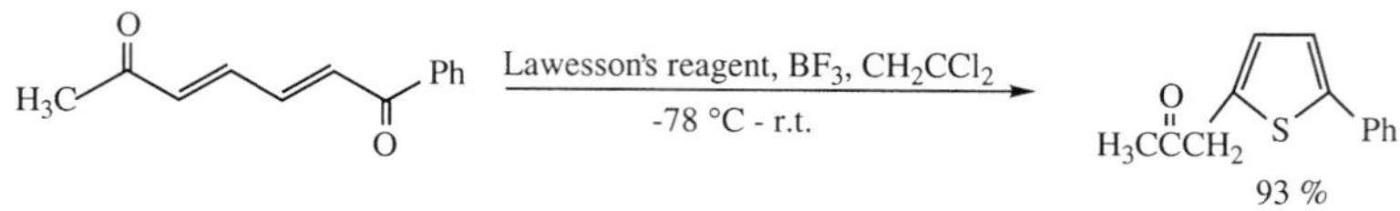

Bis(1,3-dithiolium)-4-olates react with dimethyl acetylenedicarboxylates under elimination of COS to produce 2,2′-bridged bis(thiophene) derivatives [18].

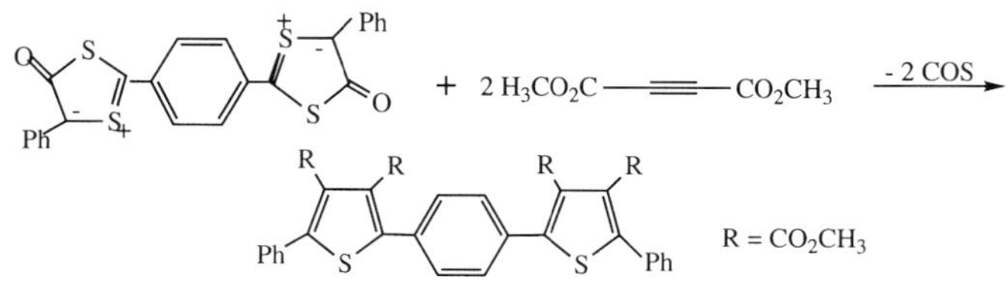

Dimethyl 2,2′-(1,4-phenylene)bis(5-phenyl-3,4-thiophenedicarboxylate) [18]

A suspension of 2,2′-(1,4-phenylene)*bis*(5-phenyl-1,3-dithiolylium)-4-olate (0.461 g, 0.997 m*M*), dimethyl acetylenedicarboxylate (0.567 g, 6.99 m*M*) and toluene (20 ml) is heated at 95–100°C. After treatment with charcoal and cooling to room temperature the precipitated crystals are filtered off. The filtrate is concentrated, diluted with methanol and kept in refrigerator giving a second crop of crystals. The combined crystal fractions are recrystallized from chloroform/petroleum ether giving 0.278 g (45%) as colorless needles mp 238°C.

It was recently found that palladium acetate catalyzes the reaction diphenylacetylene with sulfur leading to tetraphenylthiophene in 71% yield already at 160°C [19].

4D.4 ARYLTHIOPHENES THROUGH CROSS-COUPLING REACTIONS

A classical method for the preparation of aryl thiophenes is the coupling of 2-thienylcopper reagents with aryl iodides in quinoline [20,21]. At present the most useful methods for the synthesis of arylthiophenes consists in palladium-catalyzed cross-coupling reaction between either thienylmetalorganic reagents, especially boron, magnesium, zinc and tin derivatives, and aryl halides or triflates, or by the reaction of halothiophenes with the above mentioned class of arylmetalorganic reagents.

4D.4.1 From thiophene and aryl bromides

The reaction of thiophene with aryl bromides in the presence of potassium acetate and 5 mol% of palladium(0) in *N*,*N*-dimethylacetamide at 150°C in a sealed tube can be used for the preparation of 2-arylthiophenes in moderate yields [22].

2-Arylthiophenes [22]

A mixture of aryl bromides (2 m*M*), thiophene (1 ml, *ca* 12 m*M*), potassium acetate (294 mg, 3 m*M*), tetrakis(triphenylphosphine)palladium(0) (116 mg, 5 mol%) and *N,N*-dimethylacetamide (5 ml) in a sealed tube is heated at 150°C for 12 h. The solvent is evaporated *in vacuo* and the residue triturated with water (20 ml) and extracted with dichloromethane (3 × 15 ml). The combined organic phases are dried over sodium sulfate, filtered and concentrated. The crude product is purified by medium pressure liquid chromatography using hexane/diethyl ether or hexane/dichloromethane.

4D.4.2 From thienylmetalorganic reagents and aryl halides

4D.4.2.1 From zinc, copper and magnesium derivatives

2-Thiophenemagnesium bromide and 2-thiophenezinc chloride are coupled under palladium(0) catalysis with *para*-nitro, *para*-cyano, *para*-carboxy, *para*-carbobutoxy- and *para*-diethylcarboxamidoiodobenzene to give excellent yields of the corresponding arylthophenes [23] 3-Cyano-, 3-chloro- and 4-vinyl-bromobenzene have also been used in this method [24].

2-Arylthiophenes [23]

The *para*-substituted iodobenzene (5 m*M*) is added under argon to anhydrous tetrahydrofuran (5 ml) followed by tetrakis(triphenylphosphine)palladium(0) (0.5 m*M*) and 2 *M* tetrahydrofuran solution of the organometallic reagent (6 m*M*). 2-Thiophenezinc chloride is prepared by treating the Grignard reagent with anhydrous zinc chloride in tetrahydrofuran at 0°C. The reactions are performed at room temperature for 2 h and then the reaction mixture is treated with 10% aqueous hydrochloric acid. The products are extracted with dichloromethane and isolated by flash chromatography using petroleum ether and ethyl acetate as eluent.

2′-(3-Cyanophenyl)thiophene [24]

To a stirred mixture of 1-bromo-3-cyanobenzene (2.73 g, 15 mmol) and bis(triphenylphosphine)nickel dichloride (0.4 g, 0.6 m*M*, 4 mol%) in tetrahydrofuran (20 ml) is added a solution of 2-thienylzinc chloride (15 m*M*) in tetrahydrofuran (10 ml) by a double-ended needle. The reaction mixture is stirred under nitrogen for 3 h. After the usual work up and evaporation of the solvent, the residue, a black powder, is purified by chromatography using pentane/ethanol (19:1) as eluent giving 2.2 g (80%) of the title compound as light-yellow crystals mp 51–54°C.

Also less reactive bromo derivatives such as *tert*-butyldimethylsilyl protected 4-(ethylhydoxyethylamino)bromobenzene has successfully been coupled with 2-thienylzinc chloride [25].

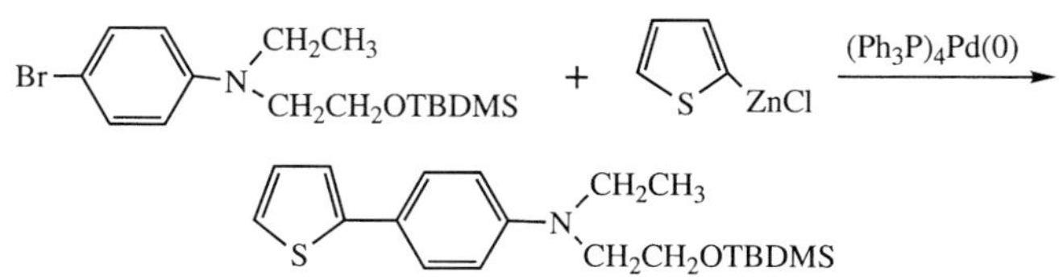

4-Thienyl(ethylhydroxyethyl)aminobenzene [25]

2-Bromothiophene (13.65 g, 83.7 m*M*) and magnesium powder (2.24 g, 92 m*M*) are added to anhydrous tetrahydrofuran (30 ml) in a 250 ml flask under nitrogen. The mixture is stirred at room temperature for 2 h and then anhydrous zinc chloride (12.55 g, 92 m*M*) is added at 0°C. The mixture is stirred for another hour and *tert*-butyldimethylsilyl protected 4-(ethylhydroxyethylamino)bromobenzene (20 g, 55.8 m*M*) in tetrahydrofuran (20 ml) is added dropwise to the resulting organozinc reagent. A catalytic amount of tetrakis(triphenylphosphine)palladium(0) (1.67 g, 1.4 m*M*) is then added and the resulting reaction mixture is refluxed overnight. After removing the solid by filtration, the reaction mixture is refluxed overnight. After removing the solid by filtration, the reaction mixture is concentrated and purified by flash chromatography using hexane as eluent giving 12.6 g (63%) of the title compound.

Two equivalents of thienylzinc chloride gave 1,4-, 1,3-di- and 1,2-di(2-thienyl)benzene upon palladium-catalyzed reaction with *para*-dibromo-, *meta*-dibromo- and *ortho*-dibromobenzene [24]. From 1,3,5-tribromothiophene and 2-thienylzinc chloride 1,3,5-tri(2′-thienyl)benzene is prepared [24].

1,3,5-Tri(2′-thienyl)benzene [24]

To a solution of 1,3,5-tribromobenzene (3.77 g, 12 mmol) and tetrakis(triphenylphosphine)palladium(0) (1.4 g, 0.12 mmol) in tetrahydrofuran (100 ml), a solution of 2-thienylzinc chloride (81 mmol) in tetrahydrofuran (100 ml) is added. To reaction mixture is stirred at 50°C for 48 h. Work up followed by purification by chromatography initially using pentane as eluent in order to remove the starting material and then pentane/chloroform (4:1). The pale-yellow crystalline material so obtained is recrystallized from ethanol giving 2.5 g (60%) of the title compound as white crystals mp 156–158°C.

1,4-Di(2-thienyl)benzene is prepared in a palladium-catalyzed reaction between *para*-dibromobenzene and 2-thienylzinc chloride [26]. Rather complicated dibromo derivatives, such as 1,4-dibromo- 2,3-*bis*((4-(dodecyloxy)phenyl)-ethynyl)benzene have successfully been coupled with 5-methyl-2-thienylzinc chloride to give 2,5-*bis*((4-(dodecyloxy)phenyl)-ethynyl)1,4-*bis*(5-methyl-2-thienyl)benzene by using dichlorobis(triphenylphosphine)palladium(II) in tetrahydrofuran as catalyst [27].

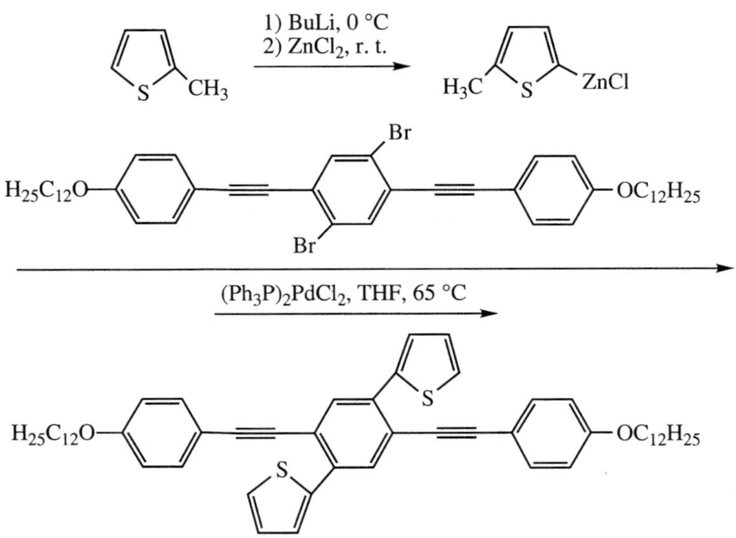

1,4-Dibromo-2,5-bis((4-(dodecyloxy)phenyl)ethynyl)benzene [27]

To a 500 ml flask charged with 1,4-dibromo-2,5-diiodobenzene (16 g, 0.0328 mol), dichlorobis(triphenylphosphine)palladium(II) (0.641 g, 0.657 m*M*), cuprous iodide (0.656 g, 3.44 mmol) and tetrabutylammonium bromide (0.634 g, 1.97 mmol), are added toluene (75 ml), diisopropylamine (75 ml) and tetrahydrofuran (50 ml). 4-(Dodecyloxy)phenylacetylene (19.74 g, 0.0689 mol) dissolved in tetrahydrofuran (25 ml) is added dropwise at 25°C generating a noticeable amount of heat. After 2 h of stirring without any external source of heat the mixture is diluted with chloroform (200 ml) and washed with 5% hydrochloric acid (3 × 50 ml), water 50 ml), 5% ammonium hydroxide (2 × 50 ml) and water (2 × 50 ml). Methanol is added to precipitate the compound as a yellow solid. After crystallization from tetrahydrofuran/methanol 19.17 g (73%) of the title compound is obtained as small colorless plates mp 108–110°C.

2,5-Bis(4-(dodecyloxy)phenyl)ethynyl),1,4-bis(5-methyl-2-thienyl) benzene [27]

To a solution of 2-methylthiophene (140 μl, 145 mmol) in anhydrous tetrahydrofuran (10 ml) cooled to 0°C is added 1.75 *M* butyllithium (0.67 ml, 1.17 mmol). After 25 min a solution of zinc chloride (0.507 g, 3.72 mmol) in tetrahydrofuran (8 ml) is added to the 2-methyl-5-lithiothiophene solution. After stirring the mixture for 20 min it is cannulated into a separate flask containing 1,4-dibromo-2,5-bis(4-(dodecyloxy)phenyl)ethynyl)benzene (0.25 g, 0.311 mmol) and dichlorobis(triphenylphosphine)palladium(II) (0.00877 g, 0.012 mmol). The reaction mixture is refluxed for 4 h and cooled to room temperature after which the reaction is carefully quenched with methanol. The so obtained solution is diluted with diethyl ether washed with 5% hydrochloric acid and dried over magnesium sulfate. The solvent is removed under reduced pressure and the product purified by chromatography on silica gel using hexane/tetrahydrofuran (15:1) as eluent, giving 0.254 g (97%) of the title compound as a yellow/orange solid mp 160–162°C.

Alternatively, Grignard reagents derived from 3-butyl-2-bromothiophene have been coupled with 1,4-dibromobenzene, using [1,3-bis(diphenyl-phosphino)propane]dichloronicke(II) as catalyst for the preparation of 1,4-di(3-butyl-2-thienyl)benzene [28]. Kumada coupling of phenylmagnesium bromide with 5-bromo-phenylthiophene using [1,3-bis(diphenylphosphino)-propane]dichloronicke(II) as catalyst is used for the preparation of 2,5-diphenylthiophene [29].

2,5-Diphenylthiophene [29]

Equimolar amounts of distilled bromobenzene (596 mg, 3.80 mmol) and magnesium (92.3 mg, 3.80 mmol) are mixed in anhydrous diethyl ether to prepare a Grignard regent. To the resulting Grignard reagent [1,3-bis(diphenyl-phosphino)propane]dichloronickel(II) (40 mg, 0.074 mmol) and 5-phenyl-2-bromothiophene (454 mg, 1.90 mmol) are added successively. After stirring overnight at room temperature, the reaction mixture is refluxed for 6 h, then cooled over an ice-water bath and subsequently hydrolyzed with 2 *M* hydrochloric acid (2 ml). The precipitate formed is collected by filtration, washed with cold methanol and recrystallized from methanol giving 157 mg (35%) of the title compound as a pale-yellow solid mp 155°C.

Kumada couplings of 2-(4-bromophenyl)thiophene with 5-phenyl-2-thio-phenemagnesium halide and *para*-diiodobenzene with 5-phenyl-2-thiophene-magnesium halides is used for the preparation of 1-(2-thienyl)-4-(5-phenylthiophene-2-yl)benzene and 1,4-bis(5-phenylthiophene-2-yl)benzene, respectively [30].

1-(2-Thienyl)-4-(5-phenylthiophene-2-yl)benzene [30]

A flask containing equimolar amounts of 2-iodo-5-phenylthiophene (286 mg, 1.00 mmol) and magnesium (24.3 mg, 1.00 mmol) is evacuated under mild heat. To this solid mixture is added diethyl ether (10 ml) to prepare a Grignard reagent. After making sure that all magnesium has disappeared, [1-3-bis(diphenylphosphino)propane]nickel(II) chloride (10 mg, 0.018 mmol) and 2-(4-bromophenyl)thiophene (120 mg, 0.50 mmol) are added successively, soon causing precipitation. After being stirred overnight at room temperature the reaction mixture is refluxed for 6 h, cooled in a ice-water bath and hydrolyzed with 2 *M* hydrochloric acid (0.5 ml). The resulting precipitate is collected by filtration, washed with methanol and Soxhlet extracted with dichloromethane. The extracted material, which precipitated in the bottom flask of the Soxhlet apparatus, is collected by filtration giving 80 mg of a crude material, which is recrystallized from acetone providing 58% of the title compound as a yellow solid mp 221°C.

Kumada coupling of 2-thiophenemagnesium iodide with 1,4-diiodobenzene and 4,4′-diiodobiphenyl gave 1,4-bis(2-thienyl)benzene and 4,4-bis(2-thienyl)-biphenyl albeit in low yields [31].

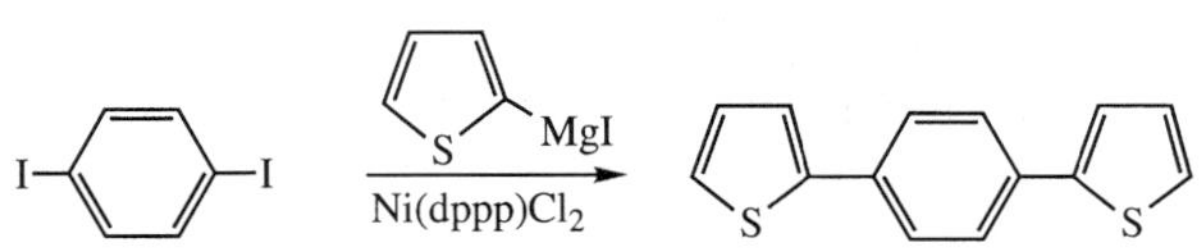

Metalation of 2,4-dimethylthiophene with butyllithium in *N,N,N′,N′*-tetramethylethylenediamine followed by reaction with zinc chloride and palladium(0)-catalyzed coupling with iodobenzene, 4-methoxy-, 4-diethyl-amino- and 4-cyano-1-iodo- or 1-bromobenzene gives 3,5-dimethyl-2-arylthiophenes in high yields [32,33].

3,5-Dimethyl-2-phenylthiophene [32]

To a solution of 2,4-dimethylthiophene (2.24 g, 20.0 mmol) and *N,N,N′,N′*-tetramethylethylenediamine (2.55 g, 22.0 mmol) in anhydrous diethyl ether (20 ml), 1.6 *M* butyllithium in hexane (13.7 ml, 22.0 mmol) is added under nitrogen at room temperature and the reaction mixture is stirred for 2 h. To this solution 1.0 *M* zinc chloride in diethyl ether (20 ml) is added and the resulting mixture stirred for 4 h. In another flask iodobenzene (4.08 g, 20.0 mmol) and tetrakis(triphenylphosphine)palladium(0) are added with stirring to anhydrous tetrahydrofuran (20 ml). To this solution the previously prepared ether solution is added dropwise at room temperature. The reaction mixture is heated at 50°C for 2 h and stirred overnight at room temperature.

After addition of water the product is extracted with ether. The combined organic phases are washed with water, dried over magnesium sulfate and evaporated. The residue is purified by chromatography on silica gel using hexane as eluent giving 3.66 g (97%) of the title compound.

1,4-Di(5′-phenyl-2′-thienyl)benzene is prepared either by coupling of 1,4-di(5-chlorozincio-2-thienyl)benzene and bromobenzene using a nickel catalyst or by an alternative strategy using two equivalents of 5-phenyl-2-thienylzinc chloride and *para*-dbromobenzene with palladium as catalyst [24].

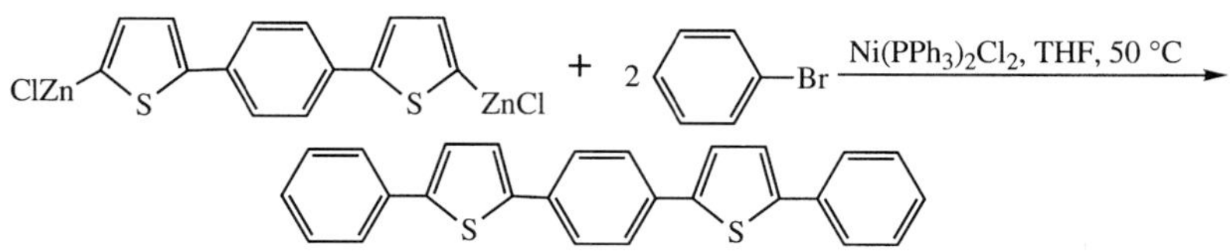

1,4-Di(5′-phenyl-2′-thienyl)benzene [24]

A solution of 2.5 *M* butyllithium in hexane (6 ml, 15 mmol) is added to a solution of 2-phenylthiophene (2.4 g, 15 mmol) in anhydrous tetrahydrofuran (30 ml) under argon at 0°C. The mixture is stirred at room temperature for 3 h, after which a freshly prepared solution of zinc chloride (2.0 g, 14.8 mmol) in anhydrous tetrahydrofuran (30 ml) is added and the mixture is stirred at room temperature for 1 h. This mixture is then added to a solution of 1,4-dibromo-benzene (1.8 g, 7.6 mmol) and tetrakis(triphenylphosphine)palladium(0) (0.06 g, 0.05 mmol, 0.7 mol %) in tetrahydrofuran (20 ml) and the reaction mixture is stirred at 50°C for 18 h. The precipitate formed is collected by filtration and washed with diethyl ether until the washings are colorless. Recrystallization from chloroform gives 1.6 g (55%) of the title compound as yellow crystals mp 301°C.

An alternating seven ring phenyl-thienyl derivative was obtained upon attempted reaction of 1,4-di(2′-thienyl-5′-zinc chloride)benzene with 1,4-dibromo-3,6-dioctylbenzene [24] instead of the polymer.

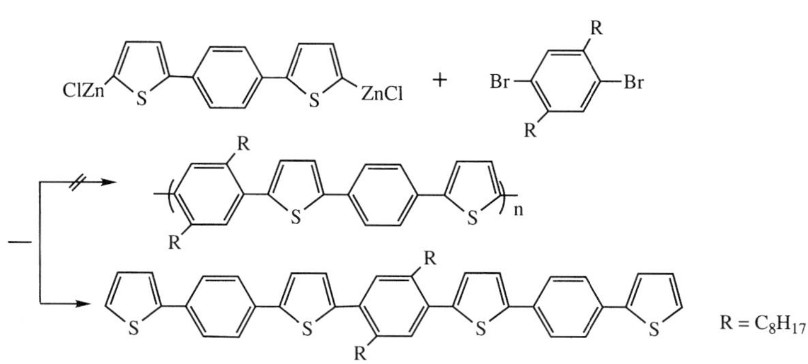

This methodology, however, can be used for the preparation of polymers with alternating thiophene and benzene rings. The nickel-catalyzed reaction of 1,4-di(5-chlorozincio-2-thienyl)benzene with 1,4-dibromobenzene and a number of 2,5-dialkyl- and 2,5-dimethoxy substituted 1,4-dibromobenzenes are used for the preparation of poly(1,4- (2′,5′-thienyl)benzenes [24]. From 2,5-dibromothiophene and 1,3-di(5-chlorozincio-2-thienyl)benzene poly(1,3(2,5′-thienyl)benzene is obtained [24].

A quite different approach is based on conversions of linked diyne units to thiophene, thiophene-1-oxide and thiophene-1,1-dioxides by zirconocene couplings. Thus the following reactions have been performed [34].

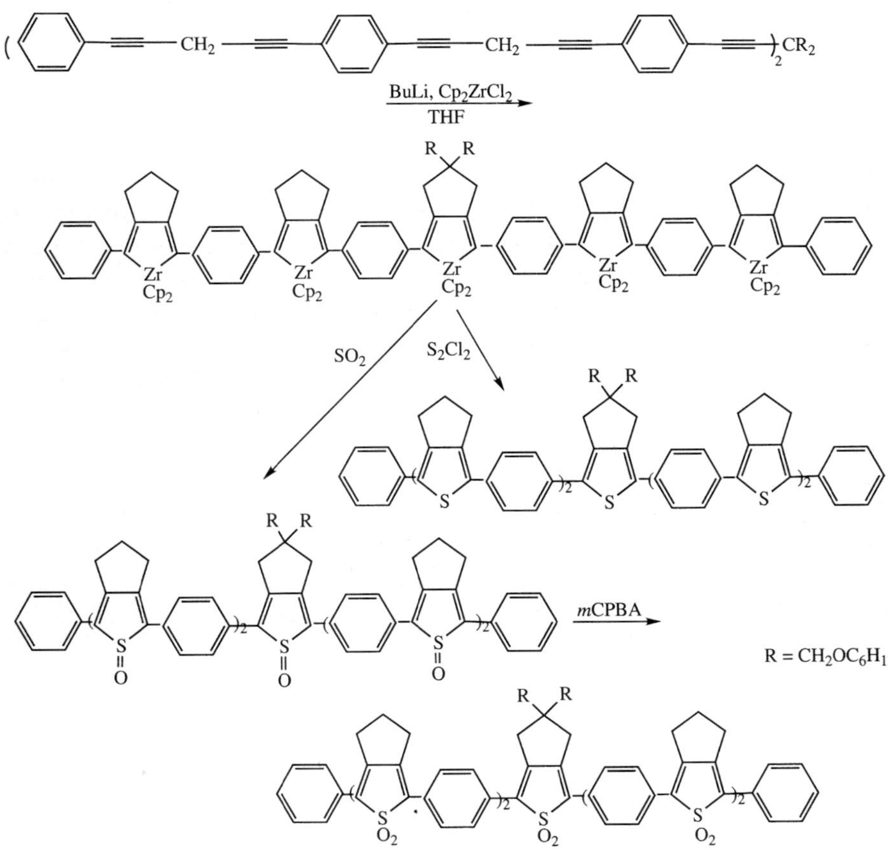

Coupling of 5-(3,5-di-*tert*-butyl-4-trimethylsilyloxyphenyl)-2-thienylzinc chloride with aryl iodide using tetrakis(triphenylphosphine)palladium(0), prepared *in situ* by treating dichlorobis(triphenylphosphine)palladium(II) with diisobutylaluminum hydride, gave 2,5-(4-alkoxyphenyl) substituted thiophenes, which could be cleaved with boron tribromide to give the hydroxyquinones, which in turn can be oxidized to the quinones [35,36].

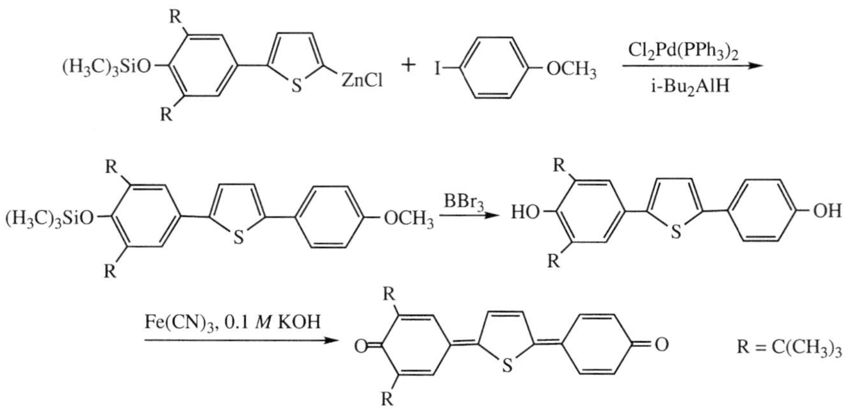

3-Thienylzinc bromide prepared from 3-thienyllithium and zinc bromide gives 3-(4-pentylphenyl)thiophene in poor yield upon reaction with 4-pentyl-1-iodobenzene using palladium(0) as catalyst [37]. 3-Thienylzinc iodide generated by the reaction of 3-iodothiophene with activated Rieke zinc has been reacted with iodobenzene and a number of substituted iodobenzenes also such containing nitro and acetyl groups to give 3-arylthiophenes [38].

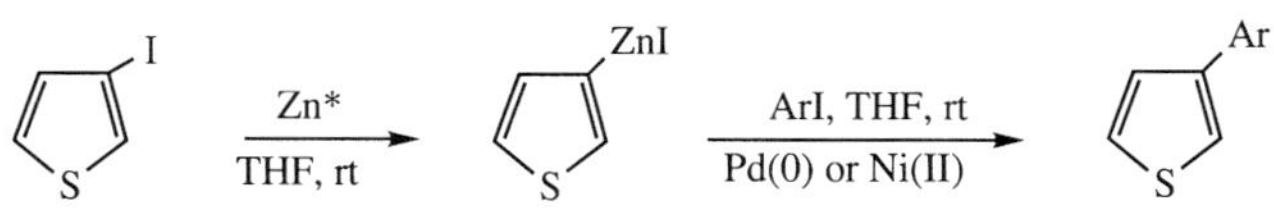

3-Thienylmanganese bromide and 4-bromo-3-thienylmanganese bromide can also be prepared from 3-bromo- and 3,4-dibromothiophene through reaction with activated Rieke manganese. These manganese derivatives do not rearrange at room temperature and upon reaction with aryl iodides a number of 3-aryl- and 4-bromo-3-arylthiophenes are obtained [38].

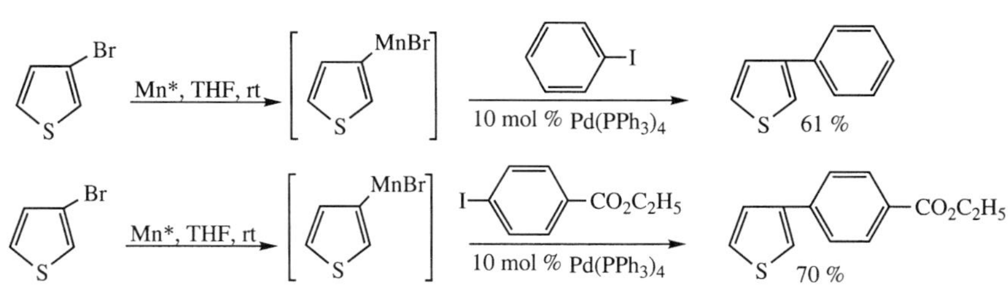

All manipulations are carried out under an atmosphere of argon on a dual manifold vacuum/argon system. The Linde purified grade argon is further purified by passage over a BASF R3-11 catalyst column at 15°C, a phosphorus pentoxide column and a column of granular potassium hydroxide. Lithium naphthalene and metal halides are weighed out and charged into reaction flasks in a Vacuum Atmospheres Company dry box. Tetrahydrofuran is distilled immediately before use from sodium/potassium alloy under an atmosphere of argon.

Preparation of highly active zinc (Rieke zinc) [38]

Active zinc is prepared by the reduction of anhydrous zinc chloride with lithium using naphthalene as an electron carrier. In a representative preparation zinc chloride (10 mmol) in tetrahydrofuran (10 ml) is transferred dropwise *via* cannula to lithium (20 mmol) and naphthalene (2 mmol) in stirring tetrahydrofuran (5 ml). The stirring is stopped when the lithium is totally consumed, and the active zinc powder is allowed to settle for 0.5 h. The supernatant is then removed *via* cannula, and freshly distilled tetrahydrofuran (10 ml) is added. The slurry is briefly stirred and then allowed to settle down for 0.5 h again, and the supernatant is subsequently removed. Freshly distilled tetrahydrofuran (15 ml) is added to newly formed and washed Zn*, which is ready for oxidative addition.

3-Phenylthiophene [38]

3-Iodothiophene (5 mmol) is added *via* syringe to active zinc (10 mmol) being stirred in tetrahydrofuran (15 ml) at room temperature. The slurry is stirred at room temperature for 8–10 h. After completion of the oxidative addition, the mixture is allowed to stand so that the excess Zn* powder could settle out of the solution. Then the newly prepared 3-thienylzinc iodide is dissolved in tetrahydrofuran, the supernatant is then transferred to a mixture of iodobenzene (4 mmol) and [1,2-bis(diphenylphosphino)ethane]dichloronicke(II) (0.4 mol%) with stirring at room temperature.

Preparation of highly active manganese (Mn) [38]*

To a mixture of lithium (20 mmol), naphthalene (2 mmol) and manganese halide (chloride, bromide and iodide) (10 mmol) freshly distlled tetrahydrofuran is added at room temperature. After stirring the mixture at room temperature for 1–3 h a black slurry is obtained, *via* syringe ready for use.

Ethyl 4-(3-thienyl)benzoate [38]

To a slurry of Mn* (5.0 mmol) in tetrahydrofuran (10 ml) under argon 3-bromothiophene (2.5 mmol) is added at room temperature. After stirring the mixture at room temperature for 5 h, 1,2-dibromethane (3.0 mmol) is syringed neat at 0°C. Then the mixture is gradually warmed to room temperature over 20 min. The resulting 3-thienylmanganese bromide is added *via* cannula to a mixture of ethyl 4-iodobenzoate (1.88 mmol) and tetrakis(triphenylphosphine)palladium(0) (0.13 mmol) in tetrahydrofuran over 40 min. The mixture is stirred at room temperature for 2 h. Aqueous hydrochloric acid (3 *M* 10 ml) is added and the product extracted with diethyl ether (2 × 20 ml). The combined organic phases are washed sequentially with saturated sodium bicarbonate solution (2 × 20 ml), saturated sodium thiosulfate solution (2 × 20 ml) and saturated sodium chloride solution (2 × 20 ml) and then dried over magnesium sulfate. Removal of solvent and flash chromatography affords the title compound in 70% yield.

Lithium 2-thienylcyanocuprate reacts with 2-methoxy-1-naphthllithium at – 125°C followed by oxygen at the same temperature to give 1-(2-thienyl)-2-methoxynaphthalene in 80% yield [39]. It was found that in certain cases the use of 2-thienylcadmium chloride is advantageous, as in the preparation of 2,5-dimethyl-1,4-di(2′-thienyl)benzene from 1,4-dibromo-2,5-dimethylthiophene. The main advantage is that cadmium chloride can be more easily made anhydrous; however it is more toxic [24].

2,5-Dimethyl-1,4-di(2′-thienyl)benzene [24]

2-Thienylcadmium chloride (23 mmol) in tetrahydrofuran (30 ml) is added to a solution of dichlorobis(triphenylphosphine)nickel(II) (0.4 g, 0.063 mmol) and 1,4-dibromo-2,5-dimethylbenzene (2.1 g, 7.5 mmol) in tetrahydrofuran (25 ml) under argon. The reaction mixture is stirred at 50°C for 16 h followed by work up and purification by chromatography using pentane as eluent. The second fraction gives 1.15 g (57%) of the title compound as a solid mp 76–78°C.

4D.4.2.2 From silicon derivatives

Ethyl (2-thienyl)difluorosilane react with aryl iodides under palladium catalyst to give 2-arylthiophenes in high yield [40].

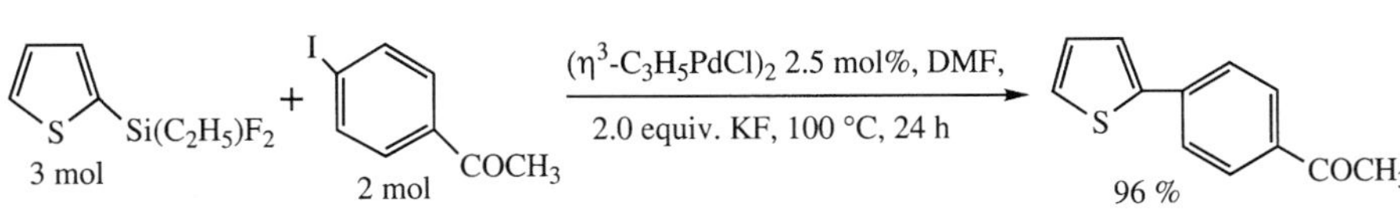

4D.4.2.3 From boron derivatives

Both 2- and 3-thiopheneboronic acid give upon Suzuki coupling with 2-bromo-1-benzoylaminomethylbenzene and palladium(0) as catalyst 2-(2-thienyl)- and

2-(3-thienyl)-1-benzoyl amino-methylbenzene in 78 and 85% yield, respective [41], when using the modification developed by Gronowitz [42], using aqueous 1,2-dimethoxyethane as solvent.

N-(2-Thienylbenzyl)benzamide [41]

N-2-(Bromobenzyl)benzamide (3 g, 10.3 mmol) and tetrakis(triphenylphosphine)palladium(0) 0.36 g, 0.3 mmol) are added to 1,2-dimethoxyethane (50 ml) with stirring under nitrogen. After 20 min a solution of anhydrous sodium carbonate (1.09 g, 10.3 mmol) in water (15 ml) and 2-thienylboronic acid (1.58 g, 12.4 mmol) are added and the reaction mixture is refluxed for 5 h. 1,2-Dimethoxyethane is evaporated and the product extracted with ethyl acetate. The extract is dried and evaporated and the residue purified by chromatography on silica using ethyl acetate/hexane (1:9) giving 2.8 g (78%) of the title compound as a white solid mp 119–120°C from ethanol.

The most recent development consists in the use of ligandless palladium and potassium fluoride. Thus the reaction of 2-iodothiophene with 3-formylphenylboronic acid gave 2-(3-formylphenyl)thiophene in 98% yield [43]. Suzuki coupling of 2-(4-bromophenyl)thiophene with phenylboronic acid and 4-biphenylboronic acid are used for the preparation of 4-(2-thienyl)biphenyl and 4-(2-thienyl)-1,1′:4′,1″ -biphenyl respectively. Coupling of two equivalents of 4-(2-thienyl)phenylboronic acid with 1,4-diiodobenzene gives 4,4″-bis(2-thienyl)-1,1′:4′,1″-terphenyl and with 2,5-diiodothiophene, 2,5-bis[4-(2-thienyl)-phenyl]- thiophene is obtained [31].

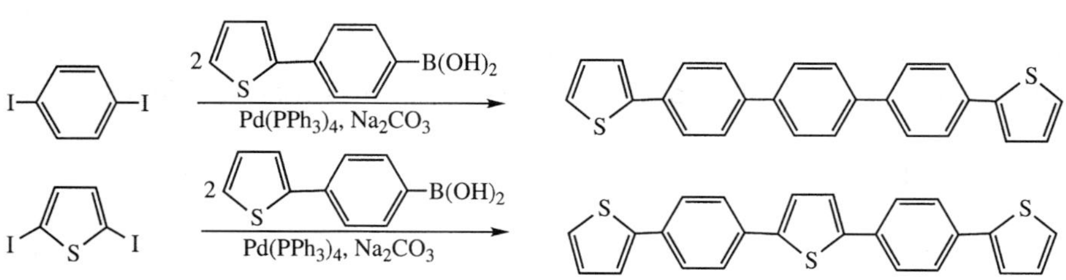

4,4″-Bis(2-thienyl)-1,1′:4′,1″-terphenyl [31]

A solution of 1,4-diiodobenzene (247 mg, 0.75 mmol), 4-(2-thienyl)phenylboronic acid (612 mg, 3.00 mmol) and tetrakis(triphenylphosphine)palladium(0) (104 mg) in benzene (60 ml) is deaerated by nitrogen for 30 min, after which a solution of sodium carbonate (636 mg, 6.00 mmol) in water (10 ml) is added. The reaction mixture is refluxed under nitrogen at room temperature for 26 h. The precipitate formed is filtered off, washed with methanol and acetone and recrystallized from 1,2,4-trichlorobenzene giving 112 mg (38%) mp 387°C.

Suzuki coupling of 3-thiopheneboronic acid with the silylated phenyl derivative gives the 3-thienyl substituted phenol, which upon oxidation with lead dioxide gives the stable phenoxy radical [44].

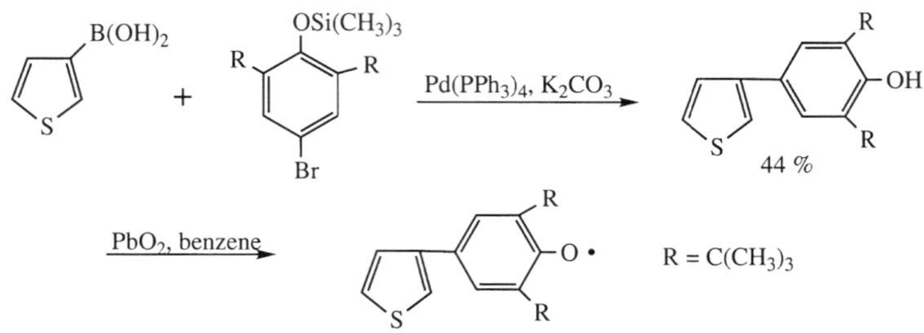

Suzuki coupling with 2-thiopheneboronic acid has been used in the preparated of 2-[4-[bis(9,9-dimethylfluorenyl)amino]phenyl]-5-dimesitylboryl)thiophene belonging to novel class of emitting amorphous molecular materials having bipolar radical formants character, allowing both stable cation and anion radicals [45].

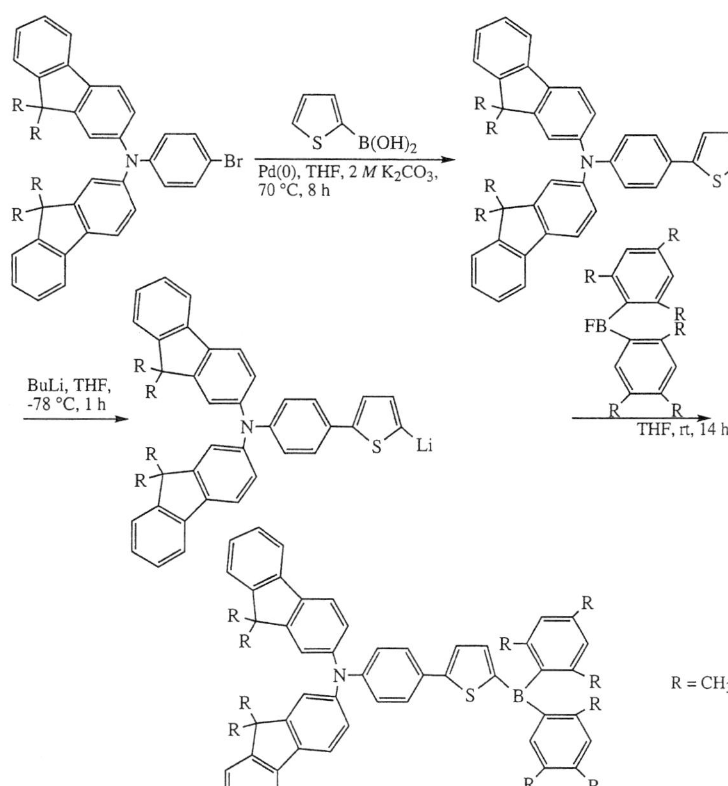

Suzuki coupling between 2-bromothiophene and 4-decyloxyphenylboronic acid is used for the preparation of 2-(4-decyloxyphenyl)thiophene [46].

2-(4-Decyloxyphenyl)thiophene [46]

A solution of crude 4-decyloxyphenylboronic acid (8.8 g, 32 mmol) in ethanol (20 ml) is added to a rapidly stirred mixture of 2-bromothiophene (5.2 g, 32 mmol), tetrakis(triphenylphosphine)palladium(0) (0.5 g, 0.4 mmol), 2 *M* aqueous sodium carbonate (30 ml) and 1,2-dimethoxyethane (30 ml). The reaction mixture is heated under reflux until all 2-bromothiophene is consumed. After cooling the product is extracted with diethyl ether, the combined organic solutions are dried over magnesium sulfate and evaporated. The residue, a brown solid, is purified by chromatography on silica gel using petroleum ether/dichloromethane (9:1) as eluent giving 6.6 g (65%) of the title compound as a pale yellow solid mp 74–77°C.

Suzuki coupling between 4,5-dibromonaphthalonitrile and 2-thiopheneboronic acid gives 4,5-di(2-thenyl)phithalonitrile in 62% yield. This is used for the preparation of a phthalocyanine derivative having eight peripheral 2-thienyl substituents, 5,5-di-(2-thienyl)-2,3-dihydro-1,3-diiminoisoindole, through refluxing with sodium methoxide in methanol, in an ammonia atmosphere, followed by heating with *N,N*-dimethylaminoethanol [47].

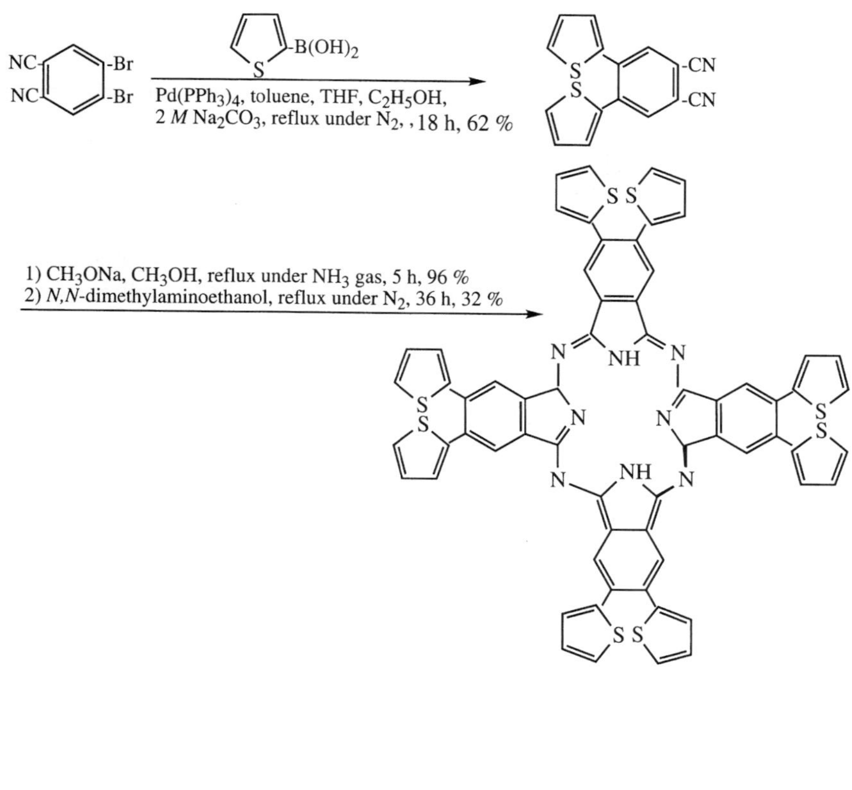

Suzuki coupling of 4-brono-3,5-dimethyl-2-thiopheneboronic and 4′-bromo-2″,4″,5″-triphenylimidazole is used for the preparation of 3-bromo-2,4-dimethyl-5-(2″,4″,5″-triphenylimidazolyl)thiophene, which by treatment with two equivalents of butyllithium and methanol is converted to 2,4-dimethyl-5-[4′-(2″,4″,5″ triphenylimidazolyl)]thiophene [48].

3-Bromo-2,4-dimethyl-5-(2″,4″,5″-triphenylimidazolyl)thiophene [48]

To a solution of 4′-bromo-2″,4″,5″-triphenylimidazole (2.0 g, 5.2 mmol), tetrakis(triphenylphosphine)palladium(0) (200 mg) and 20% aqueous sodium carbonate (6.5 ml) in tetrahydrofuran (200 ml) under argon 4-bromo-3,5-dimethyl-2-thiopheneboronic and (2.75 g, 12 mmol) is added. The reaction mixture is refluxed at 70°C for 48 h, after which 2 *M* hydrochloric acid is poured into it. The product is extracted with chloroform, the combined organic phases are dried over magnesium sulfate and evaporated. The residue is purified by chromatography on silica gel using chloroform as eluent giving 1.38 g (54%) of the title compound as a yellow powder mp 202–204°C.

2,4-Dimethyl-5-[4′-(2″,4″,5″ triphenylimidazolyl)]thiophene [48]

To a solution of 3-bromo-2,4-dimethyl-5-(2″,4″,5″-triphenylimidazolyl)thiophene (200 mg, 0.41 mmol) in anhydrous tetrahydrofuran (8 ml) under argon at − 78°C, 1.6 *M* butyllithium in hexane (0.5 ml) is added dropwise. The mixture is stirred at this temperature for 1 h followed by addition of methanol. The product is extracted with chloroform, the combined organic phases are dried over magnesium sulfate and evaporated. The residue is purified by chromatography on silica gel using chloroform as eluent giving 38 mg (23%) of the title compound as a yellow powder mp 207–208°C.

In the same way Suzuki coupling between 4-bromo-3,5-dimethyl-2-thiopheneboronic acid and the iodo-benzo-18-crown-6 was used for the preparation of 4-bromo-3,5-dimethyl-2-(benzo-18-crown-6)thiophene [49].

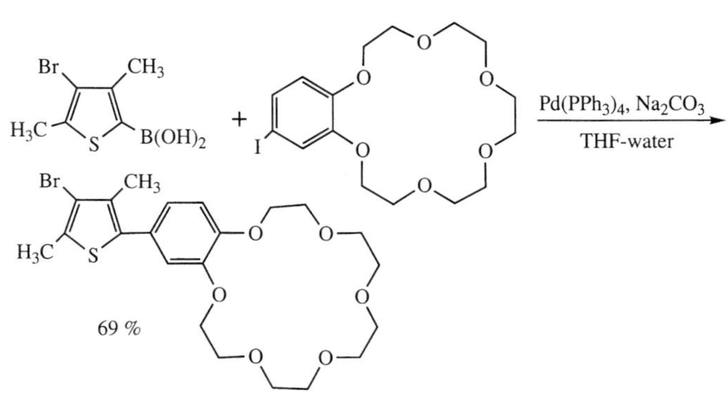

Suzuki coupling between 3-hydroxymethyl-2-thiopheneboronic acid and 4-bromobenzoic acid bound to Wang or Merrifield resin is used for the preparation of resin bound (4-(3-hydroxymethyl-2-thienyl)benzoate, which upon bromination and renewed Suzuki coupling with 3-cyclopentyloxy-4-methoxyphenylboronic acid gave polymer bound 4-(3-hydroxymethyl)-5-(3-cyclopentyloxy)-4-methoxyphenyl-2-thienul benzoate. Modification of the hydroxymethyl group by nucleophilic displacement followed by cleavage from the polymer give highly substituted 2-(4-carboxyphenyl-3-aminomethyl)-5-(3′-cyclopentyloxy-4′-methoxyphenyl) thiophenes [50].

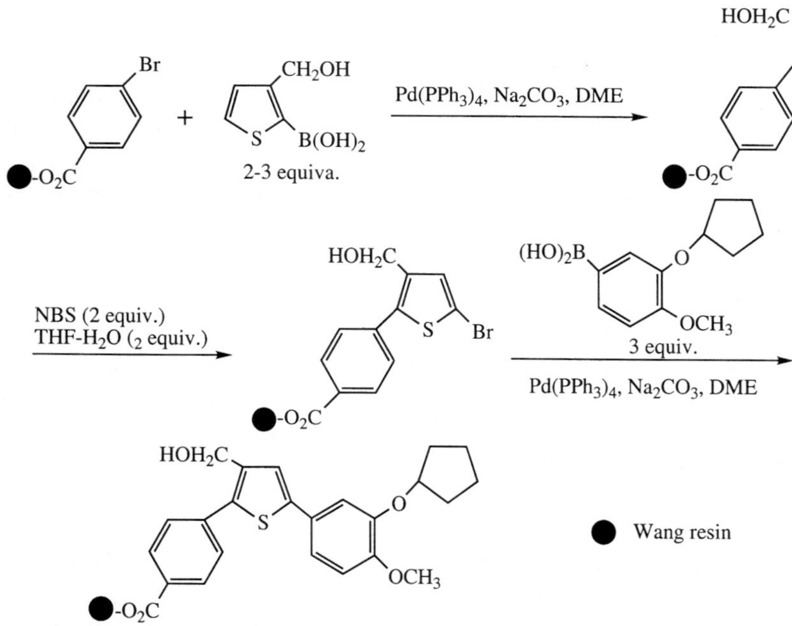

Suzuki coupling of the bromo containing resin [50]

A suspension of the resin (2.76 g), 3-cyclophentyloxy-4-methoxyphenylboronic acid (1.56 g, 6.6 mmol), tetrakis(triphenylphosphine)palladium(0) (77 mg, 0.066 mmol) and 2 *M* sodium carbonate (3 ml) in 1,2-dimethoxyethane (27 ml) is deoxygenated under a stream of nitrogen for 5 min and then heated to reflux for 7 h and poured into a 70 ml fritted polypropylene tube. The solvents are flushed out with a stream of nitrogen and the resin washed sequentially with *N*,*N*-dimethylformamide (3×), *N*,*N*-dimethylformamide/water (3×), *N*,*N*-dimethylformamide (2×), tetrahydrofuran (2×) and methanol (3×) and then dried under nitrogen overnight giving 3.0 g.

The Suzuki reaction between 3-thiopheneboronic acid has also been used with aromatic triflates in the presence of lithium chloride [51].

Buchwald and coworkers developed new catalysts consisting of mixtures of palladium acetate and *ortho*-(di-*tert*-butylphosphino)biphenyl or *ortho*-(dicyclohexylphosphino)biphenyl, which makes it possible to carry out the Suzuki coupling at room temperature with aryl bromides and aryl chlorides. The reaction of 2-bromothiophene with phenylboronic acid yielded 2-phenylthiophene in 98% yield, after 3 h using tetrahydrofuran as solvent and potassium fluoride as base [52].

2-Phenylthiophene [52]

An oven-dried resealable Schlenk flask is evacuated and backfilled with argon and charged with palladium(II) acetate (2.2 mg, 0.01 mol%), 2-(di-*tert*-butylphosphino)biphenyl (6.0 mg, 0.020 mmol) phenylboronic acid (258 mg, 1.5 mmol) and potassium fluoride (174 mg, 3.0 mmol). The flask is evacuated and backfilled with argon, after which tetrahydrofuran (1 ml) and 2-bromothiophene (163 mg, 1.0 mmol) are added through a rubber septum. The flask is sealed with a Teflon screwcap and the reaction mixture is stirred at room temperature for 3 h and then diluted with diethyl ether (30 ml) and poured into a separatory funnel. The mixture is washed with 1 *M* aqueous sodium hydroxide solution (20 ml). The phases are separated and the aqueous phase extracted with diethyl ether (20 ml). The combined organic phases are washed with aqueous sodium chloride solution (20 ml), dried over magnesium sulfate and evaporated. The crude material is purified by flash chromatography on silica gel giving 159 mg (99%) of the title compound as a white solid mp 34–35°C.

4D.4.2.4 From tin compounds

The coupling between trimethyl 2-thienylstannane and 4-nitro-iodobenzene is claimed to be best catalyzed by palladium complexes not containing phosphine ligands and can be carried out in *N,N*-dimethylformamide at room temperature [53]. However, methyl 2-iodobenzoate and 4-iodoacetoxybenzene have been coupled with tributyl 2-thienylstannane using 5 mol% of dichlorobis(triphenylphosphine)palladium(II) in refluxing tetrahydrofuran [54]. 4-Iodo nitrobenzene can be used for the preparation of 2-(4-nitrophenyl)-3-alkylthiophenes through coupling with 3-alkyl-2-trialkylstannylthiophene using palladium(0) and cuprous iodide as cocatalyst [55].

Stille coupling between 4-bromoaniline and 2-(tributylstannyl)thiophene was not successful, so instead the tin derivative was coupled with the more reactive 4-bromonitrobenzene followed by reduction of the nitrophenyl derivative with dimine, or alternatively with sodium borohydride and copper(II) acetylacetonate, which gives the desired 2-(4-aminopheyl)thiophene in high yield [55,56].

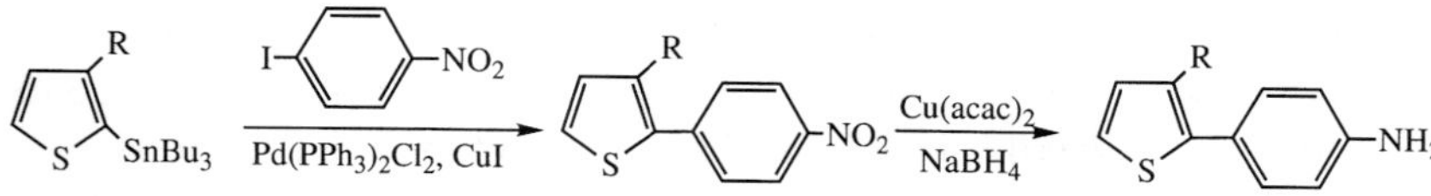

A similar approach was used for the synthesis of 3,4-ethylenedioxy-2-(4-aminophenyl)thiophene from 3,4-(ethylenedioxy)-2-trimethylstannylthiophene and 4-bromonitrobenzene [56]. The coupling of two equivalents of 4-bromonitrobenzene with one equivalent of 2,5-bis(trimethylstannyl)thiophene is used for the preparation of 2,5-bis(4′-nitrophenyl)thiophene in 66% [57].

The Stille coupling between 2-tributylstannylthiophene and dimethyl 5-iodoisophtalate using 2 mol% of bis(acetonitrile)dichloropalladium(II) in *N*,*N*-dimethylformamide gives dimethyl 5-(2-thienyl)isophtalate. A similar methodology was used for the preparation of 18-(2-thienyl)-2,3,5,6,8,9,11,12,14,15-decahydro-1,4,7,10,13,16-benzohexaoxacyclooctadecin from 2-tributylstannylthiophene and 4-bromobenzo-18-crown-6 [58].

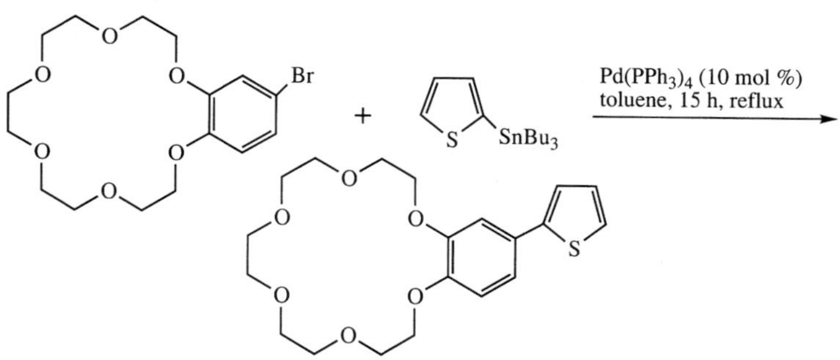

The coupling between 5-bromosalicylaldehyde and 2-(tributylstannyl)thiophene, giving 5-(2-thienyl)salicylaldehyde in 66% yield is the key step in the synthesis of cobalt salen-based conducting polymers [59].

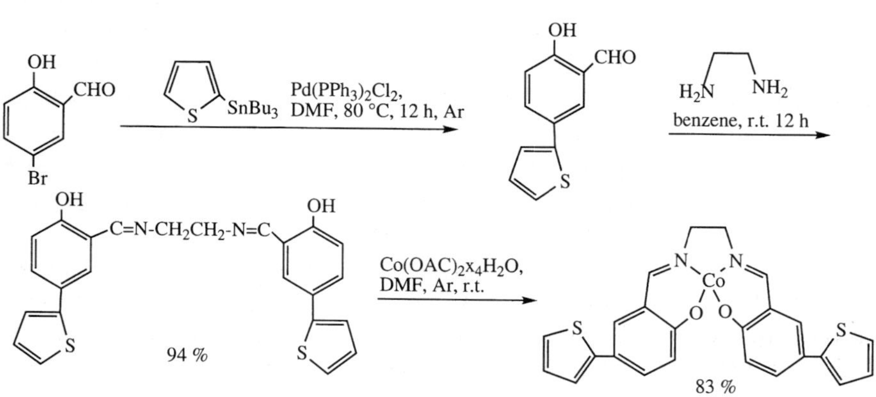

The Stille cross-coupling reaction can also be carried out on solid support. Thus 2-, 3- and 4-iodobenzoate linked to Merriefield resins has been coupled with tributyl 2-thienylstannane and after cleavage upon treatment with lithium hydroxide, 2-,3- and 4-(2-thienyl)benzoic acids were obtained in high yields [60].

The Stille reaction between 4-bromophenyl terminated dendrons and trimethylstannylthiophene gave the [G-3]-dendron, the periphery of which is fully functionalized with 2-thienyl groups [61].

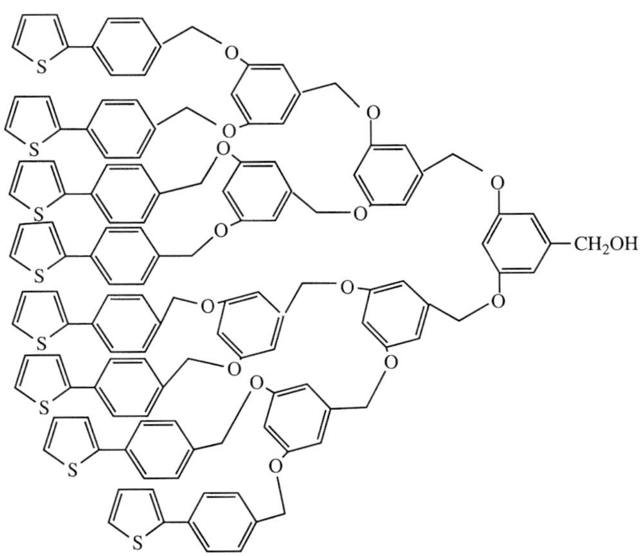

Th$_8$[G-3]-OH [61]

A 100 ml flask with gas-inlet is charged with Br$_8$-[G-3]-OH (278 mg, 0.125 mmol) and 2-trimethylstannylthiophene (525 mg, 2.13 mmol) and toluene (3 ml). The mixture is deaerated 7 times, each time saturated with argon. Then tetrakis(triphenylphosphine)palladium(0) (2 mol%) is added and the yellow mixture is refluxed constantly blanketed by argon. After 48 h the reaction is complete and the mixture is allowed to cool to room temperature. The product is extracted with dichloromethane/water giving a solid, which is purified by chromatography on silica gel (35 g) using chloroform as eluent, giving 130 mg (46%) of the title compound as a slightly yellow foam.

Cuprous iodide or manganese bromide catalyzed coupling of 2-thienyltributylstannane with *para*-iodoanisole or *para*-diiodobenzene in the presence of sodium chloride or potassium chloride in *N*-methylpyrrolidone at 90–100°C gives high yields of 2-(4-methoxyphenyl)thiophene and 1,4-di(2-thienyl)benzene [62].

The Stille coupling has been optimized using palladium on carbon with cuprous iodide as cocatalyst and from 2-tributylstannyl-3-methylthiophene and *para*-iodonitrobenzene, 3-methyl-2-(*para*-nitrophenyl)thiophene was obtained in 85% yield [63].

Various kinds of both symmetrical and asymmetrical di(thienyl)quinones can be prepared by the palladium-catalyzed cross-coupling of 2,3-dibromoquinone with tributyl-3-thienylstannane and tributyl(2-methyl-5-trimethylsilyl-3-thienyl)stannane [64].

4D.4.2.5 From lead compounds

Instead of thienyltin derivatives, 2-thienyllead triacetate can be used for the coupling with iodonium salts, using Pd_2(dibenzylideneacetone)$_3$/chloroform (5 mol%) as catalyst in the presence of two equivalents of sodium methoxide at room temperature in chloroform [65].

2-(4-Methoxyphenyl)thiophene [65]

To a stirred solution of 2-thienyllead triacetate (344 mg, 0.74 mmol) and sodium methoxide (66 mg, 1.23 mmol) in chloroform (3 ml), Pd_2(dibenzylideneacetone)$_3$/chloroform (14 mg, 5 mol%) is added, followed by *para*-methoxyphenyl(phenyl)iodonium tetrafluoroborate (245 mg, 0.61 mmol) at room temperature under nitrogen; the reaction mixture is stirred for 3 h. The product is extracted with diethyl ether (20 ml). The extract is washed three times with water, dried over magnesium sulfate and evaporated. The crude product is purified by chromatography on silica gel using hexane as eluent giving 72% of the title compound and 10% homocoupling of the lead compound.

4D.4.2.6 From thienylmetalorganic reagent and other classes of aryl compounds

The Stille coupling between 2-(trimethylstannyl)thiophene with 2-naphthyl triflate in the presence of lithium chloride gives 2-(2-thienyl)naphthalene [66].

2-(2-Thienyl)naphthalene [66]

To a solution of bis(dibenzylideneacetone)palladium (0.01 g, 5 mol%), triphenylphosphine (0.01 g, 10 mol%) and lithium chloride (0.02 g, 0.47 mmol) in tetrahydrofuran (5 ml) are added 2-naphthyl trifluoromethanesulfonate (99 mg, 0.36 mmol) and 2-(trimethylstannyl)thiophene (98 mg, 0.40 mmol).

The reaction mixture is heated at 60°C for 16 h and subsequently cooled to room temperature; the solvent is then removed under reduced pressure. The residue is dissolved in dichloromethane (10 ml). This solution is washed with 10% aqueous ammonium hydroxide solution (20 ml) and sodium chloride solution (20 ml), dried over magnesium sulfate and evaporated. The residue is purified by chromatography on silica gel using ethyl acetate/petroleum spirit (1:4) as eluent giving 53 mg (70%) of the title compound mp 103–108°C.

Using this technique 1,5-di(2-thienyl)naphthalene was prepared in high yields from 2-tributylstannylthiophene and naphthalene-1,5-ditriflate [67]. Milder conditions for the cross-coupling between organostannanes have recently been developed using iodanes at room temperature. Thus 2-thienyltributylstannane with hydroxy(tosyloxy)iodobenzene (Koser's reagent) in the presence of palladium(II) chloride (0.5 mol%) in acetonitrile/water (4:1) gave 2-phenylthiophene in 91% yield after 15 min [68].

The use of *meta*-oxobis(trifluoromethanesulfonato(phenyl)iodine, also called Zefirov's reagent, on the other hand led to a mixture of 2-phenylthiophene and 2-(4-iodophenyl)thiophene [68].

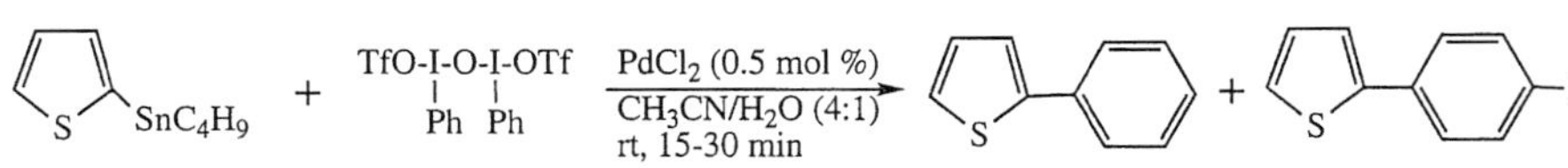

The coupling of 2-thienyl tributylstannane with diphenyliodonium tetrafluoroborate using the same catalyst and solvent system yields, after 30 minutes at room temperature, 2-phenylthiophene in 87% yield. Interestingly the unsymmetrical *para*-tolylphenylliodonium fluoroborate apparently gave only the 2-(*para*-tolyl)thiophene in 88% yield [69].

2-(para-Tolyl)thiophene [69]

To a stirred solution of *para*-tolylphenyliodonium fluoroborate (1.54 g, 4.0 mmol) and palladium(II) chloride (3.6 mg, 0.5 mol%) in acetonitrile/water (4:1) (10 ml), 2-tributylstannylthiophene (1.60 g, 4.3 mmol) is added. The reaction mixture is stirred at room temperature for 30 min and quenched with saturated aqueous ammonium chloride solution (10 ml), and the product extracted with diethyl ether (2 × 20 ml). The combined organic phases are dried over magnesium sulfate and evaporated. The residue is purified by chromatography on silica gel using hexanes as eluent, giving 0.61 g (88%) of the title compound.

Instead of palladium(II) chloride manganese dichloride hydrate (5 mol%) in *N*-methylpyrrolidone/tetrahydrofuran at 70°C can be used for coupling with diphenyl iodonium fluoroborate [70].

4D.4.2.7 From halothiophenes and arylmetal derivatives

Suzuki coupling between the benzophenone imine of (4-pinacolylborono)-phenylalanine ethyl ester and 2-bromothiophene is used for the preparation of 4-(2-thienyl)phenylalanine [71]. A modified Suzuki coupling between 2-thienyl(phenyl)iodonium tosylate and phenylboronic acid gave a quantitative yield of 2-phenylthiophene after five minutes [72].

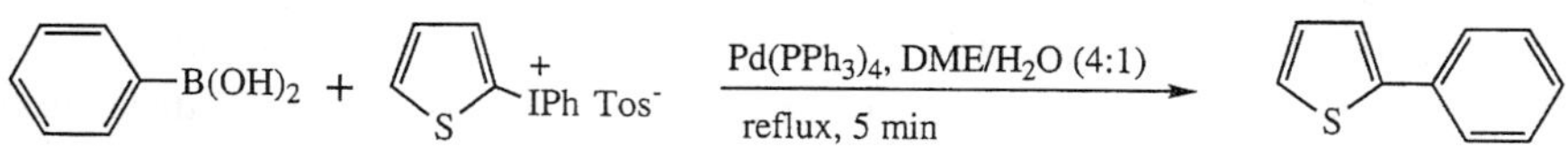

In another modification phenyl 2-thienyliodonium tetrafluoroborate is coupled with potassium *ortho*-methoxyphenyl trifluoroborate or potassium *ortho*-methylphenyl trifluoroborate to give 2-(*ortho*-methoxyphenyl)thiophene and 2-(*ortho*-methylphenyl)thiophene in excellent yield under mild conditions (60°C, 30 min, solvent 1,2-dimethoxyethane) in the presence of palladium catalyst and without added base [73].

The poly(ethylene glycol) PEG 6000 bound carboxylic ester of 5-bromo-2-thiophenecarboxylic acid were coupled with *ortho*-, *meta*- and *para*-formylphenylboronic acid to give the PEG 6000 bound carboxylic esters of 5-(formylphenyl)-2-thiophenecarboxylic acid [74]. Recently a very convenient microwave-assisted aqueous Suzuki coupling between PEG-bound 5-bromo-2-thiophenecarboxylic acid was developed and a number of arylboronic acids have been used for the preparation of 5-aryl-2-thiophenecarboxylic acid PEG esters, using palladium acetate catalysis [75].

The Stille coupling using palladium on carbon as catalyst and cuprous iodide as cocatalyst yields 2-phenylthiophene in 77% yield [76].

4D.5 BY OTHER METHODS

4D.5.1 Photochemical couplings

Photochemical coupling of 4-chloro-*N*,*N*-dimethylaniline with thiophene in acetonitrile gives 2-(4′-*N*,*N*-dimethylaminophenyl)thiophene [77].

2-(4′-N,N-dimethylaminophenyl)thiophene [77]

A solution of 4-chloro-*N*,*N*-dimethylaniline (780 mg, 5.0 mmol) and thiophene (8.4 g 100 mmol) in acetonitrile (100 ml) is irradiated in an immersion well apparatus fitted with a high-pressure mercury arc (125 W, water-cooled through a quartz jacket) after 15 min flushing with argon and maintaining a slow gas flux during the irradiation. Anhydrous potassium carbonate (2 g) is added and the solution magnetically stirred during the experiment. The irradiated solution is evaporated and the residue purified by chromatography on silica gel using cyclohexane/ethyl acetate as eluent, giving 548 mg (54%) of the title compound as a colorless solid mp 118–120°C after recrystallization from methanol.

4D.5.2 Aryne intermediates

The reaction of thiophene with different benzyne precursors yields a mixture of products, one among them being phenylthiophene [78]. Unhindered aryl triflates react with a mixture of 1.5 equivalents of lithium diisopropylamide and 8–10 equivivalents of 2-thienyllithium at – 7°C to give about 50% yield of 2-phenylthiophene [79].

4D.5.3 Trimerization of acetylthiophenes

1,3,5-Tri(2-thienyl)benzene and 1,3,5-tri(5-chloro-2-thienyl)benzene are obtained upon treatment of 2-acetyl and 5-chloro-2-acetylthiophene with silicon tetrachloride in ethanol [24,80].

1,3,5-Tri(2-thienyl)benzene [80]

2-Acetylthiophene (5.4 g, 0.04 mol) is treated with silicon tetrachloride (14.25 g. 0.08 mol) in anhydrous ethanol (40 ml) at 0°C. The reaction mixture is stirred at room temperature 6 h, after which it is poured into ice-cooled water. The product is extracted with dichloromethane, the combined organic phases are dried over magnesium sulfate and evaporated. The residue is purified by chromatography on silica gel using hexane as eluent, giving the title compound in a yield of 42%.

Cyclotrimerization of 1,2-bis(5-dodecyl-2-thienyl)acetylene using dicobalt octacarbonyl as catalyst gave hexakis(5-dodecyl-2-thienyl)benzene [81].

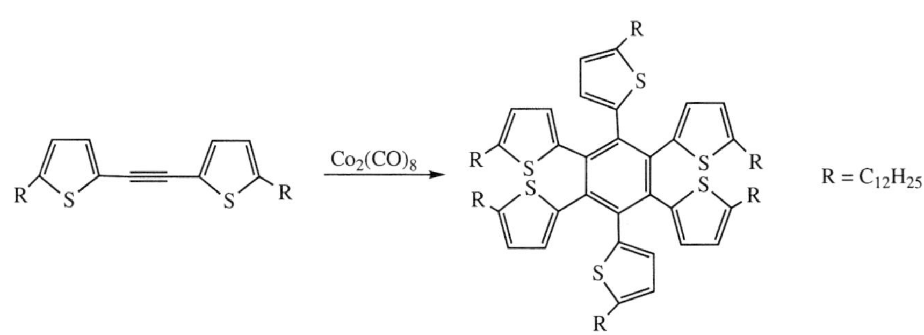

Hexakis(5-dodecyl-2-thienyl)benzene [81]

A suspension of 1,2-bis(5-dodecyl-2-thienyl)acetylene (1.50 g, 2.85 mmol) in dioxane (40.0 ml) is degassed several times with argon, after which dicobalt octacarbonyl (150 mg, 0.438 mmol) is added. The reaction mixture is refluxed for 1.5 h, cooled to room temperature and poured into water. The product is extracted three times with dichloromethane and the combined organic phases are washed in sodium chloride solution, dried over magnesium sulfate and evaporated. The residue is purified by chromatography on silica gel using petroleum ether/dichloromethane (95:5) as eluent, giving 0.09 g (61%) of the title compound as a colorless oil, which slowly formed a waxy white solid at room temperature.

The corresponding hexasubstituted benzenes carrying bithienyl and terthienyl groups were similarly obtained.

4D.5.4 Electrophilic substitution of arylthiophenes

In most cases, the higher reactivity of the thiophene than the aryl ring in electrophilic substitutions makes it possible to use such reactions for the preparation of arylthiophenes functionalized in the thiophene ring. Thus chloromethylation of 3,4-dimethyl-2-phenylthiophene and 2,4-dimethyl-5-phenylthiophene with chloromethylmethyl ether and zinc chloride in 1,2-dichloroethane give 2-chloromethyl-3,4-dimethyl-5-phenylthiophene and 3-chloromethyl-2,4-dimethyl-5-phenylthiophene, which without purification were transformed to the corresponding cyano derivatives through the reaction with tetrabutyl ammonium bromide and sodium cyanide in water in about 50% yield. These nitriles were then hydrolyzed with concentrated hydrochloric acid to the acetic acids, however, in unsatisfactory yields [82].

4D.5.5 Various methods

2-Chlorothiophene reacts easily with certain active aromatic compounds in the presence of aluminum chloride under mild conditions yielding the corresponding 2-arylthiophenes. As by-products diarylthiophenes and 5-chloro-2,2′-bithienyl are obtained. However, with 1-methoxynaphtalene an 83% yield of 2-(4-methoxy-1-naphthyl)thiophene was obtained [83].

3-Arylthiophenes can be prepared from 2,5-dichlorothiophene in two steps, as it reacts regioselectively with various aromatic compounds in the presence of aluminum chloride to give 4-aryl-2-chlorothiophene, which is easily converted to the corresponding 3-arylthiophenes by catalytic dechlorination [84].

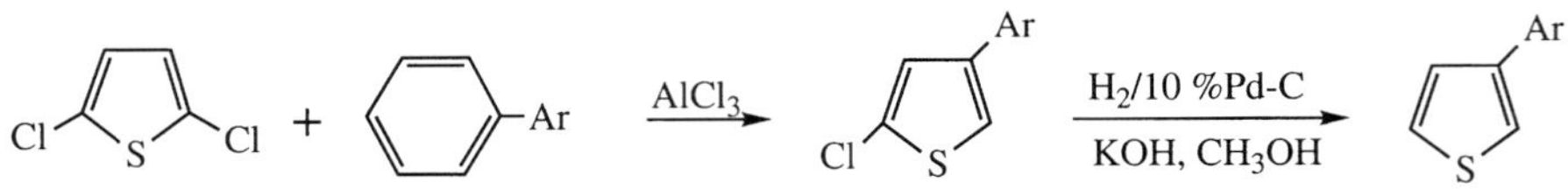

4-Phenyl-2-chlorothiophene [84]

Pulverized aluminum chloride (20 mmol) is added in portions over a 5 min period to a mixture of 2,5-dichlorothiophene (3.14 g, 20 mmol) and benzene (4.68 g, 20 mmol) in dichloromethane (10 ml) under ice cooling. An exothermic reaction occurs with the appearance of coloration and the rise of the temperature of the reaction mixture, which is stirred at 5°C for 30 min, at room temperature for 1 h and at reflux temperature for 30 min, after which it is poured into ice-water and the product extracted with chloroform (3 × 10 ml). The combined organic phases are extracted with water, 5% sodium bicarbonate solution, again with water, dried and evaporated. The residue is sublimed at 2–3 mm Hg followed by recrystallization from hexane.

REFERENCES

1. A. I. Kosak, R. J. F. Palchak, W. A. Steele and C. M. Selwitz, *J. Am. Chem. Soc.* **76**, 4450 (1954).
2. S. Gronowitz and N. Gjøs, *Acta Chem. Scand.* **21**, 2823 (1967).
3. A. G. A. Elagamey, M. A. Sofan, Z. E. Kandeel and M. H. Elnagdi, *Coll. Czech. Chem. Commun.* **52**, 1561 (1987).
4. A. Datta, H. Ila and H. Junjappa, *Tetrahedron Lett.* **29**, 497 (1988).
5. M. P. Balu, D. Pooranchand, H. Ila and H. Junjappa, *Tetrahedron Lett.* **29**, 501 (1988).
6. E. Waldvogel, *Helv. Chim. Acta* **75**, 907 (1992).
7. C. V. R. Sastry, A. K. Marwah, P. Marwah, G. S. Rao and D. R. Sridhar, *Synthesis* 1024 (1987).
8. P. Marwah, A. K. Marwah and G. Shankar, *Synth. Commun.* **19**, 2809 (1989).
9. D. R. Shridhar, M. Jogibhukta, P. S. Rao and V. K. Handa, *Synthesis* 1061 (1982).
10. P. Pouzet, I. Erdelmeier, D. Ginderow, J.-P. Mornon, P. M. Dansette and D. Mansui, *J. Heterocycl. Chem.* **34**, 1567 (1997).
11. R. M. Acheson and G. C. M. Lee, *J. Chem. Research (S)*, 380 (1986).
12. G. M. Acheson and G. C. M. Lee, *J. Chem. Soc., Perkin Trans. 1* 2321 (1987).
13. C.-D. Czogalla and F. Boberg, *Phosphorous and Sulfur* **35**, 127 (1988).
14. G. Kirsch, D. Prim, F. Leising and G. Mignani. *J. Heterocycl. Chem.* **31**, 1005 (1994).
15. K. Tsuji, K. Nakamura, T. Ogino. N. Konishi, T. Tojo, T. Ochi, N. Seki and M. Matsuo, *Chem. Pharm. Bull.* **46**, 279 (1998).
16. F. Freeman and H. Lu, *Tetrahedron Lett.* **34**, 1753 91993).
17. C. W. Ong, C. M. Chen, L. F. Wang and P. C. Shieh, *Tetrahedron Lett.* **39**, 9191 (1998).
18. H. Gotthardt, W. Phlaumbaum and P. Gutowski, *Chem. Ber.* **121**, 313 (1988).
19. T. Tsuda and A. Takeda, *Chem. Commun.* 1317 (1996).
20. M. Nilsson and C. Ullenius, *Acta Chem. Scand.* **24**, 2379 (1970).

21. M. Nilssion and C. Ullenius, *Acta Chem. Scand.* **25**, 2428 (1971).
22. A. Ohta, Y. Akita, T. Ohkuwa, M. Chiba, R. Kukunaga, A. Miyafuji, T. Nakata, N. Tani and Y. Aoyagi, *Heterocycles* **31**, 1951 (1990).
23. C. Amatore, A. Jutand, S. Negri and J.-F. Fauvarque, *J. Organomet. Chem.* **390**, 389 (1990).
24. A. Pelter, I. Jenkins and D. E. Jones, *Tetrahedron* **53**, 10357 (1997).
25. D. Yu, A. Gharavi, L. Yu, *J. Am. Chem. Soc.* **117**, 11680 (1995).
26. A. Pelter, M. Rowlands and I. Jenkins, *Tetrahedron Lett.* **28**, 5213 (1987).
27. M. B. Goldfinger, K. B. Crawford and T. M. Swager, *J. Am. Chem. Soc.* **119**, 4578 (1997).
28. S.-C. Ng, J.-M. Xu, H. S. O. Chan, A. Fujii and K. Yoshino *J. Mater. Chem.* **9**, 381 (1999).
29. S. Hotta, S. A. Lee and T. Tamaki, *J. Heterocycl. Chem.* **37**, 25 (2000).
30. S. Hotta, H. Kimura, S. A. Lee and T. Tamaki, *J. Heterocycl. Chem.* **37**, 281 (2000).
31. S. Hotta, *J. Heterocyl. Chem.* **38**, 923 (2001).
32. M. Irie, K. Sakemura, M. Okinake and K. Uchida, *J. Org. Chem.* **60**, 8305 (1995).
33. K. Uchida, Y. Kido, T. Yamaguchi and M. Irie, *Bull. Chem. Soc. Japan* **71**, 1101 (1998).
34. M. C. Suh, B. Jiang and T. D. Tilley, *Angew. Chem.* **39**, 2870 (2000).
35. K. Takahaski and T. Suzuki, *J. Am. Chem. Soc.* **111**, 5483 (1989).
36. K. Takahashi and T. Sakai, *Chem. Lett.* 157 (1993).
37. D. Melamed, C. Nuckols and M. A. Fox, *Tetrahedron Lett.* **35**, 8329 (1994).
38. R. D. Rieke, S.-H. Kim and X. Wu, *J. Org. Chem.* **62**, 6921 (19970.
39. B. H. Lipshutz, F. Kayser and N. Maullin, *Tetrahedron Lett.* **35**, 815 (1994).
40. Y. Hatanaka, S. Fukushima, and T. Hiyama, *Heterocycles* **30**, 303 (1990).
41. H. Finch, D. H. Reece and J. T. Sharp, *J. Chem. Soc., Perkin Trans. 1* 1193 (1994).
42. S. Gronowitz, V. Bobosik and K. Lavitz, *Chem. Scripta* **23**, 120 (1984).
43. G. W. Kabalka, V. Namboodiri and L. Wang, *Chem. Commun.* 775 (2001).
44. C. Xie, P. M. Lahti and C. George, *Org. Lett.* **2**, 3417 (2000).
45. Y. Shirota, M. Kinoshita, T. Noda, K. Okumoto and T. Ohara, *J. Am. Chem. Soc.* **122**, 11021 (2000).
46. A. S. Matharu, C. Grover, L. Komitov and G. Andersson, *J. Mater. Chem.* **10**, 1303 (2000).
47. T. Muto, T. Temma, M. Kimura, K. Hanabusa and H. Shirai, *Chem. Commun.* 1649 (2000).
48. K. Yagi, C. F. Soong and M. Irie, *J. Org. Chem.* **66**, 5419 (2001).
49. M. Takeshita and M. Irie, *Tetrahedron Lett.* **39**, 613 (1998).
50. Y. Han, A. Giroux, C. Lépine, F. Laliberté, Z. Huang, H. Perrier, C. I. Bayly and R. N. Young, *Tetrahedron* **55**, 11669 (1999).
51. A. Huth, I. Beetz and I. Schumann, *Tetrahedron* **45**, 6679 (1989).
52. J. P. Wolfe, R. A. Singer, B. H. Yang and S. L. Buchwald, *J. Am. Chem. Soc.* **121**, 9550 (1999).
53. N. A. Bumagin, I. C. Bumagina, and I. P. Beletskaya, *Dokl. Akad. Nauk SSSR*, **274**, 39 (1984).
54. T. R. Bailey, *Tetrahedron Lett.* **27**, 4407 (1986).
55. S. C. Ng, L. G. Xu, H. S. O. Chan, *Synthetic Metals* **94**, 185 (1998).
56. D. Nagvekar, B. Sankaran and L.-S. Tan, *Polymer Prep.* **39**, 548 (1998).
57. M. R. Kamel, S. A. Al-Taweel, M. M. El-Abadelah and K. M. Abu Ajaj, *Phosphorus, Sulfur and Silicon* **126**, 65 (1997).
58. E. Cielen, A. Tahri, K. VerHeyen G. J Hornaert, F. V. De Schryver and N. Boens, *J. Chem. Soc., Perkin Trans. 2* 1673 (1998).
59. R. P. Kingsborough and T. M. Swager, *Adv. Mater.* **10**, 1100 (1998).
60. S. Chamoin, S. Houldsworth and V. Snieckus, *Tetrahedron Lett.* **39**, 4175 (1998).
61. L. Groenendaal and J. M. J. Fréchet, *J. Org. Chem.* **63**, 5675 (1998).
62. S.-K. Kang, J.-S. Kim, S.-C. Choi, *J. Org. Chem.* **62**, 4208 (1997).
63. G. P. Roth, V. Farina, L. S. Liebeskind and E. Pena-Cabrera, *Tetrahedron Lett.* **36**, 2194 (1995).
64. S. Yoshida, H. Kubo, T. Saika and S. Katsumura, *Chem. Letters* 139 (1996).
65. S.-K. Kang, S. C. Choi and T.-G. Baik, *Synth. Commun.* **29**, 2493 (1999).
66. G. T. Crisp and S. Papadopoulos, *Aust. J. Chem.* **41**, 1711 (1988).

67. B. Sankaran, J. L. Burkett, B. Reinhardt and L.-S. Tan, *Polymer Prep.* **39**, 157 (1998).
68. S.-K. Kang, H.-W. Lee, J.-S. Kim and S.-C. Choi, *Tetrahedron Lett.* **37** 3723 (1996).
69. S.-K. Kang, H.-W. Lee, S.-B. Jang T.-H. Kim, and J.-S. Kim, *Synth. Commun.* **26**, 4311 (1996).
70. S.-K. Kang, W.-Y. Kim, Y.-T. Lee, S.-K. Ahn and J.-C. Kim, *Tetrahedron Lett.* **39**, 2131 (1998).
71. Y. Satoh, C. Gude, K. Chan and F. Firooznia, *Tetrahedron Lett.* **38**, 7645 (1997).
72. S.-K. Kang, H.-W. Lee, S.-B. Jang and P.-S. Ho, *J. Org. Chem.* **61**, 4720 (1996).
73. M. Xia and Z.-C. Chen, *Synth. Commun.* **29**, 2457 (1999).
74. C. G. Blettner, W. A. König, G. Rühter, W. Stenzel and T. Schotten, *Synlett.* 307 (1999).
75. C. G. Blettner, W.A. König, W. Stenzel, and T. Schotten, *J. Org. Chem.* **64**, 3885 (1999).
76. G. P. Roth, V. Farina, L. S. Liebeskind and E. Pena-Cabrera, *Tetrahedron Lett.* **36**, 2194 (1995).
77. B. Guizzardi, M. Mella, M. Fagnoni and A. Albini, *Tetrahedron*, **56**, 9383 (2000).
78. D. Del Mazza and M. G. Reinecke, *J. Org. Chem.* **53**, 5799 (1988).
79. K. H. Reuter and W. J. Scott, *J. Org. Chem.* **58**, 4722 (1993).
80. S. Kotha, K. Chakroborthy and E. Bramachary, *Synlett.* 1621 (1999).
81. Y. Geng, A. Fechtenkötter and K. Müllen, *J. Mater. Chem.* **11**, 1634 (2001).
82. K. Uchida, Y. Kido, T. Yamaguchi and M. Irie, *Bull. Chem. Soc. Japan* **71**, 1101 (1998).
83. T. Sone, R. Yokoyama, Y. Okuyama and K. Sato, *Bull. Chem. Soc. Japan* **59**, 83 (1986).
84. T. Sone, M. Inoue and K. Sato, *Bull. Chem. Soc. Japan* **61**, 3779 (1988).

– 4E –
Acylthiophenes

4E.1 THIOPHENE ALDEHYDES AND THEIR ALKYL DERIVATIVES

4E.1.1 Parent aldehydes

4E.1.1.1 Electrophilic substitution of thiophene and alkylthiophenes

Vilsmeyer–Haak formylation of thiophene with phosphoryl chloride and either *N,N*-dimethylformamide [1] or *N*-methylformanilide [2] is still the most useful method for the preparation of 2-thiophene aldehyde due to its exclusive selectivity for the α-position. A recent example is the preparation of 3,4-dioctyl-2-thiophene aldehyde [3].

3,4-Dioctyl-2-thiophene aldehyde [3]

A Schlenk tube under nitrogen is charged with 1,2-dichloroethane (40 ml), 3,4-dioctylthiophene (7.07 g, 23 mmol), and *N,N*-dimethylformamide (2.1 g, 28.7 mmol). The mixture is cooled to 0 °C and phosphorus oxychloride (4.4 g, 28.7 mmol) is added dropwise, after which the reaction mixture is refluxed for 3 h, cooled to room temperature and poured into 10% aqueous hydrochloric acid. This solution is stirred for 1 h and the product extracted with dichloromethane. The combined organic phases are washed several times with aqueous sodium bicarbonate solution, dried over magnesium sulfate, evaporated, and distilled giving 6.02 g (97%) of the title compound bp 165 °C/0.3 mm Hg.

3,5-Di-*tert*-butyl-2-thiophene aldehyde is obtained from 2,4-di-*tert*-butylthiophene and *N,N*-dimethylformamide [4] and 3,4-dihexyl-2-thiophene aldehyde was recently prepared from 3,4-dihexylthiophene [5]. A recent modification using trifluoromethane sulfonic anhydride/*N,N*-dimethylformamide complex might be of use for the formylation of deactivated thiophenes [6].

3,5-Di-tert-butyl-2-thiophene aldehyde [4]

Under ice-cooling, phosphorus oxychloride (5 ml, 55 mmol) is added to *N,N*-dimethylformamide (10 ml), after which the mixture is heated to 80 °C for 15 min, then cooled again and 2,4-di-*tert*-butylthiophene (8.58 g, 44 mmol) is added dropwise. The reaction mixture is heated at 100 °C for 5 h and then poured into a solution of sodium acetate. The phases are separated and the aqueous phase extracted with chloroform. The combined organic phases are washed with water until neutral reaction, dried, and evaporated. The residue is distilled *in vacuo* giving 7.24 g (74%) of the title compound as a viscous light-yellow liquid.

An alternative route for the preparation of 2-thiophene aldehyde in high yield is Rieche formylation of thiophene with dichloromethyl methyl ether, dichloromethyl butyl ether, or dichloromethyl methyl thioether using tin tetrachloride as catalyst [7,8]. From 2,5-dimethyl-3-alkylthiophenes a number of 2,5-dimethyl-3-alkyl-4-thiophene aldehydes were prepared, by the Rieche reaction with dichloromethylbutyl ether [4].

2,5-Dimethyl-3-alkyl-4-thiophene aldehydes [4]

Dichloromethyl butyl ether (0.25 mol) is added dropwise to a solution of 2,5-dimethyl-3-alkylthiophene (0.1 mol) and tin tetrachloride (0.27 mol) in dichloromethane (100 ml) under stirring at such a rate that the temperature does not exceed 5 °C. When the addition is completed the cooling bath is removed and the reaction mixture brought to room temperature for 40 min, refluxed for 3 h, and then poured into ice (500 ml) and concentrated hydrochloric acid (50 ml). This mixture is shaken in a separatory funnel until the violet color disappears. The phases are separated and the aqueous phases extracted with dichloromethane. The combined organic phases are washed with a solution of sodium carbonate and water and evaporated. The residue is steam distilled until 5 l is collected, the distillate is extracted with diethyl ether and the combined organic phases are dried over magnesium sulfate and evaporated. The residue is distilled *in vacuo*.

A mixture of the 2-formyl- and 5-formylthienylacetate is obtained upon reaction of ethyl 3-thienylacetate [9]. Other methods for the preparation of 2-thiophene aldehyde of less importance is the reaction with *N,N*-dimethylformamide triphenylphosphine dibromide [10] and with phosphorus oxychloride-1,4-diformylpiperazine [11]. 5-Methyl-2-thiophene aldehyde [12], 5-propyl-2-thiophene aldehyde [13], 5-isopropyl-2-thiophene aldehyde [14], 5-cyclopropyl-2-thiophene aldehyde [15], 5-butyl-2-thiophene aldehyde [16,17], 5-*tert*-butyl-2-thiophene aldehyde [18], 5-pentyl-, 5-isopentyl- and 5-hexyl-2-thiophene aldehyde [16], and other 5-alkyl-2-thiophene aldehydes [13,19,20] are all prepared by selective Vilsmeyer formylation in the 5-position of 2-alkylthiophenes. Also functionalized 2-alkylthiophenes such as 2-thienyl-(methylene)*n* carboxylates ($n = 3$–6) give selectively the 5-formyl derivative [21–23]. 2,3-Dialkyl-5-formylthiophenes [24–26], 3,4-trialkyl-, and 2,3,4-trialkyl-5-formylthiophenes [27] are obtained by formylation of the corresponding alkylthiophenes. 2,5-Dialkyl-3-thiophenealdehydes are obtained by the formylation of 2,5-dialkylthiophenes [14,17,28,29].

3-Alkylthiophenes are predominantly, but not selectively formylated in the 2-position [24,26,29–31]. In a recent paper a variety of methods for regioselective formylation of 3-methylthiophene have been examined. Optimal yields and regioselectivity for 2-formylation were obtained with *N*-formylpyrrolidine (11 : 1), although up to 46 : 1 ratio could be obtained with dichloromethylmethyl ether/titanium tetrachloride, albeit in lower yield. Optimal 5-formylation of 3-methylthiophene was obtained using *N*-formylindoline/oxalyl chloride [32].

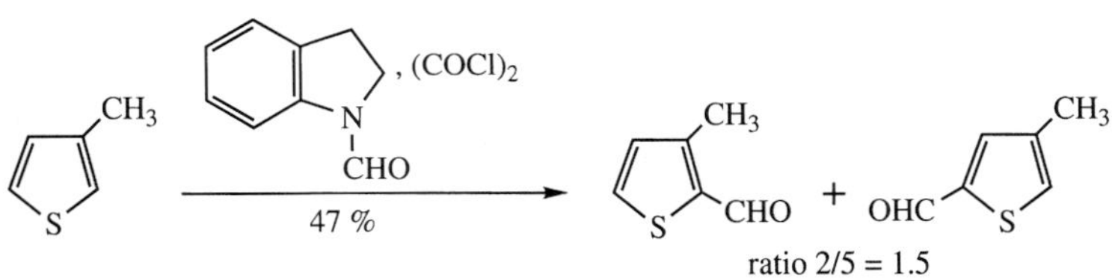

It is claimed that only 4-*tert*-butyl-2-thiophene aldehyde is obtained from the bulky 3-*tert*-butylthiophene [14]. The reaction of both 2- and 3-tributylstannylthiophene with dichloromethyl methyl ether in the presence of aluminium chloride followed by hydrolysis gives exlusively 2-thiophene aldehyde. Thus the 2-isomer gives *ipso*-substitution exclusively, while the 3-isomer reacts by electrophilic substitution in the activated 2-position [33].

2-Thiophene aldehyde [33]

A suspension of aluminium chloride in anhydrous dichloromethane (20 ml) is cooled to −78 °C under argon and a solution of dichloromethyl methyl ether and 3-tributylstannylthiophene in anhydrous dichloromethane (5 ml) is added during 15 min. After stirring at −78 °C for 4 h the mixture is hydrolyzed by saturated aqueous ammonium chloride (25 ml). The phases are separated and the aqueous phase extracted with dichloromethane (3 × 25 ml). The combined organic phases are treated with a saturated aqueous solution of potassium fluoride (15 ml), stirred vigorously for 3 h, after which the precipitated tributylstannyl fluoride is filtered off. The filtrate is extracted with dichloromethane (2 × 10 ml). The combined organic phases are dried over magnesium sulfate and evaporated. The title compound is obtained in a yield of 70%.

4E.1.1.2 From thiophenemagnesium halides and thienyllithium derivatives

The reaction of thiophenemagnesium halides and of thienyllithium derivatives with *N,N*-dimethylformamide [34,35] or other formylating reagents, such as triethyl formate [36,37] is a useful approach to thiophene aldehydes and has extensively been used during recent years. The reaction of thiophene with isopropylaminomagnesium chloride in tetrahydrofuran leads to the thiophenemagnesium derivative, which upon reaction with *N*-formylpiperidine gives 2-thiophene aldehyde [38].

2-Thiophene aldehyde [38]

Butylmagnesium chloride (0.6 *M*) in tetrahydrofuran (5 ml, 3.00 mmol) is added to a solution of diisopropylamine (0.46 ml, 3.28 mmol) in anhydrous tetrahydrofuran (12 ml) under argon at room temperature. The stirring is continued at room temperature for 24 h, after which thiophene (127 mg, 1.51 mmol) is added at room temperature. After 24 h *N*-formylpiperidine (440 mg, 3.89 mmol) is added and the stirring is continued for 27 h. The reaction mixture is diluted with saturated ammonium chloride solution (50 ml) and the product is extracted with chloroform (3 × 50 ml). The combined organic phases are dried over magnesium sulfate and evaporated. The residue is purified by chromatography on silica gel using hexane/ethyl acetate (5 : 1) as eluent giving 143 mg (84%) of the title compound as a colorless oil.

5-Ethyl-2-thiophene aldehyde was prepared from 5-ethyl-2-thiophenemagnesium bromide and triethyl formate [36]. There is generally no advantage in using the reaction of *N,N*-dimethylformamide with 2-thiophenemagnesium bromide derived from 2-bromothiophene or 2-thienyllithium obtained by metalation of thiophene over the Vilsmayer–Haak method, except in the case of 3-alkylthiophenes where the 3-alkyl group directs differently in electrophilic substitution and metalation. Thus 3-methylthiophene gives mainly the 5-formyl derivative *via* the metalation route [24,30], while Vilsmayer–Haak formylation gives, as mentioned above, predominantly the 3-methyl 2-formylthiophene [15]. However, the great synthetic usefulness is the reaction of thermally unstable thienyllithium derivatives, obtained by halogen–metal exchange at low temperatures, with *N,N*-dimethylformamide. In this way 3-thiophene aldehyde [39] and a number of alkylsubstituted 3-thiophene aldehydes are prepared [30,40–42], as well as 2-benzyl-3-thiophene aldehyde [43].

2-Benzyl-3-thiophene aldehyde [43]

2-Benzyl-3-bromothiophene (9.5 g, 38 mmol) in anhydrous diethyl ether (60 ml) is added dropwise to a solution of 1.5 *M* butyllithium in hexane (25 ml,

38 mmol) in anhydrous diethyl ether cooled to −70 °C under nitrogen. After stirring for 1 h *N,N*-dimethylformamide (3.0 g, 41 mmol) in anhydrous diethyl ether (10 ml) is added dropwise. The reaction mixture so obtained is stirred at −70 °C for 1 h, warmed to room temperature and poured onto ice. The phases are separated and the aqueous phase extracted with diethyl ether. The combined organic phases are dried over magnesium sulfate and evaporated to dryness. The residue is distilled giving 6.5 g (85%) of the title compound bp 114–116 °C/kPa

Lithiation of 2-methyl-3-thiopheneacetaldehyde acetal is used for the preparation of the aldehyde shown below [44].

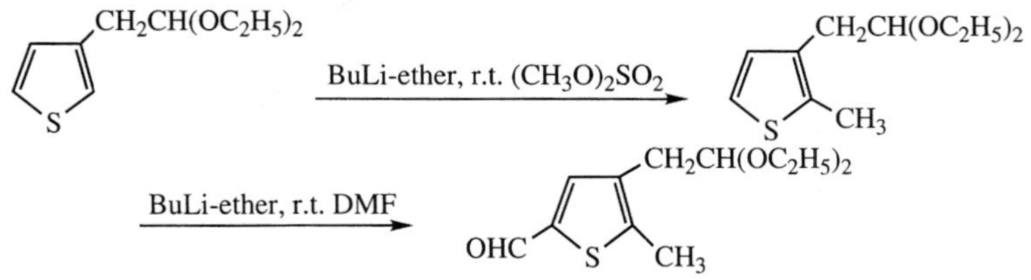

More complex derivatives can be prepared by this route. Thus bis(5-formyl-2-thienyl)methane, 2,2-bis(5-formyl-2-thienyl)butane, and 2,2-bis(5-formyl-2-thienyl)-3-methylbutane were prepared by metalation of the corresponding bis(thienyl) methane derivatives and *N,N*- dimethylformamide [45–47].

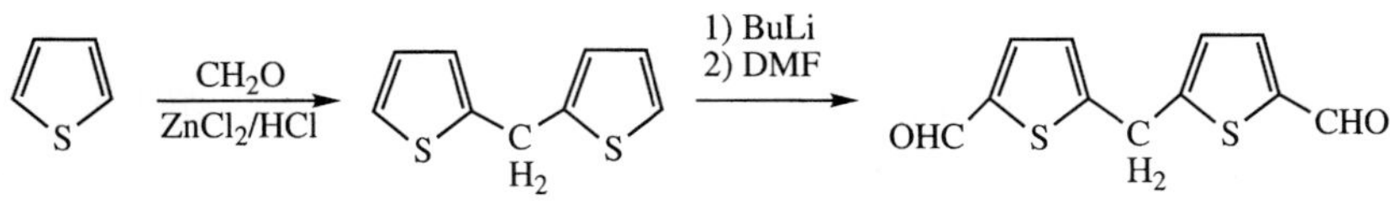

Bis(5-formyl-2-thienyl)-methane [45]

A solution of 10.1 *M* butyllithium in hexane (3.2 ml) is added dropwise to a stirred solution of 2,2′-dithienylmethane (2.64 g, 14.7 mmol) in anhydrous diethyl ether (100 ml) under nitrogen at such a rate that 20 °C is maintained. The stirring is continued at room temperature for 15 min, after which a solution of *N,N*-dimethylformamide (2.55 g) in anhydrous diethyl ether (25 ml) is added dropwise over 20 min and the stirring is continued at room temperature for 2 h. Water (100 ml) is cautiously added, the phases are separated and the organic phase is washed with water (100 ml), 2 *M* hydrochloric acid (150 ml), 5% sodium bicarbonate solution (150 ml) and water (150 ml), dried over magnesium sulfate and evaporated. The residue is recrystallized from 95% ethanol giving 2.35 g (68%) of the title compound as off-white crystals mp 93.5–95.0 °C.

Treatment of 1,1,1-tris(2-thienyl)ethane with five equivalents of butyllithium and subsequent addition of excess *N*,*N*-dimethylformamide gave 64% of the trialdehyde. Similarly the hexyl compound was obtained in 83% yield. They were used for the preparation of novel cage molecules bicapped with tris(2-thienyl)methanes by McMurry coupling [48].

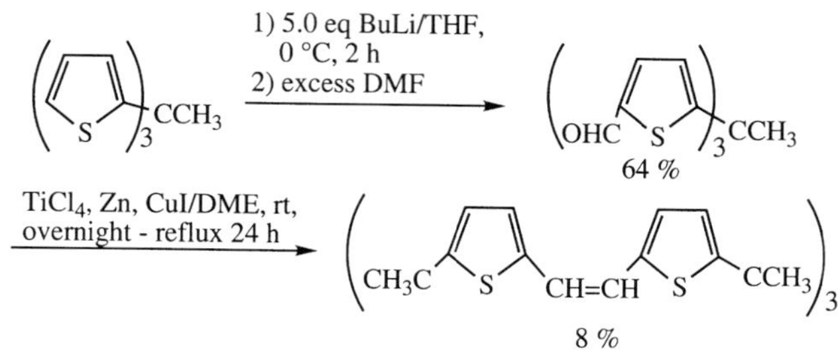

The following reaction has been performed by treatment with excess butyllithium followed by *N*,*N*-dimethylformamide [49].

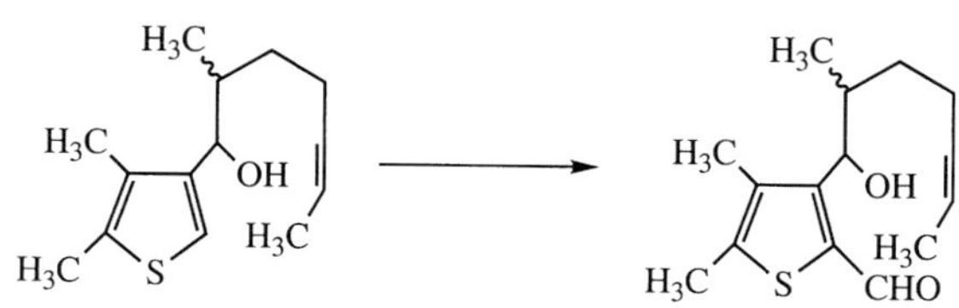

3-(1-Hydroxy-2-methyl-(E)-5-heptenyl)-4,5-dimethyl-2-formylthiophene [49]

Butyllithium (1.6 *M*) in hexane (2.5 ml, 4 mmol) is added to a solution of 4-(1-hydroxy-2-methyl-(*E*)-heptenyl)-2,3-dimethylthiophene (380 mg, 1.6 mmol) in anhydrous diethyl ether (10 ml) under nitrogen at room temperature and the mixture is refluxed for 2.5 h, after which it is treated with anhydrous *N*,*N*-dimethylformamide (1 ml, 13 mmol). After stirring the reaction mixture at room temperature for 15 h, it is hydrolyzed with 2 *M* hydrochloric acid under ice cooling. The phases are separated and the organic phase is washed with water and saturated sodium bicarbonate solution, dried over sodium sulfate and evaporated. The residue is purified by radial chromatography using cyclohexane/ethyl acetate (95 : 5) as eluent giving 69 mg (18%) as a light-yellow oil.

4E.1.1.3 From halomethyl-, hydroxymethyl- and aminomethylthiophenes

Preparation of 2-thiophene aldehyde [50] and 3-thiophene aldehyde [51] from halomethylthiophenes by the Sommelet reaction are described in

Organic Synthesis. This method has also been used for the preparation of some alkyl-substituted thiophene aldehydes, such as 2,5-dimethyl- 3-thiophene aldehyde [52]. 3-Bromomethylthiophene is oxidized to the 3-aldehyde by 2-nitropropane in ethanolic sodium ethoxide [53].

3-Thiophene aldehyde [53]

To a solution of sodium (2.3 g, 0.1 mol) dissolved in ethanol (100 ml), 2-nitropropane (8.9 g, 0.1 mol) is added. To this solution 3-bromomethylthiophene (17.7 g, 0.1 mol) is added dropwise over 10 min and the temperature is maintained at 25–35 °C. After stirring the reaction mixture at room temperature for 5 h, water is added (300 ml) and the product is extracted with diethyl ether (4 × 100 ml). The combined organic phases are washed twice with water, dried over calcium sulfate and evaporated. The residue, a red-brown oil, is distilled giving 6.2 g (56%) of the title compound bp 44–54 °C/1 mm Hg.

2-Thiophene aldehyde is obtained in a one-pot procedure, which involves successive hydrolysis to the alcohol and oxidation with dichromate [54]. 2-Thiophene aldehyde is also prepared by oxidation of the corresponding hydroxymethyl derivative with silver carbonate–celite in 94% yield [55], with acidified dichromate in 65% yield [54], dimethylsulfoxide activated by oxalyl chloride in 92% yield [56] or potassium persulfate in 43% yield [57]. Activated manganese dioxide has recently been used for preparation of the following aldehyde [58].

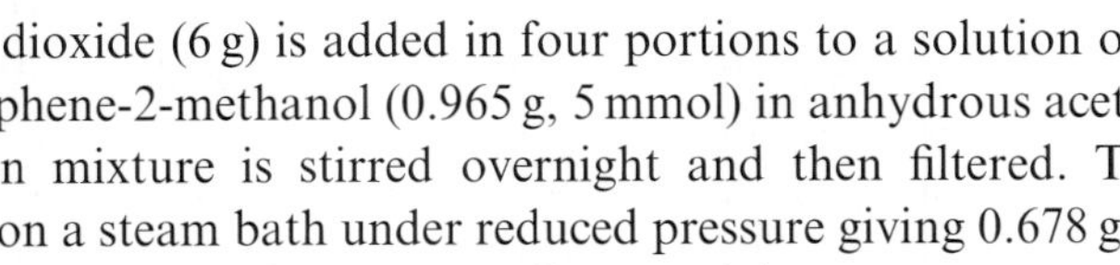

3-(Pyrrolylmethyl)-2-thiophene aldehyde [58]

Manganese dioxide (6 g) is added in four portions to a solution of 3-(pyrrolylmethyl)thiophene-2-methanol (0.965 g, 5 mmol) in anhydrous acetone (100 ml). The reaction mixture is stirred overnight and then filtered. The filtrate is evaporated on a steam bath under reduced pressure giving 0.678 g (71%) of the title compound in pure form as an oily material.

Warming 2-thenylammonium bromide with dimethylsulfoxide gives 2-thiophene aldehyde in 72% yield [59]. Another way to prepare 3-thiophene aldehyde is the side chain bromination of 3-methylthiophene to 3-(dibromomethyl)thiophene followed by hydrolysis with aqueous sodium carbonate and pyridine [60]. This approach was also used for the preparation of 3,5-di-*tert*-butyl-2-thiophene aldehyde by bromination of 2-methyl-3,5-di-*tert*-butylthiophene to the dibromomethyl derivative followed by hydrolysis [4].

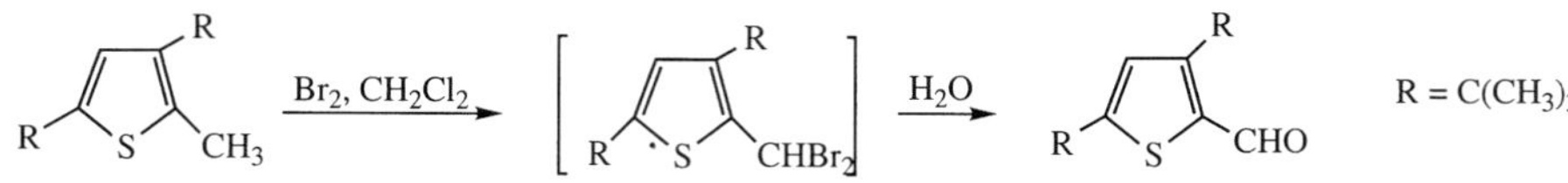

However, due to the lachrymatory properties of the halomethylthiophenes and the difficulties in obtaining selectively side chain bromination of methylthiophenes, these methods cannot seriously compete with the Vilsmeyer-Haak formylation and the route *via* lithium and magnesium compounds.

4E.1.1.4 From thiophenecarboxylic acid derivatives and cyanothiophenes

Thiophenecarboxylic acid chlorides are reduced to the corresponding aldehyde with sodium aluminium tri-*tert*-butoxide [61] or catalytically [62]. 3-Thiophene aldehyde is obtained in 80% yield by stannous chloride reduction of the iminochloride from 3-thiophenecarboxanilide (Sonn–Müller method) [63]. 3-Cyanothiophene is reduced to the aldehyde with stannous chloride/hydrochloric acid [63,64] or with sodium bis(2-methoxyethoxy) aluminium hydride [65].

4E.1.1.5 *Via* electrophilic substitution reactions of thiophene aldehydes and alkyl-substituted derivatives

The preparation of nitro-, alkoxy-, and halo-substituted thiophene aldehydes by appropriate electrophilic substitution is treated in Chapters 5, 6, and 7, respectively.

4E.1.1.5.1 Chloromethylation of thiophene aldehydes

4-Chloromethyl-2-thiophene aldehyde is selectively obtained *via* the reaction of 2-thiophene aldehyde with chloromethyl methyl ether and aluminium chloride (2.2 mol) [66,67]. If only 0.9 mol of aluminium chloride is used 5-chloromethyl-2-thiophene aldehyde is the main product [66]. 4-Chloromethyl-2-thiophene aldehyde is also obtained from 2-thiophene aldehyde by the reaction of bis(chloromethyl) ether with 1.8 mol of aluminium chloride [68]. Chloromethylation of 2-thiophene aldehyde with trioxane–zinc chloride-hydrochloric acid gives the 5-chloromethyl derivative predominantly [69]. The chloromethylation by this method gives the 4-chloromethyl derivative of 5-methyl-2-thiophene aldehyde in 79% yield [70].

4E.1.1.5.2 Other substitution reactions of thiophene aldehydes

The heptafluoropropylation of 2-thiophene aldehyde is not of preparative use as a mixture of the 5- and 3-heptaheptafluoropropyl-2-thiophene aldehyde is obtained upon reaction with bis(heptafluorobutyryl)peroxide [71]. The coupling reaction of 2-thiophene aldehydes with aromatic or aliphatic aldehydes promoted by samarium diiodide in the presence of hexamethylphosphoramide is a new method for the preparation of C-5 hydroalkylation products. The coupling reaction of 3-thiophene aldehyde occurs at C-2. Thus the reaction of 2-thiophene aldehyde with 4-methoxybenzaldehyde gave the following result in 45% yield [72].

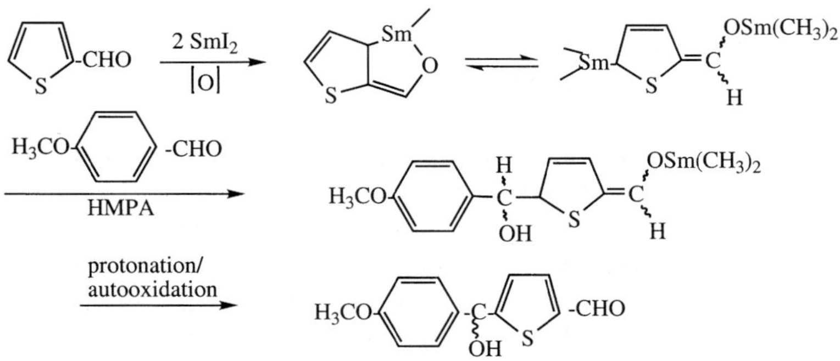

4E.1.1.5.3 Photochemical methods

Irradiation of 5-bromo-2-thiophene aldehyde in the presence of 2-ethynylthiophene leads to a mixture of 5-(2-thienylethynyl)-2-thiophene aldehyde and 2-ethynyl-5-formyl- 2,2′-bithienyl, while phenylacetylene in this reaction only gives 5-(phenylethynyl)-2-thiophene aldehyde [73–75].

Photochemical reaction of 5-bromo-2-thiophene aldehyde with 2-ethynylthiophene [74]

5-Bromo-2-thiophene aldehyde (402 mg, 2.1 mmol) in acetonitrile (300 ml) in the presence of 2-ethynylthiophene (3 g, 27.8 mmol) is purged with nitrogen for 1 h. The mixture is then irradiated in an immersion apparatus with a high-pressure mercury arc, (Helios–Italquartz) surrounded by a pyrex water-jacket. After 3 h the reaction mixture is diluted with chloroform and washed with saturated sodium chloride solution. The neutral organic phase is dried over sodium sulfate and evaporated. The residue is purified by chromatography on silica gel using chloroform/hexane (3 : 2) as eluent giving 59 mg pure 5-(2-thienylethynyl)-2-thiophene aldehyde and 10 mg 2-ethynyl-5-formyl-2,2′-bithienyl mp 78–79 °C.

Using very dilute solutions the yield of 5-(2-thienylethynyl)-2-thiophene aldehyde could be optimized. However, only small amounts could be obtained in this way [76]. When 5-bromo-2-thiophene aldehyde was used in the photochemical reaction, a higher selectivity was observed than in the reaction of the 5-iodo-2-thiophene aldehyde, but the reaction still was not synthetically useful because of a sharp decrease in reactivity [74]. However, 65% yield of 5-(5′-trimethylsilylthienylethynyl)-2-thiophene aldehyde was obtained in the photochemical reaction of 5-iodo-2-thiophene aldehyde with 2-ethynyl-5-trimethylsilylthiophene, which could be desilylated to the desired natural product 5-(2-thienylethynyl)-2-thiophene aldehyde [74].

5-(5′-Trimethylsilylthienylethynyl)-2-thiophene aldehyde [74]

A solution of 5-iodo-2-thiophene aldehyde (1 g, 4.2 mmol) and 2-ethynyl-5-trimethylsilylthiophene (2.5 g, 13.9 mmol) in acetonitrile (320 ml) is degassed with nitrogen for 1 h and then irradiated for 9 h in an immersion apparatus with a 500 W high pressure mercury arc surrounded by a Pyrex water-jacket. The reaction mixture is diluted with chloroform and washed with 0.1 *M* sodium thiosulfate solution and sodium chloride solution, dried over sodium sulfate and evaporated. The residue is purified by chromatography on silica gel using benzene as eluent giving 792 mg (65%) of the title compound.

4E.1.1.5.4 Substituted thiophene aldehydes *via* lithiation of protected aldehydes followed by electrophiles

The reaction of 2-thiophene aldehyde with lithium *N*-methyl piperazide followed by butyllithium/*N*,*N*,*N*′,*N*′-tetramethylethyleneamine and methyl iodide gives 5-methyl-2-thiophene aldehyde [77].

5-Methyl-2-thiophene aldehyde [77]

Butyllithium in hexane (3.31 mmol) is added to a solution of *N*-methylpiperazine (0.40 ml, 3.6 mmol) in anhydrous tetrahydrofuran (10 ml) at −78 °C. After 15 min 2-thiophene aldehyde (0.26 ml, 3 mmol) is added and the mixture is stirred at −78 °C for 15 min. *N*,*N*,*N*′,*N*′-tetramethylethyleneamine (1.36 ml, 9 mmol) and butyllithium in hexane (9 mmol) are then added and the mixture is stirred for 3 h at −23 °C, after which it is recooled to −78 °C and methyl iodide (1.1 ml, 18 mmol) is added dropwise. The reaction mixture is allowed to come to room temprature for 30 min and then with vigorous stirring poured into cold water (20 ml). The product is extracted with diethyl ether and the combined organic phases are washed with sodium chloride solution, dried over magnesium sulfate and evaporated. The residue is purified by radial preparative layer chromatography on silica gel using ethyl acetate/hexanes (2 : 8) as eluent, giving 290 mg (77%) of the title compound as a light yellow oil.

If lithiated *N*,*N*,*N*′-trimethylethylenediamine/lithium *N*-methyl piperazide is used, a mixture of 3-methyl-2-thiophene aldehyde and 5-methyl-2-thiophene aldehyde is obtained in the proportions of 2 : 1. 3,5-Dimethyl-2-thiophene aldehyde is obtained by the reaction of 5-methyl-2-thiophene aldehyde with lithium *N*-methyl piperazide followed by butyllithium and methyl iodide, or by the reaction of 3-methyl-2-thiophene aldehyde and lithium *N*-methyl piperazide followed by butyllithium and methyl iodide in about 90% yield [77]. 3-Thiophene aldehyde gives in the reaction with lithium *N*-methyl piperazide as the protecting group and methyl iodide 5-methyl-3-thiophene aldehyde and 2-methyl-3-thiophene aldehyde in the proportion of 83 : 17, while lithiated *N,N,N*-trimethylethylenediamine selectively gives 2-methyl-3-thiophene aldehyde. The reaction of 3-thiophenealdehyde with lithium *N*-methylpiperazine, followed by metalation with *sec*-butyl-lithium/*N,N,N′,N′*-tetramethylethyleneamine and benzaldehyde gives 2-(phenylhydroxymethyl)-4-thiophene aldehyde [78]. Finally 2,5-dimethyl-3-thiophene aldehyde is obtained from 2-methyl-3-thiophene aldehyde with lithium N-methylpiperazine as protecting group [77].

2,5-Dimethyl-3-thiophene aldehyde [77]

Butyllithium in hexane (1.8 mmol) is added to a solution of *N*-methylpiperazine (0.22 ml, 2 mmol) in anhydrous tetrahydrofuran (8 ml) at −78 °C. After 15 min 2-methyl-3-thiophene aldehyde (201 mg, 1.6 mmol) is added and the stirring is continued for 15 min. Butyllithium in hexane (3 mmol) is added and the mixture stirred at −23 °C for 3 h, after which it is recooled to −78 °C and methyl iodide (0.73 ml, 12 mmol) is added. The reaction mixture is allowed to warm to room temperature during 30 min and then under vigorous stirring poured into cold water (20 ml). The product is extracted with diethyl ether and the combined organic phases are washed with sodium chloride solution, dried over magnesium sulfate and evaporated. The residue is purified by radial preparative layer chromatography on silica gel using ethyl acetate/hexanes (2 : 8) as eluent giving 129 mg (54%) of the title compound as a light yellow oil.

2-(Acetoxyalkyl)-3-thiophene aldehydes are obtained by protecting the aldehyde function of 3-thiophene aldehyde with lithiated *N,N,N*-trimethylethylenediamine followed by lithiation of the 2-position and reaction with various aromatic aldehydes, followed by acetylation of the intermediate hydroxyalkyl derivative [79,80].

2-α-(Acetoxy-3,4-dimethoxybenzyl)thiophene aldehyde [79]

Butyllithium (33 mmol) is added to a solution of *N,N,N*-trimethylethylenediamine (4.6 ml, 36 mmol) in anhydrous tetrahydrofuran at −78 °C. After 5 min 3-thiophene aldehyde (3.33 g, 30 mmol) is added and the stirring continued for 25 min, after which diisopropylamine (4.7 ml, 33 mmol) and butyllithium

(60 mmol) are added. The flask is sealed and put in a freezer at −20 °C for 6 h. The mixture is recooled to −78 °C and 3,4-dimethoxybenzaldehyde (5 g, 30 mmol) in anhydrous tetrahydrofuran (50 ml) is added. The reaction mixture is allowed to warm to room temperature during 30 min and then under vigorous stirring poured into cold water (200 ml). The product is extracted with ethyl acetate and the combined organic phases washed with sodium chloride solution, dried, and evaporated. The residue is purified by chromatography on silica gel giving the hydroxy aldehyde, which is transformed into the acetoxy aldehyde by the usual method using acetic anhydride and triethylamine in dichloromethane. After purification by chromatography the title compound is obtained in a yield of 83% mp 91–92 °C.

Halogen–metal exchange of the neopentylglycol-protected 3-bromo-2-methyl-5-thiophene aldehyde followed by reaction with perfluorocyclopentene is used for the preparation of 1,2-bis(5-formyl-2-methylthiophene-3-yl)perfluorocyclopentene [81].

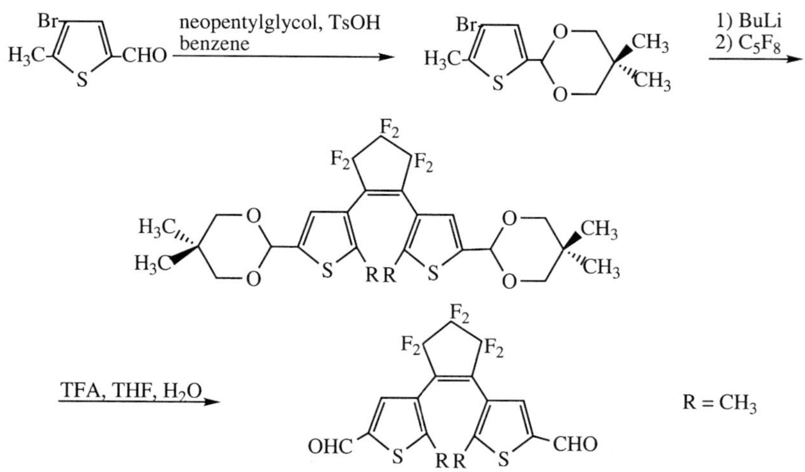

3-Bromo-2-methyl-5-thiophene aldehyde 2,2-dimethylpropan-1, 3-diyl acetal [81]

A solution of 3-bromo-2-methyl-5-thiophene aldehyde (10.0 g, 48.8 mmol), neopentylglycol (6.09 g, 58.5 mmol), and *para*-toluenesulfonic acid (1.00 g) in benzene (80 ml) is refluxed for 3 h. After cooling the solution is washed with 5% aqueous sodium bicarbonate solution (2 × 150 ml). The benzene phase is dried over sodium sulfate and evaporated giving 14.08 g (99%) of the title compound as a yellow solid mp 54.1–57.8 °C after recrystallization from hexane.

1,2Bis[5-(5,5-dimethyl-1,3-dioxacyclohex-2-yl)-2-methylthiophene-3-yl]-perfluorocyclo-pentene [81]

A solution of 3-bromo-2-methyl-5-thiophene aldehyde 2,2-dimethylpropan-1,3-diyl acetal (2.00 g, 6.85 mmol) in anhydrous diethyl ether (30 ml) under nitrogen is cooled to −78 °C and 1.6 *M* butyllithium in hexane (4.5 ml, 7.2 mmol) is added. The mixture is stirred for 2 h, after which perfluorocyclopentene (0.43 ml, 3.44 mmol) is added through a cooled syringe. The stirring is continued for 2 h and the reaction mixture is allowed to warm to room temperature. After stirring at this temperature for 2 h the reaction mixture is diluted with diethyl ether, washed with 1% hydrochloric acid, saturated aqueous sodium bicarbonate solution and water, dried over magnesium sulfate, and evaporated. The residue, a yellow–brown syrup, is purified by chromatography on silica gel using dichloromethane/hexane (1 : 1 to 3 : 1) giving 1.47 g (72%) of the title compound as a yellow solid mp 160.0–162.2 °C.

1,2- Bis(5-formyl-2-methylthiophene-3-yl)perfluorocyclopentene [81]

Trifluoroacetic acid (30 ml) is added to a solution of 1,2-bis[5-(5,5-dimethyl-1,3-dioxacyclohex-2-yl)-2-methylthiophene-3-yl]-perfluorocyclopentene (797 mg, 1.34 mmol) in tetrahydrofuran (150 ml) and water (40 ml). The reaction mixture is stirred for 50 min at room temperature and then quenched with saturated aqueous sodium bicarbonate solution. The product is extracted with diethyl ether and the combined organic phases are washed with 2% aqueous sodium bicarbonate solution, dried over sodium sulfate and evaporated. The residue is purified by chromatography on silica gel using dichloromethane/hexane (3 : 1) as eluent, giving 406 mg (71%) mp 179.8–184.5 °C.

4E.1.1.6 Palladium-catalyzed substitution reactions of thiophene aldehydes

2-Thiophene aldehydes with aryl substituents in the 4- and 5-positions are conveniently prepared by Suzuki coupling of 5-bromo- and 4-bromo-2-thiophenealdehyde with arylboronic acids in aqueous media at room temperature [82].

5-(4′-Fluorophenyl)-2-thiophene aldehyde [82]

A 50-ml round bottom flask is charged with 2-bromo-5-thiophene aldehyde (1.77 g, 6.16 mmol), 4-fluorophenylboronic acid (1.03 g, 6.78 mmol), tetrabutylammonium bromide (1.99 g, 6.16 mmol), palladium acetate (0.03 g, 2 mol%), and potassium carbonate (2.13 g, 15.4 mmol). Deionized water (10 ml) is added and the reaction mixture is stirred vigorously for 2 h, after which it is diluted with water (30 ml). The product is extracted with ethyl acetate (2 × 100 ml) and the combined organic phases are stirred over charcoal (5 g) for 30 min followed by addition of sodium sulfate. After evaporation the colorless residue is purified by medium pressure chromatography using hexane/ethyl acetate (4 : 1) as eluent, giving 1.046 g (82%) of the title compound as a yellow solid.

Palladium-catalyzed reaction of 2-thiophene aldehyde with methyl acrylate is used for the preparation of (*E*)-methyl 2-(5-formylthienyl)acrylate [83].

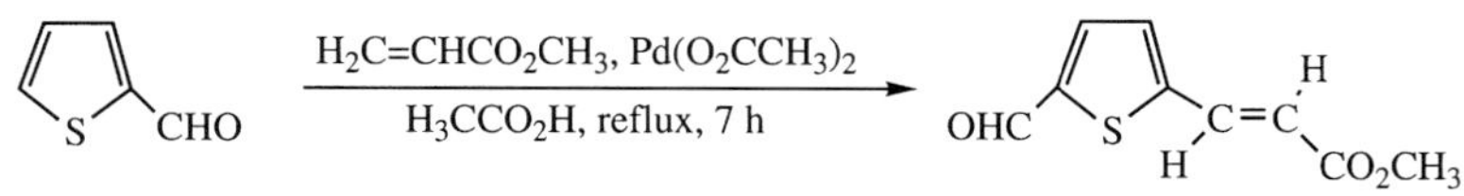

34 %, m. p. 68-69 °C

4E.2 DI-, TRI- AND TETRAFORMYLTHIOPHENES

4E.2.1 *Via* metalation of thiophene

Metalation of thiophene with two equivalents of butyllithium/*N*,*N*,*N*′,*N*′-tetramethylethyleneamine at 40 °C followed by *N*,*N*-dimethylformamide and hydrolysis is probably the best method for the preparation of 2,5-diformylthiophene [84].

2,5-Diformylthiophene [84]

A suspension of 2,5-dilithiothiophene is prepared by addition of butyllithium in hexane (0.12 mol) at room temperature to a mixture of *N*,*N*,*N*′,*N*′-tetramethylethyleneamine (13.9 g, 0.12 mol), thiophene (8.4 g, 0.10 mol), and hexane (30 ml). The temperature of the white suspension is allowed to rise to 40 °C and the conversion is completed by refluxing the mixture for 30 min. After addition of anhydrous tetrahydrofuran (120 ml) the solution is cooled to −40 °C and excess *N*,*N*-dimethylformamide (0.27 mol) is added over a period of 10 min. The temperature of the reaction mixture is gradually raised to room temperature and the stirring is continued for 30 min, after which the reaction mixture is poured into a mixture of 30% hydrochloric acid (200 g) and water (1700 ml) at −20 to −5 °C under vigorous stirring. Part of the dialdehyde may separate during this hydrolysis. Saturated sodium bicarbonate solution is slowly added until the aqueous phase has reached pH = 6. The phases are separated and the aqueous phase extracted with diethyl ether (7 × 50 ml). The combined organic phases are dried over magnesium sulfate and evaporated. The residue is crystallized from tetrahydrofuran/diethyl ether (4 : 1) giving 10.5 g (75%) of the title compound mp 109–112 °C.

Vilsmeyer formylation of (*E*)-1,2-(3,4-dihexyl-2-thienyl)ethene gave this dialdehyde shown below [5].

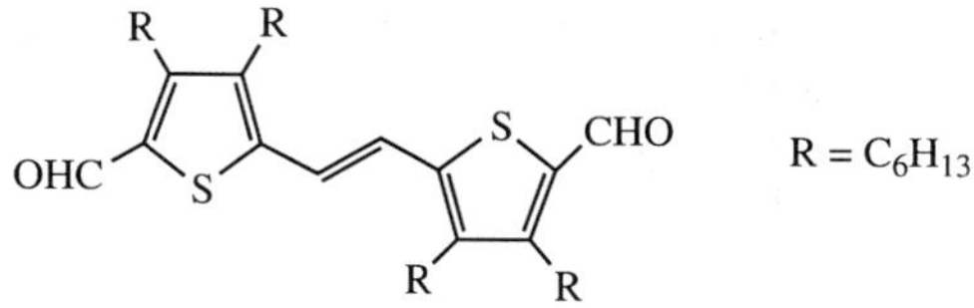

By metalation with butyllithium in tetrahydrofuran followed by *N*,*N*-dimethylformamide the following transformation has been performed [85].

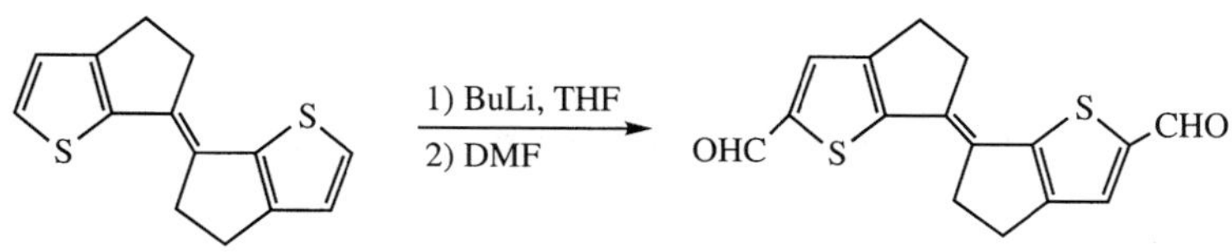

2,2′-Diformyl-6,6′-bis(4,5-dihydro-3H-cyclopenta[b]thienylidene) [85]

A round-bottomed flask equipped with a dropping funnel and nitrogen inlet is charged with 6,6′-bis(4,5-dihydro-3*H*-cyclopenta[*b*]thienylidene) (0.1 g, 0.41 mmol) in anhydrous tetrahydrofuran (20 ml). The solution is cooled to 0 °C and 1.6 *M* butyllithium in hexane (0.54 ml, 0.86 mmol) is added dropwise, after which the stirring is continued for 15 min and anhydrous *N*,*N*-dimethylformamide (0.20 ml, 2.34 mmol) is added. The reaction mixture is allowed to warm to room temperature and stirred for another 30 min. Water is added and the product extracted with diethyl ether. The combined organic phases are dried over calcium chloride and evaporated. The residue is purified by chromatography on silica gel using dichloromethane as eluent, giving 0.10 g (81%) of the title compound as a red powder mp 198–200 °C.

4E.2.2 *Via* lithiation of protected aldehydes

Acetal-protected 2-thiophene aldehyde gives 2,5-diformylthiophene [86,87] upon reaction with butyllithium followed by *N*,*N*-dimethylformamide and hydrolysis, and from 3-thiophene aldehyde, 2,3-diformylthiophene is obtained *via* metalation of the diethyl acetal followed by *N*,*N*-dimethylformamide [88,89]. The intermediate acetal can, of course, be isolated and used in carbonyl reactions. 2,3,4-Triformylthiophene and 2,3,4,5-tetraformylthiophene have likewise been prepared by metalation and reaction with *N*,*N*-dimethylformamide of 3,4-di- and 2,3,4-triacetals, respectively [90,91].

This method is used for the preparation of 2,4-diformylthiophene from 4-bromo-2-thiophene aldehyde acetal [89,91–93] and from 3-iodo-2-thiophene-aldehyde diethylacetal upon halogen–metal exchange with butyllithium, followed by *N*,*N*-dimethylformamide 3-formyl-2-thiophene aldehyde diethyl acetal is prepared [89].

4E.2.3 From dihalothiophenes *via* halogen–metal exchange and reaction with *N*,*N*-dimethylformamide

Two formyl groups are in this way introduced successively in a one-flask procedure. There is no need to specifically protect the first formyl group since the reaction intermediate itself

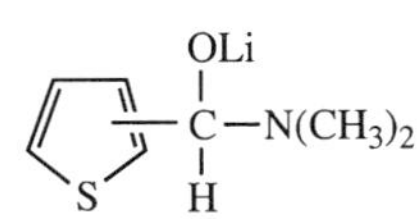

serves this purpose [94–99]. In this way 2,3-diformyl-, 3,4-diformyl-, 2,5-diformyl-, and 2,4-diformylthiophene were prepared from 2,3-dibromo-, 3,4-diiodo-, 2,4-diiodo-, and 2,5-diiodothiophene, respectively.

2,3-Diformylthiophene [98]

To a solution of butyllithium prepared from lithium (8.10 g) and butyl bromide (80 g, 0.58 mol) in anhydrous diethyl ether (350 ml) cooled to −60 °C, a solution of 3-bromothiophene (32.6 g, 0.20 mol) in anhydrous diethyl ether (60 ml) is added dropwise during 30 min. The stirring is continued for 50 min, after which a solution of *N*,*N*-dimethylformamide (45.8 g, 0.63 mol) in anhydrous diethyl ether (220 ml) is added dropwise during 2 h. The reaction mixture is left at −5 °C for 15 h and then the temperature is gradually raised to −30 °C during 3 h. The reaction mixture is poured into water and the aqueous phase is acidified. The phases are separated and the aqueous phase extracted with diethyl ether. The combined organic phases are evaporated and the residue distilled giving 11.2 g (40%) of the title compound as white crystals bp 90–95 °C/0.2 mm Hg, mp 78 °C. It is in most cases better to use the diiodo derivatives as for instance 3,4-dibromothiophene gives 4-bromo-2,3-diformylthiophene *via* a combination of halogen–metal exchange and metalation [98]. 2,4-Diformyl and 2,5-diformylthiophene have been prepared recently in good yields, by using *tert*-butyllithium in hexane at −70 °C followed by *N*,*N*-dimethylformamide [100].

2,4-Diformylthiophene [100]

tert-Butyllithium (1.7 *M*) in hexane (16.8 ml) is added dropwise to a solution of 2,4-dibromothiophene (0.97 g, 4.01 mmol) in anhydrous diethyl ether (200 ml) at −78 °C and after stirring the mixture for 5 min *N,N*-dimethylformamide (3 ml) is added. The stirring is continued for 1 h before the cooling bath is removed and the reaction mixture is allowed to warm gradually to room temperature. The yellow oil obtained by usual work-up is purified by chromatography on silica gel using hexane/dichloromethane (9 : 1) as eluent, giving 0.35 g (63%) of the title compound as a white solid mp 79–80 °C.

3,4-Diformylthiophene [101]

Butyllithium is added with stirring to a solution of 3,4-diiodothiophene (77.8 g, 0.23 mol) in anhydrous diethyl ether (200 ml) under nitrogen at −60 °C. The stirring is continued for another hour, after which *N,N*-dimethylformamide (42 g, 0.575 mol) in anhydrous diethyl ether (100 ml) is added dropwise for 1.5 h at such a rate that the temperature is kept between −60 and −55 °C. After stirring for 5 h at −40 °C the reaction mixture is hydrolyzed, the phases separated and the water phase extracted with ether. The combined organic phases give, upon evaporation, 18.9 g (58%) of the title compound as white crystals mp 81 °C.

4E.2.4 From halomethyl-, hydroxymethyl-, and aminomethylthiophenes

The Sommelet reaction of (chloromethyl)thiophene aldehydes can be used for the preparation of dialdehydes [102]. Bis(chloromethyl)thiophenes are converted to the dialdehydes by the Kröhnke procedure, *para*-nitroso-*N,N*-dimethylaniline of the bis(pyridinium) salt followed by acidic hydrolysis of the resulting nitrone [69].

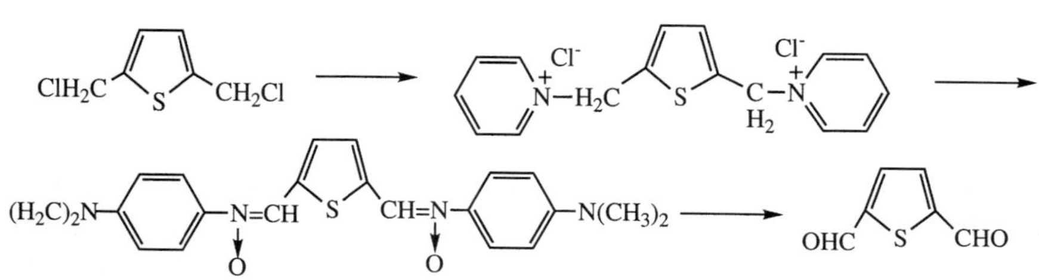

Other methods for the preparation of dialdehydes consist in the oxidation of the bis(chloromethyl) derivatives with dimethylsulfoxide/sodium bicarbonate

[103] or 2-nitropropane-potassium iodide–sodium ethylate [104]. Bis(hydroxymethyl)thiophenes are oxidized to the dialdehydes by oxidation with manganese dioxide [105–107] or chromic acid [107]. Bis(methoxymethyl)thiophenes are similarly oxidized in high yield to the dialdehydes [108].

4E.2.5 Reduction of cyano derivatives

Dicyanothiophenes are successfully reduced to the dialdehydes by the use of diisobutyl aluminium hydride [96,109–111].

4E.3 VARIOUS SUBSTITUTED THIOPHENE ALDEHYDES (OTHER THAN ALKYL SUBSTITUTED)

3-Phenyl-2-thiophene aldehyde is prepared by formylation of 3-phenylthiophene [112]. 5-(2-Pyridyl)-2-thiophene aldehyde is prepared by halogen–metal exchange of 5-(2-pyridyl)-2-bromothiophene at −70 °C followed by reaction with *N*,*N*-dimethylformamide [113] or by the reaction of 2-(2-pyridyl)thiophene with butyllithium in tetrahydrofuran at 0 °C followed by 1-formylpiperidine [114].

5-(2-Pyridyl)-2-thiophene aldehyde [114]

Butyllithium (1.6 *M*) in hexane (14 ml, 22 mmol) is added to a solution of 2-(2-pyridyl)thiophene (3.22 g, 20.0 mmol) in anhydrous tetrahydrofuran (75 ml) at 0 °C. This mixture is treated with 1-formylpiperidine (2.5 ml, 22 mmol) and the stirring is continued at 20–25 °C for 12 h, after which the reaction mixture is hydrolyzed with water (50 ml). The product is extracted with diethyl ether (3 × 40 ml), the combined organic phases are washed with concentrated ammonium chloride solution, dried over magnesium sulfate and evaporated. The residue is purified by chromatography on silica gel (200 g) using dichloromethane/ethyl acetate (1 : 1) as eluent followed by sublimation at 110 °C/0.7 mbar, giving 2.39 g (63%) of the title compound as a colorless solid mp 119–120 °C.

4E.3.1 *Via* arylation of thiophene aldehydes

5-Aryl-2-thiophene aldehyde is obtained by treatment of 2-thiophene aldehyde with a diazonium salt in acetone in the presence of copper(II) chloride [115]. The reaction of 2-thiophene aldehyde with the diazonium salt from *para*-aminobenzoic acid gave 5-(*para*-carboxyphenyl)-2-thiophene aldehyde in mediocre yield, while 3-thiophene aldehyde did not give the desired product [116].

The Heck arylation of 3-thiophene aldehyde and iodobenzene using the conditions described by Jeffery have been studied in detail. However, even under the best conditions a mixture of 2-phenyl-3-thiophene aldehyde (66%), 2,4-diphenyl-3-thiophene aldehyde (6%), and biphenyl (38%) were obtained. 2-Thiophene aldehyde is less reactive and larger amounts of biphenyl are formed, but this reaction is regiospecific and 5-phenyl-2-thiophene aldehyde can be isolated [117,118]

5-Phenyl-2-thiophene aldehyde [118]

A suspension of potassium carbonate (40 mmol), tetrabutylammonium bromide (16 mmol), and palladium acetate (0.8 mmol) in acetonitrile/water (3.7 : 0.4 ml) is stirred under nitrogen for 5 min, after which 2-thiophene aldehyde (3.58 g, 32 mmol) and iodobenzene (3.30 g, 16 mmol) are successively added. The reaction mixture is heated at 80 °C for 7 h. After cooling to room temperature water and diethyl ether are added. The phases are separated and the organic phase is washed with water, dried over magnesium sulfate and evaporated. The residue is recrystallized giving 0.77 g (30%) of the title compound as a yellow solid mp 89 °C.

Reacting 2-thiophene aldehyde with iodobenzene using a catalytic system of palladium acetate and cuprous iodide in the presence of cesium carbonate gives 5-phenyl-2-thiophene aldehyde [119].

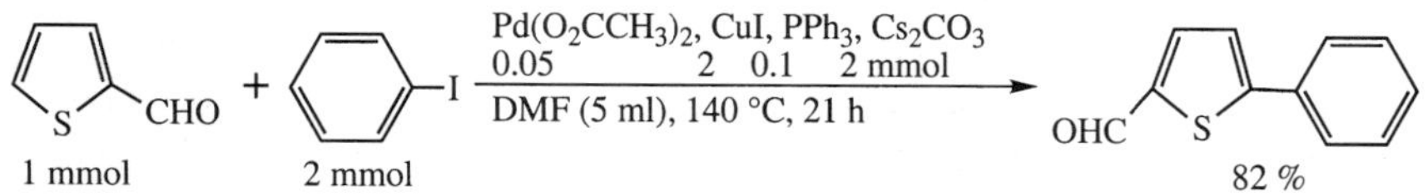

2-(2-Nitrophenyl)-3-thiophene aldehyde is prepared by palladium(0) catalyzed coupling of 2-tributylstannyl-3-thiophene aldehyde with *ortho*-bromonitrobenzene [120].

2-(2-Nitrophenyl)-3-thiophene aldehyde [120]

A 250-ml three-necked flask equipped with condenser, magnetic bar, thermometer, and nitrogen inlet is charged with 2-tributylstannyl-3-thiophene aldehyde (8.0 g, 0.02 mol), *ortho*-bromonitrobenzene (4.0 g, 0.02 mol), tetrakis(triphenylphosphine)palladium(0) (0.69 g, 0.6 mmol), and anhydrous *N*,*N*-dimethylformamide (90 ml). The reaction mixture is stirred at 100 °C for 24 h, cooled to room temperature and evaporated. The residue is diluted with water followed by extraction with diethyl ether three times. The combined organic phases are dried over magnesium sulfate and evaporated. The residue is recrystallized from ethanol giving 2.42 g (52%) of the title compound as yellow needles mp 79–80 °C.

The palladium-catalyzed reaction of 2-bromothiophene with 1,2-dimethyl-1*H*-imidazole is used for the preparation of 5-(2-thienyl)-1,2-dimethyl-1*H*-imidazole [119]. Suzuki coupling of 3-nitrophenylboronic acid with 5-bromo-2-thiophene aldehyde gives 5-(3-nitrophenyl)-2-thiophene aldehyde [121].

5-(3-Nitrophenyl)-2-thiophene aldehyde [121]

Tetrakis(triphenylphosphine)palladium(0) (346 mg, 0.3 mmol), 2 *M* sodium carbonate solution (30 ml), and ethanol (12 ml) are added to a suspension of 5-bromo-2-thiophene aldehyde (1.15 g, 6 mmol) and 3-nitrophenylboronic acid (1.2 g, 7.2 mmol) in toluene (30 ml) under nitrogen. After 5.5 h under reflux the reaction mixture is poured into dichloromethane (150 ml) and water (100 ml). The phases are separated and the aqueous phase extracted with dichloromethane (2 × 50 ml). The combined organic phases are dried over sodium sulfate and concentrated to 5 ml, which are purified by chromatography using dichloromethane as eluent. The solid obtained is recrystallized from hexane/chloroform giving 1.17 g (84%) of the title compound as beige crystals mp 147 °C.

4E.3.2 *Via* lithiation of acetal-protected formylthiophenes

2-Formyl-3-thiopheneboronic acid [122], 5-methyl-2-formyl-3-thiopheneboronic acid [123], 5-formyl-2-thiopheneboronic acid [122] 3-formyl-2-thiopheneboronic acid [124] 4-chloro-3-formyl-2-thiopheneboronic acid [123], 5-chloro-3-formyl-2-thiopheneboronic acid [123], 4-bromo-3-formyl-2-thiopheneboronic acid [125], and 3-formyl-4-thiopheneboronic acid can all be prepared from the corresponding ethylene glycol acetals by metalation or halogen–metal exchange of the bromo derivative followed by reaction with trialkyl borate.

4E.3.3 From dihalothiophenes by stepwise halogen–metal exchange and reaction with different electrophiles

2-Formyl-3-thiopheneboronic acid is most conveniently prepared from 2,3-dibromothiophene by halogen–metal exchange followed by *N*,*N*-dimethylformamide and a second halogen–metal exchange and trialkyl borate [126]. Adding first trialkyl borate and then *N*,*N*-dimethylformamide gives 3-formyl-2-thiopheneboronic acid. 5-Methyl-, 5-ethyl-, and 5-propyl-3-formyl-2-thiopheneboronic acids were all prepared from the corresponding 5-alkyl-2,3-dibromothiophenes followed by stepwise halogen–metal exchange followed by reaction with trialkyl borate and *N*,*N*-dimethylformamide [123].

2-Formyl-3-thiopheneboronic acid [126]

A solution of 2,3-dibromothiophene (72.0 g, 0.3 mol) in anhydrous diethyl ether (200 ml) is added to ethyllithium in diethyl ether (0.33 mol) at −70 °C under nitrogen with vigorous stirring. The stirring is continued for 5 min, after which *N*,*N*-dimethylformamide (23.4 g, 0.32 mol) in anhydrous diethyl ether (75 ml) is added in a slow stream. The cooling bath is removed and the mixture is stirred for 2 h, after which it is recooled to −70 °C and ethyllithium in diethyl ether (0.36 mol) is added in a slow stream followed by addition of butyl borate (97.5 g, 0.42 mol) in anhydrous diethyl ether (170 ml) in one portion. The reaction mixture is stirred at −70 °C for 4 h, the cooling bath is removed and the temperature is allowed to rise to 0 °C. After work-up and recrystallization from aqueous ethanol 16.5 g (35%) of the title compound is obtained.

3-Formyl-2-thiopheneboronic acid [127]

A solution of 2,3-dibromothiophene (96.8 g, 0.40 mol) in anhydrous diethyl ether (400 ml) is added with vigorous stirring in a slow stream under nitrogen to 1.0 *M* ethyllithium in diethyl ether (440 ml) at −75 °C. After 5 min butyl borate (110.6 g, 0.48 mol) is added in one portion through the wide-open dropping funnel and the reaction mixture is stirred at −70 °C for 3 h, followed by addition of 1.0 *M* ethyllithium in ether in a slow stream. Five minutes after the addition is complete, *N*,*N*-dimethylformamide (41.0 g, 0.56 mol) in anhydrous diethyl ether (50 ml) is added dropwise with stirring. When the temperature of the reaction mixture has risen to room temperature (*ca* 2 h), it is cooled to −10 °C, 1.0 *M* hydrochloric acid (1300 ml) is then added and the stirring continued for 1 h at room temperature. The phases are separated and the aqueous phase washed three times with ether. The combined ether phases are divided into two portions, each of these portions is extracted three times with 1 *M* sodium carbonate (200 ml) and each aqueous extract immediately acidified with 4 *M* hydrochloric acid. After cooling 35.0 g (56%) of the title compound is filtered off.

4E.3.4 Monoaddition to one formyl group in diformylthiophenes

Monoaddition to one formyl group in diformylthiophenes is an approach to substituted thiophene aldehydes. Thus 2,5-diformylthiophene reacts with the phosphonates in the following type of reaction [128].

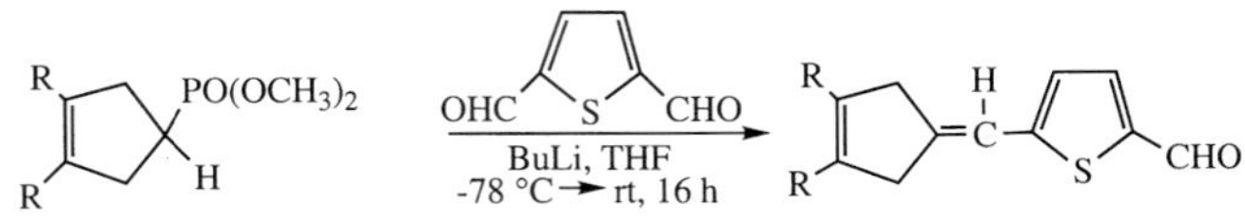

R = H, SCH_3, R,R = $S(CH_2)_2S$

The Wittig reaction of 2,5-diformylthiophene has been performed as shown below [129].

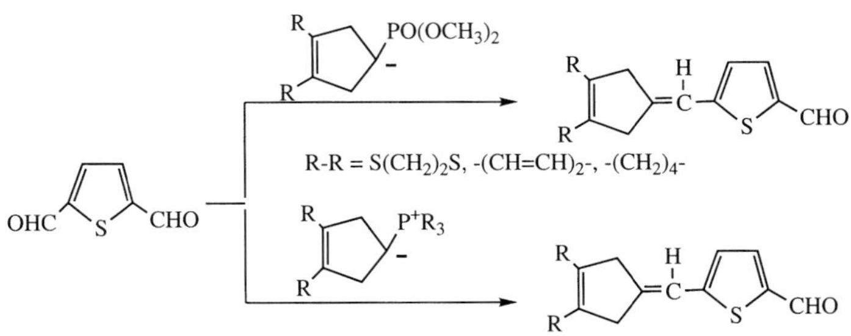

a) R = CO_2CH_3, R' = Bu b) R = H, R' = Ph c) R-R = $S(CH_2)_2S$, R' = Ph

4-Nitrobenzylphosphonate was reacted with the mono dimethylacetal of 2,5-thiophene dialdehyde, giving 2-(4-nitrostyryl)-5-thiophene aldehyde. This aldehyde was, by selective reduction of the aldehyde and reaction of the obtained alcohol with triphenylphosphonium hydrobromide, converted to the phosphonium salt, which upon Wittig reaction with the mono diethyl acetal of 2,5-thiophene dialdehyde was used for the preparation of end-functionalized oligovinylthiophenes with liquid crystal properties [130].

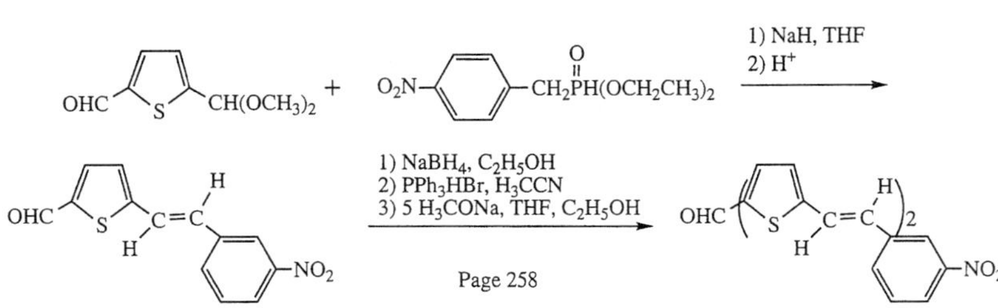

Reaction of 2,5-diformylthiophene with diphenylmethylamine and aqueous hypophosphoric acid gives, unexpectedly, the mono(α-hydroxyalkylphosphinate) derivative [131].

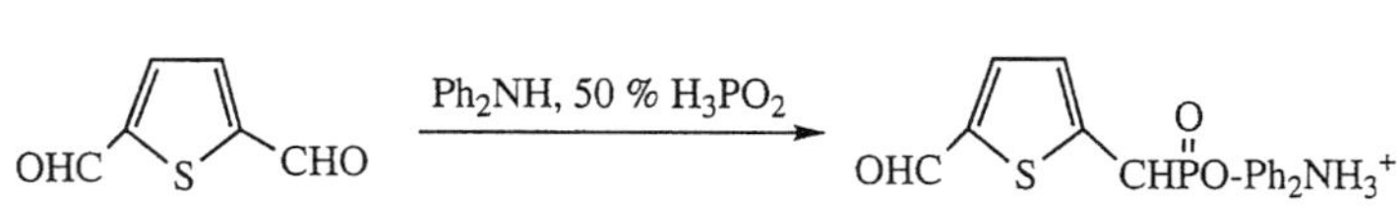

30 %

Pinacol coupling of 2,5-diformylthiophene using inexpensive aluminium powder as a reductive agent in aqueous sodium hydroxide/methanol media gives 1,2-bis(5-formylthiophene-2-yl)ethane-1,2-diol in a 5:1 mixture of racemic and meso isomers [132].

1,2-Bis(5-formylthiophene-2-yl)ethane-1,2-diol [132]

A 10% aqueous sodium hydroxide solution (6.7 ml) is added dropwise to a mixture of 2,5-diformylthiophene (700 mg, 5 mmol) and aluminium powder (0.45 g, 16.7 mmol) in methanol (15 ml) under stirring at room temperature during 10 min. The stirring is continued for 80 min, after which the reaction mixture is filtered and the filtrate extracted with ethyl acetate. The combined organic phases are washed with water, dried over magnesium sulfate and evaporated. The residue is purified by chromatography on silica gel using hexane/ethyl acetate (1 : 1) as eluent, giving a yield of 40% of a white solid as a 5: 1 mixture of racemic and meso isomers mp 107–130 °C.

4E.4 ACETALS, THIOACETALS, AND AMINALS DERIVED FROM THIOPHENE ALDEHYDES

4E.4.1 Acetals prepared from aldehydes

Acetals are important protecting groups for the carbonyl function, enabling other functional groups to be introduced at various points in the thiophene ring [133]. Dimethyl and diethyl acetals are prepared by the treatment of aldehyde with the appropriate trialkyl orthoformate in the corresponding alcohol in the presence of a trace of hydrochloric acid [134–137]. In this way the diethyl acetals of 3-ethylthio-5-methyl-2-thiophene aldehyde [138] and 2-methylthio-5-methyl-3-thiophene aldehyde were also obtained [139]. Other effective catalysts for the acetalization are ammonium nitrate [140] and lanthanoid chlorides [141].

2-Thiophene aldehyde diethylacetal [140]

A mixture of 2-thiophene aldehyde (100 g, 0.895 mol), anhydrous ethyl alcohol (75 ml), ethyl orthoformate (96.8 g, 1.1 mol) and ammonium nitrate (3.5 g) is gently refluxed for 12 h. After cooling the reaction mixture, it is filtered, sodium carbonate is added to the filtrate, which is distilled giving 120 g (75%) of the title compound bp 107 °C/23 mm Hg.

Acetals are also prepared using the tetraalkyl orthosilicate–alcohol–phosphoric acid system [86,142]. Very often cyclic acetals (1,3-dioxolanes) are used and prepared from the aldehyde by reaction with glycol in the presence of *para*-toluenesulfonic acid in toluene removing water by a Dean–Stark trap [143,144]. In this way 3-(1,3-dioxalan-2-yl)thiophene [88] and 3-bromo-2-(1,3-dioxalan-2-yl)thiophene [145] are obtained from the aldehydes.

A later spectroscopic study showed that treatment of 4- and 5-formyl-2-thiophenecarboxylic acid upon solution in methanol easily gave the corresponding acetals [146].

4E.4.2 Acetals *via* metalation or halogen–metal exchange followed by reaction with electrophiles

3-(1,3-Dioxolan-2-yl)thiophene-2-sulfonyl chloride is prepared by metalation of 3-(1,3-dioxalan-2-yl)thiophene, followed by reaction with sulfur dioxide and *N*-chlorosuccinimide [147].

4E.4.3 Thioacetals

The reaction of 2-thiophene aldehyde with propan-1,3-dithiol/hydrochloric acid is used for the preparation of 2-(2-thienyl)1,3-dithiane [148]. Alternatively trimethylsilyl chloride promoted reaction of 2-thiophene aldehyde, 5-methyl-2-thiophene aldehyde, and 5-nitro-2-thiophene aldehyde with propane-1,3-dithiol was used for the preparation of the 1,3-dithianes [149].

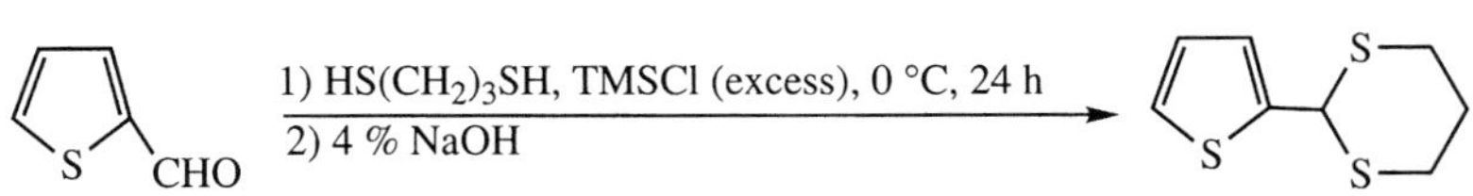

The reaction of 2-thiophene aldehyde with methyl mercaptoacetate in the presence of aluminium chloride in dichloromethane is used for the preparation of 2-[bis(methoxycarbonylmethylsulfanyl)methyl]thiophene [150].

2-[bis(Methoxycarbonylmethylsulfanyl)methyl]thiophene [150]

A solution of 2-thiophene aldehyde (11.2 g, 100 mmol) and methyl mercaptoacetate (21.24 g, 200 mmol) in dichloromethane (150 ml) is stirred at room temperature and anhydrous aluminium chloride (5.6 g, 42 mmol) is added in small portions under cooling. The reaction mixture turns turbid as the reaction proceeds. After the addition the mixture is stirred for another 10–15 min and then hydrolyzed with water (20 ml), the product is extracted with dichloromethane, the combined organic phases are washed with water, dried, and evaporated. A yellow liquid residue is purified by chromatography using ethyl acetate/light petroleum (1 : 9) as eluent giving 20.5 g (67%) of the title compound bp 90–95 °C/0.5 mm Hg.

The reaction of 2,5-diformylthiophene with two equivalents benzene-1,2-dithiol in refluxing benzene in the presence of catalytic amounts of *para*-toluenesulfonic acid gives bis(benzodithiol-2-yl)thiophene in 48% yield [151].

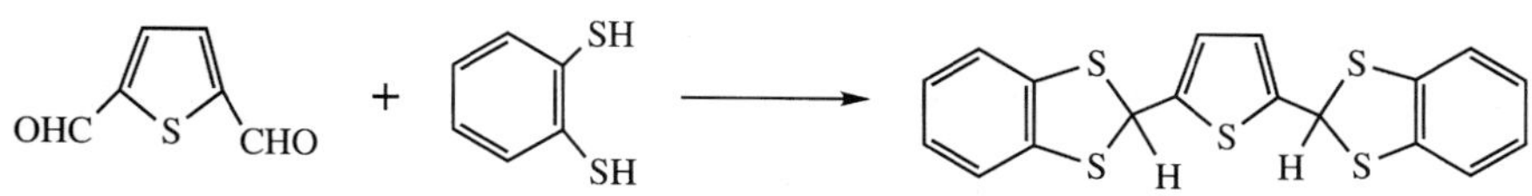

4E.4.4 Aminals derived from thiophene aldehydes

The reaction of 2-thiophene aldehyde with two equivalents of 2-aminothiazole can be used for the preparation of the aminal [152].

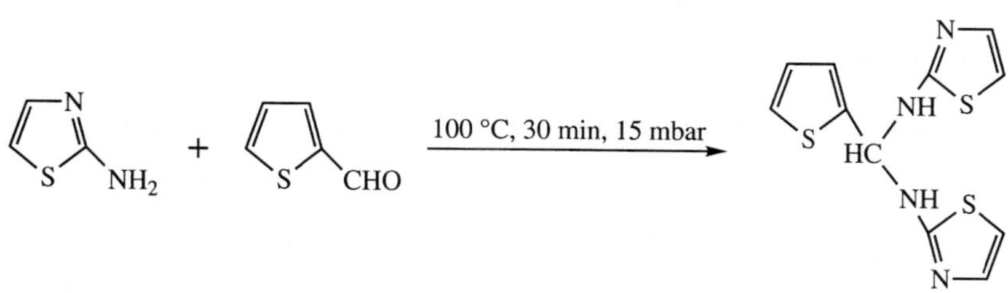

An aminal could also be prepared from 2-thiophene aldehyde and 2-aminopyridine by heating in ethanol or refluxing in acetone [153].

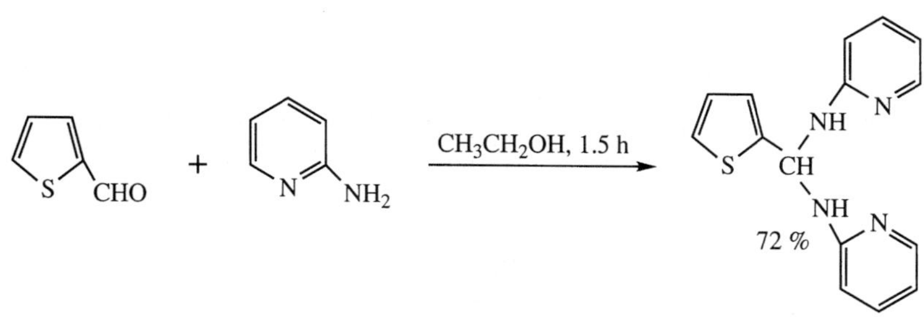

4E.4.5 Oxazolidines derived from thiophene aldehydes

The reaction of 2-thiophene aldehyde with ephedrine in benzene using a Dean–Stark trap gave *trans*- and *cis*-2-(2′-thienyl)-3,4-dimethyl-5-phenyl-oxazolidine [154].

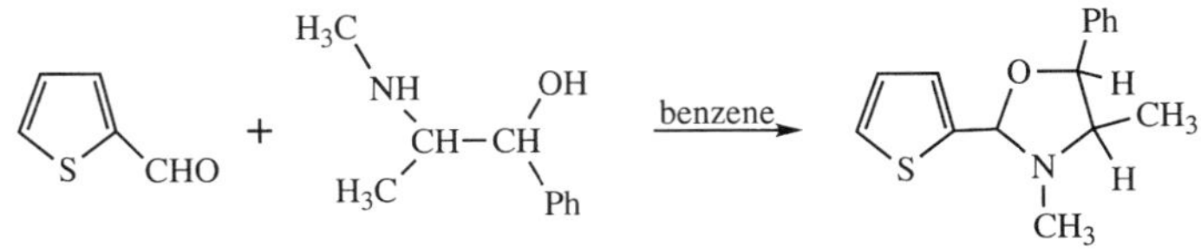

trans white crystals m. p. 82-83 °C from ethanol/water
cis cream colored m. p. 61-62 °C from heptane

The optically active oxazolidine shown below has been prepared [155] and used as a ligand in palladium-catalyzed reactions [156].

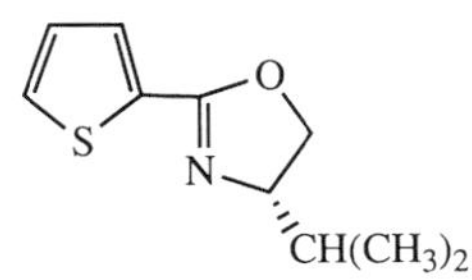

4E.4.6 Isoxazolidines derived from thiophene aldehydes

The isoxazolidines shown below are obtained by 1,3-dipolar cycloaddition of 2-thiophene aldoxime with acrylonitrile and methyl acrylate and did not give the expected [4 + 2] cycloadducts [157].

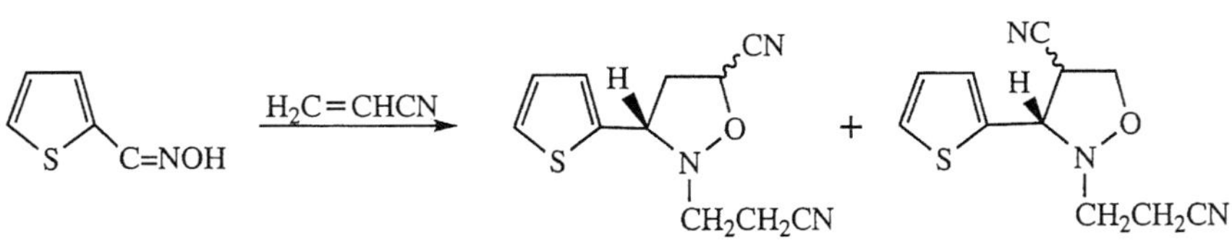

4E.5 IMINES (SCHIFF'S BASES), OXIMES, HYDRAZONES AND RELATED DERIVATIVES FROM THIOPHENE ALDEHYDES

4E.5.1 Introduction

These type of compounds have lost their importance for the characterization of thiophene aldehydes, but many are interesting starting materials for various other thiophene derivatives. The emphasis will be on such applications.

4E.5.2 Imines derived from thiophene aldehydes

Thiophene aldehydes react with a wide range of primary aromatic and aliphatic amines to give the expected imines (Schiff's bases) [37,54,158–165]. Compound such as the anils derived from 3-nitro-2-thiophene aldehyde and *para*-dimethylamino-, *para*-chloro-, *para*-methylaniline [166], as well as *para*-acetylaniline [167] are prepared by heating in ethanol.

N-(3-Nitrothiophene-2-ylidene)anilines [166]

A mixture of 3-nitro-2-thiophene aldehyde (0.79 g, 5 mmol), the appropriate aromatic amine (5.5 mmol), and ethanol (10 ml) is under stirring refluxed for 10 min and then left overnight at room tmperature. The precipitate formed is filtered off and washed with ice-cooled ethanol giving 86% of *N*-(3-nitrothiophene-2-ylidene)aniline mp 59–60 °C after recrystallization from light petroleum.

This method was used for the preparation of the Schiff's bases from 2-thiophene aldehyde and 5-methyl-2-thiophene aldehyde, which were used in a reaction with diiron nonacarbonyl [168].

N-(2-Thienylmethylidene)aniline [168]

A mixture of 2-thiophene aldehyde (3.8 ml, 40 mmol) and aniline (3.6 ml, 40 mmol) is heated at reflux in methanol for 12 h. The solvent is evaporated and the residue is distilled with a Kugelrohr distillation apparatus giving 7.02 g (94%) of the title compound as an orange-red compound bp 175 °C/ 0.1 mm Hg.

An alternative method considered to be better is the condensation of the appropriate aldehyde in toluene in the presence of molecular sieves, which was used for the preparation of the anils from 2-thiophene aldehyde and aniline and *ortho*-aminobenzonitrile [169].

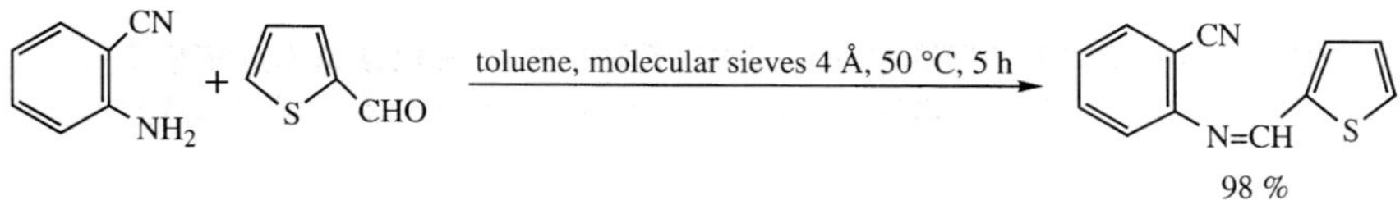

This methodology was also used in the reaction of 2-thiophene aldehyde with ethyl glycinate hydrochloride in toluene/triethylamine in the presence of 4 Å sieves at room temperature, which gave a 70% yield of ethyl 2-(2-thenylideneamino)ethanoate as a mixture of *E* and *Z* isomers (1 : 1) [170].

The imine has been prepared from 2-thiophene aldehyde and norephedrine [154].

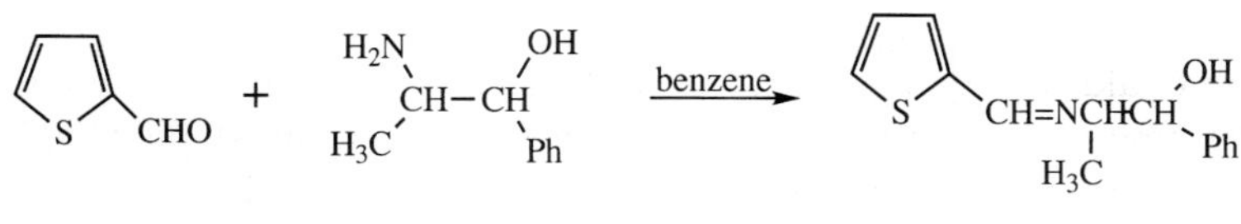

The imine prepared from 2-thiophene aldehyde and *ortho*-bromophenylethylamine has been used in a palladium-catalyzed [2+2]-cycloaddition with ketenes for the preparation of β-lactams [171].

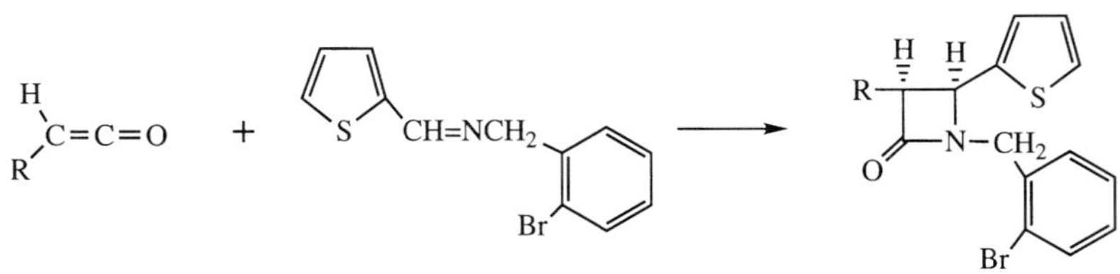

The diazadiene shown below was prepared *in situ* from *N*-trimethylsilyl-(2-thienyl)methanimine and phenylisothiocyanate and reacted with 1-morpholino-1-cyclohexene to give the morpholine derivative [172].

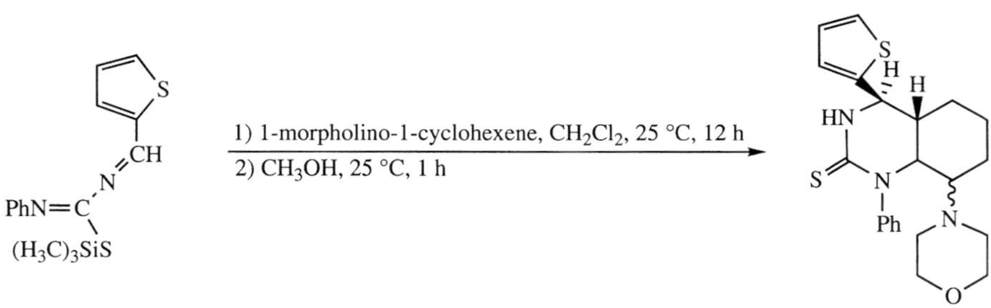

Reaction of 2,5-diformylthiophene with diphenylmethylamine gives the diimine derivative [131]. 2-Iminomethylthiophene and the analogous isopropyl and *tert*-butyl derivatives as well as 2,5-bis(methyliminomethyl)thiophene have been prepared and their conformation studied. Only the *trans* conformation was found for the iminomethyl derivatives shown below [173].

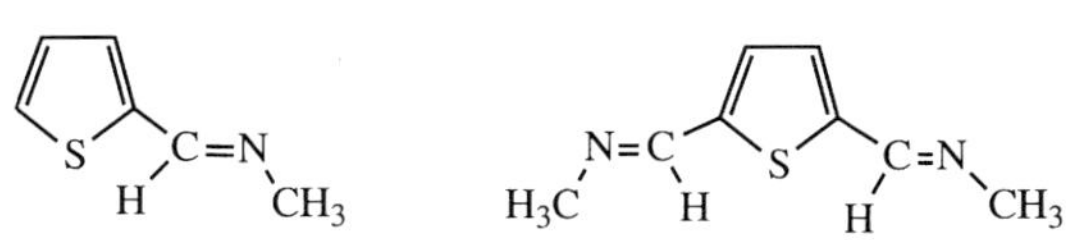

Thiophenobiscalix[4]arenes with imine linkages are prepared by condensation of 2,5-diformylthiophene with diaminocalix[4]arene in refluxing methylene chloride/methanol in the presence of magnesium sulfate and were used for binding studies with viologens [174].

Schiff's bases are used for complex formation with various transition metals. For instance the titanium complexes derived from *N*,*N*-dimethyl-*N'*-(2-thenylidene)ethylenediamine, *N*,*N'*-bis(2-thenylidene) ethylenediamine and 2-thenylidene alkyl or arylamine and its methyl derivatives have been prepared [175].

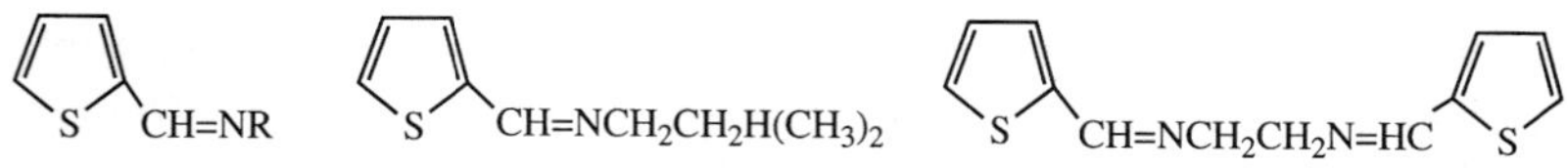

Schiff's bases have been prepared from 2-thiophene aldehyde with ethylene diamine, *ortho*-phenylenediamine or 4-methyl-*ortho*-phenylenediamine [176] and their complexes with bis(cyclopenadienyl)titanium(IV) and zirconium(IV) dichloride studied [177]. A number of Schiff's bases derived from 2-(5-formyl-2-thienylthio)benzoic acid [178] and 2-formyl-4*H*-thieno[2,3-*b*]-[1]benzothiopyran-4-one have been prepared [179]. The Schiff's bases from S-benzoyl sulphanilamide dithiocarbamate and 2-thiophene aldehyde were prepared and their lanthanide complexes studied [180].

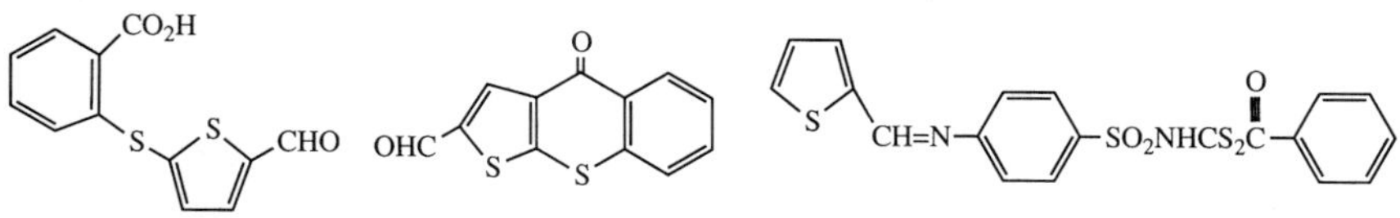

Macrocylic tetraimine Schiff's bases are prepared by non-template synthesis from bis(5-formyl-2-thienyl)methane derivatives [46].

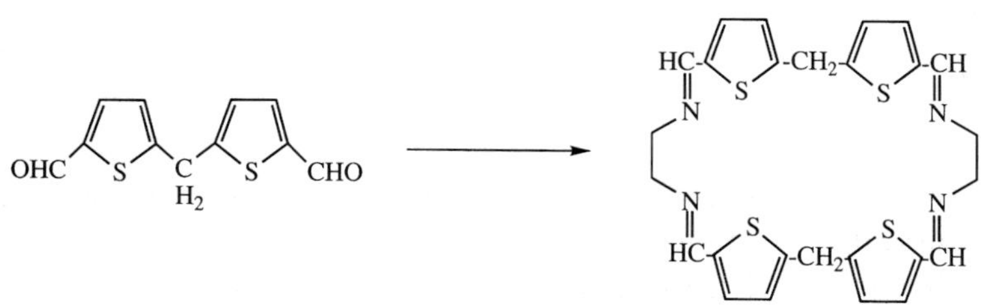

Synthesis of the macrocycle [46]

A solution of bis(5-formyl-2-thienyl)methane (0.25 g, 1.0 mmol) in chloroform (20 ml) and a solution of 1,2-ethanediamine (0.06 g, 1.0 mmol) are added simultaneously to chloroform (20 ml) over a period of 1.5 h with stirring at room temperature. The stirring is continued for 30 min, after which the solution is filtered in order to remove any insoluble material. The filtrate is dried over sodium sulfate and evaporated and the yellow powder so obtained is washed with ethanol (2 × 20 ml) giving 2.2 g (83%) of the macrocycle as a colorless powder mp 158–165 °C (decomp).

4E.5.3 Oximes derived from thiophene aldehydes

Oximes of thiophene aldehydes have during recent years been prepared as starting materials for carbonitrile-*N*-oxides and also used in 1,3-dipolar cycloaddition reactions. The condensation of 2,5-diformylthiophene with hydroxylamine leads to the formation of the following three isomers [181].

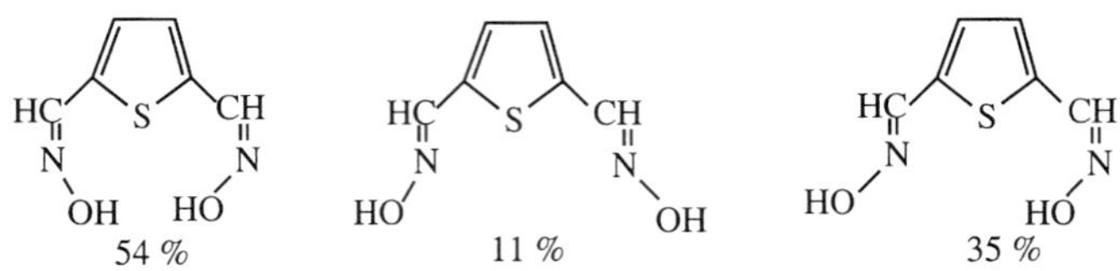

2,5-Dimethyl-4-alkylthiophene-3-aldoximes and 3,5-di-*tert*-butylthiophene-2-aldoxime [4], 2-methylthio-5-methyl-3-thiophenealdoxime [139,182] and 5-nitro-2-thiophene aldoxime [183] were prepared by the reaction of the appropriate aldehyde in ethanol with hydroxylamine hydrochloride and sodium acetate in water.

2,4,5-Trimethyl-3-thiophenealdoxime [4]

To a solution of 2,4,5-trimethyl-3-thiophene aldehyde (10.8 g , 70 mmol) in ethanol (50 ml), a solution of hydroxylamine hydrochloride (10.5 g, 140 mmol) and sodium acetate (19 g, 140 mmol) in water (50 ml) is added with stirring. The reaction mixture is heated to reflux and ethanol is added to homogenization. The refluxing is continued for 4 h, after which water is added. The precipitate formed is filtered off, washed with water, dried and dissolved in diethyl ether/hexane (1:3). After filtration through a pad of silica gel the filtrate is evaporated and the residue recrystallized from hexane giving 10.5 g (88%) of the title compound mp 101–103 °C.

The reaction of the oximes with halo derivatives in the presence of sodium alkoxide is used for the preparation of oxime ethers [183]. Using both *N*-methyl-*N*,*O*-bis(trimethylsilyl)hydroxylamine and *N*-methyl hydroxylamine hydrochloride the following reactions have been performed [184,185].

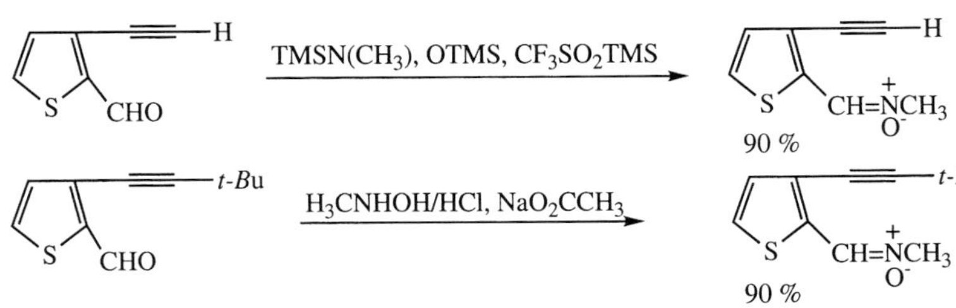

Other substituents in the thiophene ring can be modified in the presence of an acetylenic functionality. Thus 3-(3,3-dimethyl-1-butynyl)-2-thiophene aldehyde could be transformed to the nitrones by two different methods [185].

4E.5.4 Hydrazone derivatives of formylthiophenes

N,*N*-Dimethyl hydrazone of 2-thiophene aldehyde has been prepared [186,187] and its reaction with dimethyl acetylenedicarboxylate studied [188]. The sodium salts of the tosylhydrazones reacted with an equimolar amount of acrylonitrile to give the corresponding cyclopropane derivatives *via* 1,2-additions of 2- and 3-thienylmethylene [189,190].

The 2- and 3-thienylmethylenes gave stereospecific adducts with *cis* and *trans* stilbene, respectively [191]. The reaction with phenylacetylenes afforded pyrazoles [192]. The thermal decomposition of the sodium salts of the tosylhydrazones of 2- and 3-thiophene aldehyde in the presence of silver chromate gave thiophene-2- and 3-(*para*-toluenesulfonyl)methane *via* nitrogen–carbon migration of the the tosyl group [193]. The hydrazone shown below and its palladium and platinum complexes have been prepared [194].

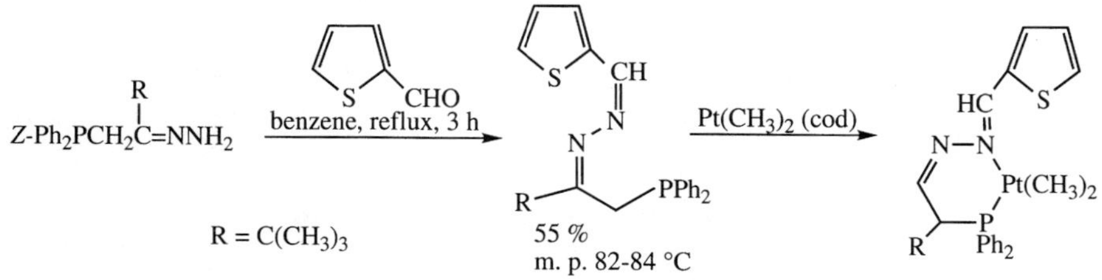

4E.5.5 Semicarbazones and thiosemicarbazones of formylthiophenes

4-Phenyl- and 4-(*ortho*-bromophenyl)semicarbazide have been used for the preparation of the semicarbazones of 2-thiophene aldehyde [195].

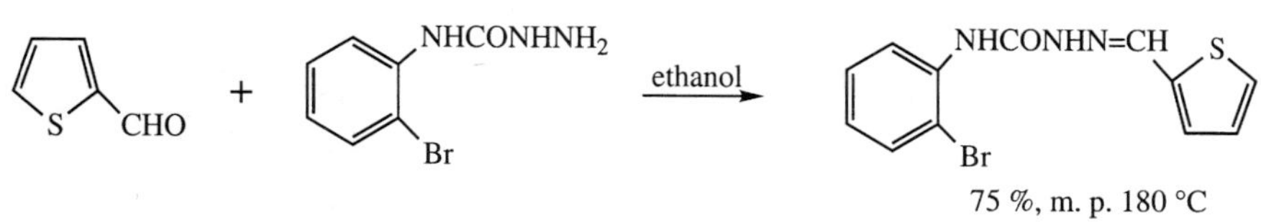

2-Thiophene aldehyde semicarbazone and its oxovanadium(V)-complexes have been prepared [196].

2-Thiophene aldehyde semicarbazone [196]

2-Thiophene aldehyde (1.12 g, 10 mmol) and semicarbazide hydrochloride (1.11 g, 10 mmol) are condensed in distilled ethanol in the presence of sodium acetate. The semicarbazone formed is recrystallized from ethanol giving 1.5 g (88%) of the title compound as a gray solid mp 223 °C.

Numerous thiosemicarbazones have been prepared in connection with studies of their tuberculostatic activity [197]. Various tin-complexes of the thiosemicarbazone of 2-thiophene aldehyde [198] were recently prepared and their crystal structure and biological activity studied [199]. The crystal structure and photochemical properties of *N′*-(2-thenylidene)benzhydrazide have been studied [200].

Chlorodiphenyl(2-thiophene aldehyde thiosemicarbazonato-N,S)tin(IV) [199]

2-Thiophene aldehyde thiosemicarbazone (0.56 g, 3 mmol) is dissolved in hot methanol (20 ml). This solution is added to a solution of diphenyltin dichloride (1.03 g, 3 mmol) in ethanol (20 ml). The reaction mixture is heated with stirring for 30 min, cooled and slowly evaporated at room temperature. The yellow crystals are filtered off and washed with ethanol giving 73% of the title compound mp 170–171 °C.

4E.6 THIOPHENE THIOALDEHYDES

2-Amino-3-thiophene thioaldehyde and 3-amino-2-thiophene thioaldehyde are prepared by reacting the corresponding *ortho*-azidoaldehydes with hexamethyldisilathiane in neat acetonitrile and/or in methanol in the presence of hydrochloric acid at room temperature [201].

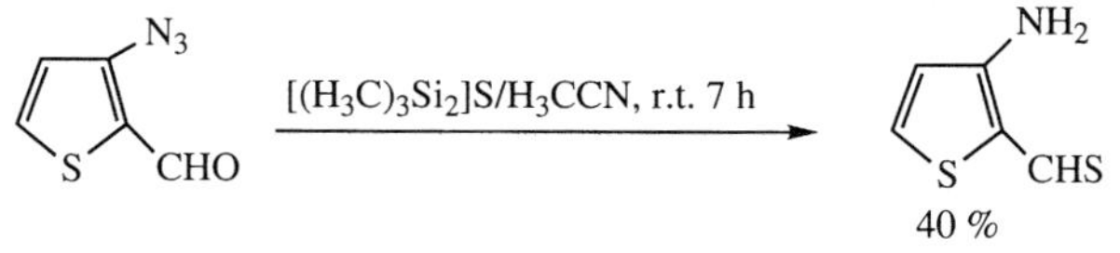

Treatment of the Schiff's bases of thiophene aldehyde 5-methyl- and 3-methyl-2-thiophene aldehyde with $[PPh_4][W(CO)_5SH]$ in the presence of an equimolar mixture of boron trifluoride-etherate and two equivalents acetic acid gave tungsten complexes of the 2-thioformylthiophenes [202].

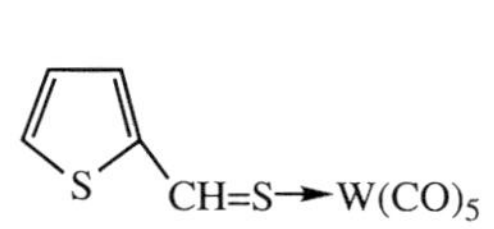

Pentacarbonyl(2-thioformylthiophene)tungsten [202]

Pure *N*-(2-thienylmethylene)aniline (0.092 g, 0.5 mmol) is dissolved in hexane (19 ml) and benzene (1 ml) after which two solutions are added, first a mixture of boron trifluoride-diethyl ether (0.072 g, 0.5 mmol) and acetic acid (0.06 g, 1 mmol) in dichloromethane (2.5 ml) and then a solution of $[PPh_4][W(CO)_5SH]$ (0.348 g, 0.5 mmol) in dichloromethane (2.5 ml). After stirring vigorously for 3 min at 0 °C the resulting reddish purple reaction mixture is pressed through a pad of silica gel. The filtrate is concentrated and diluted with cooled hexane (15 ml) giving a precipitate which is filtered off. This procedure is repeated three times giving 59% of the title compound as black needles mp 36.5 °C.

The radical anions of 2- and 3- thiophene thioaldehydes are obtained by photolysis in an alkaline medium of the corresponding thenyl thiols, prepared by reaction of the corresponding alcohols with thiourea, in the presence of di-*tert*-butyl peroxide [203].

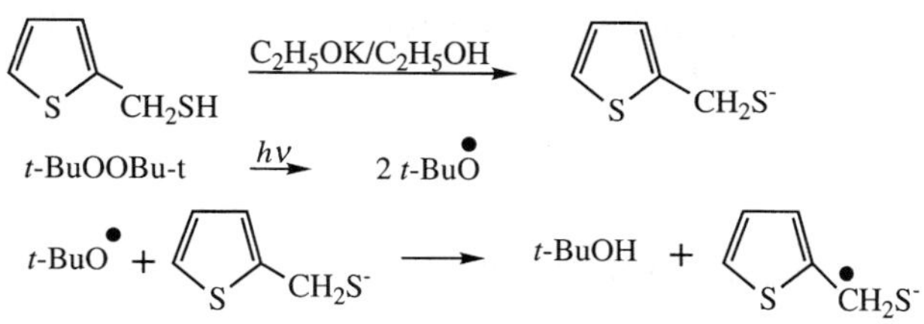

4E.7 THIENYL ALKYLKETONES AND THEIR ALKYL DERIVATIVES

4E.7.1 Friedel–Craft acylation and related reactions of thiophene and alkylthiophenes

The acylation of thiophene and its alkyl derivatives is achieved with many catalysts and many acylating agents in many solvents. Aluminum chloride is more seldom used with thiophene, as polymerization occurs giving by-products. Previously, the reaction of thiophene in benzene with acetic anhydride and tin tetrachloride as catalyst was used for the preparation of 2-acetylthiophene. However, due to the toxicity of benzene, this procedure is avoided, although this methodology was used for the preparation of 2-butanoyl and 2-heptanoylthiophene [204] and 2-phenylacetylthiophene from thiophene [205].

2-Phenylacetylthiophene [205]

Freshly distilled stannic chloride (3.4 ml, 0.029 mmol) is added to a solution of thiophene (3.4 g, 0.040 mmol) and phenylacetyl chloride (5.28 ml, 0.040 mmol) in anhydrous benzene (50 ml) at 0 °C. The greenish blue reaction mixture is stirred for 1 h, after which the intermediate formed is hydrolyzed by a slow addition of a mixture of water (18 ml) and concentrated hydrochloric acid (20 ml). The stirring is continued for 1 h and then the phases are separated, the organic phase dried over sodium sulfate and evaporated. The residue, a yellow oil, is purified by chromatography on silica gel using cyclohexane/diethyl ether (4:1) as eluent, giving the title compound as white crystals mp 49–50 °C.

These reaction conditions were also used for the preparation of 2,4,5-trimethyl-3-acetylthiophene from 2,3,5-trimethylthiophene [206], 2,5-dimethyl-3-thenyl 2,5-dimethyl-3-thienyl ketone from 2,5-dimethyl thiophene and 2,5-dimethyl-3-thiopheneacetic acid chloride [207].

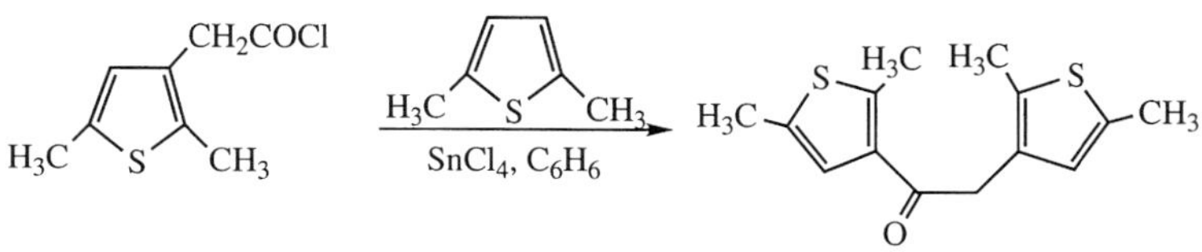

In a recent example 2-acetylthiophene was prepared using tin tetrachloride and acetyl chloride in dichloromethane [208].

2-Acetylthiophene [208]

A solution of tin tetrachloride (5.22 g, 20 mmol) and freshly distilled acetyl chloride (1.57 g, 20 mmol) in anhydrous dichloromethane (10 ml) is added dropwise at room temperature over a period of 0.5 h to a solution of thiophene (1.68 g, 20 mmol) under argon. The stirring is continued for 30 min, water (20 ml) and diethyl ether (100 ml) are added. The phases are separated and the organic phase washed with water and aqueous sodium hydrogen carbonate solution, dried over magnesium sulfate and evaporated giving 2.42 g (96%) of the title compound.

On a larger scale, however, the best method for the preparation of 2-acetylthiophene appears to be the reaction between thiophene and acetic anhydride with orthophosphoric acid as catalyst [209], which has also been published in *Organic Synthesis* [210]. This method was recently used for the preparation of 5-methyl-, 5-butyl- and 5-octyl-2-acetylthiophene [211]. Also boron trifluoride etherate [212], iodine [213], perchloric acid, [214,215] magnesium perchlorate [216,217], and heteropolyacids [218] can be used as catalysts in the reaction of thiophene with acetic anhydride. Boron trifluoride etherate, stannic chloride, and ferric chloride were previously found to be the most efficient catalysts for the acetylation of thiophene with acetic anhydride [219]. 2-Alkylthiophenes are predominantly acetylated in the 5-position although a small amount of 2-ethyl-3-acetylthiophene was formed besides 2-acetyl-5-ethylthiophene, from 2-ethylthiophene [220]. (*S*)-2-(2-methylbutanoyl)thiophene is prepared by the reaction of thiophene with (*S*)-2-methylbutanoic acid with phosporus pentoxide in benzene [221].

2-Acetyl-5-ethylthiophene [220]

A mixture of 2-ethylthiophene (448 g, 4.0 mol) and acetic anhydride (428 g, 4.0 mol) is heated to 70 °C, after which phosphoric acid (24 ml) is added with vigorous stirring. When the exothermic reaction has subsided the mixture is refluxed for 2 h. After cooling water is added, the phases separated and the organic phase washed with sodium bicarbonate solution and water, dried over magnesium sulfate and fractionated giving 474 g (77%) of a product consisting of 97% of the title compound bp 107–109 °C/8 mm Hg.

3-Alkylthiophenes generally give a mixture of 2- and 5-acyl derivatives [222,223]. 2,5-Dialkylthiophenes are acylated in the 3-position [224–226]. The reaction of di(2-thienyl)propane with acetic anhydride and 85% phosphoric acid is used for the preparation of 2,2-bis(5-acetyl-2- thienyl)propane [227].

2,2-Bis(5-acetyl-2-thienyl)propane [227]

A mixture of di(2-thienyl)propane (5.0 g, 0.024 mol) and acetic anhydride (6.0 g, 0.06 mol) is heated to 70–75 °C. The heating source is removed and 85% phosphoric acid (4 drops) is added. After the exothermic reaction has subsided the mixture is refluxed for 2.5 h, cooled, poured into water and extracted with chloroform. The combined organic phases are washed with sodium hydrogen carbonate solution and evaporated. The residue is chromatographed on silica gel using first benzene/chloroform (3 : 1) as eluent giving 1.5 g (25%) of the monoacetyl derivative. Further elution with benzene/chloroform (1 : 1) gives 4.5 g (64%) of the title compound mp 62 °C.

2-Acetyl-3-thienylacetate is prepared from ethyl 3-thienylacetate by reaction with acetyl chloride and stannic chloride in dichloromethane; minor amounts of 2-acetyl-4-thienylacetate are obtained as a by-product [9]. Trifluoroacetylthiophenes are obtained by the reaction of trifluoroacetic anhydride in 1,2-dichloroethane with thiophene and alkylthiophenes without the use of a Lewis acid catalyst [228–231]. Acylthiophenes are obtained by the reaction of thiophene with a mixture of trifluoroacetic anhydride and a carboxylic acid, which reacts as a mixed anhydride without the use of a catalyst [228,232,233]. In the case of long chain acids, the reaction is accelerated by the addition of phosphoric acid [234].

Endo-Bicyclo[2.2.1]hept-2-ene-5-yl 2-thienyl ketone is prepared from thiophene and the *endo*/*exo*-bicyclo[2.2.1]hept-2-ene-5-carboxylic anhydride in ether using zinc chloride as catalyst [235].

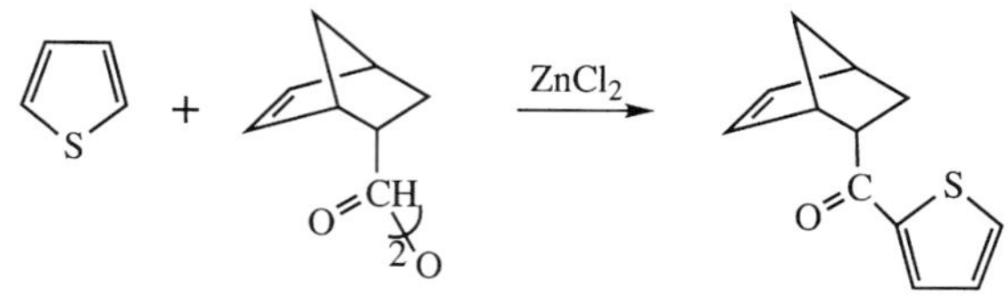

Endo-Bicyclo[2.2.1]hept-2-ene-5-yl 2-thienyl ketone [235]

Freshly fused, finely pulverized zinc chloride (3.13 g, 0.023 mol) is added to a solution of anhydrous thiophene (11.61 g, 0.138 mol) in anhydrous diethyl ether (29 ml) under a stream of anhydrous nitrogen. *endo/exo*-bicyclo[2.2.1]-hept-2-ene-5-carboxylic anhydride (6 g, 0.023 mol) is added in one portion to the stirred suspension. Reaction is indicated by the appearance of a yellow–orange color within 5 min of the anhydride addition. The reaction is quenched after 45 min by addition of saturated sodium carbonate solution (30 ml) and the stirring is continued for 5 min. The phases are separated and the organic phase washed twice with saturated sodium carbonate solution (30 ml), dried over magnesium sulfate and evaporated. The residue is redissolved in diethyl ether and this solution is washed several times with 10% sodium hydroxide solution, dried over magnesium sulfate and evaporated. The residue is purified by flash chromatography giving 1.00 g of the title compound.

2-(Trifluoromethylsulphonyloxy)pyridine is used in the reaction of thiophene with acetic acid and propionic acid in trifluoroacetic acid to give 2-acetyl and 2-propionylthiophene in quantitative yields [236].

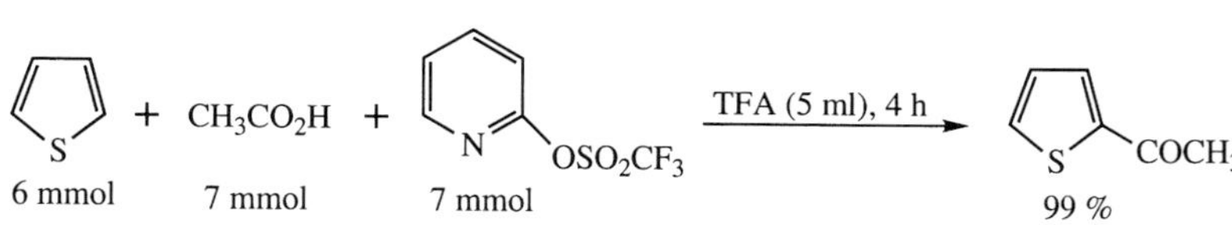

Instead of using acid chlorides or acid anhydrides in the Friedel–Crafts type reaction, the free carboxylic acids can be used in order to prepare the ketones by reacting them with thiophenes in the presence of phosphorus pentoxide, phosphoric acid, or polyphosphoric acid [237–240].

A convenient method for the acylation of thiophenes directly with a free carboxylic acid is to reflux the acid with silicon tetrachloride in benzene followed by treatment of the resulting tetraacyloxysilane with thiophene and stannic chloride [241,242]. The Friedel–Crafts reaction between thiophene and 2-alkylthiophenes with the ester acid chloride of a dicarboxylic acid using stannic chloride as catalyst is still often used for the preparation of 2-alkyl-5-thenoylalkyl acids. Alternatively an ω-ester is reacted with an acid chloride [243].

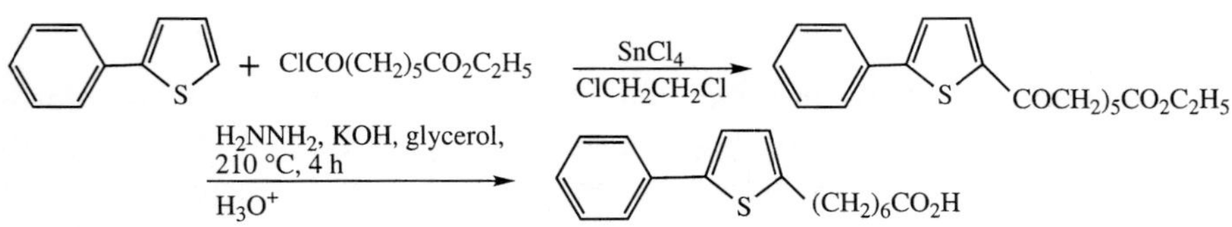

7-(5-Phenyl-2-thienyl)heptanoic acid [243]

Stannic chloride (18.7 g, 71.8 mmol) is added dropwise to a cooled solution of 2-phenylthiophene (8 g, 50 mmol) and the ester acid chloride (10.3 g, 53.6 mmol) in 1,2-dichloroethane (80 ml). During the addition the temperature is kept below 5 °C. The stirring is continued at room temperature for 1 h, after which the reaction mixture is poured into cold water (800 ml). The phases are separated and the aqueous phase extracted three times with dichloromethane. The combined organic phases are washed with water, dried over sodium sulfate and evaporated. The residue is flash chromatographed on silica gel using dichloromethane as eluent giving 9.5 g (58%) of the ethyl ester as a yellow oil. This oil (9.5 g, 28.8 mmol) is directly mixed with hydrazine monohydrate (4.3 g, 85.9 mmol) and potassium hydroxide (6.4 g, 114 mmol) in glycerol (100 ml). The reaction mixture is heated at 210 °C for 4 h. The excess of hydrazine and water are distilled off at normal pressure. After cooling to room temperature, water (100 ml) is added. The aqueous phase is acidified with concentrated hydrochloric acid and extracted three times with dichloromethane. The combined organic phases are dried over sodium sulfate and evaporated. The residue, a solid, is purified by recrystallization from toluene giving 4.95 g (34%) of the title compound as a white solid mp 135 °C.

4E.7.2 Various Friedel–Crafts like reactions

Thiophene is acetylated by the mixed sulfonic carboxylic anhydride, $H_3CCO_2SO_2CH_3$ [244] and by acetyl toluene-*para*-sulfonate [245]. 2-Acetylthiophene is also obtained from thiophene and acetic anhydride and diphosphoryl chloride [246]. The reaction of 2-thenoyl chloride with (1*E*,3*E*)-1-trimethylsilyl-1,3-nonadiene in the presence of aluminium chloride and dichloromethane is used for the preparation of (1*E*,3*E*)-1-(2-thenoyl)-1,3-nonadiene [247].

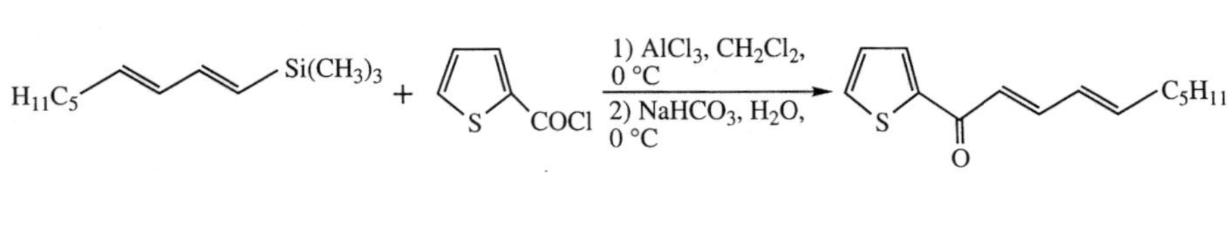

(1E,3E)-1-(2-thenoyl)-1,3-nonadiene [247]

A suspension of anhydrous aluminium chloride (1.2 g, 8.9 mmol) in dichloromethane (45 ml) is stirred for 20 min at room temperature and then cooled to 0 °C. 2-Thenoyl chloride (11.3 g, 8.9 mmol) is added and the mixture is stirred for 1 h at 20 °C and cooled to 0 °C. (1*E*, 3*E*)-1-trimethylsilyl-1,3-nonadiene (1.75 g, 9.9 mmol) is added and the reaction mixture is stirred at 0 °C for 105 min and subsequently at 20 °C for 2 h, after which it is poured into ice-water and sodium bicarbonate. The product is extracted with hexane (3 × 100 ml), the combined organic phases are worked up and evaporated. The residue is purified by medium pressure liquid chromatography on silica gel using hexane/diethyl ether (95:5) as eluent, giving 0.73 g (37%) of the title compound with chemical and stereoisomeric purity higher than 98.5%.

Acylation of 3-acetonyl- and 3-phenacyl-2,5-dimethylthiophenes with aliphatic acid anhydrides in the presence of perchloric acid directly give thieno[3,4-*c*]pyrylium perchlorates. Upon heating in ethanol, they converted to 2,5 dimethyl-3-acetyl-4-acylthiophenes [248].

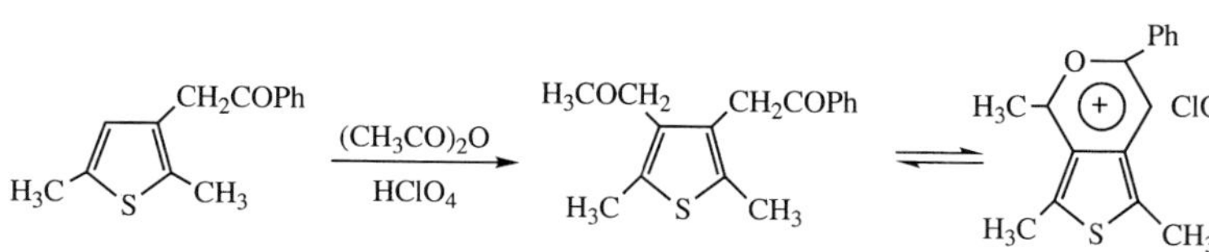

2,5-Dimethyl-3-acetyl-4-phenacylthiophene [248]

A mixture of the perchlorate (1.8 g, 5 mmol) and ethanol (340 ml) is heated to the boiling point and cooled. The colorless precipitate formed is filtered off giving 1 g (74%) of the title compound mp 179–180 °C after recrystallization from aqueous ethanol.

4E.7.3 *Via* the reaction of thienylmetalorganic reagents

4E.7.3.1 With acetic anhydrides

3-Acetylthiophene was prepared in 68% yield by converting 3-thienyllithium into the Grignard reagent with magnesium bromide followed by treatment with acetic anhydride [249].

4E.7.3.2 With nitriles

The reaction of thienyllithium derivatives with nitriles is most useful, if the nitrile does not possess an α-hydrogen atom [249]. For the preparation of *tert*-

butyl-3-thienyl ketone from 3-thiophenemagnesium bromide [250] (cf. section 46.7.3.3) used.

4E.7.3.3 With *N,N*-dimethylcarboxamides

The reaction of thienyllithia with *N*,*N*-dimethylcarboxamides is a very useful method for the preparation of acylthiophenes, provided the thienyllithium derivative does not have large *ortho* substituents [251–253].

3-Acetyl-2-ethyl-5-methylthiophene [251]

Butyllithium (1.30 *M*) in hexane (42 ml, 55 mmol) is added to a solution of 3-bromo-2-ethyl-5-methylthiophene (10.2 g, 49.7 mmol) in anhydrous diethyl ether (100 ml) cooled to −70 °C. The mixture is stirred at this temperature for 1 h, after which *N*,*N*-dimethylacetamide (4.8 g, 55 mmol) in anhydrous diethyl ether. The reaction mixture is allowed to reach room temperature, stirred overnight and poured into 5 *M* hydrochloric acid and ice. After stirring for 1 h the phases are separated and the aqueous phase extracted with diethyl ether. The combined organic phases are washed with water, dried, evaporated and distilled giving 3.5 g (42%) of the title compound bp 60–65 °C/0.6 mm Hg.

Another recent example is the following reaction starting with 5-propyl-3-bromothiophene [254].

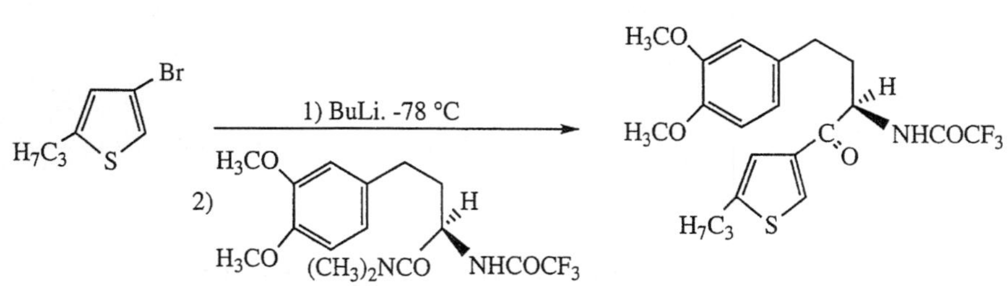

(2R)-4-(3,4-dimethoxyphenyl)-1-(2-propyl-4-thienyl)-2-((trifluoroacetyl) amino)-1-butanone [254]

Butyllithium (2.5 *M*) in hexanes is added to a solution of 5-propyl-3-bromothiophene (17.4 g, 85 mmol) in anhydrous diethyl ether (200 ml) cooled to −78 °C maintaining the temperature below −70 °C. The stirring is continued at −78 °C for 30 min, after which a precooled (−78 °C) solution of (2*R*)-*N*-methoxy-*N*-methyl-4-(3,4-dimethoxyphenyl)-2-[(trifluoroacetyl)amino]butanamide (10.7 g, 28.3 mmol) in anhydrous tetrahydrofuran (50 ml) is added at such a rate that the temperature is below −70 °C. The reaction is quenched with saturated ammonium chloride solution, stirred at room temperature and diluted with ethyl acetate (300 ml). The phases are separated and the organic phase washed with 1 *M* hydrochloric acid and saturated sodium chloride solution, dried over magnesium sulfate and evaporated giving 12.5 g (100%) of the title compound as an oil.

Another more complex example is given below [255]

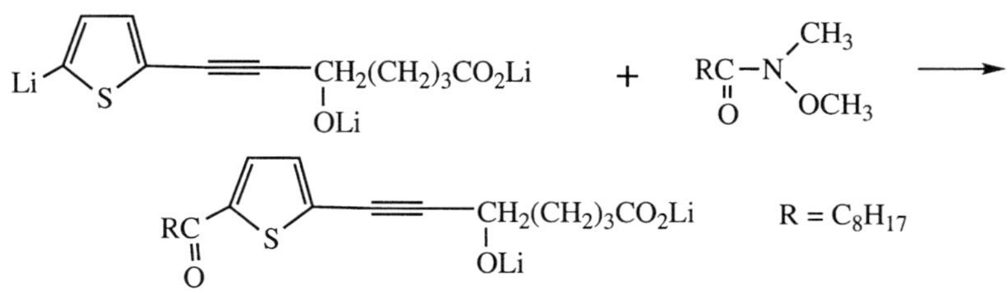

5-Hydroxy-7-{2-[5-(1-oxononyl)]thienyl}hept-6-ynoic acid [255]

Butyllithium (2.39 *M*) in hexane (0.92 ml) is added dropwise with stirring to a solution of 5-hydroxy-7-(2-thienyl)hept-6-ynoic acid (150 mg, 0.68 mmol) in anhydrous tetrahydrofuran (15 ml) at −70 °C under nitrogen. The pale yellow precipitate formed is stirred between −70 °C and −35 °C for 1 h and then recooled to −70 °C when *N*-methoxy-*N*-methylnonamide (0.18 g, 0.89 mmol) is added. The reaction mixture is left to warm to room temperature and then stirred for 18 h. After addition of saturated aqueous ammonium chloride (3 ml) followed by extraction with ethyl acetate (3 × 10 ml) the combined organic phases are reextracted with dilute aqueous sodium hydroxide solution, which is acidified to pH 1 with dilute hydrochloric acid and extracted with ethyl acetate (2 × 50 ml). The combined organic phases are washed with sodium chloride solution, dried over magnesium sulfate and evaporated giving 0.17 g (69%) of the title compound as a pale yellow solid.

tert-Butyl 2-thienylketone is prepared by the reaction of 2-thienyllithium with *N*-pivaloylaziridine [256].

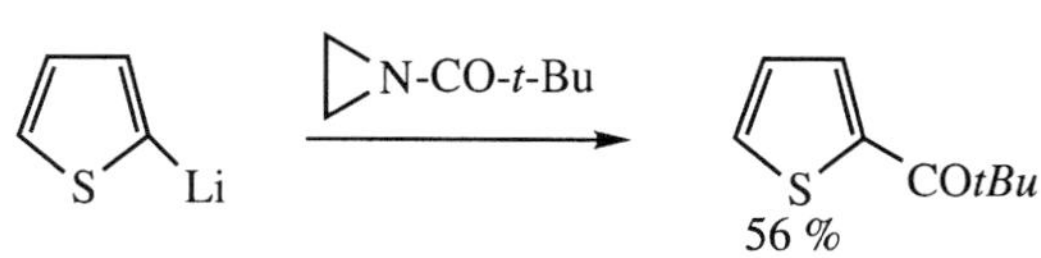

4E.7.3.4 *Via* palladium-catalyzed reactions

The palladium(0)-catalyzed reaction of 2-thiopheneboronic acid with acetyl chloride has been used for the preparation of 2-acetylthiophene [257].

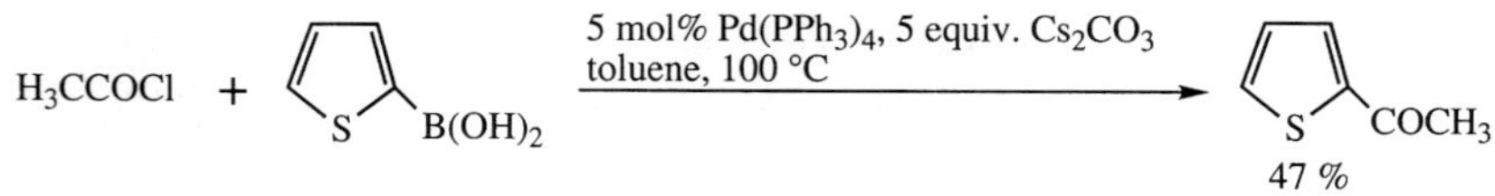

The cross-coupling reaction between 9-alkyl-9-borabicyclo[3.3.1]nonane derivatives, generated *in situ* by hydroboration of an alkene, *tert*-butyl isocyanide, and 2-iodothiophene can be carried out in dioxane at 50 °C in the presence of tripotassium phosphate and catalytic amounts of palladium(0) to give the ketone shown below in a yield of 83% [258].

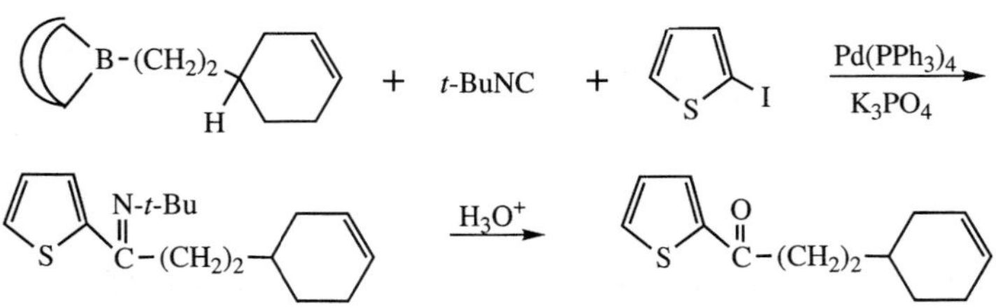

Through the use of a very efficient catalysts the Suzuki coupling between 2-acetyl-5-chlorothiophene and *ortho*-tolylboronic acid gives 2-acetyl-5-(*ortho*-tolyl)thiophene in excellent yield [259].

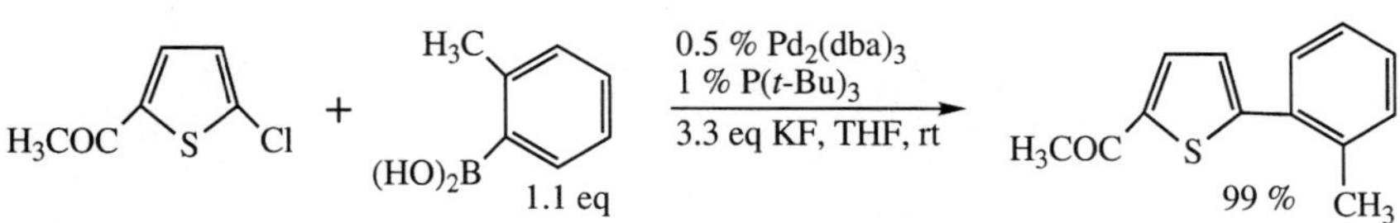

4E.7.3.5 From cyanothiophenes and alkyllithium and alkylmagnesium derivatives

The reaction of 2-cyanothiophene with butyllithium gives butyl 2-thienyl ketone in 58% yield [260]. 3-Acetylthiophene was prepared from 3-cyanothiophene and methylmagnesium iodide [261].

3-Acetylthiophene [261]

Methyl iodide (300 ml, 4.8 mol) is slowly added dropwise to magnesium (112 g, 4 mol) in diethyl ether (1000 ml) with stirring and under reflux. The stirring and reflux are continued for 1 h, after which under these conditions 3-cyanothiophene (142 g, 1.3 mol) is added dropwise. The stirring is continued until a yellow precipitate is formed. The reaction mixture is heated and then poured into ice. The phases are separated and the aqueous phase is acidified with hydrochloric acid and extracted with diethyl ether. The combined organic phases are washed with water, dried, evaporated and distilled giving 95 g (57%) bp 109 °C/16 mm Hg, mp 56 °C after recrystallization from cyclohexane.

4E.7.3.6 From formylthiophenes and Grignard or lithium reagents followed by oxidation

A recent example of this approach is the preparation of 3-pentanoylthiophene by the reaction of 3-thiophene aldehyde with butylmagnesium bromide followed by oxidation with chromic oxide [262].

3-Pentanoylthiophene [262]

A three-necked flask fitted with a condenser and a dropping funnel under nitrogen is charged with magnesium shavings (2.62 g, 108 mmol) and anhydrous diethyl ether (5 ml). A solution of 1-bromobutane (13.56 g, 99 mmol) in anhydrous diethyl ether (25 ml) is added dropwise over a period of 0.5 h. The stirring is continued for 0.5 h, after which a solution of 3-thiophene aldehyde (10.09 g, 90 mmol) in anhydrous diethyl ether (30 ml) is added over a period of 1 h. The reaction mixture is refluxed for 1 h followed by stirring at room temperature for 20 h and then poured into a mixture of ice-water (200 ml) and concentrated hydrochloric acid (30 ml). After dilution with diethyl ether (100 ml) the phases are separated and the aqueous phase extracted with diethyl ether. The combined organic phases are dried over magnesium sulfate and evaporated giving an oil corresponding to the intermediate alcohol. The latter is dissolved in acetic acid (110 ml) and a solution of chromium(VI) oxide (6.47 g, 65 mmol) in acetic acid/water (2:1) (135 ml) is added dropwise. After 1 h of stirring and further addition of water (100 ml) the reaction mixture is extracted with diethyl ether. The combined organic phases are slowly poured into a saturated aqueous solution of sodium carbonate until neutralization. The phases are separated and the organic phase washed with water, dried over magnesium sulfate and evaporated. The residue is purified by flash chromatography on silica gel using dichloromethane as eluent, giving 11.57 g (77%) of the title compound as a yellow oil.

2- and 3-Thienyl cyclohexyl ketone are prepared by a ruthenium-catalyzed reaction of 2- and 3-thiophene aldehyde with cyclohexene [263].

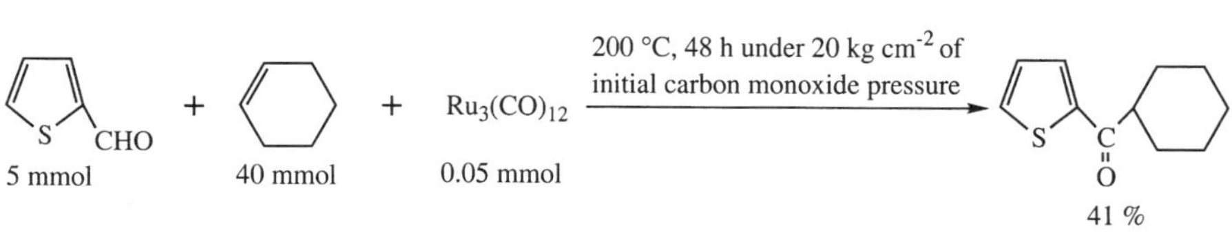

The oxidation of secondary silaacetylenic alcohols of the thiophene series with active manganese dioxide in ethereal solution is used for the preparation of the corresponding ketones [264].

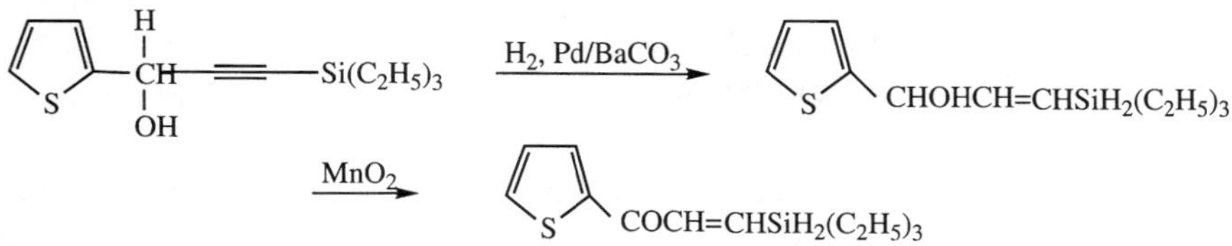

4E.7.3.7 From thienyllithium or magnesium derivatives and aldehydes followed by oxidation

Alternatively, thienylmetalorganic reagents and aliphatic aldehydes can be used. Metalation of optically active 3-(*para*-tolylsulfinyl)thiophene with lithium diisopropylamide followed by reaction with aliphatic aldehydes yields the diastereomeric carbinols in about equal amounts, which by oxidation with pyridinium chlorochromate in dichloromethane gives the ketone [265].

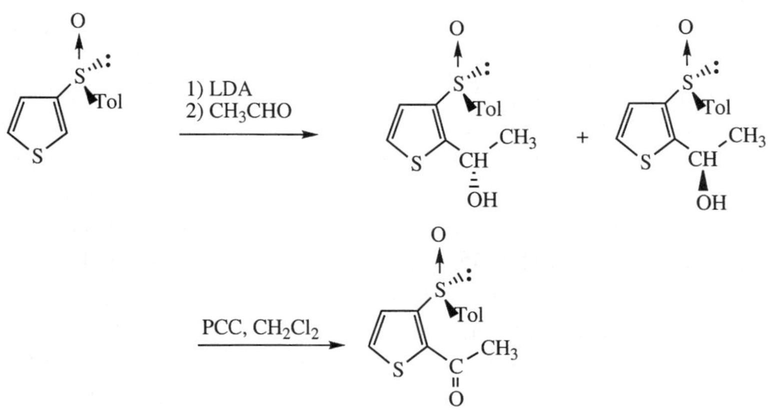

(S)-1-[3-(para-tolylsulfinyl)-2-thienyl]ethanone [265]

A solution of the diastereomeric carbinols (2.2 g, 8.3 mmol, 1:1 mixture) in anhydrous dichloromethane (15 ml) is added in one portion to a suspension of pyridinium chlorochromate (3.6 g, 16.5 mmol) and molecular sieves 4 Å (powder, 356 mg) in anhydrous dichloromethane (5 ml) at room temperature. The reaction mixture is stirred vigorously for 4 h, diluted with diethyl ether (30 ml) and filtered through a short pad of Florisil. The filtrate is evaporated and the residue purified by chromatography on silica gel using hexane/ethyl acetate (1:1) as eluent, giving 1.6 g (73%) of the title compound mp 157–158 °C after recrystallization from hexane/ethyl acetate, $[\alpha]_D^{18}$ −410.6° ($c = 1.4$, ethanol).

4E.7.3.8 From thiophenecarboxylates and alkyllithium derivatives

The reaction of methyl 2-thiophenecarboxylate with 1-bromo-2-ethoxycyclopropyllithium followed by acid-catalyzed methanolysis of the 1-bromo-2-ethoxyketones and further reaction with zinc powder and sodium decylthiolate gives 3-decylthio-substituted dimethyl acetals of 4-(2-thienyl)-4-oxobutanal [266].

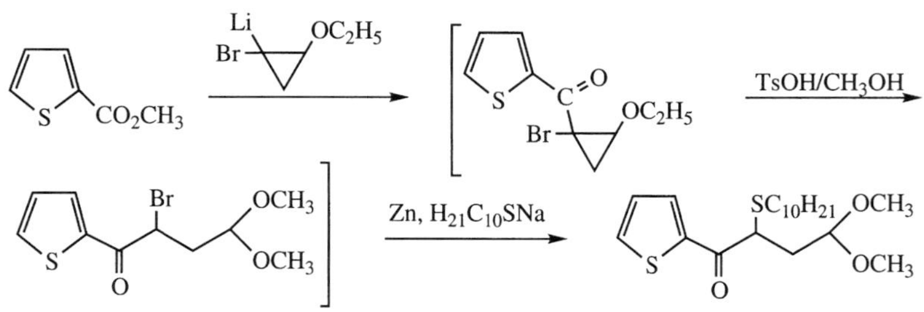

4E.7.3.9 From acylthiophenes by modifications in the thiophene ring

4E.7.3.9.1 From halo-substituted derivatives

Suzuki coupling of 5-bromo-2-acetylthiophene with phenylboronic acid can be used for the preparation of 5-phenyl-2-acetylthiophene in high yield [211].

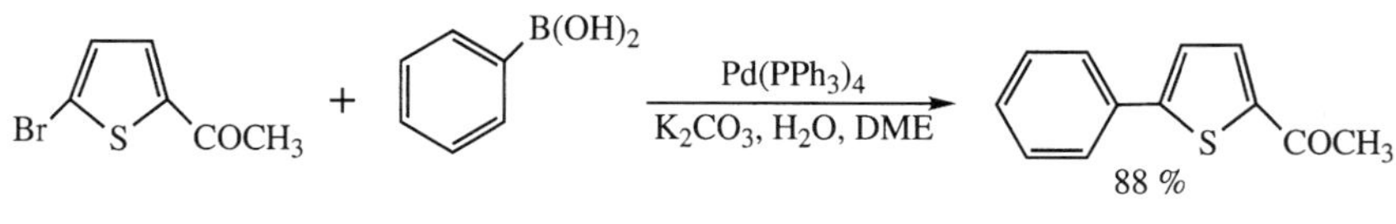

Photochemical reaction of 5-iodo-2-acetylthiophene with acrylonitrile can be used for the preparation of 2-acetyl-5-(1-iodo-2-cyanoethyl)thiophene [267].

2-Acetyl-5-(1-iodo-2-cyanoethyl)thiophene [267]

2-Acetyl-5-iodothiophene (100 mg, 0.4 mmol) is dissolved in acetonitrile (70 ml, 58 mmol) in the presence of acrylonitrile (3 g). The mixture is flashed with nitrogen for 1 h and then irradiated with a 125 W high pressure mercury arc (Helios-Italquartz). After 1 h the reaction mixture is washed with 0.1 *M* sodium thiosulfate solution, dried over sodium sulfate and evaporated. The residue is purified by chromatography on silica gel using hexane/ethyl acetate as eluent giving 73 mg (60%) of the title compound as a viscous oil.

The photochemical reaction between 5-iodo-2-acetylthiophene and 5-trimethylsilyl-2-ethynylthiophene is used for the preparation of 2-acetyl-5-(5′-trimethylsilyl-2′-thienylethynyl)thiophene [268].

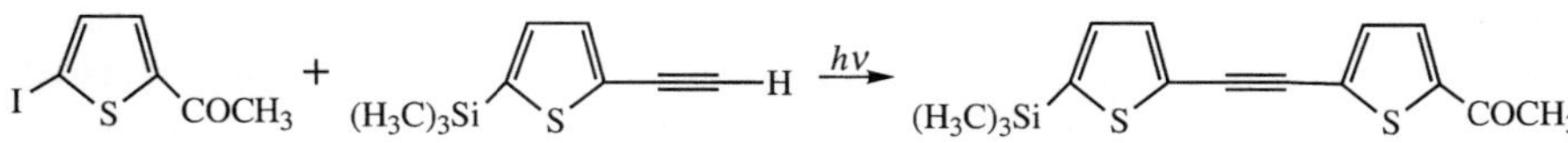

2-Acetyl-5-(5'-trimethylsilyl-2'-thienylethynyl)thiophene [268]

2-Acetyl-5-iodothiophene (1g, 3.97 mmol) and 5-trimethylsilyl-2-ethynylthiophene (2.5 g, 13.9 mmol) are dissolved in acetonitrile (320 ml). The solution is degassed with nitrogen for 1 h and then irradiated in an immersion apparatus with a 500-W high-pressure mercury arc (Helios–Italquartz) surrounded by a Pyrex water-jacket. After 6 h the reaction mixture is diluted with chloroform, washed with 0.1 *M* sodium thiosulfate solution and sodium chloride solution. The phases are separated, the organic phase dried over sodium sulfate and evaporated. The residue is purified by chromatography on silica gel using chloroform/hexane (4:1) giving 978 mg (81%) of the title compound as a dense oil.

The photochemical reaction of 2-acetyl-5-iodothiophene, 5-iodo-2-thiophene aldehyde 2-iodo-5-nitrothiophene with electron-poor olefins such as methyl acrylate, acrolein and 3-butenone gave the addition products in good yields according to the procedures described above in irradiations at 125 W [269].

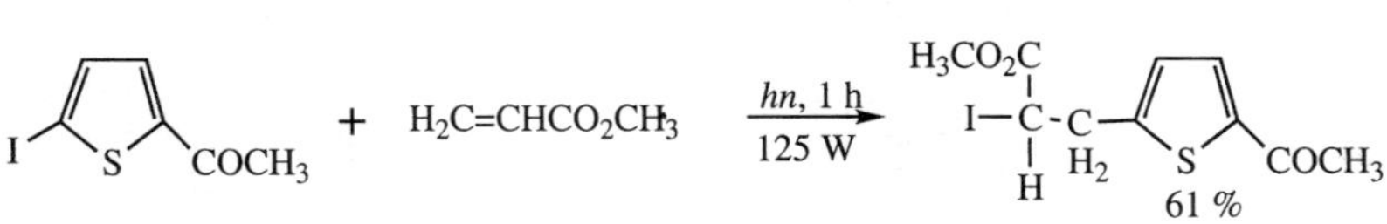

4E.7.3.9.2 Via electrophilic substitution of alkyl ketones

Usually electrophilic substitution of 2-acetylthiophene is not useful, as mixtures of 4- and 5-substituted derivatives are obtained. However, if the acetylthiophene is converted to the oxime ether, then electrophilic substitution occurs selectively in the 5-position and upon acidic hydrolysis with 25% hydrochloric acid at 70 °C for eight hours 5-substituted 2-acetylthiophenes are obtained [270]. However, 4-chloromethyl 2-acetylthiophene is prepared in 65% yield by the action of paraformaldehyde and anhydrous aluminium chloride on the corresponding carbonyl compounds [271].

4E.8 FORMYLSUBSTITUTED ACYLTHIOPHENES AND ACYLTHIOPHENEBORONIC ACIDS

4E.8.1 *Via* metalation of protected acylthiophenes and thiophene aldehydes

Metalation of the ethyleneglycol ketal of 3-acetylthiophene with ethyllithium followed by reaction with tributyl borate and hydrolysis gives 3-acetyl-2-thiopheneboronic acid, which also can be obtained by a one-pot procedure starting from 3-bromothiophene, halogen–metal exchange with ethyllithium, reaction with *N,N*-dimethylacetamide, metalation with ethyllithium and reaction with tributyl borate [123].

3-Acetyl-2-thiopheneboronic acid [123]

A solution of 2,3-dibromothiophene (28.0 g, 0.12 mol) in anhydrous diethyl ether (50 ml) under nitrogen is added to 0.8 *M* ethyllithium in diethyl ether (200 ml) cooled to −70 °C. After stirring for 20 min *N,N*-dimethylacetamide (11.0 g, 0.13 mol) in anhydrous diethyl ether (50 ml) is added in a slow stream. The cooling bath is removed and the stirring is continued for 2 h and the mixture is then recooled to −70 °C and 0.8 *M* ethyllithium in diethyl ether (250 ml) is added in a slow stream. The stirring is continued for 30 min, after which butyl borate (40.0 g, 0.18 mol) in anhydrous diethyl ether (50 ml) is added in one portion. The stirring is continued at this temperature for 4 h, the temperature is then allowed to rise to 0 °C, 1 *M* hydrochloric acid is added and the stirring is continued for 1 h. The phases are separated and the aqueous phase extracted with diethyl ether. The combined organic phases are extracted with 1 *M* sodium carbonate solution and the extract immediately acidified with 4 *M* hydrochloric acid. The precipitate formed is filtered off, dried and recrystallized from aqueous ethanol giving 7.1 g (35%) of the title compound mp 110–125 °C (decomp.).

2-Acetyl-3-thiopheneboronic acid was similarly prepared by the one-pot procedure starting with halogen–metal exchange of 2,3-dibromothiophene with ethyllithium, followed by reaction with *N,N*-dimethylacetamide, renewed halogen–metal exchange and tributyl borate. Alternatively, a two-step procedure, consisting in the preparation of the ethyleneglycol ketal of 3-bromo-2-acetylthiophene followed by halogen–metal exchange and reaction with tributyl borate can be used [123]. 3-Acetyl-2-formylthiophene is prepared by metalation of 2-methyl-(3- thienyl)-1,3-dioxolane with butyllithium followed by reaction with *N,N*-dimethylformamide and hydrolysis of the acetal group [89]. Metalation of the ethylene acetal of 3-thiophenealdehyde with butyllithium followed by reaction with *N,N*-dimethylacetamide followed by hydrolysis is used for the preparation of 2-acetyl-3-formylthiophene [98].

2-(2′-Acetyl-3′-thienyl)-1,3-dioxolane [98]

A solution of 2-(3′-thienyl)-1,3-dioxolane (30 g, 0.192 mol) in anhydrous diethyl ether (70 ml) is added to a solution of butyllithium in diethyl ether (80 ml), prepared from lithium (3.45 g) and butyl bromide (34.2 g, 0.249 mol) at −10 °C. The mixture is refluxed for 15 min and then recooled to −10 °C, after which *N,N*-dimethylacetamide (20 g, 0.274 mol) in anhydrous diethyl ether (50 ml) is added. The reaction mixture is stirred at 20 °C for 90 min and hydrolyzed. The phases are separated and the aqueous phase extracted with ether. The combined organic phases are dried, evaporated and distilled giving 15.2 g (40%) of the title compound as white crystals bp 134–136 °C/2 mm Hg, mp 60 °C.

4E.8.2 *Via* metalation and halogen–metal exchange of protected carbonyl derivatives followed by reaction with electrophiles and hydrolysis

3-Acetyl-2-formylthiophene and 2-acetyl-3-formylthiophene were prepared from the acetal-protected acetyl derivative and the acetal-protected formyl derivative by hydrolysis with 5% hydrochloric acid [89,98]. 2-Formyl-3-propanoylthiophene is prepared by halogen–metal exchange of the ethylene acetal of 3-bromo-2-thiophenealdehyde followed by propionitrile and hydrolysis [98].

4E.9 THIENYL ALKYL KETONES MODIFIED IN THE ALKYL GROUP

4E.9.1 Thenoylvinyl derivatives

4E.9.1.1 *Via* condensation of acylthiophenes with aldehydes

1-(1-Methyl-5-nitro-(2-imidazolyl)-3-(2-thienyl)-2-propene-1-one is prepared from 2-acetylthiophene and 1-methyl-5-nitroimidazole-2-thiophene aldehyde using glacial acetic acid containing traces of sulfuric acid [272].

4E.9.1.2 *Via* condensations with trichloroacetonitrile

The condensation of 2-acetylthiophene with trichloroacetonitrile in the presence of *N*-ethyl-anilinomagnesium bromide gives 3-amino-1-(2-thienyl)-4,4,4-trichloro-2-butene-1-one [273].

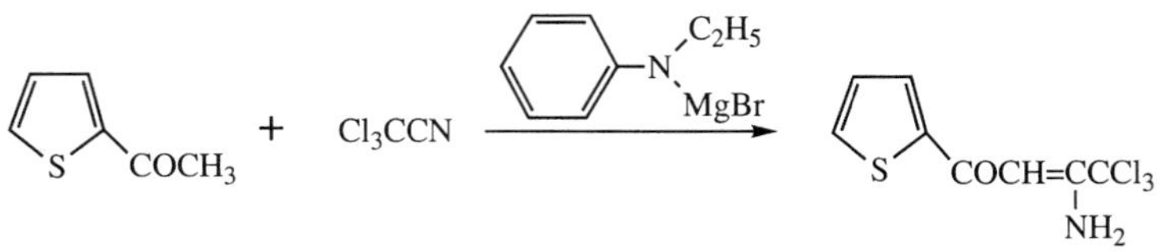

4E.9.1.3 From Mannich bases

Heating of Mannich bases to 200 °C is used for the preparation of 2- and 3-methacryloylthiophene, α-styryl-2-thienylketone, 2-acryloylthiophene and 2,5-dimethyl-3-methacryloylthiophene [237].

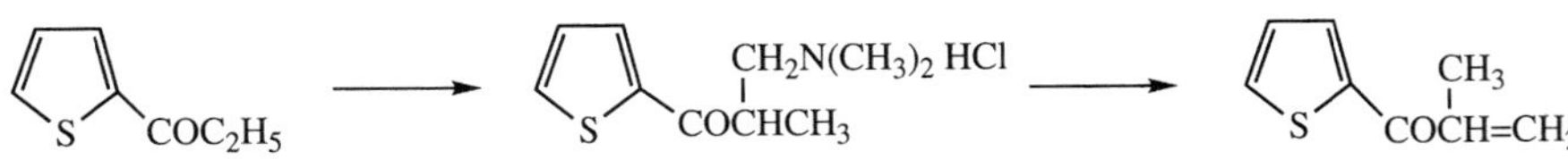

3-Dimethylamino-2-methyl-1-(2-thienyl)propanone hydrochloride [237]

A mixture of 2-propionylthiophene (140 g, 1 mol), paraformaldehyde (36 g, 1.2 mol), dimethylamine hydrochloride (98 g, 1.2 mol), concentrated hydrochloric acid (5 ml), and 95% ethanol (120 ml) is refluxed for 15 h. The ethanol is removed by evaporation and water (500 ml) is added. A small amount of oil is extracted by diethyl ether and the extract discarded. The aqueous phase is basified by ammonium hydroxide and extracted three times with diethyl ether. The combined organic phases are washed with water, dried over magnesium sulfate and anhydrous hydrogen chloride is passed through the solution. The precipate is filtered off and dried giving 216 g (95%) of the title compound as white needles mp 154–156 °C after recrystallization from ethanol.

2-Methacrylothiophene [237]

3-Dimethylamino-2-methyl-1-(2-thienyl)propanone hydrochloride and hydroquinone (0.2 g) are heated in an oil bath at 200 °C with a nitrogen bleed at 0.15 mm. The product, which distills off, is washed with water, dried, and redistilled giving the title compound (70%) bp 112–113 °C/17 mm Hg.

4E.9.2 Thenoylacetylene derivatives

N,*N*-Diphenylamino(2-thiophenecarbonyl)acetylene is prepared by the reaction of lithium diphenylamine with 2-thiophenecarbonylethynyl(phenyl)iodonium triflate [274].

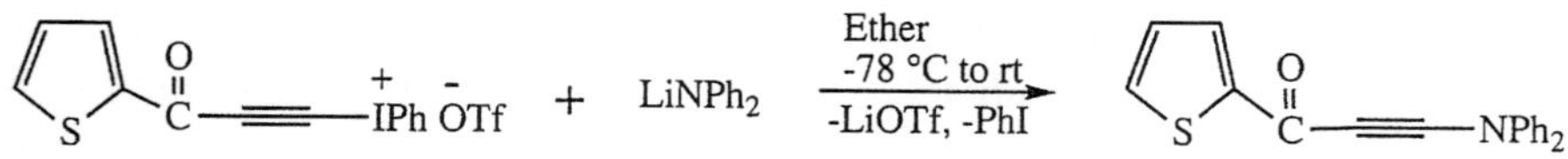

N,N-Diphenylamino(2-thiophenecarbonyl)acetylene [274]

Butyllithium (2.5 *M*, 0.41 ml, 1.0 mmol) is added to a stirred solution of diphenylamine (0.173 g, 1.00 mmol) in anhydrous diethyl ether (50 ml) at −78 °C under nitrogen, after which 2-thiophenecarbonylethynyl(phenyl)iodonium triflate (0.50 g, 1.00 mmol) is added in one portion and the stirring is continued for 5 min. The reaction mixture is then eluted through silica gel with dichloromethane and the eluate evaporated. The residue is purified by silica gel radial chromatography (200–400 mesh, 2 mm plate) using dichloromethane as eluent, giving 0.16 g (53%) of the title compound as a red oil.

4E.9.3 Thienylglyoxylic acids and other ketoacids

4E.9.3.1 Friedel–Craft type reactions with substituted aliphatic acid chlorides or anhydrides

The reaction between thiophene and various alkyl- and halo-substituted thiophenes with oxalic acid chloride ethyl ester using aluminium chloride as catalyst is used for the preparation of thienylglyoxylic ethyl esters and acids [275].

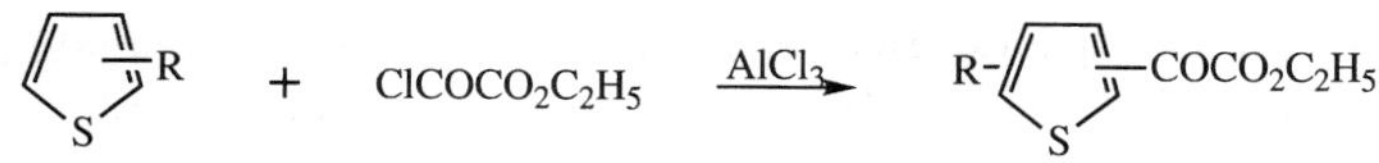

Ethyl thienylglyoxylates and thienylglyoxylic acids: General procedure [275]

A solution of aluminium chloride (10.0 g, 0.075 mol) in nitromethane (20 ml) is added to a mixture of the thiophene derivative (0.05 mol) and ethyl chlorooxoacetate (10.3 g, 0.075 mol) under cooling at such a rate that the temperature is below 10 °C. The stirring is continued at this temperature for 1 h. At room temperature the reaction mixture is poured into ice-water (200 ml). The phases are separated and the aqueous phase extracted with diethyl ether (2 × 25 ml). The combined organic phases are washed with water, 10% sodium bicarbonate solution and again with water, dried over sodium sulfate, evaporated and distilled.

Potassium hydroxide solution 20% (10 ml) is then added to the ester (0.025 mol) in 1,4-dioxane (50 ml). The reaction mixture is stirred for 24 h and then evaporated to dryness. The residue is dissolved in water and this solution is extracted with diethyl ether (2 × 20 ml). The alkaline aqueous solution is carefully acidified under ice cooling with concentrated hydrochloric acid and the precipitated acid is taken up in diethyl ether and the combined organic phases are dried over sodium sulfate and evaporated. The residue is recrystallized from diethyl ether/hexane.

5-Butyl-2-(ω-carbomethoxy)propionylthiophene and 5-heptyl-2-(ω-carbomethoxy)butanoylthiophene were prepared by Friedel-Crafts reaction of 2-butyl- and 2-heptylthiophene with β-carbomethoxypropionyl chloride and γ-carbomethoxybutyryl chloride using tin tetrachloride in benzene as catalyst [204].

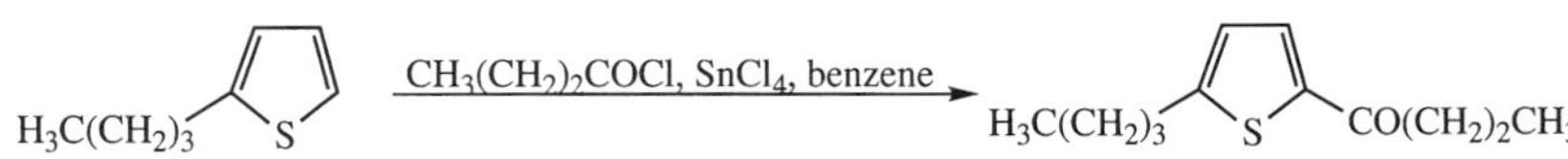

5-Butyl-2-(ω-carbomethoxy)propionylthiophene and 5-heptyl-2-(ω-carbomethoxy)butanoylthiophene [204]

Freshly distilled stannic chloride (0.2 mol) is added dropwise under stirring during a period of 30 min to a solution of the thiophene derivative and the carbonyl derivative in anhydrous benzene (100 ml). The reaction mixture is stirred at 0–5 °C for 1 h, at room temperature for 2 h and then treated with 6 *M* hydrochloric acid. The phases are separated, the organic phase is washed with 10% aqueous sodium bicarbonate solution and water, dried over sodium sulfate and evaporated. The residue is purified by chromatography on silica gel.

Friedel–Crafts reaction of thiophene with succinic anhydride in nitrobenzene gives 3-(2-thenoyl)propionic acid in 75% yield [276,277]. The Friedel–Crafts reaction of thiophene and 2-methylthiophene with glutaric anhydride with aluminium chloride in dichloromethane gave a higher yield of 5-(2-thienyl)-4-oxopentanoic acid and 5-(5-methyl-2-thienyl)-4-oxopentanoic acid than previously obtained [278].

5-(2-Thienyl)-4-oxopentanoic acid [278]

Finely pulverized anhydrous aluminium chloride (68 g, 0.5 mol) is added portionwise to a vigorously stirred solution of glutaric anhydride (20 g, 0.175 mol) in anhydrous dichloromethane (350 ml) at 0–5 °C under anhydrous conditions. The stirring is continued for 30 min, after which thiophene (14.9 g, 0.175 mol) in anhydrous dichloromethane (100 ml) is added over a period of 30 min followed by stirring for 30 min. Crushed ice (150 g) and concentrated hydrochloric acid (150 ml) are added and the mixture is warmed until the suspended materials dissolve. The phases are separated and the organic phase is washed thoroughly with water and then with sodium hydroxide (30 g) in water (250 ml). The combined alkaline extracts are acidified, the precipitate formed filtered off and crystallized from water giving 25.9 g (75%) of the title compound mp 84–85 °C.

The Friedel–Craft reaction of cyclopentane-1,1-diacetic anhydride with thiophene, 2-methylthiophene and 2-ethylthiophene in nitrobenzene, with aluminium chloride as catalyst gives 3,3-cyclopentane-5-(2-thienyl)-5-oxopentanoic acid and the analogous 5-methyl and 5-ethyl derivatives [279].

3,3-Cyclopentane-5-(2-thienyl)-5-oxopentanoic acid [279]

Aluminum chloride is added slowly to a vigorously stirred solution of thiophene (7.9 g, 0.09 mol) and cyclopentane-1,1-diacetic anhydride (13 g, 0.08 mol) in nitrobenzene (85 ml) at 0 °C. The stirring is continued for 1 h, after which the mixture is hydrolyzed with ice-hydrochloric acid followed by steam distillation in order to remove solvent. The residue is filtered and dissolved in sodium carbonate solution, filtered and the filtrate is decolorized with charcoal and acidified with concentrated hydrochloric acid. The precipitate formed is filtered off and recrystallized from light petroleum giving 14 g (72%) mp 82 °C.

Friedel–Crafts reactions between thiophene and derivatives of 1-carboxycyclohexaneacetic acid were used in connection with synthesis of thienospiran derivatives [280]. Recently another approach to thienospirans consisting in the Friedel–Crafts reaction of thiophenes with anhydrides of various unsymmetrically substituted succinic acids having substituents at the same carbon atom was developed in connection with a study of the regioselectivity of the reaction [281].

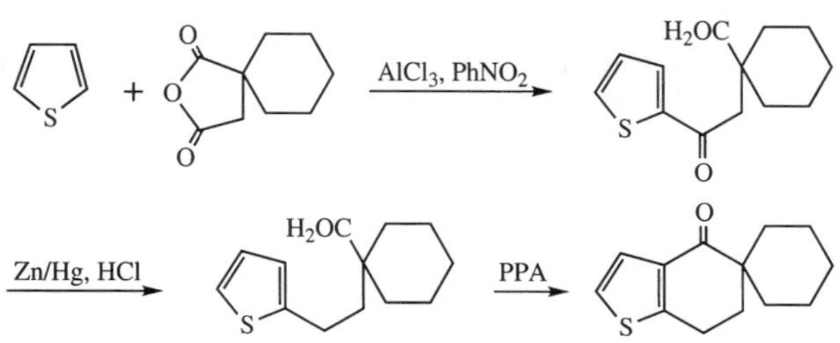

4E.9.4 ω-Halosubstituted thienyl ketones

4E.9.4.1 *Via* Friedel-Crafts reactions

16-Iodo-1-(2-thienyl)hexadecan-1-one and 12-bromo-1-(2-thienyl)dodecan-1-one are prepared by Friedel–Crafts reaction of thiophene with 16-iodohexadecanoyl chloride and 12-bromododecanoyl chloride, respectively, using stannic chloride in dichloromethane as catalyst [282].

16-Iodo-1-(2-thienyl)hexadecan-1-one [282]

A solution of 16-iodohexadecanoic acid (1.5 g, 3.9 mmol), *N*,*N*-dimethylformamide (36 mg, 0.5 mmol), and thionyl chloride (595 mg, 5 mmol) is stirred at 80 °C for 1 h. The resulting solution of 16-iodohexadecanoyl chloride is cooled to room temperature and added to a solution of thiophene (420 mg, 5 mmol) in dichloromethane (40 ml). Anhydrous stannic chloride (1.52 g, 6 mmol) is added dropwise to this mixture at 0 °C and the stirring is continued at 0 °C for 30 min and at room temperature for 2 h, followed by treatment of 6 *M* hydrochloric acid. The phases are separated and the organic phase washed with 1 *M* hydrochloric acid (2 × 50 ml), sodium hydroxide solution (2 × 50 ml) and water (2 × 50 ml), dried over sodium sulfate and evaporated. The residue is dissolved in petroleum ether (1 ml) and chromatographed on silica gel using first petroleum ether and then benzene as eluent, giving 1.19 g (66%) of the title compound as yellow crystals mp 44–46 °C after recrystallization from petroleum ether.

The reaction of thiophene with 3-chloropropionyl chloride, 4-chlorobutyryl chloride crotonyl chloride, and cinnamoyl chloride using different Friedel–Crafts catalysts, aluminium chloride/nitromethane, ferric chloride or aluminium chloride/carbon disulfide, were investigated. In most cases mixtures were obtained, except when using 4-chlorobutyryl chloride and aluminium chloride/nitromethane, which led to 75% of 4-chloro-1-(2-thienyl)-butane-1-one [283].

4E.9.5 Other ω-substituted thienyl ketones

4E.9.5.1 *Via* Friedel–Crafts reactions

The reaction of thiophene with methylthio- or phenylthioacetyl chloride in dichloromethane with stannic chloride as catalyst is a high yielding route to methylthiomethyl- and phenylthiomethyl 2-thienyl ketone [284].

The reaction of of 2-bromothiophene with β-ethylsulfanylpropionyl tetrafluoroborate gives 1-(5-bromo-2-thienyl)-3-ethylsulfanyl)-1-propanone, while with 3-bromothiophene the corresponding 6-membered sulfonium salt was obtained [285].

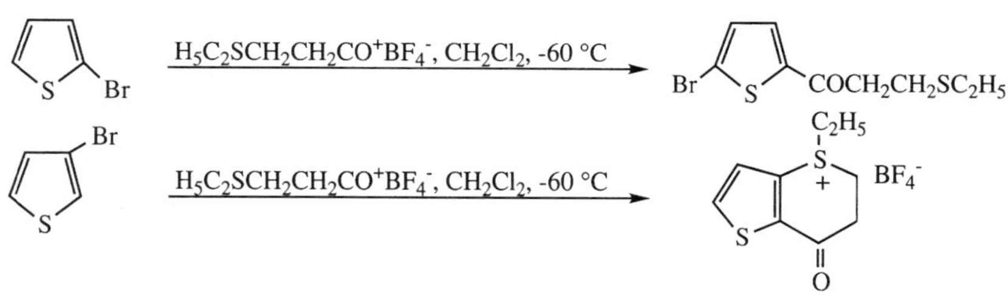

4E.9.5.2 By Claisen-like condensations

The reaction between ethyl-2-thiophenecarboxylate and dimethylsulfoxide in *tert*-butanol using potassium *tert*-butoxide as base is used for the preparation of methylsulfinylmethyl-2-thienyl ketone [286].

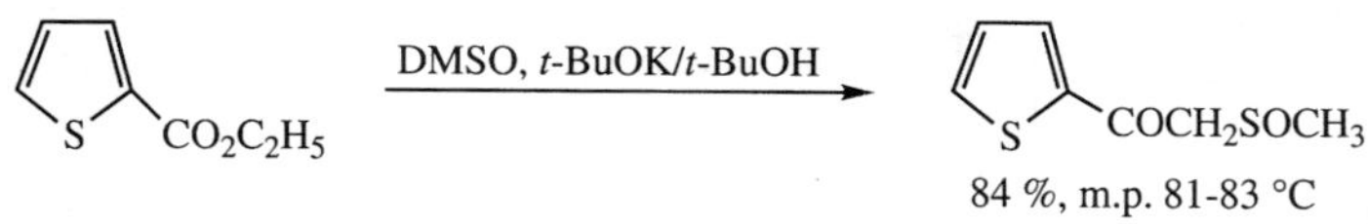

4E.10 THIENYL KETONES MODIFIED α TO THE CARBONYL GROUP

4E.10.1 Halogenation and further reactions

The chloro- and bromoacetylthiophenes are easily prepared by direct halogenation of the appropriate acylthiophenes [287], or even more conveniently *via* the enoxysilane [288].

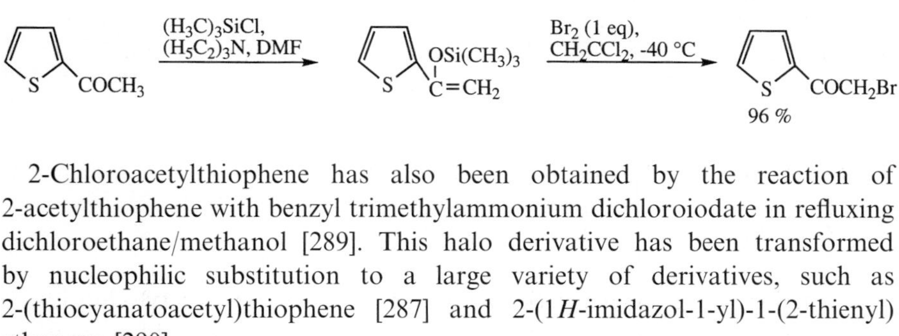

2-Chloroacetylthiophene has also been obtained by the reaction of 2-acetylthiophene with benzyl trimethylammonium dichloroiodate in refluxing dichloroethane/methanol [289]. This halo derivative has been transformed by nucleophilic substitution to a large variety of derivatives, such as 2-(thiocyanatoacetyl)thiophene [287] and 2-(1*H*-imidazol-1-yl)-1-(2-thienyl) ethanone [290].

2-(1H-Imidazol-1-yl)-1-(2-thienyl)ethanone [290]

Bromine (19.0 ml, 0.371 mol) is added dropwise to an ice-cold solution of 2-acetylthiophene (40.75 g, 0.323 mol) in diethyl ether (200 ml) under nitrogen. The stirring is continued for 90 min and the reaction is quenched with saturated aqueous ammonium chloride solution (100 ml). The phases are separated and the organic phase washed with water (100 m), dried over magnesium sulfate, and evaporated. The residue, a yellow oil, is dissolved in diethyl ether (100 ml) and added to an ice-cold solution of imidazole (89.75 g, 1.32 mol) in methanol (150 ml) under nitrogen over a period of 60 min. The stirring is continued at room temperature for 22 h and then the reaction mixture is diluted with water (1000 ml) and the product extracted with chloroform (4 × 300 ml).

The combined organic phases are dried over magnesium sulfate and evaporated. The residue is flash chromatographed on neutral silica gel using chloroform/methanol (99:1) as eluent, giving 36.10 g (58%) of the title compound mp 87–89 °C after recrystallization from ethyl acetate.

Recently the reaction of 2- and 3-bromoacetylthiophene with benzylmethylamine was used for the preparation of aminoketones, which are transformed to 4-(2- and 3-thienyl)tetrahydroisoquinolines [291].

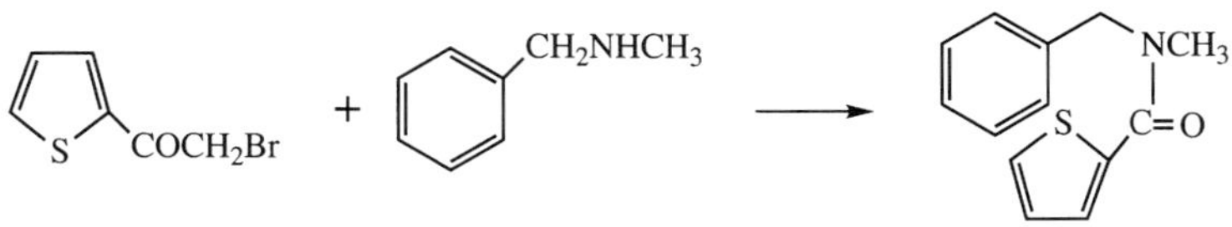

Similarly their reaction with 4-methylpyridine is used for the preparation of 2[1-(4-methylpyridinium]-1-(2-thienyl)ethanone bromide [292].

2[1-(4-Methylpyridinium]-1-(2-thienyl)ethanone bromide [292]

4-Methylpyridine (11.2 g, 0.12 mol) is added a solution of 2-bromoacetylthiophene (20.5 g, 0.1 mol) in benzene (100 ml). The reaction mixture is refluxed for 2 h. The precipitate formed is filtered off, washed with benzene and acetone and recrystallized from ethanol giving 19.6 g (69%) of the title compound as brown prisms mp 232–233 °C (dec).

The reaction of 2,5-dimethyl-3-haloacetylthiophene with sodium sulfide is used for the preparation of the ketosulfide shown below [293].

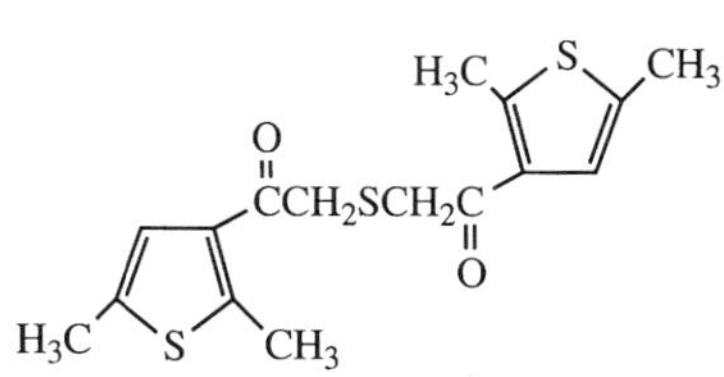

Another recent example is the preparation of 2-(2-chloro-4-nitrophenylthio)-1-(2-thienyl)ethanone from 2-chloroacetylthiophene and 2-chloro-4-nitrobenzenethiol under alkaline conditions. This compound is oxidized to the sulfone by treatment with 30% aqueous hydrogen peroxide in acetic acid [294].

2-(2-Chloro-4-nitrophenylthio)-1-(2-thienyl)ethanone [294]

A mixture of 2-chloro-4-nitrobenzenethiol (11.4 g, 60 mmol), sodium hydroxide (2.4 g, 60 mmol), and water (150 ml) is heated and stirred under nitrogen to a deep red solution. After cooling triethylbenzyl ammonium chloride (0.8 g) is added and then a solution of 2-chloroacetylthiophene (9.6 g, 60 mmol) in acetone (50 ml) is added dropwise over a period of 1 h at room temperature. The stirring is continued for 1 h, the precipitate formed is filtered off, washed with water and recrystallized from ethyl acetate giving 17.3 g (92%) of the title compound as yellow needles mp 152–153 °C.

2-(2-Chloro-4-nitrophenylsulfonyl)-1-(2-thienyl)ethanone [294]

The compound described above is dissolved in acetic acid (50 ml) and oxidized with 30 % aqueous hydrogen peroxide (10 ml) at 55 °C for 24 h. The solvent is distilled off under reduced pressure, the residue poured into ice-water and the precipitate formed is filtered off, washed with water and recrystallized from ethanol giving 10.8 g (52%) of the sulfone as pale-yellow crystals mp 142–143 °C

The reaction of 1-(2-thienyl)ethanone-2-dimethylsulfonium bromide with diazotized anilines in ethanol solution, buffered with sodium acetate or with *N*-nitrosoacetarylamides is a route to hydrazidoyl bromides [295].

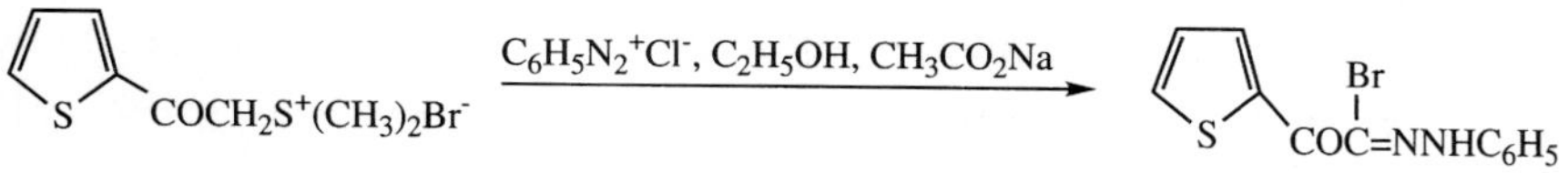

C-Thenoyl-N-phenylformhydrazidoyl bromide [295]

A solution of 1-(2-thienyl)ethanone-2-dimethylsulfonium bromide (2.67 g, 0.01 mol) in ethanol (50 ml) is stirred with sodium acetate (3 g) and the mixture is cooled to 0–5 °C. An aqueous solution of the diazonium salt (0.01 mol) cooled to 0–5 °C is added dropwise with stirring over a period of 45 min. The stirring is continued for 30 min and then the reaction mixture is left in an ice-chest for 2 h. The precipitate formed is filtered off, washed with water and recrystallized from glacial acetic acid giving 2.01 g (65%) of the title compound mp 149 °C.

Reaction of 3-acetylthiophene with lithium 1,1,1,3,3,3-hexamethyldisilazane followed by treatment with 2,2,2-trifluoroethyl trifluoroacetate and treating the resulting α-trifluoroacetylketone with methanesulfonyl azide in acetonitrile containing one equivalent of water and one and a half equivalent of triethylamine gives the diazoketone in excellent yield [296].

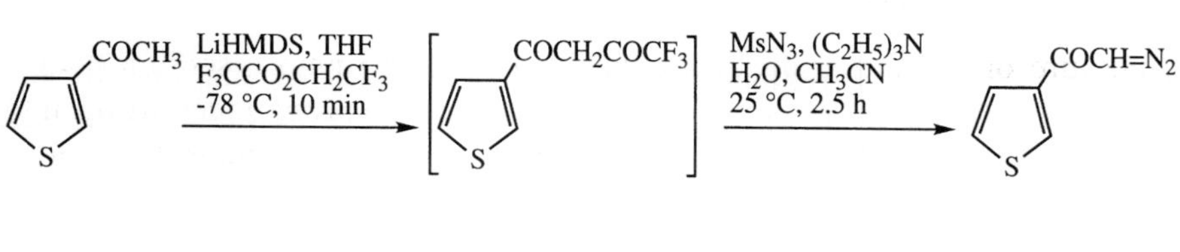

2-Diazo-1-(3-thienyl)-1-ethanone [296]

A flask equipped with rubber septum, argon inlet adapter, and pressure equalizing funnel is charged with 1,1,1,3,3,3-hexamethyldisilazane (0.98 ml, 4.71 mmol) in anhydrous tetrahydrofuran (16 ml). After cooling to 0 °C 2.20 *M* butyllithium in hexane (2.24 ml, 4.71 mmol) is rapidly added dropwise. The stirring is continued for 10 min and the mixture is cooled to −78 °C and 2-acetylthiophene (0.548 g, 4.34 mmol) in anhydrous tetrahydrofuran (16 ml) is added dropwise over a period of 15 min followed by rapid addition of 2,2,2-trifluoroethyl trifluoroacetate (1.16 ml, 8.67 mmol) by syringe in one portion. After 10 min the reaction mixture is poured into a separatory funnel containing 5% aqueous hydrochloric acid (25 ml) and diethyl ether (30 ml). The phases are separated and the aqueous phase extracted with diethyl ether (2 × 20 ml). The combined organic phases are washed with saturated sodium chloride solution (25 ml) and evaporated giving 2.2 g of a yellow oil, which is immediately dissolved in acetonitrile (15 ml). This solution is transferred to a flask equipped with rubber septum, argon inlet adapter and pressure equalizing funnel and treated with water (0.08 ml, 4.34 mmol), triethylamine (0.91 ml) and methanesulfonyl azide (0.56 ml, 6.51 mmol) in acetonitrile (30 ml) at room temperature. The stirring is continued for 2.5 h, after which the solution is concentrated to a volume of about 10 ml, diluted with diethyl ether (30 ml), washed with 10% aqueous sodium hydroxide solution (3 × 20 ml) and saturated sodium chloride solution (25 ml), dried over magnesium sulfate and evaporated. The residue is purified by chromatography on silica gel using ethyl acetate/petroleum ether (1:2) as eluent giving 0.610 g (92%) of the title compound as yellow crystals mp 55–57 °C.

4E.10.2 Dialkylaminomethylation (Mannich reaction)

The Mannich reaction is of great use in the thiophene series for dialkylaminomethylation in the α-position. 3-Dimethylamino-2-methyl-1-(2-thienyl)propanone hydrochloride was prepared in 95% yield from 2-propionylthiophene paraformaldehyde and dimethylamine hydrochloride [237]. In the same way the Mannich bases from 2-acetylthiophene, 3-propionylthiophene, 2-phenylacetylthiophene, and 2,5-dimethyl-3-propionylthiophene were also prepared [237].

4E.10.3 Alkylation reactions

Dialkylation of 3-acetylthiophene with alkyl iodides having one to four carbons is achieved in about 50% yield using a two-phase benzene-solid potassium hydroxide system in the presence of 18-crown-6 [297,298].

Isopropyl 3-thienyl ketone [297]

Finely powdered potassium hydroxide (11.2 g, 0.2 mol) is added to a solution of 3-acetylthiophene (2.52 g, 20 mmol) and 18-crown-6 (0.055 g, 0.2 mmol) in benzene (30 ml). The mixture is stirred at room temperature for 10 min, after which methyl iodide (9.97 ml) is added. The reaction mixture is stirred at room temperature for 19 h and then left for two days at room temperature. After filtration benzene and methyl iodide are evaporated at reduced pressure and the residue is distilled giving 1.6 g (52%) of the title compound bp 56–57 °C/ 0.5 mm Hg.

4E.10.4 Thienyl ketones by various reactions

2-Acetylthiophene is converted to the corresponding enaminone by the reaction with *N,N*-dimethylformamide dimethylacetal in 95% yield, which can further be transformed to the chloropropeneiminium salt by reaction with phosphorus oxychloride and further transformed to the vinamidium salt [299].

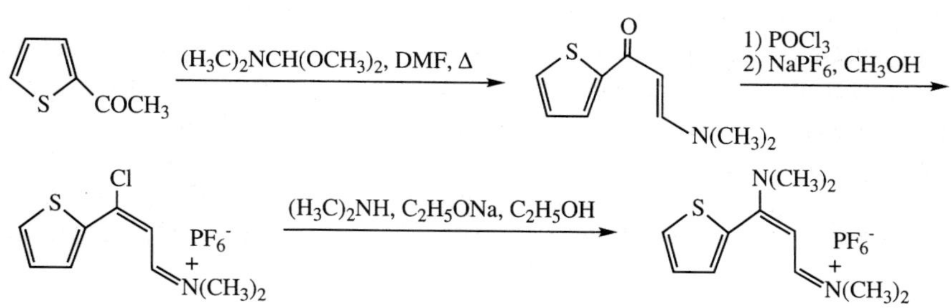

E-3-Dimethylamino-1-(2-thienyl)propenone [299]

A mixture of 2-acetylthiophene (10.0 g, 0.079 mol) and *N,N*-dimethylformamide dimethyl acetal (37.8 g, 0.317 mol) is placed in a round-bottomed flask, equipped with a reflux condenser and a magnetic bar, and stirred at reflux overnight. After cooling to room temperature and evaporation, the red–orange solid is dried *in vacuo* giving 14.3 g (95%) of the title compound, pure enough to be used in next step.

3-Chloro-3-(2-thienyl)prop-2-en-1-ylidendimethyliminium hexafluorophosphate [299]

A dry 250-ml three-necked flask is charged with anhydrous dichloromethane (100 ml) and *E*-3-dimethylamino-1-(2-thienyl)propenone (14. 0 g, 0.077 mol). Phosphorus oxychloride (11.8 g, 0.077 mol) is added dropwise under stirring and the stirring is continued for 30 min. After evaporation a cold solution of sodium hexafluorophosphate (26 g, 0.154 mol) in methanol is carefully added to the residue. The precipitate formed is filtered off giving 23.7 g (89%) of the title compound mp 177–178 °C after recrystallization from methanol.

The photo-stimulated reaction of the enolate ion of 2-acetylthiophene with iodobenzene is used for the preparation of benzyl 2-thienyl ketone [300].

2-Phenyl-1-(2-thienyl)ethanone [300]

Potassium *tert*-butoxide (6.0 mmol) and 2-acetylthiophene (63 mg, 5.0 mmol) are added to anhydrous and degassed dimethyl sulfate (40 ml) under nitrogen. After 15 min, iodobenzene (31 mg, 1.5 mmol) is added and the reaction mixture is irradiated for 3 h. The reaction is quenched with an excess of ammonium nitrate and water (120 ml). The mixture is extracted twice with dichloromethane (40 ml). The combined organic phases are washed with water, dried and evaporated. The residue is chromatographed on silica gel using petroleum ether/diethyl ether (9:1) as eluent giving a mixture of the title compound and acetylthiophene, which are separated by distillation at reduced pressure in a Kugelrohr apparatus.

The reaction of silyl enol ethers of 2-acetylthiophene with trimethylsilyl trifluoromethanesulfonate/iodosobenzene in dichloromethane is a very mild method for the preparation of the keto triflate [301].

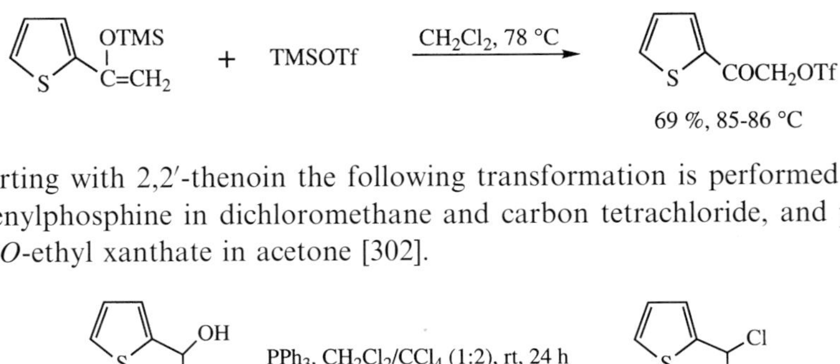

Starting with 2,2′-thenoin the following transformation is performed using triphenylphosphine in dichloromethane and carbon tetrachloride, and potassium *O*-ethyl xanthate in acetone [302].

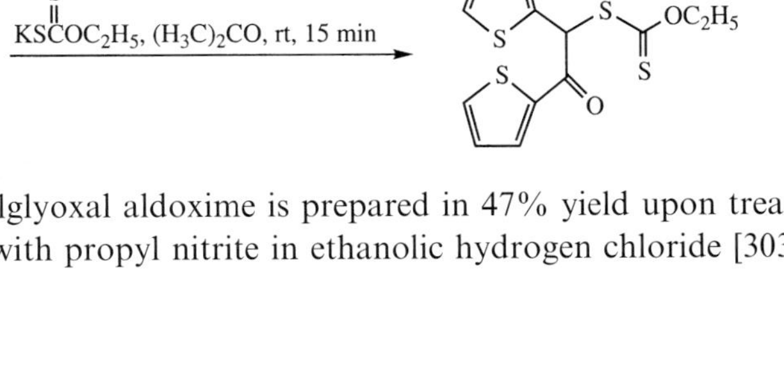

2-Thienylglyoxal aldoxime is prepared in 47% yield upon treating 2-acetylthiophene with propyl nitrite in ethanolic hydrogen chloride [303].

4E.11 THIENYL KETONES *VIA* PALLADIUM CATALYZED REACTIONS

The palladium-catalyzed reaction of 2-iodothiophene with *para*-methoxyphenylboronic acid in the presence of carbon monoxide gives methoxyphenyl 2-thienylketone [304].

para-Methoxyphenyl 2-thienylketone [304]

A flask equipped with a magnetic stirring bar, a septum inlet, and a reflux condenser is charged with dichlorobis(triphenylphosphine)palladium(II) (0.03 mmol), potassium carbonate (3 mmol), *para*-methoxyphenylboronic acid (1.1 mmol), and 2-iodothiophene (1.0 mmol). The flask is flushed with carbon monoxide, after which anisole (6 ml) is added. The reaction mixture is stirred at 80 °C for 5 h under carbon monoxide pressure (1 atm), cooled to room temperature, and diluted with benzene, washed with water, dried over magnesium sulfate and evaporated. The residue is purified by chromatography on silica gel giving 87% of the title compound.

Chemo- and regiospecific palladium-catalyzed four-component cascade process involving formation of four new bonds is used in the following reaction route initiated by oxidative addition of palladium(0) to 2-iodothiophene followed by sequential incorporation of carbon monoxide, allene and the nucleophiles [305].

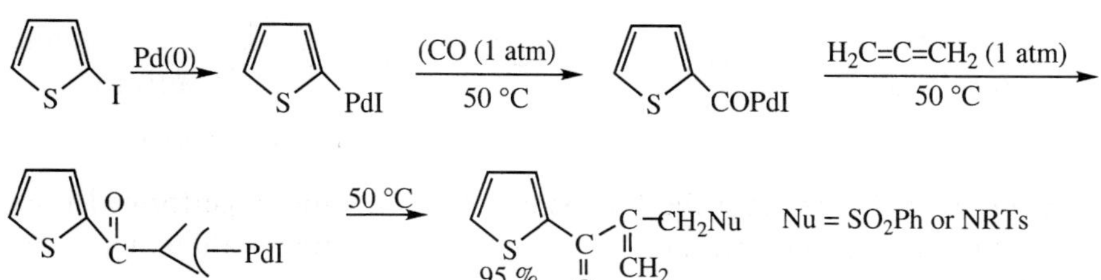

Alkylation of benzyl vinylcarbamate with anions of dimethyl malonate in the presence of *cis*-bis(acetonitrile)dichloropalladium(II) followed by cross-coupling with 2-trimethylstannylthiophene give the ketone shown below [306].

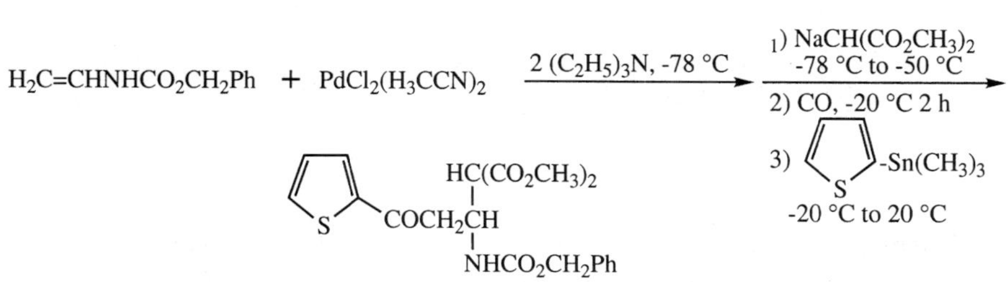

Methyl 3-[(benzyloxycarbonyl)amino]-2-(methoxycarbonyl)-5-oxo-5-(2-thienyl) pentanoate [306]

O-Benzyl-*N*-vinylcarbamate (50 mg, 0.28 mmol) is added to a stirred solution containing *cis*-bis (acetonitrile)dichloropalladium(II) (65 mg, 0.25 mmol) in tetrahydrofuran at room temperature. After 5 min the amber-red homogenous solution is cooled to −78 °C and treated with triethylamine (50 mg, 0.50 mmol) and 0.20 *M* sodium dimethyl malonate in tetrahydrofuran (1.60 ml, 0.32 mmol). The mixture is allowed to warm to −50 °C and stirred for 2 h. The reaction vessel is evacuated two times and the atmosphere replaced with carbon monoxide (1 atm). The mixture is allowed to warm to −20 °C and stirred for 2 h. The resulting black slurry is treated with 2-trimethylstannylthiophene (0.50 mmol). The reaction mixture is slowly warmed to room temperature, stirred for 15 h and filtered through a pad of silica gel using ether as eluent. After evaporation the residue is purified by radial-layer chromatography on 1 mm silica gel plate using hexane/ethyl acetate (3:2) as eluent, giving 65 mg (62%) of the title compound as a clear oil.

4E.12 THIENYL KETONES BY RING-CLOSURE REACTIONS

The reaction of 3-butyne-2-one with elemental sulfur at 205–215 °C in benzene in a stainless autoclave can be used for preparation of a mixture of 2,4-diacetyl- and 2,5-diacetylthiophene in 43% and 22% yields, respectively, which could be separated by chromatography [307]. The reaction of 2-acyleneammonium chlorides with α-mercapto derivatives can be used for the preparation of 2-acylthiophenes in good to moderate yields [308].

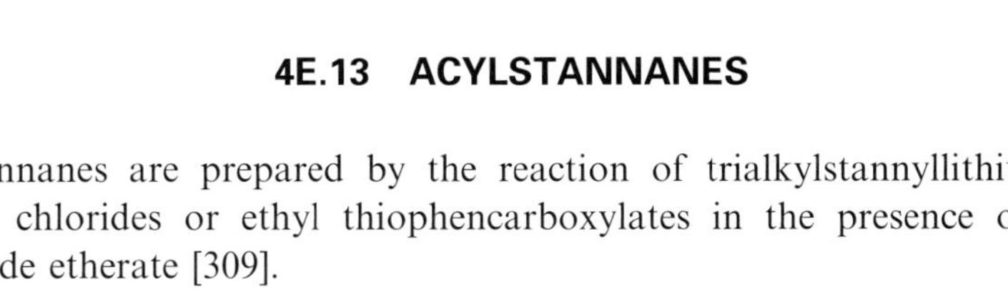

4E.13 ACYLSTANNANES

Acylstannanes are prepared by the reaction of trialkylstannyllithium with thenoyl chlorides or ethyl thiophencarboxylates in the presence of boron trifluoride etherate [309].

4E.14 THIENYLGLYOXALS

The oxidation of acetylthiophenes with selenium dioxide is an often-used method for the preparation of thienyl glyoxals [310–314].

2-Thienyl glyoxal [311]

A three-necked flask equipped with stirrer, reflux condenser, and thermometer is charged with dioxane (480 ml), water (22 ml), and powdered selenium dioxide (83 g, 0.75 mol). The mixture is stirred at 55 °C until a clear solution is obtained, after which 2-acetylthiophene (94 g, 0.75 mol) is added in one portion. The reaction mixture is stirred under reflux for 4 h and filtered hot. The filtrate is evaporated and the residue distilled giving 75 g (72%) of the title compound bp 101–102 °C/16 mm Hg.

5-Trimethylsilyl-2-acetylthiophene has successfully been oxidized to the glyoxal [315]. This method can also be used for the preparation of glyoxal derivatives from 2,2′-bithienyl and α-terthienyl [316]. Reaction of 2-acetyl-substituted thiophenes with alkyl nitrites in the presence of the corresponding aliphatic alcohols and hydrochloric acid can be used for the preparation of linear acetals of thienylglyoxals [317].

2-(2,2-Dipropoxyacetyl)thiophene [317]

Propyl nitrite (5.1 g, 0.06 mol) is gradually added to a mixture of 2-acetylthiophene (6.3 g (0.05 mol), propylalcohol (1.8 g, 0.03 mol) and hydrochloric acid (0.3 g) at such a rate that the temperature is below 50 °C. The reaction mixture is heated to 70 °C for 1.5 h, cooled and neutralized with 20% sodium hydroxide solution, washed with water, dried over magnesium sulfate and distilled giving 1 g (32%) of the title compound bp 145–150 °C/13 mm Hg.

4E.15 THENOINS (ThCOCHOHTh)

4E.15.1 By the benzoin condensation

The benzoin condensation does not work well in the thiophene series giving much lower yield than with benzaldehyde [318–320]. The reactions of 2-thiophene aldehyde and furfural have been reexamined with catalysts such as potassium cyanide, cyanide resin and a thiazolium salt. In contrast to furfural, 2-thiophene aldehyde gave both thenoins and thenils, which could partly be the reason for the low yields. On the other hand only 3,3′-thenoin was obtained in 64% yield from 3-thiophene aldehyde using the thiazolium salt as catalyst [321].

4E.15.2 From thienylglyoxals and dithienyl diketones and metalorganic reagents

The reaction of 2-thiophene aldehyde with 2-thiophenemagnesium bromide is used for the preparation of 2,2′-thenoin [320]. The reaction of di-(2-thienyl) ketone with cylopentadiene in the presence of diethylamine in methanol gives a mixture in 66% yield [322].

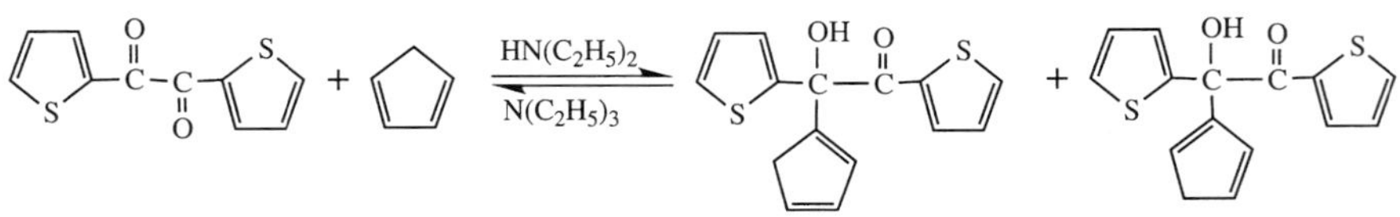

4E.16 α,ω-THENOYL ALKANES

4E.16.1 Dithenoyl methanes

These compounds are conveniently prepared by the Claisen condensation of acetylthiophenes with esters of thiophenecarboxylic acids [323-330]. From 2-acetylthiophene and ethyl 2-thiophenecarboxylate, di(2-thenoyl)methane was obtained in 75% yield using sodium amide in tetrahydrofuran for the condensation [331]. Di-(2-thenoyl)methane gave an europium complex and is applied to organic electroluminiscent devices as an emitting and electron-transporting material.

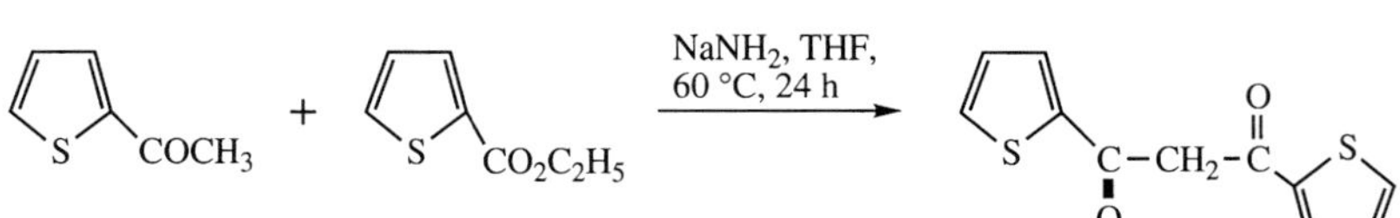

1,3-Di(2-thienyl)propane-1,3-dione [331]

Ethyl 2-thiophenecarboxylate and sodium amide are added to 2-acetylthiophene in tetrahydrofuran. The mixture is stirred at 60 °C for 24 h, after which it is concentrated and acidified with aqueous hydrochloric acid. The product is extracted with diethyl ether and purified by recrystallization from ethanol giving the title compound as yellow crystals in a yield of 75% mp 96–98 °C.

The reaction of 2-thenoyl chloride with 5-chlorooxindole is used for the preparation of 5-chloro-3-thenoyloxindole [332]. 1-(2-Thienyl)-3-(2-furyl)-3-(2,4-dithianyl)propanone is prepared by the reaction of 2-lithio-1,3-dithiane with 1-(2-thienyl)-3-(2-furyl)propenone and converted to 1-(2-thienyl)-4-(2-furyl)-4-oxabutanal on reaction with cupric oxide and cupric chloride, a suitable starting material for terheterocyclic compounds [333].

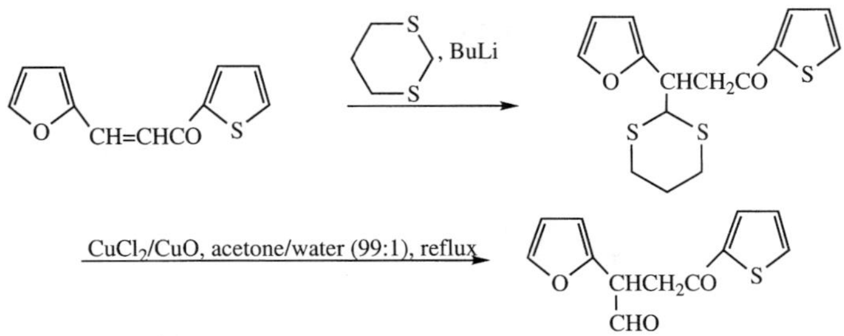

1-(2-Thienyl)-3-(2-furyl)-3-(2,4-dithianyl)propanone [333]

Butyllithium (1.6 *M*) in hexane (4.99 mmol) is added to a solution of 1,3 dithiane (500 mg, 4.16 mmol) in anhydrous tetrahydrofuran (2 ml) under argon at −78 °C. The mixture is stirred at −20 °C for 1 h and recooled to −78 °C, after which 1-(2-thienyl)-3-(2-furyl)propenone (850 mg, 4.16 mmol) is added. The stirring is continued for 5 h and the reaction quenched with saturated ammonium chloride solution. The product is extracted with diethyl ether and the combined organic phases are washed with saturated sodium chloride solution, dried over magnesium sulfate, and evaporated. The residue is purified by chromatography on silica gel using hexane/ethyl acetate (10:1) as eluent giving 875 mg (65 %) of the title compound as an oil.

1-(2-Thienyl)-4-(2-furyl)-4-oxabutanal [333]

Cupric oxide (324 mg, 4.08 mmol) and cupric chloride (274 mg, 1.02 mmol) are added to a solution of 1-(2-thienyl)-3-(2-furyl)-3-(2,4-dithianyl)propanone (330 mg, 1.01 mmol) in acetone/water (99:1) (10 ml) under argon. After refluxing for 5 h, the reaction mixture is treated with hexane/dichloromethane (1:1) and 5 *M* sulfuric acid. The phases are separated and the organic phase washed with water and saturated sodium chloride solution, dried over magnesium sulfate, and evaporated giving 180 mg of the title compound as an oil.

Recently a novel method for the preparation of di-(2-thenoyl)methane was developed, consisting in the reaction of the bis(iodozincio)methane complex of tetrahydrothiophene and 2-thenoyl chloride [334].

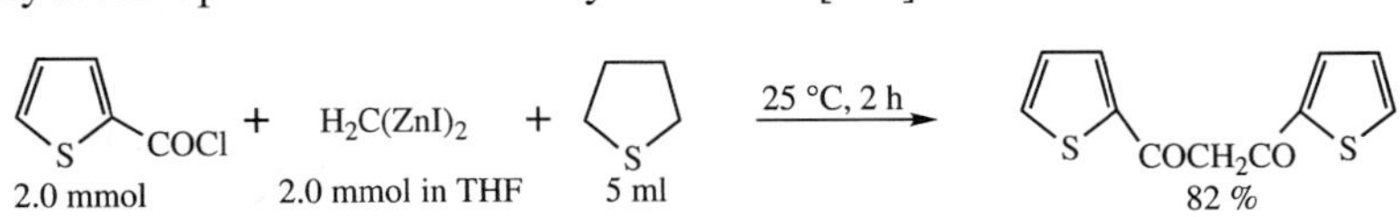

4E.16.2 1,2-Di(thenoyl)ethanes and -ethenes

Such compounds have become of great interest during recent years for the preparation of photochromic materials, polythiophenes, and other polyheterocyles. 1,2-Di(2-thenoyl)ethane is prepared by the Stetter reaction of 2-thiophene aldehyde with the Mannich base dimethylaminoethyl 2-thienyl ketone in the presence of acetic acid under azeotropic conditions [208,335].

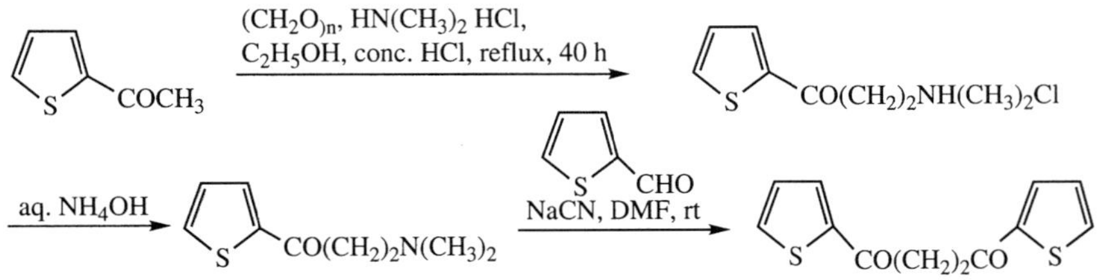

Another route, which is used for the preparation of 1,2-di(5-methyl-2-thenoyl)ethane, is the reaction of 5-methyl-2-thienyllithium with *N,N,N′,N′*-tetramethylsuccinimide [336].

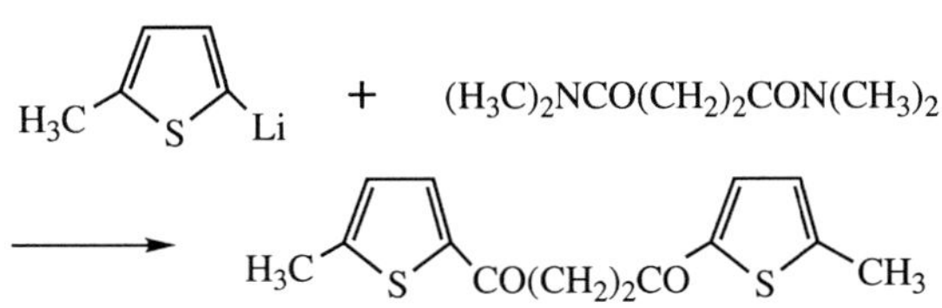

Bis(5-methyl-2-thienyl)-1,4-butadione [336]

A solution of 1.6 *M* butyllithium in hexane (82.5 ml) is slowly added under argon at −20 °C to a solution of 2-methylthiophene (11.78 g, 120 mmol) in anhydrous diethyl ether (100 ml). The stirring is continued at room temperature for 2 h. After cooling to 0 °C *N,N,N′,N′*-tetramethylsuccinimide (10 g, 58 mmol) is added in one portion and the reaction mixture is stirred at room temperature for 40 h, cooled in an ice-bath, and quenched with 10% hydrochloric acid (250 ml). The stirring is continued for 1 h, after which the phases are separated and the aqueous phase extracted with chloroform. The combined organic phases are washed with water and sodium chloride solution, dried over sodium sulfate, and evaporated. The residue is purified by chromatography on silica gel using dichloromethane as eluent, giving 5.35 g (33%) of the title compound as a yellow solid mp 179–180 °C after recrystallization from diethyl ether.

A number of 1,2-(dithenoyl)-1-thienylethanes (trithienyl-1,4-butanediones) are prepared by the Stetter reaction between 2- or 3-thiophene aldehyde and 1-(2-thienyl)-3-(3-thienyl)-2-propene-1-one and 1-(2-thienyl)-3-(2-thienyl)-2-propene-1-one [337].

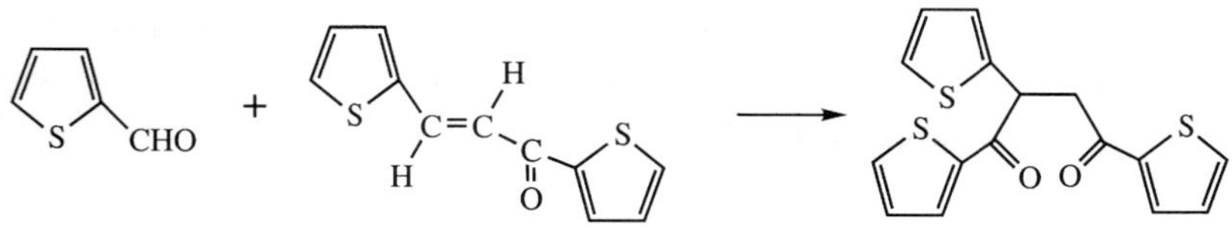

1,2,4-Tri-(2-thienyl)-1,4-butanedione [337]

A solution of 2-thiophene aldehyde (7 g, 62.6 mmol), 1,2-di-(2-thienyl)-2-propene-1-one (13.75 g, 62.6 mmol)), 3,4-dimethyl-5-(2-hydroxyethyl)thiazolium iodide (37 g, 13 mmol), and triethylamine (3.8 g, 38 mmol) in ethanol (90 ml) under argon is stirred under reflux for 12–16 h. On cooling to 0 °C the product crystallized, which is filtered off and recrystallized from methanol/acetone (2:1) giving 17.9 g (87%) as off-white crystals mp 117 °C.

The reaction of 1,2-di(thenoyl)ethane with dibromomethane and potassium *tert*-butoxide in anhydrous dimethylsulfoxide gives a double *C*-alkylation leading to 1,2-di(2-thenoyl)cyclopropane. With dichloromethane double *O*-alkylation leads to a dioxepine [338].

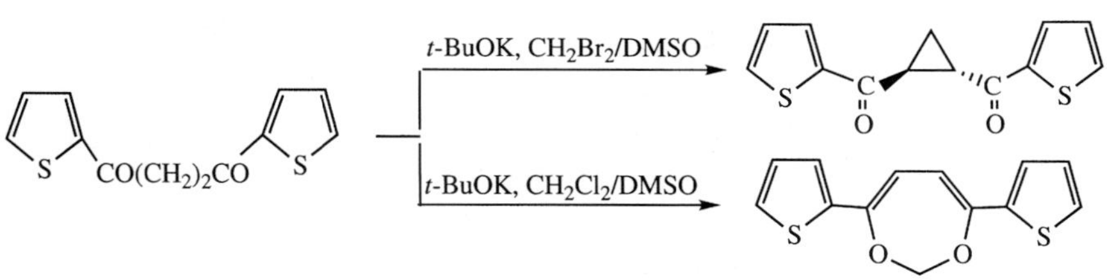

Cyclopropane-1,2-dicarbonyl-2,2′-bisthiophene [338]

A mixture of 1,2-di(thenoyl)ethane (025 g, 1 mmol) and potassium *tert*-butoxide (0.224 g, 2 mmol) in anhydrous dimethylsulfoxide (10 ml) under nitrogen is stirred for 10 min, after which dibromomethane (0.52 g, 3 mmol) is slowly added and the reaction mixture is stirred at room temperature for 30 min. After quenching with distilled water the product is extracted with dichloromethane (3 × 40 ml). The combined organic phases are washed with saturated ammonium chloride solution and sodium chloride solution, dried over sodium sulfate and evaporated. The residue is purified by chromatography on silica gel using cyclohexane/ethyl acetate (9:1) as eluent, giving 0.115 g (44%) of the title compound as a pale-yellow solid mp 110 °C.

1,3-Dioxepine-4,7-diyl-2,2′-bisthiophene [338]

This compound is prepared as described above but with dichloromethane as reactant giving 0.102 g (39%) of a dark yellow solid mp 120 °C.

From 1,4-bis(2-thienyl)but-2-yne-1,4-diol, prepared from acetylene dimagnesium bromide and 2-thiophene aldehyde, 1,4-bis(2-thienyl)but-2-yne-1,4-dione was prepared by oxidation with manganese dioxide and transformed to 4,5-bis(2-thenoyl)-1,3-dithiole-2-thione [339].

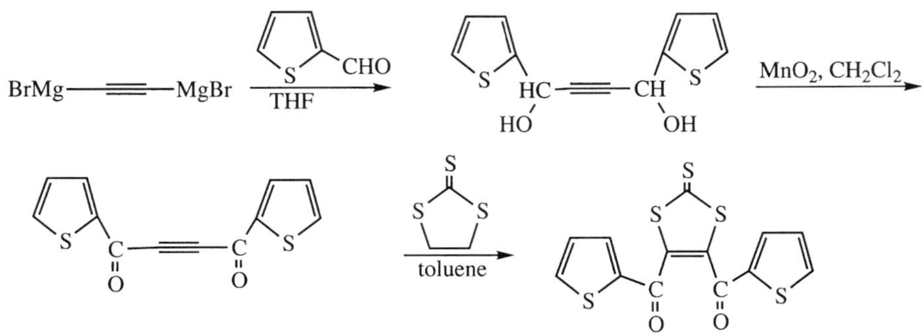

1,4-Bis(2-thienyl)but-2-yne-1,4-diol [339]

A solution of ethylmagnesium bromide is prepared from magnesium (1.42 g) and ethyl bromide (4.92 ml, 64 mmol) in tetrahydrofuran under standard Grignard conditions. Acetylene gas is purified by passing the reagent through a trap cooled to −70 °C and a second trap with concentrated sulfuric acid and then immediately transferred into the vigorously stirred Grignard solution at 50 °C over a period of 2 h and at a rate of 2–3 l/min. After this time the flow of acetylene is interrupted and 10–15 ml of the solvent is removed by distillation at normal pressure to ensure the complete decomposition of the acetylene monomagnesium bromide. 2-Thiophene aldehyde (5 ml, 54 mmol) is then added at room temperature and the stirring is continued overnight. The reaction mixture is poured into 20% aqueous ammonium chloride (200 ml) and the product extracted with ethyl acetate (2 × 30 ml). The combined organic phases are dried over magnesium sulfate and evaporated. The residue is purified by chromatography on silica gel using dichloromethane with a gradual change to dichloromethane/ethyl acetate (2:1) as eluent, giving 3.30 g (50%) of the title compound mp 110–113 °C.

1,4-Bis(2-thienyl)but-2-yne-1,4-dione [339]

Manganese dioxide (600 mg) is added to a solution of 1,4-bis(2-thienyl)but-2yne-1,4-diol (50 mg, 0.20 mmol) in dichloromethane (5 ml). The reaction mixture is stirred at room temperature for 2 min, filtered through a pad of silica gel and the pad is washed with dichloromethane (30 ml). The filtrate is evaporated giving 28 mg (55%) of the title compound mp 133–135 °C.

4E.16.3 1,3-Di(thenoyl)propanes

1,3-Di-(2-thenoyl)hexafluoropropane is prepared by the reaction of 2-thienylmagnesium bromide and perfluoroglutaryl dichloride [340].

Hexafluoro-1,3-di-(2-thenoyl)propane [340]

2-Thienylmagnesium bromide (0.16 mol), prepared with the standard Grignard technique, is added dropwise to perfluoroglutaryl dichloride (22.2 g, 0.08 mol) in anhydrous diethyl ether 200 ml) at −30 °C. The reaction mixture is refluxed for 16 h, cooled and treated with an excess of 10% ammonium chloride solution. The phases are separated and the organic phase washed with alkali and water, dried over magnesium sulfate and evaporated. The residue is taken up in hot benzene and this solution is decolorized with charcoal and evaporated giving the title compound mp 59–60 °C after recrystallization from methanol.

On the other hand the reaction of 2-thienyllithium with *N,N,N′,N′*-tetraethyl perfluoroglutarimide gave *N,N*-diethyl-2-thiophenecarboxamide and not the desired diketone [340].

4E.17 THIENYL SUBSTITUTED α-KETO ESTERS, AMIDES AND ACIDS

4E.17.1 Reaction of thiophenemagnesium halides or thienyllithium derivatives with dialkyl oxalates

The reaction of thiophenemagnesium halides or thienyllithium derivatives can be used for the preparation thienyl glyoxalate [341–344]. Reaction of 2,5-diiodothiophene and ethyl bromide (entrainment method) with magnesium followed by diethyl oxalate gave the 2,5-glyoxylic ester, which was hydrolyzed to the thiophene-2,5-diglyoxylic acid [342].

2-Thienyllithium prepared by halogen metal exchange of 2-chlorothiophene at −60 °C in tetrahydrofuran gave upon reaction with diethyl oxalate an 84% yield of ethyl 2-thienylglyoxalate [344].

4E.17.2 Friedel-Craft acylation of a thiophenes with the acid chloride $ClCOCO_2R$

An alternative useful route to thienylsubstituted α-keto esters consists of Friedel-Crafts reaction of thiophene with ethoxalyl chloride [310,345,346]. Friedel-Crafts reactions of fatty acid esters of thiophene with ethoxalyl chloride is used for the preparation of 2,5-disubstituted acids [346].

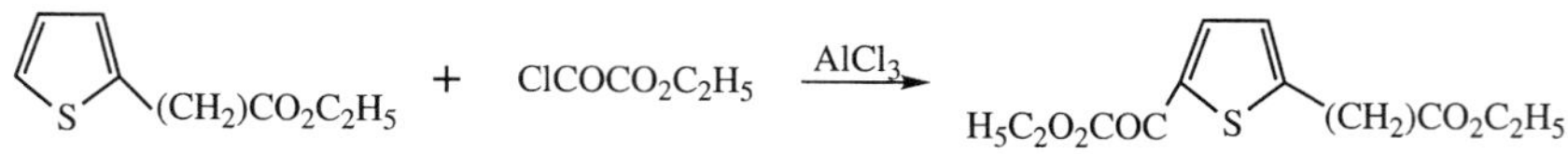

Ethyl α-oxo(5-ethoxycarbonylmethyl-2-thienyl acetate [346]

Ethoxalyl chloride (41.8 g, 0.31 mol) is added to a stirred suspension of anhydrous aluminium chloride (77 g) in dichloromethane (450 ml) cooled to −6 °C in one portion. The stirring is continued for 15 min, after which ethyl 2-thienylacetate (43.2 g, 0.25 mol) is added dropwise while the temperature is kept between −1 and −8 °C. The reaction mixture is stirred at this temperature for 30 min and poured into ice and hydrochloric acid. The phases are separated and the organic phase washed with water and sodium carbonate solution, dried over magnesium sulfate, evaporated and distilled giving 34.0 g (80%) of the title compound bp 201–203 °C/7 mm Hg.

Titanium tetrachloride-catalyzed reaction of 2-(4-chlorophenyl)thiophene with ethoxalyl chloride gives ethyl 5-(4-chlorophenyl)thien-2-glyoxylate in excellent yield [345].

Ethyl 5-(4-chlorophenyl)thien-2-glyoxylate [345]

A solution of titanium tetrachloride (3.79 g, 20 mmol) in benzene (30 ml) is added dropwise to a solution of 2-(4-chlorophenyl)thiophene (3.89 g, 20 mmol) and ethoxalyl chloride (2.73 g, 20 mmol) in benzene (40 ml) cooled to 0 °C. The stirring is continued at room temperature for 3 h and the reaction mixture is then thoroughly mixed with ice-water. The product is extracted with dichloromethane and this solution is evaporated. The residue is recrystallized from dichloromethane giving 5.43 g (92%) of the title compound.

4E.17.3 Reaction of thienyllithia with half amide, EtO_2CCONR_2 or diamide $(CONR_2)_2$

Reaction of 2-thienyllithium with tetramethyloxamide is used for the preparation of *N,N*-diethyl 2-thienyl glyoxylamide [347].

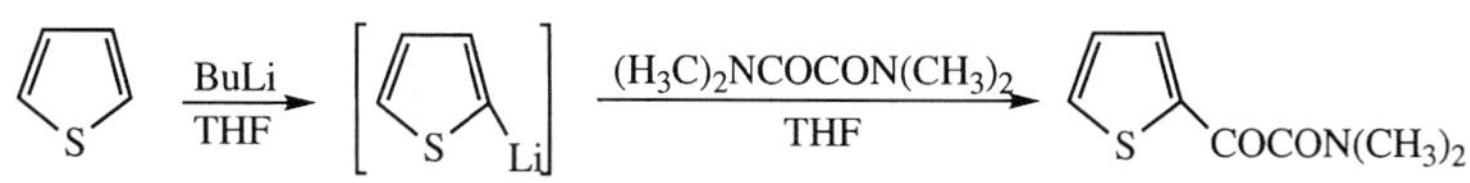

4E.17.4 α-Ketoacids by oxidation of acetylthiophenes

2-Thienylglyoxylic acid is prepared by oxidation of 2-acetylthiophene with nitric acid [348], potassium permanganate [349] or selenium dioxide in pyridine [350]. 3-Thienylglyoxylic acid as well as its 2,5-dimethyl- and 2,4,5-trimethyl derivatives were prepared by the last mentioned method [350]. This is the best method as competing further oxidation can occur with the first two oxidizing agents.

2,4,5-Trimethylthiophene-3-glyoxylic acid [350]

A stirred solution of 3-acetyl-2,4,5-trimethylthiophene (4.8 g, 28.6 mmol) in anhydrous pyridine (20 ml) is warmed to 60 °C and selenium dioxide (4.6 g, 41.5 mmol) is added in portions over a period of 30 min. The stirring is continued for 4.5 h, after which the mixture is cooled and filtered. The filtrate is evaporated and the residue dissolved in water and steam distilled in order to remove pyridine. The remaining aqueous solution is acidified with 40% phosphoric acid (10 ml). The product is extracted with diethyl ether and the combined organic phases dried over magnesium sulfate and evaporated giving 4.85 g (85%) of the title compound as a yellow solid mp 94.5–96.5 °C after recrystallization from benzene/cyclohexane.

4E.17.5 Various derivatives of α-keto acids

N-substituted 3-thienylglyoxylhydroxamic acids are prepared by the reaction of 3-thienylglyoxylic acid chloride with silylated *N*-alkylhydroxylamines [351].

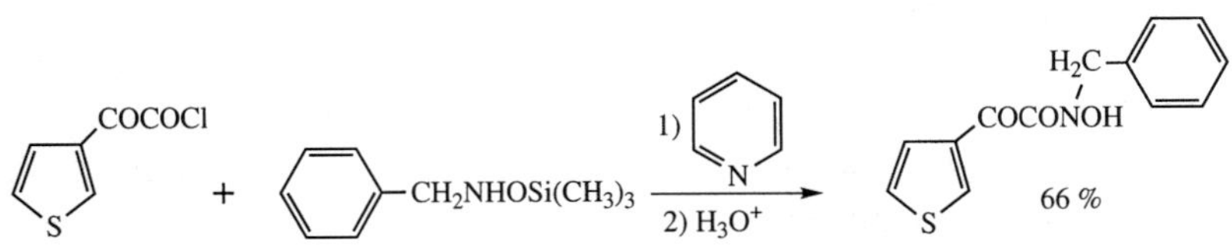

4E.18 THIENYL SUBSTITUTED β-KETO ESTERS AND ACIDS

4E.18.1 *Via* Claisen condensation of thiophenecarboxylic ester and an aliphatic ester

This method has been used for the preparation of the 2-thienyl and 5-alkyl-2-thienyl derivatives [352]. The reaction of methyl 2-thiophenecarboxylates with γ-butyrolactone with sodium methoxide in anhydrous dioxane is used for the preparation of 3-(2-thienylcarbonyl)tetrahydrofuran-2-one [353].

3-(2-Thienylcarbonyl)tetrahydrofuran-2-one [353]

Equimolar quantities of methyl 2-thiophenecarboxylate and sodium methoxide are heated with four equivalents of γ-butyrolactone in anhydrous 1,4-dioxane under nitrogen for 45 h. After addition of ice-water the pH is adjusted to 4. The product is extracted with chloroform and the combined organic phases dried, and evaporated. The residue is purified by chromatography giving 60% of the title compound as an oil.

4E.18.2 From acylthiophenes and diethyl carbonate

Base-induced condensation of acylthiophenes with diethyl carbonate is conveniently used for the preparation of thienylsubstituted β-ketoesters [354–356]. Thus the reaction of ethyl 2-thienyl ketone and dimethyl carbonate gave methyl 3-(2-thenoyl)acetate [356].

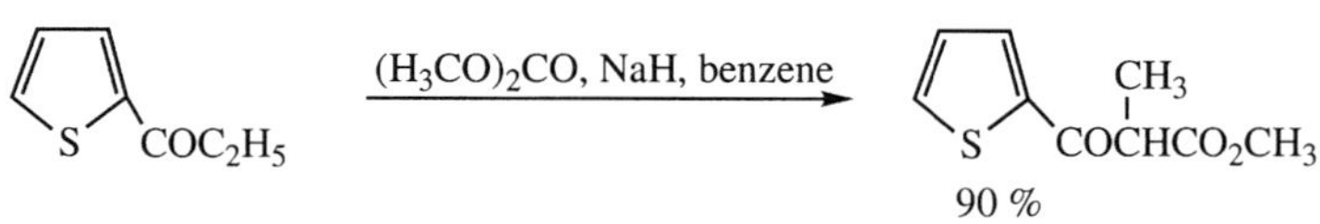

4E.18.3 Acylation of appropriate 1,3-dicarbonyl compounds with ThCOCl, followed by hydrolysis

Ethyl 2-thenoylacetate is prepared by the reaction of 2-thenoyl chloride with ethyl acetoacetate [357,358].

Ethyl 2-thenoylacetate [357]

Anhydrous ethanol (15 ml) is added to magnesium turnings (2.65 g) and the reaction is started by addition of carbon tetrachloride (0.5 ml). When the vigorous reaction has subsided anhydrous diethyl ether (100 ml) is added in portions. The stirring is continued for 2–3 h, after which the mixture is cooled to 5 °C and ethyl acetoacetate (13.0 g, 0.1 mol) in anhydrous diethyl ether (20 ml) is added dropwise. At the same temperature 2-thenoyl chloride (14.6 g, 0.1 mol) in anhydrous diethyl ether (20 ml) is added dropwise over a period of 1 h. The stirring is continued for 1 h and then the reaction mixture is left overnight and hydrolyzed with ice and dilute sulfuric acid. The phases are separated and the organic phase is washed with water, dried over sodium sulfate and evaporated. The residue is treated with 5% ethanolic ammonia solution at 0 °C for 10–12 h, part of the ethanol is distilled off and the residue treated with a mixture of ice and dilute sulfuric acid. The product is extracted with diethyl ether and the combined organic phases are washed with water, evaporated and distilled giving 12.9 g (65%) of the title compound bp 162–175 °C/12 mm Hg.

The reaction of 2-thiophenecarboxylic acid chloride with the lithium enolate of *tert*-butyl acetate gives the *β*-ketoester, which was stereoselectively alkylated at C-2 using (*S*)-methyl 2-(triflyloxy)propionate [359].

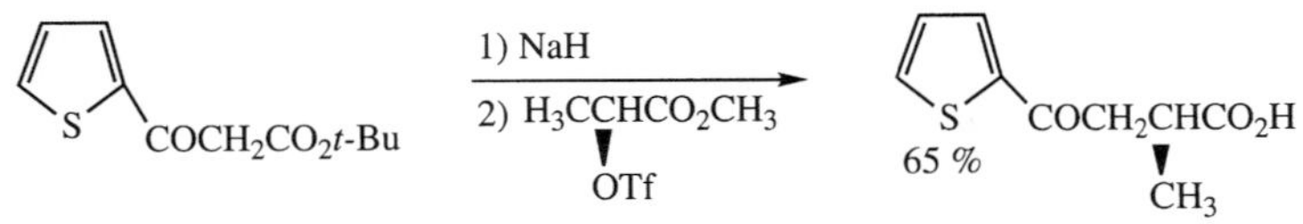

4E.18.4 From methyl thienyl ketones and acid anhydrides

Acylation of methyl thienyl ketones with an acid anhydride in the presence of ferric chloride or zinc chloride can be used for the preparation of *β*-keto esters, but the yields are low [360].

4E.18.5 *Via* oxidation of *β*-hydroxy esters

Oxidation of methyl 2-methyl-3-(2-thienyl)-3-hydroxypropionate with pyridinium dichromate is used for the preparation of methyl 2-methyl-3-(2-thienyl)-3-oxopropionate [361].

Methyl 2-methyl-3-(2-thienyl)-3-oxopropionate [361]

A mixture of 2-thiophene aldehyde (4.75 g), methyl bromoacetate (8.48 g) and activated zinc dust in anhydrous benzene (50 ml) is under stirring refluxed for 1 h and worked up. A suspension of the Reformatsky product obtained (5 g) and pyridinium dichromate (29 g) in dichloromethane (50 ml) is stirred at room temperature for 5 days. After work up the product is purified by chromatography on silica gel (200 g) using hexane/ethyl acetate (19:1) as eluent giving 4.33 g (87%) of the title compound.

4E.19 THIENYL ARYL KETONES

4E.19.1 Friedel–Crafts reactions between thiophenes and aroyl halides or anhydrides

In connection with the preparation of compounds of medicinal interest, a number of 2-thienyl arylketones were prepared. Thus 3-methoxy-4-nitrobenzoyl chloride reacted with thiophene in dichloromethane using aluminium chloride as catalyst to give 2-thienyl 3-methoxy-4-nitrophenyl ketone [362], and

2-thienyl 2-methyl-3-methoxyphenyl ketone was obtained with 2-methyl-3-methoxybenzoyl chloride, and nitrated to 2-thienyl 2-methyl-3-methoxy-4-nitro- phenyl ketone. In all these preparation the acid chlorides were prepared by the reaction with thionyl chloride and used without isolation and purification [362]. 5-Methyl-2-thienyl 4-chloro-3-nitrophenyl ketone is prepared in the same way from 2-methylthiophene 3-nitro-4-chlorobenzoyl chloride [362].

(3-Methoxy-4-nitrophenyl)(2-thienyl)methanone [362]

A suspension of 3-methoxy-4-nitrobenzoic acid (50.0 g, 0.254 mol), thionyl chloride (60.0 ml, 0.823 mol), chloroform (120 ml) and *N,N*-dimethylformamide (5 drops) is refluxed under nitrogen for several hours until a clear solution is obtained and then evaporated to dryness at 60 °C. The acid chloride so obtained is dissolved in dichloromethane (166 ml) and added in a slow stream to a suspension of aluminium chloride (33.9 g, 0.254 mol) in dichloromethane (166 ml) cooled with an ice-bath at such a rate that the temperature is below 10 °C, followed by dropwise addition of thiophene (22.4 g, 0.279 mol) at the same temperature. Stirring is continued for 0.5 h in the cold and for 2 h at room temperature, after which the reaction mixture is poured into ice-water. The phases are separated and the organic phase washed twice with 5% sodium bicarbonate solution, dried over magnesium sulfate containing decolorizing charcoal, and evaporated. The residue is recrystallized from toluene giving 28.9 g (45%) of the title compound mp 117–118 °C.

2-Thienyl 2-hydroxy-5-ethylphenyl ketone was obtained in low yield from thiophene and 2-hydroxy-5-ethylbenzoyl chloride [363]. Another example is the preparation of 2-thienyl 2,3-dimethoxyphenyl ketone from thiophene and 2,3-dimethoxybenzoyl chloride using aluminium chloride as catalyst at 0 °C. If aluminium chloride in refluxing dichloromethane is used, demethylation of the *ortho*-methoxy group occurs giving 2-thienyl 2-hydroxy-3-methoxyphenyl ketone [364].

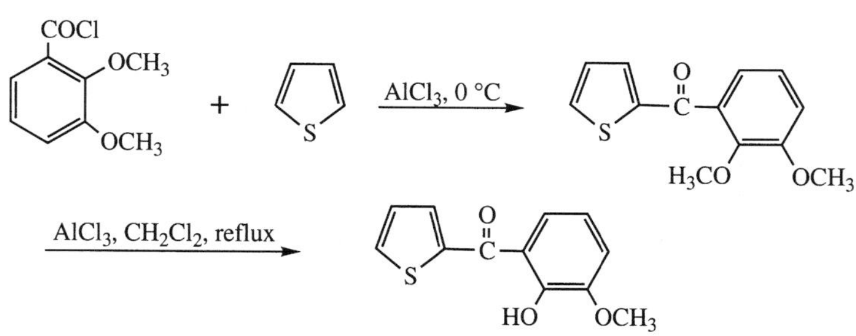

The reaction of 2-substituted thiophenes with 2-thiophenecarbonyl chloride in carbon disulfide using aluminium chloride as catalyst was recently used for the preparation of 5-substituted di(2-thienyl)ketones in connection with an investigation of their protonation equilibria [365]. Only mono-thenoylation of di(2-thienyl)methane with 2-thenoyl chloride and stannic chloride could be achieved, yielding 2-(2-thenoyl)-5-(2-thenyl)thiophene [227].

2-(2-Thenoyl)-5-(2-thenyl)thiophene [227]

Stannic chloride (1.82 g, 0.007 mol) is added dropwise to a solution of di(2-thienyl)methane (12.6 g, 0.07 mol) and thenoyl chloride (10.3 g, 0.07 mol) in benzene (150 ml) at 0 °C with stirring. The stirring is continued at room temperature for 30 min and at 40–50 °C for 30 min, after which the reaction mixture is treated with water. The phases are separated and the organic phase washed with sodium carbonate solution, dried and evaporated. The residue, a brown solid, is recrystallized from ethanol giving 15.6 g (77%) of the title compound as white needles mp 104 °C.

4E.19.2 Friedel–Crafts reactions between thiophenecarbonyl chlorides and reactive aromatics

The reaction of 2-thenoyl chloride, 3-methyl-2-thenoyl chloride, and 4-methyl-2-thenoyl chloride with anisole using aluminium chloride as catalyst is a convenient route, 3-methyl-2-thienyl- and 4-methyl-2-thienyl 4-methoxyphenyl ketones, which were easily further modified by nitration of the phenyl group [362].

(4-Methoxyphenyl)(3-methyl-2-thienyl)methanone [362]

Reaction of 3-methyl-2-thiophenecarboxylic acid (5.0 g, 0.0352 mol), thionyl chloride (7.85 ml, 0.108 mol) and chloroform (25 ml) for 5 h gives the acid chloride as an amber liquid. A solution of anisole (4.19 g, 0.0387 mol) in dichloromethane (10 ml) is slowly added to an ice-cooled suspension of aluminium chloride (7.04 g, 0.0528 mol) in dichloromethane (12 ml) at such a rate that the temperature is below 12 °C. A solution of the acid chloride (0.0352 mol) in dichloromethane (10 ml) is added at the same temperature. The stirring is continued under cooling for 25 min and then the reaction mixture is cautiously poured into ice water. The phases are separated and the organic phase washed with 5% sodium carbonate solution, dried, evaporated and distilled giving 6.0 g (73%) of the title compound bp 142–150 °C/0.3 mm Hg, mp 53–56.5 °C.

The reaction of 2-thenoyl chloride with 2-hydroxy-3,4-dichloroacetophenone in carbon disulfide and aluminium chloride as catalyst is used for the preparation of 2-thienyl 2,3-dichloro-4-hydroxy-5-acetylphenyl ketone [366].

2-Thienyl 2,3-dichloro-4-hydroxy-5-acetylphenyl ketone [366]

Aluminum chloride (16 g, 0.12 mol) is added to a stirred solution of 2-hydroxy-3,4-dichloroacetophenone (20.5 g, 0.1 mol) and 2-thenoyl chloride (14.6 g, 0.1 mol) in carbon disulfide (300 ml) under cooling. The reaction mixture is warmed on a water-bath for 1 h and evaporated. The residue is cooled and poured into ice and the product extracted with chloroform. The combined organic phases are washed with water, dried over sodium sulfate and evaporated. The residue is purified by chromatography on silica gel using dichloromethane/cyclohexane (7:3) as eluents, the first fraction containing starting material and the last the title compound 9.4 g (30%) mp 172–174 °C.

Reacting 2-thenoyl chloride with 2,4-dimethoxyacetylaminobenzene in the presence of excess aluminium chloride gives 2-thienyl 2-hydroxy-4-methoxy-5-acetylaminophenyl ketone due to demethylation [364].

(5-Acetamido-2-hydroxy-4-methoxyphenyl)(2-thienyl) methanone [364]

Aluminum chloride (10 g, 0.075 mol) is added in small portions to a suspension of 2-thenoyl chloride (7.33 g, 0.05 mol) and 2,4-dimethoxyacetanilide (9.76 g, 0.05 mol) in carbon disulfide (50 ml) at 0 °C. The stirring is continued at room temperature for 30 min, after which the reaction mixture is refluxed for 20 h and cooled. The carbon sulfide is decanted and the residue poured under vigorous stirring into crushed ice (200 g) and concentrated hydrochloric acid (60 ml). The hydrolysis continues for 1 h, after which the product is extracted with toluene. The combined organic phases are washed with 5% sodium bicarbonate and water until neutral reaction, dried over sodium sulfate and evaporated. The residue is dissolved in dichloromethane (20 ml) and purified by chromatography on silica gel using dichloromethane/ethyl acetate (80:20) as eluent, giving 5.8 g (40%) of the title compound as yellow crystals mp 187 °C.

The reaction of 2-thenoyl chloride with 2-trimetylsilylthiazole yields 2-thiazolyl-2-thienyl ketone [367].

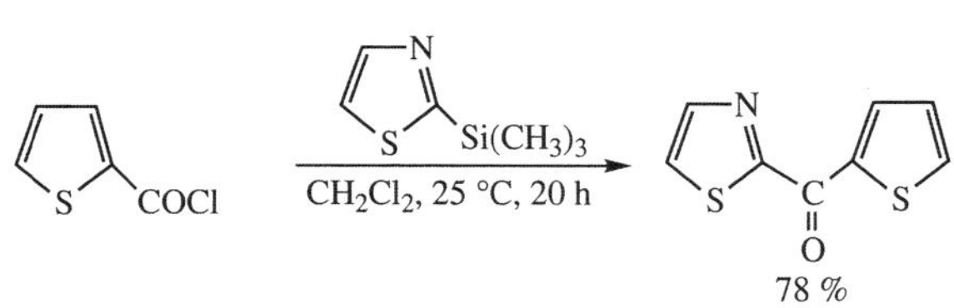

4E.19.3 From thienyllithium derivatives and carbon dioxide

Metalation of 2-methylthiophene and 2-butylthiophene with butyllithium followed by reaction with gaseous carbon dioxide is used for the preparation of the corresponding bis(5-alkylthiophene-2-yl) ketones [368].

Bis(5-methylthiophene-2-yl) ketone [368]

A three-necked 500-ml flask equipped with a reflux condenser and a thermometer is charged under nitrogen with 2-methylthiophene (9.8 g, 100 mmol) in anhydrous tetrahydrofuran. 2.5 *M* Butyllithium in hexane (40 ml) is added at −78 °C. The mixture is allowed to reach room temperature, whereupon carbon dioxide gas is introduced over a short period of time. This is repeated over a 30 min period until the temperature of the reaction mixture remained constant upon further addition of carbon dioxide. The reaction mixture is then cooled to −78 °C and chlorotrimethylsilane (12.6 ml, 100 mmol) is added. After warming to room temperature the reaction mixture is poured into 0.5 *M* hydrochloric acid (500 ml). The phases are separated and the aqeuous phase extracted twice with diethyl ether (100 ml). The combined organic phases are washed with saturated sodium carbonate solution, dried over magnesium sulfate, evaporated and distilled at reduced pressure giving 13.3 g (60%) of the title compound as light-yellow flakes mp 53–55 °C.

4E.19.4 From thienyl metalorganic reagents and acid chlorides

The preparation of di(3-thienyl) ketone *via* the reaction of 3-thienyllithium with *N,N*-dimethylcarbamyl chloride [369] has been optimized [370].

Bis(3-thienyl) ketone [370]

A solution of 3-bromothiophene (10.00 g, 61.3 mmol) in anhydrous diethyl ether (50 ml) is cooled to −78 °C and 2.4 *M* butyllithium in hexane (25.5 ml) is added dropwise. This mixture is stirred for 1 h while it is allowed to warm to −40 °C. The pale yellow suspension of thienyllithium thus obtained is cooled to −50 °C and *N,N*-dimethylcarbamyl chloride (3,3 g, 31 mmol) is added dropwise over a period of 1–2 min. The reaction mixture is allowed to warm to −40 °C and stirred for 3 h at −40 to −30 °C, then treated with 1 *M* hydrochloric acid at 0 °C. After warming to room temperature the phases are separated and the aqueous phase extracted with diethyl ether (3 × 50 ml). The combined organic phases are dried over magnesium sulfate and evaporated giving 4.64 g (78%) of the title compound. A small amount is dissolved in carbon tetrachloride and chromatographed using petroleum ether/diethyl ether (3:1) as eluent mp 73 °C after recrystallization from diethyl ether/petroleum ether.

3-Thiophenemagnesium iodide, obtained by reaction with Rieke magnesium, Mg*, upon reaction with benzoyl chloride in tetrahydrofuran at −70 °C gives 3-benzoylthiophene [371,372].

3-Benzoylthiophene [372]

3-Iodothiophene (1.05 g, 5 mmol) is added to the washed Mg* (7.5 mmol) in tetrahydrofuran with stirring at room temperature. The slurry is stirred for 5–7 min at room temperature. When the oxidative addition is completed the excess of activated magnesium is allowed to settle out of the organomagnesium solution. The supernatant is cannulated to benzoyl chloride (0.56 g, 4 mmol) in tetrahydrofuran (10 ml) at −30 °C. After stirring the reaction mixture it is quenched with saturated aqueous ammonium chloride solution (20 ml). The product is extracted with diethyl ether (2 × 20 ml). The combined organic phases are washed with sodium chloride solution (2 × 20 ml), dried over magnesium sulfate and evaporated. The residue is flash chromatographed using ethyl acetate/hexane as eluent, giving 579 mg (77%) of the title compound.

3-Thienylmanganese bromide, 3-bromo-4-thienylmanganese bromide and 4-substituted 3-thienylmanganese bromide, obtained by reaction of the corresponding bromo derivatives, followed by benzoyl halides give the corresponding aryl 3-thienyl ketones in excellent yields under mild conditions [372,373].

3-(2-Bromobenzoyl)thiophene [372]

To a slurry of Rieke manganese (10.0 mmol) in tetrahydrofuran (10 ml) under argon, 3-bromothiophene (0.82 g, 5 mmol) is added at room temperature and the stirring is continued for 5 h. Dibromoethane (6.0 mmol) is neat added to the mixture at 0 °C, which is allowed to warm to room temperature over a period of 20 min. To this organomanganese reagent 2-bromobenzoyl chloride (0.51 g, 3 mmol) is neat added at room temperature *via* syringe. The reaction mixture is stirred at room temperature for 30 min and then quenched with 3 *M* hydrochloric acid (10 ml). The product is extracted with diethyl ether (2 × 10 ml). The combined organic phases are washed with saturated sodium bicarbonate solution, sodium thiosulfate solution and sodium chloride solution, dried over magnesium sulfate and evaporated. The residue is purified by flash chromatography using ethyl acetate/hexane as eluent, giving 102 mg (34%) of the title compound.

The reaction of 2-trimethylstannylthiophene with benzoyl chloride can be used for the preparation of phenyl 2-thienyl ketone [374].

Phenyl 2-thienyl ketone [374]

Benzoyl chloride (5.8 g, 0.041 mol) and 2-(trimethylstannyl)thiophene (10.2 g, 0.041 mol) are refluxed for 6 h at 172 °C. Fractional distillation afforded trimethyltin chloride (3.45 g, 42%) and 3.5 g (45%) of the title compound bp 78–80 °C/0.04 mm Hg, mp 55–56 °C after recrystallization from hexanes.

The palladium-catalyzed reaction of 2-thiopheneboronic acid with five equivalents of cesium carbonate under anhydrous conditions with *para*-chlorobenzoyl chloride is used for the preparation of *para*-chlorophenyl 2-thienyl ketone [257].

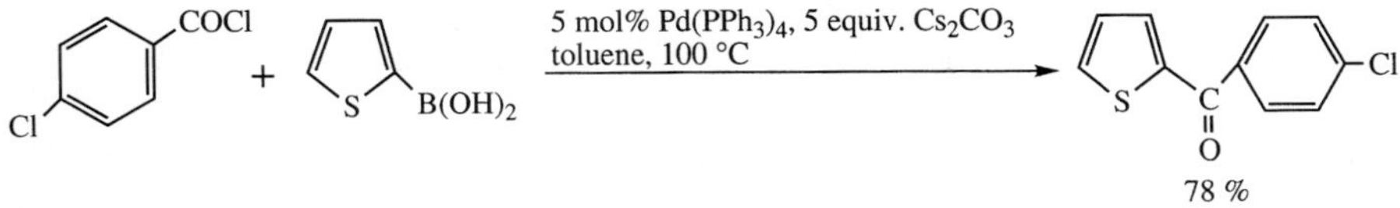

The palladium-catalyzed acyldemetalation reaction of di(2-thienyl)mercury with acid chlorides such as *para*-nitrobenzoyl chloride, 2-thenoyl chloride and 2-furoyl chloride give high yields of unsymmetrical heterocyclic ketones under mild conditions along with small amounts of homo coupled product [375].

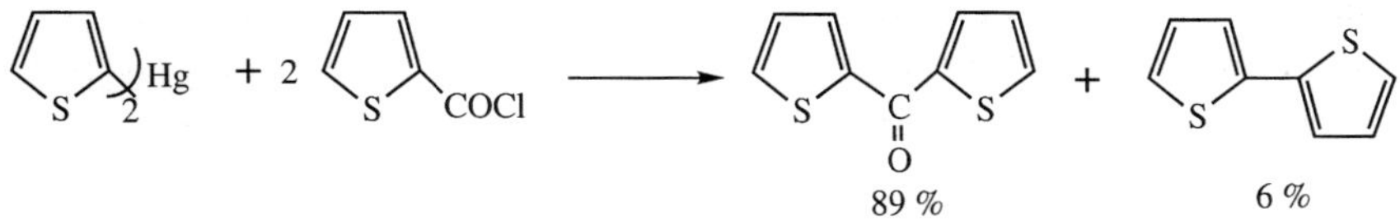

The palladium-catalyzed carbonylative coupling reaction of tributyl-2-thienylstannane with diphenyl iodonium salts gives phenyl 2-thienyl ketone and with 2-thienyl iodonium tosylate di(2-thienyl) ketone was obtained [376].

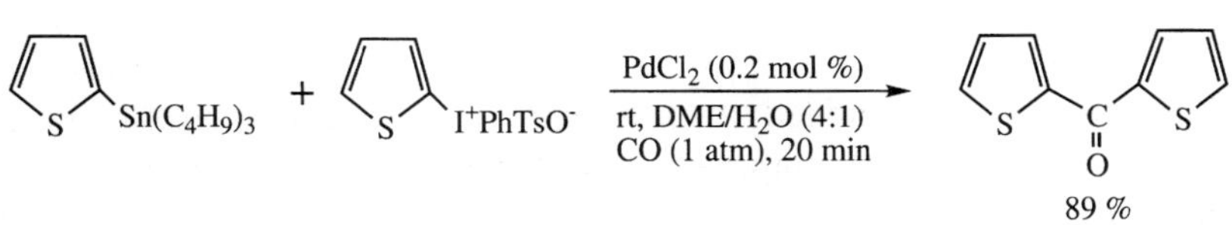

4E.19.5 From thienyllithium derivatives and aromatic nitriles

This is an often-used route to aryl thienyl ketones. Sometimes the intermediate ketimines are isolated, before hydrolysis [377]. From 3-thienyllithium and benzonitrile, 3-benzoylthiophene is prepared [249,378]. The reaction of 3-thienyllithium with 4-methyl- and 4-methoxybenzonitrile is used for the preparation of 3-thienyl 4-methylphenyl ketone and 3-thienyl 4-methoxyphenyl ketone in high yields [379,380].

3-(4-Methylbenzoyl)thiophene [379]

Butyllithium in hexane (0.031 mol) is added dropwise to a solution of 3-bromothiophene (5.0 g, 0.031 mol) in anhydrous diethyl ether (20 ml) containing anhydrous tetrahydrofuran (5 ml) under nitrogen at −78 °C. This solution is stirred for 10 min and the 4-methylbenzonitrile (3.63 g, 0.031 mol) in anhydrous tetrahydrofuran (5 ml) is added dropwise. The reaction mixture is stirred at −70 °C for 0.5 h, at −50 °C for 1 h and then allowed to warm to −10 °C over a period of 1.5 h, after which it is quenched with water (5 ml, 2 *M* hydrochloric acid 830 ml). The product is extracted with diethyl ether (2 × 50 ml). The acidic aqueous phase is refluxed for 2 h and after cooling extracted with diethyl ether (3 × 50 ml). The combined organic phases are washed with sodium chloride solution, dried and evaporated. The residue is purified by flash chromatography on silica gel using 2-propanol/hexane (5:95) as eluent, giving 5.2 g (83%) of the title compound as a clear oil.

Another use of cyanothiophene is their reaction with zirconocene complexes of substituted naphthalenes generated *in situ*, followed by hydrolysis [381].

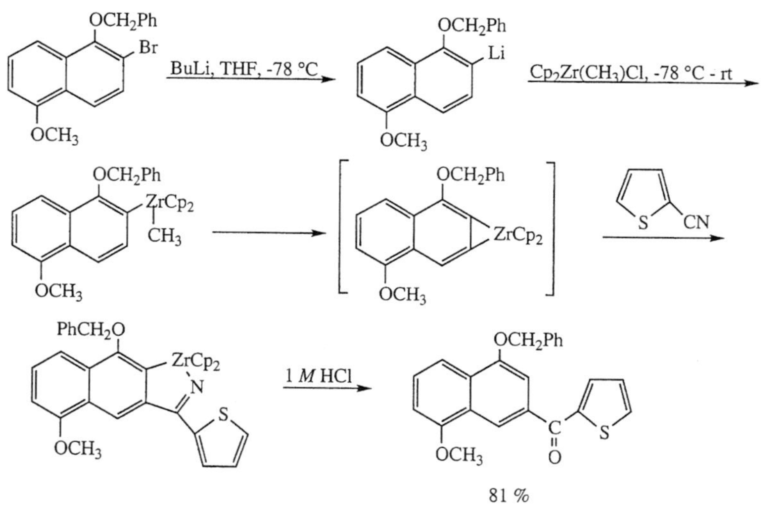

4E.19.6 From thienyllithium derivatives and aromatic aldehydes followed by oxidation

The reaction of 3-thiophene aldehyde with 3-thienyllithium followed by oxidation with chromium trioxide was first used for the preparation of 3,3′-dithienyl ketone [260]. This approach is also used for the preparation of several methylated derivatives such as di(2,3-dimethyl-4-thienyl) ketone [382].

1,2′,3,3′-Tetramethyl-4,4′-dithienyl ketone [382]

A solution of chromium trioxide (20.0 g, 0.20 mol) in water (20 ml) is slowly added to ice-cooled pyridine (400 ml). After 5 min a solution of

1,2′,3,3′-tetramethyl-4,4′-dithienyl carbinol (15.5 g, 0.061 mol) in pyridine (30 ml) is added. The reaction mixture is left for three days at room temperature, after which the precipitate formed is filtered off and the filtrate poured into an excess of water. The precipitate formed is filtered off giving 12.5 g (82%) of the title compound mp 74–75 °C.

Alternatively, reaction of two equivalents of 3-thienyllithium with one equivalent of methyl formate, immediately followed by oxidation of the resulting alcohol with pyridinium chlorochromate can be used for the preparation of 3,3′-dithienylketone [383].

4E.19.7 From thienyllithium derivatives and esters

The synthesis of substituted 2-aminophenyl 2-thienyl ketones is achieved from the corresponding anthranilic acid by treatment of 2-thienyllithium and chloro trimethylsilane [384].

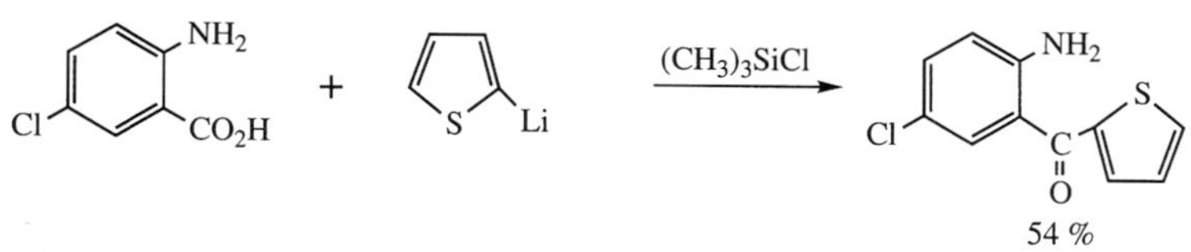

The reaction of 2-thienyllithium with ethyl trifluoroacetate is used for the preparation of trifluoromethyl-2-thienyl ketone [385].

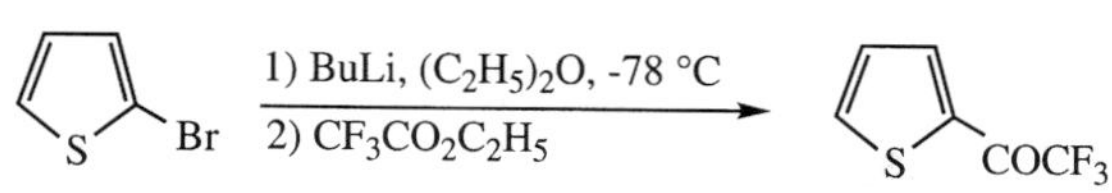

4E.19.8 From thienyllithium derivatives and *N,N*-dimethyl carbamate

5,5′-Diformyl-2,2′-dithienylmethanone is prepared by halogen–metal exchange of 2-(5-bromo-2-thienyl)-1,3-dioxalane followed by reaction with ethyl *N,N*-dimethylcarbamate and acid hydrolysis [45]. The reaction of the sterical crowded 2,5-di-tert-butyl-3-thienyllithium with 1/3 equivalent of dimethyl carbonate stopped at the ketone stage and gave di(2,5-di-*tert*-butyl-3-thienyl)ketone in 49% yield [386].

4E.19.9 From thienyllithium and -magnesium derivatives and anhydrides or lactones

A convenient route to 2-thenoylbenzoic acids is the reaction of phthalic anhydrides with 2-thienyllithium [387].

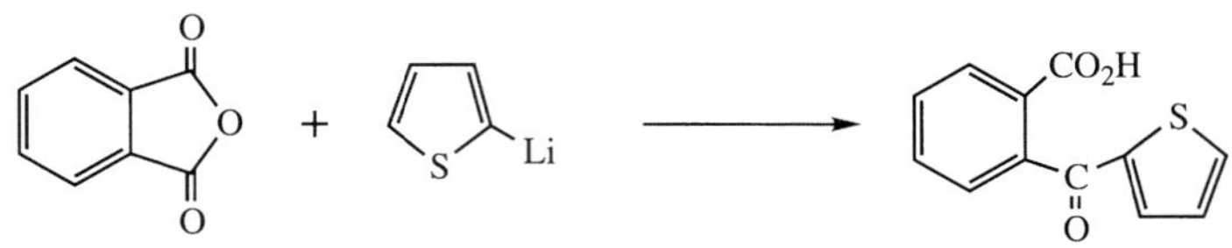

The reaction of 2-thienylphthalide conveniently prepared by the reduction of 2-(2-thenoyl)benzoic acid with sodium borohydride with 3-alkyl-2-thiophenemagnesium bromide gives an intermediate, which upon reaction with Lawesson's reagent yields 1,3-dithienylbenzo[*c*]thiophenes [388].

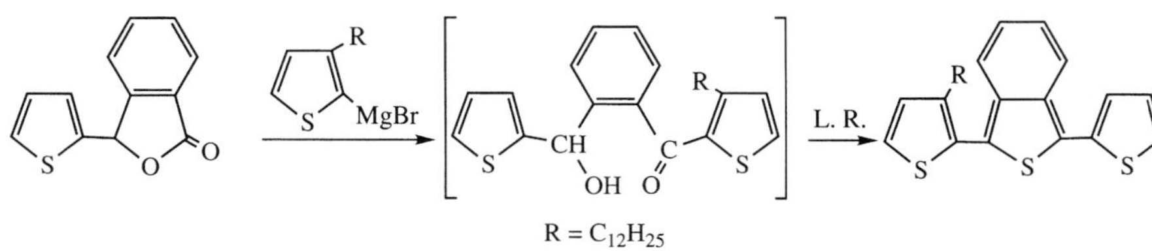

1-(3-Dodecylthienyl)-3-thienylbenzo[c]thiophene [388]

2-Bromo-3-dodecylthiophene (8 g, 0.024 mol) is added slowly to a refluxing mixture of magnesium turnings (0.705 g, 0.029 mol) and iodine (20 mg) in anhydrous tetrahydrofuran. After 6 h of reflux, the Grignard reagent formed is cooled and added slowly *via* an addition funnel to 2-thienyl phthalide in anhydrous tetrahydrofuran (50 ml) at −10 °C. The reaction mixture is stirred for 6 h and poured into ice-containing ammonium chloride. The intermediate is extracted with dichloromethane, dried over sodium sulfate and treated with Lawesson's reagent (5 g, 0.012 mol). The stirring is continued at room temperature for 6 h followed by standard work-up and filtration through a column of basic alumina giving 8.7 g (77%) of the title compound as a thick liquid.

4E.19.10 Ring-closure reactions

Hinsberg reaction of 1,2-cyclohexanedione with diphenacyl sulfide followed by treatment with thionyl chloride in pyridine gives 1,3-dibenzoyl-4,5,6,7-tetrahydrobenzo[*c*]thiophene in good yield [389].

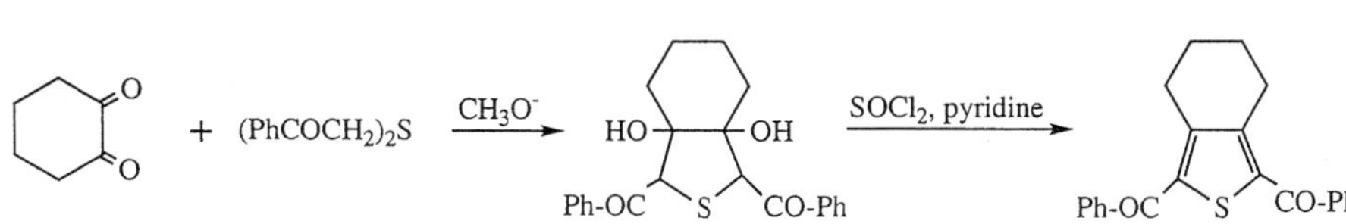

1,3-Dibenzoyl-4,5,6,7-tetrahydrobenzo[c]thiophene [389]

Thionyl chloride (5.2 ml) is slowly added to a suspension of 1,3-dibenzoyl-1,3,3a,4,5,6,7,7a-octahydrobenzo[*c*]thiophene-3a,7a-diol (1.95 g, 5.1 mmol) in chloroform (40 ml) and pyridine (5.2 ml) cooled in an ice bath. The stirring is continued at 0 °C for 1 h and at room temperature for 15 min, after which the reaction mixture is poured into ice and hydrochloric acid. The phases are separated and the aqueous phase extracted with chloroform. The combined organic phases are washed with water, dried over sodium sulfate, and evaporated. The residue is twice recrystallized from methanol/chloroform giving 1.30 g (74%) of the title compound mp 134–136 °C.

4E.19.11 Various reactions

Recently a coupling reaction of thiophene aldehydes promoted by samarium diiodide was developed. Thus, for example, the reaction of 2-thiophene aldehyde with 3-thiophene aldehyde in the presence of hexamethylphosphoramide gave the 5-hydroxyalkylation product which upon oxidation with pyridinium dichromate gave 5-(3-thenoyl)-2-thiophene aldehyde [72].

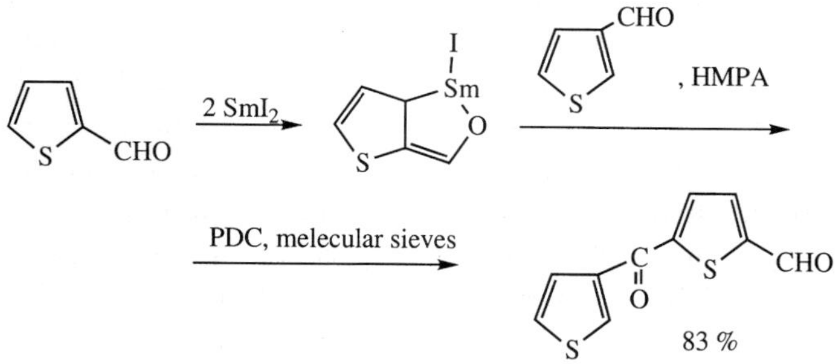

General procedure for the reactions of thiophene aldehydes with samarium diiodide [72]

Samarium metal (0.66 g, 4.4 mmol) and 1,2-diiodoethane (1.02 g, 3.6 mmol) in anhydrous tetrahydrofuran (40 ml) are stirred at room temperature under argon for 1 h to give a deep-blue solution. In most cases hexamethylphosphoramide (2.8 ml, 16 mmol) is added. Precaution should be taken as hexamethylphosphoramide is a toxic cancer suspect agent. The mixture is cooled to 0 °C and a solution of thiophene aldehyde (2 mmol) in tetrahydrofuran (2 ml) is added dropwise over a period of 2 min. The stirring is continued at 0 °C for 10 min and the mixture is then allowed to warm to room temperature. After addition of saturated aqueous ammonium chloride solution

(0.1 ml), the mixture is filtered through a pad of silica gel, which is rinsed with ethyl acetate/hexane (1:1). The filtrate is evaporated and the residue purified by chromatography on silica gel using ethyl acetate/hexane as eluent, giving the products.

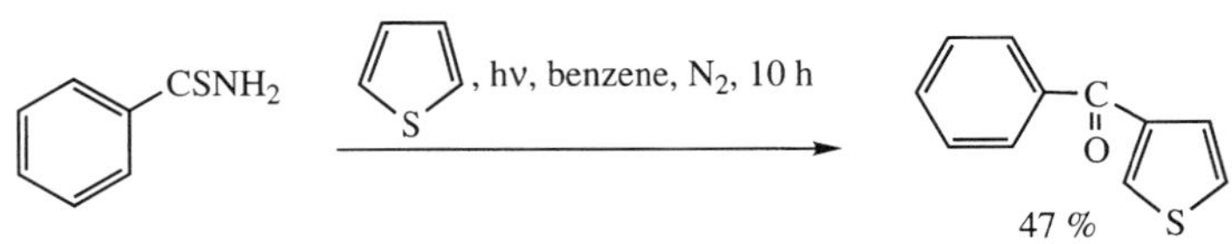

Irradiation of arenecarbothioamides with thiophenes gives 3-aroyl-substituted thiophenes [390].

4E.20 CYCLIC THIOPHENE-FUSED KETONES

4E.20.1 Five-membered cyclic ketones

4E.20.1.1 Ring-closure of 3-(thienyl)propionic acids

Five-membered cyclic ketones are important intermediates and are conveniently prepared by ring closure of 3-(3-thienyl)propionic acid derivatives, reacting the acid with polyphosphoric acid or hydrogen fluoride or the acid chloride using stannic chloride as catalyst [391]. Thus 5,6-dihydrocyclopenta[*b*]thiophene-6-one was also more recently prepared from 3-(3-thienyl)propionic acid upon reaction with a solution of phosphorus pentoxide in methanesulfonic acid [392].

5,6-Dihydrocyclopenta[b]thiophene-6-one [392]

Finely powdered 3-(3-thienyl)propionic acid (3.0 g, 19 mmol) is added to a solution of phosphorus pentoxide (15 g) and methanesulfonic acid (100 ml). The reaction mixture is stirred at room temperature for 40 min, after which it is poured into ice (150 g) and the product is extracted with dichloromethane (4 × 125 ml). The combined organic phases are washed with 5% sodium hydroxide solution (125 ml), 1 *M* hydrochloric acid (125 ml) and sodium chloride solution (125 ml), dried over sodium sulfate and evaporated. The residue is purified by chromatography using diethyl ether/petroleum ether (1:3) as eluent, giving 1.15 g (44%) of the title compound as an ivory solid mp 91–92 °C.

The same compound was also prepared in a one-pot procedure starting from 3-(3-thienyl)acrylic acid, which is reduced with sodium amalgam to the propionic acid, transformed to the acid chloride and ring closed to the desired product using aluminium chloride [262].

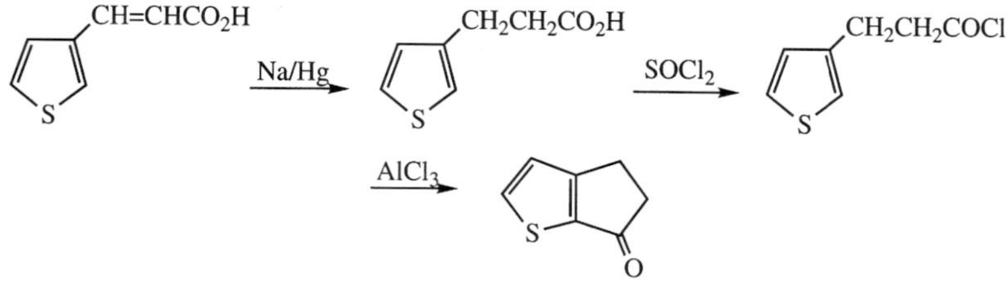

4,5-Dihydro-6H-cyclopenta[b]thiophene-6-one [262]

To a solution of 3-(3-thienyl)acrylic acid (10 g, 65 mmol) in 0.6 *M* aqueous sodium hydroxide (140 ml) 2% sodium amalgam (200 g) is added portionwise over a period of 1 h. The stirring is continued for 30 min, after which the mixture is acidified with 50% aqueous hydrochloric acid and the product extracted with diethyl ether. The combined organic phases are washed with water, dried over calcium chloride and evaporated giving 10.7 g of a pale-yellow oil that crystallized. A solution of thionyl chloride (15 ml, 210 mmol) in anhydrous diethyl ether (200 ml) is added and the mixture is refluxed for 3.5 h. The brown oil obtained after evaporation is dissolved in carbon disulfide and this solution transferred to a dropping funnel and added dropwise to a suspension of aluminium chloride (8.7 g, 65 mmol) in carbon disulfide (100 ml). The reaction mixture is stirred at room temperature for 20 h, refluxed for 2 h and then poured into a mixture of concentrated hydrochloric acid (100 ml) and ice-water (200 ml). The phases are separated and the aqueous phase extracted with dichloromethane. The combined organic phases are washed with water, dried over magnesium sulfate and evaporated. The residue is chromatographed on silica gel using petroleum ether/diethyl ether (4:1) as eluent, giving 5 g (56%) of the title compound as a white solid mp 89–90 °C.

As expected, the branched acid chloride gives upon Friedel-Crafts reaction the desired 4-butyl-4,5-dihydro-6*H*cyclopenta[*b*]thiophene-6-one, but in addition small amounts of tricyclic system is also formed probably due to a Wagner-Meerewein rearrangement of a carbocation in the presence of aluminium chloride [393].

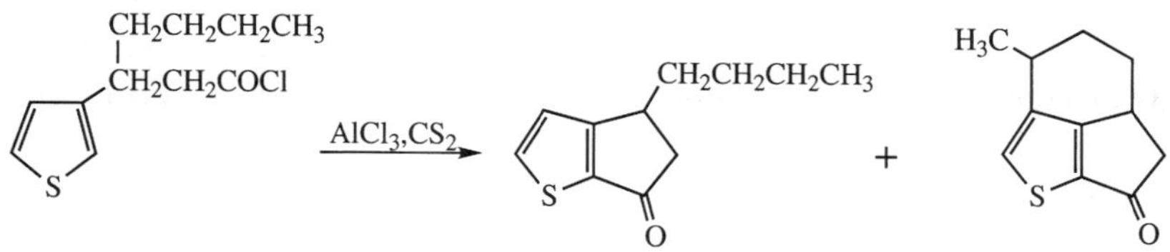

4-Butyl-4,5-dihydro-6H-cyclopenta[b]thiophene-6-one and 2-methyl-11-thiatricyclo-[7.22.0.0]undeca-1(10),8-dien-7-one [393]

The acid chloride (7.58 g, 32.88 mmol) is dissolved in carbon disulfide (100 ml) and this solution is added dropwise to a mixture of aluminium chloride (6.50 g, 48.75 mmol) in carbon disulfide (100 ml) under nitrogen. After refluxing for 4 h the reaction mixture is cooled to 20 °C and poured into a mixture of concentrated hydrochloric acid (100 ml) and ice-water (200 ml). The phases are separated and the aqueous phase extracted with dichloromethane. The combined organic phases are washed with water, dried over magnesium sulfate and evaporated. The residue is first subjected to a rapid filtration (silica gel, dichloromethane) and then chromatographed on silica gel using petroleum ether/diethyl ether (4:1) as eluent, which allowed separation of 4-butyl-4,5-dihydro-6*H*-cyclopenta[*b*]thiophene-6-one 2.63 g (40%) and 2-methyl-11-thia-tricyclo[7.22.0.0]undeca-1(10),8-dien-7-one 0.75 g (11%) as a mixture of two diastereomers (60:40).

If attempts are made to force the ring-closure to the 4-position by blocking the 2-position with halogen or *tert*-butyl, the blocking group might be lost during cyclization with aluminium chloride [394,395] or polyphosphoric acid [237]. Attempts to ring-close 3-(2,5-dibromo-3-thienyl)propionic acid derivatives give inseparable mixtures [392]. In connection with a synthesis of thianinhydrin, 5,6-dihydrocyclopenta[*b*]thiophene-6-one was brominated with *N*-bromosuccinimide followed by hydrolyses to a diketone, which was reacted with *N*,*N*-dimethylformamide dimethylacetal in dichloromethane and upon reaction with dimethyl dioxirane in acetone the desired compound was obtained [396].

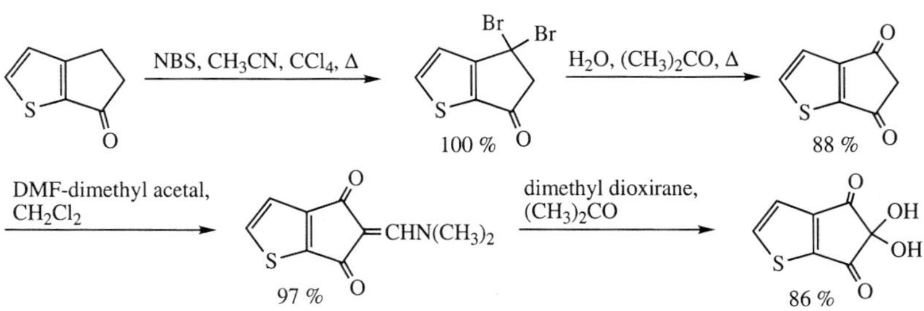

A previous route to thianinhydrin consists in the preparation of cyclopenta[*b*]thiophene-4,6-dione by diazotization of 4-aminocyclopenta[*b*]-thiophen-6-one, followed by Kornblum oxidation with bromine in dimethyl sulfoxide [397].

The oxodihydrocyclopentathiopheneacetic acids were prepared by cyclization of thienylglutaric acids. Thus treatment of the 3-isomer with 10%

phosphorus pentoxide in methanesulfonic acid gave smooth ring-closure to the 2-position. From the 2-thienylglutaric acid, ring-closure of its acid chloride to the 3-position was achieved using aluminium chloride as catalyst and 4-oxo-5,6-dihydro-4*H*-cyclopenta[*b*]thiophene-6-acetic acid was obtained. Finally ring-closure to the 4-position was obtained with the 2-methyl-3-thienylglutaric acid giving 1-methyl-4-oxo-5,6-dihydro-4*H*-cyclopenta[*c*]thiopheneacetic acid [398].

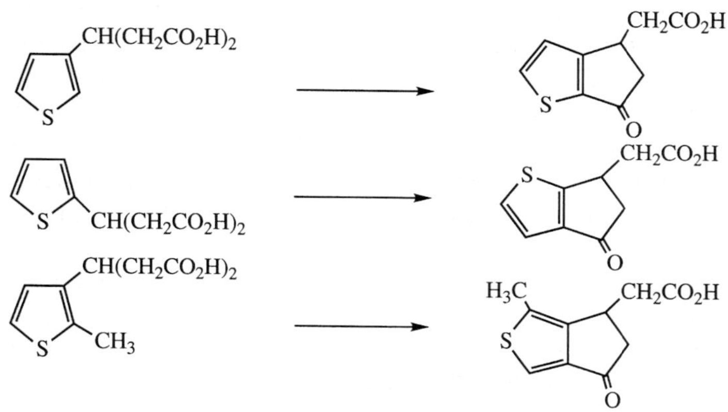

6-Oxo-4,5-dihydro-6H-cyclopenta[b]thiophene-4-acetic acid [398]

3-Thienylglutaric acid (1 g, 4.7 mmol) is quickly added to a hot solution of 10% phosphorus pentoxide in methanesulfonic acid (20 g). The reaction mixture is vigorously stirred at 100 °C for 20 min, after which it is cooled and hydrolyzed with ice. The product is extracted with diethyl ether and the combined organic phases dried over magnesium sulfate and evaporated. The residue is recrystallized from benzene giving 70% of the title compound mp 130 °C.

4-Oxo-5,6-dihydro-4H-cyclopenta[b]thiophene-6-acetic acid [398]

2-Thienylglutaric acid (2.14 g, 10 mmol) in dichloromethane is refluxed for 3 h with an excess of thionyl chloride. Then the chilled solution is treated in portions with aluminium chloride (4 g, 31 mmol) at 0 °C with vigorous stirring. The stirring is continued under cooling for 1 h and at room temperature for 2 h. After cooling the reaction is quenched by cautious addition of ice and then diluted with water. The phases are separated and the aqueous phase extracted with dichloromethane. The combined organic phases are washed with water, dried over magnesium sulfate and evaporated. The residue is recrystallized from benzene/hexane (8:2) giving 60% of the title compound as a white solid mp 75 °C.

1-Methyl-4-oxo-5,6-dihydro-4H-cyclopenta[c]thiopheneacetic acid [398]

This compound is prepared in the same way as the previous one from 2-methyl-3-thienylglutaric acid (2.28 g, 10 mmol) and recrystallized from benzene giving 40% of the title compound as a yellow solid mp 110 °C.

Several research groups have recently prepared 6-aminocyclopenta[*b*]thiophene derivatives. Thus treatment of 3-amino-3-(3-thienyl)propionic acid with a mixture of boiling trifluoroacetic acid and anhydride gave in one step 4-amino-4,5-dihydrocyclopenta[*b*]thiophene-6-one [399,400]. Reaction of its trifluoroacetylamino derivative with chlorine gave its 6-chloro derivative [401,402].

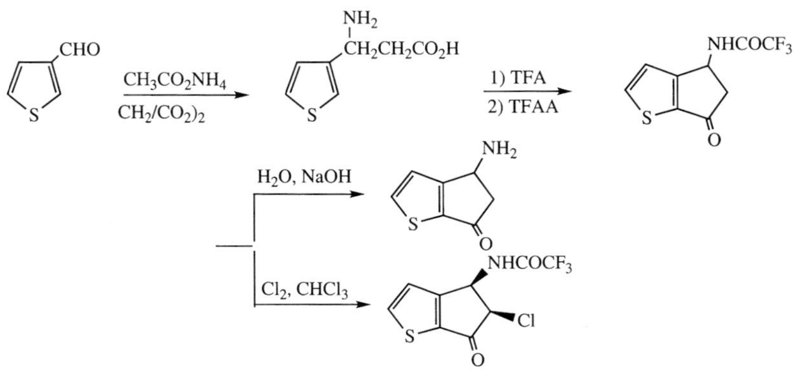

However, only low yields are obtained upon attempts to use this condition to ring-close 3-amino-3-(2-thienyl)propionic acid to the 3-position. Using Friedel-Crafts conditions the acid chloride derived from 3-trifluoroacetylamino-3-(2-thienyl)propionic acid could be cyclized in mediocre yield to 6-trifluroacetylamino-5,6-dihydrocyclopenta[*b*]thiophene using aluminium chloride in carbon disulfide, which was hydrolyzed to the amino derivative [403].

Preparation of the third thiophene-fused cyclopentanone, 6-amino-5,6-dihydro-4*H*-cyclopenta[*c*]thiophene-4-one is achieved by cyclization of amino-2,5-dihalothienylpropionic acids and subsequent dehalogenation with zinc [404].

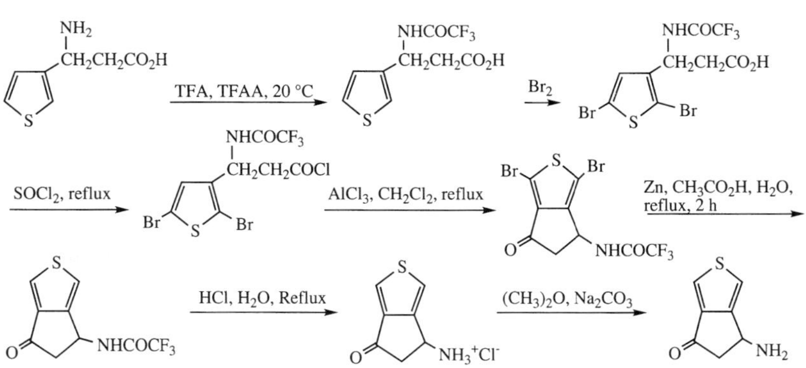

Better yields in the dehalogenation step were obtained by using diethyl phosphite and triethyl amine in tetrahydrofuran [405].

4E.20.1.2 Cyclization of α,β-unsaturated ketones

Five-membered ring ketones are very conveniently prepared by cyclization of α,β-unsaturated ketones with polyphosphoric acid [273,253,406]. Treating a thiophene directly with an α,β-unsaturated acid in the presence of polyphosphoric acid gives the same ketone, however, in low yield [237] presumably because of polymerization of the unsaturated acid under these condtions [253].

4E.20.1.3 One-step annealation on the 2,3-position

The reaction of thiophene with methacrylic acid and polyphosphoric acid in dichloromethane gave 4,5-dihydro-5-methyl-6*H*-cyclopenta[*b*]thiophene-6-one [237]. Similarly, the reaction of 2- methyl-3-phenylthiophene with methacrylic acid was recently used for the preparation of 2,5- dimethyl-3-phenyl-5,6-dihydrocyclopenta[1,2-*b*]thiophene-4-one [407].

2,5- Dimethyl-3-phenyl-5,6-dihydrocyclopenta[1,2-b]thiophene-4-one [407]

A solution of 2-methyl-3-phenylthiophene (124.7 g, 542 mmol), methacrylic acid (61.7 g, 715 mmol) in dichloromethane (200 ml) is slowly added to super polyphosphoric acid (1000 g) with stirring at 70 °C. The mixture is refluxed for 10 h and during this time additional methacrylic acid (208 g) in dichloromethane (250 ml) is added in portions of 60–75 g, after which the reaction mixture is poured onto ice. The product is extracted with hexane/dichloromethane (8:2). The combined organic phases are washed with water, saturated sodium bicarbonate solution and again with water, dried over magnesium sulfate and evaporated giving 202.9 g (82%) of the title compound as a brown oil with an isomeric ratio (3:1).

However, the corresponding reaction with β,β-dimethylacrylic acid stopped at the unsaturated ketone stage yielding 2-(β,β-dimethylacroyl)thiophene [237].

4,5-Dihydro-5-methyl-6H-cyclopenta[b]thiophene-6-one [237]

2-Methacryloylthiophene (30 g) is added dropwise to polyphosphoric acid (100 g) at 50 °C. The stirring is continued for 1 h, after which the mixture is diluted with water and the product is extracted with diethyl ether. The combined organic phases are dried over magnesium sulfate and fractionated giving 25.3 g (84%) of the title compound bp 115–119 °C/7 mm Hg.

4E.20.1.4 Intramolecular cyclization

Treatment of 1-(4-carboxy-3-thienyl)-2-(3′-thienyl)ethane and 1-(4-carboxy-3-thienyl)-2-(2′-thienyl)ethane with phophorus pentachloride and stannic chloride gave the tricyclic compounds respectively in good yields [408].

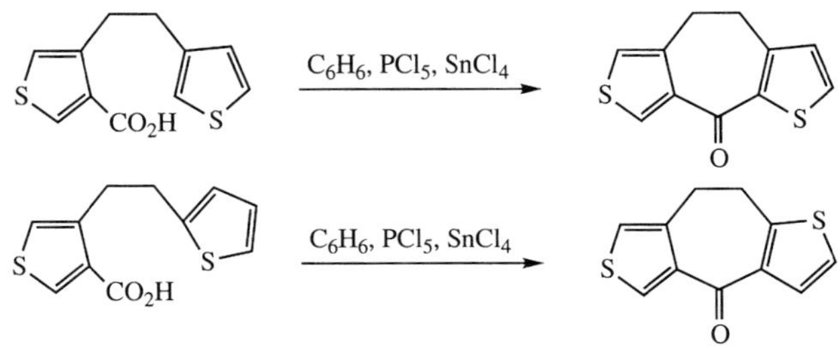

4,5-Dihydro-9H-cyclopenta[2,1-b:4,5-c′]dithiophene-9-one [408]

Phosphorus pentachloride (8.5 g) is added to a solution of 1-(4-carboxy-3-thienyl)-2-(3′-thienyl)ethane (8.1 g (0.034 mol) in benzene (200 ml). The mixture is warmed on a hot-bath until all solids are dissolved and then refluxed for 10 min. After cooling the solution is degassed by a stream of nitrogen for 15–20 min. The solution is recooled to 5–6 °C and stannic chloride (9.0 g in benzene, 50 ml) is quickly added. Efficient stirring is continued until the mixture reaches room temperature and the mixture is heated at 90 °C for 10–12 min. After cooling diethyl ether is added until a homogeneous solution is obtained. This solution is cooled in an ice-bath and 5 *M* hydrochloric acid (200 ml) followed by water (200 ml) are added through a condenser with vigorous stirring. The phases are separated and the aqueous phase diluted with water and extracted with diethyl ether. The combined organic phases are washed with 5*M* hydrochloric acid, 3 *M* hydrochloric acid, 1 *M* hydrochloric acid, water, 1 *M* sodium hydroxide solution and water, dried over calcium chloride, and evaporated. The residue is recrystallized from acetonitrile giving 6.05 g of the title compound as yellow crystals mp 110–111 °C.

4E.20.1.5 By formation of the thiophene ring

The reaction of cyclopentanone derivatives with sulfur at 210 °C gives the cyclopentanone-fused products in low yields [24,409].

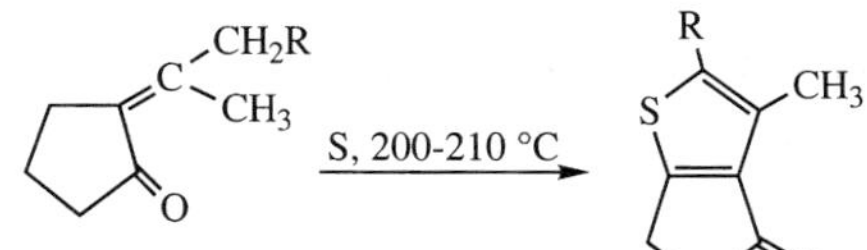

4E.20.1.6 From cyclic ketones by reaction at the α-position

4-Oxo-4*H*-5,6-dihydrocyclopenta[*b*]thiophene-5-ylideneacetic acid is prepared by the reaction of 5,6-dihydro-4*H*-cyclopenta[*b*]thiophene-4-one with glyoxylic acid and sulfuric acid in dioxane [410].

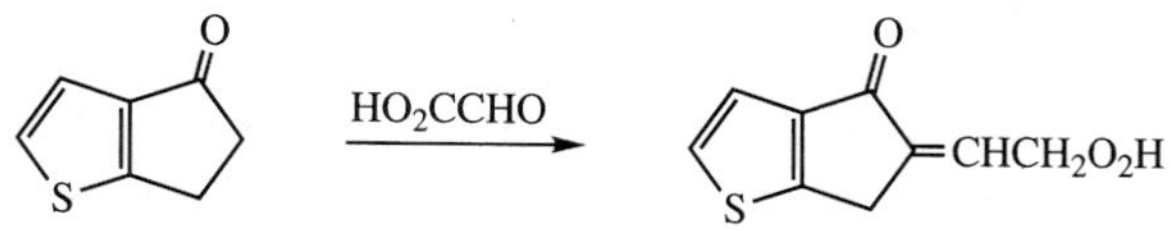

4-Oxo-4H-5,6-dihydrocyclopenta[b]thiophene-5-ylideneacetic acid [410]

A stirred mixture of 5,6-dihydro-4*H*-cyclopenta[*b*]thiophene-4-one (2 g, 14.5 mmol), glyoxylic acid (40%, 2 g, 27.6 mmol) and sulfuric acid (80%, 0.5 ml) in dioxane (20 ml) is refluxed for 3 h. After cooling the precipitated solid is collected by filtration and washed with diethyl ether giving 2.25 g (90%) of the title compound mp 224–226 °C after recrystallization from benzene.

4E.20.2 Six-membered cyclic ketones

4E.20.2.1 Ring-closure of 4-(thienyl)butyric acid derivatives

Cyclohexa[*b*]thiophene-7-one(7-oxo-4,5,6,7-tetrahydrobenzo[*b*]thiophene), 6,7-dihydrobenzo[*b*]thiophen-4(5*H*)-one, and its 2-methyl derivatives are prepared in excellent yields by cyclization of 4-(3-thienyl)butyric acid chloride in the presence of stannic chloride [411-414].

6,7-Dihydrobenzo[b]thiophene-4(5H)-one [414]

3-(2-Thenoyl)propionic acid is prepared by Friedel-Crafts succinoylation of thiophene in dichloromethane in a yield of 90%. This compound is reduced by the Huang–Minlon procedure to 4-(2-thienyl)butyric acid in a yield of 91%. This compound is cyclized with polyphosphoric acid in acetic anhydride giving the title compound in a yield of 81% mp 35–36 °C.

In contrast to the ring-closure of 3-(2-thienyl)propionic acids, the 4-(2-thienyl)butyric acids give the cyclohexa[*b*]thiophene-4-one (4-oxo-4,5,6,7-tetrahydrobenzo[*b*]thiophenes) both using the acids and polyphosphoric acid [277,415] or the acid chlorides and stannic chloride in high yields [411].

Using branched butyric acids, the 5-methyl- [411], and 6-methylcyclohexa-[*b*]thiophene-4-one [416] were prepared, as well as the 5-methyl- and 6-methylcyclohexa[*b*]thiophene-7-ones [411]. Ring-closure of 4-(3-thienyl)-butyric acid to the thiophenic 4-position is achieved when at least the 2-position or both the 2- and 5-positions are blocked with alkyl groups. Thus 4-(2-methyl-3-thienyl)butyric acid [394], 4-(2,5-dimethyl-3-thienyl)butyric acid [411,417], 4-(2,5-diethyl-3-thienyl)butyric acid [418] and 4-(2-methyl-5-*tert*-butyl-3-thienyl)butyric acid are ring closed in reasonable yield to the corresponding cyclohexa[*c*]thiophene-4-ones by reacting the acid chlorides in the presence of stannic chloride [394].

A number of six-ring ketones with a sulfur atom in the carbonyl-containing ring have been prepared. Starting from 3-(2-thenylthio)propionic acid ring-closure to 7*H*-thieno[2,3-*c*]thiopyran-4(5*H*)-one was achieved by reacting the acid chloride with stannic chloride [150,419] or the acid with phosphorus pentoxide [420].

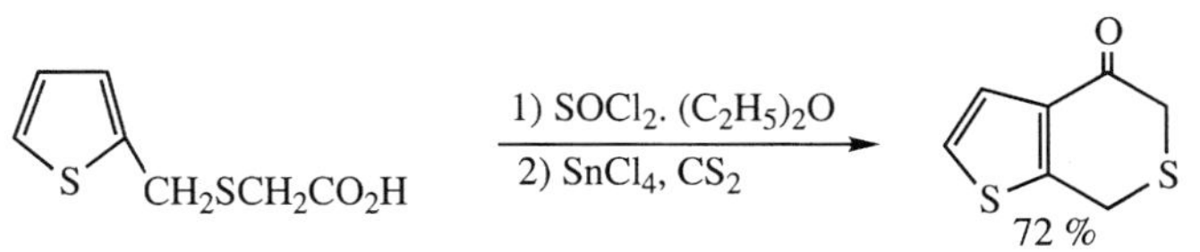

Starting from 3-(3-thienylthio)propionic acid the analogous 7-oxo ketones, 4*H*-thieno-3,2-*c*]thiopyran-7(6*H*)-one, were obtained [150,420,421].

4E.20.2.2 Ring-closure of 2-thenoylbenzoic acids

Microwave-assisted cyclization of thenoylbenzoic acids with clay in anhydrous media to thiophene-fused ketones is achieved in good yields [387].

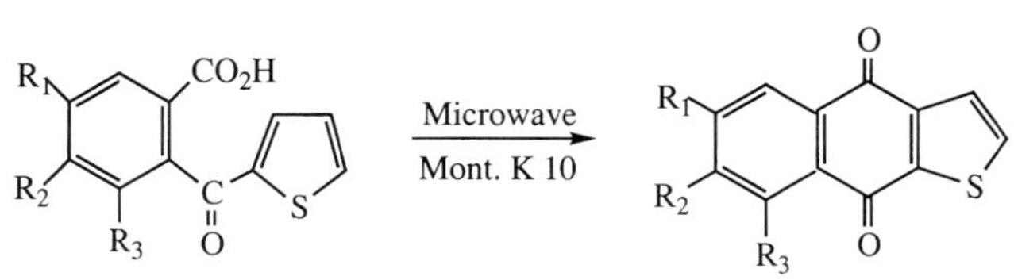

4E.20.2.3 From cyclic ketones by reaction in the α-position

Methyl 4-oxo-4,5,6,7-tetrahydrobenzo[*b*]thiophene-5-carboxylate and methyl 7-oxo-4,5,6,7-tetrahydrobenzo[*b*]thiophenecarboxylate are prepared by the reaction of the cylic ketones with dimethyl carbonate and sodium hydride in *N*,*N*-dimethylformamide [422].

Methyl 4-oxo-4,5,6,7-tetrahydrobenzo[b]thiophene-5-carboxylate [422]

A solution of 4-oxo-4,5,6,7-tetrahydrobenzo[*b*]thiophene (10 g, 0.131 mol) in anhydrous *N*,*N*-dimethylformamide (50 ml) is added dropwise to a stirred suspension of a 55% dispersion of sodium hydride in mineral oil (6.31 g, 0.115 mol) in anhydrous *N*,*N*-dimethylformamide (100 ml) under nitrogen at room temperature over a period of 30 min. The stirring is continued for 10 min, after which dimethyl carbonate (33 ml) is added at 5 °C over a period of 20 min. The reaction mixture is stirred at room temperature for 90 min and then poured into water. Extraction with ethyl acetate and evaporation gives the title compound as a solid.

Reaction of 4,5,6,7-tetrahydro-4-oxobenzo[*b*]thiophene with lithium diisopropyl amide at −70 °C followed by ethyl bromoacetate gave the 5-carboethoxy derivative in 45% yield [277]. Also the 5-cyano-6,7-dihydrobenzo[*b*]thiophen-4(5*H*)-one has been prepared by treatment of 4,5-dihydrothieno[2,3-g]-1,2-benzisoxazole with sodium methoxide [423].

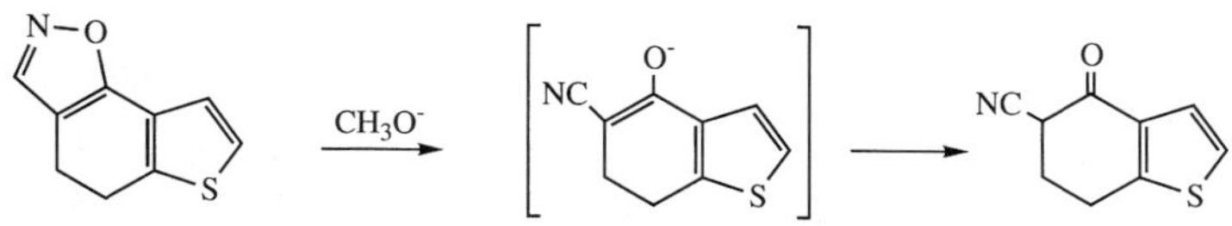

5-Cyano-6,7-dihydrobenzo[b]thiophen-4(5H)-one [423]

A cold sodium methoxide solution, prepared from sodium (1.5 g, 60 mmol) and anhydrous methanol (40 ml), is added to the isoxazole derivative (5.3 g, 30 mmol) in tetrahydrofuran (15 ml) under nitrogen over a period of 30 min. The reaction mixture is stirred under cooling for another 2 h and then 5% aqueous potassium hydroxide solution is added. After dilution with water up to a volume of 700 ml and extraction with ether the alkaline aqueous phase is acidified with dilute hydrochloric acid and extracted with diethyl ether. The combined organic phases are washed with saturated sodium chloride solution, dried over sodium sulfate, and evaporated. The residue is recrystallized from diethyl ether/petroleum ether giving 4.7 g (90%) of the title compound as white needles mp 76–77 °C.

4-Oxo-4*H*-5,6-dihydrocyclopenta[*b*]thiophen-5-yl acetic acid was prepared in the same way as the five-membered ketone derivative [410]. α-Hydroxy-α-(4-oxo-4,5,6,7-tetrahydrobenzo[*b*]thiophene-5-yl)acetic acid and α-hydroxy-α-(7-oxo-4,5,6,7-tetrahydrobenzo[*b*]thiophene-5-yl)acetic acid are prepared by the reaction of 4-oxo- and 7-oxo-4,5,6,7-tetrahydrobenzo[*b*]thiophene with glyoxylic acid monohydrate and sodium hydroxide [424].

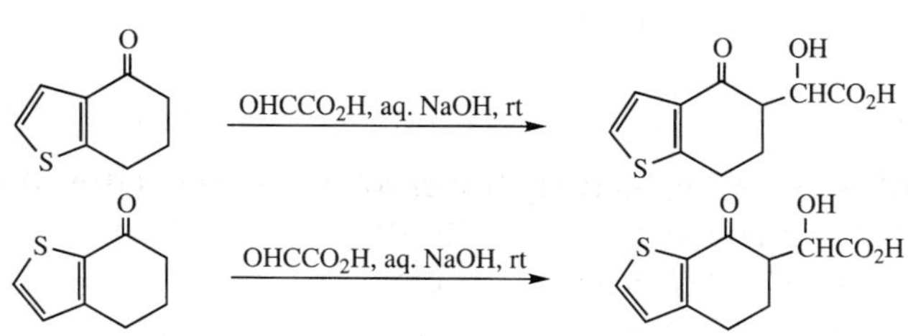

α-Hydroxy-α-(4-oxo-4,5,6,7-tetrahydrobenzo[b]thiophene-5-yl)acetic acid [424]

A solution of sodium hydroxide (2.46 g, 0.06 mol) in water (50 ml) is added to an ice-cooled mixture of the thienocyclohexanone (3 g, 0.020 mol) and glyoxylic acid monohydrate (2.07 g, 0.022 mol) vigorously stirred in water (50 ml). After stirring at room temperature for 1 h the alkaline solution is washed with diethyl ether (25 ml) and then acidified with concentrated hydrochloric acid under cooling. The stirring is continued at room temperature overnight, the precipitate formed is filtered off, washed with water and dried, giving 1.9 g (42%) of the title compound mp 159–160 °C.

4E.20.2.4 From aliphatic compounds by ring-closure reactions

6,7-Dihydrobenzo[*b*]thiophene-4(5*H*)-ones are obtained upon reaction of 3-mercaptocyclohexanone with glyoxal or a 1,2-diketone [425]. The parent compound is made by Michael addition of mercaptoacetaldehyde diethyl acetal to cyclohex-2-enone followed by cyclization [426].

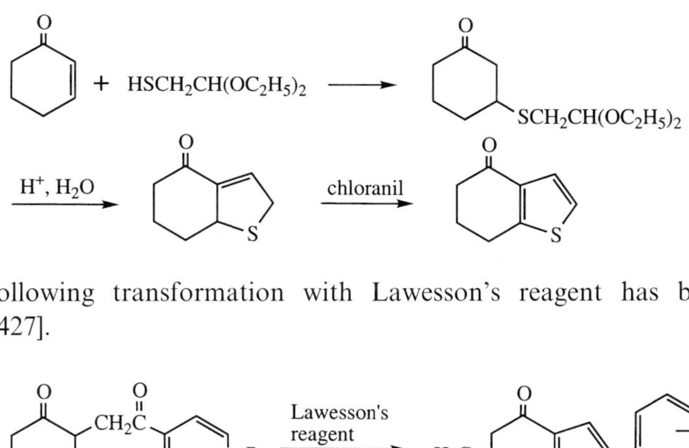

The following transformation with Lawesson's reagent has been performed [427].

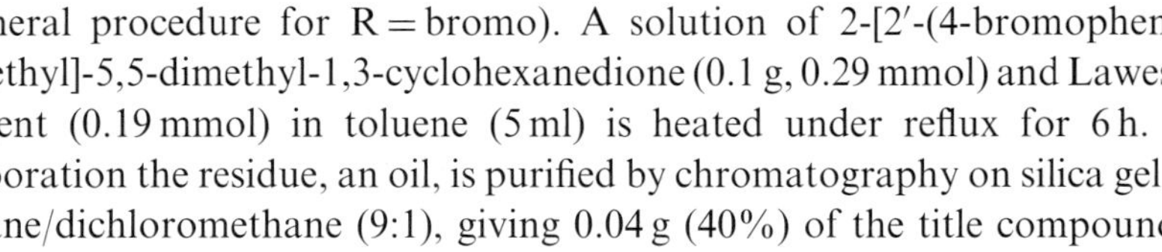

2-(4-Bromophenyl)-6,6-dimethyl-4,5,6,7-tetrahydrobenzothiophene-4-one [427]

(General procedure for R = bromo). A solution of 2-[2′-(4-bromophenyl)-2-oxoethyl]-5,5-dimethyl-1,3-cyclohexanedione (0.1 g, 0.29 mmol) and Lawesson's reagent (0.19 mmol) in toluene (5 ml) is heated under reflux for 6 h. After evaporation the residue, an oil, is purified by chromatography on silica gel using hexane/dichloromethane (9:1), giving 0.04 g (40%) of the title compound as a colorless solid mp 130–132 °C.

4E.20.2.5 *Via* ring-enlargement of five-membered cyclic ketones

4-Aminothieno[2,3-*c*]pyrid-7-one is selectively prepared by ring-enlargement of 4-amino- cyclopenta[*b*]thiophene-6-one derivatives using Schmidt and Beckman rearrangement [400].

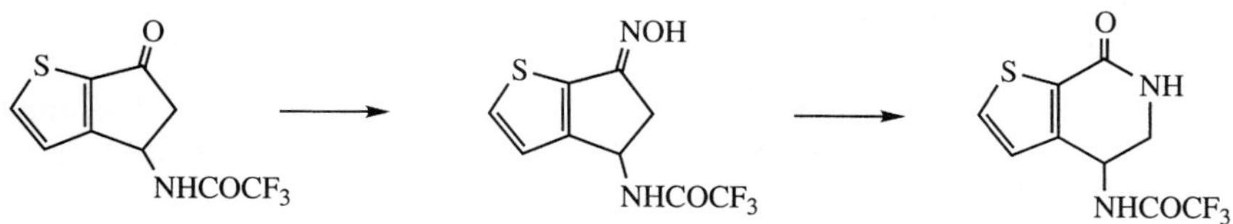

4,5,6,7-Tetrahydro-4-trifluoroacetylaminothieno[2,3-c]pyrid-7-one [400]

(a) Beckmann rearrangement
A solution of the *Z* form of the oxime (1 g, 3.7 mmol) in *tert*-butyl methyl ether (100 ml) at 0 °C phosphorus pentachloride (2.08 g, 10 mmol) is added in small portions. The reaction mixture is stirred at room temperature overnight and then poured onto ice (50 g). The phases are separated and the aqueous phase is extracted with diethyl ether (2 × 100 ml). The combined organic phases are washed with water, dried over magnesium sulfate, decolorized with charcoal, and evaporated giving 0.2 g (20%) of the title compound mp 260 °C after recrystallization from isopropanol.

(b) Schmidt rearrangement
A solution of the cyclopenta[*b*]thiophene-6-one derivative (2.5 g, 10 mmol) and sodium azide (1 g, 15 mmol) in trifluoroacetic acid (30 ml) is refluxed for 4 h. During this time additional sodium azide (0.5 g, 7.5 mmol) is added each hour. After evaporation of the solvent the residue is triturated with water (100 ml) and the precipitate formed is filtered off and dried giving 1.6 g (61%).

4E.21 SEVEN-MEMBERED CYCLIC KETONES

4E.21.1 Ring-closure of 5-(thienyl)pentanoic acid

Using high dilution the preparation of cyclohepta[*b*]thiophene-4-ones and 8-ones is achieved from the appropriate 5-(2-thienyl)pentanoic acids by the acid chloride–stannic chloride methodology [428]. Again using the appropriate branched acid chlorides the 5-methyl- [411], 6-methyl- [429], 7-methyl- [430], and 6-ethyl-cyclohepta[*b*]thiophene-4-ones [431] are prepared. In some cases aluminium chloride was used as catalyst, as in the preparation of the 6,6-diethyl derivative [432].

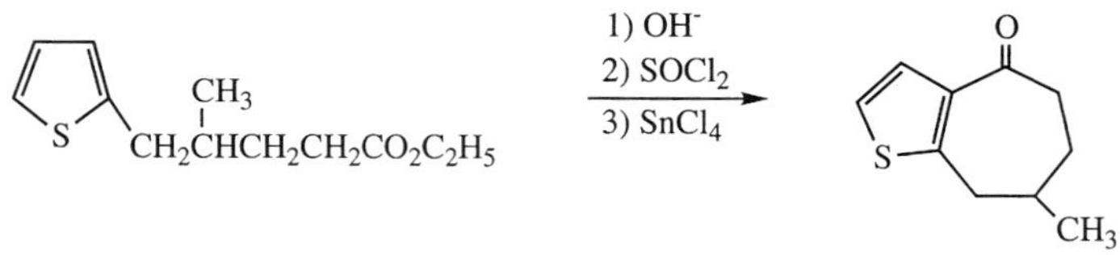

Recently 5-(2-thienyl)pentanoic acid and 5-(5-methyl-2-thienyl)pentanoic acid were prepared in high yields by Huang–Minlon reduction of the 4-oxo acids, and cyclized to 5,6,7,8-tetrahydro-4*H*-cyclohepta[*b*]thiophene-4-one and 2-methyl-5,6,7,8-tetrahydro-4*H*-cyclohepta-[*b*]-4-one by treatment with polyphosphoric acid in 11% and 75% yield, respectively [278].

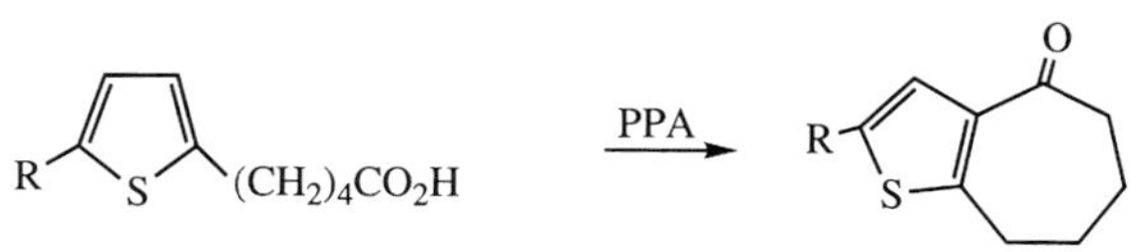

2-Methyl-5,6,7,8-tetrahydro-4H-cyclohepta[b]thiophene-4-one [278]

To vigorously stirred polyphosphoric acid (40 g) at 90 °C 5-methyl-2-thiophene-pentanoic acid (9.9 g, 50 mmol) in anhydrous toluene (60 ml) is added under anhydrous conditions. After 30 min crushed ice is added to the cooled reaction mixture. The phases are separated and the aqueous phase extracted with benzene. The combined organic phases are washed with 5% sodium hydroxide solution (3 × 50 ml) and water (3 × 50 ml), dried, evaporated and distilled giving 6.75 g (75%) of the title compound as an oil bp 120–123 °C/0.3 mm Hg.

Starting from 4,5-cycloalkyl-fused ω-2-tienylalkanoic acid chlorides, cyclic ketones, such as shown below, were prepared and by Beckmann reaction transformed to ε-caprolactams and and γ-enantholactams [433].

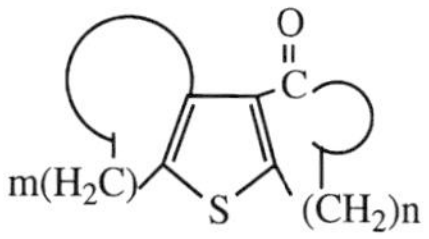

a) m = 4, n = 3; b) m = 5, n = 3; c) m = 4, n = 4; d) m = 5, n = 4

Starting from 5-(2-thienyl)pentanoic acids having methyl ethyl, propyl or longer alkyl groups in the 5-position of the thiophene ring, the corresponding 2-alkylsubstituted derivatives were prepared [428,434,435].

5,6,7,8-Tetrahydro-2,6-dimethyl-4H-cyclohepta[b]thiophene-4-one [434]

To a solution of 3-methyl-5-(5-methyl-2-thienyl)pentanoic acid (24 g) in diethyl ether (100 ml), purified thionyl chloride (18 ml) and pyridine (several drops) are added. The mixture is refluxed for 5 h, after which solvent and excess thionyl chloride is evaporated. The crude acid chloride is dissolved in anhydrous carbon disulfide (500 ml) and the solution is cooled to 0 °C. Stannic chloride is added dropwise with vigorous stirring, the reaction mixture allowed to warm to room temperature and the stirring is continued for 2.5 h. After addition of ice and hydrochloric acid the phases are separated and the organic phase dried and distilled under nitrogen giving 13.7 g (63%) of the title compound, which slowly solidified bp 121–124 °C/1.5 mm Hg.

In the preparation of the 2-butyl and 2-isobutyl derivatives titanium(IV)-chloride was used as catalyst [436]. In the same way cyclohepta[*b*]thiophene-8-ones and its 7-methyl derivative are prepared from 5-(3-thienyl)pentanoic acids [278,411]. Ring-closure of 5-(3-thienyl)pentanoic acids to the 4-position can be achieved, if the 2- and 5-positions are blocked with alkyl groups [411,418, 428,437].

Ring-closure of 3,3-cyclopentane-5-(2-thienyl)pentanoic acid chloride using stannic chloride in carbon disulfide gave 5,6,7,8-tetrahydro-6,6-spirocyclopentane-4*H*-cyclohepta[*b*]thiophene-4-one. Similarly the 5-methyl-2-thienyl, and 5-ethyl-2-thienyl substituted derivatives were prepared [279].

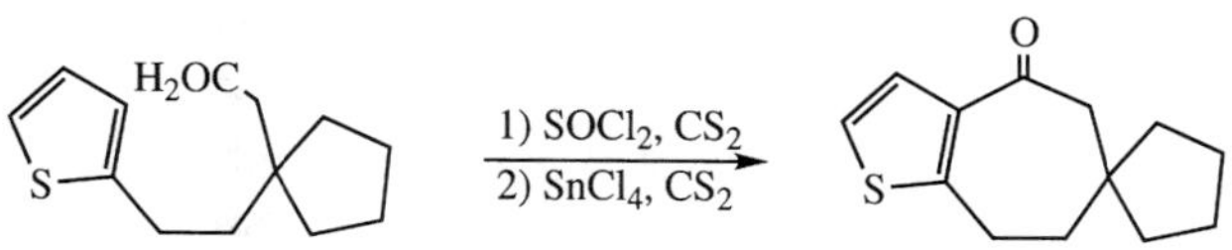

Seven-membered ketones with a sulfur atom in the carbonyl-containing ring are prepared, in much lower yields than the corresponding six-membered ketones. The 6-S-4-CO [438], the 5-S-8-CO [421], and 6-S-8-CO [438], cyclic ketones were prepared by ring-closure of the acid chlorides using tin tetrachloride.

4E.21.2 By substitution of α position to the carbonyl group in cyclic ketones

5-Hydroxymethylene-5,6,7,8-tetrahydro-4*H*-cyclohepta[*b*]thiophene-4-one and its 2-methyl derivative are prepared by formylation of the cyclic ketones with ethyl formate with sodium ethoxide in benzene [278].

5-Hydroxymethylene-5,6,7,8-tetrahydro-4H-cyclohepta[b]-thiophene-4-one [278]

Ethyl formate (0.8 g, 10 mmol) in anhydrous benzene (15 ml) is added under nitrogen to sodium ethoxide prepared from oil-free sodium hydride (20 mmol). Under external cooling 5,6,7,8-tetrahydro-4*H*-cyclohepta[*b*]thiophene-4-one

(0.83 g, 5 mmol) in anhydrous benzene (10 ml) is added over a period of 10 min. The stirring is continued at 0 °C for 5 h and at the end ice-water is added. The phases are separated and the organic phase washed with water (2 × 25 ml) and 5% sodium hydroxide solution (3 × 10 ml). The combined aqueous phases are acidified and extracted with diethyl ether and the extracts are dried, evaporated, and distilled giving 0.66 g (72%) of the title compound bp 120–123 °C/0.5 mm Hg.

4E.22 EIGHT-MEMBERED CYCLIC KETONES

It is difficult to prepare eight-membered ketones and only 8% yield was obtained upon ring-closure of 6-(2-thienyl)hexanoic acid chloride using aluminium chloride as catalyst [439]. Better yields were obtained upon ring-closure of 6-(5-ethyl-2-thienyl)hexanoic acid chloride and aluminium chloride [439]. Acceptable yield of ring-closed product was obtained when the 2- and 5-positions are blocked with methyl groups. Thus 6-(2,5-dimethyl-3-thienyl)-hexanoic acid upon treatment with polyphosphoric acid gave 60% of the following compound [411].

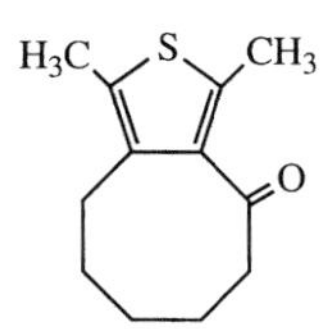

4E.23 MACROCYCLIC KETONES

4E.23.1 *Via* ring closure of ω-thienylalkanoic acid derivatives

Such compounds have been obtained by careful control of the cyclization conditions in order to obtain optimal yields. High dilution conditions are necessary and the presence of an active heterogeneous surface is also desirable. Good yields are obtained by using partially hydrolyzed aluminium chloride etherate as catalyst or by conducting the reaction in the presence of silica or alumina gel of coarse pore size [440,441]. Ring-closure of the acid chloride of ω-(2-thienyl)alkanoic acids with $n = 9$–12 proceeds smoothly to the second α-position giving in 50–64% yield [441]. For $n = 9$ the yield increased to 75% when silica gel was added to the aluminium chloride–ether–water catalyst [442].

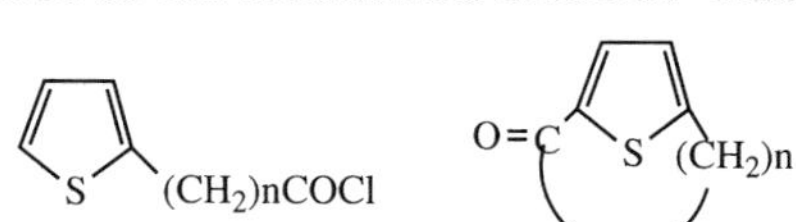

Cyclization by reacting the acids with trifluoroacetic acid anhydride-phosphoric acid in acetonitrile gave ketones with ring-size of 12–21 [443].

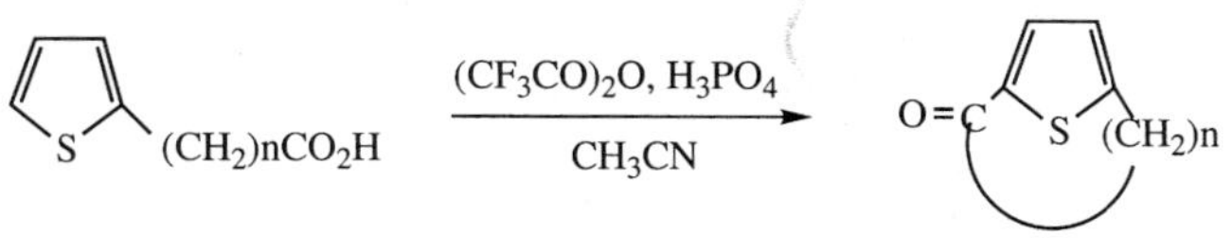

The cyclization experiment for n = 11 is representative [441]

A dried two-necked flask fitted with a reflux condenser and a calcium chloride guard tube is charged with 85% phosphoric acid (0.23 g, 2.0 mmol) and trifluoroacetic acid anhydride (1.8 ml, 12.6 mmol) dissolved in anhydrous acetonitrile 120 ml). To this solution, vigorously stirred and under light reflux, a solution of 12-(2-thienyl)dodecanoic acid (1.3 g, 4.6 mmol) in anhydrous acetonitrile (40 ml) is added over 2 h by means of a motor-driven syringe through a capillary tube dipping into the solution. During the addition the clear solution becomes cloudy. After complete addition the heating is continued for 10 min. After cooling the product is taken up in diethyl ether and the combined organic phases are washed with sodium chloride solution and water, dried, and evaporated. The residue is chromatographed on silica gel using benzene as eluent, giving 0.72 g (59%) of the title compound as an orange oil.

Cyclization of the 5-methyl-substituted acid chloride with $n = 9$ takes place entirely into the 4-position [444], for $n = 10$ or 11 cyclization occurs both to the 3- and 4-positions [444,445]. ω-(3- or 4-Methyl-2-thienyl)alkanoyl chlorides ($n = 9$ or 10) cyclize into the 5-position in very good yields for $n = 9$ or 10 [446,447].

4E.23.2 *Via* intramolecular cyclization of activated methylene group with an ω-iodomethylene group

By this method macrocyclic ketones are prepared by the intramolecular alkylation of a β-keto ester group by the 5-(ω-iodoalkyl) side chain by the use of high dilution conditions in the presence of finely powdered potassium carbonate in butanone [448-452]. Also potassium *tert*-butoxide [453] or other alkali carbonates [454] have been used as base. High yields of cyclic β-keto esters are obtained when $n = 9$ or 11 [449–453].

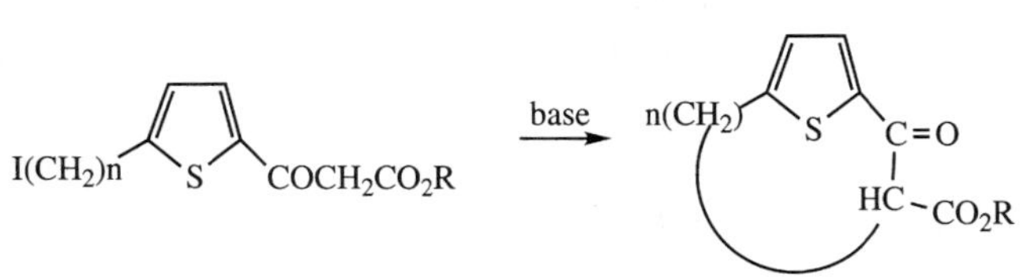

4E.24 THIOKETONES

Phenyl-, mesityl-, and *para*-chlorophenyl 2-thienyl thioketone are prepared by the reaction of the corresponding ketones with Lawesson's reagent for a study of their cycloaddition reactions [455].

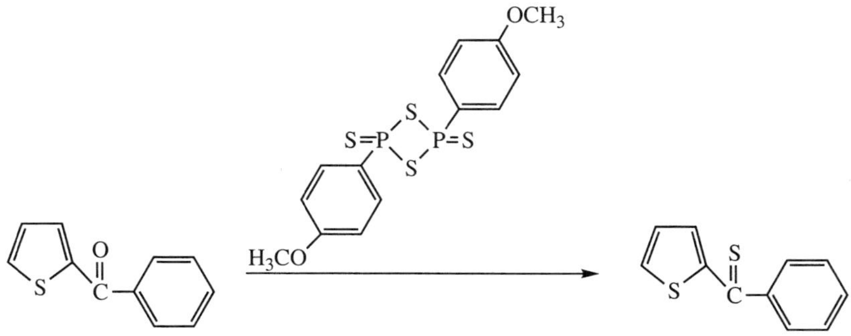

4E.25 ACETALS, THIOACETALS AND AMINALS DERIVED FROM THIOPHENE KETONES

Also in acylthiophenes as well as in thiophene aldehydes, the carbonyl group must be protected, before metalation and halogen–metal exchange reactions can be carried out, which makes acetals and thioacetals two important synthetic intermediates. Thus the acetal with ethylene glycol is prepared from 2- and 3-acetylthiophene in the usual way [89,98,456].

2-Methyl-2-(2-thienyl)-1,3-dioxolane [456]

A flask fitted with nitrogen inlet, stirrer, Vigreux column, Barrett trap, and reflux condenser is charged with 2-acetylthiophene (25 g, 1984 mol), ethylene glycol (12.3 g, 01984 mol), and *para*-toluenesulfonic acid (1 g) in anhydrous benzene (200 ml). The reaction mixture is refluxed vigorously until the theoretical amount of water is collected, after which the solution is concentrated to 40 ml and undissolved *para*-toluenesulfonic acid is filtered off. The filtrate is diluted with cyclohexane (200 ml) and left overnight at 5 °C. The crystalline material formed is collected giving 24.2 g (72%) of the title compound mp 33–34 °C after recrystallization from benzene/cyclohexane.

Upon metalation of 2-methyl-2-(2′-thienyl)-1,3-dioxolane followed by various electrophiles, various 5-substituted derivatives are obtained, which upon hydrolysis give 5-substituted 2-acetylthiophenes [140,456].

2-Acetyl-5-benzoylthiophene [456]

A refluxing etheral slurry (100 ml) of lithiated 2-methyl-2-(2-thienyl)-1,3-dioxolane (0.011 mol) is treated with freshly distilled benzonitrile (1.24 g, 0.012 mol). The reaction mixture is refluxed overnight, after which 6 *M* hydrochloric acid is added. The diethyl ether is distilled off and the acidic aqueous solution is refluxed for an additional 5 h followed by extraction with diethyl ether. The combined organic phases are evaporated and the residue, a tan solid, is recrystallized from cyclohexane/ethyl alcohol with Norite A giving 1.67 g (66%) of the title compound as a white solid mp 123 °C.

The acetal derived from 3-acetylthiophene gives upon lithiation followed by reaction with *N,N*-dimethylformamide, 2-(2′-formyl-3′-thienyl)-2-methyl-1,3-dioxalane [98].

2-(2′-Formyl-3′-thienyl)-2-methyl-1,3-dioxalane [98]

Butyllithium is added to a solution of 2-(3′-thienyl)-2-methyl-1,3-dioxalane (20 g, 0.117 mol) in anhydrous diethyl ether (50 ml) at −20 °C. The mixture is refluxed for 15 min, after which it is cooled to −10 °C and a solution of *N,N*-dimethylformamide (9.35 g, 0.128 mol) in anhydrous diethyl ether (80 ml) is added. The reaction mixture is stirred for 4 h and during this time the temperature is allowed to rise to 15 °C. After hydrolysis, extraction with diethyl ether, evaporation and distillation, 16.3 g (70%) of the title compound is obtained as white crystals bp 120–124 °C/1 mm Hg, mp 40 °C.

Metalation of 2-(3′-thienyl)-2-methyl-1,3-dioxalane with butyllithium followed by *N,N*-dimetylacetamide is used for the preparation of 2-(2′-acetyl-3′-thienyl)-1,3-dioxalane [89,98]. The ethylene acetal from 2-acetyl-3-iodothiophene was also prepared and its halogen–metal exchange at −70 °C followed by reaction with *N,N*-dimethylformamide gave the acetal protected 2-acetyl-3-formylthiophene [89].

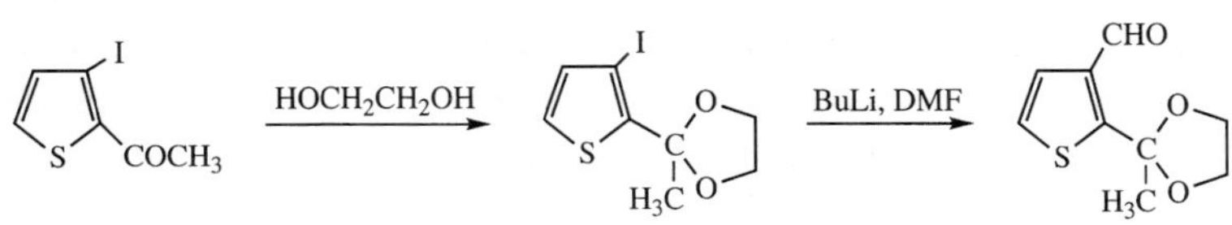

2-Methyl-2-[2-(5-trimethylsilyl)thienyl]thiazolidine is prepared from 2-acetyl-5-trimethylsilylthiophene and 2-aminoethanethiol and catalytic amounts of iodine [457].

2-Methyl-2-[2-(5-trimethylsilyl)thienyl]thiazolidine [457]

A flask equipped with a Barrett trap and condenser is charged with a solution of 2-acetyl-5-trimethylsilylthiophene (12.2 g, 62 mmol), 2-aminoethanethiol

(4.75 g, 65 mmol) and a catalytical amount of iodine in anhydrous benzene (120 mol). The reaction mixture is refluxed until the theoretical amount of water is distilled off as an azeotrop, 170 h. After cooling the benzene phase is washed twice with water, evaporated, and distilled giving 8 g (51%) of the title compound bp 116 °C/0.2 mm Hg.

4E.26 IMINES (SCHIFF'S BASES), OXIMES, HYDRAZONES AND RELATED DERIVATIVES FROM THIOPHENE KETONES

Oximes are prepared by standard procedures and upon treatment with polyphosphoric acid converted to the lactams [427].

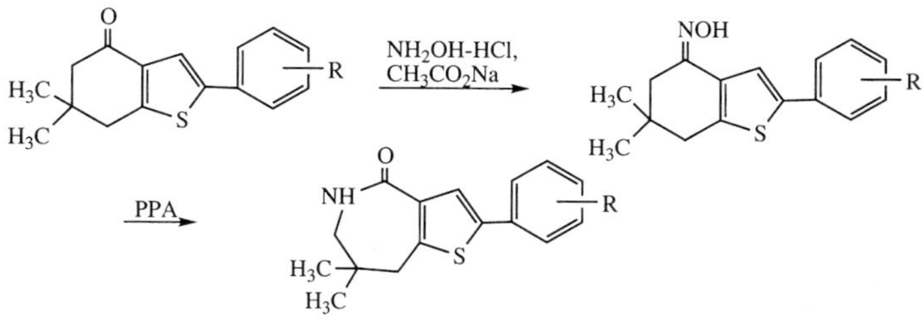

A number of 2-(5-trimethylsilyl)thienyl aryl ketimines are prepared by metalation of 5-trimethylsilylthiophene followed by reaction with aromatic nitriles [377].

2-(5-Trimethylsilyl)thienyl 4-chlorophenyl ketimine [377]

5-Trimethylsilyl-2-lithiothiophene prepared from 2-trimethylsilylthiophene (15.6 g, 0.1 mol) and 1.6 *M* butyllithium (62 ml) in diethyl ether is added dropwise under nitrogen at 0 °C to a stirred solution of *para*-chlorobenzonitrile (13.7 g, 0.1 mol) in anhydrous diethyl ether. The reaction mixture is stirred at room temperature for 6–9 h, after which a cold aqueous ammonium chloride solution (150 ml) is added rapidly and the mixture is stirred for an additional hour at room temperature. The phases are separated and the aqueous phase extracted with diethyl ether. The combined organic phases are dried over magnesium sulfate and evaporated. The residue is recrystallized from hexane giving 22 g (76%) of the title compound mp 97.5–99.5 °C.

Two different methods can be used for the preparation of oxime ethers. 2-Acetylthiophene is either reacted with *O*-alkyl hydroxylamines or the oxime of 2-acetylthiophene is alkylated on oxygen, with alkyl sulphates and sodium hydroxide in aqueous *N*,*N*-dimethylformamide. The oxime ethers give selective

5-substitution on electrophilic acetylation, nitration, chlorination, bromination, and sulfonation [270]. Acidic hydrolysis then offers a facile synthesis of 5-substituted 2-acetylthiophenes [270].

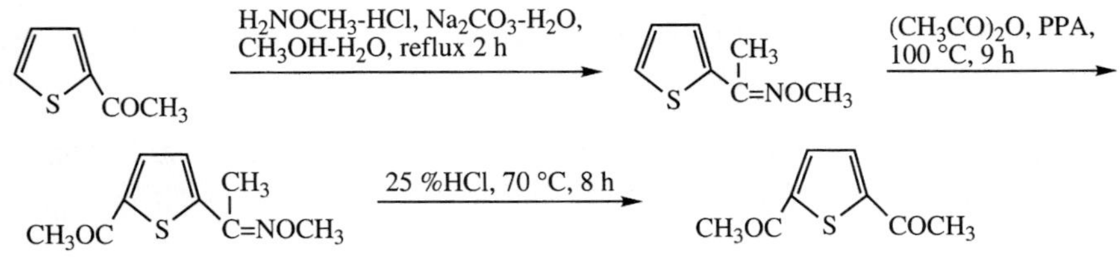

2-Thienylalkyl ketoximes were prepared in conventional ways and were alkylated with alkyl, allyl, and propargyl halides in a two-phase catalytic system [458].

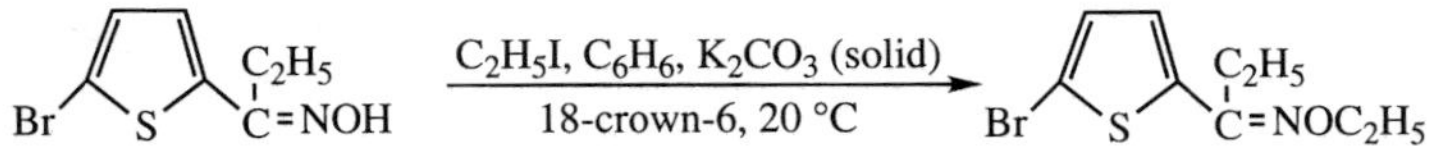

The oxime tosylate derived from trifluoromethyl-2-thienyl ketone is prepared by reaction of the ketone with hydroxylamine hydrochloride in pyridine followed by reaction with tosyl chloride, triethylamine, and 4-dimethylaminopyridine in methylene chloride [385]. Reaction of 2-thienylmethyl ketoxime with 1,2-dichloroethane in a suspension of alkali metal hydroxide and dimethylsulfoxide gives the oxime diethers [459].

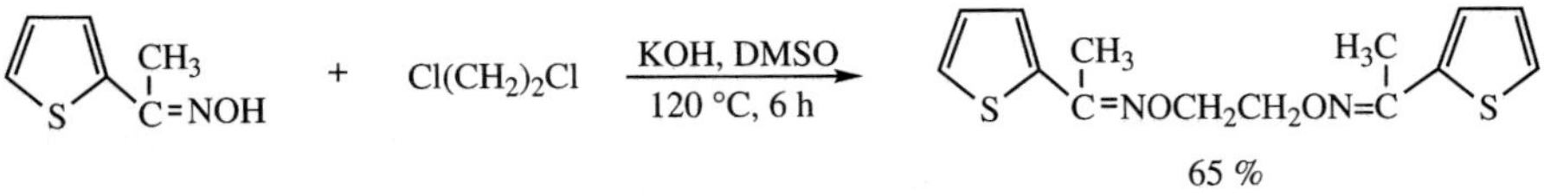

Alkyl thienyl oximes have been prepared and used for the synthesis of 2-(2-thienyl)pyroles by reaction with 1,2-dichloroethane in superbase media [460].

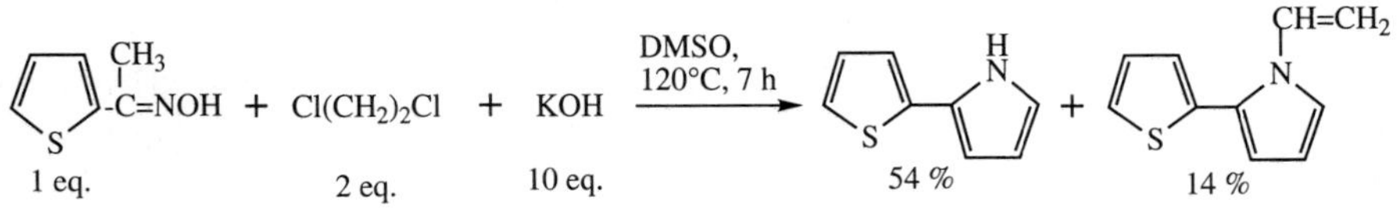

In connection with the synthesis of a series of β-aminoxypropionic acids with potential antiinflammatory activity such compounds were prepared from 2-thienyl aldoxime and ethyl acrylate [461].

Ethyl 3-thenylideneaminoxypropionate [461]

A solution of 2-thiophene aldoxime (63.5 g, 0.5 mol) in anhydrous ethanol (250 ml) is treated with ethyl acrylate (42 g, 0.42 mol) followed by a solution of 2 *M* potassium hydroxide in anhydrous ethanol (42 ml). The reaction mixture is stirred at 35 °C for 48 h and then evaporated. The residue is dissolved in diethyl ether and washed with 10% sodium hydroxide solution and water, dried, evaporated, and chromatographed on silica gel using ethyl acetate/benzene (3:2) as eluent, giving 38.6 g (34%) of the title compound as an oil.

The oximes shown below have been prepared and used in Beckmann rearrangements [400].

Guanylhydrazones of 2-acetylthiophene are prepared by reaction with aminoguanidine hydrogen carbonate and hydrochloric acid. Successive reaction with hot acetic anhydride gives the corresponding *N*,*N*′-diacetyl derivatives [462].

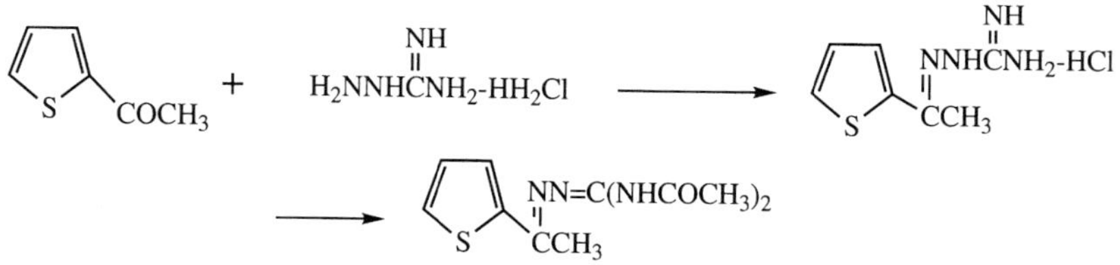

In connection with work on liquid crystals, mixed azines are prepared by the reaction of 2-acetylthiophene hydrazone with 3,4-dihydroxyacetophenone or by the reaction of 2-acetylthiophene with the hydrazone of 3,4-dihydroxyacetophenone [463].

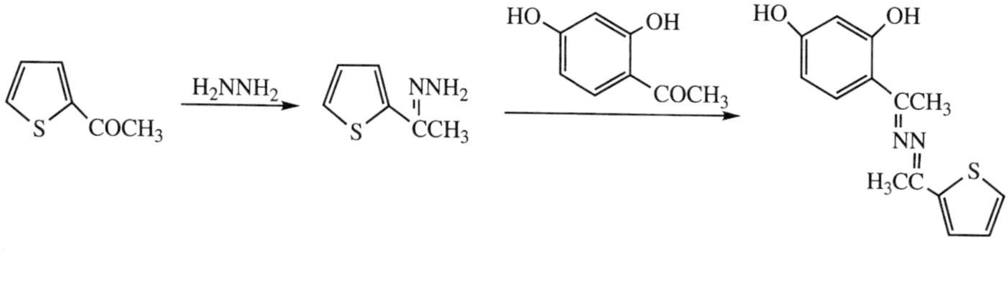

The phenylhydrazone of ethyl 2-thienylglyoxylate is prepared in the usual way in connection with a study of the rearrangement of arylhydrazones in polyphosphoric acid, which gives a mixture of ethyl 4- and 5-(4-aminoaryl)-α-thienylglyoxylate [464].

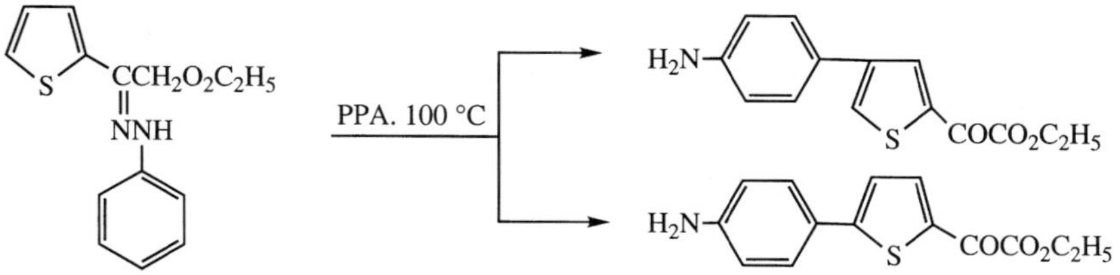

REFERENCES

1. G. V. Shishkin and V. P. Mamaev, *J. Gen. Chem USSR (Engl. Transl.)* **34**, 404 (1964).
2. A. W. Weston and R. J. Michals, *Org. Synth.* **31**, 108 (1951).
3. F. Embert, J.-P.- Lère-Porte, J. J. E. Moreau, F. Serein-Spirau, A. Righi and J.-L. Sauvajol, *J. Mater. Chem.* **11**, 718 (2001).
4. M. Krayushkin, A. A. Loktionov and L. I. Belen'kii, *Chem. Heterocycl. Compds. USSR, (Engl. Transl.)* **24**, 850 (1988).
5. I. Jestin, P. Frére, P. Blanchard and J. Roncali, *Angew. Chem. Int. Ed.* **37**, 942 (1998).
6. A. G. Martinez, R. M . Alvarez, J. O. Barcina, S. dela Moya Cerero, E. T. Vilar, A. G. Fraile, M. Hanack and R. L. Subramanian, *J. Chem. Soc., Chem. Commun.* 1571 (1990).
7. A. Rieche, H. Gross and E. Höft, *Chem. Ber.* **93**, 88 (1960).
8. H. Gross and G. Matthey, *Chem. Ber.* **97**, 2606 (1964).
9. P. M. Jackson. C. J. Moody and P. Shah, *J. Chem. Soc., Perkin Trans. 1*, 2909 (1990).
10. H. J. Bestmann, J. Lienert and L. Mott, *Liebigs Ann. Chem.* **718**, 24 (1968).
11. K. Fujii and A. Yukawa, *J. Pharm. Soc. Japan* **76**, 607 (1956).
12. E. Campaigne and W. L. Archer, *J. Am. Chem. Soc.* **75**, 898 (1953).
13. N. P. Buu-Hoï, N. D. Xuong, R. Royer and D. Lavit, *J. Chem. Soc.* 547 (1953).
14. M. Sy, N. P. Buu-Hoï and N. D. Xuong, *J. Chem. Soc.* 21 (1955).
15. R. M. Kellogg and J. Buter, *J. Org. Chem.* **36**, 2236 (1971).
16. N. P. Buu-Hoï, D. Lavit and N. D. Xuong, *J. Chem. Soc.* 1581 (1955).
17. Y. L. Gol'dfarb and P. A. Konstantinov, *Bull. Acad. Sci., Div. Chem. Soc. (Engl. Transl.)* 1013 (1956).
18. M. Sy, N. P. Buu-Hoï and N. D. Xuong, *J. Chem. Soc.* 1975 (1954).
19. N. P. Buu-Hoï, N. D. Xuong and N. Van Bac, *Bull. Soc. Chim. Fr.* 3640 (1965).
20. W. Hoek, H. Wynberg and J. Strating, *Rec. trav. Chim. Pays-Bas* **85**, 1054 (1966).
21. S. Ishimura and E. Imoto, *Bull. Chem. Soc. Japan* **35**, 432 (1962).
22. B. P. Fabrichnyi, I. F. Shalavina and Y. L. Gol'dfarb, *J. Gen. Chem. USSR (Engl. Transl.)* **34**, 3938 (1964).
23. B. P. Fabrichnyi, I. F. Shalavina and Y. L. Gol'dfarb, *J. Gen. Chem. USSR (Engl. Transl.)* **28**, 2556 (1958).
24. Y. Poirier, L. Legrand and N. Lozac'h, *Bull. Soc. Chim. Fr.* 1054 (1966).
25. N. P. Buu-Hoï and M. Khenissi, *Bull. Soc. Chim. Fr.* 359 (1958).
26. J. Lamy, D. Lavit and N. P. Buu-Hoï, *J. Chem. Soc.* 4202 (1958).
27. B. F. Crowe and F. F. Nord, *J. Org. Chem.* **15**, 1177 (1950).

28. N. P. Buu-Hoï, N. Hoán and D. Lavit, *J. Chem. Soc.* 2130 (1950).
29. A. W. Weston and T. T. Michaels, *J. Am. Chem. Soc.* **72**, 1422 (1950).
30. S. Gronowitz, P. Moses, A.-B. Hörnfeldt and R. Håkansson, *Arkiv Kemi.* **17**, 165 (1961).
31. D. J. Chadwick, J. Chambers, H. E. Hargraves, G. D. Meakins and R. L. Snowden, *J. Chem. Soc., Perkin Trans. 1* 2327 (1973).
32. O. Meth-Cohn and M. Ashton, *Tetrahedron Lett.* **41**, 2749 (2000).
33. M. Niestroj and W. P. Neumann, *Chem. Ber.* **129**, 45 (1996).
34. R. Lantz and A.-B. Hörnfeldt, *Chem. Scripta* **2**, 9 (1972).
35. B. Garrigues, B. Oussaid and C. Hubert, *Bull. Soc. Chim. Fr.* **130**, 58 (1993).
36. P. Cagniant and D. Cagniant, *Bull. Soc. Chim. Fr.* 713 (1952).
37. L. V. Dulenko, G. N. Dorofeenko, S. N. Baranov, I. G. Katts and V. I. Dulenko, *Chem. Heterocycl. Compds. (Engl. Transl.)* **7**, 294 (1971).
38. M. Shilai, Y. Kondo and T. Sakomoto, *J. Chem. Soc., Perkin Trans. 1* 442 (2001).
39. S. Gronowitz, *Arkiv Kemi.* **8**, 441 (1955).
40. Y. L. Gol'dfarb, M. A. Kalik and M. L. Kirmalova, *Bull. Acad. Sci. USSR, Div. Chem. Sci. (Engl. Transl.)* 650 (1962).
41. B. Yom-Tov and S. Gronowitz, *Chem. Scripta* **3**, 37 (1973).
42. B. Yom-Tov and S. Gronowitz, *Chem. Scripta* **3**, 165 (1973).
43. J. Skramstad and O. Eriksen, *Acta Chem. Scand.* **45**, 919 (1991).
44. A. Moradpour, *J. Chem. Soc., Perkin Trans. 1* 7 (1993).
45. Y. L. S.-T. Armiger and T. D. Lash, *J. Heterocycl. Chem.* **29**, 523 (1992).
46. T. Sone, Y. Ohba and R. Watanabe, *Bull Chem. Soc. Japan* **62**, 1346 (1989).
47. Z. Hu and M. P. Cava, *Tetrahedron Lett.* **35**, 3493 (1994).
48. H. Kurata, H. Nakaminami, K. Matsumoto, T. Kawase and M. Oda, *Chem. Commun.* 529 (2001).
49. A. Schöning and W. Friedrichsen, *Liebigs Ann. Chem.* 405 (1989).
50. K. B. Wiberg, *Org. Synth.* **29**, 87 (1949).
51. E. Campaigne, R. C. Bourgeois and W. C. McCarthy, *Org. Synth.* **33**, 93 (1953).
52. P. Cagniant, G. Merle and D. Cagniant, *Bull. Soc. Chim. Fr.* 302 (1970).
53. B. P. Das, R. T. Cunningham and D. W. Boykin, *J. Med. Chem.* **16**, 1361 (1973).
54. W. S. Emerson and T. M. Patrick, *J. Org. Chem.* **14**, 790 (1949).
55. M. Fétizon, F. Gomez-Parra and J.-M. Louis, *J. Heterocycl. Chem.* **13**, 525 (1976).
56. A. J. Mancuso, D. D. Brownfain and D. Swern, *J. Org Chem.* **44**, 4148 (1979).
57. Z. Horii, K. Sakurai, K. Tomino and T. Konishi, *J. Pharm. Soc. Japan* **76**, 1101 (1956).
58. A. Daïch, P. Ohier and B. Decroix, *J. Heterocyl. Chem.* **32**, 1731 (1995).
59. P. A. Zoretic, P. Soja, M. Jodoin and R. Levin, *Org. Prep. Proced. Int.* **8**, 33 (1976).
60. J. A. Clarke and O. Meth-Cohn, *Tetrahedron Lett.* 4705 (1975).
61. L. I. Zakharkin, D. N. Maslin and V. V. Gavrilenko, *J. Org. Chem. USSR (Engl. Transl.)* **2**, 2153 (1966).
62. E. V. Brown and J. A. Blanchette, *J. Am. Chem. Soc.* **72**, 3414 (1950).
63. S. Nishimura, R. Motoyama and E. Imoto, *Bull. Univ. Osaka Prefect., Ser. A* **6**, 127 (1958).
64. J. Srogl, M. Janda and M. Valentova, *Collect. Czech. Chem. Commun.* **35**, 148 (1970).
65. I. Stibor, M. Janda and J. Srogl, *Z. Chem.* **10**, 342 (1970).
66. L. I. Belen'kii, I, B . Karmanova and Y. L. Gol'dfarb, *J. Org. Chem. USSR (Eng. Transl.)* **7**, 1809 (1971).
67. L. I. Belen'kii, I. B. Karmanova, Y. B. Vol'kenshtein and Y. L. Gol'dfarb, *Bull. Acad. Sci, USSR, Div. Chem. Sci. (Engl. Transl.)* **20**, 878 (1971).
68. M. Nemec, M. Janda, J. Srogl and I. Stibor, *Collect. Czech. Chem. Commun.* **39**, 3527 (1974).
69. T. Sone, *Bull. Chem. Soc. Japan* **37**, 1197 (1964).
70. T. Sone, *Nippon Kagaku Zasshi* **86**, 1331 (1965).
71. M. Yoshida, T. Yoshida, N. Kamigate and M. Kobayashi, *Bull. Chem. Soc. Japan.* **61**, 3549 (1988).

72. S.-M. Yang and J.-M. Fang, *J. Org. Chem.* **64**, 394 (1999).
73. G. Piancatelli and T. Ferri, *J. Org. Chem.* **55**, 4019 (1990).
74. M. D'Auria and D. Tofani, *Tetrahedron* **48**, 9315 (1992).
75. M. D'Auria and T. Ferri, *J. Org. Chem.* **60**, 8360 (1995).
76. M. D'Auria, *Synth. Commun.* **22**, 2393 (1992).
77. D. L. Comins and M. O. Kilpack, *J. Org. Chem.* **52**, 104 (1987).
78. N. Sprutta and L. Latos-Grazy′nski, *Tetrahedron Lett.* **40**, 8457 (1999).
79. T. Kuroda, M. Takahashi, T. Ogiku, H. Ohmizu, T. Nishitani, K. Kondo and T. Iwasaki, *J. Org. Chem.* **59**, 7353 (1994).
80. T. Kuroda, K. Kondo, T. Iwasaki, A. Ohtani and K. Takashima, *Chem. Pharm. Bull.* **45**, 678 (1997).
81. A. Osuka, D. Fujikane, H. Shinmori, S. Kobatake and M. Irie, *J. Org. Chem.* **66**, 3913 (2001).
82. J. C. Bussolari and D. C. Rehborn, *Org. Lett.* **1**, 965 (1999).
83. T. Itahara and F. Ouseto, *Synthesis* 489 (1984).
84. B. L. Feringa, R. Hulst, R. Rikers and L. Brandsma, *Synthesis* 316 (1988).
85. H. Brisset, S. Le Moustarder, P. Blanchard, B. Illien, A, Riou, J. Orduna, J. Garin and J. Roncali, *J. Mater. Chem.* **7**, 2027 (1997).
86. V. I. Rogovik and Y.L. Gol'dfarb, *Chem. Heterocycl. Compds. (Engl. Transl.)* **1**, 439 (1956).
87. Y. L. Gol'dfarb, B. P. Fabrichnyi and V. I. Rogovik, *Bull. Acad. Sci. Div. Chem. Sci.* (*Engl. Transl.*) 2001 (1963).
88. D. W. H. MacDowell and T. B. Patrick, *J. Org. Chem.* **31**, 3592 (1966).
89. R. Guilard and P. Fournari, *Bull. Soc. Chim. Fr.* 1437 (1971).
90. C. Paulmier, J. Morel, P. Pastour and D. Semard, *Bull. Soc. Chim. Fr.* 2511 (1969).
91. B. Decroix, J. Morel, C. Paulmier and P. Pastour, *Bull. Soc. Chim. Fr.* 1848 (1972).
92. Y. L. Gol'dfarb, Y. B. Vol'kenshtain and B. V. Lopatin, *J. Gen. Chem. USSR, (Engl. Transl.)* **34**, 961 (1964).
93. T. M. Cresp, M. V. Sargent and P. Vogel, *J. Chem. Soc., Perkin Trans. 1* 37 (1974).
94. C. Paulmier, J. Morel, D. Semarda and P. Pastour, *Bull. Soc. Chim. Fr.* 2434 (1973).
95. F. M. Stoyanovich and B. P. Fedorov, *Chem. Heterocycl. Compds. (Engl. Transl.)* **3**, 650 (1967).
96. J. Morel and P. Pastour, *Bull. Soc. Chim. Fr.* 737 (1968).
97. M. Winn and F. G. Bordwell, *J. Org. Chem.* **32**, 1610 (1967).
98. M. Robba, B. Roques and M. Bonhomme, *Bull. Soc. Chim. Fr.* 2495 (1967).
99. R. Guilard, P. Fournari and M. Person, *Bull. Soc. Chim. Fr.* 4121 (1967).
100. T. Mitsumori, K. Inoue, N. Koga and H. Iwamura, *J. Am. Chem. Soc.* **117**, 2467 (1995).
101. M. Robba, B. Roques and M. Bonhomme, *Bull. Soc. Chim. Fr.* **7**, 2495 (1967).
102. I. B. Karmanova, Y. B. Vol'kenshtein and L. I. Belen'kii, *Chem. Heterocycl. Compds. (Engl. Transl.)* **9**, 451 (1973).
103. Y. L. Gol'dfarb, M. A. Kalik and M. L. Kirmalova, *Bull. Acad,. Sci. USSR, Div. Chem. Sci. (Engl. Transl.)* 1638 (1969).
104. A. V. El'tsov and A. A. Ginesina, *J. Org. Chem. USSR (Engl. Transl.)* **3**, 184 (1967).
105. L. Lunazzi, G. F. Pedulli, M. Tiecco, C . Vincenzi and C. A. Veracini, *J. Chem. Soc., Perkin Trans. 2*, 751 (1972).
106. Y. L. Gol'dfarb, M. A. Kalik and M. L. Kirmalova, *Bull. Acad. Sci. USSR Div. Chem. Sci.(Engl. Transl.)* 2102 (1969).
107. K. E. Schulte, J. Reisch, W. Herrmann and G. Bohn, *Arch. Pharm. (Weinheim)* **296**, 456 (1963).
108. K. Dimroth, G. Pohl and H. Follmann, *Chem. Ber.* **99**, 636 (1966).
109. P. Dubus, B. Decroix, J. Morel and P. Pastour, *Bull. Soc. Chim. Fr.* 628 (1976).
110. S. Trofimenko, *J. Org. Chem.* **29**, 3046 (1964).
111. J. Morel, C. Paulmier and P. Pastour, *Compt. rend. C* **266**, 1300 (1968).
112. H. Finch, D. H. Reece and J. T. Sharp, *J. Chem. Soc., Perkin Trans. 1* 1193 (1994).

113. M. H. Sosabowski and P. Powell, *J. Chem. Res.(S)* 12 (1997).
114. H. Nerenz, M. Meier, W. Grahn, A. Reisner, E. Schmälzlin, S. Stadler, K. Meerholz, C. Bräuchle and P. G. Jones, *J. Chem. Soc., Perkin 2*, 437 (1998).
115. R. Frimm, L. Fisera and J. Kovac, *Collect. Czech. Chem. Commun.* **38**, 1809 (1973).
116. L. Racane, V. Tralic-Kulenovic, G. Karminski-Zamola and L. Fiser-Jakic, *Monatsh. Chem.* **126**, 1375 (1995).
117. C. Gozzi, L. Lavenot, K. Ilg, V. Penalva and M. Lemaire, *Tetrahedron Lett.* **38**, 8867 (1997).
118. L. Lavenot, C. Gozzi, K. Ilg, I. Orlova, V. Penalva and M. Lemaire, *J. Organomet. Chem.* **567**, 49 (1998).
119. S. Pivsa-Art, T. Satoh, Y. Kawamura, M. Miura and M. Nomura, *Bull. Chem. Soc. Japan* **71**, 467 (1998).
120. S. Gronowitz and G. Timari, *J. Heterocycl. Chem.* **27**, 1127 (1990).
121. W. Hanefeld and M. Schlitzer, *J. Heterocycl. Chem.* **32**, 1019 (1995).
122. S. Gronowitz and A. Bugge, *Acta Chem. Scand.* **19**, 1271 (1965).
123. S. Gronowitz, T. Dahlgren, J. Namtvedt, C. Roos, G. Rosén, B. Sjöberg and U. Forsgren, *Acta Pharm. Suec.* **8**, 623 (1971).
124. S. Gronowitz and J. Namtvedt, *Acta Chem. Scand.* **21**, 2151 (1967).
125. S. Gronowitz and C. Glennow, *Chem. Scripta* **11**, 76 (1977).
126. U. Michael and S. Gronowitz, *Acta Chem. Scand.* **22**, 1353 (1968).
127. S. Gronowitz and A. Maltesson, *Acta Chem. Scand.* **25**, 2435 (1971).
128. A. I. de Lucas, N. Martin, L. Sanchez, C. Soane, J. Garín, J. Orduna, R. Alcalá and B. Villacampa, *Tetrahedron Lett.* **38**, 6107 (1997).
129. A. S. Benahmed-Gasmi, P. Frère, B. Garrigues, A. Gorgues, M. Jubeault, R. Carlier and F. Texier, *Tetrahedron Lett.* **33**, 6457 (1992).
130. C. Maertens, J.-X. Zhang, P. Dubois and R. Jerome, *J. Chem. Soc., Perkin Trans. 2* 713 (1996).
131. S. W. A. Bligh, C. F. G. C. Geraldes, M. McPartlin, M. J. Sanganee and T. M. Woodroffe, *Chem. Commun.* 2073 (1998).
132. D. A. Sahade, T. Kawaji, T. Sawada, S. Mataka, T. Thiemann, T. Tsukinoki and M. Tashiro, *J. Chem. Research (S)* 210 (1999).
133. Y. L. Gol'dfarb, Y. B. Vol'kenshtein and L. I. Belen'kii, *Angew. Chem. Int. Ed. Engl.* **7**, 519 (1968).
134. Y. L. Gol'dfarb, M. A. Kalik and M. L. Kirmalova, *Bull. Acad. Sci. USSR, Div. Chem. Sci. (Engl. Transl.)* 1578 (1964).
135. Y. L. Gol'dfarb and P. A. Konstantinov, *Bull. Acad. Sci. USSR, Div. Chem. Sci. (Engl. Transl.)* 229 (1957).
136. Y. L. Gol'dfarb, S. Ozolin, and V. P. Litvinov, *Chem. Heterocycl. Compds. (Engl. Transl.)* **3**, 737 (1967).
137. B. P. Fabrichnyi, Y. B. Vol'kenshtein, V. I. Rogovik, I. B. Karmanova and Y. L. Gol'dfarb, *Chem. Heterocycl. Compds. (Engl. Transl.)* **1**, 336 (1965).
138. Y. L. Gol'dfarb, M. A. Kalik and M. L. Kirmalova, *Chem. Heterocycl. Compds.* **3**, 45 (1967).
139. M. M. Krayushkin , M. A. Kalik and E. Y. Zvezdina, *Bull. Acad. Sci. USSR, (Engl. Transl.)* **40**, 228 (1991).
140. S. F. Thames and J. E. McCleskey, *J. Heterocycl. Chem.* **4**, 146 (1967).
141. J.-L. Luche and A. L. Gemal, *J. Chem. Soc., Chem. Commun.* 976 (1978).
142. V. B. Mochalin and N. G. Ivanova, *J. Gen. Chem. USSR (Engl. Transl.)* **32**, 1479 (1962).
143. Y. L. Gol'dfarb and P. A. Konstantinov, *Bull. Acad. Sci. USSR, Div. Chem. Sci. (Engl.Transl.)* 108 (1959).
144. S. Gronowitz, B. Gestblom and B. Mathiasson, *Arkiv Kemi* **20**, 407 (1963).
145. J. D. Prugh, G. D. Hartman, P. J. Mallorga, B. M. McKeever, S. R. Michelson, M. A. Murcko, H. Schwamm, R. L. Smith, J. M. Sondey, J. P. Springer and M. F. Sugrue, *J. Med. Chem.* **34**, 1805 (1991).

146. L. Lamartina, R. Noto and C. Arnone, *J. Heterocycl. Chem.* **25**, 1083 (1988).
147. A. P. Dantanarayana, B. DuPre, J. A. May and V. M. Lynch , *J. Heterocycl. Chem.* **36**, 65 (1999).
148. P. C. Ostrowski and V. V. Kane, *Tetrahedron Lett.* 3549 (1977).
149. M. Medarde, A. C. Ramos, R. Peláez-Lamamié de Clairac, E. Caballero and A. San Feliciano, *J. Chem. Soc., Perkin Trans. 1* 45 (1994).
150. S. S. Mandal, J. Chakraborty and A. De, *J. Chem. Soc., Perkin Trans. 1* 2639 (1999).
151. K. Takahashi, T. Nihira, K. Takase and K. Shibata, *Tetrahedron Lett.* **30**, 2091 (1989).
152. C. Hopkinson, G. D. Meakins and R. J. Purcell, *Synthesis* 621 (1991).
153. G. Musumarra and C. Sergi, *Heterocycles* **37**, 1033 (1994).
154. G. L. Eggleton, B. A. Cooper, C. L. Sturch, J. C. Trent and S. V. Mulekar, *J. Heterocycl. Chem.* **27**, 1853 (1990).
155. C. G. Frost and J. M. J. Williams, *Tetrahedron Lett.* **34**, 2015 (1993).
156. J. M. Zenner and R. C. Larock, *J. Org. Chem.* **64**, 7312 (1999).
157. R. S. Kusurkar, M. S. Wadia, D. K. Bhosale, S. S. Tavale and V. G. Puranik, *J. Chem. Res. (S)* 478 (1996).
158. R. E. Miller and F. F. Nord, *J. Org. Chem.* **16**, 1720 (1951).
159. J. Boichard, J.-P. Monin and J. Tirouflet, *Bull. Soc. Chim. Fr.* 851 (1963).
160. V. S. Egorova, V. N. Ivanova and N. I. Putokhin, *Chem. Heterocycl. Compds. (Engl. Transl.)* **3**, 654 (1967).
161. D. N. Robertson, *J. Org. Chem.* **25**, 47 (1960).
162. J. J. Pesek and J. H. Frost, *Synth. Commun.* **4**, 367 (1974).
163. B. P. Fedorov, G. I. Gorushkina and Y. Gol'dfarb, *Bull. Acad. Sci. USSR, Div. Chem. Sci. (Engl. Transl.)* 1962 (1967).
164. R. W. Drisco and H. McKennis, *J. Am. Chem. Soc.* **74**, 2626 (1952).
165. B. P. Fedorov, G. I. Gorushkina and Y. L. Gol'dfarb, *J. Gen. Chem. USSR, (Engl. Transl.)* **31**, 3668 (1961).
166. V. M. Colburn, B. Iddon, H. Suschitzky and P. T. Gallagher. *J. Chem. Soc., Perkin Trans. 1* 1337 (1979).
167. M. S. K. Youssef, *Indian J. Chem.* **19B**, 796 (1980).
168. D.-L. Wang, W.-S. Hwang, L.-C. Liang, L.-I. Wang, L. Lee and M. Y. Chiang *Organometallics* **16**, 3109 (1997).
169. L. Strekowski, M. T. Cegla, S.-B. Kong and D. B. Harden, *J. Heterocycl. Chem.* **26**, 923 (1989).
170. H. Cerecetto, R. Di Maio, M. Gonzáles and G. Seoane, *Heterocycles* **45**, 2023 (1997).
171. M. Burwood, B. Davies, I. Diaz, R. Grigg, P. Molina, V. Sridharan and M. Hughes, *Tetrahedron Lett.* **36**, 9053 (1995).
172. J. Barluenga, M. Tomás, A. Ballesteros and J. A. Lopéz, *Tetrahedron Lett.* **30**, 4573 (1989).
173. B. Oussaid, M. Ayad, B. Garrigues, J.-P. Fayet and L. T. Tran, *Can. J. Chem.* **70**, 58 (1992).
174. G. T. Hwang and B. H. Kim, *Tetrahedron Lett.* **41**, 5917 (2000).
175. A. Macias, A. Rodriguez, R. Bastida, A. De Blas, A. Sousa and D. Rodriguez, *Anales Quim.* **84**, 279 (1988).
176. R. K. Y. Ho and S. E. Livingstone, *Aust. J. Chem.* **18**, 659 (1965).
177. K. D. Mishra, R. Rai, O. P. Panday and S. K. Sengupta, *Transition Met. Chem.* **17**, 127 (1992).
178. A. S. Tantawy, H.-I. El-Subbagh, M. B. El-Ashawy and A. A. El-Enam, *Il Farmaco* **44**, 1217 (1989).
179. M. M. El-Kerdawy, A. A. El-Enam, H. I. El-Subbagh and E. Abushanab, *Monatsh. Chem.* **121**, 45 (1990).
180. K. S. Siddiqi, S. Tabassum, R. I. Kurteshi, N. H. Khan, S. A. A. Zaidi and J. Casabo, *Synth. React. Inorg. Met.-Org. Chem.* **20**, 133 (1990).
181. B. Oussaid, J.-P. Fayet, G. Pelletier and B. Garrigues, *Bull. Soc. Chim. Belg.* **101**, 969 (1992).

182. M. M. Krayushkin, M. A. Kalik, V. K. Zav'yalova and V. S. Bogdanov, *Chem. Heterocycl. Compds. (Eng. Transl.)* **24**, 689 (1988).
183. F. Delmas, M. Gasquet, P. Timon-David, N. Madadi, P. Vanelle, A. Vaille and J. Maldonado, *Eur. J. Med. Chem.* **28**, 23 (1993).
184. W. Eberbach and N. Laber, *Tetrahedron Lett.* **33**, 61 (1992).
185. J. Bussenius, N. Laber, T. Müller and W. Eberbach, *Chem. Ber.* **127**, 247 (1994).
186. R. H. Wiley and G. Irick, *J. Org. Chem.* **24**, 1925 (1959).
187. R. H. Wiley, S. C. Slaymaker and H. Kraus, *J. Org. Chem.* **22**, 204 (1957).
188. M. E. Gonzalez, P. Sancho, C. Soriano, R. Ballesteros, B. Abarca and J. Sepulvada, *Heterocycles* **27**, 1227 (1988).
189. M. Saito, H. Ishihara and K. Takahashi, *Heterocycles* **27**, 1141(1988).
190. K. Saito, T. Sato, H. Ishihara and K. Takahashi, *Bull. Chem. Soc. Japan* **62**, 1925 (1989).
191. K. Saito, H. Ishihara, T. Murase, Y. Horie and E. Maekawa, *Bull. Chem. Soc. Jpn .* **60**, 4317 (1987).
192. K. Saito, H. Fushihara, T. Sato, H. Ishihara and K. Takahashi, *Heterocycles* **29**, 1537 (1989).
193. K. Saito and H. Ishihara, *Heterocycles* **26**, 1891 (1987).
194. C. Lee, Y. Kang, S. O. Kang, J. Ko, J. W. Yoo and M. H. Cho, *J. Organomet. Chem.* **587**, 165 (1999).
195. M. R. Devi and J. M. Rao, *Synth. Commun.* **19**, 2345 (1989).
196. N. Kanoongo, R. V. Singh and J. P. Tandon, *J. Prakt. Chem.* **331**, 342 (1989).
197. E. Campaigne, P. A. Monroe, B. Arnwine and W. L. Archer, *J. Am. Chem. Soc.* **75**, 988 (1953).
198. K. Singh,, R.V. Singh and J. P. Tandon, *Synth. React. Inorg. Met.-Org. Chem.* **16**, 1341 (1986).
199. S. G. Teoh, S.-H. Ang, H.-K. Fun and C.-W. Ong, *J. Organomet. Chem.* **580**, 17 (1999).
200. S. M. Aldoshin, I. I. Chuev, L. O. Atovmyan, V. S. Nedzvetskii and A. S. Kulikov, *Bull. Acad. Sci. USSR (Engl. Transl.),* **40**, 74 (1991).
201. A. Capperucci, A. Degl'Innocenti, M. Funicello, P. Scafto, and P. Spagnolo, *Synthesis* 1185 (1996).
202. M. Muraoka, T. Yamamoto, S. Ajimi, H. Yamaguchi and T. Koinuma, *J. Chem. Soc., Perkin Trans. 1*, 667 (1994).
203. R. Borghi, M. A. Cremonini, L. Lunazzi, G. Placucci and D. Macciantelli, *J. Org. Chem.* **56**, 6337 (1991).
204. S. J. Rao, U. T. Bhalerao, B. T. Tilak, *Indian J. Chem.* **27B**, 257 (1988).
205. P. De Maria, A. Fontana and G. Cerichelli, *J. Chem. Soc., Perkin 2* , 2329 (1997).
206. K. Uchida, G. Masuda, Y. Aoi, K. Nakayama and M. Irie, *Chem. Letters* 1071 (1999).
207. T. Mrozek, H. Görner and J. Daub, *Chem. Commun.* 1487 (1999).
208. Z. Mekhalif, A. Lazarescu, L. Hevesi, J.-J. Pireaux and J. Delhalle, *J. Mater. Chem.* **8**, 545 (1998).
209. H. D. Hartough and A. I. Kosak, *J. Am. Chem. Soc.* **69**, 3093 (1948).
210. A. I. Kosak and H. D. Hartough, *Org. Synth.* **28**, 1 (1948).
211. S. Kotha, K. Chacraborty and E. Bramachary, *Synlett* 1621 (1999).
212. M. W. Farrar and R. Levine, *J. Am. Chem. Soc.* **72**, 3695 (1950).
213. G. M. Badger, H. J. Rodda and W. H. F. Sasse, *J. Chem. Soc.* 4162 (1954).
214. M. Sy and B. de Malleray, *Bull. Soc. Chim. Fr.* 1276 (1963).
215. G. N. Dorofeenko, *J. Gen. Chem . USSR (Engl. Transl.)* **31**, 918 (1961).
216. G. N. Dorofeenko, V. I. Dulenko and V. V. Baeva, J. *Gen. Chem. USSR (Engl. Transl.)* **32**, 3002 (1962).
217. G. N. Dorofeenko, V. I. Dulenko and L. M. Antonenko, *J. Gen. Chem. USSR (Engl. Transl.)* **32**, 2997 (1962).
218. K. Nomiya and M. Miowa, *Bull. Chem. Soc. Japan*, **53**, 3389 (1980).
219. M. W. Farrar, *J. Am. Chem. Soc.* **72**, 4433 (1950).

220. S. Gronowitz and J. E. Skramstad, *Arkiv Kemi* **28**, 115 (1967).
221. R. Menicagli, C. Botteghi and M. Marchetti, *J. Heterocycl. Chem.* **17**, 57 (1980).
222. E. C. Spaeth and C. B. Germain, *J. Am. Chem. Soc.* **77**, 4066 (1955).
223. N. B. Chapman, R. M. Scrowston and T. M. Sutton, J. *Chem. Soc., Perkin Trans. 1* 3011 (1972).
224. O. Dann and H. Distler, *Chem. Ber.* **84**, 423 (1951).
225. P. Cagniant and D. Cagniant, *Bull. Soc. Chim. Fr.* 62 (1953).
226. Y. L. Gol'dfarb and I. S. Korsakova, *Bull. Acad. Sci.USSR, Div. Chem. Sci. (Engl. Transl.)* 481 (1954).
227. M. Ahmed and O. Meth-Cohn, *J. Chem. Soc. (C)* 2104 (1971).
228. S. Clementi, F. Genel and G. Marino, *Ric. Sci.* **37**, 418 (1967).
229. R. K. Mackie, S. Mahtre and J. M. Tedder, *J. Fluorine Chem.* **10**, 437 (1977).
230. S. Clementi and G. Marino, *J. Chem. Soc., Perkin Trans 2* 71 (1972).
231. P. Linda and G. Marino, *J. Chem. Soc. B* 43 (1970).
232. E. J. Bourne, M. Stacey, J. C. Tatlow and J. M. Tedder, *J. Chem. Soc.* 718 (1951).
233. E. J. Bourne, M. Stacey, J. C. Tatlow and R. Worrall, *J. Chem. Soc.* 2006 (1954).
234. C. Galli, *Synthesis* 303 (1979).
235. R. R. Sauers, A. A. Hegedorn, III, S. D. Van Arnum, R. P. Gomez and R. V. Moquin, *J. Org. Chem.* **52**, 5501 (1987).
236. T. Keumi, K. Yoshimura, M. Shimada and H. Kitajima, *Bull. Chem. Soc. Japan* **61**, 455 (1988).
237. O. Meth-Cohn and S. Gronowitz, *Acta Chem. Scand.* **20**, 1577 (1966).
238. H. R. Snyder and C. T. Elston, *J. Am. Chem. Soc.* **77**, 364 (1955).
239. T. Nishiwaki, *Bull. Chem. Soc. Japan* **43**, 937 (1970).
240. H. Wynberg and A. Logothetis, *J. Am. Chem. Soc.* **78**, 1958 (1956).
241. Y. K. Yur'ev, and G. B. Elyakov, *Doklady Akad. Nauk. USSR* **86**, 337 (1952)
242. Y. K. Yur'ev, G. B. Elyakov, N. S. Zefirov and A. N. Vysokosov, *J. Gen. Chem. USSR (Engl. Transl.)* **26**, 3717 (1956).
243. R. Labeaudinière, G. Hilboll, A. Leon-Lomeli, H.-H. Lautenschläger, M. Parnham, P. Kuhl and N. Dereu, *J. Med. Chem.* **35**, 3156 (1992).
244. M. H. Karger and Y. Mazur, *J. Org. Chem.* **36**, 540 (1971).
245. S. I. Pennanen, *Heterocycles* **4**, 1021 (1976).
246. F. Effenberger, G. König and H. Klenk, *Chem. Ber.* **114**, 926 (1981).
247. B. P. Andreini, A. Carpita, R. Rossi and B. Scamuzzi, *Tetrahedron* **45**, 5621 (1989).
248. V. I. Dulenko and N. N. Alekseev, *Chem. Heterocycl. Compds. (Engl. Transl.)* 551 (1975).
249. S. Gronowitz, *Arkiv Kemi.* **12**, 533 (1958).
250. J. Hoch, *Compt. rend.* **234**, 1981 (1952).
251. S. Gronowitz and T. Frejd, *Acta Chem. Scand. B* **30**, 287 (1976).
252. G. Consiglio, S. Gronowitz, A.-B. Hörnfeldt, B. Maltesson, R. Noto and D. Spinelli, *Chem. Scripta* **11**, 175 (1977).
253. T. Frejd and O. Karlsson, *Tetrahedron* **35**, 2155 (1979).
254. P. P. Ehrlich, J. W. Ralston and M. R. Michaelides, *J. Org. Chem.* **62**, 2782 (1997).
255. S. J. Pythian, R. J. K. Taylor and J. R. Bantick, *J. Chem. Soc., Perkin Trans.* 1 194 (1990).
256. R. Fikentscher, R. Brückmann and R. Betz, *Liebigs Ann. Chem.* 113 (1990).
257. M. Haddach and J. R. McCarthy, *Tetrahedron Lett.* **40**, 3109 (1999).
258. T. Ishiyama, T. Oh-e, N. Miaura and A. Suzuki, *Tetrahedron Lett.* **33**, 4465 (1992).
259. A. F. Littke, C. Dai and G. C. Fu, *J. Am. Chem. Soc.* **122**, 4020 (2000).
260. S. Gronowitz and B. Eriksson, *Arkiv Kemi.* **21**, 335 (1963).
261. M. Janda and L. Paviensky, *Coll. Czech. Chem. Commun.* **32**, 2675 (1967).
262. P. Blanchard, H. Brisset, B. Illien, A. Riou and J. Roncali, *J. Org. Chem.* **62**, 2401 (1997).
263. T. Kondo, M. Akazome, Y. Tsuji and Y. Watanaba, *J. Org. Chem.* **55**, 1286 (1990).
264. E. N. Deryagina, A. S. Nakhmanovich, V N. Elokhina and O. G. Yarosh, *Chem. Heterocycl. Compd. (Engl. Transl.)* **6**, 891 (1970).

265. Y. Arai, A. Suzuki, T. Masuda, Y. Masaki and M. Shiro, *Chem. Pharm. Bull.* **44**, 1765 (1996).
266. A. M. Nadim, Y. N. Romashin and O. G. Kulinkovich, *J. Het. Chem. USSR (Engl. Transl.)* **27**, 918 (1991).
267. M. D'Auria, T. Ferri, G. Mauriello, A. Pesce and R. Racioppi, *Tetrahedron* **55**, 2013 (1999).
268. M. D'Auria and D. Tofani, *Tetrahedron* **48**, 9315 (1992).
269. M. D'Auria, R. Ferri, G. Poggiu, G. Mauriello and R. Racioppi, *Eur. J. Org. Chem.* 1653 (2000).
270. H. Goda, M. Sato, H. Ihara and C. Hirayama, *Synthesis* 849 (1992).
271. Y. L. Gol'dfarb, I. B. Karmanova, Y. B. Vol'kenshtein and L. I. Belen'kii, *Chem. Heterocyclic Compds. (Engl. Transl.)* **11**, 1196 (1978).
272. L. Quattara, M. Debaert and R. Cavier, *Il Farmaco Ed. Sci.* **43**, 389 (1988)
273. V. Y. Sosnovskikh and I. S. Ovsyannikov, *J. Org. Chem. USSR (Engl. Transl.)* **26**, 1801 (1990).
274. P. Murch, B. L. Williamson and P. J. Stang, *Synthesis* 1255 (1994).
275. B. Unterhalt and P. Gores, *Arch. Pharm. (Weinheim)* **322**, 839 (1989).
276. L. F. Fieser and R.G. Kennely, *J. Am. Chem. Soc.* **57**, 1611 (1935).
277. N. Garcia-Dominguez, E. Ravina, L. Santana, C. Terán and G. Garcia-Mera, *Arch. Pharm. (Weinheim)* **321**, 735 (1988).
278. A. De, S. Bhattacharya (née Mazumder), S. S. Jash, S. Mukherjee, U. Saha and P. K. Sen, *J. Heterocycl. Chem.* **29**, 1213 (1992).
279. P. K. Sen, B. Kundu, and T. K. Das, *J. Indian Chem. Soc.* **LV**, 847 (1978).
280. P. Stanetty, *J. Chem. Res. (S)* 139 (1981).
281. P. K. Sen, U. K. Saha and T. Das (neé Deb), *Indian J. Chem.* **38B**, 648 (1999).
282. M. M. Goodman, G. Kirsch and F. F. Knapp, Jr, *J. Heterocycl. Chem.* **21**, 1579 (1984).
283. A. M. El-Khawaga, M. F. El-Zohry and M. T. Ismail, *Phosphorus & Sulfur* **33**, 25 (1987).
284. H. Ishibashi, I. Takamuro, Y. Mizukami, M. Irie and M. Ikeda, *Synth. Commun.* **19**, 443 (1989).
285. M. V. Lebedev, V. G. Nenajdenko and E. S. Balenkova, *Tetrahedron* **54**, 5599 (1998).
286. P. Alcudia, I. Fernandez, J. M. Llera and F. Zorrilla, *Anal. Quimica* **84**, 333 (1988).
287. A. M. Farag, H. M. Hassaneen, I. M. Abbas, A. S. Shawaly and M. S. Algharib, *Phosphorus and Sulfur* **40**, 243 (1988).
288. J. Dubac, A. Gaset and M. Maraval, *Synth. Commun.* **21**, 11 (1991).
289. S. Kajigaeshi, T. Kakinami, M. Moriwaki, S. Fujisaki, K. Maeno and T. Okamoto, *Synthesis* 545 (1988).
290. G. B. Mullen, J. T. Mitchell, S. D. Allen and V. S. Georgiev, *J. Pharm. Sci.* **77**, 1050 (1988).
291. D. E. Tupper, T. M. Hotten and W. G. Prowse, *J. Heterocycl. Chem.* **33**, 1123 (1996).
292. K. Sato, T. Yamagishi and S. Arai, *J. Heterocycl. Chem.* **37**, 1009 (2000).
293. Z.-N. Huang, B.-A. Xu, S. Jin and M.-G. Fan, *Synthesis* 1092 (1998).
294. A.-L. Fan, S. Cao, J.-H. Xu and Z. Zhang, *J. Heterocycl. Chem.* **35**, 477 (1998).
295. A. M. Farag and M. S. Algharib, *Org. Prep. and Proc. Int.* **20**, 521 (1988).
296. R. L. Danheiser, R. F. Miller, R. G. Brisbois and S. Z. Park, *J. Org. Chem.* **55**, 1959 (1990).
297. Y. Goldberg, E. Abele and M. Shymanska, *Synth. Commun.* **20**, 2741 (1990).
298. E. M. Abele, Y. S. Gol'dberg, Y. Y. Popelis and M. V. Shimanskaya, *J. Org. Chem. USSR (Engl. Transl.)* **26**, 1545 (1990).
299. J. T. Gupton, F. A. Hicks, D. R. Wilkinson, S. A. Petrich and J. A. Sikorski, *Heterocycles* **37**, 4487 (1994).
300. M. T. Baumgartner, M. H. Gallego and A. B. Pierini, *J. Org. Chem.* **63**, 6394 (1998).
301. R. M. Moriarty, W. R. Epa, R. Penmasta and A. K. Awasthi, *Tetrahedron Lett.* **30**, 667 (1989).
302. A. Charlton, A. E. Underhill, G. Williams, M. Kalaji, P. J. Murphy, K. M. Abdul Malik and M. B. Hursthouse, *J. Org. Chem.* **62**, 3098 (1997).
303. M. Sagi, M. Amano, S. Konno and H. Yamanaka, *Heterocycles* **29**, 2249 (1989).

304. T. Ishiyama, H. Kizaki, T. Hayashi, A. Suzuki and N. Miyaura, *J. Org. Chem.* **63**, 4726 (1998).
305. R. Grigg, S. Brown, V. Sridharan and M. D. Uttley, *Tetrahedron Lett.* **38**, 5031 (1997).
306. J. J. Masters, L. S. Hegedus and J. Tamariz, *J. Org. Chem.* **56**, 5666 (1991).
307. J. Nakayama, R. Yomoda and M. Hoshino, *Heterocycles* **26**, 2215 (1987).
308. A. Alberola, J. N. Andrés, A. Gonzales, R. Pedrosa and P. Pradanos, *Synth. Commun.* **20**, 2537 (1990).
309. A. Capperrucci, A. Degl'Innocenti, C. Faggi, G. Reginato, A. Ricci, P. Dembech and G. Seconi, *J. Org. Chem.* **54**, 2966 (1989).
310. E. Profft and G. Solf, *J. Prakt. Chem.* **24**, 38 (1964).
311. S. Gronowitz, *Arkiv Kemi.* **11**, 519 (1957).
312. V. Sprio and P. Madonia, *Gazz. Chim. Ital.* **87**, 454 (1957).
313. S. Gronowitz and T. Raznikiewicz, *Arkiv Kemi* **17**, 561 (1961).
314. G. R. Pabst, O. C. Pfüller and J. Sauer, *Tetrahedron* **55**, 5047 (1999).
315. R. A. Benkeser and H. Landesman, *J. Am. Chem. Soc.* **71**, 2493 (1949).
316. O. C. Pfüller and J. Sauer, *Tetrahedron Lett.* **39**, 8821 (1998).
317. G. N. Gordeeva, S. M. Kalshnikov, Y. N. Popov, É.A. Kruglov and U. B. Imashev, *Chem. Heterocycl. Compd. (Engl. Transl.)* 638 (1987).
318. E. Campaigne and R. C. Bourgeois, *J. Am. Chem. Soc.* **75**, 2702 (1953).
319. J. H. Biel, E. P. Sprengler, H. A. Leiser, J. Horner, A. Drucker and H. L. Friedman, *J. Am. Chem. Soc.* **77**, 2250 (1955).
320. R. D. Schuetz and G. P. Nilles, *J. Org. Chem.* **36**, 2486 (1971).
321. C. K. Lee, M. S. Kim, J. S. Gong and I.-N. Han Lee, *J. Heterocycl. Chem.* **29**, 149 (1992).
322. A. Ohta, T. Kobayashi and H. Kato, *J. Chem. Soc., Perkin Trans. 1* 905 (1993).
323. J. K. Sneed and R. Levine, *J. Am. Chem. Soc.* **72**, 5219 (1950).
324. T. S. Gardner, E. Wenis and J. Lee, *J. Org. Chem.* **26**, 1514 (1961).
325. F. Clesse and H. Quiniou, *Bull. Soc. Chim. Fr.* 1940 (1969).
326. L. B. Barkley and R. Levine, *J. Am. Chem. Soc.* **73**, 4625 (1951).
327. R. A. Moore and R. Levine, *J. Org. Chem.* **29**, 1434 (1964).
328. R. A. Moore and R. Levine, *J. Org. Chem.* **29**, 1883 (1964).
329. L. B. Barkley and R. Levine, *J. Am. Chem. Soc.* **75**, 2059 (1953).
330. T. Nishiwaki, *Tetrahedron* **23**, 2979 (1967).
331. K. Okada, Y.-F. Wang, T.-M. Chen, M. Kitamura, T. Nakaya and H. Inoue, *J. Mater. Chem.* **9**, 3023 (1999).
332. K. Esses-Reiter and J. Reiter, *J. Heterocycl. Chem.* **37**, 927 (2000).
333. A. Rámila, J. Plumet and E. Camacho, *Heterocycles* **45**, 2425 (1997).
334. S. Matsubara, Y. Yamamoto and K. Utimoto, *Synlett* 1471 (1999).
335. K. Ogura, H. Yanai, M. Miokawa, M. Akazone, *Tetrahedron Lett.* **40**, 8887 (1999).
336. L. Fajarí, E. Brillas, C. Alemán and L. Juliá, *J. Org. Chem.* **63**, 5324 (1998).
337. D. M. Perrine, D. M. Bush, E. P. Kornak, M. Zhang, Y. H. Cho and J. Kagan, *J. Org. Chem.* **56**, 5095 (1991).
338. J. Doussot, A. Guy and J. Roncali, *Tetrahedron Lett.* **40**, 8811 (1999).
339. P. J. Skabara, I. M. Serebryakov, D. M. Roberts, I. F. Perepichka, S. J. Coles and M. B. Hursthouse, *J. Org.Chem.* **64**, 6418 (1999).
340. E. Jones and I. M . Moodie, *J. Chem. Soc. (C)* 1195 (1968).
341. I. I. Lapkin, Y. P. Dormidontov and T. A. Bidman, *Chem. Heterocycl. Compd. (Engl. Transl.)* **4**, 579 (1968).
342. I. I. Lapkin and Y. P. Dormidontov, *Chem. Heterocycl. Compd.* (*Engl. Transl.*) **6**, 832 (1970).
343. J. F. Bagli and E. Fernandi, *Can. J. Chem.* **53**, 2598 (1975).
344. R. G. Micetich, R. Raap and L. Mol, *Org. Pep. Proced. Int.* **3**, 167 (1971).
345. W. C. Randall, K. B. Streeter, E.L. Cresson, H. Schwamm, S. R. Michelson, P. S. Anderson, E. J. Cragoe, H. W. R. Williams, E. Eichler and C. S. Rooney, *J. Med. Chem.* **22**, 608 (1979).

346. B. P. Fabrichnyi, I. F. Shalavina and Y. L. Gol'dfarb, *J. Org. Chem. USSR* (*Engl. Transl.*) **15**, 1371 (1979).
347. E. Campaigne, G. Skowronski and R. B. Rogers, *Syn. Comm.* **3**, 325 (1973).
348. G. Combes, *Bull. Soc. Chim. Fr.* 701 (1952).
349. C. Andrieu, R. Pinel and Y. Mollier, *Bull. Soc. Chim. Fr.* 1314 (1971).
350. M. Hatanaka and T. Ishimaru, *J. Med. Chem.* **16**, 978 (1973).
351. A. Burchardt and D. Geffken, *Arch. Pharm.* (*Weinheim*) **321**, 311 (1988).
352. J. Teste and N. Lozac'h, *Bull. Soc. Chim. Fr.* 437 (1955).
353. S. S. Samanta, S. C. Gosh and A. De, *J. Chem. Soc., Perkin Trans. 1* 2683 (1997).
354. H. J. Kooreman and H. Wynberg, *Rec. trav. Chim. Pays-Bas* **86**, 37 (1967).
355. T. Kralt, *Rec. Trav. Chim. Pays-Bas* **86**, 971 (1967).
356. P. J. Kocienski, J. M. Ansell and B. E. Norcross, *J. Org. Chem.* **41**, 3650 (1976).
357. N. S. Vul'fson and V. E. Kolchin, *J. Gen. Chem. USSR,* (*Engl. Transl.*) **30**, 3392 (1960).
358. C. Trebaul and J. Teste, *Bull. Soc. Chim. Fr.* 2456 (1969).
359. R. V. Hoffman and H. -O. Kim, *J. Org. Chem.* **60**, 5107 (1995).
360. B. M. Perfetti and R. Levine, *J. Am. Chem. Soc.* **75**, 626 (1953).
361. A. Furuichi, H. Akita, H. Koshiji, K. Horikoshi and T. Oishi, *Chem. Pharm. Bull.* **32**, 1619 (1984).
362. L. L. Kruse, D. L. Ladd, P. B. Harrsch, F. L. McCabe, S.-M. Mong, L. Faucette and R. Johnson, *J. Med. Chem.* **32**, 409 (1989).
363. D. Evans, M. E. Cracknell, J. C. Saunders, C. E. Smith, W. R. N. Williamson, W. Dawson and W. J. F. Sweatman, *J. Med. Chem.* **30**, 1321 (1987).
364. M. Varache-Béranger, A. Nuhrich and G. Devaux, *Eur. J. Med. Chem.* **23**, 501 (1988).
365. R. Noto, M. Gruttadauria, P. Lo Meo and D. Spinelli, *Collect. Czech. Chem. Commun.* **64**, 1893 (1999).
366. P. Montanari, P. Da Re and P. Valenti, *J. Heterocycl. Chem.* **25**, 1277 (1988).
367. A. Dondoni, G. Fantin, M. Fogagnolo, A. Medici and P. Pedrini, *J. Org. Chem.* **53**, 1748 (1988).
368. E. Fischer, J. Larsen, J. B. Christensen, M. Fourmigué, H. G. Madsen and N. Harrit, *J. Org. Chem.* **61**, 6997 (1996).
369. U. Michael and A.-B. Hörnfeldt, *Tetrahedron Lett.* 5219 (1970).
370. P. Lucas, N. El Mehdi, H. A. Ho, D. Bélanger and L. Breau, *Synthesis* 1253 (2000).
371. X. Wu and R. D. Rieke, *J. Org. Chem.* **60**, 6658 (1995).
372. R. D. Rieke, S.-H. Kim and X. Wu, *J. Org. Chem.* **62**, 6921 (1997).
373. S.-H. Kim and R. D. Rieke, *Tetrahedron Lett.* **38**, 993 (1997).
374. J. R. Pratt, F. H. Pinkerton and S. F. Thames, *J. Organomet. Chem.* **38**, 29 (1972).
375. N. A. Bumagin, P. G. More and I. P. Beletskaya, *J. Organomet. Chem.* **365**, 379 (1989).
376. S.-K. Kang, P.-S. Ho, S.-K. Yoon, J.-C. Lee, K.-J. Lee, *Synthesis* 823 (1998).
377. F. H. Pinkerton and S. F. Thames, *J. Heterocycl. Chem.* **9**, 725 (1972).
378. D. W. H. MacDowell and A. T. Jeffries, *J. Org. Chem.* **36**, 1053 (1971).
379. G. D. Hartman and W. Halczenko, *J. Heterocycl. Chem.* **26**, 1793 (1989).
380. G. D. Hartman, W. Halczenko, R. L. Smith, M. F. Dugrue, P. J. Malergo, S. R. Michelson, W. C. Randall, H. Achwam and J. M. Sondey, *J. Med. Chem.* **35**, 3822 (1992).
381. S. L. Buchwald and S. M. King, *J. Am. Chem. Soc.* **113**, 258 (1991).
382. A. Wiersema and S. Gronowitz, *Acta Chem. Scand.* **24**, 2593 (1970).
383. R. Beyer, M. Kalaji, G. Kingscote-Burton, P. J. Murphy, V. M. S. C. Pereira, D. M. Taylor and G. O. Williams, *Synth. Met.* **92**, 25 (1998).
384. R. I. Fryer, P. Zhang and R. Rios, *Synth. Commun.* **23**, 985 (1993).
385. S. K. Richardson and R. J. Ife, *J. Chem. Soc., Perkin Trans. 1*, 1172 (1989).
386. M. Temciuc, A.-B. Hörnfeldt and S. Gronowitz, *Heterocycl. Commun.* **1**, 411 (1995).
387. A. Acosta, P. de la Cruz, P. De Miguel, E. Diez-Barra, A. de la Hoz, F. Langa, A. Loupy, M. Majdoub, N. Martin, C. Sanchez and C. Seoane, *Tetrahedron Lett.* **36**, 2168 (1995).

388. A. K. Mohanakrishnan, M. V. Lakshmikantham, C. McDougal, M. P. Cava, J. W. Baldwin and R. M. Metzger, *J. Org. Chem.* **63**, 3105 (1998).
389. W. Volz and J. Voss, *J. Prakt. Chem.* **333**, 889 (1991).
390. K. Oda and M. Machida, *Chem. Pharm. Bull.* **41**, 1299 (1993).
391. D. W. H. MacDowell, T.B. Patrick, B. K. Frame and D. L. Ellison, *J. Org. Chem.* **32**, 1226 (1967).
392. D. B. Hauze, M. Jouillié, R. Ramotowski and A. Cantu, *Tetrahedron* **53**, 4239 (1997).
393. P. Blanchard, A. Riou and J. Roncali, *J. Org. Chem.* **63**, 7107 (1998).
394. P. Cagniant, A. Reisse and D. Cagniant, *Bull. Soc. Chim. Fr.* 991 (1969).
395. J. Skramstad, *Chem. Scripta* **7**, 42 (1975).
396. D. B. Hauze and M. M. Joullié, *Tetrahedron* **53**, 4239 (1997).
397. P. Dallemagne, S. Rault and M. Robba, *Bull. Soc. Chim. Fr.* **128**, 260 (1991).
398. A. Jilale, P. Netchitaïlo, B. Decroix and D. Vegh, *J. Heterocycl. Chem.* **30**, 831 (1993).
399. P. Dallemagne, S. Rault, M. Cugnon de Sévricourt, K. M. Hassan and M. Robba, *Tetrahedron Lett.* **27**, 2607 (1986).
400. P. Dallemagne, S. Rault, M. Cugnon de Sévricourt and M. Robba, *Heterocycles* **27**, 1637 (1988).
401. P. Dallemagne, J. C. Pilo, S. Rault, M. Robba, M. Saux and A. Carpy, *Heterocycles* **34**, 1317 (1992).
402. P. Dallemagne, M.Boulouard, S. Rault and M. Robba, *J. Heterocycl. Chem.* **30**, 799 (1993).
403. P. Dallemagne, S. Rault, M. Gordaliza and M. Robba, *Heterocycles* **26**, 3233 (1987).
404. P. Dallemagne, A. Alsaïdi, M. Boulouard, S. Rault and M. Robba, *Heterocycles* **36**, 287 (1993).
405. L. P. Khanh, P. Dallemagne, A. Alsaïdi and S. Rault, *Heterocylces* **45**, 527 (1997).
406. J.-P. Maffrand, R. Boigegren, J. Courregelongue, G. Ferrand, and D. Frehel, *J. Heterocycl. Chem.* **18**, 727 (1981).
407. J. A. Ewen, M. J. Elder, R. L. Jones, A. L. Rheingold, L. M. Liable-Sands and R. D. Sommer, *J. Am. Chem. Soc.* **123**, 4763 (2001).
408. B. Yom-Tov and S. Gronowitz, *J. Heterocycl. Chem.* **15**, 285 (1978).
409. Y. Poirier and N. Lozac'h, *Bull. Soc. Chim. Fr.* 1058 (1966).
410. E. Ravina, J. Fueyo, C. F. Masaguer, J. Negreira, J. Cid, I. Loza, A. Honrubia, H. Tristán, T. G-Ferreiro, J. A. Fontenla, E. Rosa, J. M. Calleja and M. L. De Ceballos, *Chem. Pharm.Bull.* **44**, 534 (1996).
411. P. Cagniant, P. Merle and D. Cagniant, *Bull. Soc. Chim. Fr.* 322 (1970).
412. D. W. H. MacDowell and T. D. Greenwood, *J. Heterocycl. Chem.* **2**, 44 (1965).
413. C. M. Asprou, J. S. A. Brunskill, H. Jeffrey and A. De, *J. Heterocycl. Chem.* **17**, 87 (1980).
414. D. T. Drewry and R. M. Scrowston, *J. Chem. Soc. C* 2750 (1969).
415. S. Nishimura, M. Nakamura, M. Suzuki and E. Imoto, *Nippon Kagaku Zasshi* **83**, 343 (1962).
416. P. Cagniant, *Compt. Rend.* **232**, 734 (1951).
417. N. P. Buu-Hoi and N. H. Kôi, *Recl. Trav. Chim. Pays Bas* **69**, 1053 (1950).
418. P. Cagniant and D. Cagniant, *Bull. Soc. Chim. Fr.* 713 (1953).
419. P. Cagniant, D. Cagniant and A. Pancrazzi, *Bull. Soc. Chim. Fr.* 1534 (1964).
420. L. Alam and G. Thyagarajan, *Tetrahedron* **29**, 1829 (1973).
421. P. Cagniant and D. Cagniant, *Bull. Soc. Chim. Fr.* 2597 (1967).
422. Y. Amemiya, A. Terada, K. Wachi, H. Miyazawa, N. Hatakeyama, K. Matsuda and T. Oshima, *J. Med. Chem.* **32**, 1265 (1989).
423. S. Datta and A. De, *J. Chem. Soc., Perkin Trans. 1* 603 (1989).
424. M. M. Curzu, G. A. Pinna, E. Pini and L. Toma, *J. Heterocycl. Chem.* **36**, 1253 (1999).
425. R. P. Napier and C.-C. Chu, *INT: J. Sulfur Chem. Part A* **1**, 62 (1971).
426. R. P. Napier, H. A. Kaufman, P. R. Driscoll, L. A. Glick, C.-C. Chu and H. M. Foster, *J. Heterocycl. Chem.* **7**, 393 (1970).

427. R. Martinez, E. Duran, L. C. Cortés and J. G. Avila, Z, *J. Heterocycl. Chem.* **36**, 687 (1999).
428. P. Cagniant and D. Cagniant, *Bull. Soc. Chim. Fr.* 680 (1955).
429. B. P. Fabrichnyi, I. F. Shalavina, and Y. L. Gol'dfarb, *J. Gen. Chem. USSR* (*Engl. Transl.*) **31**, 1152 (1961).
430. G. Jones and M. J. Robinson, *J. Chem. Soc., Perkin Trans. 1* 505 (1977).
431. B. P. Fabrichnyi, I. F. Shalavina, S. M. Kostrova and Y. L. Gol'dfarb, *J. Org. Chem. USSR (Engl. Transl.)* **8**, 190 (1972).
432. S. S. Bhargava and G. S. Saharia, *Indian J. Chem.* **9**, 809 (1971).
433. B. P. Fabrichnyi, I. F. Shalavina, Y. L. Gol'dfarb and S. M. Kostrova, *J. Org. Chem. USSR (Engl. Transl.)* **10**, 1966 (1974).
434. G. Jones, R. K. Jones and M. J. Robinson, *J. Chem. Soc., Perkin 1* 968 (1973).
435. P. Cagniant and D. Cagniant *Bull. Soc. Chim. Fr.* 1152 (1956).
436. B. P. Fabrichnyi, I. F. Shalavina, S. M. Kostrova and Y. L. Gol'dfarb, *J. Org. Chem. USSR* (*Engl. Transl.*) **6**, 1093 (1970).
437. P. Cagniant and G. Merle, *Compt. Rend. C* **267**, 1838 (1968).
438. P. Cagniant, *Compt. Rend. C* **271**, 375 (1970).
439. Y. L. Gol'dfarb, S. Z. Taits and L. I. Belen'kii, *Bull. Acad. Sci. USSR, Div. Chem. Sci.* (*Engl. Transl.*) 2189 (1970).
440. Y. L. Gol'dfarb, S. Z. Taits and L. I. Belen'kii, *J. Gen. Chem. USSR* (*Engl. Transl.*) **29**, 3526 (1959).
441. Y. L. Gol'dfarb, S. Z. Taits and L. I. Belen'kii, *Bull. Acad. Sci. USSR, Div. Chem. Sci.* (*Engl. Transl.*) 1320 (1963).
442. S. Z. Taits, L. I. Belen'kii and Y. L. Gol'dfarb, *Bull. Acad, Sci. USSR, Div. Chem. Sci.* (*Engl. Transl.*) 1328 (1963).
443. G. Catoni, C. Galli and l. Mandolini, *J. Org. Chem.* **45**, 1906 (1980).
444. O. A. Kalinovskii, S. Z. Taits and Y. l. Gol'dfarb, *Bull. Acad. Sci. USSR, Div. Chem. Sci.* (*Engl. Transl.*) 2189 (1970).
445. S. Z. Taits, O. A. Kalinovskii, V. S. Bogdanov and Y. L. Gol'dfarb, *Chem. Heterocycl. Compds.* (*Engl. Transl.*) **6**, 1367 (1970).
446. S. Z. Taits, O. A. Kalinovskii, R. V. Lopatin and Y. L. Gol'dfarb, *Chem. Heterocycl. Compds.* (*Engl. Transl.*) **9**, 576 (1973).
447. S. Z. Taits, O. A. Kalinovskii, V. S. Bogdanov and Y. L. Gol'dfarb, *Chem. Heterocycl. Compd. (Engl. Transl.)*, **8**, 149 (1972).
448. Y. L. Gol'dfarb, S. Z. Taits and L. I. Belen'kii, *Tetrahedron* **19**, 851 (1963).
449. Y. L. Gol'dfarb, S. Z. Taits and V. N. Bulgakova, *Bull. Acad. Sci. USSR, Div. Chem. Sci.* (*Engl. Transl.*) 1181 (1963).
450. S. Z. Taits, E. A. Krasnyanskaya, Y. L. Gol'dfarb, N. F. Kononov, A. G. Pogorelov and R. F. Merzhanova, *Bull. Acad. Sci. USSR, Div. Chem. Sci.* (*Engl. Transl.*) **24**, 2422 (1975).
451. S. Z. Taits, E. A. Krasnyanskaya and Y. L. Gol'dfarb, *Bull. Acad. Sci. USSR, Div. Chem. Sci.* (*Engl. Transl.*) 735 (1968).
452. S. Z. Taits and Y. L. Gol'dfarb, *Bull. Acad. Sci. USSR, Div. Chem. Sci.* (*Engl. Transl.*) 1574 (1960).
453. S. Z. Taits, E. A. Krasnyanskaya and Y. L. Gol'dfarb, *Bull. Acad. Sci. USSR, Div. Chem. Sci.* (*Engl. Transl.*) 2098 (1970).
454. S. Z. Taits, E. A. Krasnyanskaya, A. L. Klyachko-Gurvich and Y. L. Gol'dfarb, *Bull. Acad. Sci. USSR, Div. Chem. Sci. (Engl. Transl.)* **22**, 1751 (1973).
455. H. Ohmura and S. Motoki, *Bull. Chem. Soc. Japan* **57**, 1131 (1984).
456. S. F. Thames and J. E. McCleskey, *J. Heterocycl. Chem.* **3**, 104 (1966).
457. S. F. Thames and J. E. McCleskey, *J. Heterocycl. Chem.* **4**, 371 (1967).
458. É. Abele, Y. Popelis, É. Lukevics, M. Shimanska and Y. Gol'dberg, *Chem. Heterocycl. Compds. (Engl. Transl.)* **30**, 14 (1994).

459. S. E. Korostova, L. N. Sobenina, A. I. Mikhaleva, R. N. Nesterenko, S. G. Shevchenko, V.B. Modonov and R. I. Polovnikova, *J. Org. Chem. USSR* (*Engl. Transl.*) **24**, 2291 (1987).
460. S. E. Korostova, R. N. Nesterenko, A. I. Mikhaleva. R. I. Polovnikova, and N. I. Golovanova, *Chem. Heterocycl. Compd.* (*Eng. Transl.*) **25**, 749 (1989).
461. B. Macchia, A. Balsamo, A. Lapucci, F. Maccia, A. Martinelli, S. Nencetti, E. Orlandini, M. Baldacci, G. Mengozzi, G. Soldani and P. Domiano, *J. Med. Chem.* **33**, 1423 (1990).
462. Z. Györgydéak, W. Holzer and K. Mereiter, *Monatsh. Chem.* **130**, 899 (1999).
463. J. Barbera, C. Cativiela, E. Melendez, M, Ruiz and J. L. Serrano, *Anal. Chim.* **80**, 54 (1984).
464. R. Fusco and F. Sannicolò, *J. Org. Chem.* **47**, 1691 (1982).

– 4F –

Thiophenecarboxylic Acids and Their Derivatives

4F.1 PARENT ACIDS AND ALKYL AND ARYLTHIOPHENECARBOXYLIC ACIDS

4F.1.1 Through reactions of thienylmagnesium halides, thienyl sodium or thienyllithium derivatives with carbon dioxide

4F.1.1.1 From Grignard reagents and carbon dioxide

The reaction of thiophenemagnesium bromides and iodides with carbon dioxide was first used for the preparation of thiophenecarboxylic acids. Thus already in 1915, 2-thiophene carboxylic acid was prepared from 2-iodothiophene, magnesium and carbon dioxide [1]. However, in many cases the yields are unsatisfactory, although good yields of 3-phenyl-2-thiophenecarboxylic acid and 5-phenyl-2-thiophenecarboxylic acids have been obtained from 2-bromo-3-phenylthiophene [2] and 2-bromo-5-phenylthiophene [3].

Especially with 3-bromothiophenes, the entrainment procedure has to be used (the simultaneous reaction of ethyl bromide or ethylene dibromide with magnesium) in order to obtain higher yields [4]. For these reasons the thiophenemagnesium halides have been almost completely superseded by thienyllithium derivatives in the synthesis of thiophenecarboxylic acids. This is also true with regard to the use of thienylsodium derivatives. Thus 2-thiophenecarboxylic acid has been prepared by the reaction of thiophene with sodium amalgam-chloroethane or sodium amalgam-bromobenzene and carbon dioxide.

4F.1.1.2 Metalation of thiophenes followed by reaction with carbon dioxide

The discovery of the facile metalation of thiophene by butyllithium in ether, yielding 2-thienyllithium, has revolutionized the preparation of various types of 2-substituted thiophenes by reaction with electrophiles, among them the thiophenecarboxylic acids [5]. 2-Monosubstituted thiophenes are lithiated exclusively in the 5-position, except for the few cases when the substituents have strong *ortho*-directing properties or, of course, when the substituents react with butyllithium. Good yields of 5-substituted 2-thiophenecarboxylic acids are obtained when the thienyllithium derivatives are poured onto solid carbon dioxide covered with ether. In this way 5-alkyl-, 5-alkenyl-, 5-alkynyl-, 5-aryl- and 5-heteroaryl-2-thiophenecarboxylic acids are prepared. Also 2-diarylphosphinyl-, 2-*O*-alkyl-, 2-*O*-aryl-, 2-*S*-alkyl-, 2-*S*-aryl-, 2-sulfono-, 2-dialkylsulfonamido- and 2-halo-5-thiophenecarboxylic acids are obtained by this approach, which will be treated in the appropriate chapter. A more recent example is the preparation of 5-*tert*-butyl-2-thiophenecarboxylic acid [6].

5-tert-Butyl-2-thiophenecarboxylic acid [6]

Butyllithium (0.011 mol) is added dropwise to a solution of 2-*tert*-butylthiophene (1.40 g, 0.01 mol) in anhydrous diethyl ether under argon. The mixture is stirred at room temperature for 1 h and then under argon transfered to solid carbon dioxide (0.5 mol) under anhydrous diethyl ether at −70 °C. The temperature is allowed to rise to room temperature, after which 1 *M* sodium hydroxide solution is added until a pH of about 10. Ice (40 g) and diisopropyl ether (40 ml) are added to the reaction mixture. The phases are separated, the aqueous phase acidified with 2 *M* hydrochloric acid and the product extracted with diisopropyl ether giving 1.38 g (75%) of the title compound.

The synthetic usefulness of metalation of 3-monosubstituted thiophenes depends on the nature of the substituents and the metalating agent. Thus 3-methylthiophene with butyllithium gives 61–68% of 4-methyl-2-thiophenecarboxylic acid and 19% of 3-methyl-2-thiophenecarboxylic acid [7]. A detailed investigation on the influence of the metalating agents showed that butyllithium-*N*,*N*,*N′*,*N′*-tetramethylethylenediamine gave the highest proportion of the 4-methyl-2-thienyllithium derivative, with 3-*tert*-butyllithium only 4-*tert*-butyl-2-thienyllithium was formed [8]. Only 3-methoxymethyl-2-thiophenecarboxylic acid is formed upon metalation of 3-methoxymethylthiophene followed by carbonation [9].

3-Methoxymethyl-2-thiophenecarboxylic acid [9]

To a suspension of freshly prepared ethereal solution of butyllithium (0.4 mol) in a 300 ml pressure bottle, 3-methoxymethylthiophene (38.4 g, 0.3 mol) is cautiously added in a stream of nitrogen. The flask is then filled up with anhydrous diethyl ether, stoppered and kept at 25 °C for 10 h. The mixture is slowly poured into a slurry of dry ice in anhydrous diethyl ether under a blanket of nitrogen. The reaction mixture is left at 25 °C for 3 h, after which water (375 ml) is added. The phases are separated, the ether phase washed with water (2 × 50 ml), the combined aqueous phases are washed with diethyl ether and then heated on a steam bath to expel dissolved diethyl ether. After cooling to 25 °C the solution is acidified with concentrated hydrochloric acid to pH < 2. After 3 h at 5 °C the precipitate formed is collected, washed with water, dried and recrystallized from water with charcoal giving 46 g (89%) of the title compound mp 126–127 °C.

3-Phenyl- and a variety of other 3-arylthiophenes give approximately equal amounts of 3- and 4-aryl-2-thiophenecarboxylic acids [10,11]. Other functional groups direct the metalation to the 2-position, such as methoxy- [12], *tert*-butoxy- [13], fluoro-[14], diphenylphosphinoxide [15] and protected carbonyl groups [16,17].These acids are discussed in the appropriate chapter. Metalation of 3- thiophenecarboxylic acid with two equivalents of lithium diisopropyl amide followed by methyl iodide is used for the preparation of 2-methyl-3-thiophenecarboxylic acid [18].

2,3-, 2,4- and 3,4-Disubstituted thiophenes are metalated in the free α-pos-ion. This can be used in the removal (by way of the acid) of a 2,4-dialkyl-thiophene [19] and a 2,3-dialkylthiophene [20] from 2,5-dialkylthiophene.

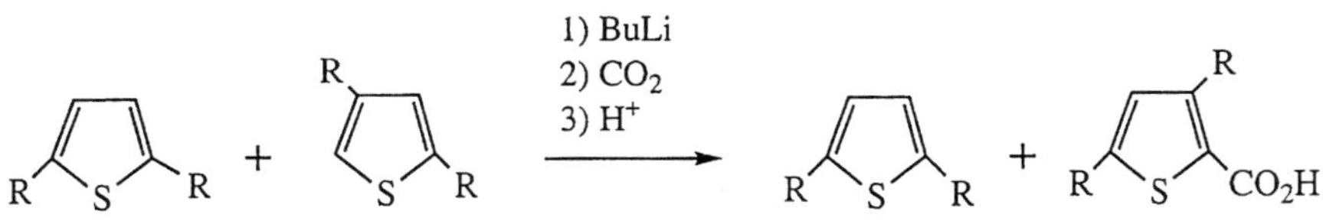

2,4-Di-tert-butyl-5-thiophenecarboxylic acid and 2,5-di-tert-butylthiophene [19]

To a mixture consisting of 75% 2,5-di-*tert*-butylthiophene and 25% 2,4-di-*tert*-butylthiophene (627 g, 3.20 mol) in anhydrous diethyl ether under nitrogen and stirring, 0.9 *M* butyllithium (1800 ml) is added. The mixture is refluxed for 4 h, cooled and poured onto solid carbon dioxide. The temperature is allowed to rise to −10 °C, after which the mixture is hydrolyzed with 2 *M* hydrochloric acid in excess. The phases are separated and the aqueous phase is extracted with diethyl ether. The combined organic phases are extracted several times with 2 *M* sodium hydroxide solution. Acidification of the combined alkaline phases gives 111 g of 2,4-di-*tert*-butyl-5-thiophenecarboxylic acid mp 197–198 °C after recrystallization from ethanol. The diethyl ether phase is dried and fractionated at reduced pressure, giving 455 g of 2,5-di-*tert*-butylthiophene bp 88–90 °C/10 mm Hg.

Metalation of 2,5-dimethylthiophene with ethyllithium-*N*,*N*,*N*′,*N*′-tetramethylethylenediamine gives both 3- and methyl lithiation, yielding 3-carboxy-5-methyl-2-thiopheneacetic acid [21]. Metalation of 2-methyl-3-thiophenecarboxylic acid with two equivalents of lithium diisopropylamide followed by methyl iodide gives 2-ethyl-3-thiophenecarboxylic acid. The use of lithium diisopropyl amide is essential, because with butyllithium as metalating agent equal amounts of this acid and 2,5-dimethyl-3-thiophenecarboxylic acid are obtained [18].

2-Ethyl-3-thiophenecarboxylic acid [18]

A solution of 2-methyl-3-thiophenecarboxylic acid (2.53 g, 17.8 mmol) in anhydrous tetrahydrofuran (5 ml) is added to a solution of lithium diisopropylamide in tetrahydrofuran (20 ml), prepared from diisopropylamine (3.60 g, 35.6 mmol) and butyllithium (36.0 mmol), under nitrogen at −30 °C. After stirring the mixture at −30 °C for 1 h a solution of methyl iodide (2.53 g, 17.8 mmol) in tetrahydrofuran (5 ml) is added and the stirring is continued at −30 °C for 2.5 h. The reaction is quenched with 10% hydrochloric acid and the product extracted with diethyl ether. The combined organic phases are washed with a small amount of sodium sulfite solution, dried over magnesium sulfate and evaporated. The residue is subjected to chromatography on silica gel using benzene/diethyl ether (5:1) as eluent, giving 2.24 g (80%) of the title compound mp 58–60 °C.

The reaction of the metalated product with butanal is used for the preparation of 2-(2-hydroxypentyl)-3-thiophenecarboxylic acid [18].

4F.1.1.3 Halogen–metal exchange of halothiophenes followed by reaction with carbon dioxide

Halogen–metal exchange, often at −70 °C, is especially useful for the preparation of 3-thienyllithium derivatives, as first demonstrated in the synthesis of 3-thiophenecarboxylic acid [22], and for the selective preparation of 2-thienyllithium derivatives. The latter case is illustrated in the synthesis of (thienylethyl)thiophenecarboxylic acids used for the synthesis of dithienotropylium ions. Thus 1-(2-carboxy-3-thienyl)-2-(3′-thienyl)ethene is prepared from *cis*-enriched 1-(2-bromo-3-thienyl)-2-(3′-thienyl)ethene at −70 °C [23] and 1-(2-carboxy-3-thienyl)-2- (3′-thienyl)ethane from 1-(2-bromo-3-thienyl)-2-(3′-thienyl)ethane [24]. This approach was also used for the preparation of 1-(3-carboxy-2-thienyl)-2-(2′-thienyl)ethene and 1-(3-carboxy-2-thienyl)-2-(2′-thienyl)ethane [24], 1-(3-carboxy-2-thienyl)-2-(2′thienyl)ethene and -ethane [23] as well as 1-(4-carboxy-3-thienyl)-2-(3-thienyl)ethane and 1-(4-carboxy-3-thienyl)-2-(2′-thienyl)ethane [25].

1-(4-Carboxy-3-thienyl)-2-(2′-thienyl)ethane [25]

To a solution of 1-(4-bromo-3-thienyl)-2-(2′-thienyl)ethane (13.6 g, 0.05 mol) in anhydrous diethyl ether (100 ml) at −70 °C 1.42 *M* butyllithium in hexane (36 ml) is rapidly added dropwise. The stirring is continued at −70 °C for 10–15 min and then the mixture is quickly poured onto crushed solid carbon dioxide covered with anhydrous diethyl ether. The mixture is allowed to warm to 2–3 °C, after which water and diethyl ether is added and it is shaken in a separatory funnel until all solid material is dissolved. The phases are separated and the ethereal phase extracted with small portions of 1 *M* sodium hydroxide solution. The combined alkaline phases are acidified with cold 5 *M* hydrochloric acid. The precipitate formed is filtered off and washed with water giving 9.5 g (80%) of the title compound mp 164 °C after recrystallization from ethanol.

Halogen–metal exchange of 3-bromo-5-methyl-2-(1-propynyl)thiophene and 3-bromo-5-methyl-2-(phenylethynyl)thiophene with butyllithium followed by crushed carbon dioxide is used for the preparation of 5-methyl-2-(1-propynyl)-3-thiophenecarboxylic acid and 5-methyl-2-(phenylethynyl)-3-thiophenecarboxylic acid, respectively [26].

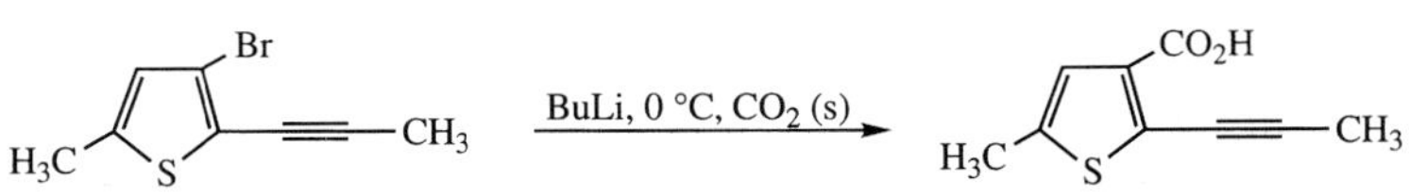

5-Methyl-2-(1-propynyl)-3-thiophenecarboxylic acid [26]

To a solution of 3-bromo-5-methyl-2-(1-propynyl)thiophene (1.0 g, 4.7 mmol) in anhydrous diethyl ether under nitrogen at 0 °C 1.52 *M* butyllithium in hexane (3.4 ml, 5.1 mmol) is added. The stirring is continued for 15 min, after which the mixture is poured onto crushed solid carbon dioxide. Workup and recrystallization from aqueous ethanol gives 0.50 g (60%) of the title compound mp 128–129 °C.

(3-Carboxy-2-thienyl)-2-(2-furyl)ethane is prepared in 77% yield from the corresponding 3-bromo derivative by halogen–metal interconversion with ethyllithium, followed by carbon dioxide [27]. Halogen–metal exchange of 3-bromothiophene followed by reaction with cyclohexanone followed by directed metalation and carbon dioxide is used for the preparation of 3-(1-hydroxy-1-cyclohexyl)-2-thiophenecarboxylic acid [28].

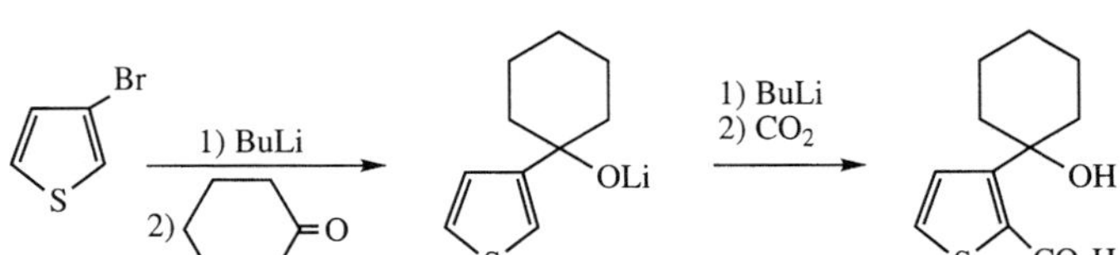

3-(1-Hydroxy-1-cyclohexyl)-2-thiophenecarboxylic acid [28]

To 0.7 *M* butyllithium in ether (45 ml, 32 mmol) under nitrogen at −80 °C 3-bromothiophene (5.2 g, 32 mmol) in anhydrous tetrahydrofuran (25 ml) is added and the stirring is continued for 10 min, after which cyclohexanone (3.3 g, 34 mmol) in anhydrous tetrahydrofuran (25 ml) is added dropwise. The mixture is kept at −80 °C and then the temperature is allowed to rise to −30 °C when another 45 ml of butyllithium is added. The mixture is stirred at −20 °C for 1.5 h, recooled to −100 °C and, under nitrogen, poured onto crushed solid carbon dioxide. Next morning the reaction mixture is poured into water, the phases are separated and the aqueous phase washed with diethyl ether and acidified with 2 *M* hydrochloric acid. The product is taken up in diethyl ether and this solution is dried over magnesium sulfate and evaporated. The residue, an oil, crystallizes when cyclohexane is added. Recrystallization from ethanol/water gives 4.9 g (68%) mp 138–139 °C.

Similarly, 4-(1-hydroxy-1-cyclohexyl)-3-thiophenecarboxylic acid is prepared in a one-pot procedure by halogen–metal exchange of 3,4-dibromothiophene followed by cylohexanone, renewed halogen–metal exchange and reaction with carbon dioxide [29].

4F.1.2 By oxidation of side chains in substituted thiophenes

4F.1.2.1 Oxidation of alkylthiophenes

Alkynyl- and alkenylthiophenes give fair yields of acids upon oxidation with potassium permanganate in acetone. 2-Thiophenecarboxylic acid is obtained from 1-(2-thienyl)propene [30] and 5-phenyl-2-thiophenecarboxylic acid was prepared both from the 5-phenyl derivatives shown below and 2,5-thiophenedicarboxylic acid from the ester shown below [31].

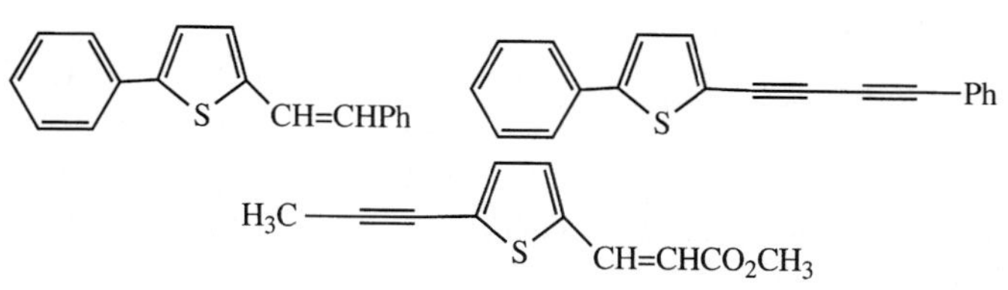

Alkaline potassium permanganate can be used for the oxidation of simple alkylthiophenes and 4-methyl-2-thiophenecarboxylic acid is obtained from 2-ethyl-4-methylthiophene [32,33]. 2,5-Thiophenedicarboxylic acid is also prepared in 40% yield by oxidation of 2,5-dimethylthiophene [31] and in 82% by oxidation of 5-methyl-2-thiophenecarboxylic acid with alkaline permanganate [9].

4F.1.2.2 Oxidation of thiophene aldehydes

The oxidation of thiophene aldehydes is a very general method for the preparation of carboxylic acids. Many other substituents, except amino groups, are tolerated. Its usefulness depended on the fact that the aldehydes were much more easily available by electrophilic reactions (Vilsmeyer and Sommelet reaction) than carboxylic acids. However, due to the development of the thienyllithium derivatives, which allows the direct preparation of thiophene-carboxylic acids, the route *via* the aldehydes has diminished in importance. The most frequently used agent for the oxidation is silver(I) oxide in sodium hydroxide, giving excellent yields. From 3-thiophenealdehyde 95–97% yield of 3-thiophenecarboxylic acid was obtained [34]. 3- Methyl-2-thiophenecarboxylic acid has been prepared by this method [35].

3 Methyl-2-thiophenecarboxylic acid [35]

To a solution of sodium hydroxide (8.40 g, 0.21 mol) in water (72 ml) silver nitrate (18.0 g (0.11 mol) is added and the suspension cooled to 5 °C, after which 3-methyl-2-thiophene aldehyde (6.30 g (0.05 mol) is added portionwise. The reaction mixture is stirred at room temperature for 1.5 h and filtered and the solid is washed with water (70 ml). The ice cooled filtrate is acidified and the precipitate formed is collected and dried giving 6.10 g (86%) of the title compound as colorless crystals mp 141–143 °C.

Numerous substituted thiophenecarboxylic acids, such as 2-cyclopropyl-[36], 2-methoxymethyl- [37], 2-phenoxymethyl- [38], 2-acetoxyethyl- 5-thiophene-carboxylic acids were obtained in this way [39]. 2-Thiophene aldehyde is also oxidized to the carboxylic acid by silver(II)oxide [40] and by a catalytic process involving silver(I)oxide–copper(II)oxide and air [41].

Thiophene aldehydes with additional protected formyl groups or keto groups can also be oxidized to the carboxylic acids by this method [42,43]. Transformations of the following type has been performed by oxidation of the corresponding aldehydes with active manganese dioxide [44].

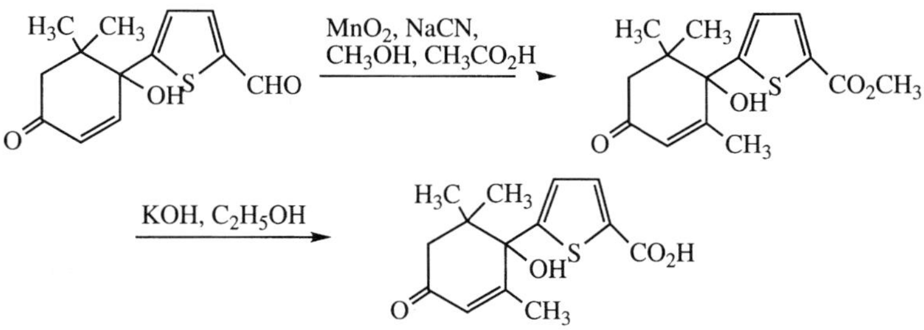

Methoxy-, alkylthio-, alkylseleno- and halogen-substituted thiophenecarboxylic acids are prepared by oxidation of the corresponding aldehydes and are treated in the appropriate chapters. Thiophene dialdehydes are oxidized without difficulty to the corresponding diacids [45–47]. Other oxidation agents than silver oxide are only rarely used. An example is potassium iodide–mercury(II) iodide, which can be used for the oxidation of 5-(methylpropinyl)-2-thiophene aldehyde (junipal), as well as 5-(1-propenyl)-2-thiophene aldehyde to the carboxylic acids [30,48]. It is also possible to use more powerful oxidation reagents for resistant thiophene aldehydes, as in the oxidation of 3-bromo-4-nitro-2-thiophenealdehyde with chromium trioxide in acetic acid [49]. Jones reagent (chromium trioxide-sulfuric acid in acetone) was utilized for the preparation of 5-methoxycarbonyl-2-thiophenecarboxylic acid from the aldehyde in 73% yield [50].

5-Methoxycarbonyl-2-thiophenecarboxylic acid [50]

A solution of methyl 5-formyl-2-thiophenecarboxylate (2.00 g, 11.75 mmol) in acetone (100 ml) is treated dropwise with Jones' reagent until a persistent orange color results. The mixture is stirred at room temperature for 1 h, the excess oxidant decomposed with isopropanol, after which the mixture is filtered. The filtrate is evaporated and the residue dissolved in ethyl acetate, dried over magnesium sulfate and evaporated to about 10 ml. Filtration gives 1.60 g (73%) of the title compound as a white solid mp 186–189 °C.

Other reagents, which have been used for the oxidation of thiophene aldehydes to thiophenecarboxylic acids include nickel peroxide [51], tetrabutylammonium permanganate [52], and mixtures of hydrogen peroxide with potassium hydroxide [53] or sulfuric acid [54].

4F.1.2.3 Hypohalite and related oxidations of acyl thiophenes

Also acyl thiophenes, like thiophene aldehydes, are more easily prepared by electrophilic subsitution reactions of thiophenes, than the carboxylic acids. So the indirect route to thiophenecarboxylic acids was previously preferred in many cases and especially the hypohalite oxidation of methyl ketones was much used. 2-Thiophenecarboxylic acid is easily prepared in this way [55]. There appears to be no preferred alkali hypohalite, although it is claimed that sodium hypochlorite gave a somewhat better yield of 5-ethyl-2-thiophenecarboxylic acid from 2-acetyl-5-ethylthiophene than sodium hypobromite [56]. Numerous 5-alkyl-2-thiophenecarboxylic acids have been prepared from the 2-acetyl derivatives [57,58] as well as also long chain 5-alkyl-2-thiophenecarboxylic acids [59], 3-alkyl- and 4-alkyl-2-thiophenecarboxylic acids [60–62] and 2-methyl- and 2-propyl-4-thiophenecarboxylic acid [63,64].

5-Decyl-2-thiophenecarboxylic acid [59]

This acid is prepared following general directions (Organic Syntheses, Coll. Vol. II, p. 428) however, using an equal volume of dioxane as a co-solvent. The crude acid (90%) is recrystalllized twice giving the title compound in pure form mp 84–85 °C.

5-Benzyl-2-thiophenecarboxylic acids [64,66] and 5-(2-methylfuryl)-2-thiophenecarboxylic acid [66] can also be prepared from the corresponding acetyl derivatives. 5-Phenyl- and other 5-aryl-2-thiophenecarboxylic acids can also be obtained in this way [68,69].

2,5-Dimethyl-3-phenyl-4-thiophenecarboxylic acid [68]

To a solution of 3-acetyl-2,5-dimethyl-4-phenylthiophene (7.5 g, 33 mmol) in dioxane (250 ml) a freshly prepared hypobromite solution obtained from sodium hydroxide (53 g) in water (250 ml) and bromine (19 ml) at 0 °C is added dropwise. When the addition is completed the reaction mixture is stirred at 55 °C for 2 h, after which sodium bisulfite (70 g) in water (750 ml) is added and the stirring is continued overnight. Bromoform is distilled off and the aqueous solution is acidified with dilute hydrochloric acid. The precipitated acid is filtered off giving 6.7 g (89%) of the title compound mp 130–131 °C after recrystallization from aqueous ethanol.

Also dicarboxylic acids, such as 2,5-dimethyl-3,4-thiophenedicarboxylic acid, have been prepared by this methodology [69].

4F.1.3 From cyanomethylthiophenes

The active methylene group of cyanomethylthiophenes can be condensed with 4-nitroso-*N*,*N*-dimethylaniline. The intermediates are hydrolyzed *via* the thenoyl cyanides to thiophenecarboxylic acids. This method can be used for the preparation of 5-ethyl-2-thiophenecarboxylic acid [70], 3-thiophenecarboxylic acid [71] and 2,5-diethyl-3-thiophenecarboxylic acid [70].

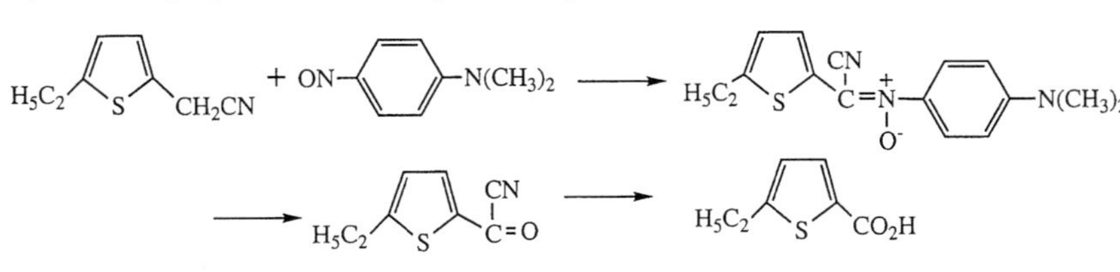

4F.1.4 Hydrolysis of cyanothiophenes

Hydrolysis of cyanothiophenes is a useful route to thiophenecarboxylic acids, as they are available from halothiophenes, especially such halothiophenes, which give poor results in halogen–metal interconversion. Full hydrolysis of cyanothiophenes is either achieved by boiling in concentrated hydrochloric acid or treatment with hot concentrated alkali solutions. 3-Thiophenecarboxylic acid is obtained in 84% yield by the acid hydrolysis of 3-cyanothiophene [73]. This method has especially been used for the preparation of nitro-2-thiophenecarboxylic acids [40,73,74].

2-(2-Formyl-1*H*-pyrrol-1-yl)thiophene-3-carboxylic acid is prepared by hydrolysis in 6 *M* sodium hydroxide in refluxing ethanol under microwave irradiation of 2-(2-formyl-1*H*-pyrrol-1-yl)-3-thiophenecarbonitrile [75].

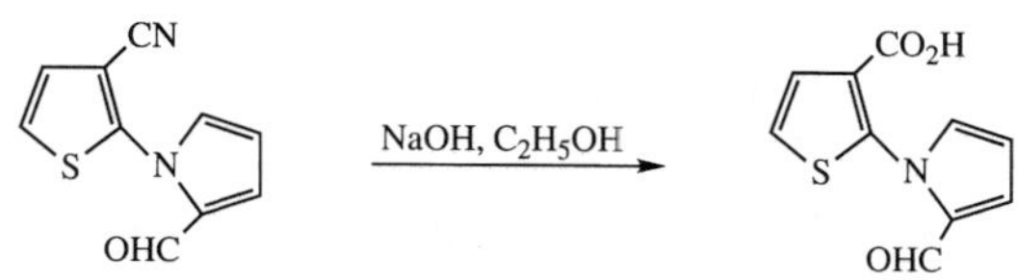

2-(2-Formyl-1H-pyrrol-1-yl)thiophene-3-carboxylic acid [75]

A 6 *M* aqueous sodium hydroxide solution (50 ml) is added to a suspension of 2-(2-formyl-1*H*-pyrrol-1-yl)-3-thiophenecarbonitrile (20 g. 0.10 mol) in ethanol (200 ml). The reaction mixture is refluxed for 15 min under microwaves. Ethanol is removed under reduced pressure. The aqueous solution is cooled, diluted with water (300 ml) and brought to pH = 1 with 35% hydrochloric acid. The precipitate formed is collected by filtration, washed with water (100 ml) and recrystallized from ethanol to give 21 g (95%) of the title compound as colorless crystals mp 208 °C.

The hydrolysis of also very highly substituted 2-(thienyl)-4,4-dimethyl-2-oxazolines is achieved with trifluoromethanesulfonic anhydride [76].

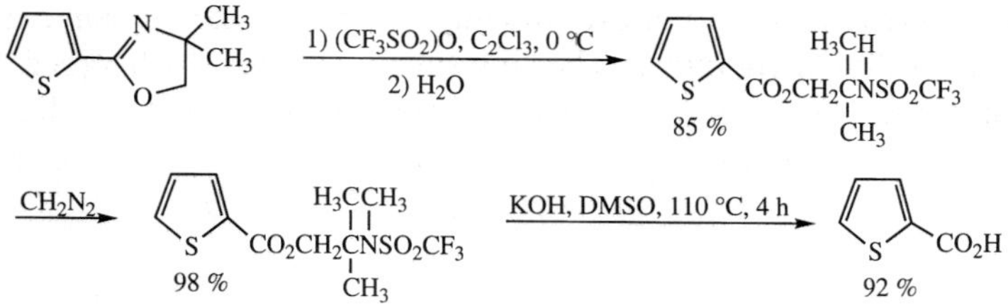

4F.1.5 Direct introduction of carboxyl groups or protected carboxyl groups in thiophenes

4F.1.5.1 *Via* palladium-catalyzed carbonylation reactions of thiophenes and halothiophenes

During recent years the carboxyl group is introduced in a thiophene ring using palladium catalysis. The reaction of thiophene with carbon monoxide at 15 atm and 100 °C in the presence of palladium(II) acetate is used for the preparation of 2-thiophenecarboxylic acid [77]. Alkoxycarbonylation of bromo- and chlorothiophenes can be achieved under atmospheric pressure of carbon monoxide in the presence of a cobalt catalyst *in situ* generated from cobalt(II) acetate under sun lamp illumination [78].

The method of carbonylation of 2-iodothiophene in aqueous media under one atmosphere carbon monoxide in the presence of palladium complexes lacking phosphine ligands and base is used for the preparation of 2-thiophenecarboxylic acid [79].

4F.1.5.2 *Via* various substitution reactions of thiophenes

There are few good methods for the direct carboxylation of thiophenes. The Gattermann reaction carried out in 1888 by Gattermann on 3-methylthiophene and some dimethyl- and trimethylthiophenes, using carbamyl chloride and aluminum chloride as catalyst gave only low yields of the amides, which could then be hydrolyzed to the free acids. It was later found that the reaction could be carried out in the absence of the catalyst, with thiophene and 2-methylthiophene with reaonable yield, however, very poor yields were obtained with 2-halothiophenes. When oxalyl chloride was used instead of carbamoyl chloride, the appropriate acid chlorides were obtained from thiophene and 2-methylthiophene, but with the 2-halothiophenes the yields were low [80]. Thiophene reacts with chloro- or fluorosulfonyl isocyanate in the absence of a catalyst and the resulting sulfonyl amide can be transformed to the carboxamide [81].

Using titanium chloride as catalyst, ethyl 2,3-dichloromethyl-5-thiophenecarboxylate was obtained upon chloromethylation of ethyl 2-thiophenecarboxylate in 98% yield [82].

Ethyl 2,3-dichloromethyl-5-thiophenecarboxylate [82]

To a mixture of ethyl 2-thiophenecarboxylate (32 g, 0.2 mol) and chloromethyl methyl ether (80 g, 1.0 mol) cooled in an ice/water bath titanium chloride (60 g, 0.3 mol) is added dropwise over a period of 40 min. The reaction mixture is stirred at room temperature for 3 h, after which dichloromethane (200 ml) is added and the solution poured into ice-water. The phases are separated and the organic phases washed with sodium chloride solution, dried over magnesium sulfate and evaporated. The residue is recrystallized from hexane giving 50 g (98%) of the title compound as colorless needles mp 60–61 °C.

Under Friedel-Crafts condition phenylisocyanate and phenylisothiocyanate condense with thiophene to give the *N*-phenyl amide and *N*-phenylthioamide in excellent yields [83]. Photochemical reaction of thiophene in ethanol gives ethyl 2-thiophenecarboxylate [84]. Friedel-Crafts reactions of thiophenes with pyrocatechol dichloromethylene acetal, using stannic chloride as catalyst followed by alkaline hydrolysis can be used for the preparation of 2-thiophenecarboxylic acid [85]. This reaction sequence has also been used with 2-bromo-3-methylthiophene to give 5-bromo-4-methyl-2-thiophenecarboxylic acid [37] and with methyl 3-thiophenecarboxylate to give 2,4-thiophenedicarboxylic acid [86].

The Haller–Bauer reaction of nonenolizable thienyl ketones, such as 2-thienyl-2,6-dichlorophenyl ketone, gives 2-thiophenecarboxylic acid when cleaved with butoxide-water reagent [87,88].

4F.1.6 By changes of substituents in thiophenecarboxylic acids and derivatives

4F.1.6.1 By palladium-catalyzed coupling reaction of halothiophenecarboxylic acid derivatives

Palladium-catalyzed coupling of methyl 3-iodo-2-thiophenecarboxylate with an *E*-alkenyl fluorosilane is used for the preparation of the 3-vinylsubstituted 2-thiophenecarboxylate [89].

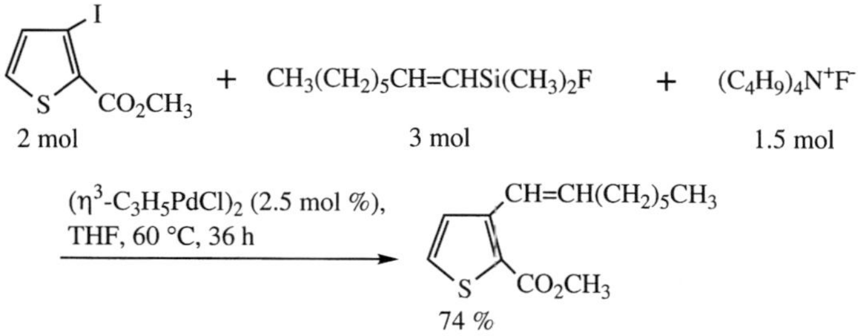

Heck coupling of 5-bromo-2-thiophenecarboxylic acid and 4,5-dibromo-2-thiophenecarboxylic acid with ethyl acrylate are used for the preparation of ethyl 3-(5-carboxy-2-thienyl)acrylate and diethyl 5-carboxy-2,3-thiophenediacrylate in 62% and 55% yield, respectively [90].

Diethyl 5-carboxy-2,3-thiophenediacrylate [90]

To a solution of palladium(II) acetate (0.07 g, 0.31 mmol) and triphenylphosphine (0.33 g, 0.26 mmol) in acetonitrile (50 ml) and triethylamine (30 ml) 4,5-dibromo-2-thiophenecarboxylic acid (2.09 g, 7 mmol) and ethyl acrylate (4.2 g, 42 mmol) are added. The reaction mixture is sealed in a glass tube and heated at 100 °C for 20 h. After cooling the content of the tube is evaporated and the residue is dissolved in water and filtered. The filtrate is refluxed with charcoal and acidified with diluted hydrochloric acid. The precipitate formed is filtered off and recrystallized from ethanol giving 1.26 g (55%) of the title compound as yellow crystals.

Microwave assisted aqueous Suzuki couplings between polyethyleneglycol 6000 bound 5-bromo-2-thiophenecarboxylate and arylboronic acids are used for the preparation of 5-aryl- 2-thiophenecarboxylates [91].

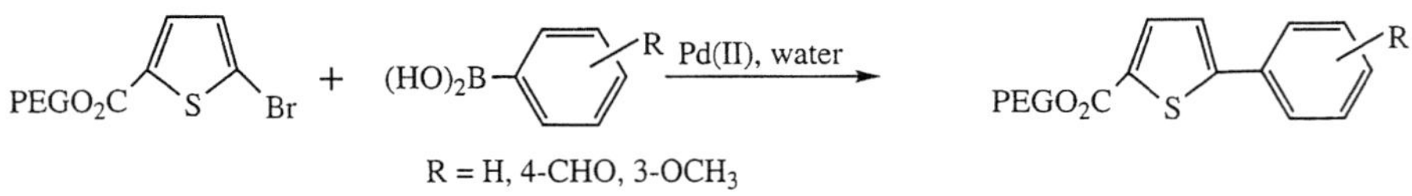

Representative cross-coupling reaction of PEG 600 bound aryl halides [91]

An argon flushed 20 ml screw cap culture tube is charged with polyethyleneglycol 6000 bound 5-bromo-2-thiophenecarboxylate (1.0 g, *ca* 0.31 mmol) and water (5 ml). 3-Methoxyphenylboronic acid (57 mg, 0.37 mmol), palladium(II) acetate (7 mg, 10 mol %) and potassium carbonate (107 mg, 0.78 mmol) are added. The reaction mixture is stirred for 2 h at 70 °C under argon. The solvent is coevaporated with toluene (400 μl) at 60 °C. The residue is dissolved in toluene (15 ml) and the mixture is centrifuged. The clear supernatant is precipitated with cold *tert*-butyl methyl ether (−18 °C) and centrifuged. The precipitate formed is dissolved in dichloromethane. This procedure is repeated twice and the product is dried under vacuum.

Benzyl 3-bromo-2-thiophenecarboxylate can be used in Suzuki coupling with the benzophenone imine of (4-pinacolylborono)phenylalanine ethyl ester to give the compound shown below [92].

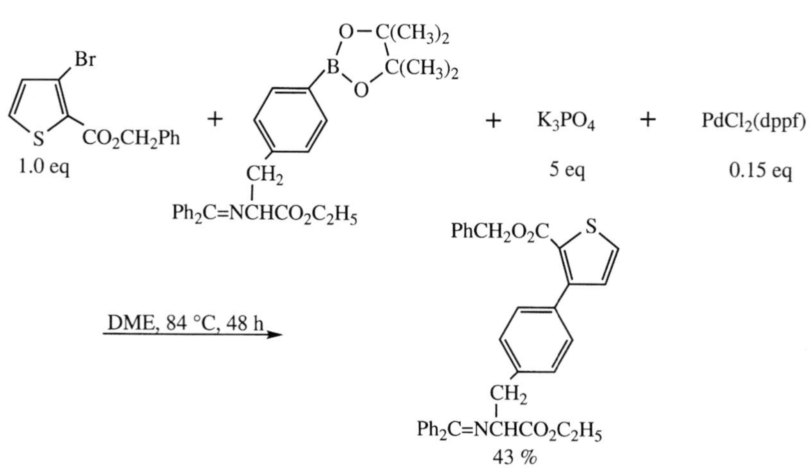

Another example of the use of Suzuki coupling is the following coupling with methyl 3-bromo-2-thiophenecarboxylate [93].

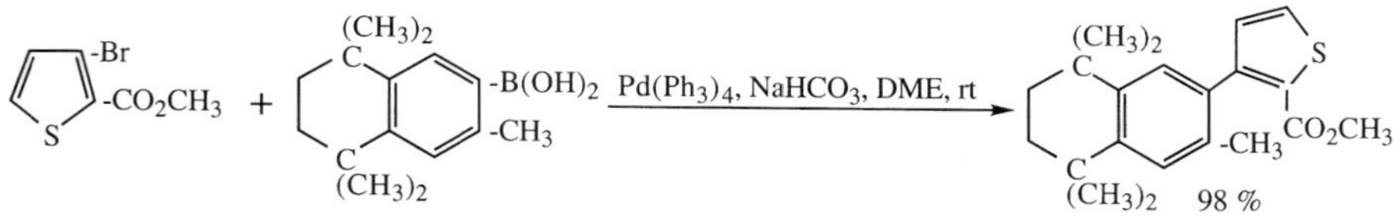

Kumada coupling of methyl-5-iodo-2-thiophenecarboxylate with the zinc derivative derived from diene is used for the preparation of the modified farnesol [94].

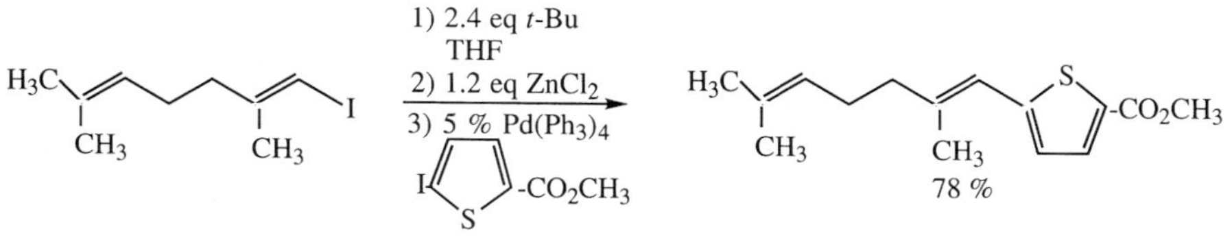

Reductive debromination of bromothiophenecarboxylic acids with Raney Cu–Al alloy in alkaline deuterium oxide is an excellent method for the preparation of deuterated 2-thiophenecarboxylic acids [95].

3-Deuterio-2-thiophenecarboxylic acid [95]

To a solution of 3-bromo-2-thiophenecarboxylic acid (620 mg, 3 mmol) in 10% sodium deuterioxide in deuteriated water (15 ml), Raney Cu–Al alloy (600 mg) is gradually added at room temperature, after which the reaction mixture is stirred at 50 °C for 1 h. The copper powder produced and unchanged Raney alloy are filtered off and washed with water (20 ml). The combined filtrate and washings are acidified with concentrated aqueous hydrochloric acid and the product taken up in diethyl ether. The diethyl ether solution is washed with water, dried over sodium sulfate and evaporated. The residue is recrystallized from water giving 271 mg (70%) of the title compound with almost the same mp as the starting material.

5-Ethylthiomethyl-2-thiophenecarboxylic acid is oxidized to the corresponding sulfone with hydrogen peroxide in acetic acid [96]. Side chain bromination of ethyl 5-methyl-2-thiophenecarboxylate gives 5-bromomethyl-2-thiophenecarboxylate, which *via* the 5-acetoxymethyl-2-thiophenecarboxylate is transformed to the 5-hydroxymethyl-2-thiophenecarboxylic acid. The same reaction sequence was applied to ethyl 4-methyl-2-thiophenecarboxylate for the preparation of 4-hydroxymethyl-2-thiophenecarboxylic acid and of

3-hydroxymethyl- 2-thiophenecarboxylic acid from ethyl 3-methyl-2-thiophenecarboxylate [61].

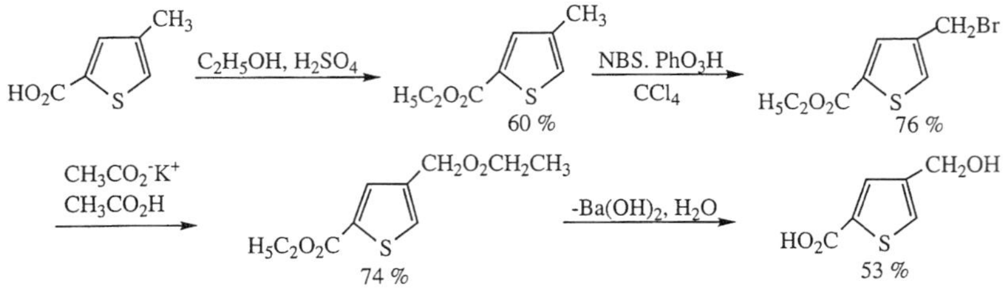

The reaction of 3,4-di(cyanomethyl)-2,5-dicarbomethoxythiophene with sulfur monochloride in the presence of triethylamine in tetrahydrofuran gave the following bicyclic compound [97].

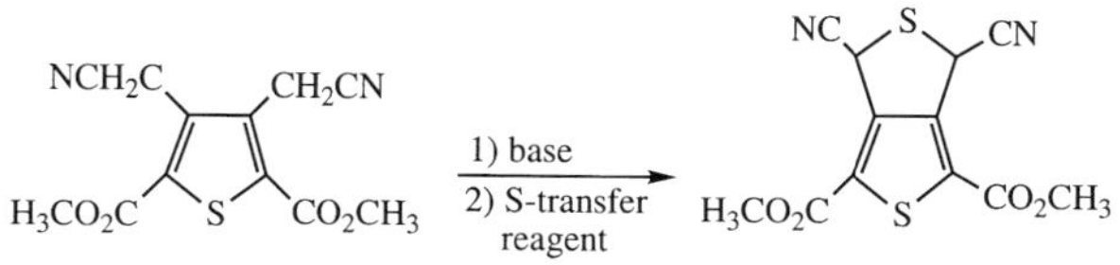

4F.1.6.2 By modifications of aldehyde and keto groups in thiophenecarboxylic acids

The reaction of methyl 3-formyl-2-thiophenecarboxylate with dimethyl L-glutamate is used for the preparation of methyl *N*-[(2-carbomethoxythiophene-3-yl)methyl]-L-pyroglutamate [98]. This compound is also prepared from crude 3-bromomethyl-2-thiophenecarboxylate, obtained by side chain bromination, followed by reaction with dimethyl L-glutamate [98].

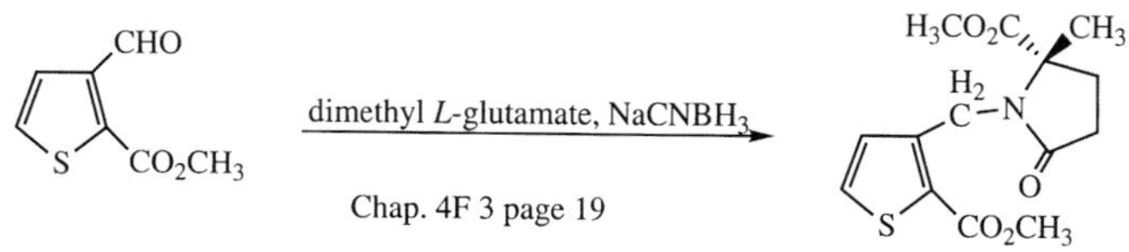

Methyl N-[(2-carbomethoxythiophene-3-yl)methyl]-L-pyroglutamate [98]

A mixture of dimethyl L-glutamate hydrochloride (0.85 g, 4.0 mmol) and methyl 3-formyl-2-thiophenecarboxylate (0.44 g, 2.6 mmol) is stirred in methanol (100 ml) and treated with triethylamine (1.01 g, 10 mmol). The stirring is continued for 15 min, after which acetic acid (4 ml, 67.6 mmol) and activated 3 Å molecular sieves (5 g) are added to the homogeneous solution. After 15 h of stirring at room temperature the mixture is heated to 50 °C and sodium cyanoborohydride (1.28 g, 20 mmol) is added. The reaction mixture is heated at 50 °C overnight and filtered. The filtrate is concentrated and the residue triturated with dichloromethane. The organic phase is separated, washed with water, dried over magnesium sulfate and evaporated. The residue is purified by chromatography on silica gel using ethyl acetate/hexane (10–70%) as eluent giving 0.12 g (36%) of the title compound as a colorless oil.

3-(2,2-Dicyanoethenyl)- and 3-(2,2-dicyano-1-methylethenyl)thiophene-2-carboxylic acid are prepared by condensation of malonitrile with 3-formyl-2-thiophenecarboxylic acid and 3-acetyl-2-thiophenecarboxylic acid using alkaline or acidic conditions [99]. Also 2-(2,2-dicyanoethenyl)thiophene-3-carboxylic acid is prepared in the same way [99].

3-(2,2-Dicyano-1-methylethenyl)thiophene-2-carboxylic acid [99]

3-Acetyl-2-thiophenecarboxylic acid (2.0 g, 12 mmol), malonitrile (0.76 g, 12 mmol) and β-alanine are suspended in benzene (50 ml). Propanol (810 ml) and acetic acid (5 ml) are added and the suspension is refluxed until no further water is collected in a Dean–Stark collector (*ca* 4 h). After addition of diethyl ether (150 ml) the suspension is filtered. The filtrate is evaporated and the remaining acetic acid is removed azeotropically with toluene. The crude product (92%) is taken up in diethyl ether and this solution is filtered through a short column of silica gel and evaporated giving the title compound mp 140–142 °C after recrystallization from benzene.

4F.1.6.3 Modifications in alkyl groups in thiophenecarboxylic acids

Side chain bromination of methyl 2-methyl-3-thiophenecarboxylate, methyl 3-methyl-2-thiophenecarboxylate and methyl 4-methyl-3-thiophenecarboxylate, is used for the preparation of the corresponding bromomethyl derivatives, which upon reaction with *ortho*-nitrophenolate and *ortho*-aminothiophenolate give the corresponding (*ortho*-nitrophenoxymethyl)- and (*ortho*-aminophenoxy methyl) carbomethoxythiophenes [100].

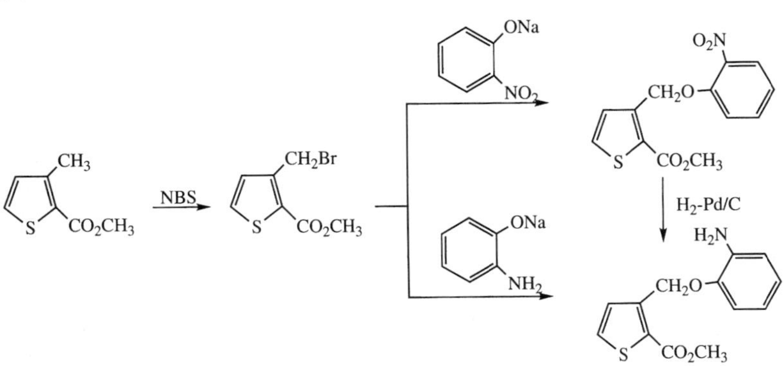

4-(1-Pyrrolylmethyl)thiophene-3-carboxylic acid is prepared by the reaction of methyl 4-bromomethyl-3-thiophenecarboxylate with pyrrole followed by alkaline hydrolysis [101]. In the same way 3-(1-pyrrolylmethyl)thiophene-2-carboxylic acid has been prepared [102].

4-(1-Pyrrolylmethyl)thiophene-3-carboxylic acid [101]

To a well-stirred suspension of the potassium salt of pyrrole, prepared from pyrrole (4.50 g, 65 mmol) and potassium metal (2.60 g) in anhydrous tetrahydrofuran under nitrogen a solution of methyl 4-bromomethyl-3-thiophenecarboxylate (14.56 g, 62 mmol) in anhydrous tetrahydrofuran (100 ml) is slowly added dropwise at room temperature. The reaction mixture is refluxed for 4 h and cooled to room temperature, after which cyclohexane (150 ml) is added and the mixture left at room temperature for 2 h. After filtration and evaporation the residue is purified by distillation giving 82% of the title compound bp 140–142 °C/0.3 mm Hg, mp 69 °C after recrystallization from hexane.

Corrresponding reactions have also been carried out with phthalimide in connection with the synthesis of complex condensed heterocyclic systems [103]. 2-Carboxy-3-thienylacetic acid [104] can be transformed to 2-acetoxy-2-(2-carboxy-5-trimethylsilylthiophen-3-yl)acetic acid [105].

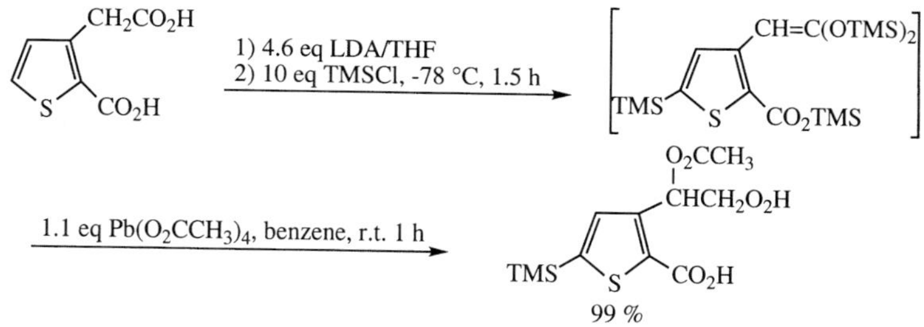

4F.1.6.4 Hydrogenation of unsaturated side chains

Vinyl groupings as in 1-(3-carboxythienyl)-2-(2-thienyl)ethene and isomeric compounds are conveniently hydrogenated to the ethane derivatives using Wilkinson's rhodium catalyst [23–25,27]. Chlorine substituents are not removed in the hydrogenation [106].

4F.1.6.5 Various methods

An alternative to the hypochlorite oxidation of acylthiophenes consists in halogenation of the α-methylene- or methyl group followed by reaction with pyridine and alkali treatment [107]. This method is especially useful for methyl ketones carrying alkylthio [108,109] and alkylseleno substituents [111,112], which are sensitive to hypochlorite oxidation.

4F.1.6.6 Thiophenecarboxylic acids *via* ring-closure reactions

4F.1.6.6.1 Alkyl- and aryl-2-thiophenecarboxylic acids

Ring-closure reactions are especially attractive for the preparation of highly substituted 2-thiophenecarboxylic acids. Mostly different modifications of the Fiesselman reaction are used. In one, often used variant, β-chlorovinyl aldehydes or ketones are reacted with an ester of thioglycolic acid in the presence of an organic base [112–118]. More recently this method is used for the preparation of ethyl 3,5-dimethyl-2-thiophenecarboxylate, ethyl 3,4,5-trimethyl-2-thiophenecarboxylate, ethyl 5-methyl-3-phenyl-2-thiophenecarboxylate and ethyl 3-methyl-5-phenyl-2-thiophenecarboxylate [119].

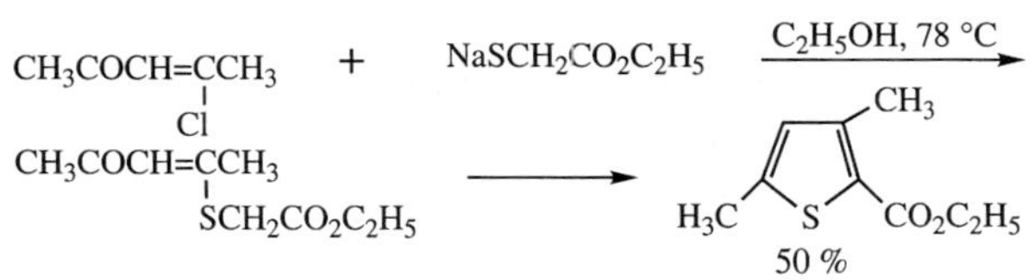

General procedure [119]

A solution of the 3-chlorovinylketone (10 mmol) and the sodium salt of ethyl thioglycolate (1.56 g, 11 mmol) is refluxed in anhydrous ethanol (25 ml). After cooling the reaction mixture to room temperature ethanol is evaporated and the residue redissolved in anhydrous tetrahydrofuran. This solution is filtered, the filtrate evaporated and the ethyl 2-thiophenecarboxylates are purified by flash chromatography on silica gel using toluene as solvent.

A double Fiesselmann reaction of 4,4′-bis(1-chloro-2-formylethenyl)phenyl ether gives 4,4′-bis(5-carbomethoxy-2-thienyl)phenyl ether, used for the synthesis of some bisthiophenecarboxamide derivatives as ditopic receptors for long chain dicarboxylic acids [120].

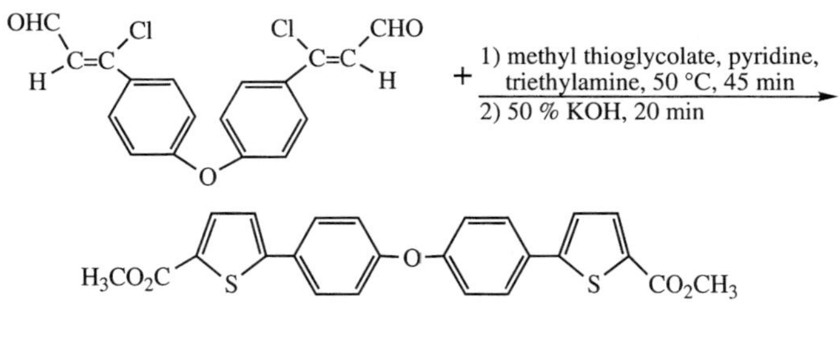

4,4′-Bis(5-carbomethoxy-2-thienyl)phenyl ether [120]

To a stirred solution of 4,4′-bis(1-chloro-2-formylethenyl)phenyl ether (0.20 g, 0.58 mmol) and methyl thioglycolate (0.11 ml, 1.27 mmol) in pyridine (3–4 ml) at 0 °C triethylamine (0.24 ml, 1.73 mmol) is added dropwise. The reaction mixture is gradually allowed to attain room temperature and then stirred at 45–50 °C for an additional 30 min. After cooling to 10–15 °C 50% aqueous potassium hydroxide solution is added. The stirring is continued at 10–15 °C for 20 min, after which the mixture is poured onto ice. The product is extracted with dichloromethane and the combined organic phases are washed with dilute hydrochloric acid and water, dried over sodium sulfate and evaporated. The residue is purified by chromatography on silica gel using benzene as eluent, giving 0.194 g (75%) mp 203–205 °C.

In another type of the Fiesselmann reaction, 1,3-dicarbonyl compounds are reacted with a thioglycolate followed by reaction with alkoxide. Unfortunately, most of these results have only been published in patents and PhD theses, which, however have been extensively discussed [121]. Methyl 4,5,6,7-tetrahydrobenzo[*c*]thiophene-1-carboxylate is obtained by acid-catalyzed condensation of 2-oxocyclohexanecarboxaldehyde and methyl thioglycolate, followed by aromatization with freshly prepared sodium methoxide in rather low yields [122].

Methyl 4,5,6,7-tetrahydrobenzo[c]thiophene-1-carboxylate [122]

To a magnetically stirred neat mixture of 2-oxocyclohexanecarboxaldehyde (3.6 g, 28 mmol) and methyl thioglycolate (6.0 g, 56 mmol) concentrated sulfuric acid (3 drops) is added. The resulting yellow solution is stirred at room temperature for 12 h, diluted with ice-water (25 ml) and extracted with dichloromethane (25 ml). The aqueous phases are extracted once more with dichloromethane (25 ml) and the combined organic phases are washed with saturated aqueous sodium chloride solution (50 ml), dried over sodium sulfate and evaporated. The residue, a viscous yellow oil, is dissolved in methanol and added dropwise over 1 h to freshly prepared sodium methoxide solution, prepared from sodium (1.7 g) and methanol (100 ml). The deep orange solution so obtained is stirred overnight, concentrated to one quarter of the volume and partitioned between dichloromethane (50 ml) and water (50 ml). The phases are separated and the aqueous phase extracted with dichloromethane (25 ml). The combined organic phases are washed with saturated aqueous sodium chloride solution (50 ml), dried over sodium sulfate and evaporated. The residue is purified by chromatography on silica gel using ethyl acetatehexane (5:95) as eluent giving 1.5 g (27%) of the title compound as a clear colorless liquid.

Various 2-thiophenecarboxylates carrying a 2-trifluoromethyl, 2-(1,1-difluoroalkyl- or 2-polyfluoroalkyl groups are recently synthesized from α-fluoroalkyl ketones or aldehydes with methyl or ethyl thioglycolates. [123]

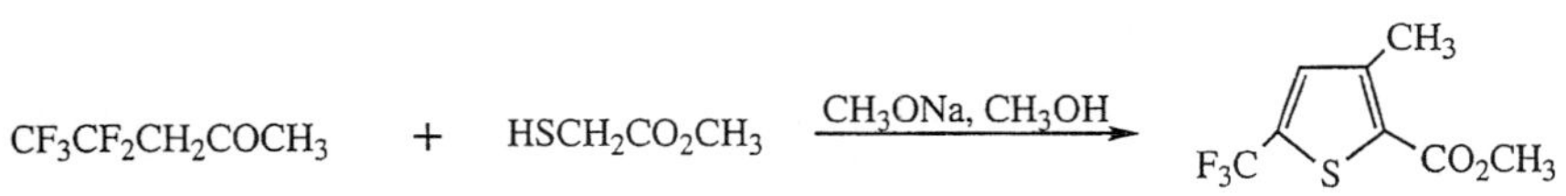

5-Methoxycarbonyl-4-methyl-2-trifluoromethylthiophene [123]

To a solution of the fluoro containing compound (0.88 g, 5 mmol) and methyl thioglycolate (0.64 g, 6 mmol) in anhydrous methanol (10 ml), sodium methoxide (20 mmol) is slowly added. The reaction mixture is stirred at room temperature for 4 h, after which it is poured into water (50 ml). The product is extracted with diethyl ether and the combined organic phases washed with ammonium chloride solution and sodium chloride solution, dried over sodium sulfate and evaporated. The residue is purified by chromatography on silica gel using light petroleum/ethyl acetate (100:0–100:6) as eluent, giving 0.85 g (76%) of the title compound.

Ring-closure with methyl thioglycolate in the presence of 1,8-diazabicyclo[5.4.0]undec-7-ene in acetonitrile gave the stable bicyclic radical [124].

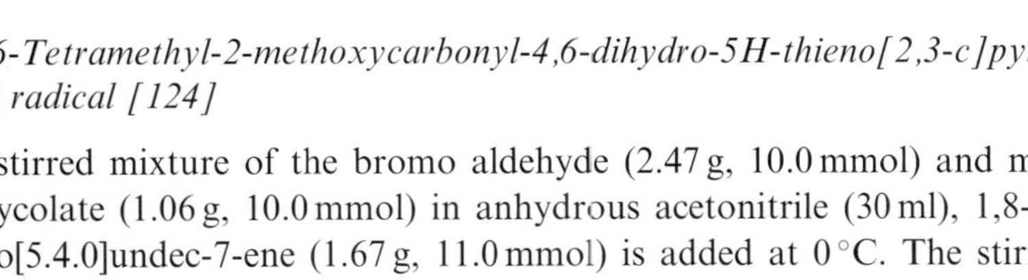

4,4,6,6-Tetramethyl-2-methoxycarbonyl-4,6-dihydro-5H-thieno[2,3-c]pyrrol-5-yloxyl radical [124]

To a stirred mixture of the bromo aldehyde (2.47 g, 10.0 mmol) and methylthioglycolate (1.06 g, 10.0 mmol) in anhydrous acetonitrile (30 ml), 1,8-diazabicyclo[5.4.0]undec-7-ene (1.67 g, 11.0 mmol) is added at 0 °C. The stirring is continued for 1 h at room temperature, after which the solvent is evaporated. The residue is dissolved in ethyl acetate (30 ml) and this solution is washed with 5% aqueous sulfuric acid, dried over magnesium sulfate and evaporated. The residue is purified by flash chromatography using hexane/diethyl ether as eluent giving 1.65 g (65%) of the title compound as a yellow crystalline solid mp 116–118 °C after recrystallization from hexane/diethyl ether (2:1).

Addition of alkyl thioglycolates to α,β-unsaturated ketones in the presence of organic base gives tetrahydrothiophene-3-ols, which are transformed to thiophenecarboxylic acids on treatment with polyphosphoric acid followed by dehydrogenation with diphenyl disulfide or chloroanil [125,126]. 5-Phenyl-2-thiophenecarboxylic acid is prepared by the reaction of 5-phenylbutynoic acid with hydrogen sulfide [127] or by oxidative cyclization of 2-mercapto-5-phenylbut-2,4-dienoic acid [118,128].

A modified Hinsberg reaction can be used for the preparation of ethyl 5-(*para*-nitrophenyl)-2-thiophenecarboxylate and ethyl 3,4-diphenyl-5-(*para*-nitrophenyl)-2-thiophenecarboxylate from glyoxal and benzil and ethyl 4-nitrobenzylthioacetate [129].

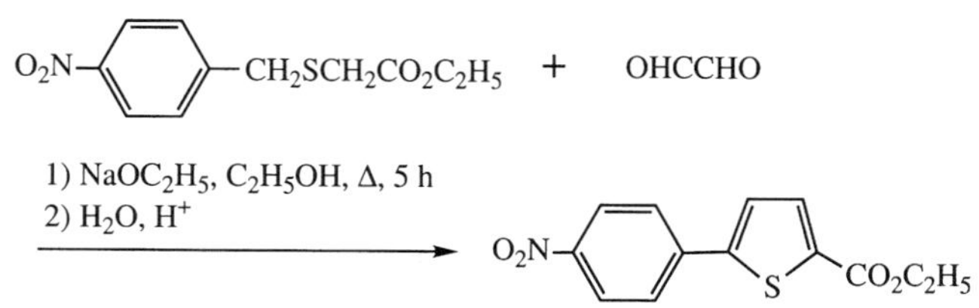

Ethyl 5-(para-nitrophenyl)-2-thiophenecarboxylate [129]

To a solution prepared from sodium (0.46 g, 0.02 mol) and absolute ethanol (100 ml) ethyl 4-nitrobenzylthioacetate (5.10 g, 0.02 mol) is slowly added with stirring at 5 °C. The stirring is continued for 15 min, after which 40% glyoxal (2.90 ml, 0.02 mol) is added. The reaction mixture is refluxed for 4–6 h, cooled to room temperature, poured into ice water and neutralized with 10% aqueous hydrochloric acid. The precipitate formed is filtered off, washed with water and recrystallized from *N,N*-dimethylformamide/ethanol (1:1) giving 2.10 g (76%) of the title compound as a light brown crystalline solid mp 227 °C.

4F.1.6.6.2 Alkyl- and aryl-substituted 3-thiophenecarboxylic acids

A useful method for the preparation of 2-aryl-4-methyl-3-thiophenecarboxylates is the reation of an α-mercaptoacetaldehyde or ketone with an acetylenic ester under the influence of catalytic amounts of sodium or potassium *tert*-butoxide [87,130]. Methyl 5-ethyl-2-phenylthiophene-3-carboxylate can be prepared from 2-mercaptobutyraldehyde and methyl phenylpropiolate [131].

A very useful and versatile route to 3-thiophenecarboxylic acids consists in the preparation of 2,5-dihydrothiophene derivatives by Michael addition of an α-mercaptocarbonyl compound to vinylphosphonium salts or phosphonates followed by spontaneous intramolecular Wittig type reaction. These dihydrothiophene are easily aromatized already by air or other mild oxidative agents [132–135]. Two examples are shown in the schemes below [136,137].

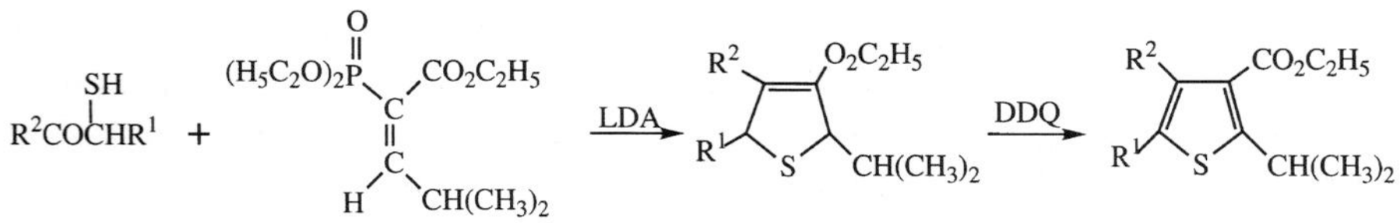

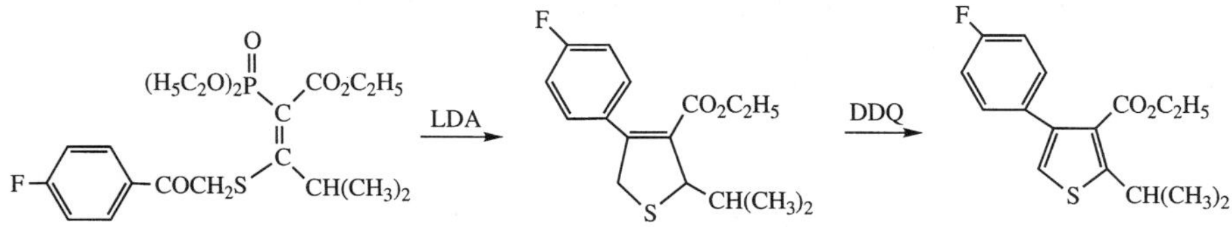

The Paal–Knorr reaction in which a 1,4-dicarbonyl compound is heated with phosphorus pentasulfide or Lawesson's reagent can be used for the preparation of 3-thiophenecarboxylic acids. In this way ethyl 5-benzylthiophene-3-carboxylate is obtained [138]. The same starting materials give thiophenes on reaction with hydrogen sulfide in ethanol [139]. The cycloaddition reactions of 1,3-dithiolium salts with propiolates give adducts, which extrude stable fragments, leading to 3-thiophenecarboxylic acids [140–142].

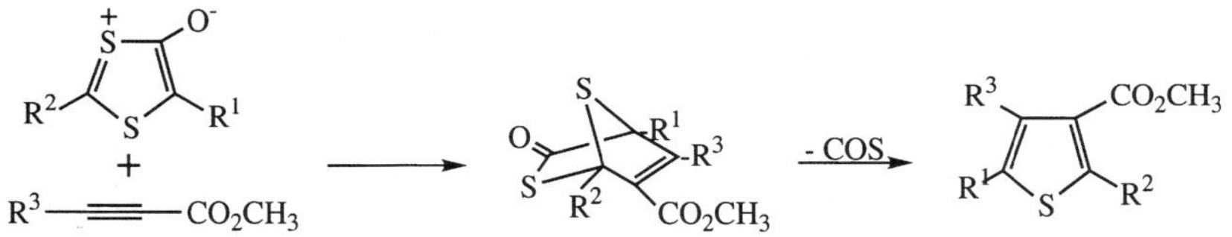

4F.2 THIOPHENEDI- AND POLYCARBOXYLIC ACIDS

4F.2.1 2,5-Thiophenedicarboxylic acids

Using excess butyllithium *N,N,N′,N′*-tetramethylethylenediamine in the metalation of 3-alkylthiophenes, gives 3-alkylthiophene-2,5-dicarboxylic acids upon reaction with carbon dioxide [143,144].

3-Octyl-2,5-thiophenedicarboxylic acid [144]

3-Octylthiophene (12.4 g, 0.0633 mol) is added to a solution of *N,N,N′,N′*-tetramethylethylenediamine (18.3 g, 0.158 mol) in anhydrous hexane (30 ml) and 1.6 *M* butyllithium in hexane (98 ml, 0.158 mol) under nitrogen at room temperature. After stirring at room temperature for 1 h the mixture is refluxed for 30 min, cooled to −70 °C and slowly poured under nitrogen to a 500 ml flask half-filled with crushed dry ice. The reaction mixture is left overnight at room temperature and then poured into 10 *M* hydrochloric acid and ice. The phases are separated and the aqueous phase extracted with diethyl ether (2 × 100 ml). The combined organic phases are extracted with 10% sodium hydroxide solution (100 ml) and the alkaline extracts acidified with 10 *M* hydrochloric acid. The product is taken up in diethyl ether (2 × 50 ml) and the ether extracts are washed with water and sodium chloride solution, dried over magnesium sulfate and evaporated. The residue is recrystallized from ethanol/water giving 14.0 g (78%) of the title compound as a white powder.

2,5-Thiophenedicarboxylic acids are conveniently prepared by the Hinsberg reaction of dialkyl diketones [145,146], diaryl ketones [147,148], phenylglyoxal [149] and pyruvic acids with diethyl thiodiglycolate. For the condensation, usually sodium ethoxide in ethanol or potassium *tert*-butoxide in *tert*-butanol are used. However, also the anion of dimethylsulfoxide can be used [146].

A very convenient preparation consists in the heating of adipic acid with thionyl chloride, followed by hydrolysis of the acid chloride formed [150]. Modest yields of dimethyl 3,4-diarylthiophene-2,5-dicarboxylates are formed upon heating of methyl arylpropiolates with sulfur in a solvent [151].

2,5-Thiophenedicarboxylic acids are obtained from 5-ethylthiomethyl-[97], 5-chloromethyl- [152] and 5-methoxymethyl-2-thiophenecarboxylic acids [153] by oxidation with potassium permanganate. Also from 2,5-bis(hydoxymethyl)thiophene, 2,5-thiophenedicarboxylic acid is obtained [154].

4F.2.2 Thiophene-2,3-dicarboxylic acids

The *ortho*-directing properties are used in the metalation of 2-thiophenecarboxylic acid with two equivalents of butyllithium in tetrahydrofuran, giving upon carbonation thiophene-2,3-dicarboxylic acid [155,156].

Thiophene-2,3-dicarboxylic acid [156]

To a solution of 2-thiophenecarboxylic acid (10 g, 78.0 mmol) in anhydrous tetrahydrofuran at −78 °C, 2.5 *M* butyllithium in hexane (68.6 ml, 171 mmol) is added. The stirring is continued at −78 °C for 30 min, after which the reaction mixture is quenched with carbon dioxide generated from dry ice. The temperature is allowed to rise to −10 °C and 2 *M* hydrochloric acid (85 ml). The phases are separated and the aqueous phase extracted with ethyl acetate (150 ml). The combined organic phases are dried over sodium sulfate and evaporated. The residue, a solid, is recrystallized from isopropanol giving 5.9 g (44%) of the title compound mp 271–271 °C.

Thiophene-2,3-dicarboxylic acid is obtained by oxidation of 3-alkyl-2-thiophenecarboxylic acids [157]. In most cases, the conditions to oxidize acetyl groups are drastic enough for the oxidation of alkyl groups and especially vinyl groups. Thus (5-acetyl-2-thienyl)ethylene gives 2,5-thiophenedicarboxylic acid [152] and 2-acetyl-3-methylthiophene gives 2,3-thiophenedicarboxylic acid upon oxidation with permanganate [158,159].

2,3-Thiophenedicarboxylic acid [158]

To a stirred mixture of 2-acetyl-3-methylthiophene (35 g, 0.25 mol) water (2 l), sodium hydroxide (280 g) and potassium permanganate (195 g) are added portionwise over a period of 90 min. The stirring is continued for 1 h at room temperature and 2 h on a steam bath, after which the mixture is cautiously acidified to about pH = 3 with 50% sulfuric acid, again heated and stirred on a steam bath for 30 min and filtered hot. The filtrate is extracted with ethyl acetate (5 × 250 ml). The combined organic phases are dried over magnesium sulfate and evaporated giving 18.7 g (43%) of the title compound mp 268–270 °C.

2,3-Thiophenedicarboxylic acid is prepared by oxidation of 3-formyl-2-thiophenecarboxylic acid [160].

2,3-Thiophenedicarboxylic acid [160]

A solution of 3-formyl-2-thiophenecarboxylic acid (3.25 g, 0.021 mol) and sodium hydroxide (2.0 g) in water (10 ml) is shaken vigorously with an aqueous suspension of silver oxide, prepared from silver nitrate (7.5 g, 0.045 mol) in water (8 ml) and sodium hydroxide (3.5 g) in water (20 ml). After a few minutes the silver is filtered off and the filtrate acidified. The precipitate formed is filtered off giving 2.1 g (69%) of the title compound as fine needles mp 271–272 °C after recrystallization from water.

However, upon oxidation of 3-acetyl-2-methyl-5-thiophenecarboxylic acid with sodium dichromate 5-methyl-2,4-thiophenedicarboxylic acid is obtained in 82% yield [161]. Alkylthiophenes functionalized in the benzylic position, such as halomethylthiophenes, hydroxy- and alkoxymethylthiophenes, as well as mercaptomethyl derivatives are conveniently oxidized.

2,3-Thiophenedicarboxylic acid can be prepared by the alkaline hydrolysis of 2,3-dicyanothiophene in ethylene glycol [162,163] and this method was also recently used for the preparation of 3,4-thiophenedicarboxylic acid from 3,4-dicyanothiophene [164], although milder conditions have been used previously for the preparation of 3,4-thiophenedicarboxylic acid [163].

Hydrolysis of 3- and 5-cyano-2-thiophenecarboxylic acid in refluxing 2 *M* sodium hydroxide solution can be used for the preparation of 2,3- and 2,5-thiophenedicarboxylic acid [165].

Similarly, 3,4-thiophenedicarboxylic acid, 2,4-thiophenedicarboxylic acid and 2,5-thiophenedicarboxylic acid were prepared by silver oxide oxidation of 3-formyl-4-thiophenecarboxylic acid, 4-formyl-2-thiophenecarboxylic acid and 5-formyl-2-thiophenecarboxylic acid, respectively [166].

2,3-Thiophenedicarboxylic acids are obtained by reaction of arylalkynethiolate with dimethyl acetylenedicarboxylate [167,168].

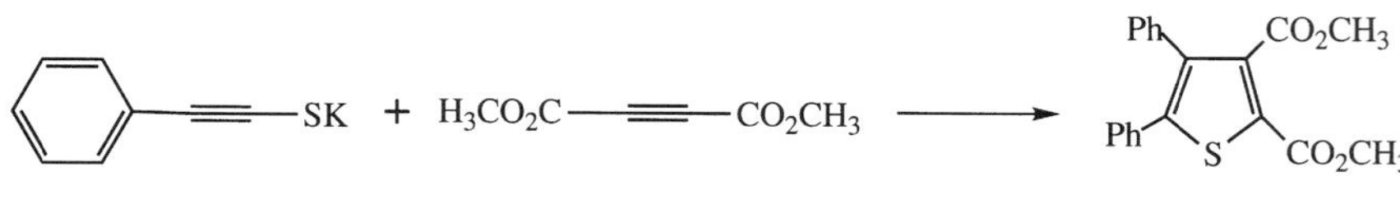

2,3-Thiophenedicarboxylic acid derivatives can be prepared by a [2 + 3]cycloaddition alkene extrusion process from dihydrothiophenes [169].

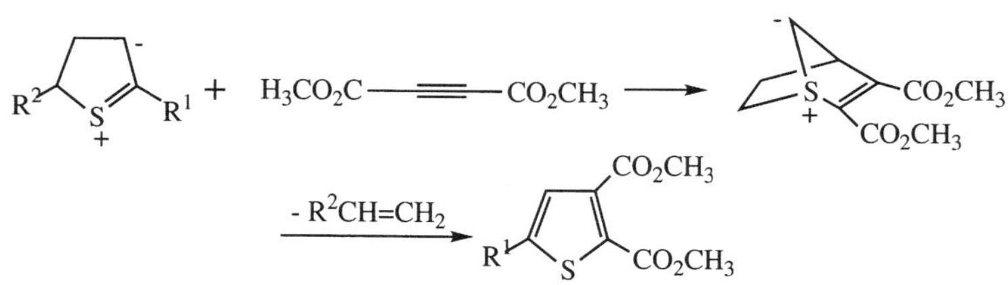

Dimethyl 4,5-bis(trifluoromethyl)thiophenedicarboxylate is prepared by reaction of dimethyl acetylenedicarboxylate with the cyclic disulfide *via* the intermediate 1,2-dithiin and sulfur extrusion [170].

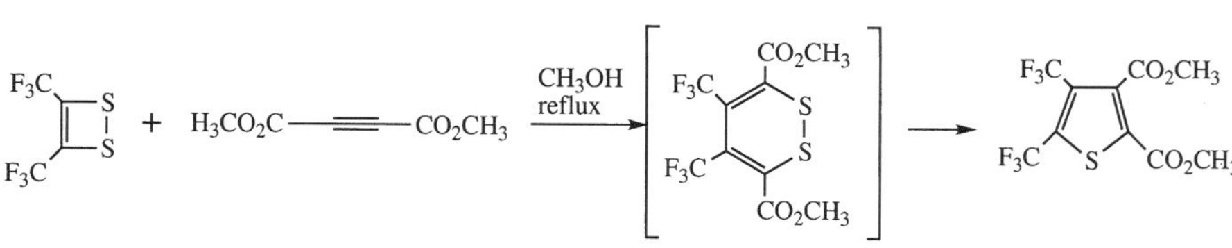

Reaction of dialkyl acetylenedicarboxylate with the Ni(II)complex derived from NiS_x and diphenylacetylene gives 4,5-diphenylthiophene-2,3-dicarboxylate in high yields, in this case probably *via* sulfur extrusion from an intermediate 1,4-dithiin [171,172].

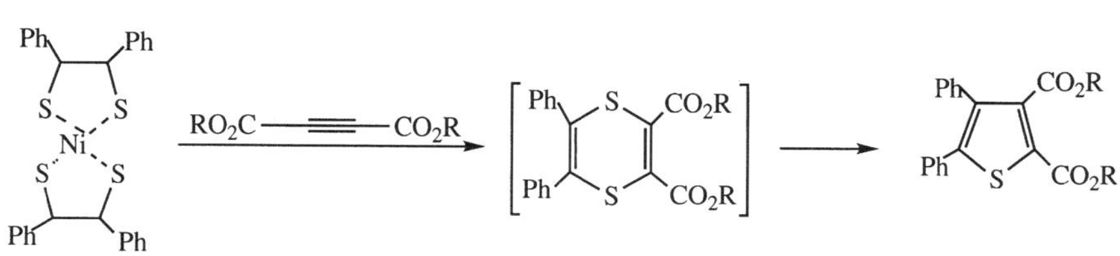

5-Methylthiophene-2,3-dicarboxylic acid can be obtained by the reaction of the diketone with sulfur [173].

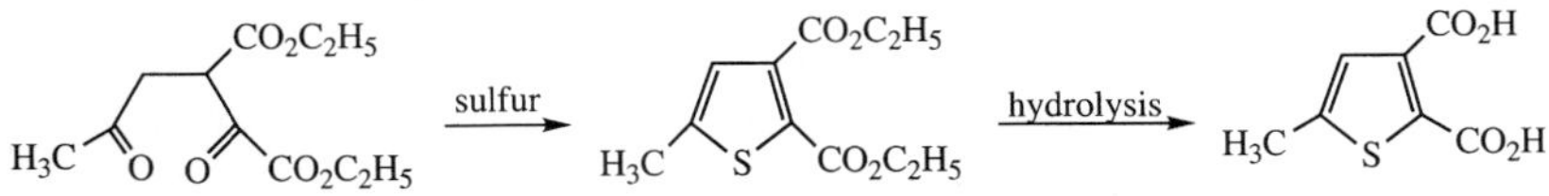

4F.2.3 Thiophene-2,4-dicarboxylic acids

Thiophene-2,4-dicarboxylic acids are not very accessible *via* ring closure reactions. The Paal synthesis applied to the ketoaldehyde ketal gives only low yields of thiophene-2,4-dicarboxylic acid [174].

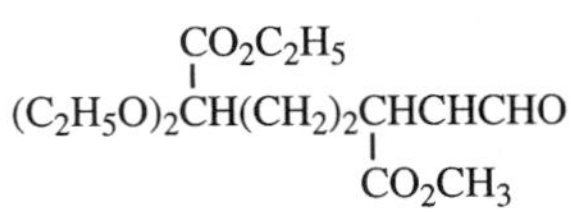

Excellent yields of 3,5-diarylthiophene-2,4-dicarboxylic acids are obtained by thermolysis of dialkyl 3,6-diaryl-1,4-dithiin-2,5-dicarboxylate [175,176].

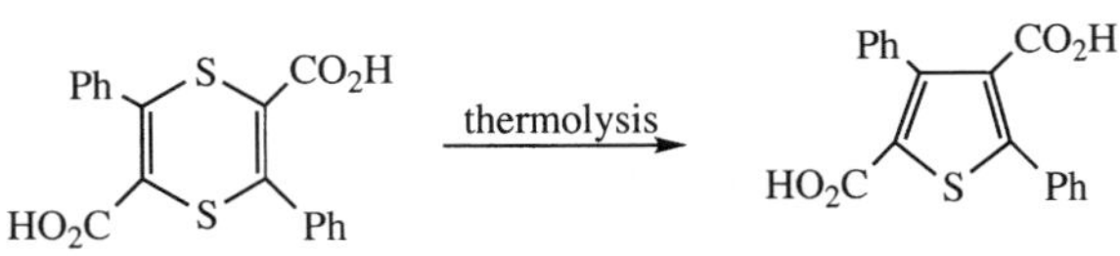

The reaction of the rhodanine condensate with sodium hydroxide is used for the following transformation [177].

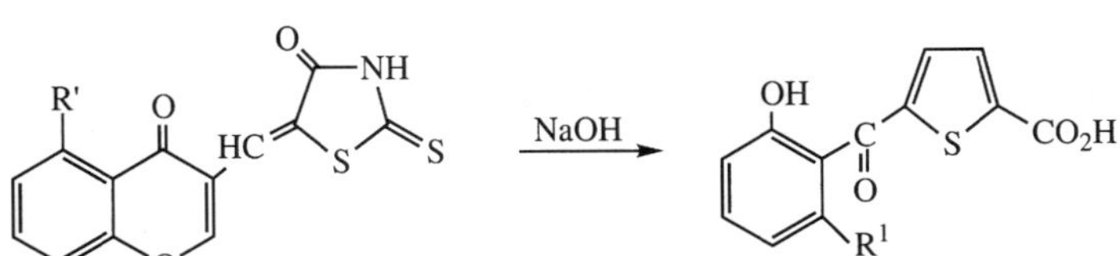

4F.2.4 Thiophene-3,4-dicarboxylic acids

The reaction of dimethyl or diethyl 2,3-diformyl succinate [178] or diethyl 1-formyl-2- diethoxymethylsuccinate with phosphorus pentasulfide in toluene gives thiophene-3,4-dicarboxylic acid [154,164].

Thiophene-3,4-dicarboxylic acid [164]

Anhydrous toluene (210 ml), diethyl 1-formyl-2-diethoxymethylsuccinate (50.0 g, 164.5 mmol) and phosphorus pentasulfide (35.48 g, 79.8 mmol) are placed in a flask fitted with a stirrer and reflux condenser. The mixture is stirred rapidly under reflux for 2 h. After cooling the toluene is decanted off from the dark resin. The resin is washed with toluene and this too is decanted off. The combined toluene solutions are washed with water (160 ml) and with ice-cold 2 *M* sodium hydroxide solution (2 × 160 ml), dried over magnesium sulfate and evaporated. The residue is distilled and the fraction boiling from 100 °C to 160 °C at 5 mm Hg is collected. This is hydrolyzed by heating with a solution of sodium hydroxide (6.46 g, 161.5 mmol) in water (16 ml) and ethanol (16 ml) for 30 min. After evaporation of the red solution the residue is treated with warm water (50 ml) and acidified with concentrated hydrochloric acid and chilled. The precipitate formed is filtered off and washed with cold diethyl ether giving 6.79 g (24%) of the title compound.

Cycloaddition extrusion reactions are quite useful for the preparation of thiophene-3,4-dicarboxylic acids. Thus the reaction of 2,2′-(1,4-phenylene)-bis(5-phenyl-1,3-dithiolylium)-4-olat with dimethyl acetylenedicarboxylate is used for the preparation of the thiophene derivative *via* the intermediate shown below [179].

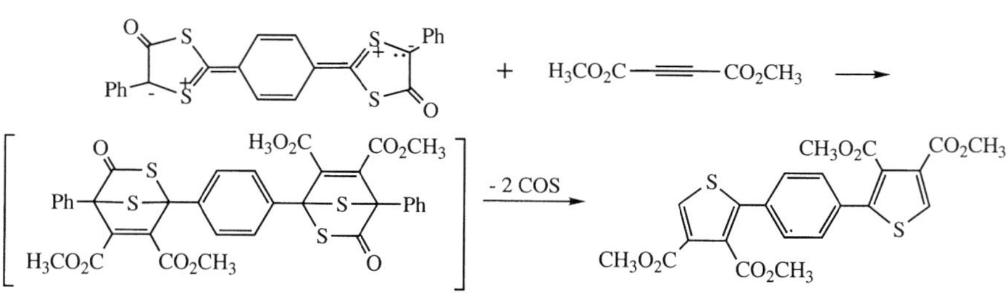

2,2′-(1,4-Phenylene)bis(5-phenyl-3,4-thiophenecarboxylic acid dimethylester [179]

A suspension of 2,2′-(1,4-phenylene)bis(5-phenyl-1,3-dithiolylium)-4-olat (0.461 g, 0.997 mmol) and dimethyl acetylenedicarboxylate (0.567 g, 3.99 mmol) in toluene (20 ml) is heated at 95–100 °C for 1.5 h. After decolorization the precipitate formed at room temperature is filtered off. The filtrate is evaporated and the residue is treated with methanol and placed in a refrigerator. The second crop of crystals are filtered off. The combined crops give 0.278 g (45%) of the title compound as small colorless needles mp 238 °C.

3,4-Thiophenedicarboxylic acid [164]

A solution of 3,4-dicyanothiophene (8.44 g, 62.9 mmol) and potassium hydroxide (23.30 g, 415.3 mmol) in ethylene glycol (90 ml) is stirred under reflux for 4 h. On cooling, the mixture is poured into water and the resulting solution is washed with diethyl ether. The aqueous phase is cooled in an ice bath and acidified with 2 *M* hydrochloric acid. The precipitate formed is filtered off, and taken up in diethyl ether, and the filtrate is thoroughly extracted with diethyl ether. The combined organic phases are dried over magnesium sulfate and evaporated giving 9.7 g (90%) of the title compound.

The [3+2] cycloaddition of suitably substituted mesoionic species to dimethyl acetylenedicarboxylate, followed by extrusion of a stable fragment can be used for the prepration of dimethyl 3,4-thiophenedicarboxylates [140,141,180–191].

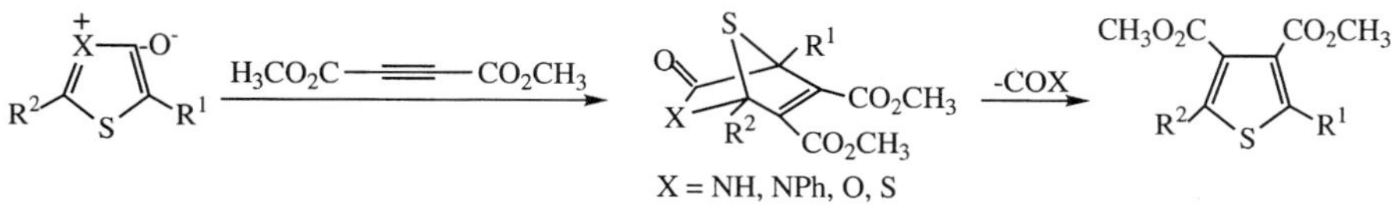

Oxidation with 30% hydrogen peroxide–acetic acid of 3,4-disubstituted 2,5- and 4,5-dihydrothiophenes with electron-withdrawing substituents at the 3- and 4-positions give excellent yields of, for instance, 3,4-dicarboxythiophenes [192]. The reaction of thiaphosphamanganacyclopentadiene complex with two equivalents of methyl propiolate gives a new complex, which upon heating loses carbon monoxide giving a third complex, which in turn gives the thiophene derivative [193].

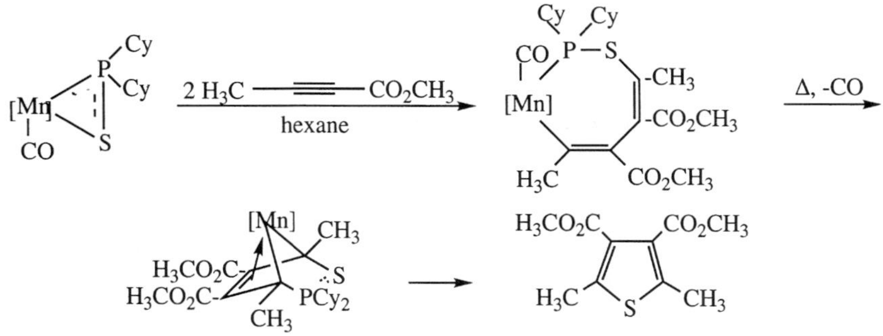

4F.2.5 Thiophenetri- and thiophenetetracarboxylic acids

Thiophene-2,3,4- and 2,3,5-tricarboxylic acids cannot be prepared easily, but have been obtained in overall low yields by the Paal–Knorr reaction from appropriate 1,4-dicarbonyl triesters [194]. Recently, 2,3,5-thiophenetricarboxylic acid trimethyl ester was prepared from a manganese complex by reaction with ammonium cerium nitrate [193].

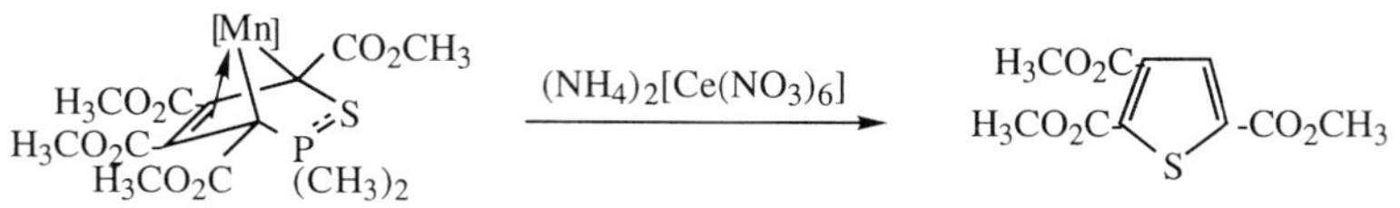

A simple method of synthesis of tetramethyl thiophenetetracarboxylate is the heating of dimethyl maleate or fumarate with sulfur at 150 to 200 °C [195]. Another route to preparation of thiophenetetracarboxylic acid consists in vigorous oxidation of tetrachlorothiophene to form the anhydride. This was transformed by reaction with hydrogen sulfide to the 1,4-dithiin, which upon treatment with nitric acid gave the desired product [196].

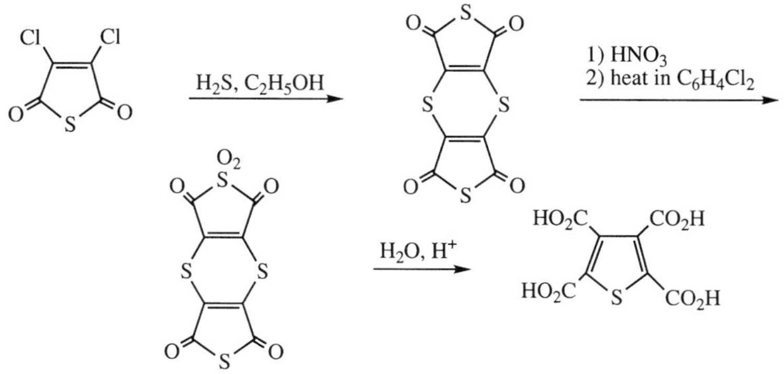

The tetramethyl ester is obtained from dimethyl acetylenedicarboxylate in a high-pressure process in which the alkyne is reacted with carbon monoxide in the presence of a manganese–phosphorus–sulfur catalyst *via* the complex structure as intermediate [197]. The tetracarboxylic acid is also prepared in excellent yield by hydrolysis with concentrated hydrochloric acid of 2,3,5,6-tetracyano-1,4-dithiin [198]. The recently discovered best method for the preparation of tetramethyl thiophenetetracarboxylate is the reaction of potassium *para*-toluenethiosulfonate with two equivalents of dimethyl acetylenedicarboxylate at room temperature in acetonitrile [199].

Tetramethyl thiophenetetracarboxylate [199]

To a stirred solution of dimethyl acetylenedicarboxylate (1.42 g, 10 mmol) in acetonitrile (20 ml) potassium *para*-toluenethiosulfonate (1.13 g, 5 mmol) is added. The reaction mixture is stirred at room temperature for 2 h and then poured into water. The product is extracted with dichloromethane and the combined organic phases are dried over magnesium sulfate and evaporated. The residue is purified by chromatography on silica gel using dichloromethane/hexane (1:2) as eluent. After recrystallization from methanol 1.20 g (76%) of the title compound is obtained mp 124–125 °C.

4F.3 FORMYL- AND ACYLTHIOPHENECARBOXYLIC ACIDS

4F.3.1 From bromothiophenecarboxylic acids *via* halogen–metal exchange and reaction with *N,N*-dimethylformamide

Both 3-and 4-formylthiophene-2-carboxylic acid were obtained in this way from the bromoacids, using two equivalents of butyllithium for the halogen–metal exchange [166].

4F.3.2 From protected thiophene aldehydes and ketones by metalation or halogen–metal exchange and carbonation

5-Formyl-2-thiophenecarboxylic acid is prepared by metalation of 2-thiophene-aldehyde diethyl acetal by butyllithium followed by carbon dioxide [43,200]. 3-Thiophenealdehyde ethylene acetal upon metalation followed by reaction with carbon dioxide gives 3-formyl-2-thiophenecarboxylic acid after hydrolysis [160]. 5-Acetyl-2-thiophenecarboxylic acid is obtained by metalation of 2-acetyl-thiophene diethyl ketal followed by carbon dioxide [201]. Metalation of the ethylene ketal of 3-acetylthiophene, followed by reaction with carbon dioxide is used for the preparation of 3-acetyl-2-thiophenecarboxylic acid [202].

3-Acetyl-2-thiophenecarboxylic acid [202]

A solution of 2-methyl-2-(3-thienyl)-1,3-dioxolane (23.3 g, 0.136 mol) in anhydrous diethyl ether (50 ml) is added dropwise at −10 °C to butyllithium in diethyl ether prepared from lithium (2.54 g) and butyl bromide (25.4 g, 0.185 mol). The mixture is refluxed for 15 min and then poured into crushed solid carbon dioxide covered by anhydrous diethyl ether. After 30 min water is added, the phases are separated, pH of the aqueous phase is adjusted to 8 and extracted with diethyl ether. The organic phase containing the acetal is treated with 5% hydrochloric acid (120 ml) for 2 h. The precipitate formed is filtered off giving 16.24 g (73%) of the title compound as white crystals mp 151 °C.

Metalation of 5-ethyl-2-thiophenealdehyde diethyl acetal occurs in the 3-position and upon carbonation and hydrolysis 5-ethyl 2-formyl-3-thiophene-carboxylic acid is obtained [203]. Halogen–metal exchange of 4-bromo-3-thiophene aldehyde ethylene acetal followed by carbonation and hydrolyis is used for the preparation of 4-formyl-3-thiophenecarboxylic acid [166]. Using the acetal derived from acetylbromothiophenes, acetylthiophenecarboxylic acids are prepared [204] and similarly 2-acetyl-5-methyl-4-thiophenecarboxylic acid is prepared from 2-acetyl-4-bromo-5-methylthiophene ethylene ketal [21]. Metalation of 2-methyl-2-(2-thienyl)-1,3- dioxolane with butyllithium followed by carbonation and hydrolysis is used for the preparation of 2-acetyl-5-thiophenecarboxylic acid in 70% yield [205].

Halogen–metal exchange of 4-bromo-2-acetylthiophene diethyl ketal followed by carbon dioxide is used for the preparation of 5-acetyl-3-thiophenecarboxylic acid in 65% yield [206]. The reaction of 3-thienyllithium with *N,N*-dimethylformamide at −70 °C gives protected 3-thiophene aldehyde and upon metalation and reaction with carbon dioxide, 3-formyl-2-thiophenecarboxylic acid is obtained in an one-pot procedure [17].

3-Formyl-2-thiophenecarboxylic acid [17]

A solution of 3-bromothiophene (16.3 g, 0.10 mol) in anhydrous diethyl ether (50 ml) is added during a few minutes with stirring to 0.74 *M* butyllithium in diethyl ether (148 ml) at −70 °C. The stirring is continued for 15 min, after which *N,N*-dimethylformamide (8.03 g, 0.11 mol) is added in one portion and the reaction mixture is allowed to warm to room temperature. After 1 h 0.74 *M* butyllithium in diethyl ether (172 ml) is added dropwise and the reaction mixture is refluxed for 135 min and then poured onto solid carbon dioxide covered with anhydrous diethyl ether. Standard work up gives 6.9 g (44%) of the title compound mp 130–131 °C after recrystallization from benzene/ligroin.

4F.3.3 From dihalothiophenes by one-pot procedures

2-Formyl-3-thiophenecarboxylic is prepared by reacting 2,3-dibromothiophene with one equivalent of butyllithium at −70 °C, followed by *N,N*-dimethylformamide, renewed reaction with butyllithium and carbonation. In the same way 2-formyl-4-thiophenecarboxylic acid is obtained from 2,4-dibromothiophene [17].

4F.3.4 By electrophilic substitution of thiophenecarboxylic acid derivatives

Ethyl 2-thiophenecarboxylate gives the 5-acetyl derivative with acetic anhydride/zinc chloride [207]. 5-Acyl-3-thiophenecarboxylic esters have also been prepared in this way [208]. The reaction of thiophene-3,4-dicarboxylic acid anhydride with benzene catalyzed by aluminum chloride is used for the preparation of 4-benzoyl-3-thiophenecarboxylic acid in 67% yield [209,210].

4F.3.5 Hydrolysis of carbonyl-substituted cyanothiophenes

All isomeric cyanothiophene aldehydes can be transformed into the methyl and ethyl esters of the formylthiophenecarboxylic acids *via* the iminoethers [211].

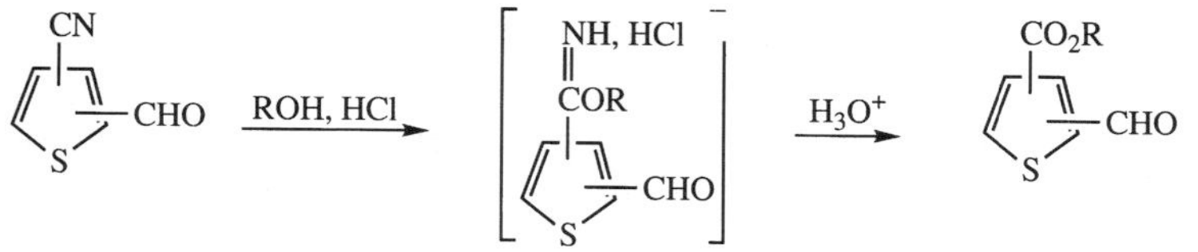

4F.3.6 Oxidation of side chains in acetylthiophenes

2-Acetyl-4-thiophenecarboxylic acid is prepared by oxidation of 2-acetyl-4-hydroxymethylthiophene [212]. 5-Acetyl-2-thiophenecarboxylic acid is obtained by oxidation of ethyl 5-acetyl-2-thenylsulfone with potassium permanganate [96] and of 5-formyl-2-acetylthiophene with silver oxide [107]. Oxidation of 5-dimethylaminomethyl-3-acetyl-2-methylthiophene with potassium permanganate is used for the preparation of 3-acetyl-2-methyl-5-thiophenecarboxylic acid [161]. Hypochlorite oxidation of 2,2′-methylene-bis(5-acetylthiophene) is used for the preparation of 2,2′-carbonylbis(5-thiophenecarboxylic acid) [213].

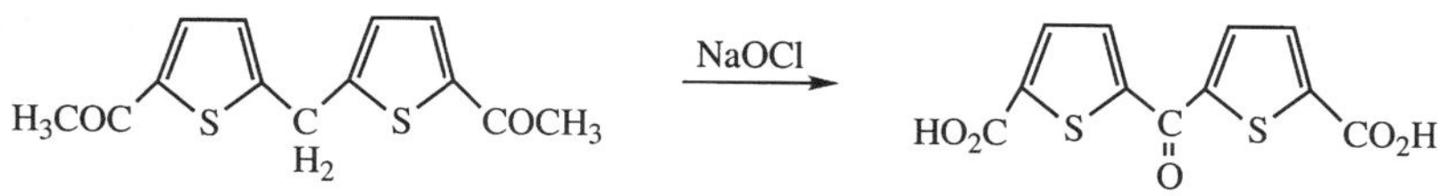

4F.3.7 Oxidation of side chains in thiophenecarboxylic acids

Oxidation of 5-hydroxymethyl-2-thiophenecarboxylic acid with 40% nitric acid is used for the preparation of 5-formyl-2-thiophenecarboxylic acid [61].

4F.4 THIOPHENECARBOXYLIC ACID HALIDES

This class of compounds is generally prepared by standard methods. 3-Thenoyl chloride is obtained by the reaction of 3-thiophenecarboxylic acid with thionyl chloride [214,215]. A more recent example is the preparation of 3-(2-nitro-1-pyrrolylmethyl)thiophene-2-acid chloride by the reaction of the acid with thionyl chloride in benzene [102], and of thiophene-2,3-dicarbonyl chloride through the reaction of the diacid with phosphorus pentachloride [163].

3-(2-Nitro-1-pyrrolylmethyl)thiophene-2-carboxylic acid chloride [102]

To a well-stirred solution of 3-(2-nitro-1-pyrrolylmethyl)thiophene-2-carboxylic acid (3 g, 11.9 mmol) in anhydrous benzene (120 ml) thionyl chloride (1.43 g, 12 mmol) in anhydrous benzene (80 ml) is cautiously added dropwise.

The stirring is continued for 15 min at room temperature and then the reaction mixture is refluxed for 1.5 h. After cooling it is evaporated and the residue recrystallized from ligroin giving 3.1 g (94%) of the title compound mp 90–91 °C.

4F.5 THIOPHENECARBOXYLIC ACID AZIDES

Many thenoyl azides, which are useful as synthetic intermediates, are prepared by standard procedures by the reaction of the acid chlorides, or a mixture of the acid ethyl chloroformate and triethylamine with sodium azide. Alternatively treatment of the acyl hydrazide with nitrous acid can be used. In this way 2-thenoyl azide [216–219], alkyl substituted 2-thenoyl azides [158,218] as well as 3-thenoyl azides are prepared [217,220] 4-(1-pyrrolylmethyl)thiophene-3-carbonyl azide and 4-[1-(2-formyl)pyrrolylmethyl]thiophene-3-carbonyl azide were prepared from the acids in anhydrous acetone in the presence of triethylamine and reaction with ethyl chloroformate [101,221]. Another alternative is the treatment of 2- and 3- thiophenecarboxylic acid as well as 5-methyl-2-thiophenecarboxylic acid with diphenylphosphoric azide [222].

4-(1-Pyrrolylmethyl)thiophene-3-carbonyl azide [101]

A solution of 4-(1-pyrrolylmethyl)thiophene-3-carboxylic acid (2.07 g, 10 mmol), anhydrous acetone (75 ml) and anhydrous triethylamine (2.73 g, 27 mmol) is cooled in an ice bath. To the well-cooled and stirred solution under nitrogen, ethyl chloroformate (3.8 g, 35 mmol) in anhydrous acetone (7.5 ml) is added over a period of 30 min. The stirring is continued at 0 °C (2.92 g, 45 mmol) and cold water is added dropwise over a period of 20 min. The reaction mixture is stirred at 0 °C for 30 min and the poured onto crushed ice. The product is extracted with carbon tetrachloride (3 × 50 ml). The combined organic phases are washed with water, dried over magnesium sulfate and evaporated at 30–35 °C. The oily residue 2.11 g (91%) of the title compound is used without further purification.

4F.6 THIOPHENECARBOXAMIDES

4F.6.1 From thiophenecarboxylic acids and acid chlorides

Thiophenecarboxylic acid chlorides are reacted with a wide range of amino compounds to give amides. A few more recent examples will be given. The reaction of 2-thiophenecarboxylic acid with thionyl chloride followed by aqueous methyl amine is used for the preparation of *N*-methyl-2-thiophenecarboxamide

[223]. The reaction of 2-thiophenecarbonyl chloride in dichloromethane with 2,4,6-trimethylaniline and triethylamine in methylene chloride is used for the preparation of *N*-(2,4,6-trimethyl)phenylthiophene-2-carboxamide [223].

N-Methyl-2-thiophenecarboxamide [223]

2-Thiophenecarboxylic acid (40.0 g, 0.31 mol) and thionyl chloride (112 ml, 1.55 mol) are heated under reflux for 5 h. The excess of thionyl chloride is removed under reduced pressure and the crude acid chloride is added dropwise to a solution of aqueous methylamine (0.31 mol) in 10% aqueous sodium hydroxide (50 ml). The stirring is continued for 12 h, after which the product is extracted with ethyl acetate (5 × 30 ml). The combined organic phases are washed with 5% aqueous hydrochloric acid (30 ml), water (2 × 30 ml) and sodium chloride solution (30 ml), dried over magnesium sulfate and evaporated. The residue is recrystallized from ethyl acetate/light petroleum giving 36.28 g (83%) of the title compound as colorless crystals mp 111–113 °C.

The reaction of 5-carboxy-2-acetylthiophene with thionyl chloride in toluene followed by treatment with aqueous ammonia gave 2-acetyl-5-thiophenecarboxamide [224].

2-Acetyl-5-thiophenecarboxamide [224]

To a suspension of 5-acetyl-2-thiophenecarboxylic acid (22 g, 0.129 mol) in toluene (200 ml) thionyl chloride is added dropwise under vigorous stirring. The mixture is gently refluxed for 2 h and then evaporated to dryness. The residue is dissolved in toluene (150 ml) and treated with aqueous ammonia below 10 °C. The white precipitate formed is filtered off, washed with water and dried giving 20 g (93%) of the title compound mp 226–228 °C after recrystallization from methanol.

N-(2-thenoyl)-L-proline, *N*-(3-thenoyl)-L-proline as well as *N*-thenoyl-2-L-thiazolidine-4-carboxylic acids are prepared by treatment of the corresponding acids with a large excess of thionyl chloride. The excess thionyl was distilled off *in vacuo* and the acid chlorides in acetone reacted with L-proline and L-thiazolidine-4-carboxylic acid [225].

The reaction of 2- and 3-thenoyl chloride with *N*-methyl(trimethylsilyl)-methylamine is used for the preparation of *N*-methyl-*N*[(trimethylsilyl)methyl]-2-thiophenecarboxamide and -3-thiophenecarboxamide [226].

N-methyl-N[(trimethylsilyl)methyl]-2-thiophenecarboxamide [226]

A mixture of 2-thiophenecarbonyl chloride (13.7 g, 93.6 mmol) and *N*-methyl-(trimethylsilyl)methylamine (10.8 g, 92.1 mmol) in 10% sodium hydroxide solution (100 ml) is stirred at room temperature for 1 h, after which it is poured into water (150 ml). The product is extracted with ethyl acetate and the combined organic phases are dried over magnesium sulfate, evaporated and distilled giving 20.1 g (96%) of the title compound bp 105–107 °C/0.4 mm Hg.

The reaction of 2,5-thiophenedicarbonyl chloride reacts with 2-mercaptoethyl amines to give *N*, *N'*-bis(substituted-2-thioethyl)2,5-thiophenedicarboxamides [227]. The crude acid chloride prepared by the reaction of 5-carbomethoxy-2-thiophenecarboxylic acid with thionyl chloride is reacted with aminoacetone hydrochloride and α-aminoacetophenone to give *N*-(5-methoxycarbonyl-2-thenoyl)-α-aminoacetone and *N*-(5-methoxycarbonyl-2-thenoyl)-α-amino-acetophenone, respectively [50].

N-(5-Methoxycarbonyl-2-thenoyl)-α-aminoacetophenone [50]

A stirred suspension of 5-carbomethoxy-2-thiophenecarboxylic (1.50 g, 8.06 mmol) in 10 ml thionyl chloride is refluxed for 1.5 h, cooled to room temperature and evaporated. The crude acid chloride is dissolved in chloroform (10 ml) and added dropwise at 0 °C to a stirred solution of α-aminoacetophenone hydrochloride (1.45 g, 8.45 mmol) and triethylamine (1.71 g, 8.45 mmol). The suspension so obtained is stirred under cooling for 15 min and evaporated. The residue, a solid, is stirred in 1 *M* hydrochloric acid for 15 min, filtered off and recrystallized from ethanol giving 2.20 g (90%) of the title compound as a light tan solid mp 170–172 °C.

3-Methyl-2-thiophenecarboxylic acid was converted to its potassium salt, which was transformed to the acid chloride upon reaction with oxalyl chloride. Reaction with dimethyl L-glutamate hydrochloride gave the desired amide [98].

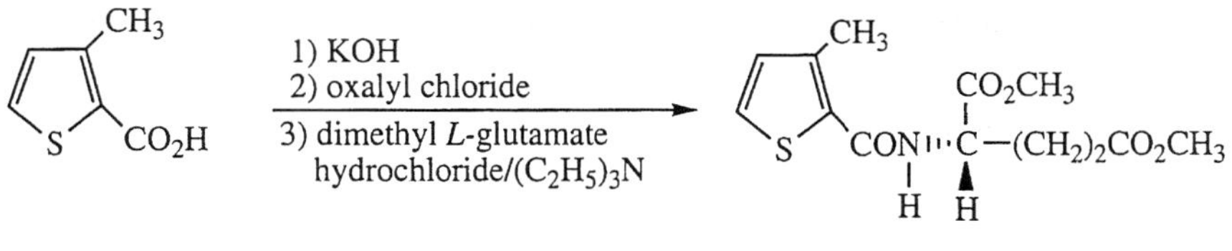

Dimethyl N-(3-methyl-2-thienylcarbonyl)-L-glutamate [98]

The potassium salt of 3-methyl-2-thiophenecarboxylic acid is prepared by stirring the acid (5.68 g, 40 mmol) with potassium hydroxide (3.15 g, 56.1 mmol) in refluxing methanol (75 ml) for 10 min. The mixture is evaporated to dryness, the residue stirred in benzene (175 ml) for 1 h, the solution is cooled in an ice bath, after which one drop of *N*,*N*-dimethylformamide is added followed by slowly addition of oxalyl chloride (36.25 g, 285 mmol) over a period of 45 min. The stirring is continued at room temperature for 1 h and the solution evaporated to dryness. The excess of oxalyl chloride is removed azeotropically by evaporation with chloroform. The resulting crude acid chloride is added to a stirred suspension of dimethyl *L*-glutamate hydrochloride (8.45 g, 40 mmol) and triethylamine (12.15 g, 120 mmol) in dichloromethane (500 ml) at 0 °C. After stirring the reaction mixture at room temperature for 20 h it is washed with aqueous sodium bicarbonate, 1 *M* hydrochloric acid and sodium chloride solution, dried over sodium sulfate and evaporated. The residue is purified by chromatography using ethyl acetate/hexane (2:8) as eluent, giving 11.05 g (92%) of the title compound as a clear yellow oil.

2-Thiophenecarboxylic acid reacts with 1,1′-oxalyldiimidazole or 1,1′-oxa--oxalyldi(1,2,4-triazole) in acetonitrile at 40 °C to give the corresponding 1-thenoylazole intermediates which promptly undergo aminolysis to give amides or dipeptides [228].

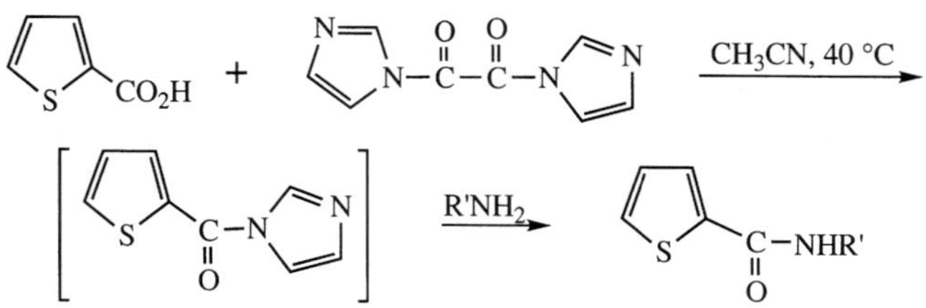

In connection with a study of ion-exchange resins for solution phase parallel synthesis of chemical libraries, nine basic ion-exchange resins were evaluated for the catalysis and purification of the benzyl amide of 2-thiophencarboxylic acid from 2-thenoyl chloride and benzyl amine [229]. *N*-(2-Hydroxy-1,1-dimethylethyl)-2-thiophenecarboxamide is prepared in a yield of 88% by the reaction of 2-thiophenecarbonyl chloride with 2-amino-2-methyl-1-propanol in anhydrous methylene chloride at 0 °C [230]. The unusual product, bis(*N*-2-thenoyl)-4,4-dimethyl-2-oxazolinylidene, is obtained by the reaction of 2-thenoyl chloride with 2-(tributylstannyl)-4,4-dimethyl-2-oxazoline [231].

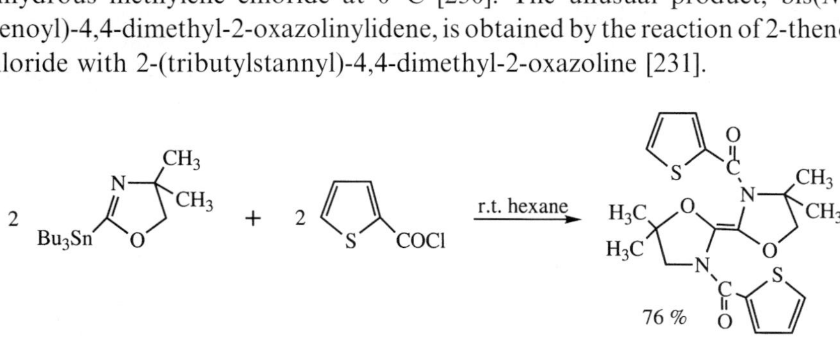

4F.6.2 By reactions of thiophenecarboxamides

The following reaction sequence has been performed from 5-bromo-2-thiophenecarboxylic acid using 1-[3-(dimethylamino)propyl]-3-ethylcarbodiimide

(EDCI) as coupling agent and a palladium-catalyzed coupling with 2-pivaloyl-6-ethynyl-5-deazapterin [232].

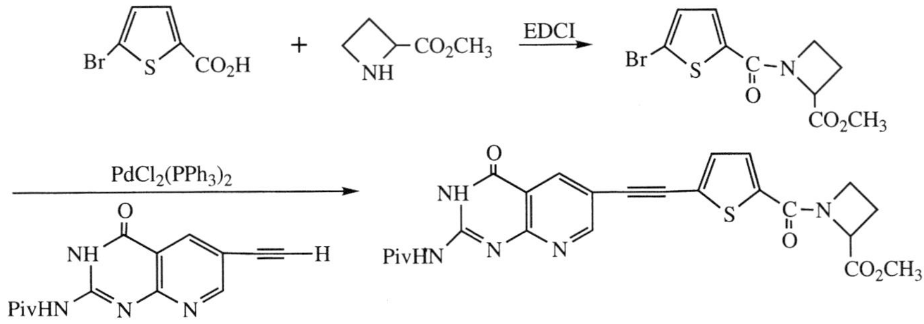

N-{5-[2-(2-Pivaloylamino-3.4-dihydro-4-oxopyrido[2,3-d)pyridin-6yl)ethynyl]thien-2-ylcarbonyl}-2-carbomethoxyazetidine [232]

A mixture of *N*-(5-bromothien-2-ylcarbonyl)-2-carbomethoxyazetidine (0.608 g, 2 mmol), 2-pivaloyl-6-ethynyl-5-deazapterin (1.08 g, 4 mmol), bis(triphenylphosphine)palladium(II) chloride (0.070 g, 0.10 mmol) and triethylamine (1.01 g, 10 mmol) in acetonitrile (50 ml) is heated under argon overnight and then cooled to room temperature. After filtration the filtrate is concentrated and the residue purified by chromatography on silica gel using methanol/dichloromethane (1–5/99–95) as eluent, giving 0.94 g (95%) of the title compound as a pale-yellow solid mp 198–200 °C.

Another example of the use of a palladium-catalyzed reaction is the Negishi cross-coupling of the zinc bromide derivative derived from 2-bromo-3-thiophenecarboxylic acid dietylamide with *ortho*-bromotoluene. On the other hand the Suzuki coupling of 3-diethylaminocarbonyl-2-thiopheneboronic acid with *ortho*-bromotoluene gave only low yields. However, excellent yields were obtained if the Suzuki coupling was applied to *ortho*-toluylboronic acid and 2-bromo-3-thiophenecarboxylic acid diethyl amide [233].

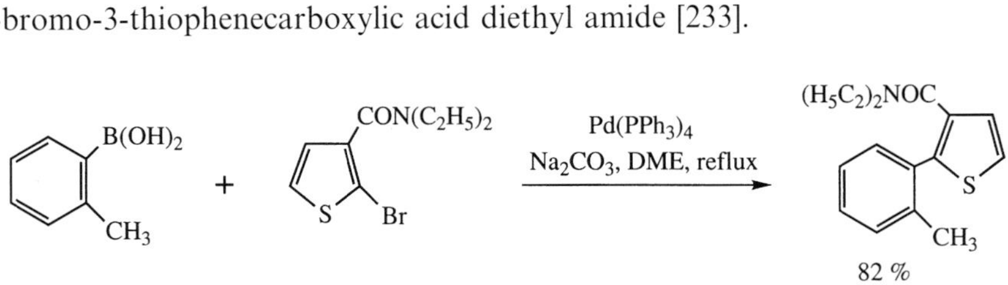

Metalation of *N*,*N*-diethylthiophene-3-carboxamide with butyllithium *N*,*N*,*N*′,*N*′-tetramethylethylenediamine at −78 °C followed by reaction with pentafluoroacetone is used for the preparation of *N*,*N*-diethyl-2-(1,1,1,3,3-pentafluoro-2-hydroxypropan-2-yl)thiophene-3-carboxamide [234].

N,N-Diethyl-2-(1,1,1,3,3-pentafluoro-2-hydroxypropan-2-yl)thiophene-3-carboxamide [234]

To a solution of 1.6 *M* butyllithium in hexane (3.8 ml, 6 mmol) and *N,N,N′,N′*-tetramethylethylenediamine (0.75 g, 6 mmol) in anhydrous diethyl ether (20 ml) at −78 °C stirred for 25 min, *N,N*-diethyl thiophene-3-carboxamide (0.915 g, 5 mmol) in anhydrous diethyl ether (10 ml) is slowly added. The stirring is continued for 1.5 h, after which pentafluoroacetone is introduced through a gas inlet tube. The reaction mixture is stirred, first at −78 °C for 7 h and then at room temperature for 12 h. Upon work up 55% of the starting material is recovered and the mixture is separated by preparative thin layer chromatography using dichloromethane as eluent, giving 101.1 mg (13.6%) of the title compound as a yellow oil.

Metalation of 2-*tert*-butylcarboxamido-3-methylthiophene with butyllithium in tetrahydrofuran gives first the lithium azaenolate, which upon further lithiation in the methyl group gives the dilithio derivative, which upon reaction with electrophiles yields 2,3-disubstituted thiophenes [235].

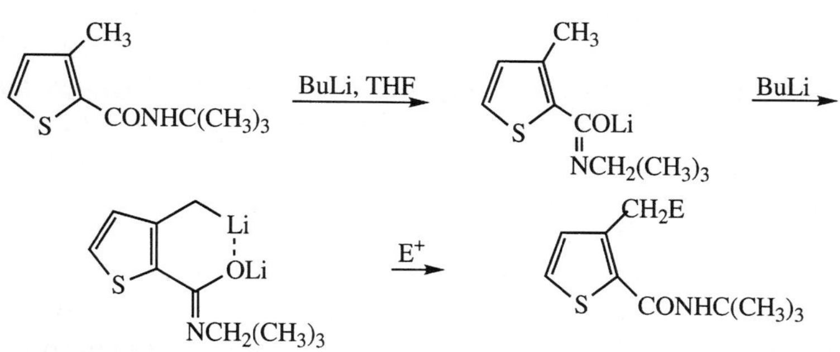

Thiophene-2-carboxanilide reacts with excess chlorosulfonic acid to give 4′-sulfonyl chloride [236].

4F.7 THENOYLSILANES

A high-yielding synthesis of a variety of functionalized acylsilane has also been applied to 2-thiophenecarbonyl chloride, which upon reaction with the copper–zinc complex $(PhMe_2Si)_2$–$CuCN(ZnCl)_2$ in tetrahydrofuran at −20 °C gave dimethylphenyl-2-thenoyl silane [237].

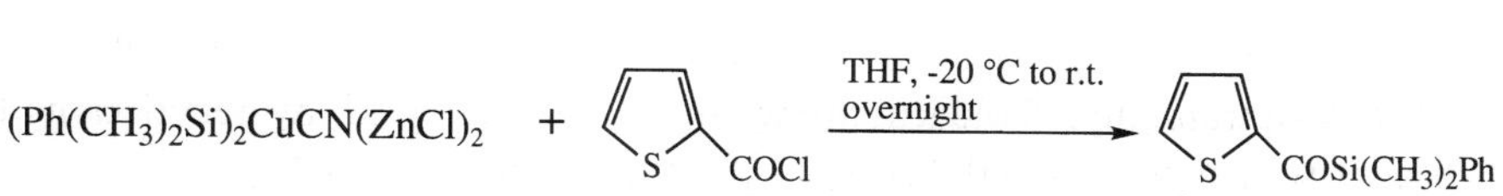

General procedure for acylsilanes [237]

A solution of dimethylphenylsilyllithium in tetrahydrofuran (5 ml, 2 mmol) is treated at −78 °C under argon with 1 *M* solution of zinc chloride in diethyl ether (2 ml, 2 mmol) and the temperature is raised to 0 °C, while the color of the solution changes from dark brown to yellow. This solution is transferred *via* cannula to a stirred suspension of cuprous cyanide (0.089 g, 1 mmol) in tetrahydrofuran (4 ml) at −78 °C. The mixture is allowed to warm to 0 °C, when a dark violet, homogeneous solution is obtained. The appropriate carboxylic acid chloride (1 mmol) in tetrahydrofuran (2 ml) is added *via* syringe to the silyl-zinc cuprate at −20 °C. After stirring the reaction mixture at 0 °C for 3 h and at room temperature overnight the reaction is quenched by addition of saturated aqueous ammonium chloride solution and the product is extracted with diethyl ether. The combined organic phases are dried over sodium sulfate and evaporated. The residue is purified by chromatography on silica gel using petroleum ether/diethyl ether (gradient from 100:1 to 10:1) giving the acylsilanes.

4F.8 THIOPHENECARBOXYLHYDRAZIDES AND -HYDRAZONES

2-Thiophenecarboxyhydrazide is prepared by refluxing an ethanolic solution of the ester with hydrazine hydrate [238–240].

2-Thiophenecarboxyhydrazide [240]

Ethyl 2-thiophenecarboxylate is heated with 95% hydrazine in 95% ethanol for 5 h. The hydrazide is treated with hot methanol and carbon black giving 75% of the title compound mp 134–136 °C.

1,2-Bis(2-thiophenecarboxy)hydrazine is obtained in 87% yield by the reaction of 2-thenoyl chloride with hydrazine hydrate [239]. 2-Thiophenecarboxyl hydrazide reacts smoothly with alkyl isocyanates to give the semicarbazide shown below [241].

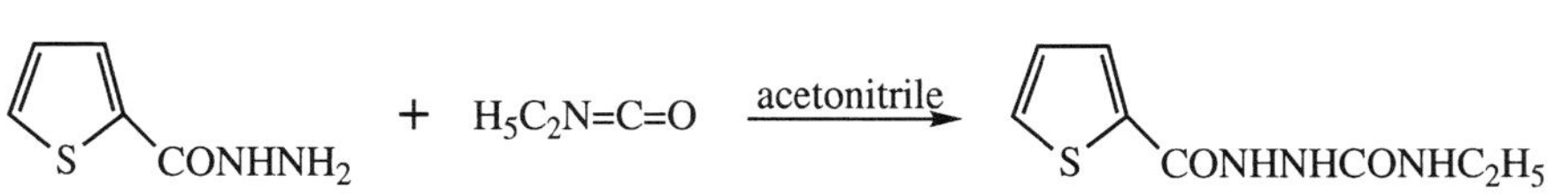

Thenoyl hydrazones of the thiophene series were prepared by the reaction of 2-thenoylhydrazines with thiophene aldehydes in connection with work on tuberculostatic compounds [241–243].

4F.9 THIENYL IMIDATES AND RELATED COMPOUNDS

This class of compounds prepared by standard methods, such as the reaction of the cyanothiophene with an alcohol and hydrogen chloride [223,244–246].

2-Thiophenecarboximidic acid ethyl ester hydrochloride [244]

A stirred solution of 2-cyanothiophene (54.6 g, 0.50 mol) and absolute ethanol (34.5 g, 0.75 mol) in anhydrous diethyl ether (300 ml) is cooled to 0 °C and is saturated with anhydrous hydrogen chloride passed into the solution at such a rate that the temperature is kept between 0 and 5 °C. The cooling bath is removed and the reaction mixture is stirred for 24 h. The crystalline precipitate formed is filtered off, rapidly washed with anhydrous diethyl ether and dried *in vacuo* over silica giving 86.2 g (91%) of the title compound mp 126–128 °C.

The imidates of 3-thiophenecarboxylic acid [244,246] as well as of various dicarboxylic acids have been obtained in this way [211,246,247]. Selective reaction of one of the cyano groups has been achieved [211]. Also the treatment of 2-cyanothiophene, but not 3-cyanothiophene with sodium borohydride in ethanol, gives the imidate [246]. Another procedure applied to 4-nitro-2-cyanothiophene, consisting in the treatment with methoxide in dimethylsulfoxide followed by acid, also gives the imidate [248]. A recent approach to thienyl imidates starts from *N*-methyl- and *N*-mesityl-2-thienylcarboxamides which by reflux with thionyl chloride gives the *N*-substituted thiophenecarboximidoyl chloride, which upon reaction with sodium methoxide and methanol in tetrahydrofuran at 0 °C gives the methyl-*N*-substituted-2-carboximidates [223]. Metalation of methyl *N*-methylthiophenecarboximidate and methyl *N*-(2,4,6-trimethyl)phenylthiophene-2-carboximidate occurs selectively in the 5-position in contrast to the 3-metalation of the oxazolidine derivative a cyclic imidate (cf below) [223].

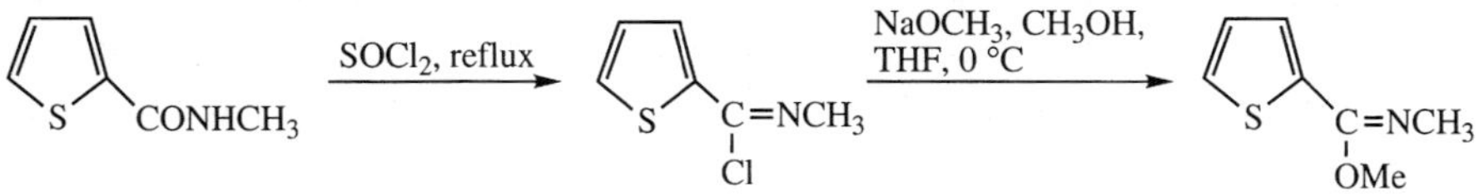

Methyl N-methylthiophene-2-carboximidate [223]

A solution of sodium methoxide (1.0 g, 18.52 mmol) in methanol (15 ml) is added dropwise to *N*-methylthiophene-2-carboximidoyl chloride (1.02 g, 6.39 mmol) in tetrahydrofuran (10 ml) at 0 °C. The reaction mixture is stirred at 0 °C for 0.5 h, allowed to reach room temperature and stirred for 12 h and then evaporated. The residue is diluted with diethyl ether (40 ml) and this solution is washed with sodium bicarbonate solution (2 × 20 ml), water (20 ml) and sodium chloride solution (20 ml), dried over magnesium sulfate, evaporated and distilled giving 0.93 g (97%) of the title compound as a colorless oil bp 60 °C/1.0 mm Hg.

4F.10 THIENYLAMIDINES AND RELATED COMPOUNDS

Thienylamidines and related compounds are normally obtained by the reaction of imidates with ammonia [244], ammonium chloride [211] or ammonium benzenesulfonate [211,246,249].

2-Thiophenecarboxamidine hydrochloride [244]

2-Thiophenecarboximidic acid ethyl ester hydrochloride (82.2 g, 0.43 mol) is added in portions to a stirred solution of ammonia (16.3 g, 0.96 mol) in absolute ethanol (200 ml), after which the mixture is stirred at room temperature for 48 h. If some ammonium chloride has precipitated, it is filtered off, the solvent is evaporated and the residue consists of 67.8 g (97%) of the title compound mp 177–178 °C.

N-substituted amidines are prepared by the use of amines in the reaction [250,251], as well as by the action of sodamide on thienylidenephenylhydrazones [252], and by the action of ethoxycarbonylthiophene-2-carboxythioamide [253]. The catalyzed direct addition of an amine to 2-cyanothiophene has also been used for direct preparation of the amidine [254]. The reaction of 2-cyanothiophene and hydroxylamine is used for the preparation of the amidoximes [255,256]. *N*-substituted derivatives are prepared from an amine and a hydroxyamyl chloride, prepared by the action of nitrosyl chloride on an oxime [257]. Addition of carbon tetrachloride to a stirred suspension of 1-phenyl-2-(2-thenoyl)hydrazine and triphenylphosphine in acetonitrile at room temperature is used for the preparation of *N*-phenyl-*C*-(2-thienyl)formohydrazidoyl chloride [258,259].

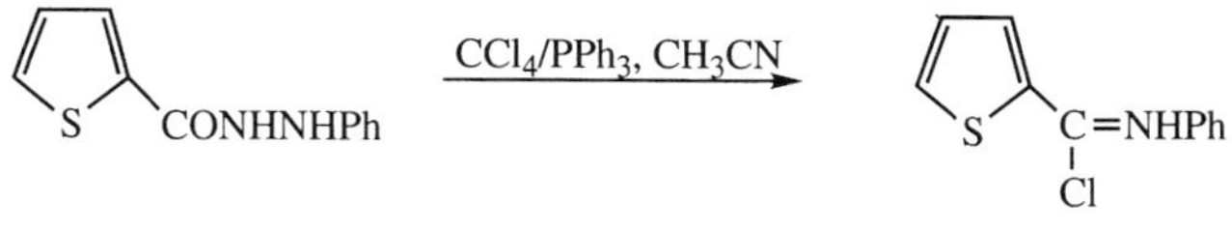

N-Phenyl-C-(2-thienyl)formohydrazidoyl chloride [258]

Carbon tetrachloride (2 ml, 20 mmol) is added to a stirred suspension of 1-phenyl-2-(2-thenoyl)hydrazine (4.3 g, 20 mmol) and triphenylphosphine (6.55 g, 25 mmol) in acetonitrile (40 ml), dried by passage through an alumina column and introduced directly from the column into the reaction flask. The stirring is continued for 10 h, after which water (100 ml) is added. The phases are separated and the organic phase dried over sodium sulfate and evaporated. The residue is crystallized from methanol giving 3.6 g (80%) of the title compound mp 87 °C.

This method was extended to the preparation of *C*-(2-thienyl)-*N*-(*para*-nitrophenyl)formohydazinoyl chloride and bromide [260]. The chloride upon reactions with equivalent amounts of nucleophiles, such as sodium cyanide, sodium phenoxide, sodium thiophenolate, sodium phenylsufinate, sodium azide as well as morpholine and phenyl hydrazine give the corresponding substitution products [260].

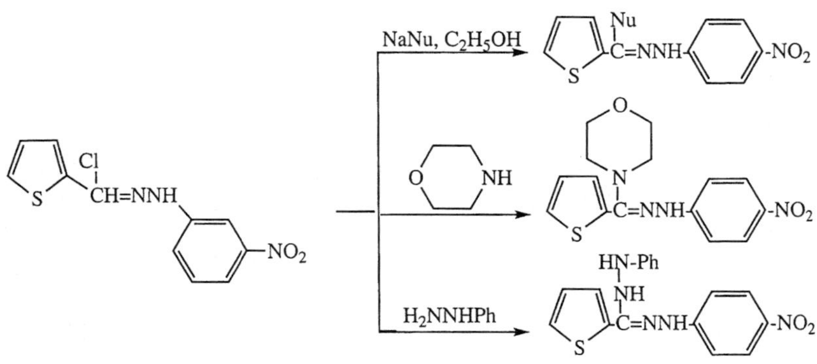

4F.11 THIENYL 2-OXAZOLINES

4F.11.1 From thiophenecarbonyl chlorides

Oxazolines derived from thiophenecarboxylic acids have become of increasing importance as directing groups in the metalation reaction. They are prepared by the reaction of thiophene carbonyl chlorides with 2-amino-2-methyl-1-propanol to *N*-(2-hydroxy-1,1-dimethylethyl)-2-thiophenecarboxamide, followed by ring closure upon treatment with thionyl chloride [230,261,262].

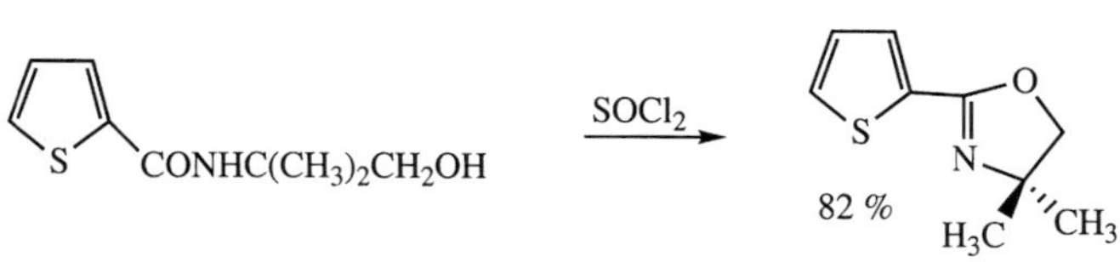

4F.11.2 From oxazolines by metalation followed by reaction with electrophiles

2-[3-(α-Hydroxybenzyl)-2-thienyl]-4,4-dimethyl-2-oxazoline is prepared by metalation in the 3-(*ortho*) position of 4,4-dimethyl-2-(2-thienyl)-2-oxazoline with butyl lithium at -78 °C followed by benzaldehyde [230].

2-[3-(α-Hydroxybenzyl)-2-thienyl]-4,4-dimethyl-2-oxazoline [230]

To a solution of 4,4-dimethyl-2-(2-thienyl)-2-oxazoline (15.0 g, 80 mmol) in anhydrous diethyl ether (320 ml) under nitrogen at −78 °C 1.6 *M* butyllithium in hexane (70 ml, 112 mmol) is added. The stirring is continued at this temperature for 15 min and after removal of the cooling bath for 20 min. The mixture is recooled to −65 °C and a solution of freshly distilled benzaldehyde (12 ml, 120 mmol) in anhydrous diethyl ether (80 ml) is added dropwise. Work up gives a yellow oil, which, under stirring, is treated with a sodium bisulfite solution for 2 h. The product is extracted with diethyl ether and the combined organic phases washed with water, dried over sodium sulfate and evaporated. The residue is filtered with diethyl ether through aluminum oxide; the filtrate is extracted with cold 1 *M* sulfuric acid, the aqueous phase made alkaline with 2 *M* sodium carbonate solution. The product is taken up in diethyl ether and the combined organic phases are washed with water, dried over sodium sulfate and evaporated giving 21.7 g (95%) of the title compound as a colorless oil.

Conversion of the 3-lithium derivative to the zinc derivative by reaction with zinc bromide followed by palladium-catalyzed coupling with iodobenzene gives 4,4-dimethyl-2-(3-phenyl-2-thienyl)oxazoline [263].

4,4-Methyl-2-(3-phenyl- 2-thienyl)oxazoline [263]

4,4-Dimethyl-2-(2-thienyl)oxazoline (0.38 g, 2.11 mmol) is treated with 1.52 *M* butyllithium in diethyl ether (30 ml, 2.15 mmol) at −78 °C for 15 min and then at 0 °C for 30 min. The solution is recooled to −20 °C and zinc bromide solution (3.16 mmol) is added. The stirring is continued at this temperature for 1 h, after which a solution of tetrakis(triphenylphosphine)palladium(0) (0.1 g, 4 mol%) and iodobenzene (0.14 ml, 2.11 mmol) in anhydrous tetrahydrofuran (25 ml) is added. The reaction mixture is allowed to warm to room temperature and then refluxed for 16 h. Work up followed by flash chromatography and kugelrohr distillation at 185 °C/0.1 mm Hg gives 0.44 g (81%) of the title compound as an oil which crystallizes to a solid mp 48–49 °C.

However, a detailed study showed that the metalation of the oxazoline by butyllithium is largely dependent on solvent, temperature and the concentration of starting material. Under kinetic control the *ortho* disubstituted product is preferred, while under thermodynamic control the 2,5-disubstituted thiophenes are obtained. Upon reaction with trimethysilyl chloride thus either 4,4-dimethyl-(5-trimethylsilyl-2-thienyl)-2-oxazoline or 4,4-dimethyl-(3-trimethylsilyl-2-thienyl)-2-oxazoline could be obtained depending upon reaction conditions [264].

Metalation of 4,4-dimethyl-2-(2-thienyl)oxazoline with butyllithium at −70 °C followed by reaction with pentafluoroacetone gives 3-(1,1,1,3,3-pentafluoro-2-hydroxypropan-2-yl)-2- (4,4-dimethyl-2-oxazoline-2-yl)thiophene, which upon treatment with acid gives a mixture of the acid and lactone [234].

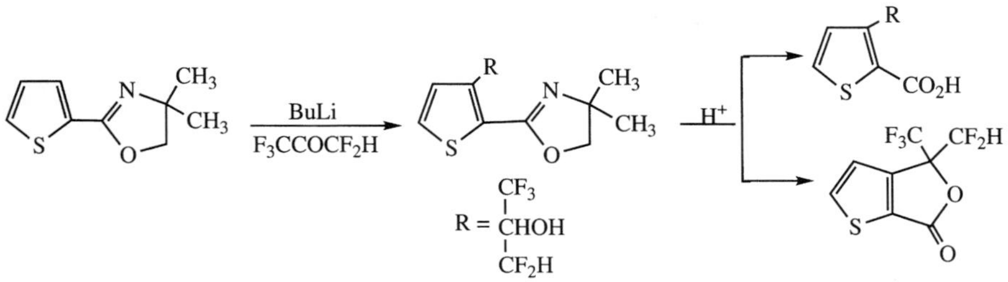

A detailed study on the effect of chiral oxazoline ligands on the palladium catalyzed reaction of aryl and vinyl iodides with 1,2-dienes was carried out [265,266].

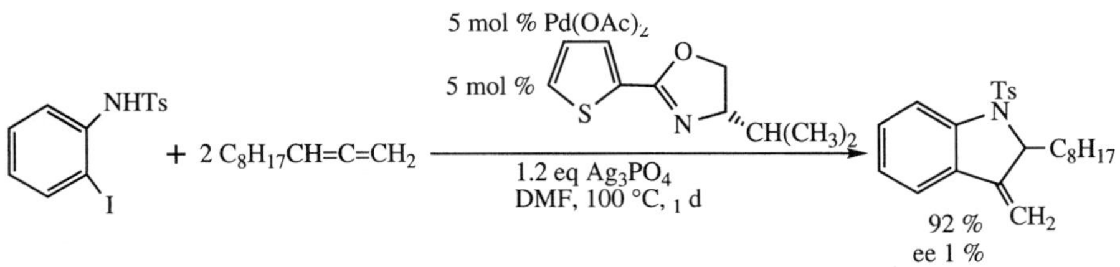

4F.12 THIOPHENECARBOXYLIC ESTERS

4F.12.1 From thiophenecarboxylic acids and alcohols

Most standard methods can be used for the esterification of thiophenecarboxylic acids. For the preparation of methyl esters, the method of Clinton and Lakovski or reaction with diazomethane can be used [215,267]. From acids containing electron withdrawing groups, such as 3-formyl-2-thiophenecarboxylic acid, the methyl ester can be obtained by refluxing in methanol-ethylene chloride using concentrated sulfuric acid as catalyst [160].

Methyl 3-formyl-2-thiophenecarboxylate [160]

A mixture of 3-formyl-2-thiophenecarboxylic acid (5.0 g, 0.032 mol), methanol (10 ml), ethylene chloride (20 ml) and concentrated sulfuric acid (1 ml) is refluxed for 16 h. The phases are separated and the organic phase washed with water, dried and distilled giving 3.7 g (68%) of the title compound, bp 140–141 °C/20 mm Hg, mp 61–62 °C after recrystallization from petroleum ether.

4F.12.2 From thiophenecarbonyl chlorides and alcohols and phenols

A number of phenol esters of 2-thiophenecarboxylic acid are prepared from 2-thiophenecarbonyl chloride and phenol and used in the Fries rearrangement to give ketones [268].

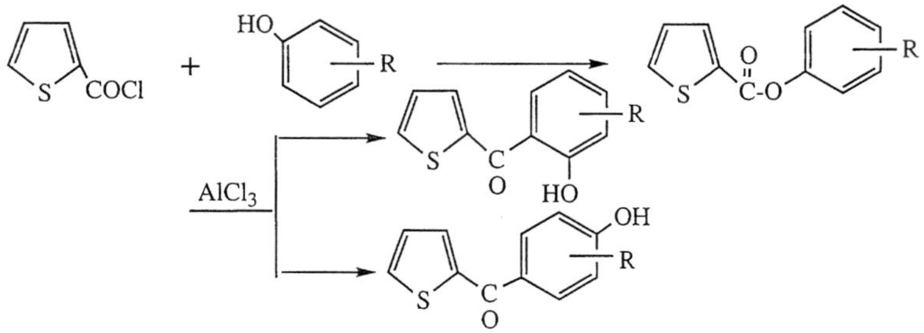

Macrocyclic esters derived from 2,5-thiophenedicarbonyl dichloride have been prepared by their reaction with the appropriate glycols [269].

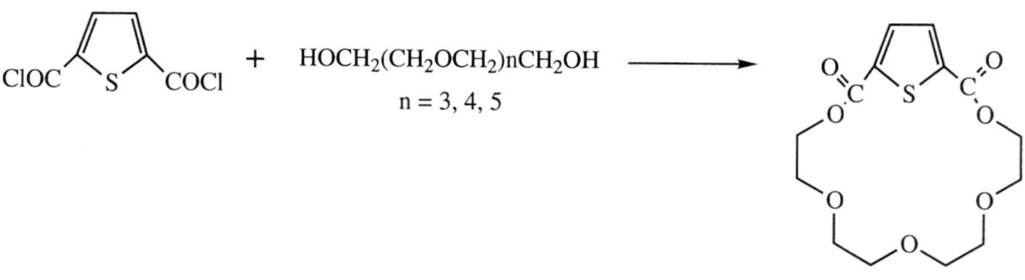

General procedure [269]

Equimolar amounts of 2,5-thiophenedicarbonyl dichloride and glycol are separately dissolved in dichloromethane and simultaneously added dropwise to rapidly stirred dichloromethane (1 l) containing an excess of triethylamine. The reaction mixture is stirred at room temperature for 48 h, after which it is concentrated. The precipitated amine salt is filtered off and the filtrate evaporated. The product is isolated by continuous liquid–liquid extraction with hot hexane followed by recrystallization from methanol or ethanol. For $n = 3$ yield 17% mp 111–113 °C; $n = 4$ yield 18% mp 96–97 °C; $n = 5$ yield 14% mp 63–64.5 °C.

A more complex example is the formation of cyclophanes by macrocyclization of the diol shown below with 2,5-thiophenedicarbonyl chloride [270].

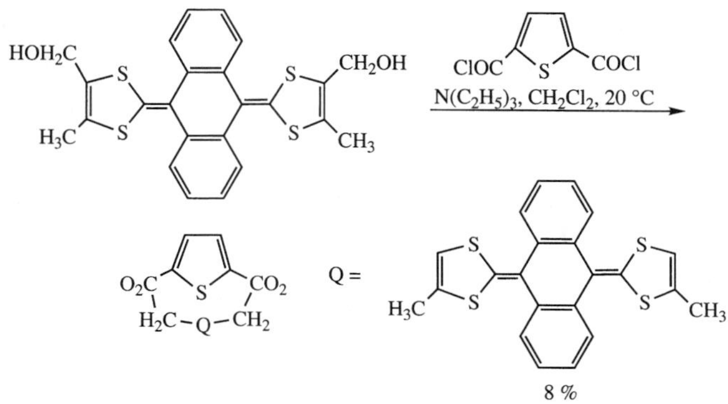

The reaction of 2-thiophenecarbonyl chloride with 2(5H)thiophenone, 2(5H)furanone and 1-methyl-2(5H)-pyrrolone with triethylamine as catalyst gives 2-thienyl, 2-furyl- and 1-methyl-2-pyrrolyl 2-thiophenecarboxylate, respectively [271].

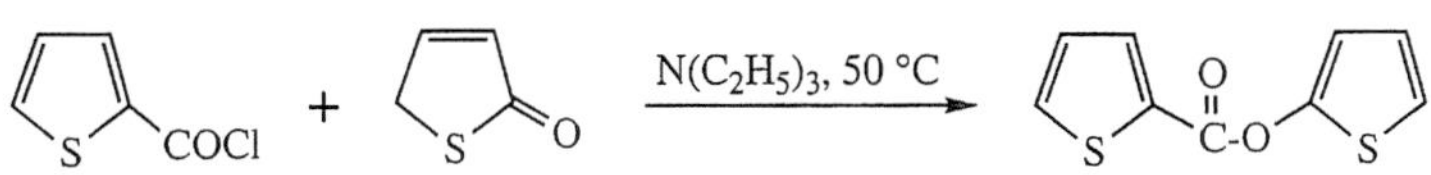

2-Thienyl 2-thiophenecarboxylate [271]

A mixture of 2(5H)thiophenone (1.40 g, 14 mmol), triethylamine (5.00 ml, 36 mmol) and 2-thiophenecarbonyl chloride (2.34 g, 16 mmol) are stirred with ice cooling for 5 min and then at 50 °C for 3 h. After cooling diethyl ether is added followed by filtration. The filtrate is evaporated and distilled giving 1.90 g (65%) of the title compound bp 110–114 °C/0.01 mm Hg.

Novel bent shaped, four ring thiophene based chiral liquid crystalline esters are prepared from 5-(4-decyloxyphenyl)thiophene-2-carboxylic acid and (*S*)-1-methylheptyl-4′-hydroxy- biphenyl-4-carboxylate using the carbodiimide esterification method [272].

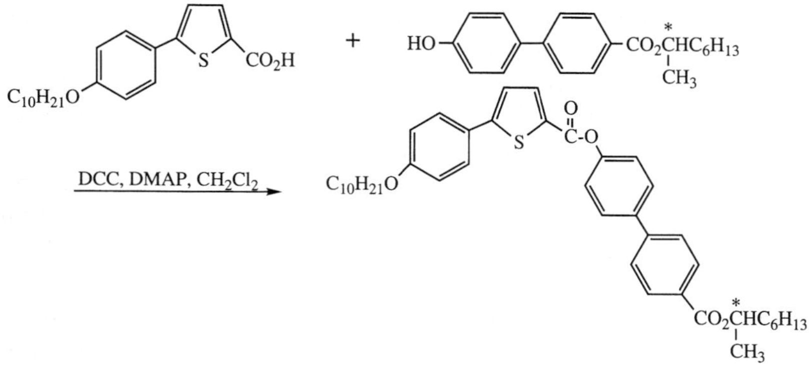

(S)-4′-(1-Methylheptyloxycarbonyl)biphenyl-4-yl-5-(4-decyloxyphenyl)thiophene carboxylate [272]

1,3-Dicyclohexylcarbodiimide (0.50 g, 2.4 mmol) and (*S*)-1-methylheptyl 4′-hydroxybiphenyl-4-carboxylate (0.36 g, 1.4 mmol) are added to 5-(4-decyloxyphenyl)thiophene-2-carboxylic acid (0.46 g,1.4 mmol) in anhydrous dichloromethane (25 ml). 4-Dimethylaminopyridine (1–2 crystals) is added and the stirring is continued at room temperature for 2 h. The white precipitate formed is filtered off and the filtrate evaporated. The residue is purified by flash chromatography using petroleum ether/dichloromethane (1:1) as eluent. Repeated recrystallizations from ethanol gave 0.9 g (64%) of the title compound as an off-white solid.

4F.12.3 From reactive thenoylamides

1-(2-Thenoyl)azole reacts with alcohols to give esters [228].

Methyl 2-thiophenecarboxylate [228]

To a rapidly stirred solution of 2-thiophenecarboxylic acid (1.28 g, 10 mmol) in acetonitrile (20 ml) 1,1-oxalyldiimidazole (2.09 g, 11 mmol) is added in a single portion. The mixture is stirred at room temperature for 15 min and then at 40 °C for 40 min. After cooling to room temperature a solution of methanol (10 ml) and methylsulfonic acid (2.1 g, 20 mmol) in acetonitrile (10 ml) is added dropwise. The reaction mixture is heated at 60 °C for 2 h, after which it is evaporated and the residue poured into ice water. The product is extracted with ethyl acetate and the combined organic phases are washed with 5% sodium bicarbonate solution and water, dried over sodium sulfate, evaporated and distilled giving 1.08 g (76%) bp 90–93 °C/11 mm Hg.

4F.12.4 Alkylation of salts of thiophenecarboxylic acids

Methyl 5-formyl-2-thiophenecarboxylate is prepared from 5-formyl-2-thiophenecarboxylic acid by reaction with sodium carbonate and methyl iodide in *N*,*N*-dimethylformamide [50].

Methyl 5-formyl-2-thiophenecarboxylate [50]

Methyl iodide (4.36 g, 30.72 mmol) is added to a stirred suspension of 5-formyl-2-thiophenecarboxylic acid (4.00 g, 25.61 mmol) and sodium carbonate (9.50 g, 89.63 mmol) in *N*,*N*-dimethylformamide (75 ml). After stirring overnight at room temperature the reaction mixture is poured into water saturated with solid sodium chloride. The product is extracted with ethyl acetate and the combined organic phases are washed with sodium chloride solution, dried over magnesium sulfate and evaporated giving 3.83 g (88%) of the title compound as a light-yellow solid mp 85–87 °C.

4F.12.5 From thienyllithium derivatives

Cyclic esters of thiophenecarboxylic acids are prepared starting from 3-thienyllithium followed by benzaldehyde and renewed treatment with butyllithium and cyclohexanone. The compound formed can also be obtained starting from 3-thiophenecarboxylic acid, which is metalated with lithium diisopropyl amide and reacted with cyclohexanone. Treatment with benzenesulfonyl chloride in pyridine gave the spiro ester [28].

2-(1-Hydroxy-1-cyclohexyl)-3-thiophenecarboxylic acid [28]

A solution of 3-thiophenecarboxylic acid (10 g, 78 mmol) in anhydrous tetrahydrofuran (100 ml) is added dropwise to a solution of lithium diisopropyl amide, prepared from diisopropylamine (17.5 g, 172 mmol) in anhydrous tetrahydrofuran (40 ml) and 0.94 *M* butyllithium in diethyl ether at (−10 °C) under nitrogen at −80 °C. The stirring is continued at −50 °C for 1 h, after which the mixture is recooled to −80 °C and cyclohexanone (9 g, 91 mmol) in anhydrous tetrahydrofuran (90 ml) is added. The reaction mixture is allowed to warm to −20 °C and poured into 2 *M* hydrochloric acid and ice. The phases are separated and the aqueous phase extracted with diethyl ether. The combined organic phases are washed with sodium chloride solution, dried and evaporated giving 14.4 g (81%) of colorless crystals mp 140–141 °C after recrystallization from methanol/water (2:1).

The corresponding spirosubstituted thieno[3,4-*c*]furans were similarly prepared [29].

4F.12.6 By reaction of thiophene esters

Metalation of ethyl 2-thiophenecarboxylate with diisopropylaminomagnesium chloride in tetrahydrofuran followed by *N*-formylpiperidine and benzaldehyde gives ethyl 5-formyl-2-thiophene carboxylate and ethyl 5-[hydroxy(phenyl)-methyl]-2-thiophenecarboxylate, respectively [273].

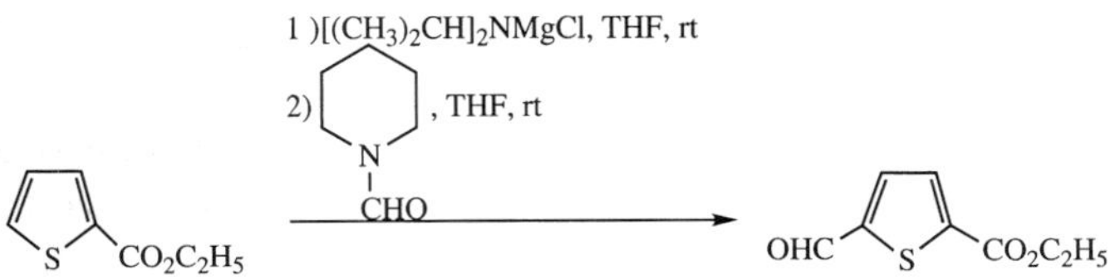

Ethyl 5-formyl-2-thiophene carboxylate [273]

To a solution of diisopropylamine (0.30 ml, 2.14 mmol) in anhydrous tetrahydrofuran (10 ml) under argon at room temperature 0.60 *M* butylmagnesium chloride (3.30 ml, 1.98 mmol) is added. The stirring is continued for 24 h, after which ethyl thiophene-2-carboxylate (159 mg, 1.02 mmol) is added at room temperature. After 10 min 1-formylpiperidine (288 mg, 2.54 mmol) is added and the stirring is continued at room temperature for 30 h. The reaction mixture is diluted with saturated aqueous ammonium chloride (50 ml), the product extracted with chloroform (3 × 50 ml), the combined organic phases dried over magnesium sulfate and evaporated. The residue is purified by chromatography on silica gel using hexane/ethyl acetate (5:1) as eluent, giving 99.8 mg (52%) of the title compound as colorless prisms mp 57–58 °C.

In connection with syntheses of a new series of mesogenic compounds with thienyl moieties, esters were prepared from methyl 2-bromomethyl-5-thiophenecarboxylate and 4-hydroxy-4′-alkoxybiphenyl and then transesterified [274].

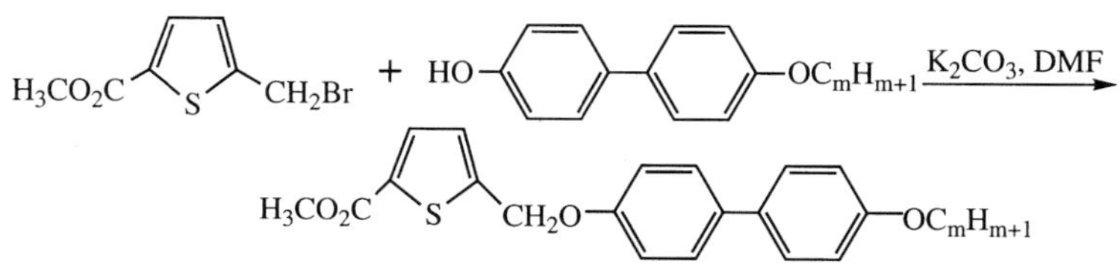

Methyl 5-(4′-alkoxybiphenyl-4-yloxymethyl)-2-carboxylate [274]

To a solution of 4-hydroxy-4′-alkoxybiphenyl (m = 12) (3.5 g, 10 mmol) in anhydrous *N*,*N*-dimethylformamide (25 ml), anhydrous potassium carbonate (2.1 g, 15 mmol), and methyl 2-bromomethyl-5-thiophenecarboxylate (2.7 g, 10 mmol) are added. The reaction mixture is stirred at 35 °C for 3 days, cooled, poured into water and extracted six times with hot chloroform. The combined organic phases are dried over magnesium sulfate and evaporated. The residue is purified by chromatography on silica gel at 40 °C in order to avoid crystallization using chloroform as eluent, giving 3.5 g (65%) of methyl 5-(4′-dodecyloxybiphenyl-4-yloxymethyl)thiophene-2-carboxylate.

4F.13 DIORGANOTIN(IV) DERIVATIVES OF 2-THIOPHENECARBOXYLIC ACID

Two series of complexes organotin(IV) derivatives of 2-thiophenecarboxylic acids have been prepared by treating 2-thiophenecarboxylic acid with diorganotin(IV) oxides in a 1:1 and 2:1 ratio [275,276].

4F.14 THIOPHENECARBOXYLIC ACID ANHYDRIDES AND DITHENOYL PEROXIDES

4F.14.1 Anhydrides

3-Thiophenecarboxylic acid anhydride has been obtained by reaction of the sodium salt of 3-thiophenecarboxylic acid in water with 3-thiophenecarbonyl chloride in acetone [54,277].

3-Thiophenecarboxylic acid anhydride [277]

A sodium hydroxide solution is added to a suspension of 3-thiophenecarboxylic acid (6.9 g, 5.4 mmol) in water until neutral to phenolphthalein. Under rapid stirring pyridine (2 drops) is added followed by 3-thiophenecarbonyl chloride (7.9 g, 5.4 mmol) in acetone (10 ml). The stirring is continued for 15 min, the precipitate formed filtered off, washed with water and a small amount of petroleum ether and dried in a desiccator giving 11.6 g (90%) of the title compound mp 53–56 °C.

2,3-Thiophene- and 3,4-thiophenedicarboxylic acid anhydride are prepared by the reaction of 2,3-thiophenedicarboxylic acid and 3,4-thiophenedicarboxylic acid in acetic anhydride followed by removal of the excess acetic anhydride [278,279].

2,3-Thiophenedicarboxylic acid anhydride [278]

A solution of 2,3-thiophenedicarboxylic acid (7.9 g, 45 mmol) in freshly distilled acetic acid anhydride (40 ml) is heated to reflux for 2 h. The excess of acetic acid anhydride is removed by evaporation and the residue recrystallized from benzene/heptane (3:2) giving 7.09 g (99%) of the title compound as light-brown plates. Sublimation gives snow-white crystals mp 138–140 °C.

Starting with 2-carboxy-3-thienylacetic acid the following reaction sequence has been performed [280].

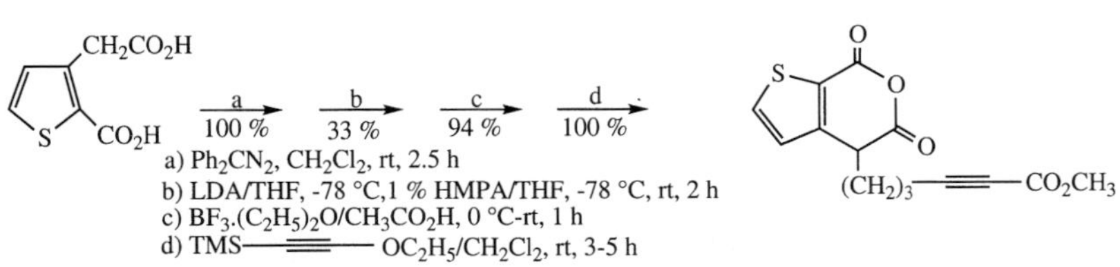

4F.14.2 Dithenoyl peroxides

Dithenoyl peroxides are prepared by reaction of the acid chlorides with sodium peroxide or with hydrogen peroxide–sodium carbonate [281,282].

4F.15 SULFUR ANALOGS OF THIOPHENECARBOXYLIC ACIDS AND THEIR DERIVATIVES

4F.15.1 Thiophenecarbothioic acids and derivatives

2-Thienyl- and 5-methyl-2-thiophenethiocarboxylic acid *O*-propyl ester are prepared by metalation of thiophene and 2-methylthiophene with butyllithium followed by magnesium bromide and *O*-propyl chlorothioformate [283].

2-Thiophenethiocarboxylic acid O-propyl ester [283]

To a solution of thiophene (252 g, 3.0 mmol) in anhydrous tetrahydrofuran (4 ml) under nitrogen at 0 °C, butyllithium (3.3 mmol) is added followed by magnesium bromide diethyl etherate (6.0 mmol). After cooling to −78 °C a solution of *O*-propyl chlorothioformate (3 mmol) in tetrahydrofuran is added in one portion. The stirring is continued for 1 h, after which anhydrous hexane (10 ml) is added and the reaction mixture is allowed to warm to room temperature and evaporated. The residue is filtered through a thin bed of silica gel using hexane as eluent, giving a yellow solution of the title compound.

4F.15.2 Thiophenecarbothioamides and selenoamides

4F.15.2.1 Addition of hydrogen sulfide to cyanothiophene

Thioamides of thiophenecarboxylic acids are readily prepared by the addition of hydrogen sulfide to cyanothiophenes. It has been applied to 2-cyanothiophene [249], 5-methyl-, 5-ethyl- and 5-propyl-2-cyanothiophene [284], 2,5-dicyanothiophene [285] and to a series of cyanothiophene acetals [286].

Thiophene selenoamides are prepared by the reaction of cyanothiophenes with hydrogen selenide [287] or in the following way [288].

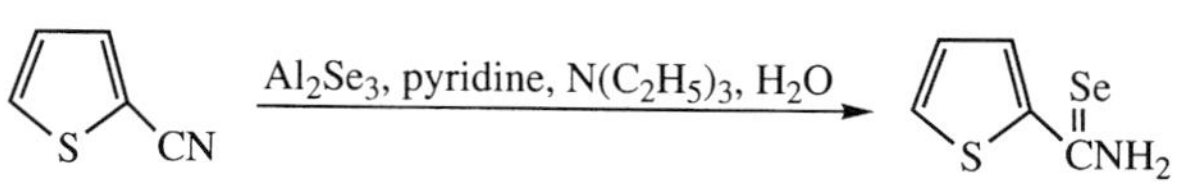

4F.15.2.2 Electrophilic substitution of thiophenes with isothiocyanates

Thiophene reacts with phenyl isothiocyanate in the presence of stannic chloride to give *N*-phenyl-2-thiophenecarbothioamide in 92% yield [83]. In the stannic chloride-catalyzed reaction of thiophene with methyl 2-isothiocyanatobenzoate, 2-(2-thienyl)-4*H*-3,1-benzothiazine-4-one is obtained, which upon treatment with sodium methoxide in methanol gives *N*-(2-methoxycarbonylphenyl)thiophene-2-thiocarboxamide [289].

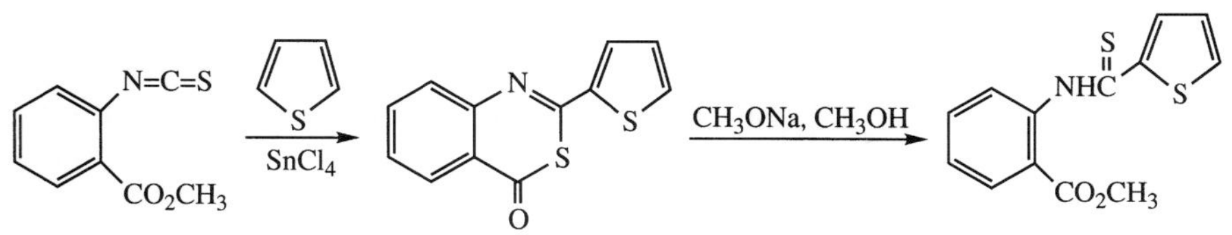

2-(2-Thienyl)-4H-3,1-benzothiazine-4-one [289]

A mixture of thiophene (2.0 g, 23.8 mmol), methyl 2-isothiocyanatobenzoate (3.41 g, 17.6 mmol) and stannic chloride (5 ml, 42.7 mmol) is stirred at room temperature for 16 h. To the dark semisolid formed ice, water and concentrated hydrochloric acid are added. The aqueous solution is decanted and the sticky solid is triturated first with hexane and then with ethanol giving 2.45 g (57%) of the title compound mp 125–127 °C. Recrystallization from ethanol gives the pure compound as buff-coloured crystals mp 127–128 °C.

N-(2-methoxycarbonylphenyl)thiophene-2-thiocarboxamide [289]

A mixture of 2-(2-thienyl)-4*H*-3,1-benzothiazi-4-one (0.50 g, 2.04 mmol), sodium methoxide (0.30 g, 5.55 mmol) and methanol (10 ml) is heated in a water bath for 5 min, after which it is diluted with water (20 ml), acidified with concentrated hydrochloric acid and filtered giving 0.55 g (97%) of the title compound mp 128–129 °C. Recrystallization from methanol gives the pure compound as buff-coloured crystals mp 129.5–131 °C.

The *N*-allyl thioamide of 2-thiophenecarboxylic acid is prepared by the reaction of thiophene with allyl isothiocyanate in nitromethane using aluminium chloride as catalyst [290].

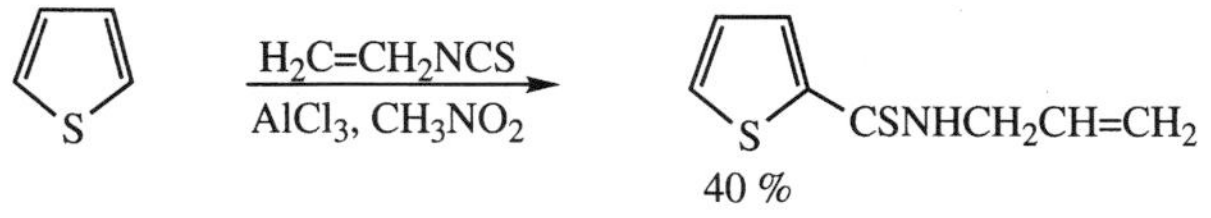

General procedure [290]

Allyl isothiocyanate (5.16 g, 0.52 mol) is added dropwise to a stirred solution of aluminium chloride (13.3 g, 0.1 mol) in anhydrous nitromethane (100 ml) at 0–5 °C. The stirring is continued at 0–5 °C for 1 h and then at room temperature for 1–2 h, after which the reaction mixture is poured into ice water. The product is extracted with ethyl acetate and the combined organic phases are dried over magnesium sulfate and distilled. The oily thiophene derivative is purified by chromatography on silica gel using benzene/ethyl acetate (8:1) as eluent.

4F.15.2.3 From thienyllithium derivatives and thiuram monosulfide

Both 2- and 3-thienyllithium react at −70 °C with *N,N′*-dimethyl-, *N,N′*-diphenyl- and tetramethylthiuram sulfide to give the thioamides in good yields [291].

General procedure [291]

The lithium compounds are prepared from the bromo derivatives (20 mmol) in anhydrous tetrahydrofuran (120 ml) under nitrogen at −70 °C and 2.0 *M* butyllithium in hexane (11 ml, 22 mmol). To these solutions the sulfide (20 mmol) is added in one portion and the stirring is continued at −70 °C for 2 h. The reaction mixture is allowed to warm to room temperature and then poured into saturated aqueous ammonium chloride solution. The phases are separated and the organic phase dried over sodium sulfate and evaporated. The residue is purified by chromatography on silica gel using heptane ethyl acetate (95:5) as eluent.

4F.15.3 Thiophenecarboxythiolic acids

Thiolesters are prepared by the reaction of thiophenecarboxylic acid chlorides and thiols [292,293]. A recent convenient method is the reaction of 1-(2-thienyl)azole with thiols in acetonitrile at 60 °C analogous to the ester described above [228]. Another method for the preparation of thiolesters consists in the palladium-catalyzed carbonylation of 2-iodothiophenetriethyltin thiolate system [294].

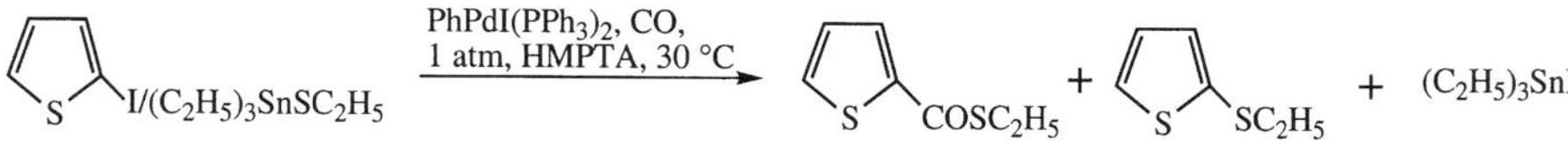

4F.15.4 Thiophenecarbodithioic acids

A convenient route is the reaction of thiophenemagnesium halides or thienyllithia with carbon disulfide, which upon alkylation give the esters [295–297].

Methyl 2-thiophenedithiocarboxylate [296]

A solution of the Grignard reagent is prepared from 2-bromothiophene (65.2 g, 0.40 mol) in tetrahydrofuran (300 ml) and a 30% excess of magnesium turnings. After decantation of the remaining magnesium the solution is heated to 60 °C and carbon disulfide (31.2 g, 0.40 mol) is added during 10 min at the same temperature. At 10–15 °C methyl iodide (64 g, 0.45 mol) is added in one portion and when the exothermic reaction has subsided the reaction mixture is heated for an additional 45 min at 50 °C. After addition of water the phases are separated and the aqueous phase extracted with diethyl ether. The combined organic phases are dried, evaporated and distilled giving 46.2 g (68%) of the title compound bp 158 °C/12 mm Hg.

Reaction of thiophene with phenyl thiochlorothioformate in the presence of stannic chloride as catalyst is used for the preparation of phenyl thiophene-2-carbodithioate [298]. Methyl 2-thiophenecarbodithioate is obtained by the reaction of 2-(2-thienyl)-4,5-dimethoxycarbonyl-1,3-dithiole with butyllithium followed by methyl iodide [299].

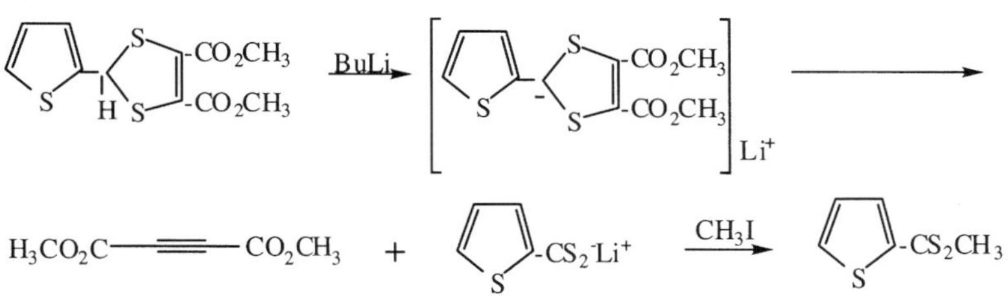

Methyl 2-thiophenecarbodithioate is also obtained by direct sulfurization of 2-chloromethylthiophene followed by alkylation [300].

4F.15.5 Various sulfur analogs of thiophenecarboxylic acid derivatives

Thiohydrazides of 2-thiophenecarboxylic acid is prepared by the reaction of methyl 2-thiophenecarbodithioate with excess hydrazine [301].

2-Thiophenethiocarboxylic acid hydrazide [301]

To a solution of methyl 2-thiophenecarbodithioate (1.74 g, 10 mmol) in ethanol (30 ml) at 30–50 °C under vigorous stirring 50% hydrazine hydrate (30 mmol) is rapidly added. The yellow solution is acidified with acetic acid and quickly cooled. After addition of water the precipitate formed is filtered off, washed with small amount of cold methanol and recrystallized from ethanol giving 0.80 g (50%) of the title compound mp 155–159 °C.

The reaction of thienyllithium derivatives with thiuram disulfide gives *N,N′*-dimethyldithiocarbamates [291].

4F.16 CYANOTHIOPHENES

4F.16.1 By direct substitution of thiophenes

2-Cyanothiophene is prepared by treatment of thiophene with chlorosulfonyl isocyanate followed by *N,N*-dimethylformamide or another amide [244,302].

2-Cyanothiophene [244]

Chlorosulfonyl isocyanate (85.0 g, 0.60 mol) is rapidly added to a solution of thiophene (168.2 g, 2.00 mol) in anhydrous diethyl ether (100 ml) and anhydrous benzene (150 ml). The reaction mixture is kept at 35 °C until crystals begin to precipitate. More chlorosulfonyl isocyanate (198.2 g, 1.40 mol) in anhydrous benzene (250 ml) is then added over a period of 1 h, during which time the temperature rises but is kept at 50 °C with a water bath. After complete addition the stirring is continued at 50 °C for another hour and the the reaction mixture is left at room temperature overnight. The flask now containing a precipitate of 2-thiophenecarboxamido-*N*-sulfochloride is placed in a cooling bath and *N,N*-dimethylformamide (308 g, 4.20 mol) is added at such a rate that the temperature is kept at 10–12 °C. The stirring is continued at 30 °C for 30 min and then the reaction mixture is poured onto ice (400 g). The phases are separated and the aqueous phase extracted twice with benzene. The combined organic phases are washed with water and sodium bicarbonate, dried over magnesium sulfate, evaporated and distilled giving 148.6 g (72%) of the title compound bp 77–78 °C/kPa.

Reaction of 2-ethylthiophene with thiocyanogen and stannic chloride gave a 30% yield of 5-ethyl-2-cyanothiophene [303].

4F.16.2 From thienylmetalic reagents and cuprous cyanide

The reaction of 2-thienylthallium bis(trifluoroacetates) with cuprous cyanide in acetonirile gives 2-cyanothiophene [304].

2-Cyanothiophene [304]

A solution of 2-thienylthallium bis(trifluoroacetate) (5.13 g, 10 mmol) and cuprous cyanide (0.91 g, 10 mmol) in anhydrous (50 ml) is under nitrogen refluxed for 15 h. After evaporation the residue is extracted with chloroform

(4 × 25 ml). The combined organic phases are washed with water (4 × 25 ml), 1 *M* hydrochloric acid (2 × 25 ml) and water (2 × 25 ml), dried over magnesium sulfate and evaporated giving 0.58 g (53%) of the title compound.

4F.16.3 Reaction of halothiophenes with cuprous cyanide

The reaction of halothiophenes with cuprous cyanide in solvents such as pyridine, quinoline or *N,N*-dimethylformamide is a widely used method for the preparation of cyanothiophenes and many substituents such as alkyl, formyl dimethoxyethyl and benzoyl are tolerated [87,209,286,305]. From 3-bromothiophene in quinoline an 86% yield of 3-cyanothiophene is obtained [306], while in *N,N*- dimethylformamide 64% was obtained [244]. Dinitriles are obtained by this method from dibromo or diiodo derivatives [162–164,279, 307,308].

3,4-Dicyanothiophene [164]

A solution of 3,4-dibromothiophene (22.36 g, 92.4 mmol) and cuprous cyanide (24.10 g, 269 mmol) in anhydrous *N,N*-dimethylformamide (25 ml) is refluxed for 4 h under stirring. The dark solution formed is poured into a solution of hydrated ferric chloride (93.0 g) in 1.7 *M* hydrochloric acid (162 ml). The temperature of the mixture is maintained at 60–70 °C for 0.5 h. After cooling the mixture is extracted five times with dichloromethane and each extract washed two times with 6.0 *M* hydrochloric acid, water, saturated aqueous sodium bicarbonate solution and water. The combined organic phases are dried over magnesium sulfate and evaporated. The residue is sublimed at 0.1 mm Hg giving 10.0 g (81%) of the title compound.

This method was recently used for the preparation of various 2,5-dialkyl-3,4-dicyanothiophenes from the corresponding dibromo derivatives, which were transformed to phtalocyanine related macrocycles [309].

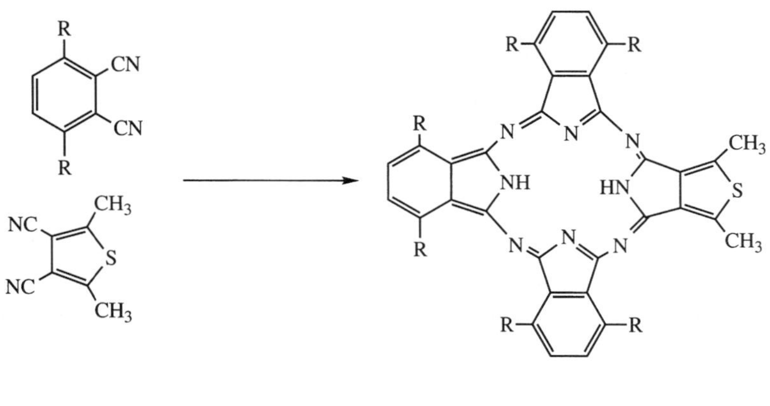

4F.16.4 Dehydration of oximes

Another very often used method for the preparation of cyanothiophenes is the dehydration of aldoximes with acetic anhydride. In this way 2-cyanothiophene [310] and 3-cyanothiophene [311] are obtained in 89% and 73% yield, respectively. Also alkyl substituted cyanothiophenes have been prepared in this way [312]. Halo methoxy and nitro substituted cyanothiophenes are obtained in these way, as mentioned in the appropriate chapters.

2-Cyanothiophene is also prepared in 93% yield from the oxime using carbodiimide and triethylamine in the presence of copper(II) ions [313].

General procedure [313]

To a solution of hydroxylammonium chloride (292 mg, 4.20 mmol) in water (1 ml) the aldehyde (4.00 mmol) and pyridine (2 ml) are added. After stirring at room temperature for 1 h hydrated cupric sulfate (200 mg, 0.80 mmol) is added followed by triethylamine (850 mg, 8.40 mmol) in dichloromethane (2 ml) and 1,3-dicyclohexylcarbodiimide (990 mg, 4.8 mmol) in dichloromethane (8 ml). The stirring is continued for 2 h, after which formic acid (0.7 ml) is added in order to destroy the excess of 1,3-dicyclohexylcarbodiimide. The reaction mixture is chromatographed on silica gel using dichloromethane as eluent. The eluate is evaporated giving the desired compound.

Dinitriles can also be prepared in this way [314]. All six cyanothiophene aldehydes are prepared by treatment of the corresponding oxime acetals with phosphorus pentachloride [286].

General procedure for preparation of cyanothiophene aldehydes [286]

To a solution of the oximes (65 mmol) in anhydrous benzene (100 ml) phosphorus pentachloride (72 mmol) is added under stirring in small portions. The reaction mixture is stirred at 50 °C for 0.5 h, cooled and poured onto ice water. The phases are separated and the aqueous phase extracted several times with chloroform. The combined organic phases are washed until neutral first with saturated aqueous sodium bicarbonate solution then with water, dried and evapoated. The residue is recrystallized from water giving the title compounds.

As an alternative to the oxime route, 2-cyanothiophene can be prepared directly from 2- thiophenealdehyde by the action of heat on its condensation product with 4-amino-1,2,4-triazole [315].

4F.16.5 Dehydration of amides

Phosphoryl chloride either alone or in pyridine or *N*,*N*-dimethylformamide is often used for the dehydration of amides to nitriles. The best method appears to be the dehydration of the amide with titanium(IV) chloride [316].

2-Cyanothiophene [316]

To anhydrous tetrahydrofuran (200 ml) at 0 °C titanium(IV) chloride (11 ml, 0.1 mol) in anhydrous carbon tetrachloride (25 ml) is added under stirring. To the precipitate formed 2-thiophenecarboxylic amide (6.35 g, 0.05 mol) in anhydrous tetrahydrofuran is added and the mixture is stirred at 0 °C for 1–2 h, after which *N*-methylmorfoline (10.2 g, 0.2 mol) in anhydrous tetrahydrofuran is added dropwise. The stirring is continued at 0 °C for 4 h and then water and diethyl ether are added to the brown suspension. The phases are separated and the aqueous phase extracted with diethyl ether (2 × 50 ml). The combined organic phases are washed with saturated sodium chloride solution (50 ml), dried over magnesium sulfate and distilled giving 4.8 g (88%) of the title compound.

4F.16.6 Direct transformation of thiophenecarboxylic acids

2-Cyanothiophene is prepared from 2-thiophenecarboxylic acid by treatment with urea/sulphamic acid mixture [317] or by treatment with chlorosulfonyl isocyanate, followed by *N*,*N*-dimethylformamide [318].

4F.16.7 By substitution reactions of cyanothiophenes

Direct arylation of 2- and 3-cyanothiophene using a Heck type reaction with palladium(II) acetate/tetrabutylammonium bromide as catalytic system has been used for the preparation of a number of 5-aryl-2-cyanothiophenes and 2-aryl-3-cyanothiophenes [319].

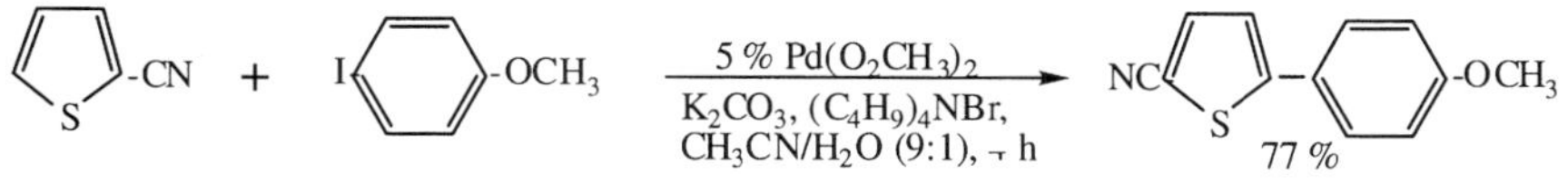

4F.16.8 By palladium-catalyzed coupling of halocyanothiophene with phenylacetylenes

Coupling of 5-bromo-2-cyanothiophene with the phenylacetylene chromium tricarbonyl complex gives the following thienyl complex [320].

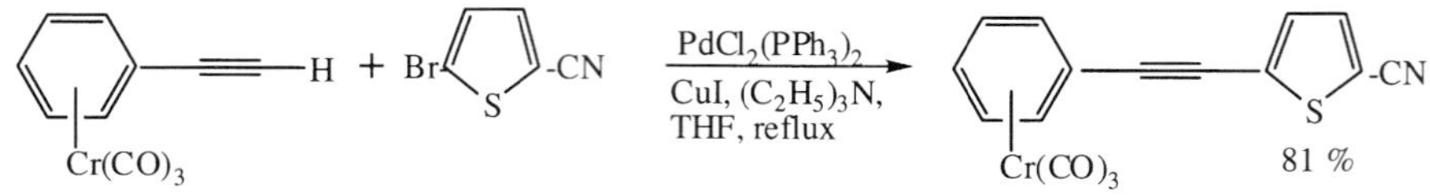

4F.16.9 By cyclization reactions

The Hinsberg reaction between 1,2-cyclohexanedione and thiodiacetonitrile gives 3-cyano-4,5,6,7-tetrahydrobenzo[*c*]thiophene in mediocre yield [321].

4F.16.10 By various reactions

An interesting high-yielding route for the preparation of 3-dimethylaminomethyl-2- and 4- thiophenecarbonitriles is a retromalonate addition reaction of the diethyl [*ortho*-bromomethylthienyl)-methylene]propanedioates with hydrazines and triethylamine in high yield [322].

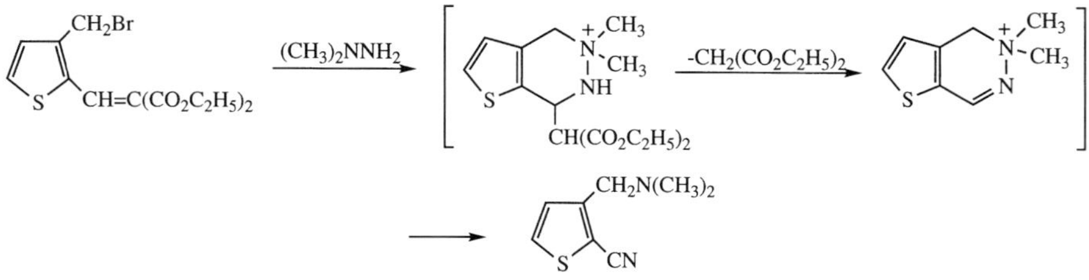

4F.17 THIOPHENE NITRILE OXIDES

Di- and trialkylsubstituted thiophenecarbonitrile oxides are prepared by the oxidation of the aldoximes with sodium hypochlorite in methylene chloride [323].

2,4,5-Trimethyl-3-nitrile oxide [323]

To a solution of 2,4,5-trimethylthiophene-3-aldoxime in dichloromethane (15 ml) 12% sodium hypochlorite (1.2 g, 16 mmol) is added with vigorous stirring at such a rate that the temperature does not exceed 0 °C. The stirring is continued for 20 min, after which the phases are separated and the aqueous phase extracted with dichloromethane. The combined organic phases are washed with water, dried over sodium sulfate and evaporated. The residue is sublimed at 60 °C/5 mm Hg giving 1.01 g (60%) of the title compound mp 72–73 °C.

Stable 2-alkylthio-and 2-alkylsulfonylthiophene 3-carbonitrile oxides containing various functional groups such as bromo, methoxy, thiomethyl and methylsulfone have been prepared similarly [324–326]. These types of compounds are also treated in chapter 6.

REFERENCES

1. W. Schlenk and W. Ochs, *Ber.* **48**, 679 (1915).
2. S. Gronowitz and T. Frejd, *Acta Chem. Scand.* **B30**, 485 (1976).
3. J. Cymerman-Craig and M. Moyle, *J. Chem. Soc.* 3907 (1963).
4. S. Gronowitz, *Arkiv Kemi* **12**, 115 (1958).
5. H. Gilman and D. A. Shirley, *J. Am. Chem. Soc.* **71**, 1870 (1949).
6. B. Garrigues, B. Oussaid and C. Hubert, *Bull. Soc. Chim. Fr.* **130**, 58 (1993).
7. V. Ramanathan and R. Levine, *J. Org. Chem.* **27**, 1667 (1962).
8. S. Gronowitz, B. Cederlund and A.-B. Hörnfeldt, *Chem. Scripta* **5**, 217 (1974).
9. K. Y. Tserng and L. Bauer, *J. Org. Chem.* **40**, 172 (1975).
10. N. Gjøs and S. Gronowitz, *Arkiv Kemi* **30**, 225 (1968).
11. P. Leardini, G. Martinelli, P. Spagnolo and M. Tiecco, *J. Chem. Soc. C* 1464 (1970).
12. S. Gronowitz, *Arkiv Kemi* **12**, 239 (1958).
13. S. Gronowitz, *Arkiv Kemi* **16**, 363 (1965).
14. S. Gronowitz and U. Rosén, *Chem. Scripta* **1**, 33 (1971).
15. J. P. Lampin and F. Mathey, *J. Organomet. Chem.* **71**, 239 (1974).
16. S. Gronowitz, J. E. Skramstad and B. Eriksson, *Arkiv Kemi* **28**, 99 (1967).
17. U. Michael and S. Gronowitz, *Acta Chem. Scand.* **22**, 1353 (1968).
18. S. Hue-Wen, K. Hirose, K. Shirai and T. Kumamoto, *Bull. Chem. Soc. Japan* **62**, 2725 (1989).
19. S. Gronowitz and G. L. Borgen, *Acta Chem. Scand.* **19**, 1180 (1965).
20. S. Gronowitz and J. E. Skramstad, *Arkiv Kemi* **28**, 115 (1967).
21. S. Gronowitz and T. Frejd, *Acta Chem. Scand.* **B29**, 818 (1975).
22. S. Gronowitz, *Arkiv Kemi* **7**, 361 (1954).
23. B. Yom-Tov and S. Gronowitz, *Chem. Scripta* **3**, 165 (1973).
24. B. Yom-Tov and S. Gronowitz, *Chem. Scripta* **3**, 37 (1973).
25. B. Yom-Tov and S. Gronowitz, *J. Heterocycl. Chem.* **15**, 285 (1978).
26. J. O. Karlsson, A. Svensson and S. Gronowitz, *J. Org. Chem.* **49**, 2018 (1984).
27. U. Michael and S. Gronowitz, *Chem. Scripta* **4**, 126 (1973).
28. F. Sauter, P. Stanetty, H. Fröhlich and W. Ramer, *Heterocycles* **26**, 2639 (1987).
29. F. Sauter, P. Stanetty and H. Fröhlich, *Heterocycles* **26**, 2637 (1987).
30. K. E. Schulte and N. Jantos, *Arch. Pharm.* **292**, 536 (1959).
31. K. E. Schulte, J. Reisch and L. Hoerner, *Chem. Ber.* **95**, 1943 (1962).
32. F. Y. Perveev and N. I. Kudryashova, *Zh. Obshch. Khim.*, **23**, 9976 (1953).
33. F. Y. Perveev and N. I. Kudryashova, *Dokl. Akad. Nauk SSSR* **98**, 975 (1954).
34. E. E. Campaigne and W. M. LeSuer, *Org. Syntheses* **33**, 94 (1953).
35. D. E. Aultz, A. R. McFadden and H. B. Lassman, *J. Med. Chem.* **20**, 456 (1977).
36. H. Wynberg and J.P. M. Houbiers, *J. Org. Chem.* **36**, 834 (1971).
37. M. Nemec, M. Janda, J. Srogl and I. Stibor, *Collect. Czech. Chem. Commun.* **39**, 3527 (1974).
38. M. Janda, J. Srogl, M. Nemec and K. Kalfus, *Collect. Czech. Chem. Commun.* **41**, 1541 (1976).
39. D. J. Chadwick, J. Chambers, G. D. Meakins and R. L. Snowden, *J. Chem. Soc. Perkin Trans. 1* 523 (1975).
40. H. Paul and H. Migulla, *Arch. Pharm. (Weinheim)* **311**, 679 (1978).
41. R. Kimura, T. Yabuuchi and Y. Tamura, *Chem. Pharm. Bull.* **8**, 103 (1960).

42. B. Decroix, J. Morel, C. Paulmier and P. Pastour, *Bull. Soc. Chim. Fr.* 1848 (1972).
43. B. P. Fabrichnyi, Y. B. Volkenshtein, V. I. Rogovik, I. B. Karmanova and Y. L. Gol'dfarb, *Khim. Geterotsikl Soedin.* 504 (1965).
44. T. Asami and S. Yoshida, *Heterocycles* **38**, 2739 (1994).
45. Y. L. Gol'dfarb, B. P. Fabrichnyi and V. I. Rogovik, *Izv. Akad. Nauk SSSR, Ser. Khim.* 2172 (1963).
46. Y. L. Gol'dfarb, Y. B. Volkenshtein and B. V. Lopatin, *Zh. Obshch. Khim.* **34**, 969 (1964).
47. T. Sone, *Bull. Chem. Soc. Japan* **37**, 1197 (1964).
48. J. H. Birkinshaw and P. Chaplen, *Biochem. J.* **60**, 255 (1951).
49. S. Gronowitz and K. Dahlgren, *Arkiv Kemi* **21**, 201 (1963).
50. C. J. Goddard, *J. Heterocycl. Chem.* **28**, 17 (1991).
51. K. Nakagawa, S. Mineo and S. Kawamura, *Chem. Pharm. Bull.* **26**, 299 (1978).
52. P. Cogolli, F. Maiolo, L. Testaferri, and M. Tingoli, *J. Chem. Soc., Perkin Trans. 2,* 1331 (1980).
53. E. Profft and G. Solf *J. Prakt. Chem.* **24**, 38 (1964).
54. D. MacKay, *Can. J. Chem.* **44**, 2881 (1966).
55. Y. Tamura, Y. Yamada and Z. Yoshida, *Tetrahedron* **35**, 39 (1979).
56. J. Teste and N. Lozac'h, *Bull. Soc. Chim. Fr.* 437 (1955).
57. N. Messina and E. V. Brown, *J. Am. Chem. Soc.* **74**, 920 (1952).
58. M. Sy, N. P. Buu-Hoi and N. D. Xuong, *J. Chem. Soc.* 1975 (1954).
59. H. Wynberg and A. Logothetis, *J. Am. Chem. Soc.* **78**, 1958 (1956).
60. Y. L. Gol'dfarb, B. P. Fabrichnyi and I. F. Shalavina, *Zh. Obshch. Khim.* **29**, 3633 (1959).
61. V. N. Gogte, B. D. Tilak, K. N. Gadekar and M. S. Sahasrabudhe, *Tetrahedron* **23**, 2443 (1967).
62. E. C. Spaeth and C. B. Germain, *J. Am. Chem. Soc.* **77**, 4066 (1955).
63. M. Sy, N. P. Buu-hoi and N. D. Xuong, *Compt. rend.* **239**, 1224 (1954).
64. M. Janda, M. Valenta and P. Holy, *Coll. Czech. Chem. Commun.* **39**, 959 (1974).
65. N. P. Buu-Hoi, M. Sy and N. D. Xuong, *Compt. rend.* **240**, 442 (1955).
66. Y. L. Gol'dfarb and Y. L. Danyushevskii, *Izvest. Akad. Nauk. SSSR, Otdel. Khim. Nauk.* 540 (1963).
67. N. P. Buu-Hoi and N. Hoan, *Rec. trav. Chim.* **69**, 1425 (1950).
68. S. Gronowitz, J. Rehnö, K. Titlestad, M. Vadzis, B. Sjöberg, P. Bamberg, B. Ekström and U. Forsgren, *Acta Pharm. Suec.* **9**, 381 (1972).
69. Y. L. Gol'dfarb and V. P. Litvinov, *Zh. Obshch. Khim.* **30**, 2719 (1960).
70. P. Cagniant and D. Cagniant, *Bull. Soc. Chim. Fr.* 713 (1952).
71. P. Cagniant and G. Merle, *Compt. rend.* **264**, 112 (1967).
72. S. Nishimura, T. Motoyama and E. Imoto, *Bull. Univ. Osaka Prefecture, Ser. A.* **6**, 127 (1958).
73. P. Fournari and J. P. Chane, *Bull. Soc. Chim. Fr.* 479, (1963).
74. P. Reynaud and L. Delaby, *Bull. Soc. Chim. Fr.* 1614 (1955).
75. M. Boulouard, S, Rault, P. Dallemagne and M. Robba, *Heterocycles* **41**, 515 (1995).
76. D. P. Phillion and J. K. Pratt, *Synth. Commun.* **22**, 13 (1992).
77. Y. Fujiwara, T. Kawauchi and H. Taniguchi, *Chem. Commun.* **1980**, 220.
78. J. Marchal, J. Bodiguel. Y. Fort and P. Caubere, *J. Org. Chem.* **60**, 8336 (1995).
79. N. A. Bumagin, K. V. Nikitin and I. P. Beletskaya, *J. Organomet. Chem.* **358**, 563 (1988).
80. G. Kühnhauss, H. Reinhardt and J. Teubel, *J. Prakt Chem.* **3**, 137 (1956).
81. R. Graf, *Liebigs Ann.* **661**, 111 (1963).
82. M. Takeshita and M. Tashiro, *J. Org. Chem.* **56**, 2837 (1991).
83. R. Jaunin, *Helv. Chim. Acta* **63**, 1542 (1980).
84. T. Akiyama, O. Ikarashi, K. Iwasaki and A. Sugimoro, *Bull. Chem. Soc. Japan* **48**, 914 (1975).
85. H. Gross, J. Rusche and M. Mirsch, *Berichte* **96**, 1382 (1963).
86. M. Janda, J. Srogl, M. Nemec and I. Stibor, *Org. Prep. Proc. Int.* **3**, 295 (1971).
87. M. Derenberg and P. Hodge, *Tetrahedron Lett.* 3825 (1971).

88. D. G. Davies, M. Derenberg and P. Hodge, *J. Chem. Soc. C* 455 (1971).
89. Y. Hatanaka, S. Fukushima and T. Hiyama, *Heterocycles* **30**, 303 (1990).
90. G. Karminski-Zamola, J. Dogan, M. Bajic, J. Blazevic and M. Malesevic, *Heterocycles* **38**, 759 (1994).
91. C. G. Blettner, W. A. König, W. Stenzel and T. Schotten, *J. Org. Chem.* **64**, 3885 (1999).
92. Y. Satoh, C. Gude, K, Chan and F. Firooznia, *Tetrahedron Lett.* **38**, 7645 (1997).
93. F.-L. Quing and X.-J. Yue, *Tetrahedron Lett.* **38**, 8067 (1997).
94. S. A. Biller, J. W. Abt, A. T. Pudzianowski, L. C. Rich, D. A. Slusarchik and C. P. Ciosek, Jr, *Organic & Medicinal Chem. Lett.* **3**, 595 (1993).
95. M. Tashiro, S. Mataka, K. Nakayama, H. Tsuzuki and T. Yonemitsu, *J. Chem. Research (S)* 136 (1988).
96. Y. L. Gol'dfarb, G. I. Gorushkina and B. P. Federov, *Izvest. Akad. Nauk SSSR, Otd. Khim Nauk.* 2021 (1959).
97. R. R. Amaresh, M. V. Lakshmikanthan, R. Geng and M.P. Cava, *Tetrahedron Lett.* **41**, 8843 (2000).
98. E. C. Taylor, L. D. Jennings, Z. Mao, B. Hu, J.-G. Jun and P. Zhou, *J. Org. Chem.* **62**, 5392 (1997).
99. P. Kolsaker, J. Arukwe, J. Barcóczy, A. Wiberg and A. K. Fagerli, *Acta Chem. Scand.* **52**, 490 (1998).
100. A. Jilale and B. Decroix, *Chem. Scripta*, **27**, 417 (1987).
101. A. Daich, J. Morel and B. Decroix, *J. Heterocycl. Chem.* **31**, 341 (1994).
102. A. Daich, P. Ohier and B. Decroix, *J. Heterocycl. Chem.* **32**, 1731 (1995).
103. P. Pigeon, M. Othman, P. Netchitailo and B. Decroix, *Tetrahedron* **54**, 1497 (1998).
104. D. E. Ames and O. Ribeiro, *J. Chem. Soc., Perkin Trans. 1* 1390 (1975).
105. Y. Tamura, M. Kirihara, J. Sekihachi, R. Okunaka, S. Mohri, T. Tsugoshi, S. Akai, M. Sasho and Y. Kita, *Tetrahedron Lett.* **28**, 3971 (1987).
106. B. Yom-Tov, S. Gronowitz, S. B. Ross and N. E. Stjernström, *Acta Pharm Suecica* **11**, 149 (1974).
107. Y. B. Vol'kenshtein and Y. L. Gol'dfarb, *Dokl. Akad. Nauk, SSSR* **138**, 115 (1961).
108. Y. L. Gol'dfarb and V. P. Litvinov, *Izvest. Akad, Nauk., SSSR, Otdel. Khim. Nauk.* 343 (1963).
109. Y. L. Go'dfarb, M. A. Kalik and M. A. Kirmalova, *Zh. Obshch. Khim.* **29**, 2034 (1959).
110. Y. L. Gol'dfarb, V. P. Litvinov and A. N. Sukiasyan, *Izvest. Akad. Nauk. SSSR, Ser. Khim* 1296 (1971).
111. A. N. Sukiasyan, V. P. Litvinov and Y. L. Gol'dfarb, *Izvest. Akad. Nauk SSSR, Ser. Khim.* 1345 (1970).
112. S. Hauptmann, M. Weissenfels, M. Scholz, E. M. Werner, H. J. Koehler and J. Weissflog, *Tetrahedron Lett.* 1317 (1968).
113. S. Hauptmann, M. Weissenfels, E. M. Werner, and J. Weissflog, *Z. Chem.* **9**, 22 (1969).
114. A. Tundo, C. M. Cammagi, R. Leardini and M. Tiecco, *J. Chem. Soc. B* 1683 (1970).
115. N. D. Trieu and S. Hauptmann, *Z. Chem.* **13**, 57 (1973).
116. S. Hauptmann and E. M. Werner, *J. Prakt. Chem.* **314**, 499 (1972).
117. J. Liebscher and H. Hartmann, *J. Prakt. Chem.* **318**, 731 (1976).
118. C. M. Beaton, N. B. Chapman, K. Clarke and J. M. Willis, *J. Chem. Soc. Perkin Trans. 1* 2355 (1976).
119. A. Alberola, J. M. Andrés, A. Gonzáles, R. Pedrosa and P. Prádanos, *Synth. Commun.* **20**, 2537 (1990).
120. J. K. Ray, S. Gupta, D. Pan and G. K. Kar, *Tetrahedron* **57**, 7213 (2001).
121. S. Gronowitz, *Chemistry of Heterocyclic Compounds, Volume 44, Part 1*, p. 88 ff.
122. E. C. Taylor and J. E. Dowling, *J. Org. Chem.* **62**, 1599 (1997).
123. H. P. Guan, B.-H. Luo and C.-N. Hu, *Synthesis* 461 (1997).
124. T. Kálai, M. Balog, J. Jekö and K. Hideg, *Synthesis* 1476 (1998).

125. B. D. Tilak, H. S. Desai and S. S. Gupta, *Tetrahedon Lett.* 1609 (1964).
126. B. D. Tilak and S. S. Gupta, *Indian J. Chem.* **7**, 9 (1969).
127. K. E. Schulte, J. Reisch and L. Hoerner, *Chem. Ber.* **95**, 1943 (1962).
128. E. E. Campaigne and R. E. Cline, *J. Org. Chem.* **21**, 29 (1956).
129. D. W. Rangnekar and and S. W. Mavlankar, *J. Heterocycl. Chem.* **28**, 1455 (1991).
130. F. Bohlmann and E. Bresinsky, *Chem. Ber.* **97**, 2109 (1964).
131. V. E. Kolchin and N. S. Vulfson, *Zh. Obshch. Khim.* **32**, 3731 (1962).
132. J. M. McIntosh and R. Steevensz, *Can. J. Chem.* **52**, 1934 (1974).
133. J. M. McIntosh, H. B. Goodbrand and G. M. Masse, *J. Org. Chem.* **39**, 202 (1974).
134. J. M. McIntosh and R. A. Sieler, *J. Org. Chem.* **43**, 4431 (1978).
135. J. M. McIntosh and R. A. Sieler, *Can. J. Chem.* **56**, 226 (1978).
136. G. M. Coppola, R. E. Damon and H. Yu, *Synlett* 1143 (1995).
137. G, M. Coppola, R. E. Damon and H.Yu, *J. Heterocycl. Chem.* **33**, 687 (1996).
138. M. Elliott, N. F. Jones and B. C. Pearson, *J. Chem. Soc. C* 2551 (1971).
139. F. Duus, *Tetrahedron* **32**, 2817 (1976).
140. H. Gotthardt and B. Christl, *Tetrahedron Lett.* 4747 (1968).
141. H. Gotthardt, C. Weisshuhn and B. Christ, *Chem. Ber.* **109**, 753 (1976).
142. K. T. Potts, S. D. Datta, J. L. Marshall, *J. Org. Chem.* **44**, 622 (1979).
143. W. Huang, W.-L. Yu, H. Meng, J. Pei, S. F. Y. Li, *Chem. Mater.* **10**, 3340 (1998).
144. H. Meng and W. Huang, *J. Org. Chem.* **65**, 3894 (2000).
145. H. Wynberg and D. J. Zwanenburg, *J. Org. Chem.* **29**, 1919 (1964).
146. D. J. Zwanenburg and H. Wynberg, *Rec. trav. Chim.* **88**, 321 (1969).
147. N. P. Buu-Hoi, M. Sy and N.D Xuong, *Rec. trav. Chim.* **75**, 463 (1956).
148. O. Dann and G.Hauck, *Arch. Pharm.* **293**, 187 (1960).
149. R. Dabard and J. Y. LeBuhan, *Bull. Soc. Chim. Fr.* 4280 (1972).
150. S. Nakagawa, J. Okumura, F. Sakai, H. Toshi, and T. Naito, *Tetrahedron Lett.* 3719 (1970).
151. M. Gindy, V. B. Baghos, E. A. Ghali and F. G. Baddar, *J. Indian Chem. Soc.* **53**, 490 (1976).
152. R. Lukes, M. Janda and K. Kefurt, *Coll. Czech. Chem. Commun.* **25**, 1058 (1960).
153. Y. Matsumoto and T. Sone, *Kogyo Kagaku Zasshi* **62**, 1559 (1959).
154. S. Oae, N. Furukawa, T. Watanabe, Y. Otsuji, and M. Hamada, *Bull. Chem. Soc. Japan* **38**, 1247 (1965).
155. D. J. Chadwick and A. Plant, *Tetrahedron Lett.* **28**, 6085 (1987).
156. R. J. Heffner and M. M. Joullié, *Synth. Commun.* **21**, 1055 (1991).
157. J. Sicé, *J. Org. Chem.* **19**, 70 (1954).
158. R. R. Baker, J. P. Joseph, R. E. Schaub, F. J. McEnvoy and J. H. Williams, *J. Org. Chem.* **18**, 138 (1953).
159. S. D. Loiwal and N. C. Jain, *Indian J. Chem.* **1**, 119 (1963).
160. S. Gronowitz, B. Gestblom and B. Mathiasson, *Arkiv Kemi* **20**, 407 (1963).
161. A. P. Yakubov, L. I. Belen'kii and Y. A. Gol'dfarb, *Zh. Org. Khim.* **7**, 525 (1971).
162. D. W. H. MacDowell and J. C. Wisowaty, *J. Org. Chem.* **37**, 1712 (1972).
163. D. W. H. MacDowell and F. L. Ballas, *J. Org. Chem.* **42**, 3717 (1977).
164. J. A. Crayston, A. Iraqi, P. Mallon and J. C. Walton, *J. Chem. Soc., Perkin Trans. 2*, 1589 (1993).
165. S. Gronowitz and B. Eriksson, *Arkiv Kemi* **21**, 335 (1963).
166. S. Gronowitz, A. Biezais and B. Mathiasson, *Arkiv Kemi* **21**, 265 (1963).
167. L. M. Petrov, V. N. Chistokletov and A. A. Petrov, *Zh. Org. Khim.* **12**, 2035 (1976).
168. L. S. Rodionova, M. L. Petrov and A. A. Petrov, *Zh. Org. Khim.* **14**, 2050 (1978).
169. K. Gollnick and S. Fries, *Angew. Chem.* **92**, 848 (1980).
170. C. G. Krespan and B. C. McKusick, *J. Am. Chem. Soc.* **83**, 3438 (1961).
171. G. N. Schrauzer and V. Mayweg, *J. Am. Chem. Soc.* **84**, 3221 (1962).
172. G. N. Schrauzer and V. Mayweg, *J. Am. Chem. Soc.* **87**, 1483 (1965).
173. R. G. Jones, *J. Am. Chem. Soc.* **77**, 4069 (1955).

174. R. G. Jones, *J. Am. Chem. Soc.* **77**, 4074 (1955).
175. F. Boberg, *Liebigs Ann.* **679**, 118 (1964).
176. F. Boberg, U. Puttins, W. Schmidt, K. F. Torges, *Phosphorus & Sulfur* **17**, 135 (1983).
177. A. O. Fitton J. R. Frost, H. Suchitzky and P. G. Houghton, *Synthesis* 133 (1977).
178. T. Kametani and H. Sunagawa, *Yakugaku Zasshi* **83**, 360 (1963).
179. H. Gotthardt, W. Pflaumbaum and P. Gutowski, *Chem. Ber.* **121**, 313 (1988).
180. A. Robert, M. Ferrey and A. Foucauld, *Tetrahedron Lett.* 1377 (1975).
181. K. T. Potts, E. Houghton, and U. P. Singh, *J. Chem. Soc. D* 1129 (1969).
182. K. T. Potts, E. Houghton and U. P. Singh, *J. Org. Chem.* **39**, 3627 (1974).
183. H. Gotthardt and C. M. Weisshuhn, *Chem. Ber.* **111**, 3178 (1978).
184. K. T. Potts, R. Ehlinger and W. M. Nichols, *J. Org. Chem.* **40**, 2596 (1975).
185. K. T. Potts, S. J. Chen, J. Kane and J. L. Marshall, *J. Org. Chem.* **42**, 1633 (1977).
186. K. Masuda, J. Adachi and K. Nomura, *Chem. Pharm. Bull.* **25**, 1471 (1977).
187. K. T. Potts and U. P. Singh, *J. Chem. Soc. D* 569 (1969).
188. K. T. Potts, D. R. Choudbury, A. J. Elliott and U. P. Singh, *J. Org. Chem.* **41**, 1724 (1976).
189. H. Gotthardt and C. M. Weisshuhn *Chem. Ber.* **111**, 2021 (1978).
190. H. Gotthardt, C. M. Weisshuhn, O. M. Huss and D. J. Brauer, *Tetrahedron Lett.* 671 (1978).
191. H. Gotthardt, O. M. Huss, C. M. Weisshuhn and C. Michael, *Chem. Ber.* **112**, 1650 (1979).
192. T. Takaya, S. Kosaka, Y. Otsuji and E. Imoto, *Bull. Chem. Soc. Japan* **41**, 2086 (1968).
193. E. Lindner, V. Käss and H. A. Mayer, *Chem. Ber.* **123**, 783 (1990).
194. R. G. Jones, *J. Am. Chem. Soc.* **77**, 4163 (1955).
195. H. Hopf and J. von der Crone, *Chimia* **13**, 107 (1959).
196. O. Scherer and F. Kluge, *Chem. Ber.* **99**, 1973 (1966).
197. E. Lindner, A. Rau and S. Hoehne, *Angew. Chem. Int. Ed.* **18**, 534 (1979).
198. H. E. Simmons, R. D. Vest, D. C. Blomstrom, J. R. Roland and T. L. Cairns, *J. Am. Chem. Soc.* **84**, 4746 (1962).
199. T. G. Kutateladze, J. L. Kice and N. D. Zefirov, *J. Org. Chem.* **57**, 5270 (1992).
200. V. I. Rogovik and Y. l. Gol'dfarb, *Izvest. Akad, Nauk SSSR, Ser. Khim.* 2178 (1963).
201. S. N. Godovikova and Y. L. Gol'dfarb, *Izvest. Akad, Nauk SSSR, Ser. Khim.* 1434 (1965).
202. M. Robba, B. Roques and M. Bonhomme, *Bull. Soc. Chim. Fr.* 2495 (1967).
203. Y. L. Gol'dfarb and M. A. Kalik, *Zh. Org. Khim.* **14**, 2603 (1978).
204. S. N. Godovikova and Y. L. Gol'dfarb, *Bull. Acad. Sci., Div. Chem. (Engl. Transl.)* **18**, 1391 (1965).
205. S. F. Thames and J. E. McCleskey, *J. Heterocycl. Chem.* **3**, 104 (1966).
206. L. I. Belen'kii, I. B. Karmanova, Y. B. Vol'kenshtein and Y. L. Gol'dfarb, *Izvest. Akad. Nauk. SSSR, Ser. Khim.* 956 (1971).
207. K. Schlögl and H. Pelöusek, *Liebigs Ann.* **651**, 1 (1962).
208. D. S. Noyce, C. Alipinski and R. W. Nichols, *J. Org. Chem.* **37**, 2615 (1972).
209. P. Pirson, A. Schonne and L. Christiaens, *Bull. Soc. Chim. Belg.* **79**, 575 (1970).
210. D. W. H. MacDowell, R. A. Jourdenais, R. W. Naylor and J. C. Wisowaty, *J. Org. Chem.* **37,** 4406 (1972).
211. B. Decroix, J. Morel and P. Pastour, *J. Chem. Research (S)* 134 (1978).
212. I. B. Karmanova, Y. B. Vol'kenshtein and L. i. Belen'kii, *Khim. Geterotsikl. Soed.* **4**, 920 (1973).
213. T. L. Cairns, B. C. McKusick and V. Weinmayr, *J. Am. Chem. Soc.* **73**, 1270 (1951).
214. E. Campaigne and W. M. LeSuer, *J. Am. Chem. Soc.* **70**, 1555 (1948).
215. R. Hoffman and S. Gronowitz, *Arkiv Kemi* **16**, 515 (1960).
216. T. Kametani, T. Katagi, T. Fujiwara, and Y. Akasawa, *Jap. J. Pharm. Chem.* **26**, 544 (1954).
217. W. C. McCarthy and L. E. Foss, *J. Org. Chem.* **42**, 1508 (1977).
218. D. Binder, G. Habison and R. Noe, *Synthesis* 255 (1977).
219. B. Stanovnik, M. Tisler, V. Golob, L. Huala and O. Nicolic, *J. Heterocycl. Chem.* **17**, 733 (1980).

220. G. Ah-Kow, C. Paulmier and P. Pastour, *Bull. Soc. Chim. Fr.* 151 (1976).
221. B. Decroix and J. Morel, *J. Heterocycl. Chem.* **28**, 81 (1991).
222. J. Reisch and H. Labitzke, *Arch. Pharm. (Weinheim)* **308**, 203 (1975).
223. R. A. Barcock, D. J. Chadwick, R. C. Storr, L. S. Fuller and J. H. Young, *Tetrahedron* **50**, 4149 (1994).
224. Y. Hara, E. Sato, A. Miyagishi, A. Aisaka and T. Hibino, *J. Pharm. Sci.* **67**, 1334 (1978).
225. D. Ladurée, J. C. Lancelot, M. Robba, E. Chenu and G. Mathé, *J. Med. Chem.* **32**, 456 (1989).
226. T. Usami, N. Shirai and Y. Sato, *J. Org. Chem.* **57**, 5419 (1992).
227. Z. Zhang, A. E. Martell, R. J. Motekaitis and L. Fu, *Tetrahedron Lett.* **40**, 4615 (1999).
228. T. Kitagawa, H. Kuroda, H. Sasaki and K, Kawasaki, *Chem. Pharm. Bull.* **35**, 4294 (1987).
229. L. M. Gayo and M. J. Suto, *Tetrahedron Lett.* **38**, 513 (1997).
230. A. Schöning, T. Debaerdemaeker, M. Zander and W. Friedrichsen, *Chem. Ber.* **122**, 1119 (1989).
231. M. Kosugi, A. Fukiage, M. Takayanagi, H. Sano, T. Migita and M. Satoh, *Chem. Lett.* 1351 (1988).
232. E. C. Taylor and B. Hu, *Heterocycles* **45**, 241 (1997).
233. M. A. F. Brandao, A. B. de Oliveiro and V. Snieckus, *Tetrahedron Lett.* **34**, 2437 (1993).
234. M. G. Reinecke and L.-J. Chen, *Acta Chem. Scand.* **47**, 318 (1993).
235. M. G. Davidson, R. P. Davies, P. R. Raithby and R. Snaith, *Chem. Commun.* 1695 (1996).
236. R. A. El-Sayed, *Khim. Geterotsikl. Soedin.* 921 (1998).
237. B. F. Bonini, M. Comes-Francchini, G. Mazzanti, U. Passamonti, A. Ricci and P. Zani, *Synthesis* 92 (1995)
238. M. Sy and B. de Malleray, *Bull. Soc. Chim. Fr.* 1278 (1963).
239. D. Evans and T. F. Gray, *J. Chem. Soc.* 3006 (1965).
240. J. L. Abernethy, D, Srulevich, and M. J. Ordway, Jr, *J. Org. Chem.* **40**, 3445 (1975).
241. H. L. Yale, K. A. Losee, F. M. Perry and J. Bernstein, *J. Am. Chem. Soc.* **76**, 2208 (1954).
242. Ng. Ph. Bu-Hoi, Ng. D. Xuong, R. Royer and D. Lavit, *J. Chem. Soc.* 547 (1953).
243. H. L. Yale, K. Loser, J. Martins, M. Holsing, F. M. Perry and J. Bernstein, *J. Am. Chem. Soc.* **75**, 1933 (1953).
244. S. Gronowitz and S. Liljefors, *Acta Chem. Scand.* **31**, 771 (1977).
245. B. Elpern and F. C. Nachod, *J. Am. Chem. Soc.* **72**, 3379 (1950).
246. B. Decroix and P. Pastour, *J. Chem. Res. (S)* 132 (1978).
247. J. W. H. Watthey, M. Desai, R. Rutledge and R. Dotson, *J. Med. Chem.* **23**, 690 (1980).
248. G. Doddi, G. Illuminati and F. Stegel, *Tetrahedron Lett.* 3221 (1973).
249. M. Bercot-Vatteroni, *Ann. Chim.* 7, 303 (1962).
250. P. M. Theuss, W. Weuffen and H. Tredt, *Arch. Pharm.* **301**, 139 (1968).
251. F. Eloy, A. Deryckere and J.P. Maffrand, *J. Med. Chem.-Chim. Ther.* **9**, 602 (1974).
252. S. Robev, *Chem. Ber.* **91**, 244 (1958).
253. E. P. Papadopoulos, *J. Org. Chem.* **39**, 2540 (1974).
254. M. Robba and R. C. Moreau, *Ann. Pharm. Fr.* **22**, 201 (1964).
255. S.-O. Lawesson, *Arkiv Kemi* **11**, 325 (1957).
256. P. Dubus, B. Decroix, J. Morel and P. Pastour, *Ann. Chim. (Paris)* **10**, 331 (1975).
257. Y. Iwakura, K. Uno, and S, Shiraishi, *Bull. Chem. Soc. Jpn.* **41**, 2954 (1968).
258. H. M. Hassaneen, H. A. H. Mousa and A. S. Shawali, *J. Heterocycl. Chem.* **24**, 1665 (1987).
259. P. Wolkoff, *Can. J. Chem.* **53**, 1333 (1975).
260. H. M. Hassaneen, H. A. H. Mousa, N. W. Abed and A. S. Shawali, *Heterocycles* **27**, 695 (1988).
261. A. J. Carpenter and D. J. Chadwick, *J. Chem. Soc., Perkin Trans. 1,* 173 (1985).
262. L. Della Vecchia and I. Vlattas, *J. Org. Chem.* **42**, 2649 (1977).
263. D. S. Ennis annd T. L. Gilchrist, *Tetrahedron* **46**, 2623 (1990).
264. P. Ribérau and G. Queguiner, *Tetrahedron* **40**, 2107 (1984).

265. J. M. Zenner and R. C. Larock, *J. Org. Chem.* **64**, 7312 (1999).
266. C. G. Frost and J. M. J. Williams, *Tetrahedron Lett.* **34**, 2015 (1993).
267. S. Gronowitz, P. Moses, A.-B. Hörnfeldt and R. Håkansson, *Arkiv Kemi* **17**, 165 (1961).
268. M. Varache-Béranger, A, Nuhrich and G. Devaux, *Eur. J. Med.* **23**, 501 (1998).
269. J. S. Bradshaw, S. L. Baxter, J. D. Lamb, R. M. Izatt and J. J. Christensen, *J. Am. Chem. Soc.* **103**, 1821 (1981).
270. N. Godbert, A. S. Batsanov, M. R. Bryce and J. A. K. Howard, *J. Org. Chem.* **66**, 713 (2001).
271. C. K. Lee, J. S. Yu, and S. H. Kim, *J. Heterocycl. Chem.* **35**, 835 (1998).
272. A. S. Matharu, C. Grover, L. Komitov and G. Andersson, *J. Mater. Chem.* **10**, 1303 (2000).
273. M. Shilai, Y. Kondo and T. Sakomoto, *J. Chem. Soc., Perkin Trans. 1* 442 (2001).
274. D. Kardas, J. Mieczkowski, D. Pociecha, J. Szydlowska and E. Gorecka, *J. Mater. Chem.* **11**, 741 (2001).
275. G. K. Sandhu and N. S. Boparoy, *Synth. React. Inorg. Met.-Org Chem.* **20**, 975 (1990).
276. V. K. Jain and V. B. Mokal, *Indian J. Chem.* **31A**, 861 (1992).
277. O. Meth-Cohn and S. Gronowitz, *Acta Chem. Scand.* **20**, 1577 (1966).
278. M. G. Reinecke, J. G. Newsom and L.-J. Chen, *J. Am. Chem Soc.* **103**, 2760 (1981).
279. A. P. Krapcho and M. E. Petry, *J. Heterocycl. Chem.* **26**, 1509 (1989).
280. Y. Kita, R. Okunaka, M. Sasho, M. Taniguchi, T. Honda and Y. Tamura, *Tetrahedron Lett.* **29**, 5943 (1988).
281. R. D. Schuetz and D. M. Teller, *J. Org. Chem.* **27**, 410 (1962)
282. M. C. Ford and D. MacKay, *J. Chem. Soc.* 4620 (1957)
283. T. Nakamura and M. Matsumoto, *Synth. Commun.* **29**, 201 (1999).
284. G. Scarpini, B. Tornetta, G. Pappalardo and A. Bernardini, *Boll. sedute Accad.Gioenia Sci. Nat. Catania* **9**, 497 (1968)
285. F. Dallacker and V. Mues, *Chem. Ber.* **108**, 576 (1975).
286. P. Dubus, B. Decroix, J. Morel and P. Pastour, *Bull. Soc. Chim. Fr.* 628 (1976).
287. P. Chauvin, J. Morel, P. Pastour and J. Martinez, *Bull. Soc. Chim. Fr.* 2079 (1974).
288. V. I. Cohen, *Synthesis* 668 (1978).
289. L. M. Deck, S. D. Turner, J. A. Deck, and E. P. Papadopoulos, *J. Heterocycl. Chem.* **38**, 343 (2001).
290. T. S. Jagodzinski, J. G. Sosnicki and B. Nowak-Wydra, *Polish J. Chem.* **67**, 1043 (1993).
291. S. Gronowitz, A.-B. Hörnfeldt and M. Temciuc, *Synthesis* 483 (1993).
292. K. Beelitz, G. Buchholtz and K. Praefke, *Liebigs Ann.* 2043 (1979).
293. M. J. S. Dewar and R. M. Riddle, *J. Am. Chem. Soc.* **97**, 6658 (1975).
294. Y. V. Gulevich, N. A. Bumagin and I. P. Beletskaya, *Zh. Org. khim. (Engl. Transl.)* **24**, 1918 (1987).
295. R. Mayer, S. Scheithauer, and D. Kunz, *Chem. Ber.* **99**, 1393 (1966).
296. T. Meijer, P. Vermeer and L. Brandsma, *Rec. Trav. chim.* **92**, 601 (1973).
297. H. Viola and R. Mayer, *Z. Chem.* **15**, 348 (1975).
298. H. Viola, S. Scheithauer and R. Mayer, *Chem. Ber.* **101**, 3517 (1968).
299. M. L. Petrov, N.I. Kuz'mina and A. A. Petrov, *Zhur Org. Khim. (Engl. Transl.)* **24**, 204 (1988).
300. W. Thiel, R. Mayer and H. Viola, *Z. Chem.* **28**, 233 (1988).
301. W. Thiel and R. Mayer, *J. prakt. Chem.* **331**, 649 (1989).
302. C. Lohaus, *Chem. Ber.* **100**, 2719 (1967).
303. F. M. Stoyanovich, G.I. Gorushkina and Y. L. Gol'dfarb, *Izvest. Akad. Nauk. SSSR, Ser. Khim*, 387 (1969).
304. E. C. Taylor, A. H. Katz and A. McKillop, *Tetrahedron Lett.* **25**, 5473 (1984).
305. A. Vecchi and G. Malone, *J. Org. Chem.* **22**, 1636 (1957).
306. D. W. H. MacDowell and T. D. Greenwood *J. Heterocycl. Chem.* **2**, 44 (1965).
307. J. Morel, C. Paulmier and P. Pastour, *Compt. rend.* **266**, 1300 (1968).
308. C. Paulmier, J. Morel, P. Pastour and D. Semard, *Bull. Soc. Chim. Fr.*, 2511 (1969).

309. M. Cook and A. Jafari-Fini, *Tetrahedron* **56**, 4085 (2000).
310. B. Östman, *Acta Chem. Scand.* **22**, 2754 (1968).
311. D. R. Arnold and R. J. Bertwell, *J. Am. Chem. Soc.* **95**, 4599(1973).
312. S. Gronowitz and B. Gestblom, *Arkiv Kemi* **18**, 513 (1962).
313. E. Vowinkel and J. Bartel, *Chem. Ber.* **107**, 1221 (1974).
314. B. E. Ayres, S. W. Longworth and J. F. W. McOmie, *Tetrahedron* **31**, 1755 (1975).
315. H. G. O. Becker and H. J. Timpe, *Z. Chem.* **4**, 304 (1964).
316. W. Lehnert, *Tetrahedron Lett.* 1501 (1971).
317. J. Luecke and R. Winkler, *Chimia* **25**, 94 (1971).
318. Höchst, Neth. Appl. 6,412, 717; *Chem. Abstr.* **63**, 11274 (1963).
319. C. Gozzi, L. Lavenot, K. Ilg, V. Penalva and M. Lemaire, *Tetrahedron Lett.* **38**, 8867 (1997).
320. J.-P. Tranchier, R. Chavignon, D. Prim, A. Auffrant, Z. F. Plyta, F. Rose-Munch and E. Rose, *Tetrahedron Lett.* **41**, 3607 (2000).
321. W. Volz and J. Voss, *J. prakt. Chem.* **333**, 889 (1991).
322. C.-K. Sha and C.-P. Tsou, *J. Chinese Chemical Soc.* **38**, 183 (1991).
323. M. M. Krayushkin, A. A. Loktionov and L. I. Belen'kii, *Chem. Heterocycl. Compd. USSR (Engl. Transl.)* **24**, 850 (1988).
324. M. M. Krayushkin, M. A. Kalik, V. K. Zav'yalova, A. A. Loktionov and V. S. Bogdanov, *Chem. Heterocycl. Compd. USSR (Engl. Transl.)* **25**, 1349 (1989).
325. M. M. Krayushkin, M. A. Kalik and A.A. Loktionov, *Chem. Heterocycl. Compds. USSR (Engl. Transl.)* **26**, 756 (1990).
326. M. S. Krayushkin, M. A. Kalik, V. K. Zav'yalova and S. Bogdanov. *Chem. Heterocycl Compd. USSR (Engl. Transl.)* **24**, 332 (1988).

– 4G –

Thiophene Derivatives Containing Silicon, Germanium, Tin and Lead

4G.1 THIENYLSILICON COMPOUNDS

4G.1.1 Thienyl hydrosilanes

4G.1.1.1 Reduction of di- and trichloro(thienyl)silanes with lithium aluminium hydride

2-(Thienyl)silane, 5-methyl- and 5-chloro-2-thienylsilane are prepared in 77–87% yield from the corresponding thienyltrichlorosilyl compounds and lithium aluminium hydride in diethyl ether [1]. Phenylvinyl-(2-thienyl)silane is prepared by reduction of phenyl vinyl(2-thienyl)chlorosilane with lithium aluminium hydride [2].

Phenylvinyl(2-thienyl)silane [2]

To a suspension of lithium aluminium hydride (1.42 g, 37.4 mmol) in anhydrous diethyl ether (75 ml) a solution of phenylvinyl(2-thienyl)chlorosilane (37.62 g, 150 mmol) in anhydrous diethyl ether (50 ml) is added dropwise under stirring. The stirring is continued and the reaction mixture is refluxed for 16 h. After cooling to 0 °C, water (25 ml) is carefully added dropwise. The phases are separated and the aqueous phase extracted with diethyl ether (3 × 50 ml). The combined organic phases are washed with water, dried over sodium sulfate, evaporated and distilled, giving 22.6 g (70%).

4G.1.1.2 From thiophenemagnesium bromide and diorganosilyl chloride

The reaction between 2-thiophenemagnesium bromide and dichlorophenylsilane and methyl phenylsilyl chloride, respectively, is used for the preparation of phenyl(2-thienyl)chlorosilane and methylphenyl(2-thienyl)silane [2,3]. Further reaction of phenyl(2-thienyl)silane with vinylmagnesium chloride is another route to phenylvinyl(2-thienyl)silane [2].

Phenyl(2-thienyl)chlorosilane [2]

To a mixture of magnesium turnings (6.56 g, 0.27 mol) and diethyl ether (140 ml) a solution of 2-bromothiophene (42.39 g, 0.26 mol) in diethyl ether (40 ml) is added over a period of 45 min. The stirring is continued at room temperature for 3 h, after which the mixture is refluxed for 1 h, cooled and slowly added at 20 °C to a solution of dichlorophenylsilane (53.1 g, 0.3 mol) in diethyl ether (500 ml). The reaction mixture is refluxed for 3 h, stirred at room temperature for 24 h and filtered. The solid magnesium salts are washed with petroleum ether (3 × 50 ml) and the combined washings filtrate is evaporated. The residue is taken up in petroleum ether (200 ml) and left for some hours and this solution is filtered through sodium sulfate and distilled giving 41.5 g (71%) of the title compound bp 70 °C/0.1 mm Hg.

2-Thienyllithium cannot be used in this type of reaction, as it gives methylphenyl bis(2-thienyl)silane [3]. Similarly diethyl-, dipropyl- and dibutyl-2-thienylvinylsilane were prepared from chlorodiethyl-, chlorodipropyl- and chlorodibutyrylvinylsilane and 2-thiophenemagnesium bromide [3]. Sonification of 2-chlorothiophene and magnesium turnings followed by dimethylchlorosilane gives 2-dimethylsilylthiophene [4].

2-Dimethylsilylthiophene [4]

A flask equipped with condenser, an ultrasound horn and a septum is charged with tetrahydrofuran (60 ml), magnesium turnings and 2-chlorothiophene (5.6 ml, 50 mmol) and then cooled in a water bath while the reaction mixture is sonicated for 1 h at a 30% power level. At this time almost all of the magnesium is consumed and dimethylchlorosilane (6.5 ml, 60 mmol) is added dropwise *via* a syringe. After 0.5 h the reaction mixture is transferred to a 500 ml separatory funnel and pentane (200 m) and water (10 ml) are added. The phases are separated and the organic phase washed with water, dried over magnesium sulfate and evaporated. The residue is distilled through a 10 cm vacuum jacketed Vigreux column giving 13 g (92%) of the title compound bp 74–75 °C/60 mm Hg.

4G.1.1.3 Reactions of thienylsilicon hydrides with tin tetrachloride

The reactions of 2-thienylsilanes with tin tetrahalides occurs selectively with the replacement of only one hydrogen atom. Thus from the above-mentioned thienylsilanes 2-thienylchloro and bromo silane as well as the 5-methyl and 5-chloro derivatives are obtained in about 40% yield [1].

4G.1.2 2-Trialkylsilylthiophenes

A review on the preparation of aryl- and heteroaryltrimethylsilanes has been published [5].

4G.1.2.1 2-Trialkylsilylthiophenes from 2-thiophenemagnesium halides and trialkylsilyl halides

The reaction of 2-thiophenemagnesium bromide and 5-bromo-2-thiophenemagnesium with trimethylsilyl chloride are used for the preparation of 2-trimethylsilylthiophene and 2-bromo-5-trimethylsilylthiophene [6]. The reaction of 2-thiophenemagnesium bromide with dialkylethynyl fluoride gives dialkylethynyl (2-thienyl)silanes [7].

Dimethylethynyl(2-thienyl)silane [7]

To a stirred solution of 2-thiophenemagnesium bromide, prepared from magnesium (1.33 g) and 2-bromothiophene (9.0 g, 56 mmol) in diethyl ether, dimethylethynyl fluoride (5.6 g, 55 mmol) in diethyl ether (25 ml) is added. The reaction mixture is refluxed for 1 h and then decomposed with saturated ammonium chloride solution. After work up and distillation 5.1 g (56%) of the title compound is obtained, bp 54 °C/4 mm Hg.

Methyl(2-thienyl)dichlorosilane is prepared in 63–73% yield by the reaction of 2-thiophenemagnesium bromide with excess methyltrichlorosilane [8]. A modification in the preparation of 2-trimethylsilyl chloride is the *in situ* reaction of 2-bromothiophene with magnesium and trimethylsilyl chloride in hexamethylphosphoramide [9].

4G.1.2.2 2-Trialkylsilylthiophenes through the reaction between 2-thienyllithium derivatives and trialkylsilyl halides

The general and most often used method consists in the reaction of 2-thienyllithium derivatives, obtained by metalation of thiophenes with butyllithium or lithium diisopropylamide or halogen–metal exhange of 2-bromothiophenes with alkyllithium derivatives, with trimethylsilyl chloride at low temperature.

In this way, 2-trimethylsilylthiophene [10–16] and 5-methyl-2-trimethylsilylthiophene [17] are prepared. In a similar way 2-thienyldimethylethylsilane, 2-thienyltriethylsilane, 2-thienyltripropylsilane, 2-tri(isopropylsilyl)thiophene

2-(*tert*-butydimethylsilyl)thiophene, 2-(trimethylsilylethoxymethyl)thiophene and 2-thienyldimethylallylsilane are prepared from 2-thienyllithium and the appropriate trialkylsilyl chlorides [15,18].

2-(tert-Butyldimethylsilyl)thiophene [18]

To a stirred solution of tert-butyldimethylchlorosilane (7.54 g, 0.05 mol) in anhydrous tetrahydrofuran (25 ml) at −78 °C 1.0 *M* thienyllithium in tetrahydrofuran (50 ml, 0.05 mol) is added *via* cannula at such a rate that the temperature is maintained below −50 °C. The stirring is continued for 1 h and then the reaction mixture is allowed to warm to room temperature. Aqueous work up affords a brown liquid, which is purified by flash chromatography using petroleum as eluent, giving 9.58 g (97%) of the title compound as a colorless liquid.

Dimetalation of thiophene and 3-methylthiophene followed by reaction with trimethylsilyl chloride gives 2,5-bis(trimethylsilyl)thiophene and 3-methyl-2,5-bis(trimethylsilyl)hiophene [19].

3-Methyl-2,5-bis(trimethylsilyl)thiophene [19]

To a solution of 3-methylthiophene (2.45 g, 25 mmol) in anhydrous diethyl ether (50 ml) 2.5 *M* butyllithium in hexane (22 ml, 55 mmol) is added at room temperature. The stirring is continued for 30 min, after which trimethylsilyl chloride (5.43 g, 50 mmol) is added. The reaction mixture is refluxed for 1 h and hydrolyzed with saturated ammonium chloride solution. The product is extracted with diethyl ether and the combined organic phases are dried over sodium sulfate, evaporated and distilled giving 4.9 g (81%) of the title compound as a colorless liquid bp 110 °C/5 mm Hg.

Metalation of α-(2-thienyl)acetaldehyde diethylacetal with butyllithium followed by trimethylsilyl chloride or dimethylphenylsilyl chloride gave silylation and elimination in one pot yielding [5-(trimethylsilyl)-2-thienyl]-β-vinylethyl ether [20].

[5-(Trimethylsilyl)-2-thienyl]-β-vinylethyl ether [20]

To a solution of α-(2-thienyl)acetaldehyde diethylacetal (11.8 g, 0.059 mol) in anhydrous diethyl ether under, nitrogen, 1.6 *M* butyllithium (0.118 mol) is added at room temperature during a period of 5 min, after which trimethylsilyl chloride (12.7 g, 0.118 mol) is added. The reaction mixture is refluxed overnight, the salts are filtered off and the filtrate is evaporated. The residue is distilled giving 10.7 g (87%) bp 72–74 °C/0.025 mm Hg.

Metalation of 2-thiophenealdehyde diethylacetal or the corresponding 1,3-dioxolane derived from 2-acetylthiophene followed by trimethylsilyl chloride gives the corresponding protected 2-(trimethylsilyl)-5-thiophene aldehyde and methyl ketone, which can be deprotected [21,22].

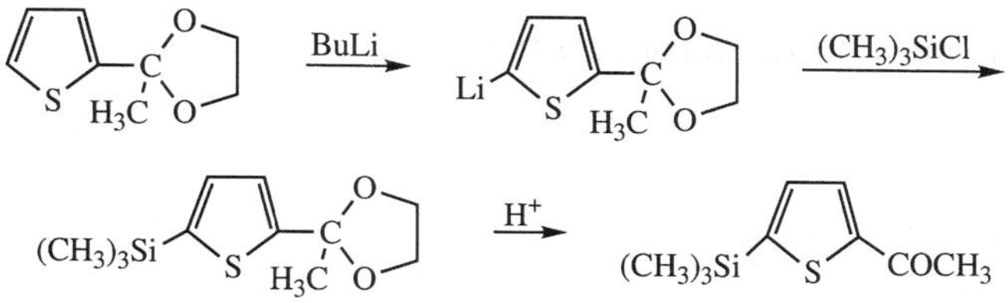

An alternative newer route to 5-trimethylsilyl-2-thiophene aldehyde consists in blocking of the aldehyde group of 2-thiophene aldehyde with lithium *N*-methylpiperazide, followed by metalation with butyllithium and reaction with trimethylsilyl chloride [23].

Metalation of 2-methylthiophenetricarbonylchromium(0) at −78 °C in tetrahydrofuran with butyllithium followed by trimethylsilyl chloride gives η^5-[2-methyl-5-trimethylsilyl)thiophene]tricarbonylchromium(0) [13].

η^5-[2-Methyl-5-trimethylsilyl)thiophene]tricarbonylchromium(0) [13]

To a solution of 2-methylthiophenetricarbonylchromium(0) (1.00 g, 4.27 mmol) in anhydrous tetrahydrofuran (17 ml) at −78 °C butyllithium (1.02 equiv.) is added over a period of 10 min. The stirring is continued for 1 h, after which trimethylsilyl chloride (1.2 equiv.) is added dropwise to the suspension and the reaction mixture is allowed to warm to room temperature. Aqueous work up affords a product which is purified by chromatography using diethyl ether/petroleum (5:95) as eluent, giving 1.26 g (96%) of the title compound as an orange crystalline solid mp 115–116 °C (decomp.).

In a similar way the triethylsilyl-, dimethylisopropylsilyl-, triisopropylsilyl- and *tert*-butyldimethylsilyl derivatives were prepared using the appropriate trialkylsilyl halide [13]. The congested nonplanar 3,4-di-*tert*-butyl-2,5-bis(tri-isopropylsilyl)thiophene is prepared in 42% yield by halogen–metal exchange of 2,5-dibromo-3,4-di-*tert*-butylthiophene with *tert*-butyllithium followed by reaction with triisopropylsilyl chloride. The less congested 3,4-di-*tert*-butyl-2,5-bis(trimethylsilyl)- and 2,5-bis(triethylsilyl)thiophene were also prepared in much better yields [24].

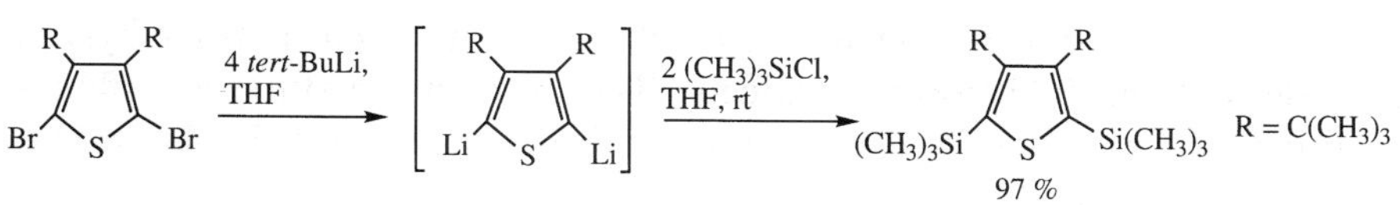

3-Methyl-2-bromothiophene has to be used for the preparation of 3-methyl-2-thienyllithium as metalation leads predominantly to 3-methyl-5-thienyllithium. Upon reaction with trimethylsilyl chloride, 3-methyl-2-trimethylsilylthiophene is obtained in 96% yield [25]. Metalation of 3-bromothiophene and 4-methyl-3-bromothiophene with lithium diisopropylamide followed by reaction with trimethylsilyl chloride gives 3-bromo-2-trimethylsilylthiophene and 3-bromo-4-methyl-2-trimethylsilylthiophene [26].

3-Bromo-2-trimethylsilylthiophene [26]

To butyllithium in hexane (0.250 mol) under nitrogen at 0 °C a solution of diisopropylamine (35.5 ml, 0.250 ml) in anhydrous diethyl ether (150 ml) is added. The stirring is continued for 30 min, after which 3-bromothiophene (40.5 ml, 0.250 mol) is added dropwise and the stirring is continued for another 30 min. The mixture is cooled to −70 °C and a solution of trimethylsilyl chloride (33.3 ml, 0.250 mol) in anhydrous diethyl ether (100 ml) is added. The reaction mixture is stirred at −70 °C for 2 h and left overnight. Dilute hydrochloric acid is added, the phases separated and the aqueous phase is extracted with diethyl ether. The combined organic phases are washed with water, dried, evaporated and distilled giving 36.9 g (63%) of the title compound bp 94–98 °C/18 mm Hg.

4G.1.2.3 By rearrangement of 3-ThCH$_2$OSiR$_3$ derivatives

2-Trialkylsilyl-3-hydroxymethylthiophenes are obtained in good yields by treatment of 3-(silyloxymethyl)thiophenes with butyllithium in hexamethylphosphoramide/tetrahydrofuran through an intramolecular [1,4]oxygen–carbon silyl migration [27,28].

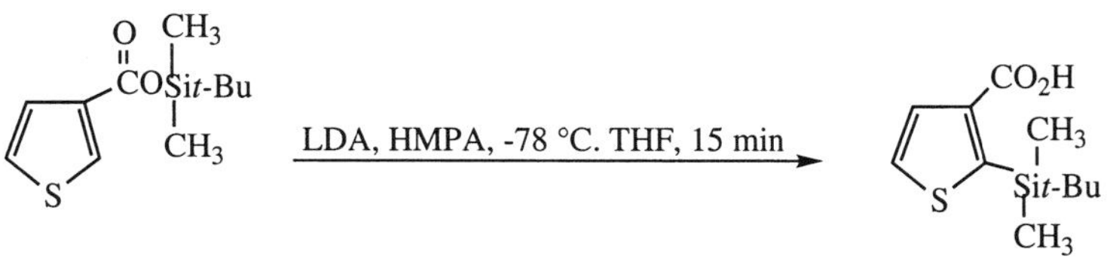

2-Trialkylsilylated-3-thiophenecarboxylic acids are prepared from silylesters of 3-thiophenecarboxylic acid upon treatment with a mixture of lithium diisopropylamide and hexamethylphosphoramide [28,29].

4G.1.2.4 From 2-trialkylsilylthiophenes by further reactions

Trialkylsilylthiophenes are rather stable and can, therefore, be modified by a variety of reactions.

4G.1.2.4.1 Metalation and halogen–metal exchange of trimethylsilylthiophenes

The reaction of trimethylsilylthiophene with butyllithium followed by solid carbon dioxide is used for the preparation of 5-trimethylsilyl-2-thiophene-carboxylic acid [11].

5-Trimethylsilyl-2-thiophenecarboxylic acid [11]

To 2-trimethylsilylthiophene (8 g, 0.05 mol) butyllithium in diethyl ether (0.05 mol) is added. The mixture is refluxed for 4 h and then quickly poured onto solid carbon dioxide. After warming to room temperature the phases are separated and the aqueous phase under vigorous stirring acidified with concentrated hydrochloric acid. The precipitate is filtered off and crystallized from ethanol/water giving 6.3 g (62%) of the title compound mp 134–135 °C.

Reaction with *N,N*-dimethylformamide is used for the preparation of 5-trimethylsilyl-2-thiophene aldehyde and with ethylene oxide β-(5-trimethyl-silyl-2-thienyl)ethanol is obtained [17].

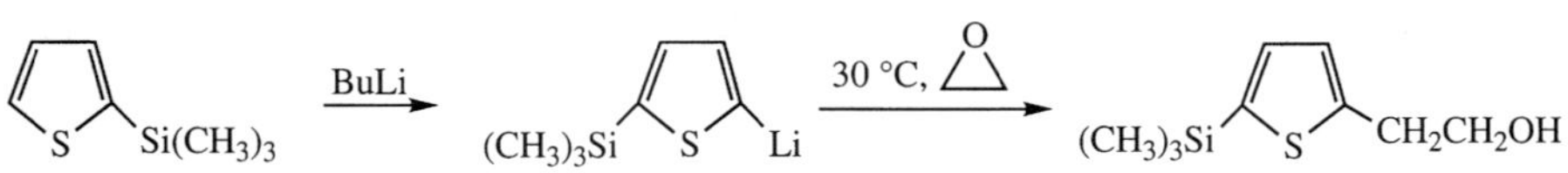

Metalation of thiophene with butyllithium/*N,N,N′,N′*-tetramethyl-1,2-ethanediamine in tetrahydrofuran followed by electrophilic quenching with diethyl carbamyl chloride and trimethylsilyl chloride is a method for the preparation of *N,N*-diethyl 5-trimethylsilylthiophene-2-carboxamide [14]. Metalation of 5-trimethylsilyl-, and 5-dimethylethylsilylthiophene with butyllithium followed by reaction with bromoacetaldehyde diethylacetal is used for the preparation of α-2-(5-trimethylsilyl)thienyl]acetaldehyde diethylacetal [20].

α-[2-(5-Trimethylsilyl)thienyl]acetaldehyde diethylacetal [20]

To a solution of 2-trimethylsilylthiophene (10.5 g, 67 mmol) in anhydrous tetrahydrofuran 1.6 *M* butyllithium (67 mmol) is added and the mixture refluxed for 3 h, after which bromoacetaldehyde diethylacetal (13.2 g, 67 mmol) is added. The reaction mixture is refluxed for 48 h, evaporated and filtered. The residue is distilled giving 5.48 g (30%) bp 77 °C/0.02 mm Hg.

The reaction of 5-trimethylsilyl-2-thienyllithium with arylnitriles gives ketimines upon hydrolysis with ammonium chloride in water. Acidic hydrolysis to the ketones is achieved without loss of the trimethylsilyl group [30].

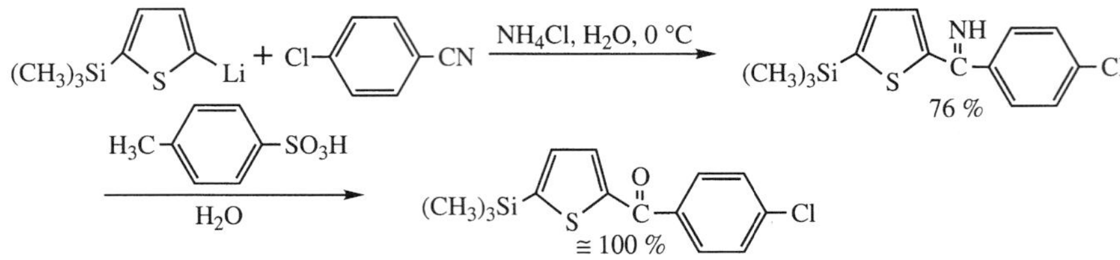

The reaction of 5-trimethylsilyl-2-thienyllithium with trichloroacetonitrile at 0 °C on the other hand gives a 45% yield of 5-trimethylsilyl-2-chlorothiophene [30]. Halogen–metal exchange of 5-trimethylsilyl-2-bromothiophene with *tert*-butyllithium in diethyl ether at −78 °C, followed by reaction with zinc chloride, or trimethyl borate is used for the preparation of 5- trimethylsilyl-2-thienylzinc chloride, and 5-trimethylsilyl-2-thiopheneboronic acid, respectively [25].

The metalation of 2,5-bis(trimethylsilyl)thiophene with butyllithium does not lead to metalation in the 3-position, but to lithium-trimethylsilyl exchange, giving 5-trimethylsilyl-2-thienyllithium (30–40%) and methylation of a silylmethyl group giving 5-trimethylsilyl-2-lithiomethyldimethylsilylthiophene in 15–20% [25].

4G.1.2.4.2 Electrophilic and other substitution of 2-trimethylsilylthiophenes

The reaction of 2-trimethylsilylthiophene with acetic anhydride in the presence of catalytic amounts of iodine gives only a small yield of 2-acetyl-5-trimethylsilylthiophene [11,12]. Surprisingly the reaction of 2-trimethylsilylthiophene with benzoyl chloride gives a low yield of 2-benzoyl-5-trimethylsilylthiophene [12]. Electrophilic substitution of 2-trimethylsilylthiophene with bis(*para*-anisyl)carbenium triflate in the presence of a proton sponge gave 2-bis(*para*-anisylmethyl)-5-trimethylsilyl)thiophene [32].

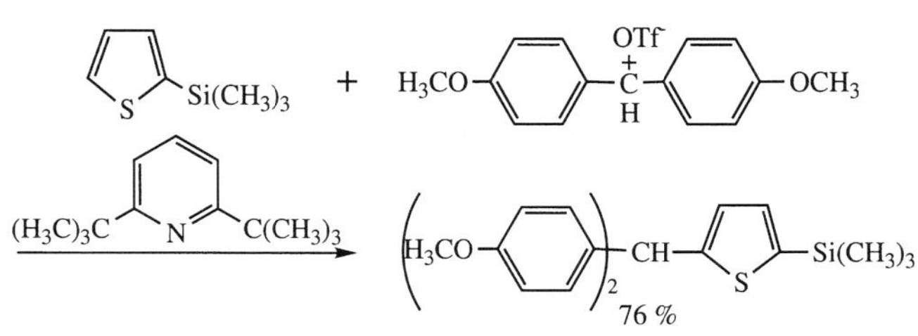

2-(Dimethyl-*tert*-butylsilyl)-5-bromothiophene is prepared by bromination of 2-(dimethyl-*tert*-butylsilyl)thiophene with *N*-bromosuccinimide in *N*,*N*-dimethylformamide at room temperature. Minor amounts of the 3-bromo derivative are formed as byproduct [33]. Heptafluoropropylation of 2-trimethylsilylthiophene with bis(heptafluorobutyryl) peroxide gives 2-heptafluoropropyl-5-trimethylsilylthiophene [34].

2-Heptafluoropropyl-5-trimethylsilylthiophene [34]

A solution of bis(heptafluorobutyryl) peroxide (0.426 g, 1.0 mmol) and 2-trimethylsilylthiophene (0.234 g, 1.5 mmol) in Freon-113 is degassed by a freeze–thaw cycle sealed in an ampoule and kept at 40 °C until the peroxide is consumed. The reaction mixture is washed two times with 5% aqueous sodium hydroxide and water. The organic phase is dried over magnesium sulfate and the product purified by chromatography on silica gel giving 0.21 g (68%) of the title compound.

From 2,5-di(trimethylsilyl)thiophene mixtures of 3-heptafluoropropyl-2,5-bis(trimethylsilyl)thiophene and 2-heptafluoropropyl-5-trimethylsilyl thiophene are obtained, the composition depending on the amounts of bis(heptafluorobutyryl) peroxide used [34]. The ligand transfer of naphthalenetricarbonylchromium(0) with 2-trialkylsilylthiophenes is an alternative route to η^5-2-(trialkylsilyl)thiophenetricarbonylchromium(0) derivatives [13].

4G.1.2.4.3 Various reactions of 2-trialkylsilylthiophenes

The carbonyl function of 2-acetyl- and 2-formyl-5-trialkylsilylthiophenes is modified to silicon-substituted thiophene derivatives of mercaptoethylamine such as 2[2-(5-trimethylsilyl)thienyl]thiazolidine of interest as possible radioprotective agents without any complication from the trialkylsilyl groups [35]. 3,4-Dimethyl-2,5-bis(trimethylsilyl)thiophene is without complications brominated in the side chain, with *N*-bromosuccinimide and dibenzoyl peroxide in carbon tetrachloride, giving 3,4-bis(bromomethyl)-2,5-bis-(trimethylsilyl)thiophene [36]. 2-Trialkylsilylthiophenes are converted to the corresponding η^5-tricarbonylchromium(0) complexes through reaction with naphthalenetricarbonyl)chromium(0) [18]. Especially the complexes of bulky 2-trialkylsilyl complexes are stable and metalation followed by an electrophile and renewed metalation and electrophile is synthetically very useful for the preparation of 2-trialkylsilyl-4,5-disubstituted thiophenes [18].

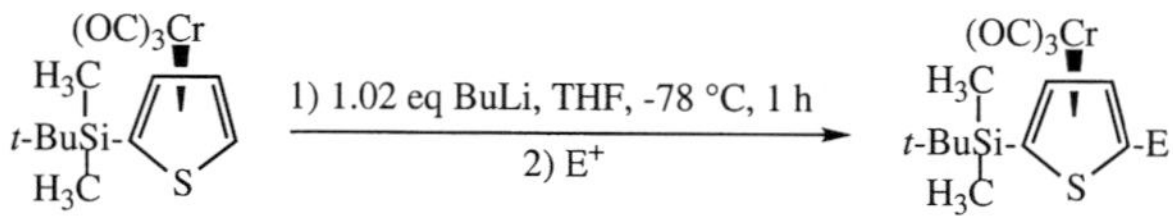

E^+	E	Yield (%)
D_2O	D	97
CH_3I	CH_3	80
$(CH_3S)_2$	SCH_3	70
PhNCO	CONHPh	44
I_2	I	28
$PhCH_2Cl$	CH_2Ph	0
CH_3COCl	$COCH_3$	0
$B(OiPr)_3$	$B(OH)_2$	0
$CH_2{=}CHCH_2Br$	$CH_2CH{=}CH_2$	72
$ClCOCH_3$	$COCH_3$	54
$CH_3COCH{=}CH_2$	$CH_2CH_2COCH_3$	0

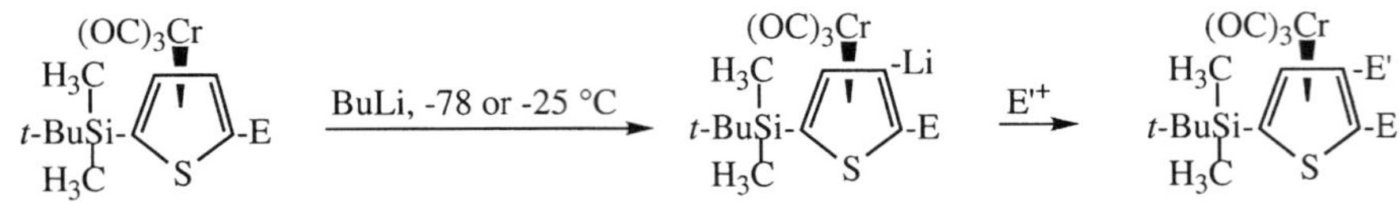

E	E'^+	E'	Yield (%)
CH_3	CH_3OD	D	74
CH_3	CH_3I	CH_3	61
CH_3	CH_3OCOCl	CO_2CH_3	60
CH_3	$(CH_3)_3Sn$	$Sn(CH_3)_3$	45
CH_3	I_2	I	34
CH_3	CH_2=$CHCH_2Br$	CH_2CH=CH_2	35
SCH_3	CH_3OD	D	37
SCH_3	CH_2=$CHCH_2Br$	CH_2CH=CH_2	39
CH_2CH=CH_2	CH_3OD	D	48

The 2-trialkylsilyl group is stable under oxidative conditions. Thus 2-acetyl-5-trimethylsilylthiophene is oxidized by selenium dioxide to (5-trimethylsilyl-2-thienyl)glyoxal monohydrate [37].

(5-Trimethylsilyl-2-thienyl)glyoxal monohydrate [37]

To a solution of freshly sublimed selenium dioxide (5 g, 45 mmol) in dioxane at 60 °C 2-acetyl-5-trimethylsilylthiophene (9 g, 46 mmol) is added. The mixture is refluxed under stirring for 2.5 h and filtered by suction through a layer of Hyflo Supercel. The filtrate is evaporated and the residue is distilled giving 7.0 g (73%) of a pale yellow oil bp 89–90 °C/0.6 mm Hg. A portion of the glyoxal obtained is heated with ten times its volume of water, boiled with Norit after solution and filtered. On cooling the filtrate gives white gelatinous flocs of the hydrate mp 102–109 °C after crystallization from dilute ethanol.

This stability under oxidative conditions also allows the preparation of trialkylsilylated thiophene-1,1-dioxide by oxidation of trialkylsilylthiophenes with *meta*-chloroperbenzoic acid [33,38].

3-Bromo-5-methyl-2-trimethylsilylthiophene-1,1-dioxide [38]

To a vigorously stirred suspension of *meta*-chloroperbenzoic acid (10.1 g, 44.0 mmol) and solid sodium bicarbonate (4.2 g, 50.0 mmol) in dichloromethane (150 ml), 3-bromo-5-methyl-2-trimethylsilylthiophene (5.0 g, 20.0 mmol) is added. The stirring is continued for 72 h, after which the 3-chlorobenzoic acid is filtered off. The filtrate is washed with saturated sodium carbonate solution and several times with water, dried over magnesium sulfate and evaporated at 0 °C giving 4.89 g (87%) of the title compound, which is stable for longer periods when stored as a cooled dichloromethane solution.

Also 2,5-bis(trialkylsilyl)thiophenes can be oxidized to the thiophen-1,1-dioxides [40].

4G.1.2.4.4 Palladium(0) and other transition metal catalyzed couplings of trialkylsilyl substituted thiophenes

During recent years there has been extensive use of such reactions in connection with synthetic work on conducting polymers and related materials [40].

4G.1.3 3-Trialkylsilylthiophenes

4G.1.3.1 From 3-thienylsodium, -lithium or -magnesium derivatives and trialkylsilyl halides

3-(Thienyl)trimethylsilane is obtained by reaction of 3-bromothiophene with trimethylsilyl chloride in the presence of a five-fold excess of ethyl bromide (entrainment method) [41].

3-(Thienyl)trimethylsilane [41]

A small amount of ethyl bromide is added to magnesium turnings (22 g) covered with diethyl ether (150 ml). When the reaction has started a mixture of 3-bromothiophene (24 g, 0.15 mol), ethyl bromide (82 g, 0.75 mol) and diethyl ether (450 ml) is added slowly under stirring, after which the mixture is refluxed for 40 h. Trimethylsilyl chloride (100 g, 0.93 mol) is added at such a rate that reflux is maintained followed by reflux for 3 h. After addition of water and dilute acid the phases are separated and the organic phase dried over sodium sulfate and distilled giving 12 g (52%) of the title compound bp 168 °C.

A later development for the preparation of 3-trimethylsilylthiophene in 80% yield consisted in the reaction of 3-bromothiophene with sodium sand in dioxane at 60 °C followed by trimethylsilyl chloride [9]. The attempted preparation of 3-trimethylsilylthiophene through the reaction of 3-bromothiophene with zinc powder and trimethylsilyl chloride in hexamethylphosphoramide at 80 °C failed as a mixture of the 2- and 3-isomers was obtained [9].

The best method for the preparation of 3-trimethylsilylthiophene therefore, is the sonochemical Grignard cross-coupling of 3-bromothiophene with trimethylsilyl chloride and magnesium in tetrahydrofuran [42].

3-Trimethylsilylthiophene [42]

A vial with 20 cm length and 1 cm internal diameter is charged with 3-bromothiophene (0.41 g, 2.5 mmol), tetrahydrofuran (2 ml), trimethylsilyl chloride (0.27 g, 2.5 mmol) and magnesium (0.06 g). The vial is placed in the middle of an ultrasonic bath (250 W, 45 kHz) in such a way that the bottom of the vial is 1 cm above the bottom of the vial. The reaction mixture is kept under ultrasonication for 8 h. After completion of the reaction the solids are filtered off and washed with tetrahydrofuran (2 ml). The combined filtrates are evaporated and distilled giving 0.29 g (75%) of the title compound bp 167–168 °C.

η^5-(2,5-Dimethylthiophene)tricarbonylchromium(0) is easily metalated by butyllithium in tetrahydrofuran/*N*,*N*,*N*′,*N*′-tetramethyl-1,2-ethanediamine at −78 °C and upon reaction with trimethylsilyl chloride, the 2,5-dimethyl-3-trimethylsilyl complex is obtained [18].

4G.1.3.2 Through further reactions of substituted 3-trimethylsilylthiophenes

3-Trimethylsilyl-4-ethynylthiophene and bis[4-(trimethylsilyl)thien-3-yl]acetylene are obtained by Stille couplings of 3-iodo-4-trimethylsilylthiophene with butylstannylacetylene and 1,2-tributylstannylacetylene, respectively. A number of 4-trimethylsilyl-3-alkyl- and arylethynyl-substituted thiophenes are prepared by Sonagashira coupling of 3-iodo-4-trimethysilylthiophene alkyl- or aryl-acetylenes [43].

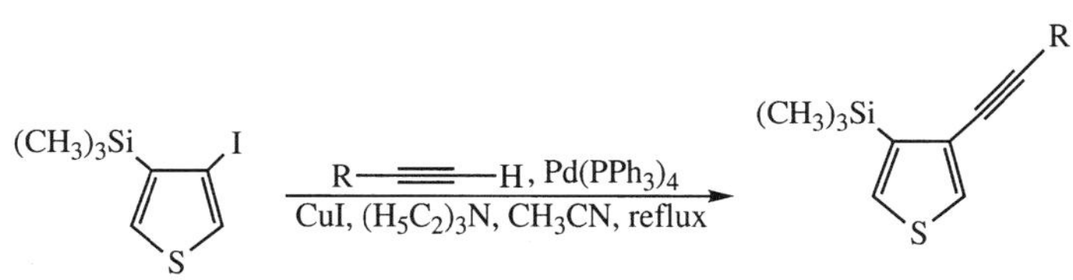

R = C_5H_{11} (88 %), C_7H_{15} (89 %), Ph (85 %), CH_2CH_2OH (92 %), $(CH_2)_4OH$ (91 %), $CH_2(CH_3)NCH_2Ph$,

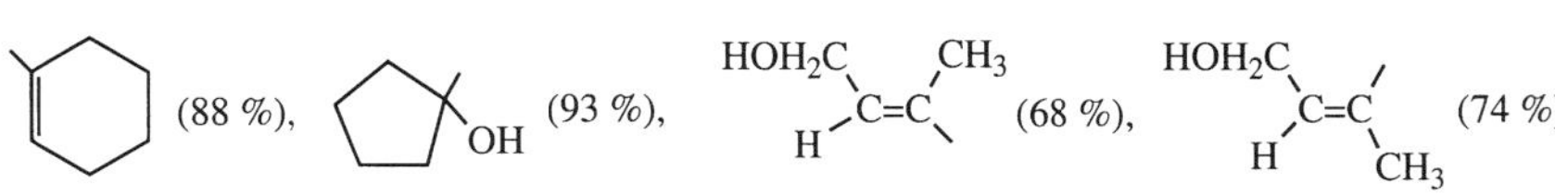

Catalytic hydrogenation of 3-alkyl- and 3-arylethynyl-4-trimethylsilylthiophene is used for the preparation of 3-alkyl-4-trimethylsilylthiophenes [43].

3-(Trimethylsilyl)-4-heptylthiophene [43]

To a solution of 3-(trimethylsilyl)-4-(heptyn-1′-yl)thiophene (500 mg, 2 mmol) in hexanes (20 ml) and triethylamine (4 ml) 10% palladium on carbon (0.11 g, 0.1 mmol) is added, after which the mixture is stirred under hydrogen atmosphere for 24 h. The reaction mixture is filtered and the filtrate evaporated giving 507 mg (100%) of the title compound as a colorless oil.

Suzuki coupling between phenylboronic acid and 3-iodo-4-trimethylsilylthiophene can be used for the preparation of 3-phenyl-4-trimethylsilylthiophene [44].

4G.1.3.3 By cycloaddition cycloreversion reactions

3-Trialkylsilylthiophenes are also obtained in good yields through the intramolecular Diels-Alder cycloaddition-cycloreversion between 4-phenylthiazole and trimethylsilylacetylene or butyltrimethylsilylacetylene in sealed tubes at 325–340 °C in the presence of 1,8-diazabicyclo[5.4.0]undecene-7. Using this method, 3-trimethylsilylthiophene and 3-butyl-4-trimethylsilylthiophene are obtained in 65 and 73% yield, respectively [43].

4G.1.4 Bis(trialkylsilyl)thiophenes

The reaction of thiophene with two equivalents of butyllithium followed by trimethylphenyldimethyl- and *tert*-butyldimethylsilyl chloride at −20 °C is used for the preparation of the corresponding 2,5-bis(trialkylsilyl)thiophenes [39].

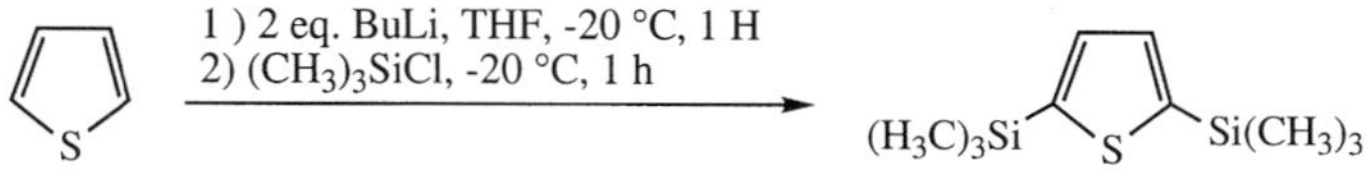

Metalation of 3,4-dimethylthiophene with a 2.6-fold excess of butyllithium in diethyl ether/hexane at reflux followed by addition of trimethylsilyl chloride at reflux for 20 h gives 3,4-dimethyl-2,5-di(trimethylsilyl)thiophene in 78% yield [36].

Several methods are available for the preparation of 3,4-bis(trimethylsilyl)thiophene. It was first prepared from 3,4-dibromothiophene *via* halogen–metal exchange with butyllithium at −70 °C and reaction with trimethylsilyl chloride to 3-bromo-4-tri(methylsilyl)thiophene followed by renewed halogen–metal exchange with two equivalents of *tert*-butyllithium and reaction with trimethylsilyl chloride [45]. Attempts to prepare 3,4-bis(trimethylsilyl)thiophene in a way similar to 3-trimethylsilylthiophene, by ultrasonification of 3,4-dibromothiophene, trimethylsilyl chloride and magnesium powder in tetrahydrofuran were less successful and gave after 7 days reaction only 30% of the desired compound [43]. Another approach to the desired compound, also giving low yield, consists in a ring-closure reaction of bis(trimethylsilylmethyl)sulfoxide with bis(trimethylsilyl)acetylene, giving 3,4-bis(trimethylsilyl)-2,5-dihydrothiophene in low yield, followed by aromatization by treatment with 2,3-dichloro-5,6-dicyano-1,4-benzoquinone [43]. Finally the best method for the preparation of 3,4-bis(trimethylsilyl)thiophene appears to be the reaction of 4-phenylthiazole with bis(trimethylsilyl)acetylene by intramolecular Diels-Alder cycloaddition–cycloreversion in sealed tubes [43].

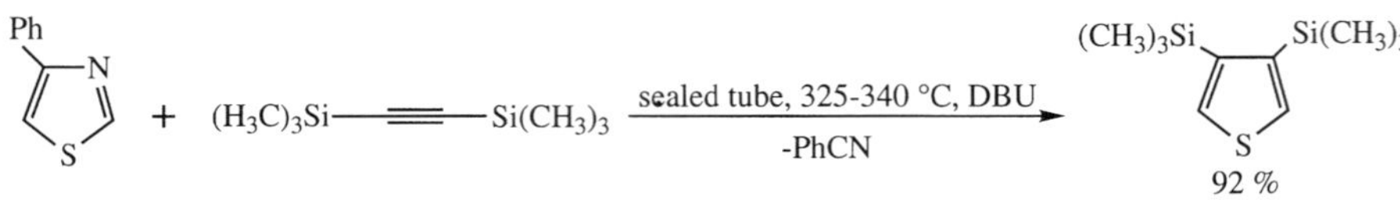

2,4-Bis(trimethylsilyl)thiophene is prepared by metalation of 3-(trimethylsilyl)thiophene with butyllithium, which occurs selectively in the 5-position, followed by reaction with trimethylsilyl chloride [45]. Further lithiation of 2,4-bis(trimethylsilylthiophene with butyllithium gives tris(2,3,5-trimethylsilyl)thiophene. 2,3-Bis(trimethylsilyl)thiophene can be prepared in moderate yield by halogen–metal exchange beween 3-bromo-2-(trimethylsilyl)thiophene at low temperature followed by the more reactive trimethylsilyl iodide in order to avoid ring opening of the 3-thienyllithium derivative [45].

4G.1.5 Di- tri- and tetra-arylsilanes

4G.1.5.1 From Grignard reagents and silyl halides

2-Thienylphenylvinylchlorosilane is obtained from 2-thiophenemagnesium bromide and phenylvinyl dichlorosilane [2].

2-Thienylphenylvinylchlorosilane [2]

To a solution of phenylvinyldichlorosilane (121.9 g, 0.6 mol) in diethyl ether (1 l) at 0 °C under vigorous stirring the Grignard solution prepared from magnesium turnings (14.6 g, 0.6 mol) and 2-bromothiophene (97.8 g, 0.6 mol) is added dropwise during 4 h. The reaction mixture is stirred at room temperature for 12 h and refluxed for 6 h, after which the magnesium salts are filtered off, the filtrate evaporated, the residue is taken up in petroleum ether and this solution is filtered through a pad of sodium sulfate. The solid is washed with petroleum ether and the combined filtrates are evaporated and distilled giving 70.7 g (47%) of the title compound as a colorless liquid bp 110 °C/0.8 mm Hg.

The ultrasonification of 2-chlorothiophene and magnesium turnings in tetrahydrofuran with dimethyl dichlorosilane gives an 80% yield of dimethyl bis(2′-thienyl)silane [4]. Phenyl-2-thienyldimethylsilane is obtained by the reaction of 2-thiophenemagnesium chloride and phenyldimethylchlorosilane as described above [4].

4G.1.5.2 From thienyllithium derivatives and arylsilyl halides

Triphenyl-2-thienylsilane, 2-thienyldimethylphenylsilane, 2-thienyldiphenylmethylsilane and 2-thienylphenyldiethylsilane are prepared from 2-thienyllithium and the appropriate silyl chlorides [15,46]. Also dimethyl-4-(2,4-dimethylphenyl)(2-thienyl)silane is prepared through the reaction of 2-thienyllithium with dimethyl-4-(2,4-dimethylphenyl)chlorosilane [47].

Dimethyl-4-(2,4-dimethylphenyl)(2-thienyl)silane [47]

To a solution of thiophene (11.7 g, 0.14 mol) in anhydrous diethyl ether (150 ml) under argon at 20 °C butyllithium in hexane (100 ml, 0.13 mol) is added over a period of 20 min. The mixture is stirred for 30 min and cooled to −20 °C, after which a solution of dimethyl-4-(2,4-dimethylphenyl)chlorosilane (27.7 g, 0.14 mol) in anhydrous diethyl ether (20 ml) is added. The temperature of the reaction mixture is allowed to rise to 20 °C over a period of 5 h and the stirring is continued for 12 h. After addition of saturated ammonium chloride solution (50 ml) the phases are separated and the organic phase dried over sodium sulfate, evaporated and distilled giving 24 g (70%) of the title compound as a colorless liquid bp 140–147 °C/4 mm Hg.

The reaction of 2-thienyllithium with dimethyldichlorosilane gives bis(2-thienyl)dimethylsilane in 82% yield [15]. Methyl tris(2-methyl-4-thienyl)silane is prepared by halogen–metal exchange of 4-bromo- 2-methylthiophene with *sec*-butyllithium at −78 °C followed by trimethoxymethylsilane [48].

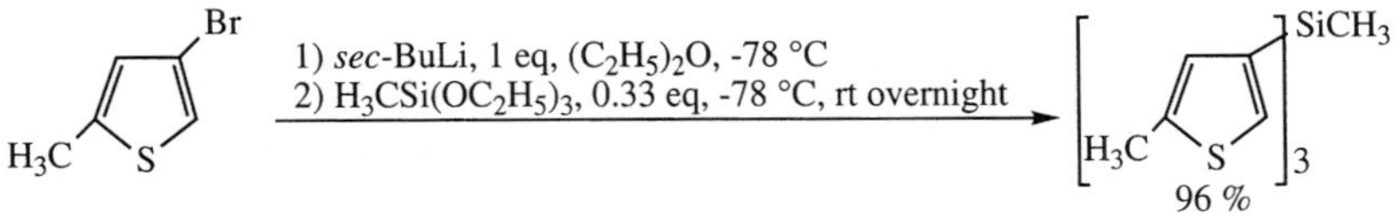

The reaction of 2,5-dilithiothiophene with 3,4-dimethylphenyltrimethylsilyl chloride is used for the preparation of 2,5-bis[dimethyl-(3,4-dimethylphenylsilyl)]thiophene. Attempted oxidation with aqueous permanganate led to cleavage of the two thienyl-silicon bonds and isolation of a siloxane [49].

When thiophene was treated with one equivalent of butyllithium and then with 0.5 equivalents of dimethyl dichlorosilane, dimethyldi(2-thienyl)silane was obtained in 74% yield [39]. The reaction of four equivalents of 2-thienyllithium with tetramethoxysilane in diethyl ether gives 70% of tetra-2-thienylsilane along with 20% of 2,5-bis[tri(2-thienyl)silyl]thiophene [50]. The latter compound is prepared in 84% yield by the reaction of 2,5-dilithiothiophene, obtained from 2,5-dibromothiophene and *tert*-butyllithium, with two equivalents of methyl tri-2-thienylsilylether. An organosilicon dendrimer composed of sixteen thiophene rings is prepared by tetralithiation of tetra-2-thienylsilane followed by reaction with methyltri-2-thienylsilyl ether [50].

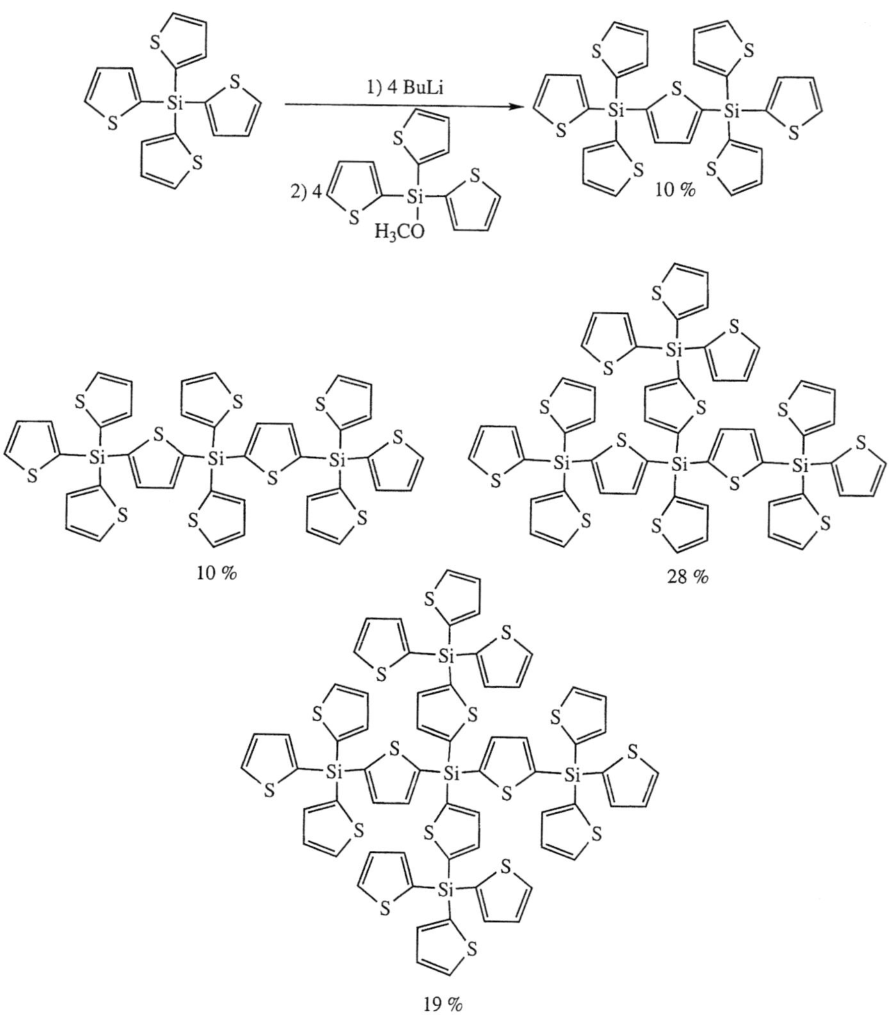

The following compound and the one starting from *cis* 1,2-di(4-bromo-3-thienyl)ethene) have been prepared [51].

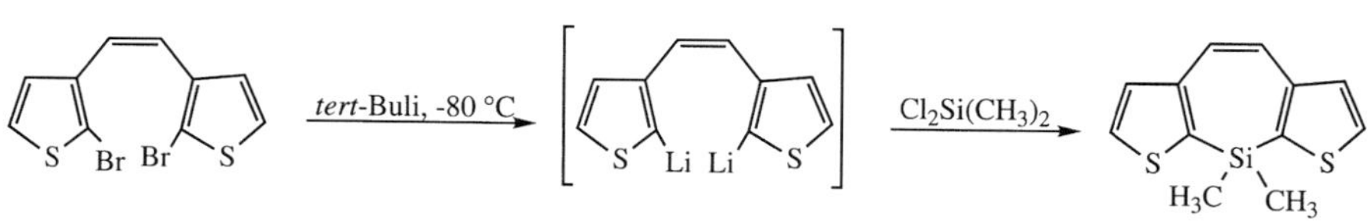

4G.1.5.3 Hydrosilylation of thienyldiorganosilanes

The reaction of 1-heptene and methylphenyl-2-thienylvinylsilane with methylphenyl-2-thinylsilane in the presence of catalytic amounts of hydrogen hexachloroplatinate(IV) is used for the preparation of methylheptylphenyl (2-thienyl)silane and 2,5-diphenyl-2,5-di-2-thienyl-2,5-disilahexane [3].

Methylheptylphenyl(2-thienyl)silane [3]

To a mixture of methylphenyl-2-thienylsilane (7.3 g, 36 mmol) and 0.1 *M* $H_2PtCl_6 \cdot 6H_2O$ in anhydrous isopropanol (0.2 ml), 1-heptene (3.5 g, 36 mmol) is added. The reaction is exothermic and after complete addition the reaction mixture is heated at 100 °C for 2 h. Upon distillation 2.3 g (22%) of the title compound is isolated.

2-(γ-Aminopropylmethylphenylsilyl)thiophene and 2,5-bis(γ-aminopropylmethylphenylsilyl)thiophene are prepared under the same conditions by hydosilylation of allylamine with 2-methylphenylsilyl)thiophene and bis(2,5-methylphenylsily)thiophene, respectively [52]. Also more complicated thiophenes such as 2,5-di-2-thienyl-3-hexyne-2,5-diol can be hydrosilylated with triethylsilane.The reaction does not stop at the addition step but complete dehydration of the adduct occurs giving 2,5-dihydro-2,5-dimethyl-2,5-di-2-thienyl-3-triethylsilylfuran [53].

4G.1.5.4 Reactions of thienyltriaryl silanes

The reaction of triphenyl-2-thienylsilane with butyllitium followed by dry ice gives triphenyl(2-thienyl-5-carboxy) silane in 45% yield [46]. The reaction of dimethyldi(2-thienyl)silane with two equivalents of lithium diisopropylamine followed by two equivalents of trimethylsilyl chloride gives dimethyl bis(5-trimethylsilyl-2-thienyl)silane, while its reaction with one equivalent of dimethylsilyl dichloride gives macrocylic compounds with four and six thiophene rings [39].

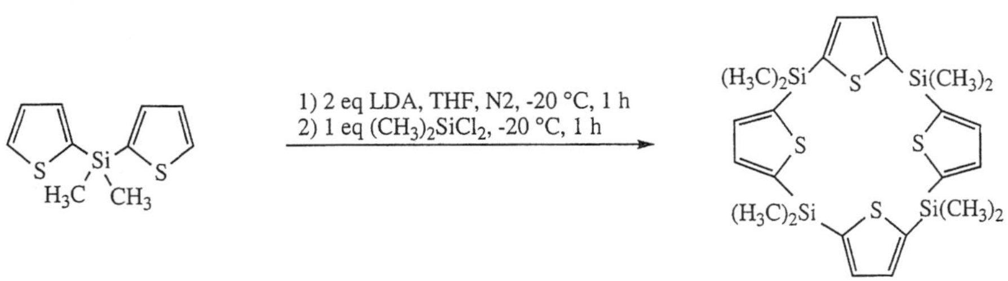

The reaction of dimethyl-di-(2-thienyl)silane with butyllithium leads to mono and dilithium derivatives which upon reaction with cupric chloride gives silathiophenoprotophanes. The reaction of the mono- and dilithium derivatives of dimethyl di(2-thienyl)silane with dichlorodimethylsilane gives the bithienyl derivatives. The reaction of the dilithio derivative of dimethyl-di-(2-thienyl)-silane, prepared in 85% yield by dilithiation at 0 °C, with one mol of dichlorodimethylsilane gave the cyclic compound shown below [54].

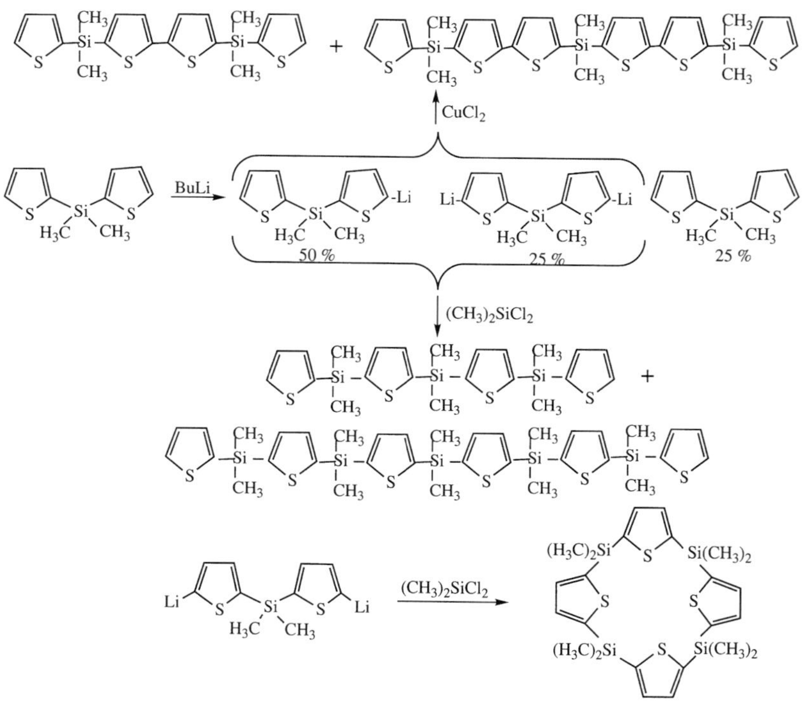

The trilithium derivatives obtained by metalation of methyl tris(2-methyl-4-thienyl)silane with three equivalents of butyllithium in diethyl ether/tetrahydrofuran at 0 °C or halogen–metal exchange of methyl tris(2-bromo-5-methyl-3-thienyl)silane with six equivalents of *tert*-butyllithium at −78 °C followed by diethyl carbonate gave 4-silathiophenetriptycene in low yield [48].

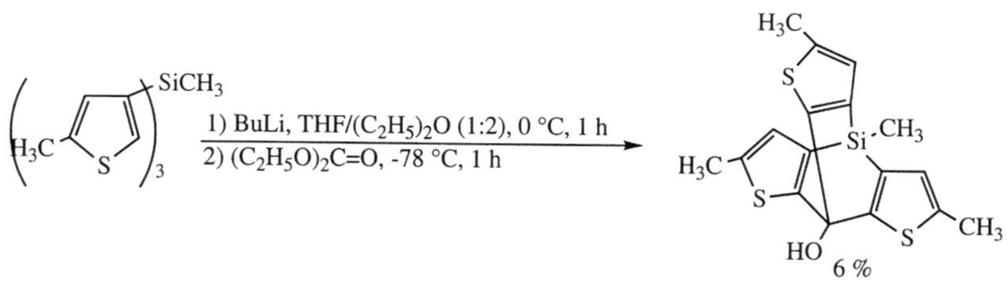

4G.1.5.5 By various reactions

Compounds such as dimethyl (dichloromethyl)(2-thienyl)silane, methyl dichloromethyl[di(2-thienyl)]silane and dichloromethyl[tri(2-thienyl)]silane are prepared in 38–66% yield by the reaction of dimethyl(2-thienyl)-, methyl[di(2-thienyl)]- and tri(2-thienyl)silane with dichlorocarbene generated from sodium trichloroacetate under solid–liquid phase transfer conditions [55].

Methyl(dichloromethyl)[di(2-thienyl)]silane [56]

To a solution of methyl[di(2-thienyl)]silane (1.1 g, 5 mmol) and 18-crown-6 (0.13 g, 0.5 mmol) in anhydrous toluene (10 ml), sodium trichloroacetate (2.78 g, 15 mmol) is added and the reaction mixture is heated under reflux for 6 h. After filtration through Celite 545 another portion of sodium trichloroacetate (2.78 g, 15 mmol) is added and heating is continued for 6 h. This procedure is repeated once again and the title compound is obtained in a yield of 43% determined by GLC.

4G.1.6 Thienylsilyl halides, alcohols and ethers

2-Thienyltrichlorosilane is prepared in 80–85% yield from 2-thiophenemagnesium bromide and silicon tetrachloride [56]. The gas phase high-temperature (600 °C) reaction of thiophene with trimethylsilyl chloride gives a low yield of 2-thienylsilyl trichloride [57]. 2-[(Methoxy)dimethylsilyl-thiophene is prepared by the reaction of 2-thienyllithium with dichlorodimethylsilane followed successively by addition of triethyl amine and methanol [58].

2-[(Methoxy)dimethylsilyl]thiophene [58]

A solution of 2-thienyllithium in diethyl ether (100 mmol) is added to a solution of dichlorodimethylsilane (19.4 g, 150 mmol) in anhydrous diethyl ether (150 ml) at 0 °C. After stirring at room temperature for 1 h triethylamine (20.3 g, 200 mmol) is added at 0 °C followed by methanol (6.41 g, 200 mmol) and the stirring is continued for 1 h. The product is extracted and the combined organic phases are washed with sodium bicarbonate solution and sodium chloride solution, dried over sodium sulfate and concentrated giving 9.1 g (63%) of the title compound bp 99 °C/80 mm Hg.

2-Thienyltrimethoxysilane is obtained by the reaction of 2-thienyltrichlorosilane in anhydrous methanol in pyridine and petroleum ether [59].

2-Thienyltrimethoxysilane [59]

To a solution of 2-thienyltrichlorosilane (435 g, 2.00 mol) and anhydrous pyridine (506 g, 6.40 mol) in petroleum ether (2.6 l), anhydrous methanol (275 ml, 6.80 mol) is added through the condenser as rapidly as possible. The precipitated pyridine hydrochloride is filtered off and thoroughly washed with petroleum ether. The combined filtrates are distilled through a Vigreux column. When the temperature of the residue attains 90 °C the distillation is continued *in vacuo* giving 339 g (83%) of the title compound bp 92 °C/9 mm Hg.

Octa- and dodeca(2-thienyl)silsesquioxane, with the latter usually predominating, are obtained by alkaline rearrangement of a polymer obtained through alkaline hydrolysis of 2-thienyltrimethoxysilane [59]. Triethoxy-3-thienylsilane is obtained in low yield upon reaction of 3-thienyllithium with tetrachlorosilane at −40 °C, followed by reaction with ethanol and pyridine [60].

2-Methoxydimethylsilylthiophene is prepared by the reaction of 2-dimethylsilylthiophene with methanol catalyzed by tris(triphenylphosphine)rhodium(I) chloride [4]. Using the same method 1,1,3,3-tetramethyl-1,3(bis(2′-thienyl) disiloxane is prepared from 2-dimethylsilylthiophene [4].

2-Methoxydimethylsilylthiophene [4]

Nitrogen is bubbled through a solution of 2-dimethylsilylthiophene (0.68 g, 4.6 mmol) in freshly distilled benzene (15 ml) and methanol (0.2 ml) for 10 min, after which tris(triphenylphosphine)rhodium(I) chloride (5 mg) is added. The flask is sealed with rubber septa and the magnetic stirring is continued overnight. The content of the flask is evaporated and distilled giving 0.6 g (72%) of the title compound bp 45–46 °C/0.5 mm Hg.

The reaction of three equivalents of 2-thienyllithium with one equivalent of tetramethoxysilane gives methyl tri-2-thienylsilyl ether in 50% yield [50]. 5-Phenyl-2-[(methoxy)dimethylsilyl]thiophene is prepared by the cupric oxide/pentafluorophenyl promoted cross-coupling of iodobenzene with 2-[methoxy) dimethylsilyl]thiophene, using 1,3-dimethyl-2-imidazolidinone as solvent [58].

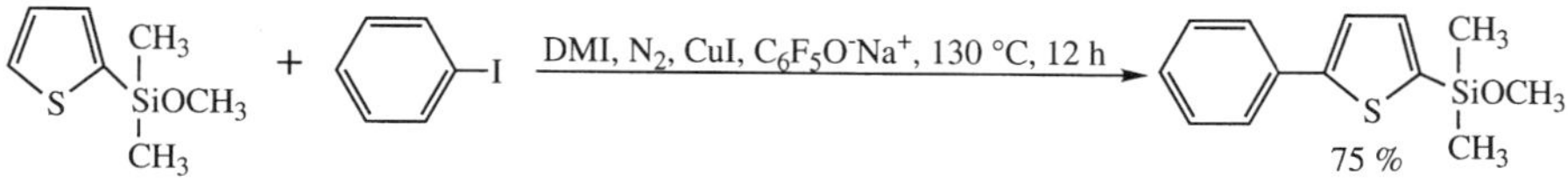

1-(5′-X-2′-Thienyl)silatranes (X = hydrogen, chloro, bromo, methyl and cyano) are prepared by transesterification of 2-thienyltrialkoxysilanes with triethanolamine in xylene, with simultaneous removal of the alcohol by distillation [61]. For a review see Ref. [62].

Another method for the preparation of this class of compounds is hydrosilylation of vinylsilatrane with dimethyl-2-thienylsilane in the presence of chloroplatinic acid giving dimethyl-1-(2′-thienylethyl)silatrane. A recent method for the preparation of silatranes is the selective cleavage of silicon–carbon bonds in heterylsilanes with triethanolamine in the presence of cesium fluoride. Thus tetra(2-thienyl)silane gives a 95% yield of 2-thienylsilatrane [63].

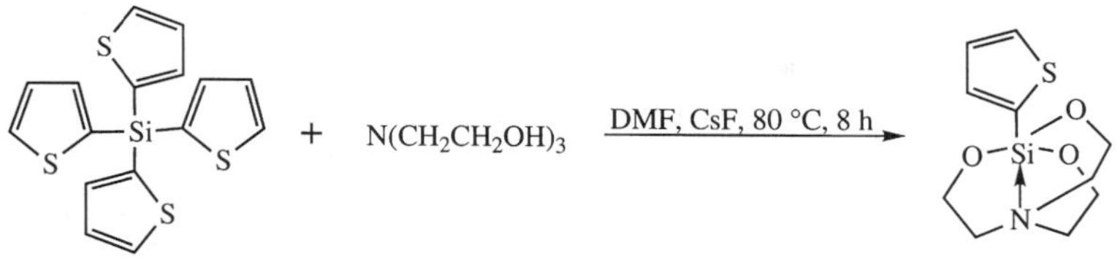

A compound containing two silatrane groups is obtained by the addition of two vinylsilatrane molecules to 2,5-bis(dimethylsilyl)thiophene [62,64]. Trimeric, tetrameric and pentameric methyl(2-thienyl)cyclosiloxanes are prepared by hydrolysis of methyl(2-thienyl)dichlorosilane with aqueous potassium carbonate or bicarbonate, extraction with ether and distillation of the crude hydrolysate [8].

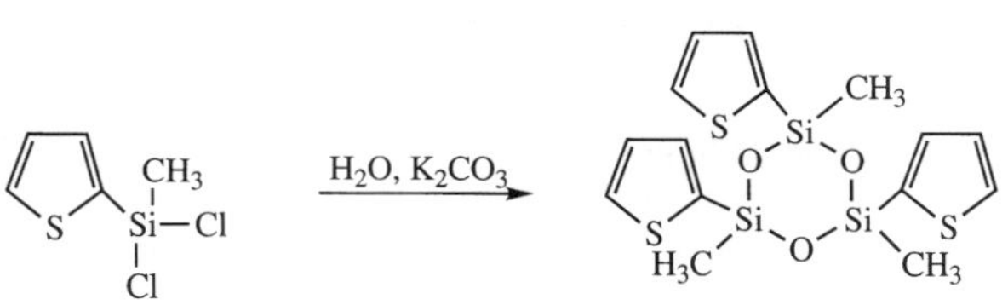

4G.1.7 Thiophenes with various silicon-containing functionalities

1,1,2,2-Tetramethyl-1,2-bis (2′-thienyl)silane is prepared in 85% yield by sonification of 2-chlorothiophene with magnesium turnings in tetrahydrofuran followed by reaction with 1,2-dichloro-1,1,2,2-tetramethyldisilane [4].

The organosilicon peroxides *tert*-(butyldioxy)-dimethyl-2-(thienyl)silane, bis(*tert*-butyldioxy)methyl(2-thienyl)silane and tris(*tert*-butyldioxy)(2-thienyl)-silane are prepared by the reaction of chlorodimethyl(2-thienyl)silane, dichloromethyl(2-thienyl)silane and trichloro-(2-thienyl)silane with tert-butyl-hydroperoxide in the presence of gaseous ammonia at 0 °C. [65,66].

The following reaction sequence has been described [67].

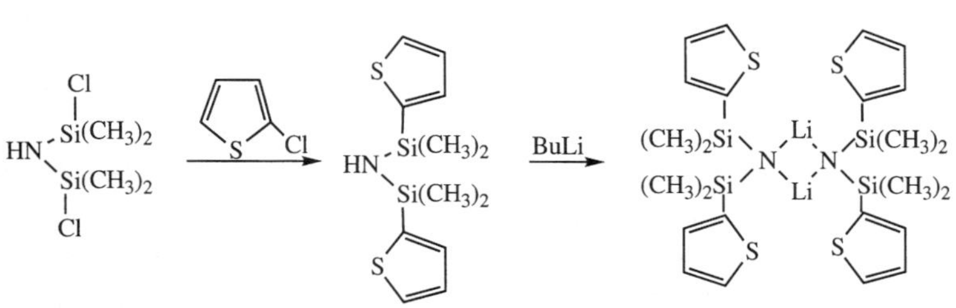

Metalation with two equivalents of butyllitium occurs quite surprisingly in the 3-position of one of the thiophene rings and is followed by ring opening. [68].

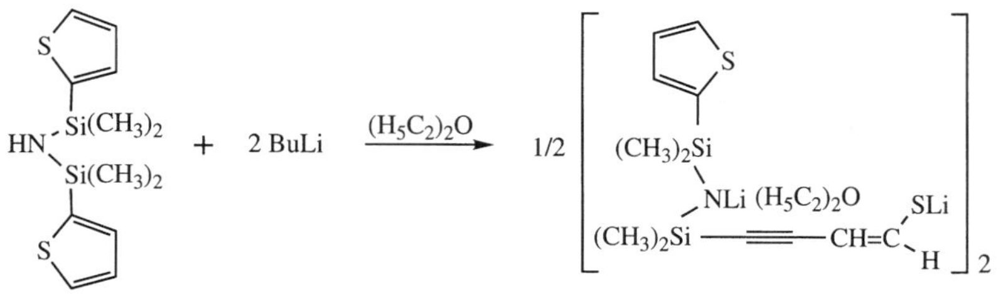

4G.2 THIENYLGERMANIUM COMPOUNDS

4G.2.1 Thienylhydrogermanes

4G.2.1.1 By reduction of ethoxygermanes

Di(2-thienyl)germane is prepared in 76% yield by the reduction of di(2-thienyl)diethoxygermane with a suspension of lithium aluminium hydride and tetrabutylammonium sulfate in pentane [69]. This method was also used for the preparation of 2-thienylgermane and tri(2-thienyl)germane [70].

2-Thienylgermane [70]

To a suspension of lithium aluminium hydride (1 g) in pentane (50 ml), triethoxy(2-thienyl)germane (6.5 g, 0.02 mol) and tetrabutylammonium hydrogen sulfate (0.74 g) are added. After stirring at room temperature for 2 h the precipitate formed is filtered off and the filtrate distilled giving 2.6 g (81%) of the title compound bp 136–138 °C.

4G.2.1.2 From thienyllithium derivatives and chlorogermanes

3-Germylthiophene is prepared by the reaction of 3-thienyllithium and chlorogermane [71].

3-Germylthiophene [71]

A vessel previously dried in an oven at 150 °C and purged with anhydrous nitrogen is charged via syringes with 3-bromothiophene (1.9 ml, 20 mmol) in anhydrous diethyl ether (15 ml). The vessel is attached to a vacuum line and the pressure is slowly reduced at room temperature to 50 mm Hg, after which the temperature is lowered to −70 °C. Butyllithium (11.2 ml, 22.4 mmol) is added by syringe. The stirring is continued for 15 min and chlorogermane (1.15 ml, 18.0 mmol) is condensed into the vessel and allowed to react under vacuum for 8 h. The reaction vessel is then allowed to warm slowly to room temperature and the reaction is quenched with water (5 ml). The phases are separated, the organic phase washed three times with water and the water phase is extracted with diethyl ether. The combined organic phases are evaporated to a volume of 5 ml and the product purified by gas chromatography giving 90% of the title compound as a colorless liquid.

2,2-Di(2-thienyl)-1,3-dioxa-6-aza-2-germacyclooctanes are obtained by the condensation of *N*-methyl or *N*-*tert*-butyl diethanolamine with di(2-thienyl)-germane without a catalyst [69].

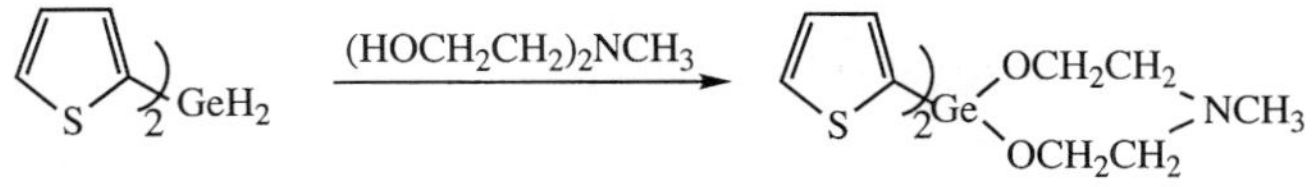

4G.2.2 Thienyl alkylgermanes

4G.2.2.1 From thiophenemagnesium compounds

Trimethyl(2-thienyl)germane is obtained by the reaction of 2-thiophenemagnesium bromide with chlorotrimethylgermane in hexane/tetrahydrofuran [72].

Trimethyl(2-thienyl)germane [72]

To a solution of chlorotrimethylgermane (17 g, 0.11 mol) in hexane (40 ml) 2-thiophenemagnesium bromide, prepared from magnesium turnings (2.65 g) and 2-bromothiophene (18 g, 0.11 mol) in diethyl ether is added. The reaction mixture is diluted with tetrahydrofuran in order to raise the solubility of the Grignard reagent and then heated for 1 h. The precipitate formed is filtered off and the filtrate evaporated giving 19.5 g (87%) of the title compound bp 71 °C/ 17 mm Hg.

This method is also used for the preparation of dibutyl- and diphenyldi(2-thienyl)germane from 2-thiophenemagnesium bromide, dibutylgermanium dichloride and diphenylgermanium dichloride, respectively [73].

Dibutyldi(2-thienyl)germane [73]

To a solution of 2-thiophenemagnesium bromide, prepared from 2-bromothiophene (5.80 g, 35.6 mmol) and magnesium (1.10 g, 45.5 mmol) in tetrahydrofuran (40 ml), dibutylgermanium dichloride (3.7 g, 14.3 mmol) is added. The reaction mixture is refluxed for 2 h, hydrolyzed and extracted with petroleum ether. The combined organic phases are dried over sodium sulfate and distilled giving 4.87 g (96%) of the title compound bp 93 °C/0.025 mm Hg.

Metalation of these germanium derivatives with butyllithium followed by reaction with trimethylsilyl chloride is used for the preparation of the 5,5′-bis(trimethylsilyl) derivatives [73]. Dimethyl(2-thienyl)germane is prepared by the reaction of 2-thiophenemagnesium bromide with dichlorodimethylgermane, followed by reduction of the mixture so obtained of chloro- and bromodimethyl(2-thienyl)germane with lithium aluminium hydride [74].

Dimethyl(2-thienyl)germane [74]

A solution of 2-thiophenemagnesium bromide, prepared from 2-bromothiophene (46 g, 0.26 mol) and magnesium (5.7 g) in diethyl ether is added to dichlorodimethylgermane (46 g, 0.26 mol). The precipitate formed is filtered off, the filtrate is evaporated and distilled giving 29 g of a mixture of chloro- and bromodimethyl(2-thienyl)germane (2:11) bp 82–84 °C/4–5 mm Hg. To this mixture (9 g) dissolved in diethyl ether at −30 °C lithium aluminium hydride 0.33 g) is added in small portions. After warming to room temperature the stirring is continued for 4 h. The precipitate formed is filtered off, the filtrated evaporated and distilled giving 1.5 g of the title compound bp 98–100 °C/32 mm Hg.

4G.2.2.2 From thienyllithium derivatives and halotrimethylgermane

3-(Trimethylgermyl)thiophene is obtained in 41% yield through the reaction of 3-thienyllithium with bromo trimethylgermane at −70 °C [71].

3-(Trimethylgermyl)thiophene [71]

A vessel previously dried in an oven at 150 °C and purged with anhydrous nitrogen is charged *via* syringes with 3-bromothiophene (1.1 ml, 12 mmol) in anhydrous diethyl ether (20 ml). The temperature is lowered to −70 °C. Butyllithium (6.6 ml, 13.2 mmol) is added through a syringe. The stirring is continued for 15 min and bromotrimethylgermane (1.5 ml, 12.0 mmol) is added. The reaction mixture is stirred at −70 °C for 2 h and then allowed to warm slowly to room temperature, after which the reaction is quenched with water (5 ml). The phases are separated, the organic phase washed three times with water and the water phase is extracted with diethyl ether. The combined organic phases are evaporated to a volume of 5 ml and the product purified by gas chromatography giving 0.97 g (41%) of the title compound as a colorless liquid.

Metalation of 3-trimethylsilylthiophene with butyllithium occurs in the 5-position and upon reaction with trimethylchlorogermane, 2-trimethylgermyl-4-trimethylsilylthiophene is obtained in 50% yield. Similarly metalation of 3-trimethylgermylthiophene followed by trimethylsilyl chloride is used for the preparation of 2-trimethylsilyl-4-trimethylgermylthiophene [75].

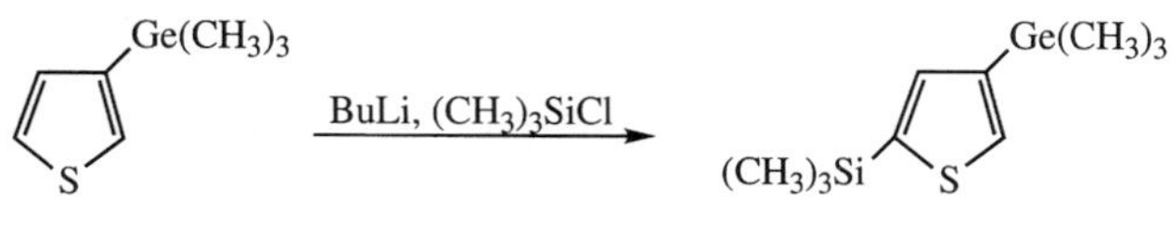

2-Trimethylsilyl-4-trimethylgermylthiophene [75]

To a solution of 3-trimethylgermylthiophene (5.1 g, 0.025 mol) in anhydrous ethyl ether (60 ml) at room temperature 2.5 *M* butyllithium in hexane (10 ml, 0.025 mol) is added. The stirring is continued for 1 h, after which trimethylsilyl chloride (2.7 g, 0.025 mol) is added dropwise. The reaction mixture is refluxed for 2 h and hydrolyzed with saturated ammonium chloride solution. The product is extracted with diethyl ether and the combined organic phases are dried over sodium sulfate, evaporated and distilled giving 4.0 g (58%) of the title compound bp 129–131 °C/15 mm Hg.

4G.2.3 Triaryl thienylgermanes

Tetra(2-thienyl)germanium is prepared by the reaction of 2-thiophenemagnesium bromide with germanium tetrachloride [76]. It is also obtained by the reaction of four equivalents of 2-thienyllithium with germanium tetrachloride in 40% yield.

4G.2.4 2,5-Di(2,5-trialkylgermanyl)thiophenes

The reaction of 2,5-dibromothiophene with *sym* dichlorotetramethyl digermane in the presence of tetrakis(triphenylphosphine)palladium(0) as catalyst gives 2,5-di(trimethylgermyl)thiophene after treatment with methyllithium. The reaction is a germylene insertion reaction leading primarily to chloro dimethylgermyl derivative [77].

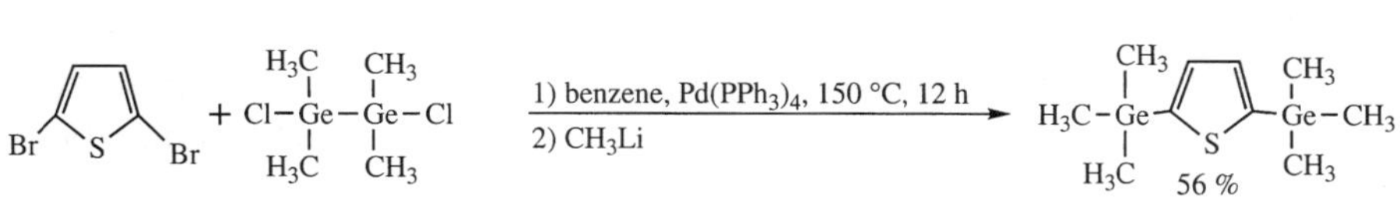

4G.2.5 Thienylalkoxygermanes

Trialkoxy 2-(thienyl)germanes are prepared by the insertion of germanium dibromide in the carbon–halogen bond of 2-bromothiophene with subsequent alcoholysis of the trihalosilyl derivative [62,78]. Alternatively heating of 2-bromothiophene with germanium tetrabromide and copper can be used [62]. Triethoxy-2-thienylgermane is obtained in low yield upon reaction of 2-thiophenemagnesium bromide with tetraethoxygermane [60]. The reaction of 2-thienyltrialkoxygermane with triethanolamine is used for the preparation of 2-thienylgermatrane [61,62,64,78]. Another route for the preparation of 2-thienylgermatranes consists in the reaction of hydroxygermatrane, obtained by germanium dioxide and triethanolamine, with thienylsilanes. No catalyst is necessary [62,79]. 3-Thienylgermatrane can also be prepared by the reaction of 3-bromothiophene with germanium(II) dioxane dibromide [80].

3-Thienylgermatrane [80]

3-Bromothiophene (3.2 g, 20 mmol) and germanium(II) dioxane dibromide (3.2 g, 10 mmol) are refluxed in a sealed ampoule at 200 °C for 4 h. Distillation gives 2.3 g (58%) of 3-thienyltribromogermatrane bp 104–106 °C/2.5 mm Hg. This compound (2.0 g, 5 mmol) is reacted with triethanolamine (0.75 g, 5 mmol) *via* the appropriate triethoxy derivative. The product is recrystallized from chloroform giving 0.7 g (46%) of the title compound mp 178–179 °C.

4G.2.6 Thienylhalogermanes

Both 2- and 3-thienyltribromogermane are prepared through the reaction of 2- and 3-bromothiophene with germanium(II) dioxane dibromide in a sealed ampoule at 200 °C [78]. Bis[2,5-(chlorodimethylgermyl)]thiophene is prepared in 65% yield by addition at room temperature of two equivalents of dimethylgermyl chloride to a hexane suspension of one equivalent of 2,5- dilithiothiophene [81]. The reaction of two equivalents of 2-thienyllithium with 1,1-dichloro-3-germacyclopentene gives 1,1-di(2-thienyl)germa-3-cyclopentene [82].

1,1-Di(2-thienyl)germa-3-cyclopentene [82]

To a solution of 2-thienyllithium (0.11 mol) in anhydrous tetrahydrofuran (100 ml) under nitrogen at −30 °C a solution of 1,1-dichloro-3-germacyclopentene (0.05 mol) in anhydrous tetrahydrofuran (50 ml) is gradually added. The reaction mixture is allowed to warm to room temperature and poured into ice water. The product is extracted with diethyl ether and the combined organic phases are washed with water, dried over magnesium sulfate and evaporated. The residue is recrystallized from ethanol or purified by chromatography on silica gel using ethyl acetate/hexane (8:2) as eluent giving 37% of the title compound as an oil.

The reaction of 2-thienyllithium with diphenyldichlorogermane is used for the preparation of diphenyldi(2-thienyl)germane [83] and the reaction with dimethyldichlorogermane gives dimethyl bis(2-thienyl)germane [81].

Dimethyl bis(2-thienyl)germane [81]

To a solution of thiophene (1.94 g, 23 mmol) and *N*,*N*,*N′*,*N′*-tetramethyl-1,2-ethanediamine (2.66 g, 23 mmol) in anhydrous diethyl ether (50 ml) under nitrogen 1.6 *M* butyllithium in hexane (14.4 ml) is added. The mixture is stirred at room temperature for 30 min, after which dimethyldichlorogermane (2.00 g, 11.5 mmol) in anhydrous diethyl ether (20 ml) is slowly added through a syringe. The stirring is continued for 1 h, the reaction mixture is then filtered, the filtrate evaporated and distilled giving 2.4 g (78%) of the title compound bp 90–91 °C/0.25 mm Hg.

[1_4]Dimethylgerma-2,5-thiophenocalixarene is obtained in up to 50% yield by the reaction in dilute media of dimethyldi(5-lithio-2-thienyl)germane with dimethylgermanyl chloride [81].

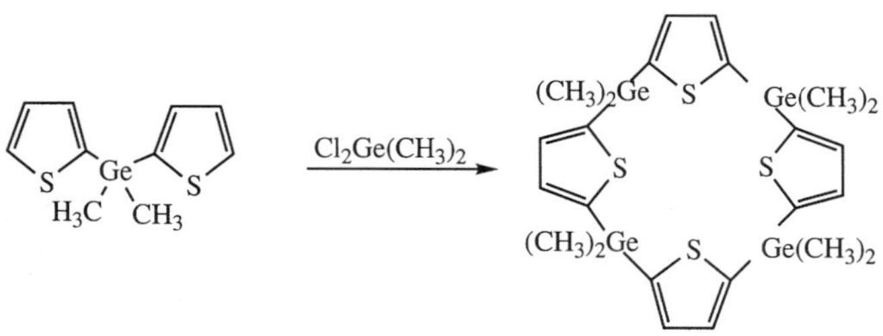

Another calixarene, [1_4]di-(2-thienyl)germa-2,5-thiopheneocalixarene is prepared by the reaction bis(2-thienyl)dichlorogermane with 2,5-dilithiothiophene [81].

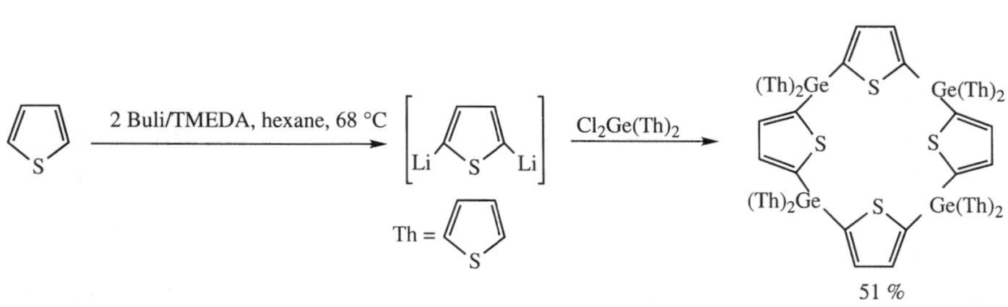

Metalation of dimethyldi(2-thienyl)germane with two equivalents of butyllithium followed by reaction with trimethylsilyl chloride gives dimethyl bis(5-trimethylsilyl-2-thienyl)germane [84].

Thermolysis at 180 °C of the calixarene led to poly(2,5-dimethylgermyl)thiophene. The reaction of 2,5-dilithiothiophene, obtained by metalation of thiophene with two equivalents of butyllithium in hexane and *N,N,N′,N′*-tetramethyl-1,2-ethanediamine, with dimethylgermanyl chloride did not lead to the desired polymer but gave only 5% of the polymer poly[2,5-(dimethylgermyl)thiophene], 20% of bis[2,5-chlorodimethylgermyl)]thiophene, the main product being the above mentioned calixarene [81].

Hydrogermylation of 3,3-dimethyl-1-butyne, phenylacetylene and methyl 2-propynyl ether with dimethyl (2-thienyl)germane in the presence of catalytic amounts of hexachloroplatinate(IV) hydrate proceeds more selectively than hydrosilylation and gives predominantly the *β-trans* isomers [74]. Also 2,5-di-2-thienyl-3-hexyne-2,5-diol has been hydrogermylated with triethylgermane giving after dehydration 2,5-dihydro-2,5-dimethyl-2,5-di-2-thienyl-3-(triethylgermyl)furan [53].

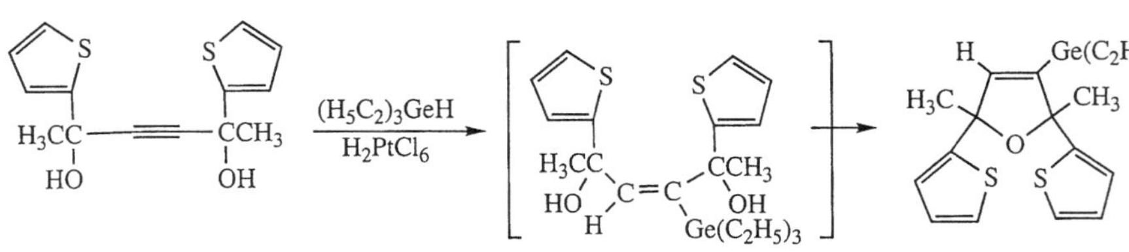

Pyrolysis in a flow-type system at 450–550 °C of a mixture of hexachlorodisilane, germanium tetrachloride and 2-chlorothiophene gives 2-thienyltrichlorogermane in 70% yield, in which germanium(II) chloride is the reactive intermediate [85].

5-Trimethylgermyl-2-thiophene aldehyde is prepared by blocking the aldehyde function of 2-thiophene aldehyde with *N*-methylpiperazide followed by metalation of the 5-position with butyllithium at −20 °C and reaction with trimethylgermyl chloride at −78 °C [23]. The compounds shown below are prepared from *cis* 1,2-di(2-bromo-3-thienyl)ethene and *cis* 1,2-di(4-bromo-3-thienyl)ethene) *via* halogen–metal exchange with *tert*-butyllithium at −80 °C followed by reaction with dichlorodimethylgermane [51].

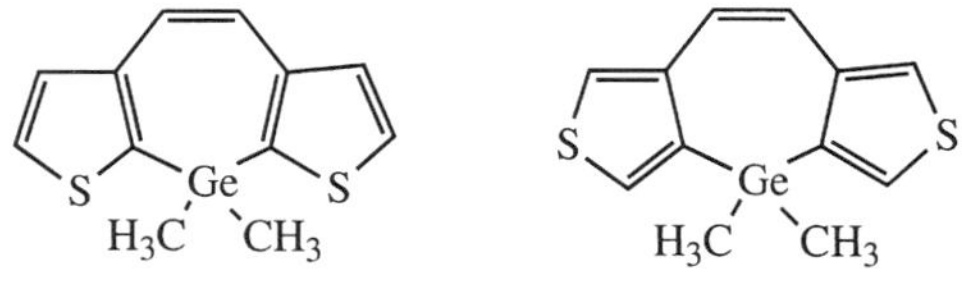

4G.3 THIENYLSTANNANE COMPOUNDS

4G.3.1 Trialkylstannylthiophenes

4G.3.1.1 2-Trialkylstannylthiophenes through the reaction between 2-thienyllithium derivatives and trialkyltin chloride

The common method for the preparation of 2-trialkylstannylthiophenes consists in the reaction between 2-thienyllithium derivatives, prepared by metalation of thiophenes with alkyllithium, and trialkyltin halides. 2-(Trimethylstannyl)thiophene is obtained by metalation of thiophene with butyllithium in hexane followed by reaction with trimethylstannyl chloride in diethyl ether [12].

2-(Trimethylstannyl)thiophene [12]

To a solution of freshly distilled thiophene (15.1 g, 0.18 mol) in anhydrous diethyl ether at 0 °C 2.37 *M* butyllithium in hexane (76 ml, 0.18 mol) is added dropwise. The stirring is continued at room temperature for 3 h, after which trimethyltin chloride (35 g, 0.18 mol) in anhydrous diethyl ether is added dropwise. The reaction mixture is stirred for 4 h and the lithium chloride is filtered off. The filtrate is washed with very dilute acetic acid and water, dried over magnesium sulfate, evaporated and distilled giving 31.5 g (71%) of the title compound bp 97–99 °C/33 mm Hg.

2-Tributylstannylthiophene is similarly obtained, through the reaction of 2-thienyllithium, prepared by metalation of thiophene with butyllithium in the presence of *N*,*N*,*N*′,*N*′-tetramethyl-1,2-ethanediamine with tributylstannyl chloride at −70 °C [86].

2-Tributylstannylthiophene [86]

To a solution of thiophene (5.0 g, 0.06 mol) and anhydrous *N*,*N*,*N*′,*N*′-tetramethyl-1,2-ethanediamine (8.6 g, 0.07 mol) in anhydrous diethyl ether (100 ml) under nitrogen, 1.42 *M* butyllithium in hexane (46 ml) is added at such a rate that moderate reflux is maintained. When the addition is completed the mixture is refluxed for another hour and then cooled to −70 °C, after which tributylstannyl chloride (19.4 g, 0.06 mol) in anhydrous diethyl ether (30 ml) is added at such a rate that the temperature does not exceed −70 °C and the stirring is continued for 4 h at −70 °C. The reaction mixture is allowed to reach room temperature and water is added. The phases are separated and the aqueous phase extracted with diethyl ether (3 × 30 ml). The combined organic phases are dried over magnesium sulfate, evaporated and distilled giving 10.2 g (46%) bp 148 °C/2 mm Hg.

2-(Trimethylstannyl)-4-octylthiophene is prepared by metalation of 3-octylthiophene with lithium diisopropylamide in tetrahydrofuran at −70 °C followed by reaction with trimethylstannyl chloride [87]. A large number of substituted trialkylstannyl compounds have been prepared by this approach. 5-(Trimethylsilyl)- and 5-(*tert*-butyldimethylsilyl-2-tributylstannyl)thiophene are thus prepared [33,40,88].

5-(Trimethylsilyl-2-tributylstannyl)thiophene [88]

To a solution of 2-(trimethylsilyl)thiophene (5.00 g, 31.99 mmol) and *N,N,N′,N′*-tetramethyl-1,2-ethanediamine (6.03 ml, 31.99 mmol) in anhydrous tetrahydrofuran (100 ml) 2.5 *M* butyllithium in hexane (15.99 ml) is added dropwise and the mixture is warmed to room temperare for 3 h. After recooling to −78 °C tributylstannyl chloride (10.41 ml, 38.39 mmol) is added dropwise. The stirring is continued for 1 h at −78 °C, the reaction quenched with 10% aqueous ammonium chloride solution (20 ml), the temperature is allowed to reach room temperature and the product extracted with diethyl ether (3 × 50 ml). The combined organic phases are dried over magnesium sulfate, evaporated and distilled giving 12.70 g (89%) of the title compound as a yellow oil bp 143–144 °C/0.15 mm Hg.

From 3-methyl-2-trimethylsilylthiophene upon metalation with lithium diisopropylamine followed by tributylstannyl chloride, 3-methyl-2-trimethylsilyl-5-tributylstannylthiophene was prepared [25].

3-Methyl-2-trimethylsilyl-5-tributylstannylthiophene [25]

To a solution of diisopropylamine (0.511 g, 5.05 mmol) in anhydrous tetrahydrofuran (5 ml) at −78 °C 2.12 *M* butyllithium in hexane (2.38 ml, 5.05 mmol) is added dropwise. The mixture is allowed to warm to 0 °C for 10 min and then recooled to −78 °C, after which 3-methyl-2-trimethylsilylthiophene (0.86 g, 5.05 mmol) in anhydrous tetrahydrofuran (5 ml) is added dropwise *via* cannula. The mixture is stirred at this temperature for 30 min and then tributyltin chloride (1.64 g, 5.05 mmol) is added dropwise. The reaction mixture is allowed to warm to room temperature and poured into water. The product is extracted with diethyl ether, the combined organic phases washed with sodium chloride solution, dried over sodium sulfate and evaporated. The residue is purified by chromatography on silica gel, washed with 10% triethylamine in hexane, using hexane as eluent giving 1.97 g (85%) of the title compound as a colorless liquid.

Halogen–metal exchange of 5-trimethylsilyl-2-bromothiophene with *tert*-butyllithium at −78 °C followed by reaction with trimethyltin chloride can also be used for the preparation of 5-trimethylsilyl-2-tributylstannylthiophene [25].

5-Trimethylsilyl-2-tributylstannylthiophene [25]

To a solution of 1.68 *M* *tert*-butyllithium in pentane (20.0 ml, 33.6 mmol) at −78 °C 5-trimethylsilyl-2-bromothiophene (3.95 g, 16.8 mmol) is slowly added and the mixture is stirred at this temperature for 1 h, after which tributyltin chloride (5.47 g, 16.8 mmol) is slowly added. The reaction mixture is allowed to warm to room temperature for 2 h before being poured into water with a few drops of 3 *M* hydrochloric acid in order to remove emulsion. The phases are separated and the aqueous phase extracted with diethyl ether. The combined organic phases are washed with sodium bicarbonate solution and sodium chloride solution, dried over sodium sulfate and evaporated. The residue is filtered through a silica plug with hexane giving 5.84 g (78%) of the title compound as a colorless liquid.

Metalation of dimethyl-5-(2-thienyl)isophtalate with the lithium derivative of 2,2,6,6-tetramethylpiperidine followed by reaction with tributylstannyl chloride gave dimethyl 5-(5-tributylstannyl-2-thienyl)isophtalate [89].

Dimethyl 5-(5-tributylstannyl-2-thienyl)isophtalate [89]

A three-necked flask equipped with septum and stopcock is charged with 2,2,6,6-tetramethylpiperidine (3.36 g, 24 mmol) in anhydrous tetrahydrofuran (50 ml). After cooling the mixture to 0 °C butyllithium (15 ml, 24 mmol) is added *via* syringe and the stirring is continued under argon at room temperature for 1 h. This mixture containing the lithiated anion is added dropwise to a solution of dimethyl-5-(2-thienyl)isophtalate (5.52 g, 20 mmol) and tributylstannyl chloride (7.82 g, 24 mmol) in anhydrous tetrahydrofuran at −78 °C. After stirring the mixture at −78 °C for 2 h and overnight at room temperature it is poured into ice cooled sodium chloride solution (200 ml) and diethyl ether (100 ml) is added. This mixture is stirred at room temperature for 15 min, the phases are separated and the aqueous phase extracted with diethyl ether. The combined organic phases are washed with sodium chloride solution and evaporated. The residue is purified by chromatography on aluminium oxide using hexane/ethyl acetate (98:2) as eluent, giving 4.66 g (41%) of the title compound as a reddish oil.

A one-pot method for the preparation of 5-(trimethylstannyl)-2-thiophene aldehyde is the reaction of 2-thiophene aldehyde with lithium *N*-methylpiperazide for protection of the formyl group, followed by butyllithium and trimethylstannyl- or tributylstannyl chloride [23]. 5-Trimethylsilyl-, 5-trimethylgermanyl and 5-trimethylstannyl-2-thiophene aldehyde are prepared by protecting 2-thiophene aldehyde with lithiated *N*,*N*,*N*-trimethylethylenediamine followed by the reaction with the appropriate trimethyl halide [23].

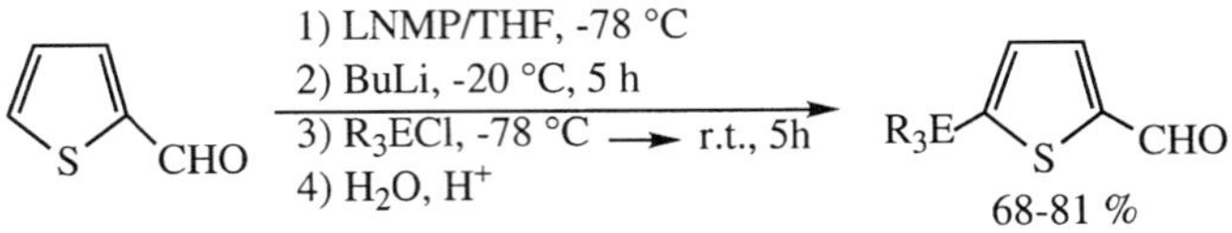

$R_3E = (CH_3)_3Si, (CH_3)_3Ge, (C_2H_5)_3Ge, (CH_3)_3Sn, (C_4H_9)_3Sn$

5-Trimethylstannyl-2-thiophene aldehyde [23]

To a suspension of lithium *N*-methylpiperazide (2.12 g, 20 mmol) [prepared from *N*-piperazine (2.0 g, 20 mmol) and 1.6 *M* butyllithium in hexane (12.5 ml, 20 mmol) at −78 °C] in tetahydrofuran (40 ml) is added 2-thiophene aldehyde (2.05 g, 18.3 mmol) at −78 °C. The reaction mixture is stirred for 15 min, after which 1.6 *M* butyllithium in hexane (12.5 ml, 20 mmol) is added and the reaction mixture stirred for 5 h at −20 °C. A solution of trimethylstannyl chloride (4.0 g, 20 mmol) in anhydrous tetrahydrofuran (10 ml) is added dropwise at −78 °C and the mixture is allowed to warm to room temperature and stirred for 5 h. The mixture is hydrolyzed by stirring with 1 *M* hydrochloric acid (120 ml) at 0 °C for 10 min and neutralized with aqueous sodium carbonate solution. The phases are separated and the aqueous phase extracted with ether. The combined organic phases are dried over sodium sulfate and concentrated. The residue is chromatographed on silica gel 60 using pentane/diethyl ether (100:3) as eluent, giving 3.93 g (78%) of the title compound mp 20–21 °C.

Metalation of acetal protected 3-thiophene aldehyde with butyllithium followed by reaction with trimethylstannyl- or tributylstannyl chloride is a good method for the preparation of 2-(2-trimethylstannyl-3-thienyl)-1,3-dioxolane and the corresponding tributylstannyl derivative [90,91].

2-(2-Trimethylstannyl-3-thienyl)-1,3-dioxolane [90]

To a stirred solution of 2-(3-thienyl)-1,3-dioxolane (20 g, 0.128 mol) in anhydrous diethyl ether under nitrogen at room temperature 1.51 *M* butyllithium in hexane (91.4 ml, 0.138 mol) is added dropwise. After refluxing the mixture for 10 min, it is cooled to −70 °C and trimethylstannyl chloride (25.5 g, 0.128 mol) in anhydrous diethyl ether (100 ml) is added dropwise. When the addition is complete, the reaction mixture is warmed to room temperature and the precipitated lithium chloride is filtered off. The filtrate is distilled giving 33.8 g (83%) of the title compound as a yellow liquid bp 116–120 °C/0.045 mbar. Chromatography on silica gel using ethyl acetate/cyclohexane (5:95) as eluent gives the title compound as a solid mp 58–59.5 °C.

Hydrolysis of the dioxolane with 1 *M* hydrochloric acid removes the protecting group, yielding 2-tributylstannyl-3-thiophene aldehyde [91].

2-Tributylstannyl-3-thiophene aldehyde [91]

A reaction mixture consisting of 2-(2-trimethylstannyl-3-thienyl)-1,3-dioxolane (11.44 g, 0.026 mol), 1 *M* hydrochloric acid (10 ml) and tetrahydrofuran (30 ml) is refluxed for 45 min, after which it is cooled to room temperature and diethyl ether (100 ml) is added. The phases are separated and the aqueous phase extracted with diethyl ether. The combined organic phases are washed with sodium bicarbonate solution and water, dried over sodium sulfate and evaporated. The residue is purified by chromatography on silica gel using ethyl acetate/hexane (95:5) as eluent, giving 9.13 g (85%) of the title compound as a colorless liquid bp 184–187 °C/5 mm Hg.

Metalation of *tert*-butyl *N*-(3-thienyl)carbamate was not successful upon attempted preparation of the 2-trimethylstannyl derivative. However, using halogen–metal exchange on *tert*-butyl *N*-(2-bromo-3-thienyl)carbamate followed by reaction with trimethylstannyl chloride at −80 °C gave *tert*-butyl *N*-(2-trimethylstannyl-3-thienyl)carbamate in high yield [92].

tert-Butyl N-(2-trimethylstannyl-3-thienyl)carbamate [92]

A solution of *tert*-butyl *N*-(2-bromo-3-thienyl)carbamate (5.85 g, (0.021 mol) in anhydrous tetrahydrofuran (50 ml) is cooled to −80 °C. At −78 to −72 °C 2.03 *M* butyllithium in cyclohexane (23 ml) is added dropwise. After complete addition the mixture is stirred at −74 °C for 45 min and trimethylstannyl chloride (4.58 g, 0.023 mol) in anhydrous tetrahydrofuran (10 ml) is added dropwise. The reaction mixture is allowed to warm to room temperature and sodium chloride is added. The product is extracted with ethyl acetate, the combined organic phases are washed with water, dried over sodium sulfate and evaporated giving 8.24 g of the title compound as a brown oil.

Also in some other cases halogen–metal exchange is preferred over metalation, as in the preparation of 3-methyl-2-thienyllithium from 2-bromo-3-methylthiophene, which upon reaction with trimethylsilyl chloride gave 3-methyl-2-trimethylsilylthiophene, which upon metalation with lithium diisopropylamide and reaction with tributylstannyl chloride gives 3-methyl-2-trimethylsilyl–5-tributylstannylthiophene in good yield [25]. 5-Trimethylsilyl-2-tributylstannylthiophene appears also best to be prepared starting from 2-bromo-5-trimethylsilylthiophene, by halogen–metal exchange at −78 °C with *tert*-butyllithium in diethyl ether followed by tributylstannyl chloride [25].

The mild conditions of the metalation reaction make it possible to start with *E*-(2-thienyl)ethenyldimethylphenylsilane, which upon reaction with lithium diisopropylamide followed by trimethylstannyl chloride gives (*E*)-2-(5-trimethylstannyl-2-thienyl)ethenyldimethylphenylsilane [93].

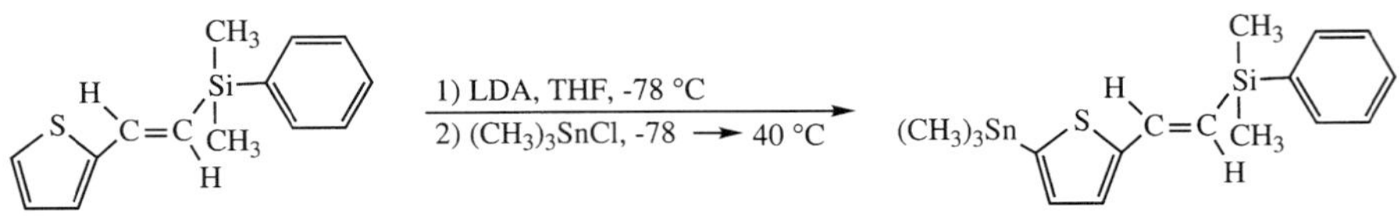

The following reaction has been investigated [94].

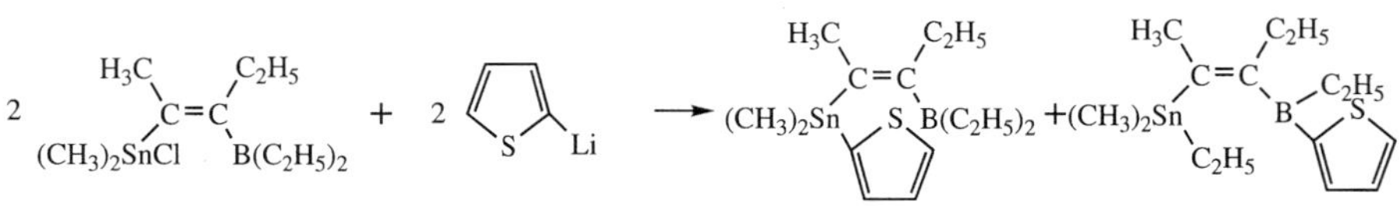

4G.3.1.2 From Grignard reagents and trialkylstannyl halides

Grignard reagents can also be used instead of 2-thienyllithium derivatives in the preparation of 2-trialkylstannylthiophenes. From the reaction between 3-alkyl-2-thiophenemagnesium bromide and tributylstannyl chloride, 3-alkyl-2-tributylstannylthiophene is obtained [95].

A simple and highly efficient method for the preparation of tributylstanylthiophene and 5-chloro-2-tributylstannylthiophene consists in the reaction of 2-bromothiophene or 5-chloro-2-bromothiophene with magnesium powder in tetrahydrofuran with bis(tributyltin)oxide *via* sonochemical Barbier type reaction in a commercial ultrasonic cleansing bath (39 kHz) [96,97].

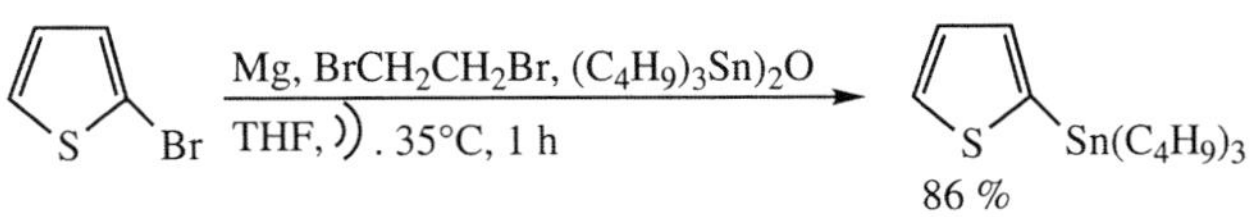

4G.3.1.3 2-Trialkylstannylthiophenes through the reaction of 2-halothiophenes with trialkylstannyllithium derivatives

Another method for the preparation of 5-(5-tributylstannyl-2-thienyl)isophthalate consists in the palladium(0)-catalyzed reaction of bis(tributyl)tin with dimethyl 5-(5-bromo-2-thienyl)isophtalate [89].

4G.3.1.4 2-Trialkylstannylthiophenes from 2,5-(trialkylstannyl)thiophenes

Transmetalation of 2,5-bis(trimethylstannyl)thiophene with methyllithium at −70 °C, followed by reaction with the appropriate electrophile is used for the preparation of 2-methyl-5-trimethylstannylthiophene [98] and 1-(5-trimethylstannylthien-2-yl)cycloalkan-1-ols through reaction with cyclopentanone and cyclohexanone [99].

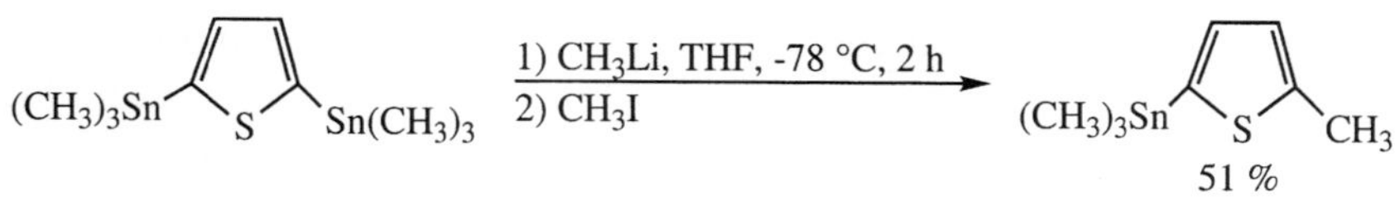

4G.3.1.5 3-(Trialkylstannyl)thiophenes through the reaction of 3-thienyllithium derivatives with trialkylstannyl derivatives

Such compounds are best obtained through the reaction of 3-thienyllithium derivatives, prepared from bromo- or iodothiophenes through halogen–metal exchange with alkyllithium derivatives. As examples can be mentioned the preparation of 3-tributylstannylthiophene [100], 5-methyl-4-tributylstannyl-2-trimethylsilylthiophene from 2,4-dibromothiophene followed by silylation and stannylation [101], and the preparation of *tert*-butyl *N*-(3-trimethylstannyl-2-thienyl)carbamate and *tert*-butyl *N*-(4-trimethylstannyl-3-thienyl)carbamate, starting from *tert*-butyl *N*-(3-bromo-2-thienyl)carbamate and tert-butyl *N*-(4-bromo-3-thienyl)carbamate, respectively [92].

3-Tributylstannylthiophene [100]

To a stirred solution of 3-bromothiophene (12.8 g, 80 mmol) in anhydrous diethyl ether (90 ml) under nitrogen at −70 °C 1.42 *M* butyllithium in hexane (62 ml, 88 mmol) is added at such a rate that the temperature is kept at −70 °C. The stirring is continued at this temperature for 30 min, after which a solution of tributylstannyl chloride (28.6 g, 88 mmol) in anhydrous diethyl ether (30 ml) is added. After stirring the reaction mixture at −70 °C for 4 h it is allowed to warm to room temperature and water (200 ml) is added. The phases are separated and the aqueous phase extracted with diethyl ether (3 × 50 ml). The combined organic phases are dried over magnesium sulfate, evaporated and distilled giving 14.9 g (50%) of the title compound bp 124 °C/0.6 mm Hg.

4G.3.1.6 3-Trialkylstannylthiophenes through palladium(0)-catalyzed reaction of 3-thienylboron derivatives with trialkylstannyl derivatives

The reaction of tris[4-(phenylethyl)thien-3-yl]boroxin with tributylstannyl chloride in methanol/toluene under palladium(0) catalysis is a good method for the preparation of 3-(phenylethyl)-4-(tributylstannyl)thiophene [43].

3-(Phenylethyl)-4-(tributylstannyl)thiophene [43]

To a stirred solution of tributylstannyl chloride (1.3 g, 4 mmol) and tetrakis-(triphenylphosphine)palladium(0) 231 mg, 0.2 mmol) in methanol/toluene (1:1) (80 ml) under nitrogen, tris[4-(phenylethyl)thien-3-yl]boroxin (428 mg, 0.67 mmol) and solid sodium methoxide (216 mg, 4 mmol) are added. The reaction mixture is refluxed for 24 h, cooled and poured into ice-water (50 ml). The product is extracted with diethyl ether (3×50 ml) and the combined organic phases are dried over magnesium sulfate and evaporated. The residue is purified by flash chromatography on silica gel using hexanes/triethylamine (99:1) as eluent, giving 595 mg (62%) of the title compound as a colorless oil.

4G.3.1.7 2,5-Di(trialkylstannyl)thiophenes through the reaction of 2,5-dilithio-thiophenes with trialkylstannyl chloride

Such compounds have during recent years become of great importance in connection with the preparation of a variety of polythiophenes through Stille coupling. 2,5-Bis(trimethylstannyl)thiophene has been prepared through the reaction of thiophene with excess butyllithium/*N*,*N*,*N*′,*N*′-tetramethyl-1,2-ethanediamine in hexane or in high yield by reaction with butyllithium in tetrahydrofuran at −78 °C followed by tributylstannyl chloride [98,102–104].

2,5-Bis(trimethylstannyl)thiophene [98]

To a solution of thiophene (2.3 ml, 28.4 mmol) and *N*,*N*,*N*′,*N*′-tetramethyl-1,2-ethanediamine (8.8 ml, 58 mmol) in anhydrous hexane (35 ml) under nitrogen at 0 °C butyllithium (37.5 ml, 58 mmol) is added dropwise. The mixture is refluxed for 30 min and cooled to 0 °C, after which a solution of trimethylstannyl chloride (11.86 g, 60 mmol) in anhydrous hexane is added dropwise. The stirring is continued at room temperature for 24 h and saturated ammonium chloride solution is added. The phases are separated and the organic phase is washed twice with aqueous copper sulfate, dried over sodium sulfate and evaporated. The residue is recrystallized from hexane giving 9.46 g (82%) of the title compound as a white solid mp 100–100.6 °C.

4G.3.2 Thienyltriorganostannane compounds

The reaction of 2-thiophenemagnesium iodide or bromide with stannic chloride in the ratio of 1:5 is a method for the preparation of tetra(2-thienyl)stannane [72,105] and in a similar way 2-thienyl triphenyltin, di(2-thienyl)diphenyltin and di-(2-thienyl) dibutyltin are prepared from diphenyltin dichloride, and triphenyltin chloride, and dibutyl tin dichloride and the appropriate amounts of 2-thiophenemagnesium iodide [105].

Tetra(2-thienyl)stannane [105]

To a stirred and cooled solution of anhydrous stannic chloride (2.71 g, 0.01 mol) in anhydrous diethyl ether (50 ml), 2-thiophenemagnesium iodide, prepared from 2-iodothiophene (10.5 g, 0.05 mol) and magnesium turnings, is added dropwise. The mixture is refluxed for 1 h and the diethyl ether is removed after addition of benzene (50 ml). The excess of Grignard reagent is destroyed by addition of aqueous saturated ammonium chloride solution (10 ml). The phases are separated and the organic phase evaporated. The gray residue is purified by chromatography on aluminum oxide using benzene as eluent, giving 4.06 g (90%) of the title compound as colorless crystals mp 153 °C.

The reaction of 2-thiophenemagnesium bromide with diphenylstannyl dichloride gives diphenyl-di(2-thienyl)stannane, which upon metalation with *tert*-butyllithium-*N,N,N′,N′*-tetramethyl-1,2-ethanediamine followed by reaction with trimethylsilyl chloride is used for the preparation of diphenyl di(5-trimethylsilyl-2-thienyl)stannane. Mixed thienyl germanium tin derivatives have also been prepared [73].

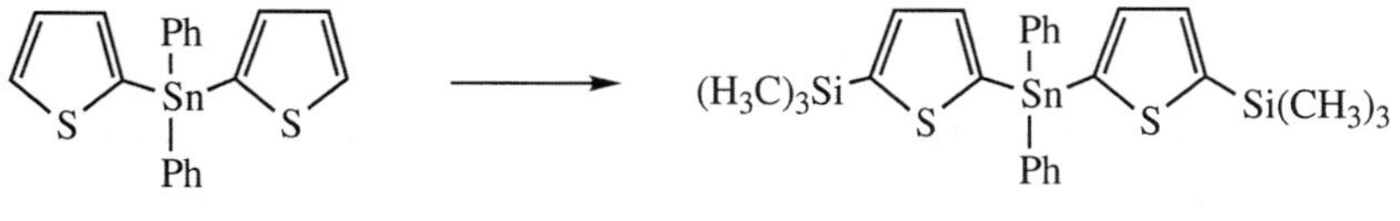

Diphenyl di(5-trimethylsilyl-2-thienyl)stannane [73]

To a solution of diphenyldi(2-thienyl)stannane (0.20 g, 046 mmol) and *N,N,N′,N′*-tetramethyl-1,2-ethanediamine (2 ml) at −95 °C 1.7 *M* *tert*-butyllithium in pentane (1.14 mmol) is added. The stirring is continued for 45 min, after which the mixture is warmed to −60 °C and treated with trimethylsilyl chloride (0.2 ml, 1.6 mmol). The reaction mixture is warmed to room temperature, refluxed for 5 min and evaporated. The residue is extracted with pentane and this solution is filtered through silica gel. The product is recrystallized from pentane/isopropanol giving 0.23 g (85%) of the title compound as white crystals mp 78 °C.

Bis[3-(2-pyridyl)-2-thienyl-*C,N*]diphenyltin(IV), the first example of a six-coordinate tetraorganotin compound, is prepared through metalation of 2-(3-thienyl)pyridine followed by reaction with diphenyltin dibromide [106,107].

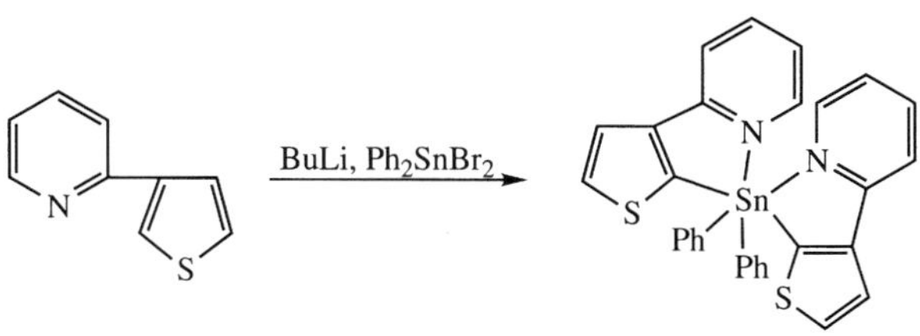

Bis[3-(2-pyridyl)-2-thienyl-C,N]diphenyltin [108]

To a solution of 2-(3-thienyl)pyridine (2 ml, 14 mmol) in anhydrous diethyl ether (30 ml) under nitrogen at room temperature 10.2 *M* butyllithium in hexane (1.5 ml) is added dropwise. The stirring is continued for 1 h followed by dropwise addition of diphenyltin dibromide (3.13 g, 7 mmol) in anhydrous diethyl ether (50 ml). The reaction mixture is refluxed for 4 h, after which lithium chloride is filtered off and the filtrate evaporated. Dropwise addition of hexane to the residue causes immediate precipitation giving 3.4 g (83%) of the title compound.

4G.3.3 Diarylthienyltin halides

4G.3.3.1 From thienyllithium, -copper and -magnesium derivatives and tin dihalides

The reaction of 2-(3-thienyl)pyridine with butyllithium followed by di(*para*-tolyl)tin dibromide is used for the preparation of 3-(2-pyridyl-2-thienyl) di(*para*-tolyl)tin bromide [108].

3-(2-Pyridyl-2-thienyl)di(para-tolyl)tin bromide [108]

To a solution of 2-(3-thienyl)pyridine (1 ml, 7.2 mmol) in anhydrous diethyl ether (30 ml) at room temperature 15% butyllithium in hexane (5 ml) is added dropwise. The stirring is continued for 2 h, after which the mixture is cooled to 0 °C and treated with dropwise addition of di(*para*-tolyl)tin dibromide

(1.6 g, 3.5 mmol) in diethyl ether. The reaction mixture is stirred at room temperature for 4 h, the lithium bromide formed is filtered off and washed with diethyl ether (2 × 20 ml). The combined filtrates are evaporated giving 1.5 g (79%) of the title compound mp 221–223 °C after recrystallization from toluene.

In the case of 2-thienyllithium and 5-methyl-2-thienyllithium the attempted monohalogenation reactions with R_2SnX_2 or R_2SnXY gave the unsymmetrical tetraorganotin derivatives. However, using the less reactive 2-thienylcopper or 5-methyl-2-thiophenemagnesium bromide in the reaction with dibutylstannyl chloroacetate, dibutyl-2-thienyltin chloride and 5-dibutyl-5-methyl-2-thienylstannyl chloride are obtained [108].

Dibutyl-2-thienyltin chloride [108]

To a solution of thiophene (3.2 ml, 42 mmol) in anhydrous diethyl ether (20 ml) at room temperature 15% butyllithium in hexane (27 ml) is added. The stirring is continued for 1 h, after which the thienyllithium is added through a syringe to a suspension of cuprous iodide (7 g, 36 mmol) in anhydrous diethyl ether (30 ml) at −70 °C. The stirring is continued at this temperature for 2 h and the mixture is treated with chlorodibutyl tin acetate (10 g, 30 mmol) in anhydrous diethyl ether (20 ml). After stirring for 18 h the cuprous acetate is filtered off, the filtrate evaporated and the residue distilled giving 5 g (47%) of the title compound bp 110–115 °C/0.06 mm Hg.

4G.3.3.2 Cleavage of tetraorganotin compounds

Tris(3-thienyl)stannyl chloride and bromide, bis(3-thienyl)stannyl) dichloride and dibromide as well as 3-thienylstannyl trichloride and tribromide have been prepared by the reaction of tetrakis(3-thienyl)stannane and tin tetrachloride or tetrabromide in appropriate proportions [109,110].

Another alternative for the synthesis of triorganotin halides consists in the controlled cleavage of the corresponding tetraorganotin compound with iodine. Thus bis(2-thienyl)phenyltin iodide was obtained from tris(2-thienyl)-phenyltin and [3-(2-pyridyl)-2-thienyl]dicyclohexyl)tin iodide from [3-(2-pyridyl)-2- thienyl]cyclohexyltin [108].

Bis(2-thienyl)phenyltin iodide [108]

A solution of tris(2-thienyl)phenyltin (0.9 g, 2 mmol) and iodine (0.5 g, 2 mmol) in diethyl ether (200 ml) is stirred at room temperature for 15 min, after which the solvent is completely evaporated. The residue, a brown solid, is dissolved in petroleum ether (20 ml) and this solution is filtered and evaporated giving a yellow oil, which upon cooling gives 0.6 g (60%) of the title compound as colorless crystals mp 72–74 °C.

4G.4 THIENYLPLUMBANE COMPOUNDS

4G.4.1 Trialkylplumbylthiophenes

3-(Trimethylplumbyl)thiophene is obtained by the reaction of 3-thienyllithium with chlorotrimethylplumbane at −70 °C [71].

3-(Trimethylplumbyl)thiophene [71]

To a solution of 3-bromothiophene (0.75 ml, 8.0 mmol) in anhydrous diethylether (10 ml) at −70 °C butyllithium (4.0 ml, 8.0 mmol) and chlorotrimethylplumbane (1.65 g, 5.7 mmol) in anhydrous tetrahudrofuran (7 ml) are added. The stirring is continued for 2 h giving 80% of the title compound as a colorless liquid.

4G.4.2 Triarylthienylplumbanes

Tetra(2-thienyl)plumbane is prepared by the reaction of 2-thiophenemagnesium bromide with lead tetrachloride [76].

4G.4.3 Thienyllead tricarboxylates

Both 2- and 3-thienyllead tricarboxylates are prepared by the reaction of 2- and 3-thienyl tributylstannane with equivalent amounts of lead tetraacetate and catalytic amounts of mercury(II) trifluoroacetate in anhydrous chloroform [111].

2-Thienyllead triacetate [111]

2-Thienyltributylstannane (4.47 g, 12 mmol) is added to a solution of lead tetraacetate 4.43 g, 10 mmol) and mercury(II) trifluoroacetate (0.21 g, 0.5 mmol) in anhydrous chloroform (50 ml). The reaction mixture is stirred in a stoppered flask for 30 min, after which acetic acid (5 ml) is added and the stirring is continued for 5 min. The solvent is evaporated and the residue is dissolved in a minimum volume of chloroform/acetic acid (19:1). This solution is slowly poured into anhydrous light petroleum (100 ml). The precipitate formed is filtered off under nitrogen and dried *in vacuo* giving 2.58 g (55%) of the title compound as cream-colored crystals mp 61–65 °C.

REFERENCES

1. N. S. Nametkin, O. V. Kuz'min, T. I. Chernysheva, V. I. Savushkina, V. Z. Anisimova and E. R. A. Chernyshev, *Doklady Akad. Nauk. SSSR* (*Engl. Transl.*) **193**, 474 (1970).
2. B. Tacke, E. Zimonyi-Hegedüs and U. Wannagat, *J. Organomet. Chem.* **172**, 21 (1979).
3. L. I. Kartasheva, N. S. Nametkin and T. I. Chernysheva, *J. Gen. Chem.* (*Engl. Transl.*) **40**, 1253 (1970).
4. S. S. Hu and P. Weber, *J. Organomet. Chem.* **369**, 155 (1989).
5. D. Häbich and F. Effenberger, *Synthesis* 841 (1979).
6. R. A. Benkeser and A. Torkelson, *J. Am. Chem. Soc.* **76**, 1252 (1954).
7. O. G. Yarosh, A. S. Nachmanovich and M. G. Voronkov, *Izvest. Akad. Nauk. SSSR* (*Engl. Transl.*) 1510 (1977).
8. W. C. Hamann, C. F. Hobbs and D. J. Bauer, *J. Org. Chem.* **32**, 2841 (1967).
9. F. Effenberger and D. Häbich, *Liebigs Ann. Chem.* 842 (1979).
10. C. Eaborn and J. A. Sperry, *J. Chem. Soc.* 4921 (1961).
11. R. A. Benkeser and R. B. Currie, *J. Am. Chem. Soc.* **70**, 1780 (1948).
12. J. R. Pratt, F. H. Pinkerton and S. F. Thames, *J. Organomet. Chem.* **38**, 29 (1972).
13. M. S. Loft, T. J. Mowlem, D. A. Widdowson and D. J. Williams, *J. Chem. Soc., Perkin Trans. 1* 105 (1995).
14. S. S. Samanta S. C. Gosh and A. De, *J. Chem. Soc., Perkin Trans. 1* 2683 (1997).
15. S. F. Thames, J. E. McClesky and P. L Kelly, *J. Heterocycl. Chem.* **5**, 749 (1968).
16. C. Eaborn and G. Seconi, *J. Chem. Soc., Perkin Trans. II*, 925 (1976).
17. P. A. Konstantinov and R. I. Shupik, *J. Gen. Chem. USSR* (*Engl. Transl.*) **33**, 1223 (1963).
18. M. S. Loft, T. J. Mowlem and D. A. Widdowson, *J. Chem. Soc. Perkin Trans. 1* 97 (1995).
19. E. Lukevics, P. Arsenyan, S. Belyakov, J. Popelis and O. Pudova, *Eur. J. Org. Chem.* 3139 (2000).
20. P. L. Kelly, S. F. Thames and J. E. McCleskey, *J. Heterocycl. Chem.* **9**, 141 (1972).
21. S. F. Thames and J. E. McCleskey, *J. Heterocycl. Chem.* **4**, 146 (1967).
22. S. F. Thames and J. E. McCleskey, *J. Heterocycl. Chem.* **3**, 104 (1966).
23. F. Denat, H. Gaspard-Iloughmane and J. Dubac, *Synthesis* 954 (1992).
24. J. Nakayama, T. Yu, Y. Sugihara, A. Ishii and S. Kumakura, *Heterocycles* **45**, 1267 (1997).
25. R. Wu, J. S. Schumm, D. L. Pearson and J. M. Tour, *J. Org. Chem.* **61**, 6906 (1996).
26. S. Gronowitz, T . Frejd, O. Karlsson, K. Lawitz, P. Pedaja and K. Pettersson, *Chem. Scripta* **18**, 192 (1981).
27. E. J. Bures and B. A. Keay, *Tetrahedron Lett.* **28**, 5965 (1987).
28. E. Bures, P. G. Spinazzé, G. Beese, I. R. Hunt, C. Rogers and B. A. Keay, *J. Org. Chem.* **62**, 8741 (1997).
29. G. Beese and B. A. Keay, *Synlett* 33 (1991).
30. F. H. Pinkerton and S. F. Thames, *J. Heterocycl. Chem.* **9**, 725 (1972).
31. F. H. Pinkerton and S. F. Thames, *J. Organomet Chem.* **29**, C4 (1971).
32. M. Herrlich, N. Hampel and H. Mayr, *Org. Lett.* **3**, 1629 (2001).
33. G. Barbarella, L. Favaretto, G. Sotgiu, M. Zambianchi, L. Antolini, O. Pudova and A. Bongini, *J. Org. Chem.* **63**, 5497 (1998).
34. M. Yoshida, T. Yoshida, N. Kamigata and M. Kobayashi, *Bull. Chem Soc. Jpn.* **61**, 3549 (1988).
35. S. F. Thames and J. E. McCleskey, *J. Heterocycl. Chem.* **4**, 171 (1967).

36. C. Hoogzand, J. Nielsen and E. H. Braye, *Chem. Commun.* 1520 (1971).
37. R. A. Benkeser and H. Landesman, *J. Am. Chem. Soc.* **71**, 2493 (1949).
38. S. Gronowitz, G. Nikitides and A. Hallberg, *Acta Chem. Scand.* **45**, 632 (1991).
39. N. Furukawa, H. Hoshiai, T. Shibutani, M. Higaki, F. Iwasaki and H. Fujihara, *Heterocycles* **34**, 1085 (1992).
40. G. Barbarella, P. Ostoja, P. Maccagnani, O. Pudova, L. Antolini, D. Casarini and A. Bongoni, *Chem. Mater.* **10**, 3683 (1998).
41. F. B. Deans and C. Eaborn, *J. Chem. Soc.* 2302 (1959).
42. Y. Goldberg, R. Sturkovich and E. Lukevics, *Synth. Commun.* **23**, 1235 (1993).
43. X.-S. Ye and H. N. C. Wong, *J. Org. Chem.* **62**, 1940 (1997).
44. X.-S. Ye and H. N. C. Wong, *Chem. Commun.* 339 (1996).
45. A. R. M. O'Donovan and M. K. Shepard, *Tetrahedron Lett.* **35**, 4425 (1994).
46. H. Gilman, R. A. Benkeser and G. E. Dunn, *J. Am. Chem. Soc.* **60**, 1410 (1950).
47. V. M. Polosin, A. A. Astakhov, V. B. A. Tafeenko and A. V. Ivashchenko. *J. Gen Chem. USSR* **60**, 1410 (1990).
48. A. Ishii, T. Tsuchiya, J. Nakayama and M. Hoshino, *Tetrahedron Lett.* **34**, 2347 (1993).
49. J. R. Pratt and S. F. Thames, *J. Org. Chem.* **38**, 4271 (1973).
50. J. Nakayama and J.-S. Lin, *Tetrahedron Lett.* **38**, 6043 (1997).
51. S. Yasuike, F. Nakashima, J. Kurita and T. Tsuchiya, *Heterocycles* **45**,1899 (1997).
52. N. S. Nametkin, T. I. Chernysheva, N. A. Pritula, M. I. Gevenyan and L. I. Kartasheva, *Izvest. Akad, Nauk. SSSR, Ser. Khim.* (*Engl Transl.*) 2097 (1969).
53. I. M. Gverdtsiteli and M. D. Chanturiya, *J. Gen. Chem.* (*Engl. Transl.*) **45**, 2309 (1975).
54. T. Kauffmann and H.-H. Kniese, *Tetrahedron Lett.* 4043 (1973).
55. E. Lukevics, R. Sturkovich, Y. Goldberg and A. Gaukhman, *J. Organomet. Chem.* **345**, 19 (1988).
56. K. Olsson and C. Axén (Grönvall), *Arkiv Kemi* **17**, 529 (1961).
57. E. A. Chernyshev and N. G. Toolstikova, *Izvest. Akad. Nauk SSSR, Ser. Khim.* 1606 (1964).
58. H. Ito, H. Sensui, K. Arimoto, K. Miura and A. Hosomi. *Chem. Lett.* 630 (1997).
59. K. Olsson and C. Axén. *Arkiv Kemi* **22**, 237 (1964).
60. É. Lukevics, O. A. Pudova and N. P. Erchak, *J. Gen. Chem. USSR* (*Engl. Transl.*) **50**, 1094 (1980).
61. M. G. Voronkov, G. I. Zelchan, V. I. Savushkina, B. M. Tabenko and E. A. Chernychev, *Chem. Heterocycl, Compds.* (*Engl. Transl.*) 646 (1976).
62. É. Lukevics and M. Ignatovich, *Chem. Heterocycl. Compds.* **75** (1992).
63. V. Gevorgyan, L. Borisova and E. Lukevics, *J. Organomet. Chem.* **527**, 295 (1997).
64. É. Lukevics, S. Germane, O. A. Pudova and N. P. Erchak, *Khim.-farm. Zh.* **13**, 52 (1979).
65. N. P. Sluchevskaya, V. A. Yablokov, N. V. Yablokova, V. I. Savushkina, and E. A. Chernyshev, *J. Gen. Chem. USSR* (*Engl. Transl.*) **47**, 213 (1977).
66. N. P. Sluchevskaya, N. V. Yablokova, V. A. Yablokov and Y. A. Aleksandrov, *J. Gen. Chem. USSR* (*Engl. Transl.*) **48**, 1037 (1978).
67. M. Veith, A. Koban, K. Fries, P. Spaniol, R. Elsässer, A. Rammo, V. Huch and U. Kleinsteuber, *Organometallics* **17**, 2612 (1998).
68. M. Veith, A. Koban and V. Huch, *Helv. Chim. Acta* **81**, 1640 (1998).
69. E. Lukevics, S. Belyakov and O. Pudova, *J. Organomet. Chem.* **523**, 41 (1996).
70. E. Lukevics, O. A. Pudova, I. Mazheika and S. Grinberga, *Chem. Heterocycl. Compds.* **33**, 171 (1997).
71. S. K. Ritter and R. E. Noftle, *Chem. Mater.* **4**, 872 (1992).
72. É. Lukevics, O. A. Pudova, Y. Popelis and N. P. Erchak, *J. Gen. Chem.* (*Engl. Transl.*) **51**, 102 (1981).
73. J. Hockemeyer, A. Castel, P. Rivière, J. Satgé, K. G. Ryder, A. Drury, A. P. Davey and W. J. Blau, *Appl. Organomet. Chem.* **11**, 513 (1997).

74. É. Lukevics, R. Y. Strukovich and O. A. Pudova, *J. Gen . Chem. USSR* (*Engl. Transl.*) **58**, 721 (1988).
75. E. Lukevics, P. Arsenian, S. Belyakov, J. Popelis and O. Pudova, *Organometallics* **20**, 2487 (2001).
76. E. Krause and G. Renwanz, *Ber.* **65**, 777 (1932).
77. N. P. Reddy, T. Hayashi and M. Tanaka, *Chem. Lett.* 677 (1991).
78. É. Lukevics, L. Ignatovich, N. Porsyurova and S. Germane, *Appl. Organomet. Chem.* **2**, 115 (1988).
79. É. Lukevics, L. Ignatovich, N. Shilina and S. Germane, *Appl. Organomet. Chem.* 1992.
80. E. Lukevics, L. Ignatovich, N. Posiurova and S. Germane, *Appl. Organomet. Chem.* **2**, 115 (1988).
81. J. Barreau, G. Rima, A. Akkari and J. Satge, *Inorg. Chim. Acta* **260**, 11 (1997).
82. N. Kakimoto, K. Sato, T. Takada and M. Akiba, *Heterocycles* **29**, 2115 (1989).
83. E. Liepins, I. Zicmane, L. M. Ignatovich and E. Lukevics, *J. Organomet. Chem.* **389**, 23 (1990).
84. B. König and M. Rödel, *Chem. Ber./Recueil* **130**, 421 (1997).
85. E. A. Chernyshev, N. G. Komalenkova and G. N. Yakovleva, *Dokl. Chem.* (*Engl. transl.*) **336**, 88 (1994).
86. D. Peters, A.-B. Hörnfeldt and S. Gronowitz, *J. Heterocycl. Chem.* **27**, 2165 (1990).
87. P. R. L. Malenfant, L. Groenendaal and J. M. J. Fréchet, *Polymer Prep.* **39**, 135 (1998).
88. L. S. Liebeskind and J. Wang, *J. Org. Chem.* **58**, 3550 (1993).
89. E. Cielen, A. Tahri, K. Ver Heyen, G. J Hornaert, F. C. De Schryver and N. Boens, *J. Chem. Soc., Perkin Trans. 2* 1573 (1998).
90. Y. Yang, A.-B. Hörnfeldt and S. Gronowitz, *Synthesis* 130 (1989).
91. S. Gronowitz, A.-B. Hörnfeldt and Y. Yang, *Chem. Scripta* **28**, 281 (1988).
92. P. Björk, T. Aakermann, A.-B. Hörnfeldt and S. Gronowitz, *J. Heterocycl. Chem.* **32**, 751 (1995).
93. R. Rossi, A. Carpita and T. Messeri, *Synth. Commun.* **21**, 1875 (1991).
94. S. Kerschl and B. Wrackmeyer, *J. Organomet. Chem.* **338**, 195 (1988).
95. S. C. Ng, L. G. Xu and H. S. O. Chan, *Synthetic Metals* **94**, 185 1998).
96. A. S.-Y. Lee and W.-C. Dai, *Tetrahedron Lett.* **37**, 495 (1996).
97. A. S.-Y. Lee and W. C Dai, *Tetrahedron* **53**, 859 (1997).
98. D. E. Seitz, S.-H. Lee, R. N. Hanson and J. C. Bottaro, *Synth. Commun.* **13**, 121 (1983).
99. S.-H. Lee, R. N. Hanson and D. E. Seitz, *Tetrahedron Lett.* **25**, 1751 (1984).
100. S. Gronowitz and D. Peters, *Heterocycles* **30**, 645 (1990).
101. S. Yoshida, H. Kubo, T. Saika and S. Katsemura, *Chem. Lett.* 139 (1996).
102. L. L. Miller and Y. Yu, *J. Org. Chem.* **60**, 6813 (1995).
103. Z. Bao, W. Chan, and L. Yu, *Chem. Mater.* **5**, 2 (1993).
104. Y. Wei, Y, Yang and J.-M. Yeh, *Chem. Mater.* **8**, 2659 (1996).
105. S. Gopinathan, C. Gopinathan and J. Gupta, *Indian J. Chem.* **12**, 623 (1974).
106. V. G. Kumar Das, L. K. Mun, C. Wei, S. Blunden and T. C. W. Mak, *J. Organomet. Chem.* **322**, 163 (1987).
107. V. G. Kumar Das, L. K. Mun and C. Wei, *Organometallics* **6**, 10 (1987).
108. K. M. Loo, S. Selvaratman, S. Weng, Ng C. Wei and V. G. K. Das. *J. Organomet. Chem.* **430**, 149 (1992).
109. D. W. Allen, D. J. Derbyshire, J. S. Brooks and P. J. Smith, *J. Organomet. Chem.* **251**, 45 (1983).
110. D. W. Allen, D. J. Derbyshire, J, S, Brooks, S. J. Blunden and P. J. Smith, *J. Chem. Soc., Dalton Trans.* 1889 (1984).
111. J. T. Pinhey and E. G. Roche, *J. Chem. Soc., Perkin Trans. 1* 2415 (1988).

– 5 –

Syntheses of Thiophenes with Group V Substituents

5.1 NITROGEN DERIVATIVES

5.1.1 Nitro derivatives

5.1.1.1 General aspects

Nitro derivatives of thiophenes are mostly prepared by direct nitration, with conditions depending upon the nature of the substituents. When the orientation rules do not lead to the desired isomers, carboxylic acid groups, chlorosulfonyl groups, or halogens are used as blocking and orienting groups, which are then removed by decarboxylation, desulfonation, or dehalogenation.

Thienyllithium derivatives can be converted to di(thienyl)iodonium salts by reaction with (*E*)-chlorovinyliodoso dichloride, which upon reaction with sodium nitrite in *N*,*N*-dimethylformamide yield nitrothiophenes. Some heavily substituted nitrothiophenes have been obtained by ring-closure reactions. Finally, other substituents can, of course, be introduced in nitrothiophenes by electrophilic substitution reactions. Which method of all these is the preferable one for the synthesis of a desired substituted nitrothiophene is dependent on the nature and position of the substituent and on the position in which one wants to introduce the nitro group.

5.1.1.2 Mono- and dinitrothiophenes

Babisinian modified the earlier methods for the nitration of thiophene itself by the use of acetic acid as cosolvent [1]. This method has appeared in *Organic Syntheses* [2]. Nitronium tetrafluoroborate [3], copper(II) nitrate [4], or aluminium nitrate [5] in acetic acid or acetic anhydride can be used for the nitration of thiophene. The main problem in the preparation of 2-nitrothiophene is that, irrespective of which of the above-mentioned methods for direct nitration is used, about 15% of the 3-nitro isomer is formed. It is extremely difficult to remove the 3-isomer by recrystallization [6–8]. The isomers can be separated with difficulty by gas liquid chromatography [8,9] or the 3-isomer can be removed by selective chlorosulfonation leaving the 2-isomer in high purity [6,10]. Also the recent nitration of thiophene with N_2O_5–SO_2 led to a mixture of 2- and 3-nitrothiophene, as well as 2,4- and 2,5-dinitrothiophene [11]. Indirect methods are therefore used sometimes for obtaining pure 2-nitrothiophene as decarboxylation of 5-nitro-2-thiophenecarboxylic acid with copper and quinoline [10,12,13]. Isomer-free 2-nitrothiophene can be obtained by the reaction of di(2-thienyl)iodonium chloride (cf. chapter 7) with sodium nitrite in *N,N*-dimethylformamide, although the yield is only 31% [14,15]. This method is definitely the best for the preparation of 3-nitrothiophene from di(3-thienyl)iodonium chloride [14].

3-Nitrothiophene [14]

A solution of di(3-thienyl)iodonium chloride (3.30 g, 0.01 mol) and sodium nitrite (3.45 g, 0.05 mol) in anhydrous *N,N*-dimethylformamide (60 ml) is stirred at 110 °C for 4–5 h. The reaction mixture is cooled, poured onto water, and extracted with ether. The combined ether phases are washed with water, dried over magnesium sulfate, and evaporated. The residue is chromatographically separated on neutral alumina using petroleum ether as eluent for 3-iodothiophene and chloroform for 3-nitrothiophene, which is recrystallized from ethanol giving 0.72 g (52%) mp 75.7–75.9 °C.

Another method is the decarboxylation of 4-nitro-2-thiophenecarboxylic acid [10,16–18]. As the carboxylic group, the chlorosulfonyl function can also be used. Thus the 4-nitro-2-chlorosulfonyl derivative formed in the purification of the crude nitrothiophene mixture obtained in the direct nitration of thiophene can be used for the preparation of about 20% of 3-nitrothiophene, upon heating with water [19–21].

Further nitration of nitrothiophenes using fuming nitric acid alone [19,20] or in the presence of sulfuric acid [1,19,22,23] always leads to isomer mixtures of dinitrothiophene and therefore is not of great preparative interest. 3-Nitrothiophene with nitric acid in trifluoroacetic acid as expected gives 2,4-dinitrothiophene (91%) predominantly. However, nitration of thiophene [1] followed by nitration of the mononitro mixture [19] gives a mixture that can be separated readily by chromatography [7,24] to give 2,5-dinitro- and 2,4-dinitrothiophene in isolated yields of 27 and 43%, respectively [7].

2,3-Dinitrothiophene can also be obtained by diazotization of 2-amino-3-nitrothiophene followed by reaction with sodium nitrite in the presence of copper salts [25]. Decarboxylation of 3,5-dinitro-2-thiophenecarboxylic acid with copper in quinoline [16] and treatment of 2-iodo-3,5-dinitrothiophene with sodium iodide in acetone and acetic acid [20] has been used for the preparation of 2,4-dinitrothiophene. The use of halogen as a blocking group is well demonstrated in the synthesis of 3,4-dinitrothiophene from 2,5-dibromo-3,4-dinitrothiophene by treatment with hypophosphorus acid [26] or with copper in boiling butyric acid [27,28]. This method was also used for the preparation of 2,3,4-trinitrothiophene from 2-bromo-3,4,5-trinitrothiophene [26]. 3,4-Dinitrothiophene has become an important synthetic starting material for many other types of compounds [29–34].

5.1.1.3 Nitrothiopheneboronic acids

Nitration of 2-thiopheneboronic acid with fuming nitric acid in concentrated sulfuric acid only leads to deboronation, yielding 2,4-dinitrothiophene in 54% yield. On the other hand 3-formyl-2-thiopheneboronic acid gives upon nitration under the same conditions a 70% yield of 3-formyl-5-nitro-2-thiopheneboronic acid, while the corresponding reaction of 2-formyl-3-thiopheneboronic acid yielded only 22% of 2-formyl-5-nitro-2-thiopheneboronic acid, due to deboronation, giving 54% of 4-nitro-2-thiophene aldehyde [35].

3-Formyl-5-nitro-2-thiopheneboronic acid [35]

3-Formyl-2-thiopheneboronic acid (1.56 g, 0.01 mol) is added to a solution of concentrated sulfuric acid (25 ml) and fuming nitric acid (0.85 ml, 0.02 mol) at −20 °C in one portion with vigorous stirring. The stirring is continued at the same temperature for 45 min, after which the reaction mixture is poured into ice (200 g). The precipitate of white crystals is filtered off, washed with water, and dried. Recrystallization from ethanol/water gives 1.4 g (70%) of the title compound mp 150–155 °C.

5.1.1.4 Alkylnitro derivatives

Not very many alkyl nitrothiophenes have been prepared. In the direct nitration of 2-alkylthiophenes isomer mixtures are often obtained, which causes separation problems and diminished yields. 2-Methyl-3-nitrothiophene is best prepared by nitration of 5-methyl-2-thiophenecarboxylic acid, followed by decarboxylation with copper in quinoline [10,36]. Another approach to 3-nitro-2-methylthiophene starts with chlorosulfonation of 2-methylthiophene followed by nitration and desulfonylation [37]. Bulkier alkyl groups give larger proportions of the 5-nitro isomer upon nitration [5,38–40]. Further nitration of 2-alkylthiophenes is of course an excellent method for the preparation of 2-alkyl-3,5-dinitrothiophenes [10,41–43].

Nitration of 2,5-dialkylthiophenes can be carried out so that either the 3-mononitro or the 3,4-dinitro product is obtained [44–47]. Nitration of 3,4-di-*tert*-butylthiophene with nitric acid in acetic acid/acetic anhydride at room temperature gives 2-nitro-3,4-di-*tert*-butylthiophene in 74% yield [48].

An alternative route to alkylsubstituted 2-nitrothiophenes is the addition of Grignard reagents to 2-nitrothiophene in tetrahydrofuran at −50 °C. Unfortunately a mixture of 3-alkyl- (predominantly) and 5-alkyl-2-nitrothiophenes is obtained upon aromatization of the primary product with 2,3-dichloro-5,6-dicyanobenzoquinone (DDQ). On the other hand 3-nitrothiophene gives 2-alkyl-3-nitrothiophene in high yield, opening a short and convenient route to this type of compounds [49].

2-Methyl-3-nitrothiophene [49]

A 100-ml three-necked round-bottom flask equipped with magnetic stirrer and a dropping funnel is charged under nitrogen with 3-nitrothiophene (0.65 g, 5.0 mmol) in anhydrous tetrahydrofuran (20 ml). The solution is cooled to −50 °C and methylmagnesium bromide (5.5 mmol) is added dropwise. 2,3-dichloro-5,6-dicyanobenzoquinone (6.0 mmol) in anhydrous tetrahydrofuran is added after 15 min to the stirred mixture. The stirring is continued for 3 h, after which the reaction mixture is quenched with saturated ammonium chloride solution. The phases are separated and the aqueous phase extracted with ether. The combined organic phases are dried over sodium sulfate, evaporated, and the residue is purified by chromatography on silica gel using cyclohexane/ethyl acetate (4:1) as eluent, giving 0.50 g (70%) of the title compound mp 61–63 °C.

Using this approach, 5-(dimethoxymethyl)-2-nitro-3-(prop-2-enyl)thiophene was prepared from 5-(dimethoxymethyl)-2-nitrothiophene and allylmagnesium bromide, followed by aromatization [50]. A reaction leading to similar results (vicarious nucleophilic substitution of hydrogen in nitrothiophenes) has been observed with many different carbanions. 2-Nitrothiophene reacts predominantly in the 3-position and 3-nitrothiophene in the 2-position [51].

5-(Dimethoxymethyl)-2-nitro-3-(prop-2-enyl)thiophene [50]

2-Nitro-5-thiophene aldehyde (314 mg, 2.0 mmol) is boiled under reflux with methanol (5 ml) and 1.0 *M* ethereal hydrochloric acid (4.0 ml) for 45 min. The solvent and excess reagent are evaporated. 1.0 *M* Allylmagnesium bromide in anhydrous tetrahydrofuran (2.0 ml, 2.0 mmol) is added to this acetal in anhydrous tetrahydrofuran (8 ml) during 15 min at −50 °C. The stirring is continued at −50 °C for 15 min, after which 2,3-dichloro-5,6-dicyanobenzoquinone (0.68 g, 3.0 mmol) in anhydrous tetrahydrofuran (10 ml) is added.

The reaction mixture is stirred at room temperature for 3 h and then poured into saturated ammonium chloride solution. The phases are separated and the aqueous phase extracted three times with ether. The combined organic phases are dried and evaporated and the residue purified by chromatography using ethyl acetate/hexane (1:4) as eluent and by radial PLC with ethyl acetate/hexane (1:5) as eluent giving 73 mg (25%) of the title compound as colorless oil.

Thiophenes with functionalized alkyl groups in the 2-position give mainly substitution in the 5-position. Bulky and/or electron-withdrawing substituents in the side chain reduce the amount of 3-nitro isomers considerably. Thus 2-benzylthiophene gives 87, 3 and 10% nitration in the 5-, 4-, and 3-position, and a pure 2-benzyl-5-nitrothiophene can be obtained in good yield [52]. From 2-thenylthiophene and its dimethyl derivative, good yields of the 5,5′-dinitro derivatives have been prepared [7]. The nitration of 2-thenyl acetate and 2-thenyl chloride gives mainly the 5-isomers, which can be easily isolated in pure form from the nitration mixture [37,53–56]. 4-Nitro-2-chloromethylthiophene has been prepared by chloromethylation of 3-nitrothiophene [56]. The nitration of 3-alkylthiophenes leads to isomer mixtures and therefore only a 46% yield of 3-methyl-2-nitrothiophene is obtained [36,57,58] from the nitration of 3-methylthiophene. Nitration of 3-methyl-2-thiophenecarboxylic acid followed by decarboxylation has been used for the synthesis of 2-nitro-4-methylthiophene [57]. Recently 2-nitro-3-thenylamine hydrochloride was prepared by this approach from 2-nitro-2-thenyl bromide [59].

3-Bromomethyl-2-nitrothiophene [59]

A mixture of ethylene dibromide (70 ml), 3-methyl-2-nitrothiophene (33g, 230 mmol), and benzoyl peroxide (0.5 g) is gently refluxed with stirring for 15 min, after which a mixture of *N*-bromosuccinimide (60 g, 339 mmol) and benzoyl peroxide (0.3 g) is added in small portions over 30 min. The red–brown solution so obtained is allowed to reflux for an additional 5 h. After cooling, carbon tetrachloride (50 ml) is added and the resulting succinimide is filtered off and washed with carbon tetrachloride (2 × 25 ml). The filtrate is washed with water (3 × 100 ml), dried, and concentrated under reduced pressure. The crude oil is dissolved in hexane (200 ml) and the unreacted starting material (8.5 g) is filtered off. After evaporation the residue is purified by distillation giving 32 g (85%) of the title compound bp 117–130 °C/1.5 mm Hg, mp 59–61 °C.

2-Nitro-3-thenylamine hydrochloride [59]

A suspension of quaternary ammonium salt from hexamethylenetetramine and 3-bromomethyl-2-nitrothiophene (3.62 g, 10 mmol) in dry ethanol (100 ml) under stirring is treated slowly with concentrated hydrochloric acid (7.5 ml), after which the reaction mixture is heated at 50–55 °C for 4 h. After filtration and cooling to 5 °C in an ice-water bath the precipitate formed is filtered off, washed with cooled ethanol (2 × 15 ml) and air dried. Recrystallization from ethanol /diethyl ether (4:1) gives 1.8 g (82%) of the title compound mp 295–297 °C (dec.).

3-Methyl-2-nitrothiophene gives with *N*-bromosuccinimide in ethylene dibromide and benzoyl peroxide, 3-bromomethyl-2-nitrothiophene [59]. 5-Nitro-2-thenyl cyanide is best prepared through the reaction of 1-(5-nitro-2-thienyl)-2-(*N*,*N*-dimethylamino)ethene with hydroxylamino-*O*-sulfonic acid [59].

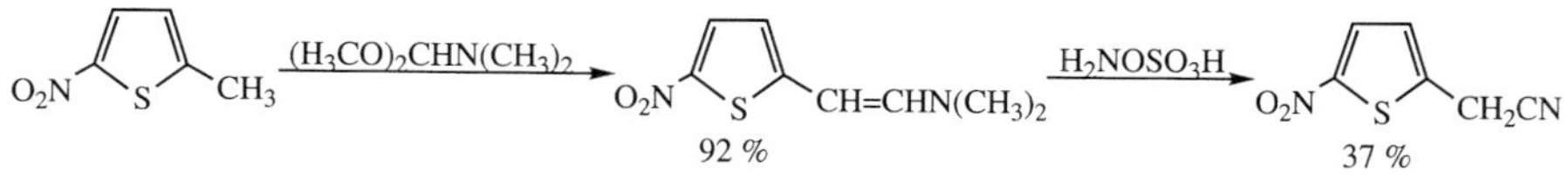

5.1.1.5 Alkenyl- and alkynylnitro derivatives

β-Aminovinylnitrothiophenes are efficiently prepared by the addition of secondary amines to 2-ethynyl-5-nitrothiophene [60].

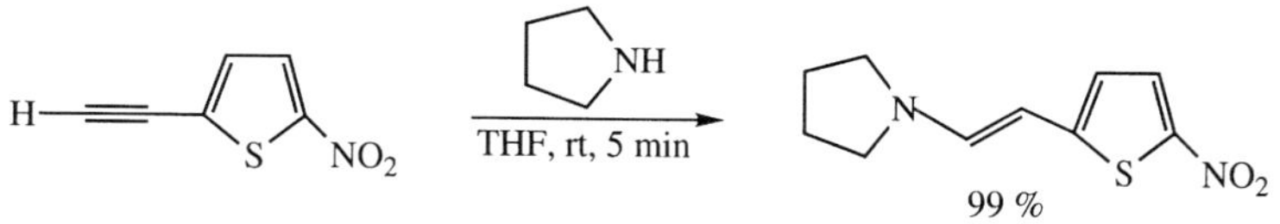

1-Aryl-3-(5-nitrothienyl)propenones are prepared through the reaction of 5-nitro-2-thiophene aldehyde with phosphoranes [61].

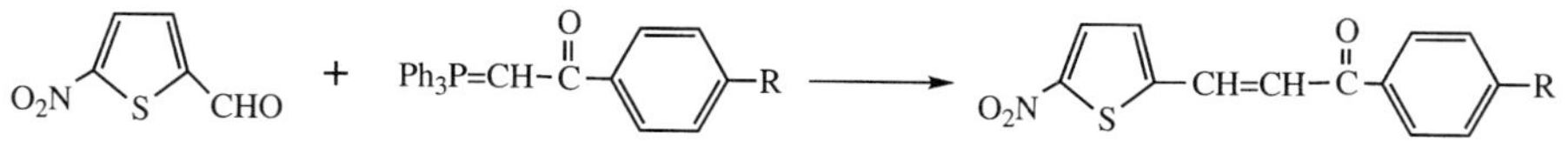

Nitration of 2-styrylthiophene is not of preparative use as under a variety of conditions attack occurs at the double bond and also lesser amounts of 5- and 3-nitration occur. On the other hand 2-thenylideneacetone [62] and *β*-(2-thienyl)acrylic acid [63] give good isolated yields of the corresponding 5-nitro derivatives. Nitration of *β*-3-thienylacrylic acid and its esters on the other hand are not synthetically useful as mixtures of about equal amounts of the 2- and 5-nitrated compounds are obtained [64,65]. 2-Nitro-5-phenylethynylthiophene is prepared in 87% yield by photochemical reaction of 2-iodo-5-nitrothiophene with phenylacetylene in acetonitrile [66].

Another approach is the condensation of nitromethylthiophenes with aromatic aldehydes. Thus condensation of 3,5-dinitro-2-methylthiophene with *O,O'*-diacetyl-4-formyl-*N*-phenyldiethanolamine using catalytic amounts of pyrolidine in tetrahydrofuran gives 2-{[4-bisacetoxyethyl)phenyl]ethenyl}-3,5-dinitrothiophene in 56% yield [67].

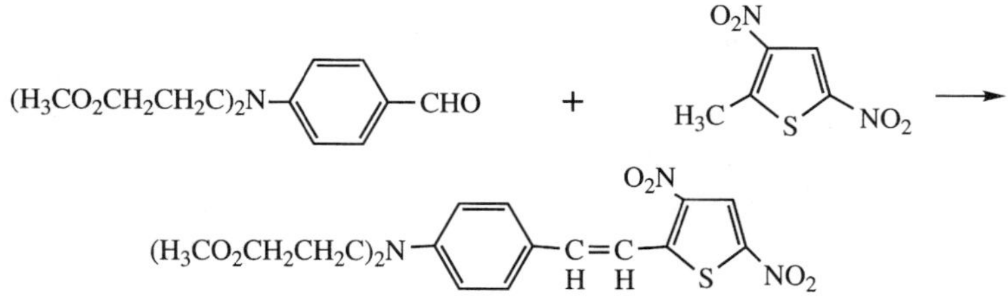

2-{[4-Bisacetoxyethyl)phenyl]ethenyl}-3,5-dinitrothiophene [67]

A freshly distilled pyrrolidine (2 drops) and 3,5-dinitro-2-methylthiophene (0.94 g, 5 mmol) are added to a solution of *O,O'*-diacetyl-4-formyl-*N*-phenyldiethanolamine (2.2 g, 7.5 mmol) in tetrahydrofuran. The reaction mixture is refluxed for 20 h, after which the solvent is evaporated and the residue recrystallized from methanol giving 1.3 g (56%) of the title compound as shining black crystals mp 169–170 °C.

5.1.1.6 Aryl-substituted nitrothiophenes

5.1.1.6.1 By nitration reactions

Nitration of 2- and 3-phenylthiophene does not occur selectively and the 3-nitro and 5-nitro isomers are obtained from the 2-isomer in the proportion of 41:59 [69,70]. However, the isomers could be separated by fractional crystallization [4,69]. The nitration of 3-phenylthiophene is useful for the preparation of 3-phenyl-2-nitrothiophene, as the 2- and 5-nitro isomers are formed in the proportion of 9:1 [69].

2-(4-Fluorophenyl)-3-[4-(methylsulphonyl)phenyl]-5-nitrothiophene is obtained upon nitration of the 2,3-diarylthiophene with nitric acid in acetic anhydride. Direct arylation of 2-nitrothiophene using a catalyst system of palladium(II) acetate and tetrabutylammonium iodide has been achieved with *para*-methoxyiodobenzene [71,72].

5.1.1.6.2 Suzuki couplings between arylboronic acids and bromonitrothiophenes

Coupling between 6-butylthio-2-naphthylboronic acid and 2-bromo-5-nitrothiophene gives 2-(6-butylthio-2-naphthyl)-5-nitrothiophene [73].

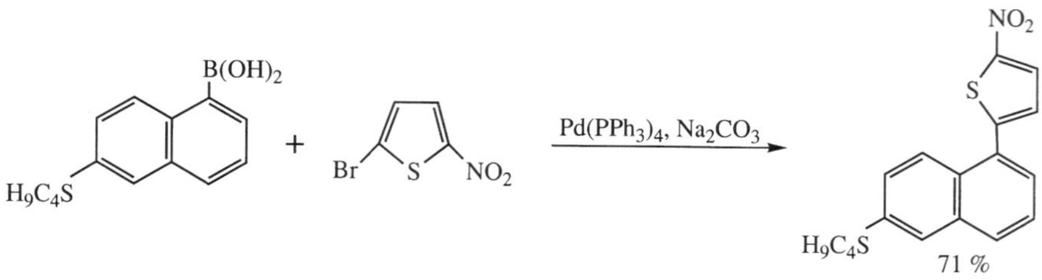

5.1.1.7 Nitrosubstituted thiophene aldehydes, ketones, nitriles and carboxylic acids and their derivatives

5.1.1.7.1 By modification of side chains in nitrothiophenes

Nitrothiophene aldehydes are converted to oximes by standard methods and converted to nitrocyanothiophenes upon treatment with acetic anhydride [74].

N-(3-nitrothiophene-2-ylidene)anilides [75]

A mixture of 3-nitro-2-thiophene aldehyde (0.79 g, 5 mmol), the appropriate aromatic amine (5.5 mmol), and ethanol (10 ml) are stirred, heated under reflux for 10 min and then cooled to ambient temperature overnight. The precipitate is filtered off, washed with ethanol (0 °C) and recrystallized giving *N-(3-nitrothiophene-2-ylidene)aniline* (86%) mp 59–60 °C from light petroleum, *N-(3-nitrothiophene-2-ylidene)dimethylaminoaniline* (90%) mp 164–165 °C from ethanol, *N-(3-nitrothiophene-2-ylidene)-4-methylaniline* (94%) mp 97.5–98.5 °C from light petroleum, *N-(3-nitrothiophene-2-ylidene)-4-chloroaniline* (67%) mp 124–125°C from light petroleum.

4-Acetyl-2-nitrothiophene can be selectively reduced to 2-nitro-4-(1-hydroxyethyl)thiophene with sodium borohydride [76].

2-Nitro-4-(1-hydroxyethyl)thiophene [76]

Sodium borohydride (4 g, 79 mmol) in methanol (30 ml) is added dropwise to a solution of 2-nitro-4-acetylthiophene (3 g, 17.5 mmol) in methanol (30 ml). The temperature is kept at 10–20 °C. The reaction mixture is then treated with water (100 ml) and extracted with dichloromethane (3 × 200 ml). The combined organic phases are dried and evaporated giving 2.0 g (66%) of the title compound as an oil.

The reaction of 2-nitro-3-thiophene aldehyde with acetonylacetone and a few drops of piperidine is used for the preparation of 2-nitro-3-thienylmethyleneacetylacetone [77].

2-Nitro-3-thienylmethyleneacetylacetone [77]

A mixture of 2-nitro-3-thiophene aldehyde (5.13 g, 32.7 mmol) and acetylacetone (3.27 g, 32.7 mmol) containing 10 drops of piperidine is allowed to stand for two days. The solid formed is filtered off and recrystallized from ethanol giving 6.99 g (52%) of the title compound as yellow prisms mp 90–91 °C.

An interesting result is obtained from the reaction of the diacetate from 5-nitro-2-thiophene aldehyde with lithium 2-nitropropan-2-ide in dimethylsulfoxide which gives the compound shown below in a yield of 70–80% [68].

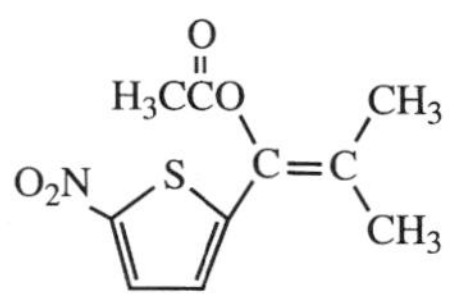

Recently it was found that the condensation of 5-nitro-2-thiophenealdehyde with glycin ester hydrochloride did not lead to the imine but to ethyl 2-amino-3-(5-nitro-2-thienyl)-2-propenoate [78].

Ethyl 2-amino-3-(5-nitro-2-thienyl)-2-propenoate [78]

Ethyl glycinate hydrochloride (50 mg, 0.36 mmol) is added in three portions to a stirred mixture of 5-nitro-2-thiophene aldehyde (50 mg, 0.32 mmol), triethylamine (0.05 ml, 0.36 mmol) and 4 Å molecular sieves (100 mg) in toluene (2 ml). The suspension is stirred at room temperature for 24 h, when the nitro compound is consumed. The reaction mixture is washed with sodium chloride solution (2 × 25 ml). The organic phase is dried over sodium sulfate and evaporated to afford the nitrothiophene derivative as an oily residue, which is purified by chromotography on silica gel using ethyl acetate in petroleum ether (0–30%) as eluent. After recrystallization from chloroform 62 mg (80%) of the title compound is obtained as green needles mp 86.5–88.5 °C.

In most cases the competition between the inherent higher reactivity of the α-position and the "meta"-directing effect of -I-M- substituents leads to the mixtures of 4- and 5-nitro derivatives in the nitration of 2-thiophene aldehydes, ketones, and carboxylic acids. This can often result in tedious separations. The proportion of the 4-nitro isomer can, however, be increased considerably by increasing the acidity of the medium. Thus the 4- and 5-nitro isomers are in the proportion of 3:1, which is increased to 4:1 using nitronium fluoroborate in 100% sulfuric acid [79]. Similar proportions are obtained in the nitration of 2-acetylthiophene [79,80] and 2-propionyl thiophene. This problem can be circumvented by first preparing the oxim-ether of 2-acetylthiophene, which due to its weaker electron-withdrawing effect (or even electron-donating effect) selectively is nitrated in the 5-position. Hydrolysis with 25% hydrochloric acid then yields 2-acetyl-5-nitrothiophene [81].

2-Acetyl-5-nitrothiophene [81]

2-Acetylthiophene (25.0 g, 0.2 mol), *O*-methylhydroxylamine hydrochloride (0.22 mol), and sodium carbonate (11.7 g, 0.11 mol) are dissolved in methanol (20 ml) and water (60 ml). Acetic acid (2.0 ml) is then added to the reaction mixture under stirring to adjust the pH to 4.5, after which the reaction mixture is refluxed for 2 h. After cooling, water (40 ml) and chloroform (80 ml) are added to this solution. The organic phase is separated, washed with water (2 × 70 ml) and dried over magnesium sulfate. The solvent is evaporated and the residue distilled bp 70–73 °C/mm Hg.

To a solution of this compound (15.5 g, 0.1 mol) and acetic acid anhydride (25.8 g, 0.25 mol), acetyl nitrate (20 ml) is added over a period of 1.6 h at 5 °C. The reaction mixture is then stirred at room temperature for 3 h. After cooling, methanol (30 ml) and water (70 ml) are added to decompose excess amounts of acetic acid anhydride and acetyl nitrate. Chloroform (50 ml) is added to this reaction mixture, the phases are separated and the organic phase washed with water (2 × 50 ml) and dried over magnesium sulfate. The solvent is evaporated and the residue recrystallized from ethanol mp 117–118 °C.

The nitrated compound (83 mmol) is mixed with 25% hydrochloric acid (100 ml) and the solution is heated at 70 °C for 8 h. The precipitate obtained is filtered off after cooling and washed with water giving the title compound mp 108–109 °C.

The nitration of 2-thiophenecarboxylic acid [10,17,22,82], 2-carbomethoxythiophene [83–85], 2-carboethoxythiophene [13], and 2-cyanothiophene [18,86] under various conditions gives higher proportions of the 5-isomers and very tedious procedures have to be used in some cases, although the 4- and 5-nitro-2-thiophenecarboxylic acids have been separated by repeated crystallization or conversion into derivatives such as methyl esters or barium salts followed by recrystallization [10,17]. 5-Alkyl-2-thiophenecarboxylic acid can be easily nitrated in the 4-position. 4-nitro-5-propyl-2-thiophenecarboxylic acid was obtained from 5- propyl-2-thiophenecarboxylic acid [87].

4-Nitro-5-propyl-2-thiophenecarboxylic acid [87]

A ground potassium nitrate (12.5 g, 124 mmol) is gradually added to a solution of 5-propyl-2-thiophenecarboxylic acid (16.9 g, 99 mmol) in concentrated sulfuric acid (170 ml) and cooled to < 5 °C. The reaction mixture is stirred at 0 °C for 1 h, after which it is poured into ice (500 g). The precipitate formed is filtered off and washed with water to neutral reaction. The crude product contains 2,4-dinitro-5-propylthiophene due to decarboxylation during the nitration. The acid is transformed to its sodium salt and the dinitro compound is removed by steam distillation. After acidification the crude acid is recrystallized from toluene/heptane giving 12.5 g (59%) of the title compound mp 106–108 °C.

A side reaction which diminishes the yields is replacement of carboxyl and acyl groups during nitration, presumably the result of *ipso*-substitution [10,17,22,82,87–89]. 2-Benzoyl 5-nitrothiophene has been prepared in 70% yield by the reaction of 2-nitrothiophene with benzoyl chloride in the presence of tin tetrachloride [90].

On the other hand the nitration of 3-thiophenecarboxylic acid [91], 3-carbomethoxythiophene [92], 3-cyanothiophene [88,93], 3-thiophene aldehyde [37,88,94], and 3-acetylthiophene [95,96] all give the 5-isomers predominantly, which are easily obtained pure by recrystallization, although small amounts of the 2- and 4-isomers are also formed [88].

2-Nitro-4-acetylthiophene [96]

To a solution of concentrated nitric acid (20 ml) and concentrated sulfuric acid (11 ml) cooled to −10 °C, 3-acetylthiophene (5 g, 40 mmol) is added at such a rate that the temperature is kept between −5 and −10 °C. When the addition is complete the reaction mixture is poured onto ice. The orange precipitate is filtered and the aqueous phase extracted with dichloromethane. After evaporation the solids are collected and purified by chromatography on silica gel using ether/hexane (1:1) as eluent, giving 4.79 g (70%) of the title compound mp 62 °C.

4-Nitrothiophene-2-carboxamide is prepared by the addition of potassium nitrate to a solution of thiophene-2-carboxamide in concentrated sulfuric acid [97].

4-Nitrothiophene-2-carboxamide [97]

Potassium nitrate (1.03 g) is slowly added to a solution of thiophene-2-carboxamide (1.3 g, 10.2 mmol) in concentrated sulfuric acid (10 ml) at 10 °C under stirring. The reaction mixture is stirred at 10 °C for 30 min and then poured into crushed ice. The precipitate formed is filtered off, washed with water, and purified by chromatography on silica gel using benzene/ethyl acetate (1:1) as eluent, giving 1.30 g (74%) of the title compound mp 152–153 °C after recrystallization from methanol.

A number of amides were prepared from 4-nitro-5-methyl-2-thiophenecarboxylic acid by reaction of its acid chloride with various amines such as

2-dimethylaminoethylamine, 4-(2-aminoethyl)morpholine, 1-(2-aminoethyl) piperidine), and 1-(3-aminopropyl)pyrrolidine [50].

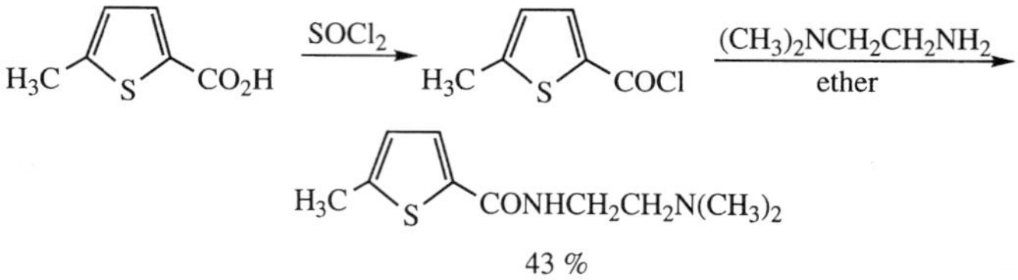

The reaction of 5-nitro-2-thiophene aldehyde with the ylides formed from the 1,3-dithian-2-yl triphenylphosphonium chloride and the corresponding diethyl phosphonates of dithiolyl and 1,3-benzodithiolyl compounds in tetrahydrofuran at −70 °C gives the donor–acceptor molecules shown below [98].

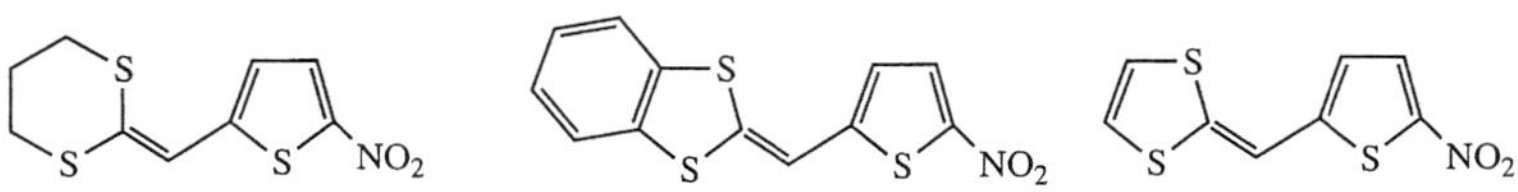

The reaction of 2-chloro-3-nitrothiophene with 2-benzylimidazoline did not give the expected S_NAr substitution product but yielded the tertiary alcohol shown below [99]. An interesting mechanism for this reaction is suggested.

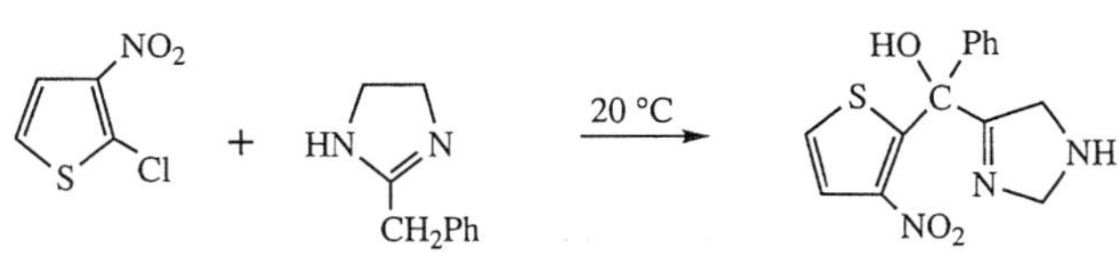

1-(4,5-Dihydro-1H-imidazol-2-yl)-1-phenyl-1-(3-nitro-2-thienyl)methanol [99]

A solution of 2-chloro-3-nitrothiophene (1.0 g, 6.0 mmol) in propionitrile (30 ml) containing molecular sieves (4 Å) is stirred and 2-benzyl-4,5-dihydro-1*H*-imidazole (1.93 g, 12.0 mmol) in propionitrile (20 ml) is added dropwise. After maintaining stirring for 24 h at ambient temperature, the solution is filtered and evaporated under vacuum to give a dark-red residue that is purified by flash chromatography on silica gel using ethyl acetate/methanol (9:1) as eluent, giving 1.13 g (62%) of the title compound as fine beige needles mp 140–142 °C after recrystallization from ethanol.

5.1.2 Amino derivatives and their *N*-alkylated, arylated and acylated derivatives

5.1.2.1 General aspects

The most important methods for the preparation of aminothiophenes are: (1) reduction of nitro derivatives; (2) conversion of acylthiophenes and derivatives of thiophenecarboxylic acids into aminothiophenes using Beckman rearrangement, Curtius rearrangement, Schmidt reaction, or Hofmann reaction; (3) reaction of halothiophenes with ammonia and amines; (4) displacement of nitrobenzenesulfonylmethoxy and *para*-nitrophenoxy groups by amino groups; (5) ring-closure reactions such as the Gewald, the Fiesselmann and the Gompper reaction. The aminothiophenes are much more reactive and less stable than the corresponding anilines and special care has to be exercised in their syntheses.

5.1.2.2 Parent aminothiophenes

5.1.2.2.1 By reduction of nitrothiophenes

A large number of reducing agents, such as tin and hydrochloric acid [17,19,100–109], tin and hydrogen chloride in ethanol [108], stannous chloride in hydrochloric acid [4,64,100,110,111] or hydrogen chloride in ethanol [64,112–114], iron and hydrochloric acid [115], iron in acetic acid [115–121], zinc in ethanolic hydrochloric acid [112], zinc and sodium hydroxide [123], sodium amalgam in methanol [124,125], aluminium amalgam in ether [13,23,126], thiosulphate in methanol [127,128], Raney-nickel [21,129–132], and catalytic hydrogenation over palladium [42,65,133–135] or platinum [136] have been used for the reduction of nitrothiophenes to aminothiophenes. A good method for the preparation of the very unstable 2- and 3-aminothiophene is the reduction with either tin and hydrochloric acid [108] or tin(II) chloride and hydrochloric acid developed by Steinkopf and Lützkendorf [19,100,130], which gives the aminothiophenes in the form of stable hexachloro stannates(IV).

A somewhat tedious route to 3-aminothiophene has also been described. 2-Formylthiophene was nitrated and the 4- and 5-nitro-2-thiophene aldehyde were separated. Oxidation of 4-nitro-2-thiophene aldehyde with sodium dichromate in sulfuric acid followed by decarboxylation with and quinoline gave 3-nitrothiophene, which was reduced to 3-aminothiophene with iron and hydrochloric acid and directly converted to the ethyl *N*-(3-thienyl)ethylaminomethylene [137]. Reduction of 2,5-dibromo-3,4-dinitrothiophene with tin in hydrochloric acid is a good method for the preparation of 3,4-diaminothiophene [138,139].

3,4-Diaminothiophene [139]

Small portions of tin (72 g) are added to a stirred suspension of 2,5-dibromo-3,4-dinitrothiophene (34 g, 102 mmol) in concentrated hydrochloric acid (700 ml) at such a rate that the temperature is maintained below 30 °C. The reaction mixture is stirred for another 4 h and kept in a refrigerator overnight. The solid formed is filtered off and washed many times with diethyl ether and acetone. The hexachlorostannate (20.3 g) so obtained is suspended in water (200 ml), the mixture is stirred after addition of diethyl ether (200 ml) and cooled to 5 °C and made basic by 4 *M* sodium hydroxide solution. The phases are separated and the water phase continuously extracted with diethyl ether for several hours. The combined organic phases are dried and concentrated under cooling to a volume of 30 ml. The precipitate is filtered off, washed with ether giving 5.9 g (51%) of the title compound as a white solid mp 96 °C after recrystallization from ether.

The reduction can also be obtained with simultaneous boc-protection [140,141].

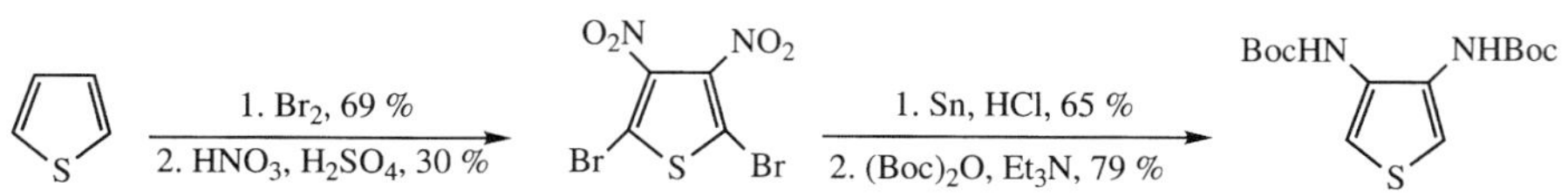

This methodology has also been used for the preparation of 3,4-diamino-2,5-dimethylthiophene [44] and 3,4-diamino-2,5-di-*tert*-butylthiophene [142]. Amides of 5-nitro-2-thiophenecarboxylic acids were reduced with sodium dithionite in aqueous ethanol to the 5-amino derivative [143].

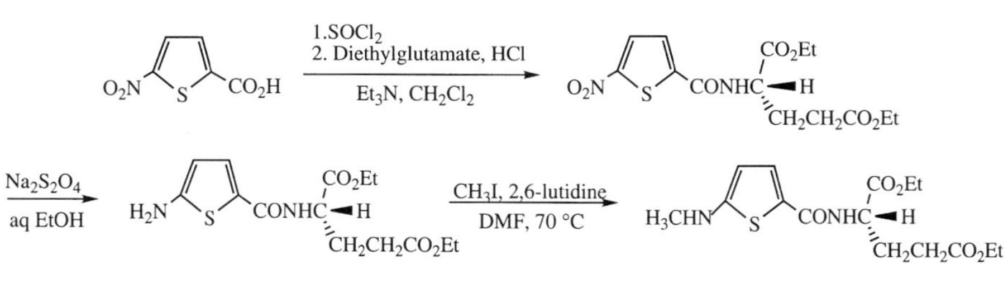

5.1.2.2.2 Alkylation of aminothiophenes

Alkylation of aminothiophenes can be used for the preparation of *N*-alkyl and *N*,*N*-dialkylaminothiophenes. Thus *N*-benzyl-*N*-dimethylaminoethylthiophene was prepared in 35% yield by alkylation of 2-aminothiophene with *β*-dimethylaminoethyl chloride in pyridine followed by alkylation with benzyl chloride [144]. Various 3-pyrrolidinothiophenes were prepared by the reactions of 3-aminothiophenes with 1,4-dibromobutane in the presence of Hünig's base [145,146]. *N*-(3-thienyl)carbamic esters have been used as starting material for the preparation of 3-methylaminothiophene and 3-dimethylaminothiophene through reduction with lithium aluminium hydride and alkylation followed by reduction respectively [147,148]. The last step in the scheme above is a methylation using methyl iodide in the presence of 2,6-lutidine in *N,N*-dimethylformamide at 70 °C [143]. 3-Thienyl trialkyl ammonium salts have been prepared by alkylation of dialkylaminothiophenes [148–150].

N-methyl-N-benzyl-3-aminothiophene methyl iodide [149]

A 50% sodium hydride suspension (4g, 0.083 mol) in anhydrous xylene (50 ml) is added to a solution of methyl *N*-(3-thienyl)carbamate (5 g, 0.038 mol) in anhydrous xylene (150 ml). The reaction mixture is refluxed for 12 h and benzyl chloride (22 g, 0.17 mol) is added after cooling. After refluxing for an additional 8 h the precipitate is filtered off and washed with benzene. The combined filtrates are evaporated and the residue distilled giving 3.32 g (45%) at 140–160 °C/0.3 mm Hg.

A solution of the benzylated urethane (2.5 g, 0.011 mol) in anhydrous tetrahydrofuran (50 ml) under nitrogen is added dropwise with stirring to lithium aluminium hydride (2 g, 0.053 mol) in anhydrous tetrahydrofuran. After refluxing for 24 h and hydrolyzing the reduction complex with water the precipitate is filtered off and washed with ether. The combined filtrates are dried over magnesium sulfate and evaporated. The oily residue is diluted with an equal volume of benzene and methyl iodide is then added. The solid product is recrystallized from methanol mp 152.5–153.5 °C.

5.1.2.2.3 Nucleophilic displacement of substituents in non-activated thiophenes by ammonia and amines

Reaction of 2-bromothiophene with potassium amide in liquid ammonia occurs with rearrangement to give 3-aminothiophene [151]. The reaction of 2-iodothiophene with trimethylsilylamido copper followed by methanol gave 45% of 2-aminothiophene [152]. The reaction of 2-iodothiophene with the anion of *N,N*-diphenylamine, prepared from *N,N*-diphenylamine and potassium hydride in dimethylpropylene urea in the presence of cuprous iodide at 80 °C, gave 2-(*N,N*-diphenylamino)thiophene [153].

2-(N,N-diphenylamino)thiophene [153]

A solution of *N,N*-diphenylamine (46 g, 275 mmol) is added dropwise to a slurry of hexane-washed potassium hydride (31 g, 35% in mineral oil, 275 mmol) in dimethylpropyleneurea (50 ml) over 1 h under nitrogen. The resulting yellow slurry is stirred for an additional hour, after which freshly prepared anhydrous copper(I) iodide (52 g, 275 mmol) is added and the stirring continued for 1 h by which time all copper(I) iodide is dissolved giving a dark-green turbid solution. This solution is heated to 80 °C and a solution of

2-iodothiophene (46 ml, 412 mmol) in dimethylpropyleneurea (75–125 ml) is added dropwise over 3 h. The dark mixture is heated until most of the *N*,*N*-diphenylamine is consumed as determined by gas chromatography. The mixture is then cooled and poured into ethyl acetate (1000 ml) and filtered, concentrated ammonium hydroxide (100 ml) is added, and the organic phase washed with 10% ammonium hydroxide solution until no blue color is observed. After drying over magnesium sulfate and evaporation a thick brown oil is obtained. The oil is subsequently subjected to molecular path distillation with the major fraction distilling at 135 °C/0.005 mm Hg. Recrystallization from ethanol gives 27 g (40%) of the title compound.

Recently, diarylaminothiophenes and 2,5-bis(diarylamino)thiophenes are prepared by tri-*tert*-butylphosphineligated palladium-catalyzed coupling of bromo- and dibromothiophenes with diarylamines [154].

2,5-Di-(diphenylamino)thiophene [154]

Diacetatopalladium(II) (44.9 mg, 0.2 mmol), 50 mg/ml tri-tert-butylphosphine in *ortho*-xylene (2.4 ml, 0.6 mmol), 2,5-dibromothiophene (2.42 g, 10 mmol), diphenylamine (3.38 g, 20 mmol), sodium *tert*-butoxide (2.11 g, 22 mmol), and *o*-xylene (40 ml) are mixed at room temperature and heated at 120 °C for 3 h under nitrogen. After addition of water, extraction with diethyl ether, and evaporation, the residue is purified by reprecipitation by adding methanol (30 ml) giving 2.39 g (57%) of the title compound mp 144–145 °C.

Reaction of 1,3,5-tris(phenylamino)benzene with 2- and 3-iodothiophene in decaline in the presence of potassium hydroxide and copper powder gave 1,3,5-tris(phenyl-2-thienylamino)benzene and 1,3,5-tris(phenyl-3-thienylamino)benzene, respectively [155].

A convenient route to dialkylaminothiophenes in 10–70% yield consists in refluxing of 2- and 3-thiophenethiol with secondary amines such as dimethylamine, diethylamine, and piperidine in toluene [156–159].

2-Piperidinothiophene [158]

2-Thiophenethiol (11.6 g, 0.1 mol) and piperidine (8.5 g, 0.1 mol) are dissolved in toluene (30 ml) at room temperature. Upon warming salt formation takes place. The reaction mixture is refluxed under nitrogen for 1 h during evolution of hydrogen sulfide. After evaporation of the solvent the residue is fractionated giving 9.2 g (55%) of the title compound bp 102–103 °C/2 mm Hg.

2-Dimethylaminothiophene and 2-pyrrolidinothiophene are best prepared by the reaction of 2-thiophenethiol with dimethylamine and pyrrolidine, respectively [160].

2-Dimethylaminothiophene [160]

A mixture of 2-mercaptothiophene (1.79 g, 15.4 mol) and toluene (5 ml) that has been saturated with dimethylamine gas is heated at 100 °C for 30 min. During this time, the solid formed on the condenser is periodically scraped back into the flask, and at aspirator pressure 0.785 g (40%) of the title compound is collected as a yellow liquid after distillation at high vacuum.

2-Dimethylamino-5-methylthiophene is obtained by heating 5-methyl-4-thiolene-2-one, a tautomeric form of 2-hydroxy-5-methylthiophene with hexamethyl phosphortriamide [161]. Reaction of 2-methoxythiophene with *N*-lithio-*N′*-methylpiperazine can be used for the preparation of 1-methyl-4-(2-thienyl)piperazine [162]. A low yield of 3-piperidinothiophene is obtained from the reaction of di-(3-thienyl) iodonium chloride with piperidine in the presence of copper(II) salts [163].

Heating of 3-aminothiophene in acetic acid/benzene can be used for the preparation of di(3-thienyl)amine in 75% yield [164]. A convenient method for the preparation of 2-alkyl-3-aminothiophenes is the acid-catalyzed reaction of 3-aminothiophene with aldehydes in the presence of selenophenol. In the absence of the reducing agent bis-(3-amino-2-thienyl)methane derivatives are obtained [165].

2-Ethyl-3-aminothiophene [165]

A solution of 3-aminothiophene (198 mg, 2.0 mmol) in dichloromethane (20 ml) at 0 °C under stirring is quickly added to acetaldehyde (88 mg, 2.0 mmol) in dichloromethane (10 ml) containing selenophenol (785 mg, 5.0 mmol) at 0 °C. A solution of *para*-toluenesulfonic acid (20 mg) in dichloromethane is then added dropwise. The stirring is continued at room temperature for 3 h, after which the reaction mixture is extracted with 1 *M* hydrochloric acid (2 × 25 ml). The phases are separated and the aqueous phase washed with ether and then basified by 4 *M* sodium hydroxide solution in the presence of ether (30 ml). After separating the phases the aqueous phase is extracted with ether (3 × 10 ml), the combined ether phases are dried and concentrated and the residue chromatographed on basic alumina using hexane/chloroform (6:4) as eluent giving 214 mg (84%) of the title compound.

Bis(3-amino-2-thienyl)methylmethane [165]

Acetaldehyde (48 mg, 1.1 mmol) is added to a solution of 3-aminothiophene (198 mg, 2.0 mmol) in dichloromethane (20 ml). A 12 *M* aqueous hydrochloric acid solution (0.5 ml) is then introduced dropwise. A precipitate appears after 2 min. The suspension is stirred at room temperature for 40 min. The solid is isolated, washed with dichloromethane (30 ml), and dissolved in water (20 ml).

In the presence of dichloromethane (30 ml) treatment with 1 *M* aqueous sodium hydroxide solution allows the transfer of the product to the organic phase which is then separated. The basic aqueous solution is extracted with dichloromethane (25 ml), the combined organic phases are dried and evaporated and the oily residue is chromatographed on basic alumina using hexane/chloroform (6:4) as eluent. After recrystallization from hexane/chloroform (95:5) 180 mg (80%) of the title compound is obtained mp 99 °C.

5.1.2.2.4 Direct and indirect electrophilic amination of thienylmetal derivatives

The reaction of thienyllithium derivatives with *N*-lithium acetamide and cuprous cyanide followed by oxidation with oxygen yields the acetamido derivatives in low yield [166], which upon acidic hydrolysis give the aminothiophenes. Higher yields are obtained upon reaction of the higher order cuprate, $(2\text{-Th})_2CuCNLi_2$, with *N*,*O*-bis(trimethylsilyl)hydroxylamine, which gives the stable [(2-trimethylsilyl)amino]thiophene, which can be easily converted to the free amine [167]. These authors have developed a general and convenient procedure for the preparation of *N*-alkylaminothiophenes, such as 2-methylamino- and 2-isopropylaminothiophene, by electrophilic amination of the above-mentioned higher order cuprates with *N*-alkylhydroxylamines [168].

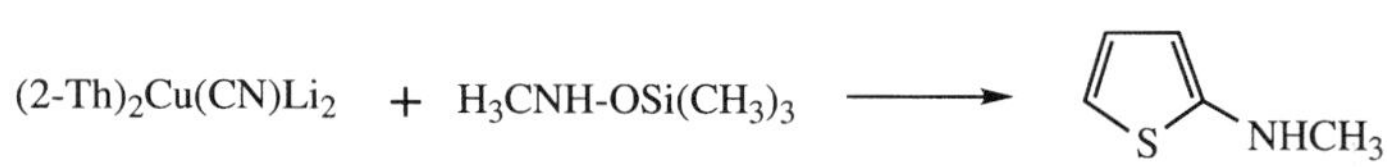

[(2-Trimethylsilyl)amino]thiophene [167]

N,*O*-bis(trimethylsilyl)hydroxylamine (0.426 ml) is added dropwise To a solution of (2-Th)2CuCNLi$_2$ (2.00 mmol) in tetrahydrofuran (6 ml), cooled to −50 °C. After 1 h the reaction mixture is warmed to room temperature, filtered through a Celite pad, and evaporated to give a dark oily residue. The crude product is purified by distillation to afford 246 mg (72%) of the title compound as a colorless liquid that slowly darkens on exposure to light bp 38–40 °C/0.3 mm Hg.

3-Amino-2-thiophene aldehyde [169]

3-Azido-2-thiophene aldehyde (6.2 g, 41 mmol) is dissolved in ethanol (50 ml) and cooled to 10 °C, after which piperidine (2 drops) is added. With continued cooling, hydrogen sulfide is bubbled through the reaction mixture whereupon an exothermic reaction takes place with evolution of nitrogen. The hydrogen sulfide inlet is at such a rate that the temperature is kept below 20 °C. After about 20 min when the evolution of nitrogen has ceased, the reaction mixture is cooled to 0 °C and the preciptiated sulfur filtered off. The filtrate is treated with ice and 4.4 g (85%) of the title compound is obtained mp 68.0–70.0 °C after recrystallization from ethanol/water.

5.1.2.2.5 Hydrolysis of acetamido- and other protected derivatives

Acid hydrolysis of acetylamino derivatives can often be used for the preparation of aminothiophenes or their salts. Recently ethyl 3-(3,4-diaminothien-2-yl)propenoate was prepared from the di-*boc*-protected derivative upon treatment with a 20% solution of hydrobromic acid in acetic acid followed by neutralization with aqueous sodium hydroxide [170].

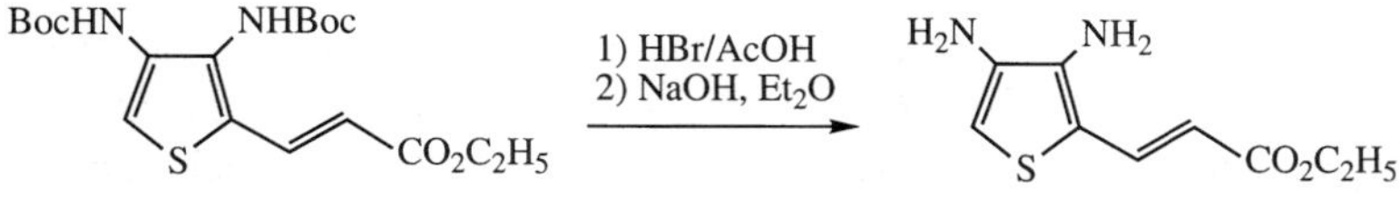

Ethyl [(3,4-diamino)thien-2-yl]propenoate [170]

Ethyl 3-[3,4-bis(*tert*-butoxycarbonylamino)thien-2-yl]propenoate (0.412 g, 1 mmol) in a 20% solution of hydrobromic acid in acetic acid (0.450 g, 1 mmol) is stirred at room temperature for 20 min and anhydrous diethyl ether (20 ml) is added. The reaction mixture is stirred for another 15 min. The salt is filtered off, washed with ether and dissolved in water (20 ml). After neutralization with diluted aqueous sodium hydroxide solution, extraction with diethyl ether (3 × 10 ml) and evaporation 0.134 g (63%) of the title compound is obtained in pure form.

5.1.2.2.6 Conversion of acylthiophenes and derivatives of thiophenecarboxylic acids into aminothiophenes

The Beckmann rearrangement of the oximes of 2-acetylthiophene and 3-acetylthiophene have been used for the preparation of 2-acetamido- [115,171] and 3-acetamidothiophene [171,172] in high yields. Upon acidic hydrolysis the parent compounds can be obtained. The Schmidt reaction with hydrazoic acid applied to 2-acetylthiophene is not a good method for the preparation of 2- acetamidothiophene as an eutectic mixture with the isomeric *N*-methyl-2-thiophenecarboxamide is formed [173], while this reaction applied to 3-acetylthiophene gives a high yield of 3-acetylaminothiophene [172].

2-Thiophenecarboxamide is hydrolyzed on attempted Hofmann reaction [174]. The first NMR spectrum of 3-aminothiophene, showing it to exist in the amino rather than the imino form was obtained on an undistilled crude sample, prepared by the Hofmann reaction [175]. It was decomposed upon attempted distillation *in vacuo* and later obtained pure by preparative gas chromatography [176].

3-Aminothiophene [176]

A sodium hypobromite solution is prepared by dropping bromine (3.7 ml) with stirring into a solution of sodium hydroxide (11.22 g) in ice-cold water (92 ml). 3-thienylacetamide (6.56 g) is added to this cold yellow solution under nitrogen. The reaction mixture is stirred for 1 h and then warmed to 70–80 °C for about 45 min. The color of the solution becomes dark red. After cooling, the product is extracted with *para*-xylene, the extract is dried over Drierite and the title compound purified by preparative gas chromatography and collected in a flask immersed in a dry ice-bath.

3-Thiophenecarboxamide and several halogenated derivatives gave 50–80% of the corresponding 3-aminothiophenes isolated as their stable *N*-acyl derivatives [172,174].

3-Aminothiophene hydrochloride has been obtained from the oxime of 3-oxotetrahydrothiophene upon treatment with ethereal hydrogen chloride [177].

The most useful of the rearrangement reactions for the preparation of derivatives of the aminothiophenes is the *Curtius* reaction. Many different methods have previously been used for the transformation of the carboxylic acid *via* the carbonyl azides to the carbamates [178–180]. The best method being a one-pot procedure introduced by Yamada *et al.* [181], consisting in the reaction of carboxylic acids with diphenyl phosphorazidate and *tert*-butyl-alcohol in the presence of triethylamine. It has been used for the preparation of halo *tert*-butyl *N*-(2-thienyl)carbamate [182] and tert-butyl *N*-(3-thienyl)carbamate [183].

tert-Butyl N-(4-iodo 3-thienyl)carbamate [183]

A 250-ml round-bottomed flask equipped with magnetic bar and condenser is charged with 4-iodo-3-thiophenecarboxylic acid (5.08 g, 0.02 mol), diphenyl phosphorazidate (6.05 g, 0.022 mol), triethylamine (2.02 g, 0.022 mol), and anhydrous *tert*-butanol (100 ml). The reaction mixture is refluxed for 15 h and the cooling water kept at 30 °C. After cooling the excess of alcohol is evaporated and the residue dissolved in dichloromethane. This solution is washed with 5% aqueous citric acid solution, water, a saturated aqueous sodium bicarbonate solution, and a saturated aqueous sodium chloride solution. After drying over magnesium sulfate and evaporation the residue is chromatographed on silica gel using ethyl acetate/cyclohexane (1:3) as eluent, giving 6.05 g (93%) of the title compound mp 58–58 °C.

The carbamates are stable in contrast to most free aminothiophenes and constitute a convenient form for storage of the amino derivatives. They give on heating, in the presence of an acid catalyst, the free amines [183,184]. 3-Aminothiophene was thus obtained in good yield by hydrolysis of isopropyl *N*-(3-thienyl)carbamate [185] and also 2,3-diaminothiophene derivatives have been prepared in this way [186]. Other carbamate esters have been hydrolyzed under basic conditions [185–187]. The Curtius reaction was also shown to be very useful for the synthesis of 3,4-diaminothiophenes from 4-amino-3-thiophenecarboxlic acid [188–190], which was of importance in connection with attempts to prepare biotin and related compounds.

An interesting synthesis of 4-dimethylamino-2-trimethylsilylthiophene in 39% yield consists in the reaction of 5-trimethylsilyl-2-thiophene sulfonamide with butyllithium/*N*,*N*,*N*′,*N*′-tetramethyl-1,2-ethanediamine at −30 °C *via* a rearrangement reaction [191].

5.1.2.2.7 Ring-closure reactions leading to simple aminothiophenes

An elegant synthesis of 2-aminothiophene consists in the ring closure of *cis* *β*-benzylthiomethyl acrylonitrile with hydrochloric acid in ether [192,193] 2,4-Diaminothiophenes have been prepared refluxing *α*-chlorothioacetanilides in methanol [194].

N,N′-Diisopropyl-N,N′-diphenyl-2,4-aminothiophene [194]

α-Chloro-*N*-isopropylthioacetanilide (13.1 g, 0.058 mol) in methanol (100 ml) is refluxed for 12 h. After cooling crystalline sulfur is removed by filtration and methanol removed by evaporation. The residue is taken up in water and the water solution extracted with ether (2 × 50 ml). The aqueous phase is neutralized with sodium bicarbonate and the resulting oil extracted with ether. The combined organic phases are dried over magnesium sulfate and evaporated giving 4.9 g (48%) of the title compound mp 74 °C after recrystallization with charcoal from cold pentane.

Self condensation of arylthioacetomorpholides in the presence of *N*,*N*-disubstituted amides and phosphorus oxychloride gives 2-aryl-3-benzyl-3-morpholinothioacrylamides, which upon treatment with ethyl glycinate unexpectedly gives 2,4-dimorpholino-3,5-diphenylthiophene in high yield [195].

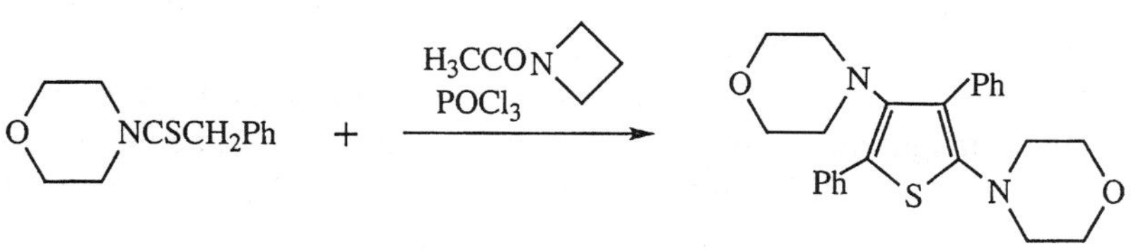

A convenient synthesis of 2,5-bis(*N*,*N*-dialkylamino)thiophenes consists in the reaction of lithiated propargylic amines with isothiocyanates followed by addition of potassium *tert*-butoxide, *tert*-butanol, and methyl iodide at −15 °C. If the temperature during base addition is kept at −40 °C 1-alkyl-2-*N*,*N*-dialkylamino-5-methylthiopyrroles are obtained instead [196].

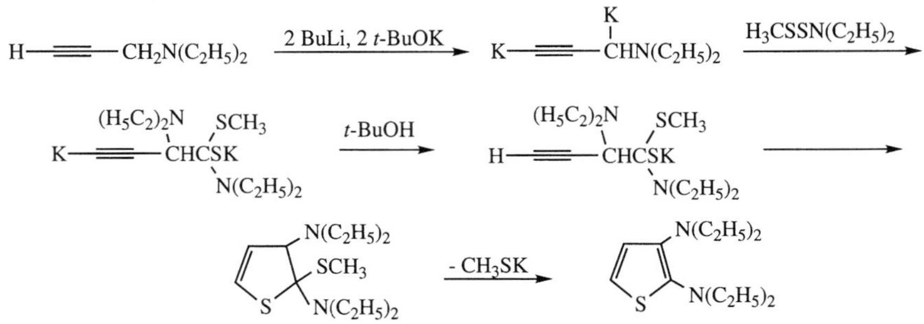

2,3-(Diethylamino)thiophene can be prepared as shown in the scheme above [197]. A convenient one-pot synthesis of 2-*N*-alkylamino-, 2-*N*-phenylamino-, 2-*N*,*N*-dialkylamino-, and 2-*N*-alkyl-*N*-phenylaminothiophenes consists in adding a solution of *tert*-butanol and potassium *tert*-butoxide in dimethylsulfoxide to a solution of the adduct from a lithiated 1-alkyne, $RCH_2C{\equiv}CLi$, or the lithiated allene, *tert*-BuCH=C=CHLi, and an isothiocyanate in tetrahydrofuran and subsequently hydrolyzing the reaction mixture or quenching with methyl iodide [198].

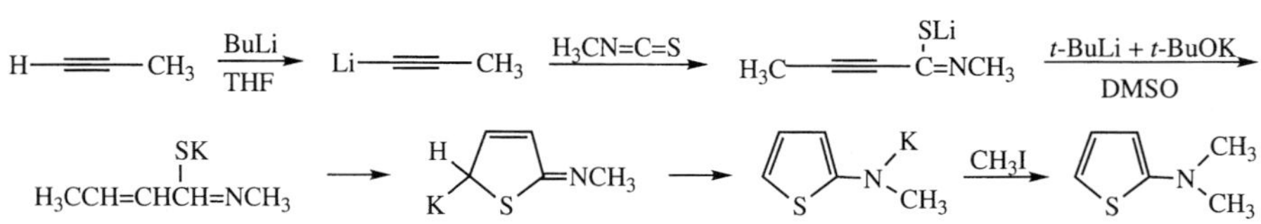

The reaction of substituted allyl benzotriazoles with butyllithium followed by isothiocyanates followed by heterocyclization by treatment with zinc bromide in dichloromethane is a convenient method for the preparation of 3-alkyl and 3-benzyl-substituted 2-aminothiophenes [199].

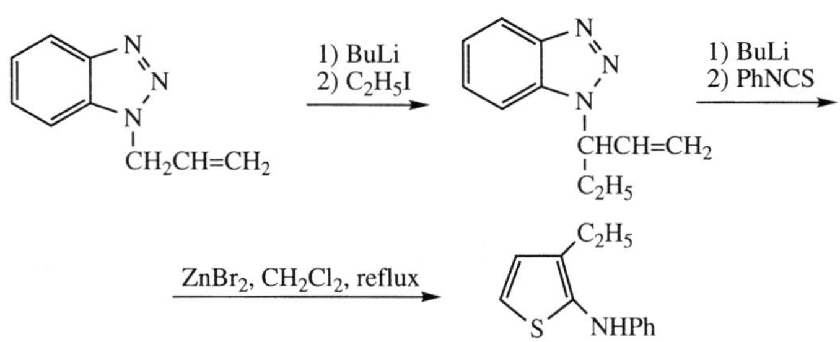

Either 2,5-bis(*N*,*N*-dialkylamino)thiophenes or 1-alkyl-2-*N*,*N*-dialkylamino-5-methylthiopyrroles can be obtained through the reaction of lithiated propargylic amines with isothiocyanates, followed by the addition of potassium *tert*-butoxide in *tert*-butanol and methyl iodide depending on the temperature during treatment with base [200].

Starting from *N*-acetylated diarylamines, alkyl 2-diarylaminothiophene-5-carboxylates are prepared *via* the 1-chlorovinamidium salts, which can be hydrolyzed and decarboxylated to the 2-diarylaminothiophenes [201].

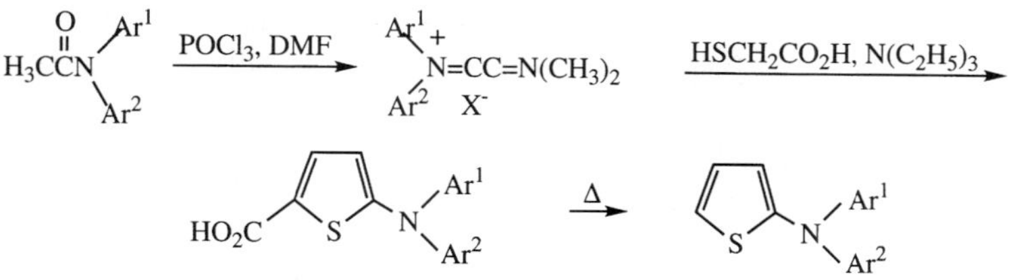

An excellent review on the synthesis of heterocyclic compounds, among them aminothiophenes, from metalated unsaturated compounds and isothiocyanates has been published recently [202].

5.1.2.3 Alkylsubstituted amino derivatives

5.1.2.3.1 Reduction of nitrothiophenes

Only very few simple alkylsubstituted aminothiophenes or their salts are prepared by reduction of nitrothiophenes. Some among them are 2-amino-5-methyl-, 2-amino- 4-methyl [197,203], and 2-amino-5-ethylthiophene.

5.1.2.3.2 From thienyllithium derivatives *via* azides

A new method for the preparation of 2-amino-5-methylthiophene is the reaction of 5-methyl-2-thienyllithium with a vinyl azide followed by hydrolysis [204].

5.1.2.3.3 From oxotetrahydrothiophenes

Among the 3-aminothiophenes the 5-methyl- [177], 5-isopropyl- [145], 5-tert-butyl- [145], 4-methyl [177], and 2-methyl derivatives [177] are prepared through amination of the corresponding oxotetrahydrothiophenes followed by aromatization. Also 2,5-dimethyl-3-aminothiophene has been synthesized.

5.1.2.3.4 By ring-closure reactions

A variety of ring-alkylated 2-*N*,*N*-dialkylaminothiophenes can be prepared through a thio-Claisen reaction of acetylenic allenic sulfides to a thioamide which is cyclized upon treatment with *tert*-butoxide in liquid ammonia [205].

5.1.2.4 Aryl- and heteroarylamino derivatives

5.1.2.4.1 Reduction of nitro derivatives

5-Amino-2-(4-fluorophenyl)-3-[4-(methylsulfonyl)phenyl]thiophene is prepared by reduction of the corresponding nitro derivative [206].

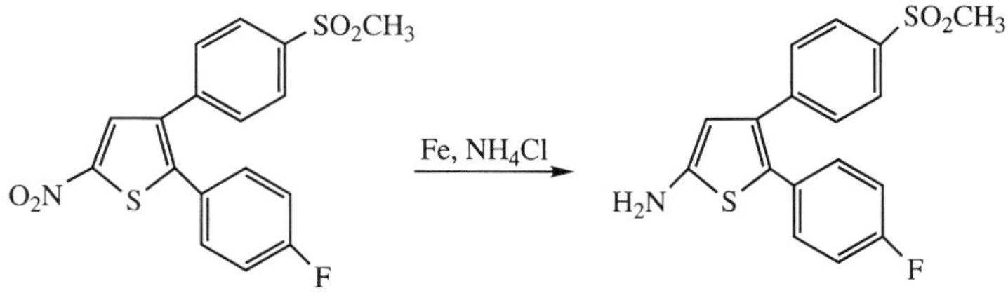

5-Amino-2-(4-fluorophenyl)-3-[4(methylsulfonyl)phenyl]thiophene [206]

A mixture of 5-nitro-2-(4-fluorophenyl)-3-[4-(methylsulfonyl)phenyl]thiophene (3.6 g, 9.55 mmol), iron powder (3.6 g), and ammonium chloride (0.36 g) in ethanol (58 ml) and water (22 ml) is stirred under reflux for 1 h. The insoluble material is removed by filtration and washed with *N*,*N*-dimethylformamide (40 ml). The filtrate is evaporated, the residue is triturated with water, the precipitate is collected and recrystallized from ethanol, giving 2.7 g (82%) of the title compound as a pale brown powder mp 207–209 °C.

5.1.2.4.2 From thienyllithium derivatives *via* azides

Aryl and hetarylsubstituted aminothiophenes are most conveniently prepared from the corresponding thienyllithium derivatives through reaction with tosyl azide giving thienyl azides which are reduced to the corresponding amino derivatives by hydrogen sulfide in ether [169]. In this way a number of isomeric amino phenylthiophenes and amino thienylthiophenes have been prepared.

5.1.2.4.3 By ring-closure reactions

Isomeric phenylbutanones [207] and also benzalacetone [207–217] and some other related compounds [214,216,218,219], upon treatment with sulfur in morfoline give 5-phenyl-2-morpholinothiophene under Willgerodt–Kindler conditions.

N,N-disubstituted 3,4-diaryl-2-aminothiophenes are readily prepared from *t*-amides of aryl tioacetic acids and phenacyl bromides followed by ring closure [156].

2-(Cycloalkylamino)-4- aryl or 3,4-annelated thiophenes are prepared by regioselective deprotonation–cyclization of acyclic and cyclic aroyl ketene *N,S*-acetals *via* dipole-stabilized carbanions in the presence of lithium diisopropylamide in tetrahydrofuran [220].

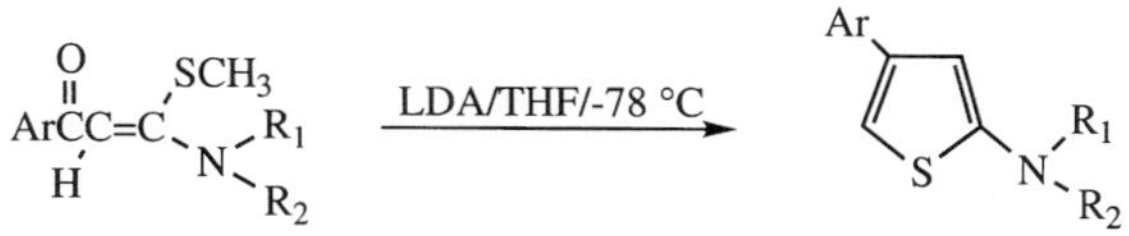

5.1.2.4.4 By various methods

2-Amino-4-phenylthiophene is also conveniently prepared by refluxing 2-amino-3-carbethoxy-4-phenylthiophene with 20% ethanolic potassium hydroxide.

5.1.2.5 Aminosubstituted thiophene aldehydes ketones, acids, esters, and nitriles

Such aminothiophenes are reasonably stable and have found use for the synthesis of many bicyclic derivatives of potential pharmaceutical interest. A great variety of synthetic useful methods are also available.

5.1.2.5.1 Through ring-closure reactions

The most useful method for the preparation of 2-aminothiophenes with carbonyl functions at the 3-position is the Gewald reaction (for reviews cf [221]). This is a cyclization reaction with three major variations. One procedure consists in the condensation of a mercapto aldehyde or ketone with an acetonitrile carrying an electron-withdrawing group, in the presence of an amine [222,223]. In simpler modifications a carbonyl derivative is reacted with sulfur and the nitrile in the presence of an amine or the Knoevenagel adduct of the carbonyl derivative and the nitrile is sulfurated in the presence of the amine [224–226]. In this way cyclopropyl-substituted crotonitriles are reacted with sulfur in the presence of a base yielding substituted 4-cyclopropyl-2-aminothiophenes [227]. Substituted acetonitriles used are esters of cyanoacetic acid [222–226,228,229], malonitrile [222,223,225,226,230–232], cyanoacetamide [222,223,225,231,234–236], *N*-substituted cyanoacetamide [223,234,237–240] as 2-aminothiophene-3-carboxanilides [241] and cyanoacetic acid [242]. A further group of nitriles that are extremely useful in Gewald syntheses are α-acylacetonitriles which yields 3-acylated 2-aminothiophenes [222–225,228,

236,243–247]. A wide range of carbonyl compounds can be used in the Gewald reaction, the only real requirement is that it undergoes the Knoevenagel condensation [248]. Ethyl 2-amino-4-methyl-5-isobutenylthiophene-3-carboxylate was prepared through the reaction of ethyl 2-cyano-3,6-dimethylhepta-2,4-dienoate with sulfur and triethylamine in absolute ethanol or from 5-methyl-2-morpholino-2,4-hexadiene, sulfur, and ethyl cyanoacetate in ethanol [249].

Ethyl 2-amino-4-methyl-5-isobutenylthiophene-3-carboxylate [249]

A mixture of 5-methyl-2-morpholino-2,4-hexadiene (544 g, 3.00 mol), sulfur (96 g, 3.00 mol), and ethyl cyanoacetate (340 g, 3.01 mol) in absolute ethanol is warmed to reflux for 72 h. The mixture is cooled and diluted with water (5 l). This solution is extracted with dichloromethane (3 × 1 l) and the combined organic phases are dried over magnesium sulfate. After evaporation a quantitative yield of the title compound is obtained as a dark oil. An analytical sample is obtained by flash chromatography using toluene as eluent. The combined clean fractions are evaporated and the residue is dissolved in boiling hexane. Upon cooling at room temperature a solid is formed, which is filtered off and rinsed with hexane giving the title compound mp 79–82 °C.

Ethyl 5-amino-2,4-bis(ethoxycarbonyl)-3-thiopheneacetate is prepared by the reaction of diethyl 3-oxopentanedioate with sulfur ethyl cyanoacetate and diethylamine in ethanol [250].

Ethyl 5-amino-2,4-bis(ethoxycarbonyl)-3-thiopheneacetate [250]

A mixture of diethyl 3-oxopentanedioate (20.2 ml, 0.1 mol), sulfur (3.2 g, 0.1 mol), ethyl cyanoacetate (11.3 ml, 0.1 mol), ethanol and diethylamine (10 ml) is stirred on a water-bath at 40–50 °C for 2 h. At the end of the reaction a reddish brown solid separates, which after cooling to room temperature is filtered off, washed with cold ethanol and dried.Upon recrystallization from ethanol 28.6 g (87%) of the title compound is obtained as shiny white needles mp 179 °C.

Ethyl 2-amino-4-phenylthiophene-3-carboxylate has been recently prepared by the reaction of arylidene ethyl cyanoacetate, sulfur, and morpholine in connection with the preparation of some new thienopyrimidine derivatives [251].

Ethyl 2-amino-4-phenylthiophene-3-carboxylate [251]

A mixture of arylidene ethyl cyanoacetate (10.5 g, 0.05 mol) and sulfur (1.3 g, 0.05 mol) in absolute ethanol (20 ml) in the presence of morpholine (10 ml, 0.011 mol) is heated at 40–60 °C for 4 h. After cooling the mixture is poured into ice-water and the solid product is filtered off and recrystallized from ethanol giving 9.88 g (80%) of the title compound as yellow crystals mp 98 °C.

A special example of application of the Gewald reaction is the reaction of the protected cyanoacethydrazide with cyclohexanone and morpholine or 2,5-dihydroxy-1,4-dithiane to give the thiophene derivative [252].

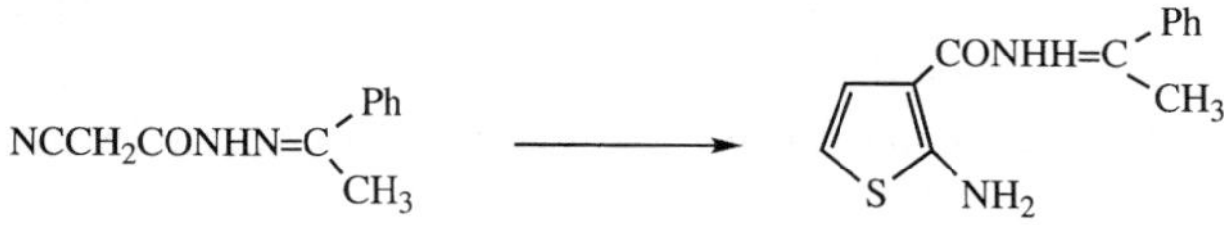

2-Amino-N′-(1-phenylethylidene)thiophene-3-carbohydrazide [252]

A mixture of *N*′-(1-phenylethylidene)cyanoacetohydrazide (12.8 g, 64 mmol), 2,5-dihydroxy-1,4-dithiane (4.85 g, 32 mmol), triethylamine (6.7 ml), and *N*,*N*-dimethylformamide (27 ml) is heated under stirring for 1.5 h. The dark reaction mixture is poured into water and the solid formed is filtered off and recrystallized from ethanol giving 12–15 g (74%) of the title compound mp 188.5–189 °C.

2-Amino-3-cyano-4-*N*-arylcarbamoyl-5-methylthiophenes are obtained by condensation of aryl amides of acetoacetic acid with sulfur and malonitrile in absolute alcohol [253].

The following reaction sequence is performed [254].

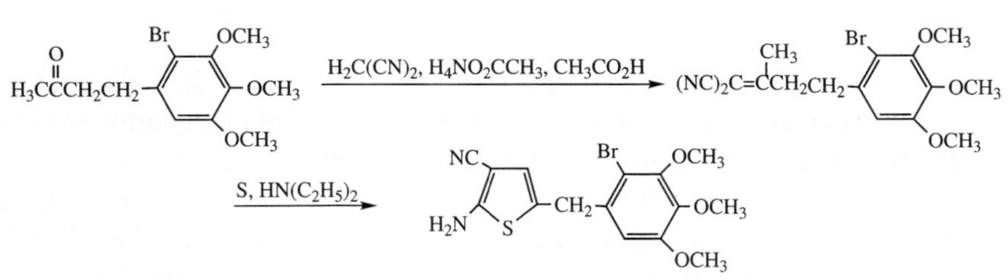

2-Amino-5-(2′-bromo-3′,4′,5′-trimethoxybenzyl)-4-methylthiophene-3-carbonitrile [254]

Diisopropylamine (0.15 ml, 1.09 mmol) is added dropwise to a stirred mixture of ylidenemalonitrile (400 mg, 1.1 mmol) and powdered sulfur (38 mg, 1.2 mmol) in 95% ethanol (4 ml). The stirring is continued at room temperature for 18 h, after which the volume is reduced to 0.5 ml by evaporation at 35 °C. A 2 *M* hydrochloric acid (20 ml) is added and the product extracted with ethyl acetate (3 × 40 ml). The combined organic phases are washed with water (30 ml), dried over magnesium sulfate, and evaporated. The residue is purified by flash chromatography on silica gel using hexane/ethyl acetate (3:1) as eluent. Recrystallization of the first fraction gives 121 mg (28%) of the title compound as orange prisms mp 124–126 °C.

4-(2-Dimethylaminovinyl)pyridine, as a surrogate for the unstable and very reactive 4-pyridineacetaldehyde, reacts in a Gewald reaction with sulfur ethyl cyanoacetate and morpholine to give 2-amino 3-carboethoxy-5-(4-pyridyl)thiophene [255].

2-Amino 3-carboethoxy-5-(4-pyridyl)thiophene [255]

A solution of 4-(2-dimethylaminovinyl)pyridine (28.2 g, 0.19 mol), cyanoacetate (21.6 g, 0.19 mol), and morfoline (5.0 ml) in absolute ethanol (250 ml) is heated at 80–85 °C under nitrogen for 3 h. The reaction mixture is then chilled in ice. The resulting crystals are collected, washed with hexane, and dried giving 38.6 g (82%) of the title compound mp 171.5–173.0 °C.

A recent example of the use of the Gewald reaction is the preparation of 4,5-disubstituted 2-amino-3-*tert*-butoxycarbonylthiophenes through the reaction of elemental sulfur with *tert*-butyl cyanoacetate, morpholine, and the appropriate ketone [256].

General procedure for the preparation of 4,5-disubstituted derivatives 2-Amino-3-tert-butoxycarbonylthiophenes [256]

Elemental sulfur (0.1 g-atom), *tert*-butyl cyanoacetate (0.1 mol), morfoline (0.1 mol), and the appropriate ketone (0.1 mol) are added to absolute ethanol (100 ml) and stirred at reflux for 18 h. The ethanol is removed *in vacuo* after cooling. The crude product is purified *via* flash chromatography using hexane/ethyl acetate (1:1) as eluent giving a straw-colored oil. 4,5-Dimethyl-2-amino-3-*tert*-butoxycarbonylthiophene is obtained after recrystallization from acetone (100 ml) in a yield of 10.9 g (48%) mp 66–67 °C.

A selective synthesis of ethyl 5-(2-acetoxyethyl)-2-amino-4-methyl-3-thiophenecarboxylate, a key intermediate for an efficient synthesis of 3-deazathiamine, by a modified Gewald reaction has been described [257].

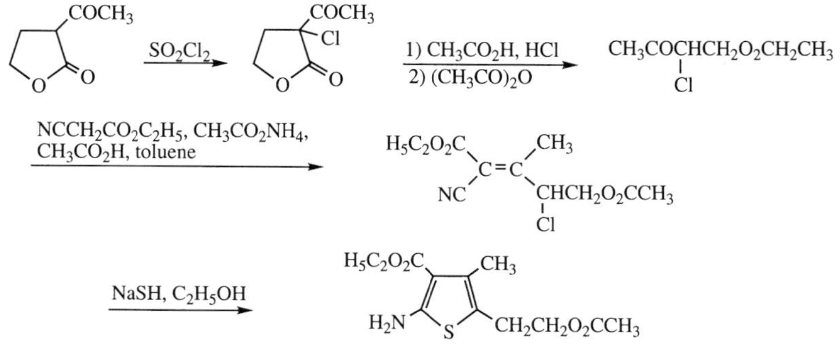

Ethyl 5-(2-acetoxyethyl)-2-amino-4-methyl-3-thiophenecarboxylate [257]

Acetic acid (9.9 g, 0.17 mol) is added dropwise to a stirred solution of sodium sulfide (13.3 g, 0.17 mol) in anhydrous ethanol (200 ml) at −40 °C followed by dropwise addition of a solution of the chloroalkene (18 g, 66 mmol) in ethanol (30 ml) over a period of 15 min. The stirring is continued for 1 h and the reaction mixture is then allowed to slowly warm to −10 °C, stirred for another hour and quenched by addition of ice-water. The precipitate formed is filtered off, washed with ice-water and dried giving 14.6 g (82%) of the title compound mp 64–65 °C.

Recently a novel four-component condensation reaction of ethyl cyanoacetate, sulfur, morpholine, and phenoxyacetone gave unexpectedly 2-amino-3-carbethoxy-4-methyl-5-morpholinothiophene, albeit in low yield. Also other cyclic secondary amines can be used [258].

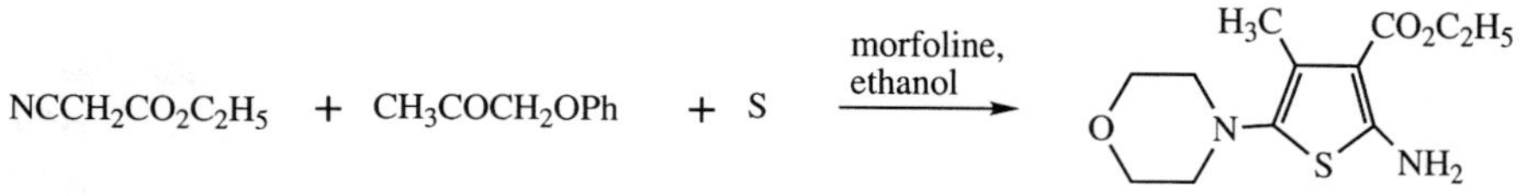

2-Amino-3-carbethoxy-4-methyl-5-morpholinothiophene [258]

To a solution of phenoxyacetone (2.0 g, 13,3 mmol) in ethanol (4 ml), ethyl cyanoacetate (1.5 g, 13.3 mmol) is added followed by sulfur (13.3 mmol) and morpholine (1.5 ml). The reaction mixture is stirred for 18 h, concentrated, and chromatographed on silica gel using dichloromethane/diethyl ether (95:5) as eluent giving 0.65 g (21%) of the title compound mp 137 °C.

The reaction of ethyl cyanobromoacetate with phenyl isothiocyanate and malonitrile in alkaline *N,N*-dimethylformamide followed by cyclization offers a method for the preparation of 4-amino-2-phenylamino-3,5-dicyanothiophenes [259].

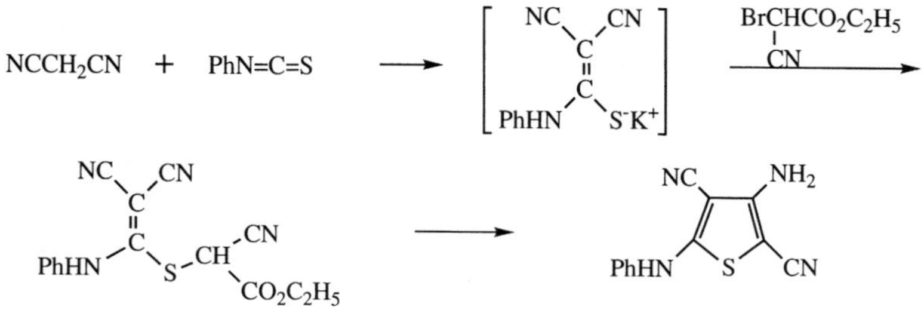

4-Amino-2-phenylamino-3,5-dicyanothiophene [259]

A solution of ethyl cyano (2,2 dicyano-1-phenylaminothio)acetate (3.12 g, 0.01 mol) in ethanol-sodium hydroxide, prepared by adding sodium hydroxide

(0.4 g, 0.01 mol) to ethanol (30 ml), is heated in a boiling water-bath for 6 h. The reaction mixture is diluted with cold water containing hydrochloric acid until pH 6, after which the solid product is collected by filtration. Crystallization from *N*,*N*-dimethylformamide gives 2.1 g (90%) of the title compound mp > 300 °C.

A series of new *N*,*N*-dialkylamino-5-thiophenecarboxylates have been prepared from acetamides using three different routes, *via* thioacetamides, 3-aminothioacrylamides, or 3-chloropropenimminium salts [260].

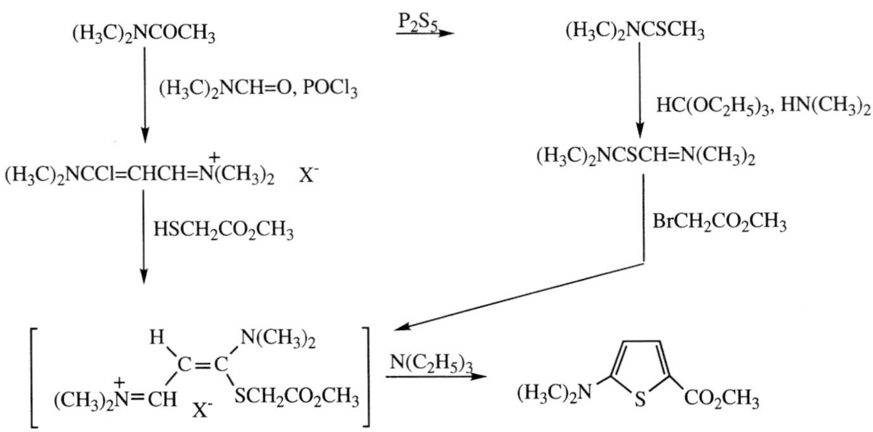

Especially the route *via* 3-aminothioacrylamides has recently been extensively used [261].

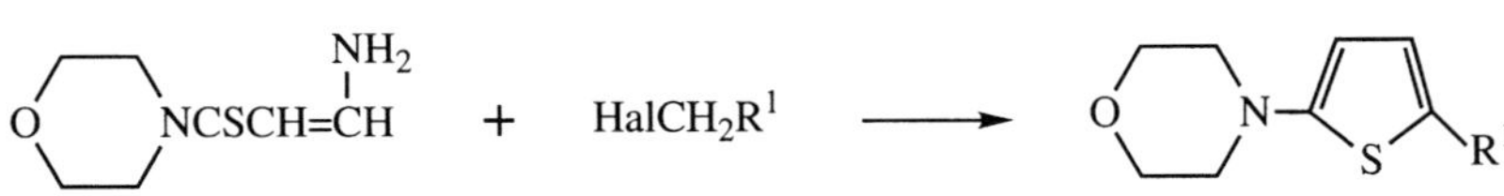

5-Acetyl-2-morpholinothiophene [261]

A mixture of 3-[dimethylamino(thioacryloyl)]morpholine (2.0 g, 10 mmol) and chloroacetone (0.92 g, 10 mmol) in acetonitrile (25 ml) is refluxed for 2 min and the triethylamine (10 ml) is added. After cooling and addition of water (5 ml) the precipitate formed is filtered off and recrystallized giving 1.7 g (80%) of the title compound mp 114–116 °C.

The reaction of 1,3-oxathiolium salts with active methylene compound is a useful route to 2-aminothiophenes having electron-withdrawing groups in the 3- and 5-positions [262–266]. Heating *N*-arylcyanothioacetamides in the presence of catalytic amounts of amines offers a route to 2-amino-3-arylthiocarbamoyl-4-cyano-5-arylaminothiophenes [267].

2-Amino-3-arylthiocarbamoyl-4-cyano-5-phenylaminothiophene [267]

A mixture of cyanothioacetanilide (1.0 g, 6.0 mmol) and triethylamine (0.15 g, 1.5 mmol) in ethanol (5 ml) is heated at 70–75 °C for 1.5–2 h. The reaction mixture is evaporated and after cooling ethanol (5 ml) is added to the residue. Rubbing with a glass rod gives a red precipitate, which is collected by filtration and washed with ethanol. After recrystallization from chloroform 0.65 g (65%) of the title compound is obtained mp 234 °C.

Treatment of 3-acylpropionamides with Lawesson's reagents gives mixtures of pyrroles and 2-aminothiophenes and therefore is not of preparative interest [268].

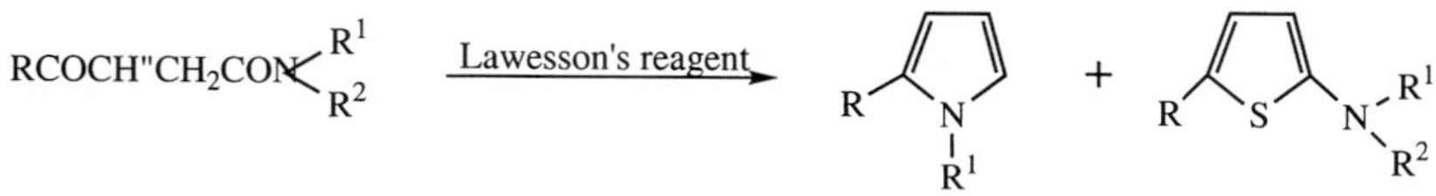

A very convenient method for the preparation of esters of 3-amino-2-thiophenecarboxylic acids is the Fiesselmann reaction (for review cf. [269]) consisting in the base-catalyzed condensation of esters of thioglycolic acid with α, β-dihalonitriles [270,271], α-halo-α, β-unsaturated nitriles [272,273], or with α, β-unsaturated olefins having suitable leaving groups in the β-position [274,275]. This reaction can be used for the preparation of methyl 3-amino-2-thiophenecarboxylate itself [275–277], as well as of various 4- or 5-alkyl or aryl substituted 3-amino-2-thiophenecarboxylates [275,276,278,279].

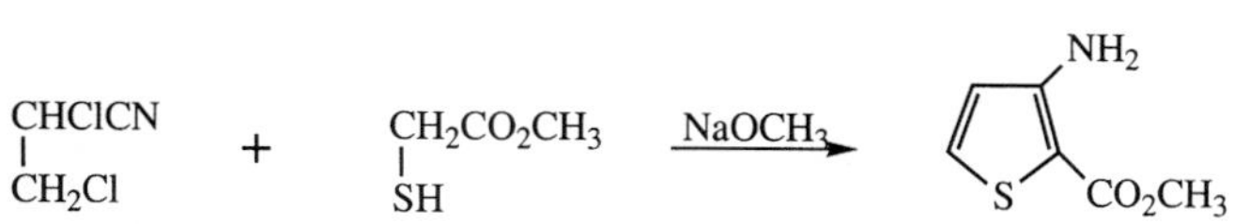

Methyl 3-amino-2-thiophenecarboxylate [277]

Methanol-free sodium hydroxide is prepared by suspending commercial sodium methoxide in a mixture of anhydrous benzene (300 ml) and diethyl ether (150 ml). This solution is evaporated to dryness and the procedure repeated twice. The residue is heated to 180 °C at 100 mm Hg for 15 min. The sodium methoxide so obtained (380 g) is suspended in anhydrous diethyl ether (2.5 l), after which hydroquinone (0.5 g) is added. The bluish suspension is cooled in an ice-bath and redistilled methyl thioglycolate (413 g, 3.90 mol) in anhydrous diethyl ether (300 ml) is added dropwise under vigorous stirring during 3 h. The color changes to pink during this operation. After stirring for an additional 0.5 h, α, β-dichloropropionitrile (322 g, 2.60 mol) in anhydrous diethyl ether (300 ml) is added dropwise during 5 h and with continued cooling. The color of the reaction mixture undergoes several changes and finally becomes brick-red. It is stirred for an additional 15 min and left overnight, after which it is diluted with water (3 l) and the pH adjusted to 5–6 by addition of acetic acid. The phases are separated and the aqueous phase is extracted three times with ether, the combined ether phases are dried, evaporated, and the excess of methyl thioglycolate is distilled off, after which the residue is distilled at reduced pressure giving 196 g (48%) of the title compound bp 100–102 °C/ 0.1 mm Hg, mp 64–65 °C from methanol containing a few drops of 2.5 *M* sodium hydroxide solution.

The bicyclic compound below prepared by Fiesselmann reaction of 2-cyano-3-[(ethoxycarbonylmethyl)thio]-*N*-methylmaleinimide is obtained by the reaction with ethyl mercaptoacetate [280].

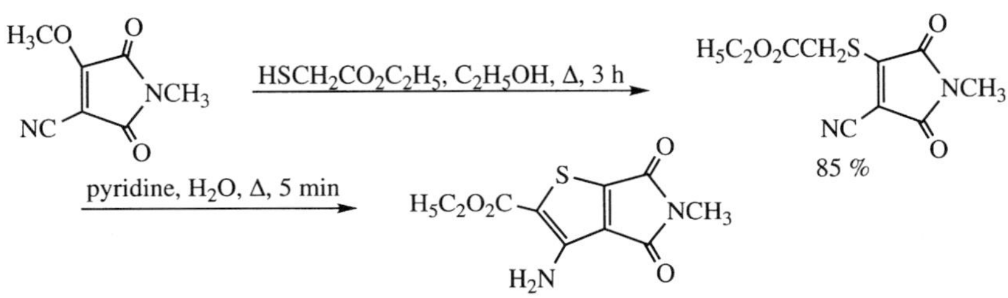

4-Amino-5-ethoxycarbonylthiophene-2,3-N-methyldicarboximide [280]

Water is added (20 ml) to a solution of 2-cyano-3-[(ethoxycarbonylmethyl)-thio]-*N*-methylmaleinimide (2.5 g, 10 mmol) in pyridine (5 ml) at room temperature. The reaction mixture is refluxed for 5 min and then cooled to room temperature. The precipitate formed is filtered off, washed with water, and dried *in vacuo* at 70 °C giving 1.1 g (43%) of the title compound mp 236 °C after recrystallization from acetic acid.

Another recent example is the reaction of the enoltosylate derived from 2-formyl-2-arylacetonitrile with methyl thioglycolate, giving good yields of 4-aryl-3-amino-2-thiophenecarboxylates [281].

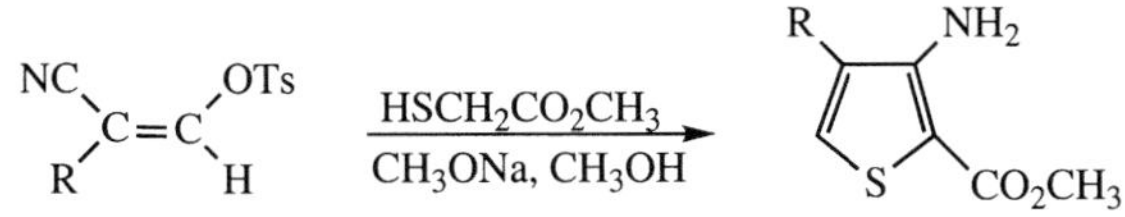

Methyl 3-amino-4-phenyl-2-thiophenecarboxylate [281]

Methyl thioglycolate (8.1 ml, 0.105 mol) is added dropwise to a 0.1 *M* solution of sodium methoxide in methanol (210 ml) at 10 °C. The stirring is continued for 30 min, after which 2-phenyl-3-(*para*-toluenesulfonato)acrylonitrile (29.1 g,

0.1 mol) is added dropwise at 10 °C. The reaction mixture is refluxed for 6 h and evaporated. The residue is recrystallized from water giving 12.6 g (54%) of the title compound mp 70 °C.

3-Amino-2-cyanothiophenes have been prepared by the use of 3-substituted propyne nitriles with α-mercaptoacetonitrile [278]. A recent modification of the Fiesselmann reaction was used for a convenient preparation of 5-aryl-3-amino-2-alkoxycarbonylthiophene by reacting β-chlorocinnamonitriles, prepared in a one-pot reaction from acetophenones, *N*,*N*-dimethylformamide, phosphoryl chloride, and hydroxylamine hydrochloride, with α-mercaptoacetic esters in the presence of a base [282].

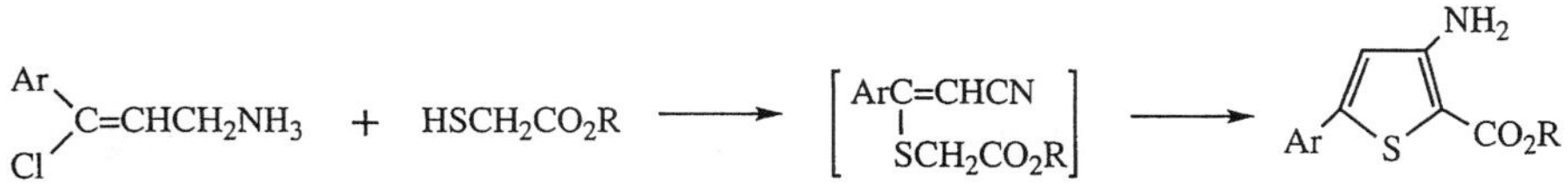

5-Phenyl-3-amino-2-ethoxycarbonylthiophene [282]

Ethyl α-mercaptoacetate (12.0 g, 0.1 mol) is added at room temperature. When the addition is complete β-chlorocinnamonitrile (16.4 g, 0.1 mol) is added to a stirred solution of sodium (2.3 g, 0.1 mol) in ethanol (100 ml). The reaction mixture is heated under reflux for 10 min, cooled, diluted with water (300 ml), and filtered. After drying and recrystallization 17.6 g (66%) of the title compound is obtained mp 101–104 °C.

A recent example is the preparation of methyl 3-amino-4-(3-phenylmethyl)thiophene-2-carboxylate from 2-benzyl-3-(*para*-toluenesulfonyl)acrylonitrile and methyl thioglycolate [283].

Methyl 3-amino-4-(3-phenylmethyl)thiophene-2-carboxylate [283]

Methyl thioglycolate (3.4 g, 31.5 mmol) is added to a solution of sodium methoxide (3.56 g, 66.0 mmol) in methanol (100 ml) at 10 °C and the reaction mixture is stirred for 15 min. 2-Benzyl-3-(*para*-toluenesulfonyl)acrylonitrile (9.39 g, 30.0 mmol) is added and the reaction mixture refluxed for 2 h. Most of the methanol is evaporated *in vacuo* and the residue is treated with water (400 ml). The red oil is taken up in ether, the ether solution is dried over sodium sulfate, and evaporated. The residue is chromatographed on silica gel (120 g) using toluene as eluent, giving 2.2 g (30%) of the title compound as colorless prisms after recrystallization from cyclohexane mp 101–102 °C.

3-Amino-2 acetylthiophenes are prepared by applying the Fiesselmann reaction to mercaptoacetone [273,274].

Fiesselmann reaction of 2-benzyl-3-(p-toluenesulfonyl)acrylnitrile with methyl thioglycolate in sodium methoxide in methanol is used for the preparation of methyl 3-amino-4-benzylthiophene-2-carboxylate. In a similar way the 4-(3-pyridinylmethyl) derivative was obtained [284].

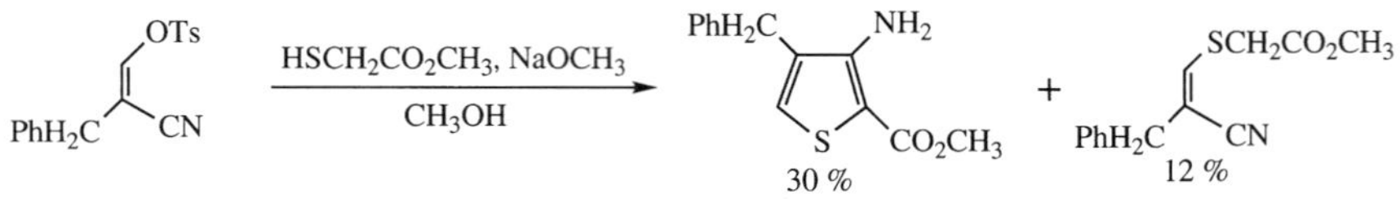

Methyl 3-amino-4-benzylthiophene-2-carboxylate [284]

Methyl thioglycolate (3.4 g, 31.5 mmol) is added to a solution of sodium methoxide (3.56 g, 66.0 mmol) in methanol (100 ml) at 10 °C and the reaction mixture is stirred for 15 min. Then tosylate (9.39 g, 30.0 mmol) is added and the mixture refluxed for 2 h. Most of the methanol is evaporated and the residue treated with water (400 ml). The red oil is extracted with ether (400 ml), the combined ether phases dried over sodium sulfate and evaporated. The residue is chromatographed on silica gel (120 g) using toluene as eluent. The product is recrystallized from cyclohexene giving 2.2 g (30%) of the title compound as colorless prisms mp 101–102 °C.

The reaction of dihydropyridine with ethyl thioglycolate in the presence of sodium carbonate gives the tricyclic compound, shown below [285].

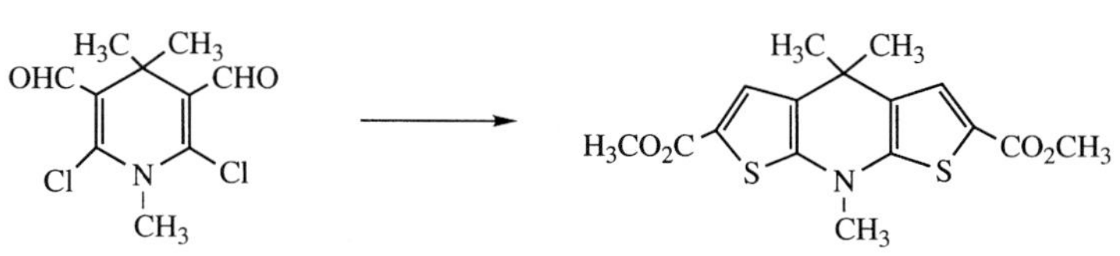

4,4,8-Trimethyl-4,8-dihydro-bis(2-ethoxycarbonylthieno [2,3-b:3′,2′-e]pyridine [285]

A solution of 2,6-dichloro-3,5-formyl-1,4,4-trimethyl-1,4-dihydropyridine (0.5 g, 2 mmol) in a minimum amount of hot ethanol is treated with ethyl thioglycolate (0.75 g, 6 mmol) and anhydrous sodium carbonate (0.64 g, 6 mmol). The reaction mixture is kept at 100 °C for 11 h. After reducing the volume to half, the mixture is left to crystallize. The solid formed is filtered off, washed with water, and recrystallized from ethanol giving 0.45 g (61%) of the title compound mp 215 °C.

2-Benzoyl-4-cyano-3-methyl-5-phenylaminothiophene is obtained in the following way [286].

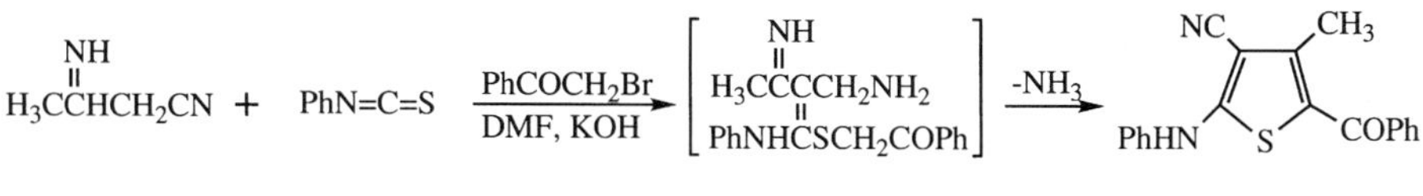

5-Substituted-4-amino-2-phenylthiophene-3-carbonitriles are prepared in one step by the reaction of β-chloro-α-cyanocinnamonitrile with α-oxthioles or successively with sodium sulfide and α-chlorocarbonyl compounds [287].

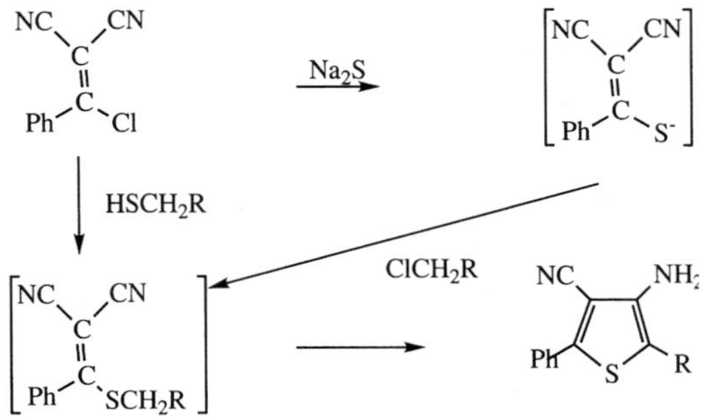

5-Alkyl and 5-arylsubstituted 3-amino-2-carbonylthiophenes can be prepared by the Gompper reaction consisting in the condensation of an activated acetonitrile with a thiocarbonyl compound such as a thioester to an enethiolate, which is alkylated and ring closed [288–291]. A modification, using benzylidene aminoacetonitrile in the condensation with thio- or dithiocarboxylic acids, followed by alkylation of the intermediate enethiolate with methyl chloroacetate and ring-closure constitutes a good method for the synthesis of methyl 3,4-diamino-2-thiophenecarboxylate [134,292]. Ring-closure reaction between 3-aminopropenethioamides and 2-bromoacetophenone in the presence of triethylamine gives 3-amino-2-benzoyl-4-cyano-5-phenylaminothiophene in 97% yield [293]. The reaction of cyanoalkenes with ethyl chloroacetate, α-haloketones, or chloroacetonitrile gives 3-amino-2-substituted thiophenes, such as 3-amino-2-acetyl-5-phenylaminothiophene [294].

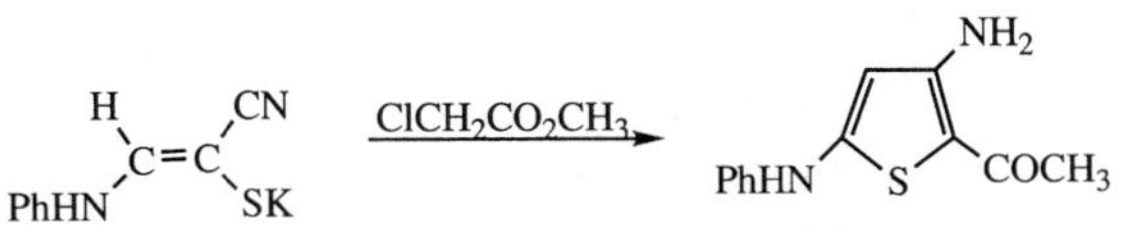

2-Acyl-3-aminothiophenes can be synthesized by the condensation of β-mercaptopropionitriles with α-haloketones followed by aromatization of the intermediate dihydrothiophene [295]. 2,4-Diamino-3,5-dicyanothiophene is conveniently prepared by the reaction of the self-condensation product of malonitrile with sulfur [296]. 2,5-Diamino-3,4-dicyanothiophene is best prepared by the reaction of tetracyanoethylene with hydrogen sulfide in the presence of pyridine [297–299].

5-Acetyl-2-amino-4-phenylthiophen-3-carbonitrile is conveniently prepared from α-cyano-β-thiocyanomethyl cinnamonitrile on refluxing in acetic acid/sulfuric acid [300,301]. Derivatives of 3-cyano-2,4-diaminothiophene are obtained upon treatment of substituted amino-2-cyano-3-methylthioacrylonitriles with lithium diisopropyl amide [302].

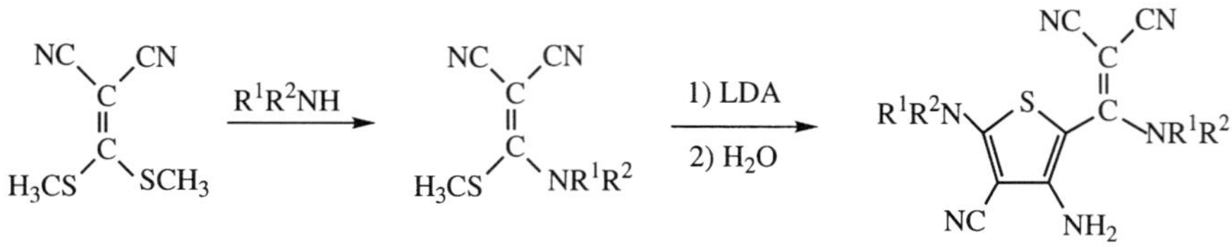

The reaction of 2,2-cyanoethenyl- and 1,2,2-tricyanoethenyl-substituted bromomethyl derivatives with 3-aminothioacrylamides is used for the preparation of 5-dicyanoethenyl- and 5-tricyanoethenyl-substituted 2-aminothiophenes [303].

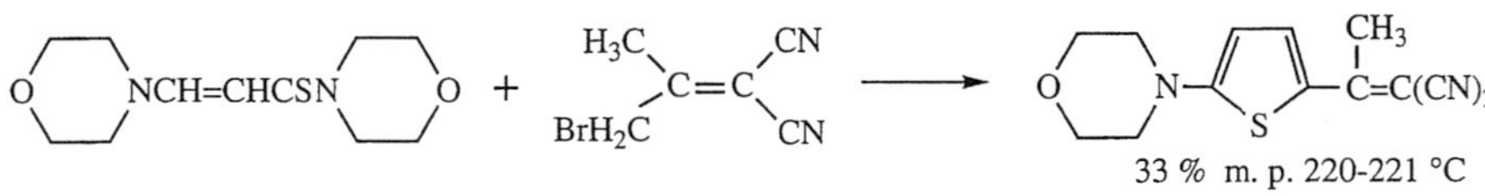

Preparation of cyanovinyl-substituted thiophenes [303]

A mixture of the cyanovinyl-substituted bromomethyl compound (0.1 mol) and 3-morphonolthioacrylamide (0.1 mol) in acetonitrile (100 ml) or methanol (100 ml) is refluxed for 10 min. After cooling triethylamine (26.0 g, 025 mol) for reactions in acetonitrile or sodium methoxide (13.5 g, 0.25 mol) for reactions in methanol is added and the mixture further refluxed for 10 min, after which it is cooled and diluted with water. The precipitate formed is filtered off.

5.1.2.5.2 Various methods

2-Aminothiophenes having electron-withdrawing groups in the 3- or 3,4-positions have been prepared by cycloaddition reactions of 2-amino-1,3-dithiolones followed by elimination of COS on heating [304,305].

The amination of 2- or 5-carboethoxy-3-oxotetrahydrothiophenes, *via* the oximes followed by aromatization has been used for the preparation of esters of 3-amino-2- and 5-carboxylic acids [306–309]. Another route to *N*-substituted 3-aminothiophenes consists in the condensation of the oxotetrahydrothiophene with amine followed by aromatization of the resulting enamine or mixtures of enamines [150].

5.1.2.5.3 Nucleophilic substitution of halothiophenes with amines

5-*N*,*N*-disubstituted thiophene aldehydes are prepared by the reaction of 5-bromo-2-thiophene aldehyde with the appropriate secondary amine [310].

Base-catalyzed reaction of weakly activated 2- or 3-bromothiophenes with dialkylamines in aqueous media is a good method for the preparation of 5 and 3-dialkylamino-2-thiophene aldehydes [311,312].

2-Diethylamino-5-thiophene aldehyde [310]

A solution of 2-bromo-5-thiophene aldehyde (3.8 g, 0.02 mol), diethylamine (2.9 g, 0.04 mol), and diethylamine hydrobromide (3.1 g, 0.02 mol) in absolute ethanol is heated in a sealed ampoule at 110 °C for 3 h. After decomposition with excess 30% sodium hydroxide solution the resulting oil is taken up in ether, the ether phase dried over sodium sulfate, and evaporated. Upon distillation the residue gives 2.6 g (69%) of the title compound bp 122 °C/ 0.1 mm Hg, mp 24 °C.

More recent work describes the preparation of 5-piperidino-2-thiophene aldehyde through the reaction of 2-bromo-5-thiophene aldehyde with piperidine and triethyl amine in toluene and of 5-piperidino-2-(2-thenoyl)thiophene from 5-chloro-2-(2-thenoyl)thiophene [313].

5-Piperidino-2-thiophene aldehyde [313]

In a 250 ml flask equipped with a condenser and under nitrogen, 2-bromo-5-thiophene aldehyde (8.4 g, 0.051 mol), triethylamine (3.7 ml, 0.051 mol), and piperidine (4.36 g, 0.051 mol) are dissolved in toluene (100 ml) and refluxed at 120 °C over a period of three days. The reaction mixture is then hydrolyzed with 0.1 *M* hydrochloric acid (5 ml), washed with 0.1 *M* hydrochloric acid, sodium bicarbonate, and water. The organic phase is dried over magnesium sulfate, evaporated and the brown solid obtained chromatographed on silica gel using hexane/ethyl acetate (8:2) as eluent giving 4.28 g (50%) of the title compound.

In connection with work on donor–acceptor substituted thiophenes, bithiophenes, and terthiophenes, of interest, for the preparation of nonlinear optical materials, a number of compounds, where the acceptor group is nitro or tricyanovinyl and the donor group various sec. amino groups, such as dimethylamino- and hydroxymethylpyrrolidine are prepared by nucleophilic subsitution with 2-iodo-5-nitrothiophene [314].

Palladium-catalyzed aminations of electron-deficient halothiophenes has recently been shown to be a useful reaction for the synthesis of a broad range of functionalized aminothiophenes. The reaction of methyl 3-bromo-2-thiophenecarboxylate has been carried out with a large number of primary and secondary amines in high yields. Also methyl 5-bromo-2-thiophenecarboxylate and 3-bromo-2-cyanothiophene give high yields of methyl 5-phenylmethylamino-2-thiophenecarboxylate and 3-butylamino-5-cyanothiophene, respectively. Also methyl 3-chloro-2-thiophenecarboxylate and the triflate of methyl-3-hydroxythiophenecarboxylate give high yields [315].

Thiophene	Amine	Product	Conditions[a]	Yields[b]
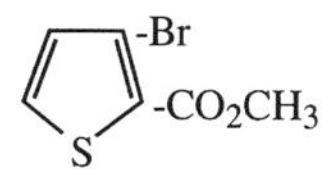	$BuNH_2$	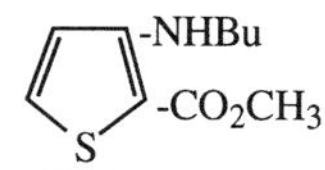	A	94 (100)
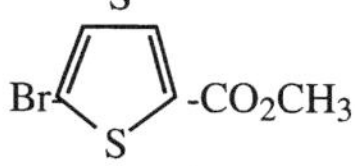	$PhNHCH_3$	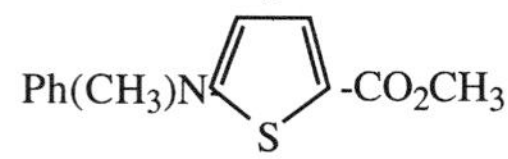	B	78 (85)
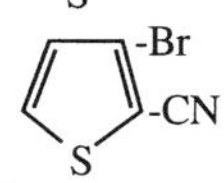	$BuNH_2$	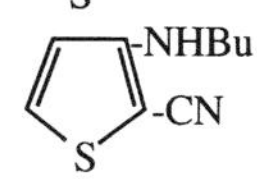	A	89 (100)
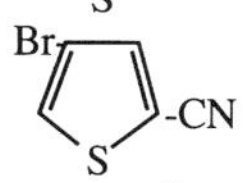	$PhNHCH_3$	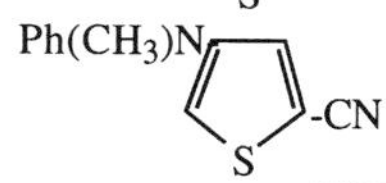	A or B	- (0)
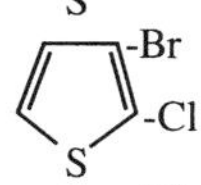	$BuNH_2$	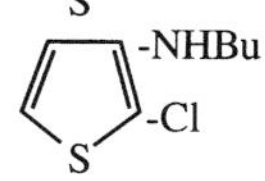	A	- (7)
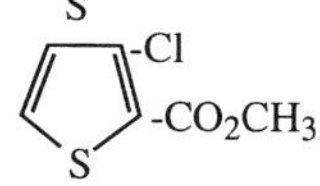	$BuNH_2$	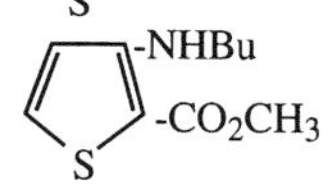	A or B	- (68)
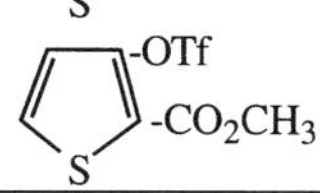	$BuNH_2$	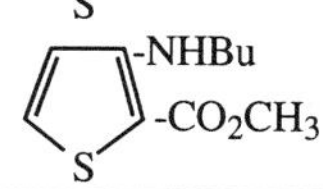	A or B	- (96)

[a]Reaction conditions: A 1.0 eq. thiophene, 1.2 eq. amine, 1.4 eq. Cs_2CO_3, 5 mol % Pd_2dba_3, 10 mol % BINAP, toluene, 110 °C, 20 h under nitrogen. B as A but 10 mol % $Pd(OAc)_2$ is used as palladium source.

[b]Isolated yields GC yields in parenthesis.

5.1.2.5.4 From aminothiophenes

3-Alkylamino-5-arylthiophenes are prepared by bubbling hydrogen gas into a methanolic solution of 2-acetyl-3-alkylamino-5-arylthiophenes [314].

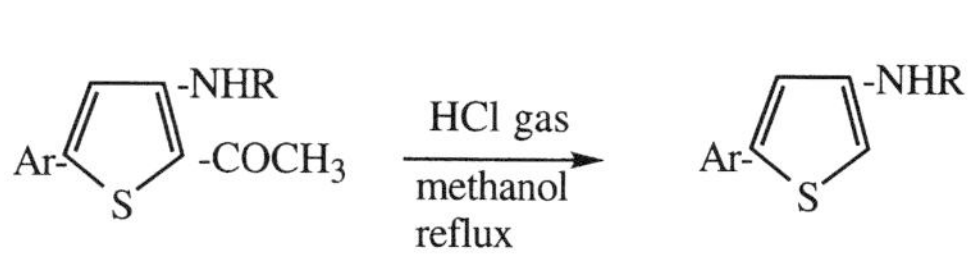

R = CH_3, C_2H_5 Ar = Ph, 4-$H_3COC_6H_4$, 3-ClC_6H_4, 3-$H_3COC_6H_4$

yields 67-99 %

3-Methylamino-5-phenylthiophene [314]

A dried hydrogen chloride gas is bubbled into a solution of 2-acetyl-3-methylamino-5-phenylthiophene (44 mg, 0.19 mmol) in anhydrous methanol

(5 ml), which is heated for 4 h at reflux. The color of the solution changes from yellow to dark blue. After cooling the reaction mixture-water (30 ml) is added and the product is extracted with dichloromethane (3 × 30 ml). The combined organic phases are dried over magnesium sulfate and evaporated. The residue is chromatographed on a silica gel column (2 × 10 cm) using ethyl acetate/hexane (1:2) as eluent giving 35 mg (98%) of the title compound mp 40–43 °C after recrystallization from dichloromethane/hexane.

5.1.2.5.5 Reduction of azidothiophenes

3-Amino-2-thiophene aldehyde, 3-amino-2-thiophenecarboxylic acid, 3-amino-2-cyanothiophene and 2-amino 3,5-diformylthiophene have been obtained by reaction of the corresponding azido derivatives with hydrogen sulfide [228].

The reaction of 3-azido-2-thiophene aldehyde with hexamethyldisilathiane in neat acetonitrile is a good method for the preparation of 3-amino-2-thiophene thioaldehyde. From 2-azido-3-thiophene aldehyde the reaction with hexamethyldisilathiane is used for the preparation of the 2-amino-3-thiophene aldehyde and 2-amino-3-thiophene thioaldehyde [316,317].

3-Amino-2-thiophene thioaldehyde [316]

3-Azido-2-thiophene aldehyde (0.14 g, 0.9 mmol) and hexamethyldisilathiane (0.38 ml, 1.8 mmol) in acetonitrile (15 ml) are stirred at room temperature for 1 h, after which more hexamethyldisilathiane (0.19 ml, 0.9 mmol) is added and the stirring is continued for 7 h. The reaction mixture is diluted with dichloromethane, washed with 10% sodium hydrogen carbonate and the solvent evaporated yielding the product, which is purified by column chromatography on neutral aluminium oxide using chloroform as eluent yielding 57.7 mg (40%) of the title compound as a red-brown solid.

5.1.2.5.6 Reduction of nitrothiophenes

Reduction of 2-nitro-4-thiophenecarboxylic acid with tin in hydrochloric acid can be used for the preparation of 2-amino-4-thiophenecarboxylic acid [96].

In connection with work on desoxynucleic acid analogs with an aromatic heterocyclic polyamide backbone, the Fmoc-protected methyl 5-amino-3-(1,2,3,4-tetrahydro-2,4-dioxopyrimidin-1-yl)-thiophene-2-carboxylate was recently prepared from the amino derivative, by reduction of methyl 5-nitro-3-(1,2,3,4-tetrahydro-2,4-dioxopyrimidin-1-yl)thiophene-2-carboxylate using hydrogen and 10% palladium on carbon as catalyst [318,319].

Methyl 5-amino-3-(1,2,3,4-tetrahydro-2,4-dioxopyrimidin-1-yl)-thiophene-2-carboxylate [318]

A 10% Pd/C (80.50 g) is added to a solution of methyl 5-nitro-3-(1,2,3,4-tetrahydro-2,4-dioxo-pyrimidin-1-yl)thiophene-2- carboxylate (0.50 g, 1.68 mmol) in degassed tetrahydrofuran/acetic acid (120 ml, 5:1), and the resulting mixture is hydrogenated with hydrogen at normal pressure. After 18 h, the mixture is filtered over *Celite*, the latter washed with methanol (200 ml) and toluene (100 ml), and the filtrate concentrated *ca* 60 ml. Then toluene (300 ml) is further added followed by evaporation to dryness. In order to remove traces of residual acetic acid, the residue is dissolved in dioxane/water (4:1) and the solution coevaporated with toluene and ethanol. This is repeated three times. One hundred milligrams of the resulting crude material (580 mg) containing 3% of acetic acid is dissolved in dioxane/water (4:1), adsorbed on silica gel and purified by flash chromatography using acetonitrile/dichloromethane/triethylamine (5:5:0.05) as eluent, giving 62 mg of the pure title compound as a yellow powder mp 254–256 °C.

Classical reduction of 2-cyanoethylcarbamoyl-5-methyl-4-nitrothiophene with ferrous sulfate heptahydrate and barium hydroxide gives 4-amino-2-cyanoethylcarbamoyl-5-methylthiophene [136].

4-Amino-2-cyanoethylcarbamoyl-5-methylthiophene [136]

A stirred mixture of ferrous sulfate heptahydrate (800 g, 2.88 mol) and barium hydroxide octahydrate (910 g, 2.88 mol) in water (3300 ml) is heated to 90 °C and a suspension of 2-cyanoethylcarbamoyl-5-methyl-4-nitrothiophene (44 g, 0.184 mol) in acetone (200 ml) is added during 5 min. The mixture is kept at 90 °C for 2 h and then filtered hot. The solid material is washed well with boiling water. The pH of the combined filtrates is adjusted to 10 with 10 *M* sodium hydroxide solution. After continuous ether extraction for 48 h, 25.5 g (73%) of the title compound is obtained as prisms from ethyl acetate mp 170–171 °C.

Reduction of a peptide of 5-nitro-2-thiophenecarboxylic acid with sodium dithionite in aqueous ethanol can be used for the preparation of the amino derivative [143]. Reduction of 2-formyl- and 2-acetyl-4-nitrothiophene with iron and hydrochloric acid can be used for the preparation of the 2-formyl and 2-acetyl-4-amino derivative, which were trapped with ethoxymethylene malonate [137].

5.1.2.5.7 Through modification of the substituents

3-Amino-2-thiophenecarbohydrazides and 2-amino-3-thiophenecarbohydrazides are prepared by the reaction of 3-amino-2-thiophenecarboxylates and 2-amino-3-thiophenecarboxylate with 85% aqueous hydrazine [252,320].

3-Aminothiophene-2-carbohydrazide [252]

A solution of methyl 3-aminothiophene-2-carboxylate (6.90 g, 44 mmol) in 95% ethanol (12.5 ml) is treated with 85% aqueous hydrazine (13 ml). The reaction mixture is refluxed for 24 h and evaporated. The residue is washed with water and air-dried giving 5.83 g (86%) of the title compound mp 157–158.5 °C.

Upon reactions with aldehydes or ketones the carbohydrazones are obtained [320]. *N*-(2-carbomethoxythienyl)imidates are prepared by the reaction of methyl 3-amino-2-thiophenecarboxylate with ortho esters [321].

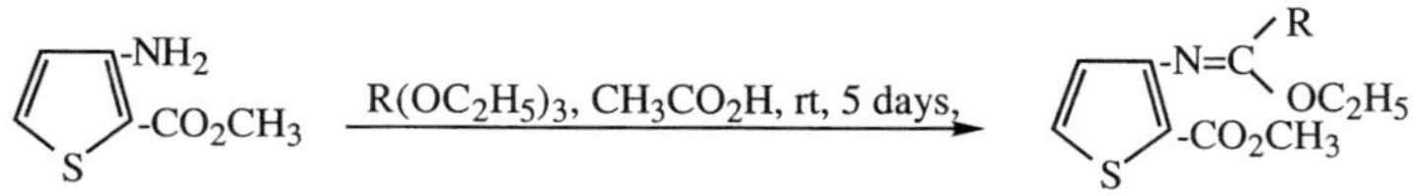

Methods for the transformation of 3-amino-2-carbomethoxythiophene to 3-amino-2-carbamoylthiophene has recently been described [322].

3-Amino-2-carbamoylthiophene [322]

A mixture of 3-amino-2-carbomethoxythiophene (1.61 g, 11.3 mmol) and 29% aqueous ammonium hydroxide (40 ml) is heated in a sealed pyrex tube (85 ml internal volume) at 160 °C for 2 h. The cooled mixture (containing one tan liquid phase) is evaporated to a volume of 2 ml or less. Addition of diethyl ether to the viscous residue produces a solid precipitate, which is collected by filtration, washed with ether, and dried in air, giving 388 mg (27%) of the title compound as white prisms mp 122–124 °C after recrystallization from 95% ethanol and sublimation 110 °C/0.03 mm Hg.

Trans 1-(2-diethylaminothienyl)-2-(2-thiazolyl)ethene is prepared through the Wittig reaction of 5-diethylamino-2-thiophene aldehyde with the 2-thiazole-methylphosphonium salt and potassium *tert*-butoxide in benzene. 5-Dibutyl-aminothiophenes with various electron-withdrawing groups in the 2-position are prepared by the condensation of 2-dibutylamino-5-formylthiophene with various CH-acidic heterocycles [323].

The reaction of 3,4-dichloro-1,2,3-dithiazolium chloride with methyl 3-amino-2-thiophenecarboxylate gives methyl [*N*-(4-chloro-5*H*-1,2,3-dithiazole-5-yliden)]-2-thiophenecarboxylate [324].

Methyl [N-(4-chloro-5H-1,2,3-dithiazole-5-yliden)]-2-thiophenecarboxylate [324]

Pyridine (702 mg, 8.87 mmol) in dichloromethane (10 ml) is added to a mixture of methyl 3-amino-2-thiophenecarboxylate (633 mg, 4.03 mmol) and

3,4-dichloro-1,2,3-dithiazolium chloride (888 mg, 4.26 mmol) in dichloromethane (100 ml) during 20 min. The reaction mixture is stirred at room temperature for 2 h, worked-up and chromatographed using hexane/ethyl acetate (3:1) as eluent giving 788 mg (67%) of the title compound, which is recrystallized from a mixture of dichloromethane and hexane and yellowish needles mp 126–127 °C are obtained.

Hydrolysis of methyl 2-amino-4,5,6,7-tetrahydrobenzo[*b*]thiophenecarboxylate with sodium hydroxide in methanol, followed by reaction with benzyl bromide in *N*,*N*-dimethylformamide at room temperature gives the benzyl ester, which is acylated with ethyl oxalyl chloride in methylene chloride in the presence of triethyl amine to give benzyl 2-(ethyloxalyl-amino)-4,5,6,7-tetrahydrobenzo[*b*]thiophene-3-carboxylate, which can be further methylated on nitrogen [325].

Benzyl 2-(ethyloxalylamino)-4,5,6,7-tetrahydrobenzo[b]thiophene-3-carboxylate [325]

A solution of ethyl oxalyl chloride (4.48 g, 25.0 mmol) in dichloromethane (40 ml) is added dropwise to a stirred solution of benzyl 2-amino-4,5,6,7-tetrahydrobenzo[*b*]thiophene-3-carboxylate (6.80 g, 23.7 mmol) and triethylamine (2.48g, 28.1 mmol) in dichloromethane (120 ml) with ice cooling. The reaction mixture is stirred at 0 °C for 0.5 h and at room temperature for 20 h. Work-up with dichloromethane furnishes 9.2 g (100%) of the title compound as a brown oil.

Acylation of methyl 3-amino-2-thiophenecarboxylates with 3-chloropivaloyl chloride and intramolecular ring closure using sodium carbonate in *N*,*N*-dimethylformamide is used for the preparation of *β*-lactam derivatives such as *N*-[2-methoxycarbonyl-4-(4-fluorophenyl)thien-3-yl]-3,3-dimethylazetidin-2-one [326].

N-[2-methoxycarbonyl-4-(4-fluorophenyl)thien-3-yl]-3,3-dimethylazetidin-2-one [326]

Sodium carbonate (1.14 g, 0.0108 mol) is added to a solution of methyl 3-(3-chloro-2,2-dimethylpropionylamino)-4-(fluorophenyl)thiophene-2-carboxylate (4.0 g, 0.0108 mol) in anhydrous *N*,*N*-dimethylformamide (40 ml). With stirring under nitrogen the reaction mixture is heated at 160 °C for 1 h, after which it is poured into ice-water (150 ml). The product is extracted with ethyl acetate (150 ml) and after drying and evaporation the residue is washed with diethyl ether/hexane giving 2.90 g (80%) as a white solid mp 84 °C.

Methyl 3-(3-benzoylthioureido)-2-thiophenecarboxylate is prepared by the reaction of methyl 3-amino-2-thiophenecarboxylate with benzoyl isothiocyanate [327].

Methyl 3-(3-benzoylthioureido)-2-thiophenecarboxylate [327]

A solution of benzoyl isothiocyanate in acetone (60 mmol in 30 ml) is added to a solution of methyl 3-amino-2-thiophenecarboxylate (4.71 g. 30 mmol) in anhydrous acetone (25 ml). The reaction mixture is refluxed for 1 h and kept at 0 °C overnight. The precipitate is filtered off giving 7.3 g (76%) of the title compound mp 144–145 °C after recrystallization from methanol.

Thieno-fused 3,1-oxazine-4-ones are prepared by the reaction of methyl 3-amino-2-thiophenecarboxylate and 2-amino-3,5-dicarbethoxy-4-methylthiophene with dichloromethylene dimethylammonium chloride [328].

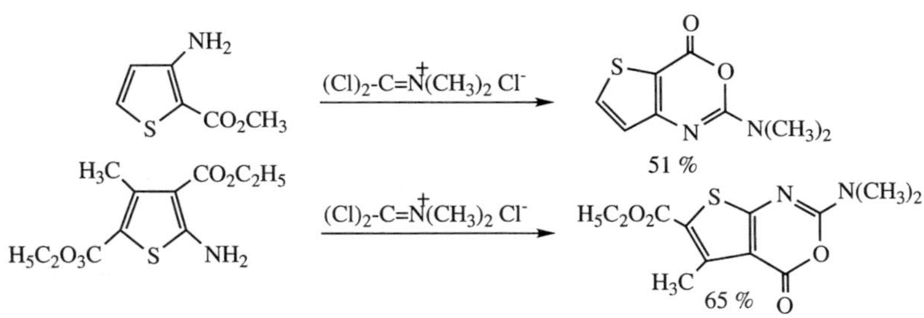

2-(Dimethylamino)-4H-thieno[3,2-d][1,3]oxazin-4-one [328]

A solution of methyl 3-amino-2-thiophenecarboxylate (0.79 g, 5.0 mmol) and dichloromethylene dimethylammonium chloride (0.90 g, 5.5 mmol) in anhydrous chloroform is stirred between −10 °C and −5 °C for 6 h. The reaction mixture is then stirred overnight at room temperature. After removal of the solvent the residue is chromatographed on silica gel using ethyl acetate/petroleum ether (1:1) as eluent giving 497 mg (51%) of the title compound mp 126–128 °C.

5.1.2.5.8 By electrophilic substitution of aminothiophenes

2-*N*,*N*-dialkylamino-5-formylthiophenes are prepared by Vilsmeier formylation of 2-*N*,*N*-dialkylaminothiophenes in 50–80% yield [156,158], while 3-*N*,*N*-dialkylaminothiophenes are formylated in lower yields in the 2-position [158].

5-Dimethylamino-2-thiophene aldehyde [158]

Phosphorus chloride (2 ml) is added to 2-dimethylaminothiophene (1.27 g, 10 mmol) in dimethylformamide (5 ml) under cooling. The yellow reaction mixture is heated at 50 °C for 30 min poured into ice-water (100 ml) neutralized with sodium carbonate. After standing for some time the precipitate is filtered off and recrystallized from hexane/benzene giving 0.93 g (60%) of the title compound mp 88–90 °C.

Stannic chloride catalyzed *C*-glycosylation of *N*-formyl- and *N*-acetyl-4-amino-3-carboalkoxythiophenes with 1-*O*-acetyl-tri-*O*-benzoyl-β-D-ribose give the nucleosides [329].

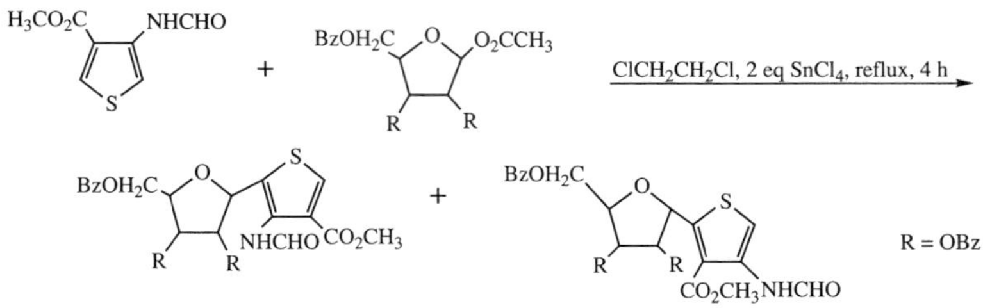

Refluxing in tetrahydrofuran of 3-methylamino-5-phenylthiophene and α β-unsaturated esters such as methyl acrylate, (*E*)-methyl crotonate, dimethyl fumarate, diethyl maleate and ethyl propiolate gives 1-methyl-3,4-dihydrothieno[2,3-*e*]pyridine-2-ones, and/or 1-methylthieno[2,3-*e*]pyridine-2-ones depending upon the structure of the esters [314].

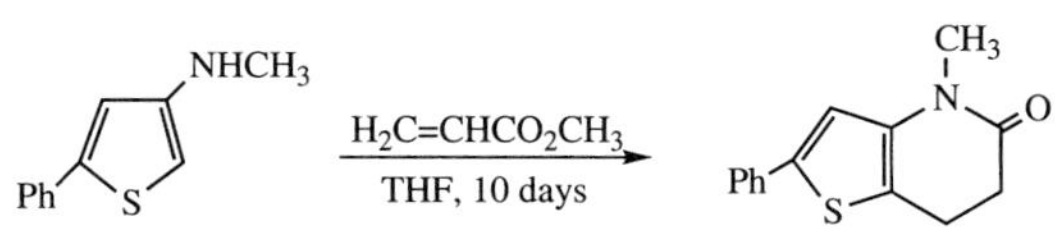

1-Methyl-3,4-dihydro-6-phenylthieno[2,3-e]pyridin-2-one [314]

Methyl acrylate (28 mg, 0.33 mmol) is added to a solution of 3-methylamino-5-phenylthiophene (51 mg, 0.33 mmol) in tetrahydrofuran (5 ml). The reaction mixture is heated at reflux for 10 days. Removal of the solvent *in vacuo* gives a residue which is chromatographed on a silica gel column (1 × 30 cm) using ethyl acetate/hexane (1:2), giving an unknown mixture. Subsequent elution with the same solvent mixture (9:1) gives 29 mg (44%) of the title compound as pale-yellow crystals mp 130–131 °C after recrystallization from ethyl acetate/hexane.

On the other hand the same reaction with α,β-unsaturated nitriles such as acrylonitrile and tetracyanoethene gives the corresponding thiophenes derivatives [314].

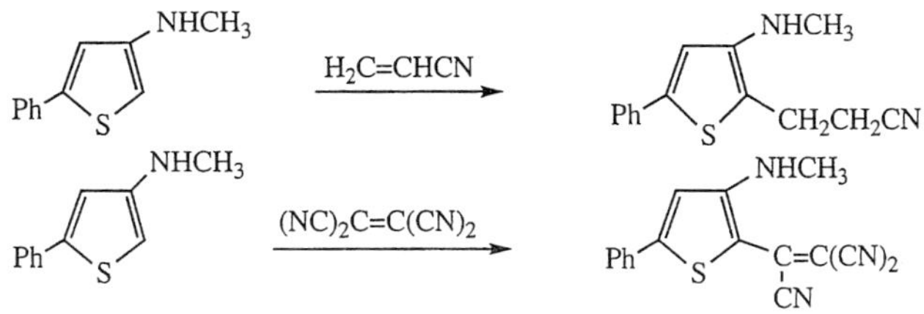

On the other hand acetylation of 2-dimethylaminothiophene in the presence of aluminium chloride only gives a low yield of the 5-acetyl derivative [330].

5-Acetyl-2-amino-3-cyano-4-phenylthiophene can be prepared from 2-acetyl-3-cyano-4-phenylthiophene upon reaction with glacial acetic acid and sulfuric acid [331]. The extremely reactive 2,4-diaminothiophene derivatives react smoothly at room temperature with isocyanates and isothiocyanates in the 5-position to give amides and thioamides, respectively. Reaction with acid chlorides in the presence of triethyl amine gives ketones [194].

3,5 Bis(N-isopropylanilino)-2-thienyl phenyl ketone [194]

N,N′-Diisopropyl-*N,N′*-diphenyl-2,4-thiophenediamine (1 g, 2.8 mmol) is added to an ether solution with an equimolar amount of triethylamine. To this solution benzoyl chloride (2.86 mmol) is added and the reaction mixture is stirred at room temperature for 4 h. The triethylamine hydrochloride is removed by filtration and the filtrate evaporated. After recrystallization of the residue from aqueous methanol, 0.9 g of the title compound is obtained.

Aminoalkylation in the 5-position of a number of 2-amino-4-aryl-3-thiophenecarboxylic esters has been achieved [332].

5.1.2.5.9 By alkylation of carbonyl-substituted aminothiophenes

In most cases 2-aminothiophenes with electron-withdrawing groups in the 3-position are alkylated only with difficulty [229] and in many cases not at all [333,334]. On the other hand if the acidic hydrogen of 3-amino-2-benzoyl-4-cyanothiophene is removed by treatment with sodium hydride alkylation can be achieved [335]. Also some 2-amino-3-carboethoxythiophenes and 2-amino-3-cyanothiophenes can be alkylated under these conditions [336].

Another way of achieving *N–C*-bonds is by the Michael type reactions. Both 3-amino-2-carbomethoxythiophene and 2-amino-3-carbomethoxythiophene react with dimethyl acetylenedicarboxylate and give upon heating in *tert*-butanol dimethyl 2-(2′-methoxycarbonyl-3′-thienylamino)-2-butenedioate and dimethyl 2-(3′-methoxycarbonyl-2′-thienylamino)-2-butenedioate in 62 and 26% yield, respectively [337].

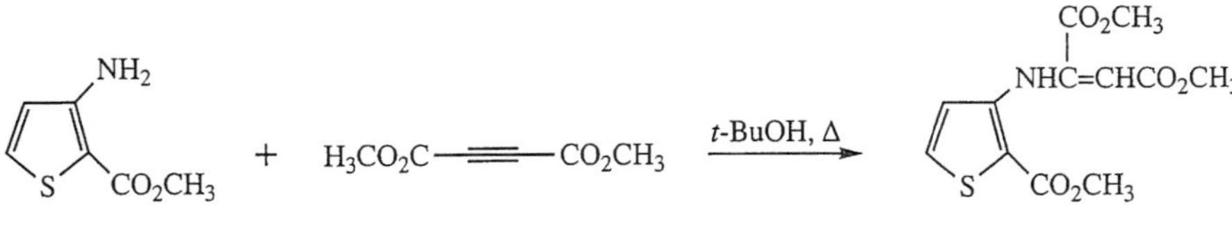

Dimethyl 2-(2′methoxycarbonyl-3′-thienylamino)-2-butenedioate [337]

A mixture of methyl 3-aminothiophene-2-carboxylate (12.60 g, 80.16 mmol) and dimethyl acetylenedicarboxylate (18.30 g, 128.7 mmol) in *tert*-butanol (220 ml) is heated under reflux for 16 h. The solvent is removed under reduced pressure and the residue is recrystallized from cyclohexane to afford 14.87 g (62%) of the title compound as yellow needles mp 83–85 °C.

Similarly allene-1,3-dicarboxylate is added to 2-amino-3-thiophenecarboxylates to give the thienylaminopent-2-enedioates [338]. The reaction of 3-amino-2-carbethoxythiophene with the potassium salt of methyl dicyanoacetate is a method for the preparation of methyl 3-[(1-amino-2-cyano-2-(methoxycarbonyl)ethenyl)amino]thiophene-2-carboxylate [339].

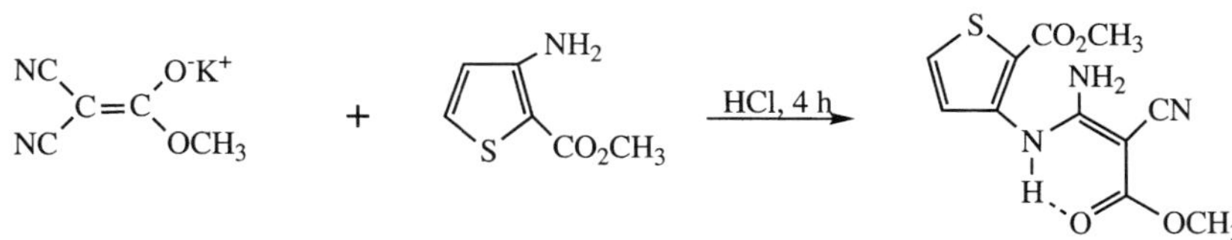

Methyl 3-[(1-amino-2-cyano-2-(methoxycarbonyl)ethenyl)amino]thiophene-2-carboxylate [339]

The potassium salt of methyl dicyanoacetate (1.62 g, 10 mmol) is added to a suspension of methyl 3-aminothiophene-2-carboxylate (1.57 g, 10 mmol) in concentrated hydrochloric acid 0.9 ml) and water (20 ml). The reaction mixture is refluxed for 4 h. After cooling, the precipitate is collected by filtration, washed with acetone and then recrystallized from acetone giving 1.17 g (42%) of the title compound as white crystals mp 265 °C (dec).

2-[Bis(methylthio)methylene amino]-3-cyanothiophene is prepared by the reaction of 4-imino-2-thioxo-2,4-dihydro-1*H*-thieno[2,2-*d*]-(1,3)-thiazine with methyl iodide in 1 *M* aqueous potassium hydroxide solution [340].

2-[Bis(methylthio)methylene amino]-3-cyanothiophene [340]

Methyl iodide (2 g) is added to a stirred solution of 4-imino-2-thioxo-2,4-dihydro-1*H*-thieno[2,2-*d*]-(1,3)-thiazine (0.509 g, 2 mmol) in 1 *M* aqueous potassium hydroxide solution (20 ml). The stirring is continued for 1 h, after which the precipitate formed is filtered off and recrystallized from methanol giving 0.51 g (90%) of the title compound mp 118 °C.

5.1.2.6 Nitroamino derivatives

5.1.2.6.1 Through the reaction of halonitrothiophenes with ammonia and amines

The reaction of *ortho*-halonitro thiophenes with ammonia [341], primary aliphatic amines [341–347], piperidine [105,341–343,348–364], and other secondary amines [105,342,343,365.370] is the most convenient method for the preparation of aminonitrothiophenes, *N*-alkylamino-, and *N*,*N*-dialkylamino-nitrothiophenes.

Nitration of methyl 3-acetylamino-2-thiophenecarboxylate gives a mixture of methyl 3-acetylamino-5-nitro-2-thiophenecarboxylate and methyl 3-acetyl-4-nitro-2-thiophenecarboxylate, which is easily separated by fractional crystallization [318].

Methyl 3-(acetylamino)-5-nitro- and 4-nitrothiophene-2-carboxylate [318]

A 100% nitric acid (15 ml) is added to a mechanically stirred solution of methyl 3-(acetylamino)thiophene-2-carboxylate (15 g, 75.3 mmol) in 95% sulfuric acid (150 ml) cooled to −35 °C. The reaction mixture is stirred at −25 °C for 45 min and allowed to warm to 0 °C. The viscous liquid is poured on ice (250 g) and the resulting aqueous phase extracted with dichloromethane (5 × 300 ml). The combined organic phases are dried over magnesium sulfate and evaporated. The residue is dissolved in hot toluene to saturation at 100 °C and the solution is left at 4 °C overnight. The crystals formed are filtered off giving 6.4 g (35%) of methyl 3-(acetylamino)-4-nitrothiophene-2-carboxylate as faint yellow needles mp 172–174 °C. Concentration of the filtrate to half volume followed by standing at 4 °C for another night yields a second crop containing both nitro compounds. The resulting filtrate is evaporated and dried *in vacuo* giving 9.35 g (51%) of methyl 3-(acetylamino)-5-nitrothiophene-2-carboxylate as a dark yellow powder mp 115–116 °C.

Triethylamine reacts with 2-bromo-5-nitrothiophene to give the corresponding trimethylammonium salt [371]. *N*-(3,5-dinitro-2-thienyl) amino acids are prepared by the reaction of 2-halo-3,5-dinitrothiophene with amino acids [345,372,373]. Also *N*-arylaminonitrothiophenes can be prepared in high yields through the reaction of aromatic amines with halo nitrothiophenes [26,342, 343,355,364,368,371]. The reaction of 5-acetyl-2-chloro-3-nitrothiophene with morpholine, pyrrolidine, and dimethylamine hydrochloride in *N*,*N*-dimethylformamide at room temperature can be used for the preparation of the corresponding dialkyl amines in excellent yields [374,375].

5-Acetyl-2-phenylamino-3-nitrothiophene [375]

2-Chloro-3-nitro-5-acetylthiophene (512 mg, 2.5 mmol) is stirred with aniline (465 mg, 5.0 mmol) in *N*,*N*-dimethylformamide (10 ml) at room temperature under argon. When tlc analysis indicates that the reaction is complete, the reaction mixture is poured into ice-water and the precipitate formed recrystallizes from 96% ethanol giving 537 mg (82%) of the title compound as orange needles mp 138–140 °C.

5.1.2.6.2 By alkylation of aminonitro derivatives

2-(*N*-alkyl-*N*-methyl)-3,5-dinitrothiophene is prepared by the reaction 2-*N*-alkyl-3,5-dinitrothiophenes with dimethyl sulfate in the presence of potassium carbonate [376], while *N*-methylation of *N*-aryl-amino-3,5-dinitrothiophenes can be achieved by treatment with diazomethane in diethyl ether [377].

5.1.2.6.3 By nitration of aminothiophenes

Free aminonitrothiophenes cannot be prepared by direct nitration of aminothiophenes, as they normally decompose upon attempted nitration, except for 3-trimethylthienyl ammonium salts, which mainly yields the 5-nitro isomer [378]. Aminonitrothiophenes are therefore in many cases best prepared by alkaline or acidic hydrolysis of 2-acetylamino-5-nitro- and 2-acetylamino-3,5-dinitrothiophene [67].

2-Amino-5-nitrothiophene [67]

2-Acetylamino-5-nitrothiophene (1.0 g, 5.38 mmol) is slowly added to aqueous potassium hydroxide (20 g) solution at 30 °C. The stirring is continued for 1 h, after which water (25 ml) is added and the reaction mixture gently neutralized at 0 °C with concentrated hydrochloric acid. The precipitate formed is filtered off and purified by chromatography on silica gel using ethyl acetate/petroleum ether (2:1) as eluent giving 0.55 g (71%) of the title compound as a yellow solid.

2-Amino-3,5-dinitrothiophene [67]

2-Acetamido-3,5-dinitrothiophene (3.0 g, 13 mmol) is added with stirring to 50% sulfuric acid. The stirring is continued at 100 °C for 4 h and after cooling the reaction mixture is poured into ice/water (300 ml) with vigorous stirring. The precipitate formed is filtered off, washed with water and dried overnight in a drying box (15 mm Hg, room temperature). The crude product is purified by chromatography on silica gel using ethyl acetate/petroleum ether as eluent giving 1.3 g (53%) of the title compound mp 179–180 °C.

5.1.2.6.4 *Via* ring-closure reactions

The reaction of nitrothioacetamides with α-haloketones in the presence of 1,8-diazabicyclo[5.4.0]undec-7-ene (DBU) is a good method for the preparation of 2-dialkylamino-3-nitro-4-substituted thiophenes [379].

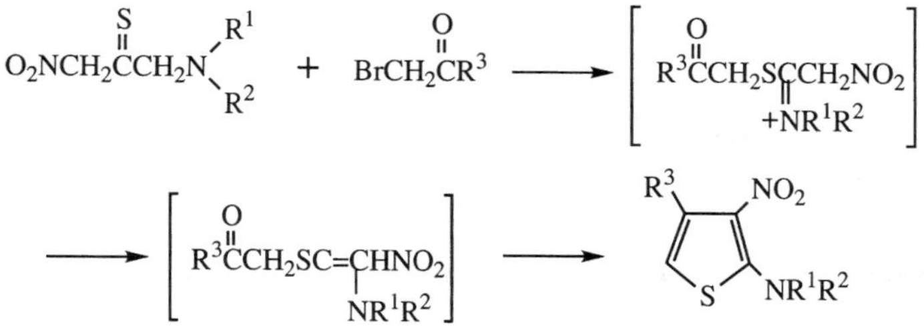

General procedure for the synthesis of 2-dialkylamino-3-nitro-4-substituted thiophenes [379]

Nitrothioacetamide (2 mmol) is solved in benzene (8 ml) and DBU (340 mg, 2.2 mmol) is added to this solution, which is stirred under nitrogen for 10 min. The α-haloketone (2.2 mmol) is dissolved in benzene (7 ml) and added to the reaction flask. The reaction mixture is stirred at 60 °C for 12 h and after evaporation, the residue is chromatographed on silica gel using benzene/petroleum ether (3:1) as eluent.

3-Nitro-4-phenyl-2-*N*-pyrrolidinylthiophene is obtained in 98% yield mp 114 °C.

A recently described general route to 2-acyl- and 2-aroyl-3-alkylamino-5-arylthiophenes and 2-ethoxycarbonyl-3-methylamino-5-arylthiophenes consists in the reaction of thioaroylketene-*S*,*N*-acetals with 1,3 dicarbonyl compounds in the presence of mercury(II) acetate [380].

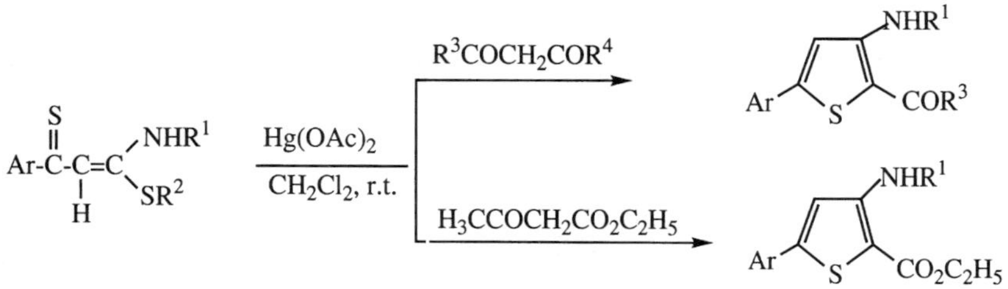

5.1.2.7 Acylaminothiophenes

As mentioned in Section 'Hydrolysis of acetamido- and other protected derivatives', acylaminothiophenes are obtained through the Beckmann, Schmidt, and Hofmann reactions as intermediates in the synthesis of aminothiophenes. Reductive acylation in the presence of acetic anhydride has been used for the preparation of acetylaminothiophene derivatives [118,381–385]. In this way the triacetyl derivative of 2,3,4-triaminothiophene was prepared from 3,4-diacetylamino-2-nitrothiophene [139].

Methyl 2-acetylamino-4-thiophenecarboxylate [385]

To a solution of methyl 2-nitro-4-thiophenecarboxylate (3.74 g, 20 mmol) in acetic acid (100 ml) and acetic anhydride (15 ml) powdered reduced iron (6 g) is added. Under vigorous stirring the reaction mixture is heated to 55–60 °C, until the beginning of a spontaneous temperature rise. A light colored precipitate soon appears, the reaction mass thickens and the temperature reaches 75–80 °C. The stirring and heating at 75–80 °C are continued for another 6–8 h. After cooling to 50–60 °C water is added until the precipitate is dissolved. The remaining iron and resinification products are filtered off, the filtrate evaporated and for complete removal of the acetic acid water (100 ml) is added for coevaporation. The dark residue is acidified with 6 *M* hydrochloric acid. This solution is left in refrigerator for several hours, the crystals formed collected and recrystallized from alcohol giving 2.5 g (69%) of the title compound mp 164.0–164.5 °C.

If the aminothiophenes are prepared by other methods, acylation of aminothiophenes is readily achieved by reaction with the appropriate acyl halide, acid anhydride or for *N*-formyl derivatives, formic acid. Steinkopf found that 2-aminothiophene is acetylated in ether solution with acetic anhydride [100], a method which has also been used later by others [130,386]. Also 3-aminothiophene [19] is readily acetylated at low temperature. Acetic anhydride in pyridine has been used in some other cases [179]. Mixed diacetyl derivatives of 2,3-diaminothiophene can be prepared from 3-acetylamino-2-nitrothiophene through reduction with iron in acetic acid followed by acetic anhydride [347]. The reaction of 2-amino-3-thiophenecarboxylates or 2-amino-3-thiophenecarboxamide with succinic anhydride is used for the preparation of *N*-(2-thienyl)succinic acid mono amides [387].

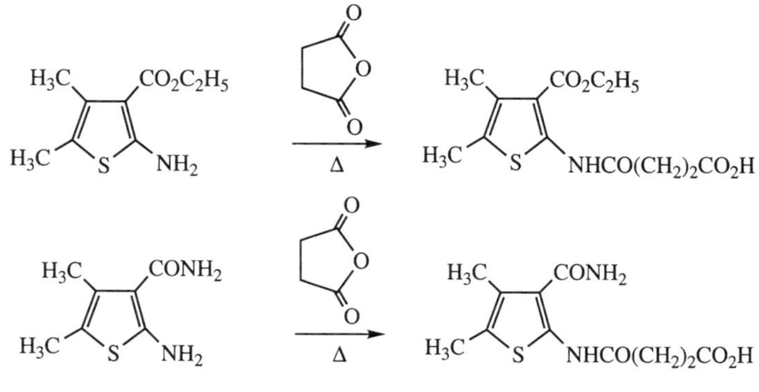

In order to prepare acylaminothiophenes with electron-withdrawing groups, more vigorous conditions have to be used, such as refluxing in acetic anhydride alone or in solvents such as dioxane or chloroform and/or in the presence of bases such as triethylamine pyridine or *N,N*-dimethylaniline. In slowly reacting cases acetyl chloride and pyridine are often preferred [236,336,388,389].

2-Acetamido-3-cyano-4,5-dimethylthiophene [389]

A solution of 2-amino-3-cyano-4,5-dimethylthiophene (45.7 g, 0.3 mol) and acetic anhydride (33.7 g, 0.33 mol) in acetic acid (50 ml) is stirred under reflux for 30 min. The reaction mixture is cooled to room temperature and poured onto crushed ice (500 g). The precipitate is filtered off, washed with water until neutral, dried, and recrystallized from ethanol to afford 43.7 g (75%) of the title compound mp 211–213 °C.

Reaction of 2-amino-3-(2-chlorobenzoyl)thiophene with phthalimidobenzoyl chloride in boiling chloroform is used for the preparation of 3-(2-chlorobenzoyl)-2-(phtalimidoacetamido)thiophene. Its hydrazinolysis in a mixture of methanol and chloroform at 60 °C stops at 2-(aminoacetamido)-2-(2-chlorobenzoyl)thiophene [390].

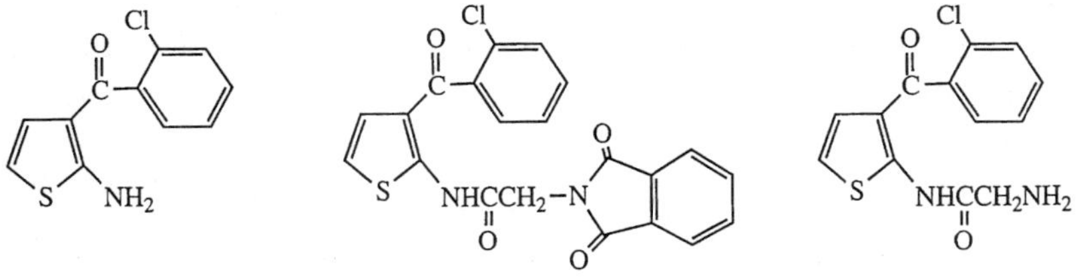

3-(2-Chlorobenzoyl)-2-(phthalimidoacetamido)thiophene [390]

A solution of 2-amino-3-(2-chlorobenzoyl)thiophene (94.3 g, 0.40 mol) in chloroform (525 ml) is stirred and treated over 5 min with phthalimidoacetyl chloride (88.0 g, 0.40 mol), slowly heated to the boiling point and refluxed for 2 h. After standing overnight a polymeric substance is filtered off and the filtrate is filtered through a column of aluminium oxide (60 g), evaporated to a volume of about 220 ml and diluted with ethanol (160 ml). After standing at room temperature for two days the product is filtered off giving 72.4 g (43%) of the title compound mp 197–200 °C after recrystallization from chloroform/ethanol.

2-(Aminoacetamido)-2-(2-chlorobenzoyl)thiophene [390]

A solution of 3-(2-chlorobenzoyl)-2-(phthalimidoacetamido)thiophene (105.7 g, 0.25 mol) in a mixture of methanol (2100 ml) and chloroform (330 ml) is stirred and treated at 45 °C over 10 min with a solution of 100% hydrazine hydrate (24.8 g, 0.5 mol) in water (110 ml) and the mixture is stirred for 1.5 h at 60 °C.

After standing overnight the precipitated phthalylhydrazine is filtered off and the filtrate evaporated *in vacuo*. The residue is extracted twice with dilute aqueous ammonia (750 ml), the product is filtered off, dissolved in chloroform (250 ml), and the solution filtered through a column of neutral aluminium oxide (150 g). The filtrate is evaporated and the residue crystallized from ethanol (210 ml) giving 42.2 g (58%) of the title compound mp 159–161 °C.

3-Amino-2-carbethoxythiophenes have been acylated with 3-chloropivaloyl chloride in pyridine and then cyclized to β-lactams, using *N,N*-dimethylformamide and anhydrous sodium carbonate [391].

Methyl 3-(3-chloro-2,2-dimethylpropionylamino)-4-(4-fluorophenyl)thiophene-2-carboxylate [391]

A solution of methyl 3-amino-4-(4-fluorophenyl)thiophene-2-carboxylate (10 g, 39.8 mmol) and 3-chloropivaloyl chloride (6.17 g, 39.8 mmol) in dioxane (100 ml) is refluxed with pyridine (3.14 g, 39.8 mmol) under nitrogen for 1 h. The solvent is evaporated and the residue dissolved in water and extracted with diethyl ether. After removal of the solvent 11 g (75%) of the title compound is obtained as a solid mp 110 °C.

N-[2-Methoxycarbonyl-4-(4-fluorophenyl)thien-3-yl]]-3,3-dimethylazetidin-2-one [391]

A solution of methyl 3-(3-chloro-2,2-dimethylpropionylamino)-4-(4-fluorophenyl)thiophene-2-carboxylate (4 g, 10.8 mmol) in *N,N*-dimethylformamide (40 ml) and sodium carbonate (1.14 g, 10.8 mmol) is heated to 160 °C with stirring under nitrogen for 1 h. This reaction mixture is poured into ice-water (150 ml) and extracted with ethyl acetate (150 ml). After evaporation the residue is washed two times with ether/hexane giving 2.90 g (80%) of the title compound mp 84 °C.

2-Amino-3-cyanothiophene exists in solution in the amino form, but undergoes protonation at the exocyclic nitrogen in dimethylsulfoxide/trifluoroacetic acid and protonation at the 5-position in trifluoroacetic acid. It reacts with other electrophiles as a typical aromatic amine giving rise to *N*-acylated products such as the bromoacetylamino [392] and the chloroacetylamino derivative [393].

Ethyl 2-Bromoacetylamino-4,5-dimethyl-3-thiophenecarboxylate [392]

A solid potassium carbonate (0.50 g, 5 mmol) is added to a solution of ethyl 2-amino-4,5-dimethyl-3-thiophenecarboxylate (1.99 g, 10 mmol) in anhydrous tetrahydrofuran (50 ml) followed by bromoacetyl bromide (2.02 g, 10 mmol). The reaction mixture is kept at 40 °C overnight, after which the solvent is evaporated and the residue washed with water. The solid obtained is collected, air dried, and recrystallized from ethanol giving 1.91 g (60%) of the title compound mp 119–121 °C.

The reaction of 3-amino-2-carbomethoxythiophene with acrylic acid chloride, methylacryclic acid chloride, and cinnamic acid chloride in benzene and triethylamine gives the corresponding amides, which are methylated on nitrogen using methyl iodide, potassium carbonate and benzyl triethylammoniuim chloride in acetone. This method is also used for the preparation of 2-methoxycarbonyl-3-(*N*-acetyl *N*-alkynyl)aminothiophenes from 3-acetylamino-2-carbomethoxythiophene and the proper alkynyl bromide [394].

Both 2-amino-3-carbomethoxy- and 4-amino-3-carbomethoxythiophene have been acylated with chloroacetyl chloride and bromoacetyl chloride using triethylamine in methylene chloride at −20 °C [395].

Methyl 2-bromoacetamidothiophene-3-carboxylate [395]

A solution of methyl 2-aminothiophene-3-carboxylate (4.0 g, 25.4 mmol) and triethylamine (3.2 g, 31.6 mmol) in dichloromethane (200 ml) is cooled to −70 °C and bromoacetyl bromide (6.2 g, 31 mmol) is added in one portion. The stirred reaction mixture is kept at −20 °C for 20 h and then allowed to warm to room temperature before being worked-up in the usual way. Evaporation of the dichloromethane phase gives a solid residue, which after recrystallization from methanol with the aid of charcoal gives 5.93 g (84%) of the title compound mp 111–113 °C.

In a few cases esters have been used for the acylations of aminothiophene as in the preparation of α-aminoacylthiophenes from ethyl glycinate [396]. In a similar fashion ethyl cyanoacetate acylates 2-amino-3-cyanothiophene [397]. *N*-Formyl aminothiophenes have also been prepared by the Curtius rearrangement of the thiophenecarbonyl azide in refluxing formic acid [172,398].

Many substituted acetylaminothiophenes have been prepared by electrophilic substitution reactions. Mercuration of 2-acetylaminothiophene with mercuric(II) chloride can be controlled by variation in the proportion of the reagent and the reaction conditions to give consecutive mercuration at the 5-, 3-, and finally the 4-position [399]. Also deactivated derivatives such as 2-acylamino-3-thiophenecarboxylic esters easily give the 5-chloromercury derivative [223].

Nitration of methyl 3-acetylamino-2-thiophenecarboxylate gives a mixture of the 4- and 5-nitro derivatives [400].

Methods for the preparation of 2-acetylamino-5-formylthiophene by different variants of the Vilsmeier reaction have been studied extensively [23,115,171,401,402]. In the beginning the reagent was prepared from *N*-methylformanilide and phosphorus oxychloride [223], but later it was found to be more convenient and less costly to use *N,N*-dimethylformamide and phosphorus oxychloride instead. The best conditions for selective preparation of 2-acetylamino-5-formyl- and 2-acetylamino-3,5-diformylthiophene use *N,N*-dimethylformamide and phosphorus oxychloride under carefully controlled reaction conditions [171,403]. Under these conditions secondary reactions such as formamidine formation are minimized [171].

General procedure for the synthesis of acetamidothiophenecarbaldehydes [403]

N,N-Dimethylformamide (2.74 g, 37.5 mmol) is cooled to 0 °C in a flask equipped with a drying tube and phosphoryl chloride (5.76 g, 37.5 mmol) is added dropwise with stirring. To this solution redistilled 1,2-dichloroethane or 1,1,2,2-tetrachloroethane (25 ml) is added followed by a solution of the acetamidothiophene (37.5 mmol) in the same chloroethane solvent (75 ml). The reaction mixture is stirred at room temperature for 15 min and then refluxed for 15 min, after which it is cooled and poured onto ice (*ca* 300 g) with stirring. Sodium acetate (8.2 g, 0.1 mol) is added and the mixture boiled for 20 min, cooled and separated. The aqueous phase is extracted with dichloromethane and the combined organic phases are washed with saturated aqueous sodium hydrogen carbonate. After drying over magnesium sulfate the solution is evaporated and chromatographed on alumina using light petroleum/chloroform (4:1) as eluent. 3-Acetamido-2-thiophene aldehyde is obtained in a yield of 88% after recrystallization from ethyl acetate mp 72 °C (decomp).

Acylations of acetylaminothiophenes proceed smoothly. 5-Acetyl-2-acetylaminothiophene has been obtained in excellent yield by the reaction of 2-acetylaminothiophene with acetic anhydride in the presence of perchloric acid [404]. 2-Acetyl-3-acetylaminothiophene is prepared by acetylation of 3-acetylaminothiophene with acetyl chloride using aluminium trichloride as catalyst [172]. Several other acylaminothiophenes have been acylated with acetyl chloride and various benzoyl chlorides in the presence of aluminium trichloride [384,405,406]. 2-Acylaminothiophene with substituents in the 5-position can be aminoalkylated in the 3-position in good yields [407,408].

5-Benzamido-2,3-tetramethylene-4-benzylaminomethylthiophene [407]

A solution of benzylamine (8.6 g, 0.08 mol) and formalin (1.2 g, 0.04 mol) in dioxane (25 ml) is refluxed for 20 min, after which a solution of 2-benzamido-4,5-tetramethylenethiophene (5.1 g, 0.02 mol) in dioxane (25 ml) is added.

The reaction mixture is then refluxed for 2.5 h and the solvents removed. The residue is recrystallized from methanol giving 4.8 g (71%) of the title compound mp 143–144 °C.

Metalation of trifluoroacetylaminothiophene with butyllithium followed by reaction with various electrophiles can be used for functionalization of the 2-position [409].

Nitration of 2-acetylaminothiophene leads to a mixture of the 3- and 5-nitro derivatives under mild conditions [67,100,410] and the 3,5-dinitro isomer under more vigorous conditions [410]. 3-Cyano-, 3-acyl-, 3-ethoxycarbonyl-, 3-carboxy, and 3-carbamoyl-2-acetylaminothiophenes smoothly give the 5-nitro derivatives in good yield [223,411].

3-Nitro-2-acetylaminothiophenes with electron-withdrawing groups in the 5-position are conveniently obtained by nitration [100,412].

Methyl 2-acetamino-4-methyl-5-nitro-3-thiophenecarboxylate [223]

A fuming nitric acid (4.5 ml) in acetic acid (12 ml) is added dropwise to methyl 2-acetamino-4-methyl-3-thiophenecarboxylate (4 g, 19 mmol) in acetic acid anhydride (40 ml) under stirring at −15 °C. After 10 min the reaction mixture is poured into ice. The precipitate is filtered off and recrystallized from toluene giving 1.3 g (26%) of the title compound mp 176–178 °C.

3-Acetylamino-2-nitrothiophene is best prepared by nitration of 3-acetylaminothiophene and is isolated in 80% yield [172,174,412]. Similarly other 3-acylamino-2-nitrothiophenes are obtained [172,178]. The nitration of 3-acylaminothiophenes with electron-withdrawing groups in the 2-position is not synthetically useful, as mixtures of isomers are obtained [134]. However, in the absence of replaceable groups at C2, 3-acetylaminothiophenes that also have substituents at C5 give the 4-nitro derivatives [43,106]. Nitration of 2-acetylamino-3-carboethoxy-4-cyclopropylthiophene with acetyl nitrate in acetic anhydride gives the 5-nitro derivative in high yield [227].

2-Acetylamino-3-carboethoxy-4-cyclopropyl-5-nitrothiophene [227]

Fuming nitric acid (6.3 g, 0.1 mol) is added to acetic anhydride (15 ml) at −50 °C. The temperature is increased to 0 °C and the mixture is kept at this temperature for 0.5 h. After having lowered the temperature to −50 °C, 2-acetylamino-3-carboethoxy-4-cyclopropylthiophene (5.1 g, 0.02 mol) in a minimum quantity of chloroform is added. The reaction mixture is agitated at −30 °C for 1 h, poured into water (200 ml), neutralized with sodium carbonate and extracted with chloroform. The combined chloroform phases are washed with water and dried over magnesium sulfate. After evaporation the residue is recrystallized or chromatographed on aluminium oxide using ether/hexane (1:3) as eluent.

From 3,4-diacetylaminothiophene the 2-nitro- and 2,5-dinitro derivative can be prepared depending upon nitration conditions [139].

5.1.2.8 Thienylisocyanates and isothiocyanates

Isocyanates can be obtained by the Curtius reaction (cf. Section 'Hydrolysis of acetamido- and other protected derivatives') of thiophenecarbonyl azides under nonhydrolytic conditions [147,149,172,179, 180,186–190,306,398,413–424]. Another modification consists in the reaction of thiophene acid chlorides with trimethylsilyl azide, which gives the carbonyl azide and upon heating leads to thienyl isocyanate [425]. The Curtius reaction was recently used for the preparation of 2-thienylisocyanate [426].

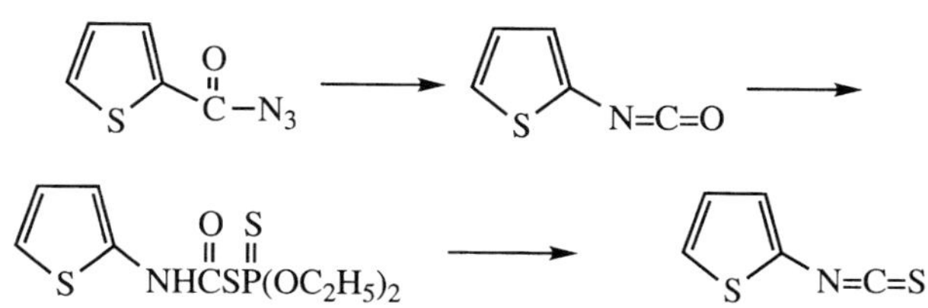

2-Thienylisocyanate [426]

A solution of 2-thiophenecarbonyl chloride (4.0 ml, 38 mmol) in acetone (20 ml) is cooled to 0 °C followed by dropwise addition of a solution of sodium azide (2.87 g, 44.1 mmol) in water (24 ml). The cooled solution is stirred until analysis by GC-MS indicates complete conversion of the chloride to 2-thiophenecarbonyl azide (3.5 h). Carbon tetrachloride (20 ml) and saturated sodium bicarbonate solution (10 ml) are added. The phases are separated and the organic phase dried over sodium sulfate and evaporated. Curtius rearrangement is performed by dissolving the above crude azide in carbon tetrachloride (25 ml) and heating in a heavy-walled glass pressure vessel sealed with a threaded teflon plug at 100 °C for 13 h. Caution: This step should be conducted behind a blast shield. After cooling to room temperature the solvent is removed *in vacuo* with protection from moisture and the residue distilled giving 1.72 g (37%) of the title compound bp 71–74 °C/1.2 mm Hg.

Another good method for the preparation of isocyanates is the reaction of aminothiophenes with phosgene, mostly described in patents [427]. 2- and 3-Thienyl isothiocyanates are prepared from the isocyanates through reaction with a sulfurating agent [428] or through the reaction of aminothiophenes with thiophosgene [429–432] as in the preparation of 2-carbomethoxy-3-isothiocyanothiophene.

Methyl 3-isothiocyanato-5-methylthiophene-2-carboxylate [432]

A solution of thiophosgene (2.5 ml) in chloroform (60 ml) and then a solution of methyl 3-amino-5-methylthiophene-2-carboxylate (5.00 g, 29 mmol) in chloroform are added dropwise to a solution of sodium hydrogen carbonate (3.7 g) in water (23 ml) at room temperature for 4 h. The phases are separated and the aqueous phase extracted with chloroform. The residue obtained from the chloroform extracts is chromatographed on silica gel using chloroform/hexane (1:1) as eluent, giving 6.80 g (93%) of the title compound mp 63–64 °C from hexane.

Catalytic hydrogenation of 2-(6-butylthio-2-naphtyl)-5-nitrothiophene gives the amino derivative which by reaction with phosgene gives the isothiocyanate [73].

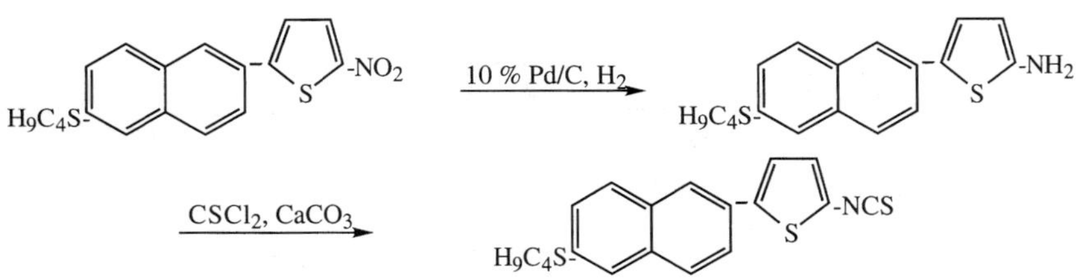

5.1.2.9 Thienylcarbamic esters, thionocarbamates, and dithiocarbamates

5.1.2.9.1 Preparation of N-thienylcarbamates by Curtius reaction

N-Thienylcarbamic esters are best prepared by the Curtius reaction [433] (cf. Section 'Hydrolysis of acetamido- and other protected derivatives') and especially useful is the one-pot procedure for *tert*-butyl esters directly from the free thiophenecarboxylic acids [181]. *N*-alkyl or *N*-aryl thiophenecarbamates are prepared by reactions of aminothiophenes with ethyl chloroformate [100,139, 234,239,429,434–438] methyl chloroformate [239,309], benzyl chloroformate [239], or phenyl chloroformate [230]. Reaction of methyl *ortho*-aminothiophenecarboxylates with 2-chloroethyl chloroformate gives carbamates, which could be hydrolyzed and transformed to cyclic carbamates [439].

Methyl 3-ethoxycarbonylamino-5-methylthiophene-2-carboxylate [438]

A solution of methyl 3-amino-5-methylthiophene-2-carboxylate (10.8 g, 63 mmol) and ethyl chloroformate (8.2 g, 76 mmol) in toluene (140 ml) is refluxed for 3 h. After evaporation the residue is chromatographed on silica gel using ethyl acetate/hexane (3:1) as eluent. Recrystallization from ethyl acetate/hexane gives 15.0 g (98%) as colorless needles mp 49–51 °C.

5.1.2.9.2 By reaction of thienylisocyanates with alcohols

Thiophenecarbamic esters have also been prepared through the reaction of thienyl isocyanates with alcohols.

5.1.2.9.3 By electrophilic substitution of *N*-thienylcarbamates

Acid-catalyzed reductive α-alkylation of the *N*-acetamides and *N*-carbamates of 3-amino and 3-methylaminothiophenes with aldehydes using catalytic amounts of *para*-toluenesulfonic acid in methylene chloride at room temperature followed by reduction with selenophenol is a good method for the preparation of the 2-alkyl-3-aminothiophene derivative. Without reduction, bis(3-amino-2-thienyl)methane derivatives are obtained. With branched aldehydes, biacetamides and biscarbamates are obtained using concentrated hydrochloric acid.

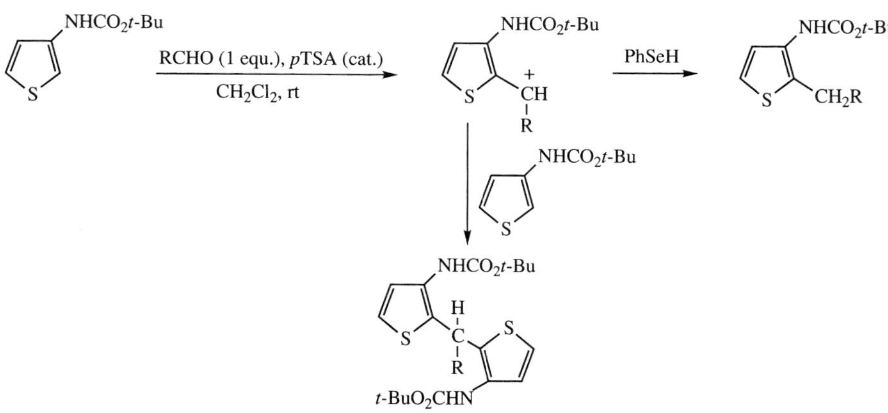

These acid-catalyzed reactions with aldehydes followed by selenophenol can also be used with the mono and dicarbamates of 3,4-diaminothiophene [440].

Ethyl (4-aminothien-3-yl)carbamate [440]

To a solution of 3,4-diaminothiophene (228 mg, 2 mmol) in dichloromethane (20 ml), ethyl chloroformate (217 mg, 2 mmol) in dichloromethane (2 ml) is quickly added. An ammonium salt is formed and triethylamine (0.5 ml) is then added dropwise. After stirring the reaction mixture for 4 h it is extracted with 1 *M* hydrochloric acid solution (2×15 ml). The phases are separated and the aqueous phase washed with dichloromethane, neutralized with 4 *M* sodium hydroxide solution and extracted with ether (2×20 ml). The combined organic phases are dried and the solvent distilled off giving 64% of the title compound as an oil.

Condensation of carbamates with α-functionalized acetals such as $ECH_2CH(OMe)_2$, (E = acetyl, carbetoxy, and cyano) gives α β-unsaturated ketones, esters, and nitriles in good yield [441].

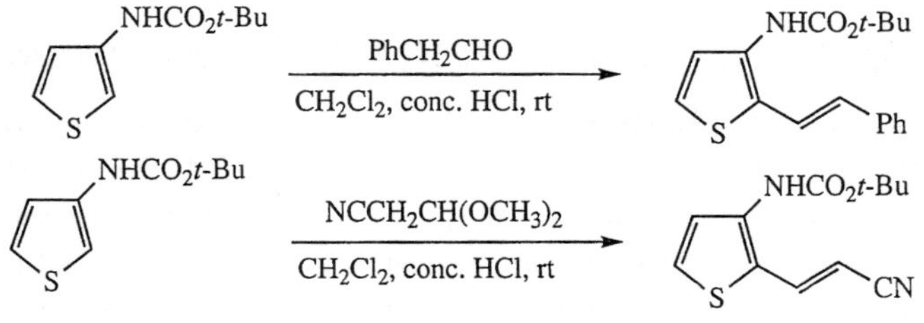

Thioamides having an enamine function can be alkylated, and the resulting product readily undergoes ring closure upon base treatment. In this way 5-aroyl-2-morpholinothiophenes are prepared [442]. 2,4-Diacyl-5-aminothiophenes are obtained by a useful and flexible synthesis based on condensation of enamines with alkyl, aryl, or acyl isothiocyanates followed by alkylation and ring closure [443–445].

5.1.2.9.4 By metalation and halogen-metal exchange of *N*-thienylcarbamates followed by reaction with electrophiles

Lithiation of boc-protected 3,4-diaminothiophene with excess lithium diisopropylamide followed by tributylstannyl chloride is used for the preparation of boc-protected 3,4-diamino-2,5-bis(tributylstannyl)thiophene. The *N*-boc-protected 3,4-diaminothiophene can be methylated in 92% yield using sodium hydroxide and dimethyl sulfate and upon lithiation with lithium diisopropylamide followed by reaction with tributylstannyl chloride, *N*,*N*′-(bis-tert-butoxycarbonyl)-*N*,*N*′-dimethyl-3,4-diamino-2,5-bis(tributylstannyl)thiophene is obtained [140].

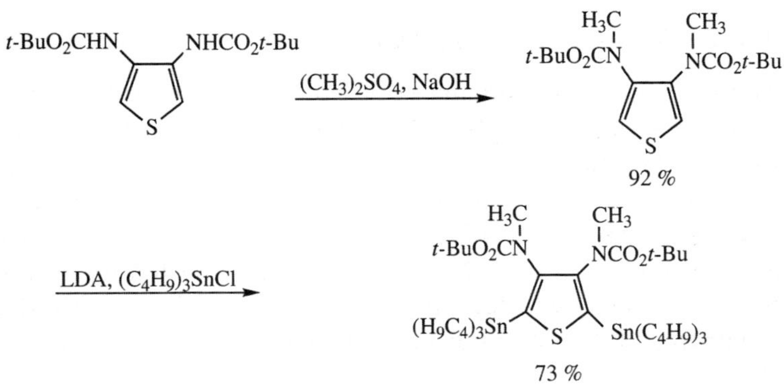

N,N′-(bis-tert-butoxycarbonyl)-N,N′-dimethyl-3,4-diamino-2,5-bis(tributylstannyl)thiophene [140]

A solution of 1.45 *M* butyllithium in hexane (19.49 ml, 28.26 mmol) is slowly added to diisopropylamine (3.70 ml, 28.26 mmol) in anhydrous diethyl ether (15 ml) at −78 °C. The solution is warmed to room temperature and then cooled to 0 °C with an ice bath. A solution of *N,N′*-(bis-*tert*-butoxycarbonyl)-*N,N′*-dimethyl-3,4-diaminothiophene (3.23 g, 9.42 mmol) in anhydrous tetrahydrofuran (8 ml) is added and the reaction mixture is warmed to room temperature for 10 min. After recooling to 0 °C chloro(tributyl)stannane (6.15 ml, 20.72 mmol) is added. After stirring the reaction mixture for 1 h, a saturated sodium chloride solution (50 ml) is added. The phases are separated and the aqueous phase extracted three times with dichloromethane. The combined organic phases are dried over sodium sulfate and filtered. Triethylamine (30 ml) is added to the filtrate and the resulting solution stirred overnight. After evaporation the residue is purified by flash chromatography on silica gel washed with neat triethylamine followed by hexane using hexane as eluent. The solvent is removed *in vacuo* giving 6.31 g (73%) of the title compound.

In later studies, the reaction of the trilithio derivative of di-*tert*-butylthiophene-3,4-diyldicarbamate with alkyl halides led to 2-alkylthiophenedicarbamates and 4-alkylthieno[3,4-*d*]imidazolones. Thus treatment with excess methyl iodide gives the trimethylated compound [170].

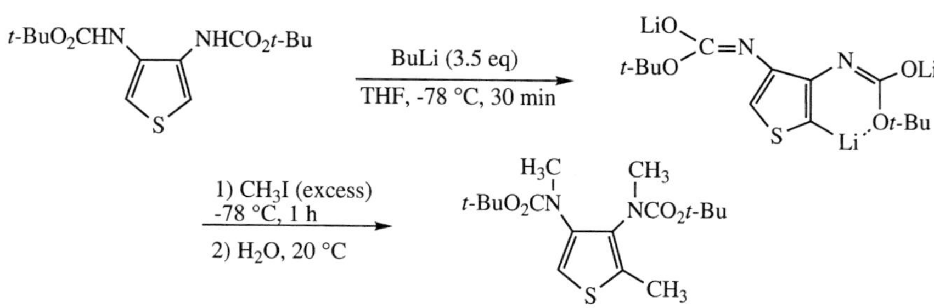

Di-tert-butyl N,N′-dimethyl (2-methylthiophene-3,4-diyl)carbamate [170]

A solution of 2.5 *M* butyllithium in hexane (1.4 ml) is added to a stirred solution of di-*tert*-butyl(thiophene-3,4-diyl)carbamate (0.314 g, 1 mmol) in tetrahydrofuran (40 ml) cooled to −78 °C under argon. The mixture is stirred at −78 °C for 30 min and methyl iodide (1.42 g, 10 mmol) is added at −10 °C. The reaction mixture is stirred for 1 h at this temperature and treated with saturated sodium chloride solution (2 ml). The phases are separated and the organic phase dried and evaporated. The residue is chromatographed on silica gel using dichloromethane/light petroleum (3:7) giving 285 mg (92%) of the title compound.

Under kinetically controlled conditions at −78 °C lithiation of boc-protected methyl 4-amino-3-thiophenecarboxylate with lithium diisopropylamide gives the boc-protected dimethyl 3-amino-2,4-thiophenedicarboxylate upon reaction with methyl chloroformate, together with small amounts of the boc-protected dimethyl 3-amino-4,5-thiophenedicarboxylate. The latter compound becomes the main product under thermodynamic conditions (−30 °C). Other electrophiles such as methyl iodiode trimethysilyl chloride and *N*,*N*-dimethylformamide give similar results [446].

Method A. Standard conditions for kinetic dianion formation [446]

To a solution of lithium diisopropylamide (2.1 equiv) in anhydrous tetrahydrofuran (1 ml) at −78 °C under nitrogen a solution of boc-protected methyl 4-amino-3-thiophenecarboxylate (58.0 mg, 0.226 mmol) in tetrahydrofuran (1 ml) is added dropwise. After 1 h at −78 °C methyl chloroformate (20 μl, 1.1 equiv.) is added to the bright yellow reaction mixture. Stirring is continued for 30 min after which the reaction mixture is quenched with saturated sodium bicarbonate solution. After warming to room temperature the reaction mixture is poured into aqueous sodium bicarbonate. The product is extracted three times with dichloromethane and the combined organic phases dried over sodium sulfate, evaporated, and flash chromatographed using hexane/ethyl acetate (5:1) as eluent, giving 71% of boc-protected methyl 4-amino-3,5-dithiophenecarboxylate and 8% of boc-protected methyl 4-amino-2,3-dithiophenecarboxylate.

Method B. Standard conditions for thermodynamic dianion formation [446]

To a solution of lithium diisopropylamide (2.1 equiv) in anhydrous tetrahydrofuran (1 ml) at −78 °C, under nitrogen, a solution of boc-protected methyl 4-amino-3-thiophenecarboxylate (124.4 mg, 0.484 mmol) in tetrahydrofuran (1 ml) is added slowly. The reaction mixture is warmed to −30 °C for 20 min, during which time the reaction color changes from yellow to amber. The mixture is cooled back to −78 °C, treated with methyl chloroformate, stirred for 30 min, allowed to warm to room temperature, quenched in saturated sodium bicarbonate solution, and extracted three times with dichloromethane. The combined organic phases are dried over sodium sulfate, evaporated, and flash chromatographed using hexane/ethyl acetate (4:1) to provide 112.4 mg (74%) of boc-protected methyl 4-amino-2,3-dithiophenecarboxylate, and 7.9 mg (5%) of boc-protected methyl 4-amino-3,5-dithiophenecarboxylate.

Metalation of 3-*tert*-butoxycarbonylaminothiophene or halogen–metal exchange of the 2-bromo derivative followed by reaction with electrophiles can be used for the preparation of 2-functionalized derivatives [409,447]. Tin-substituted derivatives of thienylcarbamic esters have been prepared from the corresponding bromo derivatives by halogen–metal exchange followed by reaction with trialkyl tin derivatives [448].

tert-Butyl N-(trimethylstannyl-3-thienyl)carbamate [448]

tert-Butyl *N*-(2-bromo-3-thienyl)carbamate (5.85 g, 0.021 mol) is dissolved in anhydrous tetrahydrofuran (50 ml) and the solution is cooled to −80 °C. At −78 to −72 °C 2.03 *M* butyllithium in cyclohexane (23 ml) is added dropwise. When the addition is completed the reaction mixture is stirred at −74 °C for 45 min, after which trimethyltin chloride (4.58 g, 0.023 mol) in anhydrous tetrahydrofuran is added dropwise. The reaction mixture is allowed to reach room temperature and sodium chloride is added. The product is extracted with ethyl acetate, the combined organic phases are washed with water, dried over sodium sulfate, and evaporated. The title compound is obtained in a yield of 8.24 g (> 100%) as a brown oil.

5.1.2.9.5 By modification of substituents in carbamates

Ethyl *N*-(2-formyl-3-thienyl)carbamate is prepared by treatment of 3-ethoxycarbonylaminothiophene-2-carboxylate with lithium aluminium hydride followed by oxidation of the alcohol with manganese dioxide or pyridinium dichromate. Ethyl *N*-(3-formyl-4-methyl-2-thienyl)carbamate is prepared similarly [449]. The reaction of these aldehydes with ethanolamine is used for the preparation of the Schiff's bases, which upon reduction with sodium borohydride give the (2-hydroxyethyl)aminomethyl-substituted carbamates [449].

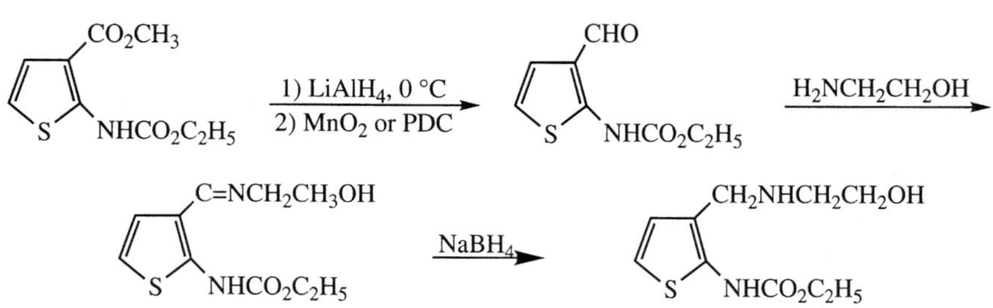

Ethyl N-(2-formyl-3-thienyl)carbamate [449]

Lithium aluminium hydride (0.5 g) is added portionwise to a solution of methyl 3-ethoxycarbonylaminothiophene-2-carboxylate (2.29 g) in anhydrous diethyl ether (40 ml) under nitrogen at 0 °C during 3 min. The mixture is stirred at the same temperature for 30 min, after which a saturated solution of ammonium chloride is carefully added. The precipitate formed is filtered through celite and the filtrate is washed with water. The aqueous phase is extracted with chloroform, the combined organic phases are dried over magnesium sulfate and filtered. The filtrate is treated with activated manganese dioxide (20.0 g) and the mixture stirred at room temperature for 1 h. Insoluble material is filtered off, the filtrate evaporated and the residue purified by chromatography on silica gel using ethyl acetate/hexane (1:9) as eluent, giving 1.47 g (74%) of the title compound as colorless leaflets mp 58–59 °C after recrystallization from ethyl acetate/hexane.

The following reaction has been performed [450].

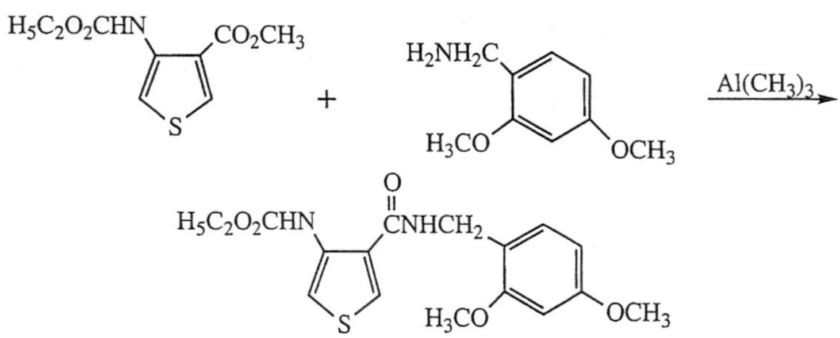

Ethyl {4-[[(2,4-dimethoxyphenylmethyl)amino]carbonyl]thiophene-3-yl} carbonate [450]

A solution of 2,4-dimethoxybenzylamine (20.9 g, 125 mmol) in toluene (300 ml) under nitrogen at 10 °C is slowly treated with 2 *M* trimethylaluminium in toluene (68.8 ml, 138 mmol). This solution is warmed to room temperature for 30 min and then recooled to 10 °C and treated with the thiophene derivative (28.6 g, 125 mmol). After stirring the reaction mixture at room temperature for three days it is cooled to 0 °C and carefully quenched with 2 *M* hydrochloric acid. The phases are separated and the aqueous phase extracted with dichloromethane. The combined organic phases are washed with 2 *M* hydrochloric acid, water and sodium chloride solution, dried over magnesium sulfate and evaporated. The residue, a white solid, is recrystallized from dichloromethane/hexane/diethylether giving 31.1 g. The mother liquid is purified by chromatography on silica gel using diethyl ether/hexane (1:1) as eluent, affording another 5.8 g and a total yield of 81% of the title compound as a white solid mp 123–126 °C.

tert-Butyl-2-allyl- and *N*-allylthienyl carbamates have been prepared and used as substrates for the preparation of thieno[3,2-*b*]pyrroles [451].

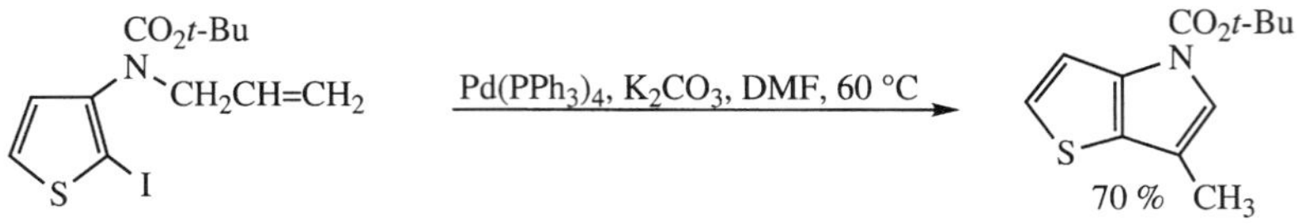

5.1.2.9.6 From halo-boc-aminothiophenes by palladium-catalyzed coupling reactions

Heck coupling of 2-bromo-3-boc-aminothiophene and di-*tert*-butyl-2-bromothiophene-3,4-diylcarbamate with methyl acrylate or methyl vinyl ketone is used for the preparation of methyl 3-(3-tert-butoxycarbonylamino-2-thien-2-yl)propenoate and 4-[3-(tert-butoxycarbonylamino)thien-2-yl]butenone in 70 and 75% yield, respectively [170].

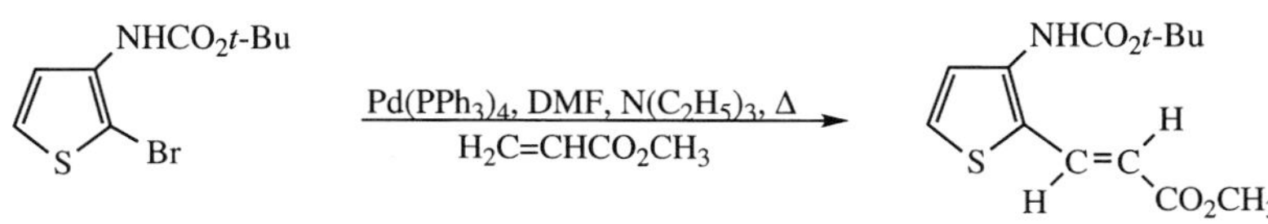

*Methyl 3-(3-*tert*-butoxycarbonylaminothien-2-yl)propenoate [170]*

To a degassed solution of *tert*-butyl(2-bromothien-3-yl)carbamate (28 mg, 1 mmol) in *N*,*N*-dimethylformamide (1 ml) tetrakistriphenylphosphine palladium(0) (0.116 g, 0.1 mmol), triethylamine (0.808 g, 8 mmol), and methyl acrylate (52 mg, 8 mmol) are added. The reaction mixture is heated at 150 °C for 10 h, diluted with water (10 ml) and extracted with diethyl ether (3 × 10 ml). The combined organic phases are dried and concentrated. The product is purified by chromatography on silica gel using dichloromethane as eluent giving 20 mg (70%) of the title compound mp 124 °C.

The same compounds are used in Heck couplings with terminal acetylenes. When iodo derivatives are used the couplings can be carried out at room temperature [170].

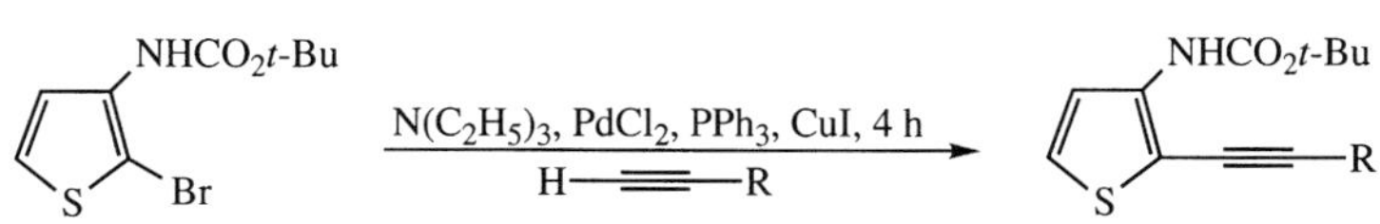

Suzuki coupling of 2-bromo-3-boc-aminothiophene and di-*tert*-butyl-2-bromothiophene-3,4-diylcarbamate with phenylboronic acid and 2-thiopheneboronic acid is used for the preparation of the 2-phenyl and 2-(2-thienyl) derivatives [170].

Thionecarbamates and dithiocarbamates are prepared through the reactions of thienylisothiocyanates with alcohols or mercaptans, respectively [428].

5.1.2.10 Thienylurea and thiourea derivatives

Thienyl urea derivatives have been prepared by the reaction of thienyl isocyanates with amines or from the reaction of aminothiophenes with isocyanates [452,453].

Methyl 4-chloroethylureathiophene-3-carboxylate [453]

Methyl 4-aminothiophene-3-carboxylate (9.65 g, 50 mmol) in toluene (100 ml) is treated with 2-chloroethyl isocyanate (4.7 ml, 55 mmol). The reaction mixture is refluxed for 3 h and after cooling two crops (12.2 g, 93%) of the title compound are obtained as white solid mp 110–112 °C from dichloromethane/ether/hexane.

The reaction of 2-amino-3-cyanothiophenes with phosgene is used for the preparation of 2,2-ureylen-bis-thiophene-3-carbonitriles [454].

2,2-Ureylen-bisthiophene-3-carbonitrile [454]

To a solution of 2-amino-3-cyanothiophene (3.1 g, 25 mmol) in dioxane (50 ml), phosgene in toluene (25 mmol) is slowly added dropwise. The stirring is continued for 10 min, after which pyridine (15 ml) in dioxane (10 ml) is added and the reaction mixture stirred for 1.5 h. Water (150 ml) is then added and the precipitate formed filtered off, washed with water and a small amount of ethanol giving 2.74 g (80%) of the title compound.

The reaction of exo-bicyclo[4.1.0]hept-3-ene-7-isocyanate with methyl 3-amino-2-thiophenecarboxylate is used for the preparation of *N*-(exo-bicyclo[4.1.0]hept-3-ene-7-yl)-*N*′-(2-methoxycarbonylthiophene-3-yl)urea [455].

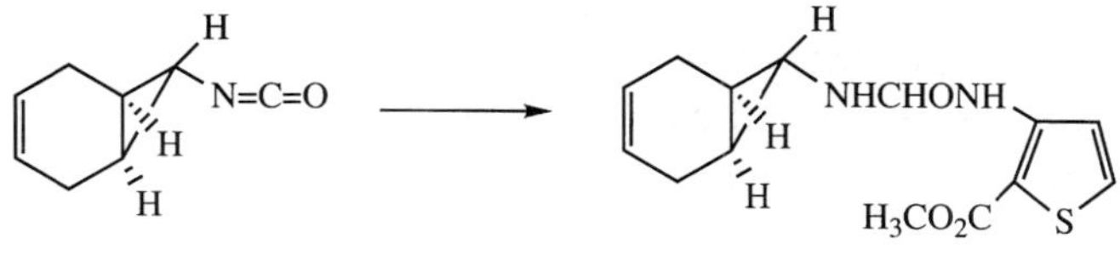

N-(exo-bicyclo[4.1.0]hept-3-ene-7-yl)-N′-(2-methoxycarbonylthiophene-3-yl)urea [455]

A solution of exo-bicyclo[4.1.0]hept-3-ene-7-isocyanate (0.1 g, 0.7 mmol) and methyl 3-amino-2-thiophenecarboxylate (0.11 g, 0.7 mmol) in diethyl ether (25 ml) is refluxed for 2 h. The precipitate formed is filtered off and the colorless solid is recrysrallized three times from diethyl ether/pentane (1:9) giving 0.17 g (84%) mp 134 °C.

The *ortho*-amino ester reacts with benzoylisothiocyanate and phenylisothiocyanate to give the urea derivative and the thiourea derivative, respectively, the *ortho*-aminonitrile reacts similarly with phenylisothiocyanate, benzoyl isothiocyanate, and ethoxycarbonylisothiocyanate to afford the thiourea and urea derivatives [456].

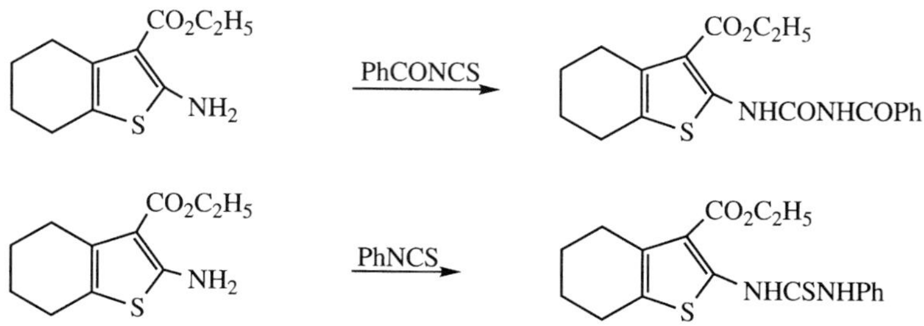

Reaction of ethyl 2-amino-4,5,6,7-tetrahydrobenzo[b] thiophene-3-carboxylate with phenyl isothiocyanate [456]

A solution of ethyl 2-amino-4,5,6,7-tetrahydrobenzo[*b*]thiophene-3-carboxylate (2.25 g, 0.01 mol) in dioxane (30 ml) is treated with phenyl isothiocyanate (0.01 mol) and the reaction mixture is refluxed for 2 h. After evaporation the residue is recrystallized from dioxane giving 2.34 g (65%) of the thiourea derivative mp 181 °C.

N,N'-dithienyl ureas are prepared by the reaction of aminothiophenes with carbonyl diimidazole as phosgene substitute [136]. *N,N*-dimethyl-*N'*-thienyl ureas are prepared from aminothiophenes and dimethylaminocarbonyl chloride [457]. Cyclic ureas are prepared by the reaction of *ortho*-diaminothiophenes with phosgene.

N-Benzyl-*N'*-(2-thienyl) urea is prepared by the reaction of 2-thienylisocyanate with benzylamine [426].

N-Benzyl-N'-(2-thienyl) urea [426]

Benzyl amine (550 μl, 5.1 mmol) is added dropwise to a stirred solution of 2-thienylisocyanate (625 mg, 5.0 mmol) in carbon tetrachloride (40 ml). A white solid is immediately formed. After stirring the reaction mixture for 15 min, it is filtered, the solid washed with additional carbon tetrachloride and dried *in vacuo* giving 1.10 g (94%) of the title compound mp 166–167 °C.

Thienylthiourea derivatives are prepared through the reaction of aminothiophenes with isothiocyanates [176,458–463] or through the reaction of thienylisothiocyanates with amines [431]. In this way 2,2'-thioureylene-bis-(4,5-dimethylthiophen-3-carbonitrile) is prepared from 4,5-dimethyl-2-amino-3-cyanothiophene and phenylisothiocyanate or by its reaction with thiophosgene in pyridine [454]. Ethyl 2-benzoylthioureidothiophene-3-carboxylate are prepared from esters of 2-amino-3-thiophenecarboxylic acids and benzoyl isothiocyanate in acetone [464].

Recent examples of the latter method are the preparation of methyl 3-(3-adamantylthioureido)thiophene-2-carboxylate [465] and methyl 3-[3-(2-hydroxy-1-methylethyl)thioureido]-2-thiophenecarboxylate [466] through the reaction of methyl 3-isothiocyanato-2-thiophenecarboxylate with 1-aminoadamantane and S-(+)-2-amino-2-propanol, respectively [103,106,133,134,188].

Methyl 3-[3-(2-hydroxy-1-methylethyl)thioureido]-2-thiophenecarboxylate [466]

To a stirred solution of methyl 3-isothiocyanatothiophene-2-carboxylate (0.45 g, 2.26 mmol) in anhydrous tetrahydrofuran (8 ml), *S*-(+)-2-amino-2-propanol (0.183 g, 2.44 mmol) is added. The reaction mixture is stirred at room temperature for 24 h. After evaporation of the solvent at reduced pressure the oily residue is cooled to 0 °C giving 0.59 g (95%) of the title compound as pale-yellow crystals mp 103–104 °C.

Another example is the reaction of ethyl 4,5-dimethyl-2-isothiocyano-3-thiophenecarboxylate with ammonium hydroxide in dichloromethane at room temperature, yielding ethyl 2-(aminothioxomethyl)amino-4,5-dimethyl-3-thiophenecarboxylate [467].

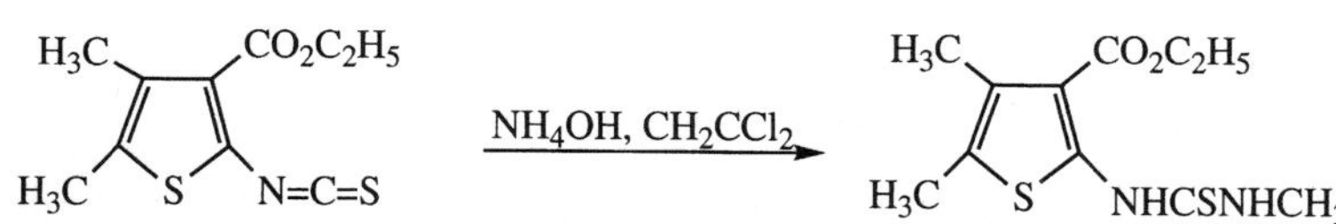

Ethyl 2-(aminothioxomethyl)amino-4,5-dimethyl-3-thiophenecarboxylate [467]

A solution of ethyl 4,5-dimethyl-2-isothiocyano-3-thiophenecarboxylate (1.76 g, 7.3 mmol) in dichloromethane (20 ml) is added to a solution of 30% ammonium hydroxide (0.8 ml) in dichloromethane (10 ml). The reaction mixture is stirred at room temperature for 0.5 h, after which the precipitate formed is filtered off, washed with ethanol, and recrystallized from ethanol/dioxane giving 1.10 g (54%) mp 214–215 °C (decomp.).

With appropriate *ortho*-substituents ring closure of the urea and thiourea derivatives to bicyclic systems is obtained [468–470]. Thus *N*-(3-carboethoxy-2-thienyl)-*N'*-allyl thiourea is formed under base catalysis [460]. 3-Cyano-2-thioureidothiophene can be prepared by treatment of its benzoyl derivative with sulfuric acid [471].

Ethyl 2-[(ethoxycarbonylmethylaminothiocarbonyl)amino]-4,7-dihydro-5*H*-thieno[2,3-*c*]thiopyran-3-carboxylate is prepared by the reaction of ethyl 2-amino-4,7-dihydro-5*H*-thieno[2,3-*c*]thiopyran-3-carboxylate with ethoxycarbonylmethyl isothiocyanate [463].

Ethyl 2-[(ethoxycarbonylmethylaminothiocarbonyl)amino]-4,7-dihydro-5H-thieno[2,3-c]thio pyran-3-carboxylate [463]

A solution of thioxycarbonylmethylisothiocyanate (0.73 g, 0.005 mol) is added to a stirred solution of ethyl 2-amino-4,7-dihydro-5*H*-thieno[2,3-c]thiopyran-3-carboxylate (1.21 g, 0.0049 mol) in absolute ethanol (10 ml). The reaction mixture is heated under reflux for 2 h. On cooling, the separated solid product is collected by filtration, dried and recrystallized from ethanol giving 1.6 g (84%) of the title compound as white crystals, mp 163–165 °C.

5.1.2.11 Other *N*-modified thiophenes

The reaction of 2- and 3-aminothiophenes with nitromalonaldehyde can be used for the preparation of 3-(thienylamino)-2-nitropropenals [472]. 3,4-Disulfinylamidothiophene can be obtained in 81% yield upon reaction of 3,4-diaminothiophene with phenylsulfinylamide in triethylamine [142]. The reaction of 3-amino-4-carbomethoxythiophene with primary and secondary chlorosulfonamides in benzene in the presence of triethylamine offers a good method for the preparation of [(amino- and isopropylamino)sulfonyl]aminomethoxycarbonylthiophenes [473].

Methyl 3-[(Isopropylamino)sulfonyl]amino-4-thiophenecarboxylate [473]

To a solution of the hydrochloride of 3-amino-4-thiophenecarboxylate (1.95 g, 10 mmol) in anhydrous benzene (30 ml), isopropylsulfamoyl chloride (1.57 g, 10 mmol) in anhydrous benzene (20 ml) is added dropwise at room temperature. The reaction mixture is refluxed until the total conversion of the starting material, observed by thin layer chromatography using methanol/chloroform (1:10) as eluent. After evaporation the residue is recrystallized from ethyl acetate/hexane giving 2.17 g (78%) of the title compound mp 95–97 °C.

The reaction of 2-amino-3-cyanothiophene with dibromotriphenyl phosphorane has been used for the preparation of 2-[triphenylphosphoranylidene)amino]-3-cyanothiophene which upon reaction with phenyl isocyanate gives the carbodiimide, *N*-phenyl-*N*′-(2-cyanothieno)carbodiimide [474]. The reaction of 3-amino-2-cyanothiophene with 2,5-dimethoxytetrahydrofuran in acetic acid provides 3-(1-pyrollo)-2-cyanothiophene [475].

3-(1-Pyrollo)2-cyanothiophene [475]

A mixture of 3-amino-2-cyanothiophene (124 g, 1.0 mol), 2,5-dimethoxytetrahydrofuran (132 g, 1.0 mol), and acetic acid (1000 ml) is heated under reflux for 30 min. The acetic acid and the ethyl acetate formed are evaporated off and the residue distilled giving 155 g (89%) (bp 105–110 °C/0.2 mm Hg) of the title compound, which crystallizes on cooling (mp 48–50 °C) from ethanol.

Amidines are formed when hydrogen chloride is led into a mixture of 2-amino-3-carbethoxythiophenes and cyanoacetate or cyanoacetanilides, which ring-close spontaneously to 3,4-dihydro-4-oxothieno[2,3-*d*]pyrimidine-2-ylacetic acid derivatives [476].

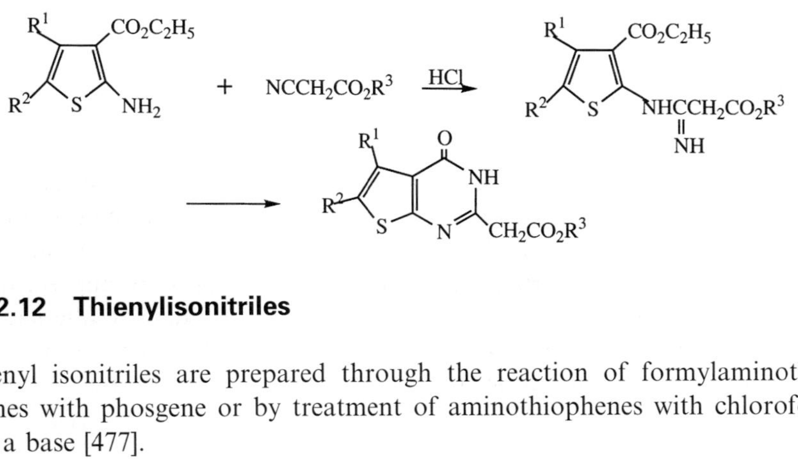

5.1.2.12 Thienylisonitriles

Thienyl isonitriles are prepared through the reaction of formylaminothiophenes with phosgene or by treatment of aminothiophenes with chloroform and a base [477].

5.1.3 Hydrazino, azo, and related derivatives of thiophene

5.1.3.1 Diazonium salts

One of the great differences between thiophene and benzene chemistry is the unavailability of simple thienyldiazonium salts prepared by diazotization of aminothiophenes. Even 2-aminothiophenes with electron-withdrawing groups but with a vacant 5-position still do not undergo diazotization reactions efficiently [478]. Only those 2-aminothiophenes with an electron-withdrawing group and a blocked 5-position can be used sucessfully, as has been demonstrated with 4,5-dialkyl-2-amino-3-cyano and -3-carbethoxythiophenes [479], (cf. scheme below) the 5-alkoxycarbonyl [414], 5-acetyl- [480], 2-amino-5-nitrothiophene and 2-amino-3,5-dinitrothiophene- [67], 5-sulfonic- [399,482] and 5-arsonic 2-aminothiophene [482].

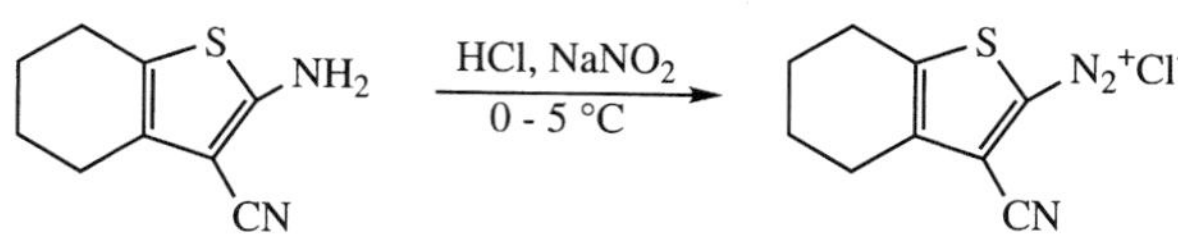

3-Aminothiophenes differ from the 2-isomers, that is derivatives without electron-withdrawing groups, but with the activated 2-position blocked give diazonium salts [483–486]. The presence of an electron-withdrawing group in the 2-position is nevertheless beneficial and 2-nitro- [25,486–489], 2-acyl- [488–492], 2-carboxy- [493], 2-alkoxycarbonyl- [270,275,279,490,494,495], and 2-aminocarbonyl-3-thienyldiazonium salts are prepared [490]. In a few cases 3-aminothiophenes with unblocked 2-positions such as 3-amino-4,5-dicarbomethoxythiophene [308,496] and 4-amino-2-thiophenecarboxylic acid are sucessfully diazotized.

5.1.3.2 Azo derivatives

Substituted 2-thiophenediazonium ions couple readily with phenols, aromatic, and heteroaromatic amines as well as active methylene derivatives to give highly colored dyes. 4-Diethanolaminobenzene-azo-2-(5-nitrothiophene) and 4-diethylaminobenzene-aza-2-(3,5-dinitrothiophene) are thus prepared by coupling of 5-nitro-2-thienyldiazonium 3,5-dinitro-2-thienyldiazonium salts with *N*-phenyldiethanolamine [67].

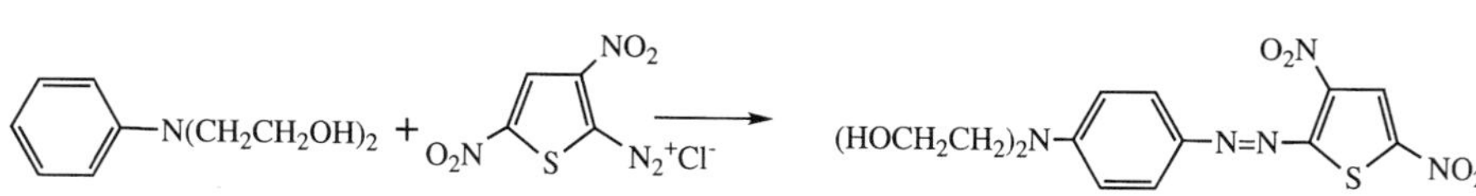

4-Diethanolaminobenzene-azo-2-(3,5-dinitro)thiophene [67]

Sodium nitrite (0.482 g, 6.96 mmol) is slowly added with stirring to concentrated sulfuric acid (5.7 ml) at ice-bath temperature. To facilitate the dissolution process, the temperature may be raised to 30 °C for a while and then lowered below 5 °C again. A mixture of propionic acid (3.2 ml) and acetic acid is added. During the addition of the acid mixture the temperature must be kept below 15 °C and then lowered to 0 °C when 2-amino-3,5-dinitrothiophene (1.2 g, 6.35 mmol) is slowly added over a period of 30 min. An aqueous solution (43 ml) of *N*-phenyldiethanolamine (1.77 g, 9.78 mmol) and concentrated hydrochloric acid (1.9 ml) cooled to 0 °C is added with stirring. The reaction is allowed to continue for another 50 min. The solid formed is filtered off, washed several times with water and dried. The crude product (1.5 g a deep blue solid) is purified by TLC on silica gel using ethyl acetate as developer.

Another example is the coupling of the diazonium salt from 2-amino-4,5,6,7-tetrahydrobenzo[*b*]thiophene-3-carbonitrile with 2-amino-1-naphthalenesulfonic acid [479].

2-(2-Amino-1-naphthyl)azo-4,5,6,7-tetrahydrobenzo[b] thiophene-3-carbonitrile [479]

To concentrated hydrochloric acid, 2-amino-4,5,6,7-tetrahydrobenzo[*b*]thiophene-3-carbonitrile (8.9 g, 0.05 mol) is added and dissolved by warming, after which the solution is cooled to 0–5 °C. With vigorous stirring sodium nitrite (3.45 g, 0.05 mol) in water (10 ml) is added gradually for about 2 h at 0–5 °C. The reaction mixture is stirred for another hour at 0–5 °C. The excess of nitrous acid is decomposed by addition of urea. The clear diazonium salt solution is slowly added to 2-amino-1-naphthalenesulfonic acid (11.15 g, 0.05 mol) in acetic acid (50 ml) at 0 °C. The pH of the reaction mixture is maintained at 4–5 throughout the coupling period by addition of sodium carbonate in portions for 2 h at 0 °C. When the addition of the diazonium salt is over the reaction mixture is stirred for a further period of 5 h and the partially separated dye is completely precipitated by neutralization. The dye is filtered, washed with water and dried. Recrystallization from ethanol gives 13.44 g (81%) of the title compound as a pink crystalline solid mp 219 °C.

5-(Heteroarylazo or arylazo)thiophenes are prepared by diazotization of ethyl 5-amino-2,4-bis(ethoxycarbonyl)-3-thiophene acetate using nitrosyl hydrogen sulfate, and coupling with suitable heterocyclic hydroxy- and *N,N*-dialkylamino-substituted arylamines [250].

Ethyl 5-[(5-hydroxy-3-methyl-1-phenyl-1H-pyrazol-4-yl)azo]2, 4-bis(ethoxycarbonyl)-3-thiopheneacetate [250]

Ethyl 5-amino-2,4-bis(ethoxycarbonyl)-3-thiophene acetate (1.64 g, 0.005 mol) is added in portions during 1 h to a cooled mixture of nitrosyl hydrogen sulfate prepared from sodium nitrite (0.38 g, 0.0055 mol) and concentrated sulfuric acid (10 ml) at 0 °C. The mixture is stirred for an additional 1 h at 0 °C, then added to an ice-water mixture under stirring, excess of nitrous acid is destroyed by the addition of urea (1 g) and the solution is filtered to obtain a clear diazonium salt. 3-Methyl-phenyl-1*H*-pyrazol (0.87 g, 0.005 mol) is dissolved in dilute sodium hydroxide. The solution is cooled by external cooling so that the temperature is at 0–5 °C. To this cooled solution the prepared diazonium salt is added slowly so that the temperature does not rise above 5 °C. The pH of the reaction mixture is maintained alkaline throughout the coupling period (1 h) by the addition of solid sodium carbonate in portions. After the addition of the diazonium salt is completed, the reaction mixture is stirred for 4 h at 0 °C, when the dye which partially separated is neutralized with dilute hydrochloric acid (5%), filtered, thoroughly washed with water and dried. The product is recrystallized from methanol giving 2.43 g (89%) of the title compound mp 246 °C.

N,*N*-disubstituted 2-aminothiophenes, such as 2-morpholino-5-acylthiophenes, couple depending on the substitution pattern at C(5); with aryldiazonium salts such as *para*-nitrophenyldiazonium and 2,4-dinitrophenyldiazonium salts, either at their C(3) or C(5) position yielding the corresponding 3-arylazo-2-morpholinothiophenes or, under elimination of the substituent at C(5) 5-arylazo-2-morpholinothiophenes [261].

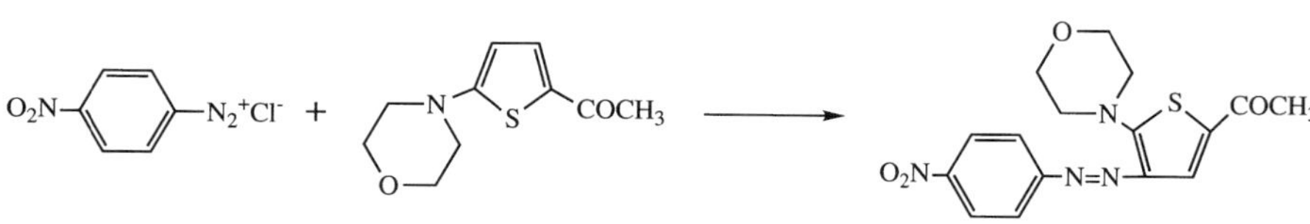

General procedure [261]

To a mixture of the appropriate 2-morpholinothiophene (10 mmol) in methanol or acetonitrile (25 ml) a solution of the nitro-substituted benzenediazonium salt, prepared by diazotization of the appropriate nitroaniline (10 mmol) dissolved in a mixture of acetic acid (50 ml) and sulfuric acid (10 ml) with sodium nitrite (0.7 g, 10 mmol) is added dropwise at room temperature. After standing at room temperature for 2 h the resulting mixture is diluted with methanol (50 ml) and water (100 ml). The precipitate formed is filtered off, the mother liquid is dried and evaporated and the residue purified by chromatography, giving 5-Benzoyl-2-morpholino-3-(4-nitrophenylazo)thiophene, 2.2 g (52%) mp 222–224 °C.

Most of the reports on the azo coupling reactions are given in the patent literature and the products are often only characterized by their color [4,114,497]. 3-Thiophenediazonium salts have especially been coupled with β-naphthol in alkaline solutions [17,43,279,483–485]. In connection with attempted diazotization of 2-aminothiophenes with free 5-positions, the formation of azo-derivatives due to "self-coupling" have been observed [248,498]. Such "self-coupling" products were also obtained as by-products in the diazotization of 4,5-dicarbomethoxy-3-aminothiophene [497]. An unusual self-coupling to 3-chloro and 3-ethoxy-2-thienyl-3-thienyl azine occurs when heating 2-carboxy-3-thienyldiazonium chloride in chloroform in the presence of trace amounts of ethanol [493].

5.1.3.3 Hydrazine derivatives

Thienylhydrazine and their derivatives have been prepared by the reduction of thiophenediazonium salts [414,494,498,499] by reaction of the anions of *N*-acylaminothiophenes with activated hydroxylamino derivatives [417,418, 500–503] or by the substitution of chlorine in 2-chloro-3-nitrothiophene with hydrazine [344,504,505].

tert-Butyl 2-(3-Thienyl)hydrazine [502]

To a suspension of sodium hydride (80% in paraffin) (40 mg, 1.3 mmol) in *N*,*N*-dimethylformamide under nitrogen, *tert*-butyl 3-thiophenecarbamate (245 mg, 1.2 mmol) is added. The resulting mixture is heated at 50–60 °C for 30 min, after which it is cooled to 10 °C and *O*-(4-nitrobenzoyl)amine (265 mg, 1.3 mmol) in *N*,*N*-dimethylformamide is added dropwise. The reaction mixture is stirred overnight and ice is then added. The product is taken up in methylene chloride and the combined organic phases are dried over magnesium sulfate. After evaporation the residue is chromatographed on silica gel using hexane/methylene chloride (7:3) as eluent, giving 180 mg (70%) of the title compound as a white solid mp 38–39 °C.

The reaction of ethyl 4-chloroquinoline-3-carboxylate with 3-hydrazino-2-thiophenecarboxylate and with 2-hydrazino-3-thiophenecarboxylate is used for the preparation of the following type of compounds [506].

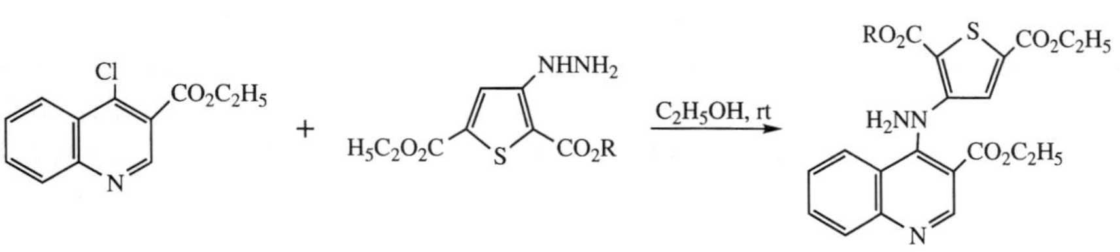

Ethyl 4-{2-[2-(methoxycarbonyl)-5-propylthien-3-yl]hydrazino} quinoline-3-carboxylate [506]

To a solution of ethyl 4-chloroquinoline-3-carboxylate (377 mg, 1.60 mmol) in ethanol (10 ml) methyl-3-hydrazino-1-propylthiophene-2-carboxylate (343 mg, 1.60 mmol) is added. The stirring is continued at room temperature for 2 h, after which the mixture is evaporated. The residue is dissolved in chloroform and this solution is washed with cold aqueous sodium carbonate solution and water, dried over magnesium sulfate, and evaporated. The residue is purified by chromatography on silica gel using toluene/methanol (3:1) as eluent, giving 616 mg (93%) of the title compound as orange crystals mp 152–153 °C.

Heating 2-amino-4-phenylthiophene with 80% hydrazine hydrate under reflux yields 2-hydrazino-4-phenylthiophene, existing as the 2(5H)-thiophenone hydrazone. Upon reaction with benzaldehyde the hydrazone is obtained [507].

4-Phenyl-2(5H)-thiophenone hydrazone [507]

4-Phenyl-2-thiopheneamine (5.26 g, 30 mmol) is heated with 80% hydrazine hydrate (40 ml) at reflux for 2 h. After cooling the precipitate is filtered off and recrystallized from ethanol giving 2.6 g (46%) of the title compound mp 122–124 °C.

Alkyl 2-hydrazino-3-thiophenecarboxylates are conveniently prepared from the corresponding alkyl 2-amino-3-thiophenecarboxylate [494,499,508]. The reaction of ethyl 2-methyl-4-hydroxythiophenecarboxylic acid with substituted hydrazines, such as *N,N*-dimethylhydrazine [509], phenylhydrazine [509,510], and various acylhydrazines [511] constitute good methods for the preparation of thienylhydrazines.

5.1.3.4 Azides

The best methods for the preparation of azidothiophenes are the reaction of thienyllithium derivatives with styryl azide [204] or tosyl azide [512–515].

2-Azidothiophene [515]

A solution of thiophene (6.72 g, 0.08 mol) in anhydrous diethyl ether (100 ml) is added with stirring under nitrogen at room temperature to 1.6 *M* butyllithium in hexane (50 ml). The reaction mixture is stirred and heated under reflux for an additional hour, after which it is cooled to −70 °C and added dropwise to a solution of tosyl azide (15.8 g, 0.08 mol) in anhydrous diethyl ether (100 ml). When the addition is completed the reaction mixture is stirred and allowed to reach 0 °C within 5 h. The pale yellow triazene salt formed is rapidly filtered off and suspended in pentane. The suspension is treated at 0 °C with a solution of tetrasodium pyrophosphate (0.08 mol) in water (200 ml). The yellow pentane phase is separated and evaporated giving a residue, which is chromatographed on "Florisil" using pentane as eluent, giving 5.5 g (55%) of the title compound as an oil.

This approach has recently been used for the preparation of 2-azido-3-methylthiophene and 2-azido-5-trimethylsilylthiophene [516]. A one-pot procedure for the preparation of isomeric acetyl- and 2,2,2-trifluoroacetylazidothiophenes from 2,3-, 3,4-, and 2,5-dibromothiophenes through stepwise halogen–metal exchange and successive reaction with *N,N*-dimethylacetamide (or *N,N*-diethyl-2,2,2-trifluoroacetamide) and tosyl azide has been developed [517].

The diazotization of aminothiophenes with electron-withdrawing groups, often conveniently available through ring-closure reactions, offers another route, as shown in the preparation of 2-acetyl-3-azidothiophene and methyl 3-azidothiophenecarboxylate from the amino derivatives by diazotization followed by reaction with sodium azide and sodium acetate [518].

Methyl 3-azidothiophene-2-carboxylate [518]

Methyl 3-aminothiophene-2-carboxylate (785 mg, 5 mmol) is dissolved in concentrated hydrochloric acid (5 ml), chilled to −5 °C and then treated dropwise during 5 min with a chilled solution of sodium nitrite (6 mmol) in water (2 ml), the temperature being kept under −4 °C. The reaction mixture is stirred for further 10 min at −5 °C, after which urea (50 mg) is added to remove the excess of nitrous acid. After 5 min the mixture is filtered to remove traces of solids and the filtrate is added in a steady stream to a stirred slurry of sodium azide (7.7 mmol) and sodium acetate (15.09 g) in ice-water (25 ml). Frothing occurs and the solid azide separates. After another 20 min 814 mg (89%) of the title compound is collected by filtration mp 67–68 °C.

A third useful way is by nucleophilic substitution of activated bromothiophenes with sodium azide. This was first applied for the preparation of 3-azido-2-thiophene aldehyde from 3-bromo-2-thiophene aldehyde using dimethylsulfoxide as solvent [169]. Better yields were obtained with hexamethylphosphoramide as solvent [519].

3-Azido-2-thiophene aldehyde [169]

3-Bromo-2-thiophene aldehyde (45.0 g, 0.236 mol) and sodium azide (45.0 g, 0.692 mol) are dissolved in dimethyl sulfoxide (600 ml). The reaction mixture is slowly stirred for 24 h at 65 °C, whereupon it is cooled and poured into water. The water solution is extracted with four portions of diethyl ether. The combined ether phases are dried over magnesium sulfate and evaporated to dryness. The crude product is recrystallized from methanol giving 17.5 g (48%) of the title compound mp 56.6–57.2 °C.

With doubly activated bromothiophenes such as 2-bromo-5-methoxycarbonyl-3-nitrothiophene, reaction with sodium azide in methanol is successful [520].

2-Azido-5-methoxycarbonyl-3-nitrothiophene [520]

A solution of sodium azide (0.9 g, 3.5 mmol) in methanol (40 ml) is treated with a solution of 2-bromo-5-methoxycarbonyl-3-nitrothiophene (0.4 g, 1.7 mmol) in methanol (40 ml). The reaction mixture is set aside in the dark until a TLC test indicates that the reaction is completed (*ca* 2 h). Upon addition of water the azido compound is precipitated and can be filtered off giving 0.2 g (58%) of the title compound mp 67–70 °C.

Modification of a substituent in the azide is also used for the preparation of other substituted azides. For instance 3-azidothiophene-2-aldoxime is transformed to the corresponding nitrile in 86% yield upon treatment with cyanuric chloride [519].

5.2 PHOSPHORUS DERIVATIVES

5.2.1 Thienylphosphines

5.2.1.1 By direct phosphorylation

The reaction of thiophene in pyridine with 0.3 mol of phosphorus tribromide at 180 °C can be used for the preparation of tris(2-thienyl)phosphine [521].

Tris(2-thienyl)phosphine [521]

Phosphorus tribromide (8.2 g, 0.03 mol) is added with stirring to a solution of thiophene (16.8 g, 0.2 mol) in pyridine (30 ml). The reaction mixture is sealed in an ampoule made of thick glass and heated for 20 h at 180 °C. After cooling the ampoule is opened and the content transferred to benzene (200 ml), this solution is filtered and successively washed with 10% aqueous sodium hydroxide solution (50 ml) and water (2 × 50 ml), dried over sodium sulfate, treated with ether (870 ml) and the precipitate filtered off. The filtrate is evaporated and the residue fractionated in a stream of argon giving 3.1 g (35%) of the title compound bp 178–179 °C/0.02 mm Hg.

5.2.1.2 *Via* thienylmagnesium or -lithium derivatives

The reaction of 2-thiophenemagnesium bromide with phosphorus trichloride yields tris(2-thienyl)phosphine [522]. Methylated tris(thienyl)phosphines are obtained in 28–66% yield through the reaction of the corresponding thienyllithium derivatives with phosphorus tribromide [523]. A better method consists in the reaction of the thienyllithium derivatives with triphenyl phosphite [524].

Tris(2-methyl-3-thienyl)phosphine [523]

To 2-methyl-3-thienyllithium, prepared from 2-methyl-3-bromothiophene (6.8 g, 0.38 mol) in anhydrous diethyl ether (125 ml) and 0.95 *M* buthyllithium in ether (370 ml, 0.35 mol), phosphorus tribromide (21.6 g, 0.08 mol) in anhydrous ether (50 ml) is added dropwise at −70 °C over a period of 30 min. The reaction mixture is stirred overnight while the cooling bath is allowed to reach room temperature. After hydrolysis with a solution of ammonium chloride and hydrochloric acid at 0 °C the organic layer is separated, washed with sodium bicarbonate solution and dried. Fractionation gives 26.5 g (64%) of the title compound as a red viscous oil, bp 158–162 °C/0.15 mm Hg. On standing overnight this oil crystallizes and recrystallization from methanol gives fine white crystals mp 81–83 °C.

The reaction of excess 2-thienyllithium with phenyl phosphonous dichloride gives phenyldi(2-thienyl)phosphine [525].

Phenyldi(2-thienyl)phosphine [525]

Butyllithium (0.4 mol) in light petroleum (bp 40–60 °C, 200 ml) is added dropwise with stirring under nitrogen to thiophene (35 g, 0.42 mol) in anhydrous ether (150 ml) during 20 min. The resulting solution is left for 1 h before being cooled in ice. Phenylphosphorous dichloride (17.9 g, 0.1 mol) in benzene (50 ml) is then added slowly, after which the reaction mixture is heated under reflux for 1 h, cooled in ice and hydrolyzed with 10% ammonium chloride solution (100 ml). The phases are separated and the organic phase dried over sodium sulfate and evaporated. The residue is distilled giving 18.4 g (67%) of the title compound bp 150 °C/0.15 mm Hg.

5.2.1.3 *Via* ring-closure reactions

Rearrangements of 1-alkynyl-2-alkynyl sulfides to thioketenes in the presence of dialkylphosphine lead upon ring-closure to 2-dialkylphosphinothiophenes [526].

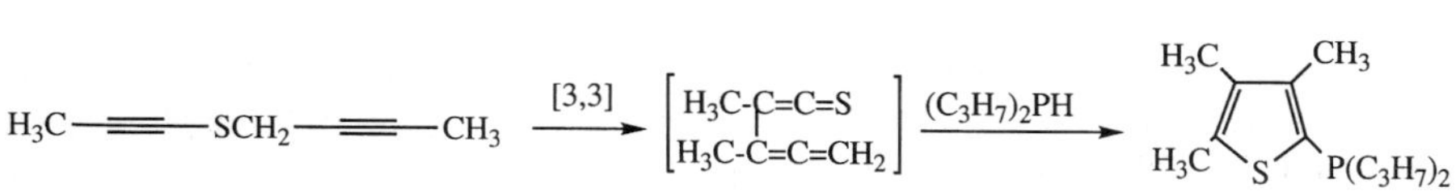

3,5-Diamino-4-phosphorylthiophenes are prepared by alkylation with α-haloketones of phosphorylated thioamides of carboxylic acids [527].

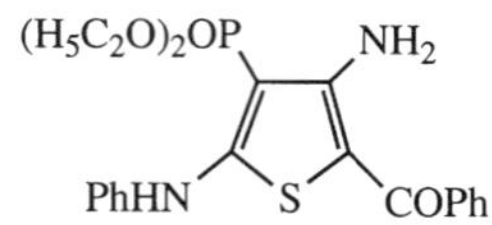

5.2.2 Thienyl phosphonium salts

Heating of tris(2-thienyl)phosphine with two equivalents of benzyl chloride in benzene is used for the preparation of tris(2-thienyl) benzylphosphonium chloride [528] and with methyl iodide, methyl tris(2-thienyl)phosphonium iodide is obtained [529]. Heating of 2-furyldiphenylphosphine, 2-bromothiophene, and anhydrous nickel bromide at 160–170 °C affords 2-furyldiphenyl-(2-thienyl)phosphonium bromide [530].

2-Furyldiphenyl-(2-thienyl)phosphonium bromide [530]

2-Furyldiphenylphosphine (1.00 g, 4 mmol), 2-bromothiophene (1.3 g, 8 mmol), and anhydrous nickel bromide (0.44 g, 2 mmol) are heated together at 160–170 °C for 5 h. On cooling the deep-green semi-crystalline mass is treated with water (10 ml) and the resulting emulsion extracted with ether (2 × 10 ml), which is discarded. The aqueous solution is then extracted with chloroform (3 × 10 ml). After evaporation 0.6 g (40%) of the title compound is obtained mp 243–246 °C (decomp.) from methyl acetate/ethanol.

5.2.3 Thienylphosphine oxides

The most convenient method for the preparation of thienyl phosphine oxide is the oxidation of thienylphosphines with hydrogen peroxide in acetone at 0 °C [525,531]. Another direct route to tris(2-thienyl)phosphine oxide consists in the reaction of 2-thiophenemagnesium bromide with diethyl chlorophosphate [532]. A number of 2,2′-disubstituted di(3-thienyl)phosphino oxides have been prepared by lithiation of 3-thienylphosphine oxides followed by reaction with electrophiles [533].

Phenyl bis(diphenylhydroxymethyl-3,3-thienyl) phosphine oxide [533]

A solution of phenyl di-3-thienyl phosphine oxide (5.50 g, 0.02 mol) in anhydrous tetrahydrofuran (150 ml) is treated with butyllithium in hexane (0.042 mol) at room temperature for about 1 h. Then benzophenone (12. 7 g,

0.07 mol) in tetrahydrofuran is added and the reaction mixture stirred at room temperature for 2–3 h, after which it is hydrolyzed by aqueous ammonium chloride. The phases are separated, the organic phase evaporated and the residue extracted with chloroform. The combined chloroform phases are dried over magnesium sulfate and evaporated. The residue is treated with acetone, the precipitate formed is filtered off and recrystallized from dioxane giving 10.2 g (80%) of the title compound mp 258 °C.

5.2.4 Thienyl phosphine sulfides

This class of compounds is conveniently prepared through the reaction of tris(thienyl)phosphines with sulfur [523].

Tris(2-methyl-3-thienyl)phosphine sulfide [523]

Tris(2-methyl-3-thienyl)phosphine (1.5 g, 4.65 mmol) in methanol (50 ml) is refluxed overnight with a slight excess of sulfur. After removal of the solvent the solid residue is dissolved in benzene and reprecipitated with petroleum ether in order to remove excess of sulfur. Recrystallization from methanol gives the title compound in quantitative yield mp 117–119 °C.

5.2.5 Thienylphosphonous dihalides and compounds derived from them

2-Thienylphosphonous dichloride can conveniently be obtained through Friedel–Crafts reaction of thiophene with phosphorus trichloride using stannic chloride as catalyst [534,535]. The reaction of thiophene with equivalent amounts of phosphorus tribromide under reflux in pyridine gives 2-thienylphosphonous dibromide, which upon reaction with diethylamine gives 2-thienylphosphonous acid bis(dimethylamide) [521].

2-Thienylphosphonous acid bis(dimethylamide) [521]

To a stirred solution of thiophene (4.2 g, 0.05 mol) in pyridine (10 ml) is added phosphorus tribromide (13.6 g, 0.05 mol) and the reaction mixture is refluxed for 48 h. It is then cooled to 20 °C and treated with hexane (20 ml). The pyridine hydrobromide formed is removed by filtration and a solution of dimethylamine in hexane (50 ml) is added to the filtrate with stirring and cooling. The reaction mixture is stirred at room temperature for 2 h, after which the dimethylamine hydrobromide is removed by filtration, the filtrate is evaporated and the residue fractionated in a stream of argon giving 6.7 g (66%) of the title compound bp 77–78 °C/0.05 mm Hg.

With 0.5 mol of phosphorus tribromide bis(2-thienyl) phosphonous bromide is obtained [521]. The reaction of 2-thienylphosphonous dichloride with tetraethylplumbane yields ethyl 2-thienylphosphinous chloride, which upon reaction with methanol in the presence of triethylamine gives methyl ethyl-2-thienylphosphonite, and upon reaction with diethylamine gives *N,N,P*-triethyl-*P*-thienylphosphinous amide [536].

N,N,P-Triethyl-P-thienylphosphinous amide [536]

A solution of ethyl 2-thienylphosphinous chloride (53.6 g, 0.3 mol) in anhydrous benzene (400 ml) is cooled with a mixture of ice and water while diethylamine (55 g, 0. 75 mol) is added at such a rate that the temperature does not rise above 8 °C. The reaction mixture is then stirred at room temperature for 6 h. Diethylamine is filtered off, the filtrate evaporated and the residue distilled giving 30 g (47%) of the title compound bp 121.122 °C/10 mm Hg.

The reaction of 2-thienylphosphonous dichloride with ethylene oxide can be used for the preparation of bis(2-chloroethyl)-2-thienylphosphonite, which upon distillation rearranges to a mixture of several compounds [537].

Bis(2-chloroethyl)-2-thienylphosphonite [537]

In a stream of argon with stirring and cooling, ethylene oxide (70 g, 1.6 mol) is pressed into a solution of 2-thienylphosphonous dichloride (105 g, 0.57 mol) in anhydrous benzene (300 ml) in such a way that the temperature of the reaction mixture does not rise above 55 °C, after which it is left overnight at room temperature. Evaporation gives 150 g (96%) of the title compound as a crude product.

(5-Chloro-2-thienyl)(1-hydroxyalkyl)phosphinic esters are prepared by the reaction of ethyl hydrogen (5-chloro-2-thienyl)phosphonite with aldehydes and ketones [538]. The reaction of 5-chloro-2-thienylphosphonous dichloride with acrylic acid gives [2-(chloroformyl)ethyl](5-chloro-2-thienyl)phosphinic chloride, which upon reaction with ethanol and triethylamine gives ethyl (5-chloro-2-thienyl)-[2-(ethoxycarbonyl)ethyl]phosphinate [538].

5.2.6 Thienylphosphonates

Low-temperature UV-photolysis of 2-iodothiophene in excess trimethyl phosphite gives a lower yield (32%) of methyl 2-thienylphosphonate than iodobenzenes [540]. Preparative electrochemical oxidation of the sodium salt of dialkylphosphorus acids in thiophene gives a 30% yield of ethyl 2-thienyl phosphonate [540].

A number of 2-trialkylsilylthiophenes are metalated by butyllithium and reacted with dialkyl chlorophosphate and dialkyl chlorothiophosphate to give 5-trialkylsilyl-substituted 2-phosphonates and thiophosphonate [541].

Diethyl 2-(5-trimethylsilyl)thienylphosphonate [541]

To a solution of thiophene (8.4 g, 0.1 mol) in anhydrous diethyl ether heated at reflux, butyllithium (0.1 mol) is added. The mixture is stirred and refluxed for 2 h, after which trimethylsilyl chloride (10.8 g, 0.1 mol) is added rapidly and the stirring is continued for 2 h. Another portion of butyllithium (0.1 mol) is added and this mixture is stirred at reflux for 3 h and added to a solution of diethyl chlorophosphate (17.2 g, 0.1 mol). The stirring is continued for 14 h, a solution of sodium bicarbonate is added and the phases separated. The organic phase is dried over magnesium sulfate, evaporated, and distilled giving 9.0 g (31%) of the title compound bp 103 °C/0.07 mm Hg.

5.3 ARSENIC, ANTIMONY, AND BISMUTH DERIVATIVES

Di(2-thienyl)chloroarsine can be prepared through the reaction of 2-thienylmercuric chloride with arsenic trichloride in toluene and was converted to di(2-thienyl)cyanoarsine upon reaction with aqueous sodium cyanide. From a mixture of di(2-thienyl)chloroarsine and 2-thienyl dichloroarsine tris(2-thienyl)arsine is obtained upon heating with an excess of 2-thienylmercuric chloride in toluene [542].

The reaction of 2-thienylmercuric chloride and phenyl dichloroarsine provides a method for the preparation of phenyl 2-thienylchloroarsine [542]. In a similar reaction 5-benzoyl-2-thienyldichloroarsine was prepared from 5-benzoyl-2-thienylmercuric chloride and arsenic trichloride. Upon treatment with sodium hydroxide this compound gives the 5-benzoyl-2-thienyl arsine-oxide. Upon oxidation of either of these compounds with 3% hydrogen peroxide in 2 *M* sodium hydroxide 5-benzoyl-2-thienylarsonic acid was obtained [543]. Tris (2-thienyl)antimon has been prepared in 83% yield through the reaction of 2-thiophenemagnesium bromide with antimon trichloride [544].

The reactions of *cis*1,2-di(2-lithio-3-thienyl)ethene and *cis*-1,2-di(4-lithio-3-thienyl)ethene with phenylantimon dichloride and phenylbismuth dichloride give the corresponding dithieno[2,3-*b*;3′,2′-]- and dithieno[3,4-*b*;3′,4′-*f*]heteropines containing group 15 elements [545].

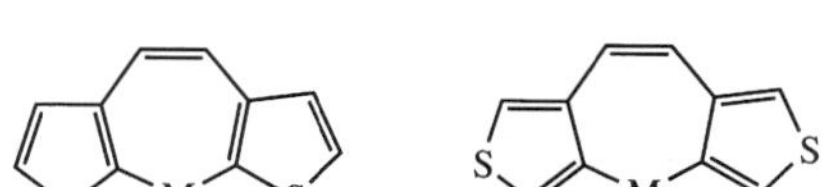

M = SbPh or BiPh

REFERENCES

1. V. S. Babasinian, *J. Am. Chem. Soc.* **50**, 2748 (1928).
2. V. S. Babasinian, *Org. Syntheses*, Coll. Volume **II**, 466 (1943).
3. G. Olah, S. Kuhn and A. Mlinko, *J. Chem. Soc.* 4257 (1956).
4. V. N. Ivanova, *J. Gen. Chem. USSR (Engl. Transl.)* **28**, 1288 (1958).
5. A. V. Zimichev and A. E. Lipkin, *Str. Svoistva Molekul.* 128 (1976); *Chem. Abstr.* **88**, 120268g (1978).
6. B. Östman, *Acta Chem. Scand.* **22**, 1687 (1968).
7. P. J. Newcombe and R. K. Norris, *Aust. J. Chem.* **31**, 2463 (1978).
8. H. Maag and B. K. Manukian, *Helv. Chim. Acta* **56**, 1787 (1973).
9. L. J. A. Martins and T. J. Kemp, *J. Chem. Soc., Faraday Trans. 1* **78**, 519 (1982).
10. L. J. Rinkes, *Recl. Trav. Chim. Pays-Bas* **51**, 1134 (1932).
11. J. M. Bakke, I. Hegbom, E. Øvreeide and K. Aaby, *Acta Chem. Scand.* **48**, 1001 (1994).
12. V. M. Zubarovskii, *Dokl. Akad. Nauk SSSR* **83**, 85 (1952).
13. O. Dann, *Ber.* **76**, 419 (1943).
14. S. Gronowitz and B. Holm, *Chem. Scripta* **2**, 245 (1972).
15. S. Gronowitz and I. Ander, *Chem. Scripta* **15**, 135 (1980).
16. I. J. Rinkes, *Recl. Trav. Chim. Pays-Bas* **52**, 538 (1933).
17. W. Steinkopf and P. J. Müller, *Liebigs Ann. Chem.* **448**, 210 (1926).
18. P. Raynaud and R. Delaby, *Bull. Soc. Chim. Fr.* 1614 (1955).
19. W. Steinkopf and T. Höpner, *Liebigs Ann. Chem.* **501**, 174 (1933).
20. A. H. Blatt, S. Bach and L. W. Kresch, *J. Org. Chem.* **22**, 1693 (1957).
21. H. Burton and W. A. Davy, *J. Chem. Soc.* 525 (1948).
22. J. Tirouflet and P. Fournari, *Bull. Soc. Chim. Fr.* 1651 (1963).
23. P. Fournari and J. P. Chane, *Bull. Soc. Chim. Fr.* 479 (1963).
24. G. Doddi, G. Illuminati, P. Mencarelli and F. Stegel, *J. Org Chem.* **41**, 2824 (1976).
25. C. Dell'Erba and G. Guanti, *Gazz. Chim. Ital.* **100**, 223 (1970).
26. A. H. Blatt, N. Gross and E. W. Tristam, *J. Org. Chem.* **22**, 1588 (1957).
27. W. T. Smith, *J. Am. Chem. Soc.* **71**, 2855 (1949).
28. W. T. Smith and L. Campanaro, *J. Am. Chem. Soc.* **75**, 3602 (1953).
29. C. Dell'Erba, A. Mele, M. Novi, G. Petrillo and S. Stagnano, *Tetrahedron* **48**, 4407 (1992).
30. C. Dell'Erba, M. Novi, G. Petrillo and P. Stagnaro, *Tetrahedron Lett.* **33**, 7047 (1992).
31. C. Dell'Erba, M. Novi, G. Petrillo and P. Stagnaro, *J. Heterocycl. Chem.* **31**, 861 (1994).
32. C. Dell'Erba, A. Mele, M. Novi, G. Petrillo and P. Stagnaro, *Tetrahedron Lett.* **31**, 4933 (1990).
33. C. Dell'Erba, M. Novi, G. Petrillo, D. Spinelli and C. Tavani, *Tetrahedron*, **52**, 3313 (1996).
34. A. Mugnoli, C. Dell'Erba, G. Guanti and M. Novi, *J. Chem. Soc., Perkin Trans. 2* 1764 (1980).
35. J. Namtvedt, *Acta Chem. Scand.* **22**, 1611 (1968).
36. H. R. Snyder, L. A. Carpino, J. F. Zack and J. F. Mills, *J. Am. Chem. Soc.* **79**, 2556 (1957).
37. C. Sone and S. Matsuki, *Nippon Kagaku Zasshi* **83**, 496 (1962); *Chem. Abstr.* **59**, 3861 (1963).
38. C. M. Cammagi, R. Leardini and G. Placucci, *J. Chem. Soc., Perkin Trans. 2* 1195 (1974).
39. R. M. Kellogg and J. Buter, *J. Org. Chem.* **36**, 2236 (1971).
40. M. Sy, N. P. Buu-Hoi and N. D. Xuong, *J. Chem. Soc.* 1975 (1954).
41. S. Nishimura, A. Sakumoto and E. Imoto, *Nippon Kagasku Zasshi* **82**, 1540 (1961); *Chem. Abstr.* **57**, 15051 (1962).
42. E. Campaigne and H. G. Grose, *J. Am. Chem. Soc.* **73**, 3812 (1951).
43. Y. L. Gol'dfarb, B. P. Fabrichnyi and I. F. Shalavina, *J. Gen. Chem. USSR (Engl. Transl.)* **6,** 1093 (1970).
44. W. Steinkopf, I. Poulsson and O. Herdey, *Liebigs Ann. Chem.* **536**, 128 (1938).
45. H. Suzuki, I. Hidaka and A. Osuka, *Chem. Lett.* 633 (1980).

46. H. Suzuki, I. Hidaka, A. Iwasa, T. Miashina and A. Osuka, *Bull. Chem. Soc. Jpn.* **54**, 771 (1981).
47. S. S. Mochalov, T. P. Surikova, F. M. Abdel'razek, V. D. Zakharova and Y. S. Shabarov, *Chem. Heterocycl. Compd. (Engl. Transl.)* **17**, 132 (1981).
48. J. Nakayama, S. Tamaoka and M. Hoshino, *Tetrahedron Lett.* **29**, 1161 (1988).
49. R. Ballini, G. Bartoli, M. Bosco, R. Dalpozzo and E. Marcantoni, *Tetrahedron* **44**, 6435 (1988).
50. M. D. Threadgill, P. Webb, P. O'Neill, M. A. Naylor, M. A. Stephens, I. J. Stratford, S. Cole, G. E. Adams and E. M. Fielden, *J. Med. Chem.* **34**, 2112 (1991).
51. M. Makosza and E. Kwast, *Tetrahedron* **51**, 8339 (1995).
52. A. Arcoria, E. Maccarone, G. Musamarra and G. Romano, *J. Heterocycl. Chem.* **9**, 849 (1972).
53. M. Bercot-Vatteroni, *Ann. Chim. (Paris)* **7**, 302 (1962).
54. C. Sone, K. Takahashi and Y. Matsuki, *Bull. Chem. Soc. Jap.* **35**, 1420 (1962).
55. C. Sone and Y. Matsuki, *Bull. Chem. Soc. Jap.* **37**, 1235 (1964).
56. C. Sone and Y. Matsuki, *Bull. Chem. Soc. Jap.* **41**, 1423 (1968).
57. I. J. Rinkes, *Recl. Trav. Chim. Pays-Bas* **52**, 1052 (1933).
58. S. Gronowitz and I. Ander, *Chem. Scripta* **15**, 20 (1980).
59. A. Daich, J. Morel and B. Decroix, *J. Heterocycl. Chem.* **30**, 675 (1993).
60. T. J. J. Müller, J. P. Robert, E. Schmälzlin, C. Bräuchle and K. Meerholz. *Org. Lett.* **2**, 2419 (2000).
61. V. N. Listvan, *Chem. Heterocycl. Compds. USSR (Engl Transl.)* 1624 (1974).
62. G. Papalardo, *Gazz. Chim. Ital.* **89**, 511 (1959).
63. R. Kimura, T. Yabuuchi and M. Hisaki, *Chem. Pharm Bull.* **10**, 1232 (1962).
64. W. J. Raich and C. S. Hamilton, *J. Am. Chem. Soc.* **79**, 3800 (1957).
65. S. Gronowitz and I. Ander, *Chem. Scripta* **15**, 145 (1980).
66. M. D'Auria, *Gazz. Chim. Ital.* **124**, 195 (1994).
67. S. Yuquan, Z. Yuxia, L. Zao, W. Jianghong, Q. Ling, L. Shixiong, Z. Jianfeng and Z. Jiayun, *J. Chem. Soc., Perkin Trans. I* 3691 (1999).
68. D. J. Freeman, P. J. Newcombe and R. K. Norris, *Aust. J. Chem.* **29**, 327 (1976).
69. S. Gronowitz and N. Gjøs, *Acta Chem. Scand.* **21**, 2823 (1967).
70. N. Gjøs and S. Gronowitz, *Acta Chem. Scand.* **26**, 1851 (1972).
71. C. Gozzi, L. Lavenot, K. Ilg, V. Penalva and M. Lemaire, *Tetrahedron Lett.* **38**, 8867 (1997).
72. L. Lavenot, C. Gozzi, K. Ilg, I. Orlova, V. Penalva and M. Lemaire. *J. Organomet. Chem.* **567**, 49 (1998).
73. A. J. Seed, K. J. Toyne, J. W. Goodby and M. Hird, *J. Mater. Chem.* **10**, 2069 (2000).
74. J. Tirouflet, *Bull. Soc. Chim. Fr.* 1066 (1958).
75. V. M. Colburn, B. Iddon, H. Suschitzky and P. T. Gallagher. *J. Chem. Soc., Perkin Trans. 1* 1337 (1979).
76. R. Benoit, J. Duflos, G. Dupas, J. Bourguignon and G. Queguiner, *J. Heterocycl. Chem.* **26,** 1595 (1989).
77. T. Kurihara, J. Sasaki, K. Santo, Y. Nakamura, R. Yoneda and S. Harusawa, *Heterocycles* **29**, 2007 (1989).
78. H. Cercetto, R. di Maio, M. Gonzales and G. Seoane, *Heterocycles* **45**, 2023 (1997).
79. Y. L. Gol'dfarb, E. I. Novikova and L. I. Belen'kii, *Bull. Acad. Sci. USSR, Div. Chem. Sci.* 1142 (1971).
80. R. P. Fabrichnyi, S. M. Kostrova, G. P. Gromova and Y. L. Gol'dfarb, *Chem. Heterocycl. Compd. (Engl. Transl.)* **9**, 1341 (1973).
81. H. Goda, M. Sato, H. Ihara and C. Hirayama, *Synthesis* 849 (1992).
82. J. Tirouflet and P. Fournari, *C. R. Hebd. Seances Acad. Sci.* **246**, 2003 (1958).
83. P. Cogolli, F. Maiolo, L. Testaferri, M. Tiecco and M. Tingoli, *J. Chem. Soc., Perkin Trans. 2* 1331 (1980).

84. K. K. Venter, M. Trushule, V. P. Litvinov, E. G. Ostrapenko and E. Liepinsh, *Chem. Heterocycl. Compd. (Engl. Transl.)* **14**, 490 (1978).
85. R. L. Elliot, P. J. O'Hanlon and N. H. Rogers, *Tetrahedron* **43**, 3295 (1987).
86. H. Paul and H. Migulla, *Arch. Pharm. (Weinhein, Ger.)* **311**, 679 (1978).
87. Y. L. Gol'dfarb, V. V. N. Bulgakova and B. P. Fabrichnyi, *Chem. Heterocycl. Compd. (Engl. Transl.)* **19**, 1283 (1983).
88. B. Östman, *Acta Chem. Scand.* **22**, 2754 (1968).
89. L. I. Belen'kii, E. I. Novikova, L. A. D'yachenko and Y. L. Gol'dfarb, *J. Org. Chem. USSR, (Engl. transl.)* **7**, 1803 (1971).
90. M. Fiorenza, A. Ricci, G. Sbrana, G. Pirazzini, C. Eaborn and J. G. Stamper, *J. Chem. Soc., Perkin Trans. 2* 1232 (1978).
91. E. E. Campaigne and R. C. Bourgeois, *J. Am. Chem. Soc.* **76**, 2445 (1954).
92. L. J. Rinkes, *Recl. Trav. Chim. Pays-Bas* **53**, 643 (1934).
93. G. Baldini, G. Doddi, G. Illuminati and F. Stegel, *J. Org. Chem.* **41**, 2153 (1976).
94. S. Gronowitz, P. Moses, A.-B. Hörnfeldt and R. Håkansson, *Arkiv Kemi* **12**, 165 (1961).
95. P. J. Newcombe and R. K. Norris, *Aust. J. Chem.* **34**, 1879 (1981).
96. R. Benoit, J. Duflos, G. Dupas, J. Bourguignon and C. Quegnier, *J. Heterocycl. Chem.* **26**, 1595 (1989).
97. C. Dell'Erba, F. Sancassan, M. Novi, D. Spinelli, G. Consiglio, C. Arnone and F. Ferroni, *J. Chem. Soc., Perkin Trans. 2* 1779 (1989).
98. A. K.-Y. Jen, V. P. Rao, K. J. Drost, K. Y. Wong and M. P. Cava, *J. Chem. Soc. Chem. Commun.* 2057 (1994).
99. R. J. Hamlyn, R. H. Jones and C. A. Ramsden, *J. Chem. Soc., Perkin 1* 1811 (2000).
100. W. Steinkopf, *Liebigs Ann. Chem.* **403**, 17 (1914).
101. W. Steinkopf and D. Jaffé, *Liebigs Ann. Chem.* **413**, 33 (1917).
102. G. Henrio, J. Morel and P. Pastour, *Bull. Soc. Chim Fr.* 265 (1976).
103. S. Nishimura and E. Imoto, *Bull. Chem. Soc. Jpn.* **35**, 432 (1962).
104. E. Campaigne and J. L. Diedrich, *J. Am. Chem. Soc.* **73**, 5240 (1951).
105. E. Profft and G. Solf, *J. Prakt Chem.* **24**, 38 (1964).
106. B. P. Fabrichnyi, I. F. Shalavina and Y. L. Gol'dfarb, *Dokl. Chem. (Engl. Trannsl.)* 447 (1965).
107. C. J. Grol, *J. Heterocycl. Chem.* **11**, 953 (1974).
108. O. Stadler, *Ber.* **18**, 1490 (1885).
109. M. J. S. Dewar and P. A. Marr, *J. Am. Chem. Soc.* **84**, 3782 (1962).
110. M. Sy and B. de Mallerray, *Bull. Soc. Chim. Fr.* 1276 (1963).
111. M. Bercot-Vatteroni, *Bull. Soc. Chim. Fr.* 234 (1964).
112. W. W. Gale, A. N. Scott and H. R. Snyder, *J. Org. Chem.* **29**, 2160 (1964).
113. A. N. Scott, B. E. Hoogenboom and H. R. Snyder, *J. Org. Chem.* **29**, 2165 (1964).
114. N. V. Stulin and N. I. Putokhin, *Chem. Heterocycl. Compd. (Engl. Transl.)* **1**, 463 (1965).
115. J. Cymerman-Craig and D. Willis, *J. Chem. Soc.* 1071 (1955).
116. J. Cymerman-Craig, G. N. Vaughan and W. K. Warburton, *J. Chem. Soc.* 4114 (1956).
117. V. Paulmier and F. Outurquin, *J. Heterocycl. Chem.* **20**, 113 (1983).
118. Y. L. Go'dfarb, B. P. Fabrichnyi and I. F. Shalavina, *Chem. Heterocycl. Compd (Engl. Transl.)* **18**, 1018 (1982).
119. O. Hromatka, D. Binder and P. Stanetti, *Monatsh. Chem.* **104**, 920 (1973).
120. L. H. Klemm and W. Hsin, *J. Heterocycl. Chem.* **12**, 1138 (1975).
121. B. M. de Malleray, *Helv. Chim. Acta* **54**, 351 (1971).
122. B. Bartho, J. Faust, R. Pohl and R. Mayer, *J. Prakt. Chem.* **318**, 221 (1976).
123. H. Schäfer and K. Gewald, *Z. Chem.* **15**, 100 (1975).
124. C. Finzi and V. Furlotti, *Gazz. Chim. Ital.* **45 II**, 290 (1915).
125. C. Finci, *Gazz. Chim Ital.* **60**, 159 (1930).
126. O. Dann and E. F. Möller, *Chem. Ber.* **80**, 23 (1947).

127. N. I. Astrakhantseva, V. G. Zhiryakov and P. I. Abramenko, *Chem. Heterocycl. Chem. (Engl. Transl.)* **12**, 1123 (1976).
128. N. I. Astrakhantseva, V. G. Zhiryakov and P. I. Abramenko, *Chem. Heterocycl. Compd. (Engl. Transl.)* **11**, 1374 (1975).
129. W. E. Parham, I. Nicholson and V. I. Traynelis, *J. Am. Chem. Soc.* **78**, 850 (1956).
130. H. Y. Lew and C. R. Noller, *J. Am. Chem. Soc.* **72**, 5715 (1950).
131. Y. L. Gol'dfarb, M. M. Polonskaya, B. P. Fabrichnyi and I. F. Shalavina, *Dokl. Chem. (Engl. Transl.)* 311 (1959).
132. D. Binder, C. R. Noe, F. Geissler and F. Hildebrand, *Arch Pharm. (Weinheim, Ger.)* **314**, 564 (1981).
133. R. Mozingo, S. A. Harris, D. E. Wolf, C. E. Hoffhine, N. R. Easton and K. Folkers, *J. Am. Chem. Soc.* **67**, 2092 (1945).
134. P. Rossy, F. G. M. Vogel, W. Hoffmans, J. Paust and A. Nürrenbach, *Tetrahedron Lett.* **22**, 3493 (1981).
135. J. K. Chakrabarti, T. M. Hotten, D. J. Steggles and D. E. Tupper, *J. Chem Res. (M)* 5101 (1978).
136. D. H. Jones and K. R. H. Woolridge, *J. Chem. Soc. C* 550 (1968).
137. G. Malicorne, J. Bompart, L. Giral and E. Despaux, *Eur. J. Med. Chem.* **26**, 3 (1991).
138. F. Outurquin and C. Paulmier, *Bull. Soc. Chim. Fr. 2*, 151 (1983).
139. F. Outurquin and C. Paulmier, *Bull. Soc. Chim. Fr. 2*, 153 (1983).
140. Q. T. Zhang and J. M. Tour, *J. Am. Chem. Soc.* **120**, 5355 (1998).
141. Q. T. Zhang and J. M. Tour, *J. Am. Chem. Soc.* **119**, 9624 (1997).
142. S. Tanaka, M. Tomura and Y. Yamashita, *Heterocycles* **37**, 693 (1994).
143. P. Marsham, *J. Heterocycl. Chem.* **31**, 603 (1994).
144. C. W. Sondern and P. J. Breivagel, *U.S. Pat. 2*, **519**, 325 (1950); *Chem. Abstr.* **45**, 666 (1951).
145. D. N. Reinhoudt, J. Geevers, W. P. Trompenaars, S. Harkema and G. J. van Hummel, *J. Org. Chem.* **46**, 424 (1981).
146. D. N. Reinhoudt, W. P. Trompenars and J. Geevers, *Synthesis* 368 (1978).
147. E. W. Brunett and W. C. McCarthy, *J. Heterocycl. Chem.* **5**, 417 (1968).
148. J. B. Sullivan and W. C. McCarthy, *J. Heterocycl. Chem.* **2**, 103 (1965).
149. J. B. Sullivan and W. C. McCarthy, *J. Org. Chem.* **30**, 662 (1965).
150. F. A. Buiter, J. H. Sperna Weiland and H. Wynberg, *Recl. Trav. Chim. Pays-Bas* **83**, 1160 (1964).
151. M. G. Reinecke, H. W. Adickes and D. C. Pyun, *J. Heterocycl. Chem.* **10**, 1067 (1973).
152. F. D. King and D. R. M. Walton, *J. Chem. Soc., Chem. Commun.* 256 (1974).
153. P. V. Bedworth, Y. Cai, A. Jen and S. R. Marder, *J. Org. Chem.* **61**, 2242 (1996).
154. M. Watanabe, T. Yamamoto and M. Nishiyama, *Chem. Commun.* 133 (2000).
155. E. Ueta, H. Nakano and Y. Shirota, *Chem. Lett.* 2397 (1994).
156. H. Hartmann, *J. Prakt. Chem.* **36**, 50 (1967).
157. S. Scheithauer, H. Hartmann and R. Mayer, *Z. Chem.* **8**, 181 (1968).
158. H. Hartmann and S. Scheithauer, *J. Prakt. Chem.* **311**, 827 (1969).
159. D. Keil, H. Hartmannn and C. Reichardt, *Liebigs Ann. Chem.* 935 (1993).
160. N. Ikemoto, I. Estevez, K. Nakanishi and N. Berova, *Heterocycles* **46**, 489 (1997).
161. E. B. Pedersen and S.-O. Lawesson, *Tetrahedron* **30**, 875 (1974).
162. W. ten Hoeve, C. G. Kruse, J. M. Luteyn, J. R. G. Thiecke and H. Wynberg, *J. Org. Chem.* **58**, 5101 (1993).
163. S. Gronowitz and B. Holm, *Chem. Scripta* **6**, 133 (1974).
164. F. Outurquin and C. Paulmier, *Tetrahedron Lett.* **34**, 5719 (1993).
165. C. Paulmier, *J. Heterocycl. Chem.* **33**, 9 (1996).
166. F. M. Stoyanovich and M. A. Marakatkina, *Chem. Heterocycl. Comp. (Engl. Transl.)* **9**, 1346 (1983).

167. A. Casarini, P. Dembech, D. Lazzari, E. Marini, G. Reginato, A. Ricci and G. Seconi, *J. Org. Chem.* **58**, 5620 (1993).
168. P. Bernardi, P. Dembech, G. Fabbri, A. Ricci and G. Seconi, *J. Org. Chem.* **64**, 641 (1999).
169. S. Gronowitz, C. Westerlund and A.-B. Hörnfeldt, *Acta Chem. Scand., Sect. B.* **29**, 224 (1975).
170. D. Brugier, F. Outurquin and C. Paulmier, *Tetrahedron* **56**, 2985 (2000).
171. O. Meth-Cohn and B. Narine, *Synthesis* 133 (1980).
172. G. Ah-Kow, C. Paulmier and P. Pastour, *Bull. Soc. Chim. Fr.* 151 (1976).
173. C. D. Hurd and J. Moffat, *J. Am. Chem. Soc.* **73**, 613 (1951).
174. E. E. Campaigne and P. A. Monroe, *J. Am. Chem. Soc.* **76**, 2447 (1954).
175. R. A. Hoffman and S. Gronowitz, *Arkiv Kemi* **16**, 515 (1960).
176. E. W. Brunett, D. M. Altwein and W. C. McCarthy, *J. Heterocycl. Chem.* **10**, 1067 (1973).
177. M. Murakami and M. Hikichi, *Jap. Pat.* **12**, 895 (1969); *Chem. Abstr.* **71**, 101702b (1969).
178. E. W. Brunett and W. C. McCarthy, *J. Pharm. Sci.* **57**, 2003 (1968).
179. D. Binder, G. Habison and C. R. Noe, *Synthesis* 255 (1977).
180. F. Garcia and C. Galvez, *Sulfur Lett.* **1**, 97 (1982).
181. K. Ninomiya, T. Shiori and S. Yamada, *Tetrahedron* **30**, 2151 (1974).
182. C. Galvez, F. Garcia, J. Garcia and J. Soldevilla, *J. Heterocycl. Chem.* **23**, 1103 (1986).
183. Y. Yang, A.-B. Hörnfeldt and S. Gronowitz, *Chem. Scripta* **28**, 275 (1988).
184. J. Reisch and H. Labitzke, *Arch. Pharm. (Weinheim)* **310**, 840 (1977).
185. S. Rault, M. Cugnon de Sévricourt and N. M. Robba, *Recl. Trav Chim. Pays-Bas* **101**, 205 (1982).
186. D. Binder, F. Hillebrand and C. R. Noe, *J. Chem. Res. (M)* 1151 (1979).
187. S. Rault, Y. Effi, M. Cugnon de Sévricourt, J.-C. Lancelot and M. Robba, *J. Heterocycl. Chem.* **20**, 17 (1983).
188. L. C. Cheney and J. R. Piening, *J. Am. Chem. Soc.* **67**, 731 (1945).
189. P. N. Confalone, G. Pizzolato and M. R. Uskokovic, *Helv. Chim. Acta* **59**, 1005 (1976).
190. P. N. Confalone, G. Pizzolato and M. R. Uskokovic, *J. Org. Chem.* **42**, 135 (1977).
191. D. W. Slocum and P. L. Gierer, *J. Org. Chem.* **38**, 4189 (1973).
192. G. W. Stacey and D. L. Eck, *Tetrahedron Lett.* 5201 (1967).
193. D. L. Eck and G. W. Stacy, *J. Heterocycl. Chem.* **6**, 147 (1969).
194. J. P. Chupp, *J. Heterocycl. Chem.* **7**, 285 (1970).
195. A. Rolfs and J. Liebscher, *J. Chem. Soc., Chem. Commun.* 1437 (1994).
196. O. A. Tarasova, N. A. Nedolya, V. Y. Vvedemsky, L. Brandsma and B. A. Trofimov, *Tetrahedron Lett.* **38**, 7241 (1997).
197. R. L. P. de Jong and L. Brandsma, *Synth. Commun.* **20**, 3427 (1990).
198. O. A. Tarasova, L. V. Klyba, V. Y. Vvedensky, N. A. Nedolya, B. A. Trofimov, L. Brandsma and H. D. Verkruisse, *Eur. J. Org. Chem.* 253 (1998).
199. A. R. Katritzky, X. Wang and A. Denisenko, *J. Org. Chem.* **66**, 2850 (2001).
200. O. A. Tarasova, N. A. Nedolya, V. Y. Vvdensky, L. Brandsma and B. A. Trofimov, *Tetrahedron Lett.* **38**, 7241 (1997).
201. H. Hartmann, P. Gerstner and D. Rohde, *Org. Lett.* **3**, 1673 (2001).
202. L. Brandsma, N. A. Nedolya, O. A. Tarasova and B. A. Trofimov, *Chem. Heterocycl. Compds.* 1443 (2000).
203. K. Gewald, *Z. Chem.* **7**, 186 (1967).
204. H. Hassner, P. Munger and B. A. Belinka, *Tetrahedron Lett.* **23**, 699 (1982).
205. J. Meijer and L. Brandsma, *Rec. Trav. Chim. Pays-Bas* **91**, 578 (1972).
206. K. Tsuji, K. Nakamura, T. Ogino, N. Konishi, T. Tojo, T. Ochi, N. Seki and M. Matsuo, *Chem. Pharm. Bull.* **46**, 279 (1998).
207. F. Asinger and A. Mayer, *Angew. Chem., Int. Ed. Engl.* **4**, 788 (1965).
208. O. Hromatka, D. Binder, P. Stanetti and G. Marischler, *Monatsh. Chem.* **107**, 233 (1976).
209. H. Schäfer and K. Gewald, *J. Prakt. Chem.* **316**, 684 (1974).

210. K. Gewald and I. Hofmann, *J. Prakt. Chem.* **311**, 402 (1969).
211. A. C. Cope, *J. Am. Chem. Soc.* **59**, 2327 (1937).
212. A. C. Cope and K. E. Hoyle, *J. Am. Chem. Soc.* **63**, 733 (1941).
213. A. C. Cope, C. M. Hofmann, C. Wyckoff and E. Hartdenbergh, *J. Am. Chem. Soc.* **63**, 3452 (1941).
214. D. Nightingale and R. A. Carpenter, *J. Am. Chem. Soc.* **71**, 3560 (1949).
215. C. D. Slater and D. L Haywood, *J. Heterocycl. Chem.* **2**, 316 (1965).
216. T. Bacchetti, A. Alemagna and B. Danieli, *Tetrahedron Lett.* 2001 (1965).
217. G. Purrello, *Gazz. Chim. Ital.* **95**, 699 (1965).
218. F. Bottino and G. Purrello, *Gazz. Chim. Ital.* **95**, 1062 (1965).
219. G. Purrello, *Gazz. Chim. Ital.* **97**, 549 (1967).
220. K. R. Reddy, M. V. B. Rao, H. Ila and H. Junjappa, *Synth. Commun.* **26**, 4157 (1996).
221. R. W. Sabnis, D. W. Ragnneker and N. D. Sonawane, *J. Heterocycl. Chem.* **36**, 333 (1999).
222. K. Gewald, *Angew. Chem.* **73**, 114 (1961).
223. K. Gewald, *Chem. Ber.* **98**, 3571 (1965).
224. K. Gewald, *Z. Chem.* **2**, 305 (1962).
225. K. Gewald, E. Schinke and H. Böttcher, *Chem. Ber.* **99**, 94 (1966).
226. K. Gewald and E. Schinke, *Chem. Ber.* **99**, 2712 (1966).
227. T. P. Surikova, V. D. Zakharova, S. S. Mochalov and Yu. S. Shabarov, *Chem. Heterocycl. Compd. (Engl. Transl.)* **24**, 860 (1988).
228. O. Meth-Cohn and B. Narine, *J. Chem. Res. (M)* 3262 (1977).
229. V. I. Shvedov and A. N. Grinev, *J. Org. Chem. USSR (Engl. Transl.)* **1**, 2269 (1965).
230. V. P. Arya, *Ind. J. Chem.* **10**, 1141 (1972).
231. V. P. Arya and S. P. Ghate, *Indi. J. Chem.* **9**, 904 (1971).
232. R. W. Sabnis and D. W. Rangnekar, *J. Heterocycl. Chem.* **29**, 1027 (1992).
233. V. P. Arya and S. P. Ghate, *Indi. J. Chem.* **9**, 1209 (1971).
234. J. D. Ramanathan, D. G. Namboothiiri, G. F. Shah, A. V. Radhakrishnan, H. J. Mehta and A. C. Padhya, *J. Ind. Chem. Soc.* **55**, 822 (1978).
235. V. P. Arya, *Ind. J. Chem.* **10**, 812 (1972).
236. M. Robba, J. M. Lecomte and M. Cugnon de Sévricourt, *Bull. Soc. Chim. Fr. Pt. 2*, 2864 (1974).
237. V. P. Arya and S. P. Ghate, *Ind. J. Chem.* **9**, 1209 (1971).
238. M. B. Devani, C. J. Shishoo, U. S. Pathak, A. V. Radhakrishnan and A. C. Padhya, *Ind. J. Chem. Sect. B* **14**, 357 (1976).
239. F. Sauter, G. Reich and P. Stanetty, *Arch. Pharm. (Weinheim, Ger.)* **309**, 908 (1976).
240. K. H. Weber and H. Daniel, *Liebigs Ann. Chem.* 328 (1979).
241. C. J. Shishoo, M. B. Devani, S. Ananthan, K. S. Jain, V. S. Bhadti, S. Mohan and L. J. Patel, *Ind. J. Chem.* **28B**, 1039 (1989).
242. K. Gewald, M. Hentschel and R. Heikel, *J. Prakt. Chem.* **315**, 539 (1973).
243. O. Hromatka and D. Binder, *Monatsh. Chem.* **104**, 704 (1973).
244. O. Hromatka, D. Binder, C. R. Noe, P. Stanetty and W. Veit, *Monatsh. Chem.* **104**, 715 (1973).
245. F. J. Tinney, J. P. Sanchez and J. A. Nogas, *J. Med. Chem.* **17**, 624 (1974).
246. M. Nakanishi, T. Tahara, K. Araki, M. Shiroki, T. Tsumagari and Y. Takigawa, *J. Med. Chem.* **16**, 214 (1973).
247. T. Hirohashi, S. Inaba and H. Yamamoto, *Bull. Chem. Soc. Jap.* **48**, 147 (1975).
248. K. Gewald, *Chem. Heterocycl. Compd. (Engl. Transl.)* **12**, 1077 (1976).
249. G. D. Madding and M. D. Thompson, *J. Heterocycl. Chem.* **24**, 581 (1987).
250. R. W. Sabnis, G. Kazemi and D. W. Ragnekar, *Bull. Chem. Soc. Jpn.* **64**, 3768 (1991).
251. Z. A. Hozien, F. M. Atta, K. M. Hassan, A. A. Abdel-Wahab and S. A. Ahmed, *Synth. Commun.* **26**, 3733 (1996).
252. M. G. Reinecke, T. A. Woodrow and E. S. Brown., *J. Org. Chem.* **57**, 1018 (1992).

253. L. G. Sharanina and S. N. Baranov, *Chem. Heterocycl. Compds. (Engl. Transl.)* 171 (1974).
254. A. Rosowsky, C. E. Mota and S. F. Queener, *J. Heterocycl. Chem.* **33**, 1959 (1996).
255. E. R. Bacon and S. J. Daum, *J. Heterocycl. Chem.* **28**, 1953 (1991).
256. M. R. Player and J. W. Sowell, Sr, *J. Heterocycl. Chem.* **32**, 1537 (1995).
257. D. Hawksley, D. A. Griffin and F. J. Leeper, *J. Chem. Soc., Perkin Trans. 1* 144 (2001).
258. I. L. Pinto, R. L. Jarvest and H. T. Serafinowska, *Tetrahedron Lett.* **41**, 1597 (2000).
259. R. M. Mohareb, S. El-Kousy and A. M. El-Torgoman, *Collect. Czech. Chem. Commun.* **57,** 1747 (1992).
260. C. Heyde, I. Zug and H. Hartmann, *Eur. J. Org. Chem.* 3273 (2000).
261. H. Hartmann and I. Zug, *J. Chem. Soc., Perkin Trans. 1*, 4316 (2000).
262. K. Hirai and T. Ishiba, *Chem. Pharm. Bull.* **19**, 2194 (1971).
263. K. Hirai and T. Ishiba, *Chem. Pharm. Bull.* **20**, 2384 (1972).
264. H. Hartmann, *Z. Chem.* **11**, 421 (1971).
265. H. Hartmann, H. Schäfer and K. Gewald, *J. Prakt. Chem.* **315**, 497 (1973).
266. Y. Tominaga, V. Matsuda and G. Kobayashi, *Heterocycles* **4**, 9 (1976).
267. I. N. Kulaeva, P. S. Pel'kis, M. O. Lozinskii and V. N. Kalinin, *J. Org. Chem. USSR (Engl. Transl.)* **20**, 106 (1981).
268. T. Nishio, *Helv. Chim. Acta* **81**, 1207 (1998).
269. S. Gronowitz, *The Chemistry of Heterocyclic Compounds, Thiophene and its Derivatives*, Vol. 44 Part 1, p. 88 John Wiley and Sons 1985.
270. H. Fiesselmann, *Angew. Chem.* **71**, 377 (1959).
271. J. Brelivet and J. Teste, *Bull. Soc. Chim. Fr.* 2289 (1972).
272. J. M. J. Trochet, O. Martin, J.-B. Zumwald, N. Le-Hong and F. Perret, *Helv. Chim. Acta* **58,** 1735 (1975).
273. P. R. Huddleston and J. M. Barker, *Synth. Commun.* **9**, 731 (1979).
274. W.-Y. Ren, M.-I. Lim, B. A. Otter and R.S. Klein, *J. Org. Chem.* **47**, 4633 (1982).
275. G. Kirsch, D. Cagniant and P. Cagniant, *J. Heterocycl. Chem.* **19**, 443 (1982).
276. R. Gompper, *Angew. Chem.* **73**, 537 (1961).
277. S. Gronowitz, J. Fortea-Laguna, S. Ross, B. Sjöberg and N. E. Stjernström, *Acta Pharm. Suec.* **5**, 563 (1968).
278. M.-I. Lim, W.-Y. Ren and R. S. Klein, *J. Org. Chem.* **47**, 4594 (1982).
279. R. Gompper, E. Kutter and W. Töpfl, *Liebigs Ann. Chem.* **659**, 90 (1962).
280. K.-H. Etzbach and H. Eilingsfeld, *Synthesis* 449 (1988).
281. F. Jourdan, D. Ladurée and M. Robba, *J. Heterocycl. Chem.* **31**, 305 (1994).
282. H. Hartmann and J. Liebscher, *Synthesis* 275 (1984).
283. P. E. Morris, A. J. Elliot and J. A. Montgomery, *J. Heterocycl. Chem.* **36**, 423 (1999).
284. P. E. Morris, Jr., A. J. Elliot and J. A. Montgomery, *J. Heterocycl. Chem.* **36**, 423 (1999).
285. A. Kurfürst and P. Sebek, *Collect. Czech. Chem. Commun.* **54**, 1705 (1989).
286. R. M. Mohareb, *Gazz. Chim. Ital.* **122**, 147 (1992).
287. K. Gewald and U. Hain, *Monatsh. Chem.* **123**, 455 (1992).
288. K. Hartke and F. Meissner, *Tetrahedron* **28**, 875 (1972).
289. K. Hartke and G. Gölz, *Liebigs Ann. Chem.* 1644 (1973).
290. G. Gölz and K. Hartke, *Arch Pharm. (Weinheim, Ger.)* **307**, 663 (1974).
291. O. Günther and K. Hartke, *Arch. Pharm. (Weinheim, Ger.)* **308**, 693 (1975).
292. K. Hartke and B. Seib, *Pharmazie* **25**, 517 (1970).
293. M. T. Cocco, C. Congiu and V. Onnis, *J. Heteocycl. Chem.* **32**, 463 (1995).
294. A. O. Abdelhamid and S. M. Al-Shehri, *J. Chem. Research (S)* 240 (1997).
295. P. A. Rossy and W. Hoffman, *Ger. Pat. 2* **700**, 215 (1978); *Chem. Abstr.* **89**, 109062k (1978).
296. K. Gewald, M. Kleinert, B. Thiele and M. Hentschel, *J. Prakt. Chem.* **314**, 303 (1972).
297. T. L. Cairns, R. A. Carboni, D. D. Coffman, V. A. Engelhardt, R. F. Heckert, E. L. Little, E. G. McGeer, B. C. McKusick and W. J. Middleton, *J. Am. Chem. Soc.* **79**, 2340 (1957).
298. W. J. Middleton, V. A. Engelhardt and B. S. Fisher, *J. Am. Chem. Soc.* **80**, 2822 (1958).

299. W. J. Middleton, *Org. Synth, Coll. Vol. IV* 243 (1963).
300. F. M. Abdelrazek and H. A. Ead, *J. Prakt. Chem.* **330**, 585 (1988).
301. F. M. Abdelrazek, *Z. Naturforsch.* **44b**, 488 (1989).
302. M. Yokoyama, M. Tophnishi, A. Kurihara and T. Imamoto, *Chem. Lett.* 1933 (1982).
303. K. Eckert, A. Schröder and H. Hartmann, *Eur. J. Org. Chem.* 1327 (2000).
304. H. Gotthardt and C. M. Weisshuhn, *Chem. Ber.* **111**, 2021 (1978).
305. H. Gotthardt and C. M. Weisshuhn, *Chem. Ber.* **111**, 2028 (1978).
306. B. R. Baker, R. E. Schaub, J. P. Joseph, F. J. McEvoy and J. H. Williams, *J. Org. Chem.* **18**, 138 (1953).
307. O. Hromatka, D. Binder and K. Eichinger, *Monatsh. Chem.* **105**, 127 (1974).
308. D. Binder and P. Stanetty, *Synthesis* 200 (1977).
309. D. Binder, C. R. Noe and M. Zahora, *Arch. Pharm. (Weinheim, Ger.)* **314**, 557 (1981).
310. F. A. Mikhailenko and L. I. Shevchuk, *Chem. Heterocycl. Compd. (Engl. Transl.)* **10**, 1151 (1974).
311. D. Prim, G. Kirsch and J. F. Nicoud, *Synlett.* **4**, 383 (1998).
312. D. Prim and G. Kirsch, *Tetrahedron* **55**, 6511 (1999).
313. C. Maertens, C. Detrembleur, P. Dubois, R. Jerome, C. Boutton, A. Persoons, T. Kogej and J. L. Bredas, *Chem. Eur. J.* **5**, 369 (1999).
314. J. S. Lee and K. Kim, *J. Heterocycl. Chem.* **37**, 363 (2000).
315. T. J. Luker, H. G. Beaton, M. Whiting, A. Mete, and D. R. Cheshire, *Tetrahedron Lett.* **41**, 7731 (2000).
316. A. Capperucci, A. Degl'Innocenti, M. Funicello, P. Scafato and P. Spagnolo, *Synthesis* 1185 (1996).
317. L. Lunazzi, A. Mazzanti, P. Spagnolo and A. Degl'Innocenti, *J. Org. Chem.* **62**, 2263 (1997).
318. G. Sauter, E. Stulz and C. Leumann, *Helv. Chim. Acta* **81**, 14 (1998).
319. G. Sauter and C. Leumann, *Helv. Chim. Acta* **8**, 916 (1998).
320. P. R. Huddleston, J. M. Barker, Y. Z. Adamczewska, M. L. Wood and D. Holmes, *J. Chem. Res. (M)* 548 (1993).
321. B. Hajjem, M. L. Ben Khoud and B. Baccar, *Bull. Soc. Chim. Belg.* **101**, 445 (1992).
322. L. H. Klemm, J. Wang and C. Hawkins, *J. Heterocycl. Chem.* **32**, 1039 (1995).
323. F. Würthner, C. Thalacker, R. Matschiner, K. Lukaszuk and R. Wortmann, *Chem. Commun.* 1739 (1998).
324. H.-S. Lee, Y.-G. Chang and K. Kim, *J. Heterocycl. Chem.* **35**, 659 (1998).
325. C. N. Johnson, R. T. Martin, H. K. A. Morgan and M. Thompson, *Synth. Commun.* **27**, 473 (1997).
326. J. C. Lancelot, C. Saturnino, H. El-Kashef, D. Perrine, C. Mahatsekake, H. Prunier and M. Robba, *J. Heterocycl. Chem.* **33**, 427 (1996).
327. M. Gütschow, *J. Heterocycl. Chem.* **33**, 355 (1996).
328. Z. Whang, R. Neidlein and C. Krieger, *Synthesis* 255 (2000).
329. S. P. Rao, K. V. B. Rao, B. A. Otter, R. S. Klein and W.-Y. Ren, *Tetrahedron Lett.* **29**, 3537 (1988).
330. N. S. Ksenzhek, L. I. Belen'kii and Y. L. Gol'dfarb, *Chem. Heterocycl. Compd. (Engl. Tansl.)* **9**, 447 (1973).
331. F. M. Abdelrazek and H. A. Ead, *J. Prakt. Chem.* **330**, 585 (1988).
332. V. I. Shvedov, V. K. Ryzhkova and A. N. Grinev, *Chem. Heterocycl. Compd. (Engl. Transl.)* **3**, 359 (1967).
333. R. F. Koebel, L. L. Needham and C. D. Blanton, *J. Med. Chem.* **18**, 192 (1975).
334. R. A. Croochet, J. T. Boatright, C. D. Blanton, C. T. Wie and W. E. Hochholzer, *J. Heterocycl. Chem.* **11**, 143 (1974).
335. K. Hirai, H. Sugimoto and T. Ishiba, *J. Org. Chem.* **45**, 253 (1980).
336. M. Wierzbicki, D. Cagniant and P. Cagniant, *Bull. Soc. Chim. Fr. Pt. 2* 1786 (1975).
337. J.-K. Luo and R. N. Castle, *J. Heterocycl. Chem.* **28**, 205 (1991).

338. G. J. S. Doad, D. I. Okor, F. Scheinmann, P. A. Bates and B. Hursthouse, *J. Chem. Soc., Perkin Trans. 1* 2993, (1988).
339. R. Neidlein and Z. Sui, *Helv. Chim. Acta* **74**, 579 (1991).
340. P. Sukumaran and K. N. Rajasekharan, *Ind. J. Chem.* **28B**, 642 (1989).
341. C. D. Hurd and K. L. Kreuz, *J. Am. Chem. Soc.* **74**, 2965 (1952).
342. G. Leandri, D. Spinelli and C. Dell'Erba, *Ann. Chim. (Rome)* **50**, 1597 (1960).
343. D. Spinelli and C. Dell'Erba, *Ann. Chim. (Rome)* **54**, 281 (1964).
344. H. Beyer and S. Melde, *J. Prakt. Chem.* **24**, 100 (1964).
345. P. N. Preston and S. K. Sood, *J. Chem. Soc., Perkin Trans. I* 80 (1976).
346. V. G. Zhiryakov, P. I. Abramenko and G. I. Prikonskikh, *Mendelev Chem J. (Engl. Transl.)* **23**, 77 (1978).
347. G. Ronsisvalle, M. S. Pappalardo, F. Vittorio, L. Pasquinucci, A. Caruso, A. Felice and M. Amico-Roxas, *Eur. J. Med. Chem.* **23**, 553 (1988).
348. D. Spinelli, C. Dell'Erba and A. Salvemini, *Ann. Chim. (Rome)* **52**, 1156 (1962).
349. C. Dell'Erba and D. Spinelli, *Tetrahedron* **21**, 1061 (1965).
350. D. Spinelli, G. Guanti and C. Dell'Erba, *J. Heterocycl. Chem.* **5**, 323 (1968).
351. D. Spinelli, G. Consiglio and A. Corrao, *J. Chem. Soc., Perkin II* 1866 (1972).
352. D. Spinelli, G. Consiglio and A. Corrao, *Tetrahedron Lett.* 4021 (1972).
353. D. Spinelli, G. Consiglio, R. Noto and A. Corrao, *J. Chem. Soc., Perkin Trans. II* 620 (1975).
354. G. Consiglio, S. Gronowitz, A.-B. Hörnfeldt, B. Maltesson, R. Noto and D. Spinelli, *Chem. Scripta* **11**, 175 (1977).
355. C. Arnone, G. Consiglio, S. Gronowitz, B. Maltesson, A.-B. Hörnfeldt, R. Noto and D. Spinelli, *Chem. Scripta* **13**, 130 (1978–79).
356. G. Consiglio, C. Arnone, D. Spinelli, R. Noto and V. Frenna, *J. Chem. Soc., Perkin Trans. 2* 621 (1982).
357. G. Consiiglio, D. Spinelli, S. Gronowitz, A.-B. Hörnfeldt, B. Maltesson and R. Noto, *J. Chem. Soc., Perkin Trans. II* 625 (1982).
358. D. Spinelli, C. Dell'Erba and G. Guanti, *Ann. Chim. (Rome)* **55**, 1260 (1965).
359. D. Spinelli, G. Consiglio, R. Noto and A. Carrao, *J. Chem. Soc., Perkin Trans. 2* 1632 (1974).
360. G. Guanti, S. Thea, C. Dell'Erba and G. Leandri, *J. Heterocycl. Chem.* **12**, 945 (1975).
361. D. Spinelli, G. Consiglio and T. Monti, *J. Chem. Soc., Perkin Trans. 2* 816 (1975).
362. D. Spinelli and G. Consiglio, *J. Chem. Soc., Perkin Trans. 2* 989 (1975).
363. D. Spinelli, G. Consiglio and R. Noto, *J. Heterocycl. Chem.* **14**, 1325 (1977).
364. G. Consiglio, C. Dell'Erba, R. Noto, M. Novi and D. Spinelli, *J. Chem. Res. (S)* 260 (1982).
365. L. H. Hellberg, C. C. Adams, R. J. Milligana and R. N. Wilke, *Org. Prep. Proced. Int.* **3**, 77 (1971).
366. L. Lunazzi, D. Macciantelli, D. Spinelli and G. Consiglio, *J. Org. Chem.* **47**, 3759 (1982).
367. G. Consiglio, C. Arnone, D. Spinelli and R. Noto, *J. Chem. Soc., Perkin Trans. 2* 721 (1982).
368. D. Spinelli and G. Consiiglio, *J. Chem. Soc., Perkin Trans. 2* 1388 (1975).
369. A. K.-Y. Jen, V. P. Rao, K. J. Drost, K. Y. Wong and M. P. Cava, *J. Chem. Soc. Chem. Commun.* 2057 (1994).
370. M. G. Hutchings, I. Ferguson, D. J. McGeein and J. O. Morley, *J. Chem. Soc., Perkin Trans. 2* 171 (1995).
371. D. Vegh, J. Kovac and M. Dandarova, *Collect. Czech. Chem. Commun.* **48**, 1885 (1983).
372. J. M. Tien and I. M. Hunsberger, *J. Org. Chem.* **25**, 2056 (1960).
373. L. H. Hellberg, M. J. Prodanovich and F. Stults, *J. Heterocycl. Chem.* **9**, 401 (1972).
374. T. Erker, *J. Heterocycl. Chem.* **30**, 187 (1993).
375. I. Puschmann and T. Erker, *Heterocycles* **36**, 1323 (1993).
376. J. Hellerbach and A. Szente, *Ger. Pat. 2* **342**, 931 (1974); *Chem. Abstr.* **80**, 146002a (1974).
377. G. Renger, B. Bbouges-Bocquet and K. H. Büchel, *J. Bioenerg.* **4**, 491 (1973), *Chem. Abstr.* **79**, 102961n (1973).

378. Y. L. Gol'dfarb, G. Zhidomirov, N. D. Chuvilkin, N. S. Ksenzhek and L. I. Belen'kii, *J. Org. Chem. USSR (Engl. Transl.)* **9**, 1536 (1973).
379. K. V. Reddy and S. Rajappa, *Heterocycles* **37**, 347 (1994).
380. B. S. Kim, K. S. Choi and K. Kim, *J. Org. Chem.* **63**, 6086 (1998).
381. L. H. Klemm and W. Hsin, *J. Heterocycl. Chem.* **12**, 1183 (1975).
382. Ya. L. Gol'dfarb, M. M. Polonskaya, B. P. Fabnrichnyi and I. F. Shalavina, *Dokl. Chem. (Engl. Transl.)* 331 (1959).
383. Ya. L. Gol'dfarb, B. P. Fabrichnyi and I. F. Shalavina, *J. Gen. Chem. USSR (Engl. Transl.)* **29**, 3596 (1959).
384. O. Hromatka, D. Binder and P. Stanetty, *Monatsh. Chem.* **104**, 920 (1973).
385. Ya. L. Gol'dfarb, V. N. Bulgakova and B. P. Fabrichnyi, *Chem. Heterocycl. Compd. (Engl. Transl.)* 1283 (1983).
386. T. Uno, K. Machida and K. Hanai, *Chem. Pharm. Bull.* **14**, 756 (1966).
387. R. Pech, E. Schleiermacher and R. Böhm, *Pharmazie* **44**, 860 (1989).
388. K. H. Weber, A. Langbein and H. Daniel. *Liebigs Ann. Chem.* 1241 (1978).
389. J. Mogensen, A. Jørgensen and E. B. Pedersen, *Chem. Scripta* **28**, 195 (1988).
390. Z. Polívka, J. Holubek, E. Svátek, J. Metys and M. Protiva, *Coll. Czech. Chem. Commun.* **49**, 621 (1984).
391. J. C. Lancelot, C. Saturnino, H. El-Kashef, D. Perrine, C. Mahatsekake, H. Pörunier and M. Robba, *J. Heterocycl. Chem.* **33**, 427 (1996).
392. A. M. Almerico, G. Cirrincione, P. Diana, S. Grimaudo, G. Dattolo, E. Aiello and F. Mingoia, *J. Heterocycl, Chem.* **32**, 985 (1995).
393. H. Schäfer and K. Gewald, *Monatsh. Chem.* **120**, 315 (1989).
394. G. Broggini, L. Garanti, G. Molteni and G. Zecchi, *J. Chem. Res. (M)* 2801 (1998).
395. S. Archer, A. Sayed-Mozaffari, E. J. Simon and T. L. Gioannini, *Eur. J. Med. Chem.* **24**, 569 (1989).
396. M. Nakanishi, M. Shiroki, T. Tahara and K. Araki, *Ger. Pat. 2* **144**, 105 (1972); *Chem. Abstr.* **76**, 140907v (1972).
397. K. Gewald, H. Schäfer and K. Sattler, *Monatsh. Chem.* **110**, 1189 (1979).
398. C. Galvez and F. Garcia, *J. Heterocycl. Chem.* **18**, 851 (1981).
399. C. D. Hurd and H. M. Priestly, *J. Am. Chem. Soc.* **69**, 859 (1947).
400. R. L. Elliott, P. J. O'Hanlon and N. H. Rogers, *Tetrahedron* **43**, 395 (1987).
401. B. M. de Malleray, *Helv. Chim. Acta* **54**, 343 (1971).
402. E. E. Campaigne and W. L. Archer, *J. Am. Chem. Soc.* **75**, 989 (1953).
403. O. Meth-Cohn, B. Narine and B. Tarnowski, *J. Chem. Soc., Perkin Trans. 1* 1531 (1981).
404. B. M. de Malleray, *Helv. Chim. Acta* **54**, 353 (1971).
405. K. Grohe and H. Heitzer, *Liebigs Ann. Chem.* 1947 (1977).
406. O. Hromatka, D. Binder and G. Pixner, *Monatsh. Chem.* **106**, 1103 (1975).
407. V. I. Shvedov, I. A. Kharzomenova, N. V. Medverda and A. N. Grinev, *Chem. Heterocycl. Compd.* 805 (1975).
408. V. I. Shvedov, I. A. Kharizomenova and A. N. Grinev, *Chem. Heterocycl. Compd. (Engl. Transl.)* **9**, 1469 (1973).
409. M. Pratts, C. Galvez, Y. Gasanz and A. Rodriguez, *J. Org. Chem.* **57**, 2184 (1992).
410. M. Bellenghi, G. Carrara, F. Fava, E. Ginoulhiac, C. Martinuzzi, A. Vecchi and G. Weitnauer, *Gazz. Chim. Ital.* **82**, 773 (1952).
411. T. Hirohashi, S. Inaba and H. Yamamoto, *Bull. Soc. Chem. Jap.* **48**, 974 (1975).
412. S. Nishimura, T. Kawasaka and E. Imoto, *Nippon Kagaku Zasshi*, **82**, 1686 (1961).
413. J. B. Press, C. M. Hofmann and S. R. Safir, *J. Org. Chem.* **44**, 3292 (1979).
414. D. Mackay, *Can. J. Chem.* **44**, 2881 (1966).
415. W. Ried and R. Christ, *Liebigs Ann. Chem.* 693 (1980).
416. D. Binder, C. R. Noe and F. Hillebrand, *Arch. Pharm. (Weinheim, Ger.)* **312**, 845 (1979).
417. D. Binder, B. C. Prager and C. R. Noe, *J. Chem. Res. (M)* 1801 (1981).

418. D. Binder, C. R. Noe and B. C. Prager, *Arch. Pharm. (Weinheim, Ger.)* **314**, 751 (1981).
419. S. Rault, M. Cugnon de Sévricourt and M. Robba, *Heterocycles* **14**, 651 (1980).
420. S. Rault, M. Cugnon de Sévricourt, N.-H. Dung and M. Robba, *J. Heterocycl. Chem.* **18**, 739 (1981),
421. J. Weinstock, *J. Org. Chem.* **26**, 3511 (1961).
422. J. Reisch and H. Labitzke, *Arch. Pharm.(Weinheim, Ger.)* **308**, 203 (1975).
423. J. D. Warren and J. B. Press, *Synth. Commun.* **10**, 107 (1980).
424. B. Stanovnik, M. Tisler, V. Golob, I. Hvala and O. Niicolic, *J. Heterocycl. Chem.* **17**, 733 (1980).
425. M. Toselli and P. Zanirato, *J. Chem. Soc., Perkin Trans. 1* 1101 (1992).
426. S. M. Graham and L. M. Ohrtman, *J. Heterocycl. Chem.* **35**, 887 (1998).
427. J. B. Kobzina, *Ger. Pat. 2* **510**, 936 (1975); *Chem. Abstr.* **84**, 59172v (1976).
428. W. C. McCarthy and L. E. Foss, *J. Org. Chem.* **42**, 1508 (1977).
429. H. Fukumi, M. Sugiyama and T. Sakomoto, *Chem. Pharm. Bull.* **37**, 1197 (1989).
430. F. Kienzle, A. Kaizer and R. E. Minder, *Helv. Chim. Acta* **66**, 1170 (1983).
431. S. Vega, J. Alonso, J. A. Diaz and F. Junquera, *J. Heterocycl. Chem.* **27**, 269 (1990).
432. M. Sugiyama, T. Sakomoto, K. Tabata, K. Endo, K. Ito, M. Kobayashi and H. Fukumi, *Chem. Pharm. Bull.* **37**, 2122 (1989).
433. B. Decroix and J. Morel, *J. Heterocycl. Chem.* **28**, 81 (1991).
434. F. Sauter, P. Stanetty, H. Potuzak and M. Baradar, *Monatsh. Chem.* **107**, 669 (1976).
435. F. Sauter, P. Stanetty and H. Potuzak, *Arch. Pharm. (Weinheim, Ger.)* **309**, 914 (1976).
436. F. Sauter, *Monatsh. Chem.* **101**, 535 (1970).
437. M. Robba, J.-M. Lecomte and M. Cugnon de Sévricourt, *Bull. Soc. Chim. Fr. Pt. 2*, 592 (1975).
438. M. Sugiyama, T. Sakomoto, K. Tabata, K. Endo, K. Ito, M. Kobayashi and H. Fukumi, *Chem. Pharm. Bull.* **37**, 2091 (1989).
439. M. Sugiyama, T. Sakamoto, H. Fukumi, *Heterocycles* **29**, 985 (1989).
440. D. Brugier, F. Outerquin and C. Paulmier, *Tetrahedron* **53**, 10331 (1997).
441. M. Berkaoui, F. Outurquin and C. Paulmier, *Tetrahedron* **54**, 9055 (1998).
442. E. J. Smutny, *J. Am. Chem. Soc.* **91**, 208 (1969).
443. S. Rajappa and B. G. Advani, *Tetrahedron Lett.* 5067 (1969).
444. S. Rajappa and B. G. Advani, *Ind. J. Chem.* **9**, 759 (1971).
445. S. Rajappa, B. G. Advani and R. Sreenivasan, *Ind. J. Chem.* **12**, 4 (1974).
446. W. A. Carroll and X. Zhang, *Tetrahedron Lett.* **38**, 2637 (1997).
447. C. Galvez, F. Garcia, A. Marzal and P. Viladoms, *J. Chem. Res. (S)* 12 (1984).
448. P. Björk, T. Aakermann, A.-B. Hörnfeldt and S. Gronowitz, *J. Heterocycl. Chem.* **32**, 751 (1995).
449. M. Sugiyama, T. Sakamoto, K. Tabata and H. Fukumi, *Chem. Pharm. Bull.* **37**, 2717 (1989).
450. R. K. Russell, C. E. van Nievelt, R. A. Rampulla and D. H. Klaubert, *Synth. Commun.* **22**, 3221 (1992).
451. D. Brugier, F. Outurquin and C. Paulmier, *J. Chem. Soc., Perkin Trans. 1* 37 (2001).
452. P. Chabrier, B. Tchoubar and S. Le Tellier-Dupré, *Bull. Soc. Chim. Fr.* 332 (1946).
453. R. K. Russell, J. B. Press, R. A. Ampulla, J. J. McNally, R. Falotico, J. A. Keiser, D. A. Bright and A. Tobia, *J. Med. Chem.* **31**, 1786 (1988).
454. K. Gewald, T. Jeschke and M. Gruner, *J. Prakt. Chem.* **333**, 229 (1991).
455. R. Neidlein and K.-F. Wesch, *Arch. Pharm.* **317**, 256 (1984).
456. L. I. Ibrahiem, G. H. Tammam and T. M. S. Abdin, *J. Chem. Soc., Pak.* **11**, 227 (1989).
457. C. A. Lundberg and J. E. Engelhart, *U.S. Pat. 3* **705**, 910 (1972); *Chem. Abstr.* **78**, 71899w (1973).
458. V. J. Ram, *Arch. Pharm. (Weinheim, Ger.)* **312**, 726 (1979).
459. F. Ishikawa and Y. Yamaguchi, *Chem. Pharm. Bull.* **28**, 3172 (1980).

460. I. V. Smolanka, A. A. Dobosh, S. M. Khripak, *Chem. Heterocycl. Compd. (Engl. Transl.)* **9**, 1169 (1973).
461. A. Cannito, M. Perrissin, C. Luu-Doc, F. Huguet, C. Gaultier and G. Narcisse, *Eur. J. Med. Chem.* **25**, 635 (1990).
462. M. Gütschow, S. Leistner and M. Pink, *J. Heterocycl. Chem.* **29**, 279 (1992).
463. E. K. Ahmed, J. Fröhlich and F. Sauter, *Collect. Czech. Chem. Commun.* **61**, 147 (1996).
464. S. Leistner, M. Gütschow, G. Wagner, R. Grupe and B. Böhme, *Pharmazie* **43**, 466 (1988).
465. U. Urleb, *J. Heterocycl. Chem.* **32**, 69 (1995).
466. U. Urleb, *J. Heterocycl. Chem.* **35**, 693 (1998).
467. M. Modica, M. Santagati and A. Santagati, *J. Heterocyclic Chem.* **38**, 973 (2001).
468. M. B. Devani, C. J. Shishoo, U. S. Pathak, S. H. Parikh, G. F. Shah and A. C. Padhya, *J. Pharm. Sci.* **65**, 660 (1976).
469. H. K. Gakhar, S. Bhardwarj and P. Baveja, *Ind. J. Chem., Sect. B* **15**, 347 (1977).
470. H. K. Gakhar, A. Madan, A. Khanna and N. Kumar, *J. Ind. Chem. Soc.* **55**, 705 (1978).
471. M. Gütschow and S. Leistner, *Z. Chem.* **30**, 23 (1990).
472. H. Schäfer, K. Gewald and M. Schmidt, *Chem. Heterocycl. Compd. (Eng. Transl.)* 1471 (1983).
473. P. Goya, J. Lissavetzky and I. Rozas, *Synthesis* 280 (1989).
474. E. C. Taylor and M. Patel, *J. Heterocycl. Chem.* **28**, 1857 (1991).
475. I. Rault, M. P. Foloppe, S. Rault and M. Robba, *Heterocycles* **36**, 2059 (1993).
476. R. Pech and R. Böhm, *Pharmazie* **44**, 790 (1989).
477. A.-G. Farbwerke Hoechst, Lucius Vormals Meister and F. R. Brüning, *Pat. 2*,018,678 (1970); *Chem. Abstr.* **74**, 141509n (1971).
478. G. V. Shishkin and V. P. Mamaev, *Izv. Sibirsk. Otd. Akad. Nauk SSSR* 112 (1962); *Chem. Abst.* **58**, 2421 (1963).
479. R. W. Sabnis and D. W. Rangnekar, *J. Heterocycl. Chem.* **27**, 417 (1990).
480. I. Lalezari, *J. Heterocycl. Chem.* **16**, 603 (1979).
481. H. Scheibler, E. Keintzel and K. Falk, *Chem. Ber.* **87**, 1184 (1954).
482. C. Finzi and V. Furlotti, *Gazz. Chim. Ital.* **45**, 290 (1915).
483. B. P. Fabrichnyi, I. F. Shalavina and Ya. I. Gol'dfarb, *J. Gen. Chem. USSR (Engl. Transl.)* **31**, 1152 (1961).
484. Ya. L. Gol'dfarb, B. P. Fabrichnyi, I. F. Shalavina and S. M. Kostrova, *J. Org. Chem. USSR (Engl. Transl.)* **11**, 2449 (1975).
485. L. C. Cheney and J. R. Piening, *J. Am. Chem. Soc.* **67**, 729 (1945).
486. C. Dell'Erba, G. Guanti and G. Garbarino, *J. Heterocycl. Chem.* **11**, 1017 (1974).
487. G. Guanti, C. Dell'Erba, and P. Macera, *J. Heterocycl. Chem.* **8**, 537 (1971).
488. C. Paulmier, G. Ah-Kow and P. Pastour, *Bull. Soc. Chim. Fr.* 1437 (1975).
489. C. Paulmier, *Bull. Soc. Chim. Fr. Pt 2* 237 (1979).
490. L. Henriksen and H. Autrup, *Acta Chem. Scand.* **26**, 3342 (1972).
491. O. Hromatka and D. Binder, *Monatsh. Chem.* **104**, 1105 (1973).
492. S. Gronowitz, C. Westerlund and A.-B. Hörnfeldt, *Chem. Scripta* **12**, 1 (1977).
493. M. G. Reinecke and R. H. Walter, *J. Chem. Soc., Chem. Commun.* 1044 (1974).
494. P. R. Huddleston and J. M. Barker and Y. Z. Adamczewska, *J. Chem. Res. (S)* 238 (1980).
495. R. Neidlein and C. M. Radke, *Helv. Chim. Acta* **66**, 2369 (1983).
496. P. Hrnciar and M. Struharik, *Chem. Zv.* **36**, 401 (1982).
497. N. V. Stulin and N. I. Putokhin, *Khimiya* 100 (1969).
498. M. Hentschel and K. Gewald, *J. Prakt. Chem.* **316**, 878 (1974).
499. A. N. Grinev, I. A. Kharizomenova, N. V. Samsonova and N. V. Kaplina, *Chem. Heterocycl. Compd. (Engl. Transl.)* **15**, 491 (1979).
500. D. Binder, C. Habison and C. R. Noe, *Synthesis* **7**, 487 (1977).
501. C. Galvez and F. Garcia, *J. Heterocycl. Chem.* **19**, 663 (1982).
502. C. Galvez and F. Garcia, *J. Heterocycl. Chem.* **21**, 393 (1984).

503. S. Vega and J. A. Diaz, *J. Heterocycl. Chem.* **30**, 1509 (1993).
504. H. Beyer, S. Melde and K. Dittrich, *Z. Chem.* **1**, 191 (1961).
505. H. Beyer and S. Melde, *J. Prakt. Chem.* **24**, 91 (1964).
506. H. Shindo, S. Takada, S. Murata, M. Eigyo and A. Matsushita, *J. Med. Chem.* **32**, 1213 (1989).
507. K. Gewald, U. Hain and M. Gruner, *Chem. Ber.* **121**, 573 (1988).
508. S. Takada, H. Shindo, T. Sasatani, N. Chopmei, A. Matsushita, M. Eigyo, K. Kawasaki, S. Murata, Y. Yakahara and H. Shintaku, *J. Med. Chem.* **31**, 1738 (1988).
509. V. I. Shvedov, V. K. Vasil'eva, A. N. Grinev and N. P. Kostyuchenko, *Chem. Heterocycl. Compd. (Engl. Transl.)* **7**, 707 (1971).
510. E. Benary and A. Baravian, *Chem. Ber.* **48**, 593 (1915).
511. V. I. Shvedov, Y. I. Trofimkin, V. K. Vasil'eva, T. F. Vlasova and A. N. Grinev, *Chem. Heterocycl. Compd. (Engl. Transl.)* **11**, 802 (1975).
512. P. Spagnolo and P. Zanirato, *J. Org. Chem.* **43**, 3539 (1978).
513. P. Spagnolo, P. Zanirato and S. Gronowitz, *J. Org. Chem.* **47**, 3177 (1982).
514. A. J. Carpenter and D. J. Chadwick, *J. Chem. Soc., Perkin Trans 1* 173 (1985).
515. D. Spinelli and P. Zanirato, *J. Chem. Soc., Perkin Trans. 2* 1129 (1993).
516. D. Davies, P. Spagnolo and P. Zanirato, *J. Chem. Soc., Perkin Trans. 1* 613 (1995).
517. P. Spagnolo and P. Zanirato, *J. Chem. Soc., Perkin Trans. 1* 963 (1996).
518. L. K. Dyall, P. M. Suffolk, W. Dehaen and G. L'abbé, *J. Chem. Soc., Perkin Trans. 2* 2115 (1994).
519. C. J. Moody, C. W. Rees and S. C. Tsoi, *J. Chem. Soc., Perkin Trans. 1* 915 (1984).
520. R. Noto, R. Rainieri and C. Arnone, *J. Chem. Soc., Perkin Trans. 2* 127 (1989).
521. A. A. Tolmachev, S. P. Ivonin, A. V. Kharchenko and E. S. Koslov, *J. Gen. Chem. USSR (Engl. Tansl.)* **61**, 778 (1991).
522. K. Issleib and A. Brack, *Z. Anorg. Chem.* **292**, 245 (1957).
523. H. J. Jakobsen, *Acta Chem. Scand.* **24**, 2661 (1970).
524. A. Ishii, R. Yoshioka, J. Nakayama and M. Hoshino, *Tetrahedron Lett.* **34**, 8259 (1993).
525. D. W. Allen and D. F. Ashford, *J. Inorg. Nucl. Chem.* **38**, 1953 (1976).
526. J. Meijer, P. Vermeer, H. J. T. Bos and L. Brandsma, *Rec. Trav. Chim.* **92**, 578 (1972).
527. V. A. Kozlov, E. V. Zheltova, V. V. Negrebetskii, A. F. Grapov and N. N. Mel'nikov, *J. Gen. Chem. USSR (Engl Transl.)* **60**, 42 (1990).
528. E. Zbiral, *Monatsh. Chem.* **98**, 1967 (1965).
529. D. W. Allen, *J. Chem. Soc. (B)* 1490 (1970).
530. D. W. Allen, S. J. Grayson, I. Harness, B. G. Hutley and W. Mowat, *J. Chem. Soc., Perkin Trans 2* 1912 (1973).
531. C. R. Griffin, R. P. Peller, K. R. Martin and J. A. Peters, *J. Org. Chem.* **30**, 97 (1956).
532. A. Burger and N. D. Dawson, *J. Org. Chem.* **16**, 1250 (1951).
533. J. P. Lampin and F. Mathey, *J. Organomet. Chem.* **71**, 239 (1974).
534. M. Bentov, L. David and E. D. Bergmann, *J. Chem. Soc.* 4750 (1964).
535. V. K. Khairullin, L. I. Nesterenko, V. I. Savushkina and E. A. Chernychev, *Izvest. Akad. Nauk. SSSR, Ser. Khim* 1846 (1974).
536. V. K. Khairullin and R. Z. Aliev, *J. Gen. Chem. USSR (Engl. Transl.)* **43**, 1907 (1973).
537. R. Z. Aliev, V. K. Khairullin and S. F. Makhmutova, *J. Gen. Chem. USSR (Engl. Transl.)* **46**, 59 (1974).
538. V. K. Khairullin and L. I. Nesterenko, *J. Gen. Chem. USSR (Engl. Transl.)* **46**, 794 (1974).
539. R. Obricki and C. E. Griffin, *J. Org. Chem.* **33**, 632 (1968).
540. E. V. Nikitin, A. S. Romakhin, O. V. Parakin, Y. A. Ignatév, G. V. Romanov, Y. M. Kargin and A. N. Pudovik, *Dokl. Akad, Nauk SSSR (Engl. Transl.)* **258**, 471 (1981).
541. S. F. Thames, L. H. Edwards, T. N. Jacobs, P. L. Grube and F. H. Pinkerton, *J. Heterocycl. Chem.* **9**, 1259 (1972).

542. L. J. Goldsworthy, W. H. Hook, J. A. John, S. G. P. Plant, J. Rushton and L. M. Smith, *J. Chem. Soc.* 2208 (1948).
543. A. W. Weitkamp and C. S. Hamilton, *J. Am. Chem. Soc.* **59**, 2699 (1937).
544. E. Krauss and G. Renwanz, *Bericht.* **65**, 777 (1932).
545. S. Yasuike, F. Nakashima, J. Kurita and T. Tsuchiya, *Heterocycles*, **45**, 1899 (1997).

– 6 –

Syntheses of Thiophenes with Group VI (Chalcogen) Substituents

6.1 OXYGEN DERIVATIVES

6.1.1 Hydroxy derivatives

6.1.1.1 General aspects

The hydroxythiophene systems are those functionalized thiophenes that differ most from their benzene analogs, the phenols. This is true both with regard to syntheses and especially to tautomeric structure and reactivity in general. The most common methods for the synthesis of phenols, such as the reaction of diazonium salts with water or fusion of sulfonic acid salts with alkali, cannot be used in the thiophene series. The hydroxythiophene systems are often very instable and air sensitive. In addition 2-hydroxythiophene itself and its simple alkyl derivatives exist as unsaturated thiolactones, as 3-thiolene-2-ones, or 4-thiolene-2-ones and not in the hydroxy form.

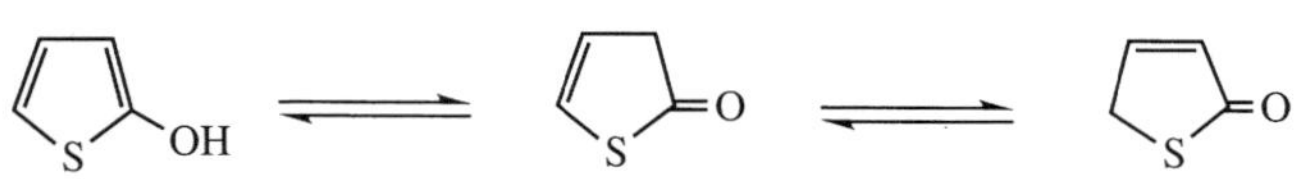

The 3-hydroxythiophene system and its alkyl derivatives exist as an equilibrium of the hydroxy form and the ketonic 3-oxo form. On the other hand, hydroxythiophene systems having electron-withdrawing and chelating substituents in the *ortho* positions exist in the phenolic form. In this chapter, for simplicity the compound will be called hydroxythiophenes, independent of their true tautomeric structure.

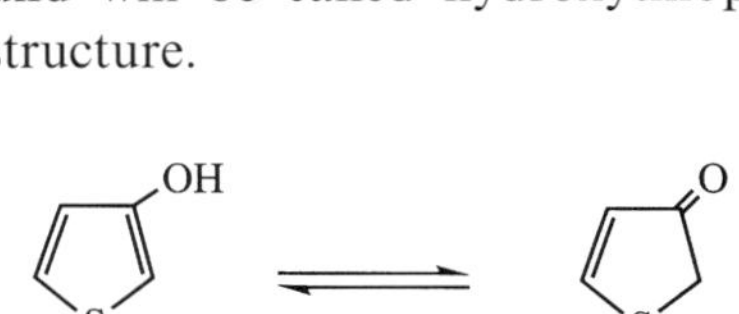

The hydroxythiophenes are best prepared (1) by the oxidation of thiopheneboronic acids, obtained via the corresponding thienyllithium derivatives by hydrogen peroxides, (2) through the dealkylation of *tert*-butoxy derivatives obtained via the corresponding magnesium derivatives and also through dealkylation of other ethers, (3) through the direct reaction of Grignard reagents or thienyllithium derivatives with oxygenating reagents and (4) through ring-closure reactions. The preferable method is of course dependent upon the desired substitution pattern.

6.1.1.2 The parent hydroxythiophenes and their alkyl derivatives

6.1.1.2.1 Dealkylation of tert-butoxythiophenes

2-Hydroxythiophene, existing as 4-thiolene-2-one, was prepared in 89–94% yield by dealkylation of 2-tert-butoxythiophene [1,2]. Dealkylation of 3-*tert*-butoxythiophene [3] only led to tar formation [4]. The 2,5-dihydroxythiophene system was first prepared by Lawesson and coworkers, through dealkylation of 2,5-di-tert-butoxythiophene, who suggested that it existed as thiosuccinic anhydride [5].

2,5-Dihydroxythiophene [6]

2,5-Di-*tert*-butoxythiophene (28 g, 0.12 mol) in a distillation flask is pyrolyzed at 150–160 °C in the presence of *para*-toluenesulfonic acid (0.1 g). The product is distilled at once under nitrogen giving a main fraction that immediately crystallized, bp 58–63 °C/0.15 mm Hg. Recrystallization from ethanol gives 11.9 g (83%) of the title compound as white crystals mp 30–31 °C.

2,4-Dihydroxythiophene and 2,4-dihydroxy-5-methylthiophene have been prepared by dealkylation of the corresponding 2,4-di-*tert*-butoxythiophenes. They exist as 4-hydroxy-3-thiolene-2-one tautomers [6]. Dealkylation of 3-methyl-2,5-di-*tert*-butoxythiophene gave the 3-methyl-2,5-dihydroxythiophene in its dioxoform [6]. 3,4-Dihydroxythiophene has been prepared by dealkylation of 3,4-di-*tert*-butoxythiophene and was found to exist as the 3-hydroxy-4-oxo tautomer.

The 2,3,4-trihydroxythiophene system can be prepared by demethylation of 3-*tert*-butoxy-2,4-dihydroxythiophene by treatment with cold trifluoroacetic acid [7] and also by demethylation of 3,4-dimethoxy-2-hydroxythiophene, existing in the 3-thiolene-2-one form, with boron tribromide [8]. This compound was used as the starting point for the preparation of thioascorbic acid.

3,4-Dihydroxy-3-thiolene-2-one [8]

A solution of 3,4-dimethoxy-2-hydroxythiophene (1.60 g, 10 mmol) in anhydrous dichloromethane (100 ml) is cooled to −78 °C under nitrogen. Boron tribromide (7.52 g, 30 mmol) in dichloromethane (30 ml) is added dropwise with stirring for 15 min, after which the reaction mixture is allowed to warm to −10 °C. After cooling again to −78 °C anhydrous diethyl ether (20 ml) is slowly added and stirring is continued for another 15 min, followed by addition of anhydrous methanol (20 ml). The reaction mixture is allowed to reach room temperature and the solvents are removed *in vacuo*. The residue is evaporated with anhydrous methanol until the distillate is free from borate. The resulting crystalline mass is dissolved in concentrated hydrobromic acid (48%, 50 ml) under nitrogen and the mixture is heated to 60 °C for 4 h. After cooling the solution is poured onto ice, the so obtained solution is saturated with sodium chloride and repeatedly extracted with warm ethyl acetate. The combined organic phases are dried over sodium sulfate and evaporated. The residue is chromatographed using 3% formic acid/diisopropyl ether as eluent giving 0.79 g (60%) of the title compound as colorless crystals mp 153 °C after recrystallization from benzene/ethyl acetate.

6.1.1.2.2 By oxidation of thiopheneboronic acids and related compounds

2-Hydroxythiophene has also been obtained by oxidation of 2-thiopheneboronic acid [9]. A number of 5-alkyl-2-hydroxythiophenes (both the 4- and 3-thiolene-2-one forms) were prepared through the boronic acid route [9–11]. In many cases the esters formed as intermediates in the reaction of thienyllithium derivatives with tributyl borate were oxidized directly without isolation and purification of the boronic acids [12].

4-Alkyl-2-hydroxythiophene was prepared similarly, but in this case only the 3-thiolene-2-one forms were obtained [9,13]. Similarly 2-alkyl-3-hydroxythiophenes [14] and 2,5-dialkyl-3-hydroxythiophenes [14–16] have been prepared; although, in order to avoid ring opening reactions the halogen–metal exchange and the reaction with the borates must be carried out at −70 °C.

6.1.1.2.3 Through the reaction of thienylmagnesium or lithium with oxygenating reagents

Another route to the 2-hydroxythiophene system exploiting 2-thienyllithium derivatives is their reaction with 2-(phenylsulfonyl)-3-phenyloxaziridine. Thus the reaction of 3,5 dimethyl-4-methoxy-2-thienyllithium with the above-mentioned reagent followed by hydrolysis with ammonium chloride gave 54% of the 2-hydroxy system in the 3-thiolene form. However, chromatographic separation from the corresponding benzoyl derivative, which was formed as a by-product, was necessary [17].

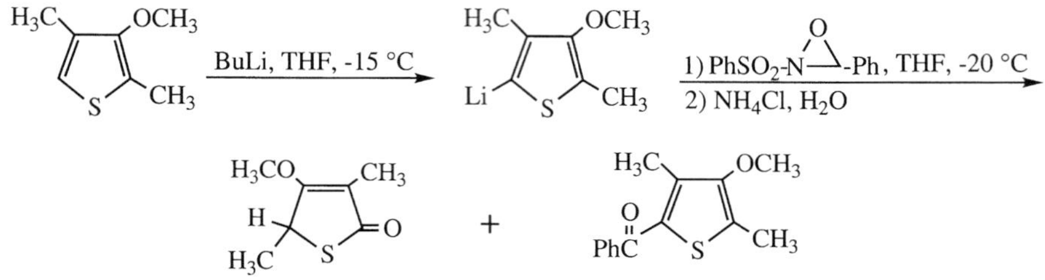

6.1.1.2.4 By ring-closure reactions

The 5-methyl-2-hydroxythiophene system historically was the first hydroxythiophene system which was synthesized by the reaction of levulinic acid with phosphorus pentasulfide [18,19], but erroneous structure assignments were made. By a now classical use of ^{1}H NMR, [20] it was shown that depending upon the workup either the 4-thiolene-2-one form or the 3-thiolene-2-one form was obtained. This discovery then led to a detailed investigation of the tautomeric equilibria and the rate and mechanism of tautomerization (for review see Ref. [21]).

4-Methyl-2-hydroxythiophene has been prepared in 40% yield through the reaction of 4-chloromethyl-4-methylazetidin-2-one with sodium hydrogen sulfide [22].

3,4-Dihydroxythiophenes have also been prepared by the reaction of α,α'-dibromo-1,2-diketones with hydrogen sulfide [23]. The most facile route to the thiosuccinic anhydride is the reaction between succinic anhydride and sodium sulfide in aqueous tetrahydrofuran. With the same methodology the 3-methyl-, 3-phenyl-, and 3,4-dimethyl-derivatives have also been prepared [24].

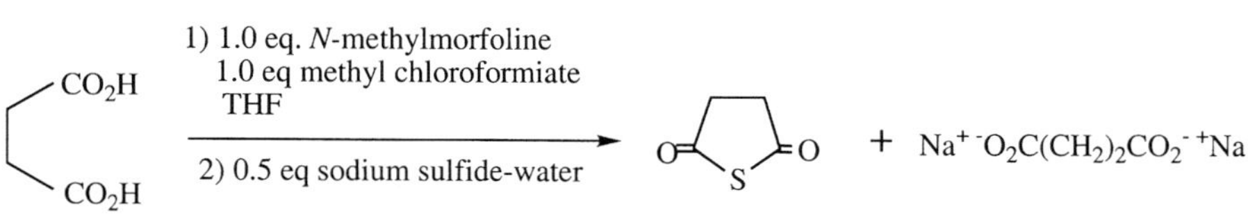

6.1.1.2.5 By various methods

The parent 3-hydroxythiophene system is best prepared by flash vacuum pyrolysis (FVP) of an appropriate alkylsulfanylmethylene-substituted derivative of Meldrum's acid, 2,2-dimethyl-1,3-dioxane-4,6-dione. 2-Substituted and 5-substituted derivatives of the 3-hydroxythiophene system such as 2-methyl- and 2-phenyl-3-hydroxythiophene as well as 5-methyl- and 5-phenyl-3-hydroxythiophene were also obtained in high yields by this technique [25–27].

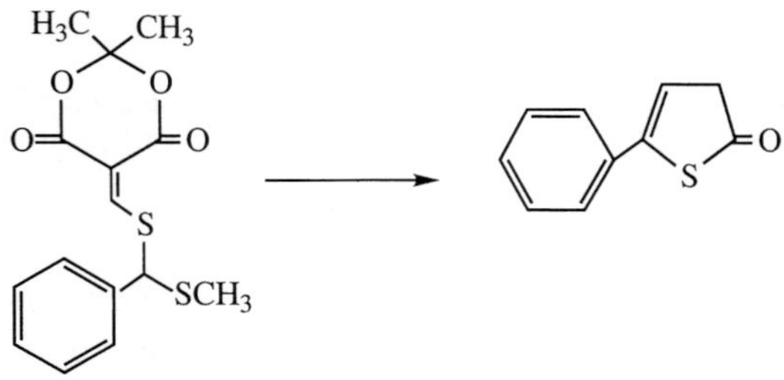

furnace temperature = 600 °C
inlet temperature = 150 °C
average pressure = 10^{-3} mm Hg
reaction time = 1 h
yield = 80 %

4-Hydroxy-2-trifluoromethylthiophene, a novel bioisoster of α,α,α-trifluoro-*meta*-cresol, is prepared by decarboxylation of 5-trifluoromethyl-3-hydroxy-2-thiophenecarboxylic acid [28].

Ethyl 4,4,4-trifluoro-3-methaxycrotonate [28]

To a mechanically stirred solution of ethyl 4,4,4-trifluoroacetoacetate (444 g, 2.41 mol) in *N,N*-dimethylformamide (2.40 l), cesium carbonate (785 g, 2.41 mol) is added. The mixture is heated to 70 °C, after which a solution of methyl *para*-toluenesulfonate (494 g, 2.65 mol) in *N,N*-dimethylformamide (890 ml) is added dropwise over a period of 1.5 h and the reaction mixture is stirred for an additional hour, cooled to room temperature and diluted with water (4.5 l). The product is extracted with diethyl ether (4 × 1 l). The combined extracts are washed with water (3 × 1 l) and saturated sodium chloride solution (1 l), dried, evaporated, and distilled through a Vigreux column giving 285 g (60%) of the title compound b.p. 40–48 °C/9 mm Hg.

Methyl 3-hydroxy-5-trifluoromethyl-2-thiophenecarboxylate [28]

A mechanically stirred solution of ethyl 4,4,4-trifluoro-3-methoxycrotonate (525 g, 2.65 mol) and methyl thioglycolate (281 g, 2.65 mol) in absolute methanol (4 l) is cooled to 5 °C. A 1 *M* methanolic solution of potassium hydroxide (3.2 l) is then added over a period of 1 h. After a small exotherm (< 30 °C) the reaction mixture is stirred at room temperature overnight. It is then poured into a mixture of ice (5 kg), water (5.2 l), and concentrated sulfuric acid (250 ml) and stirred for 15 min. The product is extracted with ethyl acetate (3 × 2.5 l). The combined extracts are washed with saturated sodium bicarbonate solution (2 × 2 l). These washings are back-extracted with ethyl acetate (2 l). The combined organic phases are washed with concentrated sodium chloride solutions (2 × 2 l), dried, evaporated and distilled through a Vigreux giving 378 g (63%) of the title compound, bp 60–62 °C/12.5 mm Hg.

3-Hydroxy-5-trifluoromethyl-2-thiophenecarboxylic acid [28]

To a stirred solution of sodium hydroxide (168 g, 4.21 mol) in water (2.1 l) a solution of methyl-3-hydroxy-5-trifluoromethyl-2-thiophenecarboxylate (238 g, 1.05 mol) in methanol (2.1 l) is added. After the initial exotherm the reaction mixture is refluxed for 3 h, cooled to room temperature, concentrated to about half the initial volume and cooled to 5 °C. Acidification with concentrated hydrochloric acid (350 ml) gives a suspension. The stirring is continued at 5 °C for 30 min, the solid collected by filtration, washed with water (3 × 1 l) and dried giving 145 g (65%) of the title compound as a white solid, mp > 100 °C (dec.)

4-Hydroxy-2-trifluoromethylthiophene [28]

3-Hydroxy-5-trifluoromethyl-2-thiophenecarboxylic acid (180 g, 849 mmol) is slowly heated under argon. Rapid evolution of gas is observed at 90 °C. Heating is continued for another 3.5 h to complete the decarboxylation. The resulting oil is vacuum distilled through a vigreux column giving 118 g (82%) of the title compound as a pale yellow liquid, bp 70–74 °C/14 mm Hg.

Bromination–dehydrobromination of saturated thiolactones has been used for the preparation of 2-hydroxy- and 3-ethyl-2-hydroxythiophene [29]. It has been claimed that 3,4-dihydroxythiophene also is obtained in 82% yield by heating *in vacuo* the 3,4-dihydroxy-2,5-dicarboxylic acid, prepared through the Hinsberg reaction [30]. Alternatively the 2,5-dihydroxythiophene system can be prepared through careful hydrolysis of the 2,5-di-(trimethylsilyloxy) derivative and it was demonstrated that rapid tautomerization from the dihydroxy form, via the 4-thiolen-2-one form, to the thiosuccinic anhydride occurred [31].

Intramolecular reaction of stabilized γ-acylphosphonium ylides derived from thiol esters with γ-lactone carbonyl groups resulted in the formation of thiotetronic acids by a Wittig-type cyclization [32]. Rhodium-catalyzed decomposition of 3-diazo-2,4-thiolanedione in *tert*-butyl alcohol at 130 °C gave the 3-*tert*-butoxy-2,4-dihydroxythiophene system [7]. Other derivatives of the 2,4-dihydroxythiophene systems (thiotetronic acids) can be prepared by α- or γ-alkylation [33,34]. The α-alkyl derivatives are most conveniently prepared by reduction of α-acyl thiotetronic acids with cyanoborohydride [33].

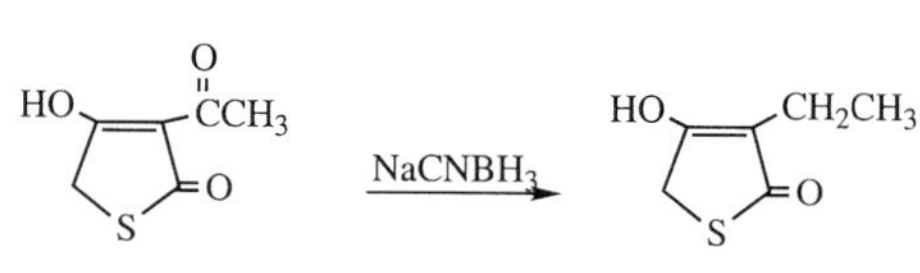

3-Ethyl-4-hydroxy-3-thiolene-2-one [33]

To a suspension of acetylthiotetronic acid (4.8 g, 30 mmol) in acetic acid (25 ml) sodium cyanoborohydride (3.75 g, 60 mmol) is added portionwise. After 1 h the reaction mixture is treated with concentrated hydrochloric acid (7.5 ml) in water (75 ml). Sodium chloride is added and the product is extracted with ethyl acetate. The combined organic phases are dried over sodium sulfate and evaporated. The residue is recrystallized from water giving 3.8 g (88%) of the title compound, mp 142 °C.

6.1.1.2.6 By metalation of alkoxythiophenes followed by reaction with electrophiles and dealkylation

Metalation of 2-butoxythiophene in the 5-position, followed by reaction with various electrophiles and dealkylation is another approach to various 5-substituted 2-hydroxythiophenes, among them 5-methyl-2-hydroxythiophene [5].

3-tert-Butoxythiophene is selectively metalated in the 2-position, which makes possible the synthesis of 2,3-di-tert-butoxythiophene and upon dealkylation the 2,3-dihydroxythiophene system, existing as 3-hydroxy-3-thiolene-2-one [6]. Metalation of 3,4-di-*tert*-butoxythiophene followed by metalation and methylation was used for the synthesis of 2-methyl-3,4-dihydroxythiophene [6].

6.1.1.4 Aryl- and hetaryl-substituted hydroxythiophenes

6.1.1.3.1 By ring-closure reactions

5-Aryl-2-hydroxythiophenes have been prepared by the reaction of β-arylpropionic acids with phosphorus pentoxide [35]. 4-(2-Furyl)-3-thiolene-2-one is prepared through the reaction of 3-bromomethyl-3-(2-furyl)-2-propenoic acid with thiourea [36].

6.1.1.3.2 By oxidation of boronic acid derivatives

2-[2-, 3- and 4-(Pyridyl)]-3-hydroxythiophenes as well as 4-(2-, 3- and 4-pyridyl)-3-hydroxythiophenes can be prepared through the hydrogen peroxide oxidation of the corresponding boronic esters [37].

General procedure for the preparation of ortho-pyridyl-3-hydroxythiophenes [37]

To a stirred solution of 3-bromo-*ortho*-(2-, 3- or 4-pyridyl)thiophene (1.20 g, 5.0 mmol) in anhydrous tetrahydrofuran (24 ml) at −100 °C under nitrogen 2.05 *M* butyllithium in hexane (2.70 ml) is added dropwise at such a rate that the temperature does not exceed −100 °C. The reaction mixture is stirred at −100 °C for 30 min and then treated with triethyl borate (0.80 g, 5.50 mmol) in anhydrous tetrahydrofuran (5.0 ml). After stirring at −100 °C for 4 h the reaction mixture is allowed to reach ambient temperature and then treated with 30% hydrogen peroxide solution (1.5 ml). The mixture is vigorously stirred at

50 °C for 2–3 h. After cooling cold water (20 ml) is added. The phases are separated and the aqueous phase is extracted with diethyl ether. The combined organic phases are washed with cold water, until the separated water does not oxidize ammonium sulfate, dried over magnesium sulfate and evaporated. The crude product is purified by column chromatography using silica gel 60.

Compound	Eluent	Solvent recrystallization	Yield (%)	Melting point (°)
3-Hydroxy-2-(2-pyridyl)-thiophene	CH_2Cl_2 C_7H_{16} (4:1)	petroleum ether	44	46.0–48.0
3-Hydroxy-2-(3-pyridyl)-thiophene	CH_2Cl_2 acetone (7:3)	methanol	47	167.0–168.0
3-Hydroxy-2-(4-pyridyl)-thiophene	ethyl acetate	methanol	34	216.0–218.0
3-Hydroxy-4-(2-pyridyl)-thiophene	CH_2Cl_2 C_7H_{16} (4:1)	petroleum ether	34	50.0–153.0
3-Hydroxy-4-(3-pyridyl)-thiophene	CH_2Cl_2 acetone (4:1)	methanol	30	150.0–153.0
3-Hydroxy-4-(4-pyridyl)-thiophene	CH_2Cl_2 methanol (92:8)	methanol	34	168.0–171.0

6.1.1.3.3 By dealkylation of alkoxythiophenes

Dealkylation of 2-methoxy-5-(2- and 3-pyridyl)thiophenes with boron tribromide resulted in a tautomeric mixture of 5-(2- and 3-pyridyl)-3-thiolene-2-ones and the corresponding 4-thiolene-2-ones [38].

6.1.1.5 Carbonyl-, carboxyl-, and cyano-substituted hydroxythiophenes

6.1.1.4.1 By ring-closure reactions

These types of hydroxythiophene derivatives are in most cases best prepared by ring-closure reactions. 5-Methyl-4-carboethoxy-4-thiolene-2-one was obtained in 54% yield by the reaction of acetylsuccinic acid with hydrogen sulfide/hydrogen chloride [39]. The classical Benary reaction [40] is useful for the preparation of 5-substituted 3-hydroxy-4-thiophenecarboxylic acids [45,41–43].

Ring-closure of S-esters of phenylpropiolic thioacids using alkali hydrides in *N,N*-dimethylformamide yields 4-phenyl-2-hydroxy-5-thiophenecarboxylic esters in very high yields [44]. The reaction of tetraketones with sulfur dichloride has been used for the preparation of 2,5-diacyl-3,4-dihydroxythiophenes [45].

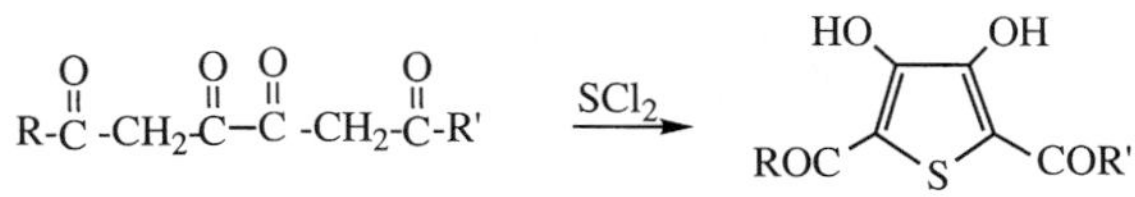

Most 3,4-dihydroxy-2,5-dicarboethoxythiophenes are best prepared by the Hinsberg reaction of diethyl thiodiacetate with diethyl oxalate [30,41,46–49].

3,4-Dihydroxy-2-carbethoxythiophene is prepared from 3,4-dihydroxy-2,5-dicarbethoxythiophene by treatment with 1 *M* sodium hydroxide, followed by 1 *M* hydrochloric acid. It was proven by NMR and X-ray that 3,4-dihydroxy-2-carboethoxythiophene exists as the dihydroxy tautomer both in the crystal state and in solution [50].

Ethyl-5-(4-nitrophenyl)-3,4-dihydroxy-2-thiophenecarboxylate was prepared through the reaction of 4-nitrobenzylthioacetate with diethyl oxalate [51,52].

5-(4-Nitrophenyl)-3,4-dihydroxy-2-thiophenecarboxylic acid [52]

To a solution prepared by addition of sodium (0.46 g, 0.02 g-atom) to absolute ethanol (200 ml), 4-nitrobenzylthioacetic acid (4.54 g, 0.02 mol) is dissolved with constant stirring for 0.5 h at 0–5 °C. To this solution diethyl oxalate (2.92 ml, 0.02 mol) is added and the reaction mixture is heated to reflux and the reflux is maintained for 5 h. After cooling to room temperature the solution is neutralized with 10% hydrochloric acid. The dark red-brown precipitate is filtered off, washed with water and dried. Recrystallization from *N,N*-dimethylformamide/ethanol (1 : 1) gives 2.5 g (89%) of the title compound as red-brown crystals, mp > 360 °C.

If α-keto esters are used esters of 4-substituted 3-hydroxy-2,5-thiophenedicarboxylic acids are obtained. The Hinsberg reaction of a diketosulfide with butyl glyoxylate, diethyl mesoxalate, and dibutyl oxalate has been used for the preparation of 2,5-dibenzoyl-3-hydroxythiophenes and 2,5-dibenzoyl-3,4-dihydroxythiophenes [54]. Even mixed sulfides can be used for preparative purposes as the condensation with butyl glyoxylate and ethyl pyruvate are regiospecific giving 5-benzoyl-2-carbomethoxy-3-hydroxythiophene and its 4-methyl derivative, respectively [54–57]. The Hinsberg reaction between the α-keto ester glycoside, methyl 2-(2,3,5-tri-O-*tert*-dimethylsilyl-(β-D-ribofuranosyl)-2-oxoacetate, and dimethyl thioacetate was the key step in the synthesis of methyl 5-carboxamido-4-hydroxy-3-(β-D-ribofuranosyl)thiophene [58].

N-(2-Hydroxyethyl)-3-hydroxy-5-methylthiophene-2-carboxamide is prepared from ethanolamine and methyl 3-hydroxy-5-methyl-2-thiophenecarboxylate [59].

N-(2-hydroxyethyl)-3-hydroxy-5-methylthiophene-2-carboxylate [59]

Ethanolamine (18.6 ml, 349 mmol) and methyl 3-hydroxy-5-methyl-2-thiophenecarboxylate (17.8 g, 103 mmol) are heated at 170 °C for 1 h. Then the mixture is cooled to room temperature, dissolved in dichloromethane and treated with sodium chloride solution saturated with 1 *M* hydrochloric acid. The phases are separated and the organic phase is washed with sodium chloride solution, dried over sodium sulfate and evaporated. The residue is purified by flash chromatography using chloroform/acetone (8 : 2) as eluent, giving 10 g (48%) of the title compound as a low melting solid.

The preparatively most important route to 3-hydroxy-2-carbonyl derivatives is most probably the Fiesselmann reaction in which β-ketoesters are reacted with two equivalents of thioglycolic acid in the presence of hydrogen chloride which quantitatively gives the thio acetals, which are esterified and ring closed with alcoholic potassium hydroxide to give 5-mono- and 4,5-disubstituted 3-hydroxy-2-thiophenecarboxylates [60–63].

Methyl-3-hydroxy-5-trifluoromethyl-2-thiophenecarboxylate is obtained by the reaction of ethyl 4,4,4-trifluoro-3-methoxycrotonate with methyl thioglycolate with methanolic potassium hydroxide, which upon base hydrolysis gave 3-hydroxy-5-trifluoromethyl-2-thiophenecarboxylic acid [28].

The Fiesselmann reaction is also used for the preparation of a number of 2-fluoroalkyl-4-hydroxy-5-methoxycarbonylthiophenes [64].

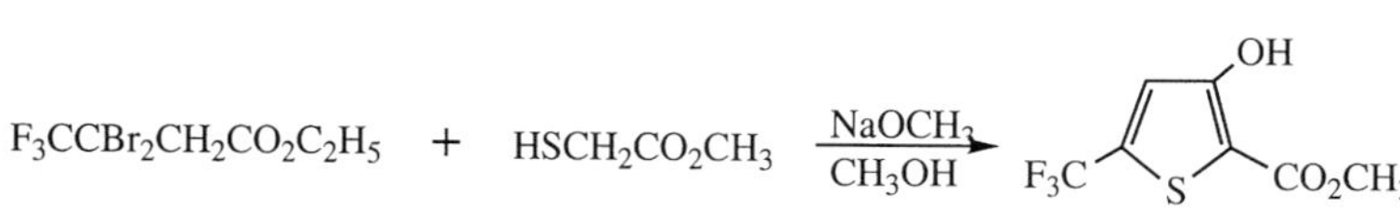

In the earliest version of the Fiesselmann reaction, esters of acetylenic carboxylic acids were used and reacted with ethyl thioglycolate and a suspension of sodium methoxide in benzene or with alcoholic potassium hydroxide. Using dimethyl acetylenedicarboxylate gives the best method for the preparation of 3-hydroxy-2,5-thiophenedicarboxylates [65–70]. However, the reaction of arylpropiolic acids gives mixtures of the 5-aryl-3-hydroxy-2- and 5-aryl-3-hydroxy-4-thiophenecarboxylic esters in the Fiesselmann reaction and are therefore preparatively less useful [63,71,72]. In another version of the Fiesselmann reaction esters of α,β-dicarboxylic acids or α-halo-α,β-unsaturated acids are used instead of β-keto esters in the synthesis of 5-substituted 3-hydroxy-2-thiophenecarboxylates [73–75]. An alternative to the Hinsberg synthesis of 3,4-dihydroxy-2,5-thiophenedicarboxylate consists in the reaction of diethyl chlorooxaloacetate with ethyl thioglycolate in pyridine followed by ring closure upon treatment with sodium ethoxide in ethanol [55]. α-Mercaptocarbonyl derivatives other than thioglycolates can be used in the various modifications of the Fiesselmann synthesis and in this way good methods for the preparation of 2-acyl-3-hydroxythiophenes are obtained [54,56]. A simple modification of the Fiesselmann reaction has been described which allows the preparation of [*b*]-condensed alkyl 3-hydroxy-2-thiophene-carboxylates in only two steps and in high yields [76].

R^1R^2	R^3	Yield (%)	MP (°C)	Solvent
$-(CH_2)_5-$	Et	86	32–34	EtOH
$-(CH_2)_6-$	Et	75	19–20	EtOH
$-CH(Me)((CH_2)_2-$	Me	91	44–46	MeOH
$-CH(Me)((CH_2)_3-$	Me	87	38–40	MeOH
$-(CH_2)_2CH(Me)CH_2$	Me	81	60–62	MeOH
$(CH_2)_2SCH_2-$	Me	46	51–53	MeOH
$-(CH_2)_3S-$	Et	54	68–70	EtOH
$-CH_2SCH_2S-$	Et	62	150–152	EtOH
$-CH_2SCH_2-$	Me	98	140–142	MeOH
$-(CH_2)_2S-$	Me	85	164–166	MeOH

General procedures for [b]-condensed alkyl 3-hydroxy-2-carboxylates [76]

A solution of the β-oxoester (0.1 mol) in absolute ethanol (500 ml) is cooled to −10 °C and a stream of gaseous hydrogen chloride is bubbled into the solution until saturation, after which thioglycolic acid (18.4 g, 0.2 mol) is added. The bubbling of hydrogen chloride is continued at −10 °C for 4 h and the reaction mixture left at room temperature for 90 h. After evaporating the solvent, the residue is neutralized with 5% sodium hydrogen carbonate solution and the product extracted with diethyl ether (200 ml). The combined organic phases are washed with water (200 ml), dried over sodium sulfate, and evaporated. The crude product is directly used in the next step.

To the above crude product is added a 2 *M* solution of the corresponding sodium alkoxide in the corresponding alcohol (175 ml). The solution so obtained is kept under nitrogen for 24 h. After evaporation the residue is treated with ice-cold water (175 ml) and this solution is acidified with ice-cold 2 *M* hydrochloric acid until pH becomes 1. Diethyl ether (175 ml) is added, the phases are separated, the organic phase washed with water, dried over sodium sulfate and evaporated. The residue is purified by chromatography on silica gel column (45 × 10 cm, 230–240 mesh) using hexane/ethyl acetate (10 : 1) as eluent and the final product is recrystallized.

Another good method for the preparation of 3-hydroxy-2-carbonylthiophenes is the so-called C_3S +C modification of the Fiesselmann reaction.

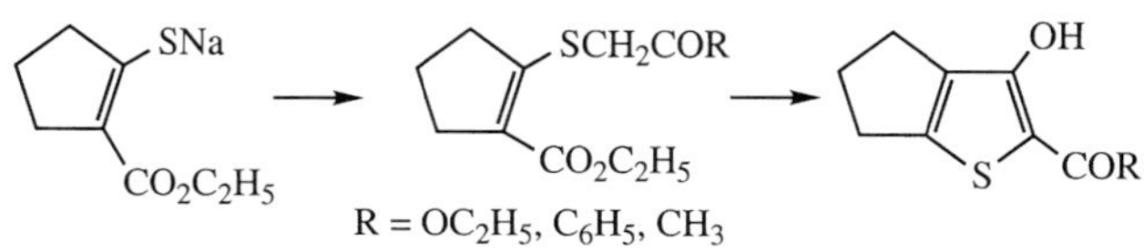

Thus the sodium salt prepared from the β-ketoester, hydrogen chloride and hydrogen sulfide were reacted with various α-halocarbonyl derivatives to yield sulfides, which in most cases without isolation were ring-closed to the 3-hydroxy-2-carbonyl derivatives upon treatment with alkoxides [56]. A good method for the preparation of esters of 3-hydroxy-2,4-thiophenedicarboxylic acids is the Fiesselmann reaction between esters of ethoxymethylene malonates and ethyl thioglycolate. Similarly the condensation of malonic acid derivatives with esters of thione or dithioacids in the presence of base gave enethiols, which upon alkylation with ethyl chlorothioacetic acid gave the 5-phenyl and 5-methyl derivative of 3-hydroxy-2,4-thiophenedicarboxylic acid [77].

2,4-Dimethyl-3-hydroxy-5-thiophenecarboxylic esters have been prepared from 3-acetoxy-substituted acetylacetone and thioglycolic acid [78].

Methyl 3-hydroxy-4-methyl-2-thiophenecarboxylate is also prepared by the reaction of methyl-3, 3-dimethoxy-2-methylpropionate and methylthioglycolate in the presence of boron trifluoride [79].

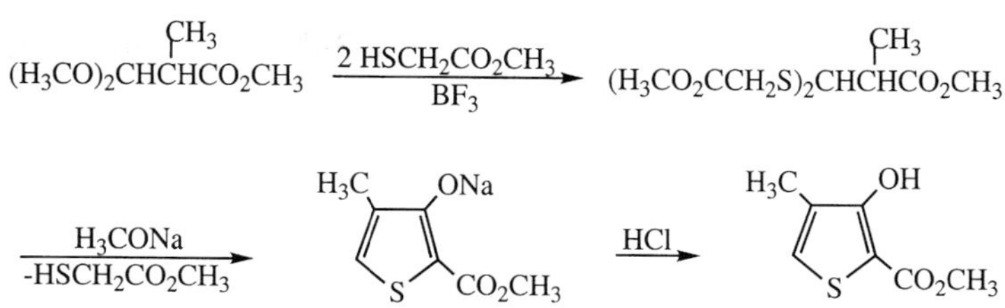

3-Hydroxythiophenes carrying electron-withdrawing groups in the 4- or 2-positions can also be prepared by direct aromatization of 3-oxotetrahydrothiophenes with bromine [45,57,80], sulfuryl chloride [81], sulfur [82], or hydrogen peroxide [83]. By this methodology methyl 3-hydroxy-4- thiophenecarboxylate [81,83,84], 2- and 4-acyl-3-hydroxythiophenes [82] and ethyl 3-hydroxy-2-phenyl-4-thiophenecarboxylate were prepared in good yields [84].

5-Substituted 3-hydroxy-2-cyanothiophenes are also conveniently prepared by the Gompper reaction starting from the enethiols derived from ketoesters, which were alkylated by chloroacetonitrile and ring-closed with sodium ethoxide [85]. 4-Cyano-3-hydroxythiophenes are prepared in high yield through the reaction of 2,3-dienonitriles with mercaptoacetates [86].

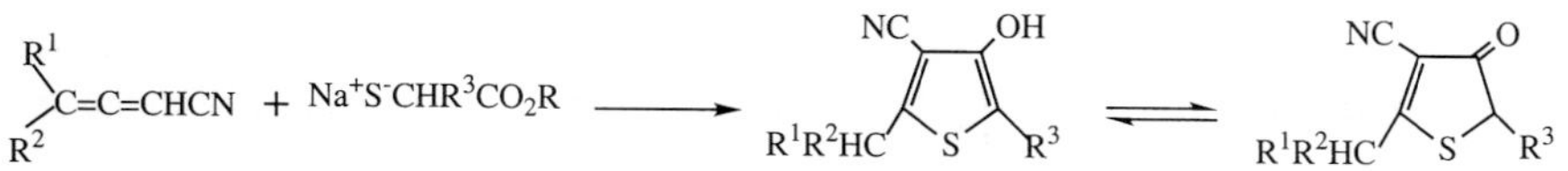

The reaction of methyl-2-chloro-3-hydroxy-5-methylthiophene-2-carboxylate with hydrogen chloride is used for the preparation of methyl-5-chloromethyl-3hydroxythiophene-2-carboxylate [79]. 4,5-Dihydro-5-(hydroxyimino)-4-oxo-3thiophenecarboxylic acid ethyl esters are prepared by nitrosation of ethyl 4-hydroxy-2-methyl-3-thiophenecarboxylate [87].

Ethyl 4,5-dihydro-5-(hydroxyimino)-2-methyl-4-oxo-3-thiophenecarboxylate [87]

To a solution of ethyl-4-hydroxy-2-methyl-3-thiophenecarboxylate (18.6 g, 100 mmol) in ethanol (27 ml) at 50 °C, isoamyl nitrite (23.43 g, 200 mmol) is added dropwise over a period of 10 min. The stirring is continued at 50 °C for 1 h, after which the reaction mixture is cooled to 0 °C. The precipitate formed is filtered off and purified by flash chromatography using hexane/ethyl acetate as eluent giving 14.9 g (75%) of the title compound as green needles after trituration with hexane mp 158 °C.

4,5-Dihydro-5-(hydroximino)-2-methyl-4-oxo-3-thiophenecarboxylic acid ethyl ester did not give the phenylhydrazone upon reaction with phenylhydrazine as previously claimed, but thiohydroxamic acid [87].

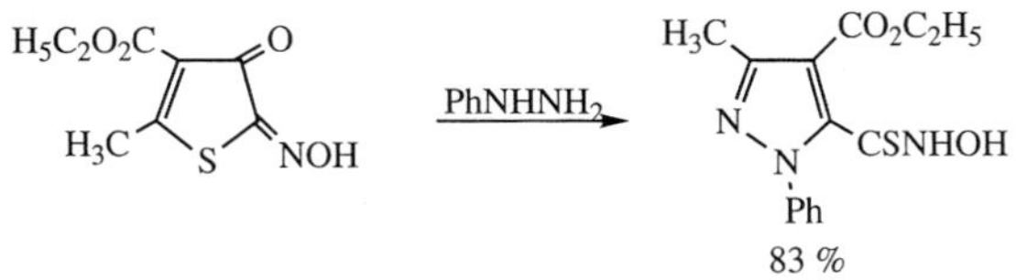

6.1.1.4.2 By metalation of alkoxythiophene followed by reaction with electrophiles and dealkylation

2-Hydroxy-5-thiophenecarboxylic acid has been prepared by metalation of 2-tert-butoxythiophene with butyllithium followed by reaction with carbon dioxide and dealkylation [5]. Starting from 3-*tert*-butoxythiophene metalation followed by reaction with electrophiles, in some cases followed by renewed metalation and reaction with electrophiles and dealkylation, has been used for the preparation of 3-hydroxy-2-acetylthiophene, 2-carboethoxy-3-hydroxythiophene, 5-carboethoxy-2-methyl-3-hydroxythiophene, and 5-carboethoxy-2,3-dihydroxythiophene [4]. 3-*tert*-Butoxy-2,5-diformylthiophene was prepared in a one-pot procedure and dealkylated to 2,5-diformyl-3-hydroxythiophene [88].

2,5-Diformyl-3-hydroxythiophene [88]

A sublimation apparatus is charged with 3-tert-butoxy-2,5-diformylthiophene (1 g, 5 mmol) and placed in an oil bath. *para*-Toluenesulfonic acid (100 mg) is added and the reaction mixture is heated to 160 °C for 1 min. Immediately the compound sublimes at reduced pressure giving 0.39 g (50%) of the title compound, mp 119 °C.

Similarly 2,5-diformyl-3,4-dihydroxythiophene was prepared [88,89].

6.1.1.4.3 By halogen–metal exchange of bromo-alkoxythiophenes followed by reaction with electrophiles and dealkylation

Halogen–metal exchange of 3-bromo-4-*tert*-butoxythiophene followed by reaction with electrophiles and dealkylation can be used for the preparation of 3-acetyl-4-hydroxythiophene [4]. By the same approach 3-acetyl-2-hydroxythiophene and 3-carboethoxy-2-hydroxythiophene were prepared from 3-bromo-2-*tert*-butoxythiophene [90]. Ethers other than *tert*-butoxy ethers can also be dealkylated.

4-Hydroxy-3-thiophenecarboxylic acid was obtained from the 4-ethoxy acid by treatment with boron tribromide in dichloromethane [84]. 3-Methoxy-2-(3-methyl-oxo-2-buten-1-yl) thiophene was similarly demethylated with boron trichloride [91].

3-Hydroxy-2-(3-methyl-oxo-2-buten-1-yl)thiophene [91]

To a solution of 1.0 *M* boron trichloride in dichloromethane (800 ml, 0.80 mol) is added a solution of 3-methoxy-2-(3-methyl-oxo-2-buten-1-yl)thiophene (52.3 g, 0.27 mol) in dichloromethane (400 ml) at −10 to 5 °C. The reaction mixture is stirred at −5 °C for 1.5 h and then poured into water (1000 ml). The phases are separated and the organic phase is dried over sodium sulfate and eluted through a pad of silica gel. After evaporation the residue, an oil, is crystallized from hexanes giving 40.0 g (82%) of the title compound, mp 32–33 °C.

Heating 3-methoxy-2-thiophene aldehyde and 2-acetyl-3-methoxythiophene with pyridine hydrochloride gives the corresponding 3-hydroxythiophene derivatives in 80–90% yield [92]. 2-Cyano-3-hydroxythiophene was prepared from 3-methoxy-2-cyanothiophene by the same method [93].

6.1.1.6 Nitro- and amino-substituted hydroxythiophenes

6.1.1.5.1 By ring-closure reactions

Methyl-5-azolyl-3-hydroxy-2-thiophenecarboxylates are prepared by treating methyl-3-hydroxy-2-thiophenecarboxylate with sulfuryl chloride, yielding crude methyl-2-chloro-3-oxo-2,3-dihydrothiophene-2-carboxylate, which upon reaction with azoles gives the desired products [94,95]. The reaction of methyl 2-chloro-3-oxo-2,3-dihydrothiophene-2-carboxylate with hydrazoic acid gives 5-azido-3-hydroxy-2-thiophenecarboxylate [94].

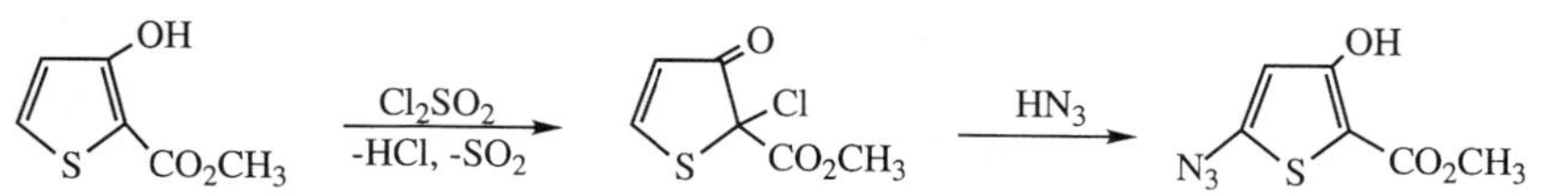

2-Amino-4-hydroxythiophenes have been prepared through the reaction of α-chloro-α-cyano ketones with sodium hydrogen sulfide [96]. Such compounds can also be prepared through the reaction of an arylacetonitrile with esters of thioglycolic acids in a refluxing organic solvent in the presence of an alkali metal alkoxide, according to a Russian patent. A similar approach to 2-amino-3-aryl-4-hydroxythiophene, described in a Japanese patent, consists in heating of the cyanosulfide.

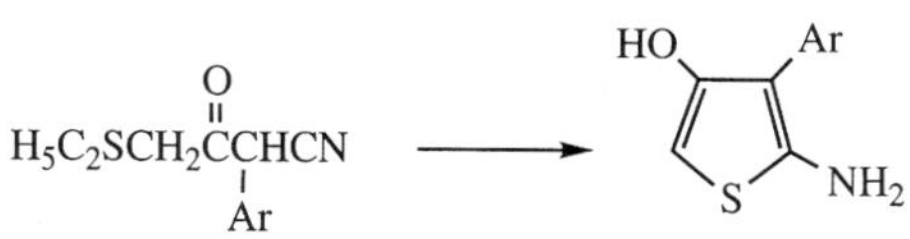

Ar = C_6H_5, p-ClC_6H_4

with 90% sulfuric acid for a few minutes [97]. 2-Amino-3-carbethoxy-4-hydroxythiophenes, existing in the tautometric ketoform are prepared by the reaction of phenylisothiocyanate or 2-isothiocyano vinylacetate with γ-chloroacetoacetate or amides of γ-bromoacetoacetic acid in the presence of sodium hydride [98–100].

Ethyl 2-anilino-4-oxo-4,5-dihydrothiophene carboxylate [98]

The sodium salt of ethyl γ-chloroacetoacetate (from 16 g, 97 mmol) of the ester and 50% sodium hydride, 4.7 g, 97 mmol) in 1,2-dimethoxyethane (75 ml) is treated with a solution of phenyl isothiocyanate (13.1 g, 97 mmol) in 1,2-dimethoxyethane (25 ml). After stirring for 5 h the reaction mixture is added to water. The product is extracted with dichloromethane, the extract is dried over magnesium sulfate and evaporated. The residue is recrystallized from ethanol giving 13.3 g (52%) of the title compound as needles, mp 146–148 °C.

The Gompper reaction has been used for the preparation of 5-amino-substituted 3-hydroxy-2-carboethoxythiophenes. Thus the reaction of diethyl monothione malonate with morpholine gave the corresponding thioamide, which was alkylated with ethyl chloroacetate and then ring closed to the 5-morpholino-3-hydroxy-2-carbethoxythiophene [101]. 2 *para*-Toluenesulfonyl-imino-3 pyridino-2,5-dihydrothiophen-4-olate can be synthesized through ring-closure reactions of a pyridinium-*N*-ylide [102].

2-para-Toluenesulfonylimino-3-pyridino-2,5-dihydrothiophen-4-olate [102]

A solution of 1-ethoxycarbonylethylpyridinium bromide (1.678 g, 4.0 mmol), N-bis(ethoxycarbonylmethylthio)toluenesulfonamide (0.984 g, 4.0 mmol) and triethylamine (4 ml) in ethanol is refluxed for 26 h. After evaporation of the solvent and the excess of triethylamine, the residue is washed with 10% hydrochloric acid and extracted with dichloromethane (30 ml). The organic phase is dried over sodium sulfate and evaporated. The residue is chromatographed on alumina using ethyl acetate/ethanol (4 : 1) as eluent giving 0.355 g (26%) of the title compound as pale-yellow crystals mp 168–170 °C after recrystallization from ethanol.

The reaction of enaminonitriles with phenyl isothiocyanate followed by cyclization with α-haloketones appears to be a good method for the preparation of 2-phenylamino-4-hydroxy-5-carbonylsubstituted thiophene derivatives [103,104].

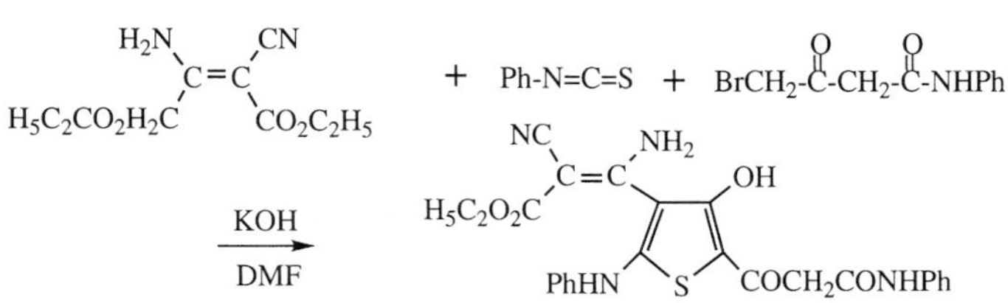

4-(3′-Amino-2′-ethoxycarbonylacrylonitrilo)-2-benzoyl-3-hydroxy-5-phenyl-aminothiophene is prepared in the same way by using phenacyl bromide [104].

4-(3′-Amino-2′-ethoxycarbonylacrylonitrilo)-2-benzoyl-3-hydroxy-5-phenylaminothiophene [104]

To a cold suspension of finely ground potassium hydroxide (1.40 g, 25 mmol) in *N,N*-dimethylfonnamide (30 ml), diethyl-3-amino-2-cyano-2-pentene-1,5-dicarboxylate (5.60 g, 25 mmol) is added followed by addition of phenyl isothiocyanate (3.38 g, 25 mmol). The mixture is stirred at room temperature for 24 h and then treated with phenacyl bromide (4.98 g, 25 mmol) at room temperature. The stirring is continued for 24 h, after which the reaction mixture is triturated with cold water (50 ml) and the pH brought to 6 with a few drops of hydrochloric acid. The precipitate formed is filtered off and washed several times with water giving 9.74 g (90%) of the title compound as orange crystals mp 222–225 °C after recrystallization from ethanol.

4-Aryl-2-arylamino-5-hydroximino-2,3,4,5-tetrahydrothiophenes are obtained through the reaction of benzoyl(thioacetanilides) with (*E*)-(*β*-nitrostyrenes. In polyphosphoric acid medium these products undergo dehydration and hydrolysis giving 4-aryl-2-arylimino-3-benzoyl-2,5-dihydro-5-oxothiophenes [105].

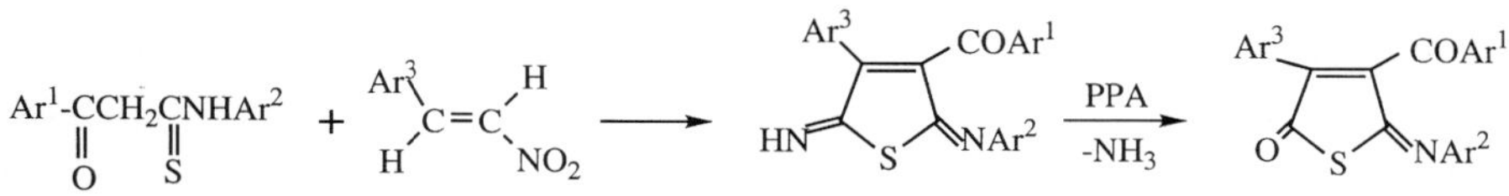

6.1.1.5.2 By dealkylation of alkoxythiophenes

4-Acetylamino-3-hydroxythiophene is prepared by dealkylation of 4-ethoxy-3-acetylaminothiophene with boron tribromide in dichloromethane [84].

4-Acetylamino-3-hydroxythiophene [84]

A solution of *N*-(4-ethoxy-3-thienyl)acetamide (2.96 g, 16 mmol) in dichloromethane (75 ml) is added to boron tribromide (8.0 g, 32 mmol) in dichloromethane. The reaction mixture is stirred overnight, after which it is quenched with water (50 ml). The phases are separated and the aqueous phase is extracted twice with diethyl ether. The combined organic phases are filtered through magnesium silicate, diluted with ethyl acetate, and evaporated giving 1.03 g (44%) of the title compound as a tan solid mp 148.5–149 °C after sublimation.

6.1.1.5.3 By nucleophilic substitution of halothiophenes

2,4-Dinitro-5-hydroxythiophene and 5-acetyl-3-nitro-2-hydroxythiophene are best synthesized through the reaction of 2-chloro-3,5-dinitrothiophene and 2-acetyl-5-chloro-4nitrothiophene with sodium formate in anhydrous methanol followed by acidification [106].

6.1.2 Acyloxy derivatives

6.1.2.1 By acylation of hydroxythiophenes

Owing to the hard nature of acyl chlorides and acyl anhydrides, *O*-acylation is always observed, even with hydroxythiophenes, which upon alkylation give large amounts of *C*-alkylation. Acetoxy derivatives have been prepared by reaction of hydroxythiophenes with acetic anhydride and sodium hydroxide or sodium acetate [49,107,108] with pyridine [109], triethylamine [110] or in the presence of catalytic amounts of sulfuric acid or *para*-toluenesulfonic acid [84,111]. Acetyl chloride has been used in many cases [9,10,16,68,112–115]. Many other esters like benzoates [9,107,116], 2,4- dinitrobenzoates [116], carbamates [86,116], and methacrylates [117] have been prepared in similar ways.

The following reaction sequence has been performed [118].

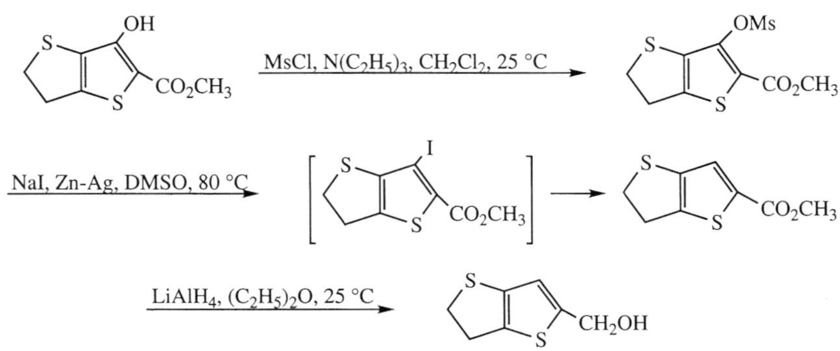

The hydroxymethyl derivative is a useful starting material for thieno[3,2-*b*]-thiophene [118].

6.1.2.2 By direct acyloxylation

Direct acyloxylation of some simple thiophenes has been achieved with diisopropyl peroxydicarbonate in acetonitrile in the presence of cupric chloride which gives thienyl isopropyl carbonates in 41–74% yields [119].

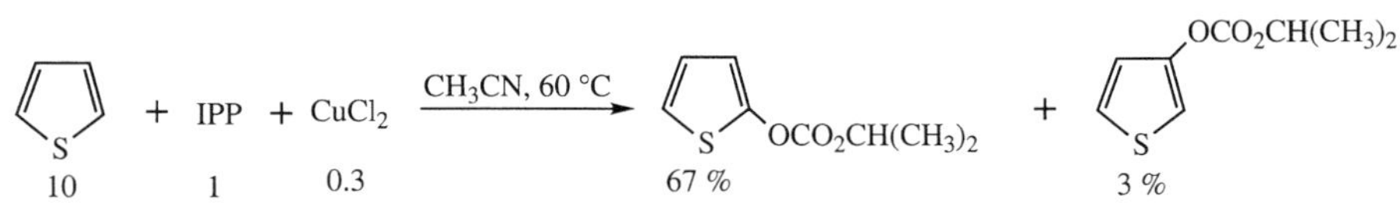

General procedure [119]

On heating and stirring, cupric chloride (1.35 g, 10 mmol) is dissolved in acetonitrile (160 ml). This solution is transferred to a 500-ml three-necked flask equipped with stirrer, Friedrich's condenser, Claisen adapter with thermometer and pressure-equalizing addition funnel, which is immersed in an oil bath at 60 °C. A tube from the condenser outlet is placed in a test tube containing a few milliliters of acetone so that evolution of gas can be observed. After addition of the thiophene derivative, diisopropyl peroxydicarbonate (6.8 g, 33 mmol) in acetonitrile (40 ml) is rapidly added. The temperature goes down to 55 °C and rises again to 60–67 °C after 5–15 min. After stirring for 2 h the reaction mixture is poured into saturated ammonium chloride solution (100 ml). The phases are separated and the organic phase washed twice with 50% ammonium chloride solution (100 ml) and three times with 10% sodium chloride solution (100 ml). The combined washings are extracted three times with diethyl ether (50 ml). The combined organic phases are dried over sodium sulfate and evaporated.

6.1.2.3 By ring-closure reactions

Acetoxy derivatives have, in a few cases, been prepared by ring-closure reactions. Thus, 2,5-diacetoxy-3,4-dichlorothiophene was prepared by treatment of dichloromaleic anhydride with acetic anhydride followed by heating with powdered zinc [120]. 4-Acetoxy-3-carbomethoxythiophene has been prepared by aromatization of methyl 2,5-dihydro-4 acetoxy-3-thiophenecarboxylate with sulfuryl chloride in dichloromethane [84] 3-Thienylcyanates have been prepared by reaction of 3-hydroxy-2-carbomethoxythiophenes with cyanogen bromide in acetone in the presence of triethyl amine at −10 to −15 °C [121].

6.1.3 Alkoxy- and aryloxythiophenes

6.1.3.1 General aspects

The most important methods for the preparation of alkoxythiophenes are: (1) copper-promoted nucleophilic aromatic substitution of halothiophenes, (2) nucleophilic substitution of activated halothiophenes, (3) preparation from thienyllithium or thiophene magnesium derivatives, (4) electrophilic substitutions of alkoxythiophenes, (5) alkylation of hydroxythiophenes and, (6) ring-closure reactions.

6.1.3.2 Parent compounds and their alkyl and aryl derivatives

6.1.3.2.1 By copper-promoted nucleophilic aromatic substitution

2-Methoxythiophene was first prepared by refluxing of 2-iodothiophene with sodium methoxide in anhydrous methanol in the presence of powdered cupric oxide for 30 hours [122]. 2-Bromothiophene can also be used if catalytic amounts of sodium iodide are added and longer reaction times are used [122–124]. Using this technique 3-methoxythiophene was prepared from

3-bromothiophene [125]. During recent years there has been renewed interest in developing efficient methods for the preparation of methoxythiophenes, which have been used for various purposes. Brandsma and coworkers found that 83% and 88% of 2- and 3-methoxythiophene was obtained when using high concentrations of methoxide and defined amounts of *N*-methyl-2-pyrrolidinone in the copper bromide-catalyzed substitution of the bromothiophenes at 100 °C [126,127].

3-Methoxythiophene [126]

A 250-ml three-necked round-bottom flask, equipped with a thermometer and a Dean Stark apparatus, is charged with solid sodium methoxide (16.2 g, 300 mmol) in methanol (30 ml) and *N*-methyl-2-pyrrolidone (50 ml). The reaction mixture is heated to 110 °C and methanol is distilled off. Subsequently 3-bromothiophene (32.6 g, 200 mmol) and copper(I) bromide (2.9 g, 20 mmol) are added. The addition of the catalyst results in a vigorous reaction and heating is interrupted until the temperature has dropped to 110 °C again. After 45 min GLC indicates that the substrate is converted with a selectivity of 98%. The reaction mixture is cooled to room temperature and a solution of sodium cyanide (5 g) in water (200 ml) is added with vigorous stirring. The product is extracted with pentane (6 × 50 ml). The combined organic phases are dried over magnesium sulfate, after which the solvent is distilled off at atmospheric pressure. Subsequent distillation *in vacuo* affords 20.1 g (88%) of the title compound, bp 81–82°C/65 mm Hg.

4-Methyl-3-methoxythiophene was also obtained in this way [128]. This compound was transformed to 3-(2-bromoethoxy)-4-methylthiophene through the reaction with bromoethanol in toluene [128,129].

Other alkoxides can also be used, and in this way 2-ethoxy- and 2-propoxythiophene and some higher alkoxy derivatives were previously prepared from 2-iodothiophene in low yields [124]. Also 2,2-dimethyl-4-[3-(4-methoxythienyl)oxymethyl]-1,3-dioxalane could be prepared through the cupric oxide-promoted reaction in the presence of potassium iodide of 3-bromo-4-methoxythiophene with the alkoxide from 1,2-*O*-isopropylideneglycerol, using the glycerol derivative as solvent [130].

2,2-Dimethyl-4-(3-(4-methoxythienyl)oxymethyl-1,3-dioxalane [130]

Sodium (6.2 g, 0.27 mol) is dissolved in 1,2-*O*-isopropylideneglycerol (139 ml). 3-bromo-4-methoxythiophene (19.3 g, 0.1 mol), potassium iodide (0.09 g, 0.54 mmol) and cupric oxide (4.0 g, 0.05 mol) are added and the reaction mixture is heated at 90°C for 24 h. Excess isopropylideneglycerol is recovered by distillation (80–81°C/11 mm Hg) and the residue treated with water and extracted with diethyl ether. The combined organic phases are dried and evaporated. The residue is recrystallized from hexane giving 18.1 g (74%) of the title compound, mp 30–31°C.

It has been claimed that the reaction of 3,4-dibromothiophene with sodium butoxide in butanol in the presence of potassium iodide after reflux for 100 h gives 3,4-dibutoxythiophene [131]. 3,4-Dialkoxy-2,5-dihydroxymethylthiophenes were recently prepared through the reduction of the diesters with lithium aluminium hydride in methanol [132].

3,4-Dibutoxythiophene [131]

Sodium (13.1 g, 57 mmol) is dissolved in butanol (240 ml), after which 3,4-dibromothiophene (23.0 g, 95 mmol), potassium iodide (0.2 g, 1.2 mmol) and copper(II) oxide (8.0 g, 100 mmol) are added. After heating under reflux for 100 h the reaction mixture is cooled and the solid material filtered off. The filtrate is poured into water (200 ml) and the product extracted with diethyl ether (3 × 50 ml). The combined organic phases are dried over magnesium sulfate and evaporated under moderately reduced pressure. The residue is distilled *in vacuo* giving 10.5 g (48%) of the title compound as a pale-yellow oil, bp 110–112°C13 mm Hg.

Under similar conditions also 3,4-dimethoxythiophene was prepared from 3,4-dibromothiophene [133]. The conditions worked out by Brandsma gave excellent yields of 3-ethoxy and 3-propoxythiophene [134].

3-Ethoxythiophene [134]

In a 100-ml three-necked flask, equipped with a thermometer, a reflux condenser, and a septum, sodium (3.5 g, 150 mmol) is dissolved under heating in 100% oxygen-free ethanol (30 ml). When all sodium has reacted the solvent is removed *in vacuo*. The remaining solid is dissolved in *N,N*-dimethylformamide. The solution is heated to 110°C, after which 3-bromothiophene (12.2 g, 75 mmol) and copper(I) bromide (1.1 g, 7.5 mmol) are added. The reaction is completed after 1 h and the product is isolated *via* the usual work-up giving 7.1 g (94%) of the title compound, bp 62–63°C/18 mm Hg.

3-Methoxy-2,5-dimethylthiophene and 3,4-dimethoxy-2,5-dimethylthiophene were prepared from 3,4-dibromo-2,5-dimethylthiophene and sodium methoxide using copper(I) bromide as catalyst [135].

Reasonable yields were obtained in the preparation of 2-methoxy-5-methyl- and 2-methoxy-5-ethylthiophene from the corresponding iodo derivatives [136]. If the reactions do not go to completion, the starting halothiophenes can be removed chemically by reaction with magnesium or butyllithium followed by hydrolysis. A 30% yield of 2,5-dimethoxy-3-methylthiophene was obtained from the diiodo derivative upon reflux for four days with sodium methoxide in methanol in the presence of cupric oxide [137]. The largest drawback of the copper-promoted nucleophilic substitution is the complications due to halogen dance that occurs when dihalothiophenes are used (cf Chapter 7). A way of avoiding this is illustrated in the synthesis of 3-bromo-4-methoxythiophene, which was prepared through bromination of methyl 3-hydroxy-2-thiophene-carboxylate to the 4-bromo derivative followed by *O*-methylation, hydrolysis, and decarboxylation [138].

6.1.3.2.2 By reaction of thiophenemagnesium halides with *tert*-butyl perbenzoate

The reaction of thiophenemagnesium halides with *tert*-butyl perbenzoate introduced by Lawesson has extensively been used for the preparation of *tert*-butoxythiophenes [1–6,88,89,92,139–148].

The thiophenemagnesium halides are prepared directly from halothiophenes and magnesium or, when they cannot be obtained in this way, from the thienyllithium derivatives through reaction with magnesium bromide. The direct use of thienyllithium derivatives with tert-butyl perbenzoate gives much lower yields. In this way 2-tert-butoxythiophene was obtained in 70–76% yield from 2-thiophenemagnesium bromide[1,2] and 3-*tert*-butoxythiophene from 3-thienyllithium and magnesium bromide [3].

3-tert-Butoxythiophene [3]

An ethereal solution of 3-thienyllithium is prepared at −70°C from 3-bromothiophene (24.0 g, 0.15 mol) and 1.5 *M* butyllithium in ether (110 ml). This solution is pressed over into a solution of magnesium bromide in anhydrous ether, obtained from magnesium (6.5 g) and bromine (29.0 g) in anhydrous ether (150 ml). After 30 min the solution is cooled to 0°C and a solution of tert-butyl perbenzoate (23.2 g, 0.12 mol) in anhydrous ether (100 ml) is added dropwise. After stirring at 0°C for an additional hour the reaction mixture is poured into ice and diluted hydrochloric acid. The phases are separated and the aqueous phase extracted with diethyl ether. The combined organic phases are extracted several times with 2 *M* sodium hydroxide solution in order to remove the benzoic acid formed. The organic phase is dried over sodium sulfate and fractionated giving 13.1 g (70%) of the title compound, bp 74–76°C/15 mm Hg.

6.1.3.2.3 By metalation of alkoxythiophenes followed by reaction with electrophiles

Di- and tri-tert-butoxythiophenes and many other derivatives have been prepared from 2- and 3-tert-butoxythiophenes and their halo derivatives (cf. Chapter 7) by metalation or halogen–metal exchange followed by appropriate electrophiles. Thus metalation of 2-*tert*-butoxythiophene followed by reaction with *tert*-butyl perbenzoate was used for the preparation of 2,5-di-*tert*-butoxythiophene and reaction with alkylating agents yielded various 2-alkyl-5-*tert*-butoxyderivatives. Metalation of 3-*tert*-butoxythiophene with butyllithium occurs selectively in the 2-position, so that reaction with *tert*-butyl perbenzoate can be used for the preparation of 2,3-di-*tert*-butoxythiophene and upon reaction with dimethyl sulfate 3-*tert*-butoxy-2-methylthiophene was obtained [4]. Further metalation of 2,3-di-*tert*-butoxythiophene followed by *tert*-butyl perbenzoate gave the 2,3,5-tri-*tert*-butoxythiophene [4].

From 5-*tert*-butoxy-2-thienyllithium and benzyl chloride 5-*tert*-butoxy-2-benzylthiophene was prepared [149].

5-tert-Butoxy-2-benzylthiophene [149]

2-*tert*-Butoxythiophene (23 g, 0.21 mol) is metalated with butyllithium (0.25 mol) in the usual way. Benzyl chloride (35 g, 0.28 mol) in anhydrous diethyl ether (50 ml) is added dropwise without cooling and the reaction mixture is gently refluxed with stirring overnight. It is then poured into water, the phases are separated and the aqueous phase extracted several times with ether. The combined organic phases are washed with water until neutral, dried over sodium sulfate and evaporated. Distillation of the residue gives 37.5 g (73%) of the title compound bp 112–114°C/0.08 mm Hg.

Dimetalation of 3,4-dimethoxy-, 3,4-dibutoxy- and 3,4-dioctoxythiophene with lithium diisopropylamide followed by tributylstannyl chloride is an excellent method for the preparation of the 2,5-di(tributylstannyl) derivatives [150]. Metalation of 3-methoxy-2-methylthiophene with butyllithium followed by reaction with tributylstannyl chloride gives 3-methoxy-2-methyl-5-tributylstannylthiophene [127].

3-Methoxy-2-methyl-5-tributylstannylthiophene [127]

To a solution of 3-methoxy-2-methylthiophene (500 mg, 3.9 mmol) in anhydrous diethyl ether (25 ml) at 0°C butyllithium (1.6 ml, 4.0 mmol) is added dropwise. After refluxing for 0.5 h the reaction mixture is cooled to room temperature and tributylstannyl chloride (1.08 ml, 4.0 mmol) is introduced. The stirring is continued at room temperature overnight and water is added. The phases are separated and the aqueous phase washed twice with diethyl ether. The combined organic phases are dried over magnesium sulfate and evaporated. The residue is purified by medium pressure liquid chromatography using hexane as eluent, giving 1.30 g (80%) of the title compound as a yellow liquid.

Metalation of 3,4-(ethylenedioxy)thiophene with butyllithium-*N*,*N*,*N*′,*N*′-tetramethylethylenediamine followed by ketones, such as di(tert-butyl) ketone, 1-adamantyl-tert-butyl ketone or di(1-adamantyl) ketone is used for the preparation of 3,4-(ethylenedioxy)-2-thienyldi(tert-alkyl)methanols, existing in two rotameric forms [151].

Alcohol synthesis [151]

To a mixture of 3,4-(ethylenedioxy)thiophene (5 mmol) and *N*,*N*,*N*′,*N*′-tetramethylethylenediamine (0.75 ml, 5 mmol) in anhydrous diethyl ether (15 ml) under argon at room temperature, 1.6 *M* butyllithium in hexane (3.2 ml, 5 mmol) is added. The stirring is continued for 30 min after which the ketone is added. The reaction mixture is stirred for 30 min and then quenched with water. The product is extracted with diethyl ether. The combined organic phases are washed with water, dried, and evaporated. The residue is purified by chromatography on silica gel using light petroleum/dichloromethane as eluent.

6.1.3.2.4 By alkylation of hydroxythiophenes

Alkylation of hydroxythiophenes can be used, although in most cases the reverse route of dealkylation of alkoxythiophenes to hydroxythiophenes is usually synthetically more important. A recent example is the synthesis of 4-, 5- and 4,5-substituted 3-methoxythiophenes from methyl-3-hydroxy-2-thiophenecarboxylates prepared by Fiesselmann ring-closure reactions, followed by methylation, hydrolysis, and decarboxylation [152].

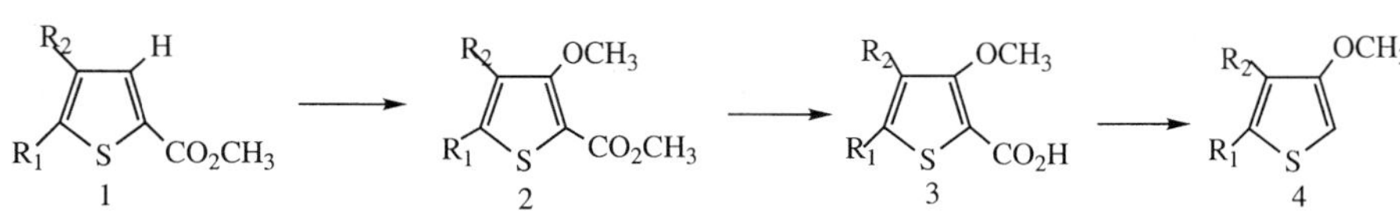

Comp. 1–4	R_1	R_2	Yield 2 (%)	mp 2 (°C)	Yield 3 (%)	mp 3 (°C)	Yield 4 (%)	bp 4 (°C)/ (mm Hg)
a	H	Me	82	Oil	94	127–129/d	87	85–87/8
b	H	SMe	88	Oil	92	137–139/c	92	120–122/8
c	H	Ph	85	93–95/a	97	150–151/c	92	98–100/8
d	Me	H	83	70–71/b	91	162–163/a	98	110–111/8
f	Me	Me	80	43–44/a	95	126–128/b	94	119–121/8
g	SMe	Me	82	oil	90	127–129/b	89	132–134/8

Solvent for recrystallization; a = methanol; b = hexane; c = toluene; d = benzene.

Methyl 3-methoxythiophene-2-carboxylates: General procedure [152]

Dimethyl sulfate (13.2 g, 0.12 mol) is added to a stirred solution of the corresponding starting material (0.1 mol) and anhydrous carbonate (16.6 g, 0.12 mol) in acetone (200 ml). The reaction mixture is heated at reflux temperature for 12 h and evaporated to dryness. The residue is treated with water and the product extracted with dichloromethane (200 ml). Evaporation of the solvent *in vacuo* gives compound **2**.

3-Methoxythiophene-2-carboxylic acids: General procedure [152]

A suspension of compound **2** (0.1 mol) in 1 *M* sodium hydroxide solution (150 ml) is heated to reflux for 30 min until total dissolution. After cooling, the reaction mixture is acidified with 1 *M* hydrochloric acid to pH 3, the solid product so formed is extracted with diethyl ether (200 ml). Evaporation of the organic solvent *in vacuo* led to compound **3**.

3-Methoxythiophenes. General procedure [152]

Compounds **3** are heated at reduced pressure (11 mm Hg) to 30°C above the melting points. After the evolution of carbon dioxide has ceased, the resulting oil is distilled giving compound **4**.

The 2,3,4-trihydroxythiophene system has been methylated to the 4-methoxy-2,3-dihydroxythiophene system with diazomethane and to the 3,4-dimethoxy-2-hydroxythiophene system with diazomethane in methanol [7]. Through methylation of ethyl-3-hydroxy-4-methyl-2-thiophenecarboxylate with diazomethane followed by reduction of the ester, 2,4-dimethyl-3-methoxythiophene was prepared [17]. Complications which can occur are due to competing *C*-alkylation, however, using hard electrophiles such as dimethyl sulfate and the ion-pair extraction method gave, predominantly and in good yields 2,5-dimethyl-3-methoxythiophene upon methylation of 2,5-dimethyl-3-hydroxythiophene [113]. Similarly 3-methyl- and 3-*tert*-butyl-2-methoxythiophene have been prepared [13,153]. This method was also used for preparation of α-(2-thienyloxy) fatty acid esters, by alkylation of 2-hydroxythiophene with a-halocarboxylates [154].

6.1.3.2.5 By ring-closure reactions

5-Aryl-2-ethoxythiophenes are prepared by the reaction of aromatic γ-keto-esters with hydrogen sulfide and hydrogen chloride [155]. A recently developed synthesis of 4-aryl-2-alkoxythiophenes is the reaction of an appropriate acylketene *O,S* acetal upon reaction with a zinc–copper couple with diiodomethane (Simmon–Smith conditions) [156].

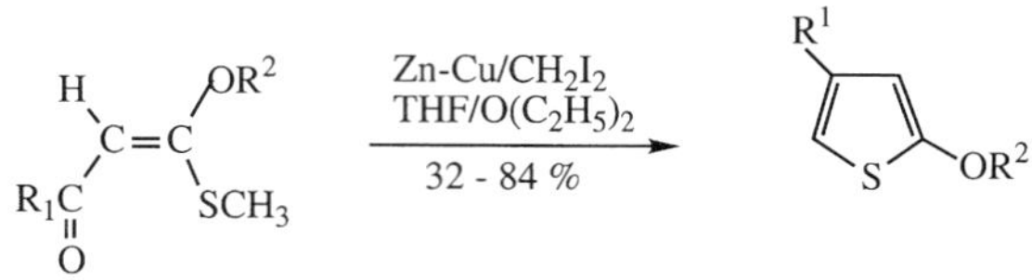

2-Alkoxy/aryloxythiophenes General procedure [156].

To a well-stirred suspension of zinc–copper couple (4.0 g, 30 mmol) in anhydrous diethylether (25 ml), under nitrogen, a small crystal of iodine and diiodomethane (6.70 g, 25 mmol) are added and the mixture is refluxed for 45 min. A solution of the appropriate acylketene *O,S*-acetal (10 mmol) in anhydrous tetrahydrofuran (15 ml) is added to the mixture, which is further refluxed with stirring for 5–8 h (monitored by TLC). The solvent is removed under reduced pressure and the residue is diluted with chloroform (150 ml) and water (200 ml). The extract is filtered to remove metal-bases residues and the residue is washed with chloroform (2 × 25 ml). The chloroform layer is separated and washed with saturated aqueous ammonium chloride (2 × 50 ml) and water (2 × 100 ml), dried over sodium sulfate, and concentrated to give the crude thiophenes that are purified by column chromatography over silica gel using hexane as eluent.

6.1.2.3.6 Vinyl- and ethynyl-substituted alkoxythiophenes

1,2-Di-(3-alkoxy-2-thienyl)ethenes were obtained by nickel-catalyzed coupling of 3-alkoxy-2-thienyllithium derivatives with 1,2-dichloroethene [157]. Another popular nickel catalyst is 1,3-bis(diphenylphosphino)propanedichloronickel(II), which is used in the coupling of 2-thiophenemagnesium bromide and 3,4-ethylenedioxy-2-thiophenemagnesium bromide with 1,2-dichloroethene to *trans*-1,2-dithienylethene [158] and 1,2-di(3,4-ethylenedioxy)thiophene, respectively [159], and with β-bromovinyl ethyl ether to 1-ethoxy-2-2-(thienyl)ethane [160]. Oxidative polymerization of the ethylenedioxy derivative gives poly [*trans*-bis(3,4- ethylenedioxythiophene)vinylene] [159].

Poly(3,4-dibutoxy-2,5-thienylenevinylene), which is more stable, can be easily prepared by titanium-induced dicarbonyl coupling of 3,4-dibutoxy-2,5-dicarbaldehyde [161].

Poly(3,4-dibutoxy-2,5-thienylene-vinylene) [161]

Titanium(IV) chloride (3.41 g, 18 mmol) is added to a suspension of zinc powder (2.36 g, 36 mmol) in tetrahydrofuran (25 ml) at −10 °C. To the green-yellow solution so obtained, 3,4-dibutoxy-2,5-formylthiophene (0.85 g, 3 mmol) in tetrahydrofuran (5 ml) is added and the reaction mixture is refluxed for 10 h. After cooling it is poured into 10% potassium carbonate solution, which is vigorously stirred for 30 min to deposit solid material. This is separated by filtration, washed with acetone using Soxhlet extractor for 24 h and extracted with chloroform for 24 h. The chloroform extract is placed under reduced pressure giving 1.1 g of the title compound as a dark blue material.

6.1.3.2.7 Change of methoxythiophenes to other alkoxythiophenes

3-(2-Bromo)ethoxy-4-methylthiophene is prepared by heating of 3-methoxy-4-methylthiophene and 2-bromo-1-ethanol in toluene in the presence of sodium bisulfate until all methanol has been distilled off and the temperature has risen to 110°C [162].

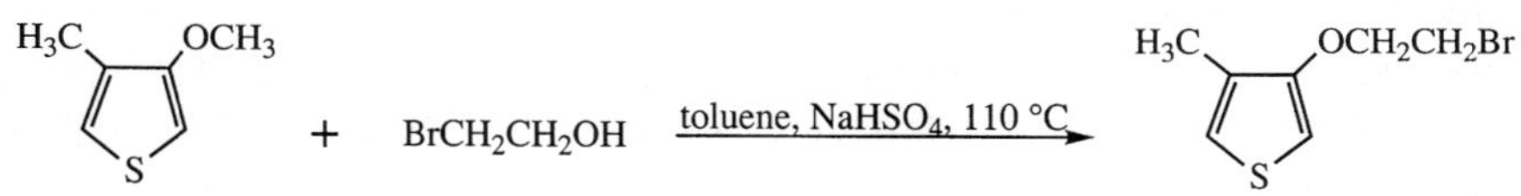

3-(2-Bromo)ethoxy-4-methylthiophene [162]

To a solution of 2-bromo-1-ethanol (8.2 g, 66 mmol) and sodium bisulfate (500 mg) in toluene (40 ml), 3-methoxy-4-methylthiophene (4.2 g, 33 mmol) is added. The reaction mixture is heated until the methanol produced is distilled off and the temperature rises to 110°C. After cooling and several washings with water the product is extracted with diethyl ether. The combined organic phases are dried over magnesium sulfate and evaporated. The residue is purified by chromatography on silica gel using hexanes as eluent, giving 4.7 g (65%) of the title compound.

In a similar way 3-(3-bromo)propoxythiophene is obtained from 3-methoxythiophene and 3-bromopropanol [163]. These bromo derivatives were used for the preparation of sodium 2-(4- methyl-3-thienyloxy)ethanesulfonate, sodium 2-(3-thienyloxy)ethanesulfonate [162] as well as sodium 3-(3′-thienyloxy)propansulfonate [163] through reaction with sodium sulfite in water. 3-(3-Bromo)-propoxythiophene was also transformed to 3-(3′-thienyloxypropyltriethylammonium bromide by reaction with triethylamine [163]. These compounds were used as starting materials for water-soluble polythiophene derivatives.

Metalation of 3-alkoxythiophenes with butyllithium can be achieved predominantly in the 2-position and upon reaction with various electrophiles such as alkyl iodides, aldehydes and ketones, 2-substituted 3-alkoxythiophenes are easily obtained. Thus the reaction of 3-methoxy-, 3-ethoxy-, and 3-isopropoxythiophene with butyllithium followed by di(1-adamantyl)ketone led to the anti[2(3-alkoxythienyl)]di(1-adamantyl)methanols [164].

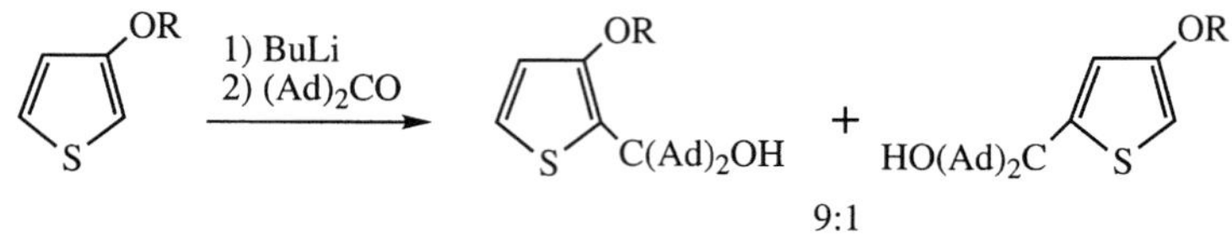

3,4-Ethylenedioxy-2,5-bis(trimethylstannyl)thiophene is prepared by dimetalation of 3,4-ethylenedioxythiophene with butyllithium followed by tributylstannyl chloride [165].

6.1.3.2.8 Modification of substituents

Reduction of methyl-3-methoxy-2-methoxythiophenecarboxylate with lithium aluminum hydride followed by neutral work up gives the very reactive 3-methoxythiophene-2 methanol. Upon treatment of this compound with acid, elimination of water and formaldehyde occurs giving bis(3-methoxy-2-thienyl)-methane [166].

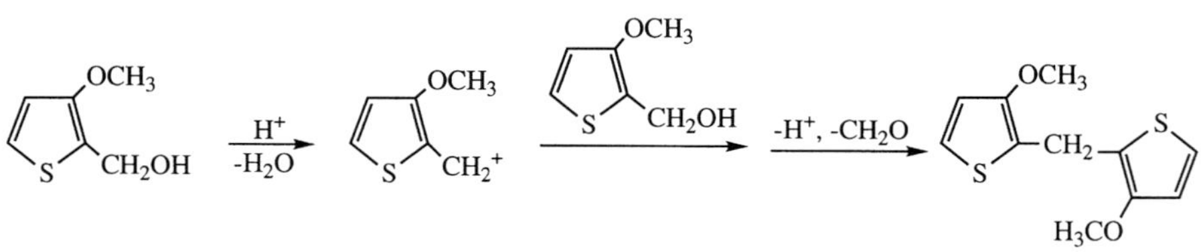

The reaction of 3,4-di-*tert*-butyl-l-[(*para*-tolyl)sulfonylimino]-1,1-dihydro-thiophene with sodium hydroxide in methanol or sodium ethoxide in ethanol gives the 2-alkoxy-3,4-di-*tert*-butylthiophene [167].

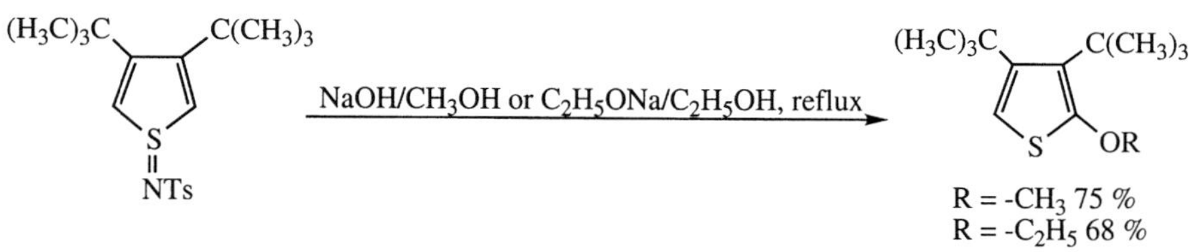

6.1.3.3 Acyl-, carboxyl-, and cyano-substituted alkoxy- and aryloxythiophenes

6.1.3.3.1 Metalation of alkoxythiophenes followed by electrophiles

Owing to the acid sensitivity and nonselectivity in electrophilic substitution of alkoxythiophenes, metalation of 2-alkoxythiophenes with organolithium compounds followed by reaction with electrophiles is in many cases the best method for the preparation of 5-substituted 2-alkoxythiophenes. In this way 5-methoxy-2-thiophene aldehyde [168], 5-methoxy-2-thiophenecarboxylic acid, 5-*tert*-butoxy-2-thiophenecarboxylic acid and its ethyl ester [5], 5-cyano-2-methoxythiophene [123] and 2,3-di-*tert*-butoxy-5-thiophenecarboxylic acid [4] have been prepared. Metalation of 3-methoxythiophene occurs predominantly [3,88,125] but not exclusively in the 2-position [169]. Metalation of 3-methoxy-2-methylthiophene with butyllithium followed by reaction with tributylstannyl chloride gives 3-methoxy-2-methyl-5-tributylstannylthiophene [127]. 3-*tert*-Butoxythiophene is metalated selectively in the 2-position and upon reaction with ethyl chloroformate and acetic anhydride, 2-carboethoxy- and 2-acetyl-3-*tert*-butoxythiophene, respectively, were obtained [4].

3-tert-Butoxy-2-thiophenecarboxylic acid [3]

To a solution of 3-*tert*-butoxythiophene (6.0 g, 38.5 mmol) in anhydrous diethyl ether (50 ml) 1.3 *M* butyllithium in diethyl ether (299 ml) is added dropwise. The mixture is refluxed for 30 min, after which it is poured into solid carbon dioxide covered with anhydrous diethyl ether. Water is added and the phases separated. The organic phase is extracted with sodium carbonate solution. The combined alkaline phases are acidified under cooling with 2 *M* hydrochloric acid. The precipitate formed is filtered off, washed with cold water and dried *in vacuo* over sulfuric acid giving 4.8 g (62%) of the title compound, mp 89.91°C, after recrystallization from aqueous ethanol.

Most selective metalation was obtained using lithium diisopropylamide [169]. 3,4-Dibutoxy-2,5-diformylthiophene has been prepared by metalation with butyllithium-*N*,*N*,*N*′,*N*′-tetramethylethylenediamine in hexane, followed by reaction with *N*,*N*-dimethylformamide [131]. 3-Phenoxy-2-thiophenecarboxylic acid can be prepared by metalation of 3-phenoxythiophene by phenyllithium followed by carbon dioxide [170].

Metalation of 3,4-ethylenedioxythiophene with butyllithium followed by reaction with *N*,*N*-dimethylformamide is used for the preparation of the 2-aldehyde and the 2,5-dialdehyde and upon reaction with pinacol borate the boronate was obtained which was of great use for the preparation of bithienyl derivatives by the Suzuki coupling [171].

3,4-Ethylenedioxy-2-thiophene aldehyde [171]

A solution of 3,4-ethylenedioxythiophene (2 g, 0.0141 mol) in anhydrous tetrahydrofuran (30 ml) is cooled to −78°C under nitrogen and treated with 2.5 *M* solution of butyllithium (6.2 ml). The temperature is slowly raised to 0°C and the mixture is stirred at the same temperature for 20 min.

The reaction mixture is recooled to −78°C and treated with anhydrous *N,N*-dimethylformamide (2 ml, 0.026 mol). The resulting mixture is then stirred at room temperature for 1 h and poured into crushed ice containing hydrochloric acid. The white precipitate of aldehyde is filtered off, washed with water and dried *in vacuo* giving 2.26 g (95%) of the title compound, mp 142°C after recrystallization from methanol.

3,4-Ethylenedioxy-2,5-thiophene dialdehyde [171]

The dilithiation of 3,4-ethylenedioxythiophene (1 g, 0.007 mol) is carried out using 2.5 *M* butyllithium (6.2 ml) and is reacted with anhydrous *N,N*-dimethylformamide (2 ml) as described above. The usual workup followed by recrystallization from methanol gives 1 g (72%) pure dialdehyde as pale brown needles, mp 141°C.

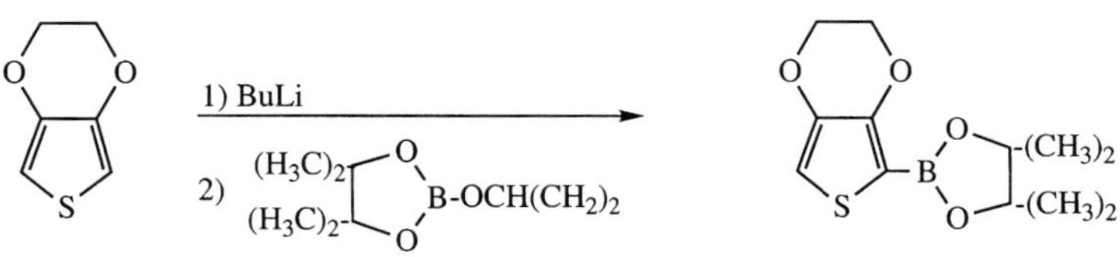

Preparation of the boronic ester [171]

Lithiation of 3,4-ethylenedioxythiophene (2 g, 0.0141 mol) is carried out as described above. To this solution 2-isopropoxy-4,4,5,5-tetramethyl-1,3,2-dioxaborolane (5.8 ml, 0.028 mol) in anhydrous tetrahydrofuran (10 ml) is added. After stirring the mixture for 6 h it is poured over crushed ice containing ammonium chloride. The product is extracted with diethyl ether (3 × 30 ml), dried and evaporated. The residue is triturated with methanol giving 2 g (71%) of the boronic ester, mp 87–88°C.

3,4-Dihexyloxythiophene gave upon lithiation and reaction with tributylstannyl chloride the 2-tributylstannyl derivative [172]. Similarly 2-tributylstannyl-3,4-ethylenedioxythiophene is prepared [173].

6.1.3.3.2 Halogen–metal exchange of bromo-alkoxythiophenes followed by electrophiles

Halogen–metal exchange of bromo-alkoxythiophenes is another alternative for the preparation of a variety of alkoxythiophenes. Thus the reaction of 4-bromo-3-*tert*-butoxythiophene with *tert*-butyllithium followed by reaction with *N*-methoxy-*N*-methyl-3,3-dimethylacrylamide was used for the preparation of 3-*tert*-butoxy-4-(3-methyl-1-oxo-2-buten-1-yl)thiophene [91].

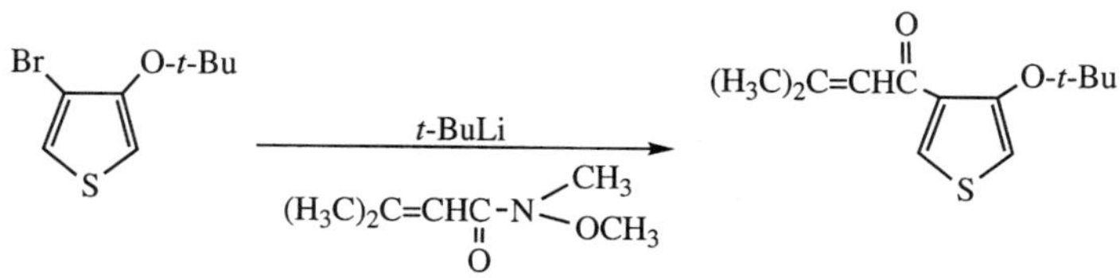

3-tert-Butoxy-4-(3-methyl-1-oxo-2-buten-1-yl)thiophene [91]

1.7 *M tert*-Butyllithium in pentanes (66 ml, 0.11 mol) is added to a solution of 4-bromo-3- *tert*-butoxythiophene (12.5 g, 0.053 mol) in anhydrous diethyl ether (500 ml) at −75°C. After stirring for 1 h *N*-methoxy-*N*-methyl-3,3-dimethylacrylamide 16.0 g, 0.112 mol) is added to the solution at −75°C. The dry ice–acetone bath is removed and the reaction mixture is stirred at ambient temperature overnight. The mixture is washed with 1 *M* hydrochloric acid and dried over magnesium sulfate. After evaporation the residue is purified by flash chromatography using diethyl ether/pentane (5:95) as eluent giving 12.3 g (97%) of the title compound as an oil.

The 3-acetyl- and 3-carboethoxy-2-*tert*-butoxythiophenes were prepared similarly from 3-bromo-2-*tert*-butoxythiophene through halogen–metal exchange at −70°C followed by acetic anhydride and ethyl chloroformate [90]. Similarly 3-acetyl-4-*tert*-butoxythiophene can be prepared from 3-bromo-4-*tert*-butoxythiophene [4].

6.1.3.3.3 By electrophilic substitution of alkoxythiophenes

In some cases electrophilic aromatic substitution is useful. 5-Alkoxy-2-thiophene aldehydes can be prepared in 40–63% yield by Vilsmeyer formylation using phosphorus oxychloride and *N*-methyl formanilide at room temperature [124]. Friedel-Crafts acylations of 2-methoxythiophene give only low yields and mixtures of 5- and 3-isomers are obtained [122,174].

Functionalization of 3,4-ethylenedioxythiophene has become of importance in connection with investigations of conducting polymers. Electrophilic substitution such as the Mannich reaction of 3,4-ethylenedioxythiophene with formaldehyde dimethylamine has been used for the preparation of 3,4-ethylenedioxy-2-dimethylaminomethylthiophene [171], which could be transformed to the quaternary ammonium salt, which upon reaction with triphenylphosphine in *N,N*-dimethylformamide gave the phosphonium salt, a useful starting material for Wittig reactions.

2-(N,N-Dimethylaminomethylene)-3,4-ethylenedioxythiophene [171]

To a stirred solution of 3,4-ethylenedioxythiophene (3.5 g, 0.0215 mol) in glacial acetic acid (5 ml) is added 37% formalin (3.49 ml, 0.043 mol) and

N,*N*-dimethylamine (4.84 ml, 0.043 mol). The mixture is then stirred at room temperature for 48 h. The brown solution is poured into water and the pH adjusted to 7 with 10% sodium hydroxide solution followed by extraction with dichloromethane. Removal of the solvent gives 3.61 g (84%) of the title compound as brown oil, which is used in the next step without further purification.

Methiodide [171]

To a stirred solution of 2-(*N*,*N*-dimethylaminomethylene)-3,4-ethylenedioxythiophene (3.60 g, 0.018 mol) in anhydrous toluene (100 ml) methyl iodide (3.6 ml, 0.058 mol) is added. The mixture is stirred at room temperature over night. The white precipitate is filtered off and washed with hot hexane giving 4.5 g (73%) of the methiodide as a white powder, mp 170–173°C.

Phosphonium salt [171]

To a solution of the methiodide (1.0 g, 2.94 mmol) in anhydrous *N*,*N*-dimethylformamide (10 ml) is added triphenyl phosphine (0.77 g, 2.94 mmol) and the resulting solution is refluxed until there is no evolution of triethylamine. The mixture is then cooled and diluted with diethyl ether and the resulting precipitate is collected by filtration and washed with diethyl ether. After recrystallization from dichloromethanelhexane, 1.35 g (84%) of the title compound is obtained as a light-brown crystalline solid, mp 205–206 °C.

Friedel–Craft conditions were found which allowed the preparation of the 2-acetyl derivative as well as the 2,5-diacetyl-3,4-ethylenedioxythiophene from the acid labile dialkoxythiophene [171].

2-Acetyl-3,4-ethylenedioxythiophene [171]

To a solution of 3,4-ethylenedioxythiophene (3.13 g, 0.022 mol) in anhydrous dichloromethane (2 ml) is added 0.25 *M* solution of acetic anhydride in anhydrous dichloromethane (114.5 ml, 0.0286 mol) and 0.25 *M* solution of stannic chloride in anhydrous acetonitrile (114.5 ml, 0.0286 mol) and the resulting mixture is stirred at room temperature under anhydrous conditions for 24 h. The wine-red solution is then poured into crushed ice containing glacial acetic acid (50 ml). The phases are separated and the aqueous phase extracted with dichloromethane (3 × 50 ml). The combined organic phases are washed with 10% sodium hydroxide solution until the washings have pH 7. Purification by chromatography on silica gel using dichloromethane/hexanes (2:1) followed by recrystallization from chloroformlhexanes gives 2.64 g (65%) of the title compound as yellow-orange needles.

2,5-Diacetyl-3,4-ethylenedioxythiophene [171]

To a solution of 3,4-ethylenedioxythiophene (0.24 g, 1.76 mmol) in anhydrous dichloromethane (1 ml) is added 0.196 *M* solution of acetic anhydride in anhydrous dichloromethane (23 ml, 4.58 mmol) and 0.1 *M* solution of stannic chloride in anhydrous acetonitrile (46 ml, 4.58 mmol) and the resulting mixture is stirred at room temperature under anhydrous conditions overnight. The mixture is worked up as described above. Chromatographic purification followed by recrystallization gives 0.277 g (70%) of the title compound as peach needles.

3-Methoxy-2-(3-methyl-l-oxo-2-buten-lyl)thiophene is obtained by reaction of 3-methoxythiophene and 3,3-dimethylacroyl chloride in dichloromethane with stannic chloride as catalyst [91].

3-Methoxy-2-(3-methyl-l -oxo-2-buten-l yl)thiophene [91]

A solution of 3-methoxythiophene (21.3 g, 0.19 mol) in dichloromethane (50 ml) is slowly added to a solution of 3,3-dimethylacroyl chloride (22 ml, 0.20 mol) and stannic chloride (23 ml, 0.20 mol) in dichloromethane (350 ml) at 0–5 °C. After stirring for 1 h the reaction mixture is poured into ice water (1 l). The phases are separated and the organic phase washed with water, dried over magnesium sulfate, and evaporated. The residue is purified by flash chromatography using dichloromethane as eluent giving 29.6 g (81%) of the title compound mp 49–51 °C.

6.1.3.3.4 By ring-closure reactions

2-*N*-Methylamino-3-methoxythiophene and 2-N-phenylamino-3-methoxythiophene are obtained by a one-pot synthesis from methyl or phenylisothiocyanate and 1-lithiomethoxyallene and are the first aminothiophenes shown to be in equilibrium with the imino tautomers. The analogous 3-methylthio derivatives prepared by a similar procedure from 1-lithiomethylthioallene exist exclusively in the amino form [175].

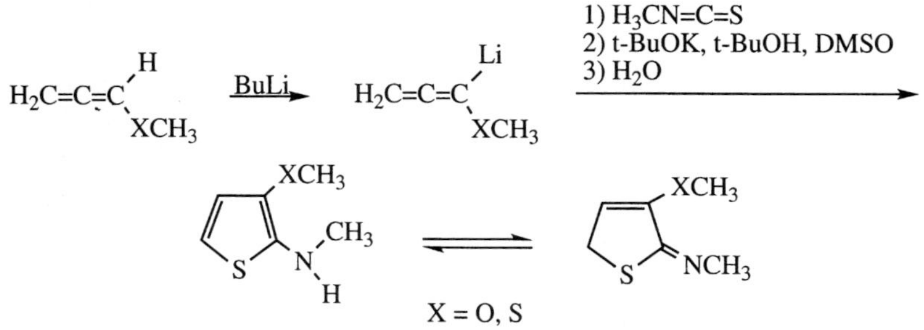

Ortho-carbonyl-substituted hydroxythiophenes, available by various ring-closure reactions are exclusively methylated on oxygen, using dimethyl sulfate and sodium hydroxide [6,176,111] or sodium carbonate, [65] diazomethane [68,86,101,177] and even by reaction with methanol and *para*-toluenesulfonic acid [84].

The following reaction sequence has been performed to give 3-ethoxy-2-methylphenylaminothiophene [178].

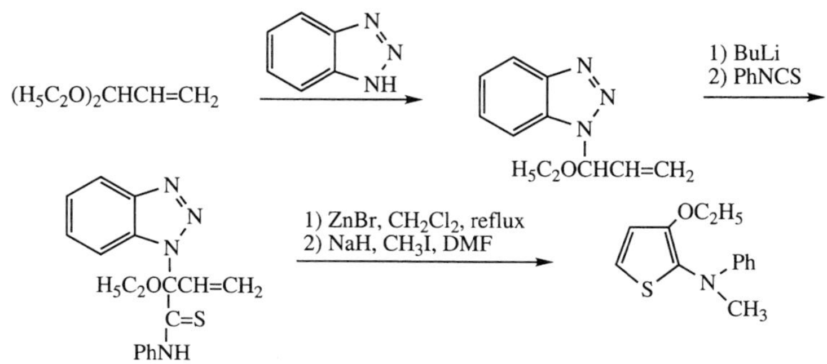

3-Ethoxy-2-methylphenylaminothiophene [178]

The crude product is dissolved in anhydrous *N,N*-dimethylformamide (30 ml) and under argon, sodium hydride (60% in mineral oil, 0.8 g, 20 mmol) is added. After 10 min methyl iodide (1.25 ml, 20 mmol) is added and the stirring is continued at room temperature for 5 h. Water (10 ml) is added and the product extracted with diethyl ether (3 × 30 ml). The combined organic phases are dried over sodium sulfate and evaporated. The residue is purified by flash chromatography on silica gel using pentane/diethyl ether (20:1) as eluent, giving the title compound as a colorless oil.

6.1.3.3.5 By alkylation of hydroxythiophenes

Diethyl 3,4-didecyloxy- and 3,4-dihexyloxy-2,5-thiophenedicarboxylate are prepared in high yields by the reaction at 100 °C of decyl bromide and hexyl bromide and potassium carbonate in *N,N*-dimethylformamide and hydrolyzed to the dicarboxylic acids with potassium *tert*-butoxide in tetrahydrofuran/water [179].

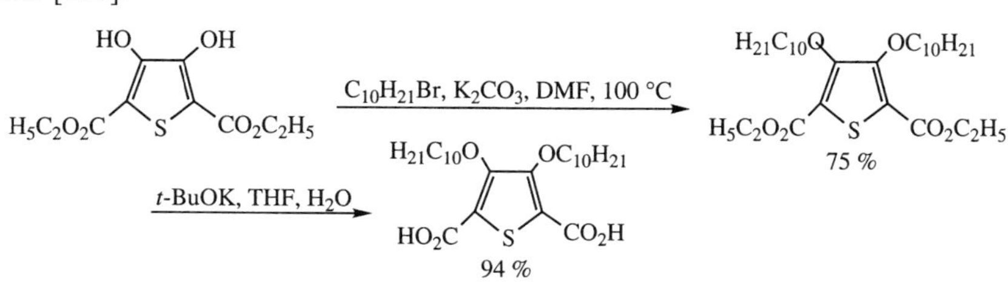

By the use of a number of dibromoalkanes cyclic 3,4-dialkylidenedioxothiophenes are prepared, which in the usual way were hydrolyzed and decarboxylated using copper chromite in quinoline [180].

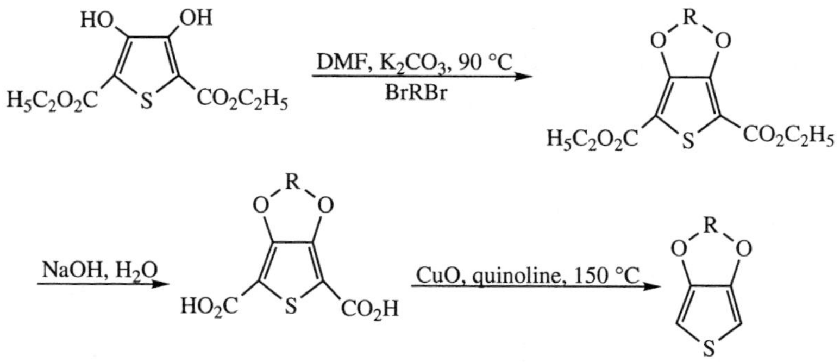

3,4-Dihydroxy-2,5-dicarbonylthiophenes have been methylated by diazomethane [49,84,181]. 3,4-Dibenzyloxy 2,5-carbonyl-substituted thiophenes have been prepared from the disodium salt by reaction with benzyl chloride in *N,N*-dimethylformamide [46] and the diethoxy derivative upon reaction with diethyl sulfate [30]. The methylenedioxy derivatives were synthesized through reaction with bromochloromethane and potassium carbonate in *N,N* dimethylformamide [47]. 3,4-Ethylenedioxythiophenes have been prepared similarly [30,182]. Upon reflux of the disodium salt of diethyl 3,4-dihydroxythiophenedicarboxylate in excess 1,2-dibromoethane a bicyclic ether and diethyl 4-(2-bromoethoxy)-2-2-bromoethyl)-3-oxo-2,3-dihydrothiophene-2,5-dicarboxylate are formed [183].

Reaction of diethyl-3,4-dihydroxythiophenedicarboxylate with 1,2-dibromoethane [183].

To a solution of sodium ethoxide (20 mmol), prepared from 460 mg sodium and 20 ml absolute ethanol, is added a solution of diethyl 3,4-dihydroxythiophenedicarboxylate (2.60 g, 10 mmol) in ethanol (40 ml), after which the reaction mixture is refluxed for 1 h. The solvent is removed *in vacuo*, and to the residual sodium salt 1,2-dibromoethane (30 ml) is added, and the mixture is heated at 110 °C for 24 h. The resulting suspension is poured into water, the phases are separated and the organic phase is successively washed with 2 *M* sodium hydroxide solution and water and evaporated in *vacuo*. The residue (1.85 g) is subjected to column chromatography. Eluting with dichloromethane gives a first component 720 mg (15%) of diethyl 4-(2-bromoethoxy)-2-2-bromoethyl)-3-oxo-2,3-dihydrothiophene-2,5-dicarboxylate as a yellow oil. Switching to ethyl acetate as mobile phase affords 710 mg (25%) of diethyl 2,3-dihydrothieno[3,4-*b*] [1,4]dioxine-5,7-dicarboxylate, which after recrystallization from ethanol gives colorless crystals, mp 151–152 °C.

Novel thiophene-based macrocycles related to azacrown-ethers are prepared from 3-hydroxy-2-carbomethoxythiophenes and α,ω-bistosylate followed by hydrolysis and decarboxylation and functionalization across the 2,2′-positions [184].

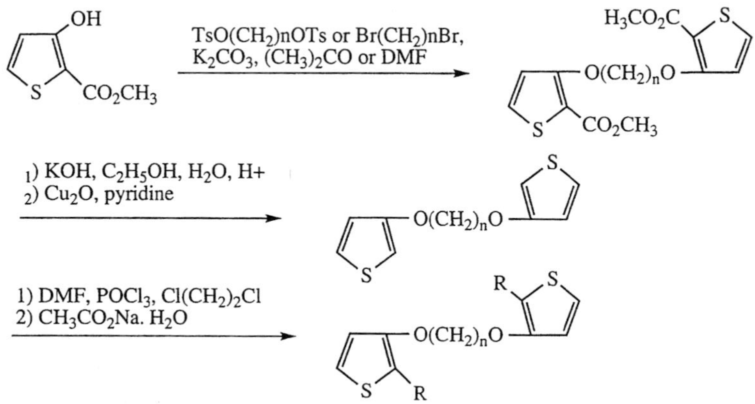

Methyl 3-hydroxy-2-thiophenecarboxylate and some halogen-substituted derivatives have been alkylated with methyl chloroacetate in dimethylsulfoxide in the presence of potassium *tert*-butoxide or by methyl bromoacetate in boiling methyl ethyl ketone in the presence of sodium carbonate [185].

α,ω-Bis(2-methoxycarbonyl-3-thienyloxy)alkanes were prepared from methyl 3-hydroxy-2-thiophenecarboxylate, anhydrous potassium carbonate, and the appropriate dibromoalkane in *N,N*-dimethylformamide and hydrolyzed to the acids under alkaline conditions [186].

1,2-Bis(2-methoxycarbonyl-3-thienyloxy)ethane [186]

A stirred solution of methyl 3-hydroxythiophene-2-carboxylate (1.58 g, 10 mmol), anhydrous potassium carbonate (0.76 g, 5.5 mmol), and 1,2 dibromoethane (1.03 g, 5.5 mmol) in anhydrous *N,N*-dimethylformamide (8 ml) is heated at 95–100 °C for 5 h. The reaction mixture is poured into ice and the resulting precipitate filtered off, washed with water, and digested in hot methanol. The residual solid is filtered off and washed with methanol giving 1.7 g (89%) of the title compound, mp 189–191 °C.

1,2-Bis(2-carboxy-3-thienyloxy)ethane [186]

A solution of 1,2-bis(2-methoxycarbonyl-3-thienyloxy)ethane (684 mg, 2 mmol) and potassium hydroxide (448 mg, 8 mmol) in 80% aqueous ethanol (14 ml) is refluxed for 2 h. The solution is filtered through glass wool into a stirred mixture of ice and slight excess of hydrochloric acid. The resulting solid is filtered off, washed with water, and dried in vacuo at 60 °C for at least three days giving 584 mg (93%), mp 219–219.5 °C.

Esters of the above-mentioned type were hydrolyzed and decarboxylated to the corresponding α,ω-bis(3-thienyloxy)alkanes and used for the preparation of macrocyclic Mannich bases, as the one shown below [147].

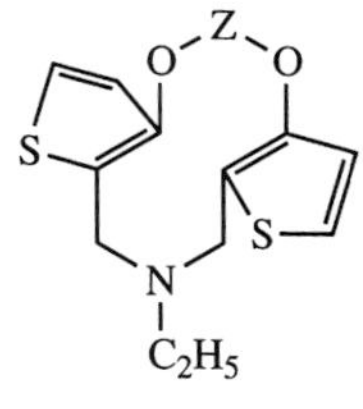

6.1.3.3.6 Modification of substituents

3,4-Dimethoxy-2,5-thiophenedicarboxylic acid is a useful starting material for preparation of the acid chloride and amides as well as for the preparation of the 2,5-diiodo-3,4-dimethoxythiophene and the 3,4-dimethoxy 2,5-diacetylthiophene [188].

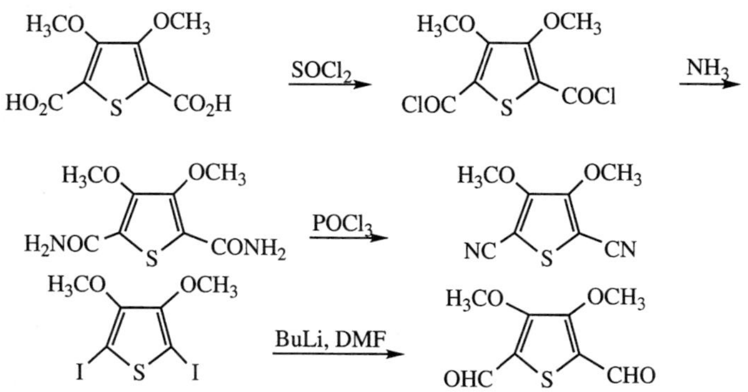

6.1.3.4 Amino- and nitro-substituted alkoxythiophenes

6.1.3.4.1 By nucleophilic aromatic substitution

Reactive dinitrohalothiophenes or nitrocarbonyl-substituted halothiophenes are useful starting materials for alkoxy- and aryloxythiophenes by nucleophilic aromatic substitution. 2-Nitro-3-phenoxythiophene was prepared in 60% yield through the reaction of 3-bromo-2-nitrothiophene with sodium phenoxide [189].

2-Methoxy-, 2-ethoxy-, and 2-phenoxy-3,5-dinitrothiophene were prepared from 2-chloro-3,5-dinitrothiophene by reaction with the corresponding alcohol and potassium hydroxide in high yields. 2-Alkoxy- and 2-aryloxynitrothiophene as well as 2-acetyl-4nitro-5-alkoxy- and 5-aryloxy derivatives were obtained from the corresponding chloro derivatives [106]. The reaction of 2-acetyl-5-chloro-4-nitrothiophene with alkoxides to 2-acetyl-4-vitro-2-alkoxythiophenes [190] and with the salt of ethyl lactate to ethyl 2-(5-acetyl-3-nitro-2-thienyloxy)propionate are good routes [191].

Ethyl 2-(5-acetyl-3-nitro-2-thienyloxy)propionate [191]

Sodium hydride (about 80% in mineral oil, 600 mg, 20 mmol) is washed three times with absolute toluene (distilled from sodium hydride) and absolute toluene (80 ml) is added under argon and a solution of ethyl lactate (2.83 g, 24 mmol) in absolute toluene (40 ml) is slowly added dropwise. After stirring this mixture for 20 min at room temperature a solution of 2-acetyl-5-chloro-4-nitrothiophene (2.05 g, 10 mmol) in absolute toluene (30 ml) is added. After 3 h a small amount of water is added to the reaction mixture, which is acidified with 2 *M* hydrochloric acid. The phases are separated and the organic phase is dried over sodium sulfate and evaporated. The residue is purified by column chromatography using toluene/ethyl acetate (9:1) as eluent. After recrystallization from 2-propanol 2.03 g (71%) of the title compound is obtained, mp 82–84 °C.

Other examples of this methodology are the synthesis of 2-methoxy-3-nitro-5-thiophenecarboxylate, [192] 3-methoxy-2-nitrothiophene, 3-methoxy-4-methyl-2-nitrothiophene, [193,194] and 3-cyano-2-methoxy-5-nitrothiophene [195] from the appropriate bromo derivatives. The reaction of 5-*tert*-butylsulfonyl-2,4-diformylthiophene with sodium alkoxides has been used for the preparation of 5-alkoxy-2,4-diformylthiophenes [196].

6.1.3.4.2 By reaction of hydroxythiophenes with activated aromatic halides

Alternatively, salts of hydroxythiophenes can be reacted with activated heterocyclic halides, like in the synthesis of 2-(2′-formyl-3′-thienyloxy)-3-nitropyridine from 2-formyl-3-hydroxythiophene and 2-chloro-3-nitropyridine [189].

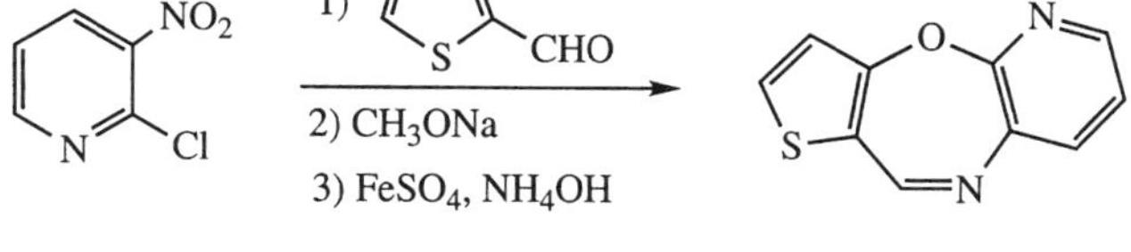

A novel type of large thiophene containing aza-crown ethers are prepared by alkylation of 2,5-dicarboethoxy-3,4-dihydroxythiophene with 5-tosyloxy-3-oxapentan-l-ol, followed by esterification with methanesulfonyl chloride, which upon reaction with methyl 3-hydroxy-2-thiophenecarboxylate gives the derivative with three thiophene rings. Hydrolysis and decarboxylation gives 3,4-bis[5′-(3″-thienyloxy)-3′oxapentyloxy]thiophene. Finally the desired product is obtained by ring closure with piperazine via a Mannich reaction [197].

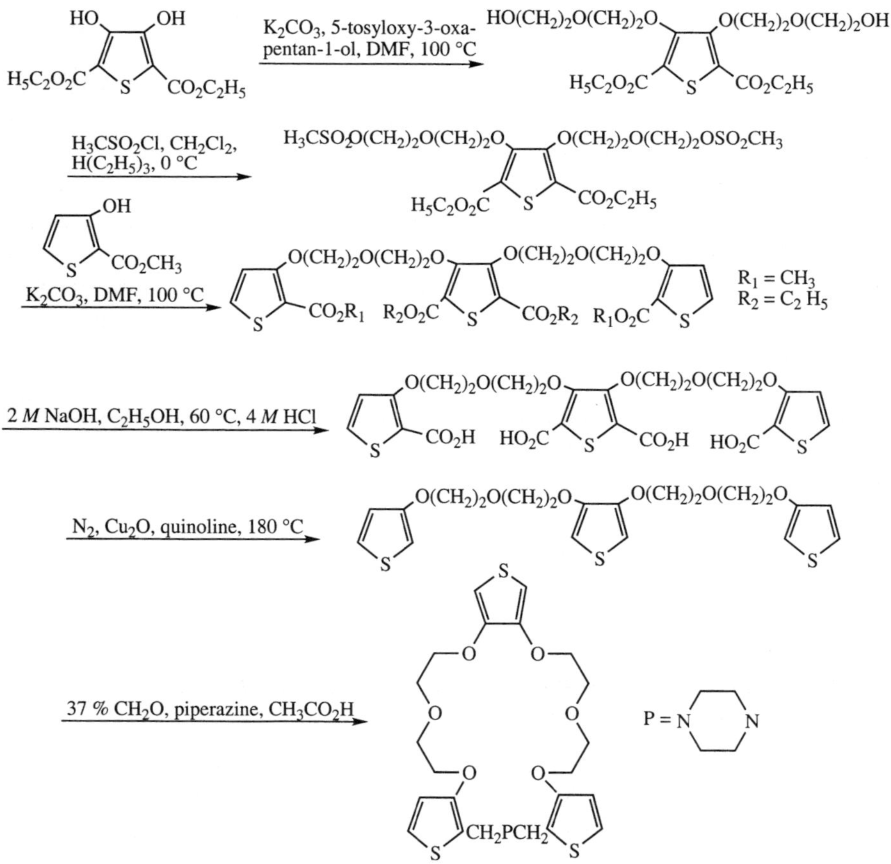

6.1.3.4.3 By various reactions

4-Methoxy-2-thiophenecarboxylic acids can, via use of the Curtius reaction, be transformed to the 4-methoxy-2-amino carbamate derivative and further transformed to the hydrazino derivative [198]. 3-Amino-4-ethoxythiophene has been prepared as shown below and is unexpectedly stable [84].

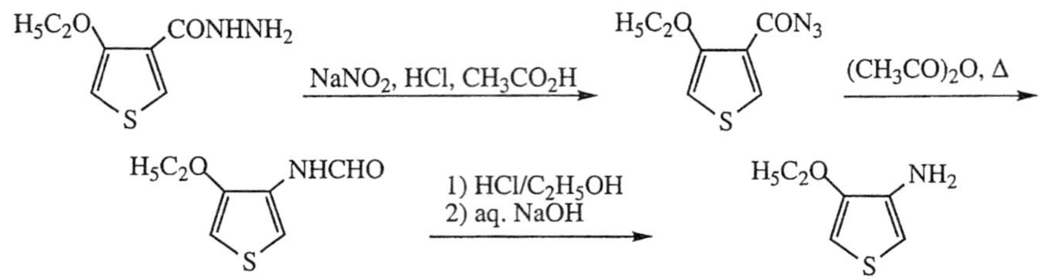

6.1.4 Silyloxy derivatives

2-(2-tert-Butyldimethylsilyloxy)thiophene can conveniently be prepared in 95 % yield through the reaction of 2-hydroxythiophene-(2(5H)thiophenone) in dichloromethane with 2,6-lutidine and *tert*-butyldimethylsilyl trifluoromethanesulfonate [199].

2-(2-tert-Butyldimethylsilyloxy)thiophene [199]

To a solution of 2(5H)-thiophenone (1.24 g, 12.4 mmol) in anhydrous dichloromethane (40 ml) are added 2,6-lutidine (4.3 ml) and *tert*-butyldimethylsilyl trifluoromethanesulfonate (4.25 g, 16.1 mol) under argon at room temperature. After stirring the reaction mixture for 30 min the solvent is evaporated and the residue purified by flash chromatography on silica gel using ethyl acetate/hexanes (1:1) as eluent, giving 2.5 g (95%) of the title compound as a pale-yellow oil.

6.2 SULFUR DERIVATIVES

6.2.1 Divalent sulfur derivatives (SH, SR, SX, SSR)

6.2.1.1 General aspects

Thiophenethiols are prepared (a) by ring closure reactions (b) by reaction of metalorganic reagents with sulfur, and (c) by modifications of thiophenes already containing a thiophene–sulfur bond. Alkylthio- and arylthiothiophenes are most conveniently prepared by (a) ring closure of aliphatic compounds, (b) by reaction of metalorganic thiophenes with appropriate sulfur reagents, (c) by alkylation or arylation of thiophenethiols, (d) by nucleophilic substitution of halothiophenes with alkyl or arylthiolates, and (e) by substitution reactions of alkyl- or arylthiothiophenes. The method of choice is, of course, dependent on the desired substitution pattern of the thiophene.

6.2.1.2 Parent thiophenethiols and alkyl-and aryl-substituted thiols

6.2.1.2.1 By reaction of thiophenemagnesium and thienyllithium derivatives with sulfur

2-Thiophenethiol is best prepared by the Grignard reagent derived from 2-bromo- or 2-iodothiophene [20,200,201] or 2-thienyllithium with sulfur. In the laboratory, 3-thiophenethiol it is best prepared through the reaction of 3-thienyllithium with sulfur [6,201,201–213]. The reaction has to be carried out rapidly at low temperatures to avoid alkylation of the 3-thiophenethiolate by the alkyl halide formed in the halogen–metal exchange [210]. 3,4-Thiophenedithiol was prepared starting from 3,4-dibromothiophene, which through halogen–metal exchange and reaction with sulfur was transformed to 3-bromo-4-thiophenethiol, which upon renewed halogen–metal exchange and reaction with sulfur yielded the desired compound [214,215]. It is even better prepared directly by a one-pot procedure from 3,4-dibromothiophene [216].

6.2.1.2.2 By dealkylation of alkylthiothiophenes

2,3-Thiophenedithiol was prepared by dealkylation of 2,3-diethylthiothiophene with four equivalents of sodium in liquid ammonia [203]. Both these thiols are of interest for the preparation of organic metals. 2,5-Thiophenedithiol was prepared similarly from 2,5-diethylthiothiophene [203]. This method was used by Goldfarb and coworkers for the preparation of a number of thiophenethiols [217–225] among them 3-aminoethyl-5-ethyl-2-thiophenethiol [222] and 5-(ethoxyethyl)-thiophenethiol [223]. The best method for the preparation of 3,4-thiophenedithiol, published recently, is the treatment of 3,4-di(isopropylthio)thiophene with pyridine and sodium at 105 °C [226].

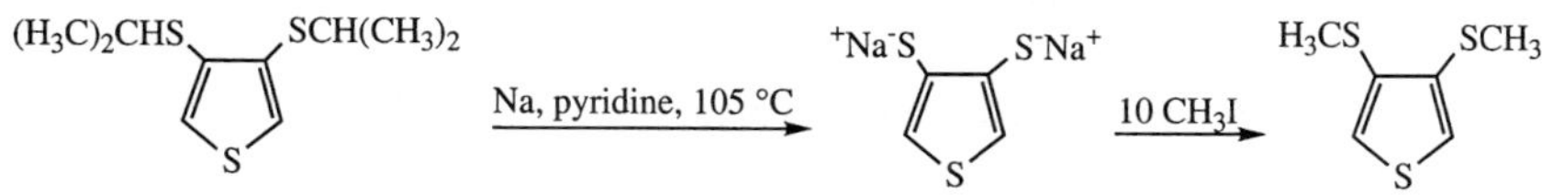

6.2.1.2.3 By reduction of sulfonyl chlorides

2-Thiophenethiol is also prepared by the reduction of 2-thiophenesulfonyl chloride with zinc dust [200,227] or lithium aluminum hydride [228]. 2,5-Dimethyl-3-thiophenethiol can also be prepared by the reduction of the sulfonyl chloride [229].

6.2.1.2.4 By ring-closure reactions

The parent 3-thiophenethiol has been prepared by pyrolysis of the thiophene tars from the commercial synthesis from butane and sulfur. 5-Isopropyl-3-methyl-2-thiophenethiol was prepared in the following way [230].

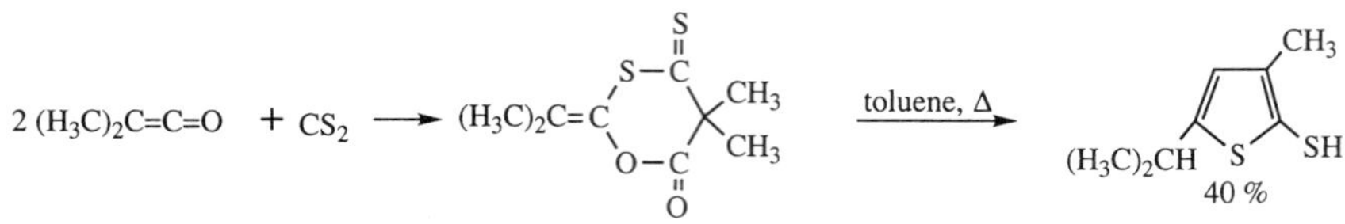

The reaction of 3-thiethanone with methanolic sodium hydrogen sulfide gives 2,4-dimethyl-3-thiophenethiol [231].

6.2.1.2.5 By thio-Claisen rearrangements

The thio-Claisen rearrangement of allyl-2-thienyl sulfides can be used for the preparation of 3-allyl-2-thiophenethiols upon heating without solvent or *N*,*N*-dimethylaniline, triethylamine or pyridine in the temperature range between 100–180°C [232,233].

6.2.1.2.6 By reaction of halothiophenes with hydrogen sulfide

2-Chlorothiophene and 5-ethyl-2-chlorothiophene react at 450–455°C with hydrogen sulfide to give the corresponding thiophenethiols in 20–40% yield [234].

6.2.1.3 Carbonyl-, carboxyl-, and cyano-substituted thiophenethiols

6.2.1.3.1 By dealkylation of alkylthiothiophenes

Dealkylation of acetal-protected *ortho*-formylalkylthiothiophenes with sodium in liquid ammonia has been used for the synthesis of 2-mercapto 3-aldimines [220,221,224], which were converted to the more stable Schiff's bases [223,225,235,236]. The corresponding *ortho*-substituted 3-ethylthio-2- and 4-aldehyde acetals gave dealkylation to the corresponding thiolates without affecting the acetal grouping [237]. 5-Formyl-2-thiophenethiol is prepared by dealkylation of the diethylacetal of 2-ethylthio-2-formylthiophene with two equivalents of sodium in liquid ammonia. Hydrolysis of 2-formyl- or 2-acetyl-3-thiocyanothiophene with aqueous sodium sulfide gave the corresponding thiols [238].

6.2.1.3.2 Ring-closure reactions followed by aromatization

Ring-closure reactions are often best used in order to reach this class of thiophenethiols. Thus the Gompper reaction [239] has been used for the synthesis of 3,5-diacetyl-4-methyl-2-thiophenethiol [240].

2-Carbomethoxy-4-methyl-3-thiophenethiol was prepared in the following way [56].

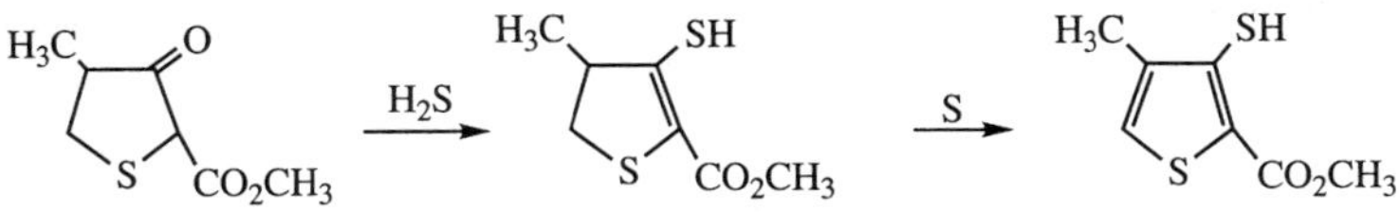

3-Mercapto-2-thiophenecarboxylic acid can be prepared in the following way [56].

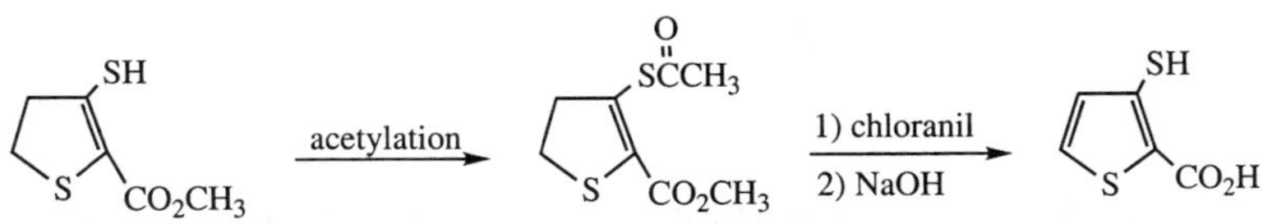

Pyridinium-1-(dithiocarboxy)methylides, prepared from the reaction of acetonyl- and phenacylpyridinium chlorides with carbon disulfide and ethyl bromoacetate in the presence of base, give upon heating with 1,8-diazabicyclo[5.4.0]undec-7-ene (DBU) in ethanol 3-(pyridinio)thiophene-2-thiolates [241].

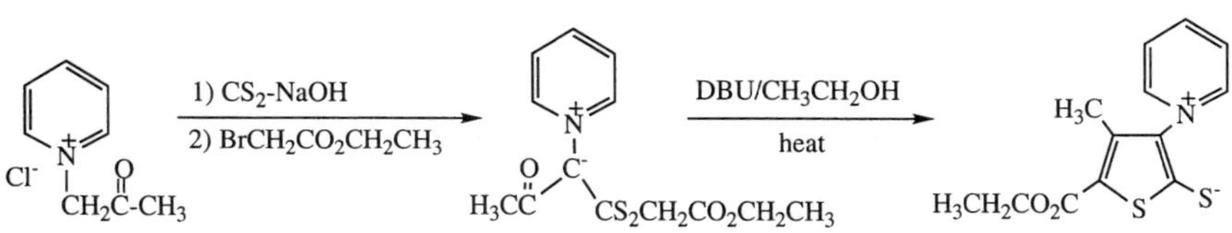

6.2.1.3.3 By nucleophilic substitution of halothiophenes with hydrogen sulfide

Ethyl 5-formyl-4-mercapto-2-methyl-3-thiophenecarboxylate was obtained in 82% yield by nucleophilic substitution of ethyl 4-chloro-5-formyl-2-methyl-3-thiophenecarboxylate with anhydrous sodium hydrogen sulfide in acetone [242]. In a similar way 5-acetyl-2-mercapto-3-nitrothiophene was obtained from 5-acetyl-2-chloro-3-nitrothiophene [190].

5-Acetyl-2-mercapto-3-nitrothiophene [190]

A solution of sodium hydrosulfide hydrate (444 mg, 6 mmol) in anhydrous methanol (15 ml) is added dropwise to a suspension of 5-acetyl-2-chloro-3-nitrothiophene (1.025 mg, 5 mmol) in anhydrous methanol (15 ml) under argon at 0 °C. The reaction mixture is stirred for 30 min, after which the precipitated product is filtered off and recrystallized from ethyl acetate/toluene (9:1) giving light brown crystals, mp 205 °C.

6.2.1.3.4 From amino-and hydroxythiophenes

3-Mercapto-2-thiophenecarboxylic acid was also obtained from 3-hydroxy-2-carbomethoxythiophene [243] and by the diazotization of methyl 3-amino-2-thiophenecarboxylate followed by reaction with potassium xantlogenate and alkaline hydrolysis [244]. 3-Mercapto-4-thiophenecarboxylic acid was prepared similarly [244].

6.2.1.4 Amino-, nitro-, hydroxy-, and alkoxythiophenethiols

6.2.1.4.1 By ring-closure reactions

2-Carbethoxy-5-methyl-3-hydroxy-4-thiophenethiol was prepared in 50% yield by the Fiesselmann reaction from ethyl 2-acetylthioacetoacetate [78]. Alkylation of the condensation product of nitromethane with carbon disulfide, with α-haloketones followed by ring closure under alkaline conditions constitutes a convenient route for the synthesis of 3-nitro-2-thiophenethiol and its 4-methyl- and 4-phenyl-derivatives [245]. 4-Substituted 3-hydroxy-2-carbomethoxy-5-thiophenethiols are prepared by the reaction of methyl 2-chloro-3-oxo-2,3-dihydrothiophenecarboxylate with thioacetic acid followed by alkaline hydrolysis [95].

6.2.1.4.2 By reduction of thiocyano derivatives

2-Amino-5-carboethoxy-4-methyl-3-thiophenethiol can easily be prepared through reduction of the corresponding thiocyano derivative either with sodium borohydride or sodium dithionite (80%) [246].

2-Amino-S-carboethoxy-4-methyl-3-thiophenethiol [246]

To a suspension of 2-amino-5-carboethoxy-4-methyl-3-thiocyanothiophene (2.4 g, 10 mmol) in ethanol (20 ml), sodium borohydride (1.1 g, 3 mmol) is added portionwise while the temperature goes up to 50 °C and the solution becomes transparent. After 10 min, water (70 ml) is added and the reaction mixture is stirred for 15 min followed by heating for 15 min. The warm solution is quickly filtered and slowly acidified with 6 *M* hydrochloric acid. Ater 2 h the precipitate is filtered off and recrystallized from ethanol giving 1.85 g (85%) of the title compound, mp 125–128 °C.

6.2.1.4.3 By reaction of thienyllithium derivatives with sulfur

Metalation of 2-tert-butoxythiophene with butyllithium followed by reaction with sulfur has been used for the preparation of 5-*tert*-butoxy-2-thiophenethiol.

6.2.1.4.4 By nucleophilic substitution of halothiophenes

Experimental conditions have been found, which permit the preparation of 3-nitro-2-thiophenethiol through the reaction of 2-chloro-3-nitrothiophene with sodium sulfide in aqueous ethanol under argon [247]. The reaction of 3-nitro-2-thiophenethiol with *trans*-(4-methoxyphenyl)glycidate under different conditions is used for the preparation of methyl threo- and methyl *erythro*-2-hydroxy-3-(4-methoxyphenyl)-3-nitro-2-thienylthio)propionate.

3-Nitro-2-thiophenethiol [247]

Sodium sulfide nonahydrate (19.58 g, 81.6 mmol) is dissolved in water (120 ml) and the mixture is stirred under argon for 20 min. A solution of 2-chloro-3-nitrothiophene (7.82 g, 48 mmol) in ethanol (50 ml) is added slowly and the stirring is continued for an additional hour. The ice-cooled mixture is acidified with 2 *M* hydrochloric acid under an argon atmosphere. The unstable precipitate is filtered off and used immediately.

6.2.1.5 Alkyl- and arylthiothiophenes and their alkyl and aryl-substituted derivatives

6.2.1.5.1 From thienyllithium derivatives and sulfur followed by alkylation

The best method for the preparation of 2-alkylthiothiophenes is the metalation of thiophenes with butyllithium, followed by reaction with sulfur and alkylation [203,248–260]. A prerequisite in the case of 3-substituted thiophenes is of course that metalation occurs regiospecifically. The reaction of 3-thienyllithia, obtained by halogen–metal exchange, with sulfur and an alkylating agent is complicated by the fact that the alkyl halide formed in the halogen–metal exchange can compete in the alkylation, causing a decrease in yield and separation difficulties [261]. However, if very reactive alkylating agents such as methyl chloroacetate or bromoacetone are used, this approach can still be applied [262–266]. Through dimetalation of 3-methoxythiophene followed by reaction with sulfur followed by 2-bromoethanol and dimethylaminopyridine and acetyl chloride, 3-methoxy-2,5-bis(acetoxyethylthio)thiophene has recently been prepared [267]. The nucleophilic ring opening of (*R*)-β-butyrolactone with 2-thiophenethiolate gives (*S*)-3-methyl-(2-thienylthio) propanoic acid [268].

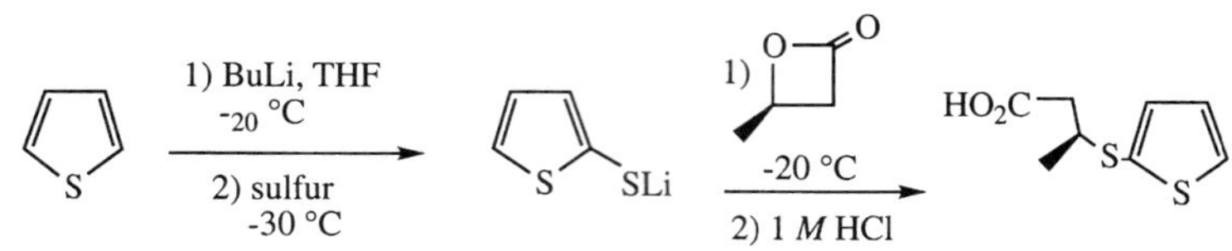

In many cases Grignard reagents derived from 2-iodo- or 2-bromothiophenes can be reacted with sulfur and then alkylated [269–271], and a detailed description for the preparation of 2-methylthiothiophene can be found in Organic Syntheses [272]. Using especially activated magnesium, it is possible to prepare 3-thiophenemagnesium iodide from 3-iodothiophene, which upon reaction with dihexyl disulfide gave 3-hexylthiothiophene in 62% yield [273].

6.2.1.5.2 From thienyllithium derivatives and disulfides

Another very useful route for the synthesis of alkylthiothiophene is the reaction of 2-thienyllithia [253,256,274,275] and especially 3-thienyllithia [226,237, 274–277] with dialkyl disulfides. 3,4-Di(isopropylthio)thiophene is prepared from 3-bromo-4-(isopropylthio)thiophene, *via* halogen–metal exchange and reaction with diisopropyl disulfide [226].

3-Bromo-4-(isopropylthio)thiophene [226]

A solution of 3,4-dibromothiophene (15 g, 0.062 mol) in anhydrous diethyl ether (125 ml) is stirred and cooled to −78 °C under argon while butyllithium (27.3 ml, 0.0682 mol) is slowly added via syringe. After stirring the mixture for 0.5 h isopropyl disulfide (10.23 g, 0.0682 mol) is added. The reaction mixture is stirred for 2 h, after which the dry ice–acetone bath is removed. The reaction mixture is allowed to warm naturally and when the temperature reached −35 °C, water (20 ml) is added. The content of the flask is poured into water (300 ml). The phases are separated and the aqueous phase extracted with diethyl ether (3 × 25 ml). The combined organic phases are washed twice with a 2 *M* potassium hydroxide solution and once with water, dried over magnesium sulfate and evaporated giving 11.77 g (80%) of the title compound as a viscous oil.

3,4-Bis(isopropylthio)thiophene [226]

3-Bromo-4-(isopropylthio)thiophene (11.77 g, 0.05 mol) is treated in the same way as 3,4-dibromothiophene giving 7.8 g (77%) of the title compound as a colorless oil after distillation, bp 150°C/15 mm Hg.

Metalation of 3-methylthiophene with butyllithium followed by reaction with dimethyl disulfide gave a mixture of 3-methyl-2-methylthio- and 3-methyl-5-methylthiothiophene, which upon renewed metalation and reaction with dimethyl disulfide gave a good yield of 2,5-bis(methylthio)-3-methylthiophene [253]. By this route also arylthio- and hetarylthiothiophenes can be obtained. In this way di-(2-thienyl)sulfide [255] and (2-thienyl)-(3-thienyl) sulfide was prepared [278]. Reaction between various thienyllithium derivatives and dimethyl disulfide can advantageously be used for the preparation of 2,5-, 2,3-, and 3,4-bis(methylthio)thiophenes as well as 2,3,4- and 2,3,5-tris-(methylthio)thiophenes [279].

2,5-Bis(methylthio)thiophene [279]

A solution of 2,5-dibromothiophene (2.42 g, 10.0 mmol) in anhydrous diethyl ether (50 ml) is cooled to −78 °C under nitrogen and treated with 2.09 *M* butyllithium in cyclohexane (10.0 ml, 20.9 mmol). The solution is stirred for 30 min and treated dropwise with a solution of dimethyl disulfide (2.07 g, 22.0 mmol) in anhydrous diethyl ether (10.0 ml) during a period of 15 min. The reaction temperature is maintained below −70 °C throughout the addition. After 1 h at −78 °C the reaction mixture is allowed to slowly reach room temperature, after which it is quenched with an ice-cold saturated ammonium chloride solution. The organic phase is separated and dried over magnesium sulfate. The solvent is evaporated *in vacuo* and the residue vacuum distilled giving 1.37 g (78%) of the title compound as a colorless oil bp 59–61°C/ 0.15 mm Hg.

6.2.1.5.3 From thienyllithium derivatives and other sulfur-containing reagents

A more convenient reagent for the conversion of thienyllithium derivatives to symmetrical dithienyl sulfides is their reaction with bis(phenylsulfonyl)sulfide [280,281]. A third method which has been used for the preparation of symmetrical sulfides, and used for the preparation of di(3-thienyl) sulfide, is the reaction of an aryllithium derivative with sulfur dichloride [282]. 2,5-Bis(mesitylthio)-3,4-ethylenedioxythiophene is prepared by dimetalation with butyllithium followed by mesitylthio chloride [165].

2,5-Bis(mesitylthio)-3,4-ethylenedioxythiophene [165]

To a solution of 3,4-ethylenedioxythiophene (0.5 g, 3.5 mmol) in anhydrous tetrahydrofuran (20 ml) at room temperature butyllithium (4.40 ml, 7.1 mmol) is added dropwise. The stirring is continued for 1.5 h, after which freshly prepared mesitylthio chloride (7.7 mmol) in hexanes (20 ml) is added. The reaction mixture is stirred at 0 °C for 1.5 h and then washed with water (3 × 50 ml). The organic phase is dried over magnesium sulfate and evaporated.

The residue is filtered through a silica frit using hexanes as eluent. The filtrate is chromatographed using hexanes/ethyl acetate (5:1) as eluent. Upon evaporation of the filtrate, 0.706 g (45%) of the title compound crystallizes mp 156–157 °C.

6.2.1.5.4 From bromothiophenes and cuprous mercaptides

Another approach to alkylthio- and arylthiothiophenes consist in the reaction of cuprous mercaptides with bromothiophenes. In this way 2-butylthio- and 2-ethylthiothiophene has been prepared in 92 and 56% yield, respectively [283]. All three isomeric dithienyl sulfides were prepared by reaction of the appropriate thiophenethiol with a bromothiophene in the presence of cuprous oxide in *N,N*-dimethylformamide [204,282], and a detailed description of 2,2-dithienyl sulfide can be found in Organic Synthesis [284]. Di(thienylthio)thiophenes have been prepared in a similar way through the reaction of 2,5- and 3,4-dibromothiophene with 2- and 3-thiophenethiol [285]. This method was recently improved through the use of freshly precipitated cuprous oxide obtained by treatment of copper(II) acetate with hydrazine. From 2,5-dibromothiophene and 2-thiophenethiol and potassium hydroxide in *N,N*-dimethylformamide 2,5-bis(2-thienylthio)thiophene was obtained [286].

2,5-Bis(2-thienylthio)thiophene [286]

A solution of 2-thiophenethiol (7.99 g, 69 mmol) in *N,N*-dimethylformamide (25 ml) is added to a stirred mixture of 2,5-dibromothiophene (8.43 g, 30 mmol), potassium hydroxide (3.89 g, 69 mmol), copper oxide(I) (4.94 g, 34 mmol) in *N,N*-dimethylformamide (50 ml). The reaction mixture is heated at 130–140°C for 16 h, cooled to room temperature and poured into 6 *M* hydrochloric acid (120 ml), and benzene (500 ml) is added. The resulting two-phase system is stirred for a while and filtered through a pad of Celite. The phases are separated and the organic phase is washed with water (10 × 100 ml), dried over magnesium sulfate and evaporated. The residue is distilled giving 6.73 g (72%) of the title compound, bp 171 °C/0.08 mm Hg.

Copper-promoted reaction between 2-iodothiophene and thiosalicylic acid was used for the preparation of 2-carboxyphenyl-2-thienylsulfide [287], which also has been obtained from 2-thiophenethiol and *ortho*-iodobenzoic acid [288].

6.2.1.5.5 From nonactivated halothiophenes and sodium thiolates in refluxing amide solvents

As an alternative to copper-promoted reactions, refluxing amide solvents such as *N,N*-dimethylformamide, *N,N*-dimethylbutyramide, or *N*-methyl-2-pyrrolidone can be used with nonactivated halo compounds. Thus 2,5-di(phenylthio)thiophene was obtained in 74% yield from 2,5-dibromothiophene and sodium thiophenolate [289]. Quite recently, however, it was found that 2-(2-arylthio)thiophenes, 2-alkylthiothiophenes, and 2-(2-thienylthio)thiophene can be prepared in high yields by simply mixing 2-iodothiophene and aryl-, alkyl-, or thiophenethiols at room temperature without solvent, base or catalyst [290].

6.2.1.5.6 From alkali salts of thiophenethiolates through alkylation, arylation and Michael addition

Many alkylthiothiophenes are prepared by alkylation or arylation of alkali salts of the thiophenethiols or by Michael addition reactions. 3-Thienylthioacetic acid [291] and 3-thienylthioacetaldehyde diethyl acetal [292] were prepared by the alkylation of 3-thiophenethiolate with chloroacetic acid and bromoacetaldehyde diethylacetal respectively in connection with synthesis of thienothiophenes. A number of thienylmethylthiothiophenes can be prepared in good yields through the reaction of the sodium salts of 2- and 3-thiophenethiols with substituted halomethylthiophenes [293]. Similarly various substituted thenyl bromides were reacted with thiophenolates and thiophenethiolates [294].

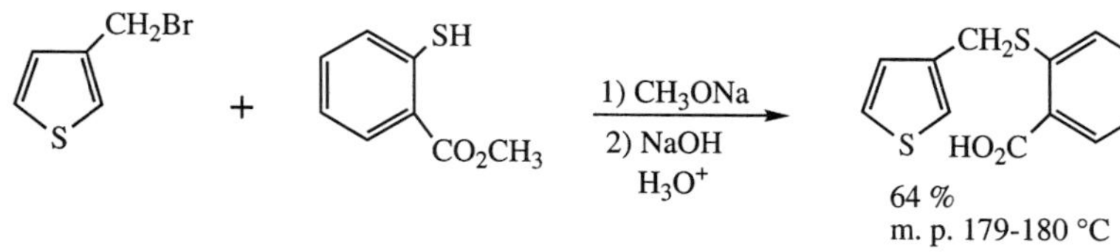

General procedure for the preparation of acids of the above type [294]

A solution of the thiol (0.05 mol) in dimethylsulfoxide (20 ml) is added to a solution of sodium methylate (0.05 mol sodium in a minimum of methanol). The mixture is stirred for 10 min and cooled to 0 °C, after which a solution of the bromo derivative (0.05 mol) in dimethylsulfoxide (10 ml) is slowly added. The reaction mixture is stirred overnight at room temperature and then poured into water. If the ester does not precipitate it is taken up in diethyl ether and this solution is dried and evaporated. The oil formed is, without further purification, dissolved in ethanol and treated with a sodium hydroxide solution, after which the reaction mixture is refluxed for 2 h. After cooling, the mixture is poured into water and acidified with hydrochloric acid. The precipitate is filtered off and recrystallized.

Reaction between 2-thiophenethiolate with ethylene chlorohydrin can be used for the preparation of 2-(β-hydroxyethylthio)thiophene [295], and with epichlorohydrin glycidol 2-thienylthioether was obtained [296]. 2-Thienylthiomethylphthalimides are synthesized by the action of chloromethylphthalide on

2- or 3-thiophenethiol using sodium methylate in methanol/dimethylsulfoxide [297].

2-(Thien-2′-ylthiomethyl)phthalimide [297]

To a mixture of 2-mercaptothiophene (1.74 g, 15 mmol) in anhydrous dimethylsulfoxide (10 ml) is slowly added 3.8 *M* sodium methylate in methanol. After 2 h of stirring, chloromethyiphthalimide (2.9 g, 15 mmol) in anhydrous dimethylsulfoxide (10 ml) is added dropwise over a period of 30 min. The reaction mixture is stirred at room temperature for 3 h and then poured into ice-water (100 ml). The precipitate is filtered off and dried. Recrystallization from methanol affords 2.33 g (66%) of the title compound as colorless crystals, mp 117 °C.

2-Thiophenethiolate can also be reacted with ethylene imine to give 2-(2-thienylthio)ethylamine [296]. Many hetaryl-thienyl sulfides have been prepared via this route [298]. Tosylates can of course be used instead of alkyl halides and methyl (*R*)-3-(*para*-toluenesulfonyloxy)butyrate was used in connection with the preparation of a topically active carbonic anhydrase inhibitor. Initial attempts to react it with lithium 2-thiophenethiolate, produced considerable elimination to methyl crotonate. Subsequent Michael addition (see below) led to the racemic product. However, using the *meta*-chlorobenzenesulfonate instead and formamide as solvent, methyl (*S*)-3-(2-thienylthio)butyrate is obtained with 98% ee [299].

Methyl (S)-3-(2-thienylthio)butyrate [299]

To a solution of thiophene (19.0 g, 226 mmol) in anhydrous tetrahydrothiophene (200 ml) at −5 °C is added 1.6 *M* butyllithium in hexane (137 ml, 219 mmol) maintaining the temperature at < 0 °C. The reaction mixture is stirred for 1 h at 0–5 °C, after which powdered sulfur (7 g, 219 mmol) is added portionwise at < 5 °C. The reaction mixture is stirred for 2.5 h at 0–5 °C and diluted with formamide (200 ml of technical grade) that has been thoroughly purged with nitrogen. To this two phase mixture solid methyl (*R*)-3-(*para*-toluenesulfonyloxy)butyrate (57 g, 209 mmol) is added and the mixture is stirred at 25 °C for 3 days. The progress of the alkylation is conveniently monitored by high-pressure liquid chromatography. The entire reaction mixture is, with stirring, poured into a vessel containing water (400 ml) and ethyl acetate (200 ml) at 25 °C. The phases are separated and the water phase extracted with ethyl acetate/hexanes (1:1, 100 ml). The combined organic phases are washed with an aqueous sodium chloride solution (200 ml). Upon evaporation viscous oil–water mixture is obtained (*ca.* 100 ml water and 60 ml oil), which is hydrolyzed directly in the next step. A small sample of the oil is chromatographed on silica gel using ethyl acetatelhexanes (1:9) as eluent for product characterization and chirality determination.

The second and main fraction $R_f = 0.52$ is identified as the title compound with chirality > 92:2 by NMR using (+)-Eu(hfc)$_3$ chiral shift reagent and monitoring the methyl doublet at 1.32 ppm.

A great variety of thienylthio derivatives with functionalized side chains can be prepared by Michael-type reactions. Thus 2-thiophenethiol gave with dimethyl maleate, in the presence of catalytic amounts dimethyl 2-thienylthio-succinate, with acrylonitrile 3-(2-thienylthio)propiononitrile and with methyl methacrylate, methyl 3-(2-thienylthio)-2-methylpropionate were obtained [296]. A large number of commercially available acrylic acids have been used in the reaction with 2-thiophenethiol for the preparation of substituted propionic acids. 3-(2-Thienylthio)-3-phenylpropionic acid have thus been obtained in almost quantitative yields [300].

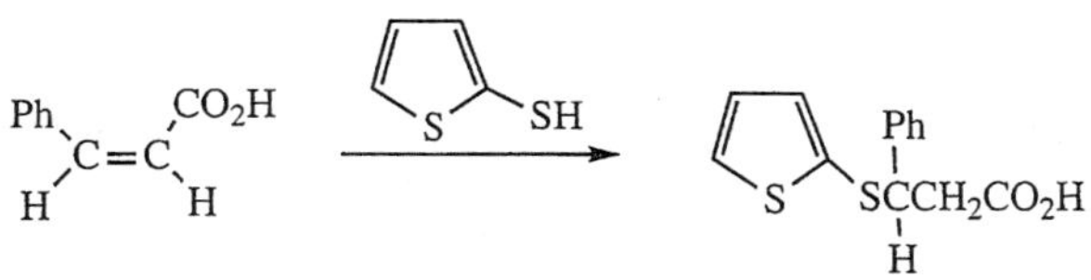

Free radical addition of 2-thiophenethiol to phenylacetyiene in benzene solution at 100 °C in the presence of azobis(isobutyro)nitrile (AIBN) gave a 70% yield of a 70:30 mixture of (*Z*) and (*E*)-2-thienyl-(*β*-styryl sulfide. With 1-hexyne the yields were not of synthetic use [301]. Upon reaction of 2-thiophenethiol with butyl vinyl ether in refluxing benzene 1-butoxy-1-(2-thienylthio)ethane can be obtained [302].

1-Butoxy-1-(2-thienylthio)ethane [302]

A degassed solution of 2-thiophenethiol (0.23 g, 2 mmol) and butyl vinyl ether (1.0 g, 10 mmol) in benzene (20 ml) is refluxed for 2 h. The reaction mixture is extracted with a 10% aqueous sodium hydroxide solution, the organic phase evaporated and the residue chromatographed giving 365 mg (85%) of the title compound.

It has been found that 2-thiophenethiol and 5-chloro-2-thiophenethiol in the presence of triethylamine or upon prolonged standing undergoes a dimerization reaction in which one molecule in its thione form acts as a Michael acceptor to give 4-(2-thienylthio)tetrahydrothiophene-2-thione [303,304].

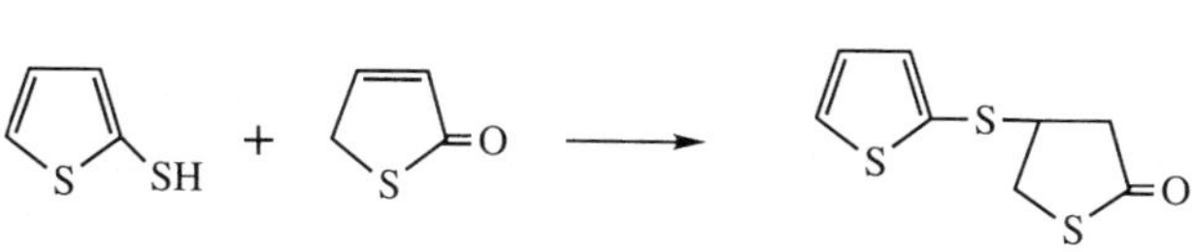

6.2.1.5.7 Through electrophilic substitution of simple thiophenes with sulfenyl chlorides

Electrophilic aromatic substitution of thiophene, 2-methylthiophene, and 2,5-dimethylthiophene with 2,4-dinitrobenzenesulfenyl chloride using tin tetrachloride as catalyst has been used for the preparation of 2,4-dinitrophenylthiothiophenes [305,306]. The tin tetrachloride-catalyzed reaction of 2,5-dimethylthiophene with perhalomethanesulfenyl chlorides can be used for the preparation of 2,5-dimethylperhalomethylthiothiophenes [307].

The reaction of 3,4-di-tert-butyl-1-[(*para*-tolyl)sulfonylimino]-1,1-dihydrothiophene gives 2-alkylthio-3,4-di-tert-butylthiophenes [167].

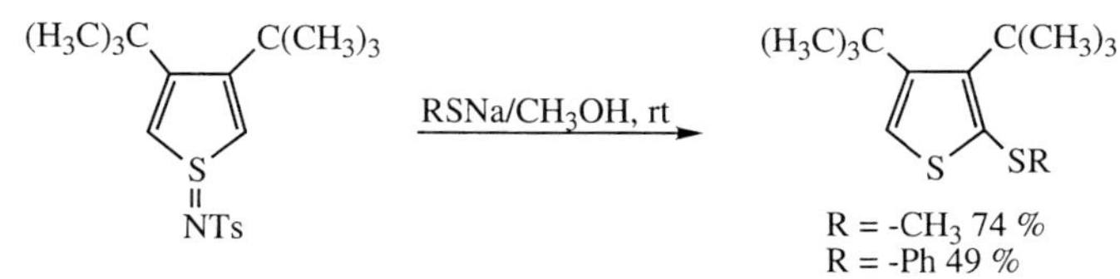

6.2.1.5.8 By ring-closure reactions

Ring-closure reactions are in many cases attractive alternatives for the synthesis of alkylthiothiophenes. The reaction of 1,3-dimetalated acetylenes with nonenolizable thiocarbonyl compounds has been used for the preparation of 2-*tert*-butyl-3-methylthiothiophene [308,309].

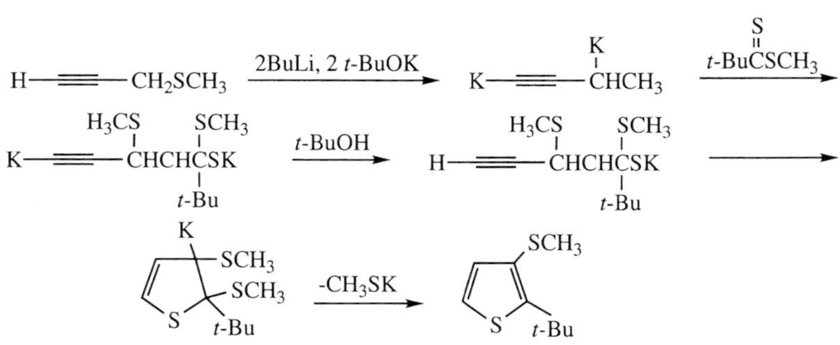

Dimetalation of acetylenic compounds [309]

A solution of butyllithium in hexane (0.22 mol) is placed in a 1-l round-bottomed, three-necked flask, equipped with a dropping funnel, combined with a gas inlet, a mechanical stirrer and a thermometer, combined with a gas outlet. A solution of the uncomplexed base potassium tert-butoxide in tetrahydrofuran (0.22 mol) is added dropwise or portionwise over 10 min, while keeping the mixture between −85 and −95 °C by occasional cooling in a bath of liquid nitrogen. Subsequently the acetylene (0.10 ml) in tetrahydrofuran (20 ml) is added dropwise over 15 min. After this addition the reaction mixture, in most cases a suspension, is stirred for an additional period of 30 min at −70 °C, for the thiomethyl derivative, and the temperature is allowed to rise to −50 °C.

Synthesis of the 2,3-disubstituted compound [309]

A mixture of the thiocarbonyl compound (0.09 mol) and tetrahydrofuran (20 ml) cooled to −75 °C is added in one portion to the vigorously stirred suspension of the dipotassiated acetylene. During this addition the temperature is between −60 and −70 °C by efficient cooling in a bath with liquid nitrogen. The stirring is continued at −70 °C for an additional 20 min and *tert*-butanol (0.22 mol) is added, after which the cooling bath is removed and the temperature allowed to rise to 10 °C. Water (300 ml) is added followed by extraction with diethyl ether. The combined organic phases are dried over magnesium sulfate and concentrated *in vacuo*. The residue is distilled through a 20 cm Vigreux column giving yields of about 95%.

Simple 2-alkylthiothiophenes, such as 3- and/or 3,4-diarylsubstituted derivatives have been prepared by regioselective deprotonation of the compounds shown below,

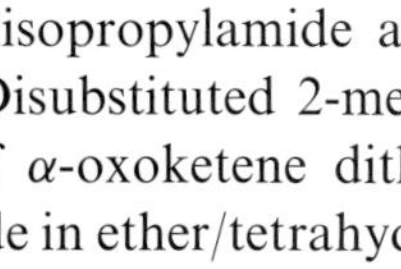

R^3	R^4
C_2H_5	C_6H_5
H	C_6H_5
H	$C_6H_4OCH_3$
OCH_3	C_6H_5
C_6H_5	H
C_6H_5	OCH_3

with lithium diisopropylamide at −78 °C in hexamethylphosphoric triamide [310,311]. 3,4-Disubstituted 2-methylthiothiophenes can be prepared through the reaction of α-oxoketene dithioacetals with the zinc–copper couple and methylene iodide in ether/tetrahydrofuran (Simmons–Smith reaction), proceeding via an intramolecular cyclocondensation of a sulfonium ylid intermediate [312]. However, ring-closure reactions are most useful for the preparation of highly substituted derivatives. Methyl 3-(3,4-disubstituted-5-alkylthio-2-thienyl)propenoates can be synthesized by *S*-alkylation of 3-oxodithioesters with methyl 4-bromocrotonoate followed by intramolecular condensation [313].

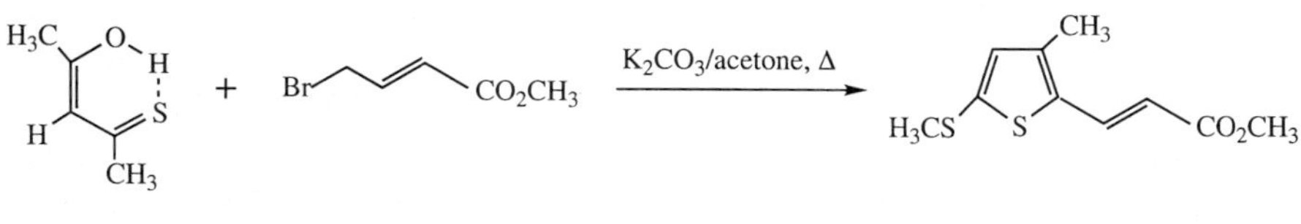

6.2.1.5.9 By modification of substituents in dithienylsulfides

Halogen–metal exchange of 4,4′-dibromo-3,3′-dithienylsulfide with cuprous cyanide produces the cyano-Gilman cuprate, which was oxidized with molecular oxygen to the cyclophane [314].

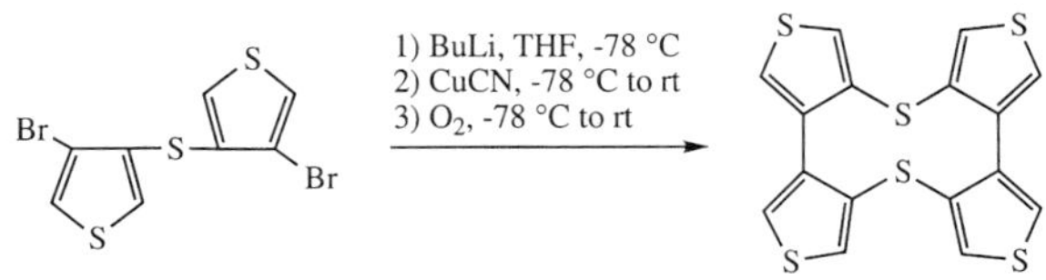

6.2.1.6 Carbonyl-, carboxyl- and cyano-substituted alkylthio- and arylthiothiophenes

6.2.1.6.1 By electrophilic substitution of alkylthio- or arylthiothiophenes

Such compounds are in most cases most conveniently prepared by electrophilic substitution or metalation and reaction with electrophiles of the appropriate alkylthio- or arylthiothiophenes. 2-Alkylthio groups direct predominantly to the 5-position in electrophilic aromatic substitution and 5-alkylthio-2-thiophene aldehydes can be prepared by Vilsmeyer formylation [222,315]. Similarly 2-acetyl-5-alkylthiothiophenes have been prepared by Friedel–Crafts type acetylation using acetyl chloride or acetic anhydride and tin tetrachloride or phosphoric acid as catalyst [269]. However, one should always be aware of the possibility that minor amounts of the 3-isomers are also formed necessitating purification.

Vilsmeyer formylation and Friedel–Crafts acetylations are used for the preparation of 3-alkylthio-2-thiophene aldehyde [263,264], and 2-acetyl-3-alkylthiothiophene [262]. 5-Alkyl-2-alkylthio-3-thiophene aldehyde, [220, 221,242] 5-alkyl-2-alkylthio-3-acylthiophenes [221,249,257] and 5-alkyl-2-alkylthio-3-chloromethylthiophene [316,317] have all been prepared by electrophilic substitution of 5-alkyl-2-alkylthiothiophenes. Care has to be exercised when trying to prepare derivatives of 2,5-di(methylthio)thiophene by electrophilic substitutions as treatment with strong acids such as trifluoroacetic acid leads to the 2,4-isomer by way of an intermolecular disproportionation reaction initiated by C-protonation [256]. Thus formylation and acetylation of 2,5-bis(methylthio)thiophene in addition to the 3-formyl and 3-acetyl derivatives yields the isomeric 3,5-bis(methylthio)-3-acylthiophenes [318].

6.2.1.6.2 Metalation of alkylthio- or aryithiothiophenes

The isomer formation in electrophilic substitution can be avoided using metalation of the 2-alkylthiothiophene, which occurs selectively in the 5-position

[206,251,291]. It has previously been believed that 3-alkylthio- or 3-arylthiothiophenes are selectively metalated in the 2-position. However, a recent investigation showed that the 2-lithio and 5-lithio derivative are formed in the proportion of 3 to 1 upon metalation of 3-methylthiothiophene [319]. Pure 3-methylthio-2-thienyllithium could, however, be prepared by halogen–metal exchange of 2-bromo-3-methylthiothiophene and was used for the preparation of 2-(1-acetoxyethyl)-3-(methylthio)thiophene [319].

6.2.1.6.3 Halogen–metal exchange of halo alkylthio- or arylthiothiophenes

Halogen–metal exchange of 2-bromo-3-methylthiothiophene with butyllithium at −70 °C followed by reaction with trimethylstannyl chloride is used for the preparation of 3-methylthio-2-(trimethylstannyl)thiophene [320].

6.2.1.6.4 Ring-closure reactions

Ring-closure reactions are very useful for the preparation of highly substituted alkylthiothiophenes. 4-Arylsubstituted 3,5-dicyano-2-methylthiothiophenes, 5-carboethoxy-3-cyano-2-methylthiothiophenes, and 5-acetyl-3-cyano-2-methylthiothiophenes have been prepared by the Gompper reaction with aroyl acetonitriles [269,321,322]. 3-Acetyl-5-carbomethoxy-4-methyl-2-methylthiothiophene was prepared in 52% overall yield from pentane-2,4-dione and carbon disulfide in the presence of aqueous potassium hydroxide, followed by successive treatment with ethyl bromoacetate, iodomethane, and alkali [242]. Another C_3S+C approach found by Smutny utilizes 3-aminodithioacrylate esters, which are reacted with α-halocarbonyl compounds, such as ethyl α-bromoacetate in the presence of triethyl amine to 5-methylthio-2-thiophenecarboxylic acid and 4-cyano-5-methylthio-2-thiophenecarboxylic acid [323]. Alkylthiothiophenes have been obtained by cyclization of allenyl dithioesters prepared by metalation of dithio esters followed by alkylation with propargyl halides on treatment with sodium ethoxide in liquid ammonia [324]. A C_3+CS approach to 3-substituted 2-methylthiothiophenes consists of the treatment of a 2-alkyne or allene derivative with butyllithium and potassium *tert*-butoxide in tetrahydrofuran followed by carbon disulfide *tert*-butylalcohol, hexamethylphosphoric triamide, and methyl iodide [308].

Applying the Fiesselmann reaction to alkylthio-β-diketones is a convenient method for the synthesis of 5-alkyl- or 5-arylsubstituted 3-methyl-4-alkylthio-2-thiophenecarboxylic acid esters [78].

Dicyanodithioacetate can successively be alkylated with derivatives of γ-bromocrotonic acid and dimethyl sulfate to thiophene derivatives [325].

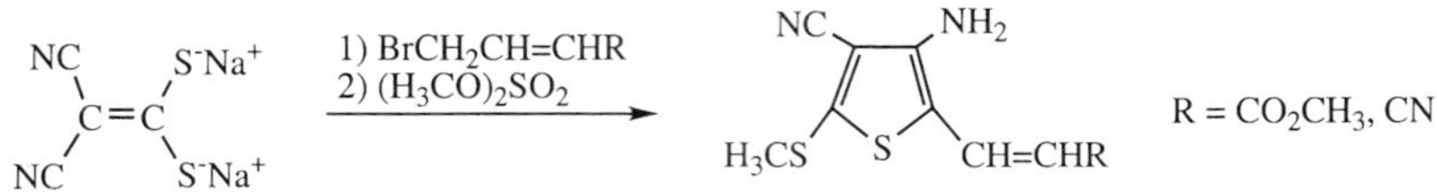

6.2.1.6.5 Various methods

Methyl 4-butylthio-3-hydroxy-5-methylthiophene-2-carboxylate is prepared by the following reaction [79].

6.2.1.6.6 Nucleophilic substitution of activated halothiophenes with alkyl thiolates

The formyl group is enough activating so that 5-ethylthio-2-thiophene aldehyde can be prepared by the reaction of 5-bromo-2-thiophene aldehyde with ethyl thiolate [326].

6.2.1.7 Amino-, nitro- and phosphoryl-substituted alkylthio- and arylthiothiophenes

6.2.1.7.1 By nucleophilic aromatic substitution of activated halothiophenes with alkyl and aryl thiolates

Nucleophilic aromatic substitution of activated halothiophenes is a convenient route to nitro-substituted alkylthio- and arylthiothiophenes. It is in most cases better to use nitrohalothiophenes and arylthiolates than to interchange the nucleophilic and electrophilic partners, as the halothiophenes are more reactive than the corresponding benzene derivatives and, in addition, thiophenolates are more stable than the corresponding thiophenethiolates. Reaction of 3-bromo-2-nitrothiophene with methyl 3-mercaptopropionate and ethyl mercaptoacetate is used for the preparation of ethyl 3-(2-nitro-3-thienylthio)-propionate and ethyl 2-nitro-3-thienylthioacetate [327].

5-Methylthio-4-nitrothiophene-2-carbonitrile is prepared by the reaction of 5-bromo-4-nitrothiophene-2-carbonitrile with sodium methanethiolate in ether [328].

5-Methylthio-4-nitrothiophene-2-carbonitrile [328]

To a stirred solution of 5-bromo-4-nitrothiophene-2-carbonitrile (0.47 g, 2 mmol) in diethyl ether (20 ml), sodium methylthiolate (0.42 g, 6 mmol) is added. The stirring is continued for 24 h, after which the solvent is distilled off and the residue purified by chromatography on silica gel using toluene as eluent giving 0.13 g (33%) of the title compound, mp 151°C after recrystallization from methanol.

Aryl thienyl sulfides have been prepared through the reaction of 2-iodo-5-nitrothiophene [329–331], 2-bromo-3,5-dinitrothiophene [332] and 3-bromo-2-nitrothiophene [333] with various substituted thiophenols in the presence of base. Methyl 4-nitro-5-[(phenylmethyl)thio]thiophene-2-carboxylate is prepared in the reaction at room temperature of methyl 5-chloro-4-nitro-2-thiophenecarboxylate with benzylmercaptan and sodium carbonate in *N*,*N*-dimethylformamide [334].

Methyl 4-nitro-5-[(phenylmethyi)thio]thiophene-2-carboxylate [334]

A solution of methyl 5-chloro-4-nitro-2-thiophenecarboxylate (14.59 g, 65.8 mmol) in anhydrous *N*,*N*-dimethylformamide (50 ml) is dropped into a cooled solution of benzylmercaptan (18.8 g, 65.8 mmol) and potassium carbonate (18.19 g, 131.6 mmol) in anhydrous *N*,*N*-dimethylformamide (20 ml). After stirring the reaction mixture for 5 h at room temperature it is poured into ice. The precipitate is filtered off, washed with cold methanol and dried *in vacuo* giving 19.2 g (94%) of the title compound as yellow crystals mp 151–152 °C after recrystallization from diisopropyl ether.

The reaction of 3-bromo-2-nitro-4-carbomethoxythiophene with benzylthiol gives 3-benzyithio-2-vitro-4-carbomethoxythiophene [335].

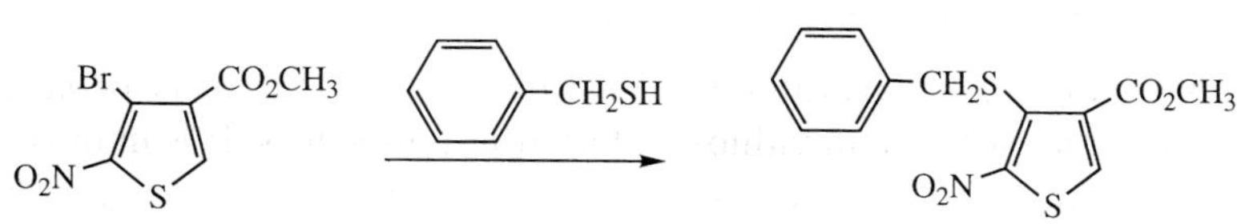

3-Benzylthio-2-nitro-4-carbomethoxythiophene [335]

A 10% solution of benzylthiol in anhydrous *N*,*N*-dimethylformamide is treated with potassium carbonate at 0 °C. To this mixture a 10% solution of 3-bromo-2-nitro-4-carbomethoxythiophene (4.50 g, 16.9 mmol) in anhydrous *N*,*N*-dimethylformamide is added maintaining the temperature below 0 °C. After hydrolysis with ice-water the product is extracted with diethyl ether. The combined organic phases are washed with 2 *M* hydrochloric acid, saturated sodium bicarbonate solution and water, dried over sodium sulfate and evaporated. The residue is recrystallized from diisopropyl ether giving 4.05 g (77%) of the title compound as yellow crystals, mp 75–77 °C.

Monosubstitution of 2,5-dibromo-3,4-dinitrothiophene cannot be achieved, but with two equivalents of thiophenolate in ethanol with triethylamine as base a good yield of 2,5-bis(arylthio)-3,4-dinitrothiophenes is obtained [336,337]. The reaction of 2,5-dichloro-3,4-dinitrothiophene, however, can be used for the preparation of either methyl 3,3′-(3,4-dinitro-2,5-thiophendiyl)dithio-bis-propionate or methyl 3-(5-chloro-3,4-dinitro-2-thienylthio)propionate [327].

Methyl 3,3′-(3,4-dinitro-2,5-thiophendiyl)dithio-bis propionate [327]

First 2,5-dichloro-3,4-dinitrothiophene is prepared. To 2,5-dichlorothiophene (15.3 g, 10 mmol) cooled to 5 °C a mixture of concentrated nitric acid (70 ml) and concentrated sulfuric acid (110 ml) cooled to 0 °C is slowly added. The temperature of the reaction mixture is not allowed to exceed 5 °C. When the addition is completed the reaction mixture is stirred at room temperature for 12 h, after which it is poured into ice-water (2l). After 30 min the precipitate is filtered off and recrystallization from ethanol gives 18.8 g (77%) of 2,5-dichloro-3,4-dinitrothiophene, mp 86 °C.

To a suspension of potassium carbonate (1.38 g, 10 mmol) and anhydrous *N*,*N*-dimethylformamide (10 ml), methyl 3-mercaptopropionate (1.34 g, 11 mmol) is added and the mixture is stirred at room temperature for 20 min. Then 2,5-dichloro-3,4-dinitrothiophene (0.972 g, 4 mmol) in anhydrous *N*,*N*-dimethylformamide (4 ml) is added under ice cooling and the reaction mixture is stirred for an additional hour, after which it is poured into ice–water. The precipitate is filtered off and after recrystallization from ethyl acetate 1.24 g (75%) of the title compound is obtained, mp 78–80 °C.

The reaction of 5-acetyl-2-chloro-3-nitrothiophene with ethyl mercapto-acetate is used for the preparation of ethyl 2-(5-acetyl-3-nitro-2-thienylsulfa-nyl)-2-methylpropionate, which upon reduction with iron/acetic acid ring closed to 6-acetyl-2,3-dihydro-3,3-dimethyl-1*H*- thieno[2,3-b][1,4]thiazinone [338].

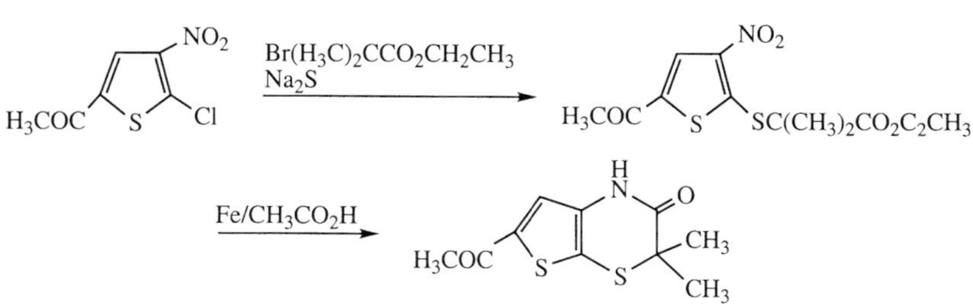

Ethyl 2-(5-acetyl-3-nitro-2-thienylsulfanyl)-2-methyl propionate [338]

To a suspension of sodium sulfide nonahydrate (5.76 g, 24 mmol) in anhydrous *N*,*N*-dimethylformamide (80 ml), 5-acetyl-2-chloro-3-nitrothiophene (4.95 g, 24 mmol) is added and the mixture is stirred under argon at room temperature.

After 1 h ethyl 2-bromoisobutyrate (9.36 g, 48 mmol) is added dropwise and the stirring is continued for 60 h at 50 °C. The reaction mixture is poured into ice-water and the precipitate is filtered off and crystallized from ethanol giving 4.77 g (63%) mp 102–104 °C.

6-Acetyl-2,3-dihydro-3,3-dimethyl-1H- thieno[2, 3-b][1,4]thiazinone [338]

To a solution of ethyl 2-(5-acetyl-3-nitro-2-thienylsulfanyl)-2-methyl propionate (4.76 g, 15 mmol) in a mixture of glacial acetic acid (110 ml), methanol (11 ml), and water (11 ml), iron powder (5.3 g) is added portionwise. The reaction mixture is heated at 60 °C for 4 h and poured into ice water. The crude product is filtered off and recrystallized from 50% ethanol giving 2.76 g (76%), mp 180–183 °C.

4-Amino-2-methylthio-3-cyano-5-carbomethoxythiophenes and related compounds are obtained by sunlight-induced conversion of thieno[2,3-c]isothiazoles with loss of elemental sulfur [339].

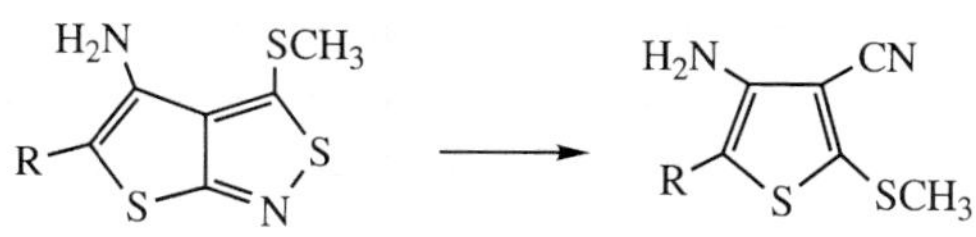

R = CN, $CO_2C_2H_5$, $CONH_2$, $COCH_3$

The reaction of 2,4-dibromo-3,5-dinitrothiophene with various substituted thiophenols in dioxane in the presence of triethylamine was used for the preparation of 2-arylthio-4-bromo-3,5-dinitrothiophene or 2,4-diarylthio-3,5-dinitrothiophene [340]. One should be aware that in some cases the nitro group and not the halogen is substituted. Thus 3,4-dibromo-2,5-dinitrothiophene yields with one equivalent of thiophenolate 2-arylthio-3,4-dibromo-5-nitrothiophene, while with two equivalents 3-bromo-2,4-diarylthio-5-nitrothiophene is obtained [336]. With benzylthiolate 2-benzylthio-3,4-dibromo-5-nitrothiophene was similarly obtained [336]. Nucleophilic substitution of 2-alkylamino- or dialkylamino-5-bromo-3,4-dinitrothiophene with arylthiolate was used for the synthesis of the corresponding 2-amino-3,4-dinitro-5-phenylthiothiophenes [332,336]. The reaction of 2,5-dibromo-3,4-dinitrothiophene with three equivalents of sodium thiophenolate in dioxane gave 2,3,5-tris(phenylthio)-4-nitrothiophene [336]. Reaction of 3-bromo-2-nitrothiophene with potassium ethyl xantogenate in ethanol offers a direct route to di(2-nitro-3-thienyl) sulfide [341] and from 5-nitro-2-iodothiophene and sodium sulfide bis(5-nitro-2-thienyl)sulfide was prepared [342]. From 4-amino-2-pyrimidinethiolate and 2-bromo-5-nitrothiophene the mixed thioether can be prepared [343].

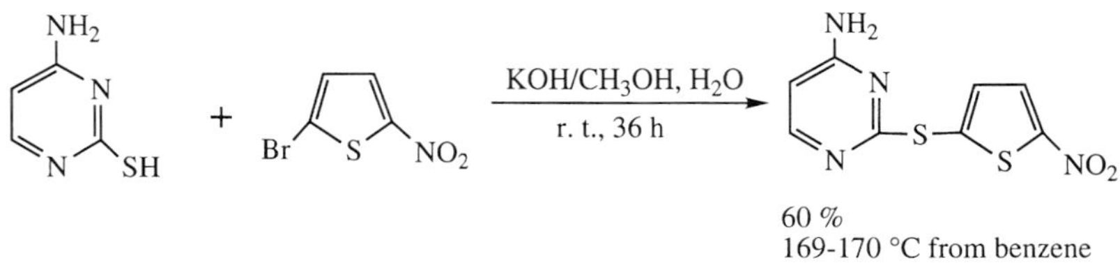

6.2.1.7.2 By nucleophilic aromatic substitution of activated haloaromatics with thiophene thiolates

In some cases the reaction of thiophenethiols with aromatic halides activated toward nucleophilic aromatic substitution has been used for the preparation of arylthiothiophenes. Most commonly used are 2,4-dinitrochloro- or bromobenzene [202,217,218,220,221,344,345], 2-chloro-3-nitrobenzene, [346,347] and 4-trifluoromethyl-2-nitrochlorobenzene [348]. Nucleophilic aromatic substitution has also been used for the preparation of mixed di(thienyl)sulfides as in the preparation of 3-thienyl (5-carbomethoxy-3-nitro-2-thienyl)sulfide from 3-thiophenethiol and methyl 2-bromo-3-nitro-2-thiophenecarboxylate [349] and 2-thienyl (5-ethyl-3-carbethoxy-2-thienyl)sulfide from 2-thiophenethiol and 5-ethyl-2-bromo-3-thiophenecarboxylate [350]. When 3-nitro- 2-thiophenethiol is heated in chlorobenzene with methyl *trans*-3-(4-methoxyphenyl)glycidate methyl *threo*-2-hydroxy-3-(4-methoxyphenyl)-3-(3-nitro-2-thienylthio)-propionate was obtained in 38% yield, while upon reaction in ethanol in the presence of sodium bicarbonate the erythro derivative was obtained in 46% yield [247].

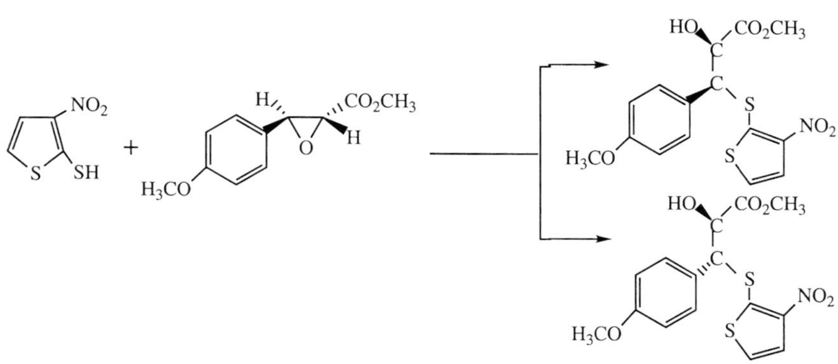

6.2.1.7.3 From activated halothiophenes sodium sulfide and alkyl or benzyl halides

5-Acetyl-2-(4-chlorobenzylthio)-3-nitrothiophene and ethyl α-(5-acetyl-3-nitro-2-thienylthio) 4-chlorophenylacetate are prepared by the reaction of

5-acetyl-2-chloro-3-nitrothiophene with sodium sulfide in ethanol followed by 4-chlorobenzyl chloride, and ethyl α-bromo-4-chlorophenyl acetate, respectively. The first-mentioned product is selectively reduced with triethylsilane to 2-(4-chlorobenzylthio)-5-ethyl-3-nitrothiophene [351].

5 Acetyl-2-(4-chlorobenzylthio)-3-nitrothiophene [351]

A mixture of 5-acetyl-2-chloro-3-nitrothiophene (410 mg, 2 mmol) and sodium sulfide nonahydrate (480 mg, 2 mmol) in ethanol (10 ml) is stirred under argon at room temperature for 1 h. After addition of 4-chlorobenzyl chloride the suspension is heated at reflux for 12 h. Then the mixture is poured into ice-water. The precipitate is filtered off and recrystallized from ethanol giving 438 mg (67%), mp 125–128 °C.

2-(4-Chlorobenzylthio)-5-ethyl-3-nitrothiophene [351]

To a solution of the compound prepared above (237 mg, 1 mmol) in trifluoroacetic acid (1.5 ml) triethylsilane (290 mg, 2.5 mmol) is added dropwise and stirred at room temperature. After 12 h the reaction mixture is poured into ice-water and neutralized with 5% aqueous sodium bicarbonate solution. The precipitate is filtered off and recrystallized from ethanol giving 291 mg (93%) mp 114–116 °C.

This methodology is also used for the preparation of 3-(4-chlorobenzylthio)-2-nitrothiophene and ethyl α-(2-nitro-3-thienylthio)-4-chlorophenylacetate from 3-bromo-2-nitrothiophene and 4-chlorobenzyl chloride and ethyl α-bromo-4-chlorophenyl acetate, respectively [351].

6.2.1.7.4 Through ring-closure reactions

Ring-closure reactions are often used for the preparation of highly substituted alkylthio derivatives. The reaction of 1,3-dimethylated acetylenes with nonenolizable thiocarbonyl compounds has been used for the preparation of 2-diethylamino-3-methylthiothiophene [309]. Thus the reaction of *bis*-(methylthio)-methylenepropanedinitrile with methyl thioglycolate in methanol/triethylamine gives methyl 3-amino-4-cyano-5-methylthiothiophene-2-carboxylate [352].

Methyl 3-amino-4-cyano-5-methylthiothiophene-2-carboxylate [352]

A mixture of bis(methylthio)methylene malonitrile (8.50 g, 50 mmol), methyl thioglycolate (5.3 g, 50 mmol), and methanol (200 ml) is refluxed for 1 h. After cooling the resulting precipitate is collected by filtration giving 9.62 g (84 %) of tan needles, mp 202–206 °C after recrystallization from methanol.

Treatment of *S*-(2- and 3-thienyl)thiophosphates with lithium diisopropylamide in tetrahydrofuran gives rearrangement to diisopropyl(2-mercapto-3-thienyl)- and (3-mercapto-2-thienyl) phosphonates [353,354].

O,O-diisopropyl S-(3-thienyl) phosphorothionate [354]

To a solution of 3-bromothiophene (1.47 g, 9 mmol) in anhydrous diethyl ether (10 ml) 9 *M* butyllithium in hexane (1 ml, 9 mmol) is slowly added at −70 °C under nitrogen. The mixture is stirred for 1 h at −70 °C after which sulfur (290 mg, 9 mmol) is added portionwise. Stirring is maintained for 1 h at −70 °C followed by addition of diisopropyl phosphorochloridate (1.8 g, 9 mmol). The reaction mixture is progressively warmed to room temperature during 16 h, and poured into a saturated aqueous solution of ammonium chloride, extracted with methylene chloride. The extracts are washed with water, dried over sodium sulfate and evaporated. Flash chromatography on silica gel using cyclohexane/ethyl acetate (90:10) as eluent gives 950 mg (38%) of the title compound as a yellow oil.

Diisopropyl (3-mercapto-2-thienyl)phosphonate [354]

A solution of the above-prepared compound (1 mmol) in anhydrous tetrahydrofuran (1 ml) is slowly added under nitrogen at −78 °C to a solution of lithium diisopropylamide (1.2 mmol) prepared from diisopropylamine (180 µl, 1.3 mmol), anhydrous tetrahydrofuran (15 ml) and 1.5 *M* butyllithium in hexane (800 µl, 1.2 mmol). The mixture is allowed to warm to 0 °C for 30 min and the stirring is continued for 30 min at 0 °C, after which the reaction mixture is, under nitrogen, added to a stirred ice-cold solution of ammonium chloride/hydrochloric acid in diethyl ether and extracted twice with diethyl ether. The combined extracts are dried over sodium sulfate and evaporated. The crude thiol is flash chromatographed on silica gel using cyclohexane/ethyl acetate (90:10) as eluent giving diisopropyl (3-mercapto-2-thienyl)phosphonate (47%) as a green oil.

2-Alkylthio- and 2-arylthio-substituted thiophenes are obtained by nucleophilic substitution of activated 2-alkylsulfinylthiophenes with alkyl- or aryl mercaptans. With thiolate anion, sulfoxide reduction is the main reaction.

6.2.1.7.5 Amino-substituted alkylthio- and arylthio derivatives by reduction of nitro derivatives

Reduction of methyl 3,3′-(3,4-dinitro-2,5-thiophendiyl)dithiobispropionate with iron powder in acetic acid is used for the preparation of the corresponding 3,4-diamino derivative, which can be cyclized to the corresponding thienodithiazepines and thieno[3,2-*b*]1,4-thiazines [327].

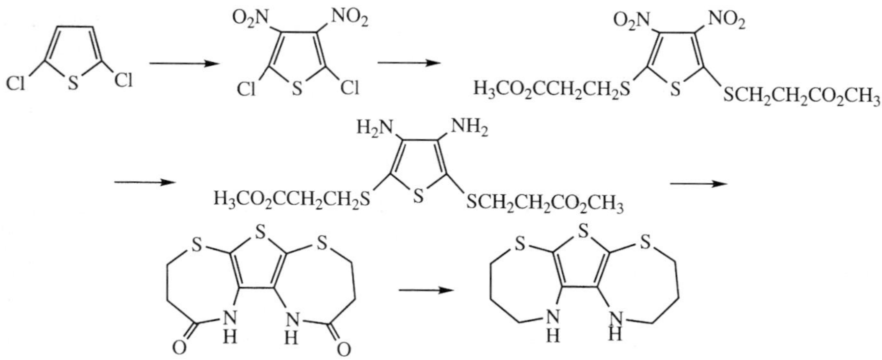

Methyl 3,3′-(3,4-diamino-2,5-thiophendiyl)dithiobispropionate [327]

Methyl 3,3′-(3,4-dinitro-2,5-thiophendiyl)dithiobispropionate (1.64 g, 4 mmol) is solved in a mixture of acetic acid (20 ml) and water (2 ml) and heated to 70 °C, after which iron powder (3.12 g, 56 mmol) is added portionwise and the heating is continued for another hour. Then the reaction mixture is poured into ice-water (500 ml). The water phase is decanted from the unreacted iron powder and extracted with dichloromethane. The combined organic phases are dried, evaporated, and column chromatographed using toluene/ethyl acetate (7:3) as eluent, giving 0.91 g (66%) of the title compound as a dark oil.

Reduction of methyl 4-nitro-[(phenylmethyl)thio]thiophene-2-carboxylate with iron in acetic acid is used for the preparation of methyl 4-amino-5-(phenylmethylthio)thiophene-2-carboxylate [334].

Methyl 4-amino-5-(phenylmethylthio)thiophene-2-carboxylate [334]

A cooled solution of methyl 4-nitro-5-(phenylmethylthio)thiophene-2-carboxylate (10.00 g, 32.32 mmol) in acetic acid (50 ml) is treated with iron powder 10.83 g, 193.9 mmol) in small portions. After stirring for 4 h at 5 °C the reaction mixture is poured into water. After neutralization with sodium hydrogen carbonate the product is extracted twice with diethyl ether. The combined organic phases are dried over sodium sulfate and evaporated giving 7.43 g (81%) of the title compound as beige crystals, mp 60–62 °C.

In the preparation of 3-amino-2-(4-chlorobenzylthio)-5-ethylthiophene from the 3-nitro derivative, stannous chloride in alcohol is used for the reduction [351].

6.2.1.7.6 Modification of substituents

From 2-alkenyl- and 2-alkynyl/thiophene aldehydes, oximes are prepared which by sodium hypochlorite are oxidized to carbonitrile oxides. These intermediates undergo spontaneous cycloaddition reactions to tricyclic systems [356–358].

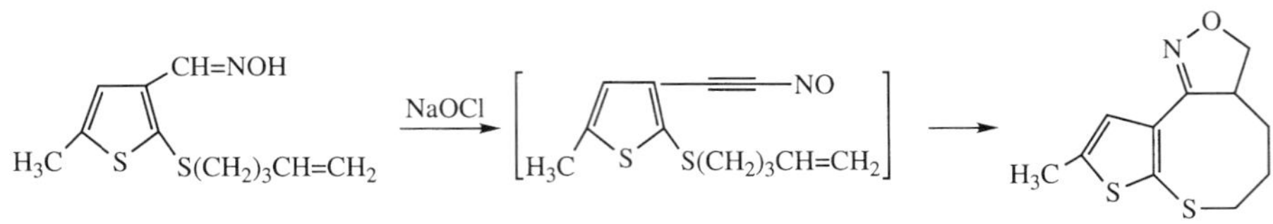

9-Methyl-3,3a,4,5-tetrahydro-6H-thieno[2,3-b]-thiocino[4,5-c]isoxazole [348]

To a solution of the oxime (0.72 g, 3.0 mmol) in dichloromethane (265 ml) at 0–5 °C an aqueous solution of sodium hypochlorite (3.3 mmol) is added. The stirring is continued for 1 h, after which the phases are separated and the organic phase is washed with water, left for 24 h at 20 °C, dried over magnesium sulfate and evaporated. The residue is dissolved in benzene and purified by chromatography on silica gel using benzene as eluent, giving 0.08 g (11%) of the title compound.

6.2.1.8 Hydroxy-, alkoxy-, and aryloxy-substituted atkylthio and arylthio derivatives

6.2.1.8.1 By ring-closure reactions

Such multifunctional derivatives are best prepared by ring-closure reactions. The reaction of 1,3-dimethylated acetylenes with nonenolizable thiocarbonyl compounds has been used for the preparation of 2-*tert*-butoxy- and 2-methoxy-3-methylthiothiophene [309]. Applying the Fiesselmann reaction to alkylthio acetoacetates is a good method for the preparation of methyl 5-alkyl- or aryl-substituted 4-alkylthio-3-hydroxy-2-thiophenecarboxylate [359]. Reaction of α-chloro-β-hydroxyacrylic acid with thioglycolic acid and esterification led to

$$H_3C{-}CO_2{-}S{-}CH_2{-}\underset{}{\overset{CO_2CH_3}{\overset{|}{C}}}H{-}CH(SCH_2CO_2CH_3)_2,$$

which upon treatment with potassium hydroxide in methanol gave 2-carboxy-3-hydroxy-4-thienylthioacetic acid [78].

The following compounds have been prepared through the reaction of bromonitromethane with vinylthiolates [360].

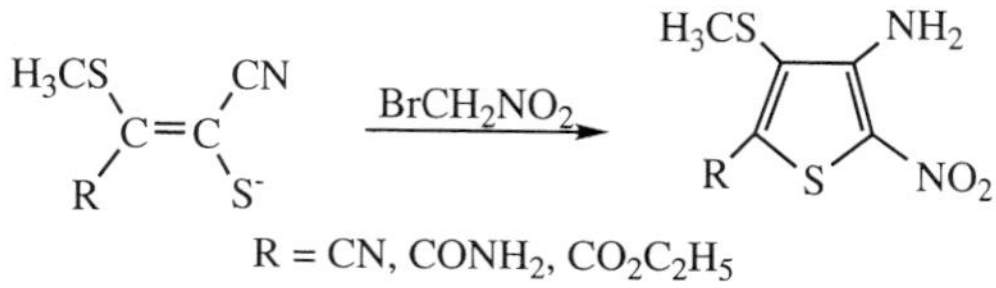

6.2.1.8.2 By aromatic substitution

Electrophilic aromatic substitution can be used for the preparation of some highly substituted and reactive arylthiothiophenes, thus 3-carbethoxy-4-hydroxy or methoxy-2-methylthiophenes and 4,5-disubstituted acylaminothiophenes react with sulfur chloride or sulfur dichloride in refluxing benzene to give the symmetrical sulfides [361].

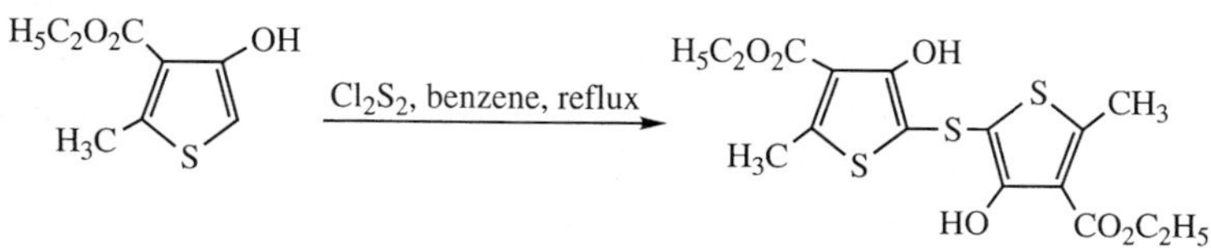

The 3-carbethoxy-4-hydroxy or methoxy-2-methylthiophene have been reacted with *ortho*-nitroarenesulfenyl chloride in dioxane to give the sulfide shown below [111].

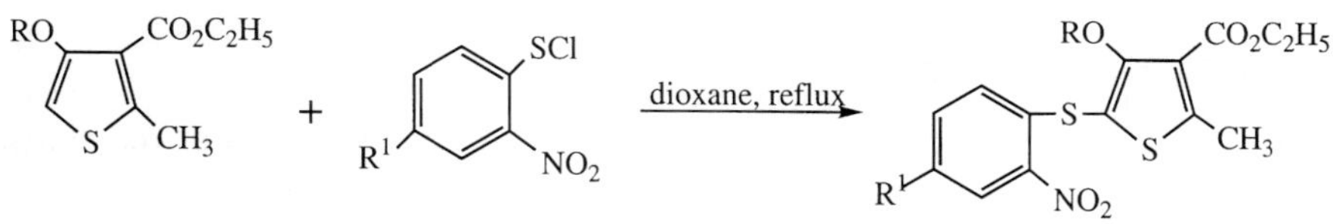

Methyl 4-benzylthio-3-thiophenecarboxylate is prepared from methyl 2,5-dihydro-4-benzylthio-3-thiophenecarboxylate through reaction with sulfuryl chloride in methylene dichloride.

Methyl 4-benzylthio-3-thiophenecarboxylate [362]

Methyl 2,5-dihydro-4-benzylthio-3-thiophenecarboxylate (4.50 g, 16.9 mmol) is treated with sulfuryl chloride giving 3.47 g (78%) of the title compound as beige crystals after recrystallization from diisopropyl ether mp 100–102°C.

The same aromatization process was used for the preparation of 4-tosyloxy-3-carbomethoxythiophene from the corresponding dihydro derivative, which could be nitrated to the 5-nitro derivative and transformed to the 4-benzylthio derivative and reduced to the 5-amino-4-benzylthio-3-carbomethoxythiophene [362].

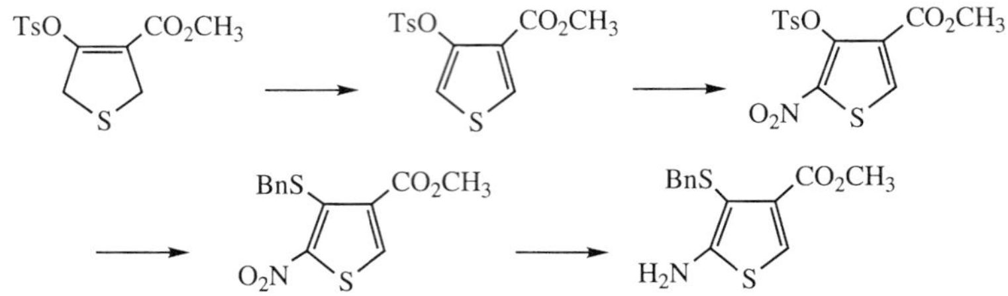

6.2.1.9 Disulfide derivatives

6.2.1.9.1 By oxidation of thiols

The most convenient method for the preparation of disulfides is, of course, the oxidation of the appropriate thiol. Air [222,246,344], iodine [202,218,237,363] potassium ferricyanide [255], and ferric chloride [224] have been used as oxidizing agents. In some cases disulfides are formed directly, if the reaction between an activated halo derivative such as 5-acetyl-2-chloro-3-nitrothiophene with sodium hydrosulfide is carried out at 80°C [190]. Another convenient method for the synthesis of bis(thienyl) disulfides is the reaction of activated halothiophenes such as 2-chloro-3-nitro- or 3-bromo-2-nitrothiophene with sodium disulfide (obtained by fusion of sodium sulfide and sulfur) in ethanol [266,364,365]. Alternatively, such activated halothiophenes are reacted with potassium hydrosulfide in methanol followed by reaction with dimethyl sulfoxide or iodine [341].

Diazotization of methyl 3-amino-2-thiophenecarboxylate followed by reaction with sulfur dioxide and zinc and hydrochloric acid is used for the preparation of di(2-carboxy-3-thienyl)disulfide which by standard methods was transformed to the acid chloride and carboxamides [366,367].

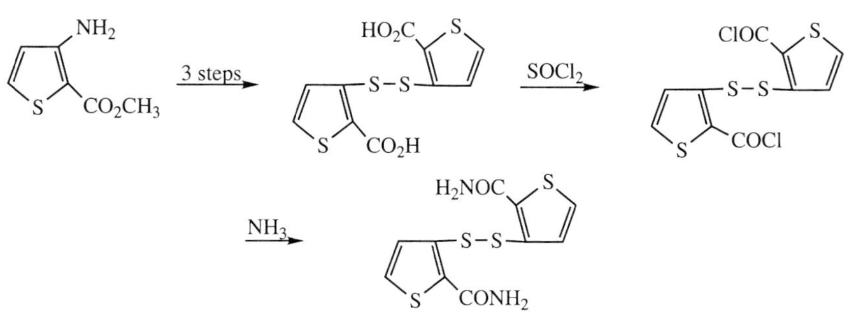

6.2.1.9.2 By various methods

The reduction of thiophenesulfonyl chlorides with hydroiodic acid in acetic acid [271] or red phosphorus in the presence of iodine [368] can also be used for the preparation of bis(thienyl) disulfides. Disulfides have also been prepared by heating 2-acylamino-3-thiocyanothiophenes in alcohol in the presence of piperidine [361]. Unsymmetrical disulfides have been prepared by the reaction of 2-thiophenethiol with arylsulfenyl chlorides [369].

Dithienyl disulfides can be prepared by condensation of dipotassium nitroethylenedithiolate and α-chlorocarbonyl derivatives followed by oxidation with iodine [370].

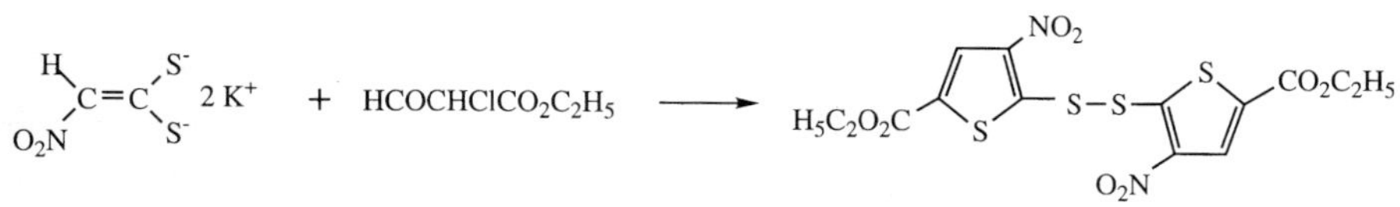

6.2.1.10 SX- (X = halogen, P(O)OR$_2$, COR, and CN) derivatives

Acyl derivatives of thiophenethiols have been prepared through reaction with benzoyl chloride [203,237,363,371], *para*-nitrobenzoyl chloride [218,219,237] acetyl chloride [371], and 2-furoyl chloride [371].

5,6-Thieno[2,3-*d*]-1,3-dithiole-2-one is prepared by aromatization of the 5,6-dihydro derivative with 2,3-dichloro-5,6-dicyano-1,4-benzoquinone and used for the preparation of gold complexes [372].

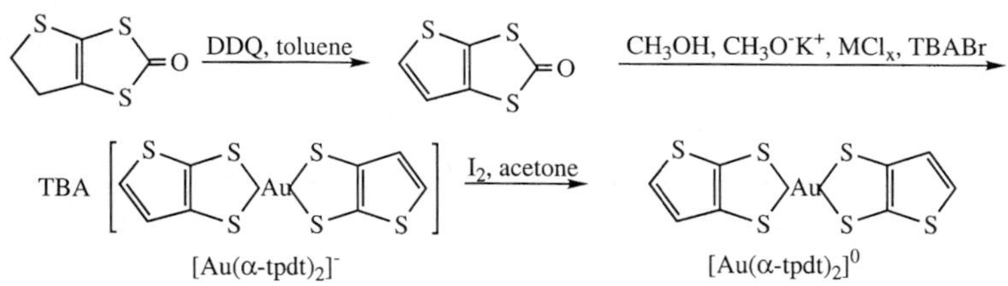

5,6-Thieno [2,3-d]-1,3-dithiole-2-one [372]

A solution of 5,6-dihydrothieno[2,3-*d*]-1,3-dithiol-2-one (0.686 g, 3.9 mmol) and 2,3-dichloro-5,6-dicyano-1,4-benzoquinone (1.869 g, 8.2 mmol) in toluene (25 ml) is stirred for 3 h at 120°C. After evaporation the residue is purified by chromatography using hexane/ethyl acetate (10:1) as eluent and crystallized giving a quantitative yield of the title compound, mp 79.5°C.

Two routes for thieno[3,4-*d*]-1,3-dithiole-2-thione have been developed from 3,4-dibromothiophene. Stepwise halogen–metal exchange followed by reaction with sulfur gives the 3,4-dithiolate, which upon reaction with carbon disulfide in aqueous sodium hydroxide gives the desired compound. Alternatively it is prepared after quenching of the dithiolate with glacial acetic acid in tetrahydrofuran at −78°C, followed by reaction with thiocarbonylbis(imidazole) [215].

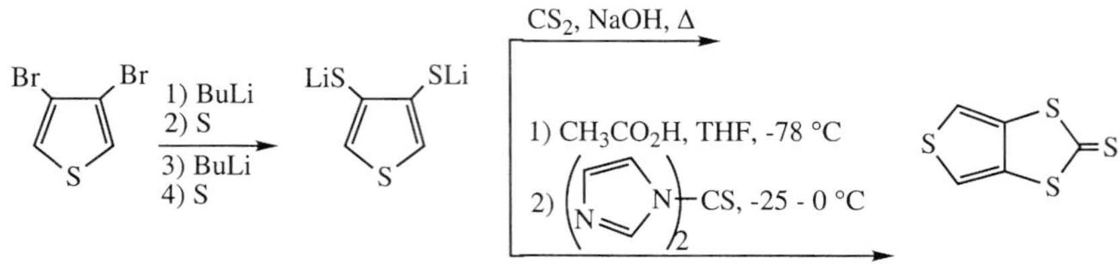

Thieno[3,4-d]-1,3-dithiole-2-thione [373]

Method A. To a solution of 3,4-dibromothiophene (4.36 g, 18 mmol) in anhydrous diethyl ether (30 ml) under argon at −78°C 1.6 *M* butyllithium in hexane (11.3 ml, 18 mmol) is added *via* syringe. The stirring is continued for 0.5 h, after which sulfur (576 mg, 18 mmol) is added. After 1 h this procedure is repeated and the stirring is continued for 1 h. A few drops of water are then added and the solution is allowed to attain room temperature and evaporated. The residue is treated with 2 *M* sodium hydroxide solution (30 ml) and carbon disulfide (12 ml). This mixture is refluxed under argon for 4 h and then left at room temperature overnight. After evaporation and filtration the residue is recrystallized from dichloromethane/hexane (1:5) giving 1.13 g (33%) of the title compound as amber needles, mp 142–143°C.

Method B. To a solution of 3,4-dibromothiophene (1.2 g, 5 mmol) in anhydrous tetrahydrofuran (40 ml) under argon at −78°C 2.3 *M* *tert*-butyllithium in pentane (4.4 ml, 10 mmol) is added *via* syringe. The stirring is continued for 0.5 h, after which sulfur (160 mg, 5 mmol) is added. After 1 h this procedure is repeated and the stirring is continued for 1 h. After quenching with glacial acetic acid (0.7 ml) in tetrahydrofuran (5 ml) thiocarbonylbis(imidazole) (890 mg, 5 mmol) in tetrahydrofuran (30 ml) is slowly added. The reaction mixture is allowed to warm to room temperature during 5 h, left overnight and treated with dilute hydrochloric acid and dichloromethane. The phases are separated and the organic phase is washed with water and sodium chloride solution, dried over magnesium sulfate and evaporated. The residue is purified by chromatography on silica gel using ethyl acetate/hexane (1:10) as eluent, giving 370 mg (39%) of the title compound.

Thieno[3,4-*d*]-1,3-dithiole-2-thione has recently been used for the preparation of gold complexes [372].

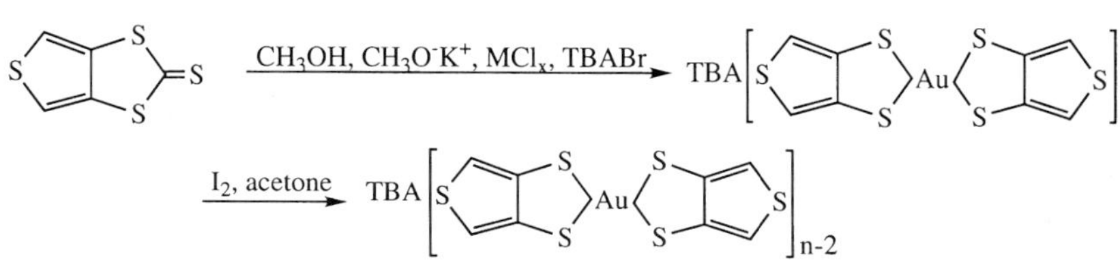

Reaction of thieno[3,4-*d*]-1,3-dithiole-2-thione with dimethoxycarbenium hexafluorophosphate in dichloromethane followed by hydrogen selenide in methanol gives thieno[3,4-*d*]-1,3-dithiole-2-selone [373].

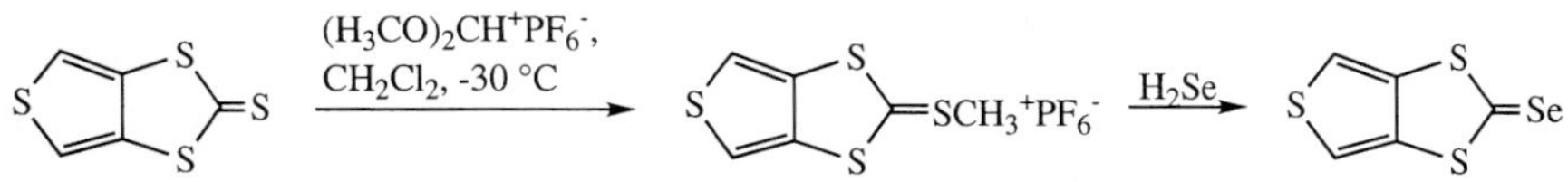

Thieno[3,4-*d*]-1,3-dithiol-2-one is prepared by heating under pressure 3,4-thiophenedithiol with selenium powder in triethylamine and tetrahydrofuran with carbon monoxide, or by reaction of the thione with mercury acetate in acetic acid [373].

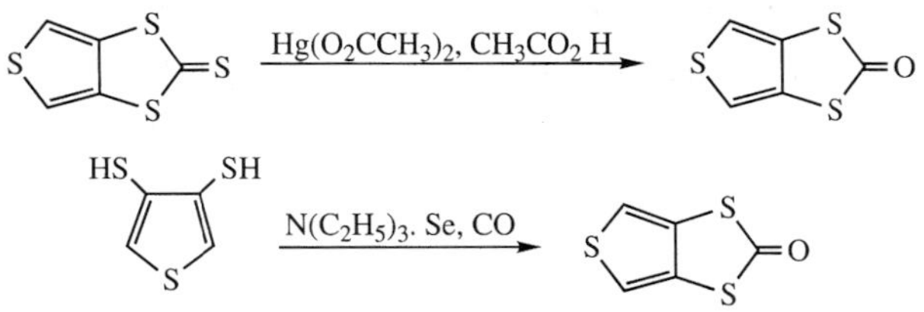

The best method for the preparation of 2-thiocyanothiophene is the reaction of thiophene with thiocyanogen, catalyzed by aluminum chloride or aluminum thiocyanide. With two equivalents of thiocyanogen, 2,5-di(thiocyano)thiophene is obtained [374]. This procedure can also be used for the preparation of alkylthiothiophenethiocyanates [375]. If the thiophene ring is activated toward electrophilic substitution, as is the case with 2-acetamidothiophene [376] 3,4-diaminothiophenes [377], acylated 3-aminothiophenes, or 3-methoxythiophene [344,378], no catalyst is necessary in the reaction with thiocyanogen, which is formed by the reaction of bromine and ammonium thiocyanate. The thiocyanogen solution can also be prepared in advance from bromine and lead thiocyanate [378].

If the thiocyanothiophenes cannot be reached by electrophilic substitution, they can be prepared by the reaction of thiophenethiolates with cyanogen bromide. In this way 3-thiocyanothiophene [210] and ethyl 4-formyl-5-thiocyano-2-thiophenecarboxylate were prepared. Another alternative is possible with activated halothiophenes and 2-nitro-3-thiocyanothiophene was obtained in 85% yield by the reaction of 3-bromo-2-nitrothiophene with potassium thiocyanate in dimethylsulfoxide at 65°C [265]. *O*,*O*-Diisopropyl-*S*-(2-thienyl)-thiophosphates are easily obtained by phosphorylation of 2-thiophenethiolate [353,354].

6.2.2 Tervalent sulfur derivatives S^+R_3, S(O)R, S(O)OR

6.2.2.1 Thienyl sulfonium salts

The most general method is the alkylation of alkylthiothiophenes in presence of mercuric iodide [379,380]. The initially formed rather unstable triiodomercurates were transformed to nitrates and perchlorates. More convenient for the preparation of dimethyl-(2-thienyl)- and dimethyl-(5-methyl-2-thienyl)sulfonium perchlorate is the reaction of thiophene or 2-methylthiophene with a mixture of dimethylsulfoxide, phosphorus oxychloride, and perchloric acid [379]. A ring-closure approach to 2-dialkylsulfonium salts consists in the condensation of malonic acid derivatives with esters of thio or dithio acids in the presence of potassium alkoxides, which was alkylated with chloromethyl methyl sulfide and converted by reaction with trimethyloxonium fluoroborate. Upon cyclization with sodium cyanide, the thiophene derivative was obtained [381].

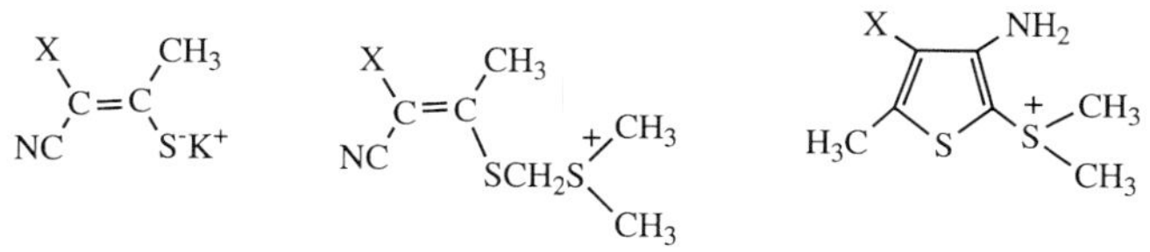

$X = CN, CO_2C_2H_5$

6.2.2.2 Thiophene sulfoxides

Sulfoxides are prepared by oxidation of sulfides with equivalent amounts of about 30% hydrogen peroxide in glacial acetic acid between 0°C and room temperature [206,237,265,270,382–385]. 4-Benzylsulfinyl-3-carbomethoxythiophene is prepared by oxidation of the benzylthio derivative with *meta*-chloroperbenzoic acid [362].

Methyl 4-benzylsulfinyl-3-thiophene carboxylate [362]

A solution of methyl 4-benzylthio-3-thiophenecarboxylate (1.50 g, 1.89 mmol) in anhydrous dichloromethane (10 ml) at −60°C is treated with *meta*-chloroperbenzoic acid (1 eq) in anhydrous dichloromethane. After completed addition the reaction mixture is warmed to room temperature and the stirring is continued for 3 h. The solution is washed with sodium bicarbonate solution, dried over sodium sulfate and evaporated giving 0.32 g (60%) of the title compound as colorless crystals after chromatography on silica gel using petroleum ether/ethyl acetate as eluent, mp 85–88°C.

No complications due to oxidation at the thiophene sulfur is observed. Oxidation can also be carried out in the presence of a thiocyano group [375,386], while formyl groups have to be protected as acetals before oxidation [385]. Chiral 2- and 3-(*para*-tolyl)sulfinylthiophene has been prepared in 74–90% yield and in almost quantitative enantiomeric excess through the reaction of 2- and 3-thiophenemagnesium bromide with -(*S*)-menthyl *para*-toluenesulfinate [387,388].

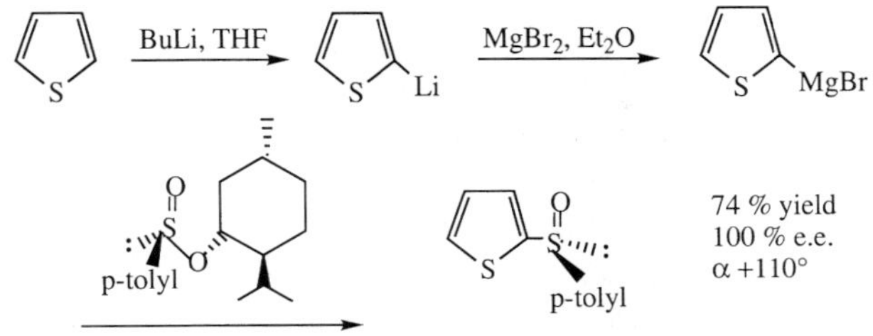

The lithium derivatives give only low e.e.-*values*. Metalation of the *para*-tolylsulfinyl thiophenes with lithium diisopropylamide followed by reaction with *N*,*N*-dimethylformamide gave S_S-5-formyl-(2-*p*-toluylsulfinyl)thiophene and S_s-2-formyl(3 *para*-toluylsulfinyl)thiophene, respectively [388]. The reaction with aldehydes lead to optically active thienyl carbinols, which were oxidized by pyridinium chlorochromate to the corresponding ketones. Stereoselective reduction of the chiral ketones is achieved with diisobutylaluminum hydride or lithium tri-*sec*-butylborohydride leading to different diastereomeric alcohols [389].

2-Methylsulfinyl 5-methylthiophene 3-carbonitrile-*N*-oxide was obtained by oxidation of 2-methylthio-5-methyl-3-thiophene aldehyde oxime with aqueous sodium hypochlorite [390].

2-Methylsulfinyl 5-methylthiophene 3-carbonitrile-N-oxide [390]

2-Methylthio-5-methyl-3-thiophene aldehyde oxime (0.94 g, 5 mmol) in dichloromethane is, under stirring, added dropwise at −5 to −8°C to an aqueous solution of sodium hypochlorite (active Cl 16.55%, sodium hydroxide 3.1%, 5.5 ml, 12.2 mmol). At the end of the addition the yellow color of the mixture changed to green and stirring is continued for 10–15 min. At 0–5°C the product is extracted with dichloromethane. The combined organic phases are washed with water, dried over calcium chloride, and the solvent is removed by distillation. The residue, a white solid, is washed with anhydrous diethyl ether mixed with a minimal amount of ethanol giving 0.9 g (89%) of the title compound, mp 87–88°C.

6.2.2.3 Thiophenesulfinic acids and derivatives

2-Thienylsulfinic acids are best prepared by the reaction of 2-thienyllithium [391] or 3-thienyllithium derivatives [261] with sulfur dioxide or by the reduction of thiophenesulfonyl chlorides with sodium sulfide solution [271,392–398].

6.2.3 Hexavalent sulfur derivatives (SO_2R, SO_3R)

6.2.3.1 Alkyl- and arylsulfonylthiophenes

6.2.3.1.1 Oxidation of sulfides

One of the best methods for the synthesis of alkyl and arylsulfonylthiophenes is the oxidation of alkyl and aryl thienyl sulfides with excess 30 to 40% hydrogen peroxide at about 100°C [202–204,248,255,261,275,277,278,285,287,307,326, 333,336,341,382,384,392,399–405].

5-Methylsulfonyl-4-nitrothiophene-2-carbonitrile is prepared in this way [238].

5-Methylsulfonyl-4-nitrothiophene-2-carbonitrile [328]

A solution of 5-methylthio-4-nitrothiophene-2-carbonitrile (1.8 g, 9 mmol) in acetic acid (7.2 ml) is treated with 30% hydrogen peroxide (3.7 ml) and the mixture is refluxed for 3 h. After removal of the acetic acid *in vacuo* the residue is taken up in water, filtered off and washed with water and purified by chromatography on silica gel using toluene/ethyl acetate as eluent giving 1.67 g (80%), mp 176°C after recrystallization from methanol/dioxane.

Oxidation of 5-methyl-2-methylthio-3-thiophene aldoxime with excess hydrogen peroxide yields 2-methylsulfonyl-5-methylthiophene-3-aldoxime [390,406]. Sterically hindered 2- or 4-alkylsulfonyl-substituted 3-thiophene-carbonitrile oxides were also prepared [407].

2-Methylsulfonyl-5-methylthiophene-3-aldoxime [390]

5-Methyl-2-methylthio-3-thiophenealdoxime (2 g, 10.5 mmol) is heated in acetic acid (15 ml) and 28% hydrogen peroxide (11 ml) at 30–40°C until the solid has dissolved completely, after which the reaction mixture is allowed to stand at 20°C for 24 h and then poured into ice water. The resulting precipitate is removed by filtration, washed with water, and dried giving 1.62 g (74%) of the title compound mp 134–136°C after recrystallization from chloroform.

In a few cases potassium permanganate in water or acetic acid has been used [329,382,404,408]. This method is useful if simultaneous oxidation of methyl or formyl groups is desired [382,410]. Chromic acid has been used for the preparation of 2,5-bis(phenylsulfonyl)thiophene from 2,5-bis(phenylthio)thiophene [383]. 3-Thienyl aryl sulfides can be oxidized to the sulfones in high yields using *meta*-chloroperbenzoic acid in chloroform at 0–10°C [411].

3-(4-Methoxylphenylsulfonyl)thiophene [411]

To a solution of 3-(4-methoxylphenylthio)thiophene (7.9 g, 0.036 mol) in chloroform (40 ml) cooled to 0–10°C is added *meta*-chloroperbenzoic acid (13.5 g, 0.079 mol) portionwise over 10 min. This suspension is stirred at 0–10°C for 2 h, after which it is washed with 1 *M* sodium hydroxide solution (2 × 50 ml) and sodium chloride solution, dried and evaporated. The residue, a tan solid, is purified by flash chromatography on silica gel using methanol/chloroform (1:99) as eluent giving 7.3 g (80%) of the title compound as a white solid, mp 137–139°C.

Methyl 2-(3-trans-2-nitrovinyl)thienyl) mercaptoacetate was oxidized to the corresponding sulfone with potassium peroxymonosulfate (oxone) in water [412].

Methyl 2-(3-trans-2-nitrovinylthienyl)sulfonylacetate [412]

Oxone (12.09 g, 19.7 mmol) in water (60 ml) is added to a solution of methyl 2-(3-*trans*-2-nitrovinyl)thienyl)mercaptoacetate (1.70 g, 6.56 mmol) in methanol (25 ml) with ice cooling. The reaction mixture is stirred at 20°C for 18 h and then heated at 60°C for 10 h, after which it is cooled, diluted with water, and extracted with ethyl acetate. The organic phases are washed with 10% sodium sulfite solution and saturated sodium chloride solution, dried over magnesium sulfate and evaporated. The residue, a yellow solid, is recrystallized from chloroform/methanol giving 1.35 g (71%) of the title compound as pale-yellow needles, mp 147–148°C.

6.2.3.1.2 Friedel–Crafts reaction of thiophenesulfonyl halides

The oxidative methods are especially important in the thiophene series since the Friedel–Crafts reaction between thiophenes and benzenesulfonyl chlorides usually leads to tar formation [393], even when catalysts weaker than aluminium chloride were used. However, reversing the reactants is more successful and 2-phenylsulphonylthiophene [393] 4-nitro-2-phenylsulfonylthiophene and 5-methyl-2-phenylsulfonylthiophene have been prepared by aluminum chloride-catalyzed reactions of 2-thiophenesulfonyl chloride, 4-nitro-2-thiophenesulfonyl chloride and 5-methyl-2-thiophenesulfonyl chloride with benzene [401].

6.2.3.1.3 Alkylation of thiophenesulfinates

Another important method for the synthesis of alkylsulfonylthiophenes consist in alkylation of thiophenesulfinates [261,394,413,414]. Trifluoromethylsulfonylthiophenes have been prepared by the reaction of thienyllithia with trifluoromethanesulfonic anhydride at −70°C.

6.2.3.1.4 Reaction of halothiophenes containing electron-withdrawing groups with alkyl or arylsulfinates

Another important method for alkyl- and arylsulfonylthiophenes with electron-withdrawing groups is the reaction of halothiophene derivatives with alkyl or arylsulfinates. It has been used for the preparation of various 5-substituted 3-nitro-2-phenylsulfonylthiophenes [416], 2,4-dinitro-5-phenylsulfonylthiophene [417,418] and of a large number of 5-nitro-2-phenylsulfonylthiophenes, carrying alkyl, carbomethoxy, amido, cyano, methylsulfonyl, and nitro groups [284,419].

6.2.3.1.5 Metalation of sulfones followed by reaction with electrophiles

Metalation of *tert*-butyl-3-thienyl sulfone with butyllithium followed by reaction with electrophiles can be used for the preparation of 2-substituted derivatives. Thus 3-*tert*-butylsulfonylthiophene-2-thiol is obtained upon reaction with sulfur [420]. This thiol could by alkylation and oxidation be transformed to 2,3-bis(*tert*-butylsulfonyl)thiophene. In similar reactions 2,5-bis-(*tert*-butylsulfonyl)thiophene could be methylated with lithium diisopropylamide in the 3-position and upon reaction with sulfur the 3-thiophenethiol can be obtained. 3-*tert*-Butylthiothiophene can be metalated to the 2,5-dilithium derivative, which upon reaction with sulfur and methyl iodide gives 3-*tert*-butylthio-2,5-di(methylthio)thiophene, which upon oxidation with hydrogen peroxide gives 3-*tert*-butylsulfonyl-2,5-di(methylsulfonyl)thiophene [420].

6.2.3.1.6 Modification of other functional groups in alkylsulfonyl or arylsulfonyl derivatives

This has been performed for the synthesis of specific compounds starting from 2-methylsulfonyl and 2-phenylsulfonyl-5-formylthiophene, the formyl group was transformed to the chloromethyl *via* the hydroxymethyl derivative and then further transformed to the phosphonate, which then through Wittig–Horner gave ethylenic derivatives suitable as second order nonlinear optics [421].

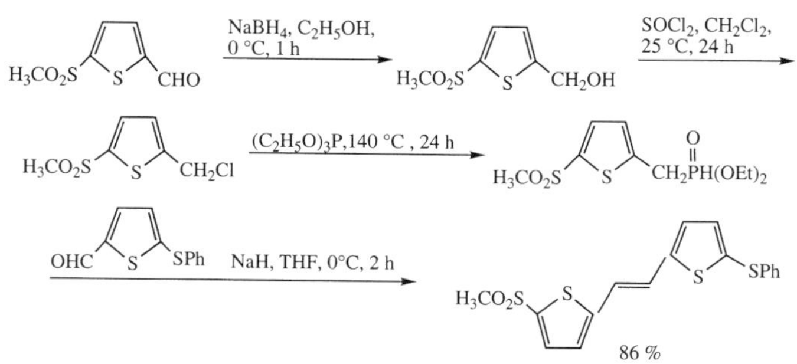

Asymmetric reduction of a cyclic ketosulfone with borane and 1,3,2-oxazaborolidine as catalyst gave more than 93% *e.e.* of the 4-(*R*)-hydroxythienopyran derivative [422].

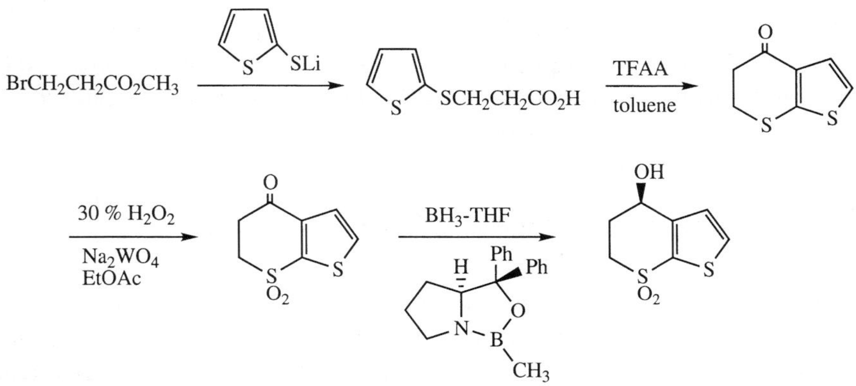

6.2.3.1.7 Ring-closure reactions

Ring-closure reactions are of importance especially for the preparation of highly functionalized alkyl and arylsulfonylthiophenes. Ring-closure reactions have recently been used for the preparation of 2-sulfonylthiophenes. 2-Sulfonylthiophenes have been prepared starting from allenylsulfones. Reaction with butyllithium and chloro trimethylsilane gave α-silylated sulfones, which added various organolithium reagents and reacted with sulfur dioxide to give various 3,4-disubstituted 2-(*para*-tosylsulfonyl)thiophenes [423].

3,4-Dimethyl-2-(para-tolylsulfonyl)thiophene [424]

A solution of 3-methyl-1-(trimethylsilyl)buta-1,2-dien-l-yl *para*-tolyl sulfone (0.29 g, 1.0 mmol) in anhydrous tetrahydrofuran (50 ml) is added dropwise to methyllithium (1.9 ml, 3.0 mmol) in anhydrous tetrahydrofuran (75 ml) kept at −78°C. After stirring for 3 h at −78°C, the reaction mixture is added by using a metal siphon to a solution of an excess sulfur dioxide in tetrahydrofuran (100 ml) kept at −78°C. After 12 h at −78°C, the volatiles are evaporated *in vacuo*, the residue is dissolved in diethyl ether and washed with a saturated aqueous ammonium chloride solution (3 × 20 ml). The organic phase is dried and evaporated. The crude product is purified by column chromatography on silica gel (60H) using petroleum ether/ethyl acetate (3:1) as eluent giving 0.197 g (75%) of the title compound, mp 125°C after recrystallization from toluene/hexane.

The reaction of phenylsulfonylacetophenone with sulfur and malonitrile in *N,N* dimethylformamide/triethylamine is used for the preparation of 2-amino-4-aryl-3-cyano-5-phenylsulfonylthiophenes [425].

2-Amino-4 phenyl-3-cyano-5 phenylsulfonylthiophene [425]

A mixture of phenylsulfonylacetophenone (2.6 g, 0.01 mol), elemental sulfur (0.32 g, 0.01 mol) and malononitrile (0.66 g, 0.01 mol) in anhydrous *N,N*-dimethylformamide (40 ml) containing anhydrous triethylamine (5 drops) is refluxed for 6 h. The reaction mixture is poured into cold water and neutralized with dilute hydrochloric acid. The precipitate formed is collected by filtration, washed with water, dried and recrystallized from dioxane giving 2.31 g (68%) of the title compound, mp 296°C.

The reaction of phenylsulfonylacetophenone with hydroxylamine hydrochloride in ammonia/methanol gives 3-amidoximo-2-amino-4-phenyl-5-phenylsulfonylthiophene [425].

3-Amidoximo-2-amino-4 phenyl-5 phenylsulfonylthiophene [425]

To a solution of 2-amino-4-phenyl-3-cyano-5-phenylsulfonylthiophene (1.7 g, 5 mmol) in methanol (40 ml), containing ammonia (2 ml), hydroxylamine hydrochloride (0.35 g, 5 mmol) is added and the reaction mixture is stirred at room temperature for 24 h and poured into water. The precipitate formed is filtered off, dried, and recrystallized from ethanol giving 0.87 g (52%), mp 274°C.

The reaction of sulfonyldiacetonitrile and methyl (cyanomethylsulfonyl)-acetate with 1,4-dithian-2,5-diol (the equivalent of mercaptoacetaldehyde) is a good method for the preparation of 2-amino-(3-cyanomethylsulfonyl)thiophene [426] and 2-amino-3-[(methoxycarbonyl)methylsulfonyl]thiophene [427] respectively.

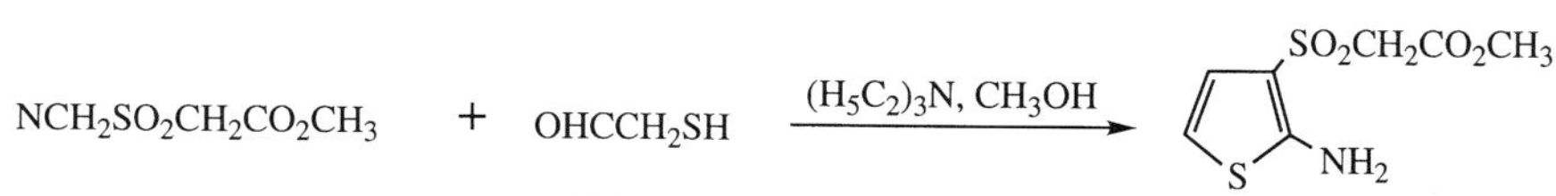

2 Amino-(3-cyanomethylsulfonyl) thiophene [426]

A suspension of sulfonyldiacetonitrile (1.44 g, 10 mmol) 1,4-dithiane-2,5-diol (4.80 g, 5.25 mmol) and piperidine (0.43 g 5 mmol) in ethanol (50 ml) is stirred overnight (*ca.* 16 h) at room temperature. The resulting orange solution is then filtered to clarify and the ethanol is removed *in vacuo*. The remaining oil is partitioned between 1 *M* hydrochloric acid and ethyl acetate. The phases are separated and the organic phase washed with sodium chloride solution, dried, and evaporated. The residue, an oil, is dissolved in a minimum amount of ethyl acetate and flushed through a short plug using ethyl acetate/hexanes (2:1) as eluent giving 1.88 g (93%) of the title compound as a light orange viscous syrup.

With two equivalents of the mercaptoacetaldehyde and triethylamine in ethanol, bis(2-amino-3-thienyl)sulfone is obtained [426]. 3-Amino-2-carbo-ethoxy-5-aryl (or methyl)sulfonyl 5-substituted thiophenes are prepared by reaction of 3-alkoxy-2-aryl (or methyl)sulfonylacrylonitriles with methyl thioglycolate in the presence of triethyl amine [428].

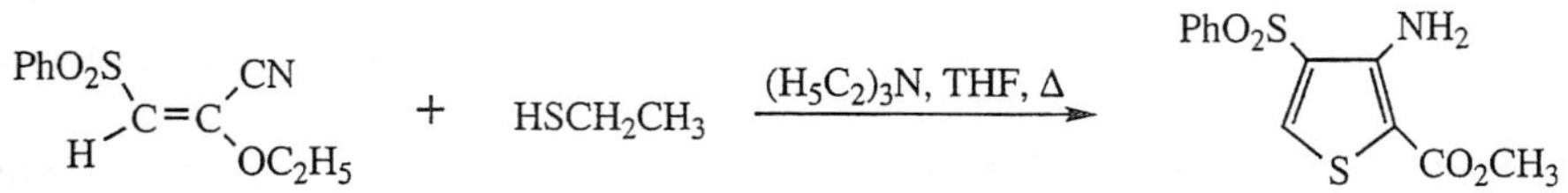

Methyl 3-amino-4-phenylsulfonyl-2-thiophene carboxylate [428]

To a solution of the nitrile (1.19 g, 5.0 mmol) in tetrahydrofuran (15 ml) methyl thioglycolate (4.58 g, 5.5 mmol) is added followed by triethylamine (0.51 g, 5.0 mmol). The reaction mixture, guarded by a calcium chloride tube, is heated to reflux until the reaction is essentially complete. After evaporation the residue is triturated with absolute methanol and recrystallized giving 1.00 g (67%) of the title compound as a white crystalline solid, mp 178–180°C.

6.2.3.2 Thiophenesulfonic acids and their salts

The most common method for the preparation of thiophenesulfonic acids is electrophilic sulfonation of thiophenes. The problem is to avoid polymerization due to the acidic conditions. A simple laboratory procedure that kept polymerization to a minimum consisted in the addition of thiophene to 95% sulfuric acid at 30–40°C [429]. The acid was purified as the calcium salt and then converted to the sodium salt in 69–76% yield [429]. An alternative is the reaction of thiophenes with chlorosulfonic acid. The primary reaction product is the thiophenesulfonyl chloride, which can be worked up as such and used for the preparation of many derivatives of the sulfonic acid or isolated after treatment with sodium hydroxide as the sodium salt of the sulfonic acid [430]. A milder sulfonation method especially useful for acid-sensitive thiophenes is the reaction with a pyridine–sulfur trioxide complex [431–433]. Reaction at room temperature with pyridine bis(sulfur trioxide) is probably the best method for the preparation of 2-thiophenesulfonic acid, isolated as the barium salt in 86% yield [434]. 3-Thiophenesulfonic acid is prepared by debromination of 2,5-dibromo-3-thiophenesulfonic acid or its sulfonyl chloride with 5% sodium amalgam [435–438]. At higher temperatures a mixture of 2,5- and 2,4-thiophenedisulfonic acid was obtained [434]. 5-Methyl-2-thiophenesulfonic acid is conveniently prepared through the reaction of 2-methylthiophene with sulfur trichloride in anhydrous ethylene chloride/dioxane below 20°C and isolated as the sodium salt after crystallization from hot ethanol [439]. 2,5-Dimethyi-3-thiophenesulfonic acid was obtained in over 90% yield upon reacting 2,5-dimethylthiophene with pyridine bis(sulfur trioxide) or dioxane/sulfur trioxide [440]. In spite of the easy de-*tert*-butylation of *tert*-butylthiophenes, it was possible to sulfonate 2,5-di-*tert*-butylthiophene and 2-*tert*-butyl-5-methylthiophene with pyridine bis(sulfur trioxide) or sulfur trioxide in dichloroethane to the 3-substituted sulfonic acids using 1,3 to 1,4 equivalents of sulfur trioxide [441]. Treatment of 3-aroylthiophenes in dichloromethane at 0–10°C with sulfuric acid in acetic anhydride is a good method for the preparation of 4-aroyl-2-thiophenesulfonic acids [442,443].

4-(4-Methylbenzoyl)thiophene-2-sulfonic acid [442]

To a solution of 3-(4-methylbenzoyl)thiophene (3.0 g, 15 mmol) in dichloromethane (25 ml) cooled to 0–10°C acetic anhydride (4.8 g, 47 mmol) followed by sulfuric acid (1.9 g, 16 mmol) are added dropwise. The resulting clear, dark solution is stirred at room temperature for 3 days and then diluted with hexane (50 ml). The suspension so obtained is stirred in an ice bath for 0.5 h, after which filtration gives 2.4 g (57%) of the title compound as pale-brown solid.

The reaction of 2-acetylaminothiophene with 100% sulfuric acid at room temperature is a good method for the preparation of 5-acetylamino-2-thiophenesulfonic acid, which was isolated as the barium salt [444]. Heating for a few minutes at 100°C with 100% sulfuric acid gives 2-acetylamino-3,5-thiophenedisulfonic acid [445].

6.2.3.3 Thiophenesulfonyl fluorides

Such compounds can conveniently be prepared through the reaction of thiophenesulfonyl chlorides with aqueous ammonium fluoride [430,435,446].

6.2.3.4 Thiophenesulfonyl chlorides

The best methods recently used by several groups for the preparation of 2-thiophenesulfonyl chlorides is the reaction of thiophene with excess chlorosulfonic acid and equivalent amounts of phosphorus pentachloride at 20°C [447–450]. Using this methodology at 55°C, a number of 4-(arylsulfonyl)-2-thiophenesulfonyl chlorides were prepared [411]. 5-Methyl-2-thiophenesulfonyl chloride has been prepared similarly [451,452].

4-(4-Methylphenylsulfonyl)-2-thiophenesulfonyl chloride [411]

To chlorosulfonic acid (1.22 g, 10.5 mmol) under nitrogen, phosphorus pentachloride is added portionwise (caution, foaming). The so obtained solution is stirred at room temperature for 10 min, after which 3-(4-methylphenylsulfonyl)thiophene (1.0 g, 4.2 mmol) is added in one portion.

The reaction mixture, a dark suspension, is heated at 55°C for 25 min and during this time foaming occurs and subsides. After pouring it onto ice the product is extracted with chloroform. The combined organic phases are filtered through a Celite pad, washed with a saturated sodium chloride solution, dried and evaporated giving 1.3 g (93%) of the title compound as a tan solid mp 118–120°C.

Various thiophenedisulfonyl chlorides have been prepared, but their structures have not been elucidated [438,449]. 2-Acetylamino-5-thiophenesulfonyl chloride [453] and 2-acetylamino-3,5-thiophenedisulfonyl chloride can be prepared in this way [438,454]. 2-Acetylaminomethyl and 2-phthalimidomethyl-5-thiophenesulfonyl chloride were also prepared in this way [455]. Even deactivated thiophenes such as 2-thiophenecarboxylic acid [449,456], 2-carbethoxythiophene [457] and 2,2′-dithienyl ketones [438] can be smoothly chlorosulfonated. It has been suggested that chlorosulfonation of these compounds occurs in the 5-position [438]. However, it was recently found that 2-thiophenecarboxylic acid yields the 4- and 5-substituted derivatives in a 4:1 ratio [449]. Only disulfonation can be obtained with *N*-benzyl-2-thiophenecarboxamide leading to *N*-(4′-chlorosulfonylbenzyl)-4-chlorosulfonyl-2-thiophenecarboxamide [458].

Electrophilic aromatic substitution is another general route to substituted thiophenesulfonyl chlorides. However, regioselectivity can often be a problem. Thus nitration of 2-thiophenesulfonyl chloride gives 4-nitro-2- and 5-nitro-2-thiophenesulfonyl chloride in a 4:1 ratio [392]. Nitration of 2-methyl-5-thiophenesulfonyl chloride is used for the preparation of 2-methyl-3-nitro-5-thiophenesulfonyl chloride [451]. Another useful method is the reaction of thienyllithium derivatives with sulfur dioxide, followed by oxidation of the intermediate sulfinate with *N*-chlorosuccinimide. In this way 3-(1,3-dioxolan-2-yl)thiophene-2-sulfonyl chloride is prepared [459].

3-(1,3-Dioxolan-2-yl)thiophene-2-sulfonyl chloride [459]

To a solution of 3-(1,3-dioxalan-2-yl)thiophene (20.0 g, 143 mmol) in anhydrous tetrahydrofuran (200 ml) at −78°C 2.5 *M* butyllithium in hexane (85.7 ml, 214 mmol) is added. After stirring at this temperature for 1 h sulfur dioxide gas is passed over the surface of the reaction mixture for 15 min. The mixture is allowed to warm to room temperature and maintained at this temperature for 2 h and evaporated. The residue is dissolved in dichloromethane (200 ml), the solution cooled to 0°C and *N*-chlorosuccinimide (30.0 g, 220 mmol) is added in portions. The reaction mixture is stirred at this temperature for 3 h, warmed to room temperature, and the stirring is continued for 2 h. The solid formed is filtered off and the filtrate evaporated giving the title compound as a brown syrup.

A few special methods might also be mentioned. Thus, 4-carbomethoxy-3-thiophenesulfonyl chloride is conveniently prepared through oxidation of di(4-carbomethoxy-3-thienyl) disulfide with chlorine in methanol [460]

Methyl 4-(chlorosulfonyl)thiophene-3-carboxylate [460]

Dimethyl 4,4′-dithiobis(thiophene-3-carboxylate) (346 g, 1 mol) is dissolved in a mixture of methanol (2 l) and water (750 ml). The solution is cooled to 0°C, chlorine gas (415 g, 5.85 mol, 1.17 equiv.) is introduced during a period of 2 h. The mixture is poured into ice-water (3 l) and the precipitated product is filtered off, dried and recrystallized from tetrachloromethane or methanol giving 433 g (90%) of the title compound, mp 70–72°C.

2-Carbomethoxy-3-thiophenesulfonyl chloride is prepared through diazotization of methyl 3-amino-2-thiophenecarboxylate followed by sulfur dioxide in the presence of cupric chloride [244]. Heating of 2,3-dihydro-3-oxo-thieno-[3,4-*d*]isothiazole-1,1-dioxide with phosphorus pentachloride and zinc chloride followed by hydrolysis gives 4-cyano-3-thiophenesulfonyl chloride in 48% yield [461]. In the same way 5-methyl-3-cyano-2-thiophenesulfonyl chloride was obtained from 2,3-dihydro-5-methyl-3-oxothieno[3,2-*d*]isothiazole-1,1-dioxide [462].

6.2.3.5 Thiophenesulfonyl bromides

2-Thiophenesulfonyl bromides are prepared from 2-thiophenesulfonyl hydrazides, potassium bromide, and potassium bromate in 10% hydrochloric acid [447].

2-Thiophenesulfonyl bromide [447]

To a solution of 2-thiophenesulfonyl hydrazide (10.7 g, 60 mmol) in 10% hydrochloric acid at 20°C, an aqueous solution (35 ml) containing potassium bromide (2.4 g, 20 mmol) and potassium bromate (6.7 g, 40 mmol) is added. The precipitate formed is filtered quickly, washed with cold water, and dried *in vacuo* giving 8.17 g (60%) of the title compound mp 48–49°C from petroleum ether.

6.2.3.6 Esters of thiophenesulfonic acids

Esters of thiophenesulfonic acids are usually prepared through the reaction of the sulfonyl chloride with alcohols in anhydrous pyridine, pyridine bases, or alkali [463,464].

6.2.3.7 Thiophenesulfonamides

A large number of thiophenesulfonamides have been prepared by the reaction of sulfonyl chlorides with aqueous or liquid ammonia for characterization purposes [81,341,430,434,436,438,445,465–470], and also because of their biological activity [342,392,411,442,452,453,455,471–474].

4-(4-Methylphenylsulfonyl)-2-thiophenesulfonamide [411]

A stream of ammonia gas is bubbled into a chloroform solution of 4-(4-methylphenylsulfonyl)-2-thiophenesulfonyl chloride (1.0 g, 2.98 mmol) at 0–10°C. The resulting suspension is then stirred at room temperature for 16 h. The solvent is evaporated and the residue purified by flash chromatography on silica gel using methanol/chloroform (5:95) as eluent, giving 0.75 g (80%) of the title compound, mp 164–166°C.

Anilides were prepared by the reaction of various anilines with thiophenesulfonyl chlorides [342,435,447,449,475,476]. Different sulfonamides can be prepared from primary alkylamines [467] secondary amines such as morpholine and piperidine [461,462] dimethylamine, cyclohexylamine, *N*-methylaniline, and various other amines [450]

N,N-Methylphenyl-2-thiophenesulfonamide [450]

2-Thiophenesulfonyl chloride (1 g, 5.5 mmol) and *N*-methylaniline are refluxed in acetonitrile for 5 h giving after recrystallization from petroleum ether, 0.85 g (60%) of the title compound, mp 105–106°C.

Metalation of *N-tert*-butylthiophene-2-sulfonamide with lithium diisopropylamide allows the selective formation of *N*-5-dilithiothiophene-2-sulfonamides from which a number of 5-substituted thiophene-2-sulfonamides are prepared.

5-(4-Methoxybenzoyl)thiophene-2-sulfonamide [477]

A solution of *N-tert*-butylthiophene-2-sulfonamide (1.00 g, 4.6 mmol) and α,α'-bipyridyl (5 mg) in anhydrous tetrahydrofuran (10 ml) is cooled to −50°C and 1.56 *M* butyllithium in hexane (55.8 ml) is added dropwise. The cooling bath is removed and the reaction mixture stirred at −10 to −15°C for 40 min. At −50°C a solution of *p*-methoxybenzonitrile (0.62 g, 4.7 mmol) in anhydrous tetrahydrofuran (5 ml) is added. The reaction mixture is allowed to warm to 10°C over a 45 min period, after which 10% hydrochloric acid (15 ml) is added. After being stirred overnight the mixture is worked up giving a yellow semisolid, which is recrystallized from butyl chloride/hexane affording 1.16 g (73%) of the title compound as a white solid, mp 120–124°C.

The dianion derived from *N-tert*-butyl thiophene-2-sulfonamide reacts with trimethylacetonitrile to give 5-(2,2-dimethylpropanoyl)thiophene-2-sulfonamide after de-*tert*-butylation by acid treatment [477].

5-(2,2-Dimethylpropanoyl)thiophene-2-sulfonamide [477]

A solution of *N-tert*-butyl thiophene-2-sulfonamide (2.36 g, 10.9 mmol) in anhydrous tetrahydrofuran (20 ml) is cooled to −30°C and 1.56 *M* butyllithium in hexane (13.6 ml) is added maintaining the temperature below −20°C. This mixture is warmed to −10°C for 25 min and then cooled to −30°C and trimethylacetonitrile (1.20 ml, 10.9 mmol) is added. The cooling bath is replaced by an ice bath and the stirring continued for 1 h, after which 10% hydrochloric acid (50 ml) is added and the mixture stirred overnight. After workup and removal of the *tert*-butyl group, 1.60 g of the title compound is obtained, mp 130–134°C after recrystallization from aqueous ethanol.

5-Trimethylsilyl-*N,N*-dimethyl-2-thiophenesulfonamide is obtained by metalation of *N,N*-dimethyl-2-thiophenesulfonamide followed by trimethylsilyl chloride [477].

5-Trimethylsilyl-N,N-dimethyl-2-thiophenesulfonamide [477]

To a solution of *N,N*-dimethyl-2-thiophenesulfonamide (5.7 g, 0.03 mol) in anhydrous tetrahydrofuran (100 ml) under argon 1.6 *M* butyllithium (22.7 ml, 0.036 mol) is added. The stirring is continued for 20 min, after which trimethylsilyl chloride (3.9 g, 0.036 mol) is added and the reaction mixture stirred for 5 h. Water is added, the phases separated and the aqueous phase extracted with diethyl ether. The combined organic phases are dried over magnesium sulfate and evaporated. The residue is recrystallized from petroleum ether giving 3.24 g (41%) of the title compound, mp 75–77°C.

Further metalation of the latter compound with butyllithium in ether/hexane occurs in the 3-position and upon reaction with solid carbon dioxide 5-trimethylsilyl-3-carboxy-2-*N,N*-dimethyl-2-thiophenesulfonamide is obtained [478]. On the other hand, reaction of 5-trimethylsilyl-*N,N* dimethyl-2-thiophenesulfonamide with butyllithium-*N,N,N′,N′*-tetramethyl-1,2-ethanediamine leads to rearrangement giving 4-dimethylamino 2-trimethylsilylthiophene in 39% yield [478].

6.2.3.8 Various other thiophenesulfonyl derivatives

N-substituted *N*-thiophenesulfonyl ureas are most conveniently prepared by the reaction of the sulfonamides with isocyanates in the presence of a base

[479–481]. They can also be prepared through the reaction of sulfonyl isocyanates with amines [404].

2-Thiophenesulfonyl azide is best prepared by the reaction of 2-thiophenesulfonyl chloride with sodium azide in aqueous acetone [449,482,483]. Thiophensulfonyl hydrazides are best prepared by the reaction of thiophene sulfonyl chlorides with 85% hydrazine hydrate in diethyl ether or tetrahydrofuran [447,482–484]. The *N*-acetylhydrazide of 2-thiophenesulfonic acid is easily prepared through the reaction of *N*-acetylhydrazine with 2-thiophenesulfonyl chloride [485]. The thiophenesulfonyl hydrazides can be reacted with various aldehydes to give hydrazones [449,484–486]. Upon reaction of the sulfonyl hydrazides with isothiocyanates, thiosemicarbazones are obtained [484,486].

Condensation of 2-thiophenesulfonamide with carbon sulfide in the presence of potassium hydroxide gave the dipotassium salt, which can be alkylated with methyl iodide [487].

Reaction of the dipotassium salt, suspended in an inert solvent such as dichloromethane, with phosgene or ethyl chloroformate and phosphorus pentachloride is a good method for the preparation of 2-thiophenesulfonyl isothiocyanate [488].

6.3 SELENIUM AND TELLURIUM DERIVATIVES

6.3.1 Divalent selenium and tellurium derivatives

Most of the methods used for the formation of thiophene–sulfur bonds can be used for the preparation of thiophene–selenium bonds. The reaction of 2-thienyllithium derivatives with elemental selenium thus gives the lithium thienylselenolates, which, in most cases, are directly transformed to alkylseleno derivatives by alkylation. In this way 2-methylseleno [252] and 2-ethylselenothiophene [489], 5-methyl-2-methylselenothiophene [252], 5-ethyl-2-ethylselenothiophene [489], 2-methylthio-5-methylseleno-, 2-methylthio-5-propylseleno-, and 2-propylthio-5-methylselenothiophene are most conveniently prepared [490]. Methyl 5-ethyl-2-thiopheneselenoacetate is prepared using methyl chloroacetate as alkylating agent [489]. This method is not very advantageous for the preparation of 3-alkylselenothiophene, as the alkyl halide formed in the halogen–metal exchange usually competes in the alkylation. Thus in an attempt to prepare the methyl 3-thienylselenoacetate through the reaction of 3-thienyllithium with selenium and ethyl chloroacetate at −70°C as much as 66% of 3-butylselenothiophene was obtained [491]. However, if the primarily formed thienylselenolate is separated from the alkylated byproduct, acidified and the selenol reacted with methyl bromoacetate in potassium hydroxide in methanol the 3-thienylselenoacetate was obtained in 50% yield [491]. Of course, if butyllithium is used in the exchange, there is no problem in the preparation of 3-butylselenothiophene [492]. For the preparation of 3-alkylselenothiophenes such as 3-methylselenothiophene the best method, in most cases therefore, is the reaction of 3-thienyllithium with dimethyl diselenide which gives the desired compound in 81% yield [492].

3-Methylselenothiophene [492]

Under nitrogen with stirring, 3-bromothiophene (8.2 g, 50 mmol) in anhydrous diethyl ether (50 ml) is added to 1.85 *M* butyllithium (27 ml, 50 mmol) at −70°C. After stirring the mixture at −70°C for 2 h, a solution of dimethyl diselenide (9.4 g, 50 mmol) in anhydrous ether (15 ml) is added at −55°C. The stirring is continued at −20°C for 1.5 h, after which a 25% ammonium chloride solution is added at −10°C. The phases are separated and the aqueous phase extracted three times with diethyl ether. The combined organic phases are washed with water and dried over calcium chloride. Distillation gives 7.2 g (81%) of the title compound, bp 95–96°C/18 mm Hg.

Another way for preparing substituted alkylselenothiophenes is electrophilic substitution. Vilsmeyer formylation and Friedel–Crafts type acylation of 2-alkylselenothiophene have been used for preparation of the 5-formyl and 5-acyl derivatives [492]. However, in the acetylation of 2-methylselenothiophene, as much as 15% of the 3-isomer was obtained as byproduct [493].

3-Acetyl-5-methyl-2-methylselenothiophene is prepared by tin tetrachloride catalyzed acetylation of 5-methyl-2-methylselenothiophene [493].

In contrast to 2-methylthiothiophene, metalation of 2-methylselenothiophene cannot be used for the preparation of 5-substituted derivatives, as lithium–methylseleno exchange occurs leading to 2-thienyllithium and butyl methyl selenide [493,494]. 3-Alkylselenothiophenes, in contrast to the 2-isomers, are metalated by butyllithium in the α-positions. Unfortunately, from the synthetic point of view the metalation is not regioselective [493]. Reduction of di(2-thienyl) diselenide with hydrazine hydrate and potassium hydroxide followed by allyl bromide is a good method to prepare allyl 2-thienyl selenide [495].

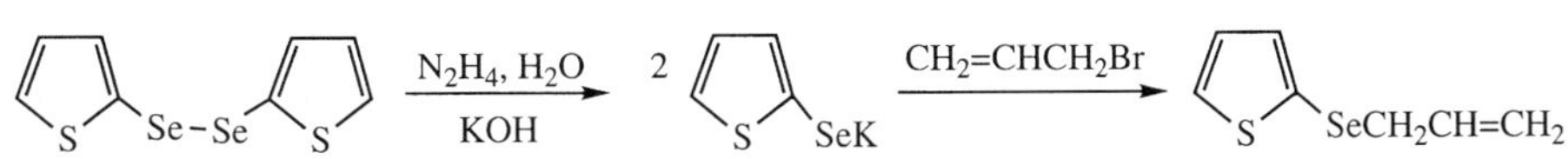

Allyl 2-thienyl selenide [495]

To a solution of potassium hydroxide (6.7 g, 0.12 mol) in hydrazine hydrate (30 ml) di(2-thienyl)diselenide (19.5 g, 0.06 mol) is added at 80–90°C. After cooling the mixture, allyl bromide (11 ml, 0.12 mol) is added dropwise. The organic phase is separated, washed with water and dried over calcium chloride. The title compound is obtained in an 86% yield.

Another convenient route to alkylselenothiophenes is starting from selenocyanates. 3-Formamido- or 3-acetylamido-2-selenocyanate reacts with aqueous sodium sulfide to give the selenolates, which upon reaction with alkyl halides or *ortho*-nitrochlorobenzene gives alkylseleno- or arylselenothiophenes [344]. Ethyl 2-nitro-3-thiopheneselenoacetate is prepared by reduction of 2-nitro-2-thiopheneselenocyanate with sodium borohydride followed by reaction with ethyl bromoacetate [265]. From the nitro derivative, the 2-acetylamino-3-thienylselenoacetate could be obtained by reductive processes [265].

Another method for the preparation of phenylselenothiophenes has been developed consisting in the generation of phenylselenyl sulfate by the oxidation of diphenyl diselenide with ammonium persulfate. This reagent effects an electrophilic aromatic substitution on thiophene and 2- and 3-methylthiophene. Under controlled conditions it is possible to introduce the desired number of phenylseleno groups in the thiophene ring [496].

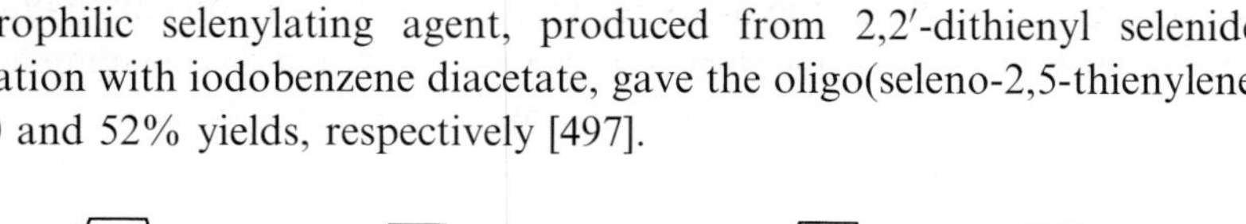

Recently, the reaction of thiophene and 2-methylthiophene with the electrophilic selenylating agent, produced from 2,2′-dithienyl selenide by oxidation with iodobenzene diacetate, gave the oligo(seleno-2,5-thienylenes) in 8, 10 and 52% yields, respectively [497].

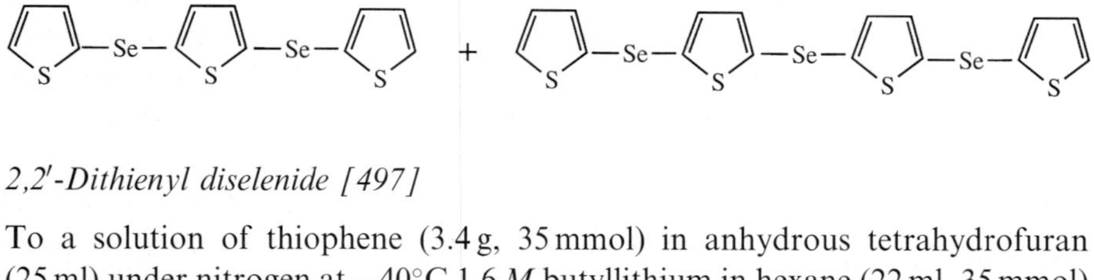

2,2′-Dithienyl diselenide [497]

To a solution of thiophene (3.4 g, 35 mmol) in anhydrous tetrahydrofuran (25 ml) under nitrogen at −40°C 1.6 *M* butyllithium in hexane (22 ml, 35 mmol) is added. After 1 h at −30°C the temperature of the reaction mixture is lowered to −70°C and powdered selenium (2.8 g, 35 mg atom) is added. After 30 min at −70°C the reaction mixture is stirred at −10°C for 1 h. The reaction is quenched with 10% aqueous ammonium chloride solution (20 ml) and potassium ferricyanide (2 g) is added. Stirring is continued overnight at room temperature. The mixture is extracted with diethyl ether and the combined extracts dried over magnesium sulfate and evaporated. The residue is purified by column chromatography on silica gel using light petroleum as eluent, giving the title compound in pure form mp 60–62°C.

Thiopheneselenocyanates are conveniently prepared by diazotization of 3-aminothiophenes with electron-withdrawing groups in the 2-position, followed by reaction with potassium selenocyanate and sodium acetate [238].

2-Formyl-3-thiopheneselenocyanate [238]

3-Amino-2-thiophene aldehyde (0.38 g, 3 mmol) is diazotated in 5 *M* hydrochloric acid. The solution of the salt is added dropwise at 0°C to a solution of potassium selenocyanate (0.35 g) and sodium acetate (10 g) in water (50 ml). The reaction mixture is warmed to 50°C and kept at this temperature for 1 h and then left overnight. The product is extracted with diethyl ether. The combined organic phases are washed several times with water, dried, and evaporated. The residue is recrystallized from petroleum ether/chloroform (90:10) giving 0.29 g (45%) of the title compound mp 111°C.

2-Thiopheneselenocyanates with electron-withdrawing groups in the 3-position are best prepared by treatment of a 3-substituted thiophene with potassium selenocyanate with bromine in methanol at 0°C [344,378]. 2-Nitro-3-thiopheneselenocyanate is prepared by the reaction of 2-nitro-3-thienyl diselenide with potassium selenocyanate [265].

Diselenides are easily formed upon acidification of thienylselenolates in the presence of air or better by the reaction of 2-thienyllithium with powdered selenium and potassium ferricyanide [497].

2,2′-Dithienyldiselenide [497]

To a solution of thiophene (3 g, 35 mmol) in anhydrous tetrahydrofuran (25 ml) under nitrogen at −40°C, 1.6 *M* butyllithium in hexane (22 ml, 35 mmol) is added. The stirring is continued at −30°C for 1 h, after which the temperature is lowered to −70°C and powdered selenium (2.8 g, 35 mmol) is added. After stirring the mixture at −70°C for 30 min and at −10°C for 1 h the reaction is quenched with 10% aqueous ammonium chloride solution (20 ml) and potassium ferricyanide (2 g) is added. The stirring is continued at room temperature overnight and the product is extracted with diethyl ether. The combined extracts are dried over sodium sulfate and evaporated. The residue is purified by chromatography using light petroleum as eluent, giving 6.16 g (55%) of the title compound, mp 60–62°C.

Reaction of 2-nitro-3-selenocyanothiophene with sodium borohydride followed by ethyl bromoacetate is used for the preparation of ethyl 2-nitro-3-thienylselenoacetate, which can further be reduced to the 2-acetylamino derivative [265]. Di(2-nitro-3-thienyl) diselenide can also be prepared through the reaction of 3-bromo-2-nitrothiophene with sodium diselenide in ethylene glycol at 50°C [265]. 4-Formyl-3-methylselenothiophene gives, upon reaction with methyl bromoacetate, a selenonium salt, which could be used as starting material for the preparation of the selenolo[3,2-*c*]thiophene system [498].

Di(Thienyl)ditelluride is prepared in 74% yield through the reaction of 2-thienyllithium with elemental tellurium in tetrahydrofuran followed by air oxidation [499]. In similar ways different di(3-thienyl) ditelluride and 3-alkyltellurothiophenes were obtained [500]. The insertion of further equivalents of chalcogen into lithium 2-thienylselenolate or -tellurolate can be used for the preparation of di(thienyltellurenyl)selenide and diselenide [501].

Stepwise halogen–metal exchange of 3,4-dibromothiophene followed by addition of selenium, quenching with acetic acid and reaction with thiocarbonylbis(imidazole) is used for the preparation of thieno[3,4-d]-1,3-diselenole-2-thione, which upon reaction with dimethoxycarbenium tetrafluoroborate in dichloromethane followed by hydrogen selenide in methanol gives thieno[3,4-d]-1,3-diselenol-2-selone. A similar reaction sequence using tellurium instead of selenium can be used for the preparation of thieno[3,4-d]-1,3-ditellurol-2-thione [215].

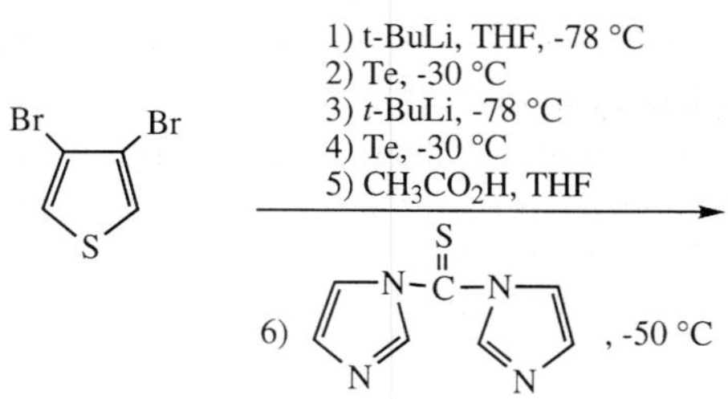

Thieno[3,4-d]-1,3-ditellurol-2-thione [215]

To a solution of 3,4-dibromothiophene (725 mg, 3 mmol) in anhydrous tetrahydrofuran (50 ml) under argon at −8 °C 2.3 *M* *tert*-butyllithium in pentane (2.7 ml, 6 mmol) is added *via* syringe. The stirring is continued for 30 min, after which tellurium powder (381 mg, 3 mmol) is added. The mixture is stirred at −30 °C for 1.5 h, recooled to −78 °C and the procedure is repeated. After quenching with acetic acid (0.6 ml) in tetrahydrofuran (5 ml) thiocarbonylbis(imidazol) (588 mg, 3.3 mmol) in tetrahydrofuran(25 ml) is added. The reaction mixture is allowed to warm slowly to −25 °C, stirred at this temperature for 3 h and then at room temperature overnight. Dichloromethane and dilute hydrochloric acid are added, the phases separated and the organic phase washed with water and sodium chloride solution, dried over magnesium sulfate, and evaporated. The residue is immediately purified by chromatography on Florisil using hexane/diethyl ether (5:1) as eluent, giving 160 mg of the title compound as a reddish brown solid.

REFERENCES

1. C. Frisell and S.-O. Lawesson, *Org. Synth.* **43**, 55 (1961).
2. S.-O. Lawesson and C. Frisell, *Arkiv Kemi* **17**, 393 (1961).
3. S. Gronowitz, *Arkiv Kemi* **16**, 363 (1960).
4. H. J. Jakobsen and S.-O. Lawesson, *Tetrahedron* **21**, 3331 (1965).
5. H. J. Jakobsen, E. H. Larsen and S.-O. Lawesson, *Tetrahedron* **19**, 1867 (1963).
6. J. Z. Mortensen, B. Hedegaard and S.-O. Lawesson, *Tetrahedron* **27**, 3839 (1971).
7. H. D. Stachel, H. Poschenrieder, J. Redlin, J. Schachtner and K. Zeitler, *Lieb. Ann. Chem.* 129 (1994).
8. H. Stachel, J. Schachtner and H. Lotter, *Tetrahedron* **49**, 4871 (1993).
9. A.-B. Hörnfeldt and S. Gronowitz, *Arkiv Kemi* **21**, 239 (1963).
10. A.-B. Hörnfeldt, *Arkiv Kemi* **22**, 211 (1964).
11. B. Cederlund and A.-B. Hörnfeldt, *Acta Chem. Scand.* **25**, 3324 (1971).
12. A.-B. Hörnfeldt, *Acta Chem. Scand* **21**, 1952 (1967).
13. B. Cederlund and A.-B. Hörnfeldt, *Chem. Scripta* **8**, 140 (1975).
14. A.-B. Hörnfeldt, *Acta Chem. Scand* **19**, 1249 (1965).
15. C. Lemarie-Retour, M. Staveaux and N. Lozac'h, *Bull. Soc. Chim. Fr. Part* 2, 1659 (1973).
16. A.-B. Hörnfeldt and P.-O. Sundberg, *Acta Chem. Scand.* **26**, 31 (1972).
17. R. Cruz-Almanza, T. Hemandez-Quiros, L. J. Brena-Valle and F. Pérez-Flores. *Tetrahedron Lett.* **38**, 183 (1997).
18. W. Kues and C. Paal, *Bericht.* **19**, 555 (1886).
19. W. Steinkopf and F. Thormann, *Lieb. Ann. Chem.* **540**, 1 (1939).
20. S. Gronowitz and R. A. Hoffman, *Arkiv Kemi* **15**, 499 (1960).
21. A.-B. Hörnfeldt, *Svensk Kem. Tidskr.* **80**, 343 (1968).
22. K. Clauss, *Tetrahedron Lett.* 1271 (1974).
23. B. King, E. Demole and A. F. Thomas, (Firmenich et Cie). Ger. Offen. 2,214, 540. *Chem. Abstr.* **78**, 29615a (1973).
24. M. J. Kates and J. H. Schauble, *J. Heterocycl. Chem.* **32**, 971 (1995).
25. A. Ben Cheikh, H. Dhimane, J. C. Pommelet and J. Cuche, *Tetrahedron Lett.* **29**, 5919 (1988).
26. G. A. Hunter and H. McNab, *J. Chem. Soc., Chem. Commun.* 375 (1990).
27. G. A. Hunter and H. McNab, *J. Chem. Soc., Perkin Trans.* 1, 1209 (1995).
28. G. M. Karp, D. Samant, S. Mukhopadhyay, M. Condon and A. Kleemann, *Synthesis* 1078 (2000).
29. H. Y. Aboul-Enein, A. A. Al-Badr, S. E. Ibrahim and M. Ismail, *Pharm. Act. Helv.* **55**, 228 (1980).
30. V. N. Gogte, L. G. Shah, B. D. Tilak, K. N. Kadekar and M. B. Sahasrabudhe, *Tetrahedron* **23**, 2437 (1967).
31. B. Capon and Z.-P. Wu, *J. Org. Chem.* **54**, 1211 (1989).

32. J. Brenan and P. J. Murphy, *Tetrahedron Lett.* **29**, 2063 (1988).
33. H. D. Stachel and A. Fendl, *Arch. Pharm. (Weinheim)* **321**, 439 (1988).
34. H. D. Stachel and A. Fendl, *Arch. Pharm. (Weinheim)* **321**, 563 (1988).
35. A.I. Kosak, R. J. F. Palchak, W. A. Steele and C. M. Selwitz, *J. Am. Chem. Soc.* **76**, 4450 (1954).
36. K. Spirkova, J. Kovác, V. Konecny, M. Dandarova and M. Cernayova, *Cold. Czech Chem. Commun.* **45**, 142 (1980).
37. S. Gronowitz, Y. Zhang and A.-B. Hörnfeldt, *Acta Chem. Scand.* **46**, 654 (1992).
38. Y. Zhang, A.-B. Hörnfeldt and S. Gronowitz, *J. Heterocycl. Chem.* **32**, 771 (1995).
39. F. Duus, *J. Chem. Soc., Perkin Trans. 1*, 292 (1978).
40. E. Benary and A. Baravian, *Bericht.* **48**, 593 (1915).
41. A. Courtin, E. Class and H. Erlenmeyer, *Helv. Chim. Acta* **47**, 1748 (1964).
42. Ng. Ph. Buu-Hoi, Ng. Hoan, Ng. H. Koi and Ng. D. Xuong, *J. Org. Chem.* **4**, 802 (1949).
43. V. I. Shvedov, V. K. Vasiléva and A. N. Grinev, *Chem. Heterocycl. Compd. USSR* 1495 (1973).
44. W. D. Rudorf and R. Schwarz, *Z. Chem.* **28**, 101 (1988).
45. K. Balenovic, A. Deljac, B. Gaspert and Z. Stefanac, *Monatsh. Chem.* **98**, 1344 (1967).
46. D. J. Chadwick, J. Chambers, G. D. Meakins and R. L. Snowden, *J. Chem. Soc., Perkin Trans. 1*, 2079 (1972).
47. F. Dallacker and V. Mues, *Chem. Ber.* **108**, 569 (1975).
48. R. H. Eastman and R. M. Wagner, *J. Am. Chem. Sac.* **71**, 4089 (1949).
49. C. G. Overberger and J. Lal, *J. Am. Chem. Soc.* **73**, 2956 (1973).
50. W. Hada, F. W. Rusek, J. Bordner and L. S. Melvin, Jr, *Tetrahedron Lett.* **34**, 8229 (1993).
51. D. W. Rangnekar and S. V. Mavlankar, *J. Heterocycd. Chem.* **28**, 1455 (1991).
52. D. W. Rangnekar and S. V. Mavlankar, *J. Heterocycd. Chem.* **28**, 1449 (1991).
53. P. P. Paranjpe and G. Bagavant, *Indian J. Chem.* **11**, 313 (1973).
54. E. Rose, Dissertation Friedrich-Alexander-Universität, Erlangen-Nürnberg, 1961.
55. S. Kroll, Dissertation, Friedrich-Alexander-Universität, Erlangen-Nürnberg, 1961.
56. H. Pfeiffer, Dissertation, Friedrich-Alexander Universität, Erlangen-Nürnberg, 1961.
57. H. Schadler, Dissertation, Friedrich-Alexander-Universität, Erlangen-Nürnberg, 1960.
58. L. Huybrechts, D. Buffel, E. Freyne and G. Hornaert, *Tetrahedron* **40**, 2479 (1984).
59. F. Benedini, G. Bertolini, F. Ferrario, A. Motti, A. Sala and F. Somenzi, *J. Heterocycl. Chem.* **32**, 103 (1995).
60. H. Scheibler and B. Frenz, *J. Prakt. Chem.* **2**(4), 127 (1955).
61. J. Brelivet, P. Apriou and J. Teste, *Bull. Soc. Chem. Fr.* 1344 (1971).
62. J.-P. Maffrand, D. Frehel, M. Miquel and M. Roc, *Bull. Soc. Chim. Fr.* **II**, 48 (1978).
63. P. Appriou, J. Brelivet and J. Teste, *Bull. Soc. Chim. Fr.* 1497 (1970).
64. H. P. Guan, B.-H. Luo and C.-M. Hu, *Synthesis* 461 (1997).
65. F. Bohlmann, K.-M. Kleine and H. Bornowski, *Chem. Ber.* **95**, 2934 (1962).
66. J. B. Hendrickson, B. Rees and J. F. Templeton, *J. Am. Chem. Soc.* **86**, 197 (1964).
67. H. Fiesselmann and P. Schipprak, *Chem. Ber.* **87**, 835 (1954).
68. H. Fiesselmann, P. Schipprak and L. Zeitler, *Chem. Ber.* **87**, 841 (1954).
69. H. Fiesselmann and P. Schipprak, *Chem. Ber.* **89**, 1897 (1956).
70. H. Fiesselmann and W. Böhm, *Chem. Ber.* **89**, 1902 (1956).
71. E. Larsson, *J. Prakt. Chem.* **325**, 328 (1983).
72. F. Bohlmann and E. Bresinsky, *Chem. Ber.* **97**, 2109 (1964).
73. P. R. Huddleston and J. M. Barker, *Synth. Commun.* **9**, 731 (1979).
74. F. Memmel, Dissertation, Friedrich-Alexander-Universität, Erlangen-Nürnberg, 1956.
75. R. Knerr, Dissertation Friedrich-Alexander-Universität, Erlangen-Nürnberg, 1958.
76. B. Donoso, P. Jordan de Urries and J. Lissavetzky, *Synthesis* 526 (1992).
77. K. Hartke and G. Gölz, *Lieb. Ann. Chem.* 1644 (1973).
78. F. Schweigert, Dissertation, Friedrich-Alexander-Universität, Erlangen-Nürnberg, 1964.

79. C. Corral, J. Lissavetzky and I. Manzanares, *J. Heterocycl. Chem.* **27**, 315 (1990).
80. U. Tergall, Dissertation Friedrich-Alexander-Universität, Erlangen-Nürnberg, 1962.
81. P. A. Rossy, W. Hoffmann and N. Mullet, *J. Org. Chem.* **45**, 6617 (1980).
82. R. Rippel, Dissertation, Friedrich-Alexander-Universität, Erlangen-Nürnberg, 1965.
83. A. P. Stoll and R. Süess, *Helv. Chim. Acta* **57**, 2487 (1974).
84. J. B. Press, C. M. Hofmann and S. R. Safer, *J. Org. Chem.* **44**, 3292 (1979).
85. B. Hedegaard, J. Z. Mortensen and S.-O. Lawesson, *Tetrahedron* **27**, 3853 (1971).
86. I. T. Kay and N. Punja, *J. Chem. Soc.* 2409 (1970).
87. R. L. Robey, C. A. Alt and E. E. van Meter, *J. Heterocycl. Chem.* **34**, 413 (1997).
88. C. Paulmier, J. Morel, D. Semard and P. Pastour, *Bull. Soc. Chim. Fr. Part* 2, 2434 (1973).
89. K. Takahashi, T. Sakae and K. Takase, *Chem. Lett.* 179 (1980).
90. V. I. Chernov and A. V. Mashkina, *Kinet. Katal.*, **10**, 307 (1969). *Chem.* Abstr. **71**, 38180n (1969).
91. J. J. NcNally, P. J. Sanfilippo, L. Fitzpatrick and J. B. Press, *J. Heterocycl. Chem.* **29**, 247 (1992).
92. G. Henrio, J. Morel and P. Pastour, *Bull. Soc. Chim. Fr. Part 2,* 265 (1976).
93. P. Netchitalo, J. Morel and J.-C. Halle, *J. Chem. Res. (M)* 2222 (1977).
94. C. Corral, M. B. El-Ashwamy, J. Lissavetzky and I. Manzanares. *J. Heterocycl. Chem.* **24**, 1301 (1987).
95. C. Corral, J. Lissavetzky and I. Manzanares, *Anal. Quim.* **86**, 89 (1990).
96. Yu. M. Volovenko, and F. S. Babichev, *Chem. Heterocycl. Compd. USSR* 1146 (1978).
97. S. Umio, K. Kariyone and K. Tanaka, Fujisawa Pharmaceutical Co Ltd, Japan 19090 ('67); *Chem.* Abstr. **69**, 10352e (1968).
98. A. W. Faull and R. Hull, *J. Chem. Soc., Perkin Trans. 1*, 1078 (1981).
99. A. W. Faull and R. Hull, *J. Chem. Research (S)* 240 (1979).
100. N. S. Ibrahim, K. U. Sadek, S. I. Aziz and M. H. Elnagdi, *Z. Naturforsch.* **40b**, 129 (1955).
101. R. Raap, *Can. J. Chem.* **46**, 2255 (1968).
102. Y. Tominaga, H. Ueda, K. Ogata and S. Kobra, *J. Heterocycl. Chem.* **29**, 209 (1992).
103. R. M. Mohareb, *Monatsh. Chem.* **123**, 341 (1992).
104. R. M. Mohareb, *Gazz. Chim. Ital.* **122**, 147 (1992).
105. K. Bogdanowicz-Szwed, J. Grochowski, A. Palasz, B. Rys, P. Serda and D. Soja, *Liebigs Ann.* 1457 (1996).
106. C. D. Hurd and K. L. Kreuz, *J. Am. Chem. Soc.* **74**, 2965 (1952).
107. C. D. Hurd and K. L. Kreuz, *J. Am. Chem. Soc.* **72**, 5543 (1950).
108. A.-B. Hörnfeldt and S. Gronowitz, *Acta Chem. Scand.* **16**, 789 (1962).
109. J. F. Bagli and E. Fedinandi, *Can. J. Chem.* **53**, 2598 (1975).
110. K. Gewald, H. Jablokoff and M. Hentschel. *J. Prakt. Chem.* **317**, 861 (1975).
111. V. I. Shvedov, O. B. Romanova, V. K. Vasil'eva, V. P. Pakhomov and A. N. Grinev, *Chem. Heterocycl. Compd. USSR* 679 (1973).
112. A.-B. Hörnfeldt, *Arkiv Kemi* **29**, 427 (1968).
113. R. Lantz and A.-B. Hörnfeldt, *Chem. Scripta* **10**, 126 (1976).
114. R. Lantz and A.-B. Hörnfeldt, *Chem. Scripta* **2**, 9 (1972).
115. T. Kralt, *Trav. Chim. Pays-Bas* **86**, 971 (1967).
116. H. C. Ford and D. Mackay, *J. Chem. Soc.* 4985 (1956).
117. R. T. Hawkins, *J. Heterocycl. Chem.* **11**, 291 (1974).
118. H.-H. Tso, J.-S. Wang, C.-Y. Wu and H.-C. Lin, *New. J. Chem.* 771 (1998).
119. A. P. Manzara and P. Kovácic, *J. Org. Chem.* **39**, 504 (1974).
120. O. Scherer and F. Kluge, Farbwerke Hoechst A.-G.; Ger 1,246,756, *Chem. Abstr.* **68**, 59589f (1968).
121. M. Hedayatullah, J. Pailler and L. Denivelle, *Bull. Soc. China. Fr. Part 2,* 2161 (1974).
122. J. Sicé, *J. Am. Chem. Soc.* **75**, 3697 (1953).
123. M. Bercot-Vatteroni, *Ann. Chim.* **7**, 303 (1962).

124. E. Profft, *Lieb. Ann.* **622**, 196 (1959).
125. S. Gronowitz, *Arkiv Kemi* **12**, 239 (1958).
126. M. A. Keegstra, T. H. A. Peters and L. Brandsma, *Synth. Comet.* **20**, 213 (1990).
127. L. L. Miller and Y. Yu, *J. Org. Chem.* **60**, 6813 (1995).
128. K. Faid and M. Leclerc, *Chem. Commum.* 2761 (1996).
129. G. Zotti, M. C. Gallazi, G. Zerbi and S. V. Meille, *Synth. Meth.* **73**, 217 (1995).
130. C. Corall, M. B. El-Ashmawy, J. Lissavetsky, A. Basilio and A. Giraldez, *Eur. J. Med. Chem.* **22**, 251 (1987).
131. S. Iwatsuki, M. Kubo and Y. Itoh, *Chem. Lett.* 1085 (1993).
132. H. Cheng and L. R. Eisenbaumer, *J. Chem. Soc. Chem. Commun.* 1451 (1995).
133. F. Goldoni, B. M. W. Langeveld-Voss and E. W. Meijer, *Synth.* Commun. **28**, 2237 (1998).
134. M. A. Keegstra, T. H. A. Peters and L. Brandsma, *Tetrahedron* **48**, 363 (1992).
135. L. D. Peeters, S. G. Jacobs, W. Eevers and H. J. Geise, *Tetrahedron* **50**, 11533 (1994).
136. Ya. L. Gol'dfarb and M. A. Kalik, *Chem. Heterocycl, Compds, USSR* 155 (1971).
137. A. Hallberg, T. Frejd and S. Gronowitz, *J. Chem. Soc., Perkin Trans. 1*, 1390 (1980).
138. C. Corall, M. B. El-Ashmawy, J. Lissavetsky, A. Basilio and A. Giraldez, *Eur. J. Med. Chem.* **22**, 251 (1987).
139. E. B. Pedersen and S.-O. Lawesson, *Tetrahedron* **26**, 2050 (1975).
140. H. J. Jakobsen, *Tetrahedron* **23**, 3737 (1967).
141. F. M. Stoyanovich, Ya. L. Gol'dfarb and G. B. Chermanova, *Bull. Acad. Sci. USSR, Div. Chem. Sci.* 2228 (1973).
142. J.-M. Meunier and P. Fournari, *Bull. Soc. Chim. Fr. 3343* (1971).
143. K. Takahashi, K. Nishijima, K. Takase and T. Shimozawa, *Chem. Lett.* 875 (1982).
144. K. Takahashi, K. Nishijima, K. Takase and S. Katagiri, *Tetrahedron Lett.* **24**, 205 (1983).
145. H. J. Jakobsen and S.-O. Lawesson, *Tetrahedron* **23**, 871 (1967).
146. E. B. Pedersen and S.-O. Lawesson, *Tetrahedron* **27**, 3861 (1971).
147. H. J. Jakobsen, E. H. Larsen and S.-O. Lawesson, *Rec. Trav. Chim. Pays-Bas* **82**, 791 (1963).
148. A. Mac Eachern, C. Soucy, L. C. Leitch, J. T. Amason and P. Morand, *Tetrahedron* **44**, 2403 (1988).
149. H. J. Jakobsen, E. H. Larsen and S. O. Lawesson, *Tetrahedron* **19**, 1867 (1963).
150. Q. T. Zhang and J. M. Tour, *J. Am. Chem. Soc.* **120**, 5355 (1998).
151. J. S. Lomas, *J. Chem. Soc., Perkin Trans.* **2**, 754 (2001).
152. J. Lissavetzky and I. Manzanares, *Heterocycles* **43**, 1767 (1996).
153. B. Cederlund, A. Jesperson and A.-B. Hörnfeldt, *Acta Chem. Scand.* **25**, 3656 (1971).
154. S. Gronowitz, R. Svenson, G. Bondesson, O. Magnusson and N. Stjernström, *Acta Pharm. Suec.* **15**, 361 (1978).
155. J. Brunet, D. Pacquer and P. Rioult, *Phosphorus and Sulfur* **3**, 377 (1977).
156. L. Bhat, H. IIa and H. Junjappa, *Synthesis*, 959 (1993).
157. J. Kwan-Yue, H. Eckhardt, T. R. Jaw, L. W. Shacklette and R. I. Elsenbaumer, *J. Chem. Soc., Chem. Commun.* 215 (1988).
158. J. P. Montheard and J. C. Dubois, *J. Heterocycl. Chem.* **22**, 719 (1985).
159. G. A. Sotzing and J. R. Reynolds, *J. Chem. Soc., Chem. Commun.* 703 (1995).
160. Z. V. Todres, N. G. Furmanova, S. P. Avagyan, Y. T. Struchkov and D. N. Kursanov, *Phosphor, Sulfur Relat. Elem.* **5**, 309 (1979).
161. S. Iwatsuki, M. Kubo and Y. Itoh, *Chem. Lett.* 1085 (1993).
162. M. Chayer, K. Faïd and M. Leclerc, *Chem. Mater.* **9**, 2902 (1997).
163. J. Lukkari, M. Salomäki, A. Viinikanoja, T. Ääritalo, J. Paukkunen, N. Kocharova and J. Kankare, *J. Am. Chem. Soc.* **123**, 6083 (2001).
164. J. Lomas, E. Vauthier and J. Vaissermann, *J. Chem. Soc., Perkin Trans. 2* 1399 (2000).
165. R. G. Hicks and M. B. Nodwell, *J. Am. Chem. Soc.* **122**, 6746 (2000).
166. R. A. Aitken and A. N. Garnett, *J. Chem. Soc., Perkin Trans.* **1**, 3020 (2000).
167. T. Otani, Y. Sugihara, A. Ishii and J. Nakayama, *Chem. Letters* 744 (2000).

168. S. Gronowitz, A. Hallberg and C. Glennow, *J. Heterocycl. Chem.* **17**, 171 (1980).
169. S. Gronowitz, B. Cederlund and A.-B. Hörnfeldt, *Chem. Scripta* **5**, 217 (1974).
170. J. W. H. Watthey and M. Desai, *J. Org. Chem.* **47**, 1755 (1982).
171. A. K. Mohanakrishnan, A. Hucke, M. A. Lyon, M. V. Lakshmikantham and M. P. Cava, *Tetrahedron*, **55**, 11745 (1999).
172. S. Akoudad, P. Frére, M. Mercier and J. Roncali, *J. Org. Chem.* **64**, 4267 (1999).
173. J.-M. Raimundo, P. Blanchard, H. Brisset, S. Akoudad and J. Roncali, *Chem. Commun.* 939 (2000).
174. N. S. Kzenzhek, L. I. Belen'kii and Ya. L. Gol'dfarb, *Khim. Geterosikl. Soedin.* 486 (1973).
175. L. Brandsma, V. Y. Vvedensky, N. A. Nedolya, O. A. Tarasova and B. A. Trofimov, *Tetrahedron Lett.* **39**, 2433 (1998).
176. F. Bohlmann, H. Bornowski and D. Kramer, *Chem. Ber.* **96**, 584 (1963).
177. R. Jaunin, *Helv. Chim. Acta* **63**, 1542 (1980).
178. A. R. Katritzky, X. Wang and A. Denisenko, *J. Org. Chem.* **66**, 2850 (2001).
179. B. A. Reinhardt, L. L. Brott, S. J. Clarson, A. G. Dillard, J. C. Bhatt, R. Kannan, L. Yuan, G. S. He and P. N. Prasad, *Chem. Mater.* **10**, 1863 (1998).
180. A. Kumar, D. M. Welsh, M. C. Morvant, F. Piroux, K. A. Abboud and J. R. Reynolds, *Chem. Mater.* **10**, 896 (1998).
181. F. Cappellina, *Ann. Chim. (Rome)* **48**, 535 (1958).
182. B. C. Jain, B. H. Iyer and P. C. Guha, *J. Ind. Chem. Soc.* **24**, 177 (1947).
183. W. Holzer, E. Schmid and C. Slatin, *Monatsh. Chem.* **125**, 1287 (1994).
184. J. M. Barker, J. D. E. Chaffin, J. Halfpenny, P. R. Huddleston and P. F. Tseki, *J. Chem. Soc., Chem Commun.* 1733 (1993).
185. C. Corral and J. Lissavetzky, *Synthesis* 847 (1984).
186. J. Halfpenny and Z. Sloman, *J. Chem. Soc., Perkin Trans 1* 1877 (2000).
187. J. D. E. Chaffin, J. M. Barker and P. R. Huddleston, *J. Chem. Soc., Perkin Trans. 1* 1398 (2001).
188. J. Morel and P. Pastour, *Bull. Soc. Chim. Fr.* 737 (1968).
189. G. Marchand, B. Decroix and J. Morel, *J. Heterocycl. Chem.* **21**, 877 (1984).
190. I. Puschmann and T. Erker, *Heterocycles* **36**, 1323 (1993).
191. I. Puschmann and T. Erker, *Monatsh. Chem.* **125**, 927 (1994).
192. G. Consiglio, C. Arnone, D. Spinelli and R. Noto, *J. Chem. Soc., Perkin Trans. 2*, 642 (1981).
193. C. Dell'Erba and G. Guanti, *Gazz. Chim. Ital.* **100**, 223 (1970).
194. C. Consiglio, D. Spinelli and R. Noto, *J. Chem. Res. (S)* 242 (1979).
195. G. Baldini, G. Doddi, G. Illuminati and F. Stegel, *J. Org. Chem.* **41**, 2153 (1976).
196. B. P. Fedorov and F. M. Stoyanovich, *Chem. Abstr.* **62**, 10413a (1965).
197. J. Halfpenny, P. B. Rooney and Z. S. Sloman, *Tetrahedron Lett.* **41**, 6223 (2000).
198. D. Binder, B. C. Prager and C. R. Noe, *J. Chem. Research (S)* 140 (1981).
199. G. Rassu, P. Spanu, L. Pinna, F. Zanardi and G. Casiraghi, *Tetrahedron Lett.* **36**, 1941 (1995).
200. W. H. Houff and R. D. Schuetz, *J. Am. Chem. Soc.* **75**, 6316 (1953).
201. S. Gronowitz, P. Moses and A.-B. Hörnfeldt, *Arkiv. Kemi.* **17**, 237 (1961).
202. Ya. L. Gol'dfarb, M. A. Kalik and M. L. Kirmalova, *J. Gen. Chem. USSR* **32**, 216 (1962).
203. Ya. L. Gol'dfarb and M. A. Kalik, *Chem. Heterocycl. Compds.* **4**, 571 (1968).
204. E. Jones and F. Moodie, *Tetrahedron* **21**, 2413 (1965).
205. B. Gestblom, S. Gronowitz, R. A. Hoffman, B. Mathiasson and S. Rodmar, *Arkiv. Kemi.* **23**, 501 (1965).
206. B. P. Fedorov and F. M. Stoyanovich, *J. Org. Chem. USSR* **1**, 189 (1965).
207. L. Brandsma and H. J. T. Bos, *Rec. Trav. Chim. Pays-Bas* **88**, 732 (1969).
208. E. Jones and I. M. Moodie, *Org. Synth.* **50**, 104 (1970).
209. E. Jones And I. M. Moodie, *Tetrahedron* **21**, 1333 (1965).
210. S. Gronowitz and R. Håkansson, *Arkiv Kemi* **16**, 309 (1960).

211. M. J. Janssen and J. Bos, *Angew. Chem.* **81**, 579 (1969).
212. P. I. Abramenko and V. G. Zhiryakov, *Chem. Heterocycl. Compds.* **12**, 860 (1976).
213. B. Gestblom, S. Gronowitz, R. A. Hoffman, B. Mathiasson and S. Rodmar, *Arkiv. Kemi.* **23,** 483 (1965).
214. S. Gronowitz and P. Moses, *Acta Chem. Scand.* **16**, 105 (1962).
215. L.-*Y.* Chiang, P. Shu, D. Holt and D. Cowan, *J. Org. Chem.* **48**, 4713 (1983).
216. P. Shu, L. Chiang, T. Emge, D. Hikt, T. Kistenmacher, M. Lee, J. Stokes, T. Poehler, A. Bloch and D. Cowan, *J. Chem. Soc., Chem. Commun.* 920 (1981).
217. Ya. L. Gol'dfarb, M. A. Kalik and M. L. Kirmalova, *Bull. Acad. Sci. USSR, Div. Chem. Sci.* 1572 (1960).
218. Ya. L. Gol'dfarb, M. A. Kalik and M. L. Kirmalova, *Bull. Acad. Sci. USSR, Div. Chem. Sci.* 1578 (1964).
219. Ya. L. Gol'dfarb, M. A. Kalik and M. L. Kirmalova, *Bull. Acad. Sci. USSR, Div. Chem. Sci.* 2107 (1969).
220. Ya. L. Gol'dfarb, M. A. Kalik and M. L. Kirmalova, *Bull. Acad. Sci. USSR, Div. Chem. Sci.* 650 (1962).
221. Ya. L. Gol'dfarb, M. A. Kalik and M. L. Kirmalova, *Bull. Acad. Sci. USSR, Div. Chem. Sci.* 1653 (1963).
222. Ya. L. Gol'dfarb, M. B. Ibrarmigova and and O. A. Kahnovskii, *Bull. Acad Sci. USSR, Div. Chem. Sci.* 1029 (1962).
223. Ya. L. Gol'dfarb, M. A. Kalik and Z. G. Kozlova, *Chem. Heterocycl. Compds.* **16**, 1007 (1980).
224. Ya. L. Gol'dfarb, M. A. Kalik, M. L. Kirmalova and M. M. Pölonskaya, *Bull. Acad. Sci. USSR, Div. Chem. Sci.* 857 (1966).
225. Ya. L. Gol'dfarb and M. A. Kalik, *Chem. Heterocycl. Compds.* **6**, 1232 (1970).
226. J. L. Reddinger and J. R. Reynolds, *J. Org. Chem.* **61**, 4833 (1996).
227. A. Biedermann, *Bericht.* **19**, 1615 (1886).
228. W. Tardel, Dissertation, Friedrich-Alexander-Universität Erlangen-Nürnberg, 1966.
229. O. Dann and W. Dimmling, *Chem. Ber.* **87**, 373 (1954).
230. J. C. Martin, R. D. Burpitt, P. G. Gott, M. Harris and R. H. Meen, *J. Org. Chem.* **36**, 2205 (1971).
231. B. Föhlisch and B. Czauderna, *Phosphorus and Sulfur* **4**, 167 (1978).
232. A. V. Anisimov, V. F. Ionova, V. K. Govorek, V. S. Babaitsev and E. A. Victorova, *Dokl. Chem.* **244**, 29 (1979).
233. A. V. Anisomov, V. F. Ionova and E. A. Victorova, *J. Org. Chem. USSR*, **13**, 2624 (1977).
234. M. G. Voronkov, É. N. Deryagina, A. S. Nakhmmanovich and L. G. Klochkova, *Chem. Heterocycl. Compds.* 620 (1974).
235. Ya. L. Gol'dfarb and M. A. Kalik, *Chem. Heterocycl. Compds.* **3**, 799 (1967).
236. L. G. Bogdanova, V. V. Zelentsov, M. A. Kalik, Ya. Gol'dfarb and V. A. Petukhov, *Bull. Acad. Sci. USSR, Div. Chem. Sci.* 1472 (1973).
237. Ya. L. Gol'dfarb, M. A. Kalik and M. L. Kirmalova, *Chem. Heterocycl. Compds.* **3**, 45 (1967).
238. C. Paulmier, *Bull Soc. Chim. Fr. II* 237 (1979).
239. R. Gompper and H. Schaefer, *Chem. Ber.* **100**, 591 (1967).
240. L. Dalgaard, L. J. Jensen and S.-O. Lawesson, *Tetrahedron* **30**, 93 (1974).
241. A. Kakehi, S. Itoh, Y. Ohno, S. Shiba and S. Kamata, *Bull. Chem. Soc. Japan* **60**, 3713 (1987).
242. K. Clarke, W. R. Fox and R. M. Scrowston, *J. Chem. Soc., Perkin Trans. 1* 1029 (1980).
243. C. Corral, J. Lissavetzky and A. M. Valdeolmillos, *Synthesis* 172 (1984).
244. C. Corral, J. Lissavetzky, A. S. Alvarezinsua and A. M. Valdeolmillos, *Org. Prep. Proc. Int.* **17**, 163 (1985).
245. L. Henriksen and H. Autrup, *Acta Chem. Scand.* **24**, 2629 (1974).

246. K. Gewald, U. Hain and H. Madlenscha, *J. Prakt. Chem.* **330**, 866 (1988).
247. I. Puschmann and T. Erker, *Heterocycles* **41**, 709 (1995).
248. Ya. L. Gol'dfarb, M. A. Kalik and M. L. Kirmalova, *Bull. Acad. Sci. USSR, Div. Chem. Sci.* 1638 (1969).
249. Ya. l. Gol'dfarb and V. P. Litvinov, *Bull. Acad. Sci. USSR, Div. Chem. Sci.* 310 (1963).
250. Ya. L. Gol'dfarb and V. P. Litvinov, *Bull. Acad Sci. Div. Chem. Sci.* 1479 (1963).
251. Ya. L. Gol'dfarb, M. A. Kalik and M. L. Kirmalova, *Dokl. Chem. (Engl. Transl.)* 3592 (1959).
252. Ya. L. Gol'dfarb, V. P. Litvinov and A. N. Sukiasyan, *Dokl. Chem. (Engl. Transl.)* **182**, 802 (1968).
253. C. M. Camaggi, I. Lunazzi and G. Piacucci, *J. Chem. Soc., Perkin Trans. 2* 1491 (1973).
254. S. Gronowitz and P. Moses, *Acta Chem. Scand.* **16**, 155 (1962).
255. B. P. Fedorov and F. M. Stoyanovich, *J. Gen. Chem. USSR* **33**, 2194 (1963).
256. A. P. Yakubov, N. V. Gregoreva and L. I. Belen'kii, *J. Org. Chem. USSR* **14**, 593 (1978).
257. A. Bugge, *Acta Chem. Scand.* **25**, 27 (1971).
258. V. P. Litvinov, T. V. Shchedrinskaya, P. A. Konstantinov and Ya. I. Gol'dfarb, *Chem. Heterocycl. Compds.* **11**, 431 (1975).
259. Y. Mazaki and K. Kobayashi, *Tetrahedron Lett.* **30**, 3315 (1989).
260. J. D. Prugh, G. D. Hartman, P. J. Mallorga, B. M. McKeever, S. R. Michelson, M. A. Murcko, H. Schwam, R. L. Smith, J. M. Sonday, J. P. Springer and M. F. Sugrue, *J. Med. Chem.* **34**, 1805 (1991).
261. S. Gronowitz, *Arkiv Kemi* **13**, 269 (1958).
262. Ya. L. Gol'dfarb and V. P. Litvinov, *Bull. Acad. Sci. USSR, Div. Chem. Sci.* 317 (1963).
263. A. Bugge, *Acta Chem. Scand.* **22**, 63 (1968).
264. S. Gronowitz, M. Herslöf, R. Svenson, G. Bondesson, O. Magnusson and N. Stjernström, *Acta Pharm. Suec.* **15**, 368 (1978).
265. C. Paulmier and F. Quturquin, *J. Heterocycl. Chem.* **20**, 113 (1983).
266. Ya. L. Gol'dfarb, V. P. Litvinov and S. A. Ozolin, Bull. *Acad. Sci. USSR, Div. Chem. Sci.* 486 (1965).
267. I. Tabakovic, T. Maki, L. L. Miller and Y. Yu, *Chem. Commum.* 1911 (1996).
268. O. Tempkin, T. J. Blacklock, J. A. Burke and M. Anastasia, *Tetrahedron Asymm.* **7**, 2721 (1996).
269. Ya. L. Gol'dfarb, M. A. Kalik and M. L. Kirmalova, *J. Gen. Chem. USSR* **29**, 2003 (1959).
270. M. Rajsner, E. Svátek, J. Metys and M. Protiva, *Coil. Czech. Chem. Commun.* **39**, 1366 (1974).
271. J. Cymerman-Craig and J. W. Loder, *J. Chem. Soc.* 237 (1954).
272. J. Cymerman-Craig and J. W. Loder, *Org. Synth.* **35**, 85 (1955).
273. X. Wu and R. D. Rieke, *J. Org. Chem.* **60**, 6658 (1995).
274. S. Gronowitz, P. Moses, A.-B. Hörnfeldt and R. Håkansson, *Arkiv Kemi* **17**, 165 (1961).
275. S. Gronowitz and U. Rosén, *Chem. Scripta* **1**, 33 (1971).
276. S. Gronowitz, P. Moses and R. Håkansson, *Arkiv Kemi* **16**, 267 (1960).
277. Ya. L. Gol'dfarb, M. A. Kalik and V. K. Zavylova, *Chem. Heterocycl. Compds.* **17**, 126 (1981).
278. F. M. Stoyanovich and B. P. Fedorov, *J. Org. Chem.* USSR **1**, 1296 (1965).
279. S. Gronowitz, M. Temciuc and A.-B. Hörnfeldt, *J. Heterocycl. Chem.* **30**, 1111 (1993).
280. F. de Jong and M. J. Janssen, *J. Org. Chem.* **36**, 1645 (1971).
281. P. Meunier and G. Pfister-Guillouzo, *Can. J. Chem.* **55**, 3901 (1977).
282. J. Ashby, M. Ayad and O. Meth-Cohn, *J. Chem. Soc., Perkin Trans. 1*, 1104 (1973).
283. R. Adams, A. Ferretti, *J. Am. Chem. Soc.* **81**, 4927 (1959).
284. E. Jones and I. M. Moodie, *Org. Synth.* **50**, 75 (1970).
285. E. Jones and I. F. Moodie, *J. Chem. Soc.* 7018 (1965).
286. J. Nakayama, N. Katano, Y. Shimura, Y. Sugihara and A. Ishii, *J. Org. Chem.* **61**, 7608 (1996).

287. O. Mayor and N. Rasanu, *Rev. Roum. Chim.* **21**, 745 (1976).
288. United Pharmaceutical Works Brit. 1,081,360, SPOFA; *Chem. Abstr.* **68**, 95849 (1968).
289. J. R. Campbell, *J. Org. Chem.* **29**, 1830 (1964).
290. S. B. Lee and J.-I. Hong, *Tetrahedron Lett.* **36**, 8439 (1995).
291. F. Challenger and J. L. Holmes, *J. Chem. Soc.* 1837 (1953).
292. V. V. Ghaisas and B. D. Tilak, *Proc. Indian Acad. Sci.* **39A**, 14 (1954).
293. G. Marchand, B. Decroix and J. Morel, *Chem. Scripta* **23**, 80 (1984).
294. A. Jilale, B. Decroix and J. Morel, *Chem. Scripta* **27**, 421 (1987).
295. K. I. Sadykhov, S. M. Aliev and M. M. Seidov, *Chem. Heterocycl. Compds.* 299 (1975).
296. K.I. Sadykhov, S. M. Aliev and M. M. Seidov, *J. Org. Chem. USSR* **11**, 2191 (1975).
297. P. Netchitailo, M. Othman and B. Decroix, *J. Heterocycl. Chem.* **34**, 321 (1997).
298. B. Iddon and M. Nicholas, *J. Chem. Research (S)* 512 (1996).
299. T. J. Blacklock, P. Sohar, J. W. Butcher, T. Lamanec and E. J. J. Grabowski, *J. Org. Chem.* **58**, 1672 (1993).
300. G. S. Ponticello, M. R. Freedman, C. N. Habecker, M. K. Holloway, J. S. Amato, R. S. Conn and J. J. Baldwin, *J. Org. Chem.* **53**, 9 (1988).
301. L. Benati, L. Capella, P. C. Montevecchi and P. Spagnolo, *J. Org. Chem.* **60**, 7941 (1995).
302. P. C. Montevecchi and M. L. Navacchia, *J. Org. Chem.* **60**, 6455 (1995).
303. G. S. Ponticello, C. N. Habecker, S. L. Varga and S. M. Pitzenberger, *J. Org. Chem.* **54**, 3223 (1989).
304. Y. Y. Vvedenskii, E. D. Shtefan, M. V. U'lev, A. R. Zhnikin, T. A. Shkarupa, É. N. Deryagina and M. G. Voronkov, *J. Org. Chem. USSR* **28**, 176 (1992).
305. C. M. Buess and N. Kharasch, *J. Am. Chem. Soc.* **72**, 3529 (1950).
306. R. D. Schuetz and W. L. Fredericks, *J. Org. Chem.* **27**, 1301 (1962).
307. A. Haas and V. Hellvig, *Chem. Ber.* **109**, 2475 (1976).
308. R. L. P. de Jong and L. Brandsma, *J. Organomet. Chem.* **238**, C17 (1982).
309. R. L. P. de Jong and L. Brandsma, *Synth. Comm.* **20**, 3427 (1990).
310. J. P. Marino and J. L. Kostusyk, *Tetrahedron Lett.* 2489 (1979).
311. J. P. Marino and J. L. Kostusyk, *Tetrahedron Lett.* 2493 (1979).
312. A. Thomas, G. Singh, H. Ila and H. Junjappa, *Tetrahedron Lett.* **30**, 3093 (1989).
313. A. Datta, H. Ila and H. Junjappa, *Synthesis* 556 (1988).
314. S. M. H. Kabir and M. Ioda, *Chem. Commun.* 2329 (2000).
315. B. P. Fedorov and F. M. Stoyanovich, *Bull. Acad. Sci., USSR, Div. Chem. Sci.* 1705 (1960).
316. Ya. L. Gol'dfarb and M. L. Kirmalova, *Chem. Heterocycl. Compd.* **5**, 360 (1969).
317. R. Helmers, *Liebigs Ann.* 181 (1973).
318. Ya. L. Gol'dfarb, M. A. Kalik, N. A. Shults and L.I. Belen'kii, *J. Org. Chem. USSR* **15**, 1150 (1979).
319. E. C. Taylor and D. E. Vogel, *J. Org. Chem.* **50**, 1002 (1985).
320. U. Folli, D. Iarossi, M. Montorsi, A. Mucci and L. Schenetti, *J. Chem. Soc., Perkin Trans. 1* 537 (1991).
321. K. A. Jensen and L. Henriksen, *Acta Chem. Scand.* **22**, 1107 (1968).
322. T. Liljefors and J. Sandström, *Acta Chem. Scand.* **24**, 3109 (1970).
323. E. J. Smutny, *J. Am. Chem. Soc.* **91**, 208 (1969).
324. D. Schuijl-Laros, P. J. W. Schuijl and L. Brandsma, *Rec. Trav. Chim. Pays-Bas* **88**, 1343 (1969).
325. D. Wobig, *Liebigs Ann. Chem.* 115 (1990).
326. Ya. L. Gol'dfarb, G. P. Lokhil and L. I. Belen'kii, *J. Gen. Chem. USSR* **37**, 2541 (1967).
327. T. Erker, *Monatsh. Chem.* **129**, 679 (1998).
328. C. Dell'Erba, F. Sancassan, M. Novi, D. Spinelli, G. Consiglio, C. Arnone and F. Ferroni, *J. Chem. Soc., Perkin Trans. 2* 1779 (1989).
329. O. Dann and E. F. Möller, *Chem. Ber.* **82**, 76 (1949).
330. P. A. van Zwieten and H. O. Huisman, *Rec. Trav. Chim. Pays-Bas* **81**, 554 (1962).
331. P. A. van Zwieten, J. Meltzer and H. O. Huisman, *Rec. Trav. Chim. Pays-Bas* **81**, 616 (1962).

332. R. D. Schuetz and C. O. Okafor, *Chim. Ther.* 289 (1968).
333. A. J. Boulton and D. Middleton, *J. Org. Chem.* **39**, 2956 (1974).
334. P. Stanetty and M. Kremslehner, *Heterocycles* **48**, 259 (1998).
335. P. Stanetty, E. Gömer and M. D. Mihovilovic, *J. Heterocycl. Chem.* **36**, 761 (1999).
336. G. Leandri, D. Spinelli and C. Dell'Erba, *Ann. Chim. (Rome)* **50**, 1597 (1960).
337. G. Consiglio, D. Spinelli, S. Gronowitz, A.-B. Hörnfeldt, B. Maltesson and R. Noto, *J. Chem. Soc., Perkin Trans. 2* 625 (1982).
338. M. E. Schreder and T. Erker, *J. Heterocyclic. Chem.* **37**, 349 (2000).
339. A. Corsaro, F. Guerrera, M. C. Sarvá and M. A. Siracusa, *Heterocycles* **27**, 2539 (1988).
340. D. Spinelli and C. Dell'Erba, *Ann. Chim. (Rome)* **54**, 281 (1964).
341. G. Ronsisvalle and G. Pappalardo, *Farm. Ed. Sci.* **32**, 678 (1977).
342. H. Y. Lew and C. R. Noller, *J. Am. Chem. Soc.* **72**, 5715 (1950).
343. G. Ronsisvalle, M. S. Pappalardo, F. Vittorio, L. Pasquinucci, E. Bousquet, S. Olivieri and E. Cammarata, *Il Farmaco* **44**, 383 (1989).
344. C. Paulmier, *Bull. Soc. Chim. Fr. II*, 592 (1979).
345. Z. V. Todres, F. M. Stoyanovich, Ya. L. Gol'dfarb and D. N. Kursanov, *Chem. Heterocycl. Compds*, **9**, 583 (1973).
346. C. J. Grol, *J. Heterocycl. Chem.* **11**, 953 (1974).
347. M. Rajsner and M. Protiva, *Collect. Czech. Chem. Commun.* **33**, 1846 (1968).
348. C. J. Grol and H. Rollema, *J. Med. Chem.* **18**, 857 (1975).
349. C. J. Grol and J. S. Faber, *Rec. Trav. Chim. Pays-Bas* **89**, 68 (1970).
350. M. Rajsner, F. Miksik, J. Metisová and M. Protiva, *Collect. Czech. Chem. Commun.* **44**, 2997 (1979).
351. T. Erker, *J. Heterocycl. Chem.* **35**, 1521 (1998).
352. Y. Tominaga, J-K. Luo and R. N. Castle. *J. Heterocycl. Chem.* **31**, 771 (1994).
353. S. Masson, J. F. Saint-Clair and M. Saquet, *Tetrahedron Lett.* **35**, 3083 (1994).
354. S. Masson, J.-R. Saint-Clair, A. Dore and M. Saquet, *Bull. Sac. Chim. Fr.* **133**, 951 (1996).
355. J. M. Luteijn, H. Dolman and H. C. Wals, *Tetrahedron* **44**, 5921 (1988).
356. M. M. Krayushkin, M. A. Kalik, E. Y. Zvezdina and V. S. Bogdanov, *Bull. Acad. Sci. USSR (Engl. Transl.)* **40**, 2474 (1991).
357. M. Krayushkin, M. A. Kalik, E. Y. Zvedina and M. G. Kurella, *Mendelleev Commun.* 114 (1993).
358. M. M. Krayushkin, M. A. Kalik, L. G. Vorontsova, E. Y. Zvezdina and M. G. Kurella, *Russian Chem. Bull. (Engl. Transl.)* **42**, 1204 (1993).
359. Ya. L. Gol'dfarb, G. P. Pokhil and L. I. Belen'kii, *Dokl. Chem. (Engl. Transl.)* **167**, 385 (1966).
360. B. R. Fishwick, D. K. Rowles and C. J. M. Stirling, *J. Chem. Soc., Chem. Commun.* 834 (1983).
361. V. I. Shvedov, I. A. Kharizomenova, O. B. Romanova, V. K. Vasileva and A. N. Grinev, *Chem. Heterocycl. Compds.* **11**, 799 (1975).
362. P. Stanetty and M. D. Mihovilovic, *Monatsh. Chem.* **130**, 573 (1999).
363. Ya. L. Gol'dfarb, M. A. Kalik and M. L. Kirmalova, *Chem. Heterocycl. Compds.* **3**, 50 (1967).
364. N. I. Astrakhantseva, V. G. Zhiryakov and P. I. Abramenko, *Chem. Heterocycl. Compds.* **11**, 1364 (1975).
365. N.I. Astrakhantseva, V. G. Zhiryakov and P.I. Abramenko, *Chem. Heterocycl. Compds.* **12**, 1123 (1976).
366. A. Daich and B. Decroix, I. *Heterocycl. Chem.* **28**, 1881 (1991).
367. C. Corral, J. Lissavetzky, S. Alvarez-Insua and A. M. Valdeolmillos, *Org. Prep. Proceed. Int.* **17**, 163 (1985).
368. K. Kawahara, *Yakugaku Zasshi* **77**, 963 (1957).
369. F. Runge, A. Jumar and P. Held, *J. Prak Chem.* [4] **8**, 44 (1959).

370. G. Ronsisvalle, Il Farmaco **35**, 341 (1980).
371. J. W. Brooks, E. G. Howard and J. J. Wehrle, *J. Am. Chem. Soc.* **72**, 1289 (1950).
372. D. Belo, H. Alves, E. B. Lopes, M. T. Duarte, V. Gama, R. T. Henriques, M. Almeida, A. Pérez-Benítez, C. Rovira and J. Venciana, *Chem. Eur. J.* **7**, 511 (2001).
373. L.-Y. Chiang, P. Shu, D. Holt and D. Cowan, *J. Org. Chem.* **48**, 4713 (1983).
374. E. Söderbäck, *Acta Chem. Scand.* **8**, 1851 (1954).
375. F. M. Stoyanovich, G. I. Gorushkina and Ya. L. Gol'dfarb, *Bull. Acad. Sci. USSR, Div. Chem. Sci.* 1742 (1970).
376. V. I. Shvedov, I. A. Kharizomenova and A. N. Grinev, *Chem. Heterocycl. Compeds.* **10**, 1045 (1974).
377. F. Outurquin and C. Paulmier, *Bull. Soc. Chim. Fr. II* 153 (1983).
378. C. Paulmier, *Tetrahedron Lett.* 1797 (1978).
379. Ya. L. Gol'dfarb, N. S. Ksenzhek and L. I. Belen'kii, *Chem. Heterocycl. Compds.* **8**, 144 (1972).
380. Ya. L. Gol'dfarb, N. S. Ksenzhek and L. I. Belen'kii, *Bull. Acad. Sci., USSR, Div. Chem. Sci.* 1124 (1973).
381. G. Gölz and K. Hartke, *Arch. Pharm., Weinheim* **307**, 663 (1974).
382. Ya. L. Gol'dfarb, M. A. Kalik and M. L. Kirmalova, *J. Gen. Chem. USSR* **30**, 1026 (1960).
383. T. S. Murthy, L. J. Pandya and B. D. Tilak, J. *Sci. Ind. Res.* **20B**, 169 (1961).
384. G. Consiglio, S. Gronowitz, A.-B. Hörnfeldt, R. Noto, K. Pettersson and D. Spinelli, *Chem. Scripta* **13**, 20 (1978–79).
385. E. Kesler and S. Gronowitz, *Monatsh. Chem.* **111**, 119 (1980).
386. Z. Polívka, M. Rajsner, J. Metys, J. Holubek, E. Svátek, M. Ryska and M. Protiva, *Coll. Czech. Chem. Commun.* **48**, 2970 (1983).
387. L. D. Girodier, C. Maignan and F. P. Rouessac, *Tetrahedron Asymm.* **3**, 857 (1992).
388. Y. Arai, A. Suzuki, T. Masuda, Y. Masaki and M. Shiro, *J. Chem. Soc., Perkin Trans. 1* 2913 (1995).
389. Y. Arai, A. Suzuki, T. Masuda, Y. Masaki and M. Shiro, *Chem. Phar. Bull.* **44**, 1765 (1996).
390. M. M. Krayushkin, M. A. Kalik, V. K. Zav'yalova and V. S. Bogdanov, *Chem. Heterocycl. Compels.* **24**, 332 (1988).
391. W. E. Truce and E. Wellisch, *J. Am. Chem. Soc.* **74**, 5177 (1952).
392. J. Cymerman-Craig, G. N. Vaughan and W. K. Warburton, *J. Chem. Soc.* 4114 (1956).
393. H. Burton and W. A. Davy, *J. Chem. Soc.* 525 (1948).
394. J. Cymerman and J. L. Lowe, *J. Chem. Soc.* 1666 (1949).
395. N. H. Nilsson, C. Jacobsen, O. Nøregaard-Sørensen, N. K. Haunsøe and A. Senning, *Chem. Ber.* **105**, 2854 (1972).
396. T. P. Forrest and D. E. Ryan, *Can. J. Chem.* **36**, 1674 (1958).
397. J. H. Uhlenbroek, M. J. Koopmans and H. O. Huisman, *Rec. Trav. Chim. Pays-Bas* **76**, 129 (1957).
398. C. Dell'Erba and D. Spinelli, *Tetrahedron* **21**, 1061 (1965).
399. D. Spinelli, G. Guanti and C. Dell'Erba, *J. Chem. Soc., Perkin Trans. 2* 441 (1972).
400. D. Spinelli, C. Consiglio and T. Monti, *J. Chem. Soc., Perkin Trans. 2* 816 (1975).
401. C. Dell'Erba, D. Spinelli and G. Leandri, *Gazz. Chim. Ital.* **99**, 535 (1969).
402. Ya. L. Gol'dfarb, F. M. Stoyanovich and G. B. Chermanova, *Bull. Acad. Sci. USSR, Div. Chem. Sci.* 2233 (1973).
403. F. M. Stoyanovich and B. P. Fedorov, *Chem. Heterocycl. Compds.* **3**, 650 (1967).
404. M. R. C. Gerstenberger, A. Haas and H. Pauling, *Helv. Chim. Acta* **65**, 490 (1982).
405. L. Skattebøl, B. Boulette and S. Solomon, *J. Org. Chem.* **33**, 548 (1968).
406. M. M. Krayushkin, M. A. Kalik, V. K. Zav'yolova, A. A. Loktionov and V. S. Bogdanv, *Chem. Heterocyl. Compds. USSR* **25**, 1349 (1989).
407. M. M. Krayushkin, M. A. Kalik and A. A. Loktionov, *Chem. Heterocycl. Compds.* USSR **26**, 756 (1990).

408. F. Kipnis and J. Ornfelt, *J. Am. Chem. Soc.* **71**, 2270 (1949).
409. B. P. Fedorov and F. M. Stoyanovich, *J. Gen. Chem., USSR* **32**, 1504 (1962).
410. B. P. Fedorov and F. M. Stoyanovich, *J. Gen. Chem., USSR* **31**, 222 (1961).
411. G. D. Hartman and W. Halczenko, *J. Heterocycl. Chem.* **27**, 127 (1990).
412. T. M. Williams, R. J. Hudcosky, C. A. Hunt and K. L. Shepard, *J. Heterocycl. Chem.* **28**, 13 (1991).
413. D. Spinelli, G. Consiglio and A. Carrao, *J. Chem. Soc., Perkin Trans. 2* 1866 (1972).
414. A. Jurá, J. Kovác, H. Krutosiková and M. Hrdina, *Collect Czech. Chem. Commun.* **37**, 3144 (1972).
415. P. Hurtel, B. Decroix, J. Morel and F. Terrier, *J. Chem. Res. (S)* 58 (1983).
416. D. Spinelli and G. Consiglio, *J. Chem. Soc., Perkin Trans. 2* 989 (1975).
417. D. Spinelli and G. Consiglio, *J. Chem. Soc., Perkin Trans. 2* 1388 (1975).
418. C. Arnone, G. Consiglio, S. Gronowitz, B. Maltesson, A.-B. Hörnfeldt, R. Noto and D. Spinelli, *Chem. Scripta* **13**, 130 (1978–79).
419. G. Consiglio, S. Gronowitz, A.-B. Hörnfeldt, B. Maltesson, R. Noto and D. Spinelli, *Chem. Scripta* **11**, 175 (1977).
420. Ya. L. Gol'dfarb, F. M. Stoyanovich, G. B. Chermanova and E. D. Lubuzh, *Bull. Acad. Sci. USSR, Div. Chem. Sci.* 2462, (1977).
421. S.-S. P. Chou, D.-J. Sun, J.-Y. Huang, P.-K. Yang and H.-C. Lin, *Tetrahedron Lett.* **37**, 7279 (1996).
422. I. Shinkai, *J. Heterocycl. Chem.* **29**, 6627 (1992).
423. S. Braverman, P. F. T. M. van Asten, J. B. van der Linden and B. Zwanenburg, *Tetrahedron Lett.* **32**, 3867 (1991).
424. J. B. van der Linden, P. F. T. M. van Asten, S. Braverman and B. Zwanenburg, *Rec. Trav. Chim. Pays-Bas* **114**, 51 (1995).
425. S. M. Sherif and A. M. Hussein, *Monatsh. Chem.* **128**, 687 (1997).
426. C. E. Stephens and J. W. Sowell, Sr, *J. Heterocycl. Chem.* **35**, 927 (1998).
427. C. E. Stephens and W. Sowell, Jr, *J. Heterocycl. Chem.* **35**, 933 (1998).
428. C. E. Stephens, M. B. Price and J. W. Sowell, Sr, *J. Heterocycl. Chem.* **36**, 659 (1999).
429. W. Steinkopf and W. Ohse, *Lieb. Ann.* **437**, 14 (1924).
430. W. Steinkopf and T. Höpner, *Lieb. Ann.* **501**, 174 (1924).
431. A. P. Terentyev and L. A. Kazitsina, *Proc. Acad. Sci. USSR* **55**, 631 (1947).
432. A. P. Terentyev and L. A. Kazitsina, *J. Gen. Chem. USSR* **18**, 723 (1948).
433. A. P. Terentyev and L. A. Kazitsina, *J. Gen. Chem. USSR* **19**, 531 (1949).
434. A.-P. Terentyev and G. M. Kadatsky, *J. Gen. Chem. USSR* **22**, 189 (1952).
435. A. Arcoria, E. Maccarone, G. Musumarra and G. A. Tomaselli, *J. Org. Chem.* **39**, 1689 (1974).
436. F. Challenger, S. A. Miller and G. M. Gibson, *J. Chem. Soc.* 769 (1948).
437. J. Langer, *Bericht.* **18**, 1114 (1883).
438. A. Buzas and J. Teste, *Bull. Soc. Chim. Fr.* 793 (1960).
439. W. E. Truce and M. F. Amos, *J. Am. Chem. Soc.* **73**, 3013 (1951).
440. A.-P. Terentyev and G. M. Kadatsky, *J. Gen. Chem. USSR* **23**, 263 (1954).
441. Y. L. Gol'dfarb, I. V. Antik and P. A. Konstantinov, *Izv. Akad. SSSR, Otdel. Khim. Nauk* 624 (1956).
442. G. D. Hartman and W. Halczenko, *J. Heterocycl. Chem.* **26**, 1793 (1989).
443. G. D. Hartman, W. Halczenko, R. L. Smith, M. F. Dugrue, P. J. Malergo, S. R. Michelson, W. C. Randall, H. Achwam and J. M. Sondey, *J. Med. Chem.* **35**, 3822 (1992).
444. C. D. Hurd and H. M. Priestley, *J. Am. Chem. Soc.* **69**, 859 (1947).
445. H. Scheibler, E. Keintzel and K. Falk, *Chem. Ber.* **87**, 1184 (1954).
446. A. Arcoria, F. P. Ballistreri, G. Musumarra and G. A. Tomaselli, *J. Chem. Soc., Perkin Trans. 2*, 221 (1981).
447. E. Maccarone, G. Musumarra, G. A. Tomaselli, *J. Org. Chem.* **39**, 3286 (1974).

448. W. E. Truce, B. Vangemert and W. W. Brand, *J. Org. Chem.* **43**, 101 (1978).
449. R. J. Cremlyn, K. H. Goulding, F. J. Swinbourne and K.-M. Yung, *Phosphorus Sulfur* **10**, 111 (1981).
450. C. A. Obafemi, *Phosphorus and Sulfur*, **8**, 197 (1980).
451. C. Sone and Y. Matsuki, *Nippon Kagaku Zasshi* **83**, 496 (1962).
452. G. deStevens, A. Halamandaris, S. Ricca Jr and L. H. Werner, *J. Med. Pharm. Chem.* **1**, 565 (1959).
453. V. S. Yakovlev and N. I. Putokhin, *Dokl. Akad. Nauk SSSR* **96**, 539 (1954).
454. C. D. Hurd and J. Moffat, *J. Am. Chem. Soc.* **73**, 613 (1951).
455. J. Cymerman and D. Faiers, *J. Chem. Soc.* 165 (1952).
456. M. Lora-Tamayo, R. Madronero, J. del Rio and G. Alonso, *An. R. Espan. Fis. Quim. (Madrid, Ser. B.)* **62**, 187 (1966).
457. W. O. Foye, H. M. Kosak and J. Hefferren, *J. Am. Pharm. Assoc.* **XLI**, 273 (1952).
458. R. Cremlyn, L. Ellis and A. Pinney, *Phosphorus, Sulfur and Silicon* **44**, 167 (1989).
459. A. P. Dantanarayana, B. DuPre, J. A. May and V. M. Lynch, *J. Heterocycl. Chem.* **36**, 65 (1999).
460. P. A. Rossy, W. Hoffman and N. Müller, *J. Org. Chem.* **45**, 617 (1980).
461. B. Unterhalt, F. Bodinka and F. Brunisch, *Arch. Pharm. (Weinheim)* **320**, 854 (1987).
462. B. Unterhalt and F. Bodinka, *Arch. Pharm. (Weinheim)* **320**, 1181 (1987).
463. R. A. Earl and R. E. Harmon, *J. Org. Chem.* **9**, 776 (1966).
464. R. N. Boyd and R. H. Hansen, *J. Am. Chem. Soc*, **75**, 3737 (1953).
465. W. Steinkopf, H. Jacob and H. Penz, *Lieb. Ann.* **512**, 136 (1934).
466. E. Profft and A. Kubat, *Lieb. Ann.* **634**, 185 (1960).
467. R. H. Cundiff and R. R. Estes, *J. Am. Chem. Soc.* **72**, 1424 (1950).
468. A. P. Terentyev and G. M. Kadatsky, *J. Gen. Chem. USSR* **21**, 1667 (1951).
469. J.-M. Bastian, A. Ebnnöther, E. Jucker, E. Rissi and A. P. Stoll, *Helv. Chim. Acta* **49**, 214 (1966).
470. J. Sicé and M. Mednick, *J. Am. Chem. Soc.* **75**, 1628 (1953).
471. J. Bulkacz, M. A. Apple, J. C. Craig and A. R. Naik, *J. Pharm. Sci.* **57**, 1017 (1968).
472. A. Buzas, J. Frossard and J. Teste, *Arch. Pharm. Fr.* **19**, 449 (1961).
473. G. D. Hartman, W. Halczenko, R. L. Smith, M. F. Sugrue, P. J. Mallorga, S. R. Michelson, W. C. Randall, H. Schwam and J. M. Sonday, *J. Med. Chem.* **35**, 3822 (1992).
474. S. Vega, R. Madronero, J. A. Diaz, F. Junquera, L. Alonso, V. Darias, L. Bravo and S. Abdala, *Eur. J. Med. Chem.* **23**, 329 (1988).
475. A. Arcoria, E. Maccarone, G. Musumarra and G. A. Tomaselli, *J. Org. Chem.* **38**, 2457 (1973).
476. H. Scheibler and K. Falk, *Chem. Ber.* **87**, 1186 (1954).
477. S. Graham and T. H. Scholz, *J. Org. Chem.* **56**, 4260 (1991).
478. D. W. Slocum and P. L. Gierer, *J. Org. Chem.* **38**, 4189 (1973).
479. W. Stoll, H. Dietrich, J. R. Geygy and A.-G. Ger, 1,139,507; *Chem. Abstr.* **68**, 10173a (1963).
480. M. M. El-Kerdawy and H. A. Selim, *J. Drug Res.* **4**, 135 (1972).
481. M. M. El-Kerdawy and H. A. Selim, *J. Drug Res.* **5**, 135 (1973).
482. J. H. Bowie, D. H. Williams, S.-O. Lawesson, J. Ø. Madsen, C. Nolde and G. Schroll, *Tetrahedron* **22**, 3515 (1966).
483. C. A. Obafemi, *Phosphorus and Sulfur* **13**, 119 (1982).
484. A. A. Munshi and J. P. Trivedi, *J. Indian Chem. Soc.* **40**, 1039 (1963).
485. R. J. Cremlyn, F. J. Swinbourne and K. M. Yung, *J. Heterocycl. Chem.* **18**, 997 (1981).
486. A. A. Munshi, N. M. Shah and J. P. Trivedi, *Indian J. Chem.* **1**, 318 (1963).
487. W. O. Foye, J. M. Kauffman, J. J. Lanzillo and E. F. LaSala, *J. Pharm. Sci.* **64**, 1371 (1975).
488. K. Dickore and E. Kuehle, *Angew. Chem.* **77**, 429 (1965).
489. Ya. L. Gol'dfarb and V. P. Litvinov, *Bull. Acad Sci. USSR, Div. Chem. Sci.* 1987 (1964).

490. O. S. Chizkov, B. M. Zolotarev, A. N. Sukiasyan, V. P. Litvinov and Ya. L. Gol'dfarb, *Org. Mass. Spectrometri* **3**, 1379 (1970).
491. V. P. Litvinov, Ya. L. Gol'dfarb, V. S. Bogdanov, I. P. Konjaeva and A. N. Sukiasyan *J. Prakt. Chem.* **315**, 850 (1973).
492. Ya. L. Gol'dfarb, V. P. Litvinov and A.N. Sukiasyan, *Bull. Acad. Sci. USSR, Div. Chem. Sci.* 1199 (1971).
493. A. N. Sukiasyan, V. P. Litvinov and Ya. L. Gol'dfarb, *Bull. Acad. Sci. USSR, Div. Chem. Sci.* 1270 (1970).
494. Ya. L. Gol'dfarb, V. P. Litvinov and A. N. Sukiasyan, *Bull. Acad. Sci. USSR, Div. Chem. Sci.* 2470 (1967).
495. N. A. Korchevin, É. N. Sukhomazova, N. V. Russovskaya, L. P. Turcahninova, M. V. Sigalov, L. V. Kyba, É. N. Deryagina and M. G. Voronkov, *Chem. Heterocycl. Compds.* **27,** 1049 (1991).
496. M. Tiecco, L. Testaferri, M. Tingoli, F. Marini and S. Marrigio, *Tetrahedron* **50**, 10549 (1994).
497. M. Tiecco, L. Testaferri, L. Bagnoli, F. Marini, A. Temperini, C. Tomassini and C. Santi, *Tetrahedron* **56**, 3255 (2000).
498. V. P. Lltvlnov, A. N. Sukiasyan, Ya. L. Gol'dfarb and L. V. Bogacheva, *Bull. Acad. Sci. USSR, Div. Chem. Sci.* 1498 (1971).
499. L. Engman and P. Cava, *Organometallics* **1**, 470 (1982).
500. M. Renson and N. Dereux, *J. Organometal. Chem.* **258**, 163 (1983).
501. C. Kollemann, D. Obendorf and F. Sladky, *Phosphorus and Sulfur* **38**, 69 (1988).

– 7 –

Syntheses of Thiophenes with Group VII Substituents

7.1 FLUORINE DERIVATIVES

7.1.1 Direct fluorination of thiophenes

The reaction of thiophene in chloroform with a dilute (5% v/v) solution of elemental fluorine in helium at −63 °C gives a mixture of 2-fluorothiophene (66 ± 3%) and 3-fluorothiophene (34 ± 3%) and is thus the least selective of the electrophilic halogenation reactions [1].

7.1.2 From Thienyllithium Derivatives

The best method for the preparation of fluorothiophenes consists in the reaction of thienyllithium derivatives prepared by metalation of thiophenes (2-substituted derivatives) or halogen–metal exchange (3-substituted derivatives) with perchloryl fluoride [2–6].

General procedure for the reaction of thienyl derivatives with perchloryl fluoride [2]

Ethereal ethyllithium solution is placed in a nitrogen swept wide-neck reaction vessel with a five socket flat flange lid, fitted with stirrer, a thermometer, nitrogen inlet, perchloryl fluoride inlet, and a reflux condenser protected with a calcium chloride tube. The thiophene to be metalated is diluted with anhydrous ether and added at such a rate that gentle reflux is maintained. The mixture is then refluxed for 20 min, cooled to −15 °C and perchloryl fluoride bubbled into the reaction mixture at a moderate rate, while the temperature is kept at −15 °C. When 20% excess of perchloryl fluoride has been bubbled into the solution, the cooling bath is removed and nitrogen is bubbled through for 60 min. The reaction mixture is then poured into saturated sodium carbonate solution, the organic layer separated and washed once with soda solution. Most of the ether is distilled and the residue is steam distilled. The distillate is extracted with ether, dried and fractionated through an efficient column.

When halogen–metal exchange is performed, the ethyllithium solution is cooled to $-40\,^{\circ}C$ before the appropriate bromo derivative is added and perchloryl fluoride is introduced at the same temperature [4].

In this way all four isomeric difluorothiophenes are obtained [4]. Also 2,3,5-trifluorothiophene can be prepared through halogen–metal exchange of 3-bromo-2,5-difluorothiophene followed by reaction with perchloryl fluoride [4]. The reaction is very general. However, unintentional hydrolysis of the intermediate lithium derivatives gives thiophenes as byproducts, which in many cases can be separated from the desired fluorothiophenes only by preparative gas liquid chromatography. In contrast to the benzene series, formation of thienyl chlorates has not been observed. The drawback with this method is the explosive nature of the perchloryl fluoride, which has made this difficult to obtain from commercial sources.

During recent years other electrophilic fluorinating reagents have been developed (for review see Ref. [7]). 2-Fluorothiophene has been obtained in low yield through reaction with *N*-fluoroquinuclidinium fluoride [8].

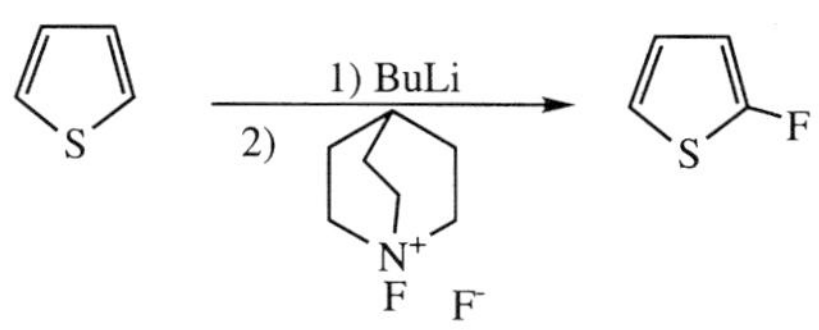

The best reagent appears to be *N*-fluoro-*N*-(phenylsulfonyl)benzenesulfonamide. The reaction of 3,4-dibromo-2,5-bis(trimethylsilyl)thiophene with butyllithium followed by the above-mentioned fluorinating reagent provided a mixture of the starting dibromo derivative and 3,4-difluoro-2,5-bis(trimethylsilyl)thiophene. Further lithiation and fluorination were repeated, without isolation of the mixture, giving the 3,4-difluorothiophene derivative [9].

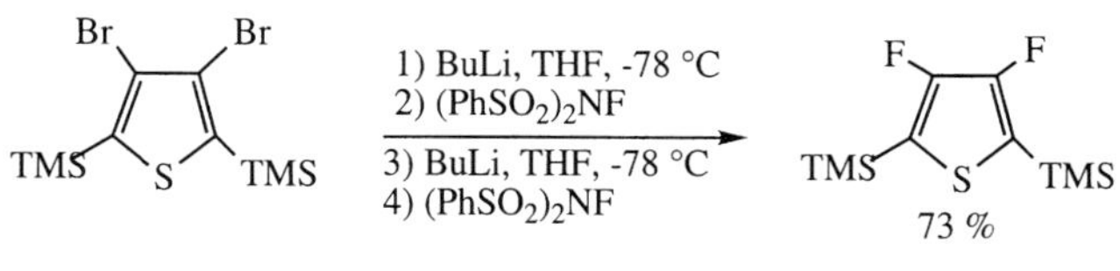

7.1.3 Nucleophilic substitution of halothiophenes

Direct substitution of 2-iodothiophene with antimony trifluoride was the first method used for the preparation of 2-fluorothiophene [10]. However, the yield was low. The reaction of tetrafluorothiophene with sodium methoxide is a method for the preparation of 2-methoxy-3,4,5-trifluorothiophene [11–13].

7.1.4 From fluorothiophenes

7.1.4.1 By electrophilic substitution

Nitration of 2-fluorothiophene and 3-fluorothiophene with nitric acid in acetic anhydride are good methods for the preparation of 2-fluoro-5-nitrothiophene [5,14] and 3-fluoro-2-nitrothiophene [5].

Nitration of 2- and 3-fluorothiophene [5]

The mixture obtained upon fluorination (10 g) is dissolved in acetic anhydride (34 ml). At 0 °C fuming nitric acid (8 g) in acetic acid (60 ml) is added dropwise during 45 min. The reaction mixture is stirred at 0 °C for 24 h and then poured into ice. 2-Fluoro-5-nitrothiophene is obtained as a red-yellow oil, which is separated and the aqueous phase is extracted with ether. The combined organic phases are washed with sodium hydroxide solution and water and dried over magnesium sulfate. Fractionation yields 9.1 g of the product bp 50–60 °C/ 1 mm Hg. 3-Fluoro-2-nitrothiophene is obtained as crystals, which are filtered off, washed with ice-water and dried, yielding 5.0 g, mp 58 °C, after recrystallization from petroleum ether.

Vilsmeyer formylation of 2-fluorothiophene and 3-fluorothiophene is a good method for the preparation of 5-fluoro-2-thiophene aldehyde and 3-fluoro-2-thiophene aldehyde, respectively [5]. Vilsmeyer formylation of methyl 2-fluoro-4-thienylthioacetate can be used for the preparation of 2-fluoro-5-formyl-4-thienylthioacetate [15]. Friedel-Crafts acylations of 2- and 3-fluorothiophene with acetyl chloride using tin tetrachloride as catalyst leads to 2-acetyl-5-fluoro- and 2-acetyl-3-fluorothiophene, respectively [5].

7.1.4.2 Metalation and halogen–metal exchange

Since fluorine does not interfere in the metalation or in the halogen–metal exchange reactions, an alternative route to prepare substituted 2- and 3-fluorothiophenes is possible. The substituents so introduced can be modified [5].

4-(1-Cyclohexenyl)-2-fluorothiophene [5]

To 0.85 *M* butyllithium (130 ml) cooled to −70 °C the mixture, obtained upon fluorination of 2,4-dibromothiophene (18.1 g), is added in a slow stream with vigorous stirring. After 5 min cyclohexanone (9.8 g, 1.10 mol) in anhydrous ether (25 ml) is added in a slow stream and the reaction mixture stirred at −70 °C for 1 h. The cooling bath is removed and the stirring continued for 4 h. After cooling to −10 °C 6 *M* hydrochloric acid is added dropwise. The phases are separated and the water phase extracted with ether. The combined ether phases are washed with water and dried over magnesium sulfate. Distillation yields 13.4 g of the product bp 121–123 °C/12 mm Hg, which crystallizes in the condenser mp 34–35 °C.

2-Fluoro-4-phenylthiophene [5]

To a solution of 2,3-dichloro-5,6-dicyano-1,4-benzoquinone (22.7 g, 0.10 mol) in toluene (200 ml) crude 4-(1-cyclohexenyl)-2-fluorothiophene (8.9 g) in toluene (20 ml) is added, after which the reaction mixture is refluxed for 2 h. After cooling, the precipitate formed is filtered off and the organic phase is washed several times with sodium hydroxide solution followed by 1 *M* hydrochloric acid, sodium bicarbonate solution and water. After drying over magnesium sulfate the solvent was evaporated and the crystalline residue was separated on a chromatographic column of alumina using hexane as eluent. Recrystallization of the product so obtained gives 5.5 g (64%) of colourless crystals mp 91–92 °C.

Bromination of 3,4-difluoro-2,5-bis(trimethylsilyl)thiophene with one equivalent of *N*-bromosuccinimide in acetic acid at 80 °C gave 2-bromo-3,4-difluoro-5-trimetylsilylthiophene which upon halogen–metal exchange followed by reaction with tributyltin chloride gave 3,4-fluoro-2-tributylstannyl-5-trimethylsilylthiophene [9]. A one-pot procedure was used for the preparation of 2,3,4-trifluoro-5-tributylstannylthiophene in 85% yield by the reaction of 2,5-dibromo-3,4-difluorothiophene with two equivalents of butyllithium followed by *N*-fluoro-*N*-(phenylsulsulfonyl)benzenesulfonamide and tributyltin chloride [9].

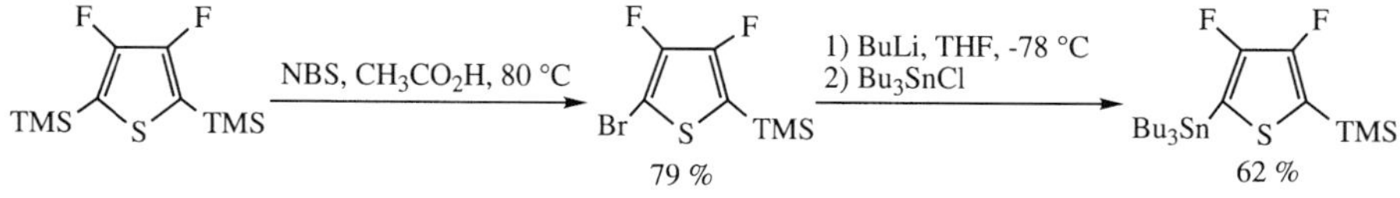

7.1.5 Other methods

The claim that methyl 3-fluoro-2-thiophene carboxylate is obtained by a Schiemann reaction of the 3-diazonium salts in xylene [16] is not correct. Instead Gomberg coupling products with *ortho*-xylene were formed. However, carrying out the Schiemann reaction in sand gave a 30% yield of the desired methyl 3-fluoro-2-thiophenecarboxylate, which could be transformed to the hydrazide [17].

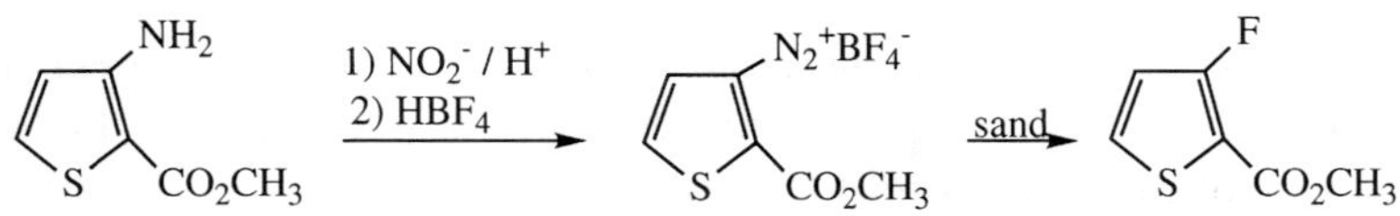

Methyl 3-fluorothiophene-2-carboxylate [17]

Methyl 3-amino-2-thiophenecarboxylate (9.4 g, 60 mmol) is added gradually to a vigorously stirred 6 *M* hydrochloric acid solution (25 ml). The reaction mixture is stirred at room temperature for 30 min and then cooled to 0 °C, after which sodium nitrite (4.2 g, 60 mmol) in water (10 ml) is added. The resulting diazonium salt is stirred at 0 °C for 1 h and then treated with a 36.6% tetrafluoroboric acid solution (18 ml). The solid diazonium salt is filtered, washed with 6 *M* hydrochloric acid, methanol and ether and dried *in vacuo* over sulfuric acid, giving methyl 3-diazothiophene-2-carboxylate tetrafluoroborate (68%) mp 142–143 °C. A mixture of methyl 3-diazothiophene tetrafluoroborate (7.68 g, 0.03 mol) and 40 g sand is heated in a round-bottomed flask at 160 °C under vacuum distillation conditions (1-5-1 mm Hg). When the head thermometer shows 75–80 °C a pale-yellow liquid distills and solidifies in the condenser. Distillation is stopped when a dark-yellow liquid begins to distill. The solid is then scraped out of the condenser and allowed to remain in the hood for a few hours to get rid of boron trifluoride gas. This compound is then recrystallized from methanol/water to give 1.54 g (32%) of the title compound mp 51–53 °C.

Direct electrophilic fluorination of 3-acetylamino- and 2-acetylaminothiophene to 3-acetylamino-2-fluoro- and 2-acetylamino-3-fluorothiophene was achieved with Selectfluor^{TM} [18], albeit in very low yields [17]. When this reaction was performed with ethyl 2-amino-3-thiophenecarboxylate, the diazonium salt coupled with the starting material, giving the azo compound [16].

Polyfluorothiophenes can be prepared from polyfluorothiolanes. Thus tetrafluorothiophene, 2,5-difluorothiophene and 3-methoxy-2,4,5-trifluorothiophene were obtained in low yield by chlorination–dechlorination reaction of fluorochlorothiolenes [11–13]. Some fluorothiophenes have also been prepared by ring-closure of aliphatic compounds or polyfluorothiolanes.

4-Fluoro-2-thiophenecarboxylic acids such as 3-methyl-, 3,5-dimethyl-, 5-methyl- and 5-methyl-3-phenyl-4-fluoro-2-thiophenecarboxylic acid are prepared by the reaction between two equivalents of methyl thioglycolate anion and α-fluoro-β-(phenylthio)enones (or -enals) in dimethylsulfoxide at 70 °C in yields ranging between 41 and 85% [19].

7.2 CHLORINE DERIVATIVES

7.2.1 Direct chlorination

The reaction of thiophene with chlorine in the absence of solvents and catalysts is complicated and leads to a mixture of substitution and addition products, depending on reaction and work-up conditions [20–22]. The ratio of chlorine to thiophene in the direct chlorination is important for the direction of the chlorination, giving mainly 2-chloro- [21], 2,5-dichloro- [21], 2,3,5- trichloro- [21] and tetrachlorothiophene [21]. These procedures are useful when large amounts of these compounds are desired and good equipment for fractional distillation is available. Some old controversies in the literature with regard to the structures of trichlorothiophenes have recently been elucidated, and it was found that the best method for the preparation of 2,3,5-trichlorothiophene is the direct chlorination of thiophene in presence of catalytic amounts of ferric chloride [23].

2,3,5-Trichlorothiophene and tetrachlorothiophene [23]

Chlorine (400–450 g) is bubbled to thiophene (42 g, 0.5 mol) under magnetic stirring. The temperature is maintained at 50 °C by means of a water bath for 8.5 h. A vigorous stream of nitrogen is then bubbled through the reaction mixture for 20 min. The reaction mixture is diluted with 75 ml of anhydrous carbon tetrachloride and ferric chloride (0.81 g, 5.00 mmol) is added. The stirring is continued at 35–40 °C for 3 h, whereupon nitrogen is bubbled through the reaction mixture for 10 min. The reaction mixture is filtered, washed with two 100 ml portions of water, dried over magnesium sulfate and evaporated. The residue is distilled *in vacuo*.

2,3,5-Trichlorothiophene (21.6 g, 23%) bp 77–79 °C/8 mm Hg. ^{13}C NMR ($CDCl_3$): δ 121.3 (d, $J = 179.6$ Hz), 123.0, 126.0, 126.8.

Tetrachlorothiophene (53.3 g, 48%) bp 61–64 °C/1.2 mm Hg, mp 29.0–30.0 °C (methanol).

Also thiophenes with more complicated side chains can be chlorinated by elemental chlorine in chloroform at room temperature, as is demonstrated by the recent preparation of 3-trifluoroacetylamino-3-(2,5-dichlorothien-3-yl) propionic acid from the parent trifluoroacetylaminothienylpropionic acid [24].

3-Trifluoroacetylamino-3-(2,5-dichlorothien-3-yl)propionic acid [24]

A solution of trifluoroacetylaminothienylpropionic acid (5.0 g, 19 mmol) in chloroform (100 ml) is bubbled for 10 min at room temperature with a chlorine flow. The reaction mixture is then refluxed for 15 min and left at room temperature. The precipate formed is filtered off giving the title compound (4.2 g, 70%) mp 162 °C (diethyl ether).

The greater reactivity of thiophene when compared to benzene in electrophilic substitution makes possible the direct chlorination in acetic acid of 3-thiophenecarboxylic acid, to 2-chloro-4-thiophenecarboxylic acid and upon further chlorination 2,5-dichloro-3-thiophenecarboxylic acid is obtained [25]. With strongly deactivated thiophenes aluminium chloride as catalyst is necessary. 2-Nitrothiophene gave in the presence of 0.6 mol of aluminium chloride, 4,5-dichloro-2-nitrothiophene in quantitative yields, via a mixture of two monochloro nitro derivatives [26]. 3-Nitrothiophene was much more reactive and in the presence of 0.0075 mol of aluminium chloride, 3-nitro-2,4,5-trichlorothiophene was obtained in 93% yield [27]. Different isomer distributions can be obtained, with carbonyl derivatives using catalytic amounts of aluminium chloride or excess aluminium chloride (swamping catalyst conditions). Thus 2-thiophene aldehyde gives 4-chloro-2-thiophene aldehyde under the latter conditions in good yield [28,29]. Thus from 2-chloro-3-thiophene aldehyde, 2,5-dichloro-3-thiophene aldehyde can be prepared [29], but ethyl 2-thiophenecarboxylate gives a less satisfactory yield of ethyl 4-chloro-2-thiophenecarboxylate [30]. 3-Thiophene aldehyde gives 2,3-dichloro-3-thiophene aldehyde under swamping catalyst conditions [31]. 5-Chloro-2-(α-alkoxyamino)-ethylthiophene is obtained from 2-(α-alkoxyamino)ethylthiophene upon reaction with chlorine in the presence of iron [32].

Instead of chlorine, crystalline benzyltrimethylammonium tetrachloro iodate, which can be easily prepared by bubbling chlorine gas into a mixture of benzyl trimethylammonium chloride (1 mol) and iodine (0.5 mol) in dichloromethane, can also be used [33]. Thus 2,4,5-trichloro-3-methylthiophene can be obtained from 3-methylthiophene [34].

3-Methyl-2,4,5-trichlorothiophene [34]

Benzyltrimethylammonium tetrachloro iodate (6.5 g, 15.5 mmol) is added to a solution of 3-methylthiophene (0.49 g, 5.00 mmol) in acetic acid (50 ml) and the reaction mixture is stirred at 70 °C for 24 h. During this time the color of the solution turns black and a yellow precipitate is formed. The reaction mixture is cooled to room temperature and the precipitate is filtered off. The filtrate is concentrated *in vacuo* and the residue is treated with 5% sodium bisulfite (10 ml) and 5% sodium bicarbonate (10 ml) and then extracted with dichloromethane (3 × 50 ml). The combined black organic phases are dried over magnesium sulfate. The traces of acetic acid and the impurities are removed by column chromatography on alumina. Concentration of the eluate gives 0.64 g (63%) of the title compound as a colorless oil bp 219 °C.

7.2.2 Chlorination with sulfuryl chloride

Sulfuryl chloride chlorinates thiophene and has been considered the best reagent for the preparation of 2-chloro- and 2,5-dichlorothiophene [35]. By this method 2-benzyl-5-chlorothiophene can be prepared from 2-benzylthiophene [36].

2-Benzyl-5-chlorothiophene [36]

Sulfuryl chloride (7.8 g) is added to 2-benzyl thiophenes (10 g) over 30 min, with constant stirring, at room temperature. A vigorous evolution of gas is observed and the stirring is continued until the gas evolution ceases, after which it is taken up in dichloromethane (50 ml). The dichloromethane solution is washed with water (2 × 25 ml), and saturated aqueous sodium hydrogen carbonate (25 ml) and dried over magnesium sulfate. After evaporation of the solvent the residue is dried over calcium hydride before distillation, giving the title compound (8.5 g, 71% bp 94–96 °C/0.2 mm Hg).

2,3,5-Trichlorothiophene can be conveniently obtained in good yield through chlorination with thionyl chloride and sulfuryl chloride, using aluminium chloride as catalyst [23].

The reaction of methyl 3-hydroxy-2-thiophenecarboxylate with sulfuryl chloride in chloroform or *N*-chlorosuccinimide in acetic acid yields 2-chloro-2-methoxycarbonyl-3-oxo-2,3-dihydrothiophene, which upon treatment with hydrochloric acid gives 5-chloro-3-hydroxy-2-methoxycarbonylthiophene [37].

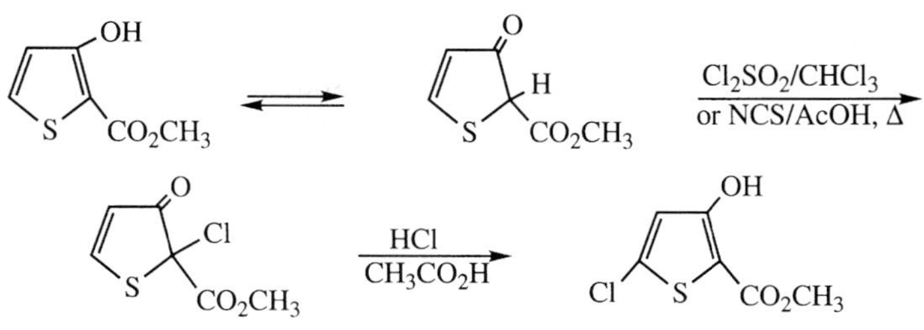

2-Chloro-2-methoxycarbonyl-3-oxo-2,3-dihydrothiophene [37]

Sulfuryl chloride (8.9 ml, 0.11 mol) is added to a stirred solution of 3-hydroxy-2-methoxycarbonylthiophene (15.8 g, 0.10 mol) in anhydrous chloroform (50 ml). There is a rapid evolution of gas. The reaction mixture is allowed to stand at room temperature for 4 h with protection from moisture. The solution is then evaporated to dryness at reduced pressure and the solid residue is used as crude product in the next step.

5-Chloro-3-hydroxy-2-methoxycarbonylthiophene [37]

Hydrogen chloride is bubbled into the crude product (5.8 g, 0.03 mol) in acetic acid (20 ml) until saturation. The reaction mixture is allowed to stand for 3 days, and the acetic acid is evaporated. The oily residue is distilled giving the title compound as the main fraction (3.99 g) bp 69 °C/0.5 mm Hg. The yellow liquid so obtained crystallizes spontaneously mp 42–44 °C (acetic acid/water). The overall yield is 69%.

These tandem reactions can be applied to 4- and 5-alkyl or aryl derivatives with minor modifications. However, the second step takes another route [38].

7.2.3 Chlorination with *N*-chlorosuccinimide

The experience of Gronowitz and co-workers is, however, that the use of *N*-chlorosuccinimide in acetic acid is the reagent of choice for the chlorination of many thiophene derivaives. It can be used for the preparation of 2-chloro-2,5-dichlorothiophene [39] and 2,3,5-trichlorothiophene [40], 3-acetamido-2-chloro- and 3-acetamido-2,5-dichlorothiophene [41], 2-chloro-5-methoxythiophene [42] 3-chloro-2,5-dimethoxythiophene [43]. 3,4-Di-*tert*-butyl-2,5-dichlorothiophene can be prepared in 91% yield using *N*-chlorosuccinimide in excess in acetic acid/dichloromethane under reflux [44].

2-Chloro-5-methoxythiophene [42]

To 2-methoxythiophene (16.0 g, 0.140 mol) in glacial acid (200 ml) *N*-chlorosuccinimide (18.7 g, 0.140 mol) is added in one portion under stirring. After 2 h the red reaction mixture is poured into sodium hydroxide (150 g) in water (1000 ml) at 0 °C. This solution is extracted three times with ether and the combined ether solutions are washed with 20% sodium chloride solution, dried over magnesium sulfate and evaporated. The residue is distilled giving 17.5 g (84%) of the title compound as a light-yellow liquid bp 83–86 °C/48 mm Hg.

The use of *N*-chlorosuccinimide for chlorination is especially useful for the preparation of mixed halothiophenes [35] since *trans* halogenation reactions are minimized [45]. A very useful modification, using biphasic chlorination with *N*-chlorosuccinimide in hexane or carbon tetrachloride and catalytic amounts (.01–1 mol%) of 70% perchloric acid has recently been described. Thiophene gave 90% yield of 2-chlorothiophene with one equivalent of *N*-chlorosuccinimide and 82% of 2,5-dichlorothiophene with two equivalents [46].

7.2.4 Chlorination with hypochlorous acid

Chlorination with hypochlorous acid has been used for the chlorination of thiophene and some alkylthiophenes [47], but appears not to be of the same importance as the methods discussed above, except maybe for the chlorination of thiophenecarboxylic acids either directly [48] or as generated *in situ* through the haloform reaction of the corresponding acetylthiophene [48,49].

7.2.5 From thienyllithium derivatives

Chlorothiophenes not available through direct chlorinations can conveniently be prepared through the reaction of thienyllithium derivatives especially with hexachloroethane [6,43,50–52].

4-Bromo-2-chlorothiophene [6]

To a solution of 2,4-dibromothiophene (97.6 g, 0.40 mol) in anhydrous diethyl ether under nitrogen at −70 °C 1.64 *M* butyllithium in hexane (244 ml) is added dropwise over a period of 1 h. A solution of hexachloroethane (95.0 g, 0.40 mol) in anhydrous diethyl ether (300 ml) is then added during 1 h. The stirring is continued for 0.5 h at −70 °C then the reaction mixture is allowed to reach room temperature and poured into water. After neutralization with 5 *M* hydrochloric acid the phases are separated and the combined organic phases are washed with water, dried over magnesium sulfate, evaporated and distilled bp 191–195 °C. The distilled product contains traces of hydrochloric acid and is purified before further use by dissolving in diethyl ether. This solution is washed with aqueous sodium bicarbonate solution, dried over magesium sulfate and evaporated giving 46.2 g (59%) of the title compound.

In some cases also chlorine [5,53] and trichloroacetonitrile [54] were used in the reaction with thienyllithia. 3-Chlorothiophene has also been obtained by deomposing 3-thienyl iodonium salts in *N*,*N*-dimethylformamide in the presence of copper sulfate [50] although the yield was only 53%.

7.2.6 From chlorothiophenes by metalation, halogen–metal exchange and electrophilic substitution

Like fluorothiophenes, many chlorothiophenes are best prepared by metalation or halogen–metal exchange of appropriate chlorothiophenes or by electrophilic substitution followed by reaction with electrophiles.

7.2.6.1 Metalation and halogen–metal exchange of chlorothiophenes

A recent example is the preparation 1-(3-chloro-2-thienyl)-2-butanol in good yield through metalation of 3-chlorothiophene followed by reaction with butylene oxide [55].

1-(3-Chloro-2-thienyl)-2-butanol [55]

Butyllithium (1.6 *M* solution in hexanes, 44.8 ml, 71.7 mmol) is added to a stirred solution of 3-chlorothiophene (8.50 g, 71.7 mmol) in anhydrous tetrahydrofuran (100 ml) at 0 °C. The solution is stirred for 1.5 h at 0 °C and then butene oxide (5.68 g, 78.9 mmol) was added dropwise. The mixture is stirred at 0 °C for 4 h and then quenched with saturated ammonium chloride (5 ml). Brine (50 ml) is added and the layers are separated. The organic phase is dried over magnesium sulfate, filtered and the solvent is removed *in vacuo* giving 9.40 g (69%) of a light-yellow oil.

The reverse addition of trichloro-2-thienyllithium to methyl dichlorosilane is an excellent method for the preparation of bis(trichloro-2-thienyl)methylsilane [56].

Bis(trichloro-2-thienyl)methylsilane [56]

To a solution of methyldichlorosilane (5.75 g, 0.05 mol) in anhydrous diethyl ether at −30 °C trichloro-2-thienyllithium (0.10 mol) in anhydrous diethyl ether (250 ml) is added. The stirring is continued for 1 h, after which the reaction mixture is filtered while cold to remove lithium chloride. Distillation of the filtrate gives 19.1 g (92%) of the title compound bp 164–165 °C/0.06 mm Hg, mp 55–56.5 °C.

Similarly tris(trichloro-2-thienyl)silane is obtained from trichlorosilane and trichloro-2-thienyllithium [56]. The reaction of 3,4-dichloro-2,5-dilithiothiophene, prepared by reaction of tetrachlorothiophene with a little more than two equivalents of butyllithium, followed by chlorodimethylsilane gives an 82% yield of 3,4-dichloro-2,5-bis(dimethylsilyl)thiophene [57].

2-Chlorothiophene and sodium amalgam sand [58]

A solution of 2-chlorothiophene (118 g, 1.0 mol) in anhydrous diethyl ether (700 ml) is added dropwise with stirring under nitrogen over a period of 2 h to a suspension of sodium amalgam, containing sodium (35 g, 1.5 mol) and mercury (25 g, 0.125 mol), in refluxing anhydrous diethyl ether. When the addition is completed, the reaction mixture is refluxed for 2 h, after which it is cooled to 25 °C and freshly crushed dry ice is added. The temperature goes up momentarily (but is not permitted to rise over 30 °C), and goes down rapidly when the carbonation is completed. Ethanol (100 ml) is added dropwise to destroy the unreacted sodium, after which water (350 ml) is added cautiously. The phases are separated and the aqueous phase is acidified with concentrated hydrochloric acid (174 ml) giving 150 g (92%) of 5-chloro-2-thiophenecarboxylic acid as white needles after recrystallization from water mp 153.0–153.5 °C.

2,5-Dichloro-3,4-diiodothiophene, upon halogen–metal exchange, gives 5-dichloro-3,4-dilithiothiophene, which when reacted with various electrophiles gives sulfur, selenium, tellurium and titanium heterocycles [59].

Halogen–metal exchange of 3-bromo-2-chlorothiophene with butyllithium at −70 °C followed by reaction with dry ice is a good method for the preparation of 2-chloro-3-thiophenecarboxylic acid [60].

2-Chloro-3-thiophenecarboxylic acid [60]

A solution of 3-bromo-2-chlorothiophene (134.3 g, 0.69 mol) in anhydrous ether (200 ml) is added slowly under nitrogen at −70 °C to butyllithium (500 ml 1.42 *M* solution in hexane diluted with 250 ml anhydrous ether). After stirring for 15 min the reaction mixture is poured into crushed carbon dioxide covered with anhydrous ether. At 0 °C saturated sodium bicarbonate solution is added carefully. The phases are separated and the ether phase extracted several times with sodium carbonate solution. The combined aqueous phases are acidified with 5 *M* hydrochloric acid. The precipitate is filtered off and recrystallized from aqueous ethanol (1:1) giving 75.5 g (68%) of the title compound mp 162–163 °C.

Another route used in connection with the preparation of photochromic diarylperfluorocyclopentenes is the preparation of 1,3-di(5-chloro-2-methyl-3-thenoyl) hexafluoropropane, by the reaction of 3-lithio-5-chloro-2-methylthiophene with the ethyl ester of hexafluoroglutaric acid [61].

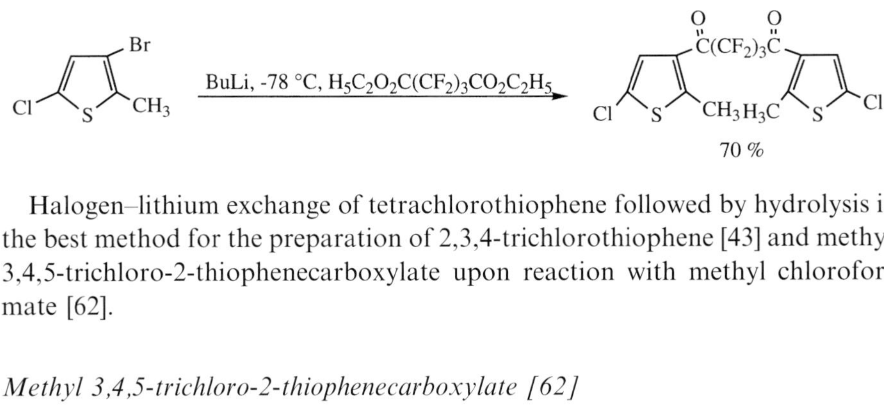

Halogen–lithium exchange of tetrachlorothiophene followed by hydrolysis is the best method for the preparation of 2,3,4-trichlorothiophene [43] and methyl 3,4,5-trichloro-2-thiophenecarboxylate upon reaction with methyl chloroformate [62].

Methyl 3,4,5-trichloro-2-thiophenecarboxylate [62]

To a solution of tetrachlorothiophene (2.22 g, 10 mmol) in anhydrous diethyl ether (75 ml) at −78 °C 2.0 *M* butyllithium in cyclohexane (6.0 ml) diluted with anhydrous diethyl ether is added. The stirring is continued for 1 h after which methyl chloroformate (1.89 g, 20 mmol) in anhydrous diethyl ether is added. After stirring the reaction mixture at −78 °C for 30 min it is allowed to warm to room temperature and poured into saturated ammonium chloride solution. The phases are separated and the aqueous phases extracted with diethyl ether

(2 × 20 ml). The combined organic phases are dried over magnesium sulfate and evaporated giving 1.59 g (65%) of the title compound as colorless crystals mp 74–75 °C after recrystallization from hexane.

Metalation of 2,3,5-trichlorothiophene with lithium diisopropylamide, followed by methyl chloroformate is used for the preparation of methyl 2,4,5-trichloro-3-thiophenecarboxylate. Upon reaction with dimethyl disulfide 3-methylthio-2,4,5-trichlorothiophene is obtained [62].

Also the hydrolysis of trichloro-2-thienylmagnesium chloride prepared through halogen–metal exchange between ethylmagnesium bromide and tetrachlorothiophene gives 2,3,4-trichlorothiophene in excellent yield [63]. Using this approach, trichloro-2-thienyl derivatives of titanocen and niobocene [64], as well as methyl 3,4,5-trichloro-2-thiophenecarboxylate can be prepared [62].

Trichloro-2-thienyllithium prepared through halogen–metal exchange between tetrachlorothiophene and *tert*-butyllithium in tetrahydrofuran at −70 °C followed by reaction with trimethylsilyl chloride is a good method for the preparation of trichloro-2-thienyltrimethylsilane [65].

Halogen–metal exchange of tetrachlorothiophene with butyllithium followed by reaction with butyl borate and oxidation with hydrogen peroxide gave the 2-hydroxy-3,4,5-trichlorothiophene system, existing in the 2,5-dihydrothiophene-2-one form [40].

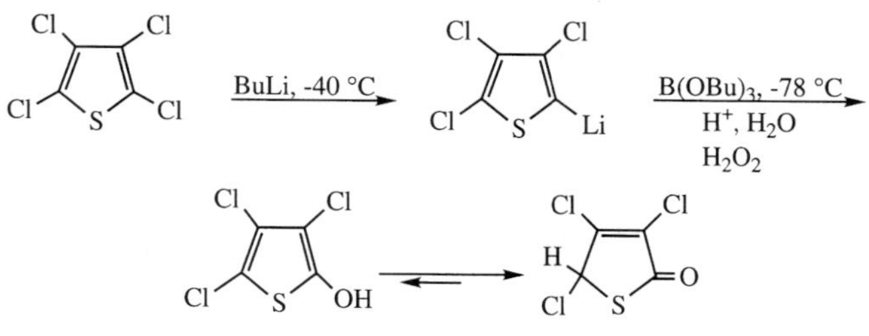

Bromine–lithium exchange of 3-bromo-2,4,5-trichlorothiophene followed by reaction with butyl borate and oxidation gave the 3-hydroxy-2,4,5-trichlorothiophene system, existing in an equilibrium of the hydroxy form and the keto form in the proportion of 1:3.2. In the work up of the trihalogenated hydroxythiophenes, it is important to wash the ethereal solution with water only, as polymerization started immediately if base is added [40].

Reaction of these chlorinated hydroxythiophenes with diazomethane and acetyl chloride, respectively is used for the preparation of the methoxy and acetoxy derivatives [40].

2-Methoxy-3,4,5-trichlorothiophene [40]

Diazomethane is generated by adding *N*-methyl-*N*-nitrosotoluene-*para*-sulfonamide (16.4 g, 76 mmol) in diethyl ether (96 ml) to a solution of potassium hydroxide (3.8 g) in water (6 ml) and ethanol (21 ml) at 65 °C. The generated diazomethane in diethyl ether is distilled into an Erlenmeyer flask cooled in ice. To this flask 2-hydroxy-3,4,5-trichlorothiophene (1.5 g, 7.5 mmol) in diethyl ether (25 ml) is added with a Pasteur pipette. An instantaneous reaction can be observed with the evolution of nitrogen bubbles. The solution is washed with water, dried and evaporated. The residue is purified by sublimation giving 1.51 g (94%) of the title compound as colorless needles mp 45 °C.

The reaction of 3,4-dichloro-2,5-dilithiothiophene with trimethylsilyl chloride gives 3,4-dichloro-2,5-di-(trimethylsilyl)thiophene [66]. The reaction of 3,4,5-trichloro-2-thienylcopper with 4-iodoanisole is used for the preparation of 2-(4-methoxyphenyl)trichlorothiophene [67].

2(4-Methoxyphenyl)trichlorothiophene [67]

A 0.20–0.25 *M* solution of trichloro-2-thienylcopper in diethyl ether (80 ml) and pyridine (80 ml) are mixed in a reaction flask and slowly heated in an oil bath under a stream of nitrogen. When the ether has evaporated and the temperature of the reaction mixture has reached 115 °C, 4-iodoanisole (3.74 g, 16.0 mmol) and 2-methoxynaphthalene (0.20 g) are added. After 130 min at 115 °C the reaction is interrupted by cooling and diethyl ether (300 ml) is added. The aqueous and organic phases are separated; the organic phase is washed with hydrochloric acid and water, dried over sodium sulfate and evaporated. The residue is chromatographed on silica gel giving 3.04 g (65%) of the title compound after recrystallization from ethanol mp 110–111 °C.

7.2.6.2 Electrophilic substitution of chlorothiophenes

Chloromethylation of chlorinated thiophenes is very useful for the preparation of various chloromethylated thiophenes. From 2-chlorothiophene, 2-chloro-5-chloromethylthiophene can conveniently be prepared [68]. 2,5-Dichloro-3-thenyl chloride was obtained by using equivalent amounts of 2,5-dichlorothiophene and chloromethyl methyl ether in carbon disulfide using stannic chloride as catalyst [69].

2,5-Dichloro-3-thenyl chloride [69]

2,5-Dichlorothiophene (26.0 g, 0.17 mol) and chloromethyl methyl ether (16.0 g, 0.20 mol) are dissolved in carbon disulfide (50 ml). The reaction mixture is cooled in an ice–salt bath, after which stannic chloride (15.0 g, 0.06 mol) is added dropwise. After 1 h the cooling bath is removed and the reaction mixture containing two phases is allowed to stand at room temperature for 2 h and then poured into ice water. The phases are separated and the organic phase washed several times with water and dried. The solvent is removed and the residue fractionated *in vacuo* giving 19 g (55%) of the title compound bp 107–110 °C/14 mm Hg.

2,3,5-Trimethyl-4-chloromethylthiophene [70] was prepared through the chloromethylation of 2,3,5-trimethylthiophene with chloromethyl ether in carbon disulfide using tin tetrachloride as catalyst. Recently, albeit in low yield, 2,3,4,5-tetrachloromethylthiophene was prepared from zinc chloride, chloromethyl methyl ether and thiophene in the proportions of 5:1 [71]. Using a three-fold excess of chloromethyl methyl ether in carbon disulfide [72], or an excess of chloromethyl methyl ether as solvent [73], 2,5-dichloro-3,4-dichloromethylthiophene, which is a very useful starting material for 3,4-*c*-fused bicyclic derivatives, is obtained.

2,5-Dichloro-3,4-dichloromethylthiophene [73]

A solution of 2,5-dichlorothiophene (76.5 g, 0.50 mol) in chloromethyl methyl ether (500 ml) is cooled to −5 °C in a bath of an ice–salt mixture. Stannic chloride (130 g, 0.50 mol) is added over 1.5 h at 0–5 °C. Then the reaction mixture is stirred at 0 °C for 1 h, at room temperature for 5.5 h and refluxed for 1 h. The dark solution is cooled to room temperature and poured with stirring into crushed ice (1500 g). The aqueous and organic phases are separated and the organic phase washed with water, dried over sodium sulfate, concentrated and distilled giving 94 g (75%) of the title compound bp 85 °C/0.2 mm Hg, mp 40–41 °C (petroleum ether).

Chloromethylation of 2,3,4-trichlorothiophene gave 2,3,4-trichloro-5-chloromethylthiophene, which was used as starting material for 2-vinyl-3,4,5-trichlorothiophene [74]. Friedel-Craft's acylation of 2,5-dichlorothiophene with acetyl chloride can be used for the preparation of 3-acetyl-2,5-dichlorothiophene [75]. With the acid chloride of the monomethyl ester of succinic acid in chloroform, using aluminium chloride as catalyst, methyl β-(2,5-dichloro-3-thenoyl)propionate is obtained [76].

Methyl β-(2,5-dichloro-3-thenoyl)propionate [76]

To a solution of 2,5-dichlorothiophene (17.0 g, 0.11 mol) and the acid chloride of monomethyl ester of succinic acid (20.4 g, 0.14 mol) in chloroform (100 ml) anhydrous aluminium chloride (33 g) is gradually added at 2–4 °C. The reaction mixture is stirred at 20 °C for 3 h, after which it is poured onto ice and allowed to stand overnight. The aqueous and organic phases are separated and the organic phase washed with dilute hydrochloric acid solution, saturated sodium chloride solution, sodium bicarbonate solution, salt solution and dried. The solvent is removed and the residue distilled giving 84% of the title compound bp 165–167 °C/4 mm Hg, mp 65.0–65.5 °C (hexane).

The reaction of 2-chlorothiophene with 5-chlorovaleryl chloride under Friedel-Crafts conditions gives the 5-chlorovaleryl thienyl ketone, which undergoes nucleophilic displacements, for instance, with ethyl 4-hydroxy-benzoate in the presence of potassium carbonate [77].

In another case 5-chloro-2-thenoyl chloride was prepared by reaction of the acid with oxalyl chloride in refluxing dichloromethane and reacted with benzene and aluminium chloride to 5-chloro-2-thienyl phenyl ketone [78]. An approach for the preparation of 1,3-di(2-methyl-5-chloro-3-thenoyl)-propane consists in the Friedel-Crafts reaction of 2-chloro-5-methylthio-phene with glutaryl chloride in carbon disulfide with aluminium chloride as catalyst [79].

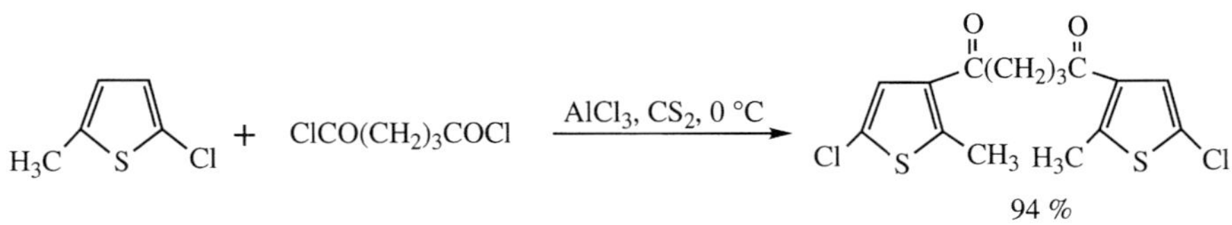

Starting from 2-chloro-, 3-chloro- or 2,5-dichlorothiophene and carrying out the Friedel-Crafts reaction with ethyl oxalyl chloride using nitromethane as solvent and aluminium chloride as catalyst gave ethyl 5-chloro-2-, 3-chloro-2- and 2,5-dichloro-3-thienylglyoxylate [80].

General procedure for preparation of ethyl 5-chloro-2-, 3-chloro-2- and 2,5-dichloro-3-thienylglyoxylate [80]

To a mixture of the thiophene derivative (0.05 mol) and ethyl oxalyl chloride (10.30 g, 0.075 mol) a solution of aluminium chloride (10.00 g, 0.075 mol) in nitromethane (20 ml) is added dropwise, under cooling and stirring, at such a rate that the temperature remains below 10 °C. After stirring at this temperature for 1 h, the reaction mixture is poured into ice-water (200 ml). The aqueous and organic phases are separated and the water phase extracted two times with ether (25 ml). The combined organic phases are washed with water, sodium bicarbonate solution, once again with water and dried over sodium sulfate. After removal of the solvent the residue is distilled *in vacuo*. *Ethyl 5-chloro-2-thienylglyoxylate* (71%) bp 103 °C/0.3 mm Hg, mp 43 °C. *Ethyl 3-chloro-2-thienylglyoxylate* (65%) bp 106 °C/0.4 mm Hg. *Ethyl 2,5-dichloro-3-thienylglyoxylate* (77%) bp 103 °C/0.5 mm Hg.

4-Aryl-2-chlorothiophenes are easily prepared through the reaction of 2,5-dichlorothiophene with various aromatic compounds in the presence of aluminium chloride [81, 82].

General procedure for 4-aryl-2-chlorothiophenes [81]

To a solution of 2,5-dichlorothiophene (20 mmol) and the aromatic compound (60 mmol) in dichloromethane (20 ml), pulverized $AlCl_3$ (20 mmol) is added in portions under ice-cooling over a period of 5 min. Exothermic reactions occur with the appearance of a coloration and the rise of the temperature of the reaction mixture. This is stirred at 5 °C for 30 min, at room temperature for 1 h, and then refluxed for an additional 30 min. The reaction mixture is poured into ice-water (70 ml). The phases are separated and the water phase extracted with chloroform (3 × 10 ml). The combined organic phases are washed with water, sodium bicarbonate, water and dried. After removal of the solvent and starting materials, the residue is sublimed under reduced pressure (2–3 mm Hg).

Aryl-3-arylmethyl-2-chlorothiophenes can be prepared through the reaction of 2,5-dichloro-3-chloromethylthiophene with benzene and alkylbenzenes in the presence of aluminium chloride [83].

7.2.7 Rearrangements of chlorothiophenes

3-Chlorothiophene is prepared by heating 2-chlorothiophene on a zeolite H-ZSM-5 fixed bed catalyst at 300–450 °C at atmospheric pressure. Similarly 2,4-dichlorothiophene can be obtained from 2,5-dichlorothiophene, which can be further isomerized to 3,4-dichlorothiophene [84].

7.2.8 From hydroxythiophenes

The reaction of ethyl 2-anilino-4-oxo-4,5-dihydrothiophene-3-carboxylate with the Vilsmeyer reagent gives ethyl 2-anilino-4-chloro-5-formyl-3-thiophene-carboxylate [85].

Ethyl 2-anilino-4-chloro-5-formyl-3-thiophenecarboxylate [85]

The Vilsmeyer reagent is prepared by adding phosphorus chloride (100 ml, 1.10 mol) in tetrachloroethane (100 ml) to dimethylformamide (36.5 ml, 0.47 mol) in tetrachloroethane (400 ml) under cooling and stirring. After stirring for an additional 0.5 h the cooling bath is removed and a solution of ethyl 2-anilino-4-oxo-4,5-dihydrothiophene-3-carboxylate (43.7 g, 166 mmol) in 1,1,2,2-tetrachloroethane (400 ml) is added dropwise with stirring. The stirring is continued at room temperature for 16 h, after which the reaction mixture is poured into water. The water solution is treated with dilute sodium hydroxide solution until faintly acidic. The phases are separated and the organic phase dried over magnesium sulfate. The oil remaining after evaporation crystallized from ethanol to give the title compound (40 g, 78%) as needles mp 117–119 °C.

The reaction of a 2,5-diaryl-3,4-dihydroxythiophene with phosphorus pentachloride at 180 °C gave the 3,4-dichloro derivative [86].

5-(4-Nitrophenyl)-2-(benzoxazol-2-yl)-3,4-dichlorothiophene [86]

A mixture of 5-(4-nitrophenyl)-2-(benzoxazol-2-yl)-3,4-dihydroxythiophene (35.4 g, 0.10 mol) and an excess of phosphorus chloride (207.5 g, 1.00 mol) is heated in an oil bath so that the reaction mixture refluxes gently. This is continued for 8 h, and after cooling the reaction mixture is slowly added to ice-water with constant stirring. The light yellowish-brown solid formed is filtered and recrystallized from dimethyl formamide/ethanol (1:1) to yield 24.6 (63%) of the title compound as a creamy solid mp >360 °C.

7.2.9 Modification of existing side chains in chlorothiophenes

As the chlorine in thiophenes is not a very reactive functionality, existing side chains can easily be modified. Thus 5-chloro-3-thiophenecarboxylic acid can by a modified Curtius reaction conveniently be transformed to 5-chloro-2-*N*-(*tert*-butoxycarbonylamino)thiophene [87].

2-Acetyl-3,4,5-trichlorothiophene is reduced with sodium borohydride and dehydrated by treatment with potassium bisulfate in the presence of hydroquinone [88].

3,4,5-Trichloro-2-vinylthiophene can also be synthesized in high yield by the Wittig reaction of 3,4,5-trichlorothenyl triphenylphosponium chloride and formaldehyde [88].

2-(α-Hydroxyethyl)-3,4,5,-trichlorothiophene [88]

To a solution of 2-acetyl-3,4,5-trichlorothiophene (18.4 g, 81 mmol) in methanol (200 ml) containing potassium hydroxide (0.2 g), sodium borohydride is added with stirring in portions at such a rate that the temperature of the exothermic reaction does not exceed 40 °C. The stirring was then continued for 2 h. The reaction mixture is neutralized with 18% hydrochloric acid, after which it is made weakly alkaline with potassium hydroxide solution. The methanol is removed by evaporation, the residue extracted with ether, the combined ether phases dried over magnesium sulfate and evaporated. The residue is distilled giving 16.2 g (87.5%) of the title compound bp 107–107 °C/1 mm Hg, mp 50–51 °C.

2-Vinyl-3,4,5-trichlorothiophene [88]

In a distillation flask 2-(α-Hydroxyethyl)-3,4,5,-trichlorothiophene (11.6 g, 50 mmol) is placed together with potassium bisulfate (0.11 g) and a small amount of hydroquinone. The dehydration is conducted at 180–200 °C/10–15 mm Hg. The distillate is extracted with ether, the combined ether phases dried over magnesium sulfate and evaporated. The residue is distilled giving 8.9 g (83%) of the title compound bp 70–72 °C/1 mm Hg.

1-(2-Carboxy-5-chloro-3-thienyl)-2-(3′-thienyl)ethane is prepared by hydrogenation of the corresponding ethene using Wilkinsons rhodium catalyst [89].

1-(2-Carboxy-5-chloro-3-thienyl)-2-(3′-thienyl)ethane [89]

A solution of 1-(2-carboxy-5-chloro-3-thienyl)-2-(3′-thienyl)ethene (4.9 g, 18 mmol) in absolute ethanol (60 ml) is degassed by passing a stream of nitrogen through it for 5 min. After adding tris(triphenylphosphine)rhodium (0.2 g) the solution is again degassed for 3 min. This mixture is hydrogenated in a Parr apparatus at about 60 °C and a hydrogen pressure between 60 and 30 psi for 30 h. The solvent is evaporated and the residue dissolved in 0.5 *M* sodium hydroxide solution. This solution is filtered through a plug of cotton wool and acidified. After 2 h the precipitate formed is filtered off and dried under vacuum giving 4.7 g of the title compound mp 110–111 °C after recrystallization from ethanol/water.

Starting from (5-chloro-2-thienyl)phosphonous dichloride a number of compounds with modified phosphorus functionalities have been prepared. 3-(2,5-Dichloro-3-thienyl)-3-oxopropanoate can be prepared from 3-acetyl-2,5-dichlorothiophene through condensation with diethyl carbonate in the presence of sodium hydride [91]. The reaction of 2-chloro-3-bromomethylthiophene with succinimide in toluene containing solid sodium carbonate yields 1-(2′-chloro-3′-thienylmethyl)succinimide in 72% yield [92]. From 2-acetyl-3,4,5-trichlorothiophene the corresponding carbinol is obtained upon sodium borohydride reduction without loss of chlorine [93].

7.2.10 Various methods

Because of the instability of many aminothiophenes, diazotization and the application of the Sandmeyer reaction is of little synthetic use in the thiophene series, except for the preparation of methyl 3-chloro-2-thiophenecarboxylate and 4-chloro-3-thiophenecarboxylate [16].

Methyl 3-chloro-2-thiophenecarboxylate [16]

Methyl 3-amino-2-thiophenecarboxylate (9.4 g, 60 mmol) is added gradually to a vigorously stirred 6 *M* hydrochloric acid solution (25 ml). The reaction mixture is stirred at room temperature for 30 min and then cooled to 0 °C, after which sodium nitrite (4.2 g, 60 mmol) in water (10 ml) is added. The resulting diazonium salt is stirred at 0 °C for 1 h and then poured at once into a well-stirred solution of cuprous chloride (60 mmol in concentrated hydrochloric acid, 25 ml) cooled to 0 °C. The reaction mixture is heated to 60 °C and kept at this temperature until evolution of nitrogen has ceased. After cooling, the reaction mixture is extracted several times with ether. The combined ethereal phases are washed with water, dried over magnesium sulfate and evaporated to dryness. Distillation of the residue gives the title compound (65%) bp 113–114 °C/14 mm Hg, mp 34–35 °C.

2-Chloro 5-tributylstannylthiophene can be prepared through the reaction of 2-bromo-5-chlorothiophene with magnesium powder and bis(tributyltin)oxide *via* sonochemical Barbier type reaction in a commercial ultrasonic cleaning bath [94].

2-Chloro 5-tributylstannylthiophene [94]

A solution of 2-bromo-5-chlorothiophene (198.5 g, 1.0 mol), magnesium powder (2.3 mol) and bis(tributyltin)oxide (1.0 mol) in anhydrous tetrahydrofuran (4 ml) is sonicated for 1 h in a commercial ultrasonic cleaning bath (Cest-575-D, 39 kHz) at about 45 °C. When the reaction is complete (monitored by TLC), water (15 ml) is added and the reaction mixture is extracted with ethyl acetate (3 × 10 ml). The combined organic phases are washed with brine (15 ml), dried over sodium sulfate, filtered and evaporated. Flash chromatography of the residue on silica gel using ethyl acetate/hexane as eluent gives the title compound (80%).

A preparatively very useful method for the synthesis of chlorothiophenes starts from the more easily available analogous bromothiophenes, which are reacted with cuprous chloride/pyridine in *N,N*-dimethylformamide or dimethylsulfoxide [95].

General procedure for the preparation of chlorothiophenes from bromothiophenes [95]

A stirred mixture of the corresponding bromothiophene (0.2 mol) and anhydrous copper(I) chloride (0.3 mol for each atom of bromine to be substituted) in anhydrous *N,N*-dimethylformamide (150 ml) is heated for 12–18 h under reflux in a nitrogen atmosphere. After being cooled the reaction mixture is poured into a flask containing water (1500 ml) and the chlorothiophene produced is codistilled with water from the flask. The phases of the distillate are separated and the organic phase dried over calcium chloride and redistilled.

Derivative	Yield (%)	Boiling point (°C)	Purity (GLC%)
2-Chloro	95	128	98
3-Chloro	98	137–139	97
2,5-Dichloro	85	162	99
2,4-Dichloro	80	167–169	85
2,3-Dichloro	90	173–174	86
3,4-Dichloro	95	182–185	98
2,3,5-Trichloro	60	198	96
2,3,4-Trichloro	85	210–212	99
Tetrachloro	30	235	87

This method should definitely be the best synthetic method for the preparation of 3-chloro- and 2,4-dichlorothiophene.

β-Deuterated chlorothiophenes are conveniently prepared by zinc dust reduction in acetic anhydride, as exemplified by the preparation of 2,5-dichloro-3,4-dideuteriothiophene from 2,5-dichloro-3,4-diiodothiophene [96].

2,5-Dichloro-3,4-dideuteriothiophene [96]

Under argon 2,5-dichloro-3,4-diiodothiophene (89.0 g, 0.22 mol) is mixed with zinc dust (296.0 g, 4.53 mol). A mixture of freshly distilled acetic anhydride (170.0 g, 1.66 mol) and heavy water (100.0 g, 5.00 mol) is added all at once. After heating gently the reaction begins spontaneously and vehemently. Subsequently the reaction mixture is refluxed for 2 h and cooled. The liquid phase is decanted and the remaining zinc dust is washed three times with methanol and five times with ether. The combined organic phases are washed with 3 *M* aqueous sodium hydroxide. This procedure is complete when the aqueous phase gives a basic reaction. Subsequently the organic phase is washed with water and dried over sodium sulfate. After removal of the solvent the residue is distilled giving 17.9 g (52.4%) of the title compound as a colorless liquid (bp 48 °C/13 mm Hg).

In the laboratory tetrachlorothiophene is best prepared by the reaction of hexachlorobutadiene with sulfur [51,97,98].

Tetrachlorothiophene [51]

In a 1-l three-necked flask equipped with a mechanical stirrer, distillation head, thermometer and a condenser is placed hexachloro-1,3-butadiene (400 ml, 2.55 mol), which is heated to 160 °C, whereupon sulfur (245 g, 7.66 mol) is added. The mixture is heated to 220–225 °C and after about 1.5 h sulfur monochloride begins to distill over. Heating is continued for 7 h and the reaction mixture is left overnight at 200 °C after which it is again heated to 220–225 °C for 7 h. By this time the distillation of sulfur monochloride ceases and 206 ml of the product can be collected. This is purified by a simple distillation at about 90 °C/6–7 mm Hg to remove tars, and 438 g of distillate is obtained. More advanced distillation gives 350 g (62%) of tetrachlorothiophene (bp 93–94 °C/7.7 mm Hg, mp 29 °C).

7.3 BROMINE DERIVATIVES

7.3.1 Direct bromination

Bromine-substituted thiophenes are in most cases, best prepared by direct bromination with bromine in chloroform or acetic acid. Starting from thiophene, 2-bromo-, 2,5-dibromo-, 2,3,5-tribromo- [99], and tetrabromothiophene [100] can be obtained depending on the amounts of bromine used and on the reaction conditions. Overbromination to 2,5-dibromothiophene is often the case upon attempted synthesis of 2-bromothiophene, which is best overcome by using dioxane dibromide as brominating agent [101,102], thallium acetate [103] or an inert solvent [104]. Recently convenient modifications making fast bromination possible have been introduced. Thus 2-bromothiophene is obtained in high yields by adding a solution of bromine in 48% hydrobromic acid or a 35% aqueous solution of hydrogen peroxide to a mixture of thiophene, diethyl ether and 48% hydrobromic acid. 2,5-Dibromothiophene was obtained in good yield from thiophene and bromine in hydrobromic acid [105].

2,5-Dibromothiophene [105]

A three-necked round-bottomed flask, equipped with a mechanical stirrer, a dropping funnel and a thermometer is charged with thiophene (42.0 g, 0.50 mol), 48% aqueous hydrobromic acid (150 ml) and diethyl ether (100 ml). A mixture of bromine (160.0 g, 1.00 mol) and 48% aqueous hydrobromic acid (150 ml) is added dropwise with vigorous stirring at −10 °C. After the addition the stirring is continued for 30 min at +10 °C. The phases are separated and the aqueous phase extracted with methylene chloride (4 × 50 ml). The combined organic phases are washed with water (1 × 200 ml), dried over magnesium sulfate and concentrated at reduced pressure. The remaining liquid is distilled through a 20 cm Vigreux column giving 110–113.7 g (91–94%) of the title compound (bp 76–80 °C/10 mm Hg).

2,4-Dibromo-5-ethylthiophene is prepared by bromination of 2-ethylthiophene with two equivalents of bromine in acetic acid [106].

2,4-Dibromo-5-ethylthiophene [106]

To a solution of 2-ethylthiophene (25 g, 0.22 mol) in acetic acid (500 ml) bromine (25 ml, 0.49 mol) is slowly added. The stirring is continued at room temperature overnight, after which the reaction mixture is neutralized and extracted with diethyl ether. The combined extracts are dried over magnesium sulfate and evaporated. The residue is purified by chromatography on silica gel using hexane as eluent, giving 39 g (65%) of the title compound as a colorless oil.

3,4-Dibromo-2,5-dichlorothiophene and 3-bromo-2,4,5-trichlorothiophene were recently obtained by bromination with bromine in acetic acid [40].

3-Bromo-2,4,5-trichlorothiophene [40]

To a solution of 2,3,5-trichlorothiophene (44 g, 0.23 mol) in acetic acid (150 ml) bromine (73.6 g, 0.46 mol) is added. The mixture is stirred at 60 °C for 2 days, after which more bromine (18.4 g, 0.12 mol) is added and the stirring is continued for another 2 days. After cooling, water saturated with sodium chloride (100 ml) is added and the product is extracted five times with dichloromethane. The combined organic phases are vigorously washed first with sodium thiosulfate solution until all bromine is reduced and then with sodium bicarbonate solution, dried, evaporated and distilled giving 33.2 g (54%) of the title compound bp 110 °C/11 mm Hg, mp 42 °C after recrystallization from ethanol.

2,5-Dibromo-3,4-difluorothiophene was recently prepared in 86% yield from the reaction of 3,4-difluoro-2,5-bis(trimethylsilyl)thiophene with bromine in dichloromethane [9].

With three equivalents of bromine 80% of 2,3,5-tribromothiophene can be quickly obtained [107]. A mixture of 30% hydrogen peroxide in ether to which 48% hydrobromic acid is added was used for the preparation of tris(5-bromo-2-thienyl)methane from tris(2-thienyl)methane [108].

Tris(5-bromo-2-thienyl)methane [108]

To a mixture of tris(2-thienyl)methane (7.86 g, 0.30 mol) in diethyl ether (70 ml) and 30% hydrogen peroxide (9.2 ml) cooled to −15 °C, 47% hydrobromic acid (30 ml) is added dropwise during 1 h. When the addition is completed the stirring is continued for another 2 h and the temperature is kept below −5 °C. At room temperature the phases are separated and the organic phase is washed with saturated sodium chloride solution, 10% sodium hydroxide solution and water. After drying over magnesium sulfate and evaporation the residue is recrystallized first from petroleum ether (80–100 °C) and then from ethanol. The title compound is obtained as colorless crystals in a yield of 8.1 g (54%) mp 81–82 °C.

2-Alkylthiophenes can be brominated in good yields, giving 2-alkyl-5-bromo derivatives on monobromination [109–111] and the 2-alkyl-3,5-dibromo derivatives upon dibromination [110,112,113] except with bulky substituents. Thus from 2-*tert*-butylthiophene the 4,5-dibromo derivative is obtained [114]. 3-Alkyl-2-bromothiophenes are obtained predominantly on bromination of 3-alkylthiophenes, not having too bulky groups. With the isopropyl and *tert*-butyl derivatives increasing amounts of the 5-substituted product are obtained [115]. A number of 3-alkylthiophenes ranging from 3-butyl- to 3-dodecylthiophenes have been brominated with bromine in acetic acid to give the corresponding 2-bromo-3-alkylthiophenes [115–117].

General procedure for bromination of 3-alkylthiophenes [116]

3-Alkylthiophene (0.1 mol) in acetic acid (125 ml) is cooled in an ice bath and nitrogen is bubbled through the solution. Under stirring, bromine (0.1 mol) in acetic acid (40 ml) is added slowly for 25 min, after which the stirring is continued for another 30 min and the reaction mixture poured into ice water. The organic material is taken up in chloroform and after separation of the phases the aqueous phase is extracted two times with chloroform. The combined chloroform phases are washed with 2 *M* sodium hydroxide solution until neutral reaction, then with water and dried over magnesium sulfate. After evaporation the residue is distilled at reduced pressure.

Alkyl group	Yield (%)	Boiling point °C/mm Hg
Methyl [115]	65	63–65/10
Ethyl [115]	71	78–82/10
Propyl [116]	64	96–101/15
Butyl [117]	48	80/1.2
Hexyl [116]	58	135–140/10
Octyl [117]	40	123/1.1

3-Alkyl-2,5-dibromothiophenes are obtained in good yields upon dibromination. [110,115] In this way 3-(2,5-dibromothiophene-3-yl)propionic acid was recently prepared from 3-thiophene-3-yl-propionic acid [118].

3-(2,5-Dibromothiophene-3-yl)propionic acid [118]

A solution of bromine (3.8 ml, 0.07 mol) in chloroform (50 ml) is added to 3-thiophene-3-yl-propionic acid (5.724 g, 0.0367 mol) in chloroform (75 ml) in an ice bath over 0.5 h. The reaction mixture is stirred at room temperature for 2 h and then poured onto ice (100 g). The chloroform phase is separated, dried over sodium sulfate and evaporated. The residue is recrystallized from hexane giving 8.30 g (70%) of the title compound as a white solid mp 87–88 °C.

Bromination of methyl tris(2-methyl-4-thienyl)silane with bromine in dichloromethane at −78 °C is used for the preparation of methyl tris(2-bromo-5-methyl-3-thienyl)silane [119]. 2-Bromo-3-thienylacetic acid is prepared by bromination of 3-thienylacetic acid with bromine in acetic acid [120].

2-Bromo-3-thienylacetic acid [120]

To a solution of 3-thienylacetic acid (3.0 g, 21.1 mmol) in acetic acid (40 ml) bromine (1.1 ml, 21.4 mmol) in acetic acid (10 ml) is added. The reaction mixture is refluxed for 2 h, after which a small amount of sodium bisulfite is added, and it is poured into ice-water (100 ml). After filtration the product is extracted with chloroform. The combined organic phases are evaporated and the residue dried giving 3.01 g (65%) of the title compound as a yellowish oil.

The best method for the preparation of 2-bromo-5-methoxythiophene appears to be the direct bromination of 2-methoxythiophene with bromine in ether [121].

5-Bromo-2-methoxythiophene [121]

A solution of 2-methoxythiophene (5.20 g, 0.046 mol) in diethyl ether (50 ml) at 0 °C is treated dropwise with bromine (2.3 ml, 0.046 mol). After stirring for 2 h, the reaction mixture is poured into ice-cold aqueous saturated sodium bicarbonate solution (50 ml). Ether is added (50 ml) and the phases separated. The organic phase is dried over sodium sulfate and evaporated giving 8.0 g (90%) of the title compound.

Bromination of some activated thiophenes is reversible [122–124] and can be directed to give either the kinetic- or thermodynamic- controlled product. Thus 3-phenylthiophene yields, upon reaction with bromine, in the presence of hydrogen bromide mainly the thermodynamically controlled product, 2-bromo-4-phenylthiophene [123,124]. Bromination of 3-phenylthiophene with *N*-bromosuccinimide or dibromo dimethylhydantoin in chloroform/acetic acid (1:1) gave almost exclusively the kinetic controlled product, 2-bromo-3-phenylthiophene [122]. Bromination of trifluoroacetylaminothienylpropionic acid with bromine in chloroform gives 3-trifluoro-acetylamino-3-(2,5-dibromothien-3-yl)propionic acid [24].

3-Trifluoroacetylamino-3-(2,5-dibromothien-3-yl)propionic acid [24]

To a solution of trifluoroacetylaminothienylpropionic acid (3.00 g, 0.011 mol) in chloroform (60 ml) bromine (1.8 ml, 0.033 mol) is added. The reaction mixture is then refluxed for 15 min and allowed to stand at room temperature. The precipitate formed is filtered off giving 3.5 g (75%) the title compound mp 180 °C after recrystallization from diethyl ether.

tert-Butyl *N*-(3-thienyl) carbamate upon bromination yields *tert*-butyl *N*-(2-bromo-3-thienyl)carbamate [125]. Bromination of methyl 5-substituted 3-hydroxy-2-thiophenecarboxylates with bromine in acetic acid is a good method for the preparation of methyl 5-alkyl 4-bromo- 3-hydroxy-2-thiophenecarboxylates [126].

Methyl 5-methyl-4-bromo-3-hydroxy-2-thiophenecarboxylate [126]

A solution of methyl 5-methyl-3-hydroxy-2-thiophenecarboxylate (5.0 g, 29 mmol) in acetic acid (25 ml) is, under stirring, treated with bromine 4.64 g (29 mmol). The stirring is continued for 16 h at room temperature and then stirred into ice-water (200 ml). The precipitate is isolated by filtration, washed with water, 5% sodium thiosulfate and again water. After drying it is recrystallized from methyl *tert*-butyl ether giving 4.8 g (66%) of the title compound mp 96–97 °C.

5-Bromo-2-(α-alkoxyamino)ethylthiophene is obtained from 2-(α-alkoxyamino)ethylthiophene upon reaction with bromine in the presence of iron [32]. Bromination in acetic acid has been used for the preparation of 2-(2′-bromothien-5′-ylmethyl)phtalimide from 2-(2′-thienylylmethyl)phtalimide [127].

Also deactivated thiophenes can be brominated with molecular bromine, the products being dependent on the reaction conditions. From 5-methyl-2-thiophene aldehyde, 4-bromo-5-methyl-2-thiophene aldehyde is prepared [113]. From 2-thiophene aldehyde with one equivalent of bromine in chloroform 5-bromo-2-thiophene aldehyde is obtained in 66% yield [128,129], and with two equivalents of bromine, 4,5-dibromo-2-thiophene aldehyde is obtained. When strong acid media such as concentrated sulfuric acid are used in the presence of silver sulfate, 4-bromo-2-thiophene aldehyde can be prepared in 91% yield [130]. In reaction using the swamping catalyst method (excess aluminium chloride) the yields are lower [131,132]. It should be mentioned that the best method for preparing 4-bromo-2-thiophene aldehyde is the halogen–metal exchange of 2,4-dibromothiophene with butyllithium at −70 °C, followed by reaction with *N*,*N*-dimethylformamide [133].

Acetylthiophenes and other ketones having α-hydrogens undergo acid-catalyzed enolization followed by side chain bromination especially in aprotic solvents such as carbon tetrachloride [134] and diethyl ether [135,136]. This can be supressed by using acetic acid [136,137], especially containing sodium acetate [138] as solvent. Under such conditions 2-acetylthiophene gives 2-acetyl-5-bromothiophene in 60% yield [138]. Using the swamping catalyst conditions 4-bromo-2-acetylthiophene can be prepared in 75% yield [139] and with two equivalents of bromine 2-acetyl-4,5-dibromothiophene is obtained [140]. 3-Acetylthiophene gives under both conditions 3-acetyl-5-bromothiophene [138,141,142]. 4-Bromo-2-cyanothiophene can also be prepared by swamping catalyst bromination of 2-cyanothiophene [143]. 2-Propionyl-4-bromothiophene was recently obtained from 2-propionylthiophene by using two equivalents of aluminium chloride in chloroform [144].

1-(4-Bromo-2-thienyl)-1-propanone [144]

To a solution of 2-propionylthiophene (47.3 g, 338 mmol) in chloroform (225 ml) aluminium chloride (101.3 g, 760 mmol) is added under stirring. To this mixture bromine (57.5 g, 360 mmol) in chloroform (375 ml) is added. The reaction mixture is stirred at room temperature for 18 h and then poured into ice-water (500 ml). The phases are separated and the organic phase washed with water (2×200 ml), dried over magnesium sulfate and evaporated giving 81 g of the title compound mp 40–41 °C.

Brominated thiophenecarboxylic acids are conveniently prepared by bromination with molecular bromine. Bromodecarboxylation is in some cases a side reaction. 2-Thiophenecarboxylic acid gives upon monobromination 5-bromo-2-thiophenecarboxylic acid [145,146]. Further bromination leads to 4,5-dibromo-2-thiophenecarboxylic acid [145,147]. Tribromination cannot be achieved, instead bromodecarboxylation occurs leading to tetrabromothiophene [148]. Upon bromination of 3-thiophenecarboxylic acid, 2-bromo-4-thiophenecarboxylic acid is first obtained. Further bromination gives the 2,5-dibromo-3-thiophenecarboxylic acid [25,149]. In contrast to 2-thiophenecarboxylic acid, the 3-isomer upon treatment with excess bromine gives 2,4,5-tribromothiophenecarboxylic acid in good yield [149,150].

2,4,5-Tribromo-3-thiophenecarboxylic acid [150]

3-Thiophenecarboxylic acid (0.50 g, 4.0 mmol) is treated with excess of bromine. The bromine is evaporated overnight and the white residue recrystallized from toluene giving 1.2 g (80%) of the title compound mp 203–205 °C.

From methyl 3-methyl-2-thiophenecarboxylate, 4,5-dibromo-3-methyl-2-thiophenecarboxylic acid can be prepared [151]. The use of one equivalent of bromine in water is claimed to give the 4-bromo-3-methyl-2-thiophenecarboxylic acid [151]. An alternative to using free bromine is to use benzyl trimethylammonium tribromide in the presence of zinc chloride. With this reagent 2,5-dibromothiophene [34], as well as methyl 3-bromo-4,5,6,7-tetrahydrobenzo [*c*]thiophene-1-carboxylate was prepared [152].

2,5-Dibromothiophene [34]

Benzyl trimethylammonium tribromide (6.50 g, 15.5 mmol) and zinc chloride (about 1 g) are added to a solution of thiophene (0.42 g, 5.0 mmol) in acetic acid

(30 ml). The reaction mixture is stirred until the orange colour disappears, 2 h, after which aqueous sodium hydrogen sulfate (5%, 10 ml) is added. This mixture is extracted with hexane (3 × 50 ml). The phases are separated and the organic phase dried over magnesium sulfate and passed through a short aluminia column. The eluent is concentrated and distilled giving 0.65 g (54%) of the title compound as a colourless oil bp 68 °C/2 mm Hg.

Methyl 3-bromo-4,5,6,7-tetrahydrobenzo [c]thiophene-1-carboxylate [152]

To a solution of methyl 4,5,6,7-tetrahydrobenzo[*c*]thiophene-1-carboxylate (0.40 g, 2.0 mmol) in acetic acid (5 ml) benzyl trimethylammonium tribromide (0.84 g, 2.1 mmol) is added followed by zinc chloride (1.1 g, 8.0 mmol). The reaction mixture, a suspension, is stirred at room temperature for 5 h, after which dichloromethane (25 ml) and water (25 ml) are added. The phases are separated and the organic phase is washed with sodium chloride solution (25 ml), dried over sodium sulfate and evaporated. The residue is purified by radial chromatography on a 2 mm plate using ethyl acetate/hexanes (5:95) as eluent giving 0.47 g (85%) of the title compound as a white solid mp 85–85 °C after recrystallization from hexanes.

β-(2-Bromo-3-thienyl)acrylic acid is obtained upon bromination of β-3-thienylacrylic acid with one equivalent of bromine [153]. On the other hand ethyl 3-(5-nitro-2-thienyl)acrylate is brominated at the ethylenic bond giving ethyl 2-bromo-3-(5-nitro-)2-thienylacrylate [154].

7.3.2 Bromination with *N*-bromosuccinimide

N-Bromosuccinimide in acetic acid with chloroform or carbon tetrachloride as cosolvent can, in most cases, be used instead of molecular bromine and is the best method for the preparation of 2,5-dibromothiophene from thiophene [122]. Its greatest advantage is that *trans* halogenation does not occur in the bromination of iodothiophenes [4,35,155,156]. 2-Bromo-3-methylthiophene and 2,3-dibromo-4,5-dimethylthiophene are conveniently obtained by from 3-methylthiophene and 2,3-dimethylthiophene, respectively, with *N*-bromosuccinimide in acetic acid in the presence of hydroquinone to avoid radical side chain bromination [157].

2-Bromo-3-methylthiophene [157]

To a suspension of *N*-bromosuccinimide (98.0 g, 0.55 mol) and hydroquinone (10 mg) in acetic acid (250 ml) 3-methylthiophene (54.0 g, 0.55 mol) is added rapidly with vigorous stirring and ice cooling. The reaction mixture is poured into water and the organic material is taken up in ether. The phases are separated and the organic phase washed with sodium hydrogen carbonate solution and water, and dried over magnesium sulfate. The solvent is evaporated and the residue fractionated giving 83.2 g (85%) of the title compound bp 63–73 °C/15 mm Hg.

N-Bromosuccinimide is also the preferred brominating agent for thiopheneboronic acids [158]. Thus bromination of 2- and 3-thiopheneboronic acid with *N*-bromosuccinimide in carbon tetrachloride gave 5-bromo-2-thiopheneboronic acid and 5-bromo-3-thiopheneboronic acid in 75% and 69% yields respectively. Lower yields were obtained with bromine in chloroform at −40 °C, most probably due to deboronation [158].

Alkoxythiophenes, even sensitive ones like 3,4-ethyleneoxythiophene, can be dibrominated to 2,5-dibromo-3,4-diethylenoxythiophene with *N*-bromosuccinimide in tetrahydrofuran/glacial acetic acid in quantitative yields [159].

2,5-Dibromo-3,4-diethylenoxythiophene [159]

To a solution of 3,4-ethyleneoxythiophene (5.0 g, 35.2 mmol) in tetrahydrofuran (20 ml) and glacial acetic acid (20 ml) *N*-bromosuccinimide is added in small portions. The stirring is continued at room temperature for 2 h, after which the deep-red reaction mixture is poured into water (100 ml). The precipitate formed is filtered off giving a quantitative yield of the title compound as a light-yellow solid.

The most serious side-reaction with alkylthiophenes is due to bromination in the benzylic position. This can be suppressed by adding hydroquinone as a radical chain inhibitor. However, bromination of 2-benzylthiophene with *N*-bromosuccinimide in *N*,*N*-dimethylformamide gives 2-benzyl-5-bromothiophene [36].

2-Benzyl-5-bromothiophene [36]

To a solution of 2-benzylthiophene (10.0 g, 58 mmol) in anhydrous *N*,*N*-dimethylformamide a solution of *N*-bromosuccinimide (10.2 g) in dimethylformamide (25 ml) is added. The reaction mixture is stirred at room temperature for 24 h after which it is poured into water (500 ml) and extracted with ether (3 × 100 ml). The combined ether phases are washed with water (4 × 100 ml), dried over magnesium sulfate and evaporated. The crude product is dried over and distilled from calcium hydride giving 12.6 g (86%) of the title compound bp 98–102 °C/0.2 mm Hg.

2-Bromo-3-bromomethylthiophene is obtained in a one-pot procedure from 3-methylthiophene upon *N*-bromosuccinimide bromination in anhydrous benzene in the presence of benzoyl peroxide [160].

2-Bromo-3-bromomethylthiophene [160]

To a solution of 3-methylthiophene (9.0 g, 92 mmol) and benzoyl peroxide (0.1 g, 0.41 mmol) in anhydrous benzene (500 ml), a mixture of *N*-bromosuccinimide (20 g, 112 mmol) and benzoyl peroxide (0.1 g, 0.41 mmol) is gradually added under reflux during 1 h. When the addition is completed the reaction mixture is refluxed for an additional hour and cooled. The precipitate is filtered off and the filtrate evaporated *in vacuo* giving 15.6 g of the title compound as an orange oil.

Bromination of 3-hexylthiophene with *N*-bromosuccinimide in *N,N*-dimethylformamide in the dark can be used for the preparation of 2-bromo-3-hexylthiophene and of 2,5-dibromo-3-hexylthiophene, if two equivalents of *N*-bromosuccinimide are used [161].

2-Bromo-3-hexylthiophene [161]

A solution of 3-hexylthiophene (3.0 g, 0.18 mol) in dimethylformamide (50 ml) is protected from light and cooled to −20 °C. A solution of *N*-bromosuccinimide (3.18 g, 0.18 mol) in *N,N*-dimethylformamide (50 ml) is slowly added and the reaction mixture is stirred at −20 °C for 30 min and then at room temperature for 5 h, after which the reaction mixture is poured into ice (50 g). This solution is extracted with dichloromethane (3 × 100 ml). The combined organic extracts are dried over sodium sulfate, the solvent evaporated and the residue fractionated giving 2.31 g (52%) of the title compound bp 95–96 °C/0.1 mm Hg.

Due to steric hindrance 3,4-di-*tert*-butylthiophene cannot be dibrominated in good yield. However, the monobromo derivative is obtained in 68% yield with one equivalent of *N*-bromosuccinimide [44]. Changing the solvent from a mixture of acetic acid and dichloromethane to *N,N*-dimethylformamide made it possible to obtain the 2-bromo derivative and with two equivalents the 2,5-dibromo derivative in good yields [162].

In spite of addition of radical initiators such as benzoyl peroxide or 2,2′-azobisisobutyronitrile, 2-bromo-3-methyl-5-phenylthiophene is obtained in good yield upon reaction of 2-phenyl-4-methylthiophene with *N*-bromosuccinimide in carbon tetrachloride [163].

2-Bromo-3-methyl-5-phenylthiophene [163]

To a solution of 4-methyl-2-phenylthiophene (5.1 g, 29 mmol) in anhydrous carbon tetrachloride (50 ml) *N*-bromosuccinimide (5.34 g, 30 mmol) and benzoyl peroxide (0.5 g) are added. The reaction mixture is irradiated with a 200-W lamp and refluxed for 1 h, then cooled and filtered to remove succinimide. The filtrate is evaporated giving a brown solid with an irritating odor, which is taken up in chloroform (50 ml). This solution is refluxed with hexamethylene tetramine for 1 h. Chloroform is removed under reduced pressure and the residue triturated with ether (2 × 100 ml). The combined ether extracts are washed with water, dried over sodium sulfate and evaporated. The residue is recrystallized from methanol giving 4.54 g (76%) of the title compound mp 62–63 °C.

However, continued reaction with *N*-bromosuccinimide in the presence of radical initiators led to 2-bromo-3-bromomethyl-5-phenylthiophene, which by the Sommelet reaction was transformed to 2-bromo-3-formyl-5-phenylthiophene [163].

Di-*tert*-butyl(thiophene-3,4-diyl)dicarbamate is brominated with one equivalent of *N*-bromosuccinimide in carbon tetrachloride to the 2-bromo derivative and with two equivalents of bromine to the 2,5-dibromo derivative [164].

*Di-*tert*-butyl(2-bromothiophene-3,4-diyl)dicarbamate [164]*

N-Bromosuccinimide (0.178 g, 1 mmol) is added to a solution of di-*tert*-butyl (thiophene-3,4-diyl)dicarbamate (0.314 g, 1 mmol) in carbon tetrachloride (20 ml). The reaction mixture is stirred overnight, filtered, washed with water, dried and evaporated. The residue is purified by chromatography using dichloromethane as eluent giving 0.335 g (93%) of the title compound.

5-Bromo-2-styrylthiophene is prepared by the reaction of 2-styrylthiophene with one equivalent of *N*-bromosuccinimide [165]. Reaction of 3-bromo-4-methylthiothiophene with *N*-bromosuccinimide in carbon tetrachloride occurred selectively *ortho* to the methylthio group yielding 2,4-dibromo-3-methylthiothiophene in 71% yield [166]. Bromination of 2,5-bis-(methylthio)thiophene and 2,3-bis(methylthio)thiophene in the same way gave 3-bromo-2,5-bis(methylthio)- and 5-bromo-2,3-bis(methylthio)thiophene in 86% and 81% yield, respectively. Two equivalents of *N*-bromosuccinimide can conveniently be used for the preparation of 3,4-dibromo-2,5-bis-(methylthio)- and 2,5-dibromo-3,4-bis(methylthio)thiophene. 4-Bromo-2,3-bis(methylthio)thiophene gave upon *N*-bromosuccinimide bromination 2,3-dibromo-4,5-bis(methylthio)thiophene in 94% yield. 3-Bromo-2,4,5-tris-(methylthio)thiophene was also prepared in this way in high yield [166].

3-Bromo-2,4,5-tris(methylthio)thiophene [166]

To a solution of 2,3,5-tris(methylthio)thiophene (3.60 g, 16.2 mmol) in carbon tetrachloride (40 ml) *N*-bromosuccinimide (2.88 g, 16.2 mmol) is added and the reaction mixture is stirred overnight at room temperature. After filtration the filtrate is washed with 10% sodium bicarbonate, dried over magnesium sulfate and evaporated. The residue, a colourless solid, is recrystallized from hexane and then from ethanol giving 4.0 g (82%) of the title compound as white crystals mp 52–53 °C.

A very useful modification, using biphasic bromination with *N*-bromosuccinimide in hexane or carbon tetrachloride and catalytic amounts (0.1–1 mol%) of 70% perchloric acid, has recently been described. 3-Bromothiophene gave 93–99% yield of 2,3-dibromothiophene and from 2-chlorothiophene, 2-bromo-5-chlorothiophene was prepared [46].

2-Bromo-5-chlorothiophene [46]

To a suspension of *N*-bromosuccinimide (1.78 g, 10 mmol) in hexane (15 ml) 2-chlorothiophene (1.19 g, 10 mmol) and 70% perchloric acid (14 μl, 1 mol%) are added. The reaction mixture is stirred at room temperature until the starting material has disappeared (GLC control, 7 h). Potassium carbonate is added, the solids are filtered and the solvent is evaporated at room temperature. The residue is distilled giving 1.49 g (75%) of the title compound bp 45–46 °C/1 mm Hg.

2-Bromo-5-chloro-3-methoxythiophene is prepared in a one-pot procedure by bromination with *N*-bromosuccinimide followed by chlorination with *N*-chlorosuccinimide [167].

2-Bromo-5-chloro-3-methoxythiophene [167]

N-Bromosuccinimide (7.79 g, 43.8 mmol) is added in parts to a solution of 3-methoxythiophene (5.0 g, 43.8 mmol) in dichloromethane (200 ml) at 0 °C. The stirring is continued at 0 °C for 0.5 h and *N*-chlorosuccinimide (5.84 g, 43.8 mmol) is added in parts, after which the stirring is continued for 1.5 h at 0 °C. The resulting solution is washed with water, dried over magnesium sulfate and evaporated giving 8.44 g (85%) of the title compound as a brown oil.

2-Bromo-3-(2-thienyl)acrylophenones can be prepared by the reaction of the corresponding acrylophenone with tetrabromo-*ortho*-benzoquinone [168].

7.3.3 From thienyllithium derivatives

Bromothiophenes have also been prepared from thienyllithium derivatives, obtained by metalation with butyllithium, followed by reaction with carbon tetrabromide or molecular bromine. In this way isomers, which are not available through electrophilic bromination such as 2-bromo-4-methylthiophene [169] can be synthesized.

2-Bromo-4-methylthiophene [169]

Anhydrous tetramethyletylenediamine (64.0 g, 0.55 mol) and 1.45 *M* butyllithium in hexane (380 ml, 0.55 mol) are added dropwise under nitrogen to 3-methylthiophene (49.0 g, 0.50 mol) at such a rate that gentle reflux is maintained. After the addition is completed the reaction mixture is refluxed for 30 min and then cooled to −70 °C. A solution of carbon tetrabromide (110.6 g, 0.33 mol) in anhydrous ether (200 ml) is added dropwise and the reaction mixture is stirred overnight at −70 °C, after which it is poured into ice-water. The phases are separated and the water phase extracted with ether. The combined organic phases are washed with 6 *M* hydrochloric acid, sodium bicarbonate solution, 20% sodium chloride solution and dried over magnesium sulfate. The solvent is evaporated and the residue fractionated *in vacuo* giving 60.2 g (68%) of the title compound bp 60–62 °C/10 mm Hg.

Metalation of the diethyl acetal of 3-thiophene aldehyde followed by reaction with bromine and hydrolysis can be used for the preparation of 2-bromo-3-thiophene aldehyde [170].

2-Bromo-3-thiophene aldehyde [170]

The diethyl acetal of 3-formylthiophene (21.5 g, 0.12 mol) is dissolved in anhydrous ether and, under nitrogen, cooled to −45 °C, and butyllithium (0.13 mol) is added dropwise at this temperature. When the addition is completed, the reaction mixture is stirred at −45 °C for an additional 30 min. The so obtained organolithium solution is transferred to a solution of bromine (12 g) in anhydrous ether (500 ml) at −70 °C. After stirring at this temperature for 2 h the temperature of reaction mixture is allowed to rise to 0 °C and then poured into ice-water. The phases are separated and the aqueous phase is acidified with hydrochloric acid and extracted with ether. The combined ether phases are washed with sodium bicarbonate, dried and evaporated. The residue is immediately hydrolyzed in refluxing anhydrous ethanol (85 ml) and concentrated hydrochloric acid (6 ml) for 30 min. After addition of ether (150 ml) and water (150 ml) the phases are separated and the ether phase washed with sodium bicarbonate solution, dried and evaporated. Upon distillation, a fraction boiling at 108–110 °C/15 mm Hg is obtained. The yield of the title compound after recrystallization from hexane (75 ml) is 5.0 g (22%) mp 34 °C.

Halogen–metal exchange of 3-bromo-2-iodo-4-thiophenecarboxylic acid with two equivalents of butyllithium followed by reaction with carbon dioxide is used for the preparation of 3-bromo-2,4-thiophenedicarboxylic acid [171].

3-Bromo-2,4-thiophenedicarboxylic acid [171]

To a solution of 3-bromo-2-iodo-4-thiophenecarboxylic acid (13.3 g, 40 mmol) in anhydrous diethyl ether (300 ml) under nitrogen at −70 °C 1.15 *M* butyllithium in diethyl ether (80 ml, 92 mmol) over a period of 5 min. The stirring is continued for 15 min, after which the reaction mixture is poured onto crushed solid carbon dioxide covered with anhydrous diethyl ether. At −10 °C the mixture is hydrolyzed. The phases are separated, the water phase is washed with diethyl ether and acidified with hydrochloric acid. The precipitate formed is filtered off and dried giving 6.4 g (64%) of the title compound mp 305–310 °C (decomp.) after recrystallization from water.

4-Bromo-2-methylthiophene is prepared from 4-bromo-2-thienyllithium and dimethyl sulfate [172].

4-Bromo-2-methylthiophene [172]

To butyllithium in diethyl ether under nitrogen at −70 °C 2,4-dibromothiophene (241 g, 1.0 mol) in anhydrous diethyl ether is added in a slow stream. The temperature is allowed to rise to −40 °C and dimethyl sulfate (170 g, 1.4 mol) in an equal volume of anhydrous diethyl ether is added at such a rate that the temperature is kept below −30 °C. After warming to 0 °C the reaction mixture is stirred for 2 h with concentrated aqueous ammonium hydroxide in order to destroy the excess of dimethyl sulfate. The phases are separated and the organic phase is washed with water, dried over calcium chloride and distilled giving 164 g (93%) of the title compound bp 61–62 °C/10 mm Hg.

Other metallorganic derivatives, which upon reaction with bromine give brominated products [173–175], are mercuriacetoxy derivatives, obtained by selective α-mercuration.

2-Benzoyl-5-bromothiophene is obtained in quantitative yield in this way [176].

7.3.4 Selective debromination of di- and polybromothiophenes

Selective debromination of, especially, tri- and tetrabromo substituted thiophenes with butyllithium followed by hydrolysis, by zinc in acetic acid or by other reduction methods gives access to many bromothiophenes not accessible by direct bromination methods. In addition, the intermediate thienyllithium derivatives can be reacted with many different electrophiles to give a great variety of functionalized bromothiophenes. The success of this approach is due to the stability of *o*-bromothienyllithium derivatives, which in contrast to *o*-bromophenyllithium do not split lithium bromide to give dehydrothiophene. Thus partial reduction of 2,3,5-tribromothiophene with zinc in acetic acid is the best method for the preparation of 3-bromothiophene [177–181] and has completely superseded earlier methods such as reaction of 2,3,5-tribromothiophene with two equivalents of butyllithium [182], and the hydrolysis of the Grignard reagent by the entrainment method followed by hydrolysis [99,145,149,183,184]. 4-Bromo-2,3-dimethylthiophene is obtained upon reduction of 2,3-dibromo-4,5-dimethylthiophene with zinc powder in acetic acid [157].

4-Bromo-2,3-dimethylthiophene [157]

A mixture of water (600 ml), zinc powder (29 g, 0.44 mol) and acetic acid is heated to reflux with vigorous stirring and 2,3-dibromo-4,5-dimethylthiophene (40.0 g, 0.148 mol) is added. After refluxing for 5 h the reaction mixture is steam distilled. The organic layer is taken up in diethyl ether. The ether phase is washed with sodium bicarbonate solution and dried over magnesium sulfate. Fractionated distillation gave 20.0 g (71%) of the title compound, bp 88–92 °C/15 mm Hg.

However for the preparation of 3,4-dibromothiophene [185] and 2,3,4-tribromothiophene the entrainment method is still quite useful [149], although comparable yields of 3,4-dibromothiophene are obtained upon treatment of tetrabromothiophene with two equivalents of butyllithium followed by water [149] or zinc in acetic acid [100,178]. Reaction of 2,3,5-tribromothiophene with one equivalent of butyllithium at −70 °C followed by hydrolysis is the best method for the preparation of 2,4-dibromothiophene [146,186,187].

2,4-Dibromothiophene [187]

To a solution of 2,3,5-tribromothiophene (80.2 g, 0.25 mol) in anhydrous ether (800 ml), under argon at −78 °C, 2.5 *M* butyllithium in hexane (100.0 ml, 0.25 mol) is added slowly and the temperature is not allowed to rise above −70 °C. When the addition is completed the stirring is continued for another 5 min, after which the reaction mixture is poured into cold water. The phases are separated and the organic phase is washed with water, dried over magnesium sulfate, evaporated and distilled giving 47.9 g (79%) of the title compound bp 81.5–96.5 °C/10 mm Hg.

An excellent method for the preparation of 3-bromothiophene in 81% yield, recently developed, consists in debromination of 2,3,5-tribromothiophene with sodium borohydride and tetrakis(triphenylphosphine)palladium(0) as catalyst, using half the amount of sodium borohydride to give 2,3-dibromothiophene in 83% yield [188]. The same methodology was applied to tetrabromothiophene for the preparation of 2,3,4-tribromothiophene and 3,4-dibromothiophene in 82% and 86% yield, respectively [188].

3-Bromothiophene [188]

An oxygen-free mixture of 2,3,5-tribromothiophene (6.42 g, 20 mmol) and tetrakis(triphenylphosphine)palladium(0) (0.231 g, 0.2 mmol) in acetonitrile

(100 ml) was stirred at 70 °C for about 10 min, sodium borohydride (2.5g, 6.6 g) is added in portions under 3 h with stirring at 70 °C. Upon completion of reaction (monitored by TLC) the solvent was removed and the crude compound distilled *in vacuo* giving 2.64 g (81%) of the title compound bp 50–52 °C/20 mm Hg.

Selective debromination of polybrominated thiophenes can be achieved in high yield by electrochemical methods. In this way 3,4-dibromothiophene was obtained from tetrabromothiophene [189].

3,4-Dibromothiophene [189]

In the cathode part of an electrochemical cell 2,3,4,5-tetrabromothiophene (10.0 g, 25 mmol) is dissolved in methanol (100 ml) and dichloromethane (25 ml) and cadmium dichloride (0.3 g, 2 mmol) are added. The electrolysis takes place at 35 °C and a voltage of 13.5–11.5 V with 1 A (83 mA/cm$^{2)}$ for 5 h. The reaction mixture is treated with water (150 ml) and the phases separated. The aqueous phase is extracted with dichloromethane (2 × 50 ml). The combined organic phases are washed with water, dried over magnesium sulfate, evaporated and distilled giving 5.4 g (89%) of the title compound bp 75 °C/ 3 mm Hg.

However, upon electrochemical reduction of dihalothiophenes at carbon cathodes in *N,N*-dimethylformamide halogen dance occurs and from either 2,3-dibromo- or 2,4-dibromothiophene a mixture of 3-bromo- and 3,4-dibromothiophene was obtained [190].

7.3.5 From halothiophenes

7.3.5.1 *Via* metalations

Metalation of 2-bromothiophene with lithium diisopropylamide followed by trimethyl silyl chloride is used for the preparation of 2-bromo-5-trimethylsilylthiophene. Upon reaction with magnesium bromide [191,192] followed by *tert*-butyl perbenzoate 2-bromo-5-*tert*-butoxythiophene was obtained.

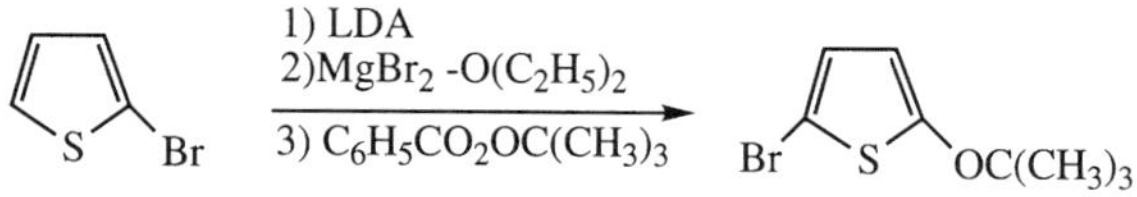

2-Bromo-5-trimethylsilylthiophene [192]

To a solution of 1.62 *M* butyllithium in hexanes (7.41 ml, 12 mmol) diluted with anhydrous tetrahydrofuran (10.0 ml) at −78 °C diisopropylamine (1.42 g,

14 mmol) is added. The mixture is warmed to 0 °C for 5 min and then recooled to −78 °C.

2-Bromothiophene (1.63 g, 10 mmol) is added and the mixture is allowed to warm to 0 °C for 5 min. After recooling to −78 °C trimethylsilyl chloride (1.30 g, 12 mmol) is added in one portion and the reaction mixture is allowed to warm to room temperature for 30 min, poured into water with a few drops of 3 *M* hydrochloric acid in order to remove the emulsion. The phases are separated and the aqueous phase extracted with diethyl ether. The combined organic phases are washed with sodium bicarbonate solution and sodium chloride solution, dried over sodium sulfate, evaporated and distilled and evaporated giving 1.8 g (77%) of the title compound as a colorless liquid bp 70 °C/2 mm Hg.

Metalation of 3-bromothiophene with lithium diisopropylamide at −78 °C followed by reaction with trimethyltin chloride gives 3-bromo-2-trimethylstannylthiophene in 75% yield [193]. Also metalation with phenyllithium can be used synthetically [194,195].

3-Bromo-2-trimethylstannylthiophene [193]

To a solution of 1.6 *M* butyllithium (85.3 ml, 0.14 mol) in anhydrous tetrahydrofuran (85 ml) at −70 °C is added a solution of diisopropylamine (16.1 g, 0.16 mol) in tetrahydrofuran (15 ml). The temperature is raised to −60 °C for 10 min and then the reaction mixture is cooled to −70 °C before a solution of 3-bromothiophene (18.5 g, 0.11 mol) in anhydrous tetrahydrofuran (5.0 ml) is rapidly added. The reaction mixture is stirred at this temperature for 30 min, then a solution of trimethyltin chloride (36.2 g, 0.18 mol) in anhydrous tetrahydrofuran (37 ml) is added dropwise. The temperature is maintained between −40 and −30 °C. The reaction mixture is allowed to warm to room temperature overnight and then it is poured into ice. The phases are separated and the aqueous phase extracted with ether (200 ml). The combined organic phases are dried over magnesium sulfate and evaporated. Distillation of the residue under reduced pressure gives 27.9 g (75%) of the title compound bp 77–78 °C/0.9 mm Hg.

Metalation of 2-bromo-3-octylthiophene with lithium diisopropylamide in tetrahydrofuran at −40° followed by reaction with tributylstannyl chloride gives 2-bromo-3-octyl-5-tributylstannylthiophene [196].

2-Bromo-3-octyl-5-tributylstannylthiophene [196]

To a solution of 2-bromo-3-octylthiophene (8.5 g, 30 mmol) in anhydrous tetrahydrofuran (30 ml) at −40 °C 2 *M* lithium diisopopylamide is added dropwise. The stirring is continued at −40 °C for 30 min, after which chlorotributylstannane (9.49 ml, 35 mmol) is slowly added at the same temperature. The mixture is allowed to warm to room temperature stirred for 12 h and hydrolyzed. The product is taken up in diethyl ether and the combined organic phases are dried over magnesium sulfate, evaporated and distilled bp 165 °C/0.015 torr. The crude product is purified by cromatography on aluminia using chloroform as eluent giving 13.4 g (79%) of the title compound.

Metalation of 3-bromothiophene and 4-methyl-3-bromothiophene with lithium diisopropylamide followed by reaction with trimetylsilyl chloride gives 3-bromo-2-trimethylsilylthiophene and 3-bromo-4-methyl-2-trimethyl-silylthiophene [197].

3-Bromo-2-trimethylsilylthiophene [197]

To a solution of butyllithium in hexane (173 ml, 0.25 mol) under nitrogen at 0 °C diisopropylamine (35.5 ml, 0.25 mol) in anhydrous diethyl ether (150 ml) is added. The stirring is continued for 30 min, after which 3-bromothiophene (40.5 g, 0.25 mol) is added dropwise. The mixture is stirred for 30 min and cooled to −70 °C. A solution of trimethylsilyl chloride (33.3 ml, 0.25 mol) in anhydrous diethyl ether (100 ml) is added and the reaction mixture is stirred at this temperature for 2 h, left overnight and then dilute hydrochloric acid is added. The phases are separated and the aqueous phase extracted with diethyl ether. The combined organic phases are washed with water, dried, evaporated and distilled giving 36.9 g (63%) of the title compound bp 94–98 °C/18 mm Hg.

Metalation of 4-bromo-2-methylthiophene with lithium diisopropylamide followed by dry ice or 2-methylcyclohexanone are good methods for the preparation of 3-bromo-5-methyl-2- thiophenecarboxylic acid [198] and 3-bromo-5-methyl-2-(2-methyl-1-cyclohexenyl)thiophene respectively [199].

3-Bromo-5-methyl-2-(2-methyl-1-cyclohexenyl)thiophene [199]

Butyllithium 1.5 *M* in hexane (16 ml, 24 mmol) is added to a diisopropylamine (2.4 g, 24 mmol) solution in anhydrous ether (50 ml). After 10 min 4-bromo-2-methylthiophene (3.5 g, 20 mmol) in anhydrous ether (100 ml) is added. The mixture is stirred for 45 min at room temperature and 2-methylcyclohexanone (2.2 g, 20 mmol) in anhydrous ether is then added. The reaction mixture is stirred for 3 h and then hydrolyzed with 5 *M* hydrochloric acid. The organic phase is separated, washed with water, dried over magnesium sulfate and evaporated. The remaining liquid is refluxed for 1.5 h with *p*-toluenesulfonic acid (0.5 g) in toluene in a flask provided with a water separator. The mixture is diluted with water and the phases are separated. The organic phase is washed with water, aqueous sodium bicarbonate and water. After drying over magnesium sulfate and evaporation the residue is chromatographed (silica, hexane) giving 3.6 g (67%) of the title compound.

Metalation of 2,3-dibromothiophene with lithium diisopropylamide under the same conditions followed by reaction with trimethylsilyl chloride leads to rearrangement yielding 3,5-dibromo-2-trimethylsilylthiophene [193,200].

3,5-Dibromo-2-trimethylsilylthiophene [193]

To a solution of 1.5 *M* butyllithium (5.0 ml, 7.4 mmol) in anhydrous tetrahydrofuran (22 ml) at −70 °C is added a solution of diisopropylamine (0.81 g, 8.1 mmol) in anhydrous tetrahydrofuran (8 ml). The temperature is raised to −60 °C for 10 min. To the reaction mixture cooled to −78 °C is added rapidly a solution of 2,3-dibromothiophene (1.5 g, 6.2 mmol) in anhydrous tetrahydrofuran (8 ml), and stirring is continued for 30 min at the same temperature. The intermediate thus formed is trapped with trimethylsilyl chloride (0.67 g, 6.2 mmol) in anhydrous tetrahydrofuran (4 ml), and stirring is continued for 30 min at −80 °C. The reaction mixture is then poured into 1 *M* hydrochloric acid (40 ml), followed by extraction with ether (80 ml). The phases are separated and the organic phase washed with water (15 ml) and dried over magnesium sulfate. The solvent is removed and the residue is distilled under reduced pressure giving 1.42 g (73%) of the title compound bp 88–89 °C/1.2 mm Hg.

Rearrangement also takes place in the same reaction with 2,5-dibromothiophene [201].

7.3.5.2 *Via* halogen–metal exchange

The four dibromothiophenes (2,3-dibromothiophene obtained by direct bromination of 3-bromothiophene) [185,202,203] constitute the most useful starting materials for the preparation of numerous thiophene derivatives through selective halogen–metal exchange followed by reaction with an electrophile. 3-Bromo-2-, 4-bromo-2-, and 5-bromo-2-thiophene aldehyde as well as 4-bromo-3-thiophene aldehyde are successfully prepared in this way [133,170]. Similarly 3-bromo-4-methyl-2-thiophene aldehyde was obtained from 2,3-dibromo-4-methylthiophene [204].

3-Bromo-4-methyl-2-thiophene aldehyde [204]

2,3-Dibromo-4-methylthiophene (7.7 g, 30 mmol) in anhydrous ether (60 ml) at −78 °C under nitrogen is treated with 1.6 *M* butyllithium (20 ml, 32 mmol) and stirring is continued at this temperature for 30 min, after which

N,*N*-dimethylformamide (5.0 ml, 64 mmol) is added. The reaction mixture is stirred at −78 °C for 3 h and at room temperature for 15 h, treated with hydrochloric acid 2 *M* and extracted with ether. The combined organic phases are washed with water followed by saturated sodium bicarbonate solution and dried over sodium sulfate. After evaporation the residue is distilled at 0.6 mm Hg giving a forerun 29 °C of 3-bromo-4-methylthiophene (50 mg) and a main fraction at 87–88 °C, which turns to a solid. Recrystallization from pentane gives 4.3 g (70%) of the title compound mp 62 °C.

Halogen–metal exchange of 2,4-dibromo-5-methylthiophene with butyllithium at −78 °C followed by reaction with trimethylsilyl chloride gives 4-bromo-5-methyl-2-trimethylsilylthiophene in 60% yield [205].

3-Methyl-2-bromothiophene has to be used for the preparation of 3-methyl-2-thienyllithium as metalation leads predominantly to 3-methyl-5-thienyllithium. Upon reaction with trimethylsilyl chloride, 3-methyl-2-trimethylsilylthiophene is obtained in 96% yield [192].

A one-pot procedure for the preparation of 2-benzyl-3-bromothiophene consists in halogen–metal exchange of 2,3-dibromothiophene followed by reaction with benzaldehyde followed by reduction with lithium aluminium hydride [206].

2-Benzyl-3-bromothiophene [206]

Butyllithium 1.5 *M* in anhydrous diethyl ether (300 ml, 0.45 mol) is cooled to −70 °C under nitrogen. A solution of 2,3-dibromothiophene (109 g, 0.45 mol) in anhydrous diethyl ether (100 ml) is added dropwise. The resulting solution is stirred for 1 h, after which a solution of freshly distilled benzaldehyde (48 g, 0.45 mol) in anhydrous diethyl ether (50 ml) is added over a period of 1 h at −70 °C. The reaction mixture is stirred for 1 h and warmed to room temperature and then poured onto ice. The phases are separated and the aqueous phase extracted with ether. The combined ether phases are washed with water and dried over magnesium sulfate. Evaporation of the solvent gives 2-(α-hydroxybenzyl)-3-bromothiophene (121 g) as a crude oil, which is used without further purification. The oil (121 g) is dissolved in anhydrous ether (100 ml) and added slowly to an ice-cooled solution of lithium aluminium hydride (26 g, 0.69 mol) and anhydrous aluminium chloride (90 g, 0.67 mol) in anhydrous ether (350 ml). The reaction mixture is stirred at room temperature for 10 h and then poured onto ice. After separation of the layers the aqueous phase is extracted several times with ether. The combined ether phases are washed with aqueous sodium bicarbonate and dried over magnesium sulfate. Evaporation of the solvent and distillation of the crude product afforded 96 g (84% from 2,3-dibromothiophene) of the title compound as a colorless oil bp 100–102 °C/1.5 mm Hg.

Both the metalation of 2,5-dibromothiophene and 2,3-dibromothiophene with lithium diisopropylamide in tetrahydrofuran at −70 °C occurs with rearrangement, yielding 3,5-dibromo-2-trimethylsilylthiophene [201,207] upon reaction with trimethylsilyl chloride. In a recent reinvestigation it was found that lithiation of 2,5-dibromothiophene with lithium diisopropylamide followed by triethylsilyl chloride gives 3,5-dibromo-2-trimethylsilylthiophene or 3,4- dibromo-2,5-bis(trimethylsilylthiophene depending on the ratio of the reagents [208].

The very labile 4-bromo-2,5-dichloro-3-hydroxythiophene system, existing as a 1:2 mixture of the keto and hydroxyforms, is prepared by halogen–metal exchange of 3,4-dibromo-2,5-dichlorothiophene followed by reaction with tributyl borate followed by oxidation with hydrogen peroxide. Reaction with diazomethane and acetyl chloride, respectively, leads to 4-bromo-2,5-dichloro-3-methoxythiophene and 3-acetoxy-4-bromo-2,5-dichlorothiophene [40].

3-Acetoxy-4-bromo-2,5-dichlorothiophene [40]

To a stirred solution of 3-hydroxy-4-bromo-2,5-dichlorothiophene (1.44 g, 5 mmol) in acetyl chloride (2 ml) excess of triethylamine is carefully added. The product is formed spontaneously. After addition of diethyl ether the phases are separated and the organic phase washed with water and dilute hydrochloric acid, dried over magnesium sulfate and evaporated. The residue is purified by preparative thin-layer chromatography on silica gel using diethyl ether/petroleum ether (1:10) as eluent giving 0.66 g (61%) of the title compound as a colorless liquid, which becomes reddish after standing under nitrogen for some days.

Halogen–metal exchange of 2,5-dibromo-3,4-ethylenedioxythiophene with one equivalent of butyllithium followed by mesitylthio chloride gives 2-bromo-5-mesitylthio-3,4-ethylenedioxythiophene [159].

2-Bromo-5-mesitylthio-3,4-ethylenedioxythiophene [159]

To a solution of 2,5-dibromo-3,4-ethylenedioxythiophene (4.99 g, 16.6 mmol) in anhydrous tetrahydrofuran (20 ml) at −78 °C, butyllithium (17.4 mmol) is added dropwise. The stirring is continued at −78 °C for 1.5 h, after which a solution of mesitylthio chloride (16.6 mmol) in hexanes (20 ml) is slowly added. This solution is stirred at −78 °C for 30 min, warmed to room temperature and stirred for an additional 30 min. After washing with water (3 × 50 ml) the organic phase is dried over magnesium sulfate and evaporated giving 4.5 g (73%) of the title compound as a red solid mp 95–96 °C.

Halogen–metal exchange of 2,4-dibromo-3-(methylthio)thiophene with butyllithium followed by reaction with dimethyl disulfide offers a convenient route to 4-bromo-2,3-bis(methylthio)thiophene [166].

4-Bromo-2,3-bis(methylthio)thiophene [166]

A cooled solution of 2,4-dibromo-3-(methylthio)thiophene (4.32 g, 15.0 mmol) in anhydrous diethyl ether (100 ml) at −78 °C under nitrogen is treated dropwise with 2.02 *M* butyllithium in cyclohexane (8.0 ml, 16.0 mmol), maintaining the temperature below −70 °C. The resulting mixture is stirred at this temperature for 30 min and then a solution of dimethyl disulfide (1.41 g, 15.0 mmol) is added slowly. The reaction mixture, after stirring at −78 °C for 3 h, is allowed to warm to room temperature, and then poured into ice-cold saturated ammonium chloride solution. The phases are separated and the organic phase washed with 1 *M* sodium hydroxide solution and dried over magnesium sulfate. After removal of the solvent, column chromatography using light petroleum for elution affords 2.44 g (64%) of the title compound (bp 88–90 °C/0.16 mm Hg).

Some other recent examples are the halogen–metal exchange of 3,4-dibromothiophene to 4-bromo-3-thienyllithium followed by reaction with isopropyl disulfide to give 3-bromo-4-isopropylthiothiophene [209] or the reaction with sulfur and methyl chloroacetate to give methyl 4-bromo-3-thienylthioacetate [210]. Another recent example is the preparation of 4-bromo-3,5-dimethyl-2-thiopheneboronic acid by halogen–metal exchange of 2,4-dibromo-3,5-dimethylthiophene with butyllithium followed by reaction with tributyl borate [211].

4-Bromo-3,5-dimethyl-2-thiopheneboronic acid [211]

To a stirred solution of 2,4-dibromo-3,5-dimethylthiophene (15.0 g, 55.5 mmol) in anhydrous diethyl ether (300 ml) at −70 °C 1.6 *M* butyllithium in hexane (40 ml, 64 mmol) is added dropwise. The stirring is continued for 30 min at −70 °C, before tributyl borate (10 ml, 81 mmol) is added. The reaction mixture is stirred for 2 h and then allowed to attain room temperature, after which 4% hydrochloric acid (100 ml) is added. The phases are separated and the organic phase extracted two times with 4% sodium hydroxide (100 ml). The combined aqueous phases are acidified with 10% hydrochloric acid. The white precipitate is filtered off, washed with water and dried *in vacuo* giving 10.3 g (80%) of the title compound.

Bromination of 3-formyl-2-thiopheneboronic acid and 2-formyl-3-thiopheneboronic acid with bromine in chloroform gives 5-bromo-3-formyl-2-thiopheneboronic acid and 5-bromo-2-formyl-3-thiopheneboronic acid in 46% and 70%, respectively [212].

4-Bromo-3-formyl-2-thiopheneboronic acid [212]

To a solution of butyllithium in hexane (0.14 mol) and anhydrous diethyl ether (100 ml) under nitrogen, freshly distilled diisopropylamine (14 g, 0.14 mol) in anhydrous diethyl ether (25 ml) is added. The mixture is cooled to −70 °C, after which 2-(4-bromo-3-thienyl)-1,3-dioxolane is added dropwise and the stirring is continued at −70 °C for 2.5 h, whereupon butyl borate (36 g, 0.16 mol) in anhydrous diethyl ether (100 ml) is added and the stirring is continued for another 3 h. When the temperature has risen to 0 °C, 5 *M* hydrochloric acid is added. The phases are separated and the organic phase extracted with sodium carbonate solution. The combined aqueous phases are acidified with 2 *M* hydrochloric acid, the precipitate formed is filtered off and recrystallized from aqueous ethanol giving 19.5 g (65%) of the title compound (mp 112–120 °C).

3-Bromo-2-ethyl-5-phenylthiophene was prepared by halogen–metal exchange of 2,4-dibromo-5-ethylthiophene followed by reaction with tributyl borate and palladium(0)-catalyzed coupling of the intermediate 3-bromo-2-ethyl-5-thiopheneboronic acid with iodobenzene [106].

3-Bromo-2-ethyl-5-phenylthiophene [106]

To a solution of 2,4-dibromo-5-ethylthiophene (20g, 74 mmol) in anhydrous diethyl ether (340 ml) under nitrogen at −78 °C 15% butyllithium in hexane (80 mmol) is added. After stirring at this temperature for 1 h tributyl borate (30 ml, 0.11 mol) is slowly added. The stirring is continued at −78 °C for 2 h and then at room temperature overnight. The reaction mixture is neutralized with concentrated hydrochloric acid and extracted with diethyl ether. The combined organic phases are extracted with sodium hydroxide solution and the alkaline phase is acidified with concentrated hydrochloric acid. The precipitate formed is filtered off and dissolved in tetrahydrofuran containing 20% aqueous sodium carbonate solution (100 ml), a palladium catalyst, and iodobenzene (9.6 g, 47 mmol). The reaction mixture is refluxed for 5 h at 70 °C. The product is extracted with diethyl ether and the combined organic phases are dried over magnesium sulfate and evaporated. The residue is purified by chromatography on silica gel using hexane as eluent giving 9.1 g (46%) of the title compound as a colorless oil.

Reaction of 2,4-dibromo-5-methylthiophene with butyllithium at −70 °C followed by tributyl borate and Suzuki coupling with 4-bromoanisole and 4-bromopyridine gives 3-bromo-2-methyl-5-(4-methoxyphenyl)thiophene [213–215] and 3-bromo-2-methyl-5-(4-pyridyl)thiophene, respectively [215,216].

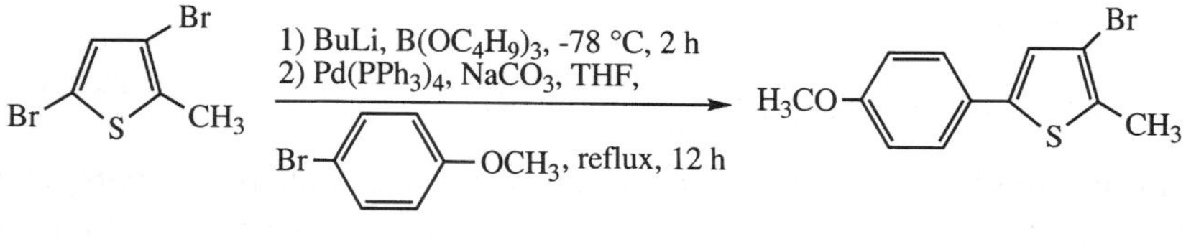

Interestingly the presence of bromine in 3-bromo-2-methyl-5-thiopheneboronic acid does not disturb in the Suzuki coupling with iodobenzene, *para*-alkylsubstituted iodobenzenes [217] or 1-iodo-4-trimethylsilylethynylbenzene, which gives a 73% yield of 3-bromo-2-methyl-5-(4-trimethylsilylethynylphenyl)thiophene [218].

It is interesting to note that halogen–metal exchange of 3,4-dibromothiophene in ether followed by reaction with cyclohexanone or *N*-methyl-4-piperidone gives excellent yields of 1-(4-bromo-3-thienyl)-cyclohexan-1-ol and 4-(4-bromo-3-thienyl)-1-methyl-4-piperidinol, respectively. Only mediocre yields are obtained if the reactions are carried out in tetrahydrofuran [219].

1-(4-Bromo-3-thienyl)-cyclohexan-1-ol [219]

3,4-Dibromothiophene (2.8 g, 11 mmol) in anhydrous ether (15 ml) at −80 °C under nitrogen is added to butyllithium (0.79 *M*) in ether (14.5 ml, 11.0 mmol) followed by cyclohexanone (1.1 g, 11 mmol) in anhydrous ether (10 ml). After stirring the reaction mixture for 1 h, it is poured into water, extracted with ether, dried over magnesium sulfate and evaporated. The residue affords 2.7 g (90%) of the title compound as colorless crystals mp 49–51 °C.

Halogen–metal exchange of 3,5-dibromo-2-phenylthiophene with butyllithium at −70 °C, followed by reaction with dimethyl sulfate gives 3-bromo-5-methyl-2-phenylthiophene [198].

3-Bromo-5-methyl-2-phenylthiophene [198]

To 3,5-dibromo-2-phenylthiophene (50.0 g, 0.157 mol) in anhydrous ether (250 ml) is added butyllithium (1.40 *M*) in hexane (115 ml, 0.161 mol) at −70 °C followed by dimethyl sulfate (20.5 g, 0.161 mol) in anhydrous ether (100 ml). During the addition the temperature is kept under −65 °C and when the addition is completed the reaction mixture is stirred at −70 °C for 4 h, after which it is allowed to reach room temperature. Concentrated ammonium hydroxide is added, the phases separated and the organic phase washed with hydrochloric acid (2 *M*) and water. After drying and evaporation distillation gives 33.3 g (56%) of the title compound bp 116–120 °C/1.0 mm Hg.

The isomeric 3-bromo-2-methyl-5-phenylthiophene can be prepared by halogen–metal exchange of 3,5-dibromo-2-methylthiophene with butyllithium, followed by reaction with cyclohexanone and aromatization of the intermediate cyclohexenyl derivative with 2,3-dichloro-4,5-dicyanobenzoquinone [198]. From 2,4-dibromothiophene, 4-bromo-2-thiophenecarboxylic acid is obtained in 79% yield upon reaction with butyllithium at −70° C followed by carbon dioxide [187]. By using *sec*-butyllithium followed by reaction with triphenyl phosphite, tris(3- bromothienyl)- and tris(5-methyl-3-bromothienyl)-phosphine can be prepared [220].

Recently 3-bromo-4-*tert*-butoxythiophene was also prepared from 4-bromo-3-thienyllithium with magnesium bromide followed by *tert*-butyl perbenzoate [121]. Halogen–metal exchange of 3,5-dibromo-2-methylthiophene followed by reaction with tributyl borate is a useful route to 4-bromo-5-methyl-2-thiopheneboronic acid [113].

Halogen–metal exchange of 2,3-dibromothiophene with butyllithium at −70 °C, followed by reaction with *N*-formylpiperidine is a recent modification of the synthesis of 3-bromo-2-thiophene aldehyde [221].

3-Bromo-2-thiophene aldehyde [221]

2,3-Dibromothiophene (24.2 g, 0.10 mol) in anhydrous ether (100 ml) is treated with 2.3 *M* butyllithium (44.3 ml, 0.102 mol). The addition is slow and the temperature kept between −70 and −64 °C. The reaction mixture is stirred at this temperature for 15 min, after which *N*-formylpiperidine (11.7 ml, 0.105 mol) is slowly added over a period of 20 min and the stirring continued for 30 min. The cooling bath is removed and stirring continued at room temperature until the internal temperature becomes 0 °C. Hydrochloric acid (3 *M* 50 ml) is added while keeping the internal temperature at 0 °C. The mixture is transferred to a separatory funnel, more hydrochloric acid (50 ml) is added and the funnel is shaken vigorously. The phases are separated and the aqueous phase extracted with ether. The combined ether phases are first washed with water and then with saturated sodium bicarbonate solution. After drying over magnesium sulfate and evaporation, 18.7 g (98%) of the title compound is obtained.

The reaction of 3-bromo-2-thienyllithium with tropylium chloride in tetrahydrofuran at −70 °C was recently used for the preparation of 3-bromo-2-cycloheptatrienylthiophene [222]. Halogen–metal exchange of *tert*-butyl 4,5-dibromo-2-thiophenecarboxylate at −80°C followed by reaction with *N*,*N*-dimethylformamide can be used for the preparation of *tert*-butyl 4-bromo-5-formyl-2-thiophenecarboxylate without any complication due to the ester function [223].

tert-Butyl 4-bromo-5-formyl-2-thiophenecarboxylate [223]

To a solution of *tert*-butyl 4,5-dibromo-2-thiophenecarboxylate (118 mg, 0.35 mmol) in tetrahydrofuran (4 ml) under argon cooled to −80 °C, butyllithium (1.68 *M*) in hexane (0.21 ml, 0.35 mol) is added. After stirring at this temperature for 4 min *N*,*N*-dimethylformamide (54 µl, 0.70 mmol) is added and the stirring continued for 20 min, after which the reaction is quenched with an aqueous saturated ammonium chloride solution. Usual work-up gives a residue, which is separated by column chromatography over silica gel using hexane/ethyl acetate (13:1) giving 74.5 mg (74%) of the title compound as a colorless syrup.

7.3.5.3 *Via* other metalorganic reagents

Grignard reagent, followed by reactions with trialkylsilyl halides, is also a possibility for the preparation of trialkylsilyl halides as illustrated by the preparation of 2-bromo-5-trimethylsilylthiophene [224].

2-Bromo-5-trimethylsilylthiophene [224]

To a solution of 2,5-dibromothiophene (136 g, 0.569 mol) in anhydrous diethyl ether (300 ml) under nitrogen, magnesium turnings are added in small portions at such a rate that the mixture is gently refluxing. When the addition is completed trimethylchlorosilane (62.0 g, 0.569 mol) is added and the reaction mixture refluxed and poured into cold dilute hydrochloric acid solution. The phases are separated and the organic phase washed with saturated sodium carbonate solution and water. The solvent is distilled off and the residue distilled at reduced pressure giving 66.5 g (50%) of the title compound as a clear liquid bp 92–96 °C/10 mm hg.

The reaction of 3,4-dibromothiophene with Rieke manganese leads to 4-bromo-3-thiophenemanganese bromide, which upon reaction with acid chlorides such as benzoyl chloride gives 3-benzoyl 4-bromothiophene and with iodobenzene gives 3-bromo-4-phenylthiophene [225].

3-Benzoyl-4-bromothiophene [225]

To a slurry of Rieke manganese (10.0 mmol) in tetrahydrofuran (10 ml) under argon is added 3,4-dibromothiophene (5.0 mmol) at room temperature, and the mixture is stirred at room temperature for 5 h. 1,2-Dibromoethane (6.0 mmol) is added neatly to the reaction mixture at 0 °C and the reaction mixture allowed to warm to room temperature over 20 min. To the resulting organomanganese reagent is added benzoyl chloride (3.0 mmol) neatly at room temperature *via* syringe. The resulting mixture is stirred at room temperature for 30 min. The mixture is quenched with 3 *M* hydrochloric acid (10 ml) and extracted with ether (2 × 10 ml) and the combined organic layers are sequentially washed with saturated sodium bicarbonate, sodium thiosulfate, and sodium chloride solutions, dried over magnesium sulfate, and concentrated. Flash chromatography using ethyl acetate/hexane as eluent affords the title compound in 80% yield.

7.3.5.4 *Via* electrophilic substitution

A large number of substituted bromothiophenes can be prepared using various types of electrophilic substitution reactions. Halogen-dance can occur upon attempted chloromethylation of halothiophenes. Thus reaction of 2,5-dibromothiophene with chloromethyl methyl ether in carbon disulfide in the presence of tin tetrachloride gives a 21% yield of 2,4-dichloromethyl-3,5-dibromothiophene [226], which has also been prepared by chloromethylation of 2,4-dibromothiophene [227]. Other conditions for the dichloromethylation have also been developed. Thus 3,4-dibromothiophene gives the 2,5-dichloromethyl derivative using formaldehyde, hydrochloric acid and zinc chloride in tetrachloromethane. 3,5-Dimethylthiophene and 3-bromo-5-methylthiophene gave the 2,4-chloromethyl derivatives, when chloromethyl methyl ether and titanium chloride in carbon disulfide was used as chloromethylating reagent. Chloromethylation of 2,5-dibromothiophene with chloromethyl methyl ether in carbon disulfide using tin tetrachloride as catalyst occurs with rearrangement yielding 2,4-dibromo-3,5-dichloromethylthiophene. Although the yield is low, this product is useful for the preparation of otherwise difficultly accessible compounds [228].

2,4-Dibromo-3,5-dichloromethylthiophene [228]

To a stirred mixture of 2,5-dibromothiophene (35.0 g, 0.25 mol), chloromethyl methyl ether (64.0 g, 0.80 mol) and carbon disulfide (190 ml) is added dropwise a mixture of tin tetrachloride (77.0 g, 0.30 mol) and carbon disulfide (15 ml) over a period of 2 h. The reaction mixture is then stirred for 2.5 h at 0 °C, for 1 h at room temperature and poured into ice-water. The organic phase is separated, washed with sodium chloride solution and dried over magnesium sulfate. The solvent is evaporated and the residue treated with hot hexane (100 ml). This solution affords, upon cooling, 18 g (21%) of the title compound as colorless needles mp 86.0–88.0 °C.

Nitration of 2,5-dibromothiophene with fuming nitric acid in fuming sulfuric acid is an excellent method for the preparation of 2,5-bromo-3,4-dinitrothiophene [229].

5-Bromo-4-nitrothiophene-2-carbonitrile is prepared by nitration of 5-bromothiophene-2-carbonitrile with fuming nitric acid in acetic anhydride at 20 °C [230].

5-Bromo-4-nitrothiophene-2-carbonitrile [230]

To a solution of fuming nitric acid (17 ml) in acetic anhydride (34 ml) at 20 °C 5-bromothiophene-2-carbonitrile (1.9 g, 10 mmol) is added dropwise. The reaction mixture is stirred at 25–30 °C for 2 h and then heated to 60 °C on a water bath. After cooling it is poured onto crushed ice. The precipitate formed is filtered off and washed with water giving 1.5 g (65%) of the title compound as a white solid mp 123 °C after recrystallization from methanol.

Chlorination of 2-bromo-3-methylthiophene with *N*-chlorosuccinimide in acetic acid gives 2-bromo-5-chloro-3-methylthiophene [89].

2-Bromo-5-chloro-3-methylthiophene [89]

To 2-bromo-3-methylthiophene (88.5 g, 0.050 mol) in acetic acid (300 ml) *N*-chlorosuccinimide (66.8 g, 0.50 mol) is added in one portion. The reaction mixture is stirred at 40 °C for 2 h and then refluxed for 1 h, after which it is poured into water (1500 ml) and extracted with ether (3 × 500 ml). The combined ether phases are washed with 1 *M* sodium hydroxide solution (3 × 300 ml) followed with water until neutral, dried over sodium sulfate, evaporated and distilled, giving 95.5 g (90%) of the title compound bp 79 °C/10 mm Hg.

Friedel–Crafts reaction of 2-bromothiophene [231] and 2,4-dibromothiophene with benzoyl chloride are good methods for the preparation of 2-bromo-5-benzoylthiophene and 2-benzoyl-3,5-dibromothiophene, respectively [232].

2-Benzoyl-3,5-dibromothiophene [232]

To a mixture of 2,4-dibromothiophene (10.0 g, 41.3 mmol) and benzoyl chloride (6.4 g, 45.5 mmol) in anhydrous dichloroethane (100 ml), aluminium chloride (8.3 g, 63 mmol) is added under stirring at such a rate that the temperature is kept at 25 °C. When the addition is completed the stirring is continued for 1 h at room temperature, after which the reaction mixture is poured into 2 *M* hydrochloric acid. The phases are separated and the aqueous phase extracted several times with dichloromethane. The combined aqueous phases are washed with 2 *M* sodium hydroxide solution and water, dried over sodium sulfate/charcoal and evaporated giving 12.0 g (84%) of the title compound as a yellow oil.

3-[2-(5-Bromothienyl)phthalide] is prepared by the aluminium chloride-catalyzed reaction of 2-bromothiophene with phthalic anhydride in dichloromethane [233].

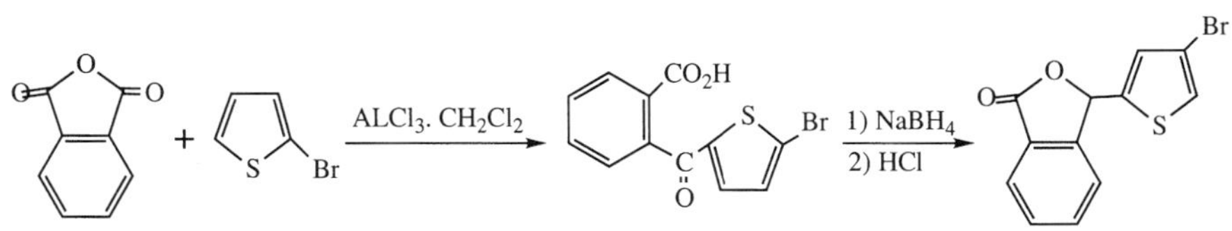

2,5-Dibromo-3,4-di(4-butylbenzoyl) thiophene is prepared by the reaction of 2,5-dibromo-3,4-dicarboxythiophene with oxalyl chloride, followed by butylbenzene and aluminium chloride [234].

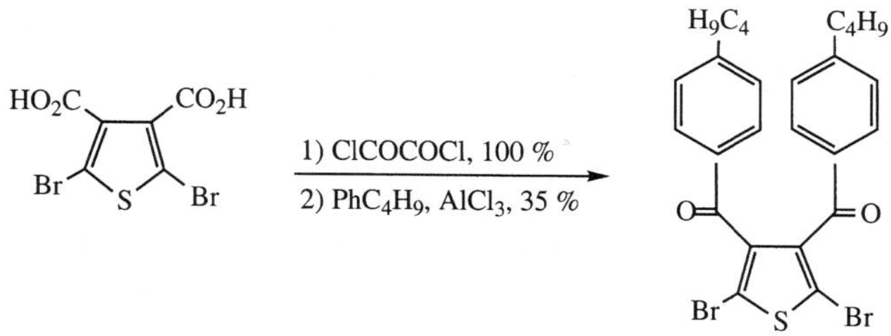

From 4-bromo-2-thenoyl chloride and anisole, 4-bromo-2-thienyl 4-methoxyphenyl ketone is prepared [187].

(4-Bromo-2-thienyl)(4-methoxyphenyl)methanone [187]

4-Bromo-2-thiophenecarboxylic acid (13.40 g, 0.0647 mol) is converted to the acid chloride by reaction with thionyl chloride (27 ml, 0.37 mol) in chloroform (67 ml). The mixture is refluxed for 2 h and concentrated to an oil, which is dissolved in dichloromethane (67 ml). This solution is added under ice cooling to anisole (7.70 g, 0.0712 mol) and aluminium chloride (9.49 g, 0.0712 mol) in dichloromethane (134 ml). The stirring is continued under cooling for 30 min and at room temperature for 30 min, then the reaction mixture is cooled with an ice-bath while water is slowly added. The phases are separated and the organic phase washed with water and 5% sodium bicarbonate solution, dried and evaporated giving a quantitative yield of an oil, which solidifies upon standing.

Its oxime as well as the oxime of 2,5-dichloro-3-acetylthiophene and 3-bromo-2-acetyl- thiophene were prepared as starting materials for thienoisoxazoles [232].

7.3.5.5 *Via* Heck-couplings

Heck-coupling of 3-bromo-2-iodo-3-methylthiophene with the appropriate acetylenzinc derivative gives the acetylenic derivative shown below [199].

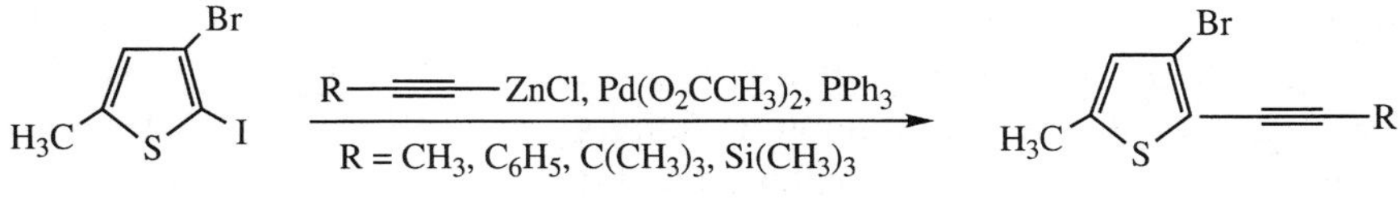

7.3.6 *Via* isomerization of bromothiophenes

3-Bromothiophene can be prepared on a large scale by treatment of 2-bromothiophene with an excess of sodamide in liquid ammonia and subsequent quenching of the reaction mixture with solid ammonium chloride [235].

3-Bromothiophene [235]

A 3-l three-necked, round-bottomed flask, equipped with a dropping funnel, an efficient mechanical stirrer and a thermometer, is charged with thiophene (420 g, 5.0 mol), 48% aqueous hydrobromic acid (1000 g, 6.0 mol), and diethyl ether (250 ml). This mixture is cooled to −20 °C, after which 35% aqueous hydrogen peroxide (400 g, 4.0 mol) is added dropwise. Initially the temperature of the reaction mixture is kept between −15 and −20 °C, but it is gradually raised to between 0 and +5 °C in the last stage of the addition. Efficient stirring and cooling is necessary during the addition, which is carried out over 90 min. When after the addition, the exothermic reaction has subsided, the temperature of the mixture is allowed to rise gradually to +20 °C. The reaction is finished when the upper layer no longer turns brown. After addition of 500 ml of water the layers are separated. The aqueous layer is extracted with diethyl ether (3 × 50 ml). The first organic phase and the combined extracts are washed separately with water, and subsequently mixed and dried over magnesium sulfate.

In a 10-l wide-necked (*ca* 5 cm diameter) round-bottomed flask, anhydrous liquid ammonia (5–6 l), is introduced *via* a plastic tube from a cylinder. The water-content is controlled before starting the preparation of sodamide: small pieces of sodium (*ca* 0.1 g each) are cut from a larger piece and are introduced into the ammonia immediately after cutting. The quality is satisfactory if less than 3 g of pieces give a persisting, uniformly blue solution. After this control the flask is equipped with an efficient mechanical stirrer, which is placed centrally, the neck being left open. Stirring is started and iron(III)nitrate (1 g) is added. After a few seconds 5 g of freshly cut (under light petrol, bp 40–60 °C) pieces of sodium (*ca* 1 g each) are introduced. After 5–10 min a grey solution is formed. The remaining 195 g of the amount of 200 g sodium is then added over a few min. After about half an hour the blue color has disappeared and a greyish suspension is formed. The mirror of sodium on the upper part of the wall is rinsed into the solution, either by very vigorous stirring or by allowing a small amount of ammonia from the cylinder to run down the wall. The ethereal solution of crude 3-bromothiophene is then cautiously added (by pouring) over 5 min, while stirring at a moderate rate (but nevertheless efficiently). After 5 min the finely powdered ammonium chloride (500 g) is introduced with the aid of a spoon (10 g portions). After an additional 5 min the stirrer is removed, a plug of cotton wool is placed on the flask and the ammonia is allowed to evaporate overnight. After addition of water (3 l) to the remaining slurry, and dissolution of the salts, (by mechanical stirring or vigorous swirling by hand) the organic (lower) layer is separated as completely as possible. The aqueous phase is extracted two to five times with small portions of pentane. The combined organic phases are, if necessary, washed with diluted hydrochloric acid and dried over magnesium sulfate. The solvent is removed by distillation through an efficient column. Subsequently very careful distillation through a 40 cm Widmar column gives 66–72% of 3-bromothiophene bp 47 °C/15 mm Hg. The higher boiling fraction gives after redistillation *ca* 80 g of 3,4-dibromothiophene bp 92 °C/15 mm Hg.

3,4-Dibromothiophene can be prepared by treating 2-bromothiophene with a mixture of equimolar amounts of sodamide and potassium *tert*-butoxide in liquid ammonia [236]. Lithiation of 2,5-dibromothiophene by lithium diisopropylamide followed by silylation gives 3,5-dibromo-2-trimethylsilylthiophene or 3,4-dibromo-2,5-bis(trimethylsilyl)thiophene depending upon the ratio of the reagents in accordance with the halogen-dance mechanism [208].

7.3.7 *Via* side chain modification

Side chain bromination of 2-bromo-3-methylthiophene [69,237] and 2,5-dibromo-3-methylthiophene [238] with *N*-bromosuccinimide using benzoyl peroxide as initiator is an excellent method for the preparation of 2-bromo-3-bromomethylthiophene and 2,5-dibromo-3-bromomethylthiophene in high yields.

2,5-Dibromo-3-bromomethylthiophene [238]

A mixture of 2,5-dibromo-3-methylthiophene (43.3 g, 0.169 mol), *N*-bromosuccinimide (30.1 g, 0.169 mol), carbon tetrachloride (100 ml), benzoyl peroxide (0.5 g) and four drops of water is refluxed with stirring for 2 h. The mixture is filtered and the residue washed well with carbon tetrachloride. The filtrate is evaporated giving 56.6 g (100%) of the title compound as an orange oil.

The compound so obtained was further modified to 3-(2,5-dibromothiophen-3-yl)propionic acid [238]. 3,4,5-Tribromo-2-thenylalcohol has also been prepared from the corresponding aldehyde by the Cannizzaro reaction [239,240].

3-Bromo-2-(*N,N*-dimethylaminomethyl)-5-methylthiophene is prepared by reduction of 3-bromo-2-(*N,N*-dimethycarboxamido)-5-methylthiophene with lithium aluminium hydride [198].

3-Bromo-2-(N,N-dimethylaminomethyl)-5-methylthiophene [241]

To lithium aluminium hydride (1.05 g, 0.0277 mol) in anhydrous ether (25 ml) 3-bromo-2-(*N,N*-dimethylcarboxamido)-5-methylthiophene (10.0 g, 0.0403 mol) in anhydrous ether (25 ml) is added. When the addition is completed the reaction mixture is refluxed for 3 h, cooled and water (1.0 ml) is added. The reaction mixture is filtered and the filter washed with ether. The filtrate is dried over potassium hydroxide pellets, evaporated and distilled giving 5.0 g (53%) of the title compound bp 69–73 °C/0.6 mm Hg.

Similarly 2,5-dibromo-3,4-bis(bromomethyl)thiophene is prepared from 2,5-dibromo-3,4-dimethylthiophene [242,243].

3,4-Dibromo-2-bromomethyl-5-methylthiophene is obtained upon reaction of 3,4-dibromo-2,5-dimethylthiophene with one equivalent of bromine in dichloromethane, while two equivalents yield 3,4-dibromo-2,5-bis(bromomethyl)thiophene in high yields [244].

3,4-Dibromo-2,5-dimercaptomethylthiophene has been recently obtained by the reaction of 2,5-di(chloromethyl)-3,4-dibromothiophene with thiourea [245].

3,4-Dibromo-2,5-dimercaptomethylthiophene [245]

A solution of 2,5-dichloromethyl-3,4-dibromothiophene (6.8 g, 20 mmol) and thiourea (3.8 g, 50 mmol) in dimethylsulfoxide (50 ml) is stirred under nitrogen at room temperature for 14 h. The reaction mixture is poured into 5% aqueous sodium hydroxide solution (100 ml) and stirred for 1 h. This solution is acidified with 10% hydrochloric acid and extracted with chloroform. The combined organic phases are washed twice with sodium chloride solution, dried over magnesium sulfate and evaporated. The residue is recrystallized from hexane to afford 5.6 g (83%) of the title compound as colorless prisms mp 51.5–54.5 °C.

The *meta*-cyclothiophenophane derivative shown below is prepared from 1,3-bis(bromomethyl)benzene and 2,4-bis(mercaptomethyl)-3,5-dibromothiophene [246].

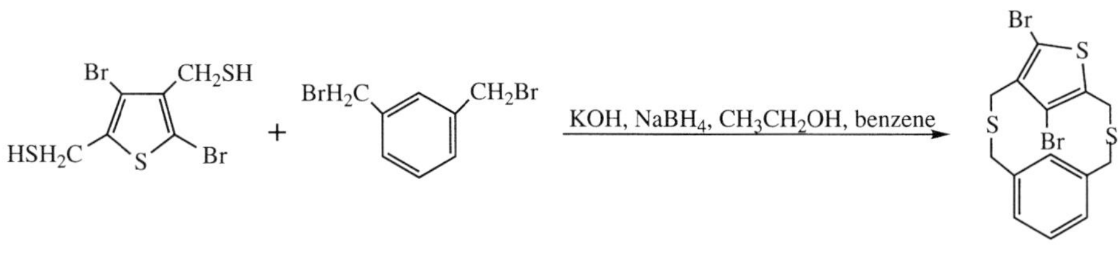

14,17-Dibromo-2,11-dithia [3]metacyclo [3](2,4)thiphenophane [246]

To a refluxing solution of potassium hydroxide (4.3 g, 80 mmol) and sodium boron hydride (380 mg, 10 mmol) in ethanol a solution of

2,4 dibromo-3,5-mercaptomethylthiophene (6.7 g, 20 mmol) and 1,3-bis(bromomethyl)benzene (5.3 g, 20 mmol) in ethanol/benzene (1:1) (200 ml) is added over 23 h. The solvents are removed by distillation and the residue poured into ice-water. The product is extracted with dichloromethane and the combined organic phases are washed with sodium chloride solution, dried over magnesium sulfate and evaporated. The residue is purified by chromatography on silica gel using hexane/dichloromethane (2:1) as eluent giving 4.8 g (55%) of the title compound and colorless prisms after recrystallization from ethanol/benzene mp 137–138 °C:

The reaction of 1,4-bis(methylmercapto)-2,5-dimethoxybenzene with 2,5-bis(chloromethyl)-3,4-dibromothiophene is used for the preparation of the paracyclothiophenophane described below [247].

14,15-Dibromo-6,9-dimethoxy-2,11-dithia[3]paracyclo[3]-(2,5)-thiophenophane [247]

A solution of 1,4-bis(methylmercapto)-2,5-dimethoxybenzene (4.5 g, 20 mmol) and 2,5-bis(chloromethyl)-3,4-dibromothiophene (6.8 g, 20 mmol) in ethanol/benzene (1:1) is added dropwise to a refluxing solution of 80% potassium hydroxide (4.5 g, 80 mmol) and sodium hydroxide (380 mg, 10 mmol) in ethanol (4 l) for 48 h. After work-up the residue is subjected to column chromatography on silica gel using hexane/chloroform (2:1) as eluent. From the eluate the solvents are removed by evaporation and the residue is recrystallized from ethanol/hexane (3:1) giving 5.0 g (51%) of the title compound as colorless needles mp 172.0–173.0 °C.

In contrast to α-bromine substituents [248], the β-bromine does not interfere in Wolff–Kishner reduction of 4-bromo-2-propionylthiophene to 4-bromo-2-propylthiophene [144]. Wolf–Kishner reduction of a mixture of 4,5-dibromo-3-methyl-2-thiophene aldehyde and 4-bromo-3-methyl-2-thiophene aldehyde gives 4-bromo-2,3-dimethylthiophene [156].

4-Bromo-2-propylthiophene [144]

To a solution of 4-bromo-2-propionylthiophene(53 g, 242 mmol) in ethylene glycol (210 ml) is added hydrazine monohydrate (30 ml, 617 mmol). The resulting solution is then heated to 160 °C for 45 min under stirring. After cooling the reaction mixture to 35 °C potassium hydroxide (42 g, 750 mmol) is added. Under stirring the mixture is heated to 160 °C and the stirring continued for 1.5 h. At room temperature water (450 ml) is added, the mixture is acidified with concentrated hydrochloric acid and the product extracted with pentane (3 × 200 ml). The combined organic phases are dried over magnesium sulfate and evaporated. The crude product (39.7 g) can either be flash chromatographed on silica gel eluting with pentane to afford 33 g (65%) or distilled bp 120–125 °C/7 mm Hg.

The reaction of 2-bromo-5-chloro-3-thenyl bromide with triethyl phosphite is used for the preparation of the halogenated thenyl phosphonate [41].

Diethyl (2-bromo-5-chloro-3-thenyl) phosphonate [89]

A mixture of 2-bromo-5-chloro-3-thenyl bromide (130 g, 0.43 mol) and triethylphosphite is heated under nitrogen at 120 °C for 1 h and at 150 °C for another 2 h. After distillation *in vacuo* 142.6 g (91%) of the title compound is obtained bp 128–134 °C/0.1 mm Hg.

A modified Curtius reaction of 3-bromo-2- and 4-bromo-3-thiophenecarboxylic acid can conveniently be used for the preparation of *tert*-butyl *N*-(3-bromo-2-thienyl)carbamate and *tert*-butyl *N*-(4-bromo-3-thienyl)carbamate [249].

tert-Butyl N-(3-bromo-2-thienyl)carbamate [249]

A 250-ml round-bottomed flask equipped with magnetic stirrer and condenser is charged with 3-bromo-2-thiophenecarboxylic acid (4.14 g, 0.02 mol), diphenyl phosphorazidate (0.022 mol), triethylamine (0.022 mol) and anhydrous *tert*-butanol (100 ml). The reaction mixture is refluxed for 15 h and the cooling water kept at 30 °C. After cooling, the excess of alcohol is evaporated and the residue dissolved in dichloromethane (50 ml). This solution is washed with 5% aqueous citric acid, water, saturated sodium bicarbonate solution, and saturated sodium chloride solution. After drying and removal of the solvent, the residue is chromatographed on silica gel, using ethyl acetate/cyclohexane (1:3) as eluent giving 4.50 g (81%) bp 104–106 °C/0.03 mm Hg.

2-(2-Nitropropen-1-yl)-5-bromothiophene is best prepared through the reaction of 5-bromo-2-thiophene aldehyde with ammonium acetate in nitroethane [250]. It can be reduced by borohydride supported on an ion-exchange resin, without removal of the bromine, to 2-(2-nitropropyl)-5-bromothiophene in 82% yield [251]. 4-Bromo-2-thienyl aryl carbinols and 4-bromo-2-thienyl aryl ketones can be prepared *via* the reaction of 4-bromo-2-thiophene aldehyde with phenylmagnesium derivatives [252].

In connection with the preparation of ionic liquid crystalline polythiophenes having viologen side chains from 2,5-dibromo-3-(2-oxa-12-bromodecyl)thiophene [253] through reaction with equimolar amounts of 1-decyl-4-(4-pyridyl)pyridinium bromide in *N*,*N*-dimethylformamide at 70 °C and successive work-up of ion-exchange from aqueous ethanol containing excess *para*-toluenesulfonic acid [254].

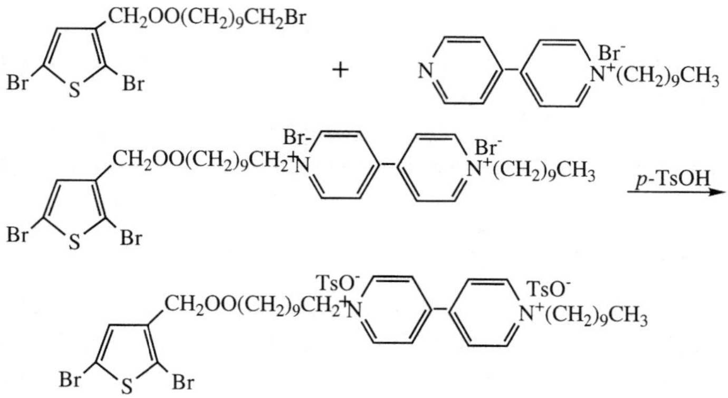

3-Bromo-2-isopropylidene-5-methylthiophene is prepared from 3-bromo-2-methyl-5-thiophene aldehyde and isopropylidene phosphorane in good yield [199].

3-Bromo-5-methyl-2-(2-methyl-1-propenyl)thiophene [199]

To a solution of isopropyltriphenylphosphonium iodide 4.3 g (10 mmol) in anhydrous tetrahydrofuran (100 ml), butyllithium (1.5 *M*) in hexane (7.0 ml, 10 mmol) is added. After 5 min, 3-bromo-5-methyl-2-thiophenealdehyde in anhydrous tetrahydrofuran (50 ml) is added. The reaction mixture is stirred for 1 h and then poured into ice-water and acidified with hydrochloric acid (5 *M*). This solution is extracted with diethyl ether. The combined organic phases are washed with sodium bicarbonate and water, dried over magnesium sulfate and evaporated. The residue is chromatographed on silica gel using hexane as eluent, giving 1.4 g (61%) of the title compound.

The isomeric *ortho*-bromothiophene aldehydes, when reacted with *ortho*-bromothenyl phosphonates or *ortho*-bromothenyltriphenylphosphoranes led to *trans* or *cis* 1,2-di(bromothenyl)ethenes in satisfactory yields [255–258]. Also bromo-chloro substituted 1,2-dithienylethenes are prepared by this approach [89,259]. 1-(Bromothienyl)-2-(bromofuryl)ethenes are prepared in a similar way [260].

General procedure for the Wittig olefin synthesis of 1,2-dithienyl ethenes [256]

To a well-stirred suspension of a phosphonium salt in anhydrous *N,N*-dimethylformamide (100 ml per 0.1 mol of phosphonium salt) an excess of sodium methoxide (1.5 equiv.) suspended in anhydrous *N,N*-dimethylformamide (30 ml per 0.1 mol) is added at 0–5 °C under nitrogen. The phosphorane intermediate (from tan to intensive orange in color), which is formed almost instantaneously, is stirred for an additional 30 min and then an equivalent amount of aldehyde dissolved in anhydrous *N,N*-dimethylformamide is added at such a rate that the reaction temperature does not exceed 30 °C. A few minutes after the addition is completed the cooling bath is removed, and the reaction mixture is stirred at room temperature for 2 h. It is then poured into ice-water and acidified with aqueous hydrochloric acid to pH 3–4, extracted with diethyl ether, washed, dried over magnesium sulfate and concentrated to about 100 ml/0.1 mol. By storing this cold concentrate overnight, most of the triphenyl phosphine oxide precipitates. After filtration the filtrate is concentrated and the oily residue is freed from the remaining triphenyl phosphine oxide by chromatography on a silica gel column using benzene as eluent. The eluent is washed with 1 *M* sodium bisulfite solution and water and dried over calcium chloride. The *cis–trans* ratio is determined by GLC. Pure *trans*-isomer can usually be obtained by fractional precipitation by hexane of the oily residue obtained after evaporation. The *cis*-isomer may be obtained as a solid from the hexane solution.

cis-1,2-Di(4-bromo-3-thienyl)ethene [258]

To a suspension of 4-bromo-3-thenyltriphenylphosphonium chloride (71.0 g, 0.15 mol) in anhydrous *N,N*-dimethylformamide (150 ml) under nitrogen, cooled to 0 °C, sodium ethoxide (11.9 g, 0.22 mol) suspended in anhydrous *N,N*-dimethylformamide (70 ml) is added. After stirring for 15 min 3-bromo-4-formylthiophene (28.5 g, 0.15 mol) in anhydrous *N,N*-dimethylformamide (130 ml) is added and the reaction mixture stirred at room temperature for 1 h. It is then poured into ice-water and dilute hydrochloric acid is added until the solution is slightly acidic. Ether is added and the phases are separated and the organic phase washed with sodium bicarbonate solution and water and dried over magnesium sulfate. After evaporation the viscous oil obtained sometimes contains triphenyl phosphine oxide. Silica gel is added in such an amount that a dry powder is obtained. A large suction funnel is used as a short chromatography column and packed with silica gel and hexane. The powder is placed on top and the ethenes diluted with hexane. The triphenylphosphine oxide remains on the silica gel. About 35.6 g (68%) of a mixture of *cis*- and *trans*-1,2-di(4-bromo-3-thienyl)ethene in a 81:19 ratio is obtained. Fractional distillation *in vacuo* gives the *cis*-isomer bp 135–140 °C/0.5 mm Hg, mp 35–37 °C as white needles after recrystallization from hexane. From the distillation residue the *trans*-isomer is obtained through extraction and recrystallization from chloroform mp 126 °C.

1-(2-Bromo-5-chloro-3-thienyl)-2-(3′-thienyl)ethene [89]

A mixture of diethyl (2-bromo-5-chloro-3-thenyl)phosphonate (21.2 g, 0.061 mol) and 3-thiophene aldehyde (7.8 g, 0.07 mol) dissolved in anhydrous

N,*N*-dimethylformamide (70 ml) is added dropwise to a suspension of sodium methoxide (9.1 g) in anhydrous *N*,*N*-dimethylformamide (30 ml), cooled in an ice-bath under nitrogen. When the addition is complete the reaction mixture is stirred for another 10 minutes and the cooling bath is removed. Stirring is continued for a further 1.5 h at room temperature and the mixture is then poured into ice-water (500 ml). The phases are separated and the organic phase extracted with diethyl ether. The combined organic phases are washed with 1 *M* aqueous sodium bisulfite and water, dried over calcium chloride and evaporated giving 16.0 (86%) of the title compound. Recrystallization from petroleum ether gives the pure *trans*-isomer mp 73–74 °C.

Other solvent-base systems which can be used are sodium hydride in 1,2-dimethoxyethane [261–264] or in *N*,*N*-dimethylformamide [265,266].

The following reaction route has been performed and used for the preparation of amine-functionalized polythiophenes [267].

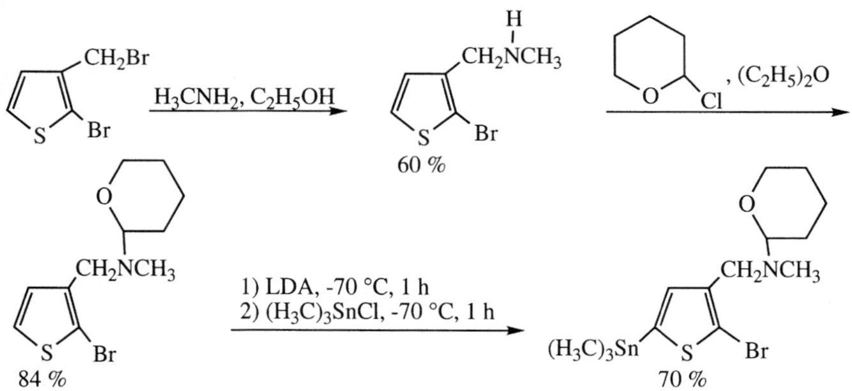

The condensation of acetylthiophenes with thiophene aldehydes is an often-used method for the preparation of compounds containing the ThCOC=C-grouping. Thus treatment of 2-acetyl-3-bromothiophene with 4-bromo-3-thiophenealdehyde with sodium hydroxide in aqueous ethanol gives 1-(3-bromo-2-thienyl)-3-(4-bromo-3-thienyl)-2-propene-1-one [268].

1-(3-Bromo-2-thienyl)-3-(4-bromo-3-thienyl)-2-propen-1-one [268]

To a solution of sodium hydroxide (10.0 g, 0.25 mol) in water (10 ml) and ethanol (40 ml), 3-bromo-4-thiophene aldehyde (21.9 g, 0.11 mol) and 3-bromo-2-acetylthiophene are added with stirring at 0 °C. After stirring at 0 °C for 3 h, the reaction mixture is kept at −15 °C overnight. The water phase is separated and the organic phase taken up in ether. The ether solution is washed with diluted hydrochloric acid and water, dried over magnesium sulfate and evaporated. The semisolid residue is crystallized from ethanol giving 20.0 g (50%) of the title compound mp 96.0–96.5 °C.

A two-step procedure can be used for the reduction of brominated thiophenic chalcones. Thus the double bond is first reduced with triethylsilane and trifluoroacetic acid to the saturated ketone, which is then reduced in the usual way with sodium borohydride [268].

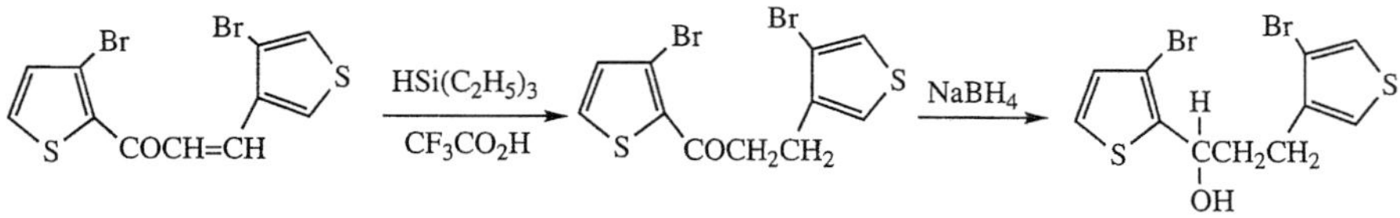

Acetals are prepared also from halogenated thiophene aldehydes [221].

3-Bromo-2-(1,3-dioxalan-2-yl)thiophene [221]

To a mixture of 3-bromo-2-thiophene aldehyde (18.7 g, 97.9 mmol) and ethylene glycol (22 ml, 400 mmol) in anhydrous toluene (100 ml) pyridinium tosylate (1 g, 4 mmol) is added. The reaction mixture is refluxed and the water removed by a Dean–Stark trap. After 1 h the mixture is cooled to room temperature and water (100 ml) and diethyl ether (100 ml) are added. The phases are separated and the organic phase washed with water (3 × 50 ml) and a saturated solution of sodium bicarbonate (25 ml), dried over magnesium sulfate and evaporated. The residue is distilled giving 17.6 g (75%) of the title compound bp 86–87 °C/0.5 mm Hg.

In some cases dimethyl acetals, such as 4,4′-dibromo-2,2′-bis(dimethoxymethyl)-3,3′-bithienyl, are obtained by dissolving the corresponding aldehyde in acidified hot methanol, and upon cooling the desired acetal crystallizes out [269–271].

Condensation of 5-bromo-2-thiophene aldehyde with (1*R*,2*R*)-1,2-diaminocyclohexane, followed by reduction with sodium borohydride of the intermediate imine gives the desired chiral (1*R*,2*R*)-*N*,*N*′-bis(5-bromothiophene-2-ylmethyl),1,2-diaminocyclohexane [271].

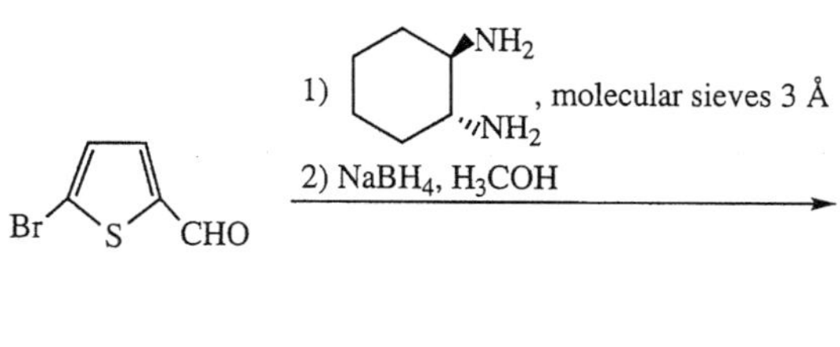

7.3.8 Various Methods

5-Bromo-3-hydroxythiophene-2-carbonitrile is prepared in the following way [232].

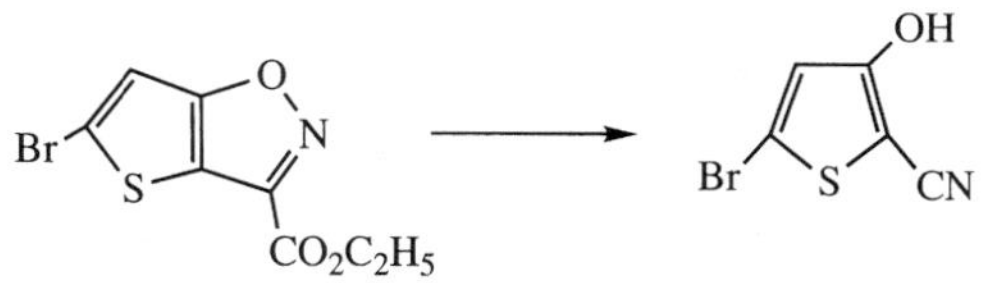

5-Bromo-3-hydroxythiophene-2-carbonitrile [232]

To a solution of ethyl 5-bromo-3-thieno[2,3-*d*]isoxazylcarboxylate (0.5 g, 1.81 mmol) in anhydrous *N*,*N*-dimethylformamide (10 ml), sodium methoxide (0.2 g, 6.6 mmol) is added and the reaction mixture stirred for 10 min. 2 *M* Hydrochloric acid is added, the phases separated and the aqueous phase extracted three times with diethyl ether. The combined organic phases are washed several times with water, dried over sodium sulfate/charcoal and evaporated. The residue is recrystallized from ethanol giving 0.32 g (87%) of the title compound as colorless crystals mp 122–125 °C.

The reaction of 3,5-dibromo-2-thiophene aldehydes with secondary amines such as morpholine, piperidine, dimethylamine, and benzylmethylamine in the presence of triethylamine yields a mixture of 3-bromo-5-dialkylamino-2-thiophene aldehyde and 5-bromo-3-dialkylamino-2-thiophene aldehyde. Chromatography gave pure 3-bromo-5-dialkylamino-2- thiophene aldehydes [272].

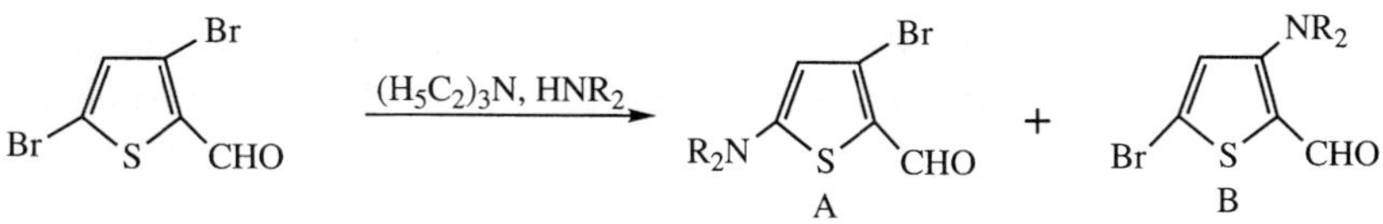

Amine	Ratio A:B	Yield of A (%)
morpholine	71:29	42
piperidine	74:26	31
dimethylamine	81:19	28
butylmethylamine	80:20	46

Bromodecarboxylation of 3,4-dimethylthiophene-2,5-dicarboxylic acid prepared *via* Hinsberg ring closure is one of the best methods for the preparation of 2,5-dibromo-3,4-dimethylthiophene [243].

An elegant method for the preparation of 4-bromo-3-methoxythiophene starts with methyl 3-hydroxy-2-thiophenecarboxylate, which is brominated to methyl 4-bromo-3-hydroxythiophenecarboxylate, which is then methylated, hydrolyzed and decarboxylated to the desired 4-bromo-3-methoxythiophene [273].

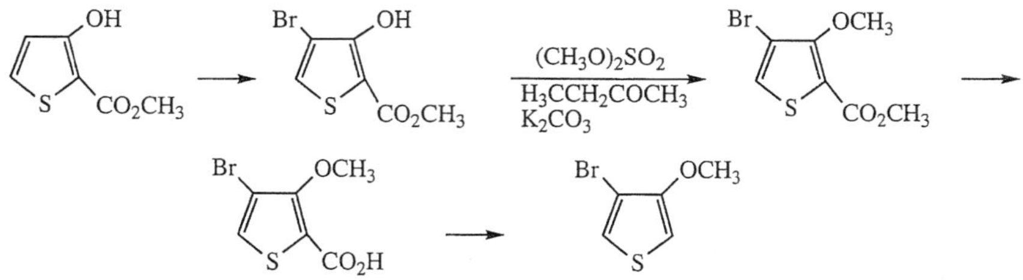

Heating 3,4-dibromothiophene and trimethylsilylacetylene in refluxing diethylamine in the presence of catalytic amounts of palladium dichloride, triphenylphosphine, and copper(I) iodide gives an 80% yield of 4-bromo-3-(trimethylsilylacetyleno)thiophene [274].

Palladium-catalyzed coupling of 3-bromo-2-iodo-5-methylthiophene with propynylzinc chloride gives 3-bromo-5-methyl-2-(1-propynyl)thiophene [199].

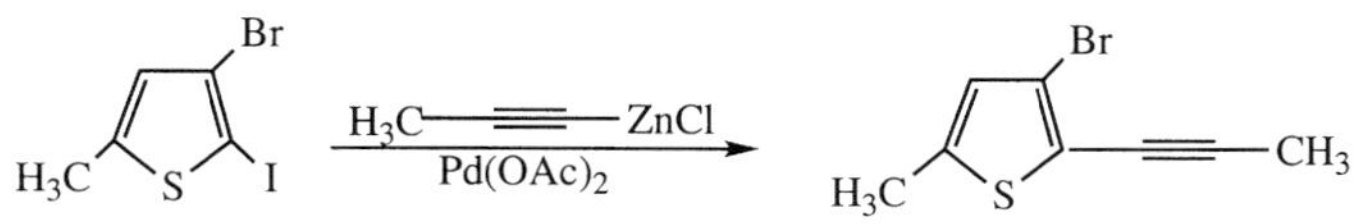

Various compounds substituted in the 3-position with groups derived from active methylene groups, such as 5-bromo-2-nitro-(3-thienyl)methyl *tert*-butylsulfonamide or *N*,*N*-dimethyl (5-bromo-2-nitro-3-thienyl)methanesulfonamide can be prepared in high yields by vicarious substitution of 5-bromo-2-nitrothiophene with anions from active methylene compounds [275]. (3-Bromo-2-nitro-5-thienyl)methyl phenylsulfone is obtained from the reaction of 3-bromo-2-nitrothiophene with the anion from chloromethyl phenylsulphone [275].

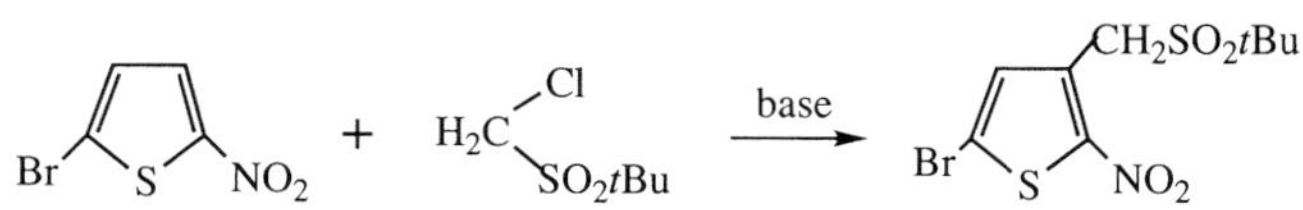

2-Bromo-5-heptafluoropropylthiophene is obtained through the reaction of 2-bromothiophene with bis(heptafluorobutyryl) peroxide [276,277]. Treatment of tritylthiomethylsubstituted acetylenic ketones with hydrobromic acid in acetic acid gives 2-substituted 4-bromothiophenes in very high yield. Acetylenic acetals substituted with tritylthiomethyl groups yield 2-substituted 3-bromothiophenes [278].

3-Bromo-2-*tert*-butyldimethylsilyl-5-tert-butyldimethylsilylethynyl-4-butylsulfanylthiophene is obtained in 70% yield by first treating 2,3,5,6-tetrabromo [3,2-*b*]thiophene successively with two equivalents of butyllithium and *tert*-butyldimethylsilyl chloride, then the resulting solution of 3,6-dibromo-2,5-bis(tert-butyldimethylsilylyl)thieno[3,2-*b*]thiophene is cooled to −78 °C and a further equivalent of butyllithium is added. The resulting mixture is allowed to warm to room temperature and is quenched with ammonium chloride [279].

7.4 IODINE DERIVATIVES

7.4.1 Direct iodination

The major problem in the direct iodination of thiophenes is the reversibility of the iodination and that the hydroiodic acid formed polymerizes thiophene and ruptures the ring with evolution of hydrogen sulfide. Both problems are solved by scavenging the hydroiodic acid. The classical method was to use mercuric oxide which reacts with the hydroiodic acid, precipitating mercuric iodide. In this way 2-iodothiophene was obtained in 75% yield [280]. A number of 2-alkyl-5-iodothiophenes are also prepared by this approach [281–285]. Diiodination is difficult to achieve. The most serious drawback of this method is the inefficient use of the iodine. In order to avoid this, iodination is carried out in the presence of an oxidant, which will convert the hydroiodic acid back to iodine. The first and cheapest reagent which has been used is nitric acid [286,287], which also catalyzes the iodination [288]. This method was recently used for the preparation of 2-benzyl-5-iodothiophene [36].

2-Benzyl-5-iodothiophene [36]

A mixture of 2-benzylthiophene (10.0 g, 0.057 mol) and iodine (4.32 g) is stirred at room temperature. 8 *M* Nitric acid (1.5 ml) is added to initiate and a further quantity of 8 *M* nitric acid (5.0 ml) is added over 30 min, after which the reaction mixture is heated and refluxed for 30 min. The organic phase is separated from the residual nitric acid, aqueous 40% sodium hydroxide solution is added and the mixture heated under reflux for 30 min. The phases are separated and the organic phase washed with water and dried over calcium chloride. Column chromatography over silica using light petroleum/ethyl acetate as eluent gives 4.6 g (43%) of the title compound as an oil.

The yields of 2-chloro-5-iodo- and 2,5-diiodothiophene prepared in this way are among the best obtained by direct iodination methods [173,289]. The only potential side reactions are nitration [286] and oxidation of the substrate. 2,4-Diiodo-5-trifluoroacetylaminothiophene and 2-iodo-3-trifluoroacetylaminothiophene can be obtained by iodination of 2- and

3-trifluoroacetylaminothiophene, according to a method previously used for iodination of aniline, in 56 and 72% yield, respectively, when iodine in a two-phase system of aqueous sodium bicarbonate and dichloromethane is used [290].

It has been found that the 2-iodo-5-methylthiophene is most conveniently prepared by the iodination of 2-methylthiophene with *N*-iodosuccinimide containing equimolar amounts of glacial acetic acid in methanol [291].

2-Iodo-5-methylthiophene [291]

To a solution of 2-methylthiophene (9.82 g, 0.10 mol) and *N*-iodosuccinimide (27.0 g, 0.12 mol) in methanol (240 ml), acetic acid (6.87 ml. 0.12 mol) is added. The reaction mixture is stirred at 0 °C for 3 h, after which water (240 ml) and diethyl ether (360 ml) are added. The phases are separated and the organic phase washed with water, 10% aqueous sodium hydroxide solution (4 times) and water (2 times) dried over calcium chloride and evaporated giving 15.4 g (69%) of the title compound as a pale-yellow liquid, which becomes colorless after distillation.

Bromination and iodination of 2-phenylthiophene with *N*-bromosuccinimide and *N*-iodosuccinimide, respectively, is used for the preparation of 2-bromo- and 2-iodo-5-phenylthiophene, respectively [292].

2-Iodo-5-phenylthiophene [292]

2-Phenylthiophene (8.01 g, 50 mmol) and *N*-iodosuccinimide (16.87 g, 75 mmol) are dissolved in methanol (50 ml). To this solution acetic acid (4.29 ml, 75 mmol) is added and soon a white precipitate is formed. The reaction flask is put into a refrigerator to ensure complete generation of the precipitation. The precipitate is collected by filtration, washed with water/methanol (1:1), dried and recrystallized from methanol/water (20:3) giving 9.50 g (68%) of the title compound.

A more convenient reagent is iodic acid, introduced by Wirth *et al.* [293], which has the advantage that the only by-product of the oxidation is water. In addition small amounts of sulfuric acid have a catalytic effect on the iodination. Using this method 2-iodo-[293], 2,5-diiodo-[294], 2,3,5-triiodo-[294,295], and tetraiodothiophene [296] as well as alkyl [114,157,199,297–304] and halogen-substituted iodothiophenes [5,35,155,170,199,294,305] were synthesized. It has also been used for the preparation of 3-iodo-2,5-dimethyl-4-phenylthiophene [198].

2,3,5-Triiodothiophene [295]

A solution of thiophene (10.5 g, 0.125 mol), acetic acid (80 ml), water (36.9 ml), carbon tetrachloride (30 ml), sulfuric acid (2.1 ml), iodine (38.1 g), and iodic acid (13.5 g) is refluxed for 100 h. Water and carbon tetrachloride are added and the phases separated. The organic phase is washed with water, 0.1 *M* sodium thiosulfate and water again, dried over sodium sulfate and evaporated. The residue is crystallized from ethanol giving 44.5 g (77%) of the title compound mp 83–85 °C.

3-Iodo-2,5-dimethyl-4-phenylthiophene [198]

A mixture of 2,5-dimethyl-3-phenylthiophene (5.00 g, 26.6 mmol), iodine (3.6 g, 14 mmol), iodic acid (1.2 g, 6.8 mmol), acetic acid (15 ml), water (20 ml), tetrachloromethane (15 ml), and concentrated sulfuric acid (0.1 ml) is stirred vigorously at 60 °C for 4.5 h. After cooling, the reaction mixture is poured into aqueous sodium thiosulfate. The phases are separated and the aqueous phase extracted with tetrachloromethane. The combined organic phases are washed with water, dried and evaporated. The crystalline residue (7.5 g, 90%) is recrystallized from hexane in the cold (−25 °C) giving 6.5 g (79%) of the title compound mp 53–56 °C.

The directing effect of the carboxyl group of 4-bromo-3-thiophenecarboxylic acid leads to the exclusive formation of 2-iodo-3-bromo-4-thiophenecarboxylic acid upon iodination with iodine–iodic acid. Using two equivalents of iodination reagent gives 3-bromo-2,5-diiodo-4-thiophenecarboxylic acid [171].

3-Bromo-2,5-diiodo-4-thiophenecarboxylic acid [171]

To a mixture of 3-bromo-4-thiophene carboxylic acid (31.0 g, 0.15 mol), acetic acid (175 ml), heated to 90 °C, sulfuric acid (3 ml), and iodic acid (6.5 g, 37 mmol) in water (25 ml), iodine (15.2 g, 60 mmol) is added in five portions over a period of 3 h. The stirring is continued for 1 h, after which the reaction mixture is allowed to attain room temperature overnight giving 42 g (84%) of the title compound as white crystals mp 204–206 °C after recrystallization from ethyl acetate.

Iodination of 3-thiophene aldehyde by this technique gives 2-iodo-4-thiophene aldehyde [170] and from methyl 2-thiophenecarboxylate, methyl 5-iodo-2-thiophene carboxylate as well as methyl 4,5-diiodothiophenecarboxylate are obtained, depending on the amount of iodinating reagent used [294]. Also 2,5-dimethyl-3-thiophene aldehyde is smoothly iodinated in the 4-position to give 2,5-dimethyl-4-iodo-3-thiophene aldehyde in 83% yield [157]. It should be noted that this compound is not conveniently obtained through halogen–metal exchange of 2,5-dimethyl-3,4-diiodothiophene followed by reaction with *N*,*N*-dimethylformamide [157]. Also 2,2′,3,3′-tetramethyl-4,4′-dithienyl ketone can be iodinated with iodine–iodic acid to 2,2′-diiodo-4,4′,5,5′-tetramethyl-3,3′-dithienyl ketone in excellent yield [157]. Also periodic acid, which has a higher efficiency than iodic acid in oxidizing hydroiodic acid, has been used for the iodination of thiophene, 2-chloro-, 2,5-dichloro-, 2-bromo-, and 2,5-dibromothiophene, as well as to some alkylthiophenes [95,96,306,307]. However, this reagent is more expensive.

2-Chloro-3,4-diiodo-5-methylthiophene [96]

A mixture of 2-chloro-5-methylthiophene (24.4 g, 0.18 mol), iodine (39.4 g, 0.15 mol), periodic acid (17.8g, 78 mmol), acetic acid (115 ml), water (24 ml), and concentrated sulfuric acid (3.2 ml) is stirred at 90 °C. During the reaction the product is precipitated, after cooling it is filtered off and stirred for 2 h in a saturated aqueous sodium dithionite solution at 60 °C. The product is separated, washed with water and ethanol and recrystallized from ethanol, giving 45.4 g (64%) as pale-yellow crystals mp 64 °C.

tert-Butyl (2,5-diiodothiophene-3,4-diyl)carbamate is prepared from the dicarbamate by refluxing with two equivalents of *N*-iodosuccinimide in carbon tetrachloride [164].

tert-Butyl (2,5-diiodothiophene-3,4-diyl)carbamate [164]

N-iodocuccinimide (1.35 g, 6 mmol) is added to a solution of di-*tert*-butyl (thiophene-3,4-diyl)carbamate (0.314 g, 1 mmol) in carbon tetrachloride (50 mml). The reaction mixture is heated under reflux for 10 h, filtered, washed with water and evaporated. The title compound obtained in a yield of 50% is used without further purification mp 151 °C.

It has been shown that in contrast to what has previously been claimed [34], that benzyl trimethylammonium dichloro iodate is not a selective iodination reagent. Depending upon the substituents on the thiophene ring more or less complicated mixtures of chloro and iodo derivatives are obtained [308].

2-Bromo-5-iodo-3-(11-diethylphosphorylundecyl)thiophene has recently been prepared in connection with work on polythiophenes, by bromination with *N*-bromosuccinimide followed by iodination with iodine and mercury(II) oxide of 3-(11-diethylphosphorylundecyl)thiophene [309].

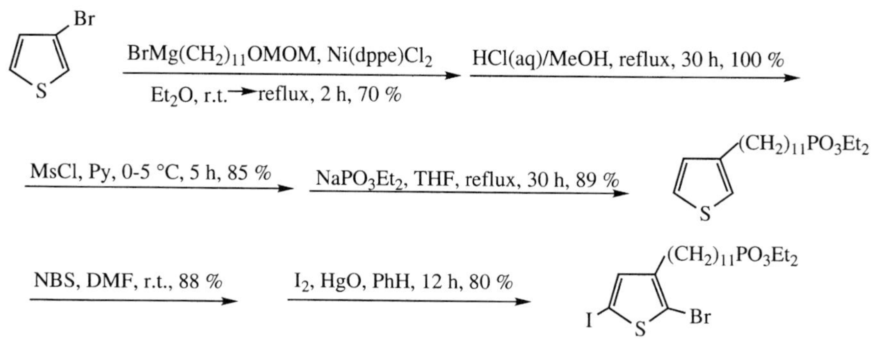

7.4.2 From thienylmetalorganic reagents

7.4.2.1 From thienylmercury derivatives

Previously, thienylmercury derivatives have extensively been used for the preparation of iodo derivatives through reaction with iodine or preferably potassium triiodide. As mentioned in Chapter 2 thienylmercuric chlorides and thienylmercuric acetates can be prepared by direct reaction of thiophenes with mercuric chloride or mercuric acetate. The chloride derivatives are less soluble than the acetates and are used when monomercuration is desired, while the acetate is used when polymercuration is desired, as in the preparation of tetraacetoxymercurythiophene, which was used for the classical synthesis of tetraiodothiophene [310] and 2-methyl-3,4,5-triiodothiophene [311]. 5-Ethyl-2-iodothiophene has been prepared by this approach [312]. As also nitrothiophenes are easily mercurated, these compounds can be used for the preparation of iodinated nitrothiophenes [313–315]. In some cases thiophenecarboxylic acids give mercurydecarboxylation leading after reaction with iodine to iodo derivatives [316]. Iodination in the presence of mercuric oxide can be used for the preparation of 2-iodo-3-methoxythiophene [317].

2-Iodo-3-methoxythiophene [317]

To a solution of 3-methoxythiophene (684 mg, 6.0 mmol) in benzene (10 ml) at 0 °C is added in small portions mercuric oxide (1.32 g, 6.1 mmol yellow crystals) and iodine (1.57 g, 6.2 mmol). The reaction mixture is stirred at room temperature for 0.5 h, after which the precipitate is filtered off and washed with diethyl ether. The filtrate and washings are combined, washed with aqueous sodium thiosulfate and dried over sodium sulfate. After evaporation the residue is purified by medium pressure liquid chromatography using 100% hexane as eluent, giving 1.40 g (97%) of the title compound as a light-yellow liquid.

Selective iodination of 2-silylated 3-phenylthiophene shown below, without any desilylation, is achieved by treatment first with mercury acetate in chloroform, followed by iodine, while treatment with *N*-bromosuccinimide gives bromodesilylation (89%) and substitution (11%) and treatment with halogen electrophiles gives a mixture of 2-iodo-3-phenylthiophene and the desired product [318].

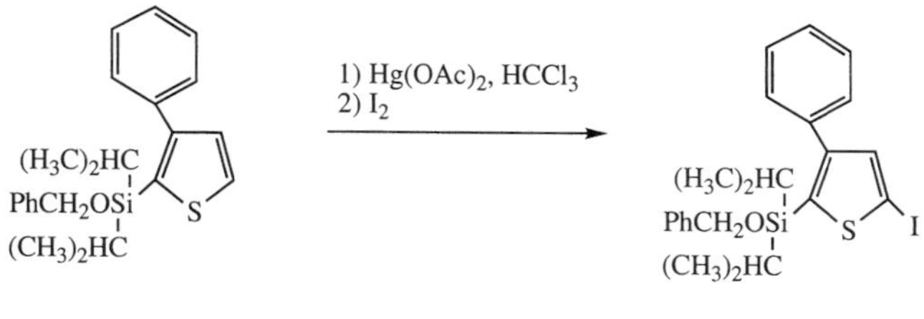

Selective metalation of ethyl-2-thiophenecarboxylate and ethyl-3-thiophenecarboxylate with diisopropylmagnesium amide chloride in tetrahydrofuran can be used for the preparation of ethyl 5-iodo-2-thiophenecarboxylate and ethyl 3-iodo-2-thiophenecarboxylate [318].

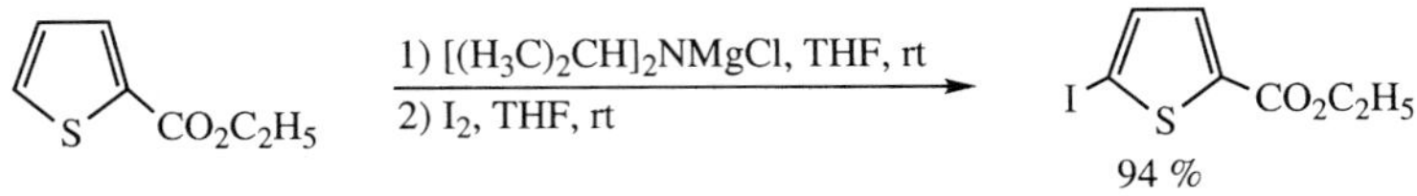

7.4.2.2 From thienyllithium derivatives

Thienyllithium derivatives prepared either by metalation or by halogen–metal exchange of bromothiophenes are especially useful for the preparation of iodothiophenes, through the reaction with iodine. Usually the yields are acceptable [5,42,170,198,258,320–324]. Recently, it has been claimed that 2-iodothiophene can be obtained from 2-thienyllithium and iodine at 0 °C [198] and 2-iodo-5-dodecylthiophene is prepared by metalation of 2-dodecylthiophene with butyllithium followed by iodine [325].

2-Iodothiophene [326]

To a solution of thiophene (1.00 g, 11.9 mmol) in anhydrous diethyl ether (20 ml) butyllithium (1.57 *M*) in hexane (7.57 ml, 11.9 mmol) is added dropwise with stirring under argon at 0 °C. The stirring is continued at room temperature for 1 h, after which a solution of iodine (3.02 g, 11.9 mmol) in anhydrous diethyl ether (40 ml) is added dropwise over a period of 10 min at 0 °C. The reaction mixture is stirred at room temperature for 1 h, poured into ice-water and extracted with ether. The combined ether phases are washed successively with saturated aqueous sodium bisulfite and sodium chloride, dried over sodium sulfate and evaporated. The oily residue is chromatographed on silica gel using pentane as eluent. 2.20 g (88%) of the title compound is obtained as a colorless oil.

2-Iodo-5-dodecylthiophene [325]

To a solution of 2-dodecylthiophene (3.00 g, 11.9 mmol) in anhydrous tetrahydrofuran (20 ml) at 0 °C butyllithium (1.6 *M*)in hexane (7.4 ml, 11.9 mmol) is added. After stirring the mixture at room temperature for 20 min it is cooled to −78 °C and iodine (3.02 g, 11.9 mmol) in tetrahydrofuran is added dropwise. The reaction mixture is allowed to warm to room temperature and then poured into petroleum ether (200 ml). This solution is washed with an aqueous solution of sodium chloride and sodium thiosulfate, dried over magnesium sulfate and evaporated. The residue is passed through a short column of silica gel giving 4.15 g (92%) of the title compound as a light-brown liquid.

Upon reaction of 5-trimethylsilyl-2-thienyllithium with iodine an 87% yield of 2-iodo-5-trimethylsilylthiophene is obtained [192].

It may be mentioned that 3-iodo-2-thiophene aldehyde can be prepared in 25% yield in an one-pot procedure from 2,3-dibromothiophene [258]. The only convenient way to prepare 2-bromo-4-iodothiophene is halogen–metal exchange between 2,4-diiodothiophene and butyllithium followed by reaction with bromine [323]. Halogen–metal exchange of 3-bromo-5-methyl-2-phenylthiophene with butyllithium at −70 °C followed by reaction with iodine gives 3-iodo-5-methyl-2-phenylthiophene. Similarly, 3-iodo-2-methyl-5-phenylthiophene has been prepared in a yield of 60% [198].

3-Iodo-5-methyl-2-phenylthiophene [198]

To a solution of 3-bromo-5-methyl-2-phenylthiophene (8.38 g, 0.033 mol) in anhydrous diethyl ether (100 ml) 1.40 *M* butyllithium (25 ml, 0.035 mol) is added dropwise at −70 °C followed by iodine (8.90 g, 0.035 mol) in anhydrous diethyl ether (100 ml). The reaction mixture is allowed to reach room temperature, after which it is poured into aqueous sodium thiosulfate. The phases are separated, the organic phase washed with water until neutral reaction, dried and evaporated. The residue (8.7 g) gives upon distillation 5.40 g (54%) of the title compound bp 118–118 °C/0.001 mm Hg.

2-Formyl-3-iodothiophene is prepared in 71% yield through halogen–metal exchange of 2-(3-bromo-2-thienyl)-1,3-dioxalane with butyllithium at −70 °C, followed by reaction with iodine, while 3-formyl-2-iodothiophene was prepared in 68% yield by metalation of 2-(3-thienyl)-1,3-dioxalane with butyllithium at room temperature followed by reaction with iodine at −70 °C [258]. Metalation of 3-thiophenecarboxylic acid with two equivalents of lithium diisopropylamide followed by iodine can be used for the preparation of 2-iodo-3-thiophenecarboxylic acid in excellent yield [327].

2-Iodo-3-thiophenecarboxylic acid [327]

To a solution of lithium diisopropylamide (105 mmol), prepared from butyllithium (1.6 *M*) in hexane (66 ml) and diisopropylamine (10.6 g) in anhydrous tetrahydrofuran (40 ml), is added dropwise a solution of 3-thiophenecarboxylic acid (6.4 g, 50 mmol) in anhydrous tetrahydrofuran (40 ml) at −78 °C. The mixture is stirred at the same temperature for 20 min, after which a solution of iodine (12.7 g, 50 mmol) in anhydrous ether (100 ml) is added dropwise. The resulting reaction mixture is poured into water (300 ml), the phases are separated and the water phase washed with ether (100 ml). The aqueous phase is acidified with 10% hydrochloric acid and extracted with ethyl acetate (3 × 100 ml). The combined phases are dried over magnesium sulfate and evaporated. The residue is crystallized from ethanol/water giving 10.7 g (84%) of the title compound mp 176–177 °C.

Lithiation of the diethylacetal of 3-thiophene aldehyde followed by reaction with iodine and hydrolysis is a good method for the preparation of 2-iodo-3-formylthiophene [170]. Using the technique applied for the dimetalation of phenylacetylene [328], treating 3-ethynylthiophene with potassium *tert*-butoxide and butyllithium in hexane and converting the dilithiated product to the magnesium derivative by reaction with magnesium bromide etherate and then reacting with iodine gives 2-iodo-3-ethynylthiophene and 2-iodo-4-ethynylthiophene in a 3:1 ratio [329]. Metalation of silyl-protected 2-aminomethylthiophene with butyllithium followed by reaction with iodine is used for the preparation of the 5-iodo derivative [330]. 3-Dodecyl-2-iodothiophene is prepared by reacting the Grignard reagent from 2-bromo-3-dodecylthiophene with iodine [331].

3-Dodecyl-2-iodothiophene [331]

A solution of 2-bromo-3-dodecylthiophene (10.0 g, 30 mmol) in anhydrous tetrahydrofuran (100 ml) is added to magnesium turnings (0.8 g, 33 mmol) and refluxed until the magnesium is dissolved. The resulting mixture is cooled to −60 °C and a solution of iodine (7.4 g, 30 mmol) in anhydrous tetrahydrofuran (100 ml) is slowly added. The reaction mixture is stirred for 2 h and then allowed to warm to room temperature. After dilution with diethyl ether (50 ml) the mixture is hydrolyzed with aqueous hydrochloric acid. The phases are separated and the aqueous phase extracted twice with diethyl ether (50 ml). The combined organic phases are washed with aqueous sodium thiosulfate and dried over sodium sulfate. After evaporation the residue is distilled *in vacuo* to afford 9.40 g (83%) of the title compound as a slightly yellow liquid bp 125 °C/8 × 10^{-4} mm Hg.

7.4.2.3 From thienylthallium derivatives

Thienylthallium derivatives are the most recently developed reagents for iodine introduction. Thallation with thallium trifluoroacetate in acetonitrile regiospecifically gives the unisolated α-thallium derivatives, which upon reaction with potassium iodide, yields the iodothiophenes [332,333]. The yields are generally good and the presence of aldehyde function is tolerated [334].

2-Methyl-5-iodothiophene [333]

2-Methylthiophene (4.90 g, 0.05 mol) solved in acetonitrile is treated with solid thallium(III) trifluoroacetate (19 g, 0.05 mol) and an aqueous solution of potassium iodide (*ca* 0.12 mol) all at once at room temperature. After this the reaction mixture is stirred for 20 min, sodium metabisulfite is added to destroy free iodine and the mixture is stirred for a further 10–15 min followed by basification with 4 *M* sodium hydroxide solution. Ether is added and the precipitated thallium(I) iodide is removed by filtration. The ethereal layer is separated and the aqueous phase extracted with ether. The combined ether phases are dried over sodium sulfate and concentrated. The crude title compound is purified by standard techniques.

Since this method is most conservative of iodine, it is particularly well suited for preparing radioiodine compounds [335], such as 2-(2(*RS*-aminopropyl)-5-[^{125}I]iodothiophene [250]. The drawback is the toxicity of thallium compounds. 17-[5-iodo-(2-thienyl)]heptadecanoic acid and 13-[5-iodo-(2-thienyl)]tridecanoic acid were prepared by this technique in connection with the preparation of radioiodine labeled compounds [335].

Other metalorganic reagents, such as thienylboron [336] thienyltin [250,337,338] and thienylcopper derivatives [339,340] give iodo derivatives upon reaction with iodine, but as these reagents mostly are prepared from thienyllithium derivatives, they seldom offer synthetic advantages.

7.4.3 Rearrangement and disproportionation reactions

Iodine rearrangement or disproportionation reactions are sometimes of synthetic use. Thus treatment of 2-iodothiophene with potassium *N*-methylanilide gives 3-iodothiophene and constitutes one of the best methods for this compound [112].

3-Iodothiophene [112]

To potassium *N*-methylanilide (87 g, 0.6 mol) in liquid ammonia at −33 °C 2-iodothiophene (21 g, 0.1 mol) is added as rapidly as possible. After 15 min sufficient ammonium chloride is added to destroy excess amide. When the ammonia is evaporated with the aid of a lukewarm water bath, water is added to the residue and the organic layer separated. Any insoluble tars are removed by filtration. The aqueous solution is extracted with three portions of ether and the combined extracts are washed with four portions of 1 *M* hydrochloric acid and one of water, dried over calcium chloride, evaporated and distilled giving 14.7 g (77%) of the title compound.

7.4.4 Partial reduction of di- and polyiodothiophenes

Like in the synthesis of especially bromothiophenes partial deiodination of polyiodothiophenes is of synthetic use. Thus 2,4-diiodothiophene has been prepared by treatment of 2,3,5-triiodothiophene with one equivalent of butyllithium followed by hydrolysis [341]. Treatment of the triiodo derivative with zinc in acetic acid gives 3-iodothiophene [294].

3-Iodothiophene [294]

Water (100 ml), zinc powder (70 g) and acetic acid (40 ml) are introduced to a flask equipped with a reflux condenser, efficient stirrer and arranged for downward distillation. The stirred mixture is heated to boiling, after which 2,3,5-triiodothiophene (110 g, 0.238 mol) is added continuously through the reflux condenser. When the distillate no longer contains any 3-iodothiophene (1 h), the organic layer is separated, dried over calcium chloride and fractionated. After a forerun of about 10 g of thiophene 25.5 g (52%) of the title compound is obtained bp 66–68 °C/9 mm Hg.

Treatment of the triiodothiophene with zinc powder in acetic anhydride and deuterium oxide gives 2,5-dideutero-3-iodothiophene [296]. Similarly 3,4-diiodothiophene has been prepared from the reaction of tetraiodothiophene with two equivalents of butyllithium [296,341,342]. Other examples of selective deiodinations are the preparations of 4-fluoro-2-iodothiophene from 2,5-diodo-3-fluorothiophene [5], 2,3-dimethyl-4-iodothiophene from 2,3-dimethyl-4,5-diiodothiophene [157], 2,4-dimethyl-3-iodothiophene from 2,4-dimethyl-3,5-diiodothiophene [299] 2-*tert*-butyl-4-iodothiophene from 2-*tert*-butyl-4,5-diiodothiophene [114] and 2-*tert*-butyl-3-iodothiophene from *tert*-butyl-3,5-diiodothiophene [114].

7.4.5 From iodothiophenes

Substituted iodothiophenes, in many cases, can be most conveniently prepared by electrophilic substitution reactions or by halogen–metal exchange of di- or polyiodothiophenes followed by reaction with the appropriate electrophile.

7.4.5.1 Electrophilic substitution

Nitration of 2-iodothiophene with fuming nitric acid in acetic anhydride yields 70% of 2-iodo-5-nitrothiophene [343]. 2-Acetyl-5-iodothiophene is prepared by acetylation of 2-iodothiophene with acetic anhydride using phosphoric acid as catalyst [344].

Biphasic halogenation of 2-iodothiophene with *N*-chloro- and *N*-bromosuccinimide catalyzed by perchloric acid yields 2-chloro-5-iodo- and 2-bromo-5-iodothiophene in 70–82% yield [46].

Nitration of 2,3,5-triiodothiophene with nitric acid in acetic anhydride gives a mixture of 2,4-diiodo-5-nitrothiophene and 2,3-diiodo-5-nitrothiophene, which are separated by preparative thin-layer chromatography [295].

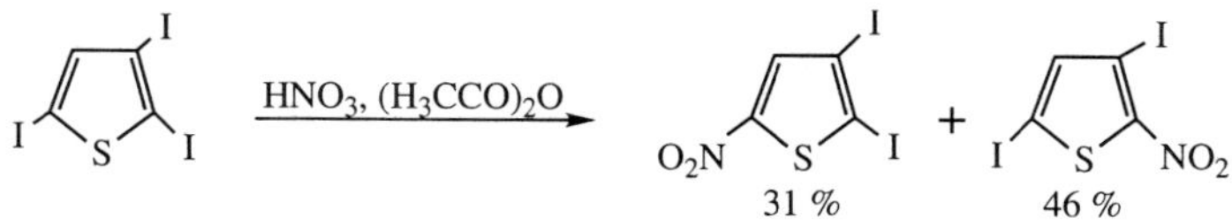

7.4.5.2 Metalation of iodothiophenes

Metalation of 3-iodothiophene with lithium diisopropylamide followed by reaction with carbon dioxide gives 3-iodo-2-thiophenecarboxylic acid [249].

3-Iodo-2-thiophenecarboxylic acid [249]

A 1-l three-necked flask, equipped with stirrer, dropping funnel, condenser with drying tube and nitrogen inlet, is charged with butyllithium (1.48 *M*) in hexane (111.5 ml) and anhydrous diethyl ether (100 ml), followed by addition of diisopropylamine (16.7 g, 0.165 mol) in anhydrous diethyl ether (50 ml) at room temperature. After the addition is completed the reaction mixture is cooled to −70 °C and 3-iodothiophene (31.5 g, 0.15 mol) in anhydrous diethyl ether (100 ml) is added dropwise, after which the stirring is continued at this temperature for 2 h. The reaction mixture is then poured onto an excess of carbon dioxide covered with anhydrous diethyl ether. At 0 °C the reaction mixture is hydrolyzed with hydrochloric acid (2 *M*). The phases are separated and the ether phase extracted with 2 *M* sodium hydroxide solution. The combined aqueous phases are treated with 2 *M* hydrochloric acid and the precipitate is filtered off, giving 24.0 g (63%) of the title compound as a white solid, mp 201–203 °C after recrystallization from water/ethanol.

The metalation of 2-iodothiophene with lithium diisopropylamide in tetrahydrofuran at −40 °C to −10 °C followed by reaction with various electrophiles is used for the preparation of 5-formyl-, 5-acetyl-, 5-carboxy-, 5-tributylstannyl-2-iodothiophene and several other derivatives [312,345].

2-Acetyl-5-iodothiophene [345]

A solution of lithium diisopropylamide (1.05 mmol) in tetrahydrofuran (2 ml) under nitrogen is cooled to −40 °C and 2-iodothiophene (210 mg, 1.0 mmol) is added with vigorous stirring. After 10 min the mixture is warmed to −10 °C and stirred for 10–20 min until a dark red color indicative of the complete formation of 2-lithio-5-iodothiophene develops. The reaction mixture is recooled to −40 °C and *N*-methyl-*N*-methoxyacetamide (113 mg, 1.1 mmol) is added in one portion. The reaction mixture is allowed to warm slowly to 0 °C and saturated ammonium chloride solution (2 ml) is added. After extraction with diethyl ether (3 × 10 ml), the combined organic phases are dried over magnesium sulfate and evaporated giving a brown solid. Purification by preparative centrifugal chromatography using petroleum ether/ ethyl acetate (6:1) as eluent gives 182 mg (72%) of the title compound mp 128–129 °C.

Metalating 3-hexyl-2-iodothiophene with lithium diisopropylamide and reacting with tributylstannyl chloride at −80 °C gives 2-iodo-3-hexyl-5-tributylstannylthiophene in 86% yield [346].

7.4.5.3 Halogen–metal exchange of iodothiophenes

Halogen–metal exchange of 2,3-, 2,4-diiodo-, 3,4-diiodo-, and 2,5-diodothiophene with butyllithium at −70 °C followed by reaction with *N*,*N*-dimethylformamide are among the best methods for the preparation of 3-iodo-2-, 4-iodo-2-, 4-iodo-3-, and 5-iodo-2-thiophene aldehyde [341]. Using two equivalents of butyllithium dilithiation is achieved, which opens a route to 2,5- and 3,4-diformylthiophene [341]. Starting from 2,3,5-triiodothiophene selective halogen–metal exchange takes place and reaction with *N*,*N*-dimethylformamide gives 2,4-diiodo-5-thiophene aldehyde and using two equivalents of butyllithium gives 3-iodo-2,5-diformylthiophene [341].

2,3-Dimethyl-4-iodothiophene is prepared in 76% yield by halogen–metal exchange of 2,3-dimethyl-4,5-diiodothiophene with butyllithium at −70 °C followed by hydrolysis [157], while reaction with *N*,*N*-dimethylformamide yields 4,5-dimethyl-3-iodo-2-thiophene aldehyde [157].

4,5-Dimethyl-3-iodo-2-thiophene aldehyde [157]

To a stirred solution of 2,3-dimethyl-4,5-diiodothiophene (72.8 g, 0.20 mol) in anhydrous diethyl ether (750 ml) 0.75 *M* ethyllithium (200 ml) is added under nitrogen at −70 °C. After stirring at this temperature for 1 h a solution of *N*,*N*-dimethylformamide (22 g, 0.30 mol) in anhydrous diethyl ether (25 ml) is added. The cooling bath is removed and when the temperature of the reaction mixture has risen to 0 °C dilute hydrochloric acid is slowly added. The phases are separated and the ether phase is washed with dilute hydrochloric acid, 10% sodium bicarbonate solution and water. After drying over magnesium sulfate and evaporation the residue is recrystallized from hexane giving 47.4 g (89%) of the title compound mp 107.5–109 °C.

4,4′-Diiodo-2,2′,3,3′-tetramethyl-5,5-dithienylcarbinol is prepared in 80% yield through the reaction of 2,3-dimethyl-4-thienyllithium with 4,5-dimethyl-3-thiophene aldehyde [157]. In a similar way 3,3′-diiodo-2′4,5,5′-tetramethyl-2,4′-dithienyl carbinol can be prepared from 3-iodo-2,5-dimethyl-4-thiophene aldehyde and 3-iodo-4,5-dimethyl-3-thienyllithium [157].

7.4.6 Modifications of side chains in iodothiophenes

3-Iodo-4-trimethylsilylthiophene is prepared through the reaction of 3,4-bis(trimethylsilyl)thiophene with iodine and silver trifluoroacetate in tetrahydrofuran at −78 °C [147].

3-Iodo-4-trimethylsilylthiophene [347]

To a solution of 3,4-bis(trimethylsilyl)thiophene (1.82 g, 8 mmol) in tetrahydrofuran (120 ml) under nitrogen at −78 °C silver trifluoroacetate (3.54 g, 16 mmol) is added. The stirring is continued for 5 min, after which iodine (4.06 g, 16 mmol) in tetrahydrofuran (60 ml) is added dropwise and the reaction mixture is stirred at −78 °C for 8 h, diluted with diethyl ether (150 ml) and filtered through Celite. The filter cake is washed with diethyl ether (50 ml). The combined filtrates are washed with sodium thiosulfate solution (2 × 80 ml), dried over magnesium sulfate and evaporated. The residue is purified by chromatography on silica gel using hexanes as eluent giving 2.16 g (96%) of the title compound as a colorless oil.

Reduction of iodothiophene aldehydes with sodium borohydride in ethanol yields the corresponding iodothenyl alcohols in good yields [258].

3-Iodo-2-thenyl alcohol [258]

To a suspension of sodium borohydride (1.90 g, 0.05 mol) in absolute ethanol (50 ml) 3-iodo-2-thiophene aldehyde (23.8 g, 0.10 mol) in absolute alcohol (200 ml) is added dropwise. During the addition the temperature is kept below 25 °C, when this is completed the reaction mixture is heated at 50 °C for 1 h, after which most of the ethanol is distilled off. The white solid residue is treated with water and acidified with 5 *M* hydrochloric acid, followed by extraction with ether. The combined ether phases are washed with sodium carbonate solution and water and dried over magnesium sulfate. After evaporation the residue is recrystallized from petroleum ether (65–71 °C) giving 18.7 g (78%) of the title compound mp 49.0–49.5 °C.

7.4.7 Various methods

The bis(trifluoroacetoxy)iodobenzene-iodine system is a good iodinating reagent for thiophene giving 5-substituted 2-iodo derivatives also with electron-withdrawing substituents [348]. 3-Iodo-4-trimethylsilylthiophene is prepared in 96% yield by mono-*ipso*-substitution of 3,4-bis(trimethylsilyl)-thiophene with iodine and silver trifluoroacetate in tetrahydrofuran at −78 °C [347,349]. This method was also used for the preparation of 3-iodo-4-phenyl-thiophene and 3-iodo-4-(phenylethynyl)thiophene from 4-phenyl-3-(trimethyl-silyl)thiophene and 3-(trimethylsilyl)-4-(phenylethynyl)thiophene [347].

3-Iodo-4-phenylthiophene [347]

4-Phenyl-3-trimethylsilylthiophene (58 mg, 0.25 mmol) in tetrahydrofuran/methanol (5:1, 8 ml) under nitrogen is cooled to −78 °C. Silver trifluoroacetate (111 mg, 0.5 mmol) is added, and the mixture stirred for 5 min to ensure complete dissolution. Then iodine (127 mg, 0.5 mmol) in tetrahydrofuran/methanol (5:1, 3 ml) is added dropwise. The reaction mixture is stirred at −78 °C for 4 h and then at room temperature for 20 h, after which it is diluted with ether and filtered through Celite. The filter cake is washed with ether and the filtrates are washed with 50% sodium thiosulfate solution, dried over magnesium sulfate and evaporated. The residue is purified by chromatography on silica gel using hexanes as eluent giving 48 mg (67%) of the title compound as colorless crystals mp 50–52 °C.

Palladium-catalyzed and cuprous iodide-promoted coupling of tetraiodo-thiophene with trimethylsilylacetylene leads to an 84% yield of 2,5-bis[(trimethylsilyl)ethynyl]-3,4-diiodothiophene [350]. Another example of such a Heck coupling is the preparation of 2-iodo-5-(1-propynyl)thiophene from 2–5-diiodothiophene and 1-propyne [351].

2,5-Bis(trimethylsilyl)ethynyl-3,4-diiodothiophene [350]

To a solution of tetraiodothiophene (0.85 g, 1.45 mmol) in diisopropylamine (50 ml) dichlorobis(benzonitrile)palladium(II) (27 mg, 0.07 mmol), triphenyl-phosphine (37 mg, 0.14 mmol) and cuprous iodide (14 mg, 0.07 mmol) are added. The clear yellow solution is cooled in ice and the flask flushed with argon. (Trimethylsilyl)acetylene (0.44 ml, 3.19 mmol) is added dropwise *via* a syringe. The color of the solution changes from yellow to green and finally to black with the formation of a heavy white precipitate. The reaction mixture is stirred at 0 °C for 1 h and then allowed to warm to room temperature. After 12 h at room temperature it is heated to reflux for 1 h, cooled and filtered. After evaporation of the solvent the residue is dissolved in dichloromethane and this solution is washed with dilute aqueous hydrochloric acid and water, dried over magnesium sulfate and evaporated. The product, a dark-brown solid, is purified by chromatography on silica using hexane as eluent giving 640 mg (84%) of the title compound as white needles mp 148–150 °C.

2-Iodo-5-(1-propynyl)thiophene [351]

A deaerated solution of 2,5-diiodothiophene (19.0 g, 35.5 mmol) in benzene (30 ml) is added to a mixture of tetrakis(triphenylphosphine)palladium(0) (1.62 g, 1.4 mmol), cuprous iodide (0.67 g, 3.5 mmol), and triethylamine (11.7 ml, 84.3 mmol) in benzene (70 ml). The mixture is cooled to 0 °C and treated during 5 min with a solution of 1-propyne (2.25 g, 56.2 mmol) in benzene (45 ml). After stirring at room temperature for 8 h, the reaction mixture is added to a large excess of a saturated aqueous ammonium chloride solution and the resulting mixture after stirring for 0.5 h is extracted with ether. The organic extract is washed with water, dried and concentrated under reduced pressure. The residue is diluted with hexane and filtered through Celite. The filtrate is concentrated under reduced pressure and the residue purified by medium pressure liquid chromatography on silica gel, using hexane as eluent, giving 6.0 g (43%) of the title compound in a purity of 98.5%.

Irradiation of 2,3-diiodo-5-nitrothiophene in benzene gives 2-phenyl-3-iodo-5-nitrothiophene [295].

2-Phenyl-3-iodo-5-nitrothiophene [295]

A solution of 2,3-diiodo-5-nitrothiophene (100 mg, 0.26 mmol) in benzene (100 ml) is degassed with nitrogen for 10 min and then irradiated with a 125 W high pressure mercury arc (Helios–Italquartz surrounded by a Pyrex water-jacket). After 24 h the solvent is evaporated giving 85 mg (98%) of the title compound.

7.5 IODONIUM DERIVATIVES

7.5.1 From thiophenes

In connection with a study of their reactivity, 2- and 3-(diacetoxyiodo)thiophene have recently been prepared by the reaction of 2- and 3-iodothiophene with sodium perborate in acetic acid [352].

2-(Diacetoxyiodo)thiophene [352]

To a solution of 2-iodothiophene (2.1 g, 10 mmol) in acetic acid (100 ml) under argon sodium perborate tetrahydrate (100 mmol) is added over a period of

30 min. The solution is then heated at 50–60 °C until the 2-iodothiophene is consumed and evaporated. Water is added to the residue and the product extracted with chloroform (3 × 30 ml). The combined organic phases are dried over sodium sulfate and evaporated giving the title compound mp 120–122 °C (decomp.).

Di(2-thienyl) iodonium salts can be prepared by direct oxidation of thiophene with potassium iodate or iodine trifluoroacetate [353]. A number of aryl-2-thienyl iodonium salts have been prepared by the reaction of thiophene with iodobenzeneacetic anhydride [353] or aryliodoso diacetates [353–356] or, to avoid the necessity of strong acid catalysts, [hydroxy(tosyloxy)iodo]arenes [357,358].

Reaction of 3,4-bis(trimethylsilyl)thiophene with iodobenzene diacetate in the presence of trifluoromethanesulfonic acid in methylene chloride gives phenyl [(trimethylsilyl)thien-3-yl] iodonium triflate [347,359].

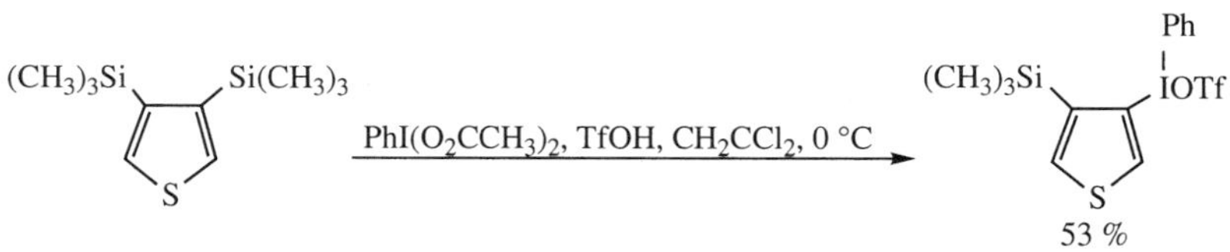

Similarly the reaction of 2,5-bis(trimethylsilyl)thiophene with iodosyl triflate in dichloromethane at −78 °C, is a method for the preparation of bis [2-(5-trimethylsilyl)-thienyl] iodonium triflate. However, the reaction of di(cyano) iodonium triflate, prepared *in situ* from iodosyl triflate and cyanotrimethylsilane, with tributylstannylthiophene and 5-bromo-2-tributystannylthiophene, respectively is the preferred method for the preparation of bis(2-thienyl)iodonium triflate and bis [2-(5-bromo-2-thienyl]iodonium triflate [360].

Bis[2-(5-bromo)thienyl]iodonium triflate [360]

Cyanotrimethylsilane (0.54 ml, 4.0 mmol) is added to a stirred suspension of iodosyl triflate (0.58 g, 4.0 mmol) in dichloromethane (15 ml) at −78 °C under nitrogen. The mixture is allowed to warm to −20 °C and stirred at this temperature for 10–15 min until the clear solution of di(cyano)iodonium triflate is obtained. This solution is cooled to −78 °C and transferred to a cold solution of 2-tributylstannyl-5-bromothiophene (1.8 g, 4.0 mmol) in dichloromethane (15 ml). The reaction mixture is allowed to warm to room temperature and crystallized by the addition of anhydrous hexane (20–30 ml). The precipitate of the iodonium triflate is filtered under nitrogen, washed with anhydrous ether (30 ml) and dried *in vacuo*. After recrystallization from acetonitrile 0.91 g (76%) of the title compound is obtained as a white microcrystalline solid mp 141–143 °C (dec).

Diaryl iodonium salts can be prepared by reacting aryl trialkylstannanes with hydroxy(tosyloxy)iodobenzene [361] and in this way 2-thienylphenyliodonium tosylate is prepared [362]. However, instead of using toxic organostannanes in the synthesis of radioligands, thiopheneboronic acids were reacted with diacetoxyiodobenzene and trifluoromethanesulfonic acid [363].

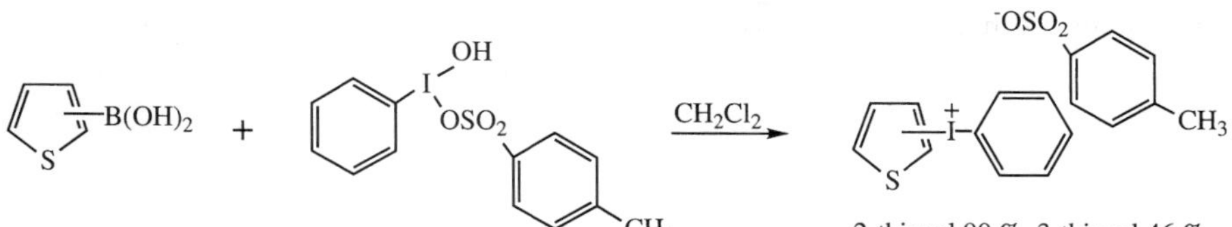

Typical procedure for the formation of diaryliodonium tosylates [363]

The boronic acid (64 mg, 5 mmol) is added to a stirred solution of hydroxy(tosyloxy)iodobenzene (Koser's reagent) (1.96 g 5 mmol) in dichloromethane (5 ml). The resulting mixture is stirred overnight. The solvent is then removed *in vacuo* to give the crude product.

7.5.2 From thienyllithium derivatives

The reaction of the thienyllithium derivatives with *trans*-chlorovinyliodoso dichloride yields the symmetric dithienyl iodonium salts [50,364–367] *via* an intermediate chlorovinyl thienyl iodonium salt [364]. Reaction of an aryl iodoso dichloride [365] a chlorovinyl aryl iodonium salt [364] or an aryl (*tert*-butylethynyl) iodonium tosylate with a thienyllithium derivative, yields aryl thienyl iodonium salts [357]. Starting from 3-thienyllithium derivatives di(3-thienyl) iodonium salts can also be prepared by this method.

Symmetrical dithienyliodonium chlorides [50]

To a solution of *trans*-chlorovinyliodoso dichloride (0.10 mol) in anhydrous toluene (200 ml) cooled to −70 °C, a thienyllithium solution is transferred through a rubber tube. The thienyllithium solution is prepared from thiophene or bromothiophene (0.20 mol) in anhydrous diethyl ether (100 ml) and butyllithium in hexane (0.21 mol). The mixture is stirred at −70 °C for 2–3 h, after which the cooling bath is removed and the temperature is allowed to rise to 0 °C. The reaction mixture is then poured into water, the precipitated iodonium salt filtered off, washed with water, acetone and ether and dried. The salts should, in order to minimize decomposition, be stored at about 0 °C, if not used within a few days.

REFERENCES

1. G. Cerichelli, M. E. Crestoni and S. Fornarini, *Gazz. Chim. Ital.* **120**, 749 (1990).
2. S. Rodmar, B. Rodmar, M. K. Sharma, S. Gronowitz, H. Christiansen and U. Rosén, *Acta Chem. Scand.* **22**, 907 (1968).
3. R. D. Schuetz, D. D. Taft, J. P. O'Brian, J. L. Shea and H. M. Mork, *J. Org. Chem.* **28**, 1420 (1963).
4. H. Christiansen, S. Gronowitz, B. Rodmar, S. Rodmar, U. Rosén and M. K. Sharma, *Arkiv Kemi* **30**, 561 (1969).
5. S. Gronowitz and U. Rosén, *Chem. Scripta* **1**, 33 (1971).
6. S. Gronowitz, R. Svenson, G. Bondesson, O. Magnusson and N. Stjernström, *Acta Pharm. Suec.* **11**, 211 (1974).
7. G. S. Lal, G. P. Pez, and R. G. Syvret, *Chem. Rev.* **96**, 1737 (1996).
8. R. E. Banks, R. A. DuBoisson and E. Tsiliopoulis, *J. Fluorine Chem.* **32**, 461 (1986).
9. Y. Sakamoto, S. Komatsu and T. Suzuki, *J. Am. Chem. Soc.* **123**, 4643 (2001).
10. R. T. Van Vleck, *J. Am. Chem. Soc.* **71**, 3256 (1949).
11. J. Burdon, J. Campbell and J. C. Tatlow, *Chem. Commun.* 27 (1969).
12. J. Burdon, J. G. Campbell, I. W. Parsons and J. C. Tatlow, *J. Chem. Soc. (C)* 352 (1971).
13. J. Burdon and I. W. Parsons, *J. Fluorine Chem.* **13**, 159 (1979).
14. R. D. Schuetz and G. P. Nilles, *J. Org. Chem.* **36**, 2188 (1971).
15. S. Gronowitz, M. Herslöf, R. Svenson, G. Bondesson, O. Magnusson and N. Stjernström, *Acta Pharm. Suec.* **15**, 368 (1978).
16. C. Corall, A. Lasso, J. Lissavetzky, A. S. Alvarez-Insua and A. M. Valdeolmillos, *Heterocycles* **23**, 1431 (1985).
17. F. Kobarfard, J. M. Kauffman and W. J. Boyko, *J. Heterocycl. Chem.* **36**, 1247 (1999).
18. R. E. Banks, *J. Fluorine Chem.* **87**, 1 (1998).
19. D. F. Andrès, E. G. Laurent and B. S. Marquet, *Tetrahedron Lett.* **38**, 1049 (1997).
20. W. Steinkopf and W. Kohler, *Liebigs Ann. Chem.* **532**, 250 (1937).
21. H. L. Coonradt, H. D. Hartough and G. C. Johnson, *J. Amer. Chem. Soc.* **70**, 2564 (1948).
22. H. L. Coonradt and H. D. Hartough, *J. Amer. Chem. Soc.* **70**, 1158 (1948).
23. M. Temciuc, A.-B. Hörnfeldt and S. Gronowitz, *J. Heterocycl. Chem.* **32**, 791 (1995).
24. P. Dallemagne, A. Alsaïdi, M. Boulard, S. Rault and M. Robba, *Heterocycles* **36**, 287 (1993).
25. E. Campaigne and R. C. Bourgeois, *J. Amer. Chem. Soc.* **76**, 2445 (1954).
26. O. Hromatka, D. Binder and P. Stanetti, *Monatsh. Chem.* **104**, 920 (1973).
27. O. Hromatka, D. Binder and G. Pixner, *Monatsh. Chem.* **106**, 1103 (1975).
28. J. Iriarte, E. Martinez and J. M. Muchowski, *J. Heterocycl. Chem.* **13**, 393 (1976).
29. M. Moreno-Manas, R. Cuberes, C. Palacin, M. Raga, J. M. Castello and J. A. Ortiz, *Eur. J. Med. Chem.* **23**, 477 (1988).
30. S. Conde, R. Madronero, M. P. Fernandez-Tome and J. del Rio, *J. Med. Chem.* **21**, 978 (1978).
31. S. Gronowitz and I. Ander, *Tetrahedron* **32**, 1403 (1976).
32. H. Goda, M. Sato, H. Ihara and C. Hirayama, *Synthesis* 849 (1992).
33. S. Kajigeshi, Y. Shinmasu, S. Fujisaki and T. Kakinami, *Chem. Lett.* 415 (1989).
34. T. Okamoto, T. Kakinami, H. Fujimoto and S. Kajigaeshi, *Bull. Chem. Soc. Japan* **64**, 2566 (1991).
35. S. Gronowitz and B. Holm, *Acta Chem. Scand. B* **30**, 423 (1976).
36. T. Bowles, R. Jones, A. E. A. Porter, J. A. Rechka, H. S. Rzepa and D. J. Williams, *J. Chem. Soc., Perkin Trans. 1* 1023 (1988).
37. C. Corral and J. Lissavetzky, *Synthesis* 487 (1984).
38. C. Corall, J. Lissavetzky and I. Manzanares, *J. Heterocycl. Chem.* **27**, 315 (1990).
39. A. Bugge, *Chem. Scripta* **2**, 137 (1972).

40. J. Skramstad, A. Lunde, H. Hope, V. Björnstad and P. Frøyen, *J. Chem. Soc., Perkin Trans. 2* 1453 (2000).
41. E. Campaigne and P. A. Monroe, *J. Amer. Chem. Soc.* **76**, 2447 (1954).
42. S. Gronowitz, A. Hallberg and C. Glennow, *J. Heterocycl. Chem.* **17**, 171 (1980).
43. A. Hallberg, T. Frejd and S. Gronowitz, *Chem. Scripta* **13**, 186 (1978–79).
44. J. Nakayama, S. Yamaoka and M. Hoshino, *Tetrahedron Lett.* **29**, 1161 (1988).
45. H. Ulrich, E. Kober, R. Ratz, H. Schroeder and C. Grundmann, *J. Org. Chem.* **27**, 2593 (1962).
46. Y. Goldberg and H. Alper, *J. Org. Chem.* **58**, 3072 (1993).
47. A. Kergomard and S. Vincent, *Bull. Soc. Chim. Fr.* 2197 (1967).
48. J. F. Bunnett, D. M. Bachmann, L. P. Snipper and J. H. Maloney, *J. Amer. Chem. Soc.* **71**, 1493 (1949).
49. J. Teste and N. Lozac'h, *Bull. Soc. Chim. Fr.* 437 (1955).
50. S. Gronowitz and B. Holm, *J. Heterocycl. Chem.* **14**, 281 (1977).
51. S. Gronowitz and T. Frejd, *Acta Chem. Scand. B* **29**, 818 (1975).
52. A. Almqvist and R. Håkansson, *Chem. Scripta* **11**, 57 (1977).
53. W. Steinkopf and A. Otto, *Liebigs Ann.* **424**, 61 (1921).
54. F. H. Pinkerton and S. F. Thomas, *J. Heterocycl. Chem.* **9**, 725 (1972).
55. C. A. Fink, A. P. Spada, D. Colussi, L. Rivera, and L. Merkel, *Nucleosides & Nucleotides* **11**, 1077 (1992).
56. R. D. Howells and H. Gilman, *J. Organomet. Chem.* **77**, 177 (1974).
57. M. R. Smith, Jr and H. Gilman, *J. Organomet. Chem.* **42**, 1 (1972).
58. J. W. Schick and H. D. Hartough, *J. Am. Chem. Soc.* **70**, 286 (1948).
59. M. J. Earle, A. G. Massey, A.-R. Al-Soudani and T. Zaidi, *Tetrahedron* **23**, 2817 (1989).
60. B. Dafgård, S. Gronowitz, G. Bondesson, O. Magnusson and N. E. Stjernström, *Acta Pharm. Suec.* **11**, 309 (1974).
61. L. N. Lucas, J. van Esch, R. M. Kellog and B. L. Feringa, *Tetrahedron Lett.* **40**, 1775 (1999).
62. S. Gronowitz, M. Temciuc and L. Eberson, *J. Heterocycl. Chem.* **32**, 65 (1995).
63. M. T. Rahman, *J. Indian Chem. Soc.* **LVIII**, 21 (1981).
64. V. A. Knizhnikov, V. I. Potkin, R. V. Kaberdin and Y. A. Ol'dekop, *J. Gen. Chem. USSR (Engl. Transl.)* **61**, 1243 (1991).
65. I. Haiduc and H. Gilman, *Rev. Romain. Chim.* **16**, 305 (1971).
66. M. R. Smith, Jr and H Gilman, *J. Organomet. Chem.* **42**, 1 (1972).
67. M. Nilsson and C. Ullenius, *Acta Chem. Scand.* **25**, 2428 (1971).
68. T. L. Cairns and B. C. McKusick, *J. Org. Chem.* **15**, 790 (1950).
69. S. Gronowitz, *Arkiv Kemi* **8**, 441 (1955).
70. M. Irie and M. Mohri, *J. Org. Chem.* **53**, 803 (1988).
71. M. Takeshita, M. Koike, H. Tsuzuki and M. Tashiro, *J. Org. Chem.* **57**, 4654 (1992).
72. M. Winn and F. G. Bordwell, *J. Org Chem.* **31**, 1610 (1967).
73. D. J. Zwanenburg and H. Wynberg, *J. Org. Chem.* **34**, 333 (1969).
74. S. G. Kon'kova, A. A. Safaryan and A. N. Akopyan, *Zh. Org. Khim. USSR (Engl. Transl.)* **14**, 1978 (1980).
75. G. B. Bachman and L.V. Haisey, *J. Am. Chem. Soc.* **70**, 2378 (1949).
76. B. P. Fabrichnyi, I. F. Shalavina, S. E. Zurabyan, Ya. L. Gol'dfarb and S. M. Kostrova, *Zhur. Org. Khim. USSR (Engl. Transl.)* **4**, 663 (1968).
77. S. Massa, M. Artico, A. Mai, R. Ragno, F. Corelli, A. Pani, M. E. Morangiu, E. Tramontano and P. La Colla, *Biorg. and Med. Chem. Lett.* **1**, 575 (1991).
78. M. Moreno-Manas, R. Cuberes, S. Palacín, M. Raga, J. M. Castello and J. A. Ortiz, *Eur. J. Med. Chem.* **23,** 477 (1988).
79. L. N. Lucas, J. van Esch, R. M. Kellogg and B. L. Feringa, *Chem. Commun.* 2314 (1998).
80. B. Unterhalt and P. Gores, *Arch. Pharm. (Weinheim)* **322**, 839 (1989).
81. T. Sone, M. Inoue and K. Sato, *Bull. Chem. Soc. Japan* **61**, 3779 (1988).

82. T. Sone, Y. Umetsu and K. Sato, *Bull. Chem. Soc. Japan* **64**, 864 (1991).
83. T. Sone, H. Kawasaki, S. Nagasawa, N. Takahashi, K. Tate and K. Sato, *Chem. Lett.* 399 (1981).
84. U. Dettmer, K. Eichler, K. Kühlein, E. I. Leupold and H. Litterer, *Angew. Chem. Int. Ed. Engl.* **26**, 468 (1987).
85. A. W. Faull and R. Hull, *J. Chem. Soc., Perkin Trans. 1* 1078 (1978).
86. D. W. Rangnekar and S. V. Mavlankar, *J. Heterocycl. Chem.* **28**, 1449 (1991).
87. S. Takada, H. Shindo, T. Sasatani, N. Chopmei, A. Matsushita, M. Eigyo, K. Kawasaki, S. Murata, Y. Yakahara and H. Shintaku, *J. Med. Chem.* **31**, 1738 (1988).
88. A. N. Akopyan, A. A. Saakyan, V. B. Gavallyan, A. G. Smbatyan and E. G. Darbinyan, *Zhur Organ. Khim. (Engl. Transl.)* **24**, 1521 (1988).
89. B. Yom-Tov, S. Gronowitz, S. Ross and N. E. Stjernström, *Acta Pharm. Suec.* **11**, 149 (1976).
90. V. K. Khairullin and L. I. Nesterenko, *J. Gen. Chem. USSR (Engl. Transl.)* **46**, 794 (1974).
91. M. M. El-Abdalahh, S. S. Sabri and A. Al-Ashqar, *Heterocycles* **45**, 255 (1997).
92. A. Mamouni, A. Daïch and B. Decroiz, *J. Heterocycl. Chem.* **33**, 1251 (1996).
93. A. N. Akopyan, A. A. Saakyan, V. B. Gavalyan, A. G. Smbatyan and E. G. Darbinyan, *J. Org. Chem. (USSR)* **24**, 1370 (1988).
94. A. S.-Y. Lee and W.-C. Dai, *Tetrahedron Lett.* **37**, 495 (1996).
95. S. Conde, C. Corral, R. Madronero and A. Sanchez Alvares-Insua, *Synthesis* 412 (1976).
96. G. Engelmann, R. Stösser, G. Kossmehl, W. Jugelt and H.-P. Weizel, *J. Chem. Soc., Perkin Trans. 2* 2015 (1996).
97. E. J. Geering, *J. Org. Chem.* **24**, 1128 (1959).
98. M. J. Raasch, *J. Org. Chem.* **45**, 856 (1980).
99. C. Troyanowsky, *Bull. Soc. Chim. Fr.* 424 (1955).
100. S. Gronowitz, P. Moses and R. Håkansson, *Arkiv Kemi* **16**, 267 (1960).
101. Y. Tamaru, Y. Yamada and Z. Yoshida, *Tetrahedron* **35**, 329 (1979).
102. A. P. Terenteyev, L. I. Belen'kii and A. L. Yanovskaya,. *J. Gen. Chem. USSR (Engl. Transl.)* **24**, 1151 (1954).
103. A. M. McKillop, D. Bromley and E. C. Taylor, *J. Org. Chem.* **37**, 88 (1972).
104. H. V. van der Plas and C. J. Persoons, *Rec. Trav. Chim. Pays-Bas* **83**, 701 (1964).
105. M. A. Keegstra and L. Brandsma, *Synthesis* 890 (1988).
106. S. Kobatake, K. Shibata, K. Uchida and M. Irie, *J. Am. Chem. Soc.* **122**, 12135 (2000).
107. L. Brandsma and H. D. Verkruiisse, 18, 1763 (1988).
108. M. Temciuc, A.-B. Hörnfeldt and S. Gronowitz, *Heterocycl. Comm.* **1**, 411 (1995).
109. M. G. Reinecke, H. W. Adickes and C. Pyun, *J. Org. Chem.* **36**, 3820 (1971).
110. M. Nemec, M. Janda and J. Srogl, *Collect. Czech. Chem. Commun.* **38**, 3857 (1973).
111. Y. L. Gol'dfarb, M. A. Kalik and M. A. Kirmalova, *J. Gen. Chem. USSR, (Engl. Transl.)* **29**, 2003 (1959).
112. M. G. Reinecke, H. W. Adickes and C. Pyun, *J. Org. Chem.* **36**, 2690 (1971).
113. S. L. Gilat, S. H. Kawai and J.-M. Lehn, *J. Chem. Soc. Chem. Commun.* 1439 (1993).
114. A.-B. Hörnfeldt, *Acta Chem. Scand.* **21**, 1952 (1967).
115. S. Gronowitz, B. Cederlund and A.-B. Hörnfeldt, *Chem. Scripta* **5**, 217 (1974).
116. G. Consiglio, S. Gronowitz, A.-B. Hörnfeldt, B. Maltesson, R. Noto and D. Spinelli, *Chem. Scripta* **11**, 175 (1977).
117. R. D. McCullough, R. D. Lowe, M. Jayaraman and D. L. Andersen, *J. Org. Chem.* **58**, 904 (1993).
118. D. B. Hauze, M. M. Joullié, R. Ramotoski and A. Cantu, *Tetrahedron* **53**, 4239 (1997).
119. A. Ishii, T. Tsuchiya, J. Nakayama and M. Hoshino, *Tetrahedron Lett.* **34**, 2347 (1993).
120. J. Heerklotz, A. Linden and M. Hesse, *Tetrahedron* **56**, 7205 (2000).
121. J. J. McNally, P. J. Sanfilippo, L. Fitzpatrick and J. B. Press, *J. Heterocycl. Chem.* **29**, 247 (1992).

122. R. M. Kellogg, A. P. Schaap, E. T. Harper and H. Wynberg, *J. Org. Chem.* **33**, 2902 (1968).
123. S. Gronowitz, N. Gjøs and R. M. Kellogg, *J. Org. Chem.* **32**, 463 (1967).
124. N. Gjøs and S. Gronowitz, *Acta Chem. Scand.* **21**, 2893 (1967).
125. E. W. Brunett and W. C. McCarthy, *J. Pharmaceutical Sci.* **57**. 2003 (1968).
126. M. D. Mullican, R. J. Sorenson, D. T. Connor, D. O. Thueson, J. A. Kennedy and M. C. Conroy, *J. Med Chem* **34**, 2186 (1991).
127. P. Pigeon and B. Decroix, *J. Heterocycl. Chem.* **33**, 129 (1996).
128. S. Gronowitz, *Arkiv Kemi* **8**, 87 (1955).
129. N. P. Buu-Hoi and D. Lavit, *J. Chem. Soc* 1721 (1958).
130. Y. L. Gol'dfarb, Y. B. Volkenshtein and B. V. Lopatin, *J. Gen. Chem. USSR (Engl. Transl)* **34,** 961 (1964).
131. L. I. Belen'kii, G. P. Gromova and Y. L. Gol'dfarb, *Izv. Akad. Nauk SSSR Ser. Khim (Engl. Transl.)* 1137 (1971).
132. D. J. Chadwick, J. Chambers, G. D. Meakins and R. l. Snowden, *J. Chem. Soc. Perkin Trans. 1* 1766 (1973).
133. S. Gronowitz, P. Moses, A.-B. Hörnfeldt and R. Håkansson, *Arkiv Kemi* **17**, 165 (1960).
134. F. Kipnis, H. Soloway and J. Ornfelt, *J. Amer. Chem. Soc.* **71**, 10 (1949).
135. J. Sam and G. G. Advani, *J. Pharm. Sci.* **54**, 753 (1965).
136. D. T. Drewy and R. M. Scrowston, *J. Chem. Soc. C* 2750 (1969).
137. T. J. Broxton, L. W. Deady, J. D. McCormack, L. C. Kam and S. H. Toh, *J. Chem. Soc., Perkin Trans. 1* 1769 (1974).
138. O. Karlsson, *Syn. Comm*, **11**, 29 (1981).
139. A. S. Alvarez-Insua, S. Conde and C. Corall, *J. Heterocycl. Chem.* **19**, 713 (1982).
140. Y. B. Volkenshtein, V. P. Lopatin and V. A. Petukhov, *Izvest. Akad, Nauk, SSSR Otdel Khim Nauk (Engl Transl.)* 1752 (1961).
141. S. Gronowitz, T. Dahlgren, J. Namtvedt, C. Roos, G. Rosén, B. Sjöberg and U. Forsgren, *Acta Pharm. Suec.* **8**, 623 (1971).
142. M. T. Rahman and H. Gilman, *J. Ind. Chem. Soc.* **53**, 582 (1976).
143. Y. L. Gol'dfarb, G. P. Gromova and L.I. Belen'kii, *Izv. Akad. Nauk. SSSR, Ser. Khim. (Engl. Transl.)* 2191 (1974).
144. P. P. Ehrlich, J. W. Ralston and M. R. Michaelides, *J. Org. Chem.* **62**, 2782 (1997).
145. W. Steinkopf, H. Jacob and H. Penz, *Liebigs Ann.* **512**, 136 (1934).
146. S. O. Lawesson, *Arkiv Kemi* **11**, 317 (1957).
147. S. Gronowitz, J. Rehnö, K. Titlestad, M. Vadzis, B. Sjöberg, P. Bamberg, B. Ekström and U. Forsgren, *Acta Pharm. Suec.* **9**, 381 (1972).
148. J. Marcusson, *Ber.* **26**, 2457 (1893).
149. S.-O. Lawesson, *Arkiv Kemi* **11**, 325 (1957).
150. S. Gronowitz and A. Rosenberg, *Arkiv Kemi* **8**, 22 (1955).
151. W. Steinkopf and W. Nitschke, *Liebigs Ann. Chem.* **536**, 135 (1938).
152. E. C. Taylor and J. E. Dowling, *J. Org. Chem.* **62**, 1599 (1997).
153. E. Campaigne, L. Fedor and R. E. Johnson, *Heterocycl. Chem.* **1**, 242 (1964).
154. R. Kimura, T. Yabuuchi and M. Hisaki, *Chem. Pharm. Bull.* **10**, 1232 (1962).
155. S. Gronowitz and T. Frejd, *Acta Chem. Scand. B* **30**, 439 (1976).
156. S. Gronowitz and B. Holm, *Acta Chem. Scand.* **23**, 2207 (1969).
157. A. Wiersema and S. Gronowitz, *Acta Chem. Scand.* **24**, 2593 (1970).
158. S. O. Lawesson, *Arkiv Kemi* **11**, 387 (1957).
159. R. G. Hicks and M. B. Nodwell, *J. Am. Chem. Soc.* **122**, 6746 (2000).
160. K. Sasaki, O. Tokuda, T. Hirota, J.-K. Luo and R. N. Castle, *J. Heterocycl. Chem.* **33**, 847 (1996).
161. P. A. Chaloner S. R. Gunatunga and P. B. Hitchcock, *J. Chem. Soc., Perkin Trans. 2* 1597 (1997).
162. J. Nakayama, Y. Ting, Y. Sugihara and A. Ishii, *Heterocycles* **44**, 75 (1997).

163. E. Campaigne and R. L. White Jr, *J. Heterocycl. Chem.* **25**, 367 (1988).
164. D. Brugier, F. Outurquin and C. Paulmier, *Tetrahedron* **56**, 2985 (2000).
165. F. Bottari, G. Lippi and B. Macchia, *Gazz. Chim. Ital.* **99**, 762 (1969).
166. S. Gronowitz, M. Temciuc and A.-B. Hörnfeldt, *J. Heterocycl. Chem.* **30**, 1111 (1993).
167. I. J. Turchi, J. B. Press, J. J. McNally, M. P. Bonner and K. L. Sorgi, *J. Org. Chem.* **58**, 4629 (1993).
168. N. Latif, N. Mishriky and N. S. Girgis, *Indian J. Chem., Sect. B*, **15**, 118 (1977).
169. C. Consiglio, D. Spinelli, S. Gronowitz, A.-B. Hörnfeldt, B. Maltesson and R. Noto, *J. Chem. Soc., Perkin Trans. 2* 625 (1982).
170. P. Fournari, R. Guilard and M. Person, *Bull. Soc. Chim. Fr.* 4115 (1967).
171. R. Håkansson and A. Svensson, *Chem. Scripta* **7**, 186 (1975).
172. R. Håkansson, A. Ask and A. Almquist, *Chem. Scripta* **2**, 72 (1972).
173. J. M. Barker, P. R. Huddleston and S. W. Shutler, *J. Chem. Soc., Perkin Trans. 1* 2483 (1975).
174. D. Spinelli, G. Guanti and C. Dell'Erba, *J. Heterocycl. Chem.* **5**, 323 (1968).
175. D. Spinelli, C. Consiglio, R. Noto and A. Corrao, *J. Chem. Soc., Perkin Trans 2* 620 (1975).
176. A. E. Weitkamp and C. S. Hamilton, *J. Amer. Chem. Soc.* **59**, 2699 (1937).
177. D. W. H. MacDowell and T. D. Greenwood, *J. Heterocycl. Chem.* **2**, 44 (1965).
178. S. Gronowitz, *Acta Chem. Scand.* **13**, 1045 (1959).
179. S. Gronowitz and T. Raznikiewicz, *Org. Syn.* **44**, 9 (1964).
180. Y. L. Gol'dfarb, V. P. Litvinov and S. A. Ozolin, *Izv. Akad. Nauk. SSSR, Ser. Khim. (Engl. Transl)* 486 (1965).
181. R. D. Schuetz, F. M. Gruen, D. R. Byrne and R. L. Brennan, *J. Heterocycl. Chem.* **3**, 184 (1966).
182. S.-O. Lawesson, *Acta Chem. Scand.* **10**, 1020 (1956).
183. C. D. Hurd and D. L. Kreuz, *J. Amer. Chem. Soc.* **74**, 2965 (1952).
184. S. Gronowitz, *Arkiv Kemi* **7**, 267 (1954).
185. S.-O. Lawesson, *Arkiv Kemi* **11**, 373 (1957).
186. Y. I. Shapiro, L. I. Belen'kii, I. A. Romanskii, F. M. Stoyanovich, Y. L. Gol'dfarb and A. I. Shatenstein, *J. Gen. Chem. USSR (Engl. Transl.)* **38**, 1938 (1968).
187. D. L. Ladd, P. B. Harrsch and L. I. Kruse, *J. Org. Chem.* **53**, 417 (1988).
188. Y. Xie, B.-M. Wu, F. Xue, S.-V. Ng, T. C. W. Mak and T. S. A. Hor, *Organometallics* **17**, 3988 (1998).
189. S. Dapperheld, M. Feldhues, H. Litterer, F. Sistig and P. Wegener, *Synthesis* 403 (1990).
190. M. S. Mubarak and D. G. Peters, *J. Org. Chem.* **61**, 8074 (1996).
191. G. M. Davies and P. S. Davies, *Tetrahedron Lett.* 3507 (1972).
192. R. Wu, J. S. Schumm, D. L. Pearson and J. M. Tour, *J. Org. Chem.* **61**, 6906 (1996).
193. L. Antolini, F. Goldoni, D. Iarossi, A. Mucci and L. Schenetti, *J. Chem. Soc., Perkin Trans. 1* 1957 (1997).
194. D. C. Harrowven, *Tetrahedron Lett.* **34**, 5653 (1993).
195. F. Sauter, P. Stanetty, H. Fröhlich and W. Ramer, *Heterocycles* **26**, 2639 (1987).
196. J.-P. Lère-Porte, J. J. E. Moreau and C. Torreilles, *Eur. J. Org. Chem.* 1249 (2001).
197. S. Gronowitz, T. Frejd, O. Karlsson, K. Lawitz. P. Pedaja and K. Pettersson, *Chem. Scripta* **18**, 192 (1981).
198. S. Gronowitz and T. Frejd, *Acta Chem. Scand.* **30**, 485 (1976).
199. J. O. Karlsson, A. Svensson and S. Gronowitz, *J. Org. Chem.* **49**, 2018 (1984).
200. F. Sauter, H. Fröhlich and W. Kalt, *Synthesis* 771 (1989).
201. H. Fröhlich and W. Kalt, *J. Org. Chem.* **55**, 2993 (1990).
202. M. Nemec, P. Vopatrna, M. Janda, J. Strogl and I. Stibor, *Synthesis* 545 (1972).
203. S. Gronowitz and K. Dahlgren, *Arkiv Kemi* **21**, 201 (1963).
204. A. Schöning and W. Friedrichsen, *Z. Naturforsch.* **44b**, 825 (1989).
205. S. Yoshida, H. Kubo, T. Saika and S. Katsemura, *Chem. Lett.* 139 (1996).
206. J. Skramstad and O. Eriksen, *Acta Chem. Scand.* **45**, 919 (1991).

207. L. Antolini, F. Goldoni, D. Iarossi, A. Mucci and L. Schenetti, *J. Chem. Soc., Perkin Trans. 1* 1957 (1997).
208. E. Lukevics, P. Arsenyan, S. Belyakov, J. Popelis and O. Pudova, *Tetrahedron Lett.* **42**, 2039 (2001).
209. J. L. Reddinger and J. R. Reynolds, *J. Org. Chem.* **61**, 4833 (1996).
210. Y. Mazaki and K. Kobayashi, *Tetrahedron Lett.* **30**, 3315 (1989).
211. M. Takeshita and M. Irie, *J. Org. Chem.* **63**, 6643 (1998).
212. S. Gronowitz and C. Glennow, *Chem. Scripta* **11**, 76 (1977).
213. S. H. Kawai, S. L. Gilat amd J.-M. Lehn, *J. Chem. Soc., Chem. Commun.* 1011 (1994).
214. S. H. Kawai, S. L. Gilat, R. Ponsinet and J.-M. Lehn, *Chem. Eur. J.* **1**, 285 (1995).
215. A. Fernandez-Acebes and J.-M- Lehn, *Chem. Eur. J.* 3285 (1999).
216. S. L. Gilat, H. Kawai and J.-M. Lehn, *Chem. Eur. J.* **1**, 275 (1995).
217. M. Irie, T. Lifka, S. Kobatake and N. Kato, *J. Am. Chem. Soc.* **122**, 4871 (2000).
218. A. Osuka, D. Fujikane, H. Shinmori, S. Kobatake and M. Irie, *J. Org. Chem.* **66**, 3913 (2001).
219. F. Sauter, P. Stanetty and H. Fröhlich, *Heterocyles* **26**, 2657 (1987).
220. A. Ishii, R. Yoshioka, J. Nakayama and M. Hoshino, *Tetrahedron Lett.* **34**, 8259 (1993).
221. J. D. Prugh, G. D. Hartman, P. J. Mallorga, B. M. Mckleever, S. R. Michelson. M. A. Murcko, H. Schwam, R. L. Smith, J. M. Sonday, J. P. Springer and M. F. Sugrue, *J. Med. Chem.* **34**, 1805 (1991).
222. N. Kusuhara, Y. Sugano, H. Takagi, H. Miyake and K. Yamamura, *Chem. Commun.* 1951 (1997).
223. H. Muratake, K. Okabe, M. Takahashi, M. Tonegawa and M. Natsume, *Chem. Pharm. Bull.* **45**, 799 (1997).
224. R. A. Benkeser and A. Torkelson, *J. Am. Chem. Soc.* **76**, 1252 (1954).
225. S. H. Kim and R. D. Rieke, *Tetrahedron Lett.* **38**, 993 (1996).
226. M. Takeshita and M. Tashiro, *J. Org. Chem.* **57**, 746 (1992).
227. M. Hori, T. Kataoka, H. Shimiz and M. Yoshimura, *Yakugaku Zasshi* **94**, 1429 (1974).
228. M. Takeshita and M. Tashiro, *J. Org. Chem.* **57**, 746 (1992).
229. M. Pomerantz, B. Chaloner-Gill, L. O. Harding, J. J. Tseng and W. J. Pomerantz, *J. Chem. Soc., Chem. Commun.* 1672 (1992).
230. C. Dell'Erba, F. Sancassan, M. Novi, D. Spinelli, G. Consiglio, C. Arnone and F. Ferroni, *J. Chem. Soc., Perkin Trans. 2* 1779 (1989).
231. R. Noto, M. Gruttadauria, P. Lo Meo and D. Spinelli, *Collect. Czech. Chem. Commun.* **64**, 1893 (1999).
232. D. Binder, C. R. Noe, K. Baumann and W. Holzer, *Arch. Pharm. (Weinheim)* **321**, 391 (1988).
233. A. K. Mohanakrishnan, M. V. Lakshmikantham, C. McDougal, M. P. Cava, J. W. Baldwin and R. M. Metzger, *J. Org. Chem.* **63**, 3105 (1998).
234. Q. T. Zhang and J. M. Tour, *J. Am. Chem Soc.* **119**, 9624 (1997).
235. L. Brandsma and R. L. P. de Jong, *Synth. Commun.* **20**, 1697 (1990).
236. L. Brandsma and H. D. Verkruijsse *Synth. Commun.* **20**, 2119 (1990).
237. H. Finch, D. H. Recce and J. T. Sharp, *J. Chem. Soc., Perkin Trans. 1*, 1193 (1994).
238. O. Meth-Cohn and S. Gronowitz, *Acta Chem. Scand*, **20**, 1577 (1966).
239. W. Steinkopf, *Ann.* **513**, 293 (1934).
240. J. Cymerman-Craig and J. W. Loder, *J. Chem. Soc.* 237 (1954).
241. S. Gronowitz and T. Frejd, *Acta Chem. Scand.* **B 30**, 485 (1976).
242. D. J. Zwanenburg and H. Wynberg, *Rec. Trav. Chim. Pays-Bas* **88**, 321 (1969).
243. K. J. Stone, M. M. Greenberg, S. C. Blackstock and J. A. Berson, *J. Am. Chem. Soc.* **111**, 3659 (1989).
244. J. Nakayama, T. Kawamura, K. Kuroda, and A. Fujita, *Tetrahedron Lett.* **34**, 5725, (1993).
245. M. Takeshita and M. Tashiro, *J. Org. Chem.* **56**, 2837 (1991).
246. M. Takeshita and M. Tashiro, *J. Org. Chem.* **57**, 746 (1992).

247. M. Takeshita, M. Tashiro and A. Tsuge, *Chem. Ber.* **124**, 1403 (1991).
248. S. Gronowitz and H. Frostling, *Acta Chem. Scand*, **16**, 1127 (1962).
249. Y. Yang, A.-B. Hörnfeldt and S. Gronowitz, *Chem. Scripta* **28**, 275 (1988).
250. M. M. Goodman, G. W. Kabalka, R. C. Marks, F. F. Knapp Jr, J. Lee and Y. Liang, *J. Med. Chem.* **35**, 280 (1992).
251. N. M. Goudgaon, P. P. Wadgaonkar and G. W. Kabalka, *Synth. Commun.* **19**, 805 (1989).
252. G. E. Boswell, R. W. McNutt, D. G. Bubacz, A. O. Davis and K. J. Chang, *J. Heterocycl. Chem.* **32**, 1801 (1995).
253. I. Osaka, S. Shibata, R. Toyoshima, K. Akagi and H. Shirakawa, *Synth. Met.* **102**, 1437 (1999).
254. M. Kijima, K. Setoh and H. Shirakawa, *Chem. Lett.* 936 (2000).
255. B. Yom-Tov and S. Gronowitz, *Chem. Scripta* **3**, 37 (1973).
256. B. Yom-Tov and S. Gronowitz, *Chem. Scripta* **3**, 165 (1973).
257. B. Yom-Tov and S. Gronowitz, *J. Heterocycl Chem.* **15**, 285 (1978).
258. S. Gronowitz and T. Dahlgren, *Chem. Scripta* **12**, 57 (1977).
259. S. Gronowitz and L. Svensson, *Chem. Scripta* **15**, 169 (1980).
260. U. Michael and S. Gronowitz, *Chem. Scripta* **4**, 126 (1973).
261. M. L. Tedjamulia, J. G. Stuart, Y. Tominaga, R. N. Castle and M. L. Lee, *J. Heterocycl. Chem.* **21**, 1215 (1984).
262. W. Hinz, R. Jones and T. Anderson, *Synthesis* 620 (1986).
263. R. M. Kellogg, M. B. Groen and H. Wynberg, *J. Org. Chem.* **32**, 3093 (1967).
264. R. Pratap, Y. Tominaga, M. L. Lee and R. N. Castle, *J. Heterocycl. Chem.* **18**, 973 (1981).
265. Y. Tominaga, M. L. Tedjamulia, R. N. Castle and M. L. Lee, *J. Heterocycl. Chem.* **20**, 487 (1983).
266. M. L. Tedjamulia, Y. Tominaga, R. N. Castle and M. L. Lee, *J. Hetrocycl. Chem.* **20**, 1143 (1983).
267. P. C. Ewbank, G. Nuding, H. Suenaga, R. D. McCullough and S. Shinkai, *Tetrahedron Lett.* **42**, 155 (2001).
268. P. Pedaja and S. Gronowitz, *Chem. Scripta* **20**, 53 (1982).
269. E. Wiklund and R. Håkansson, *Acta Chem. Scand.* **24**, 341 (1970).
270. E. Wiklund and R. Håkansson, *Chem. Scripta* **6**, 174 (1974).
271. J.-P. Lère-Porte, J. J. E. Moreau, F. Serein-Spirau and S. Wakim, *Tetrahedron Lett.* **42**, 3073 (2001).
272. E. Migianu, D. Prim and G. Kirsch, *Synlett* 459 (2000).
273. C. Corral, M. B. El-Ashmawy, J. Lissavetzky, A. Basilio and A. Giraldez, *Eur. J. Med. Chem.* **22**, 251 (1987).
274. L. Brandsma and H. F. Verkruijsse, *Synth. Commun.* **20**, 2275 (1990).
275. M. Makosza and E. Kwast, *Tetrahedron* **51**, 8339 (1995).
276. H. J. Sawada, M. Yoshida, H. Hagii, K. Aoshima and M. Kobayashi, *Bull. Chem. Soc. Jpn* **59**, 215 (1986).
277. M. Yoshida, T. Yoshida, N. Kamigata and M. Kobayashi, *Bull. Chem. Soc. Jpn* **61**, 3549 (1988).
278. T. Masquelin and D. Obrecht, *Tetrahedron Lett.* **35**, 9387 (1994).
279. L. Fuller, B. Iddon and K. A. Smith, *Chem. Commun.* 2355 (1997).
280. M. Rajsner, J. Metysova and M. Protiva, *Collect. Czech. Chem. Commun.* **35**, 378 (1970).
281. S. Gronowitz and B. Gestblom, *Arkiv Kemi* **18**, 513 (1962).
282. R. Gonzales and E. V. Brown, *J. Org. Chem.* **17**, 698 (1952).
283. Y. L. Gol'dfarb and M. A. Kalik, *Chem. Heterocycl. Compds. (Engl. Transl.)* **7**, 155 (1971).
284. M. Sy, N. P. Buu-Hoi and N. D. Xuong, *J. Chem. Soc.* 1975 (1954).
285. R. M. Kellog and J. Buter, *J. Org. Chem.* **36**, 2236 (1971).
286. Y. Lew and C. R. Noller, *Org. Syn.* **30**, 53 (1950).
287. J. M. Barker, P. R. Huddleston and M. L. Wood, *Syn. Commun.* **5**, 59 (1975).

288. A. R. Butler and A. P. Sanderson, *J. Chem. Soc., Perkin Trans 2* 1214 (1974).
289. T. Sone, T. Sakai and K. Kuroda, *Bull. Chem. Soc. Japan* **43**, 1411 (1970).
290. M. Prats and C. Gálvez, *Heterocycles* **34**, 149 (1992).
291. S. Hotta, *J. Heterocycl. Chem.* **38,** 923 (2001).
292. S. Hotta, S. A. Lee and T. Tamaki, *J. Heterocycl. Chem.* **37**, 25 (2000).
293. H. O. Wirt, O. Konigstein and W. Kern, *Liebigs Ann.* **634**, 84 (1960).
294. S. Gronowitz and V. Vilks, *Arkiv Kemi* **21**, 191 (1963).
295. M. D'Auria, C. Distefano, F. D'Onofrio, G. Mauriello and R. Racioppi, *J. Chem. Soc., Perkin Trans. 1* 3513 (2000).
296. G. de Luca, G. Martelli, P. Spagnolo and M. Tiecco, *J. Chem. Soc C* 2504 (1970).
297. S. Gronowitz and S. Hagen, *Arkiv Kemi* **27**, 153 (1967).
298. A. Hallberg, T. Frejd and S. Gronowitz, *J. Chem. Soc., Perkin Trans 1* 1390 (1980).
299. E. Wiklund and R. Håkansson, *Chem. Scripta* **3**, 320 (1973).
300. S. Gronowitz and R. Beselin, *Arkiv Kemi* **21**, 349 (1963).
301. S. Gronowitz and J. Skramstad, *Arkiv Kemi* **28**, 115 (1968).
302. S. Gronowitz and T. Frejd, *Acta Chem. Scand. B* **30**, 287 (1976).
303. T. Frejd, J. O. Karlsson and S. Gronowitz, *J. Org. Chem.* **46**, 3132 (1981).
304. T. Frejd and J. O. Karlsson, *Acta Chem. Scand. B* **37**, 895 (1983).
305. P. Meunier and G. Pfister-Guillouzo, *Can. J. Chem.* **55**, 3901 (1977).
306. B. E. Ayres, S. W. Longworth and J. F. W. McOmie, *Tetrahedron* **31**, 1755 (1975).
307. J. W. Terpstra and A. W. van Leusen, *J. Org. Chem.* **51**, 230 (1986).
308. M. D'Auria and G. Mauriello, *Synthesis* 248 (1995).
309. J. Kowalik and L. M Tolbert, *Chem. Commun.* 877 (2000).
310. W. Steinkopf and W. Hanske, *Liebigs Ann.* **527**, 247 (1937).
311. G. N. Jean and F. F. Nord, *J. Org. Chem.* **20**, 1363 (1955).
312. M. D'Auria, *Synth. Commun.* **22**, 2393 (1992).
313. G. Consiglio, C. Arone, D. Spinelli, R. Noto and V. Frenna, *J. Chem. Soc., Perkin Trans. 2* 621 (1982).
314. D. Spinelli, G. Consiglio, R. Noto and A. Corrao, *J. Chem. Soc., Perkin Trans. 2* 1632 (1974).
315. D. Spinelli, G. Consiglio and T. Monti, *J. Chem. Soc. Perkin Trans. 2* 816 (1975).
316. J. Morel and P. Pastour, *Bull. Soc. Chim. Fr.* 737 (1968).
317. L. L. Miller and Y. Yu, *J. Org. Chem.* **60**, 6813 (1995).
318. C. A. Briehn, T. Kirschbaum and P. Bäuerle, *J. Org. Chem.* **65**, 352 (2000).
319. M. Shilai, Y. Kondo and T. Sakomoto, *J. Chem. Soc., Perkin Trans. 1* 442 (2001).
320. A. J. Carpenter and D. J. Chadwick, *Tetrahedron Lett.* **26**, 1777 (1985).
321. S. Gronowitz and R. Håkansson, *Arkiv Kemi* **16**, 309 (1960).
322. P. Spagnolo, P. Zanirato and S. Gronowitz, *J. Org. Chem.* **47**, 3177 (1982).
323. S. Gronowitz and B. Holm, *Acta Chem. Scand. B* **30**, 505 (1976).
324. Ya. L. Gol'dfarb, M. A. Kalik and V. K. Zav'yalova, *Chem. Heterocycl. Copmpds., USSR (Eng. Transl.)* **17**, 126 (1981).
325. Y. Geng, A. Fechtenkötter and K. Müllen, *J. Mater. Chem.* **11**, 1634 (2001).
326. K. Takahashi and S. Tarutani, *Heterocycles* **43**, 1927 (1996).
327. M. Takahashi, T. Kuroda, T. Ogiku, H. Ohmizu, K. Kondo and T. Iwasaki, *Heterocycles* **36**, 1867 (1993)
328. L. Brandsma, H. Hommesa, H. D. Verkruisse, and R. L. P. de Jong, *Rec. Trav. Chim Pays-Bas* **104**, 226 (1985).
329. D. Sooloki, J. D. Bradshaw, C. A. Tessier and W. J. Youngs, *Organometallics* **13**, 451 (1994).
330. H. Muguruma, T. Saito, S. Sasaki, S. Hota and I. Karube, *J. Heterocycl. Chem.* **33**, 173 (1996).
331. T. Kirschbaum, R. Azumi, E. Mena-Ostreritz and P. Bäuerle, *New J. Chem.* 241 (1999).
332. A. McKillop, J. S. Fowler, M. J. Zelesko, J. D. Hunt, E. C. Taylor and G. M. McGillivray, *Tetrahedron Lett.* 2427 (1969).

333. A. McKillop, J. D. Hunt, M. J. Zelesko, J. C. Fowler, E. C. Taylor, G. McGillivray, F. Kienzle, *J. Amer. Chem. Soc.* **93**, 4841 (1971).
334. D. J. Chadwick, J. Chambers, H. E. Hargraves, G. D. Meakins and R. L. Snowden, *J. Chem. Soc., Perkin Trans 1* 2327 (1973).
335. M. M. Goodman, G. Kirsch and F. F. Knapp Jr, *J. Heterocycl. Chem.* **21**, 1579 (1984).
336. R. D. Brown, A. S. Buchanan and A. A. Humffray, *Aust. J. Chem.* **18**, 1527 (1965).
337. D. E. Seitz, S. H. Lee, R. N. Hansson and J. C. Bottavo, *Syn. Commun.* **13**, 121 (1983).
338. S.-H. Lee, R. N. Hanson and D. E. Seitz, *Tetrahedron Lett.* **25**, 1751 (1984).
339. M. T. Rahman and H. Gilman, *J. Ind. Chem. Soc.* **51**, 1018 (1974).
340. M. R. Smith and H. Gilman, *J. Organometal. Chem.* **42**, 1 (1972).
341. R. Guilard, P. Fournari and M. Person, *Bull. Soc. Chim. Fr.* 4121 (1967).
342. G. Wittig and M. Rings, *Liebigs Ann.* **719**, 127 (1968).
343. M. D'Auria, *Gazz. Chim. Ital.* **124**, 195 (1994).
344. F. Bohlmann and J. Kocur, *Chem. Ber.* **107**, 2115 (1974).
345. P. T. De Sousa, Jr and R. J. K. Taylor, *Synlett.* 755, (1990).
346. A. Iraqi and G. W. Barker, *J. Mater. Chem.* **8**, 25 (1998).
347. X.-S. Ye and H. N. C. Wong, *J. Org. Chem.* **62**, 1940 (1997).
348. M. D'Auria and G. Mauriello, *Tetrahedron Lett.* **36**, 4883 (1995).
349. X. S. Ye and H. N: C. Wong, *Chem. Comm.* 339 (1996).
350. T. X. Neenan and G. M. Whitesides, *J. Org. Chem.* **53**, 2489 (1988).
351. R. Rossi, A. Carpita and T. Messeri, *Synth. Commun.* **21**, 1875 (1991).
352. H. Togo, T. Nabana and K. Yamaguchi, *J. Org. Chem.* **65**, 8391 (2000).
353. F. M. Beringer, H. E. Bachofner, R. A. Falk and M. Leff, *J. Amer. Chem. Soc.* **80**, 4279 (1958).
354. Y. Yamada and M. Okawara, *Bull. Chem. Soc. Jpn*, **45**, 2515 (1972).
355. Y. Yamada and M. Okawara, *Bull. Chem. Soc. Jpn.* **45**, 1860 (1972).
356. J.-P. Battioni, D. Dupré, M. Delaforge M. Jaouen and D. Mansui, *J. Organomet. Chem.* **358**, 389 (1988).
357. A. J. Margida and G. F. Koser, *J. Org. Chem.* **49**, 3643 (1984).
358. G. F. Koser and R. H. Wettach, *J. Org. Chem.* **45**, 1542 (1980).
359. X. Shan, W.-K. Lee and H. N. C. Wong, *J. Am. Chem. Soc.* **118**, 2511 (1996).
360. P. J. Stang, R. Tykwinski and V. V. Zhdankin, *J. Heterocycl. Chem.* **29**, 815 (1992).
361. V. W. Pike, F. Butt, A. Shah, D. A. Widdowson, *J. Chem. Soc. Perkin Trans. 1* 245 (1999).
362. S. Martin-Santamaria, M. A. Carrol, C. M. Carroll, C. D. Carter, V. W. Pike, H. S. Rzepa, D. A. Widdowson, *Chem. Commun.* 649 (2000).
363. M. A. Carroll, V. W. Pike and D. A. Widdowson, *Tetrahedron Lett.* **41**, 5393 (2000).
364. F. M. Beringer and R. A. Nathan, *J. Org. Chem.* **35**, 2095 (1970).
365. S. Gronowitz and B. Holm, *Tetrahedron* **33**, 557 (1977).
366. S. Gronowitz and I. Ander, *Acta Chem. Scand. B* **29**, 513 (1975).
367. S. Gronowitz and I. Ander, *Chem. Scripta* **15**, 135 (1980).

– 8 –

Bi-, ter- and oligothienyls

8.1 INTRODUCTION

In this final chapter, methods of preparation of bithienyls, terthienyls, quaterthienyls and compounds containing even more connected thiophene rings up to polythienyls will be treated in a systematic way. These types of compounds are also called bithiophenes, terthiophenes, etc., or as groups, oligothiophenes and polythiophenes. However, we consider the "thienyl" nomenclature to be more correct, as these compounds are not polymers of thiophene $(C_4H_4S)_n$.

In connection with work on conducting polymers, electro-optic devices such as organic light-emitting diodes, optical switches and in other fields of materials science, there has been a great interest in the preparation of such compounds in recent years. A Handbook of Oligo- and Polythiophenes with D. Fichou as editor has recently been published, which contains a chapter by Peter Bäuerle on the synthesis of oligothiophenes of interest in materials science [1]. In the present chapter, synthetic methods for bithienyl terthienyls etc. will be described systematically, although this results in some repetition. We will also give examples of some very recent experimental procedures.

8.2 SYMMETRICAL BITHIENYLS

8.2.1 Coupling of thienyllithium or thiophenemagnesium halides with cupric chloride or iron(III) acetylacetonate

The first practical synthesis of 2,2′-bithienyl was described by Steinkopf and Roch, who obtained a 44% yield by the reaction of 2-thiophenemagnesium bromide with cupric chloride [2]. A more convenient route was discovered in the reaction of 2-thienyllithium, obtained by metalation of thiophene with butyllithium followed by cupric chloride [3]. Alternatively halogen–metal exchange between bromo- or iodothiophenes with butyllithium at −70 °C can be used for the preparation of the thienyllithium derivatives. In this way a large number of substituted 2,2′-bithienyls, such as 4,4′,5,5′-tetramethyl-2,2′-bithienyl [4,5], 4,4′,5,5′-tetraphenyl-2,2′-bithienyl [5], 3,3′-dimethyl-2,2′-bithienyl [6,7], 4,4′-dimethyl- [7], and 5,5′-dihexadecyl-2,2-bithienyl [8] were prepared from 4,5-dimethyl-2-thienyllithium, 3-methyl-2-thienyllithium, 4-methyl-2-thienyllithium, and 5-hexadecyl-2-thienyllithium, respectively.

This method is still very much in use and was recently applied to the preparation of 5,5′-bis(mesitylthio)-2,2′-bithienyl and 5,5′-bis(mesitylthio)-2,2′-bis(3,4-ethylendioxythienyl) from 5-mesitylthio-2-thienyllithium and 5-mesitylthio-3,4-ethylenedioxythienylmagnesium bromide and cupric chloride [9].

5,5′-Bis(mesitylthio)-2,2′-bithienyl [9]

A solution of 2-bromo-5-mesitylthiothiophene (0.61 g, 1.95 mmol) in anhydrous tetrahydrofuran is cooled to − 78 °C and treated with butyllithium (1.28 ml, 2.05 mmol). Upon addition of butyllithium the slightly purple solution turns dull yellow. The solution is stirred for 1.5 h and solid cupric chloride (0.262 g, 1.95 mmol) is added. The brown solution so obtained is allowed to warm to room temperature, stirred for 10 min and then heated to 50 °C for 2 h. The black solution is cooled to room temperature and quickly passed through a silica frit using chloroform as eluent, giving a deep-purple solution. The solvent is evaporated and the residue recrystallized from benzene/methanol giving 0.30 g (66%) of the title compound as purple crystals mp 145–147 °C.

4,4′,5,5′-Tetraphenyl-2,2′-bithienyl [5]

A solution of 2,3-diphenylthiophene (2.36 g, 10 mol) in anhydrous diethyl ether (40 ml) is added to 1.6 *M* butyllithium (6.9 ml, 11 mmol) and anhydrous diethyl ether and the mixture is heated under reflux for 2 h. After cooling to − 70 °C and addition of anhydrous cupric chloride (5 g, 15 mmol) under argon the reaction mixture is stirred at this temperature for 4 h. The mixture is carefully hydrolyzed at 0 °C with 50% hydrochloric acid (50 ml). The resulting powder and the evaporation residue of the ether phase are recrystallized from ethanol/benzene (2:1) giving 1.74 g (74%) of the title compound as deep-yellow leaflets mp 209–210 °C.

From 3-cyclopentylthiophene, 4,4′-dicyclopentyl-2,2′-bithienyl is obtained in mediocre yield upon metalation with butyllithium and *N*,*N*,*N*′,*N*′-tetramethyl-1,2-diaminoethane followed by coupling with cupric chloride at − 78 °C [10]. In a similar way 4,4′-dioctyl-2,2′-bithienyl was prepared from 3-octylthiophene [11]. More complex 5,5′-substituted 2,2′-bithienyls such as the following have also been prepared [12].

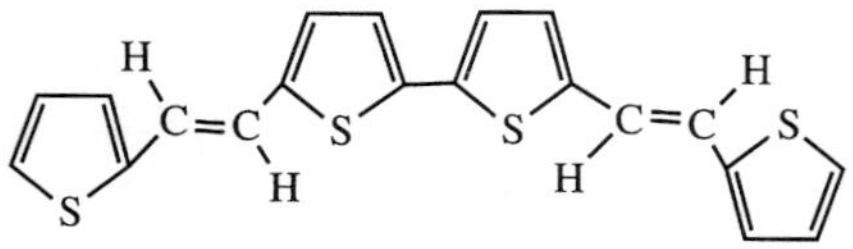

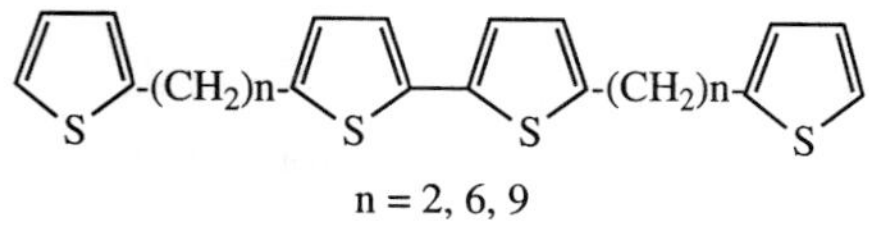

Halogen–metal exchange of 2,3-dibromothiophene with butyllithium at −70 °C followed by reaction with cupric chloride gives, 3,3′-dibromo-2,2′-bithienyl in 61% yield [13,14]. Similarly 4,4′-dibromo- and 5,5′-dibromo-, 3,3′, 4,4′-tetrabromo-, 3,3′5,5′-tetrabromo- and hexabromo-2,2′-bithienyl are prepared [13].

Metalation of 3-pentoxythiophene with lithium diisopropylamide, surprisingly occurs in the 5-position and upon coupling with cupric chloride 4,4′-dipentoxy-2,2′-bithienyl is claimed to be obtained [15].

The other symmetrical bithienyls, 3,3′-bithienyls, can in the same way be obtained through halogen–metal exchange at −70 °C of 3-halothiophenes followed by reaction with cupric chloride. In this way 3,3′-bithienyl [3], 4,4′-dibromo-3,3′-bithienyl [14,16], 4,4′-dichloro- [17], 5,5′-dichloro- [17], 2,2′,5,5′-tetrachloro- [18], 4,4′-diiodo- [18], 2,2′,5,5′-tetraethyl- [19], 2,2′-dimethyl- [20], 4,4′-dimethyl- [20], 5,5′-dimethyl- [21–23], 2,2′,4,4′-tetramethyl- [24], 2,2′,5,5′-tetramethyl- [25], 4,4′,5,5′-tetramethyl- [4], and 5,5′-diphenyl 3,3′-bithienyl [26] are obtained in yields of about 50–60%. However, for the preparation of 4,4′,5,5′-tetramethyl-3,3′-bithienyl, the halogen–metal exchange has to be carried out with 2,3-dimethyl-4-iodothiophene and not the bromo derivative, as in the latter case rearrangement of the lithium derivative occurs [4]. 2,3,4,5-Tetramethyl-7*H*-cyclopenta[1,2-*b*;4,3-*b*′]dithiophene is prepared in low yield through the reaction 4,4′-diiodo-2,2′3,3′-tetramethyl-5,5′-dithienylmethane with butyllithium at −70 °C followed by cupric chloride [4]. Compounds containing two nitronyl nitroxide or iminyl nitroxide radicals bound to 2,2′-bithienyl were prepared from 3,3′-, 4,4′-, and 5,5′- dibromothiophene, prepared from the dibromothiophene by halogen–metal exchange followed by coupling with cupric chloride and further transformation to the dialdehydes [27].

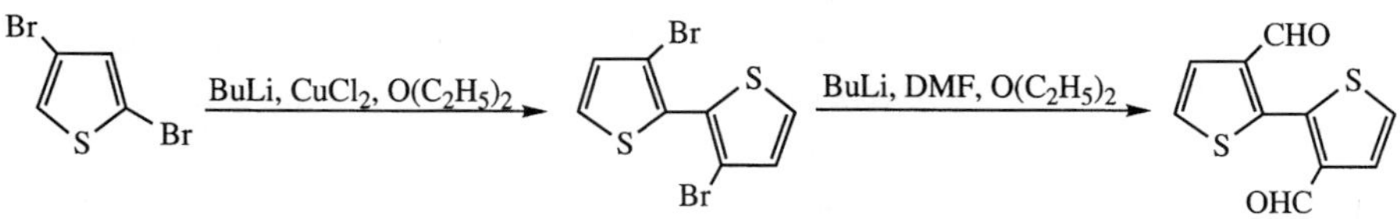

Treatment of 3-methoxy- and 4-methoxy-5-methyl-2-thienyllithium with iron(III) acetylacetonate in terahydrofuran are good methods for the preparation of 3,3′-dimethoxy- and 4,4′-dimethoxy-5,5′-dimethyl-2,2′-bithienyl [28,29].

3,3′-Dimethoxy-2,2′-bithienyl [28]

At 0 °C butyllithium (4.1 ml, 10.1 mmol) is added dropwise to a solution of 3-methoxythiophene (1.19 g, 10.1 mmol) in anhydrous tetrahydrofuran (20 ml). The mixture is stirred at 0 °C for 2 h before it is transferred *via* cannula to a solution of iron(III) acetylacetonate (3.56 g, 10.1 mmol) in anhydrous tetrahydrofuran (70 ml). After refluxing at 80 °C for 2 h the reaction mixture is allowed to cool to room temperature. The red precipitate is filtered off and washed with diethyl ether. The combined organic phases are treated with saturated aqueous ammonium chloride solution, dried over sodium sulfate and evaporated. The crude product is purified by flash chromatography using hexane as eluent, giving 928 mg (79%) as yellow crystals.

4,4′-Dimethoxy-2,2′-bithienyl is prepared by oxidative coupling of 2-trimethylsilyl-3-methoxy-5-thienyllithium followed by desilylation [29]. Halogen–metal exchange of 2,5-dibromo-3-hexylthiophene at − 80 °C in tetrahydrofuran followed by iron(III) acetylacetonate gives only a 55% yield of 5,5′-dibromo-4,4′-dihexyl-2,2′-bithienyl [30].

8.2.2 Symmetrical bithienyls through Ullman-type coupling of iodothiophenes

The conventional Ullmann reaction of 2-iodothiophene, heating with copper powder without solvent at about 200 °C, has not been successful. However, this conditions are used for the preparation of 3,3,′4,4′-tetraehyl-2,2′-bithienyl from 2-iodo-3,4-diethylthiophene [10].

3,3,′4,4′-Tetraethyl-2,2′-bithienyl [10]

A mixture of 2-iodo-3,4-diethylthiophene (1.0 g, 3.7 mmol) and copper powder (2.56 g) is heated at 200 °C for 20 min. After cooling, the content of the flask is extracted with hot chloroform, the solution is evaporated and the residue dissolved in hexane and passed through a silica gel column. Evaporation to dryness gives 0.19 g (36.5%) of the title compound mp 30–32 °C after recrystallization from methanol.

However, by reaction of 2-iodothiophene with copper powder in *N,N*-dimethylformamide, a 67% yield of 2,2′-bithienyl is obtained [31]. Tritium-labeled 2,2′-bithienyl was prepared by this method from tritium-labeled 2-iodothiophene [32]. Also 3-bromothiophene can be used for the preparation of 2,2′-bithienyl by this method [33]. Other examples of the preparation of symmetrical 2,2′-bithienyls by the Ullman reaction are the preparations of hexachloro-2,2′-bithienyl [34], 5,5′-dimethyl- [35], 5,5′-dibenzyl- [37],

5,5′-di(triphenylmethyl)- [38], 5,5′-di-*tert*-butyl-2,2′-bithienyl [39], and 3,3′-diformyl-2,2′-bithienyl from 2-iodo-3-formylthiophene [40].

3,3′-Diformyl-2,2′-bithienyl [40]

A mixture of 2-iodo-3-thiophene aldehyde (2.38 g, 10 mmol) and copper (63.5 g, 1.00 mol) in *N*,*N*-dimethylformamdie (30 ml) is stirred at 130 °C for 17 h and then filtered through a Celite filter. The filtrate is treated with water (100 ml) and extracted with benzene. The extract is dried over magnesium sulfate and evaporated. The residue is recrystallized from benzene giving 878 mg (79%) of the title compound as brown plates mp 158 °C.

Ullmann reactions of 3-iodothiophenes have, in most cases, been unsuccessful or give very low yields [18]. Strongly electron-withdrawing groups facilitate the Ullman reaction and 2-bromo-5-nitro-and 2-bromo-3-nitrothiophene give 5,5′-dinitro- [41,42] and 3,3′-dinitro-2,2′-bithienyl [42] in good yields. In highly activated cases chloro derivates have also been used and from 3,5′-dinitro-, 5-acetyl-3-nitro- and 5-carbomethoxy-3-nitro-2-chlorothiophene, 3,3′,5,5′-tetranitro-, 5,5′-diacetyl-3,3′-dinitro-, and 5,5′-dicarbomethoxy-3,3′-dinitro-2,2′-bithienyl are obtained in 40–50% yield [43].

In connection with the synthesis of 3,3′-bithienyl, it was found that the quality of the copper powder is of great importance and in some cases reduction of the halogen is observed as a serious side reaction. Using electrolytically prepared copper powder in *N*,*N*-dimethylformamide, 2,2′-dicarbomethoxy-4,4′-dinitro-3,3′-bithienyl is obtained in 73% yield [44]. However, even with *N*,*N*-dimethylformamide as solvent and with the same kind of copper powder up to 50% reduction to the dimethyl ester of 2,4-thiophenedicarboxylic acid is observed upon attempted coupling of 2,4-dicarbomethoxy-3-bromothiophene. Yet in one experiment 53% of pure 2,2′, 4,4′-tetracarbomethoxy-3,3′-bithienyl was obtained [45]. 3-Iodo-4-nitrothiophene in aqueous ammonia gives an 86% yield of 4,4′-dinitro-3,3′-bithienyl, upon addition to a mixture of cupric sulfate and copper powder in acetone/acetonitrile at room temperature with subsequent gentle heating [46], while conventional Ullmann coupling of 3-iodo-4-nitrothiophene fails [47]. Reaction of 4,4′-diiodo-2,2′,3,3′-tetramethyl-5,5′-dithienyl ketone with copper bronze in *N*,*N*-dimethylformamide gives 2,3,4,5-tetramethyl-7H-cyclopenta[1,2-*b*;4,3-*b*′]dithiophene-7-one in 76% yield, which upon treatment with lithium aluminium hydride and aluminium chloride gives the desired thiophene analogue of fluorene in more satisfactory yield than mentioned above [4].

2,3,4,5-Tetramethyl-7H-cyclopenta[1,2-b;4,3-b]dithiophene-7-one [4]

A mixture of 4,4′-diiodo-2,2′,3,3′-tetramethyl-5,5′-dithienyl ketone (8.0 g, 16.0 mmol), copper bronze (4.0 g, 63.0 mmol), and anhydrous

N,*N*-dimethylformamide (100 ml) is refluxed with stirring for 7 h. The red reaction mixture is poured into water. The precipitate is filtered off and the filtrate extracted with chloroform. The combined chloroform extracts are washed once with water, dried over magnesium sulfate and evaporated. The residue is recrystallized from hexane giving 3.0 g (76%) of the title compound as red crystals, mp 166–167 °C.

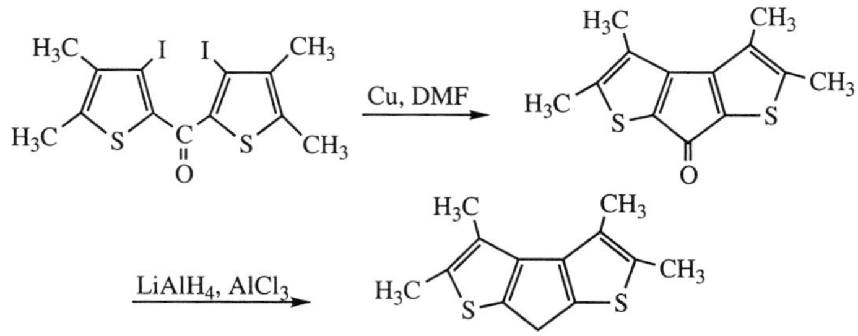

2,3,4,5-Tetramethyl-7H-cyclopenta[1,2-b:4,3-b-]dithiophene [4]

A solution of 2,3,4,5-tetramethyl-7*H*-cyclopenta[1,2-*b*;4,3-*b*-]dithiophene-7-one (3.0 g, 12.1 mmol) in anhydrous diethyl ether (1000 ml) is added to a mixture of lithium alumium hydride (8.1 g, 0.21 mol) and aluminium chloride (9.5 g, 0.071 mol) in anhydrous diethyl ether (25 ml). After the addition is completed water is carefully added and the phases separated. The aqueous phase is extracted twice with ether and the combined organic phases are washed with water, dried over magnesium sulfate and evaporated. The residue is dissolved in carbon tetrachloride/petroleum ether (10:1). This solution is chromatographed on neutral alumina using the same solvent mixture as eluent. After evaporation of the solvent the residue is sublimed and recrystallized from ethanol giving 1.85 g (65%) of the title compound mp 176–177 °C.

Similar approaches have been used for the synthesis of 2,3,5,6-tetramethyl-4*H*-cyclopenta[2,1-*b*:3,4-*b*′]dithiophene and 2,3,4,6-tetramethyl-7*H*-cyclopenta-[1,2-*b*:3,4-*c*′]dithiophene [4]. 2-Thienylcopper prepared from 2-thienyllithium or the corresponding Grignard reagent with copper(I) iodide or bromide followed by reaction with iodothiophenes in pyridine or quinoline at temperatures between 0 and 115 °C gives 2,2′-bithienyl in 42% yield [48]. A recently described Ullmann-like reduction coupling of 2-iodothiophene at room temperature, promoted by copper(I)-thiophene-2-carboxylate in *N*-methyl-pyrrolidinone, gives after 48 h, 2,2′-bithienyl [49].

2,2′-Bithienyl [49]

Copper(I)-thiophene-2-carboxylate (1.14 g, 6.0 mmol) is added in one portion to a solution of 2-iodothiophene (0.420 g, 2.0 mmol) in *N*-methylpyrrolidinone

(8 ml) under nitrogen. After being stirred at room temperature for 48 h the mixture is diluted with ethyl acetate (15 ml) and the slurry so obtained is passed through a plug of silica gel using ethyl acetate (150 ml) as eluent. The solvents are evaporated and the residue distilled *in vacuo*. Residual *N*-methylpyrrolidinone is removed under vacuum overnight giving 256 mg (77%) of the title compound.

8.2.3 Various dimerization reactions

Palladium-catalyzed reaction of 2-bromo-3-octylthiophene with hexabutyldistannane in *N*,*N*-dimethylformamide gives 3,3′-dioctyl-2,2′-bithienyl [50].

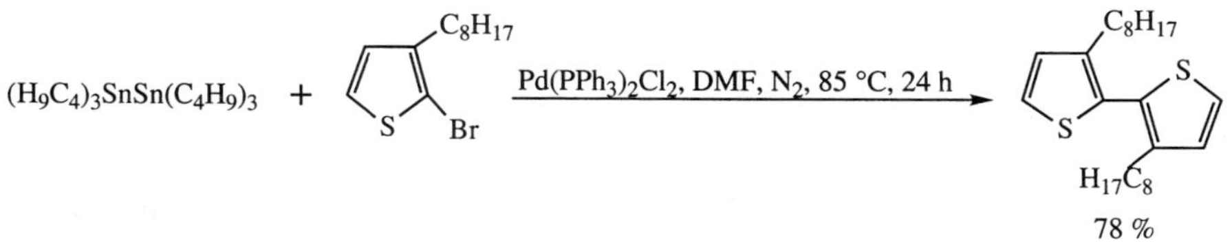

2,2′-Bithienyls can be prepared by oxidative dimerization of thiophenes containing electron-withdrawing groups, such as 2-formyl-, 2-nitro-, 2-formyl-3-methyl, and 3-methyl-2-nitrothiophene with palladium(II) acetate in acetic acid at room temperature in about 70% yield [51].

Halothiophenes containing electron-withdrawing groups, such as 5-bromo-2-thiophene aldehyde, 5-bromo-2-acetylthiophene, 5-chloro-2-acetylthiophene, 2-bromo-5-nitrothiophene, and also 2-bromo-5-chlorothiophene and 3-methyl-2-iodothiophene give, upon reaction with palladium(II) acetate and diisopropylethyl amine as base at 110 °C in toluene, the homo-coupled bithienyls together with minor amounts of reduction products. Tetraalkylammonium salts accelerate the coupling [52].

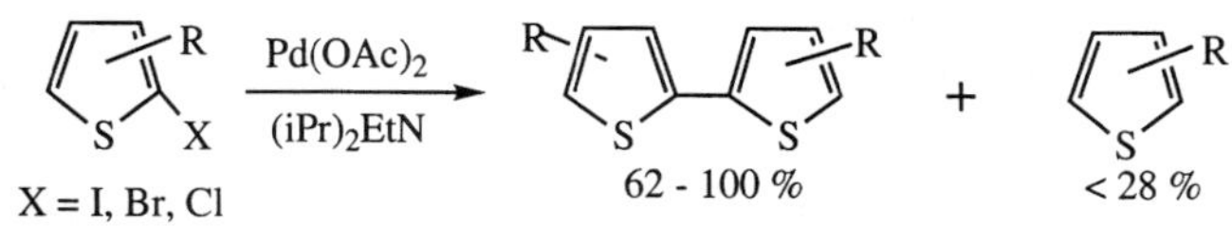

Chlorination of the 5-position of the thiophene ring of 1-(2-thienyl)naphthalene with one equivalent of *N*-chlorosuccinimide followed by self-coupling catalyzed by tris(triphenyl)phosphinenickel gives 5,5′-di(1-naphthyl)-2,2′-bithienyl in good yields [53]. By the same approach the following transformation takes place by monochlorination of 1,8-di(2-thienyl)naphthalene followed by self-coupling [53].

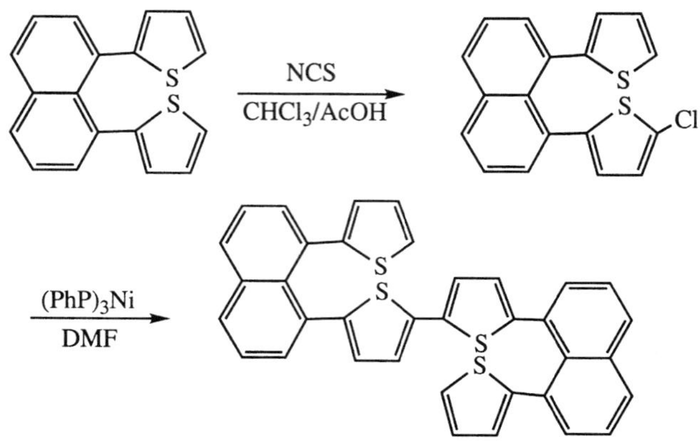

3,3′-Dipentoxy-22′-bithienyl appears to be prepared from 2-iodo-3-pentoxythiophene by the [1,3-bis(diphenylphosphine)propane]nickel(II) chloride-catalyzed Grignard method, using half an equivalent of magnesium [15].

An *in situ* Suzuki cross-coupling reaction between 2-iodothiophene and 0.5 equivalent of butyllithium followed by 1.5 equivalents of trimethyl borate followed by palladium(0) catalyst in sodium bicarbonate/toluene gives 73% of 2,2′-bithienyl. This methodology, however has hardly any advantage over the coupling of 2-thienyllithium with cupric chloride [54]. An electrochemical synthesis of 5,5′-[bis(4-bromophenyl)aminophenyl)]-2,2′-bithienyl by dimerization of bis(4-bromophenyl)-4-(2-thienyl)phenylamine proceeds in 71% yield [55].

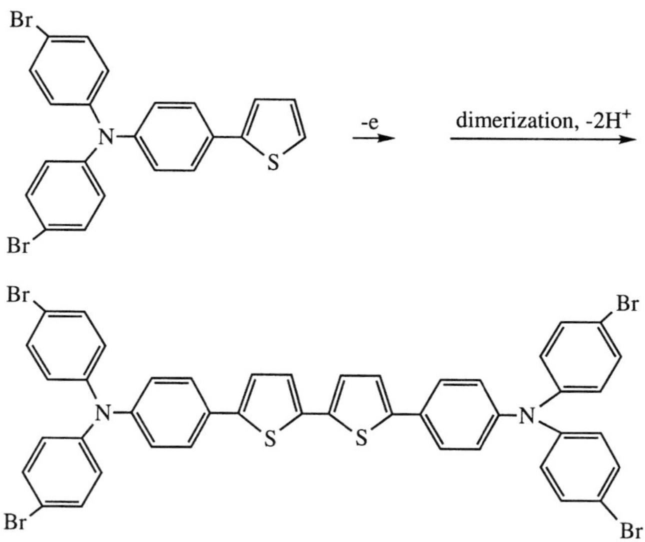

Electrochemical synthesis of 5,5′-[bis(4-di-(4-bromophenyl)amino phenyl)]-2,2′-bithienyl [55]

Into an anodic compartment of a divided cell equipped with a platinum gauze anode (3 × 5 cm) and a graphite cathode, filled with 0.1 *M* solution of lithium perchlorate in acetonitrile (40 ml) bis(4-bromophenyl)-4-(2-thienyl)phenylamine (241 mg) is added. The anode potential is maintained at 1.0 V and electrolysis is continued until 1.5 F/mol has been passed. The potential is adjusted 0.0 V vs. SCE and the reduction is carried out until 0.5 F/mol has been passed. After the electrolysis, the dark solution is evaporated and 40% aqueous hydrazine solution (70 ml) is added. The reaction mixture is sonicated for 15 min and extracted with dichloromethane (3 × 30 ml). The combined organic phases are dried over sodium sulfate and evaporated. The residue is purified by flash chromatography on silica gel using hexane/dichloromethane (9:1) as eluent, giving 170 mg (71%) of the title compound as a yellow solid mp 224–227 °C.

In a study of the indophenine reaction, it was found that the sulfuric acid-promoted condensation of *N*-heptylisatin with thiophene gives all six possible geometric isomers [56].

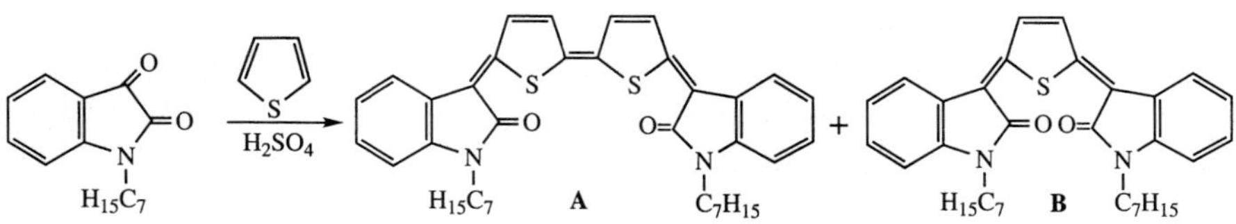

*Preparation of **A** and **B** [56]*

Concentrated sulfuric acid (0.1 ml) is added dropwise to a rapidly stirred solution of *N*-heptylisatin (0.50 g, 2.04 mmol) and thiophene (0.34 g, 4.08 mmol) in benzene (15 ml). After the solution is stirred for 3 h at room temperature, water (50 ml) is added and the product is extracted with chloroform (2 × 20 ml). The combined organic phases are washed with water (4 × 10 ml), dried over calcium chloride and evaporated. The residue is chromatographed on silica gel using methanol as eluent, giving 0.11 g (20%) of compound **B** as a brown amorphous solid. Further elution with tetrahydrofuran/hexane (1:6) gradually increases the tetrahydrofuran to (1:2) giving 0.41 g (65%) of compound **A** as a deep blue amorphous solid mp 193 °C.

4,4′-Diaryl-2,2′-bithienyls are prepared by the dimerization of 2-chloro-4-arylthiophenes with nickel(II) chloride, triphenylphosphine/zinc in *N,N*-dimethylformide [57]. The reaction of 2-chloro-4-arylthiophene with aluminium chloride in dichloromethane gives 5-chloro-3,3′-bis(aryl)-2,2′-bithienyl, which upon hydrogenation over 105 palladium and charcoal in potassium hydroxide/methanol gives 3,3′-bis(aryl)-2,2′-bithienyls [57].

8.3 UNSYMMETRICAL BITHIENYLS

8.3.1 Transition metal-catalyzed cross-coupling reactions

8.3.1.1 From thienylmagnesium derivatives and halothiophenes, Kumada coupling

3,3′-Bithienyl and 2,3′-bithienyl are prepared in 88 and 77% yield, respectively, by the coupling of 3-thiophenemagnesium bromide with 2- and 3-bromothiophene in the presence of [1,3-bis(diphenylphosphine)propane]nickel(II) chloride [58]. A number of dialkyl-2,2′-bithienyls [59,60] and 4,5,5′-trimethyl, 3,5,5′-trimethyl-, 4,5,4′,5′-tetramethyl-, 3,4,3′,5′-tetramethyl-,3,4,5,3′,5′-pentamethyl-2,2′-bithienyl were also prepared by this methodology [61].

4,5,5′-Trimethyl-2,2′-bithienyl [61]

Magnesium turnings (0.95 g, 39.3 mmol) in anhydrous diethyl ether (10 ml) and several iodine crystals are placed in a 100 ml three-necked flask fitted with a reflux condenser, drying tube, dropping funnel and gas inlet. Under an inert gas atmosphere a solution of 2-bromo-4,5-dimethylthiophene (5.00 g, 26.2 mmol) in anhydrous diethyl ether (20 ml) is slowly added dropwise. After the addition is completed the mixture is refluxed for 2 h with stirring and then transferred into a dropping funnel under an inert gas to separate the unreacted magnesium.

This Grignard solution is then added to 2-bromo-5-methylthiophene (4.63 g, 26 mmol) in anhydrous diethyl ether (50 ml) and [1,3-bis(diphenylphosphine)propane]nickel(II) chloride (460 mg, 0.85 mmol) at such a rate that gentle reflux is maintained. The reaction mixture is refluxed for 2 h and then poured into 3 *M* hydrochloric acid (100 ml) at 0 °C. The product is extracted with ether (3 × 100 ml). The combined organic phases are washed three times with water, dried over calcium chloride and evaporated. The residue is purified by chromatography on silica gel using hexane as eluent giving 2.38 g (44%) of the title compound as colorless crystals, mp 56 °C, after recrystallization from methanol.

3-Dodecyl-2,2′-bithienyl is prepared from 2-thiophenemagnesium bromide and 2-bromo-3-dodecylthiophene [62]. 5-Dodecyl-2,2′-bithienyl is better prepared by the reaction of 2-thiophenemagnesium bromide with 5-bromo-2-dodecylthiophene than by the reaction of 5-dodecyl-2-thiophenemagnesium bromide with 2-bromothiophene (60%) [63].

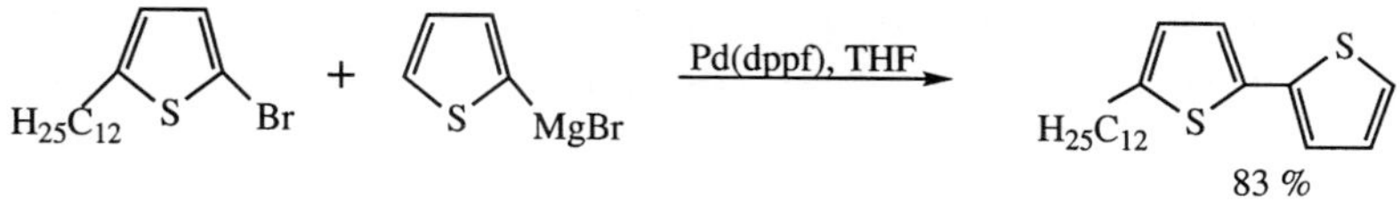

5-Dodecyl-2,2′-bithienyl [63]

From 2-bromothiophene (2.61 g, 16 mmol) and magnesium turnings (0.41 g, 17 mmol) in diethyl ether (25 ml) the Grignard reagent is prepared and coupled with 5-bromo-2-dodecylthiophene (3.31 g, 10 mmol) and [1,3-bis(diphenylphosphine)propane] nickel(II) chloride (55 mg, 0.1 mmol) in diethyl ether (25 ml). The reaction mixture is refluxed for 6 h, and after cooling to room temperature is hydrolyzed with cold 0.5 *M* hydrochloric acid. The product is extracted several times with ether and the combined organic phases washed successively with sodium bicarbonate and water, dried over sodium sulfate and evaporated. The residue is purified by chromatography on silica gel using hexane/dichloromethane (2:1) as eluent. Recrystallization from methanol/diethyl ether gives 2.90 g (87%) of the title compound as a bright-yellow solid, mp 38 °C.

3-Butylthio-2,2′-bithienyl is obtained in 90% yield from 2-thiophenemagnesium bromide and 2-bromo-3-butylthiothiophene [64]. Also bis(diphenylphosphine)ferrocenepalladium(II) chloride [65] can be used as a catalyst, as in the preparation of 5-bromo-2,2′-bithenyl from equivalent amounts of 2-thiophenemagnesium bromide and 2,5-dibromothiophene [66].

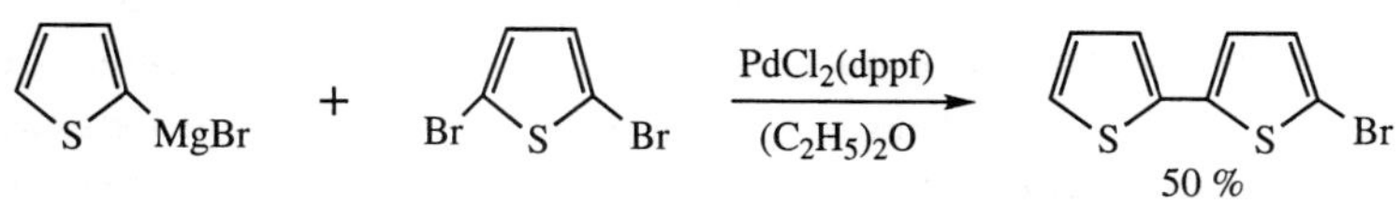

The Kumada reaction between 2-bromo-5-phenylthiophene and 5-phenyl-2-thiophenemagnesium bromide is recently used for the preparation of 5,5′-diphenyl-2,2′-bithienyl [67].

5,5′-Diphenyl-2,2′-bithenyl [67]

A flask containing 2-bromo-5-phenylthiophene (478 mg, 2.00 mmol) and magnesium (48.6 mg, 2.00 mmol) is evacuated under mild heat. To this solid mixture anhydrous diethyl ether (15 ml) is added to prepare the Grignard reagent. When all magnesium has disappeared [1,3-bis(diphenylphosphine)propane]nickel(II) chloride (20 mg, 0.037 mmol) and additional 2-bromo-5-phenylthiophene (383 mg, 1.60 mmol) are added successively. Precipitation occurs immediately. After stirring the reaction mixture overnight, it is refluxed for 6 h, cooled over an ice-water bath and hydrolyzed with 2 *M* hydrochloric acid (1 ml). The precipitate formed is collected by filtration, washed with methanol, and recrystallized from toluene giving 439 mg (69%) of the title compound as a bright-yellow solid mp 238 °C.

In some cases, it is necessary to prepare the Grignard reagent through the reaction of the lithium derivative with magnesium bromide as is the case in the preparation of 3-methylthio-2-thiophenemagnesium bromide. Upon [1,3-bis(diphenylphosphine)propane]nickel(II) chloride-catalyzed coupling with 4-methylthio-2-bromothiophene, 3,4′-bis(methylthio)-2,2′-bithienyl is obtained [68].

3,4′-Bis(methylthio)-2,2′-bithienyl [68]

To a solution of 2-bromo-3-methylthiothiophene (1.25 g, 6 mmol) in anhydrous diethyl ether (10 ml) at − 78 °C 1.3 *M* butyllithium (4.6 ml, 6 mmol) is added dropwise. The temperature of the mixture is raised to − 30 °C. A mixture of freshly prepared magnesium bromide* and diethyl ether is added with vigorous stirring at a temperature between − 30 and − 20 °C, the reaction mixture is then stirred at 0 °C for 30 min. This solution of the Grignard reagent is slowly added (30 min) *via* a cannula to a solution of 2-bromo-4-methylthiothiophene (1.05 g, 5 mmol) and [1,3-bis(diphenylphosphine)propane]nickel(II) chloride (0.068 g, 0.125 mmol) in diethyl ether (3 ml) with cooling in an ice bath. After 20 h under reflux the reaction mixture is hydrolyzed by addition to 2 *M* hydrochloric acid (20 ml) and diethyl ether in a separatory funnel. The phases are separated and the aqueous phase extracted with ether. The combined organic phases are washed with aqueous sodium bicarbonate and water, dried over magnesium sulfate and evaporated. The residue is short-path distilled giving 0.58 g (45%) of an oily product bp 145–152 °C/0.1 mm Hg as a mixture of the title compound (90%) and an isomer (10%). Purification by chromatography on silica gel using light petroleum as eluent followed by distillation gives 0.36 g of the title compound bp 148–151 °C/0.1 mm Hg.

The magnesium bromide* and diethyl ether mixture is prepared by dropwise addition of 1,2-dibromoethane (1.34 g, 7.2 mmol) and magnesium turnings (0.35 g, 14.4 mmol) in diethyl ether (5 ml) and reflux temperature during the addition. After another 30 min the oily underlayer and the supernatant layer of diethyl ether are transferred *via* a cannula to the solution of the lithium derivative, the excess of magnesium being discarded.

The same methodology is used in the synthesis of highly congested bithienyls, such as 3,4,3′,4′-tetra-*tert*-butyl-2,2′-bithienyl, which is obtained in 33% yield from 3,4-di-*tert*-butyl-2-bromothiophene and its Grignard reagent prepared by halogen–metal exchange with butyllithium followed by magnesium bromide [69].

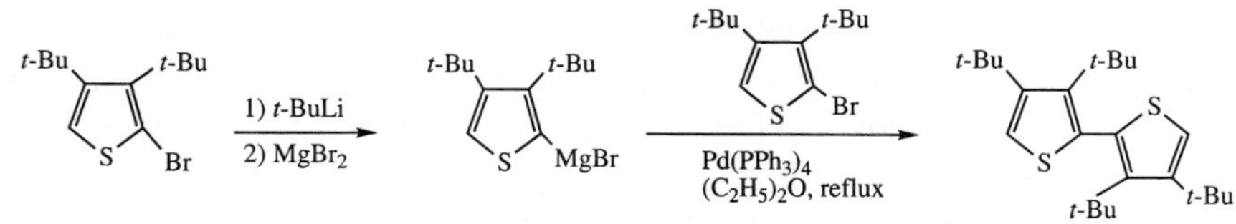

1,2-Bis(3′-octyl-2,2′-bithiophene-5-yl)ethane is prepared by conventional nickel(II) catalyzed Kumada coupling of 1,2-bis(5-bromo-2-thienyl)ethane and the Grignard reagent from 2-bromo-3-octylthiophene [70].

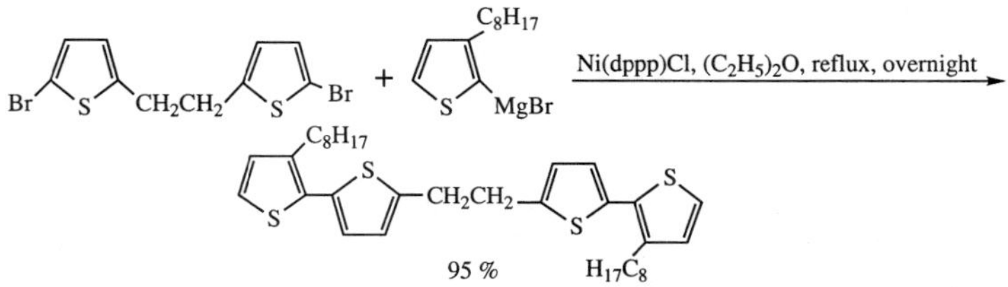

Kumada coupling of 3,4-ethylenedioxy-2-thiophenemagnesium bromide with 2-bromo-3-(11-(tetrahydropyranylox)undecylthiophene is used for the preparation of the highly oxygenated bithienyl, which is polymerized *via* 1,3-bis(diphenylphosphino)propanenickel(II) chloride-catalyzed polymerization of the 5-bromo-5′-thiophenemagnesium bromide [71].

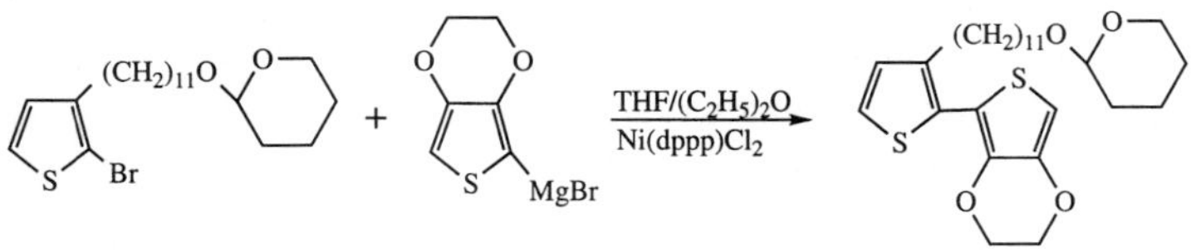

8.3.1.2 From thienylzinc derivatives and halothiophenes

Metalation of 3-octyl-2-trimethylsilylthiophene with butyllithium and zinc chloride followed by palladium(0)-catalyzed coupling with 2-bromothiophene gives 3-octyl-5-trimethylsilyl-2,2′-bithienyl [72]. 5-(Pyrolidin-1-yl)-2,2′-bithiophene is conveniently prepared by palladium(0)-catalyzed coupling of 2-iodothiophene with 5-(pyrolidin-1-yl)-2-thienylzinc chloride [73].

5-(Pyrolidin-1-yl)-2,2′-bithiophene [73]

The palladium catalyst is prepared from dichlorobis(triphenylphosphine)-palladium (2.0 g, 2.85 mmol) and 1.0 *M* diisobutylaluminium hydride in hexane in tetrahydrofuran (65 ml) at 0 °C. To this solution are added 2-iodothiophene (6.3 ml, 57.0 mmol) and a solution of 5-(pyrrolidin-1-yl) 2-thienylzinc chloride [prepared by treatment at room temperature for 30 min in tetrahydrofuran

(115 ml) of 5-(pyrrolidin-1-yl)-2-thienyllithium (57.7 mmol) with zinc chloride (7.9 g, 57.9 mmol)]. The reaction mixture is stirred overnight, the tetrahydrofuran evaporated and the residue hydrolyzed and extracted with ether. From the black solution the product is taken up in hot hexane. After filtration and evaporation the crude product is obtained. Crystallization from pentane affords 6.4 g (48%) of the title compound mp 88 °C.

Similarly metalation of 2-(pyrrolidin-1-yl)thiophene with butyllithium and zinc chloride is used for the preparation of 5-(pyrrolidino)-5′-cyano-2,2′-bithienyl by coupling with 5-cyano-2-bromothiophene [74]. In the same way metalation of 2-dimethylaminothiophene with butyllithium followed by reaction with zinc chloride and palladium(0)-catalyzed coupling with 5-nitro-2-iodo- and 5-cyano-2-bromothiophene are used for the preparation of 5-dimethylamino-5′-nitro-2,2′-bithienyl and 5-dimethylamino-5′-cyano-2,2′-bithienyl, respectively [74,75].

5-Dimethylamino-5′-cyano-2,2′-bithienyl [75]

To a solution of dimethylaminothiophene (254 mg, 2.00 mmol) in anhydrous tetrahydrofuran (2 ml) cooled over an ice bath 1.6 *M* butyllithium in hexane (1.62 ml) is added over 7 min. After 10 min the ice bath is removed and the mixture stirred at room temperature for 2 h and then transferred *via* cannula to a solution of anhydrous zinc chloride (299 mg, 2.20 mmol) in anhydrous tetrahydrofuran (2 ml). This mixture is stirred at room temperature for 1 h. Tetrakis(triphenylphosphine)palladium(0) (115 mg, 0.10 mmol) is added followed by 5-bromo-2-cyanothiophene (0.233 ml, 2.00 mmol) and the reaction mixture is stirred at room temperature for 18 h. After being diluted with dichloromethane and washed with saturated aqueous ammonium chloride solution, the organic phase is dried over magnesium sulfate and evaporated. The residue is purified by flash chromatography on silica gel using hexane/ethyl acetate (10:1 to 5:1) as eluent, giving 234 mg (50%) of the title compound as an orange–brown solid.

5,5′-Bis(trimethylsilyl)-2,2′-bithienyl is obtained in 56% yield from the Grignard reagent from 2-iodo-5-trimethylsilylthiophene and 2-iodo-5-trimethylsilythiophene using [1,3-bis(diphenylphosphine)propane]nickel(II) chloride as catalyst. The Grignard reagent from 2-iodo-3-methylthiophene and 2-iodo-5-trimethylsilylthiophene give 3-methyl-5′-trimethylsilyl-2,2′-bithienyl [76] by using the same catalyst.

3-Methyl-5′-trimethylsilyl-2,2′-bithienyl [76]

To magnesium turnings (0.82 g, 33.75 mmol) in diethyl ether (20.0 ml), 2-iodo-3-methylthiophene (5.04 g, 22.5 mmol) is added dropwise at 0 °C. The mixture is stirred at room temperature for 2 h and then transferred to a solution of 2-iodo-5-(trimethylsilyl)thiophene (4.26 g, 15.0 mmol) and [1,3-bis(diphenylphosphine)propane]nickel(II) chloride (0.183 g, 0.34 mmol) in diethyl ether (10 ml) at 0 °C. The reaction mixture is stirred at room temperature overnight and then poured into water and filtered through Celite. The phases are separated and the aqueous phase extracted with ether. The combined organic phases are washed with aqueous sodium chloride solution, dried over magnesium sulfate and evaporated. The residue is purified by chromatography on silica gel using hexane as eluent, giving 2.67 g (70%) of the title compound as a yellow liquid.

2-Bromo-4-methylthiothiophene is transformed to 4-methylthio-2-thienylzinc bromide by direct metalation with a highly reactive form of zinc prepared from zinc dichloride by reduction with lithium naphthalenide, which upon coupling with 2-bromo-4-methylthiothiophene gives 4,4′-bis(methylthio)-2,2′-bithienyl [68].

4,4′-Bis(methylthio)-2,2′-bithienyl [88]

Finely cut lithium (0.14 g, 20 mmol) is added under a stream of argon to a solution of naphthalene (2.86 g, 21 mmol) in anhydrous tetrahydrofuran (10 ml) and the suspension is stirred for 2 h at room temperature. The metal is slowly consumed to afford a dark-green solution of lithium naphthalenide. In a second vessel anhydrous zinc chloride (1.36 g, 10 mmol) under a stream of argon is fused for a few minutes, cooled and dissolved in anhydrous tetrahydrofuran (10 ml). The solution so obtained is added dropwise *via* cannula to the lithium naphthalenide solution and a precipitate of activated zinc is formed. After stirring for 15 min at room temperature a solution of 2-bromo-4-methylthiothiophene (1.046 g, 5 mmol) in anhydrous tetrahydrofuran (4 ml) is introduced with a syringe causing a slightly exothermic reaction. The suspension is stirred at 25–30 °C for 2 h to complete the formation of the organozinc bromide. The solution is left for 3 h to allow the excess of zinc to settle. The supernatant solution is carefully transferred *via* a cannula to a third reaction vessel containing a solution of 2-bromo-4-methylthiothiophene (1.046 g, 5 mmol) and [1,3-bis(diphenylphosphine)propane]nickel(II) chloride, (0.136 g, 0.25 mmol) in anhydrous tetrahydrofuran (10 ml). After being stirred at room temperature overnight the reaction mixture is poured into a saturated aqueous ammonium chloride solution and the product is extracted with dichloromethane. The combined organic phases are dried over magnesium sulfate and evaporated. The residue is distilled *in vacuo* in a short-path apparatus. Initially naphthalene sublimes followed by an oily compound, which readily turn into a solid. 0.99 g (77%) of the title compound is obtained bp 151–154 °C/0.1 mm Hg; mp 101–103 °C.

The [1,3-bis(diphenylphosphine)propane]nickel(II) chloride-catalyzed coupling of 2-thiophenemagnesium bromide with 2-bromothiophene is also being used recently for the preparation of 2,2′-bithienyl, albeit only a moderate yield was obtained [77].

2,2′-Bithienyl [77]

Magnesium turnings (22 g, 192 mmol) are heated with a hot air gun under nitrogen with rapid stirring and after cooling covered with anhydrous tetrahydrofuran (300 ml). 2-Bromothiophene (120 g, 0.74 mol) is slowly added *via* a dropping funnel, after which the mixture is refluxed for 1 h, cooled and transferred into a dropping funnel by decanting the solution from excess of magnesium. To a solution of 2-bromothiophene (100 g, 0.61 mol) in anhydrous tetrahydrofuran (300 ml), [1,3-bis(diphenylphosphine)propane]nickel(II) chloride (1.5 g, 3.6 mmol) is suspended followed by addition of the Grignard reagent from the dropping funnel. During the addition the temperature is kept at room temperature with the aid of a water-bath. After complete addition the reaction mixture is stirred for another 5 h at 70 °C and then kept overnight. The excess Grignard reagent is quenched with water and the product extracted with diethyl ether. The combined organic phases are washed with sodium chloride solution, dried over magnesium sulfate and evaporated. The residue, a black oil, is distilled giving 56.6 g (46%) bp 84 °C, mp 32–33 °C.

Kumada coupling of 1,3-di-(5-bromo-2-thienyl)benzene with two equivalents of thienylzinc chloride is used for the preparation of 1,4-di-(5′-(2″-thienyl(thienyl)benzene [78].

8.3.1.3 From thiopheneboronic acids and halothiophenes, Suzuki coupling

In 1983, Gronowitz and coworkers modified the Suzuki reaction to make it useful for the preparation of bithienyls from thiopheneboronic acids and bromothiophenes [79,80]. They used aqueous sodium carbonate or bicarbonate in glycol dimethylether as solvent, which increased the rate of coupling and diminished deboronation, especially of the 2-thiopheneboronic acids. Tetrakis(triphenylphosphine)palladium(0) was used as a catalyst. In this way 3,3′-bithienyl and 2,3′-bithienyl were obtained in 70% yield from 3-bromothiophene and 3- and 2-thiopheneboronic acid, respectively [79]. The great advantage of boronic acids in contrast to magnesium or zinc derivatives is that almost all functional groups, such as formyl and nitro groups are tolerated. Thus six of the nine *ortho*, *ortho*′-formylnitrobithienyls were prepared from the three *ortho*-bromonitrothiophenes with 4-formyl-3-thiopheneboronic acid and 2-formyl-3-thiophenebornonic acid [81]. 3-Formyl-2-thiopheneboronic acid deboronated too fast under the usual reaction conditions and also other reaction conditions were not successful [81]. 3,4′-Didodecyl-2,2′-bithienyl has been recently prepared by the coupling of sodium 4-dodecyl-2-thienylboronate with 3-dodecyl-2-iodothiophene, following the Gronowitz modification [82].

3,4′-Didodecyl-2,2′-bithienyl [82]

Tetrakis(triphenylphosphine)palladium(0) (0.33 g, 10 mmol) is added to a solution of 3-dodecyl-2-iodothiophene (4.00 g, 10 mmol) in 1,2-dimethoxyethane (50 ml) and the mixture is carefully degassed. After addition of sodium 4-dodecyl-2-thienylboronate (3.30 g, 10 mmol) the mixture is heated to reflux and a saturated aqueous solution of sodium bicarbonate (1.80 g, 20 mmol) is slowly added. After 3 h the coupling reaction is completed and the mixture is diluted with diethyl ether (50 ml). The phases are separated and the organic phase washed with diluted hydrochloric acid and repeatedly with saturated aqueous sodium chloride solution, dried over sodium sulfate and evaporated. The residue is purified by chromatography on silica gel using cyclohexane as eluent, giving 3.9 g (85%) of the title compound as a slightly yellow oil.

Suzuki couplings between 4-alkyl-3-thiopheneboronic acid derivatives and 3-iodo-4-trimethylsilylthiophene are also used for the preparation of 4-(trimethylsilyl)-4′-heptyl-3,3′-bithienyl [83].

The coupling of 2-bromo-4,6-dihydrothieno[3,4-*b*]thiophene-5,5′-dioxide with 2-thiopheneboronic acid under these conditions gives 2-(2-thienyl)-4,6-dihydrothieno[3,4-*b*]thiophene-5,5′-dioxide in 68% yield and from 5-(9-anthryl)-2-thiopheneboronic acid, prepared *in situ*, and the same bromo derivative 2-[5′-(9-anthrylmethyl)-2′-thienyl]-4,6-dihydrothieno[3,4-*b*]thiophene-5,5′-dioxide as prepared [84].

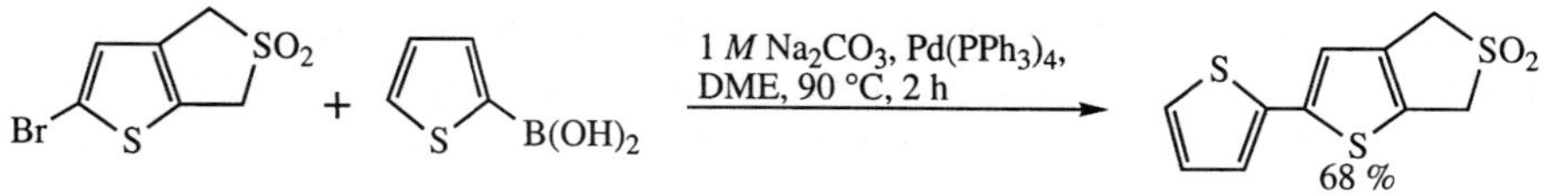

2-(2′-Thienyl)-4,6-dihydrothieno[3,4-b]thiophene dioxide [84]

To a stirred solution of 2-bromo-4,6-dihydrothieno[3,4-*b*]thiophene-5,5′-dioxide (1.42 g, 5.61 mmol) and tetrakis(triphenylphosphine)palladium(0) (5 mol%) in dimethoxyethane (50 ml) at 90 °C, a solution of 2-thiopheneboronic acid (790 mg, 6.17 mmol) in 1 *M* aqueous sodium carbonate solution (20 ml) is added dropwise within 5–10 min. The reaction mixture is stirred at 90 °C for another 2 h and hydrolyzed with water (50 ml). The phases are separated and the aqueous phase extracted with ethyl acetate (150 ml) or with dichloromethane (150 ml). The combined organic phases are washed with water, dried over sodium sulfate and evaporated. The black residue is chromatographed on silica gel using hexane/ethyl acetate (1:1) as eluent, giving 978 mg (68%) of the title compound as light-grey crystals, mp 180 °C (dec.) after recrystallization from boiling acetone.

If boronic acids sensitive to protodeboronations are obtained, they are not isolated, but immediately allowed to react with 2,2-dimethyl trimethyleneglycol, to give a stable boronate. Thus from 3-octyl-2-bromothiophene, [1′3′-(2,2′-dimethylpropylene)]-3-octyl-2-thienyl boronate was obtained *via* the Grignard reagent. Suzuki coupling with 3-octyl-2-chloro-5-iodothiophene gives 3,4′-dioctyl-5′-chloro-2,2′-bithienyl in good yield [85].

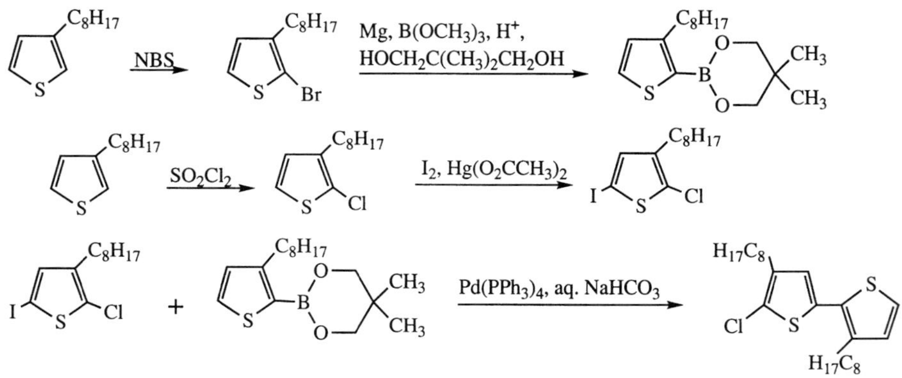

1′,3′-(2,2′-Dimethylpropylene)]-3-octyl-2-thienylboronate [85]

To magnesium (1.34 g, 55 mmol) in anhydrous tetrahydrofuran (50 ml) heated to maintain a gentle reflux, 2-bromo-3-octylthiophene (10.00 g, 36.3 mmol) is added dropwise. When the addition is completed the mixture is refluxed for 1 h and then transferred *via* cannula to a solution of trimethyl borate (16 ml, 143 mmol) in tetrahydrofuran at − 78 °C. The reaction mixture is allowed to warm to room temperature, stirred for 30 min and poured into 10% hydrochloric acid (50 ml). The phases are separated and the aqueous phase extracted with diethyl ether. The combined organic phases are dried first over sodium sulfate and then over molecular sieves in the presence of 2,2-dimethyl-1,3-propanediol (3.78 g, 35, 5 mmol). Upon evaporation a slightly yellow liquid contaminated with white crystals is obtained. This crude product is dissolved in hexane, filtered through Celite and heated for one day at 100 °C under vacuum giving 8.32 g (74%) of the title compound as a thick slightly yellow liquid.

3,4′-Dioctyl-5′-chloro-2,2′-bithienyl [85]

A solution of 5-iodo-2-chlorothiophene (3.0 g, 8.4 mmol) in 1,2-dimethoxyethane (45 ml) is carefully degassed and tetrakis(triphenylphosphine)palladium(0) (240 mg, 0.52 mmol) is added. After stirring the mixture at room temperature for 10 min 1′,3′-(2,2′-dimethylpropylene)]-3-octyl-2-thienylboronate (3.05 g, 9.9 mmol) and 1 *M* aqueous sodium bicarbonate solution (25 ml) are successively added. The reaction mixture is then refluxed with vigorous stirring until complete consumption of the iodo compound, generally 4 h, occurs, and poured into water. The phases are separated and the aqueous phase extracted with diethyl ether. The combined organic phases are washed with water and aqueous sodium chloride solution, dried over sodium sulfate and evaporated. The residue is heated overnight (100 °C) under vacuum and purified by chromatography on silica gel using hexane as eluent, giving 2.85 g (80%) of the title compound as a slightly yellow liquid.

8.3.1.4 From stannylthiophenes and halothiophenes, Stille coupling

The palladium(0)-catalyzed coupling of thienyltin derivatives with halothiophenes is very useful for the preparation of unsymmetrical bithienyls. Like the boronic acids, they tolerate a large number of functional groups, which are incompatible with Grignard or zinc reagents. The tin compounds have the advantage over boronic acids, that no base needs to be present, but the disadvantage is that they are toxic.

Thus 2-tributylstannylthiophene is coupled with 2-iodo-, 2-iodo-5-acetoxymethyl-, 2-iodo-5-hydroxymethyl-, and 2-iodo-5-thiophene aldehyde and more complex 5-substituted 2-iodothiophenes to give i.e. 2,2′-bithienyl, 5-acetoxymethyl-2,2′-bithienyl, 5-hydroxymethyl-, and 5-formyl-2,2′-bithienyl in 60–70% yield. 5 mol% of bis(Triphenylphosphine)palladium(II) chloride is used as catalyst in tetrahydrofuran at 60 °C [86]. 2,2′-Bithienyl, 5-nitro-, 5-formyl, 5-acetyl-, and 3,5-dinitro-2,2′-bithienyl was similarly prepared from 2-trimethylstannylthiophene and the corresponding substituted bromothiophenes in 65–85% yield [87]. A larger number of donor–acceptor substituted 2,2′-bithienyls are prepared by reaction of trimethylstannylthiophenes with acceptor-substituted halothiophenes [88]. The palladium(0)-catalyzed reaction of 3-methyl-2-bromothiophene with 3-methyl-2-trimethylsilyl-5-tributylstannylthiophene gives 3,4′-dimethyl-5′-trimethylsilyl-2,2′-bithienyl in good yield [89].

3,4′-Dimethyl-5′-trimethylsilyl-2,2′-bithienyl [89]

A flame-dried flask is charged with 3-methyl-2-trimethylsilyl-5-tributylstannylthiophene (1.89 g, 4.11 mmol), 2-bromo-3-methylthiophene (0.728 g,

4.11 mmol), bis(triphenylphosphine)palladium(II) chloride (0.144 g, 0.20 mmol), triphenylphosphine (0.052 g, 0.20 mmol), and toluene (20.0 ml). The reaction mixture is stirred at room temperature for 4 h, warmed to 100 °C for 4 h and then poured into saturated aqueous ammonium chloride solution. The phases are separated and the aqueous phase extracted with diethyl ether. The combined organic phases are washed with aqueous sodium chloride solution, dried over sodium sulfate and evaporated. The residue is purified by chromatography on silica gel using hexane as eluent, giving 0.727 g (66%) of the title compound as a light-green liquid.

The presence of the bromine in 3-bromo-2-trimethylstannylthiophene does not hinder coupling with 3,5-dibromo-2-trimethylsilylthiophene yielding 3′4-dibromo-5-trimethylsilyl-2,2′-bithienyl albeit only in 40% yield [90].

3,4-Dibromo-5′-trimethylsilyl-2,2′-bithienyl [90]

A solution of 3,5-dibromo-2-trimethylsilylthiophene (1.44 g, 4.60 mmol) in anhydrous toluene (10 ml) is added dropwise during 4 h to a stirred solution of 5-bromo-2-trimethylstannylthiophene (1.50 g, 4.60 mmol) and tetrakis(triphenylphosphine)palladium(0) (0.42 g, 0.36 mmol) in anhydrous toluene (4 ml) at 105 °C. The reaction mixture is stirred at this temperature for three days, after cooling it is transferred into a separatory funnel, and diluted with diethyl ether (40 ml). The solution is washed with saturated aqueous sodium bicarbonate and water, dried over magnesium sulfate and evaporated. The dark oily residue is purified by chromatography on silica gel using light petroleum as eluent, giving 0.72 g (40%) of the title compound.

3,3′4,4′,5-Pentafluoro-5′-trimethylsilyl-2,2′-bithienyl has recently been prepared by the coupling of 3,4,5-trifluoro-2-tributylstannylthiophene with 2-bromo-3,4-difluoro-5-trimethylsilylthiophene [91].

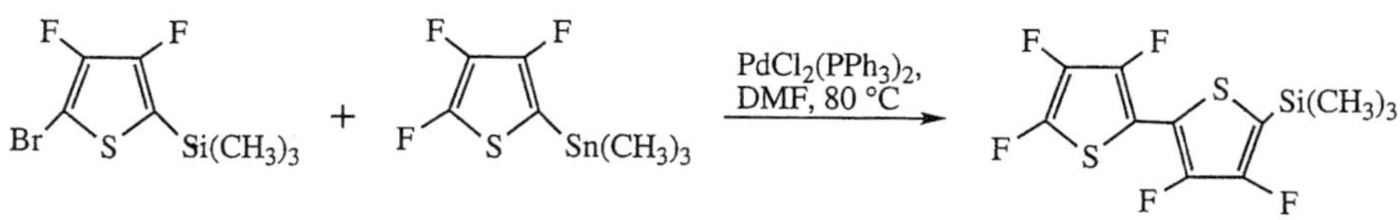

3,3′-Dimethylthio-2,2′-bithienyl is obtained in 78% from 2-bromo-3-methylthiothiophene and 3-methylthio-2-trimethylstannylthiophene [68].

4,4′-Dimethoxy-2,2′-bithienyl is prepared by coupling of 2-isopropylsilyl-5-bromo-3-methoxythiophene and 2-trimethylstannyl-3-methoxythiophene in the presence of palladium(0) and subsequent destilylation [29]. In connection with the preparation of polyhydroxy oligothiophenes, 3,3′-di[2-(tetrahydropyranyloxy)ethyl]- and 3,4′-di[(tetrahydropyranyloxy)ethyl]-2,2′-bithienyl were obtained by the coupling of 2-trimethyltin-3-[2-(tetrahydroxypyranyloxy)-ethylthiophene and 5-trimethyltin-3-[2-(tetrahydropyranyloxy)ethyl]thiophene, respectively with 2-bromo-3-[2-tetrahydropyranyloxy)ethyl]thiophene [92].

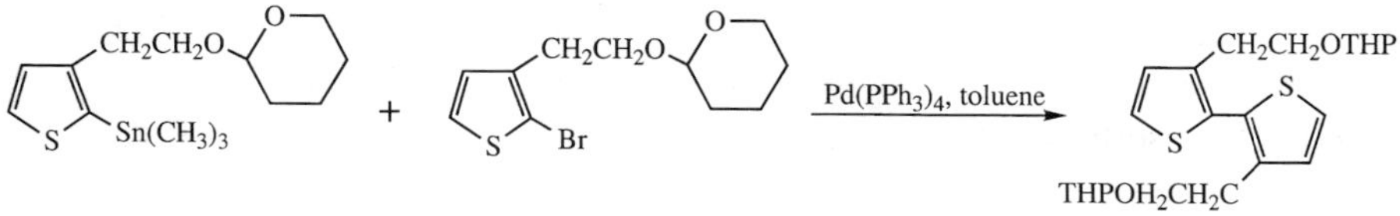

3,3′-Di[2-tetrahydropyranyloxy)ethyl]-2,2′-bithienyl [92]

To a solution of 2-trimethyltin-3-[2-(tetrahydropyranyloxy)ethyl]thiophene (1.79 g, 4.78 mmol) in toluene (50 ml) 2-bromo-3-[2-tetrahydropyranyl-oxy)ethyl]thiophene (1.39 g, 4.78 mmol) and tetrakis(triphenylphosphine)-palladium(0) (55 mg, 0.048 mmol) dissolved in toluene (30 ml) are added. The reaction mixture is refluxed overnight, hydrolyzed with 2 *M* hydrochloric acid, neutralized with aqueous sodium bicarbonate solution, washed with aqueous sodium chloride solution, dried over magnesium sulfate and evaporated. The residue is purified by chromatography on silica gel using dichloromethane/ethyl acetate (95:5) as eluent, giving 1.4 g (69%) of the title compound.

A very detailed study was undertaken in order to find the best method for the preparation of 5,5′-diphenyl-2,2′-bithienyl. The palladium(0)-catalyzed coupling of 2-bromo-5-phenylthiophene did not give satisfactory yields, also many different catalyst systems were used. However, the use of the Stille coupling between 5-phenyl-2-tributylstannylthiophene with 2-iodo-5-phenylthiophene gave the desired 5,5′-diphenyl-2,2′-bithienyl in good yield [78].

5,5′-Diphenyl-2,2′-bithienyl [78]

A solution of 2-tributylstannyl-5-phenyl thiophene (3.1 mmol) in *N,N*-dimethylformamide (10 ml) is added to a reduced palladium catalyst prepared from dichlorobis(triphenylphosphine)palladium(II) and diisobutylaluminium hydride (0.093 mmol, 3 mol%) in tetrahydrofuran (2 ml). A solution of 5-phenyl-2-iodothiophene in *N,N*-dimethylformamide (10 ml) is then added and the reaction mixture stirred at 60 °C for 3 h, within 20 min the product begins to precipitate. After cooling the reaction is quenched with 3 *M* aqueous hydrochloric acid (20 ml), the solid product collected by filtration, washed with methanol and diethyl ether and air dried giving 0.85 g (85%) of the title compound as a golden yellow solid mp 240–242 °C.

Reaction of four equivalents of 2-tributylstannylthiophene with 1,4-dibutoxy-2,5-di(2′-iodo-5′-thienyl)benzene is used for the preparation of 2,5-di(5′-(2′,2″-thienyl)thienyl)-1,4-dibutoxybenzene. The seven ring polyaryl below was prepared by Stille coupling of three equivalents of 5-phenyl-2-tributylstannylthiophene with 1,4-dibutoxy-2,5-di(2′-iodo-5′-thienyl)benzene [78].

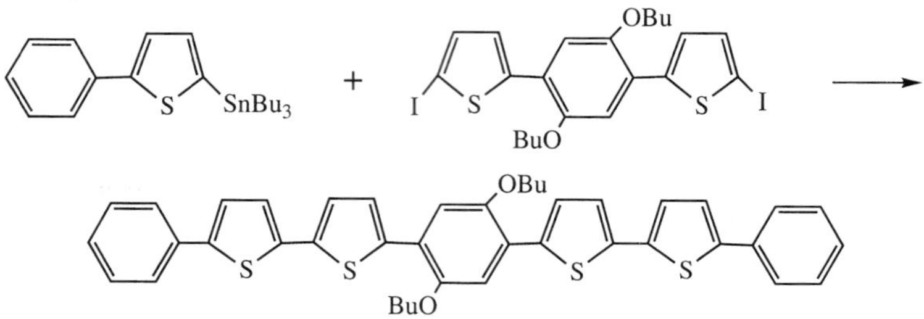

1,4-Dibutoxy-2,5-di(5-phenyl-5″-di-2,2″-thienyl)benzene [78]

To a stirred solution of 1,4-dibutoxy-2,5-di(5′-iodo-2′-thienyl)benzene (3.0 g, 4.7 mmol) in *N*,*N*-dimethylformamide (100 ml) is added a solution of 2-tributylstannyl-5-phenylthiophene (13.8 mmol) in tetrahydrofuran (50 ml) followed by addition of reduced palladium catalyst (0.13 mmol, 2.8 mol%) in tetrahydrofuran (10 ml). The reaction mixture is heated under reflux for 2 h and then cooled to room temperature, after which 3 *M* aqueous hydrochloric acid is added (100 ml). The precipitate formed is collected by filtration giving 1.6 g (48%) of the title compound mp 228–232 °C.

Stille coupling of 2,5-dibromo-3,4-dinitrothiophene with 2-tributylstannyl-3,4,-ethylenedioxythiophene unfortunately gives 3,4-dinitro-5-bromo-3′,4′ ethylenedioxy-2,2-bithienyl in only 17% yield [93]. Stille coupling of 2-tributylstannyl-3,4-ethylenedioxythiophene and 3,4-ethylenedioxy-5′-mesitylthio-2,2′-bithienyl is used for preparation of 3,4-ethylenedioxy-5′-mesitylthio-2,2′bithienyl [9].

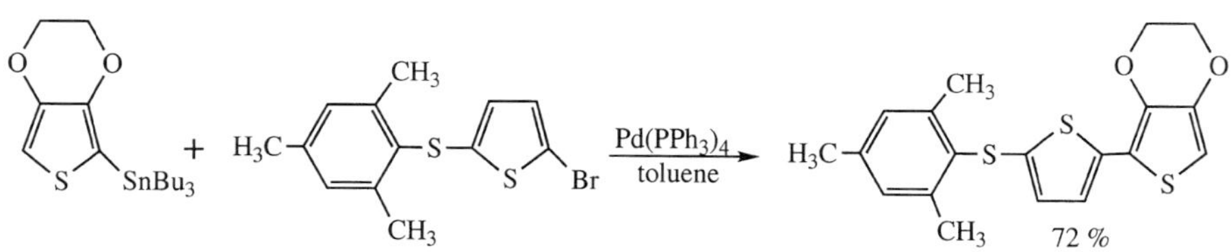

Coupling of 2-(trimethylstannyl)-5-*N*-piperidinylthiophene with 5-bromo-2-thiophene aldehyde is used for the preparation of 5-formyl-5′-*N*-piperidinyl-2,2′-bithienyl [94].

5-Formyl-5′-N-piperidinyl-2,2′-bithienyl [94]

To a degassed solution of 5-bromo-2-thiophene aldehyde (10.5 g, 55 mmol) and 2-(tributylstannyl)-5-*N*-piperidinylthiophene (22.3 g, 55 mmol) in *N*,*N*-dimethylformamide (100 ml) is added bis(triphenylphosphine)palladium(II) chloride (1.4 g, 2 mmol). The mixture is heated to 80 °C for 10 min and then cooled to 40 °C and stirred under nitrogen for 5 h. After cooling the solution of 0 °C for 10 h it is filtered and washed with cold *N,N*-dimethylformamide giving 10.44 g (80%) of the title compound.

Coupling of 4-dodecyl-2-trimethylstannylthiophene with the oxazolidine protected 3-(2-bromothienyl)propionic acid is used for the preparation of the bithienyl shown below, using tribenzylideneacetonepalladium as catalyst and cupric oxide as cocatalyst in *N,N*-dimethlformamide as solvent [95].

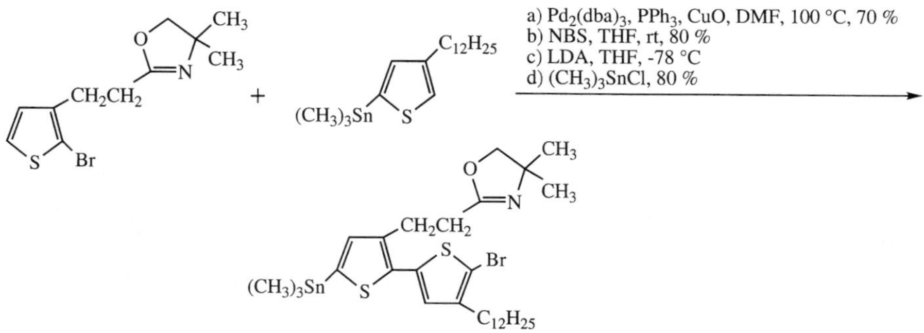

Quite crowded bithienyls can be made through the Stille reaction. Thus coupling of 2-iodo-5-(1-propynyl)thiophene with (*E*)-2-(5-trimethylstannyl-2-thienyl)ethenyldimethylphenylsilane followed by desilylation with tetrabutylammonium fluoride in tetrahydrofuran is used for the synthesis of the natural phototoxin 5-ethenyl-5′-(1-propynyl)-2,2′-bithienyl in good yield [96,97].

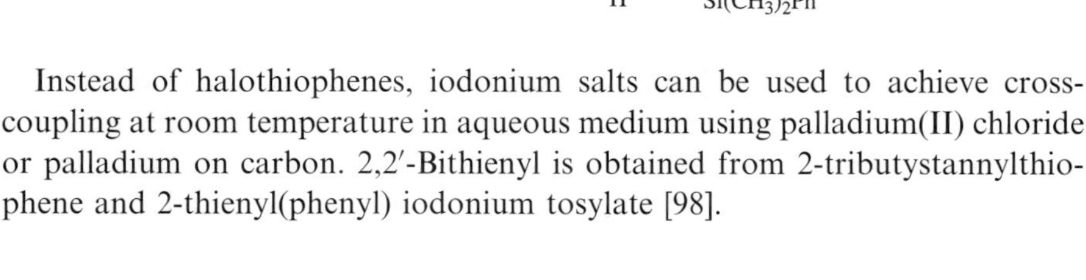

Instead of halothiophenes, iodonium salts can be used to achieve cross-coupling at room temperature in aqueous medium using palladium(II) chloride or palladium on carbon. 2,2′-Bithienyl is obtained from 2-tributystannylthiophene and 2-thienyl(phenyl) iodonium tosylate [98].

2,2-Bithienyl [98]

To a stirred solution of 2-thienyl(phenyl) iodonium tosylate (1.50 g, 4.0 mmol) and palladium(II) chlorides (3.6 mg, 0.5 mol%) in acetonitrile/water (4:1)

(10 ml) 2-tributylstannylthiophene (1.60 g, 4.3 mmol) is added. The reaction mixture is stirred at room temperature for 20 min, quenched with saturated ammonium chloride solution (10 ml) and the product is extracted with diethyl ether (2 × 20 ml). The combined organic phases are dried over magnesium sulfate and evaporated. The residue is purified by chromatography on silica gel using hexanes as eluent giving 0.62 g (94%) of the title compound.

8.3.2 Various coupling methods

The coupling of lithium 2-thienylcyanocuprate with 5-methyl, 5-ethyl- and 5-benzyl-2-thienyllithium in 2-methyltetrahydrofuran at − 125 °C followed by oxidation with oxygen is an excellent method for the preparation of 5-methyl-, 5-ethyl- and 5-benzyl-2,2′-bithienyl in high yields [99].

8.3.3 Substitution reactions of bithienyls

Due to the interest in polythiophenes in material sciences, intensive investigations on the functionalization of bithienyls have been carried out during recent years. For this purpose electrophilic substitution, metalation with organolithium derivatives as well as halogen–metal exchange has proved useful in many cases, when selectivity can be achieved. In most cases, however, the above discussed coupling of two suitably substituted thiophenes is superior.

8.3.3.1 Acylation of bithienyls

Vilsmeyer formylation of 2,2′-bithienyl using phosphorus oxychloride and *N,N*-dimethylformamide in 1,2-dichloromethane at 60 °C gives 5-formyl-2,2′-bithienyl [100–102].

5-Formyl-2,2′-bithienyl [100]

Phosphorus oxychloride (12.5 ml, 134 mmol) is added to a solution of 2,2′-bithienyl (20 g, 120 mmol) and *N,N*-dimethylformamide (10.5 ml, 135 mmol) in 1,2-dichloroethane (200 ml) cooled in an ice-bath. The solution is then warmed to room temperature and heated to reflux. After refluxing overnight the yellow suspension is cooled to room temperature and poured into a saturated aqueous sodium acetate solution (500 ml). This solution is stirred for several hours to complete the hydrolysis. The product is extracted with dichloromethane (3 × 300 ml). The combined organic phases are washed with water (1 × 100 ml), dried over magnesium sulfate and evaporated. The residue is purified by chromatography on silica gel using hexane/ethyl acetate (15:1) as eluent, giving 22 g (94%) of the title compound mp 57–59 °C.

5-[(2-Thienyl)ethynyl]thiophene-2-aldehyde is prepared in 50% yield by Vilsmeyer–Haak formylation of 2,2′-bithienyl acetylene [103].

5[(2-Thienylethynyl]-2-thiophene aldehyde [103]

2,2′-Bithienylacetylene (1 g, 5.3 mmol) is added to a mixture of *N*,*N*-dimethylformamide (424 mg) and phosphoryl chloride (890 mg). The reaction mixture is stirred at 60 °C for 0.5 h. After cooling, an aqueous solution of sodium acetate is added and the mixture is extracted with diethyl ether. The neutral combined organic phases are dried over sodium sulfate and evaporated. The residue is chromatographed on silica gel using chloroform/hexane (4:1) as eluent giving 450 mg (50%) of the title compound.

5-Methyl-2,2′-bithienyl is formylated in the 5′-position and 5,5′-dimethyl-3,3′-bithienyl in the 3-position [101]. 5-Acetyl-2,2′-bithienyl is prepared in 79% yield using tin tetrachloride as catalyst [104]. 5,5′-Diacetyl-2,2′-bithienyl was obtained from 2,2′-bithienyl and excess acetic anhydride in 85% polyphosphoric acid [31]. The reaction of 2,2′-bithienyl with propionyl chloride and titanium tetrachloride is used for the preparation of 5-propionyl-2,2′-bithienyl [105] and with propionic anhydride in phosphoric acid 5,5′-dipropionyl-2,2′-bithienyl is obtained [105]. The Friedel–Crafts reaction between 2,2′-bithienyl and stearic anhydride using boron trifluoride etherate as catalyst yields 5-octadecanoyl-2,2′-bithienyl [106].

5-Octadecanoyl-2,2′-bithienyl [106]

A mixture of 2,2′-bithienyl (3.5 g, 21 mmol) and stearic anhydride (9.5 g, 18 mmol) is heated to the melt and then cooled until resolidification begins. Boron trifluoride etherate (0.25 ml) is added with rapid swirling. The exothermic reaction is moderated with an ice bath and the temperature goes up to 70 °C. Heat is then applied to melt all solids attaining a maximum temperature of 100 °C. Dioxane (100 ml) is added and the reaction mixture is heated to reflux after which water (2 ml) is added to decompose any unreacted anhydride. Upon cooling to room temperature the precipitating crystals are collected and washed with dioxane and diethyl ether and dried in vacuum giving 6.7 g of purely monosubstituted solid bithienyl with a 20% impurity of stearic acid.

5-Acetyl-3,3′-diiodo-2,2′-bithienyl is prepared by refluxing of 3,3′-diiodo-2,2′-bithienyl with acetyl chloride in benzene for 15 h, using stannic chloride as catalyst [107]. Acylation in the free β-position of 2,2′,5,5′-tetramethyl- [108] and 2,2′,5,5′-tetraethyl-3,3′-bithienyl [19] with acetyl chloride in benzene using stannic chloride as catalyst is used for the preparation of the corresponding 4,4′-diacetyl derivatives. Acylation of 2,3′-bithienyl with acetic anhydride and phosphoric acid at room temperature gives only 39% of 5-acetyl-2,3′-bithienyl [109].

8.3.3.2 Nitration of bithienyls

Mononitration of 2,2′-bithienyl with nitric acid in acetic acid or acetic anhydride gives 40% of 5-nitro- and 9.5% of 3-nitro-2,2′-bithienyl [42]. Further nitration of 5-nitro-2,2′-bithienyl gives 5,5′-dinitro-2,2′- and 3,5′-dinitro-2,2′-bithienyl in 52 and 24% yields, respectively. Further nitration of these dinitro isomers gives selectively 3,5,5′-trinitro-2,2′-bithienyl and further nitration gave 3,3′,5,5′-tetranitrothiophene [42]. Also nitration with copper nitrate and acetic anhydride of 2,2′-bithienyl is not selective as a mixture of 5,5′-dinitro-(25%) and 3,5′-dinitro-2,2′-bithienyl (41%) is obtained [41]. Also nitration of the deactivated 5-acetyl- and 5-formyl-2,2′-bithienyl is not selective and gives mixtures of 5′- and 3′-nitrated products [110,111]. Nitration of a number of 3,3′-dialkyl-2,2′-bithienyls (alkyl = butyl, hexyl, octyl, decyl, and dodecyl) with nitric acid in acetic acid gives the 5,5′-dinitro derivatives in 50–60% yield [112].

3,3′-Dialkyl-5,5′-dinitro-2,2′-bithienyl [112]

A solution of 3,3′-dialkyl-2,2′-bithienyl (13.5 mmol) in chloroform (65 ml) is added over 1.5 h at 10 °C (50 °C for alkyl = decyl and dodecyl) to a mixture of 70% nitric acid (95 ml), glacial acetic acid (52.5 ml), and chloroform (10 ml). The mixture is stirred at room temperature until complete consumption of the dialkylated compound, about 1 h. Concentrated sulfuric acid (13 ml) is then added dropwise at 10 °C. The mixture is stirred at room temperature until complete nitration is achieved and then poured into ice-water. After removing the chloroform the residue is washed with saturated aqueous sodium bicarbonate solution, taken up in diethyl ether, washed again successively with deionized water and saturated aqueous sodium chloride solution, dried over magnesium sulfate and evaporated. The residue is recrystallized giving 50–60% of the title compounds.

Mononitration of 3,3′-bithienyl with fuming nitric acid in acetic anhydride, on the other hand, is selective and gives 2-nitro-3,3′-bithienyl in almost quantitative yield [113].

2-Nitro-3,3′-bithienyl [113]

To a suspension of 3,3′-bithienyl (10 g, 0.060 mol) in acetic acid anhydride (50 ml) a mixture of fuming nitric acid (5.3 g, 0.084 mol) and acetic acid anhydride (16 ml) is added at 0 °C. After stirring the reaction mixture for 1 h it is poured into ice. The precipitate is filtered off giving 13 g (98%) of the title compound mp 90–91 °C from methanol.

Further nitrations are not selective. Thus nitration of 2-nitro-3,3′-bithienyl leads to a mixture of 70% of 2,5′-dinitro- and 25% of 2,2-dinitro-3,3′-bithienyl [113]. 4,4′-Dicarboxy-5,5′-dinitro-3,3′-bithienyl is obtained in 63% yield upon nitration of 4,4′dicarboxy-3,3′-bithienyl [114]. Nitration of 2,3′-bithienyl with fuming nitric acid in acetic acid is the most unselective one and gives a mixture of 3-nitro- (38.7%), 3′-nitro-(34.8%) and 5-nitro-2,3′-bithienyl (26.5%) [115].

The different mononitro isomers of 2,2,- and 2,3′-bithienyls should therefore best be prepared by palladium(0)-catalyzed coupling reactions between bromonitrothiophenes and thiopheneboronic acids or thienylstannyl derivatives.

8.3.3.3 Chlorination of bithienyls

Selective monochlorination of 2,2′-bithienyl is difficult and 5-chloro-2,2′-bithienyl is best obtained by chlorination with sulfuryl chloride of 5-carboethoxy-2,2′-bithienyl followed by hydrolysis and decarboxylation [105]. Direct chlorination of 2,2′-bithienyl with sulfuryl chloride leads to complex mixtures [116]. However, 5,5′-dichloro-2,2′-bithienyl has been obtained in this way in 69% yield [117]. Chlorination of 3,3′-bithienyl with *N*-chlorosuccinimide in refluxing acetic acid gives 2,2′-dichloro-3,3′-bithienyl [17].

2,2′ Dichloro-3,3′-bithienyl [17]

N-chlorosuccinimide (3.2 g, 12 mmol) is added to a solution of 3,3′-bithienyl (2.0 g, 12 mmol) in glacial acetic acid (100 ml). The mixture is refluxed for 30 min and then cooled to room temperature. Water and chloroform are added and the phases separated. The aqueous phase is extracted twice with chloroform. The combined organic phases are washed with aqueous sodium bicarbonate solution and water to neutral reaction, dried over magnesium sulfate and evaporated. The residue is dissolved in hot ligroin (80–110 °C) and treated with active neutral aluminium oxide. This suspension is filtered and evaporated giving 2.2 g (78%) of the title compound mp 52–54 °C.

Direct chlorination of 3,3′-bithienyl with gaseous chlorine or *N*-chlorosuccinimide gives 2,2′,5,5′-tetrachloro-3,3′-bithienyl in 70% yield [17]. Another route to chlorinated 3,3′-bithienyls consists in the reaction of bromo-3,3′-bithienyls with chlorine. Thus hexachloro-3,3′-bithienyl is convenient prepared in 99% yield by reacting 2,2′,4,4′-tetrabromo-3,3′-bithienyl with excess chlorine for twenty hours [17].

8.3.3.4 Bromination of bithienyls

8.3.3.4.1 Bromination of 2,2′-bithienyls

Bromination with *N*-bromosuccinimide, using *N*,*N*-dimethylformamide, tetrahydrofuran, acetic acid or a mixture of chloroform and acetic acid as solvents are the most convenient methods for the preparation of brominated 2,2′-bithienyls. It is difficult to obtain selective monobromination of 2,2′-bithienyl. However, at −20 °C with *N*-bromosuccinimide in *N*,*N*-dimethylformamide results in a mixture of 9% starting material, 81% of 5-bromo-2,2′-bithienyl and 10% of 5,5′-dibromo-2,2′-bithienyl. Upon fractional distillation at reduced pressure followed by recrystallization pure 5-bromo-2,2′-bithienyl is obtained [63].

5-Bromo-2,2′-bithienyl [63]

In the absence of light *N*-bromosuccinimide (44.5 g, 0.25 mol) is added portionwise at −20 °C to a solution of bithienyl (41.5 g, 0.25 mol) in *N*,*N*-dimethylformamide (150 ml) and the reaction mixture is stirred for 4 h and poured into ice. The product is extracted several times with dichloromethane. The combined organic phases are washed with water, dried over sodium sulfate and evaporated. Fractional distillation under reduced pressure gives 42.8 g (70%) of a yellowish solid mp 33–34 °C. After recrystallization from methanol 33.7 g (55%) of the title compound is obtained in pure from mp 34 °C.

Bromination with *N*-bromosuccinimide of 5-dodecyl-2,2′-bithienyl gives a 99% yield of 5-bromo-5′-dodecyl-2,2′-bithienyl [118]. 5,5′-Dibromo-2,2′-bithienyl is prepared in 96% yield upon reaction of 2,2′-bithienyl with *N*-bromosuccinimide in a 1:1 mixture of chloroform and acetic acid [13,119] and in 88% yield upon reaction in *N*,*N*-dimethylformamide at 20 °C [63,120].

5,5′ Dibromo-2,2′-bithienyl [63]

In the absence of light *N*-bromosuccinimide (26.6 g, 0.15 mol) is added portionwise at 20 °C to a solution of bithienyl (12.3 g, 0.074 mol) in *N*,*N*-dimethylformamide (100 ml). The reaction mixture is stirred for 3 h and poured onto ice. The white precipitate is filtered off and washed several times with water, dried over phosporus pentoxide and recrystallized from absolute ethanol giving 21.2 g (88%) of the title compound as a white solid mp 146 °C.

4,4′,5,5′-Tetrabromobithienyl is obtained in 94% yield upon bromination with *N*-bromosuccinimide of 4,4′-dibromo-2,2′-bithienyl [13]. In order to achieve bromination in the *β*-positions, it is sometimes better to use elemental bromine when the *α*-positions are blocked, as in the preparation of 3,3′-dibromo-4,4′,5,5′-tetramethyl-2,2-bithienyl [4,5] and 3,3′-dibromo-4,4′,5,5′-tetraphenyl-2,2′-bithienyl [5].

3,3′-Dibromo-4,4′,5,5′-tetraphenyl-2,2′-bithienyl [5]

A solution of bromine (1.6 g, 10 mmol) in chloroform (20 ml) is added dropwise with ice cooling over 20 min to a solution 4,4′,5,5′-tetraphenyl-2,2′-bithienyl (2.25 g, 5 mmol) in chloroform (50 ml). After further stirring for 15 min with ice cooling and for 45 min at room temperature chloroform (100 ml) is added. The solution obtained is washed with sodium bicarbonate solution and water, and evaporated. The residue is recrystallized from benzene/ethanol (2:1) giving 2.6 g (83%) of the title compound as pale-yellow needles mp 271 °C.

For polybromination of 2,2′-bithienyl elemental bromine is used and with four equivalents with chloroform/acetic acid as solvent 3,3′,5,5′-tetrabromo-2,2′-bithienyl is obtained in 97% yield [112]. Hexabromo-2,2′-bithienyl is prepared from 2,2′-bithienyl in 61% yield [13].

Reaction of pentafluoro-5-trimethylsilyl-2,2′-bithienyl with *N*-bromosuccinimide in acetic acid at 80 °C gives 5-bromo-3,3′,4,4′,5-pentafluoro-2,2′-bithienyl in 74% yield [91]. 1,2-bis(5′-bromo-3′-octyl-2,2′bithiophene-5-yl)ethane is prepared in 97% yield by bromination of 1,2-bis(3′-octyl-2,2-bithiophene-5-yl)ethane with *N*-bromosuccinimide in chloroform/acetic acid at room temperature [70].

Interestingly, bromination of 3-butylthio-2,2′-bithienyl occurs selectively in the unsubstituted ring giving 5′-bromo-3-(butylthio)-2.2′-bithiophene in 91% yield [64]. Bromination of 1,8-bis(2,2′-bithienyl-5-yl)naphthalene with *N*-bromosuccinimide in chloroform/acetic acid is used for the dibrominated derivative [121].

2-(5-Bromo-2-thienyl)-4,6-dihydro[3,4-*b*]thiophene-5,5′-dioxide is obtained upon bromination with *N*-bromosuccinimide in *N,N*-dimethylformamide in 78% yield [84]. Bromination of 3,3′-hexyl-2,2′-bithienyl and 4,3′-dihexyl-2,2′-bithienyl selectively gives 5-bromo-3,3′-dihexyl-2,2′-bithienyl and 5-bromo-3′,4-dihexyl-2,2′-bithienyl with *N*-bromosuccinimide in acetic acid/chloroform (1:1) in 60% and 100%, respectively [122,123]. 5,5′-Dibromo-3,4′-dihexyl-2,2′-bithienyl is obtained upon bromination of the monobromo derivative with *N*-bromosuccinimide in chloroform/acetic acid [30]. 5-Bromo-3,3′-dihexyl-5′-methoxy-2,2′-bithienyl and 5-bromo-3,4′-dihexyl-5′-methoxy-2,2′-bithienyl were similarly prepared from the nonbrominated compounds [122,123].

5-Bromo-3,3′-dihexyl-5′-methoxy-2,2′-bithienyl [123]

To a solution of 3,3′-dihexyl-5′-methoxy-2,2′-bithienyl (466 mg, 1.28 mmol) in a mixture of chloroform and acetic acid (10:1, 165 ml) is added *N*-bromosuccinimide (171 mg, 0.96 mmol) in small portions at room temperature. The reaction mixture is stirred for 30 min and poured into water. The product is extracted with chloroform. The combined organic phases are washed with aqueous potassium hydroxide and aqueous sodium chloride solution and evaporated. The residue is purified by chromatography on alumina (3.2 × 18 cm) using hexane as eluent, giving 520 mg (49%) of the title compound as a pale-yellow oil.

Bromination of 2,2′-bithienyl-5-carboxylic acid with *N*-bromosuccinimide in *N,N*-dimethylformamide at − 10 °C gives 5-bromo-2,2′-bithienyl-5′-carboxylic acid [124]. 2-Bromo-5′-formyl-2,2′-bithienyl is prepared by refluxing of 5-formyl-2,2′-bithienyl with bromine and sodium bicarbonate for four hours [100,102].

2-Bromo-5′-formyl-2,2′-bithienyl [100]

To a solution of 5-formyl-2,2′-bithienyl (9 g, 46 mmol) in chloroform (100 ml) sodium bicarbonate (4.28 g, 51 mmol) is added followed by a dropwise addition of bromine (8.16 g, 51 mmol) in chloroform (100 ml) over a period of 1 h. The reaction mixture is refluxed for 4 h and then cooled and filtered. The filtrate is washed with water (2 × 100 ml), dried over magnesium sulfate and evaporated. The residue is purified by chromatography on silica gel using hexane/ethyl acetate (15:1) as eluent giving 11.2 g (89%) of the title compound as a yellow solid mp 141–145 °C (dec.)

From 5-acetyl-2,2′-bithienyl, 5-acetyl-5′-bromo-2,2′-bithienyl was obtained in 90% yield [125]. Benzyl (2,2′-bithienyl)-5-carboxylate is brominated quantitatively in the remaining *α*-position by *N*-bromosuccinimide in *N,N*-dimethylformamide [126,127]. 2,2′-Bithienyl-5-carboxylic acid, bound to solid support, is brominated by *N*-bromosuccinimide in *N,N*-dimethylformamide in the 5′-position. Using an alternating sequence of bromination and Stille couplings affords oligomers up to the pentamer in excellent yields [128,129]. 2-Bromo-3,3′-bis[2-(tetrahydropyranyloxy)ethyl]-2,2′-bithienyl [130] and the oxaxolidine protected 3-(4-dodecyl-5-bromo-2,2′-bithienyl)propionic acid [95] were obtained upon reaction with *N*-bromosuccinimide in *N,N*-dimethylformamide and tetrahydrofuran, respectively.

2-Bromo-3,3′-bis[2-(tetrahydropyranyloxy)ethyl]-2,2′-bithienyl [130]

To a solution of 3,3′-bis[2-(tetrahydropyranyloxy)ethyl]-2,2′-bithienyl (1.0 g, 2.37 mmol) in anhydrous *N,N*-dimethylformamide (50 ml) at − 20 °C, *N*-bromosuccinimide (0.42 , 2.37 mmol) is added stepwise. The reaction mixture is allowed to warm to room temperature, stirred overnight and quenched with ice. The phases are separated and the aqueous phase extracted with dichloromethane. The combined organic phases are washed with saturated aqueous sodium chloride solution, dried over sodium sulfate and evaporated. The residue is purified by flash chromatography using dichloromethane/ethyl acetate (85: 15) as eluent giving 1.07 g (90%) of the title compound as a pale-yellow oil.

5-Bromo-4-octyl-2,2′-bithienyl is prepared from 5-trimethylsilyl-4-octyl-2,2′-bithienyl by treatment with hydrochloric acid in tetrahydrofuran, followed by bromination with *N*-bromosuccinimide [72]. Bromodesilylation with bromine in ether at −90 °C is another route for the preparation of 5,5′-dibromo-2,2′-bithienyl derivatives [131].

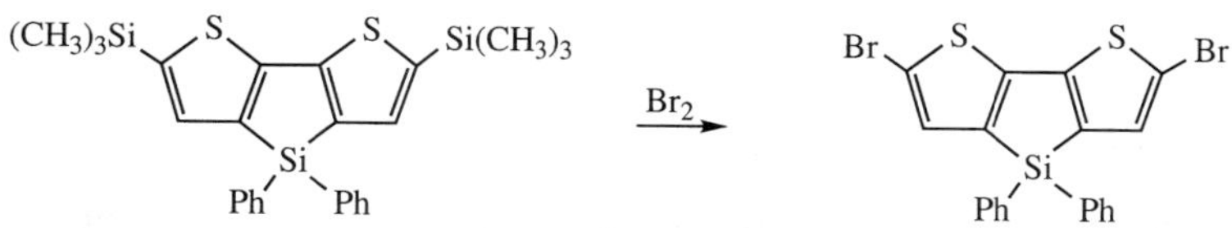

5,5′-Dibromo-4,4-diphenyl-dithieno[3,2-b:2′,3′-d]silole [131]

To a solution of the disilylated compound (6.15 g, 12.5 mmol) in diethyl ether (400 ml) bromine (1.32 ml, 25.6 mmol) is added dropwise at −90 °C. The stirring is continued at this temperature for 2 h, after which the solvent is removed under reduced pressure, 15 mm Hg. The residue is washed with hexane giving a white solid, which is recrystallized from chloroform giving 4.17 g (66%) of the title compound as a white solid mp 262–266 °C.

8.3.3.4.2 Bromination of 3,3′-bithienyls

Brominated 3,3′-bithienyls are of importance in the preparation and study of atropisomeric 3,3′-bithienyls [132]. In order to avoid rearrangements induced by hydrogen bromide, which occurs upon bromination of 3-arylthiophenes and 3,3′-bithienyls, *N*-bromosuccinimide in acetic acid is used [119,133]. The very unstable 2-bromo-3,3′-bithienyl is obtained in 90% yield upon reaction of 3,3′-bithienyl with one equivalent of *N*-bromosuccinimide in carbon tetrachloride in the presence of benzoyl peroxide [134,135]. 2,2′-Dibromo-3,3′-bithienyl is obtained in quantitative yield upon bromination of 3,3′-bithienyl with two equivalents of *N*-bromosuccinimide in chloroform/acetic acid and further bromination with *N*-bromosuccinimide in carbon tetrachloride gives 2,2′,5,5′-tetrabromo-3,3′-bithienyl in only 34% yield [119]. All these brominated 3,3′-bithienyls are unstable at room temperature and decompose under evolution of hydrogen bromide. They can be stored at −20 °C for some time [119]. The tendency for decomposition seems to be characteristic for 2,2′-dibromo-3,3′-bithienyl without deactivating groups in at least one of the rings [136].

2,2′,4,4′-Tetrabromo-5,5′-dichloro-3,3′-bithienyl, 2,2′,5,5′-tetrabromo-4,4′-dichloro-3,3′-bithienyl and 5,5′-dibromo-2,2′,4,4′-tetrachloro-3,3′-bithienyl are prepared by bromination of the appropriate chloro derivatives with elemental bromine [17]. Hexabromo-3,3′-bithienyl is easily prepared by the reaction of 3,3′-bithienyl with excess bromine [16].

Hexabromo-3,3′-bithienyl [16]

Excess bromine is cautiously added portionwise to 3,3′-bithienyl (38 g, 0.23 mol) in a large glass vessel. After the excess bromine has evaporated the remaining crystalline mass is ground in a mortar and boiled with ethanol. The insoluble part is filtered off and recrystallized from carbon tetrachloride or ligroin, bp 80 °C, giving 140 g (95%) of the title compound mp 181–182 °C.

Bromination of 3,3′-bithienyl with two equivalents of *N*-bromosuccinimide in *N,N*-dimethylformamide gives 2,2′-dibromo-3,3′-bithienyl which upon further chlorination with *N*-chlorosuccinimide gives 2,2′-dibromo-5,5′-dichloro-3,3′-bithienyl, which upon treatment with one equivalent of butyllithium followed by hydrolysis gave 2-bromo-5,5′-dichloro-3,3′-bithienyl. Renewed halogen–metal exchange followed by tributylstannyl chloride led to 5,5′-dichloro-2-tributylstannyl-3,3′-bithienyl [137].

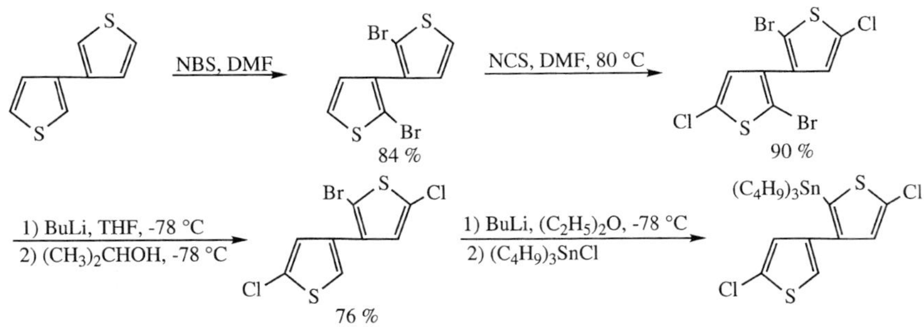

This bromination can also be carried out with optically active materials. Thus (−)-2,2′,4,4′-tetrabromo-3,3′-bithienyl gives (−)-hexabromo-3,3′-bithienyl [138] and *R*-(+)-4,4′, 5,5′-dibromo-2,2′-dicarboxy-3,3′-bithienyl [139], in this way achieving stereochemical correlation.

R-(+)-4,4′, 5,5′-tetrabromo-2,2′-dicarboxy-3,3′-bithienyl [139]

Excess bromine is poured onto R-(+)-4,4′-dibromo-2,2′-dicarboxy-3,3′-bithienyl (0.41 g, 1.0 mmol, $[\alpha]_D^{25} = +44°$). After evaporation of the bromine the remaining solid is dissolved in aqueous sodium hydroxide solution followed by reprecipitation with hydrochloric acid giving 0.52 g (91%) of the title compound, $[\alpha]_D^{25} = +53°$. Recrystallization from ethanol/water gives 0.3 g, $[\alpha]_D^{25} = +56°$. The crystals started to melt at 215 °C and new crystals, probably the racemic form, are formed mp 275–277 °C.

2,2′-Dibromo-5,5′-dimethyl-3,3′-bithienyl is obtained upon reaction with *N*-bromosuccinimide in acetic acid/chloroform [23,136]. In the same way

2,2′-dibromo-4,4′,5,5′-tetramethyl-3,3′-bithienyl and 4,4′-dibromo-2,2′,5,5′-tetramethyl-3,3′-bithienyl were obtained [136].

2,2′-Dibromo-5,5′-dimethyl-3,3′-bithienyl [136]

To a solution of 5,5′-dimethyl-3,3′-bithienyl (2.5 g, 13 mmol) in acetic acid/chloroform (1:1) cooled in ice water, hydroquinone (2 mg) is added followed by *N*-bromosuccinimide (4.6 g, 26 mmol) added in portions. The reaction is completed within a few min and the reaction mixture is diluted with water. The phases are separated and the organic phase washed with water, sodium carbonate solution, and again with water, dried over calcium chloride and evaporated at room temperature, giving 4.4 g (97%) of the title compound mp 90–91 °C after recrystallization from ligroin, bp 65–75 °C.

4,4′,5,5′-Tetrabromo-3,3′-bithienyl is prepared by similar bromination of 4,4′-dibromo-3,3′-bithienyl [136]. Reaction of 5,5′-dimethyl-3,3′-bithienyl with excess bromine gave 2,2′,4,4′-tetrabromo-5,5′-dimethyl-3,3′-bithienyl [23]. From deactivated 3,3′-bithienyls, such as 4,4′-diformyl-3,3′-bithienyl the 2,2′-dibromo derivative is prepared upon reaction with bromine in methylene chloride in the presence of aluminium chloride [140].

2,2′-Dibromo-4,4′-diformyl-3,3′-bithienyl [140]

To an ice-cooled solution of 4,4′-diformyl-3,3′-bithienyl (20 g, 90 mmol) in anhydrous dichloromethane (600 ml), aluminium chloride (60 g, 0.45 mmol) is added with stirring. The mixture is heated to reflux and bromine (30 g, 190 mmol) in anhydrous dichloromethane (100 ml) is added. When the addition is completed the reaction mixture is refluxed for 8 h and then poured onto a mixture of crushed ice and concentrated hydrochloric acid. The phases are separated and the aqueous phase extracted several times with ethyl acetate. The combined organic phases are washed with aqueous sodium bicarbonate solution and water, dried and evaporated. The residue is recrystallized from ethyl acetate giving 26 g (76%) of the title compound mp 155.0–155.5 °C.

Bromination of 4,4′-dicarboxy-3,3′-bithienyl on the other hand gives only a low yield of 2,2′-dibromo-4,4′-dicarboxy-3,3′-bithienyl, so it is preferentially prepared by permanganate oxidation of the 2,2′-dibromo-4,4′-diformyl-3,3′-bithienyl. Upon its reaction with excess bromine 2,2′,5,5′-tetrabromo-4,4′-dicarboxy-3,3′-bithienyl is obtained in 83% yield. The reaction can also be carried out with optically active 2,2′-dibromo-4,4′-dicarboxy-3,3′-bithienyl [140].

8.3.3.4.3 Bromination of 2,3′-bithienyls

2′,5-Dibromo-2,3′-bithienyl is prepared in quantitative yield, upon reaction of two equivalents of *N*-bromosuccinimide in chloroform/acetic acid with

2,3′-bithienyl [119]. Further bromination gives 2′,5,5′-tribromo-2,3′-bithienyl. Also these bromo derivatives are unstable and decompose, sometimes violently, at room temperature. 2′,3-Dibromo-2, 3′-bithienyl is prepared by bromination of 3-bromo-2, 3′-bithienyl in chloroform/acetic acid [141].

8.3.3.5 Iodination of bithienyls

5-Iodo-2,2′-bithienyl is prepared in good yields through mercuration–iodination of 2,2′-bithienyl [142, 143]. 5,5′-Diiodo-2,2′-bithienyl is prepared in 82% yield by the reaction of 2,2′-bithienyl with yellow mercuric oxide in benzene at room temperature overnight [28], or by reaction with *N*-iodosuccinimide in acetic acid [67].

5,5′-Diiodo-2,2′-bithienyl [67]

2,2′-Bithienyl (831 mg, 5.00 mmol) and *N*-iodosuccinimide (2.81 g, 12.5 mmol) are dissolved in methonal (75 ml). To this solution acetic acid (0.72 m, 12.5 mmol) is added and soon precipitation starts. The reaction mixture is put into a refrigerator to ensure complete generation of the precipitation. The precipitate is filtered off, washed with methanol, dried and recrystallized from acetone giving 1.01 g (48%) of the title compound as colorless crystals.

Two iodination methods were used for the preparation of 5′-iodo-3-(butylthio)-2,2′-bithienyl from 3-(butylthio)-2,2′-bithienyl in high yields. Either the reaction with iodine and mercuric oxide in benzene or preferably its reaction with iodine chloride and sodium acetate in methanol [64].

5′-Iodo-3-(butylthio)-2,2′-bithienyl [64]

To a stirred solution of 3-(butylthio)-2,2′-bithienyl (2.54 g, 10 mmol) and sodium acetate (0.98 g, 12 mmol) in methanol (24 mil), a solution of iodine monochloride (1.94 g, 12 mmol) in methanol (6 ml) is added dropwise during 1 h at room temperature. The reaction mixture is stirred for 1 h and then poured onto ice (20 mil). The phases are separated and the aqueous phase extracted with chloroform (100 ml). The combined organic phases are washed with aqeuous sodium thiosulfate solution, dried over magnesium sulfate and evaporated. The residue is purified by chromatography on silica gel, neutralized with 2% triethylamine solution, using light petroleum/diethyl ether (98:2) as eluent, giving 3.3 g (86%) of the title compound.

4-Iodo-2,2′,5,5′-tetramethyl-3,3′-bithienyl and 4,4′-dibromo-5,5′-diiodo-3,3′-bithienyl are obtained in 81% yield and 53% yield, respectively using the iodine–iodic acid method on 2,2′,5,5′-tetramethyl-2,2′-bithienyl [4,144], and 4,4′-dibromo-3,3′-bithienyl [136]. Also 4,4′-diiodo-2,2′,5,5′-tetraethyl-3,3′-bithienyl is prepared by this iodination method [19]. Iodination of 3,3′-bithienyl with excess iodine/iodic acid stopped after formation of 2,2′, 5,5′-tetraiodo-3,3′-bithienyl [145]. Hexaiodo-3,3′-bithienyl could be obtained *via* the mercury derivatives [145]. 4,4′-Dicarboxy-2,2′-diiodo-3,3′-bithienyl is prepared by iodination of 4,4′-dicarboxy-3,3′-bithienyl with ammonium peroxydisulfate and iodine in a mixture of water and carbon tetrachloride giving 51% of the desired product [45].

4,4′-Dicarboxy-2,2′-diiodo-3,3′-bithienyl [45]

A solution of ammonium peroxydisulfate (7.2 g, 32 mmol) in water (30 ml), iodine (5.8 g, 23 mmol) and carbon tetrachloride (25 ml) are added to 4,4′-dicarboxy-3,3′-bithienyl (6.0 g, 23 mmol) in acetic acid (250 ml). The reaction mixture is stirred for 36 h at 80 °C and then poured onto ice (500 g). The crystals are filtered off, washed with aqueous sodium bisulfate solution and water and recrystallized from ethanol/water giving 4.5 g (38%) of the title compound as brown crystals mp 237–239 °C.

8.3.3.6 Various electrophilic substitution reactions

5,5′-Di-*tert*-butyl-2,2′-bithienyl is obtained by aluminium chloride-catalyzed Friedel-Crafts reaction using excess of tert-butyl chloride in dichloromethane [146].

5,5′-Di-tert-butyl-2,2′-bithienyl [146]

Aluminum chloride (1.0 g, 7.5 mmol) is added in one portion to a vigorously stirred solution of 2,2′-bithienyl (0.830 g, 5.0 mmol) and *tert*-butyl chloride (4.0 ml, 60 mmol) in dichloromethane (75 ml). The reaction mixture is stirred at room temperature for 24 h and then poured into an excess of 10% aqueous hydrochloric acid. The product is extracted with diethyl ether. The combined organic phases are washed with water, dried over sodium sulfate and evaporated. The residue is purified by chromatography on silica gel using hexane as eluent, giving 1.05 g (78%) of the title compound mp 91–92 °C, after recrystallization from methanol.

Reaction of 4,4′-dibromo-2,2′-bithienyl with β-ethylsulfanylpropionyl tetrafluoborate led to a mixture of 2-(4-bromo-2-thienyl)-4-ethyl-7-oxo-6, 7-dihydro-5*H*-thieno[3,2-*b*]thiopyran-4-ium tetrafluoroborate and bis (4-ethyl-7-oxo-6,7-dihydro-5*H*-(thieno[3,2-b]thiopyran-4-iumyl) ditetrafluoroborate in the proportion of 3:1, from which the monoacylated product is easily obtained pure. For 4,5′-dibromo-2,2′-bithienyl only reaction of the β-bromothiophene part is observed [147].

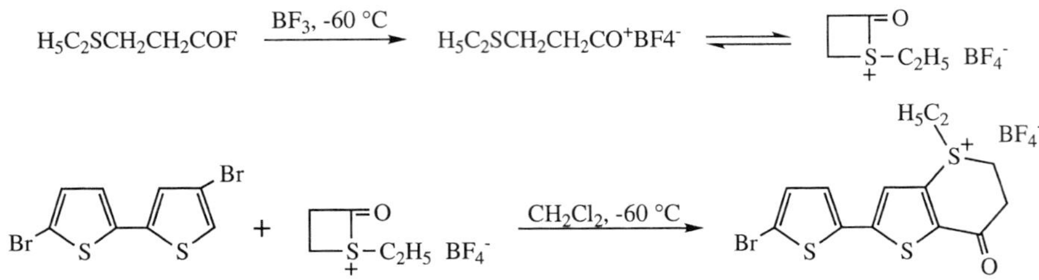

4-Ethyl-7-oxo-6,7-dihydro-5H-thieno[3,2-b]thiopyran-4-ium tetrafluoroborate [147]

A well-stirred solution of β-ethylthiopropionyl fluoride (2.7 g, 0.02 ml) in dichloromethane (40 ml) is saturated by gaseous boron trifluoride at $-60\,^\circ$C. A solution of 4,5′-dibromo-2,2′-bithienyl (6.5 g, 0.02 mol) in dichloromethane (10 ml) is added dropwise. After stirring for 15 min at $-40\,^\circ$C the temperature is allowed to rise to $0\,^\circ$C and the reaction mixture is stirred at this temperature for 2 h, after which it is poured into anhydrous diethyl ether (80 ml). The precipitated sulfonium salt is filtered off, washed with anhydrous diethyl ether (20 ml), cold methanol (10 ml), anhydrous diethyl ether (30 ml) and dried *in vacuo* giving 2.95 g (33%) of the title compound mp 123–125 °C (dec.).

8.3.4 Substituted bithienyls from Grignard reagents of halobithienyls

The reaction between the Grignard reagent from 5-bromo-2,2′-bithienyl and trimethylstannyl chloride in tetrahydrofuran is used for the preparation of 5-trimethylstannyl-2,2′-bithienyl in 97% yield [66]. This approach has been used for the preparation of 5-(trimethylstannyl)-[2-3,3′-(tetrahydropyranyloxy) ethyl]2,2′-bithienyl from the 5-bromo derivative [148].

The Grignard reagent from 4-bromo-2,2′-bithienyl is obtained by the entrainment method using the reaction with magnesium and 1,2-dibromomethane in ether and upon reaction with carbon dioxide and ethylene oxide 4-carboxy- and 4-hydroxyethyl-2,2-bithienyl were obtained in 78% and 44% yield, respectively, as halogen–metal exchange with butyllithium followed by reaction with ethylene oxide was not successfull [149].

4-Hydroxyethyl-2,2′-bithienyl [149]

Magnesium turnings (5.0 g, 0.206 mol) and 4-bromo-2,2′-bithienyl (10.0 g, 0.0410 mol) are combined in diethyl ether (200 ml). About 5 ml of a solution of dibromoethane (14.0 ml, 0.163 mol) in diethyl ether (100 ml) is added to initiate the reaction. Once the reaction is initiated, addition is continued dropwise over 1 h. The flask with the Grignard reagent is fitted with a mechanical stirrer and cooled to 0 °C. Ethylene oxide (*ca* 1 ml, *ca* 0.02 mol) is slowly transferred to the Grignard solution *via* cannula, inducing precipitation. Additional aliquots (1 ml) of ethylene oxide are added until precipitate is no longer formed. The reaction mixture is heated at reflux for 1 h, the reaction quenched with saturated aqueous ammonium chloride solution. The phases are separated and the aqueous phase extracted three times with diethyl ether. The combined organic phases are washed at least five times with 10% hydrochloric acid, twice with water, dried over sodium sulfate and evaporated. The residue is purified by Chromatotron using hexane/ethyl acetate (6:1) as eluent. The third band is collected and concentrated giving 3.8 g (44%) of the title compound as a light-yellow oil.

The reaction of 2,2′-bithienyl-5-magnesium bromide with the aldehyde was a key step in the preparation of an intermediate in the synthesis of steroid-bridged anthryloligothienylporphyrins [150].

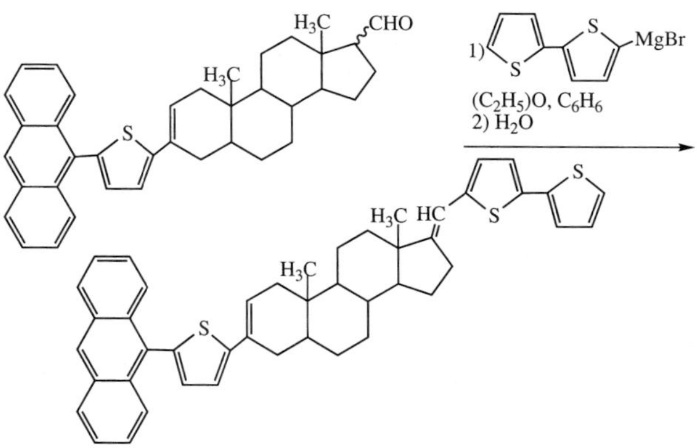

3-[5-(9-Anthryl)-2-thienyl)]-17-[2,2′-bithienyl-5-yl)methylidene]-5α-androst-2-ene [150]

A solution of 2,2′-bithienyl-5-magnesium bromide, prepared from magnesium (0.296 g, 12.19 mmol) in diethyl ether (5 ml) and a solution of 5-bromo-2,2′-bithienyl (1.87 g, 7.62 mmol) in diethyl ether/benzene (5:3), is added dropwise at room temperature within 30 min to a solution of 3-[5-(9-antryl)-2-thienyl]-17ξ-formyl-5α-androst-2-ene (4.15 g, 7.62 mmol) in diethyl ether/benzene (1:1). The reaction mixture is stirred for 2 h, hydrolyzed with ice water and extracted with dichloromethane. Then the aqueous phase is acidified and extracted with dichloromethane. Then the aqueous phase is acidified and extracted for the last time with dichloromethane. The combined organic phases are washed with water until neutral, dried over sodium sulfate and evaporated. The residue is chromatographed on silica gel using dichloromethane as eluent. The alcohols obtained are dissolved in dichloromethance/methanol (340: 160 ml). Concentrated hydrochloric acid (4 ml) is added and the reaction mixture stirred at room temperature for 4 h followed by workup as described above. Chromatography on silica gel using hexane/dichloromethane (2:1) as eluent and recrystallization from ethyl acetate gives 2.75 g (52%) of the title compound as fine-yellow crystals mp > 206 °C (sintering), 224 °C (dec.).

The Grignard reagent derived from 5-bromo-3′,4-dithexyl-2,2′-bithienyl is used for the preparation of 3,4-dihexyl-2,2′-bithienyl-5-boronic acid propanediyl ester and 3′,4-dihexyl-2,2′-bithienyl-5-carboxylic acid, which are key compounds in the solid phase synthesis up to dodecamers, by iodination and cross-coupling reactions. The acid is the anchoring group to the solid phase [151].

3′,4-Dihexyl-2,2′-bithienyl-5-carboxylic acid [151]

To a solution of 5-bromo-3′,4-dihexyl-2,2′-bithienyl (5.20 g, 12.5 mmol) in anhydrous tetrahydrofuran (100 ml), magnesium turnings (0.36 g, 15 mmol) are added. The mixture is heated under reflux until the magnesium has disappeared, after which it is cooled to 0 °C and excess of solid carbon dioxide is added. After reaching room temperature the mixture is hydrolyzed with 1 *M* hydrochloric acid. The phases are separated and organic phases washed several times with water, dried over sodium sulfate and evaporated. The residue is purified by chromatography on silica gel using dichloromethane/ethyl acetate as eluent, giving 4.2 g (88%) of the title compound as a yellow solid mp 62 °C.

8.3.5 Substituted bithienyls through metalation followed by reaction with electrophiles

Due to the α-selectivity in the metalation of 2,2′-bithienyl with alkylithium derivatives, many 5-substituted and 5,5′-disubstituted bithienyls are conveniently prepared by reaction of such lithium derivatives with suitable electrophiles. This method is used for the preparation of 5-carboxy-2,2′-bithienyl [109,124,127] and its benzyl ester [127], 5-fluoro-, 5-chloro-, 5-iodo-, 5-methyl-2,2′-bithienyl [122] and 5-trimethylstannyl-2,2′-bithienyl [87,153].

5-Trimethylstannyl-2,2′-bithienyl [153]

Butyllithium (1.6 *M* in hexane, 1.6 ml) is added to 2,2′-bithienyl in anhydrous tetrahydrofuran (10 ml) at 70 °C. The stirring is continued for 30 min, after which trimethyltin chloride (0.49 g, 2.47 mmol) in anhydrous tetrahydrofuran (15 ml) is added and the stirring is continued at 70 °C for 45 min and at ambient temperature overnight. The yellow solution is poured into water. The phases are separated and the aqueous phases extracted with diethyl ether. The combined organic phases are washed with water, dried over magnesium sulfate and evaporated giving 0.69 g (85%) of the title compound as a green oil.

Tetra-5-(2,2′-bithienyl)silane is obtained from 5-lithio-2,2′-bithienyl and tetramethoxysilane in 36% yield [154]. 5-Trimethylsilyl-2,2′-bithienyl [11], 2,2′-bithienyl-5-boronic acid [155], 5-acetyl-2,2′-bithienyl [156], and 5-(dimesitylboryl)-2,2′-bithienyl [157] are similarly prepared by reaction of the monolithium reagent with the appropriate electrophile. 2-[2-(5-(2,2′-bithienyl)) ethyl] 1,3-dioxolane is obtained through the reaction of the 5-lithium derivative with 2-(2-iodoethyl)-1,3-dioxolane [158].

2-[2-(5-(2,2′-bithienyl))ethyl]1,3-dioxolane [158]

Butyllithium (1.6 *M*, 45 ml) is added dropwise to a stirred solution of 2,2′-bithienyl (10 g, 0.06 mol) in anhydrous diethyl ether (150 ml) at − 78 °C under nitrogen. After completed addition, the cooling bath is removed and stirring is continued for 2 h. The mixture is again cooled to − 78 °C and 2-(2-iodoethyl)-1,1-dioxolane (17 g, 0.07 mol) in anhydrous diethyl ether (60 ml) is added dropwise. The reaction mixture is then refluxed for 12 h and quenched with saturated aqueous ammonium chloride solution. The phases are separated and the aqueous phase extracted with diethyl ether. The organic phases are dried over sodium sulfate and evaporated. The residue is purified by chromatography on silica gel using hexane/ethyl acetate (1:10) as eluent giving 6.1 g (40%) of the title compound mp 61.0–61.5 °C.

The reaction of 2,2′-bithienyl with butyllithium and cuprous iodide in tetrahydrofuran followed by diethyl (iodomethyl) phosphonate in dimethylsulfoxide gave the bithienylmethylphosphonate, which upon Wittig reaction with *para*-dibutylaminobenzaldehyde followed by reaction with butyllithium and tetracyanoethylene gave donor–acceptor substituted phenylethenyl bithienyls as highly efficient and stable non-linear optical chromophores [159].

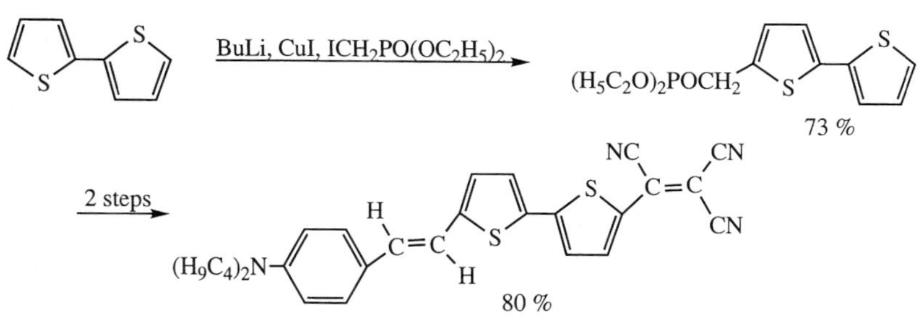

The reaction of 5-lithio-2,2′-bithienyl with dichlorobis(tributylphosphine) platinum(II) and dichlorobis (tributylphosphine) palladium(II) gives bithienyl derivatives of the two transition metals [160].

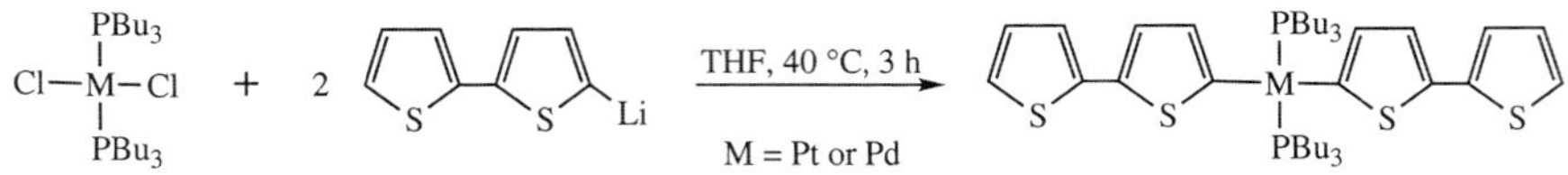

The reaction of four equivalents 5-lithio-2,2-bithienyl with tetramethoxysilane gives tetrakis(2,2′-bithienyl-5-yl)silane in 36% yield [161]. Reaction of 5-lithio,2,2′-bithienyl with 2,4,6-trichloro-1,3,5-triazine gives a star-shaped thiophene oligomer [156].

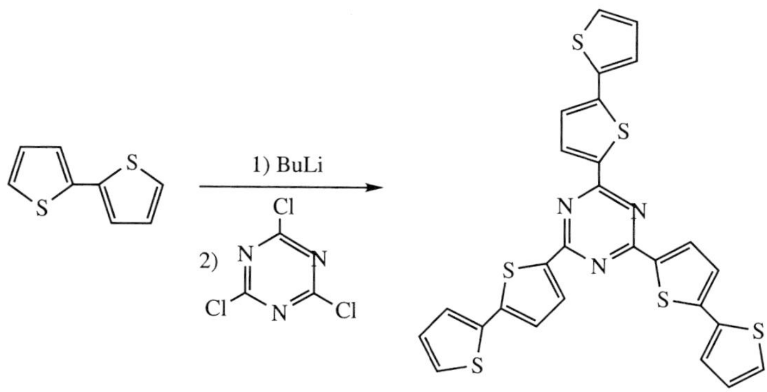

Metalation of 2,2′-bithienyl followed by reaction with 1,12-dibromododecane is used for the preparation of 5-(12-bromododecyl)-2,2′-bithienyl [162].

5-(12-Bromododecyl)2,2′-bithienyl [162]

To a solution of 2,2′-bithienyl (5.32 g, 20 mmol) and *N*,*N*,*N*′,*N*′-tetramethylethylenediamine (2.55 g, 22 mmol) in anhydrous tetrahydrofuran (100 ml) under argon 2.5 *M* butyllithium in hexane (8.8 ml, 22 mmol) is slowly added dropwise. The stirring is continued at room temperature for 1 h and then under inert conditions transferred to a dropping funnel and added to a solution of 1,12-dibromododecane (9.84 g, 30 mmol) in refluxing anhydrous tetrahydrofuran. The reaction mixture is refluxed for 8 h, hydrolyzed with a minimum amount of water and filtered through a pad of silica gel. The filtrate is diluted with water (100 ml) and the product is extracted with chloroform. The combined organic phases are dried over magnesium sulfate, evaporated and distilled giving 5.13 g (62%) of the title compound bp 154 °C/0.1 mm Hg.

Metalation of 5-hexyl-2,2′-bithienyl with butyllithium in tetrahydrofuran at – 30 °C, followed by reaction with sulfur and acetyl chloride at – 78 °C is used for the preaparation of 2-acetylthio-5-hexyl-2,2′-bithienyl [163]. This compound on reaction with sodium ethylate in ethanol gives the thiolate derivative, which upon reaction with hexafluorobenzene in dimethylimidazole gives the hexakis(5-hexyl-2,2′-bithienyl)benzene [163].

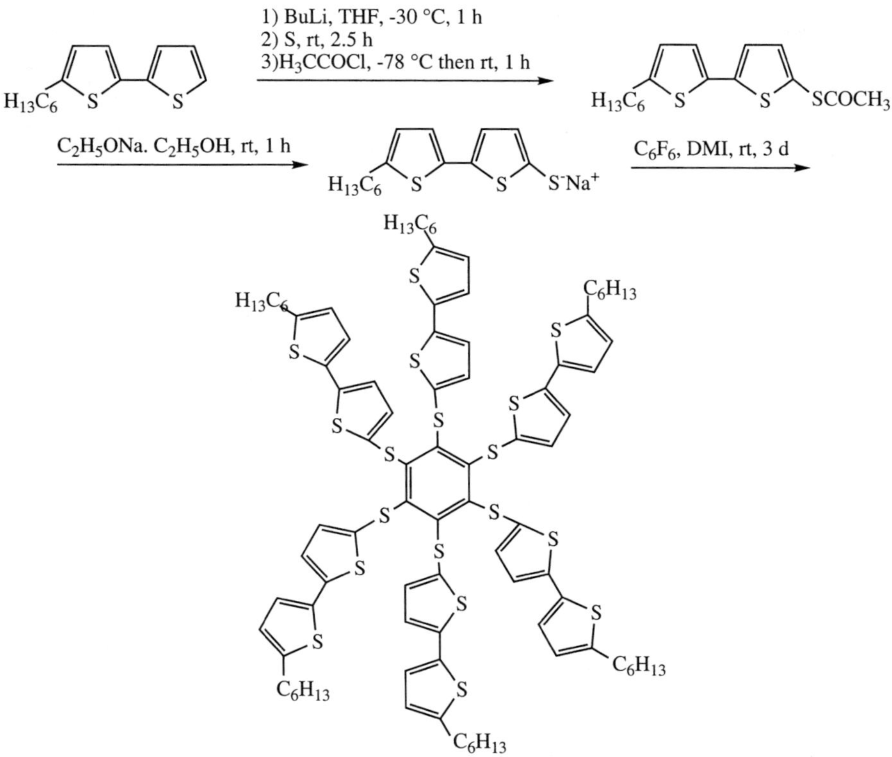

The metalation of 5-hexyl- and 5-octadecyl-2,2′-bithienyl with butyllithium followed by tributylstannyl chloride is used for the synthesis of the corresponding 5-alkyl-5′-tributylstannylbithienyls [106].

5-Octadecyl-5′-tributylstannyl-2,2′-bithienyl [106]

5-Octadecyl-2,2′-bithienyl (1.79 g, 4.3 mmol) is dissolved in anhdryous tetrahydrofuran (150 ml) and treated with butyllithium (1 equiv) at − 70 °C. After the solution is warmed to 0 °C and recooled to − 70 °C tributyltin chloride (1.4 g, 4.3 mmol) is added. The reaction mixture is warmed to room temperature and partitioned between diethyl ether and aqueous sodium chloride solution. The title compound is isolated in more than 90% yield.

5-(Dimesitylboryl)-5′-formyl-2,2′-bithienyl [157]

To a solution of the 5′-lithium derivative of 5-(dimesitylboryl)-2,2′-bithienyl, prepared from (5.7 g, 13.75 mmol) in anhydrous tetrahydrofuran (80 ml) at $-25\,^{\circ}C$ for 2 h, freshly distilled *N,N*-dimethylformamide (6.0 ml) is added at $-60\,^{\circ}C$. The reaction mixture is stirred overnight and then the solvent is evaporated. The residue is hydrolyzed and the product extracted with diethyl ether. The crude product is purified by chromatography on silica gel using pentane, which is progressively enriched with dichloromethane as eluent. The product so obtained is dissolved in a minimum amount of dichloromethane and then hot methanol is added. This solution is filtered and concentrated giving 1.7 g (28%) of the title compound as a yellow solid.

The reaction of 3-butyl-5-trimethylsilyl-2,2′-bithienyl with lithium diisopropylamide, followed by tributylstannyl chloride is used for the preparation of 3-butyl-5-tributylstannyl-5′-trimethylsilyl-2,2′-bithienyl [89]. Similarly metalation of 3′,4-dimethyl-5-trimethylsilyl-2,2′-bithienyl with lithium diisopropylamide followed by reaction with tributylstannyl chloride gives 3′,4-dimethyl-5-trimethylsilyl-5′-tributylstannyl-2,2′-bithienyl [89].

3′,4-Dimethyl-5-trimethylsilyl-5′-tributylstannyl-2-′-bithienyl [89]

To a solution of diisopropylamine (0.536 g, 5.3 mmol) in anhydrous tetrahydrofuran (4.0 ml) 2.12 *M* butyllithium (2.5 ml, 5.3 mmol) is added dropwise at $-78\,^{\circ}C$. The mixture is warmed to $0\,^{\circ}C$ for 10 min and recooled to $-78\,^{\circ}C$. 3′,4-Dimethyl-5-trimethylsilyl-2,2′-bithienyl (1.42 g, 5.3 mmol) in anhydrous tetrahydrofuran (4.0 ml) is added dropwise *via* cannula and the mixture is stirred at this temperature for 30 min. Tributyltin chloride (1.73 g, 5.3 mmol) is added dropwise, the reaction mixture is allowed to warm to room temperature for 2 h and poured into water. The phases are separated and the aqueous phase extracted with diethyl ether. The combined phases are washed with aqueous sodium chloride solution, dried over sodium sulfate and evaporated. The residue is filtered through a plug of silica gel washed with 10% triethylamine using hexane as eluent, giving 2.72 g (93%) of the title compound as a light-green liquid.

Trialkylstannyl derivatives of bithienyl and terthienyls are of great importance for the preparation of polythiophenes of technical interest. Thus 5-trimethylstannyl-2,2′-bithienyl [87,164] and 5′-trimethylstannyl-5-phenyl-2,2′-bithienyl can be prepared in excellent yield by metalation of 2,2′-bithienyl and 5-phenyl-2,2′-bithienyl with one equivalent of butyllithium at $-70\,^{\circ}C$ followed by trimethylstannyl chloride [165]. Similarly 5-tributylstannyl-2,2′-bithienyl [166], 2,5-tributylstannyl-3-butyl-5′-trimethylsilyl-2,2′-bithienyl [89], 5-octadecyl-5′-tributylstannyl-2,2′-bithienyl [167], 5-(dimethyl-*tert*-butylsilyl)-5′-(tributylstannyl)-2,2′-bithienyl [168], 5-(hexyl)-5′-(tributylstannyl)-2,2′-bithienyl [168], and

5-trimethylstannyl-3,3′-di(methylthio)-2,2′-bithienyl are prepared in high yields [169].

Lithiation of 3-alkyl-5′-trimethylsilyl-2,2′-bithienyl with lithium diisopropylamide followed by reaction with iodine is an excellent method for the preparation of 3-alkyl-5-iodo-5′-trimethylsilyl-2, 2′-bithienyls [89].

3-Methyl-5-iodo-5′-trimethylsilyl-2, 2′-bithienyl [89]

To a solution of diisopropylamine (0.401 g, 3.96 mmol) in anhydrous tetrahydrofuran 2.48 *M* butyllithium in hexane (1.60 ml, 3.96 mmol) is added dropwise at − 78 °C. The mixture is warmed to 0 °C for 5 min and recooled to − 78 °C. 3-Methyl-5′-trimethylsilyl-2,2′-bithienyl (1.00 g, 3.96 mmol) is added dropwise and the solution is stirred at − 78 °C for 1 h. Iodine (1.01 g, 3.96 mmol) in anhydrous tetrahydrofuran (5 ml) is added dropwise and the reaction mixture allowed to warm to room temperature for 30 min, after which it is poured into water. The phases are separated and the aqueous phase extracted with diethyl ether. The combined organic phases are washed with aqueous sodium chloride solution, dried over sodium sulfate and evaporated giving 100% of the title compound in pure form.

Metalation of 5-dodecyl-2,2′-bithienyl followed by iodine is used for the preparation of 5-dodecyl-5′-iodo-2,2′-bithienyl [118].

5-Dodecyl-5′-iodo-2, 2′-bithienyl [118]

To a solution of 5-dodecyl-2,2-bithienyl (1.88 g, 5.64 mmol) in anhydrous tetrahydrofuran (10.0 ml) at 0 °C 1.6 *M* butyllithium in hexane (3.5 ml, 5.64 mmol) is added. The mixture is stirred at room temperature for 20 min and then cooled to − 78 °C, after which iodine (1.43 g, 5.64 mmol) in anhydrous tetrahydrofuran (5.0 ml) is added dropwise. The reaction mixture is allowed to warm to room temperature and then poured into petroleum ether (100 ml). This solution is washed with dilute sodium chloride/sodium thiosulfate solution, dried over magnesium sulfate and evaporated. The residue is recrystallized from hexane giving 2.43 g (92%) of the title compound as white crystals.

In many cases dilithiation of 2,2′-bithienyls with alkyllithia has been used for the preparation of 5,5′-disubstituted 2,2′-bithienyls *via* reaction with appropriate electrophiles. This has been the method of choice for the preparation of 5,5′-dimethyl-, 5,5′-diisopropyl, 5,5′-di-*tert*-butyl- and 5,5′-di(trimethylsilyl)-2,2′-bithienyl [146], 5,5′-bis(trimethylsilyl)-2,2′-bithienyl-3,3′-d_2 [146], 5,5′-di-(triethyloxysilyl)-2,2′-bthienyl [170,171] 5,5′-bis(tributylstannyl)-2,2′-bithienyl [28,100,172,173], a number of 5,5′-dialkylthio-2,2′-bithienyls and 5,5′-bis-(phenylhydroxymethyl)-2,2′-bithienyl [174].

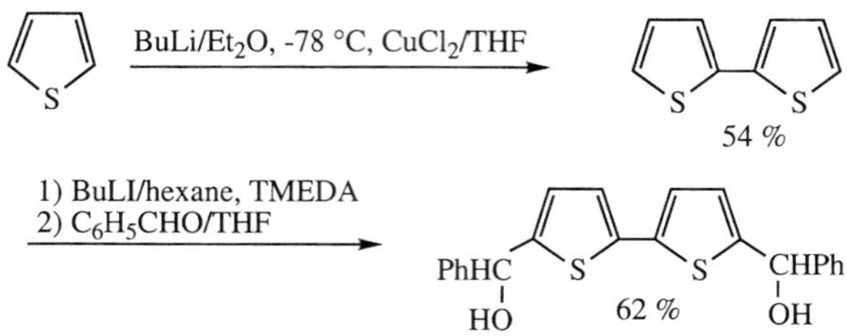

General procedure for 5,5′-di(alkylthio)-2,2′-bithienyls [175]

A dry flask under nitrogen is charged with 2,2′-bithienyl (1 g, 6 mmol) in anhydrous diethyl ether (10 ml). After cooling to 0 °C 1.6 *M* butyllithium in hexane (7.6 ml, 12 mmol) is added slowly and the stirring is continued for 1 h at 0 °C followed by addition of dialkyl disulfide. The reaction mixture is stirred for 10 min and allowed to warm to room temperature for 1 h. After addition of water the phases are separated and the aqueous phase extracted. The combined organic phases are washed, dried over sodium sulfate and evaporated. The residue is purified by chromatography on silica gel using petroleum ether as eluent giving 50–70% of the title compounds.

Reaction of 5,5′-dilithio-2,2′-bithienyl with bromine is used for the preparation of 5,5′-dibromo-2,2′-bithienyl [176]. Similarly the reaction of 5,5′-dilithio-2,2′-bithienyl with aroyl trimethylsilylacetylene gives the product shown below [177].

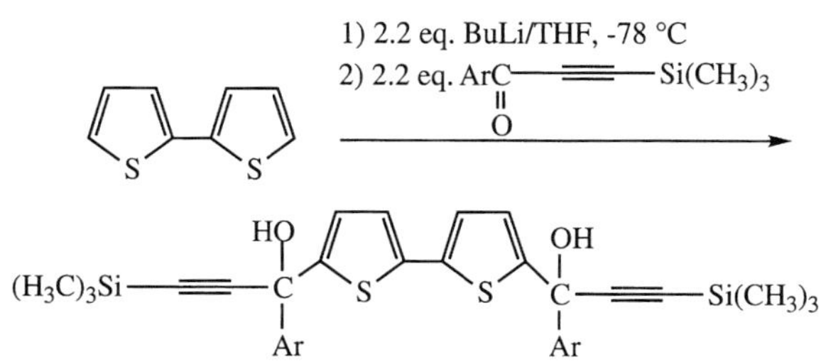

3,3′-Dimethoxy-2,2′-bithienyl, upon metalation with butyllithium followed by reactions with sulfur and 2-bromoethanol and then with 4-dimethylaminopyridine and acetyl chloride, gives 5,5′-bis(acetoxyethylthio)-2,2′-bithienyl, which gives persistent 3-alkoxy-2,5-bis(alkylthio)thiophene cation radicals [178].

Dimetalation of the bridged 2,2′-bithienyl with butyllithium and conversion of the dilithium derivative to the dizinc derivative, followed by palladium(0)-catalyzed coupling of the iodo derivative gives the product shown below, which was used in the synthesis of a dicyanoheterotriquinone methide [179].

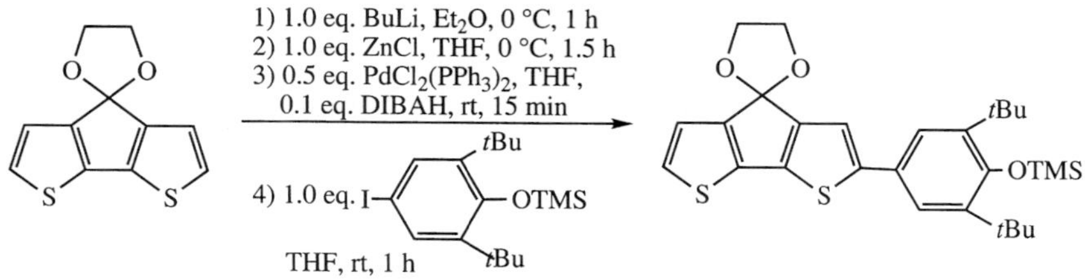

Metalation of 2,2′-bithienyl and 3,4-dihexyl-2,2′-bithienyl with butyllithium followed by reaction with copper(I) chloride and iodomethylphosphonate is a facile synthesis for the corresponding 2,2′-bithienylmethylphosphonates [180].

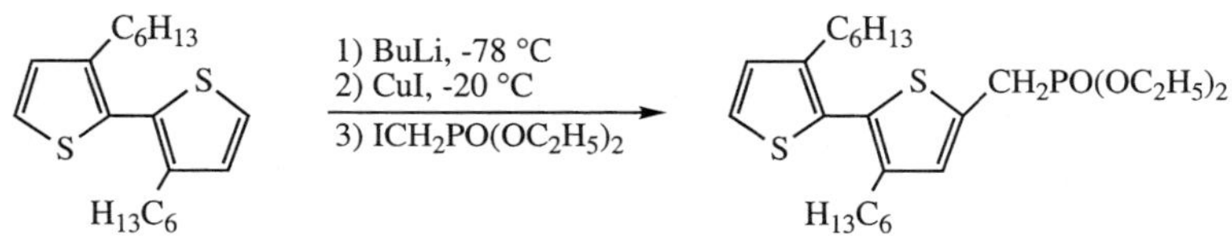

8.3.6 Substituted bithienyls through halogen–metal exchange of halobithienyls followed by electrophiles

2-Carboxy-3,3′-bithienyl is prepared from 2-bromo-3,3′-bithienyl upon halogen–metal exchange and reaction with carbon dioxide [135]. Halogen–metal exchange between especially β-substituted halobithienyls and alkyllithia, followed by electrophiles is an excellent method for the preparation of β-substituted bithienyls. 3,4′-Bis(methylthio)- and 3,4′-bis(butylthio)-2,2′-bithienyl are prepared by halogen–metal exchange of 3,4′-dibromo-2,2′-bithienyl with butyllithium at $-70\,^{\circ}$C in tetrahydrofuran followed by reaction with dimethyl disulfide and dibutyl disulfide, respectively [90].

3,4′-Bis(methylthio)-2,2′-bithienyl [90]

1.4 *M* butyllithium (4.8 ml, 6.8 mmol) is added to anhydrous tetrahydrofuran (5 ml) at $-70\,^{\circ}$C and a solution of 3,4′-dibromo-2,2′-bithienyl (1.00 g, 3.1 mmol) in anhydrous tetrahydrofuran (3 ml) is added dropwise over 5 min at $-75\,^{\circ}$C. The stirring is continued at this temperature for 40 min followed by addition of dimethyl disulfide (0.69 g, 7.4 mmol) in one portion. The temperature is allowed to rise to $-20\,^{\circ}$C and the stirring is continued at this temperature for 40 min before water (20 ml) is added. The product is extracted with diethyl ether (4×10 ml) and the combined organic phases are washed successively with 10% sodium hydroxide solution (15 ml) and water (15 ml) and dried over magnesium sulfate. The volatile materials are carefully removed under reduced pressure and the residue is purified by chromatography on silica gel using light petroleum as eluent, giving 0.58 g (47%) of the title compound.

Thus halogen–metal exchange of 4,4′-dibromo-3,3′-bithienyl with ethyllithium in tetrahydrofuran at −70 °C followed by reaction with *N,N*-dimethylformamide gives 4,4′-diformyl-3,3′-bithienyl [140].

4,4′-Diformyl-3,3′-bithienyl [140]

To a stirred solution of 4,4′-dibromo-3,3′-bithienyl (88 g, 0.27 mol) in tetrahydrofuran distilled over lithium aluminium hydride (1000 ml) 0.91 *M* ethyllithium in ether (650 ml) is added in a slow stream at −70 °C. After stirring for an additional 15 min the reaction mixture is allowed to attain room temperature during 2 h. Under cooling the mixture is hydrolyzed with 5 *M* hydrochloric acid. The phases are separated and the aqueous phase extracted several times with chloroform. The combined organic phases are washed with water, dried over calcium chloride and evaporated. The solid residue is recrystallized from ethyl acetate giving 35 g (58%) of the title compound mp 169.5–170 °C.

4,4-Diphenyl- and 4,4′-di-*para*-toluyl-2,6-bis(trimethylsilyl)dithieno[3,2-b:2′,3′-*d*]silole are prepared in 73% and 52% yield by the reaction of 3,3′-dibromo-5,5′-bis(trimethylsilyl)-2,2′-bithienyl with two equivalents of butyllithium followed by dichloridiphenylsilane and dichloro-di-*para*-tolysilane, respectively. Similarly 2,4,6-tris(trimethylsilyl)dithienosilyles were obtained in somewhat lower yields from the reaction of dilithiobis(trimethylsilyl) with the respective 1,1-difluorosilanes. The 1,1-dichlorosilanes gave in these cases unsatisfactory yields [131].

Cyclic compounds containing a disilyl functional group such as 2,7-bis(trimethylsilyl)-dithieno[3,2-*c*:2′,3′-*e*]disilacyclodiene are obtained by the reaction of 3,3′-dilithio-5,5′-bis(trimethysilyl)-2,2′-bithiophene with 1,2-dichlorodisilane [131].

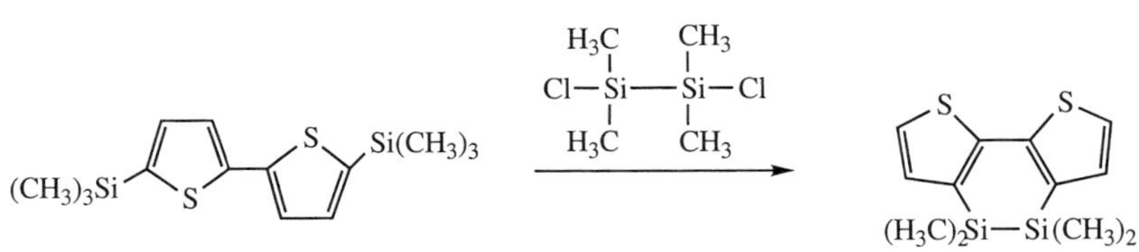

4,4′-Dicarboxy-3,3′-bithienyl was obtained with butyllithium in ether at −70 °C followed by carbon dioxide [14]. However, the reaction of 4,4′-dibromo-3,3′-bithienyl with one equivalent of butyllithium or ethyllithium at −70 °C led to metalation giving 4,4′-dibromo-5-carboxy-3,3′-bithienyl [136]. These complications are observed when free *α*-positions are present. Thus the reaction of 4,4′-dibromo-5,5′-dimethyl-3,3′-bithienyl with two equivalents of alkyllithium gave 4,4′-dicarboxy-3,3′-bithienyl as the minor product, the main product being 4-bromo-2,4′-dicarboxy-5,5′dimethyl-3,3′-bithienyl [136]. From

2,2′-dibromo-3,3′-bithienyl, 2,2′-dicarboxy-3,3′-bithienyl is prepared in 88% yield [136]. A very interesting observation was made when halogen–metal exchange was studied with optically active 2,2′,4,4′-tetrabromo-3,3′-bithienyl or 4,4′-dibromo-2,2′,5,5′-tetramethyl-3,3′-bithienyl. When these compounds were treated with two equivalents of alkyllithium at −70 °C for less than two minutes followed by carbonation, completely racemic dicarboxylic acids were obtained. However, optically active monocarboxylic acids could be isolated using one equivalent of alkyllithium, implying an achiral conformation in the intermediate dilithium compound [138,181].

The reaction of 2,2′-dibromo-4,4′-dichloro-3,3′-bithienyl with butyllithium at −70 °C, followed by hexachloroethane is used for the preparation of 2,2′,4,4′-tetrachloro-3,3′-bithienyl [17]. The selective halogen–metal exchange of α-halogens over β-halogens is utilized in the preparation of 4,4′-dibromo-2,2′-diformyl-3,3′-bithienyl [45,182] and 4,4′-dibromo-2,2′-dicarboxy-3,3′-bithienyl [183,184] from 2,2′4,4′-tetrabromo-2,2′-bithienyl and 2,2′,4,4′-tetrabromo-5,5′-dicarboxy-2,2′-bithienyl from hexabromo-3,3′-bithienyl [16] utilizing the appropriate electrophiles. The dimethyl acetal from 2,2′-diformyl-4,4′-dibromo-3,3′-bithienyl is used in the preparation of 2,2′-diformyl-4,4′-dicarboxy-3,3′-bithienyl [45].

4,4′-Dicarboxy-2,2′-diformyl-2,2′-bithienyl [45]

4,4′-Dibromo-2,2′-bis(dimethoxymethyl)-3,3′-bithienyl (39.0 g, 82 mmol), obtained as crystals after dissolving 4,4′-dibromo-2,2′-diformyl-3,3′-bithienyl in acidified hot methanol and subsequent cooling of the solution, is suspended in anhydrous diethyl ether (250 ml) and the mixture is cooled to − 70 °C under nitrogen. Ethyllithium 0.9 *M* in diethyl ether (270 ml) is added and the mixture is stirred. After 2 h a homogeneous solution is obtained. This solution is poured into crushed carbon dioxide ice covered with anhydrous diethyl ether and at − 10 °C it is hydrolyzed. The phases are separated and the aqueous phase is acidified with 5 *M* hydrochloric acid and heated in order to hydrolyze the acetal. Water is added to make a volume about 1000 ml to dissolve the precipitated crystals. After filtration and cooling to room temperature 23.1 g (91%) of the title compound is obtained as yellow crystals mp 225–229 °C.

Halogen–metal exchange of 3,3′,5,5′-, 4,4′,5,5′-tetrabromo-2,2′-bithienyl and hexabromo-2,2′-bithienyl with butyllithium in ether at − 78 °C followed by *N*,*N*-dimethylformamide gave the corresponding brominated 5,5′-diformyl-2,2′-bithienyls in 65–72% yield [13]. Halogen–metal exchange of 3,3′,5,5′-tetrabromo-2,2′-bithienyl with two equivalents of butyllithium in hexane/diethyl ether followed by reaction with trimethylsilyl chloride gives 3,3′-dibromo-5,5′-bis(trimethylsilyl)-2,2′-bithienyl in 72% yield [131].

3,3′-dibromo-5,5′-dibenzoyl-2,2′-bithienyl and 3,3′-dibromo-5,5′-diacetyl-2,2-bithienyl are prepared by halogen–metal exchange of 3,3′5,5′-tetrabromo-2,2′-bithienyl with butyllithium in tetrahydrofuran at − 70 °C, followed by reaction with benzonitrile and *N,N*-dimethylacetamide, respectively [185].

5,5′-Dibenzoyl-3,3′-dibromo-2,2′-bithienyl [185]

To a solution of 3,3′,5,5′-tetrabromo-2,2′-bithienyl (2.41 g, 5 mmol) in anhydrous tetrahydrofuran (150 ml) 2.5 *M* butyllithium in hexane (4.4 ml, 11 mmol) is added dropwise at − 78 °C during 30 min. The mixture is stirred at the same temperature for 1 h and anhydrous benzonitrile (1.1 ml, 11 mmol) is added in one portion after which the stirring is continued at the same temperature for 2 h. At room temperature the mixture is slowly added to ice-water (300 ml) acidified with 10 *M* hydrochloric acid (50 ml) under vigorous stirring for 30 min followed by neutralization with potassium carbonate and dilution with chloroform (200 ml). The phases are separated and the aqueous phase extracted with chloroform (100 ml). The combined organic phases are washed with water, dried over magnesium sulfate and evaporated. The residue is purified by chromatography using chloroform as eluent and recrystallized from toluene giving 1.38 g (65%) of the title compound as yellow needles mp 189–191 °C.

Interesting differences between ethyllithium and butyllithium were observed in the reaction of various dibromo-3,3′-bithienyls with one equivalent of the alkyllithium followed by carbon dioxide. 2,2′-Dibromo-3,3′-bithienyl gave mainly 2-bromo-2′-carboxy-3,3′-bithienyl while ethyllithium gave 60% of 2,2′-dicarboxy-3,3′-bithienyl and 40% of the monocarboxylic acid [136]. Halogen–metal exchange of 3,3′-dibromo-4,4′,5,5′-tetramethyl-2,2′-bithienyl with butyllithium at − 70 °C followed by carbon dioxide is used for the preparation of 3,3′-dicarboxy-4,4′,5,5′-tetramethyl-2,2′-bithienyl in 69% yield [4].

Recently 4,4′-bis(diphenylphosphinyl)-2,2′5,5′-tetramethyl-3,3′-bithienyl was prepared from 4,4′dilithio-2,2′,5,5′-tetramethylthiophene and diphenylphosphinous chloride and resolved by fractional crystallization of the diasteromeric adducts with (+)- and (−)-dibenzoyltartaric acids. The complexes with ruthenium(II) and ruthenium(I) were used as catalysts in homogeneous stereoselective reactions [186].

(+/−)4,4′-Bis(diphenylphosphinyl)-2,2′5,5′-tetramethyl-3,3′-bithienyl [186]

To a solution of 4,4′-dibromo-2,2′5,5′-tetramethyl-3,3′-bithienyl (1.15 g, 3.0 mmol) in anhydrous tetrahydrofuran (30 ml) at − 60 °C under nitrogen 1.6 *M* butyllithium in hexane (3.8 ml, 6.08 mmol) is added dropwise. After stirring for 10 min diphenylphosphinous chloride (1.1 ml, 6.13 mmol) is added. The stirring is continued for 1 h and the temperature is allowed to rise to room temperature. The reaction mixture is concentrated under reduced pressure and the residue treated with water and extracted with dichloromethane. The combined organic phases are treated with 35% hydrogen peroxide (10 ml) at 0 °C. The mixture is stirred for 1 h at 0 °C and for 1 h at room temperature and then diluted with water. The organic phase is separated, dried over sodium sulfate and evaporated. The residue is purified by chromatography on silica gel using ethyl acetate/dichloromethane/triethylamine (3:7:0.1), giving 1.83 g (98%) of the title compound mp 140 °C.

Resolution. A mixture of the above compound (12.5 g) and (−)-2,3-*O*, *O*-dibenzyoltartaric acid (7.6 g) is dissolved in tetrahydrofuran, refluxed for a few min and allowed to stand at room temperature for 24 h. The adduct of the (+)-form is collected and the filtrate stored for recovery of the (−)-form. The adduct is dissolved in warm tetrahydrofuran (450 ml) and 8.76 g of the pure adduct is collected mp 159–172 °C (dec.), $[\alpha]_D^{25} = -36.9$ ($c = 0.49$, benzene). The adduct is treated with 0.75 *M* sodium hydroxide solution and the mixture is extracted exhaustively with dichloromethane. The combined organic phases are dried over sodium sulfate and evaporated and the residue is recrystallized from hexane/benzene (1:1) giving 2.1 g of the (+)-form mp 236 °C, $[\alpha]_D^{25} = +62$ ($c = 0.49$, benzene). The mother liquors from the first resolution step are concentrated to dryness giving a solid. This solid is treated with a 0.75 *M* sodium hydroxide solution and dichloromethane is extracted. The combined organic phases are washed with water, dried over sodium sulfate and evaporated and the residue recrystallized from hexane/benzene (1:1) giving 1.2 g of the (−)-form mp 236 °C $[\alpha]_D^{25} = +62$ ($c = 0.49$, benzene).

8.3.7 Substituted bithienyls through various modifications of substituents in bithienyls

8.3.7.1 From halo substituted bithienyls

2-Perfluorohexyl-2,2′-bithienyl is prepared by treatment of 5-bromo-2,2′-bithienyl with copper bronze and perfluorohexyl iodide in dimethylsulfoxide and transformed to 5-bromo-5′-perfluorohexyl-2,2′-bithienyl by bromination with *N*-bromosuccinimide in *N*, *N*-dimethylformamide at room temperature [187].

2-Perfluorohexyl-2,2′-bithienyl [187]

A suspension of copper bronze (3.38 g, 53.16 mmol) in anhydrous dimethylsulfoxide (30 ml) is heated at 125 °C for 15 min under nitrogen. Perfluorohexyl iodide (11.34 g, 25.43 mmol) is added at such a rate that the temperature is maintained under 135 °C. After 45 min a solution of 5-bromo-2,2′-bithienyl

(2.50 g, 10.17 mmol) in anhydrous dimethylsulfoxide (7.0 ml) is added dropwise. The reaction mixture is stirred at 125 °C for 16 h, after which it is quenched with cold water (100 ml) and diethyl ether (200 ml) is added. The mixture is shaken and filtered through Celite. The phases are separated and the aqueous phase extracted with diethylether (100 ml). The residue is purified by chromatography on silica gel using hexane as eluent, giving 2.90 g (59%) of the title compound as a yellow solid.

Kumada coupling of 5-iodo-2,2′-bithienyl with 4-*tert*-butylbenzenemagnesium bromide gives 5-(4-*tert*-butylphenyl)-2,2′-bithienyl [188].

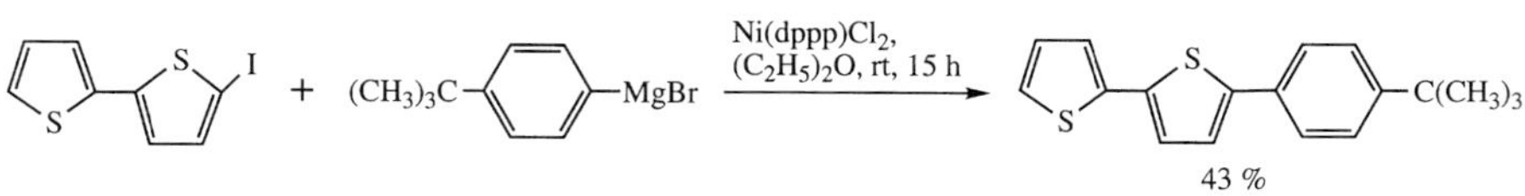

3,3′-Dialkyl-2,2′-bithienyls were prepared in 90–95% yield through the reaction of 3,3′-dibromo-2,2′-bithienyl with alkylmagnesium bromide using [1,3-bis(diphenylphosphine) propane] nickel(II) chloride as catalyst [112].

3,3′-Dialkyl-2,2′-bithienyl [112]

1-Bromoalkane (58.1 mmol) in anhydrous diethyl ether 65 ml) is reacted with magnesium (1.70 g, 69.9 mmol) activated with iodine in anhydrous atmosphere at room temperature for 1 h and cooled. The resulting Grignard reagent is then added to an ice cooled solution of 3,3′-dibromo-2,2′-bithienyl (6.30 g, 19.4 mmol) in anhydrous diethyl ether (55 ml) containing [1,1-bis(disphenylphosphine)propane]nickel(II) chloride (0.11 g, 0.20 mmol). The reaction mixture is stirred at room temperature overnight, cooled in ice and treated with an aqueous saturated ammonium chloride solution. The phases are separated and the organic phase dried over magnesium sulfate and evaporated. The residue is purified by chromatography on silica gel using hexane as eluent giving 90–95% of the title compounds as yellow liquids or semisolids.

3,4′-Dibutoxy-, 3,3′-dibutoxy- and 4,4′-dibutoxy-2,2′-bithienyl are obtained in high yields from the 4,4′-dibromo derivatives by reaction with sodium butylate in butanol using cupric oxide and potassium iodide as catalysts [189].

3,4′-Dibutoxy-2,2′-bithienyl [189]

To a solution of sodium (0.40 g, 17.40 mmol) in butanol (25 ml), copper(II) oxide (0.25 g, 3.14 mmol) and potassium iodide (0.04 g, 0.24 mmol) are added followed by addition of 3,4′-dibromo-2,2′-bithienyl (1.00 g, 3.09 mmol). The reaction mixture is stirred at 100 °C for 3 days and more potassium iodide (0.04 g, 0.24 mmol) is added. Stirring is continued for 2 days at the same temperature and the mixture is filtered. The butanol solution is poured into water and the product is extracted with diethyl ether. The combined organic phases are washed with water, dried over magnesium sulfate and evaporated. The residue is purified by chromatography on silica gel using petroleum ether/chloroform (4:1) as eluent, giving 0.86 g (90%) of the title compound.

In connection with work on the synthesis of photochromic 3H-napthto-[2,1-*b*]pyrans the Sonogashira reaction was used under phase transfer conditions with 5-bromo-2,2′-bithienyl. When 5,5′-diiodo-2,2′-bithienyl was used in the same way the disubstituted bithienyl was prepared [190].

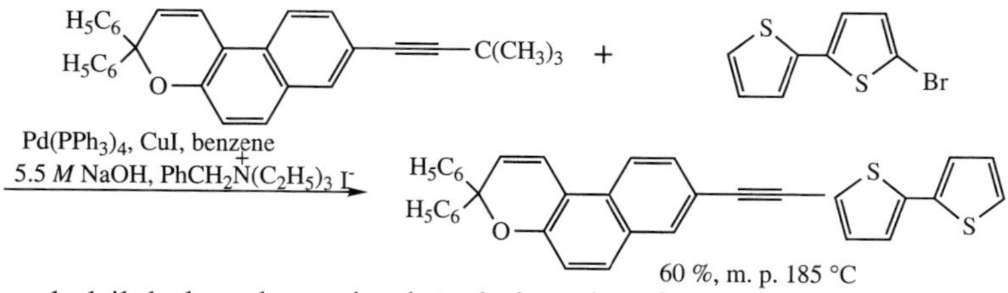

Trimethylsilylethynyl- and phenylethynyl substituted 2,2′-bithienyls are prepared by palladium-catalyzed Heck coupling of 3,3′-, 4,4′ and 5,5′-dibromo-2,2′-bithienyls [13, 185].

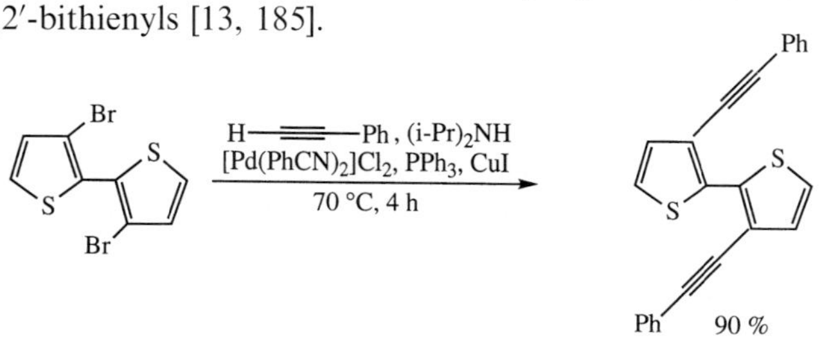

Heck coupling of 5-bromo-5′-hydroxymethyl- and 5-bromo-5′-formyl-2,2′-bithienyl with propyne and copper(I) iodide using palladium(0)-catalyst and triethylbenzylammonium chloride as base gives 5-bromo-5′-propynyl derivatives [191].

Coupling of 3,3′-dibromo- and 4,4′-dibromo-5,5′-diformyl-2,2′-bithienyl with trimethylsilylacetylene can be used for the preparation of 3,3′-bis[(trimethylsily)ethynyl]-5,5′-diformyl-2,2′-bithienyl and carbonate related compounds, which could easily be desilylated upon treatment with potassium in methanol [13].

Chloro substituents, which are used as protecting and directing groups in bithienyl chemistry, can be removed by hydrogenation over palladium on carbon or by reaction with tributyl tinhydride and azo-bisisobutyronitrile [85].

General procedure for the dechlorination reactions [85]

A mixture containing the chlorinated oligomer (0.15–0.30 mmol), excess tributyltin hydride and azo-bisisobutyronitrile (5 mg) in anhydrous tetrahydrofuran

(2 ml) is refluxed overnight before being poured into water. The aqueous phase is extracted with hexane and the combined organic phases are washed with aqueous saturated sodium chloride solution, dried over sodium sulfate and evaporated. The residue is purified by semipreparative HPLC on reverse phase using chloroform/acetonitrile mixtures as eluents.

Suzuki coupling of 3-bromo-2,2′-bithienyl with 3,6-dialkoxyphenylboronic acid is used for the preparation of 3-[2″,5″-(bis (1‴,4‴,7‴-trioxaoctyl)phenyl]-2,2′-bithienyl [192].

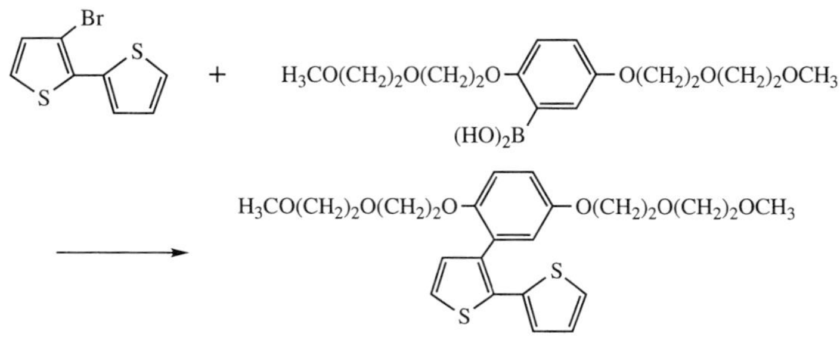

3-[2″,5″-(Bis (1‴,4‴,7‴-trioxaoctyl)phenyl]-2,2′-bithienyl [192]

To a solution of 3-bromo-2,2′-bithienyl (0.93 g, 3.4 mmol) in 1,2-dimethoxyethane (20 ml) tetrakis(triphenylphosphine)oalladium(0) (0.2 g, 0.17 mmol) is added and the mixture is stirred for 10 min. A solution of 2,5-di-trioxaoctylphenylboronic acid (1.5 g) in 1,2-dimethoxyethane (10 ml) is added followed by 1 *M* sodium bicarbonate solution (15 ml). The reaction mixture is heated to reflux overnight, cooled to room temperature, filtered and evaporated. Water is added to the residue and the so obtained solution is extracted with diethyl ether. The combined organic phases are washed with water, dried over sodium sulfate and evaporated. The residue is purified by chromatography on silica gel using diethyl ether as eluent, giving 0.8 g (50%) of the title compound.

5-Bromo-5′-phenyl-2,2′-bithienyl was prepared by Suzuki coupling of 5,5′-dibromobithienyl with phenylboronic acid [193].

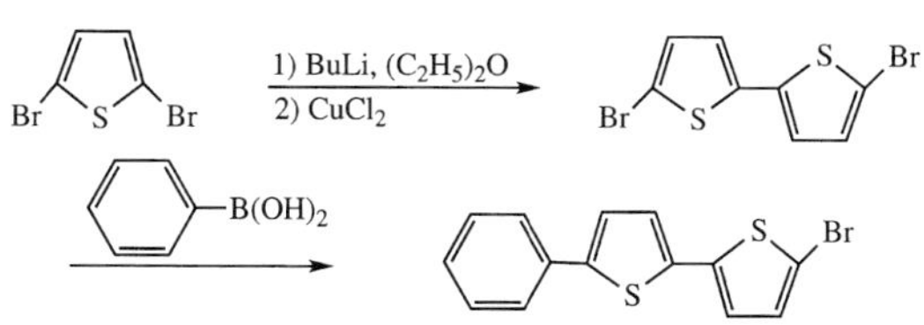

5,5-Bis(para-tolyl)-2,2′-bithienyl [67]

2,5-Diiodothiophene (418 mg, 1.00 mmol), *para*-tolylboronic acid (544 mg, 4.00 mmol) and tetrakis(triphenylphosphine)palladium(0) (139 mg, 0.12 mmol) are dissolved in benzene (80 ml) and this reaction solution is purged with nitrogen for 30 min to remove dissolved oxygen. An aqueous solution of sodium carbonate (5 ml, 848 mg, 8.00 mmol) is added to the reaction solution. The reaction mixture is stirred under nitrogen at room temperature for two days and nights. Under cooling with an ice-water bath, 30% aqueous hydrogen peroxide (1 ml) is added and the precipitate formed is collected by filtration. It is then washed with methanol followed by Soxhlet extraction with dichloromethane. The extracted material is collected by filtration and recrystallized from chloroform giving 100 mg (38%) of the title compound as a bright-yellow solid mp 252 °C.

Suzuki coupling of 5,5′-diiodo-2,2′-bithienyl with 4-biphenylylboronic acid gives 5,5′-bis(4-biphenyl)-2,2′-bithienyl in 57% yield [194]. Among a series of 2,5-diaryl-3,4-diphenylsilinoles of interest as new materials for organic electroluminiscent devices, the following compound was prepared by palladium catalyzed coupling of 5-bromo-2,2′-bithienyl with 3,4-diphenyl-2,5-dilithio-1,1-dimethylsilole [195].

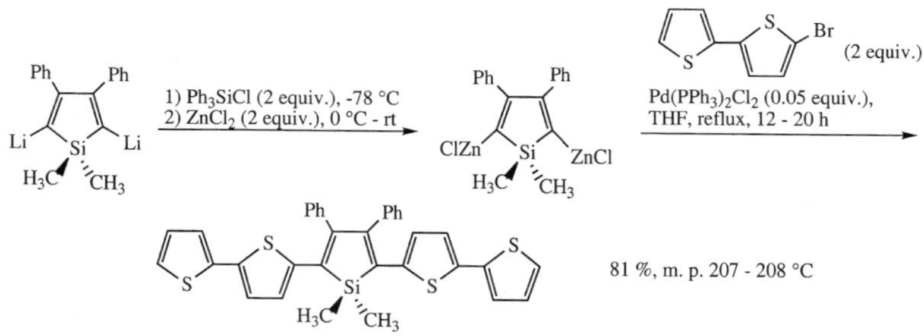

A chiral oligothiophene, 6,6′-bis(5-phenyl-2,2′-bithiophen-5′-yl)-bis(decyloxy)-1,1′-binaphthyl) possessing in-chain chirality is prepared from (*R*), (*S*) and racemic dibromo-2,2′-bis(decyloxy)-1,1′-binaphthyl by Kumada coupling with 5-phenyl-2,2′-bithienyl-5′-zinc chloride [193].

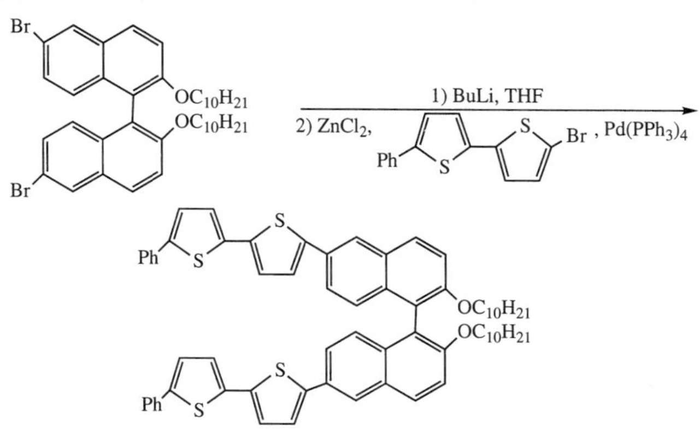

1,4-Di(thienyl)diacetylenes using an organoborane-mediated approach are useful for the preparation of unsymmetrically substituted diacetylenes. The lithium salt of each acetylenic reagent is introduced successively onto a boron compound to form a complex, which is coupled to the diacetylene by iodine oxidation. In this way, 1-(2-thienyl)-4-(2,2′-bithienyl-5)-diacetylene is prepared [196].

1-(2-Thienyl)-4-(2,2′-bithienyl-5-yl)-1,3-butadiyne [196]

9-Borabicyclo[3.3.1]nonane (0.244 g, 2 mmol) and anhydrous freshly distilled tetrahydrofuran (5 ml) are placed in a dry three-necked 250 ml flask fitted with a gas-inlet tube and a septum inlet. Under nitrogen and with magnetic stirring, dry methanol (0.064 g, 2 mmol) is added dropwise at room temperature and the solution is stirred for an additional hour. To a separate 100 ml flask containing 5-ethynyl-2,2′-bithienyl (0.380 g, 2 mmol) in anhydrous tetrahydrofuran (10 ml) kept at − 78 °C butyllithium (0.128 g, 2 mmol) is added dropwise and stirring at this temperature is continued for 1 h. The first flask is then cooled to −78 °C for 30 min and freshly distilled boron trifluoride etherate (0.376 g, 3.3 mmol) is added with a syringe. The mixture is stirred at − 78 °C for 20 min and then warmed to room temperature. The flask is again cooled to − 78 °C and 2-(lithioethynyl)thiophene, prepared by reacting butyllithium (0.128 g, 2 mmol) with 2-ethynylthiophene (0.216 g, 2 mmol) in tetrahydrofuran (10 ml) is added. The mixture is stirred at − 78 °C for 30 min, iodine (0.508 g, 2 mmol) dissolved in tetrahydrofuran (5 ml) is added dropwise and stirring is continued for 30 min at − 78 °C. After warming to room temperature, the solution is washed twice with 3 *M* sodium hydroxide solution (2 × 0.6 ml), 3 *M* sodium hydroxide solution (0.8 ml) is added followed by 30% hydrogen peroxide (0.8 ml) at such a rate that the temperature remains below 50 °C. After saturation with potassium carbonate the phases are separated and the aqueous phase extracted with dichloromethane (2 × 10 ml). The combined organic phases are dried over anhydrous potassium carbonate and concentrated. The residue is recrystallized from tetrahydrofuran/water giving 0.410 g (69%) of the title compound mp 120–121 °C.

8.3.7.2 From alkylbithienyls

Side chain bromination of methylated bithienyls is used for the preparation of bromomethylbithienyls. From 4,4′-dimethyl-3,3′-bithienyl 4,4′-bis(bromomethyl)-3,3′-bithienyl was obtained [60,197] and from 3,3′-dimethyl-2,2′-bithienyl, 3,3′-bis(bromomethyl)-2,2′-bithienyl was obtained using *N*-bromosuccinimide in carbon tetrachloride and azo-bisisobutyronitrile as catalyst [6].

4,4′-Bis(bromomethyl)-3,3′-bithienyl [197]

From a suspension of *N*-bromosuccinimide (48.0 g, 0.27 mol) in carbon tetrachloride (700 ml) 50 ml of solvent is distilled off. To the refluxing suspension azo-bisisobutyronitrile (.6 g) is added and after 2 min followed by addition of 4,4′-dimethyl-3,3′-bithienyl (26.1 g, 0.134 mol) in one portion. After a few minutes a vigorous reaction started, which subsided after 10 min. The reaction mixture is refluxed for an additional 30 min and cooled. The succinimide is filtered off and the filtrate evaporated. The residue, a brown oil, crystallizes upon standing, after which hexane (250 ml) is added and the mixture stirred at room temperature for 2 h and filtered giving 25 g (55%) of the title compound mp 61.0–62.0 °C after recrystallization from hexane.

The previous method for the preparation of 5,5′-dibromomethyl-2,2′-bithienyl by side chain bromination of 5,5′-dimethyl-2,2′-bithienyl was modified [198].

5,5′-Bis(bromomethyl)-2,2′-bithienyl [198]

A suspension of *N*-bromosuccinimide (2.77 g, 15.6 mmol) in carbon tetrachloride (60 ml) is refluxed for 4 min. The heating is stopped and azo-bis-isobutyronitrile (45 mg) is added and after 3 min a solution of 5,5′-dimethyl-2,2′-bithienyl (1.39 g, 7.8 mmol) in carbon tetrachloride (20 ml) is added. The reaction mixture is gently heated to reflux and when an exothermic reaction begins the external heating is removed. The stirring is continued for 20 min. After 10 min the reaction mixture is gently heated. After cooling to 40 °C the reaction mixture is filtered, and the filtrate cooled further. The product collected by filtration is washed with cold diethyl ether giving 1.69 g (59%) of the title compound as flat almost transparent crystals, which decompose within hours.

In some cases, it is necessary to prepare the Grignard reagent through the reaction of the lithium derivative with magnesium bromide as is the case in the preparation of 3-methylthio-2-thiophenemegnesium bromide. Upon [1,3-bis(diphenylphosphine)propane]nickel (II) chloride catalysed coupling with 4-methylthio-2-bromothiophene 3,4′-bis(methylthio)-2,2′-bithienyl is obtained [68].

This was also the case in the synthesis of highly congested bithienyls, such as 3,4,3′,4′-tetra-*tert*-butyl-2,2′-bithienyl, which was obtained in 33% yield from 3,4-di-*tert*-butyl-2-bromothiophene and its Grignard reagent prepared by halogen–metal exchange with butyllithium followed by magnesium bromide [69].

8.3.7.3 From hydroxymethylbithienyls

Reaction of hydroxymethylbithienyls with phosphorus tribromide is extensively used for the preparation of bromomethyl derivatives [182–184]. The hydroxymethyl carboxy-3,3′-bithienyls were oxidized to the dicarboxylic acids with potassium permanganate and reduced to the dihydroxymethyl derivatives [182]. 4,6-Dihydrodithieno[2,3-*c*:3′,2′-*e*]oxepin [199] and 1,9-dibromo-4,6-dihydro-2,8-dimethyldithieno[2,3-*c*:3′,2′-*e*]oxepin [23] are prepared from 2,2′-bis(hydroxymethyl)-3,3′-bithienyl and 4,4′-dibromo-2,2′-bis(hydroxymethyl)-5,5′-dimethyl-3,3′-bithienyl by heating with *para*-toluenesulfonic acid in toluene under water separation.

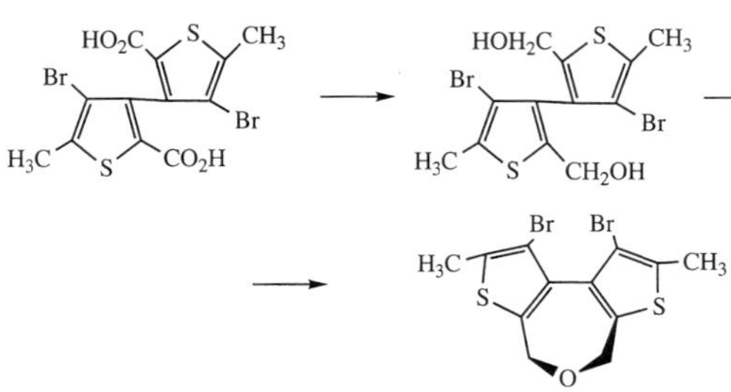

1,9-Dibromo-4,6-dihydro-2,8-dimethyldithieno[2,3-c:3′,2′-e]oxepin [23]

A mixture of 4,4′-dibromo-2,2′-bis(hydroxymethyl)-5,5′-dimethyl-3,3′-bithienyl (0.50 g, 1.2 mmol) and *para*-toluenesulfonic acid (10–20 mg) in toluene (250 ml) is refluxed for 30 min in a 500 ml bottle provided with a water separator. The cooled reaction mixture is washed with aqueous sodium bicarbonate solution and water followed by evaporation giving 0.43 g (90%) of the title compound mp 169–171 °C after recrystallization from ligroin.

Core modified mesoaryl sapphyrins are prepared from 5,5′-bis(phenylhydroxymethyl)-2,2′-bithienyl through reaction with 5,10-diphenyl-16-thiatripyrane in methylene chloride/tetrahydrofuran followed by chloranil [200].

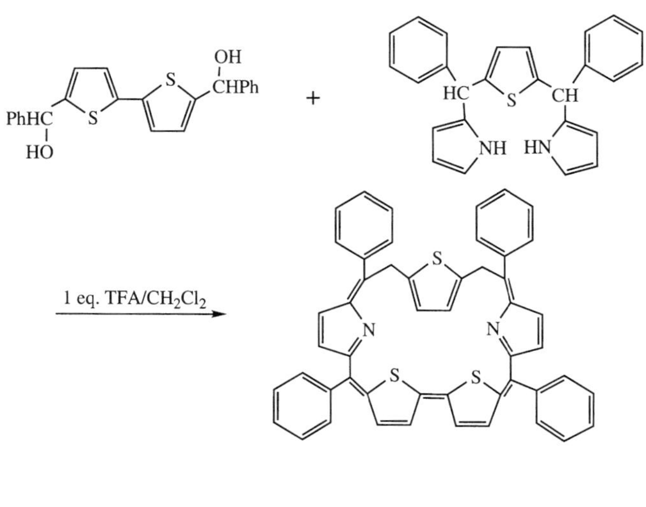

5,10,15,20-Tetraphenyl-25,27,29-trithiasapphyrin [200]

A mixture of 5,5′-bis(phenylhydroxymethyl)-2,2′-bithienyl (150 mg, 0.397 mmol) and 5,10-diphenyl-16-thiatripyrane (156 mg, 0.397 mmol) in anhydrous dichloromethane (150 ml) is stirred under nitrogen for 15 min at room temperature. Trifluoroacetic acid (0.03 ml, 0.0397 mmol) is added and the stirring is continued for 1 h under dark conditions. The solution is opened to air and chloranil (293 mg, 1,191 mmol) is added; the reaction mixture is heated to reflux in a preheated oil bath at 50 °C for 1 h. After removal of the solvent the residue is purified by chromatography on basic alumina using dichloromethane/ethyl acetate (95:5), giving 196 mg (36%) of the title compound as a green lustrous solid.

3,4′- and 3,3′-Di(hydroxyethyl)-2,2′-bithienyl were converted to the corresponding sodium sulfonate salts in the conventional way, by reaction with methanesulfonyl chloride, sodium iodide and sodium sulfite, indicating that polyhydroxy oligothiophenes are useful precursors for the preparation of self-doping, water-soluble oligothiophenes [92].

The bridged thiepin could be ring closed to the doubly bridged compound, shown below, by reaction with *para*-toluenesulfonic acid [184].

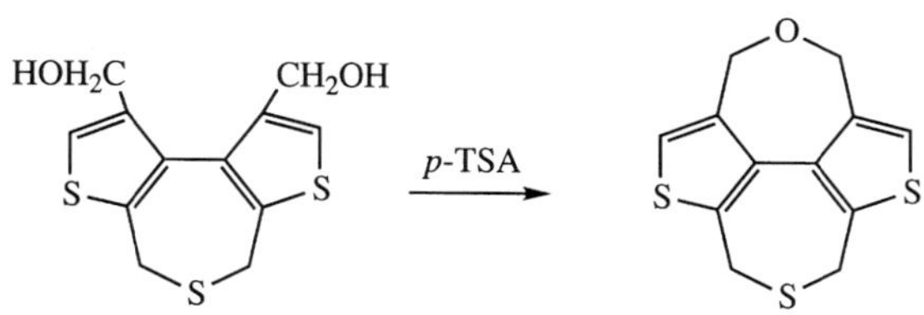

8.3.7.4 From halomethyl derivatives

Reduction of such compounds with sodium borohydride is used for the preparation of the dimethyl derivatives [182]. In order to obtain 3,3′-bithienyls with locked conformation 2,2′-and 4,4′-bridged derivatives were prepared from the bromomethyl derivatives. Thus the reaction of 2,2′-(dibromomethyl)-3,3′-bithienyl and 4,4′-dibromo-2,2′-(dibromomethyl)-3,3′-bithienyl with sodium sufide was used for the preparation of the thiepin derivatives 4,6-dihydrodithieno[2,3-*c*:3′,2′-*c*]thiepin [199] and its 1,9-dibromo derivative respectively [182].

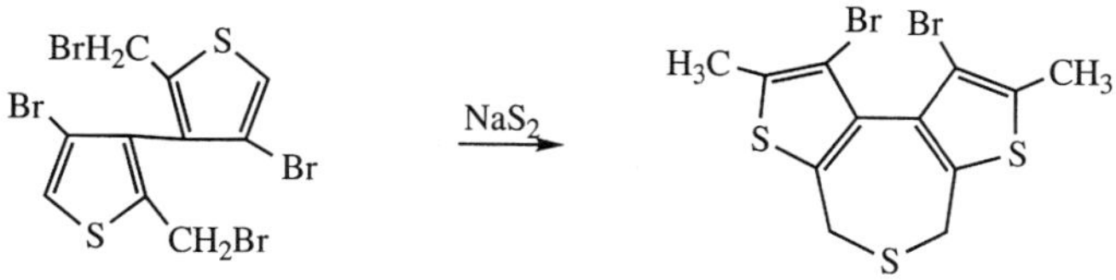

1,9-Dibromo-4,6-dihydrodithieno [2,3-c:3′,2′-e]thiepin [182]

To a solution of sodium sulfide nonahydrate (0.30 g, 1.3 mmol) in hot ethanol (50 ml) 4,4′-dibromo-2,2′-bis (bromomethyl)-3,3′-bithienyl (0.20 g, 0.39 mmol) is added. The reaction mixture is refluxed for 3 h and evaporated. The residue is treated with water in order to dissolve inorganic salts, the insoluble material is filtered off and purified by recrystallization from ligroin giving 0.10 g (67%) of the title compound mp 155–156 °C.

From 2,2′-dibromo-4,4′-dibromomethyl-3,3′-bithienyl, 1,9-dibromo-4,6-dihydrodithieno[3,4-*c*:3′,4′-*e*] thiepin was prepared [140].

1,9-Dibromo-4,6-dihydrodithieno[3,4-c:3′,4′-e]thiepin[190]

A solution of 2,2′-dibromo-4,4′-dibromomethyl-3,3′-bithienyl (0.20 g, 0.39 mmol) and sodium sulfide nonahydrate (0.30 g, 1.3 mmol) in ethanol (50 ml) is refluxed for 3 h. After cooling in ice-water the mixture is evaporated. The residue is recrystallized from 1,2-dichloroethane with a small amount of ethanol giving 0.10 g (67%) of the title compound mp 181–183 °C.

Another example of bridging of 3,3′-bithienyls is the preparation of diethyl 5*H*-4,6-dihydrocyclohepta[1,2-*c*:3,4-*c*′]dithiophene-5,5′-dicarboxylate from 4,4′-bis (bromomethyl)-3,3′-bithienyl and diethyl malonate [197].

Diethyl 5H-4,6-dihydrocychlohepta [1,2-c:3,4-c′] dithiophene-5,5′-dicarboxylate [197]

To a suspension of sodium hydride (1.7 g, 0.07 mol) in toluene (300 ml) diethyl malonate (9.7 g 0.06 mol) is added followed, after 10 min, by 4,4′-bis (bromomethyl)-3,3′-bithienyl (21.0 g, 0.06 mol) in toluene (100 ml). The reaction mixture is heated to 50–60 °C for 10 min and then another portion of sodium hydride (1.7 g, 0.07 mol) is added. After stirring for 30 min at 50–60 °C, the reaction mixture is cooled and the precipitated sodium bromide filtered off. The organic phase is washed with dilute hydrochloric acid and water and evaporated. The residue, an oil, is crystallized from cold ethanol giving 12.2 g (58%) of the title compound as slightly yellow crystals mp 61–61.5 °C.

The crown-ethers shown below are prepared through the reaction of bis (3,3′-bromomethyl)-2,2′-bithienyl with α,ω-dialkoxides derived from polyethylene glycols [60].

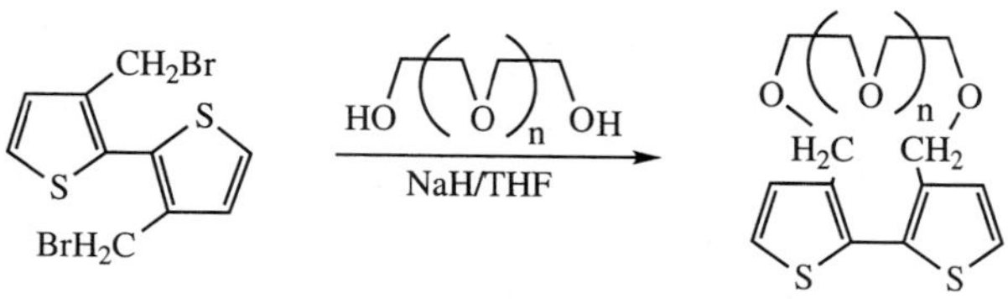

8.3.7.5 From carbonyl-containing derivatives

Formyl-substituted bithienyls undergo most of the reactions of aromatic aldehydes. Various formyl and acetyl derivatives have been reduced by sodium borohydride to the hydroxymethyl derivatives [100,191,201].

4,4′-Dibromo-2,2′-bis(dimethoxymethyl)-3,3′-bithienyl [202]

A solution of 2,2′-dibromo-4,4′-diformyl-3,3′-bithienyl (40.0 g, 0.11 mol) in methanol (350 ml) containing a few drops of concentrated hydrochloric acid is refluxed for 20 min; all material dissolves and a solid starts to separate. Sodium hydroxide (solid) is added until a slightly alkaline reaction is noticed and the heating is interrupted. The methanol solution is allowed to attain room temperature and then placed in a refrigerator and filtered giving 40.0 g (81%) of the title compound mp 140–141 °C.

5-Bromo-5′-formyl-2,2′-bithienyl has been transformed to the corresponding 5-bromo-5′-dodecanoyl-2,2′-bithienyl through reaction with dodecylmagnesium bromide followed by oxidation of the intermediate carbinol with pyridinium dichromate [100].

5-Bromo-5′-dodecanoyl-2,2′-bithienyl [100]

To a suspension of 5-bromo-5′-formyl-2,2′-bithienyl (1 g, 3.9 mmol) in tetrahydrofuran (30 ml) at − 78 °C under nitrogen 1.0 *M* solution of dodecylmagnesium bromide in diethyl ether (4.0 ml, 4 mmol) is added. The reaction mixture is slowly warmed to room temperature and stirred for 4 h under nitrogen. A saturated aqueous solution of ammonium chloride (20 ml) is added and the product extracted with diethyl ether (3 × 50 ml). The combined organic phased are washed with water (2 × 10 ml), dried over sodium carbonate and evaporated giving the alcohol as a yellowish solid. After drying under vacuum the solid is dissolved in anhydrous dichloromethane (50 ml). To this solution pyridinium dichromate (5.5 g, 14.6 mmol) is added under nitrogen, followed by oven-dried magnesium sulfate (5 g). This mixture is stirred at room temperature for 5 h, after which diethyl ether (200 ml) is added and the solution so obtained is filtered through Celite. The filtrate is dried over magnesium sulfate and evaporated. The residue is purified by chromatography on silica gel using hexane/ethyl acetate (20:1) as eluent giving 1.0 g (58%) of the title compound as an off-white waxy solid mp 96–99 °C.

2,2′-Diformyl-4,4′-dibromo-3,3′-bithienyl, 2,2′-dibromo-4,4′-diformyl-3,3′-bithienyl and 2,2′-diformyl-4,4′-dicarboxy-3,3′-bithienyl are transformed to 4,4′-dibromo-2-carboxy-2′-hydroxymethyl-3,3′-bithienyl [140], 2,2-dibromo-4-carboxy-4′-hydroxymethyl-3,3′-bithienyl [182] and 2,4,4′-tricarboxy-2′-hydroxymethyl-3,3′-bithienyl [45], respectively, through an intermolecular Cannizzaro reaction. Upon heating in toluene in the presence of *para*-toluenesulfonic acid all hydroxymethyl carboxylic acids gave ε-lactones [203]. The bridged *N*-methyl- and *N*-benzyl-5,6-dihdro-4*H*-dithieno [2,3-*c*:3′,2′-*c*] azepine and the 1,9-dibromo derivatives are prepared by the reaction of 2,2′-diformyl-3,3′-bithienyl and 4,4′-dibromo-2,2′-diformyl-3,3′-bithienyl with methyl amine and benzylamine in the presence of sodium dithionite [183, 199].

N-methyl-5,6-dihydro-4H-dithieno[2,3-c:3′2′-c]azepin [183]

4,4′-Dibromo-2,2′-diformyl-3,3′-bithienyl (1.0 g, 2.6 mmol) and 33% aqueous solution of methylamine (0.5 g, 16 mmol) are dissolved in absolute ethanol (60 ml). The mixture is heated until the starting material has dissolved, after which a solution of sodium dithionite (3.3 g, 19 mmol) in water (60 ml) is added all in one portion. The reaction mixture is stirred at room temperature for 2.5 h, refluxed for 1 h and then left overnight. After evaporation the remaining solid is collected by filtration, washed with water and recrystallized from ligroin giving 0.51 g (69%) of the title compound mp 113–115 °C.

The ylide from methyl triphenyl phosphonium bromide generated by treatment with sodium hydride in dimethylsulfoxide upon reaction with 5-bromo-5′-formyl-2,2′-bithienyl gives 5-bromo-5′-vinyl-2,2′-bithienyl [191,204].

The reaction with the phosphorane derived from ferrocenyl (methyl) triphenylphosphonium bromide was used in connection with the preparation of ferrocene and capped palladium(II) and platinum(II) complexes with thiophene spacers [205].

5′Bromo-5(2-ferrocenylvinyl)-2,2′-bithienyl [205]

To a suspension of (ferrocenylmethyl) triphenylphosphonium bromide (2.70 g, 5.00 mmol) in tetrahydrofuran potassium *tert*-butoxide (0.63 g, 5.66 mmol) is added through a sidearm with vigorous stirring. After 30 min 5-bromo-5′-formyl-2,2′-bithienyl (1,36 g, 5.00 mmol) is added in one portion. The red reaction mixture is refluxed for 6 h, cooled, poured into crushed ice and extracted with dichloromethane. The combined organic phases are washed with aqueous sodium chloride solution, dried over magnesium sulfate and evaporated. The crude product is separated from triphenylphosphine oxide by chromatography on silica gel using hexane/dichloromethane (2:3) as eluent, of which 200 ml is collected and evaporated. The residue is recrystallized from hexane giving 1.3 g (57%) of the *trans*-isomer of the title compound.

The reaction of 5-formyl-5-propynylthiophene with zinc, carbon tetrachloride and triphenylphosphine followed by butyllithium and water is used for the preparation of 5-ethinyl-5′propynyl-2,2′-bithienyl [191]. 2-Formylbithienyls undergo Michael addition to Mannich bases and in this way 1-(2-thienyl)-4-(5:2,2′-bithienyl)-1,4-butandione was obtained from the cyanohydrin of 5-formyl-2,2′-bithienyl and the free Mannich base 3-dimethylamino-1-(2-thienyl) propanone in anhydrous *N*,*N*-dimethylformamide [206].

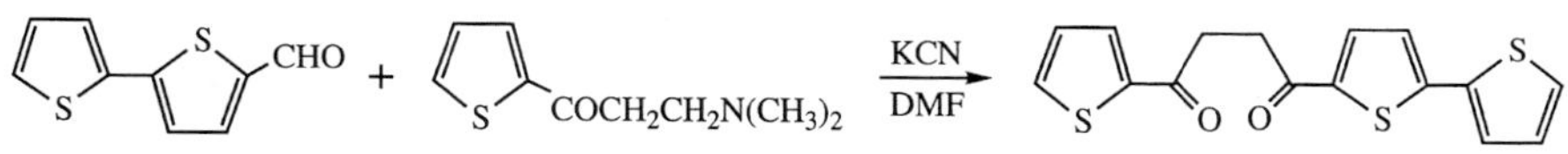

1-2(2-Thienyl)-4-(5:2,2′bithienyl)-1,4-butandione [206]

A solution of 5-formyl-2,2′-bithienyl (3.08 g, 1589 mmol) in anhydrous *N*,*N*-dimethylformamide (10 ml) is added dropwise under nitrogen during 10 min to a suspension of potassium cyanide (516 mg, 7.93 mmol) in anhydrous *N*,*N*-dimethylformamide (4 ml). After stirring the reaction mixture for 15 min, 3-dimethylamino-1-(2-thienyl) propanone (1.1 g, 6 mmol) in anhydrous *N*,*N*-dimethylformamide (10 ml) is added over a period of 30 min and the reaction mixture is stirred overnight. After addition of water (20 ml) the tan precipitate is filtered off, thoroughly washed with diethyl ether and recrystallized from acetone giving 3.82 g (94%) of the title compound mp 163.5–164.6 °C.

The condensation of 5-formyl-2,2′-bithienyl with aromatic or heterocyclic methyl ketones gives, as expected, α,β-unsaturated ketones [207]. 5-Acetyl- and 5,5′-bis(acetyl)-2,2′-bithienyl are successfully oxidized with selenium oxide to 2,2′-bithienyl-5-glyoxal hydrate and 2,2′-bithienyl-5,5′-bisglyoxal hydrate in 80 and 48% yield, respectively [208].

A number of macrocyclic compounds containing bithienyl groups have been prepared. Condensation of 5,5′-diformyl-2,2′-bithienyl with a diacid tripyrrane gives the dithiasapphyrin, shown below [209].

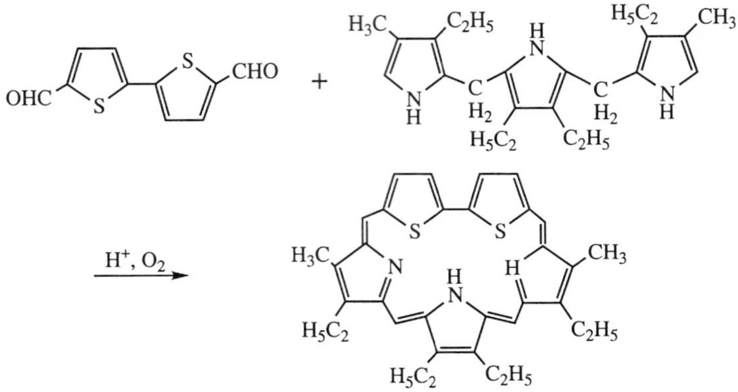

Brominated tetrasulfido [20] annulenes are prepared by McMurry coupling of brominated 2,2′-bithienyl 5,5′-dialdehydes [210].

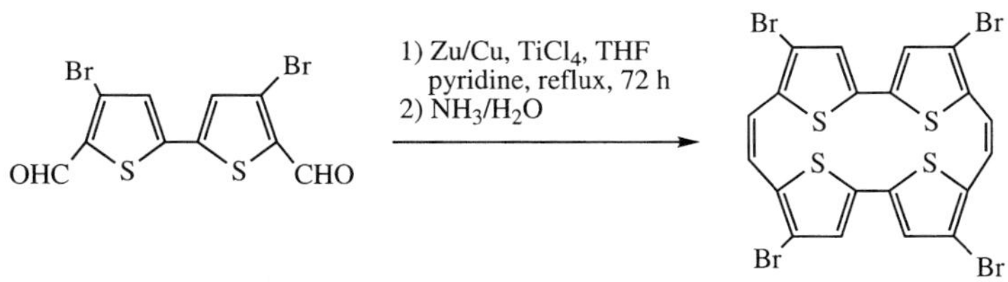

2,7,12,17-Tetrabromo-1,4;5,8:11,14:15,18-tetrasulfido[20]annulene [210]

To a suspension of zinc/copper pair (4.0 g, 0.06 mol) in tetrahydrofuran (200 ml) titanium tetrachloride (3.3 ml, 0.03 mol) is added *via* syringe at 0 °C for 30 min, after which pyridine (1.5 ml) is added and the suspension is refluxed for 2 h. To this freshly prepared and gently refluxing McMurry slurry 4,4′-dibromo-5,5′-formyl-2,2′-bithienyl (1.14 g, 3 mmol) in tetrahydrofuran (600 ml) is added dropwise in three portions. Each portion is added for 14 h and the reaction mixture is then refluxed for 10 h. After 72 h of constant stirring the red mixture is allowed to cool to room temperature and filtered through Celite. The filtrate is treated with 25% aqueous ammonia (200 ml) and diluted with chloroform (750 ml). The phases are separated and the organic phase is washed with water, dried over magnesium sulfate and evaporated. The residue is purified by chromatography on silica gel using hexane as eluent and recrystallized from benzene giving 271 mg (26%) of the title compound as yellow crystals mp 225–226 °C.

The reaction of 5,5′-diaroyl-2,2′-bithienyl with a diacetylide has been used for the preparation of the cyclic compound shown below, which upon treatment with stannous chloride in diethyl ether followed by triethylamine was used for the preparation of polydiene [177].

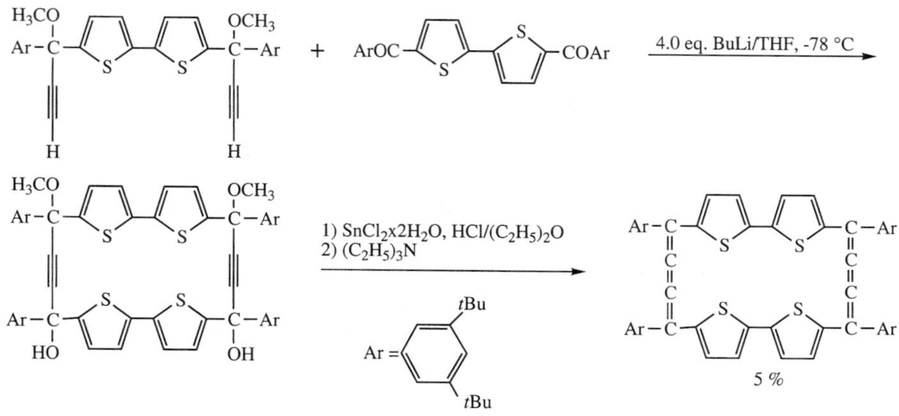

8.3.7.6 From carboxylic acids

3,3′-Bithienyldicarboxylic acids can be converted to many other 3,3′-bithienyl derivatives. Reduction with lithium aluminium hydride and aluminium chloride has extensively been used for the preparation of hydroxymethyl derivatives [140,183,184]. The reaction of 2,2′-bithienyl-5-carboxylic acid chloride has been reacted with third generation [G3] alcohol for the synthesis of dendrimer supported oligothiophenes [211].

The bisbromobithienyl macromer shown below is prepared by the reaction of 5-bromo-2,2′-bithienyl-5′-carbonyl chloride with poly(oxyethylene) in toluene/pyridine [77].

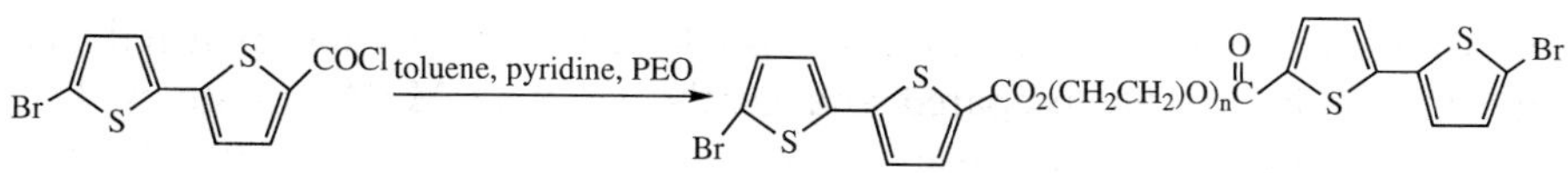

α-(5-Bromo-2,2′-bithienyl-5′-carbonyl)-ω-(5-bromo-2,2′-bithienyl-5′-carbonyloxy)-poly(oxyethane) [77]

Poly(oxyethylene) (10.8 g, 5.4 mmol) is heated to 90 °C and the melt stirred under vacuum for 0.5 h. Toluene (50 ml) and pyridine (1.5 g, 16.3 mmol) are added to the degassed polymer melt and the mixture is stirred for about 30 min, until a clear solution is obtained, after which 5-bromo-2,2′-bithienyl-5′-carbonyl chloride (5 g, 16.3 mmol) is added as a solid to the hot polymer solution. When the addition is completed the reaction mixture is stirred at 90 °C for 3 h, at room temperature for 3 days and then filtered through Celite. The filtrate is added to hexane and the precipitate formed collected by filtration and dried under vacuum. The crude products are dissolved in toluene and precipitated in hexane. The pale-yellow waxy solid is recovered by filtration, dried under vacuum for 1 day giving 10 g (73%) of the title compound.

8.3.7.7 From cyano derivatives

5-Dimethylamino-5′cyano-2,2′-bithienyl is transformed, by alkaline hydrolysis, to the 5-dimethylamino-2,2′-bithienyl-5′-carboxylic acid and by reduction with diisobutylamuminum hydride to 5-dimethylamino-2,2′-bithienyl-5′-aldehyde [74].

8.3.7.8 From various bithienyls

Reduction of nitro-substituted bithienyls such as 5-bromo-3,4-dinitro-3′,4′-ethylenedioxy-2,2′-bithienyl with stannous dichloride and hydrochloric acid is used for the preparation of 3,4-diamino-3′,4′-ethylenedioxy-2,2′-bithienyl which was condensed with 1,2-diketones [93]. A large number of 5-acetylamino-2,2′-bithienyls are prepared by reduction of nitro derivatives [212, 213]. The following Wittig reaction below has been carried out [214].

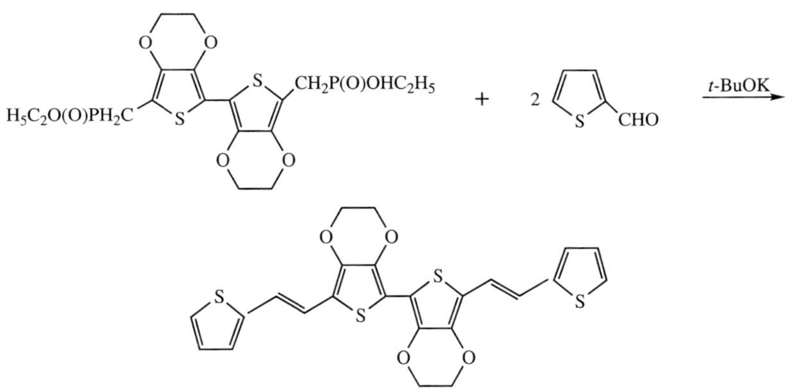

Acid-catalyzed condensation of 5,5′-bis(tolylhydroxymethyl)-2,2′-bithienyl with 16-oxa-5,10,15,17-tetrahydrotripyrrin followed by oxidation with chloranil gives 5,20-ditolyl-27-oxa-25,29-dithiasapphyrin [215].

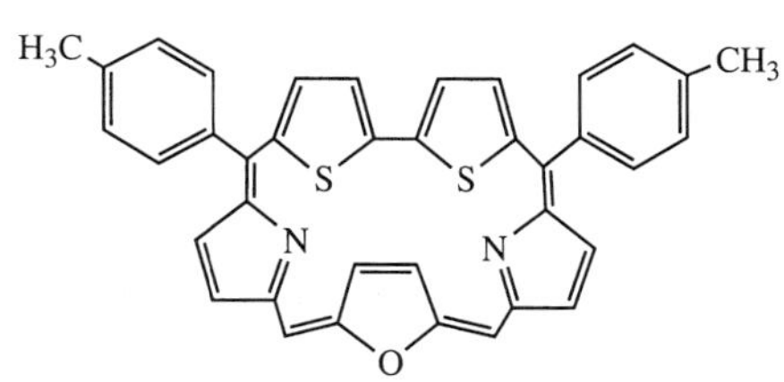

8.3.8 Preparation of bithienyls by photochemical reactions

The photochemical arylation of halosubstituted thiophenes bearing electron-withdrawing substituent has been extensively studied during recent years by D'Auria and coworkers [216]. Irradiation of 5-iodo-2-thiophene aldehyde in the presence of 2-bromothiophene or 2-methylthiophene gives 5-bromo- and 5-methyl-2,2′-bithienyl-5′-aldehyde in 99 and 69% yield respectively [191].

5-Bromo-5′-formyl-2,2′-bithienyl [191]

5-Iodo-2-thiophene aldehyde (1 g, 4.2 mmol) and 2-bromothiophene (3 ml) are dissolved in acetonitrile (300 ml) and the solution is degassed with nitrogen for 1 h, after which it is irradiated for 3 h in an immersion apparatus with a 500-W high-pressure mercury arc surrounded by a Pyrex water jacket. The mixture is diluted with chloroform and washed with 0.1 *M* sodium dithionite solution and sodium chloride solution, dried over sodium sulfate and evaporated. The residue is purified by chromatography on silica gel using chloroform/hexane (3:2) as eluents giving 1.1 g (99%) of the title compound.

The photochemical reaction of 5-iodo-2-formylthiophene and 2-iodo-5-acetylthiophene with 2-(3-acetoxyprop-1-enyl) thiophene gives the following product [217].

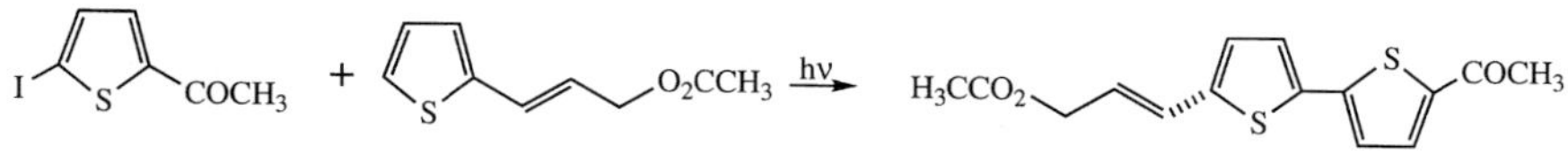

(E,Z)-5-(3-Acetoxyprop-1-enyl)-5′-acetyl-2,2′bithienyl [217]

5-Iodo-2-actylthiophene (1 g, 4 mmol) and 2-(3-acetoxyprop-1-enyl) thiophene (5 g, 27 mmol) are dissolved in acetonitrile (300 ml) and the solution is purged with nitrogen for 1 h. It is then irradiated in an immersion apparatus with a 500 W high-pressure mercury arc surrounded by a Pyrex water jacket. After 5 h the mixture is diluted with chloroform (500 ml), washed with 0.05 *M* aqueous sodium dithionite, dried over sodium sulfate and evaporated. The residue is purified by chromatography on silica gel using hexane/ethyl acetate (4:1) as eluent, giving 800 mg (66%) of the title compound as a very dense oil.

Irradiation of methyl 2-thienylacetate and methyl 2-(2-thienyl) propionate with 5-iodo-2-acetylthiophene gave the corresponding coupling products [217]. Photochemical reaction of 5-iodo-2-thiophene aldehyde with 2-(3-acetoxy-1-propynyl) thiophene gives 5-(3-acetoxy-1-propynyl)-5′-formyl-2,2′-bithienyl [218]. Photochemical rearrangement of 2,2′-bithienyl gives a low yield of 2,3′-bithienyl [219].

Photochemical reaction of 2,3-diiodo-5-nitrothiophene with thiophene, 2-chloro- and 2-bromothiophene in acetonitrile is used for the preparation of 3-iodo-5-nitro-2,2′-bithienyl, 5′-chloro-3-iodo-5-nitro-2,2′-bithienyl and 5′-bromo-3-iodo-5-nitro-2,2′-bithienyl [220].

3-Iodo-5-nitro-2,2′-bithienyl [220]

2,3-Diiodo-5-nitrothiophene (100 mg, 0.26 mmol) is dissolved in acetonitrile (100 ml) in the presence of thiophene (3 ml). The reaction mixture is degassed for 10 min and then irradiated with a 125-W high-pressure mercuric arc (Helios-Italqartz) surrounded by a Pyrex water-jacket for 20 h. After evaporating the solvent 87 mg (99%) of the title compound is obtained in pure form.

8.3.9 Preparation of bithienyls by ring-closure reactions

4-Bromo-2,2′-bithienyl and 4-bromo-2,3′-bithienyl are prepared in high yield by the treatment of the acetylenic ketones with 33% hydrobromic acid/acetic acid in dichloromethane [221].

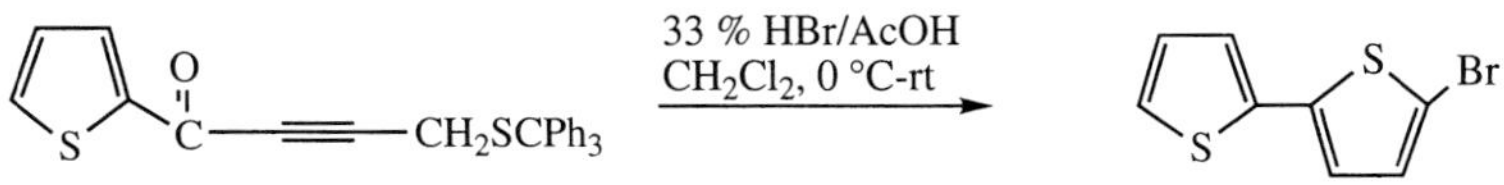

The disulfides derived form 2-mercapto-5-(2-thienyl)-, 2-mercapto-5-(2-chloro-3-thienyl)- and 2-mercapto-5-(2,5-dichloro-3-thienyl)-2,4-pentadienoic acid gave upon oxidative cyclization with iodine in ethanol 2,2′-bithienyl-5-carboxylic acid (72%), 2-chloro-2,3′-bithienyl-5-carboxylic acid (52%) and 2′,5′-dichloro-2,3′-bithienylcarboxylic acid (34%), respectively [222].

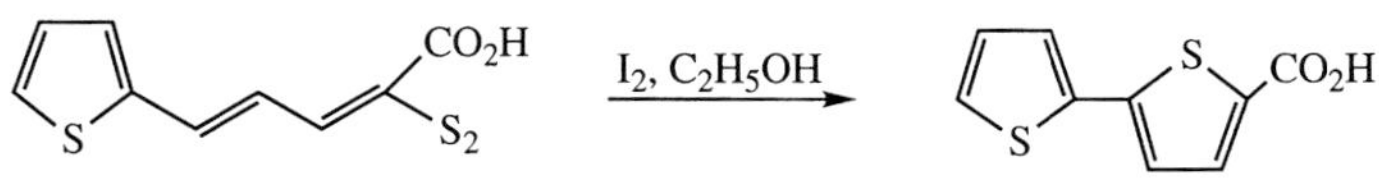

2,2′-Bithienyl-5-carboxylic acid [222]

A solution of iodine (5.08 g, 20 mmol) in ethanol is added dropwise during 2 h to a cold stirred solution of 2-mercapto-5-(2-thienyl)-2,4-pentadienoic acid (4.25 g, 20 mmol) in ethanol (50 ml). The stirring is continued for 1 h, after which saturated sodium bisulfite solution is added and the solution obtained poured into ice-water (500 ml) followed by acidification with hydrochloride acid giving 3.85 g (91%) of the crude disulfide mp 168–170 °C. Without purification this compound (1.64 g, 3.9 mmol) is dissolved in 1,2-dimethoxyethane (50 ml) and iodine (0.99 g, 3.9 mmol) in 1,2-dimethoxythane (50 ml) is added rapidly. The reaction mixture is refluxed for 3 h, cooled and poured into ice-water (500 ml). This solution is decolorized with sodium bisulfite solution and acidified with hydrochloric acid. The precipitate formed is collected, dried and sublimed at 140 °C/2 mm Hg giving 1.18 g (72%) of the title compound mp 177–179 °C.

2,3′-Bithienyl was first prepared in a practical way by the reaction of 2-thienyllithium with 3-ketotetrahydrothiophene followed by aromatization of the resulting tertiary alcohol with chloranil [223]. 3-Bromo-2-3′-bithienyl was prepared in an analogous way [134] and 5,5′-diphenyl-2,3′-bithienyl was obtained from 5-phenyl-2-thienyllithium and 2-phenyl 4-ketotetrahydrothiophene [26]. A recently used new method consists in the palladium-catalyzed ring closure of 2-[1-(prop-2-ynylthio)prop-2-ynyl]thiophene using carbon monoxide and air in methanol followed by triethylamine [224].

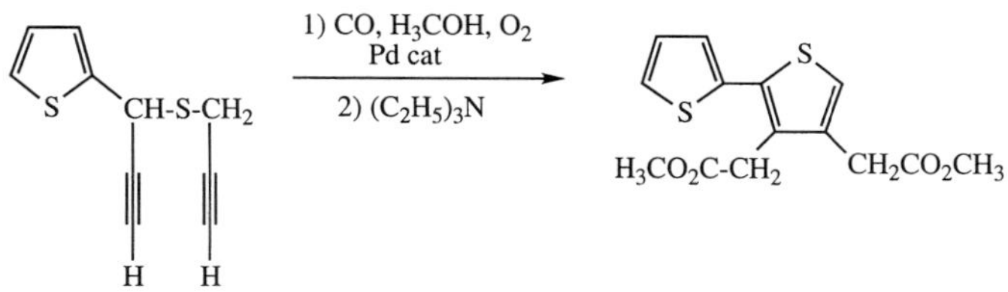

3,4-Bis[(methoxycarbonyl)methyl]-2,2′-bithienyl [224]

A stainless steel autoclave with magnetic stirring is charged in the presence of air with palladium(II) iodide (70 mg, 0.194 mmol), potassium, iodide (323 mg, 1.95 mmol) and 2-[1-(prop-2-ynylthio)prop-2-ynyl]thiophene (1.12 g, 5.8 mmol) dissolved in methanol (58 ml). The autoclave is pressurized with carbon monoxide (15 atm) and air up to 20 atm of total pressure. The reaction mixture is stirred and heated at 70 °C for 2 h, after which it is cooled and the methanol is removed *in vacuo*. The residue is taken up in dichloromethane and this solution is added to triethylamine (235 mg, 2.32 mmol). The resulting solution is stirred at 25 °C for 2 h. Chromatography on silica gel using hexane/ethyl acetate (7:3) gives 810 mg (45%) of the title compound as a colorless oil.

A new method for the preparation of 5-nitro-5′-dimethylamino-2,2′-bithienyl according to the following scheme has recently been described [255].

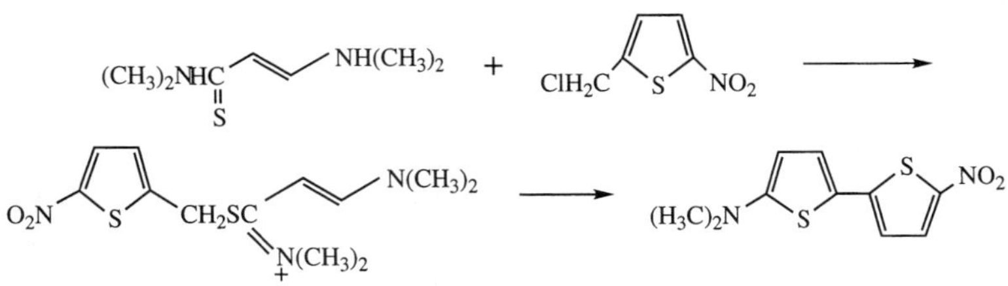

5-Nitro-5′dimethylamino-2,2′bithienyl [255]

A solution of 1,3-bis(dimethylamino) propene-3-thione (1.58 g, 10 mmol) and 5-chloromethyl-2-nitrothiophene (1.78 g, 10 mmol) in methanol (50 ml) is heated to reflux for a short time. After cooling 25% sodium methylate solution (50 ml) is added and the reaction mixture is briefly warmed again followed by cooling and addition of water (10 ml). The precipitate formed is filtered off giving 2.36 g (93%) of the title compound mp 174–176 °C.

8.3.10 Dioxides from bithienyls

Oxidation of 5,5′-bis (dimethyl-*tert*-butylsilyl)-2,2′-bithienyl with *meta*-chloroperbenzoic acid in dichloromethane gives 47% of the bis sulfone and 31% of the monosulfone [226]. Another approach consists in the Stille reaction between 2-bromo-5-hexyl- and 2-bromo-5-(dimethyl-*tert*-butylsilyl)thiophene-1,1-dioxide and 2-tributylstannylthiophene which gives monosulphones in 83% and 36% yield respectively [227].

8.4 TERTHIENYLS

All fourteen different isomers of terthienyl, shown below have been prepared by various coupling reactions or ring-closure reactions. The linear, 2,2′:5′,2″ -terthienyl is usually called α-terthienyl.

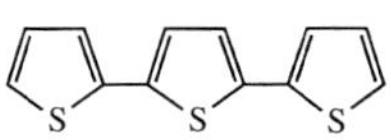
2,2':5',2"-terthienyl

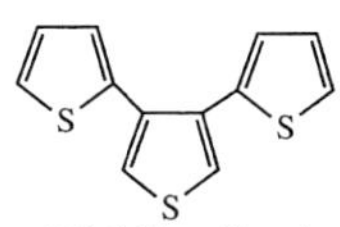
2,3':4',2"-terthienyl

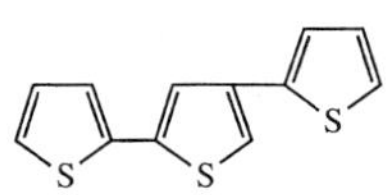
2,2':4',2"-terthienyl

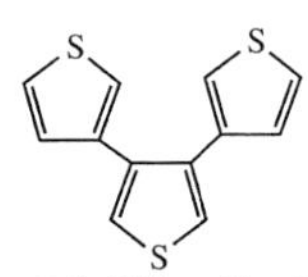
3,3':4',3"-terthienyl

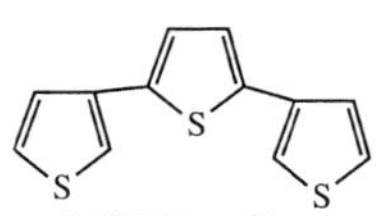
3,2':5',3"-terthienyl

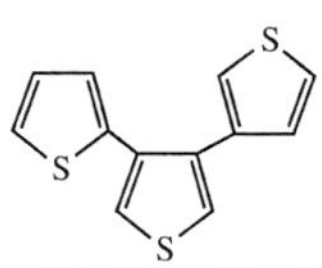
2,3':4',3"-terthienyl

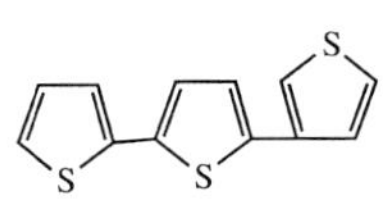
2,2':5',3"-terthienyl

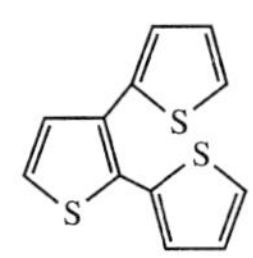
2,2':3',2"-terthienyl

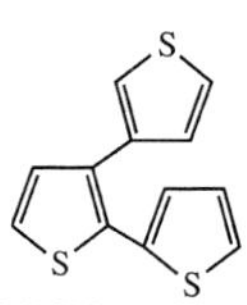
2,2':3',3"-terthienyl

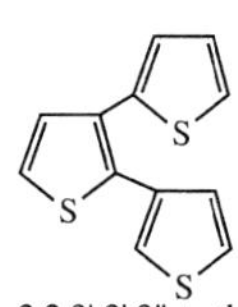
2,2,3':2',3"-terthienyl

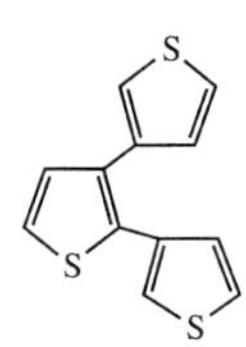
3,2':3',3"-terthienyl

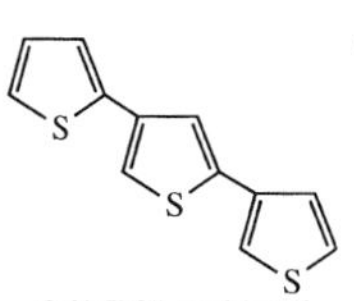
2,3':5',3"-terthienyl

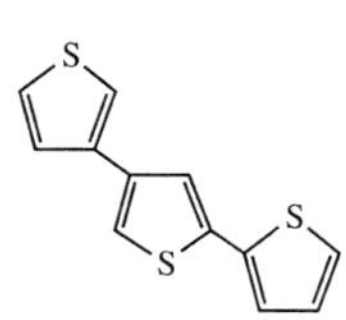
2,2':4',3"-terthienyl

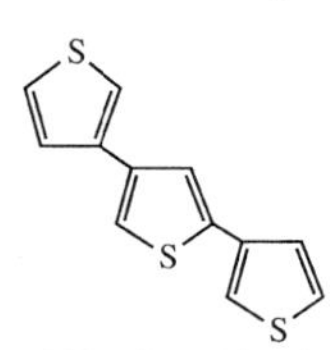
3,2':4',3"-terthienyl

8.4.1 Preparation of terthienyls through ring-closure reactions

8.4.1.1 From butadiymes

α-Terthienyls are conveniently prepared from 1,4-di(thienyl)butadiynes and hydrogen sulfide [228–231]. This method is also successfully used for the preparation of the 3,3″-dimethyl-, 5,5″-dimethyl- and 5,5″-dibutyl derivatives [230,232]. This method was also used for the preparation of 3,2′:5′:″-terthienyl from 1,4-di(3-thienyl)-1,3-butadiyne [230] and 2,2′:5′,3″-terthienyl from 1-(2-thienyl)-4-(3-thienyl)-1,3-butadiyne [233].

8.4.1.2 From 1,4-diketones and similar compounds

Another ring-closure route to α-terthienyls consists in the ring closure of 1,4-dithienyl substituted 1,4-diketones upon treatment with hydrogen chloride and hydrogen sulfide [234,235]. Alternatively phosphor pentasulfide or Lawesson's reagent can be used for the cyclization [236], as was the case in the preparation of 3,2:5′,3″-terthienyl from 1,4-di(3-thienyl)-1,4-butanedione in 75% yield [237]. Its 5,5″-dichloro derivative was obtained from the reaction of 1,4-di(2-chloro-4-thienyl)-1,4-butanedione and hydrochloric acid/hydrogen sulfide [234]. This method is also used for the preparation of 2,2′:5′,3″-terthienyl from 1-(2-thienyl)-4-(3-thienyl)-1,4-butanedione [236]. 1,3-Di-(2-thienyl)benzo-[*c*]thiophene is prepared by ring closure of 1,2-di(2-thenoyl)benzene with phosphor pentasulfide in tetrahydrofuran [238]. The reaction of 1,4-bis(2-thienyl)-2,3-dimethyl-1,4-butanedione with Lawesson's reagent gives a 76% yield of 3′,4′-dimethyl-α-terthienyl [206].

3′,4′-Dimethyl-2,2′:5′,2″-terthienyl [206]

A solution of 1,4-bis(2-thienyl)-2,3-dimethyl-1,4-butanedione (300 mg, 1.079 mmol) and Lawesson's reagent (262 mg, 0.65 mmol) in anhydrous toluene (2 ml) is refluxed under argon for 15 min. The reaction mixture is purified by flash chromatography on silica gel using dichloromethane/petroleum ether (1:1) as eluent. The tan solid obtained is recrystallized from methanol giving 228 g (77%) mp 126–128 °C.

Using 1,4-bis(5-bromo-2-thienyl)-1,4-butandione gives 5,5′-dibromo-α-terthienyl in 70% yield [215]. Recently 3′-(3-cyanopropyl)-α-terthienyl has been prepared by reaction with Lawesson's reagent [239].

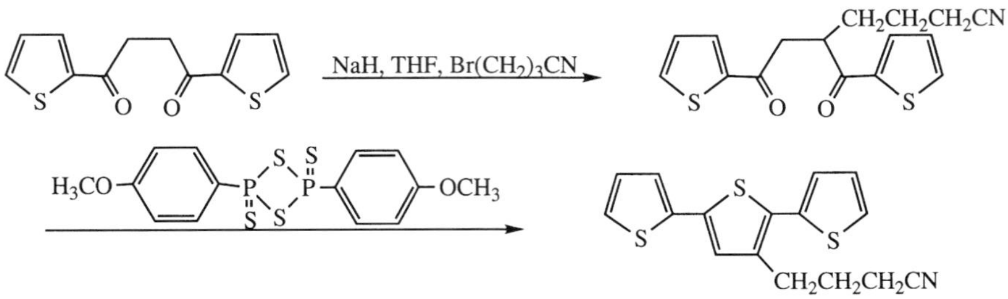

3′-(3-Cyanopropyl)-2,2′:5′,2″-terthiophene [239]

A mixture consisting of 1,4-di(2-thienyl)-2-(3-cyanopropyl)butane-1,4-dione (0.0350 g, 1.10 mmol), Lawesson's reagent (0.539 g, 1.32 mmol) and anhydrous toluene (15 ml) is slowly heated and maintained at reflux for 6 h. The orange solution is allowed to cool down to room temperature and poured into water (10 ml). The product is extracted with diethyl ether (3 × 15 ml), the combined organic phases are washed with water, dried over magnesium sulfate and evaporated. The residue, a brown oil, is purified by chromatography on silica gel using hexane/ethyl acetate (6:1) as eluent giving 0.284 g (82%) of the title compound as a yellow solid mp 260 °C.

The following reaction has been performed [240].

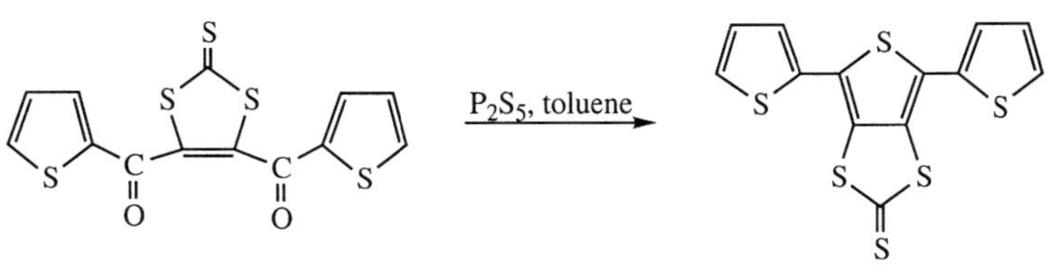

4,6-Bis(2-thienyl)thieno[3,4-d]-1,3-dithiole-2-thione [240]

A mixture of the ketone (1.0 g, 2.80 mmol), phosphorus pentasulfide (3 g, 13 mmol) and sodium bicarbonate (1 g) in 1,4-dioxane (20 ml) is stirred under nitrogen while the temperature is increased from 60 to 100 °C over a period of 1 h, after which it is poured into water (150 ml). CAUTION! Evolution of hydrogen sulfide and carbon dioxide. The suspension formed is refluxed for 15–20 min, the precipitate is filtered off, dried and purified by chromatography on silica gel using toluene as eluent. The volume of the eluent is reduced to 15–20 ml and diluted with petroleum ether (50 ml) giving 0.70 g (70%) of the title compound as an orange powder mp 195–196 °C.

A special type of α-terthienyls having a *c*-fused benzo ring on the middle thiophene ring, such as 1,3-bis(2-thienyl)benzo[*c*]thiophene, are prepared by the reaction of 2-thienylphtalide and 2-thiophenemagnesium bromide or 3-alkylthiophenemagnesium bromide giving an intermediate, which upon thionation with Lawesson's reagent gives the desired compounds. The bromo derivative 1-(5-bromo-2-thienyl)-2′-thienylbenzo[*c*]thiophene was prepared similarly from 3-[2-(5-bromothienyl)]phthalide [241].

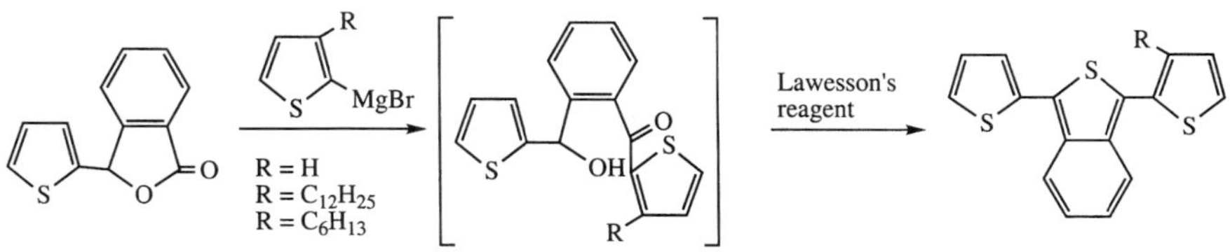

1-(3-Dodecylthienyl)-3-thienylbenzo[c]thiophene [241]

To a refluxing mixture of magnesium turnings (0.705 g, 0.029 mol) and iodine (20 mg) in anhydrous tetrahydrofuran, 2-bromo-3-dodecylthiophene (8 g, 0.024 mol) is added slowly. After 6 h of reflux the Grignard reagent is formed; it is cooled and slowly added *via* an addition funnel to a solution of 2-thienyl phthalide in anhydrous tetrahydrofuran (50 ml) at $-10\,^{\circ}C$. The reaction mixture is stirred at room temperature for 6 h and then poured into ice-containing ammonium chloride. The intermediate compound is extracted with dichloromethane, the combined organic phases are dried over sodium sulfate and treated with Lawesson's reagent (5 g, 0.012 mol). The reaction mixture is stirred at room temperature for 6 h followed by standard workup and filtration through a column of basic alumina giving 8.7 g (77%) of the title compound as a thick liquid.

The reaction of α-(2-thienyl)naphthalene with phosphorus oxychloride and *N*,*N*-dimethylformamide led to the formyl derivative, which upon reaction with divinylsulfone catalyzed by 3,4-dimethyl-5-(2-hydroxyethyl)thiazolium iodide gave the 1,4-diketone, which upon reaction with Lawesson's reagent gave the terthienyl in reasonable yields [242].

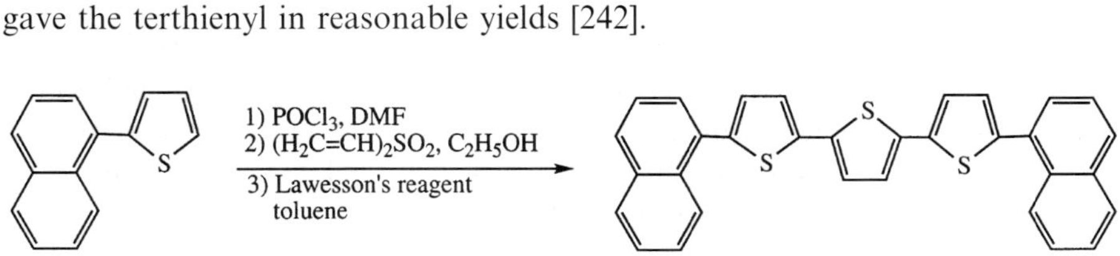

8.4.1.3 From diketo sulfides

The following reaction using Lawesson's reagent gives 2,6-di(2-thienyl)dithiin, which upon refluxing in *ortho*-dichlorobenzene gives a 13:1 mixture of α-terthienyl and 2,3′4′,2″-terthienyl [243,244] from which α-terthienyl could be obtained in pure form. However, a high yield of the latter compound was obtained upon reductive coupling of the sulfide with titanium tetrachloride/zinc followed by acid-catalyzed dehydration and aromatization [245].

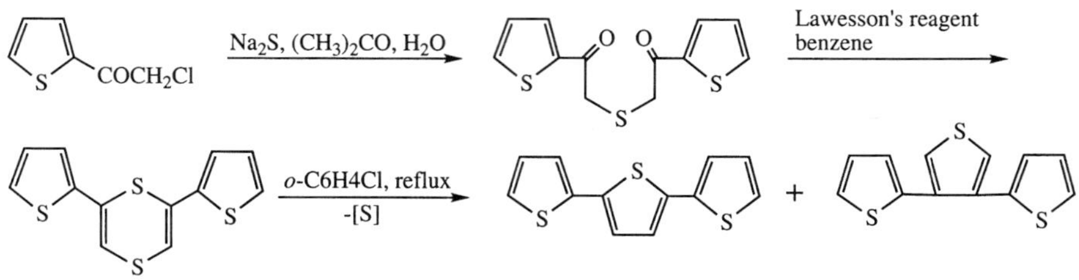

8.4.1.4 By the Hinsberg reaction

The reaction of the 2,3′-thenil with diethyl thiodiacetate gave 2′,5′-dicarboethoxy-5-chloro-2″,5″-dimethyl-2,3′:4′,3″ -terthienyl in 77% yield [234].

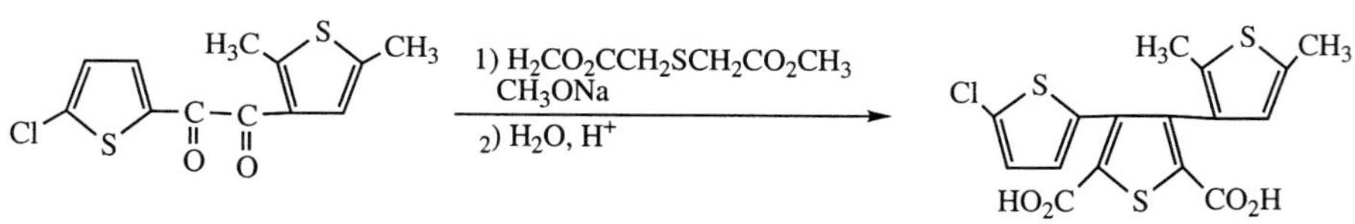

The Hinsberg reaction of 2,2′-dithenil with diethyl thiodiacetate gives 2′,5′-dicarboethoxy-2,3′:4′,2″-terthienyl, which was hydrolyzed and decarboxylated to the parent terthienyl [105].

8.4.1.5 From 3-ketotetrahydrothiophene

The reaction of the Grignard reagent from 5-iodo-2,2′-bithienyl with 3-keto-tetrahydrothiophene gave the tertiary alcohol in 80% yield. However, the subsequent aromatization gave only 12% of the desired 2,2′:5′,3″-terthienyl [223]. Also 2,2′:4′,2″-terthienyl is prepared by this approach, reacting 2-thiophenemagnesium bromide and the ketone shown below [223].

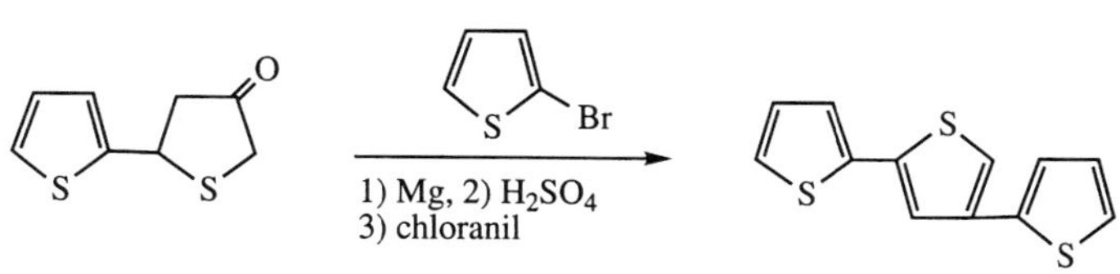

8.4.1.6 Various ring-closure methods

The palladium-catalyzed reaction of 2,5-bis[(prop-2-ynylthio)prop-2-ynyl]thiophene with carbon monoxide in methanol followed by treatment with methanol gives 3,4,3″,4″ -tetra[(methoxycarbonyl)methyl]-2,2′:5′,2″ -terthienyl in moderate yields [224].

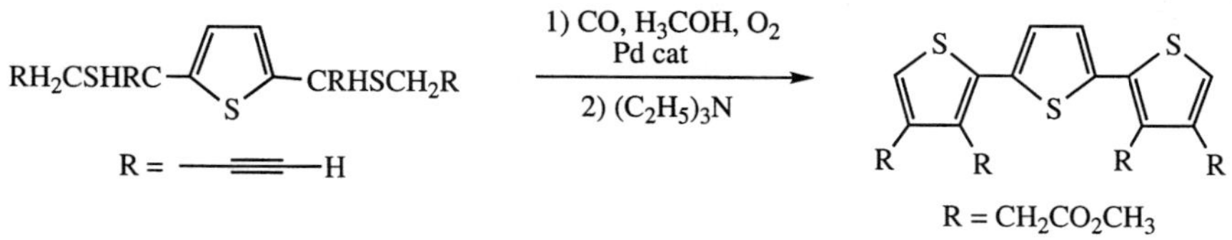

3,4,3″,4″ -Tetra[(methoxycarbonyl)methyl]-2,2′:5′,2″ -terthienyl [224]

A stainless steel autoclave with magnetic stirring is charged in the presence of air with palladium(II) iodide (180 mg, 0.50 mmol), potassium iodide (830 mg, 5.0 mmol) and 2,5-bis[(prop-2-ynylthio)prop-2-ynyl]thiophene (0.75 g, 2.5 mmol) dissolved in 1,2-dimethoxyethane/methanol (1:1) (50 ml). The autoclave is pressurized, with carbon monoxide (15 atm) and air, up to 20 atm of total pressure. The reaction mixture is stirred and heated at 70 °C for 15 h, after which it is cooled and the methanol is removed *in vacuo*. The residue is taken up in dichloromethane and this solution is added to triethylamine (253 mg, 2.5 mmol). The resulting solution is stirred at 25 °C for 5 h. Chromatography on silica gel using hexane/ethyl acetate (7:3) gives 322 mg (24%) of the title compound as a yellow solid mp 101–102 °C.

Donor–acceptor substituted α-oligothiophenes such as α-terthienylfullerenes with C_{60}-fullerene as acceptor and anthracene as donor were synthesized *via* [4 + 2] cycloaddition with the heteroanalogous *ortho*-quinodimethanes of anthryloligothiophenes [84].

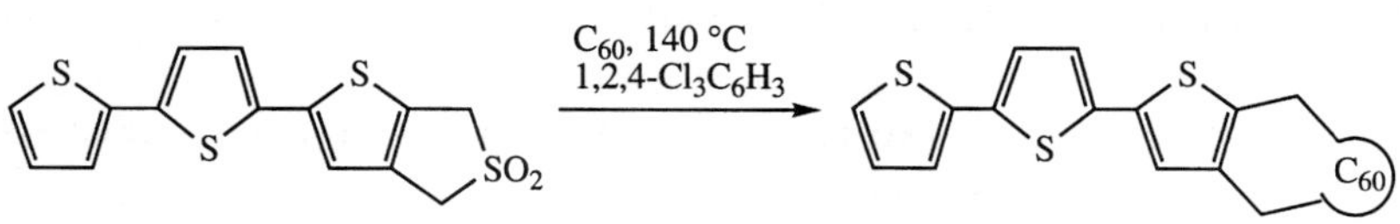

A route starting from 2-acetylthiophene, *via* 3,6-di(2-thienyl)1,4-dithiin-2,5-dicarboxylic acid dimethyl ester can be used for the preparation of 2,4′:2′,2″-terthienyl-3,5′-dicarboxylic acid dimethyl ester [246].

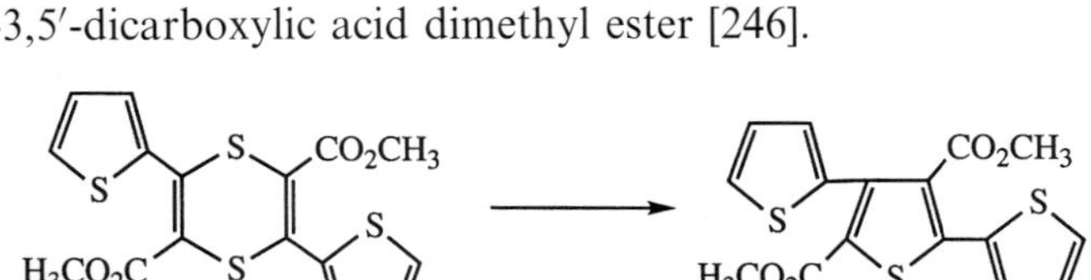

2,3′:2′,2″ -Terthienyl,3,5′-dicarboxylic acid dimethyl ester [246]

3,6-Di(2-thienyl)1,4-dithiin-2,5-dicarboxylic acid dimethyl ester (10 g, 25 mmol) in benzene (200 ml) is refluxed for 1 h. The red hot solution is filtered and evaporated giving a red oil, which is dissolved in hot ethanol (250 ml). After 6 days at room temperature the solution is decolorized and sulfur is formed in almost quantitative yield and can be filtered off. The filtrate is left for 1 h and the crystals formed are filtered off, washed with a small amount of ethanol and recrystallized from petroleum ether giving 7.3 g (80%) of the title compound mp 86 °C.

8.4.2 Palladium and nickel-catalyzed couplings of Grignard reagents and dihalothiophenes and halobithienyls

8.4.2.1 From thiophenemagnesium or zinc halide and dihalothiophenes

The reaction of 2-thiophenemagnesium bromide with 2,5-dibromothiophene using bis(diphenylphosphino)ferrocenepalladium(II) dichloride as catalyst is the best method for the preparation of α-terthienyl in good yield [66].

2,2′:5′,2″-Terthiophene [66]

To a stirred mixture of bis(diphenylphosphino)ferrocenepalladium(II) dichloride (753 mg, 1.03 mmol) and 2,5-dibromothiophene (15.2 g, 60 mmol) a 0.6 *M* solution of 2-thiophenemagnesium bromide in diethyl ether (150 mmol) is added. The reaction mixture is kept at 0 °C for 0.5 h, at 20 °C for 12 h and at 35 °C for 1 h, after which it is quenched by addition of a saturated aqueous ammonium chloride solution and diluted with water. The product is extracted with diethyl ether and the combined organic phases are filtered through Celite and evaporated. The residue is dissolved in hot hexane containing charcoal, filtered through Celite, partly concentrated and the crude product is recrystallized from hexane giving 14.4 g of the compound (67%) mp 93–94 °C.

Coupling of ω-halofunctionalized 3-alkyl-2,5-dihalothiophenes with 2-thiophenemagnesium bromide followed by transformation of the halides to thiol groups is used for the preparation of the terthienyl shown below [162].

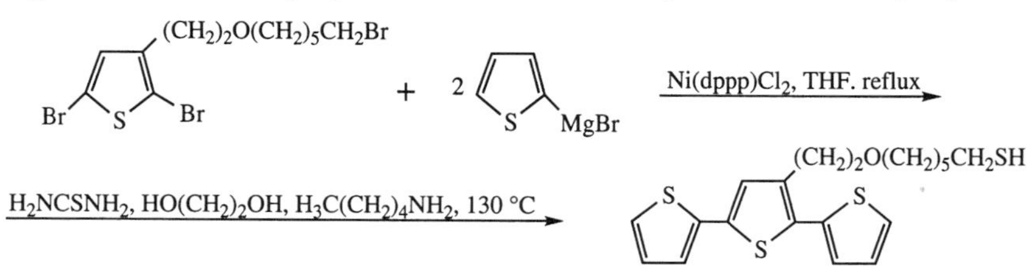

The coupling of 2-thiophenemagnesium bromide with 2,5-dibromo-3,4-dibutylthiophene is used for the preparation of 3′,4′-dibutyl-α-terthienyl [247]. From 3-pentyl-2-thiophenemagnesium bromide and 2,5-dibromothiophene, 3,3″-pentyl-α-terthienyl is prepared in 62% yield [84]. Coupling of 2,5-diiodothiophene with the Grignard reagent from 2-iodo-3-pentoxythiophene is used for the preparation of 4,4″-dipentoxy-α-terthienyl [15], and with two equivalents of 5-phenyl-2-thiophenemagnesium bromide, 5,5″-diphenyl-α-terthienyl is obtained [67].

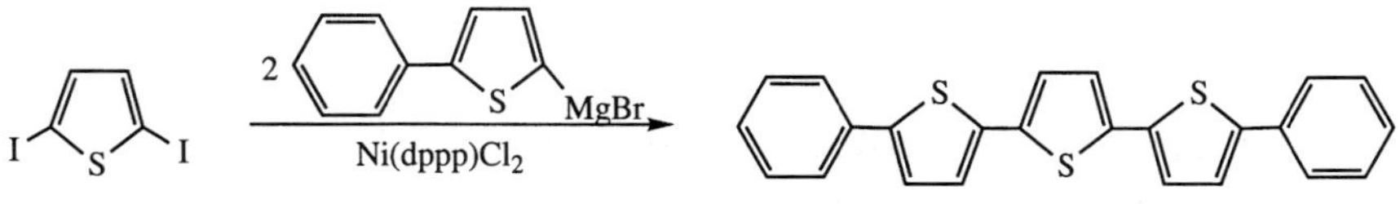

3,3″,4,4′-Tetrabutyl-α-terthienyl is prepared in 82% yield by nickel catalyzed cross-coupling of the Grignard reagent of 2-bromo-3,4-dibutylthiophene with 2,5-dibromothiophene [248]. The bis(diphenylphosphino)ferrocene-palladium(II) dichloride-catalyzed coupling of the Grignard reagent from 2-bromo-4-trimethylsilylthiophene is used in the preparation of 4,4‴-bis(trimethylsilyl)-α-terthienyl [249]. Coupling of two equivalents of 5-[3,5-bis(*tert*-butyl-4-trimethylsilyloxy)phenyl]thienyl-2-zinc chloride with 2,5-dibromothiophene has been performed [250].

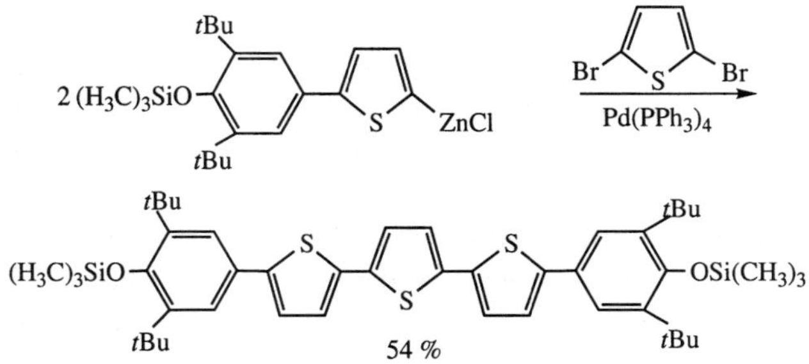

Several phenyl substituted α-terthienyls were recently prepared by Kumada couplings of 2-thiophenemagnesium bromide or phenyl-substituted 2-thiophenemagnesium bromides with phenyl-substituted 2,5-dibromothiophenes. By this approach 3,3″-diphenyl-α-terthienyl, 3,3′,3″-triphenyl-α-terthienyl, 2′-phenyl-α-terthienyl and 3′,4′-diphenyl-α-terthienyl are conveniently obtained [251].

3,3″-Diphenyl-2,2′:5′,2″-terthienyl [251]

A solution of 2-(3-phenylthienyl)magnesium bromide freshly prepared from 2-bromo-3-phenylthiophene (2.51 g, 10.5 mmol) is transferred *via* cannula to a mixture of 1,3-bis(diphenylphosphino)propanenickel(II) chloride (27 mg, 0.05 mmol) in anhydrous diethyl ether (10 ml). The mixture is refluxed for 2 h and during this period 1,3-bis(diphenylphosphino)propanenickel(II) chloride (9 mg, 0.017 mmol) is added each over 30 min. The reaction mixture is carefully hydrolyzed with water (30 ml) and the phases separated. The aqueous phase is extracted with diethyl ether (3 × 30 ml). The combined organic phases are dried over magnesium sulfate and evaporated. The residue is purified by using chromatotron (2 × 4 mm) and petroleum ether giving 0.69 g (38%) of the title compound as a light-green powder mp 139–140 °C.

Using this approach 3′-phenyl-α-terthienyl and 3′,4′-diphenyl-α-terthienyl were recently prepared from 2-thiophenemagnesium bromide and 2,5-dibromo-3-phenylthiophene [252].

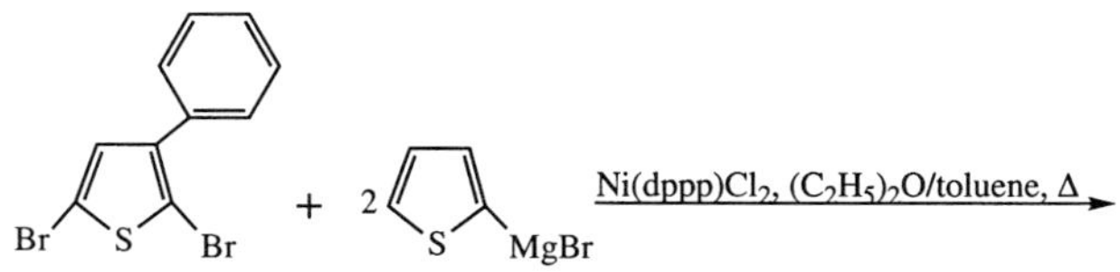

3′-Phenyl-2,2′:5′,2″-terthiophene [252]

A solution of 2-bromothiophene (3.34 g, 2.05 mmol) in anhydrous diethyl ether (20 ml) is added dropwise to magnesium chips (598 mg, 24.6 mmol) in boiling anhydrous diethyl ether (10 ml). The resulting mixture is refluxed for 3 h, allowed to cool to room temperature and transferred by means of a syringe to the dropping funnel of a second apparatus. The Grignard solution is added dropwise to a suspension of 2,5-dibromo-3-phenylthiophene (2.49 g, 7.83 mmol) and 1,3-bis(diphenylphosphino)propane nickel(II) chloride (42 mg, 0.079 mmol) in anhydrous diethyl ether (25 ml). The reaction mixture is refluxed for 72 h, cooled to room temperature, acidified with 2 *M* hydrochloric acid (50 ml) and extracted with dichloromethane. The combined organic phases are washed with water, saturated sodium bicarbonate solution and again with water, dried over magnesium sulfate and evaporated. 2,2′-Bithienyl, due to homocoupling, is removed by distillation and the residue is purified by chromatography on silica gel using petroleum ether/dichloromethane (5:1) as eluent giving 1.35 g (53%) of the title compound as an amorphous yellow solid mp 85–86 °C.

The palladium-catalyzed Kumada reaction between 2-thiophenemagnesium bromide and 2,5-diiodo-3-methoxythiophene is best used for the preparation of 3′-methoxy-α-terthienyl in 84% yield [66].

From 3-methyl-2-thiophenemagnesium bromide and 2,5-diodo-3,4-dimethylthiophene, 3,3′,4′,3″-tetramethyl-α-terthienyl is obtained [66]. 3,3′:4′3″-Terthienyl is prepared by the reaction of 3-thiophenemagnesium bromide with

3,4-dibromothiophene using bis(diphenylphosphino)ferrocene palladium(II) dichloride [65] or 1,3-bis(diphenylphosphine)propane] nickel(II) chloride [58] as catalyst. Similarly 3,2′:3′,3″-terthienyl was prepared from 3-thienylmagnesium bromide and 2,3-dibromothiophene [58], and finally 3,2′:4′,3″-terthienyl from 3-thienylmagnesium bromide and 2,4-dibromothiophene [253]. A number of 3′-alkyl-2,2′:5′,2″-terthienyls were prepared from 2-thiophenemagnesium bromide and 3-alkyl-2,5-dibromothiophenes, such as the 3′-hexyl derivative [254,255], 3′-dodecyl [62] or 3′-methyl derivative [66]. The use of 3-thiophenemagnesium bromide in the reaction with 2,5-dibromo-3-hexylthiophene led to 3,3″-dihexyl-2,2′:5′,2″-terthiophene, while the Grignard reagent from 2-bromo-3-hexylthiophene prepared by the entrainment reaction, followed by coupling with 2,5-dibromothiophene gave 3,3″-dihexyl-2,2′:5′,2′-terthienyl [254].

3′-Hexyl-2,2′:5′,2″-terthiophene [254]

A solution of 2-bromothiophene (22.5 g, 0.138 mmol) in anhydrous diethyl ether (30 ml) is added to magnesium turnings (3.4 g, 0.138 mol) covered with anhydrous diethyl ether (30 ml) at such a rate that gentle reflux is maintained. The refluxing is continued for another 20 min, after which [1,3-bis-(diphenylphosphine)propane]nickel(II) chloride (0.15 g, 0.2 mol%) is added and a solution of 2,5-dibromo-3-hexylthiophene (15 g, 0.046 mol) in anhydrous diethyl ether (80 ml) is added dropwise. The reaction mixture is refluxed for 15 h, cooled to room temperature and poured into a mixture of crushed ice (100 g) and 2 *M* hydrochloric acid (20 ml). The product is extracted with dichloromethane (5 × 30 ml). The combined organic phases are washed to neutrality with saturated sodium hydrogen carbonate and water (2 × 50 ml), dried over magnesium sulfate and evaporated. The residue is purified by flash chromatography using hexane/dichloromethane (40:1) as eluent and chromatographed again using only hexane as eluent, giving 6.81 g (45%) of the title compound as an intensely yellow-green oil.

The reaction of the Grignard reagent derived from 2-iodo-3-[(*S*)-(+)-2-methylbutyl]thiophene and 2,5-dibromothiophene in anisole with [1,3-bis-(diphenylphosphine)propane]nickel(II) chloride as catalyst is used for the preparation of 3,3″-di[(*S*)-(+)-2-methylbutyl]-α-terthienyl as precursor to chiral regioregular (poly)thiophenes [256]. Various deuterated terthienyls such as 4,4″-dideuterio-5,5″-dimethyl-α-terthienyl, 5,5″-dimethyl-3,3″,4,4″-tetradeuterio-α-terthienyl and 3′,4′-dideuterio-5,5″-dimethyl-α-terthienyl were prepared by this method through the reaction of 4-deuterio-5-methyl-2-thiophenemagnesium bromide and 3,4-dideuterio-5-methyl-2-thiophenemagnesium bromide with 2,5-dibromothiophene and 5-methyl-2-thiophenemagnesium bromide with 3,4-dideuterio-2,5-dichlorothiophene using [1,3-bis(diphenylphosphine)propane]nickel(II) chloride as catalyst. These compounds were used in a study of their radical cations in order to obtain unambiguous assignments of the β-hyperfine coupling constants [257].

3′,4′-Dideuterio-5,5″-dimethyl-2-2′:5′,2″-terthiophene [257]

2-Bromo-5-methylthiophene (6.0 g, 34 mmol) is reacted with magnesium turnings (1.24 g, 51 mmol) and a small amount of iodine in anhydrous diethyl ether (10 ml) to form the corresponding Grignard reagent, which is transferred to a dropping funnel by pressing with an inert gas in order to separate from unreacted magnesium. The Grignard solution is added dropwise to a solution of 2,5-dichloro-3,4-dideuteriothiophene (2.23 g, 14.4 mmol) and [1,3-bis(diphenyl-phosphine)propane]nickel(II) chloride (690 mg, 1.28 mmol) in anhydrous diethyl ether (40 ml). After refluxing for 6 h with stirring the reaction mixture is poured into ice-water. The product is extracted with diethyl ether (3×100 ml) and the combined organic phases are dried over calcium chloride. The solvent is distilled off in vacuo and the residue purified by chromatography on silica gel using hexane as eluent, giving 1.3 g (33%) of the title compound as yellow crystals mp 94 °C.

8.4.2.2 From thiophene dimagnesium or dizinc halides and halothiophenes

Due to sterical crowding the palladium(0)-catalyzed reaction of 3,4-di-*tert*-butyl-2,5-thiophenedizinc chloride and 2-bromothiophene gave a mixture of the expected 3′, 4′-di-*tert*-butyl-α-terthienyl (43%) and 3,4-di-*tert*-butyl-2,2′-bithienyl (42%) [69].

8.4.2.3 From thiophenemagnesium bromide and bromo- or iodobithienyls

2,3′;5′,3″-Terthienyl is prepared from 2-thienylmagnesium bromide and 4-bromo-2,3′-bithienyl and 2,2′:4′,3″-terthienyl from 3-thienylmagnesium bromide and 4-bromo-2,2′-bithienyl [258]. 2,3′:4′,3″-Terthienyl is prepared by [1,3-bis(diphenylphosphine)propane]nickel(II) chloride-catalyzed coupling of 2-thienylmagnesium bromide with 4-bromo-3,3′-bithienyl [258], 2,2′:3′,3″-terthienyl from 2-thienylmagnesium bromide and 2-bromo-3,3′-bithienyl, and finally 2,3′:2′,3″-terthienyl from 2-thienylmagnesium iodide with 5-bromo-2,2′-bithienyl. The coupling of 5-methyl-2-thiophenemagnesium iodide with 5-bromo-2,2′-bithienyl is used for the preparation of 5-methyl-α-terthienyl [66]. Coupling with 5-iodo-3-methyl-2-[5-trimethylsilyl)thienyl]thiophene is used for the preparation of 3′,3″-dimethyl-5-trimethylsilyl-α-terthienyl [76].

3′,3″-Dimethyl-2-(trimethylsilyl)-5,2′:5′,2″-terthiophene [76]

To a solution of 5-iodo-3-methyl-2-[5-trimethylsilyl)thienyl]thiophene and [1.3-bis(diphenylphosphine)propane]nickel(II) chloride (0.054 g, 0.1 mmol) in anhydrous diethyl ether (5.0 ml) the Grignard reagent prepared from 2-iodo-3-methylthiophene (1.49 g, 6.66 mmol) is added dropwise at 0 °C. The reaction mixture is allowed to warm to room temperature overnight, poured into water and filtered through Celite. The phases are separated and the aqueous phase extracted with diethyl ether. The combined organic phases are washed with sodium chloride solution, dried over magnesium sulfate and evaporated. The residue is purified by chromatography on silica gel using hexane as eluent, giving 1.04 g (67 %) of the title compound as a thick light-green liquid.

Palladium(0)-catalyzed Kumada coupling of 2-thiophenemagnesium bromide with 5-bromo-5′-dodecyl-2,2′-bithienyl is used for the preparation of 5-dodecyl-terthienyl [18]. The Grignard reagents from 3-methyl-and 3-butyl-2-iodothiophenes were coupled with 3-methyl-and 3-butyl-5-iodo-5′-trimethylsilyl-2,2′-bithienyl using [1,3-bis(diphenylphosphine)propane nickel(II) chloride as catalyst to give 3,3′-dimethyl- and 3,3′-dibutyl-5-trimethylsilyl-2,2′:5′,2″-terthienyl [89].

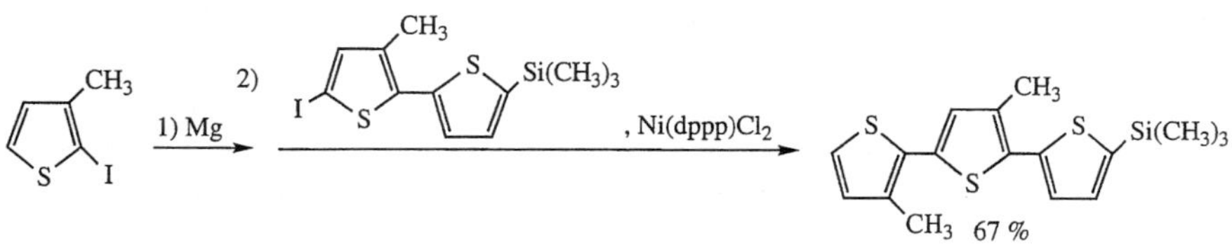

The Grignard reagent from 5-dodecyl-2-bromothiophene prepared by the entrainment method gave upon [1,3-bis(diphenylphosphine)propane]nickel(II) chloride-catalyzed coupling with 5-bromo-2,2′-bithienyl, 5-dodecyl-α-terthienyl [63]. The [1,3-bis(diphenylphosphine)propane]nickel(II) chloride-catalyzed coupling of 5-*tert*-butyl-2-thiophenemagnesium bromide, 5-*tert*-butoxy-2-thiophenemagnesium bromide and 5-trimethylsilylthiophenemagnesium bromide with 5-iodo-2,2′-bithienyl is used for the preparation of 5-*tert*-butyl-, 5-*tert*-butoxy-, and 5-trimethylsilyl-α-terthienyl, respectively [25].

5-tert-Butyl-2,2′:5′,2″-terthienyl [259]

To a solution of 2-*tert*-butylthiophene (0.458 g, 326 mmol) in anhydrous diethyl ether (10 ml) under nitrogen, 1.6 *M* butyllithium in hexane (2.04 ml) is added via syringe at room temperature. The mixture is stirred at this temperature for 5 h and then refluxed for 1 h, after which magnesium bromide etherate (0.95 g, 3.67 mmol) is added at 0 °C. The solution obtained is stirred at room temperature for 30 min and 5-iodo-2,2′-bithienyl (0.96 g, 3.3 mmol) in anhydrous diethyl ether is added along with [1,3-bis(diphenylphosphine)propane]nickel(ll) chloride (10 mg). The reaction mixture is stirred for 15 h followed by the usual workup giving 0.45 g (45%) of the title compound.

8.4.2.4 From bithienylmagnesium or zinc halides and halothiophenes

The bis(diphenylphosphine)ferrocenepalladium(II) chloride-catalyzed coupling between 2,2′-bithienyl-5-zinc chloride and 5-iodo-2-acetoxymethyl thiophene, gave only 24% of the expected 5-acetoxymethyl-α-terthienyl and 11% of 5-(2,2′-bithienyl-5-yl)methyl-α-terthienyl [66]. The [1,3-bis(diphenylphosphine)- propane]nickel(II) chloride-catalyzed coupling of 2,2′-bithienyl-5-magnesium bromide with 1,8-bis(5-bromothien-2-yl)naphthalene gives 1,8-bis(5,2′:5′,2″-terthienyl-2-yl)naphtalene in 62% yield [121]. The reaction of 1-(5-bromo-2-thienyl)naphthalene with 2,2′-bithienyl-5-magnesum bromide gives 1-(1-napthyl)-5-α-terthienyl [121,260]. 5-(3″-Methyl-2″-butenyl)-2,2′:5′,2″-terthienyl was prepared by [1,3-bis(diphenylphosphine)propane]-nickel(II) chloride-catalyzed coupling of 2,2′-bithienyl-5-magnesium iodide and 2-bromo-5-(3′-methyl-2-butenyl)thiophene [261].

5-(3′-Methyl-2″-butenyl)-2,2′:5′,2″-terthienyl [261]

To a solution of the Grignard reagent prepared from 5-iodo-2,2′-bithienyl (397 mg, 1.35 mmol) and magnesium turnings (33 mg, 1.35 mmol) in diethyl ether (10 ml) [1,3-bis(diphenylphosphine)propane]nickel(II) chloride (10 mg) is added followed by dropwise addition during 2 h of a solution of 2-bromo-5-(3′-methyl-2′-butenyl)thiophene (314 mg, 1.35 mmol) in diethyl ether (2 ml). The reaction mixture is stirred overnight at room temperature and acidified with 2 *M* hydrochloric acid until pH about 2. The product is extracted with dichloromethane (4 × 25 ml) and the combined organic phases are dried over calcium chloride and evaporated. The residue is purified by chromatography on silica gel using pentane as eluent, giving 48 mg (11%) of the title compound as a yellow solid mp 71–72 °C.

5-(Dimesitylboryl)-2,2′:5′,3″-terthienyl is prepared by palladium(0)-catalyzed coupling of 3-bromothiophene with the 5′-lithium derivative of 5-(dimesitylboryl)-2,2′-bithienyl.

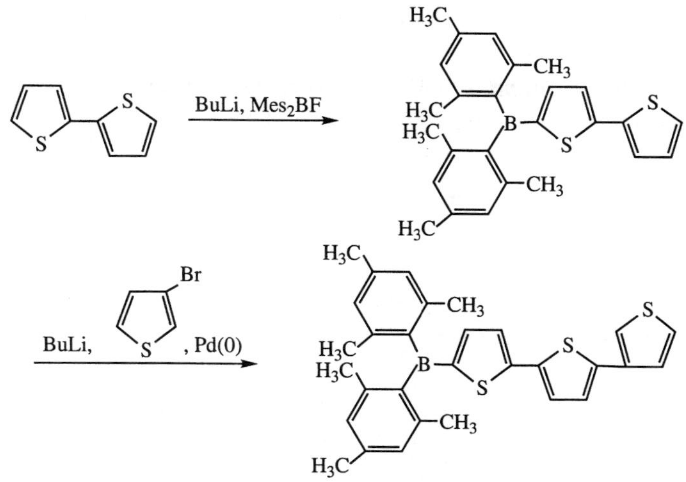

8.4.2.5 Through Kumada coupling of thienylzinc derivatives and halobithienyls

The palladium-catalyzed Kumada coupling of 5-methyl-2-thienylzinc chloride with 5-bromo-5′-iodo-2,2′-bithienyl occurs selectively at the iodine yielding 5-bromo-5′-methyl-α-terthienyl [188].

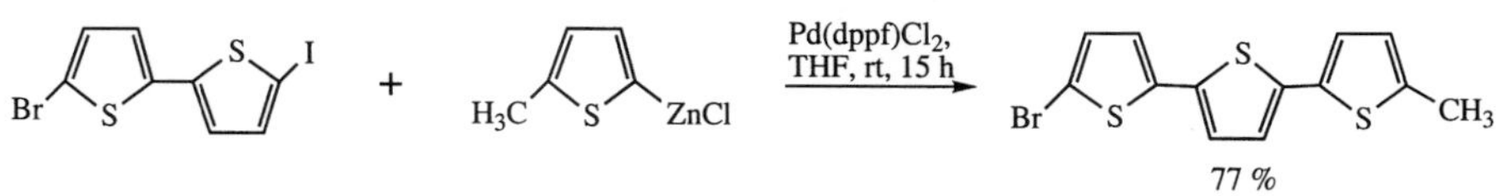

8.4.2.6 Through Suzuki coupling of thiopheneboronic acids with dihalothiophenes

8.4.2.6.1 From dihalothiophenes and two equivalents of thiopheneboronic acids

From 2.4 equiv. of 2-thiopheneboronic acid and 1 equiv. of 2,5-dibromothiophene a 40% yield of α-terthienyl is obtained. With 3-thiopheneboronic acid and 2,5-dibromothiophene, 3,2′:5′3″-terthienyl is obtained in 48% yield [262]. Coupling of 3,4-diiodo-2,5 dimethylthiophene with 2- and 3-thiopheneboronic acid gives 2,5-dimethyl-3,4-di(2-thienyl)- and 2,5-dimethyl-3,4-di(3-thienyl)thiophene in 44–49% yield [262]. Attempts to use the dibromo derivative were not successful. Suzuki coupling between 2,5-diiodothiophene and 4-pentoxy-2-thiopheneboronic acid gives a satisfactory yield of

4,4″-dipentoxy-α-terthienyl [15]. The coupling of 2-trimethylsilyl-5-thiopheneboronic acid with 2,5-diiodo-3,4-dimethylthiophene using palladium(0) catalyst in dimethoxyethane and sodium carbonate gave 5,5″-bis(trimethylsilyl)-3′,4′-dimethyl-α-terthienyl [76].

5,5″-Bis(trimethylsilyl)-3′,4′dimethyl-2,2′:5′,2″-terthienyl [76]

A 25-ml round bottomed flask is charged with 2-trimethylsilyl-5-thiopheneboronic acid (0.63 g, 3.15 mmol), 2,5-diiodo-3,4-dimethylthiophene (0.364 g, 1.0 mmol), tetrakis(triphenylphosphine)palladium(0) (0.046 g, 0.04 mmol), 2 *M* sodium carbonate solution (2.0 ml) and 1,2-dimethoxyethane (4.0 ml). The reaction mixture is stirred overnight at 85–90 °C and poured into water. The product is extracted with diethyl ether and the combined organic phases washed with sodium chloride solution, dried and evaporated. The residue is purified by flash chromatography giving 0.309 (73%) of the title compound as light-green crystals.

8.4.2.6.2 From thiophenediboronic acids and two equivalents of halothiophenes

The best method for the preparation of 3,3″-dipentoxy-α-terthienyl, due to the ease of purification, is the Suzuki coupling between 2,5-thiophenediboronic acid and 2-iodo-3-pentoxythiophene [15].

3,3″-Dipentoxy-2,2′:5′,2″-terthienyl [15]

A solution of 2-iodo-3-pentoxythiophene (8.2 g, 27.8 mmol) and tetrakis(triphenylphosphine)palladium(0) (841 mg, 0.73 mmol) in 1,2-dimethoxyethane (95 ml) is stirred under argon for 15 min. 2,3-Thiophenediboronic acid (2 g, 11.9 mmol) is added immediately followed by 1 *M* sodium bicarbonate solution (71 ml). The reaction mixture is refluxed for 4 h with vigorous stirring. After cooling to room temperature the mixture is filtered and the solid washed with hexane. The filtrate is extracted with diethyl ether and the combined organic phases are washed with water, dried and evaporated. The residue, a brown oil, is purified by flash chromatography on silica gel using hexane/chloroform in proportion 9:1 as eluent to start with and then in the proportion 8:2 giving 2.5 g (50%) of the title compound.

8.4.2.6.3 From halobithienyls and thiopheneboronic acid

2-Thiopheneboronic acid and 5-bromo-2,2′-bithienyl has been used in the preparation of the terthienyl shown below [84].

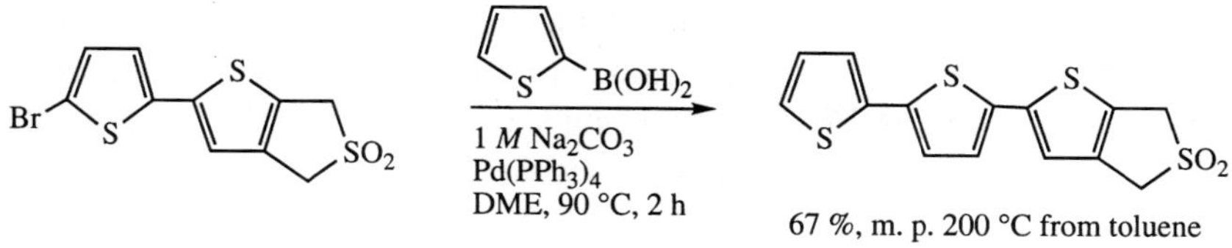

The coupling of 2,3,5-triiodothiophene with 2-thiopheneboronic acid is used for the preparation of 3′-iodo-2,2′:5′,2″-terthienyl [263]. Palladium(0)-catalyzed coupling of 3-octyl-2-thiopheneboronic acid ester with 5-chloro-5′-iodo-3,3′-octyl-2,2′-bithienyl gives 5″-chloro-3,3′,3″-tris(octyl)-α-terthienyl in 80% yield [85,264].

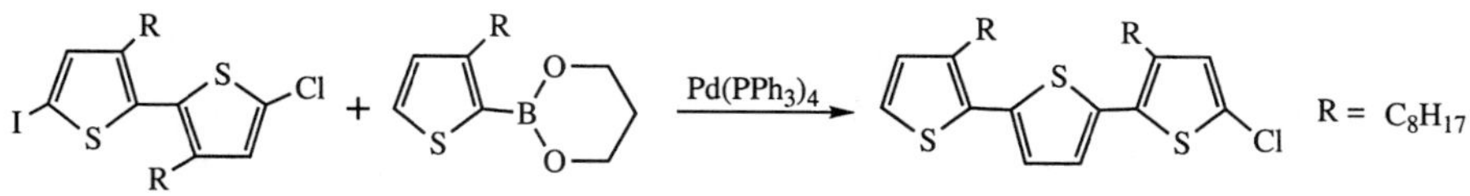

8.4.2.7 Through Stille couplings

8.4.2.7.1 From halothiophenes and trialkystannyl bithienyls

Stille couplings between 5-trimethylstannyl-2,2′-bithienyl and 2-bromothiophene, 2-bromo-5-nitrothiophene, 2-bromo-5-thiophenealdehyde, 2-acetyl-5-bromothiophene and 2-bromo-3,5-dinitrothiophene give α-terthienyl, 5-nitro, 5-formyl-, 5-acetyl- and 3,5-dinitro-α-terthienyl, respectively, in 65–86% yield [87].

Stille coupling between 2-acetoxymethyl-5-iodothiophene and 5-trimethylstannyl-2,2′-bithienyl has been used for the preparation of 5-acetoxymethyl-α-terthienyl in only 47% yield, as 15% of α-quaterthienyl was obtained as byproduct [66].

8.4.2.7.2 From halobithienyls and trialkylstannylthiophenes

Stille coupling of 5-bromo-2,2′-bithienyl with 2-tributylstannyl-5-trimethylsilylthiophene is used for the preparation of 5-trimethylsilyl-α-terthienyl [89].

5-(Trimethylsilyl)2,2′:5′,2″-terthienyl [89]

A flame-dried flask is charged with 2-tributylstannyl-5-trimethylsilylthiophene (4.41 g, 9.9 mmol), 5-bromo-2,2′-bithienyl (1.66 g, 6.6 mmol), tetrakis(triphenylphosphine)palladium(0) (0.38 g, 0.33 mmol) and toluene (30 ml). The reaction mixture is heated at 100 °C overnight under nitrogen, after which it is poured into water containing a few drops of 3 *M* hydrochloric acid to destroy the emulsion. The phases are separated and the aqueous phase extracted with diethyl ether. The combined organic phases are washed with sodium bicarbonate solution and sodium chloride solution, dried over sodium sulfate and evaporated. The residue is purified by chromatography on silica gel using hexane as eluent giving 1.78 g (84%) of the title compound as bright-yellow crystals.

Coupling of 3,4-difluoro-2-tributylstannyl-5-trimethylsilylthiophene with 5-bromopentafluoro-2,2′-bithienyl yields heptafluoro-5-trimethylsilyl-α-terthienyl, which upon treatment with *N*-bromosuccinimide gives the perfluoro-5-bromo-α-terthienyl [91].

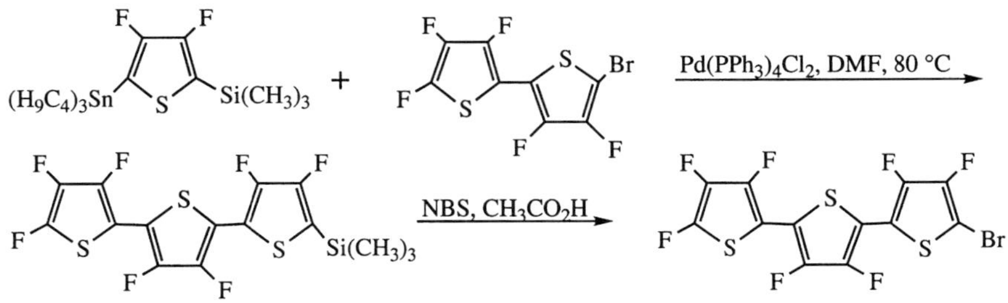

Stille coupling between 2-trimethylstannyl-4-octylthiophene and 5-bromo-5′-carbobenzyloxy-2,2′-bithienyl is used for the preparation of 5-carbobenzyloxy-4″-α-terthienyl in good yield [127]. This was used as a model reaction for the preparation of the polymer-bound 4-octyl-5″-carboxy-α-terthienyl from polymer-bound 5-bromo-2.2′-bithienyl-5′-carboxylic acid and 2-(trimethylstannyl)-4-octylthiophene [129]. Combined with bromination these constitute a method for solid phase synthesis of oligothiophenes up to pentamers. 5-Bromo-3″-alkylthio-α-terthienyl is formed in low yield by Stille reaction between two equivalents of 3-alkythio-2-trimethylstannylthiophene and 5,5′-dibromo-2,2′-bithienyl. The main product was a sexithiophene (see below) [265].

8.4.2.7.3 From dihalothiophenes and two equivalents of alkylstannylthiophenes

α-Terthienyl can also be prepared by the Stille coupling of 1 mol of 2,5-dibromothiophene and 2 mol of 2-trimethylstannylthiophene, but in only 42% yield [87]. Palladium-catalyzed coupling of ethyl 2,5-dibromothiophene-3-acetate with 2-tributylstannylthiophene gives ethyl α-terthienyl-3′-acetate [266,267].

Ethyl(2,2′:5′,2″-terthienyl)-3′-acetate [267]

Ethyl 2,5-dibromothiophene-3-acetate (3.2 g, 10 mmol) and 2-tributylstannylthiophene (7.45 g, 20 mmol) are added to a tetrahydrofuran solution of bis(triphenylphosphine)palladium(II) chloride (0.46 g, 0.4 mmol). The reaction mixture is heated at 80 °C for 24 h and the solvent removed under vacuum. The residue is taken up in dichloromethane and this solution is washed with water, dried over magnesium sulfate and evaporated. The crude product is purified by chromatography on silica gel using toluene as eluent, giving 1.68 g (50%) of the title compound.

Reaction of 2,5-dibromo-3,4-dinitrothiophene with 3-alkyl-2-tributylstannylthiophene is used for the preparation of 3,3″-dialkyl-3′,4′-dinitro-α-terthienyl [268,269].

3′,4′-Dinitro-2,2′:5′2″-terthienyl [269]

To a round-bottomed flask containing freshly prepared bis(triphenylphosphine)palladium(II) chloride 2-tributylstannylthiophene (3.51 g, 9.4 mmol), 2,5-dibromo-3,4-dinitrothiophene (1.53 g, 9.2 mmol) and anhydrous toluene (60 ml) are added. The reaction mixture is refluxed for 12 h, after which the solvent is removed by evaporation. The residue is dissolved in dichloromethane and this solution is added to an aqueous solution of potassium fluoride, which is filtered in order to remove insoluble tributyltin fluoride. The phases are separated and the organic phase washed three times with water, dried over magnesium sulfate and evaporated. The residue, an orange/brown solid, is purified by chromatography on silica gel using hexane/dichloromethane (6:4) as eluent giving 1.15 g (74%) of the title compound as a gold/orange solid mp 148–150 °C.

Similarly 3′,4′-bis[methoxycarbonyl)methyl]-α-terthienyl is prepared in 88% yield from 2,5-dibromo-3,4-bis[(methoxycarbonyl)methyl]thiophene and 2-tributylstannylthiophene [224]. Stille coupling between *N-tert*-butoxycarbonyl protected 3-amino-2-tributylstannylthiophene and 2,5-dibromo-3,4-dinitrothiophene and 3,4-(N-butylimido)-2,5-dibromothiophene using the recently developed tri(dibenzylideneacetone)palladium(0), copper(I) and trithienylarsine as catalyst system, gives excellent yields of the substituted α-terthienyls [270]. An α-terthienyl, carrying crown ether subunits across the 3′,4′-position is prepared by palladium(0)-catalyzed coupling of 2-(trialkylstannyl)thiophene and a 2,5-dichlorothiophene with a crown-ether across the 3,4-position [271].

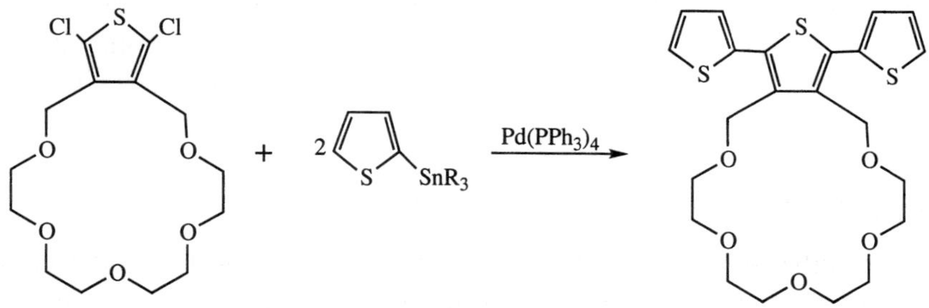

Stille coupling of 2,5-dibromo-2,4-dihexylthiophene-1,1-dioxide with two equivalents of 2-tributylstannylthiophene using tri(dibenzylideneacetone)-palladium(0) and triphenylarsine as catalyst gave 3′,4′-dihexyl-α-terthienyl-1′,1′-dioxide [272].

3′,4′-Dihexyl-2,2′:5′,2″-terthienyl-1′,1′-dioxide [272]

To a solution of tri(dibenzylideneacetone)palladium(0) (0.027 g, 0.026 mmol) and triphenylarsine (0.032 g, 0.104 mmol) in toluene (10 ml) 2,5-dibromo-3,4-dihexylthiophene-1,1-dioxide (0.77 g, 1.74 mmol) in toluene (5 ml) is added. The mixture is heated to reflux and 2-tributylstannylthiophene (1.10 ml, 3.5 mmol) is added. The reaction mixture is refluxed for 1 h, after which the solvent is removed by evaporation. The residue is purified by chromatography on silica gel using hexane/ethyl acetate (95:5) as eluent giving 500 mg (71%) of the title compound in yellow microcrystalline form.

8.4.2.7.4 From bis(trialkylstannyl)thiophenes and two equivalents of halothiophenes

3,3″-Dipentoxy-α-terthienyl is also obtained from 2,5-bis(trimethylstannyl)thiophene and 2-iodo-3-pentoxythiophene. However, in the Stille reaction the highest amounts of higher oligomers are formed, which is attributed to halogen–tin exchange [15]. 3,3″-Dimethoxy-α-terthienyl is obtained by Stille coupling of 2,5-bis(tributylstannyl)thiophene with 2-iodo-3-methoxythiophene and with 2-iodo-4-methoxy-5-methylthiophene, 4,4″-dimethoxy-5,5″-dimethyl-α-terthienyl is obtained [28].

3,3″-Dimethoxy-2,2′:5′,2″-terthienyl [28]

A flask is charged with 2-iodo-3-methoxythiophene (1 g, 4 mmol), 2,5-bis(tributylstannyl)thiophene(1.35 g, 2.04 mmol), tetrakis(triphenylphosphine)-palladium(0) (200 mg) and toluene (15 ml). The reaction mixture is first purged with argon for 20 min and then heated at 100–110 °C overnight, after which it is poured into saturated ammonium chloride solution. The phases are separated and the aqueous phase extracted with diethyl ether. The combined organic phases are washed with sodium chloride solution, dried over sodium sulfate and evaporated. The residue is purified by flash chromatography using ethyl acetate/hexane (1:99) as eluent giving 320 mg (51%) of the title compound as yellow crystals.

Stille coupling between 2,5-bis(tributylstannyl)thiophene and 2-bromo-5-mesitylthiophene or 2-bromo-3,4-ethylenedioxythiophene gives 2,5″-bis(mesitylthio)-α-terthienyl and 3,4,3″,4″-bis(ethylenedioxy)-2,5″-(mesitylthio)-α-terthienyl, respectively [9]. A series of mixed thiophene/3,4-ethylenedioxythiophene containing oligothiophene were also prepared. Thus coupling of 2,5-bis(tributylstannyl)-3,4-ethylenedioxythiophene with two equivalents of 2-bromo-5-mesitylthio-3,4-ethylenedioxythiophene gives 5,5″-bis(mesitylthio)-2,2′:5′,2″-ter(3,4-ethylenedioxy)thienyl [9]. 3,4-Ethylenedioxy-2,5″-bis(mestylthio)-α-terthienyl is obtained by Stille coupling of 2-bromo-5-mesitylthiophene and 3,4-ethylenedioxy-2,5-bis(tributylstannyl)thiophene [9].

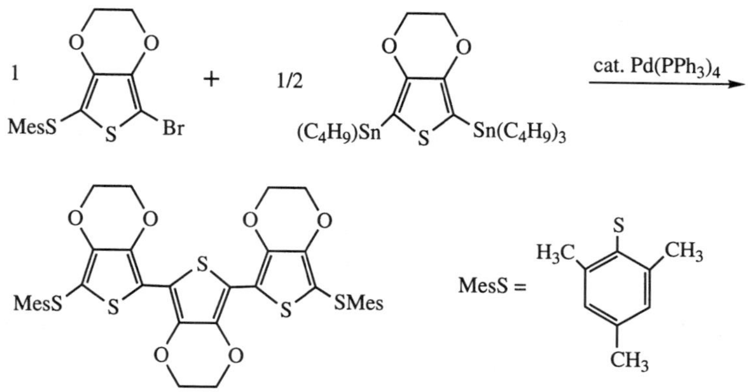

5,5″-Bis(mesitylthio)-2,2′:5′,2″-ter(3,4-ethylenedioxythienyl) [9]

A solution of 2-bromo-5-mesitylthio-3,3-ethylenedioxythiophene (2.00 g, 5.38 mmol), 2,5-bis(tributylstannyl)-3,4-ethyleneoxythiophene (0.77 g, 2.5 mmol) and tetrakis(triphenylphosphine)palladium(0) (75 mg) in toluene (40 ml) is refluxed for 18 h. After cooling the reaction mixture to room temperature and reducing the solvent volume, hexane (150 ml) is added and a shiny gold precipitate is obtained, which is filtered off. This material is purified by chromatography using hexane/ethyl acetate (5:1) as eluent, giving 28% of the title compound.

Stille coupling between 2,5-bis(tributylstannyl)thiophene and two equivalents of 5-bromo-2-thiophene aldehyde is used for the preparation of 5,5″-diformyl-α-terthienyl [100] and with 2,5-dibromo-3,4-bis[(methoxycarbonyl)methyl]thiophene an excellent yield of 3,4,3″,4″-tetra[(methoxycarbonyl)methyl]-α-terthienyl is obtained [224].

5,5″-Diformyl-2,2′:5′,2″-terthienyl [100]

A solution of 2,5-bis(tributylstannyl)thiophene (1 g, 1.51 mmol) and 2-bromo-5-thiophene aldehyde (578 mg, 3.02 mmol) in *N,N*-dimethylformamide (20 ml) is deaerated twice with argon before tetrakis(triphenylphosphine)palladium(0) (19 mg, 0.016 mmol) is added. The reaction mixture is heated at 65–70 °C for 6 h under argon. The resultant reddish brown suspension is concentrated under reduced pressure. The solids collected by filtration are washed with hexane (400 ml), diethyl ether (200 ml) and dichloromethane (200 ml) giving 380 mg (83%) of the title compound mp 222–224 °C after recrystallization from tetrahydrofuran.

8.4.2.8 Various metal-catalyzed reactions

A special example of regioselective oligomerization of 3-alkylthiothiophenes consists in their treatment with four equivalents of ferric chloride in dichloromethane giving 3,3′,3″-tris(methylthio)-α-terthienyl [273].

3,3′,3″-Tris(methylthio)-2,2′:5′,2″-terthienyl [273]

Anhydrous ferric chloride (2.49 g, 15.3 mmol) is dissolved in freshly distilled dichloromethane (60 ml) and the mixture is stirred at room temperature for 30 min. Then 2-methylthiophene (0.5 g, 3.84 mmol) in dichloromethane (20 ml) is added dropwise over a period of 20 min. After 1 h methanol (200 ml) is added and as no precipitate is formed the solution is evaporated and the solid residue taken up in chloroform. The chloroform solution is washed with water and 5% hydrazine solution, dried over sodium sulfate and evaporated. The residue, a red solid, is purified by chromatography on silica gel neutralized with a 2% triethylamine solution using cyclohexane/dichloromethane (85:15, 75:25, 65:35) as eluent, giving 0.31 g (62%) of the title compound as a brown oil.

8.4.3 Electrophilic substitutions of terthienyls

8.4.3.1 Alkylation of terthienyls

The reaction of α-terthienyl and *tert*-butyl chloride with aluminium chloride in dichloromethane is claimed to give 5,5″-di(*tert*-butyl)-α-therthienyl [259].

5′,5″-Di-tert-butyl-2,2′:5,2″-terthienyl [259]

Aluminum chloride (2.0 g, 15 mmol) is added to a vigorously stirred solution of 2,2′:5,2″-terthienyl (2.0 g, 15 mmol) and *tert*-butyl chloride (8.0 ml, 120 mmol) in dichloromethane (150 ml). The reaction mixture is stirred for 24 h at room temperature and poured into cold water (150 ml). After the usual workup a brown solid is obtained, which is recrystallized from ethanol giving 2.6 g (72%) of the title compound as tan crystals mp 134–135 °C.

8.4.3.2 Formylation and acylation of terthiophenes

Vilsmeyer monoformylation of α-terthienyl derivatives has been carried out using *N,N*-dimethylformamide and phosphorus oxychloride [100,206,274] or N-methylformanilide and phosphorus oxychloride [206,259].

5-Formyl-2,2′:5′,2″-terthienyl [274]

To a solution of *N,N*-dimethylformamide (1.2 ml, 15 mmol) and anhydrous dichloromethane (5 ml), phosphorus oxychloride (1.4 ml, 15 mmol) is added dropwise with stirring at 0 °C. The mixture is removed from the ice bath and warmed to 40 °C until a clear pale-yellow solution is obtained. The Vilsmeier reagent thus prepared is added dropwise to a solution of 2,2′:5′,2″-terthienyl (2.5 g, 10 mmol) in anhydrous dichloromethane (15 ml) at 0 °C. After 12 h at room temperature the solvent is evaporated, cold 1 *M* aqueous sodium hudroxide solution is added until pH 8–9 and the mixture heated on a steam bath for 2 h. The precipitate is filtered off, washed with water and dried. The yellow powder obtained is flash chromatographed on silica gel using hexane/dichloromethane (1:1) as eluent giving 2.67 g (96%) of the title compound as yellow crystals mp 134 °C.

Diformylation was difficult to achieve, but was forced by the reaction with three equivalents of reagents and then an additional equivalent at room temperature to give 5,5″-diformyl-α-terthienyl in 52% yield [259]. Monoformylation of 1,3-bis(2-thienyl)benzo[*c*]thiophene and its 3-dodecyl-2-thienyl and 3-hexyl-2-thienyl derivatives are also successfully carried out [241].

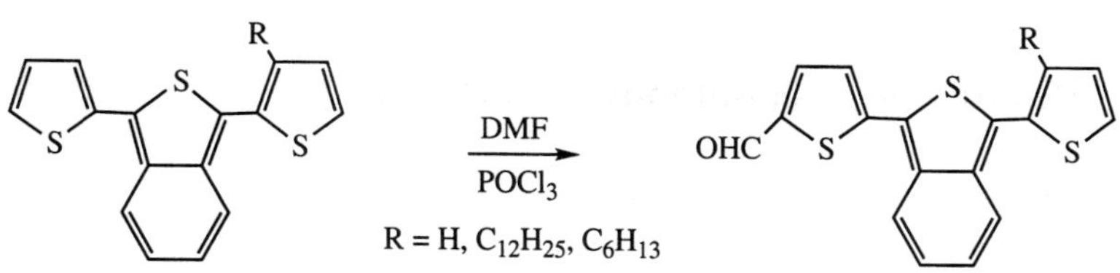

1-(2-Thienyl)-3-(5′-formyl-2′-thienyl)benzo[c]thiophene [241]

Phosphorous oxychloride (2.96 ml, 0.032 mol) is slowly added to a mixture of dichloromethane (40 ml) and *N,N*-dimethylformamide (2.45 ml, 0.032 mmol) at 0 °C. When the addition is completed the mixture is stirred until a pale-yellow solution is obtained. This solution is added to 1-(2-thienyl)-3-(2′-thienyl)benzo[*c*]thiophene (9 g, 0.03 mol) dissolved in dichloromethane (50 ml) at 0 °C. The reaction mixture is stirred at room temperature for 10 h, after which the solvent in completely removed. Aqueous sodium hydroxide solution is added, the mixture heated on a steam bath for 1 h. After cooling, the product is filtered off, air dried and passed through a column of aluminium oxide using chloroform/hexane (1:1) as eluent giving 8 g (83%) of the title compound mp 112 °C after recrystallization from methanol.

Acylation of α-terthienyl with hexanoyl chloride is used for the preparation of 5-hexanoyl-α-terthienyl [275]. 5,5″-Diacetyl-4′-alkyl-α-terthienyl is prepared by treatment of 4′-alkyl-α-terthienyl with acetic anhydride and phosphoric acid at 100 °C [255].

8.4.3.3 Nitration of terthienyls

Nitration of α-terthienyl with fuming nitric acid/acetic acid in dichloromethane/acetic anhydride is claimed to give almost selectively 5-nitro-α-terthienyl. Trace amounts of the 3′-nitro and 3-nitro derivatives were also isolated [261].

5-Nitro-2,2′:5′,2″-terthienyl [261]

To solution of 2,2′:5′,2″-terthienyl (1.24 g, 5 mmol) in dichloromethane (10 ml) and acetic anhydride (10 ml) at 5 °C 3.5 ml of a solution composed of fuming nitric acid (3 ml) in glacial acetic acid (30 ml) is added under stirring. The reaction mixture is stirred at room temperature for 6 h. Hydrolysis with water (60 ml) gives a red precipitate, which is collected by filtration, washed with water and air dried. The crude product is purified by flash chromatography on silica gel using a gradient hexane/ethyl acetate (5:95) as eluent, giving 1.20 g (82%) of the title compound mp 150–156 °C.

8.4.3.4 Bromination of terthienyls

Bromination with *N*-bromosuccinimide of α-terthienyl in *N,N*-dimethylformamide is the best method for the preparation of 5,5″-dibromo-α-terthienyl [63,276].

5,5″-Dibromo-2,2′:5′,2″-terthienyl [63]

In the absence of light *N*-bromosuccinimide (3.74 g, 21 mmol) is added portionwise at 20 °C to a solution of 2,2′:5′,2″-terthienyl (2.48 g, 10 mmol) in *N,N*-dimethylformamide (130 ml). The reaction mixture is stirred for 3 h and poured into ice. The precipitate formed is filtered off, washed several times with water, dried over phosphorus pentoxide and recrystallized from toluene/hexane giving 3.42 g (84%) of the title compound mp 159–160 °C.

Bromination with *N*-bromosuccinimide has also been used for the preparation of 5,5′-dibromo-3′-phenyl- and 5,5′ dibromo-3′,4′-diphenyl-α-terthienyl [252].

Bromination of 3′-dodecyl- [62], 3,3′-diphenyl-α-terthienyl and 3,3″-dihexyl-α-terthienyl with N-bromosuccinimide in *N,N*-dimethylformamide is used both for the preparation of the 5,5″-bromo derivative and the 5-bromo derivative [84,244,254,277].

5,5″-Dibromo-3,3″-dihexyl-2,2′:5′,2″-terthienyl [254]

To a solution of 3,3″-dihexyl-2,2′:5′,2″-terthienyl (0.50 g, 1.20 mmol) in *N,N*-dimethylformamide 20 ml) at 0 °C *N*-bromosuccinimide (0.384 g, 2.20 mmol) in *N,N*-dimethylformamide (20 ml) is added dropwise with stirring in the dark. The reaction mixture is stirred for 4 h, allowed to reach room temperature and stirred for another 36 h. It is poured into ice (100 g), the product is extracted with diethyl ether (5 × 50 ml), dried over magnesium sulfate and evaporated. The residue is purified by flash chromatography using petroleum ether as eluent, giving 0.60 g (49%) of the title compound as a yellow oil.

For the bromination of 5-heptadecyl-α-terthienyl, 1,3-dibromo-5,5-dimethylhydantoin in tetrahydrofuran, containing small amounts of azo bisisobutyronitrile, is also used [274]. Bromination with *N*-bromosuccinimide of 5-benzyloxycarbonyl-4″-octyl-α-terthienyl gives selectively a quantitative yield of the 5″-bromo derivative [126,127]. Bromination of 3,4:3″,4″-tetra-[(methoxycarbonyl)methyl]-α-terthienyl with *N*-bromosuccinimide in *N,N*-dimethyl formamide gives the 5,5″-dibromo derivative in quantitative yield [224]. 5-Bromo-5″-formylthiophene is prepared by bromination of 5-formyl-α-terthienyl with *N*-bromosuccinimide in dichloromethane/acetic acid [274] or in a somewhat better yield with bromine in chloroform–sodium bicarbonate [100].

5-Bromo-5″-formyl-2,2′:5′,2″-terthienyl [100]

To a solution of 5-formyl-2,2′:5′,2″-terthienyl (1.77 g, 6.42 mmol) in chloroform (100 ml) sodium bicarbonate (1.5 g, 17.8 mmol) is added follwed by dropwise addition of a solution of bromine (1.1 g, 6.88 mmol) in chloroform (10 ml) over a period of 1 h. The reaction mixture is stirred for an additional 5 h and then filtered. The filtrate is washed with water (2 × 20 ml), dried over magnesium sulfate and evaporated. The solid residue is recrystallized from dichloromethane giving 1.9 g (85%) of the title compound as golden brown needle-shaped crystals mp 158–160 °C.

8.4.3.5 Iodination of terthiophenes

5-Iodo-α-terthienyl was prepared in 45% yield using mercuric oxide and iodine [259]. 5-iodo-3,4′,4″-trioctyl-5″-chloro-α-terthienyl is prepared by reaction of 3,4′,4″-α-terthienyl with mercuric acetate and iodine in a mixture of chloroform/acetic acid in 96% yield [85]. 5,5″-Diiodo-α-terthienyl is prepared by reaction with *N*-iodosuccinimide in dichloromethane/acetic acid [67].

5,5″-Diiodo-2,2′:5′,2″-terthienyl [67]

2,2′:5′,2″-Terthienyl (994 mg, 4.00 mmol) and *N*-iodosuccinimide (1.98 g, 8.80 mmol) are dissolved in dichloromethane (60 ml). To this solution acetic acid (0.50 ml, 8.8 mmol) is added. Soon precipitation starts and the reaction mixture is magnetically stirred for 2 h under ice/water cooling. The precipitate is filtered off, washed with methanol and recrystallized from 2-butanone giving 1.26 g (63%) of the title compound as a yellow solid.

Recently 5,5″-diiodo-3,3″,4,4″-tetrabutyl-α-terthienyl was prepared by reaction of 3,3″,4,4′-tetrabutyl-α-terthienyl with iodine and mercuric oxide.

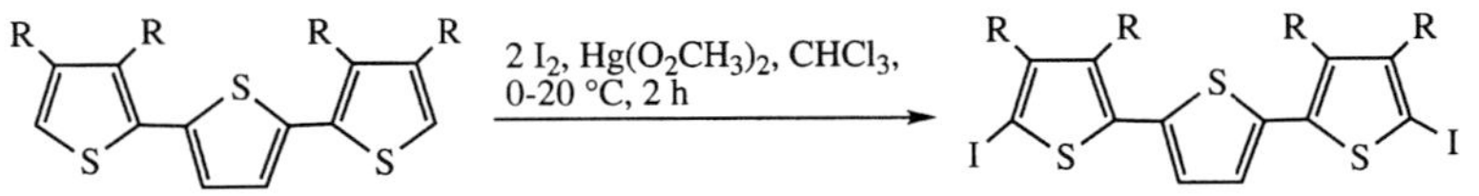

$R = C_4H_9$

8.4.3.6 Various electrophilic substitutions

Bromination with *N*-bromosuccinimide of 5-iodo-2,2′-bithienyl is used for the preparation of 5-bromo-5′-iodo-2,2′-bithienyl [188]. Treatment of α-terthienyl with one equivalent of chlorosulfonyl isocyanate at −30 °C in dichloromethane generated the *N*-chlorosulfonyl derivative of α-terthienyl-5-carboxamine, which on addition of *N*,*N*-dimethylformamide gave 5-cyano-α-terthienyl [261]. With excess chlorosulfonyl isocyanate the 5,5″-dicyano derivative is prepared [261].

5-Cyano-2,2′:5′,2″-Terthienyl [261]

To a solution of 2,2′:5′,2″-terthienyl (2 g, 8.05 mmol) in dichloromethane (40 ml) chlorosulfonyl isocyanate (5.25 ml, 8.5 mmol) in dichloromethane (10 ml) is added at −30 °C over a period of 10 min. The *N*-chlorosulfonyl amide is separated as a minute crystalline solid and the mixture is stirred until the temperature reaches 10 °C. After recooling to −30 °C *N*,*N*-dimethylformamide (17.05 ml, 7.9 mmol) is added over a period of 5 min and the reaction mixture is stirred at −30 °C for 1 h, after which it is gradually allowed to rise to room temperature and cubes of ice are added. When the ice has melted water (100 ml) is added, the phases are separated and the aqueous phase is extracted with dichloromethane (4 × 100 ml). The combined organic phases are washed with water (3 × 15 ml), dried and evaporated. The residue is purified by chromatography on a silica gel chromatotron plate using dichloromethane as eluent giving 1.65 g (75%) of the title compound as an orange solid mp 104–105 °C.

8.4.4 Substituted terthienyls through metalation followed by electrophiles

Metalation of α-terthienyls with one equivalent of butyllithium in hexane followed by methyl iodide gives a separable mixture of 7% of α-terthienyl, 69% of 5-methyl-α-terthienyl and 23% of 5,5″-dimethyl-α-terthienyl. Using excess butyllithium gave 82% yield of the 5,5″-dimethyl derivative [259]. Preparative gas liquid chromatography was necessary to obtain pure products. Depending upon the amount of butyllithium used in the metalation of terthienyl followed by *tert*-butyldimethylsilyl chloride the mono- and disilylated derivatives are obtained [277]. Metalation of α-terthienyl with butyllithium followed by *tert*-butyldimethylsilyl chloride is used for the preparation of the 5-*tert*-butyldimethylsilyl derivative and the 5,5″-disubstituted derivative [226,277]. In a similar way 5,5″-di(tributylstannyl)-α-terthienyl was prepared [28].

5,5″-Bis(dimethyl-tert-butylsilyl)-2,2′:5′,2″-terthienyl [226]

To a solution of 2,2′:5′,2″-terthienyl (1.2 g, 4.8 mmol) in anhydrous tetrahydrofuran butyllithium in hexane (3.85 ml, 96 mmol) is added dropwise. After 1 h dimethyl-*tert*-butylsilyl chloride in anhydrous tetrahydrofuran (10 ml) is added. The reaction mixture is strirred at room temperature overnight, after which water is added. The phases are separated and the aqueous phase is extracted twice with diethyl ether. The combined organic phases are dried over magnesium sulfate and evaporated. The residue is refluxed with methanol for about 5 min. The fraction soluble in methanol gives 0.46 g (26%) of 5-dimethyl-*tert*-butylsilyl-2,2′:5′,2″-terthienyl as a green solid mp 82 °C. The fraction not soluble in methanol gives 0.68 g (29%) of the title compound as a yellow solid mp 115–116 °C.

5,5″-Bis(tributylstannyl)-2,2′:5′,2″-terthienyl [28]

At −78 °C butyllithium (3.3 ml, 8.3 mmol) is added dropwise to a solution of 2,2′:5′,2″-terthienyl (1.02 g, 4.1 mmol) in anhydrous tetrahydrofuran (15 ml).

The mixture is stirred at 0 °C for 0.5 h and then recooled to −78 °C. Tributyltin chloride is introduced and the reaction mixture is allowed to warm to room temperature and stirred for an additional hour before water is added. The phases are separated and the aqueous phase extracted three times with diethyl ether. The combined organic phases are washed with saturated cupric sulfate solution, dried and evaporated giving the title compound in pure form.

Similarly dimetalation of 2,2′:5′,2″-terthienyl- [28], and 3,3″-dimethoxy-2,2′:5′,2″-terthienyl [28] followed by tributylstannyl chloride have been used for the preparation of di(tributylstannyl) derivatives. Metalation of 4′-alkyl-α-terthienyl with lithium diisopropylamide at −78 °C, followed by *N*-methylformanilide is used for the preparation of the 5,5″-dialdehyde [255]. Upon reaction with bromine, 5,5″-dibromo-α-terthienyl is obtained [176]. Metalation of α-terthienyl with butyllithium in tetrahydrofuran at −70 °C, followed by treatment with magnesium bromide and dodecanoyl chloride gives 5-dodecanoyl-α-terthienyl [106]. In the same way 5-hexanoyl-α-terthienyl is prepared [278].

5-Hexanoyl-2,2′:5′,2″-terthienyl [278]

To a solution of 2,2′:5′,2″-terthienyl (4.0 g, 16 mmol) in anhydrous tetrahydrofuran at −78 °C butyllithium in hexane (16 mmol) is added over a period of 10 min after which the stirring is continued for 20 min and magnesium bromide diethyl etherate (4.0 g, 15.5 mmol) is added. The cooling bath is removed and when the suspended solids have dissolved this solution is added to a second solution containing hexanoyl chloride, freshly distilled from *N*,*N*-diethylamine, (2.2 g, 16 mmol) and lithium tetrachloromanganate(II) (1 mmol) in tetrahydrofuran (40 ml) at 0 °C over a period of 30 min. The reaction mixture is kept at 0 °C for 1 h and then allowed to warm to room temperature overnight. Diethyl ether or toluene is added and the solution obtained washed with dilute hydrochloric acid and dilute sodium bicarbonate solution, dried and evaporated. The residue is purified by chromatography on silica gel (70 g) using hexane/dichloromethane as eluent, giving 2.0 g (36%) of the title compound as a bright, fluorescent yellow solid.

Metalation of α-terthienyl with butyllithium at − 70 °C followed by cerium trichloride gave the cerium intermediate, which upon further reaction with tetrahydrofuran gave the desired spiro compound [279].

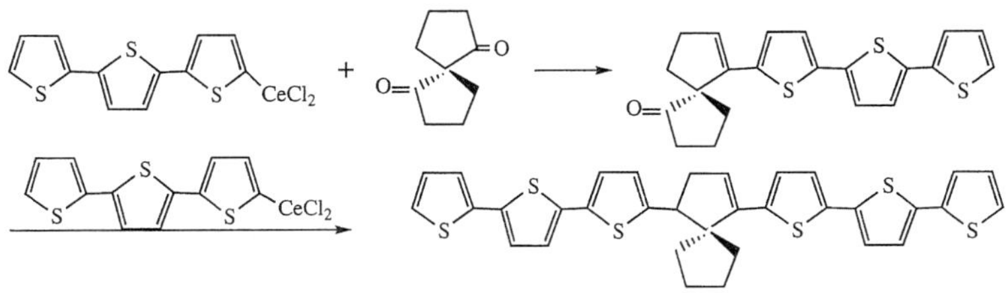

Metalation of 3,3″-dimethoxy-α-terthienyl with two equivalents of butyllithium in tetrahydrofuran followed by dimethyl sulfate gives 5,5″-dimethyl-3,3″-dimethoxy-α-terthienyl in 85% yield and with tributylstannyl chloride 3,3″-dimethoxy-5-methyl-α-terthienyl was obtained [28]. Dimetalation of 3,3″-dimethoxy-α-terthienyl with butyllithium followed by dimethylphenylsilyl chloride gives 5,5″-bis(dimethylphenylsilyl)-3,3″-dimethoxy-α-terthienyl [280].

5,5″-Bis(dimethylphenylsilyl)-3,3″-dimethoxy-2,2′:5′,2″terthienyl [280]

To a solution of 3,3″-dimethoxy-2,2′:5′,2″-terthienyl (136 mg, 0.44 mmol) in anhydrous tetrahydrofuran (20 ml) at −78 °C butyllithium (0.36 ml, 0.09 mmol) is added dropwise. The mixture is stirred at 0 °C for 30 min and then recooled to −78 °C, after which dimethyl phenylsilyl chloride (154 mg, 0.9 mmol) is added. The reaction mixture is stirred at −78 °C for 3 h, warmed to room temperature for 1 h and poured into water. The phases are separated and the aqueous phase extracted with diethyl ether. The combined organic phases are washed with sodium chloride solution, dried over sodium sulfate and evaporated. The residue is purified by flash chromatography giving 214 mg (85%) as yellow crystals.

Metalation of 5-trimethylsilyl-α-terthiophene with lithium diisopropylamide followed by trimethylstannylchloride gives a quantitative yield of 5-trimethylsilyl-5″-trimethylstannyl-α-terthienyl [89].

5-Trimethylsilyl-5″-trimethylstannyl-2,2′:5′2″-tertienyl [89]

To a solution of diisopropylamine (0.304 g, 3.0 mmol) in anhydrous tetrahydrofuran (2.0 ml) at −78 °C 2.24 *M* butyllithium in hexane (1.4 ml, 3.0 mmol) is added dropwise. The mixture is allowed to warm to 0 °C for 10 min and then recooled to −78 °C. To this solution 5-trimethylsilyl-2,2′:5′,2″-terthienyl (0.641 g, 2 mmol) in anhydrous tetrahydrofuran (2.0 ml) is added dropwise via cannula. After stirring this mixture at the same temperature for 2 h trimethyltin chloride (0.73 g, 3.66 mmol) in anhydrous tetrahydrofuran (2.0 ml) is added dropwise via cannula. The reaction mixture is allowed to warm to room temperature for 1 h and then poured into sodium chloride solution. The phases are separated and the aqueous phase extracted with diethyl ether. The combined organic phases are dried over sodium sulfate and evaporated giving a quantitative yield of the title compound as deep green crystals, which can be confirmed to be pure by NMR analysis. The compound darkens during workup and decomposes to starting material upon chromatography on silica gel using hexane as eluent.

Similarly metalation of 3,4′-dimethyl-5″-(trimethylsilyl)-α-terthienyl followed by trimethylstannyl- or tributylstannyl chloride gives 3,4′-dimethyl-5″-(trimethylsilyl)-5-(trimethylstannyl) or (tributylstannyl)-α-terthienyl in 96%

and 97% yield, respectively [76]. Using iodine in tetrahydrofuran as the electrophile gave 5-iodo-3,4′-dimethyl-5″-trimethylsilyl-α-terthienyl [76].

5-Iodo-3,4′-dimethyl-5″-trimethylsilyl-2,2′:5′,2″-terthienyl [76]

To a solution of diisopropylamine (0.203 g, 2.01 mmol) in anhydrous tetrahydrofuran (2.0 ml) at −78 °C 2.44 *M* butyllithium in hexane (0.82 ml, 2.01 mmol) is added dropwise. The mixture is warmed to 0 °C for 5 min and then recooled to −78 °C. To this solution 3′,3″-dimethyl-2-(trimethylsilyl)-5,2′:5′,2″-terthienyl (0.70 g, 2.01 mol) in anhydrous tetrahydrofuran (2.0 ml) is added dropwise *via* cannula. The reaction mixture is allowed to warm to room temperature for 20 min and poured into water. The phases are separated and the aqueous phase extracted with diethyl ether. The combined organic phases are washed with sodium chloride solution, dried over magnesium sulfate and evaporated. The residue is purified by flash chromatography on silica gel using hexane as eluent giving 0.616 g (66%) of the title compound as a light-green liquid.

Metalation of α-terthienyl with butyllithium, followed by magnesium bromide and phthalic anhydride followed by methylation leads to 5-(2‴-carbomethoxybenzoyl)-α-terthienyl. From the dilithium derivative similarly the 5,5″-bis(2‴-carbomethoxybenzoyl) derivative is prepared [261]. The reaction of dilithiated α-terthienyl with dimesitylboron fluoride is used for the preparation of 5,5″-bis(dimesithylboron)-α-terthienyl [281]. Both 5-methylthio- and 5,5″-bis(methylthio)-α-terthienyl were obtained by metalation of α-terthienyl with butyllithium in hexane followed by reaction with dimethyl disulfide [259].

Dimetalation of 1,3-di(2-thienyl)benzo[c]thiophene with two equivalents of butyllithium/*N*,*N*,*N*′,*N*′-tetramethylethylenediamine in tetrahydrofuran gives a high yield of the 5,5′-dicarboxaldehyde [241].

1,3-Di(2-thienyl)benzo[c]thiophene-5,5′-dicarboxaldehyde [241]

A solution of 1,3-di(2-thienyl)benzo[*c*]thiophene (5 g, 0.0168 mol) and *N*,*N*,*N*′,*N*′-tetramethylethylenediamine (10.1 ml, 0.067 mol) in anhydrous tetrahydrofuran (150 ml) is cooled to −78 °C under nitrogen and treated with 2 *M* butyllithium (50 ml). The temperature is slowly raised to 0 °C and the mixture is stirred at that temperature for 1 h, after which it is recooled to −78 °C and treated with *N*,*N*-dimethylformamide (45 ml, 0.58 mol). The reaction mixture is stirred at room temperature for 12 h and the poured into ice-containing hydrochloric acid. After stirring for 2 h the precipitate is filtered off, washed with water and dried *in vacuo* giving 5.2 g (88%) of the title compound mp 241 °C.

Similarly the 5,5′-diformyl derivatives were prepared from the 1-(3-dodecyl-2-thienyl) and 1-(3-hexyl-2-thienyl) derivatives [241]. The bisborolane shown below is prepared upon dimetalation of 1,3-di(2-thienyl)benzo[c]thiophene as described above followed by reaction with pinacolboronate [241].

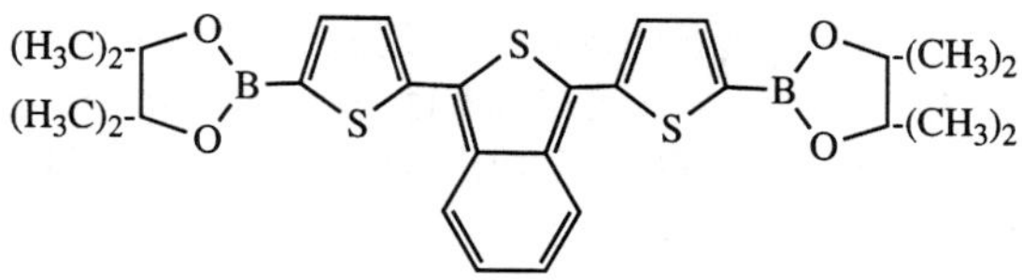

Metalation of 5-hexyl-α-terthienyl with butyllithium in tetrahydrofuran at − 30 °C, followed by reaction with sulfur and acetyl chloride at − 78 °C is used for the preparation of 5-acetylthio-5″-hexyl-α-terthienyl [171]. Reaction of α-terthienyl with butyllithium in tetrahydrofuran at − 70 °C followed by sulfur and hexanoyl chloride is used for the preparation of 2-hexanoylthio-α-terthienyl [278].

5-Hexanoylthio-2,2′:5′,2″-terthienyl [278]

To a solution of 2,2′:5′,2″-terthienyl (4.0 g, 16 mmol) in tetrahydrofuran (150 ml) at − 78 °C butyllithium in hexane (16 mmol) is added over 10 min. The stirring is continued at − 78 °C for 20 min, after which the mixture is added to a suspension of sulfur (0.51 g) in tetrahydrofuran (40 ml) at − 78 °C and treated with freshly distilled hexanoyl chloride (2.2 g). After warming the reaction mixture to room temperature overnight it is partitioned between diethyl ether and dilute hydrochloric acid. The phases are separated and the organic phase washed with dilute sodium bicarbonate, dried and evaporated. The residue is purified by chromatography on silica gel using dichloromethane/hexane as eluent, giving 2.6 g (43%) of the title compound.

Reacting this compound with sodium ethylate in ethanol gave the thiolate derivative, which upon reaction with hexafluorobenzene in DMI gives the hexakis(5″-hexyl-5-α-terthienylthio)benzene [171].

Metalation of α-terthienyl with butyllithium followed by reaction with cuprous iodide and iodomethylphosphonate is a facile synthesis of 2-(α-terthienyl)methyl phosphonate [180].

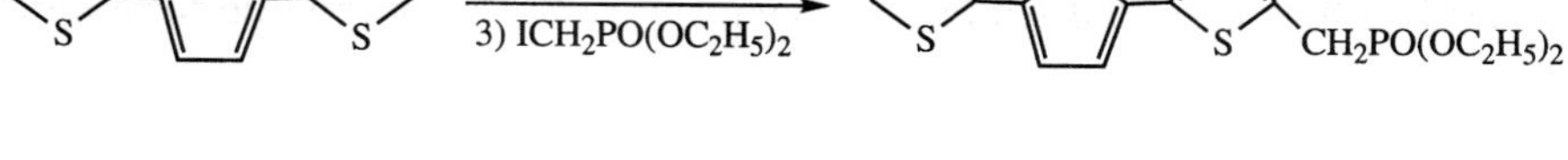

8.4.5 Substituted terthienyls via halogen–metal exchange of haloterthienyls followed by reaction with electrophiles

Halogen–metal exchange of 5-bromo-3,3″-dipentyl-α-terthienyl butyllithium followed by trimethyl borate is used for the preparation of 3,3″-dipentoxy-5-α-ter(thienyl)boronic acid [84].

8.4.6 Substituted terthienyls through various modifications of substituents

8.4.6.1 From nitroterthienyls

Reduction of 3′,4′-dinitro-α-terthienyl with tin in hydrochloric acid gives the 3′,4′-diamino derivative [268,269]. Also the corresponding 3,3″-dimethyl derivative is prepared similarly [269]. Alternatively, stannous chloride in ethanol and hydrochloric acid can be used for the reduction of 3,4-dinitrothiophene derivatives [93,268].

3′,4′-Diamino-2,2′:5′,2″-terthienyl [269]

To a mixture of 3′,4′-dinitro-2,2′:5′,2″-terthienyl (1.10 g, 0.0033 mol) in absolute ethanol (50 ml) and concentrated hydrochloric acid (50 ml) tin metal (2.5 g/atoms) is added in small portions. The reaction mixture is stirred for 24 h. The yellow solid formed is collected on a coarse frit and washed with ethanol (25 ml), after which it is suspended in diethyl ether (100 m) and water (100 ml). Sodium hydroxide solution (4 *M*) is added to basify the solution and stirring is continued for 4 h. The phases are separated and the water phases stirred with additional diethyl ether for 1 h. After repeating this procedure the combined organic phases are washed twice with water, dried over magnesium sulfate and evaporated giving 0.57 g (85%) of the title compound as a brown/orange solid mp 95–97 °C.

Hydrogenation with palladium on carbon as catalyst has been used for the preparation of 5-amino-α-terthienyl from the 5-nitro derivative [261].

5-Amino-2,2′:5′,2″-terthienyl [261]

To a solution of 5-nitro-2,2′:5′,2″-terthienyl (100 mg, 0.34 mmol) in ethyl acetate (15 ml) 10% palladium on carbon (20 mg) is added. The hydrogenation is carried out under a positive pressure of hydrogen and the reaction consumes 90 ml of hydrogen over a period of 6.5 h. The catalyst is filtered off, the solvent removed and the brown residue purified by chromatography on silica gel using, ethyl acetate as eluent, giving 59 mg (66%) of the title compound as a green solid mp 85–89 °C.

2-Aryl-4-(carboxyaryl)-3-thiophenecarboxylic acids or esters are formed from 4,5-annulated methyl 2-aryl-6-oxo-6H-cyclopenta[b]thiophene-3-carboxylate by cleavage of the cyclopentadienone ring with methanolic potassium hydroxide [282].

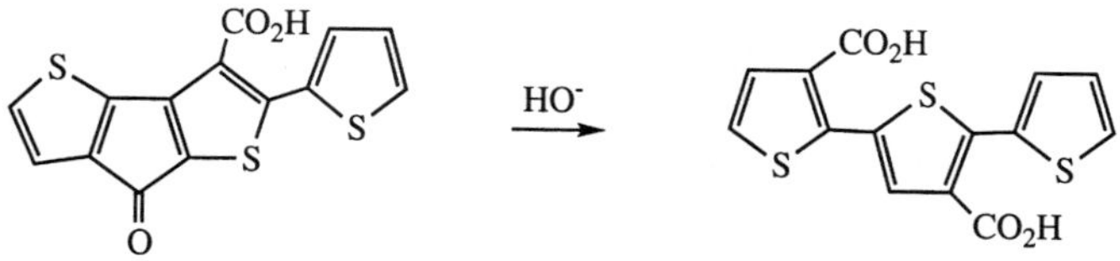

8.4.6.2 From haloterthienyls

Palladium(0)-catalyzed and cuprous iodide promoted Heck coupling of 5-iodo-2,2′:5′,2″-terthiophene with *N*-trimethylsilylacetylene in tetrahydrofuran and diisopropylamine gives 1-(trimethylsilyl)-2-[5-(2,2′:5′,2″-terthienyl)]acetylene in 88% yield, which was desilylated with anhydrous potassium carbonate in methanol/chloroform to the free acetylene derivative, which was transformed to the tributylstannyl derivative through reaction with di(isopropylamino)tributylstannane and was used for the preparation of ruthenium oligothienylacetylide complexes [283].

1-(Trimethylsilyl)-2-(5-(2,2′:5′,2″-terthienyl))acetylene [283]

A suspension of 5-iodo-2,2′:5′,2″-terthienyl (0.64 g, 1.7 mmol), diisopropylamine (0.33 g, 3.3 mmol), dichlorobis(triphenylphosphine)palladium(II) (0.14 g, 0.20 mmol) and cuprous iodide (0.022 g, 0.12 mmol) in anhydrous tetrahydrofuran (60 ml) is degassed for 2 min under nitrogen. (Trimethylsilyl)acetylene (0.33 g, 3.4 mmol) is added to the suspension via syringe. The reaction mixture is stirred at room temperature overnight while it turns dark green. The reaction is quenched by adding distilled water (20 ml). The phases are separated and the aqueous phase extracted with dichloromethane (2 × 20 ml). The combined organic phases are washed with sodium chloride solution (30 ml) and water, dried over magnesium sulfate and evaporated. The residue, a brown solid, is purified by flash chromatography on silica gel using hexanes as eluent, giving 0.52 g (88%) of the title compound.

Terthiophene bearing pyrenes at the terminal α-positions have been prepared as novel emitting materials in organic electroluminescence devices by Stille coupling of 5,5″-bis(tributylstannyl)-α-terthienyl with 1-bromopyrene [284].

Sonagashira coupling of 5″-dodecyl-5-iodo-α-terthienyl with trimethylsilylacetylene gives, after removal of the silyl group, 5-ethynyl-5″-dodecyl-α-terthienyl in excellent yield [118].

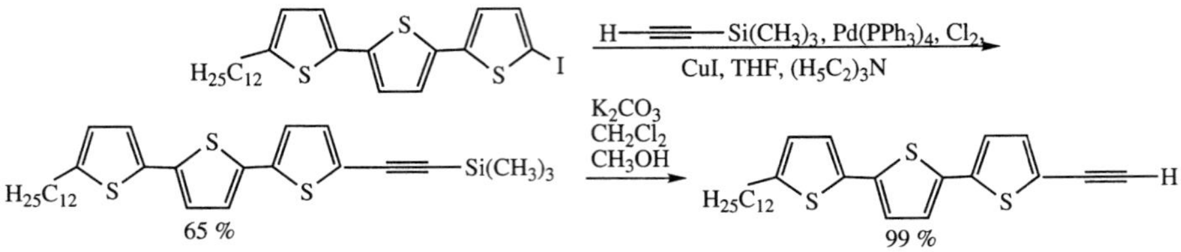

In connection with work on the synthesis of photochromic 3*H*-naphtho-[2,1-*b*] pyrans, the Sonogashira reaction was used under phase transfer conditions between the protected 8-ethynyl-3,3-diphenyl-3*H*-naphtho[2,1-*b*]pyran and 5-iodo-α-terthienyl as shown below [190].

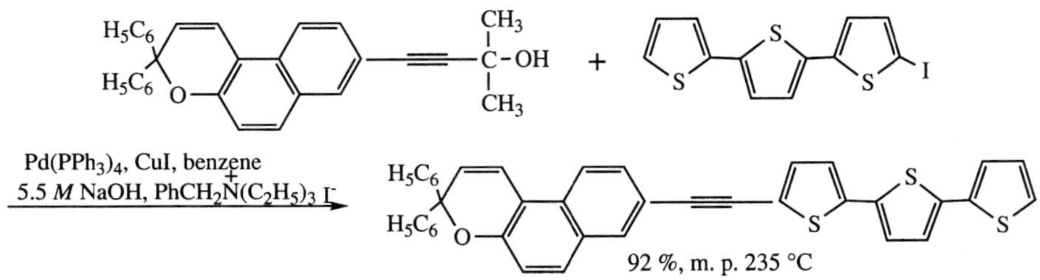

The reaction of 5,5″-dibromo-α-terthienyl with tetracyanoethylene oxide gives the highly conductive tetracyanoquinodimethane derivative [276,285].

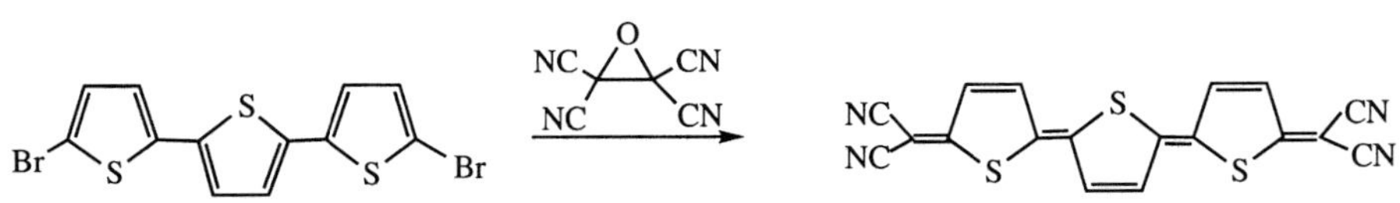

The reaction of 3′-iodo-α-terthienyl with sodium methoxide and cupric oxide in methanol is a good method for the preparation of the 3′-methoxy-derivative [262].

8.4.6.3 From hydroxymethyl derivatives of α-terthienyl

Reaction of 5-hydroxymethyl- and 5-bromo-5″-hydroxymethyl-α-terthienyl with sodium hydride and hexadecyl bromide is used for the preparation of 5-hexadecyloxymethyl- and 5-bromo-5″-hexadecyloxymethyl-α-terthienyl [274].

5-Bromo-5″-hexadecyloxymethyl-2,2′:5′,2″-terthienyl [274]

To a sodium hydride dispersion (50% paraffin) (0.27 g, 5.6 mmol), washed twice with anhydrous hexane, 5-bromo-5″-hydroxymethyl-2,2′:5′,2″-terthienyl

(1 g, 2.8 mmol) in anhydrous *N,N*-dimethylformamide (40 ml) is added. The reaction mixture is stirred at room temperature for 30 min before hexadecyl bromide (0.86 g, 2.8 mmol) is added. After 10 h the reaction mixture is diluted with water (200 ml), the stirring is continued for 3 h and the phases are separated. The organic phase is passed through a short column using dichloromethane as eluent. The solvent is removed by evaporation and the residue treated with methanol giving 1 g (63%) of the title compound as a pale yellow powder mp 101 °C after recrystallization from hexane/benzene.

Treatment of 5-hydroxymethyl-α-terthienyl with lithium aluminium hydride and aluminium chloride in diethyl ether is an unconventional method for the preparation of 5-methyl-α-terthienyl [259].

8.4.6.4 From carbonyl derivatives of terthienyls

5-(1-Hydroxyheptadecyl)-α-terthienyl is prepared from 5-formyl-α-terthienyl and heptadecylmagesium bromide. Similarly the 5-bromo-5″-(1-hydroxyheptadecyl) derivative was prepared from 5-bromo-5″-formyl-α-terthienyl [274]. 5-Hydroxymethyl-α-terthienyl and 5-bromo-5″-hydroxymethyl-α-terthienyl are prepared by sodium borohydride reduction of the 5-formyl derivative [259] and 5-bromo-5″-formyl derivative [274].

5-(1-Hydroxyheptadecyl)-2,2′:5′,2″-terthienyl [274]

To a flame dried, loosely capped 500-ml round-bottomed flask containing magnesium turnings (1.59 g, 65 mmol) and anhydrous diethyl ether (75 ml), 1,2-dibromoethane (0.7 ml, 7.6 mmol) is added. The mixture is warmed gently until constant bubbling is observed at room temperature, after which hexadecyl bromide (13.26 g, 43 mmol) is added producing a vigorous reaction to form the alkyl Grignard reagent. This is refluxed for 1 h and 5-formyl-2,2′:5′,2″-terthienyl in anhydrous diethyl ether (20 ml) is added in one portion. After refluxing for 1.5 h the reaction mixture is cooled to room temperature. Cold saturated ammonium chloride solution is added and the stirring continued for 30 min. The yellow precipitate formed is taken up in hexanes/diethyl ether (1:1) and the combined organic phases are washed with water, dried over sodium sulfate and evaporated. The residue, a yellow waxy solid, is treated with hexane (250 ml) and this mixture is stirred in a sonicator for 2 h. The precipitate is filtered off and washed with hexane giving 5 g (93%) of the title compound as a yellow powder mp 96 °C after recrystallization from hexane/benzene.

5-Methyl- and 5,5″-dimethyl-α-terthienyl is best obtained by Wolf–Kishner reduction of the corresponding aldehydes [259]. Wolff-Kishner reduction of

5-hexanoyl-α-terthienyl is used for the preparation of 5-hexyl-α-terthienyl [275]. Alternatively reduction of acyl derivatives with lithium aluminium hydride and aluminium chloride is used, as in the preparation of 5-dodecyl-α-terthienyl from the 5-dodecyl-α-terthienyl from the 5-dodecanoyl derivative [106,278].

5-Hexyl-2,2″:5′2″-terthienyl [278]

A solution of 5-hexanoyl-2,2′:5′2″-terthienyl (1.06 g, 31. mmol) in anhydrous toluene (30 ml) is added to a stirred suspension of aluminium chloride (0.79 g) and lithium aluminium hydride (0.91 g) in anhydrous diethyl ether (200 ml). After 1 h the excess of reagents are quenched by addition of ethyl acetate (12 ml) and concentrated aqueous hydrochloric acid (4 ml). The so obtained solution is filtered, concentrated, dissolved in warm toluene and chromatographed on silica gel (20 g) using hexane as eluent, giving 0.81 g (80%) of the title compound as a light-yellow solid.

5-(1-Hydroxyheptadecyl)-α-terthienyl and 5-bromo-5″-(1-hydroxyheptadecyl)-α-terthienyl are reduced to 5-heptadecyl-α-terthienyl and 5-bromo-5″-heptadecyl-α-terthienyl with sodium cyanoborohydride and zinc iodide in dichloromethane [274]. The best method for the preparation of *trans*-5-(heptadec-1-enyl)-α-terthienyl is treatment with *para*-toluenesulfonyl chloride in pyridine [274].

For the preparation of 5-bromo-5″-(heptadece-1-enyl)-α-terthienyl, treatment of 5-bromo-5″-(1-hydroxyheptadecyl)-α-terthienyl with a catalytic amount of *para*-toluenesulfonic acid in chloroform is used [274].

Trans-5-bromo-5″-(heptadece-1-enyl)-2,2′:5′,2″-terthienyl [274]

To a stirred solution of 5-bromo-5″-(1-hydroxyheptadecyl)-2,2′:5′,2″-terthienyl (1 g, 1.7 mmol) in chloroform (100 ml) a catalytic amount of *para*-toluenesulfonic acid (0.2 g) is added. The reaction mixture is kept at room temperature for 16 h, after which water is added. The phases are separated and the aqueous phases extracted with chloroform. The combined organic phases are washed with a dilute aqueous solution of sodium carbonate and water, dried over sodium sulfate and evaporated. The residue is purified by chromatography on silica gel using hexane/dichloromethane (12:1) as eluent followed by recrystallization from a mixture of hexanes and benzene giving 0.71 g (73%) of the title compound as yellow microcrystals mp 127–130 °C.

Selenium oxide oxidation of 5-acetyl-α-terthienyl and 5,5″-diacetyl-α-terthienyl is used for the preparation of the glyoxal derivatives [208]. A number of aldol condensation products from 5-formyl- and 5,5′-diformyl-1,3-di(2-thienyl)benzo[*c*]thiophene have been prepared by conventional methods by reaction with malonitrile, 2-thiopheneacetonitrile, 1,3-dithiolyl-4,5-dicarbomethoxyphophonium tetrafluoroborate and (2-thienylmethyl)triphenylphosphonium bromide [241].

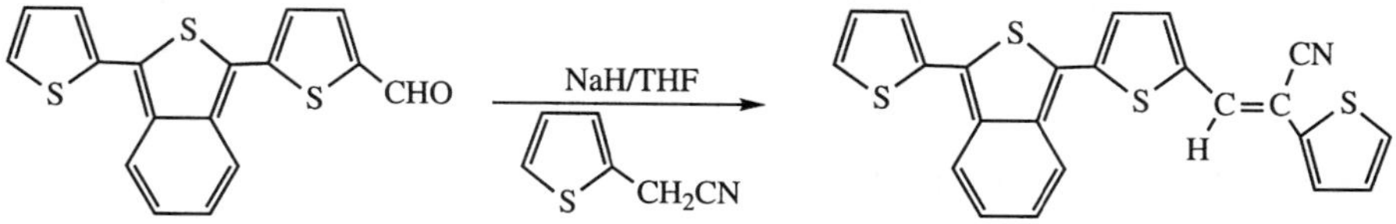

Olefination of a number of 4′-subsituted 5,5″-diformyl and diacetyl-α-terthienyls was successfully carried out with a phosphonate anion carrying the 1,3-dithiole-2-ylidene moiety [255]. The reaction between methyl(triphenylphosphoranylidene)acetate and 5-formyl-α-terthienyl is used for the preparation of methyl α-terthienyl-5-[(E)-3‴]acrylate, which was successfully hydrogenated to the corresponding propionate using 10% palladium on carbon as catalyst [261].

Methyl 2,2′:5′,2″-terthienyl-5-[(E)-3‴]-acrylate [261]

To a solution of 5-formyl-2,2′:5′,2″-terthienyl (715 mg, 2.58 mmol) in tetrahydrofuran (200 ml) methyl (triphenylphosphoranylidene)acetate (1.28 g, 3.87 mmol) is added. The reaction mixture is refluxed for 13 h, after which it is cooled to room temperature and evaporated. The residue is dissolved in chloroform (20 ml) and the insoluble material is filtered off, washed with chloroform (2 ml) giving 794 mg (92%) of the title compound as a yellow solid mp 199–200 °C.

8.4.6.5 From carboxylic acid derivatives

Alkaline hydrolysis of dimethyl 2,4′:2′,2″-terthienyl-3′,5′-dicarboxylate with potassium hydroxide in aqueous methanol is used for the preparation of 2,4′:2′,2″-terthienyl-3′,5′-dicarboxylic acid. However, selective hydrolysis can also be achieved leading to 3′-methoxycarbonyl-2,4′:2′,2″-terthienyl-5-carboxylic acid [286]. Heating of this and other carboxylic acids to 190 °C led to decarboxylation giving methyl 2,4′:2′,2″-terthienyl-3′-carboxylate [286]. Acid chlorides of the carboxylic acids were prepared in the usual way and ring-closed to cyclopenta[*b*]- and cyclopenta[*c*]thiophene derivatives using Friedel-Crafts conditions [286].

3′-Methoxycarbonyl-2,4′2′,2″-terthienyl-5-carboxylic acid [286]

Dimethyl 2,4′:2′,2″-terthienyl-3′,5′-dicarboxylate (4.3 g, 11.8 mmol) is treated with a solution of potassium hydroxide (0.7 g) in methanol (22 ml) and water (4 ml). The reaction mixture is refluxed for 3 h and upon addition of water (30 ml) the starting material forms oil drops, which upon cooling can be filtered off. The filtrate is diluted with water (50 ml), washed with diethyl ether (50 ml) and acidified with 2 *M* hydrochloric acid. The precipitate is filtered off, washed three times with water and recrystallized from methanol/water (1:1) giving 3.4 g (82%) of the title compound as yellow needles mp 186 °C.

8.4.6.6 From various derivatives

5-Hexylthio-α-terthienyl is prepared by hydrolysis of 5-hexanoylthio-α-terthienyl tetrabutylammonium hydroxide followed by alkylation with hexyl iodide [278].

5-Hexylthio-2,2′:5′,2″-terthienyl [278]

To a solution of 5-hexanoylthio-2,2′:5′,2″-terthienyl (1.05 g, 3.0 mmol) in tetrahydrofuran (60 ml) and water (1.5 ml) 55% tetrabutylammonium hydroxide in water (3 ml) is added. After 15 min hexyl iodide (0.6 ml) is added to the red mixture. The product is taken up in diethyl ether and after evaporation the residue is purified by chromatography on silica gel using hexane as eluent, giving 0.86 g (85%) of the title compound.

The diiodo derivative shown below gave upon palladium-catalyzed coupling with trimethylsilyl acetylene the trimethylsilyl protected terminal acetylene, which was deprotected nearly quantitatively under mild basic conditions to the α-terthienyldiyne [248].

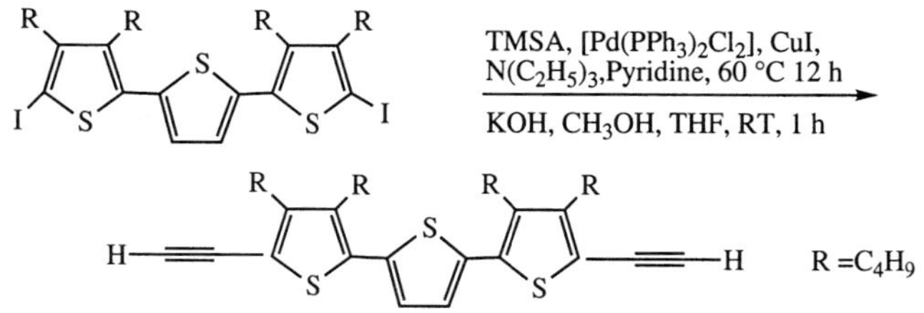

Ruthenium complexes of 3′-(diphenylphosphinoethyl)-5,5″-dimethyl-α-terthienyl are prepared as redox-switchable ligands for oxidation state dependent molecular uptake and release [287,288].

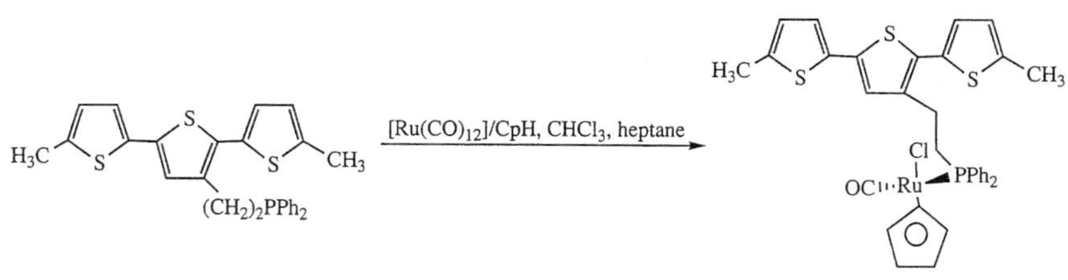

8.4.6.7 Oxidation of terthienyls to 1,1-dioxides and cation radicals

Oxidation of 5,5″-bis(dimethyl-*tert*-butylsilyl)-α-terthienyl with *meta*-chloroperbenzoic acid in dichloromethane at room temperature gave 17% of the 1,1-dioxide and 54% of the 1,1,1″,3″-tetradioxide [226]. An alternative route is the Stille reaction of 2-bromo-5-dimethyl-*tert*-butylsilyl- or 2-bromo-5-hexylthiophene-1,1-dioxide with 2,5-bis(tributylstannyl)thiophene giving 1,1′,1″-tetradioxide of 5,5″-bis(dimethyl-tert-butylsilyl)-α-terthienyl, while the reaction with 5-tributylstannyl-(5′-dimethyl-*tert*-butylsilyl)-2,2′-bithienyl gives the 1,1′-dioxide of the α-terthienyl [227].

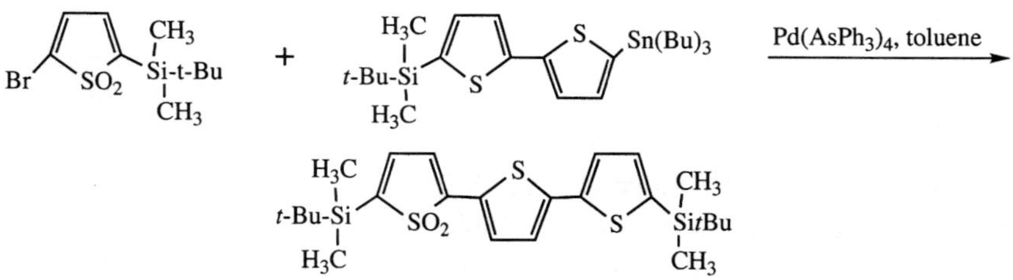

3′,4′-Dibutyl-5,5″-diphenyl-α-terthienyl is oxidized electrochemically to the stable cation radical hexafluorophosphate salt [289, 290].

8.5 CYCLIC TERTHIENYLS (BENZOTRITHIOPHENES)

8.5.1 Preparation by photochemical reactions of terthienyls

Terthienyls in which thienyl groups are attached to the 2- and 3-positions of the central thiophene ring undergo oxidative photocyclization to benzotrithiophenes. A single unsymmetrical benzotrithiophene is obtained from 2,2′:3′,2″- and 2,2′:3′3″-terthienyls. 3,2′:3′.3″-Terthienyl gives a mixture indicating that a rearrangement occurred [291].

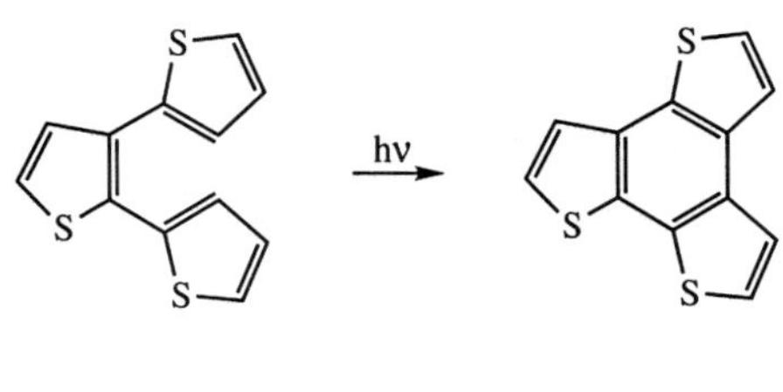

Benzo[1,2-b:3,4-b′:6,5-b″]trithiophene [291]

A solution of 2,2′:3′,2″-terthienyl (1.50 g, 0.006 mol) and iodine (15 mg) in toluene (500 ml) is irradiated in a quartz vessel placed in a Rayonet apparatus equipped with 15 tubes RPR-3500, emitting between 320 and 400 nm, with a maximum at 350 nm. During the irradiation a slow stream is allowed to bubble through the solution, which is stirred magnetically. After 25 h of photolysis the solution is concentrated, the residue extracted twice with boiling hexane (200 ml) and the extracts are filtered under pressure through a column of silica gel (2 cm). The filtrate is concentrated and cooled giving 1.10 g (74%) of pale-yellow crystals of the title compound, mp 161–166 °C after recrystallization from hexane.

8.5.2 By reactions of benzotrithiophenes

Bromination of benzotrithiophene with three equivalents of *N*-bromosuccinimide in *N,N*-dimethylformamide led to a mixture of the tribromo derivatives, which were converted to the tris(trimethylsilyl)acetylene) derivatives in the proportion of 1:5, by palladium(0)-catalyzed reaction with trimethylsilylacetylene in the presence of triethylamine and cuprous iodide. Flash-vacuum pyrolysis then gave the bowl-shaped molecule shown below [292,293].

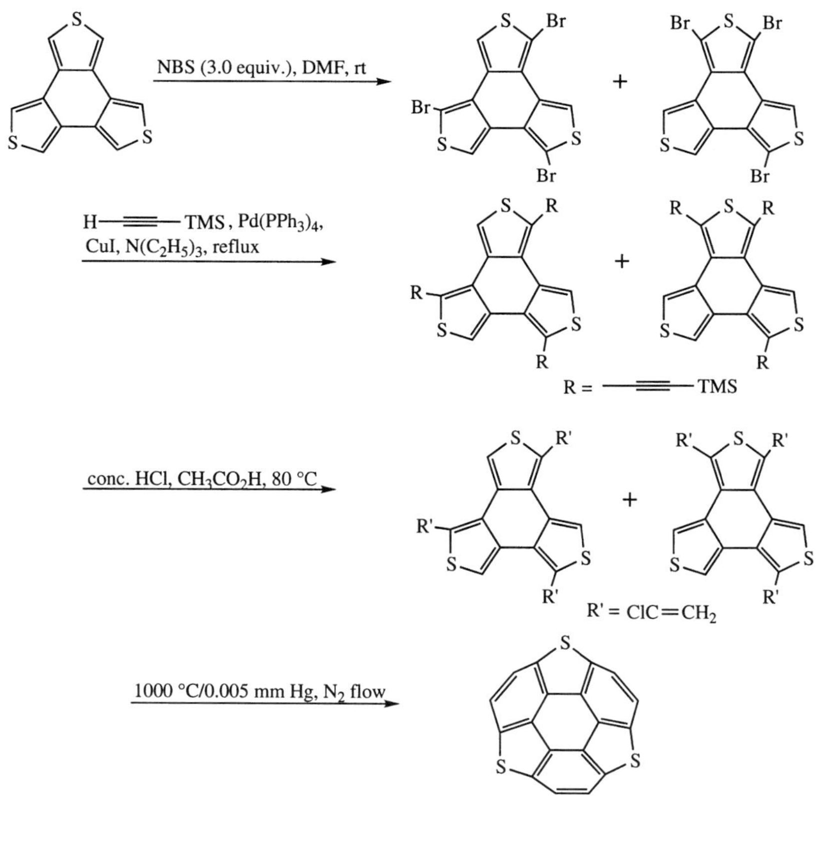

8.5.3 By other methods

New mesoaryl-30π heptaphyrins containing a terthienyl grouping, shown below have been synthesized [294].

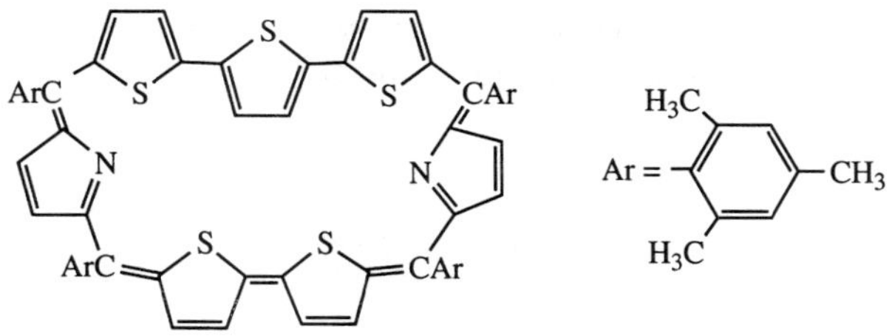

8.6 QUATERTHIENYLS

8.6.1 Introduction

Interest in the preparation of quaterthienyls has increased dramatically during recent years in connection with investigations of their use as starting materials for polythienyls of interest in material science. The development was made possible, as in the case of bithienyls and terthienyls, by the use of transition metal-catalyzed coupling reactions.

8.6.2 Symmetrical quaterthienyls through the coupling of bithienyllithium derivatives with cupric chloride or ferric chloride

The reaction of 5-lithio-2,2′-bithienyls with cupric chloride is a useful route to quaterthienyls and α-quaterthienyl itself was obtained in 85% yield through the reaction of two equivalents of 2,2′-bithienyl with one equivalent of lithium diisopropyl amide and then one equivalent of cupric chloride [295].

This method is superior to the older method consisting in the reaction of 5-iodo-2,2′-bithienyl with cuprous acetate in pyridine [296]. Also α-substituted bithienyls work well in this reaction. From 5-hexyl-5′-lithio-, 5-dodecyl-5′-lithio- and 5-3-(butoxypropyl)-5′-lithio-2,2′-bithienyl, the corresponding 5,5‴-substituted α-quaterthienyls were obtained in reasonable yields [296,297].

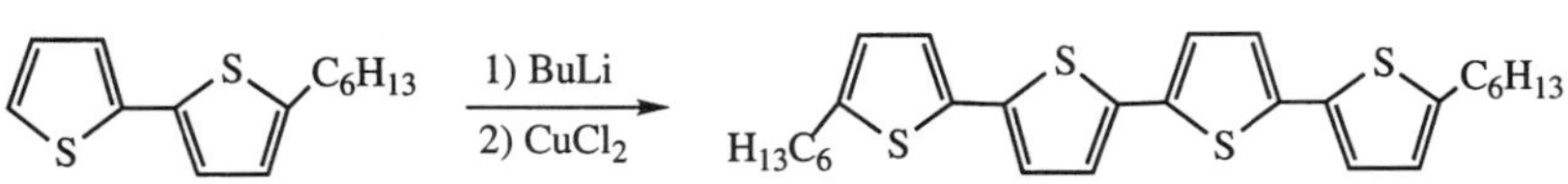

Mixed conjugated quaterthienyls of thiophene and 3,4-diethylenedioxythiophene are prepared by halogen–metal exchange of 5-bromo-3,4-ethylenedioxy-2,2′-bithienyl followed by cupric chloride [214,298].

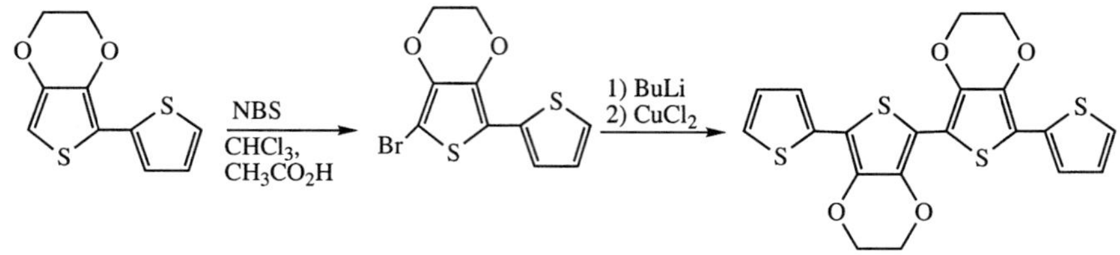

3,3′;2,2″;3″,3‴-Quaterthienyl is obtained only in 40% yield upon reaction of 2-lithio-3,3′-bithienyl, prepared by halogen–metal exchange of the 2-bromo derivatives followed by cupric chloride [299]. The quaterthienyl is obtained from the bithienyl by lithiation with lithium diisopropyl amide followed by homo coupling with cupric chloride [250].

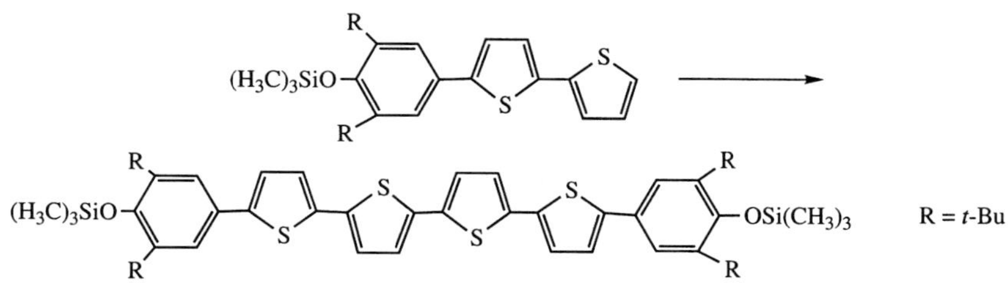

Interring bridged quaterthienyls have recently been prepared. Thus reaction of 4,4-dibutylcyclopenta[2,1-*b*;3,4-*b*′]bithiophene with lithium diisopropyl amide and cupric chloride is used for the preparation of the compound shown below [300].

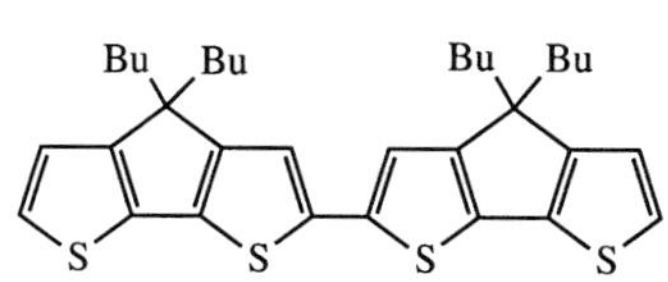

The nonbutylated compound was prepared by mercuration of cyclopenta[2,1-*b*;3,4-*b*′]bithiophene followed by reaction with copper powder and palladium(II) chloride [300]. Metalation of 5-*tert*-butyldimethylsilyl- and 5-hexyl-2,2′-bithienyl with butyllithium followed by ferric chloride or ferric acetylacetonate gives 5,5‴-bissubstituted-α-quaterthienyls [226,227].

5,5‴-Bis(tert-butyldimethylsilyl)-2,2′-:5′,2″:5″,2‴-quaterthienyl [226]

To a solution of 5-(dimethyl-*tert*-butylsilyl)-2,2′-bithienyl (1.26 g, 4.5 mmol) in anhydrous tetrahydrofuran (20 ml), butyllithium in hexane (1.8 ml, 4.5 mmol) is added dropwise. After 1 h the mixture is cooled to $-60\,^{\circ}\mathrm{C}$ and ferric acetylacetonate (1.6 g, 4.5 mmol) is added stepwise. Then the reaction mixture is allowed to warm to room temperature and 3 *M* hydrochloric acid (35 ml) is added. The phases are separated and the aqueous phases extracted with dichloromethane. The combined organic phases are washed with sodium chloride solution, dried over magnesium sulfate and evaporated. The residue is washed with diethyl ether giving 0.95 g (76%) of the title compound as a green solid mp 163–164 °C.

Similarly, halogen–metal exchange of 5-bromo-3,3′dihexyl-2,2′-bithienyl with butyllithium followed by ferric acetylacetonate gave 3,3′,3″,3‴-tetrahexyl-α-quaterthienyl [301].

3,3′,3″,3‴-Tetrahexyl-2,2′:5′,5″:2″,2‴-quaterthienyl [301]

To a solution of 5-bromo-3,3′dihexyl-2,2′-bithienyl (600 mg, 1.45 mmol) in anhydrous tetrahydrofuran (20 ml), 1.6 *M* butyllithium in hexane (1.0 ml) is added dropwise at $-80\,^{\circ}\mathrm{C}$ under argon. When the addition is completed the mixture is stirred at $-80\,^{\circ}\mathrm{C}$ for 10 min, after which ferric acetylacetonate (512 mg, 1.45 mmol) is added in portions over 15 min and the stirring is continued at $-80\,^{\circ}\mathrm{C}$ for 30 min. The reaction mixture is diluted with water (30 ml) and neutralized with 3 *M* hydrochloric acid (40 ml) and the product extracted with dichloromethane. The combined organic phases are washed with sodium chloride solution. After evaporation the residue is purified by chromatography on silica gel using hexane as eluent giving 292 mg (60%) of the title compound as a colorless oil.

3,3′,4″,3‴-Tetra(methylthio)-α-quaterthienyl is prepared by treatment of 3,3′-bis(methylthio)-2,2′-bithienyl with ferric chloride in chloroform [302].

3,3′,4″,3‴Tetra(methylthio)-2,2′:5′,2″:5″,2‴-quaterthienyl [302]

To a solution containing ferric chloride (1.25 g, 7.7 mmol) in chloroform (40 ml) 3,3′-bis(methylthio)-2,2′-bithienyl (0.5 g, 1.9 mmol) in chloroform (20 ml) is added dropwise. The mixture is stirred overnight, after which methanol (200 ml) is added. The brown solid formed is filtered off and dissolved in chloroform. The chloroform solution is washed twice with 2 *M* hydrochloric acid, dried over sodium sulfate and evaporated. The residue is purified by chromatography on silica gel using cyclohexane/dichloromethane (85:15) as eluent giving 0.28 g (59%) of the title compound as an orange solid mp 152 °C.

3,4′-Bis(methylthio)- and 3,3′-bis(methylthio)-2,2′-bithienyl give upon reaction with ferric chloride in chloroform the following quaterthienyls [273].

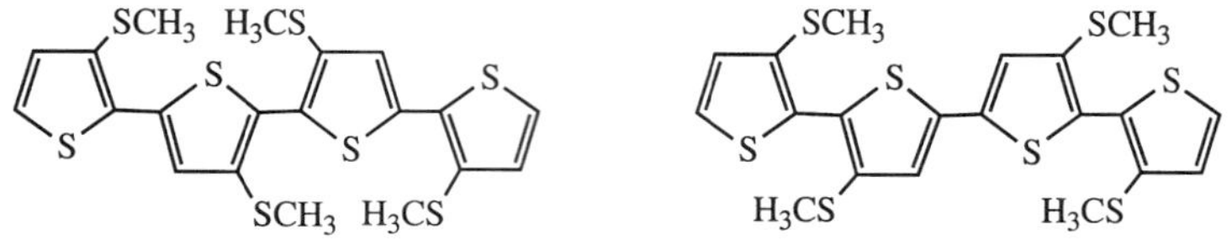

76 % 24 %

Coupling of 3-butylthio-2,2′-bithienyl with ferric chloride gives 3,3‴-bis(butylthio)-α-quaterthienyl, however this compound is difficult to purify [64]. Treatment of 3,4-ethylenedioxy-5′-mesitylthio-2,2′-bithienyl with butyllithium followed by ferric acetylacetonate gives 3′,4′3″,4″-bis(ethylenedioxy)-5,5‴-bis(mesitylthio)-α-quaterthienyl [9].

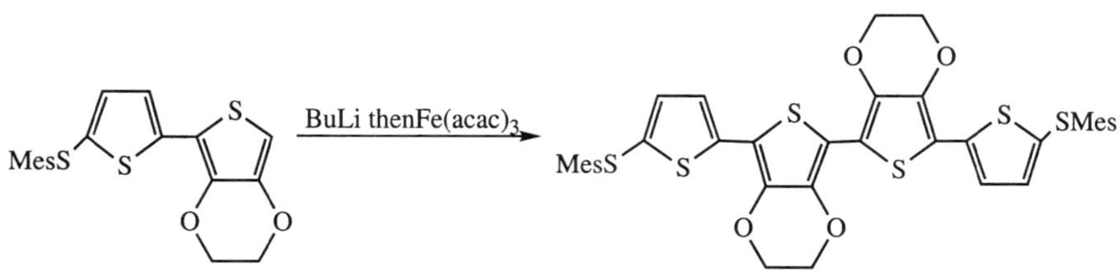

3′,4′3″,4″-Bis(ethylenedioxy-5,5‴-bis(mesitylthio)-1,2′:5′,2″:5″,2‴ quaterthienyl [9]

A solution of 3,4-(ethylenedioxy)-5′-mesitylthio-2,2′-bithienyl (0.297, 0.793 mmol) in tetrahydrofuran (10 ml) is cooled to 0 °C and treated with butyllithium (0.5 ml, 0.8 mmol). The dirty-green solution is stirred for 45 min at 0 °C and then transferred via cannula to a refluxing solution of ferric acetylacetonate (0.28 g, 0.793 mmol) in tetrahydrofuran (20 ml). The reaction mixture is refluxed for 44 h. After evaporating the bright red residue is purified by chromatography on silica gel using chloroform/hexane (1:1) as eluent, giving 0.35 g (59 %) of the title compound as a bright orange-red solid mp 296–298 °C after recrystallization from tetrahydrofuran.

8.6.3 Symmetrical quaterthienyl by other coupling methods

α-Quaterthienyl is prepared by reductive coupling of 5-bromo-2,2′-bithienyl using nickel chloride–zinc and triphenylphosphine in *N*,*N*-dimethylformamide [244]. 5,5‴-Diformyl-α-quaterthienyl is prepared by refluxing 5-bromo-5′-formyl-2,2′-bithienyl with nickel chloride and zinc and triphenylphosphine in *N*,*N*-dimethylformamide for six hours [102].

The following reactions using tris(triphenyl)phosphine nickel in *N,N*-dimethylformamide have been performed [53].

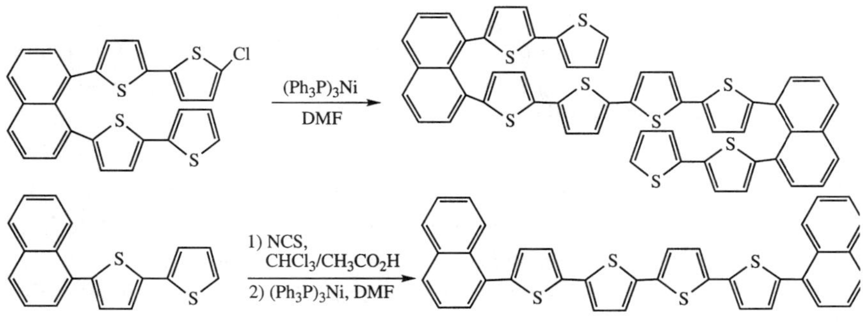

8.6.4 Transition metal-catalyzed couplings according to Kumada

8.6.4.1 From 5,5′-dihalobithienyls and two equivalents of thiophenemagnesium halides

1,3-Bis(diphenylphosphino)propanenickel(II) dichloride-catalyzed coupling of two equivalents of 3-pentoxy-2-thiophenemagnesium iodide and 5,5′-diiodo-2,2,-bithienyl gives 3,3‴-dipentoxy-α-quaterthienyl [15].

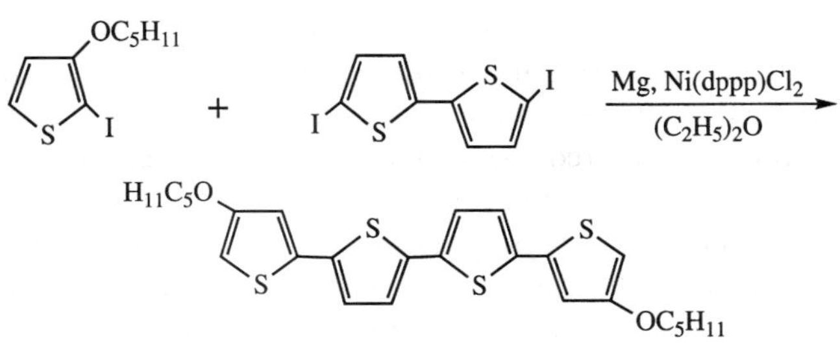

Similarly, the coupling of 5-dodecyl-2-thiophenemagnesium bromide and 3-octyl-2-thiophenemagnesium bromide with 5,5′-dibromo-2,2′-bithienyls is used for the preparation of 5,5‴-didodecyl-α-quaterthienyl [63] and 3,3‴-octyl-α-quaterthienyl [303], respectively.

5,5‴-Dodecyl-2,2′:2″,2‴-quaterthienyl [63]

From 5-bromo-2-dodecylthiophene (3.98 g, 12 mmol) and magnesium turnings (0.34 g, 14 mmol) in diethyl ether (10 ml) the corresponding Grignard reagent is prepared under heating to reflux for 2 h, and with the aid of an ultrasonic bath and a few drops of dibromoethane as entrainer. The Grignard solution is transferred to the dropping funnel of a second apparatus via cannula and is added dropwise through a frit to an ice-cooled suspension of 5,5-dibromo-2,2′-bithienyl (1.62 g, 5 mmol) and 1,3-bis(diphenylphosphino)propanedichloronickel(II) (54 mg, 0.1 mmol) in diethyl ether (50 ml). The reaction mixture is refluxed for 24 h, cooled to room temperature and hydrolyzed with 1 *M* hydrochloric acid. The phases are separated and the organic phase neutralized, washed with water, dried over sodium sulfate and evaporated. The residue is recrystallized from hexane/toluene (1:2) giving 2.35 g (70%) of the title compound as an orange solid mp 163 °C.

This method has also been used for the preparation of 3,3‴-dipropyl-,3,3‴-dihexyl- and 3,3‴-didodecyl-α-quaterthienyl from the corresponding 3-alkyl-2-thiophenemagnesium bromides and 5,5′-dibromo-2,2′-bithienyl in 44–88% yield [304–306]. 3,4‴-Bis(trimethylsilyl)-α-quaterthienyl is obtained in 83% yield upon 1,3-bis(diphenylphosphino)propanedichloronickel(II)-catalyzed coupling of 5,5′-dibromo-2,2′-bithienyl with two equivalents of 4-trimethylsilyl-2-thiophenemagnesium bromide [249]. Recently the Grignard reagent from 5-phenyl-2-thiophenemagnesium iodide was coupled with 5,5-dibromo-2,2′-bithienyl to give 5,5‴-diphenyl-α-quaterthienyl in 41% yield [67].

5,5‴-Diphenyl-2,2′:5′,2″:5″,2‴-quaterthienyl [67]

The Grignard reagent is prepared from 5-phenyl-2-iodothiophene (544 mg, 1.90 mmol) and magnesium (46.2 mg, 1.90 mmol) and the diethyl ether is evaporated with a stream of anhydrous nitrogen and subsequently replaced with anisole (6 ml). To this mixture 1,3-bis(diphenylphosphino)propanedichloronickel(II) (10 mg, 0.018 mmol) is added followed by 5,5′-dibromo-2,2′-dithienyl (334 mg, 0.80 mmol) in anisole (4 ml). The reaction mixture is stirred at room temperature overnight, then at 100 °C for 6 h, cooled in an ice-water bath and hydrolyzed with 2 *M* hydrochloric acid (1 ml). The precipitate formed is collected by filtration, washed in turns with acetone and dichloromethane and recrystallized from 1,2,4-trichlorobenzene giving 466 mg (41%) of the title compound as an orange solid mp 336 °C.

8.6.4.2 From halo-2,2′-bithienyls and 5-bithienylmagnesium halides

The best method for the preparation of α-quaterthienyl is the reaction of 5-bromo-2,2′-bithienyl with zinc powder and bis(triphenylphosphine)dichloronickel(II) and tetrabutylammonium iodide in tetrahydrofuran giving a good yield of the desired compound [276].

2,2′:5′,2″:5″2‴-Quaterthienyl [276]

A mixture of zinc powder (785 mg, 12.0 mmol), bis(triphenylphosphine)dichloronickel(II) (393 mg, 0.600 mmol) and tetrabutylammonium iodide (2.22 g, 6.00 mmol) in tetrahydrofuran (15 ml) is stirred at room temperature under nitrogen for 30 min. To the resulting reddish brown solution a solution of 5-bromo-2,2′-bithienyl in tetrahydrofuran (5 ml) is added. The reaction mixture is stirred at 50 °C for 2 h, after which water is added the precipitate formed is filtered off. This is purified by chromatography on silica gel using carbon disulfide as eluent followed by recrystallization from chloroform giving 867 mg (87%) of the title compound as yellow leaflets mp 216.5–218 °C.

The 1-napthtylsubstituted quaterthienyl has been prepared as shown below [121].

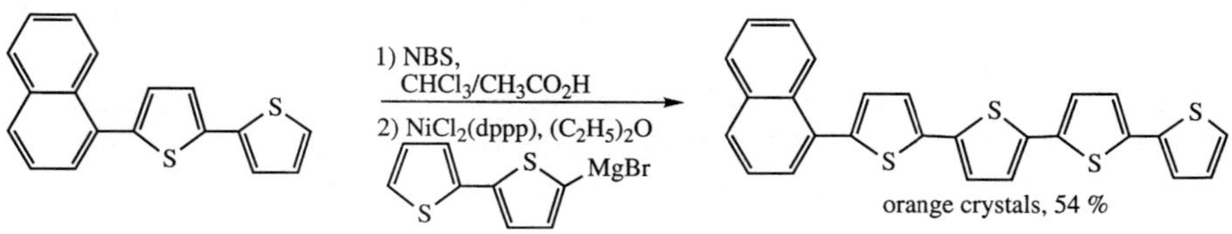

Kumada coupling of 5-bromo-5′-(12-bromodedecyl)-2,2′-bithienyl with the Grignard reagent from 5-bromo-2,2′-bithienyl gives 5-(12-bromododecyl)-α-quaterthienyl, which by standard reaction with thiourea was transformed to the thiol [162].

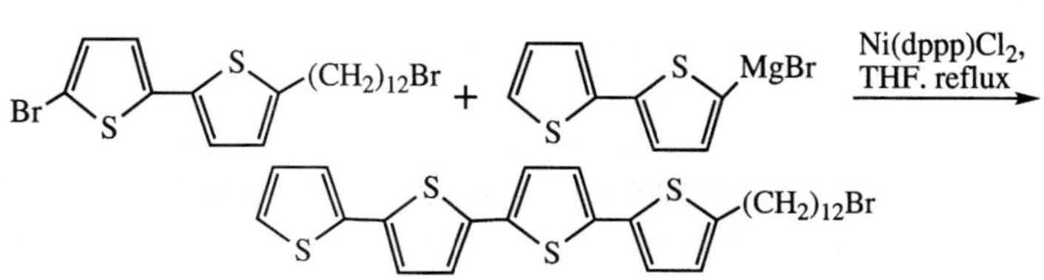

5-(12-Bromododecyl)-2,2′:5′,2″:5″2‴-quaterthienyl [162]

A solution of 5-bromo-2,2′-bithienyl (2.4 g, 10 mmol) in anhydrous tetrahydrofuran (30 ml) is under inert atmosphere added dropwise to magnesium turnings (0.24 g, 10 mmol) in anhydrous tetrahydrofuran (10 ml). The mixture is refluxed until the magnesium disappears, then transferred to a dropping funnel and slowly dropped to a solution of 5-bromo-5′-(12-bromododecyl)-2,2′-bithienyl (2.8 g, 9.5 mmol) and 1,3-bis(diphenylphosphino)-propanenickel(II) dichloride (1 mol%) in anhydrous tetrahydrofuran (40 ml). The reaction mixture is refluxed overnight and hydrolyzed with 1 *M* hydrochloric acid. The product is extracted with chloroform and the combined organic phases are dried over magnesium sulfate and evaporated. The product was purified by silicagel chromatography (70–200 mesh) with heptane until the elution of 5-(12-bromododecyl)-2,2′-bithienyl and quaterthienyl. Then, ethyl acetate is progressively added to the heptane to elute the desired product, giving 3.3 g (59%) of the title compound as a red-brown powder mp 122 °C.

1,3-Bis(diphenylphosphino)propanenickel(II) dichloride-catalyzed coupling of 3-butylthio-2,2′-bithienyl-5′-magnesium bromide with 3-butylthio-5′-iodo-2,2′-bithienyl is the better method for the preparation of 3,3‴-bis(butylthio)-α-quaterthienyl [64].

3,3‴-Bis(butylthio)-2,2′:5′,2″:5″,2‴-quaterthienyl [64]

The Grignard reagent generated from 3-butylthio-5′-iodo-2,2′-bithienyl (2.2 g, 5.7 mmol) is slowly added via cannula to a solution of 3-butylthio-5′-iodo-2,2′-bithienyl (2.0 g, 5.2 mmol) and 1,3-bis(diphenylphosphino)propanenickel(II) dichloride (0.07 g, 0.13 mmol) in diethyl ether (6 ml) cooled in an ice-bath. The reaction mixture is refluxed for 20 h, after which the reaction is quenched with hydrochloric acid/ice-water. The phases are separated and the aqueous phase extracted with diethyl ether (100 ml). The combined organic phases are washed with aqueous sodium bicarbonate solution and water and evaporated. The residue, a brown oil, is purified by chromatography on silica gel using first tetrachloromethane and then petroleum ether/diethyl ether (2:1) as eluents. The later fraction gives a yellow solid, which is stirred with pentane. About 1.4 g (53%) of the title compound is collected mp 54–55 °C.

The best method for the preparation of 3,3′;2′2″,3″,3‴-quarterthienyl is the reaction of 3,3′-bithienyl-2-magnesium bromide with 2-bromo-3,3′-bithienyl [299].

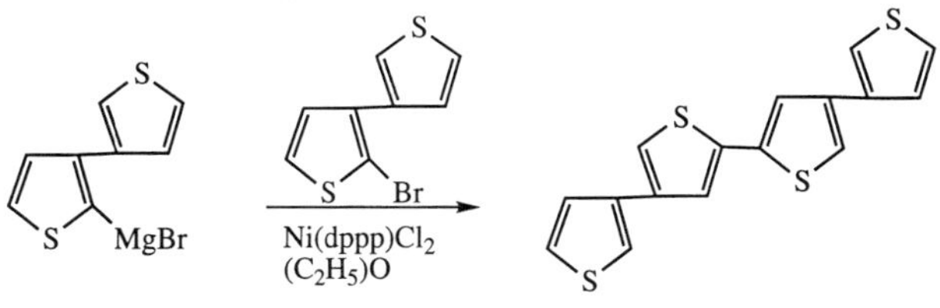

3,3′;2′2″;3″,3‴-Quaterthienyl [299]

A mixture of 2-bromo-3,3′-bithienyl (280 mg, 1.14 mmol), 1,2-dibromoethane (300 mg, 1.6 mmol) and magnesium turnings (300 mg, 12.3 mmol) in diethyl ether is refluxed for 2.5 h. The Grignard reagent is added to a mixture of 2-bromo-3,3′-bithienyl (200 mg, 0.82 mmol) and 1,3-bis(diphenylphosphino)-propanedichloronickel(II) (5 mg) in diethyl ether (10 ml). The reaction mixture is refluxed for 48 h, after which the reaction is quenched with 2 *M* hydrochloric acid (10 ml). Dichloromethane is added until the solid formed is dissolved. The phases are separated and the organic phase washed with saturated sodium chloride solution, dried and evaporated. The residue is dissolved in boiling ligroin and filtered through a short silica gel column. The filtrate is concentrated and cooled and the solid formed is collected and recrystallized from ethanol giving 226 mg (84%) of the title compound mp 166–170 °C.

5,5‴-Bis[*N*,*N*-bis(trimethylsilyl)aminomethyl]-α-quaterthienyl is prepared from 5,5′-dibromo-2,2′-bithienyl and 2-[*N*,*N*-bis(trimethylsilyl)aminomethyl]-5-thiophenemagnesium bromide prepared via metalation with butyllithium followed by magnesium bromide etherate [307].

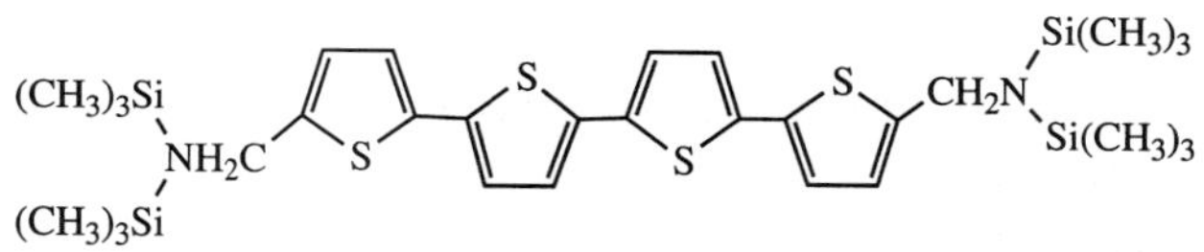

5,5‴-Bis[N,N-bis(trimethylsilyl)aminomethyl]-2,2′:5′,2″:5″,2‴-quaterthienyl [207]

To a solution of 2-(aminomethyl)thiophene (12.4 g, 0.11 mol) in diethyl ether (80 ml) 3 *M* ethylmagnesium bromide diethyl ether solution (78.7 ml, 0.236 mol) is added. The mixture is refluxed for 2 h, after which chloromethylsilane (25.7 g, 0.236 mol) in diethyl ether (50 ml) is added and the refluxing is continued for 4 h. The resulting solution is washed, dried over potassium carbonate and evaporated. The residue, a pale-yellow liquid, is distilled at 85 °C/2–3 mm Hg giving 15.4 g (55%) of 2-[*N*,*N*-bis(trimethylsilyl)aminomethyl]thiophene. A solution of this compound (5.1 g, 20 mmol) in diethyl ether (80 ml) is cooled to 0 °C under nitrogen, after which 1.6 *M* butyllithium in hexane (13.75 ml, 22 mmol) is added dropwise, and after 30 min followed by magnesium bromide etherate (5.68 g, 22 mmol). The mixture is allowed to warm to room temperature and stirred for 30 min, when 1,3-bis(diphenylphosphino)propane- dichloronickel(II) (0.16 g, 0.033 mmol) and 5,5′-dibromo-2,2′-bithienyl (2.6 g, 8 mmol) are added. The reaction mixture is stirred overnight and refluxed for 4 h. The precipitate formed is filtered off and recrystallized from hexane giving 1.9 g (36%) of the title compound as yellow, thin, blade-shaped crystals.

8.6.4.3 From halo-α-terthienyl and thiophenemagnesium halides

Coupling of 5-bromo-3,4″-bihexyl-α-terthienyl and 2-thiophenemagnesium bromide is used for the preparation of 3,3″-dihexyl-α-quaterthienyl [254].

3,3″-Dihexyl-2,2′:5′,2″:5″,2‴-quaterthienyl [254]

To a solution of 2-thiophenemagnesium bromide (3.50 mol) in diethyl ether (20 ml) 1,3-bis(diphenylphosphino)propane-dichloronickel(II) (0.1 g, 5 mol %) is added followed by a solution of 5-bromo-3,3″-dihexyl-2,2′:5′,2″-terthienyl (0.585 g, 2 mmol) in diethyl ether (20 ml). The reaction mixture is then refluxed under stirring for 18 h, after which it is allowed to cool to room temperature, poured into crushed ice and 2 *M* hydrochloric acid. The product is extracted with dichloromethane and the combined organic phases are washed to neutrality with saturated sodium bicarbonate and then with water, dried over magnesium sulfate and evaporated. The residue is recrystallized from hexane giving 0.364 g (62%) of the title compound as a yellow solid.

The reaction between 5-bromo-α-terthienyl and 5-dodecyl-2-thiophenemagnesium bromide yields 5-dodecyl-α-quaterthienyl [63].

5-Dodecyl-2,2′:5′,2″:5″,2‴-quaterthienyl [63]

From 2-bromo-5-dodecylthiophene (2.83 g, 8.6 mmol) and magnesium turnings (0.22 g, 9 mmol) in diethyl ether (15 ml) the corresponding Grignard reagent is prepared under heating to reflux for 2 h and with the aid of an ultrasonic bath and a few drops of dibromoethane as entrainer. The Grignard reagent is transferred to the dropping funnel of a second apparatus via cannula and added dropwise through a frit to an ice-cooled suspension of 5-bromo-2,2′:5′,2″-terthienyl (1.0 g, 3.1 mmol) and 1,3-bis(diphenylphosphino)propane-dichloronickel(II) (16.6 mg, 0.03 mmol) in diethyl ether/benzene (2:1) (25 ml). The reaction mixture is refluxed for 64 h, cooled to room temperature and hydrolyzed with 1 *M* hydrochloric acid. The phases are separated, the aqueous phase extracted with diethyl ether and the combined organic phases are washed with sodium bicarbonate and water and dried over sodium sulfate and evaporated. The residue is purified by chromatography on silica gel using first hexane and then dichloromethane, giving 0.91 g (59%) of the title compound as an orange-red solid mp 156–158 °C after recrystallization from benzene.

8.6.5 Transition metal-catalyzed couplings according to Stille

8.6.5.1 From 5,5′-bis(tributylstannyl)-2,2′-bithienyl and two equivalents of halothiophenes

3,3‴-Dimethoxy-α-quaterthienyl and 4,4‴-dimethoxy-5,″-dimethyl-α-quaterthienyl are prepared in acceptable yield by palladium(0)-catalyzed coupling of 5,5′-(tributylstannyl)-2,2′-bithienyl and 2-iodo-3-methoxythiophene and 2-iodo-4-methoxy-5-methylthiophene, respectively [28,171].

3,3‴-Bis(methoxy)2,2′:5′,2″:5″,2‴-quaterthienyl [171]

A solution of 2-iodo-3-methoxythiophene (5 g, 20.8 mmol), 5,5′-bis(tributyltin)-2,2′-bithienyl (7.6 g, 10.2 mmol) and tetrakis(triphenylphosphine)palladium(0) (700 mg) in toluene (10 ml) is first bubbled with argon and then heated to 110–115 °C under nitrogen. The reaction mixture is refluxed overnight, after which water is added. The precipitate formed is filtered off and washed with dichloromethane (100 ml). The water phase is extracted with dichloromethane (3 × 30 ml). The combined organic phases are treated with charcoal, dried over magnesium sulfate and evaporated. The residue is flash chromatographed using hexane, 3% ethyl acetate in hexane and 8% ethyl acetate in hexane successively as eluents giving 1.74 g (44%) of the title compound as orange crystals.

Quaterthienyls bearing pyrenes at the terminal α- or β-positions have been prepared as novel emitting materials in organic electroluminescence devices by Stille coupling of 5,5‴-bis(tributylstannyl)-α-quaterthienyl with 1-bromopyrene and by the coupling of 3-(1-pyrenyl)-2-bromothiophene with 5,5′bis(tributylstannyl)-2,2′-bithienyl [284].

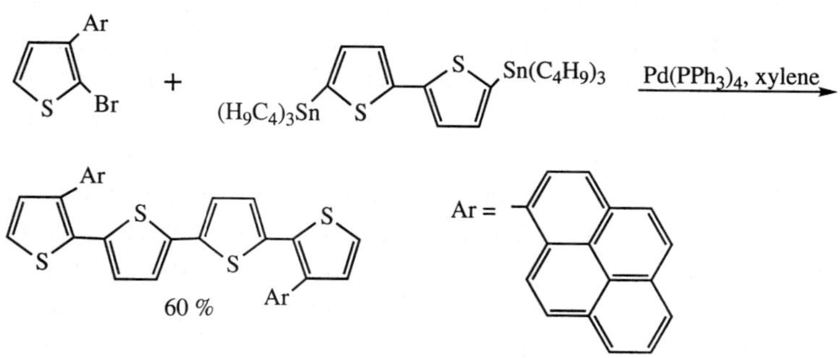

Coupling of 5,5′-bis(tributylstannyl)-2,2′-bithienyl and two equivalents of 5-bromo-2-thiophenealdehyde or 2-bromo-5-mesitylthiophene is used for the preparation of 5,5‴-diformyl-α-quaterthienyl [91,100] and 5,5‴-bis(mesitylthio)-α-quaterthienyl [9], respectively.

5,5‴-Diformyl-2,2′:5′,2″:5″,2‴-quaterthienyl [100]

A solution of 5,5′-bis(tributylstannyl)-2,2′-bithienyl (0.98 g, 1.3 mmol) and 2-bromo-5-formylthiophene (503 mg, 2.63 mmol) in *N,N*-dimethylformamide (20 ml) is deaerated twice with argon. After addition of tetrakis(triphenylphosphine)palladium(0) (15 mg, 0.013 mmol) the reaction mixture is heated at 70–75 °C for 6 h under argon. The bright-red suspension formed is concentrated under reduced pressure and filtered. The solid collected is washed first with hexane (400 ml) and then with diethyl ether (200 ml) giving 450 mg (90%) of the title compound as a bright-red solid. Further purification is carried out by boiling the red solid (200 mg) in tetrahydrofuran (400 ml) and filtered while hot. The filtrate is concentrated to 100 ml, cooled to room temperature and filtered giving 100 mg of the title compound as bright-red crystals mp 270–275 °C (decomp.).

Upon Stille coupling of 2,5-bis(tributylstannyl)-2,2′-bithienyl with two equivalents of 2-bromo-5-mesitylthio-,3,4-ethylenedioxythiophene, 3,4,3″,4‴-bis(ethylenedioxy)-5,5‴-bis(mesitylthio)-α-quaterthienyl is obtained [9].

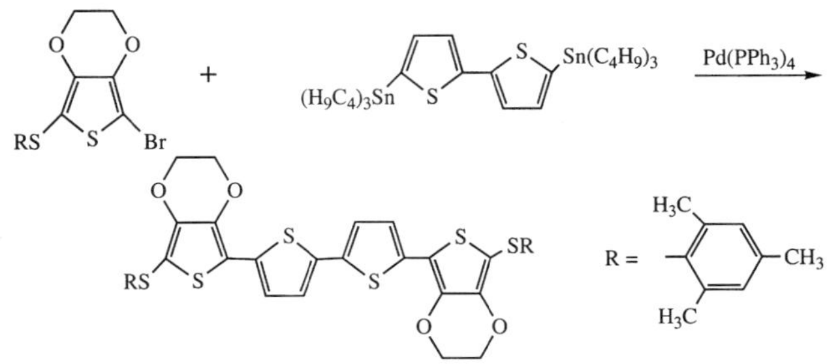

3,4,3‴, 4‴-Bis(ethylenedioxy)-5,5‴-bis(mesitylthio)-2,2′;5′,2″:5″,2‴-quaterthienyl [9]

A solution of 2-bromo-5-mesitylthio-3,4-ethylenedioxythiophene (0.56 g, 1.51 mmol), 2,5-bis(tributylstannyl)-2,2′-bithienyl (0.51 g, 0.680 mmol) and tetrakis(triphenylphosphine)palladium(0) (0.140 g, 8.0 mol%) in toluene (20 ml) is refluxed for 22 h. After cooling, the reaction mixture is poured into hexane (100 ml) resulting in a brick-red precipitate. The precipitate is filtered off and recrystallized from benzene giving 0.37 g (73%) of the title compound as a brick-red solid mp 245–247 °C.

8.6.5.2 From 5-halo-2,2′-bithienyls and 5-trialkylstannyl-2,2′-bithienyls

3′,4″-Dibutyl-5-(trimethylsilyl)-α-quaterthienyl is prepared by Stille coupling of 3-butyl-5-tributylstannyl-5′-trimethylsilyl-2,2′-bithienyl with 3-butyl-5-iodo-2,2′-bithienyl, using palladium(0) and cuprous iodide as catalytic system and *N,N*-dimethylformamide as solvent [89].

3′,4″-Dibutyl-5-(trimethylsilyl)-α-quaterthienyl [89]

A round-bottom flask is charged with 2-(5′-trimethylsilylthienyl)-3-butyl-5-(tributylstannyl)thiophene (3.18 g, 5.45 mmol), 3-butyl-5-iodo-2,2′-bithienyl (2.11 g, 6.06 mmol), copper(I) iodide (0.023 g, 0.123 mmol) and tetrakis (triphenylphosphine)palladium(0) (0.094 g, 0.082 mmol). The mixture is then degassed *in vacuo* followed by introduction of nitrogen and *N*,*N*-dimethylformamide (5 ml), stirred at room temperature overnight and heated at 75–80 °C overnight. The reaction mixture is poured into saturated ammonium chloride and the product is extracted with diethyl ether. The combined organic phases are washed with sodium chloride solution, dried over magnesium sulfate and evaporated. The residue is purified by chromatography on silica gel using hexane as eluent, giving 1.74 g (62%) of the title compound as a light-yellow liquid.

3′,3″-Dimethyl-5,5‴-(trimethylsilyl)-α-quaterthienyl is prepared from 3-methyl-5-tributylstannyl-5′-trimethylsilyl-2,2′bithienyl and 3-methyl-5-iodo-5′-trimethylsilyl-2,2′-bithienyl using palladium(0) as catalyst [76].

3′,3″-Dimethyl-5,5‴-(trimethylsilyl)-2,2′:5′,2″:5″,2‴-quaterthienyl [76]

A flask is charged with 3-methyl-5-tributylstannyl-5′-trimethylsilyl-2,2′-bithienyl (0.541 g, 1.0 mmol), 3-methyl-5-iodo-5′-trimethylsilyl-2,2′-bithienyl (0.378 g, 1.0 mmol), tetrakis(triphenylphosphine)palladium(0) (0.035 g, 0.03 mmol) and toluene (2.0 ml). The reaction mixture is heated at 100–105 °C overnight before being poured into water. The product is extracted with diethyl ether. The combined organic phases are washed with sodium chloride solution, dried over magnesium sulfate and evaporated. The residue is purified by flash chromatography giving 0.209 g (42%) of the title compound as yellow-orange crystals.

3,3′,4″3‴-Tetrakis[2-(tetrahydropyranyloxy)ethyl]-α-quaterthienyl is obtained by Stille coupling of 5-bromo-3,3′-[2-(tetrahydropyranyloxy)-ethyl]2,2′-bithienyl with 5-(trimethylstannyl)-3,3-[2-tetrahydropyranyloxy)-ethyl]-2,2′-bithienyl. The protecting groups are removed by treatment with 10% hydrochloric acid [130].

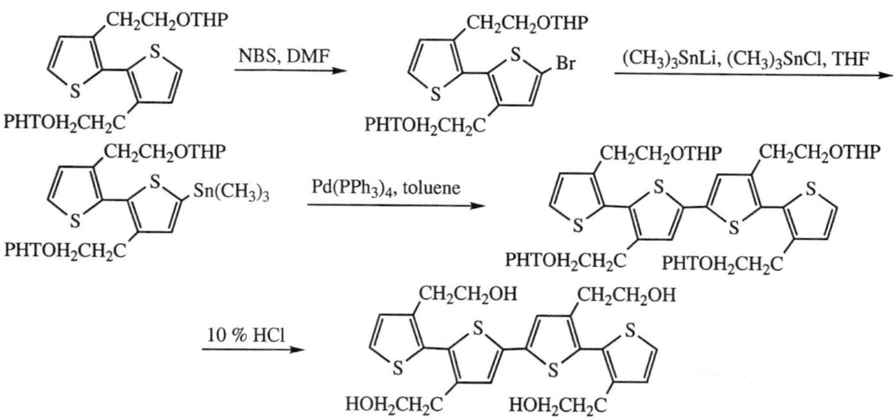

5-(Trimethylstannyl)-3,3′-[2-(tetrahydropyranyloxy)ethyl]-2,2′-bithienyl [130]

To a flask immersed in a bath at −10 °C and containing lithium wire (0.88 g, 0.12 mmol) in anhydrous tetrahydrofuran (10 ml), trimethyltin chloride (2.5 g, 0.012 mmol) dissolved in anhydrous tetrahydrofuran (40 ml) is added dropwise. The mixture is stirred at room temperature overnight. This trimethyltin lithium solution (24 ml, 6.0 mmol) is added to a solution of 5-bromo-3,3′-bis[2-(tetrahydropyranyloxy)ethyl]-2,2′-bithienyl (1.0 g, 2.0 mmol) in anhydrous tetrahydrofuran (20 ml). After stirring at room temperature for 2 h the solution is cooled to 0 °C and trimethyltin chloride (0.4 g, 2.0 mmol) in anhydrous tetrahydrofuran (20 ml) is added dropwsie. The reaction mixture is stirred at ambient temperature for 1 h and then quenched with an aqueous saturated solution of ammonium chloride. The product is extracted with diethyl ether and the combined organic phases washed with sodium chloride solution, dried over magnesium sulfate and evaporated. The residue, 0.97 g of a yellow-green liquid, contains 60% of the title compound and 40% of 3,3′-bis[2-(tetrahydropyranyloxy)ethyl]-2,2′-bithienyl.

8.6.5.3 From 5,5′-dihalo-2,2′-bithienyls and two equivalents of trialkylstannylthiophenes

The palladium(0)-catalyzed reaction of 5,5′-dibromo-3,3′-bis(methylthio)-2,2′-bithienyl with 2-trimethylstannyl-3-methylthiothiophene gives 40% of 3,4′,3″,3‴-tetra(methyltio)-α-quaterthienyl [302]. However, the Stille reaction between 5,5′-dibromo-2,2′-bithienyl and 3-alkylthio-2-trimethylstannylthiophene is rather complex due to exchange of functionalities between the reagent and unexpectedly a sexithienyl was obtained as main product. The desired product 3,3‴-bis(butylthio)-α-quaterthienyl was obtained only in 21% yield [265].

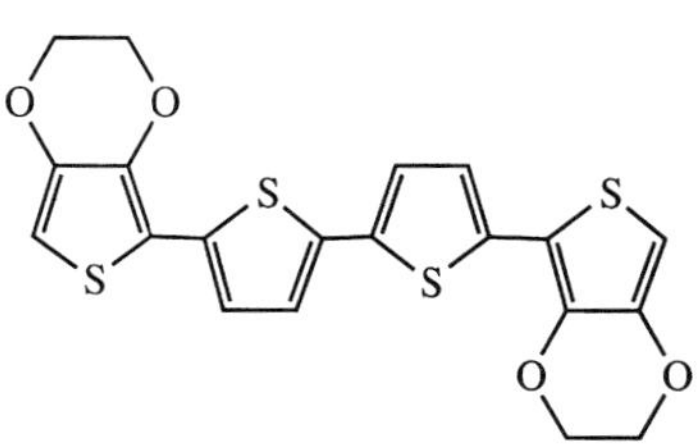

The compound shown above is best prepared by Stille coupling of 5,5′-dibromo-2,2′-bithienyl with two equivalents of 3,4-methylenedioxy-2-tributylstannylthiophene. A better method for the preparation of the other isomer is the following [214].

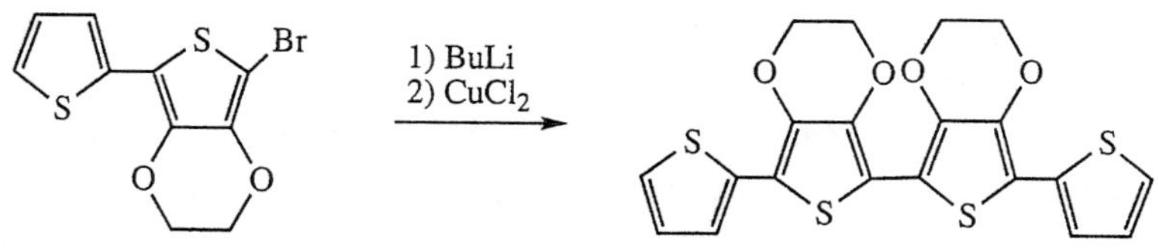

8.6.5.4 From haloterthienyls and trialkylstannylthiophenes

The reaction of polymer-bound 5-bromo-4-octyl-α-terthienyl-5″-carboxylate and 2-(trimethylstannyl)-4-octylthiophene is used for the preparation of polymer-bound 4,3′-dioctyl-α-quaterthienyl-5‴-carboxylate [126,129].

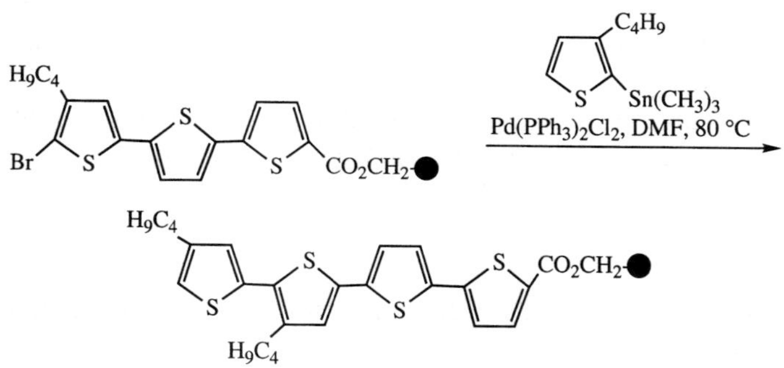

8.6.6 Transition metal-catalyzed couplings according to Suzuki

8.6.6.1 From thiopheneboronic acids and 5,5′-dihalobithienyls

Coupling of 4-pentoxy-2-thiopheneboronic acid with 5,5′-diiodobithienyl is used for the preparation of 4,4‴-dipentoxy-α-quaterthienyl [15]. An interesting modification of the Suzuki coupling is used for the preparation of 4,3′,3″3‴ -tetradodecyl-α-quaterthienyl-5-carboxylic acid benzyl ester from 4,3′-didodecyl-2,2′-bithienyl-5-boronic acid propane-1,3-diol ester and 3,3′-didodecyl-5′-iodo-2,2′-bithienyl-5-carboxylic acid benzyl ester using palladium(0) and cesium fluoride as catalyts [82].

4,3′,3″3‴-Tetradodecyl-2,2′:5′,2″:5″,2‴-quaterthienyl-5-carboxylic acid benzyl ester [82]

To a solution of 3,3′-didodecyl-5′-iodo-2,2′-bithienyl-5-carboxylic acid benzyl ester (1.00 g, 1.30 mmol) in anhydrous tetrahydrofuran (30 ml), tetrakis(triphenylphosphine)palladium(0) (0.07 g, 0.065 mmol) and cesium fluoride (1.00 g, 6.00 mmol) are added. The mixture is refluxed under inert gas atmosphere, after which 4,3′-didodecyl-2,2′-bithienyl-5-boronic acid propane-1,3-diol ester (1.16 g, 2.00 mmol) in anhydrous tetrahydrofuran is added dropwise. The reaction mixture is refluxed for 3 h and evaporated. The residue is purified by chromatography on silica gel using hexane/dichloromethane (8:2) giving 1.13 g (77%) of the title compound as an orange solid mp 55 °C.

The quaterthienyl, bis(3,4-ethylenedioxothienyl)-2,2′-bithienyl, has been obtained through a Suzuki reaction [308].

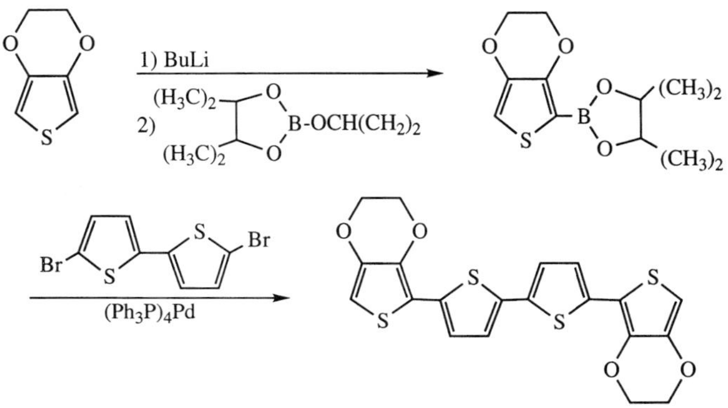

Preparation of the boronic ester [308]

Lithiation of 3,4-ethylenedioxythiophene (2 g, 0.0141 mol) is carried out as described previously. To this solution 2-isopropoxy-4,4,5,5-tetramethyl-1,3,2-dioxaborolane (5.8 ml, 0.028 mol) in anhdryous tetrahydrofuran (10 ml) is added. After stirring the mixture for 6 h it is poured over crushed ice containing ammonium chloride. The product is extracted with diethyl ether (3 × 30 ml), dried and evaporated. The residue is triturated with methanol giving 2 g (71%) of the boronic ester mp 87–88 °C.

Bis(3,4-Ethylenedioxothienyl)-2,2′-bithienyl [308]

A solution of 5,5′-dibromo-2,2′-bithienyl (325 mg, 1 mmol), the boronic ester (805 mg, 3 mmol), tetrakis(triphenylphosphine)palladium(0) (230 mg, 0.2 mmol) and anhydrous potassium phosphate (200 mg) in *N*,*N*-dimethylformamide (15 ml) is heated at 100 °C for 16 h under nitrogen. The reaction mixture is poured into water and extracted with dichloromethane. The phases are separated and the organic phase washed several times with saturated aqueous sodium chloride solution followed by water and evaporated. The residue is purified by chromatography using hexanes/dichloromethane (3:2) as eluent giving 90 mg (20%) of the title compound as an orange solid mp 188–191 °C.

8.6.6.2 From haloterthienyls and thiopheneboronic acid derivatives

3,4′,4″,4″-Tetraoctyl-5‴-chloro-α-quaterthienyl is prepared from 5-chloro-5′-iodo-α-terthienyl and [1′,3′-(2′,2′-dimethylpropylene)]-3-octyl-2-thienylboronate [85,264].

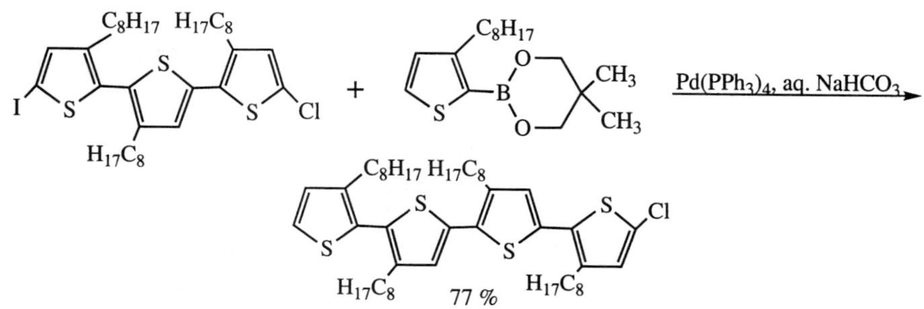

The first regular head-to-tail coupled oligothiophene, a quater(3-aryl)-thiophene was prepared on solid support according to the following scheme [309].

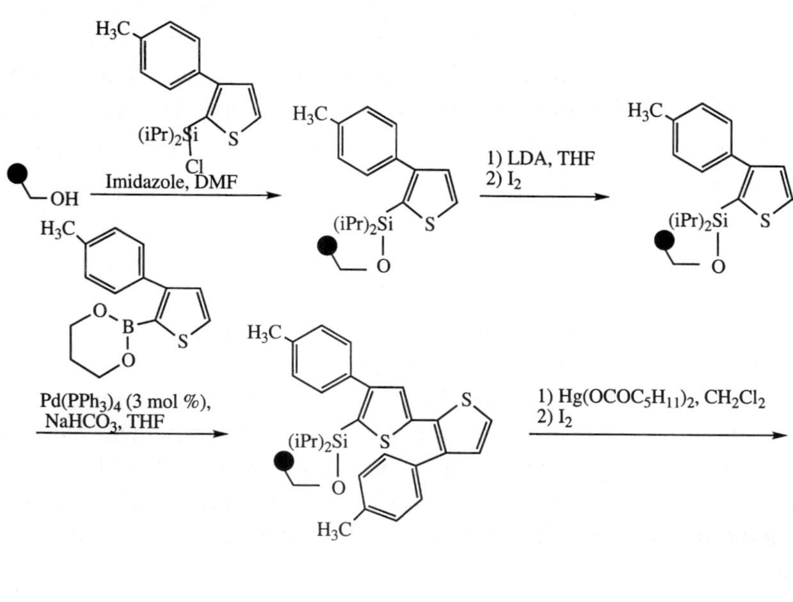

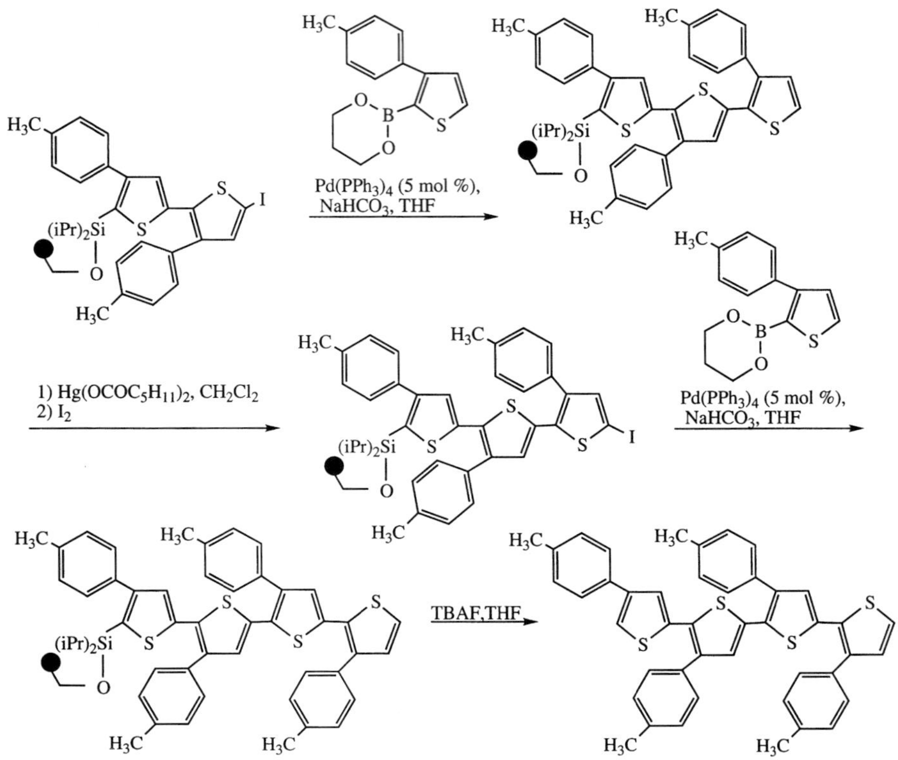

2-[Chloro(diisopropyl)silyl)-3-para-tolylthiophene [309]

To a solution of 2-bromo-3-*para*-tolylthiophene (2.53 g, 10.0 mmol) in anhydrous diethyl ether (40 ml) is added 1.6 *M* butyllithium in hexane (11 mmol) at − 70 °C. The mixture is warmed to room temperature and added *via* cannula to a solution of dichloro(diisopropyl)silane (15 mmol) in anhydrous diethyl ether (40 ml) at − 30 °C. After the mixture is stirred at ambient temperature for 1 h, the solvent and the excess of dichloro(diisopropyl)silane are evaporated and the residue distilled giving 1.94 g (60%) of the title compound bp 146 °C/0.34 mm Hg.

2-[Benzyloxy(diisopropyl)silyl]-3-phenylthiophene [309]

To a solution of 2-[chloro(diisopropyl)silyl]-3-phenylthiophene 0.50 g, 1.62 mmol) in *N*,*N*-dimethylformamide (0.5 ml) benzyl alcohol (0.44 g, 4.05 mmol) and imidazole (0.28 g, 4.05 mmol) are added. The reaction mixture is stirred at ambient temperature for 18 h, poured into water and extracted with cyclohexane. The combined organic phases are washed with sodium chloride solution, dried over sodium sulfate and evaporated. The crude product is purified by flash chromatography on aluminium oxide using pentane as eluent, giving 0.52 g (85%) of the title compound as a colorless solid mp 38 °C.

8.6.6.3 From terthienyl-5-boronic acids and bromothiophenes

The α-quaterthienyl shown below has been prepared by this approach [84].

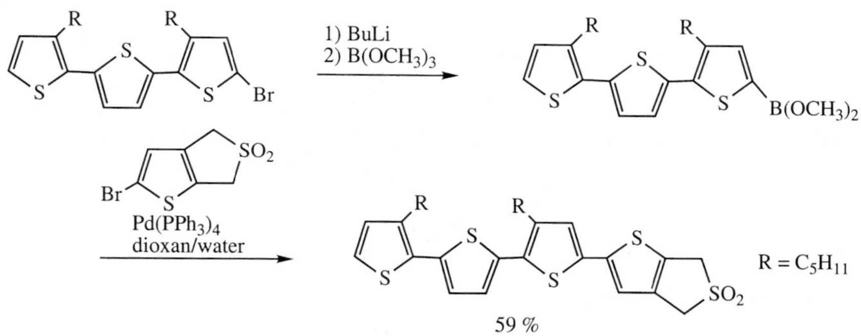

8.6.6.4 By oxidation of ate-complexes

α-Quaterthienyl is prepared by reaction of 5-lithio-2,2′-bithienyl with the β-methoxy derivative of 9-borabicyclo[3.3.1]nonane and the ate-complex neutralized with boron trifluoride etherate, followed by a second equivalent of 5-lithio-2,2′-bithienyl and oxidation with iodine and worked up by alkaline hydrogen peroxide. Alternatively 2-thienyllithium and 5-lithio-α-terthienyl can be used for the preparation of α-quaterthienyl [310].

8.6.7 Preparation of quaterthienyls through ring-closure reactions

8.6.7.1 *Via* butadiynes

1-(2-Thienyl)-4-(2,2′-bithienyl-5-yl)-1,3-butadiyne gives upon reaction with sodium sulfide nonahydrate in tetrahydrofuran a quantative yield of α-quaterthienyl [311].

1-(2-Thienyl)-4-(2,2′-bithienyl-5-yl)-1,3-butadiyne [311]

A solution of bromo(2-thienyl)acetylene (380 mg, 2 mmol) in methanol (10 ml) is added at room temperature to a well-stirred solution of cuprous chloride

(11.2 mg), a small crystal of hydroxylamine hydrochloride, 70% aqueous ethylamine (0.277 mmol) and 5-ethynyl-2,2′-bithienyl (380 mg, 2 mmol) in methanol (5 ml) and diethylether (5 ml). After stirring the reaction mixture for 1 h, sodium cyanide (4.5 mg) in water (5 ml) is added and the mixture is extracted with dichloromethane (4 × 15 ml). The combined organic phases are washed with water (3 × 10 ml), dried over magnesium sulfate and evaporated. The residue is recrystallized from tetrahydrofuran/water giving 566 mg (96%) of the title compound mp 120–121 °C.

2,2′:5′,2″:5″,2‴-Quaterthienyl [311]

A mixture of the compound described above (200 mg, 0.6 mmol) and sodium sulfide nonahydrate (2.4 mmol) in tetrahydrofuran (40 ml) is refluxed overnight. After cooling and evaporation the residue is recrystallized from 95% ethanol giving 200 mg (100%) of the title compound mp 211–212 °C.

8.6.7.2 From 1,4-diketones

The reaction of 1-(2-thienyl)-4-(2,2′-bithienyl-5-yl)-1,4-butanedione with Lawesson's reagent gives α-quaterthienyl in 93% yield [206].

2,2′:5′,2″:5″,2‴-Quaterthienyl [206]

A mixture of 1-(2-thienyl)-4-(2,2′-bithienyl-5-yl)-1,4-butanedione (3 g, 9.36 mmol) and Lawesson's reagent (2.19 g, 5.42 mmol) in anhydrous toluene (180 ml) is refluxed under argon for 1 h. The yellow-orange precipitate is filtered off and recrystallized from 95% ethanol giving 2.77 g (93%) of the title compound mp 212–213 °C.

Four of the 16 possible isomers of branched quaterthienyls (thienylterthienyls) are prepared from 1-(thienyl)-2-(thienyl)-4-(thienyl)-1,4-butanedione and phosphorus pentasulfide in acetonitrile and dichloromethane [312].

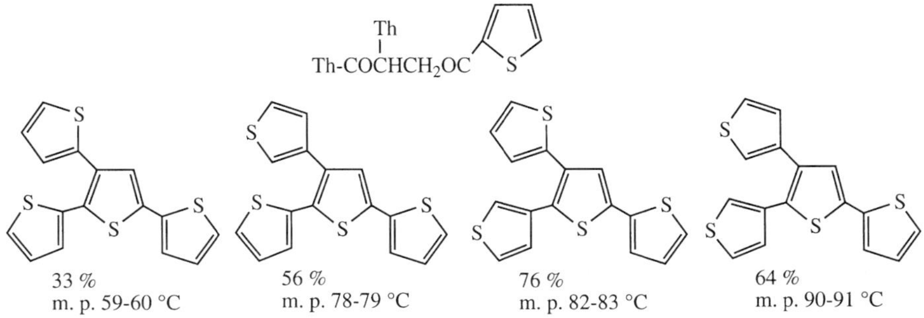

8.6.7.3 Various ring-closure reactions

Reaction of 3-(2-bithienyl)phthalide with 2-thiophenemagnesium bromide followed by ammonium chloride and Lawesson's reagent gives 1-bithienyl-3-thienylbenzo[*c*]thiophene [241].

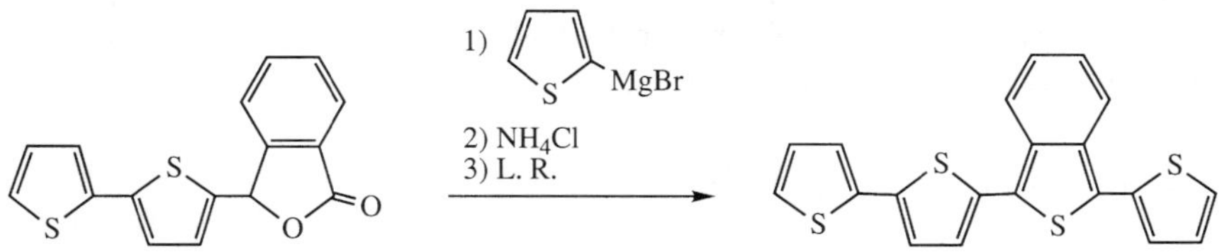

8.6.8 Electrophilic substitutions of quaterthienyls

8.6.8.1 Nitration of quaterthienyls

Both mononitration and dinitration of 3,3′,3″,4″-tetrahexyl-α-quaterthienyl can be achieved by reaction with nitric acid in acetic anhydride [123]. Attempted coupling of 3,4‴-bis(trimethylsilyl)-α-quaterthienyl with ceric ammonium nitrate gave instead the 5-nitro derivative in 43% yield [249].

8.6.8.2 Acylation and formylation of quaterthienyls

Vilsmeyer formylation of 3,3‴-dihexyl-α-quaterthienyl with *N*,*N*-dimethylformamide and phosphorus oxychloride in dichloroethane at 60 °C gives the 5-formyl derivative in 72% yield [305].

8.6.8.3 Bromination of quaterthienyls

With *N*-bromosuccinimide in *N*,*N*-dimethylformamide 5,5‴-dibromoquaterthienyl is obtained in good yield [63].

5,5‴-Dibromo-α-quaterthienyl [63]

A solution of *N*-bromosuccinimide (0.356 g, 2 mmol) in *N*,*N*-dimethylformamide (10 ml) is rapidly dropped to a stirred solution of quaterthienyl (0.330, 1 mmol) in *N*,*N*-dimethylformamide (80 ml) at 80 °C. After 2 h the reaction mixture is cooled and poured onto ice. Workup gives 0.37 g (76%) of the title compound as an orange solid mp 263–264 °C.

Dibromination of substituted quaterthienyls is achieved in almost quantitative yield using two equivalents of *N*-bromosuccinimide in a mixture of chloroform/acetic acid or *N*-bromosuccinimide in *N*,*N*-dimethylformamide. In this way 5,5‴-dibromo-3,3‴-octyl-α-quaterthienyl [303] and 5,5‴-dibromo-3,3′,3‴,4″-tetrahexyl-α-quaterthienyl are obtained [276,301], as well as 5,5‴-dibromo-3,3′,4″,3‴-tetramethylthio-α-quaterthienyl and 5,5‴-dibromo-3,4′,3″,3‴-tetramethylthio-α-quaterthienyl [302].

Monobromination is difficult to achieve cleanly and a mixture of 62% of the monobromo- and 15% of the dibromo derivative were obtained from 3,3′,3‴,4″-tetrahexyl-α-quaterthienyl and were separated by chromatography on silica gel [301]. Bromination of polymer-bound 3′4-dioctyl-α-quaterthienyl-5‴-carboxylate is used for the preparation of the polymer-bound 5-bromo-derivative as a step in the solid phase synthesis of oligothiophenes [126,129].

8.6.8.4 Iodination of quaterthienyls

The 5‴-iodo derivative of 4,3′,3″,3‴-tetradodecyl-α-quaterthienyl-5-carboxylic acid benzyl ester is prepared by iodination with mercuric caproate and iodine in chloroform/acetic acid [82].

5‴-Iodo-4,3′,3″,3‴-tetradodecyl-2,2′:5′,2″:5″,2‴-quaterthienyl-5-carboxylic acid benzyl ester [82]

To a solution of 4,3′,3″,3‴-tetradodecyl-2,2′:5′,2″:5″,2‴-quaterthienyl-5-carboxylic acid benzyl ester (0.91 g, 0.80 mmol) in chloroform/acetic acid (95:5), mercuric capronate (0.34 g, 0.84 mmol) is added, after which the solution is cooled down to 0 °C. A solution of iodine (0.21 g, 0.84 mmol) in the same solvent is added slowly dropwise over a period of 2 h. The reaction mixture is allowed to warm to room temperature and then washed several times with aqueous sodium bicarbonate solution, aqueous thiosulfate solution and water. The organic phase is dried over sodium sulfate and evaporated. The residue is purified by chromatography on silica gel using petroleum ether/dichloromethane (75:25) as eluent, giving 0.96 g (95%) of the title compound as an orange solid mp 71 °C.

8.6.9 Substituted quaterthienyls *via* metalation reactions followed by electrophiles

Metalation of 3,3‴-dimethoxy-α-quaterthienyl with butyllithium followed by reaction with dimethyl sulfate gives 5,5″-dimethyl-3,3‴-dimethoxy-α-quaterthienyl [28] and with carbon dioxide the dicarboxylic acid is obtained [171].

3,3‴-Dimethoxy-5,5‴dimethyl-2,2′:5′,2″:5″,2‴-quaterthienyl [28]

At $-78\,^{\circ}C$ butyllithium (0.36 ml, 0.90 mmol) is added dropwise to a solution of 3,3‴-dimethoxy-2,2′:5′,2″,5″,2‴-quaterthienyl (172 mg, 0.44 mmol) in anhydrous tetrahydrofuran (15 ml). The mixture is warmed with stirring to $0\,^{\circ}C$ for 30 min and then recooled back to $-78\,^{\circ}C$ followed by addition of dimethyl sulfate (113 mg, 0.9 mmol). The reaction mixture is stirred at $-78\,^{\circ}C$ for 3 h, warmed to room temperature for 1 h and poured into water. The product is extracted with diethyl ether and the combined organic phases washed with sodium chloride solution, dried over sodium sulfate and evaporated. The residue is purified by flash chromatography using ethyl acetate/hexane (5:95) as eluent, giving 156 mg (85%) of the title compound as orange crystals.

Metalation of 3′,4″-dibutyl-5-(trimethylsilyl)-α-quaterthienyl with lithium diisopropylamide followed by tributylstannyl chloride gives 3′,4″-dibutyl-5-trimethylsilyl-5‴-tributylstannyl-α-quaterthienyl [89].

8.6.10 Substituted quaterthienyls *via* halogen–metal exchange reactions followed by electrophiles

3,3′,3‴,4″-Tetrahexyl-α-quaterthienyl 5,5‴-dicarboxylic acid is prepared from 5,5‴-dibromo derivative by halogen–metal exchange with butyllithium followed by reaction with carbon dioxide and upon reaction with *N,N*-dimethylformamide the 5,5‴-dicarbaldehyde is obtained in 50% yield [301]. Halogen–metal exchange of the 5-bromo derivative followed by reaction with carbon dioxide is used for the preparation of the 5-carboxylic acid and upon reaction with *N,N*-dimethylformamide the 5-carbaldehyde is obtained [123]. An α-quaterthienyl with push–pull substituents, as in 3,3′,3‴,4″-tetrahexyl-5‴-methoxy-α-quterthienyl-5-carbaldehyde was prepared *via* halogen–metal exchange of the 5-bromo-5‴-methoxy derivative [123].

3,3′,3‴,4″-Tetrahexyl-5‴-methoxy-2,2′:5′2″:5″,2‴-quatrthienyl-5-carbaldehyde [123]

Reaction of 3,3′,3‴,4″-Tetrahexyl-5-bromo-5‴-methoxy-2,2′:5′,2″:5″,2‴-quaterthienyl (204 mg, 0.26 mmol), *N,N,N′;N′*-tetramethylethylenediamine (33 mg, 0.28 mmol), 1.6 *M* butyllithium (0.17 ml) and *N,N*-dimethylformamide (0.1 ml, 1.31 mmol) in anhydrous tetrahydrofuran (10 ml) gives 40 mg (21%) of the title compound as a yellow oil.

8.6.11 Substituted quaterthienyls *via* various modifications of substituents

8.6.11.1 From haloquaterthienyls

5,5‴-Dicyano-3,3′,3″,3‴-tetrahexyl-α-quaterthienyl is obtained *via* the reaction of the 5,5‴-dibromo derivative and cuprous cyanide in *N*-methyl-2-pyrrolidone [123].

5,5‴-Dicyano-3,3′,3″,3‴-tetrahexyl-2,2′:5′,5″:2″,2‴-quarterthienyl [123]

A mixture of 5,5‴-dibromo-3,3′,3″,3‴-tetrahexyl-2,2′:5′,5″,2″,2‴-quaterthienyl (616 mg, 0.75 mmol), cuprous cyanide (269 mg, 2.97 mmol) and *N*-methyl-2-pyrrolidone (20 ml) is refluxed for 1 h, after which it is poured into 25% aqueous ammonia. The product is extracted with benzene and the combined organic phases are washed with sodium chloride solution and evaporated. The residue is purified by chromatography on silica gel using benzene/hexane (4:6) as eluent, giving 76% of the title compound as pale-yellow microcrystals mp 85–86 °C after recrystallization from hexane.

From the 5-bromo derivative using the same method the 5-cyano derivative was obtained [123].

3,3′,3‴,4′-Tetrahexyl 5-methoxy-α-quaterthienyl is prepared in low yield from the 5-bromo derivative, cupric oxide, sodium methoxide and sodium iodide in methanol [123]. From the 5,5‴-dibromo derivative the 5,5‴-dimethoxy derivative as well as the 5-bromo-5‴-methoxy compound were obtained, and were separated by chromatography [123]. This compound was converted to the 5-cyano-5‴-methoxy derivative with cuprous cyanide in *N*-methyl-2-pyrrolidone and in this way α-quaterthienyl with push–pull substituents at the ends of the quaterthienyl was obtained [123]. Treatment of 3,4′,4″,4‴-tetraoctyl-5‴-chloro-α-quaterthienyl with tributyltin hydride and aza-isobutyronitrile gives 3,4′,4″,4‴-tetraoctyl-α-terthienyl in 70% yield [85].

8.6.12 Preparation of 1,1-dioxides of α-quaterthienyls

A mixture of 5,5‴-bis(dimethyl-*tert*-butylsilyl)-α-quaterthienyl-1,1-dioxide and 1′,1′-dioxide is formed upon oxidation of 5,5‴-bis(dimethyl-tert-butylsilyl)-α-quaterthienyl with *meta*-chloro-perbenzoic acid in dichloromethane in addition with small amounts of 1,1,1‴,1‴-tetradioxide [226]. Alternatively defined 1,1-dioxides are prepared by Stille coupling of 2-bromo-5-hexyl- and 2-bromo-5-*tert*-butyldimethylsilylthiophene-1,1-dioxide with 5,5′-bis(tributyl-stannyl)-2,2′-bithienyl giving 5,5‴-dihexyl- and 5,5‴-bis(butyldimethylsilyl)-α-quaterthienyl-1,1,1‴,1‴-tetraoxide [227].

5,5‴-Bis(hexyl)-2,2′:5′,2″:5″,2‴-quaterthienyl-1,1-dioxide [227]

To a solution of tetrakis(triphenylarsine)palladium(0) (0.012 mmol) in toluene (9 ml), 5-(hexyl)-5″-(tributylstannyl)-2,2′:5′,2″-terthienyl (260 mg, 0.42 mmol) and 2-hexyl-5-bromothiophene-1,1-dioxide (120 mg, 0.42 mmol) in toluene (9 ml) are added dropwise. The reaction mixture is refluxed overnight and then treated with a saturated solution of ammonium chloride. The phases are separated and the organic phase is washed with water, dried over magnesium sulfate and evaporated. The residue is chromatographed on silica gel using pentane/ethyl acetate (9:1) as eluent, giving 110 mg (50%) of the title compound as an orange powder mp 172 °C.

5,5‴-Hexyl-and 5,5‴-bis(butyldimethylsilyl)-α-quaterthienyl-1,1-dioxide is obtained by coupling of 5-hexyl-and 5-butyldimethylsilyl-5″-tributylstannyl-α-terthienyl with 2-bromo-5-hexyl- and 2-bromo-5-butyldimethylsilyl-thiophene-1,1-dioxide [227].

8.7 QUINQUETHIENYLS

8.7.1 Introduction

As this class of compounds contains an odd number of thiophene rings dimerization reactions cannot be used. Most of the defined quinquethienyls are prepared *via* transition metal-catalyzed cross-coupling reactions. In a few cases ring-closure reactions forming one or several of the thiophene rings are also used.

8.7.2 Transition metal-catalyzed couplings according to Kumada

8.7.2.1 Coupling between two equivalents of thiophenemagnesium halides and 5,5″-dibromo-α-terthienyls

4,4⁗-Di(trimethylsilyl)-α-quinquethienyl is prepared by [1,3-bis(diphenyl-phosphine)propane]nickel(II) chloride-catalyzed coupling of 4-trimethylsilyl-2-thiophenemagnesium bromide with 5,5′-dibromo-α-terthienyl 4′,3‴-Dihexyl-α-quinquethienyl is similarly prepared from 2-thiophenemagnesium bromide and 5,5″-dibromo-3,3′-dihexyl-α-terthenyl [254].

4′3‴-Dihexyl-2,2′:5′,2″:5″,2‴:5‴,2⁗-quinquethienyl [254]

To a solution of 2-thiophenemagnesium bromide (2.79 mmol) prepared in diethyl ether (20 ml) [1,3-bis(diphenylphosphine)propane]nickel(II) chloride

(0.1 g, 2 mol%) is added, after which a solution of 5,5″-dibromo-3,3″-dihexyl-2,2′:5′,2″-terthienyl (0.53 g, 0.90 mmol) in diethyl ether (20 ml) is added dropwise. The reaction mixture is refluxed under stirring for 18 h and then cooled and poured into a mixture of crushed ice (100 g) and 2 *M* hydrochloric acid (20 ml). The product is extracted with dichloromethane (5 × 30 ml). The combined organic phases are washed to neutrality with saturated sodium hydrogen carbonate solution, then with water (2 × 50 ml), dried over magnesium sulfate and evaporated. The residue is purified by flash chromatography using hexane/dichloromethane (40:1) as eluent followed by recrystallization from hexane giving 0.423 g (79%) of the title compound as an orange solid.

In connection with work on the synthesis of anthryloligothienylporphyrins as energy transfer and light-harvesting systems, 3,3⁗-pentyl-α-quinquethienyl is prepared from 3-pentyl-2-thiophenemagnesium bromide and 5,5″-dibromo-α-terthienyl [313].

3,3⁗-Dipentyl-2,2′:5′,2″:5″,2⁗-quinquethienyl [313]

From 2-bromo-3-pentylthiphene (2.91 g, 12.5 mmol) and magnesium (0.32 g, 13 mmol) in diethyl ether (15 ml) the Grignard reagent is prepared under heating to reflux for 2 h using 1,2-dibromoethane as entrainer. The Grignard solution is transferred to a dropping funnel of a second apparatus and added dropwise to a solution of 5,5″-dibromo-2,2′:5′,2″-terthienyl (2.03 g, 5 mmol) and [1,3-bis(diphenylphosphine)propane]nickel(II) chloride (0.027 g, 0.05 mmol) in diethyl ether/benzene (4:1) (75 ml). The reaction mixture is stirred under reflux for 20 h and then hydrolyzed with cold 0.5 *M* hydrochloric acid. The phases are separated and the aqueous phase extracted several times with dichloromethane. The combined organic phases are washed with a 1 *M* sodium bicarbonate solution followed by water, dried over sodium sulfate and evaporated. The residue is purified by chromatography on silica gel using hexane/dichloromethane (10:1) as eluent giving 2.49 g (90%) of the title compound as a yellow-orange solid mp 81–82 °C.

3″,4″-Dibutyl-α-quinquethienyl is obtained by coupling of 3′,4′-dibutyl-α-terthienyl with two equivalents of 2-thiophenemagnesium bromide [314].

5,5⁗-Diphenyl-α-quinquethienyl has recently been prepared by Kumada coupling of 5,5″-dibromo-α-terthienyl with two equivalents of 5-phenyl-2-thiophenemagnesium bromide [67].

Nickel-catalyzed coupling of the Grignard reagent of 2-bromo-3,4-dibutylthiophene with 5,5″-dibromo-3′,4′-dibutyl-2,2′:5′,2″-terthienyl gives 3,3″,3⁗,4,4″,4‴-tetrabutyl-2,2′:5′,2‴:5‴,2⁗-quinquethienyl in 76% yield [248].

Kumada coupling of 4,5,6,7-tetrahydrobenzo[*b*]thiophen-2ylmagnesium bromide with 5,5′-dibromo-3′-phenyl-α-terthienyl, 5,5′-dibromo-3′,4′-diphenyl-α-terthienyl and 1,3-bis(5-bromo-2-thienyl)benzo[*c*]thiophene is used for the preparation of 5,5″-bis(4,5,6,7-tetrahydrobene[*b*]thiophene-2-yl)-3″-phenyl-α-terquethienyl, 5,5″-bis(4,5,6,7-tetrahydrobenzo[*b*]thiophene-2-yl)-3′,4′-diphenyl-α-terthienyl and 1,3-bis[5-(4,5,6,7-tetrahydrobenzo[*b*]thiophene-2-yl)-2-thienyl]benzo[*c*]thiophene, repectively [252].

8.7.3 Transition metal-catalyzed couplings according to Stille

8.7.3.1 Coupling between two equivalents of 5-trialkylstannyl-2,2′-bithienyl and dihalothiophenes

The Stille coupling of 2,5-diiodothiophene with two equivalents of 3-methyl-5′-trimethylsilyl-5-trimethylstannyl-2,2′-bithienyl is used for the preparation of 5,5⁗-bis(trimethylsilyl)-3′,4‴-dimethyl-α-quinquethienyl [76].

5,5⁗-Bis(trimethylsilyl)-3′,4‴-dimethyl-2,2′:5′,2″:5‴,2⁗-quinquethienyl [76]

An oven-dried test tube, washed with ammonium hydroxide, is charged with 3-methyl-5-(tributylstannyl)-5′-trimethylsilyl)-2,2′-bithienyl (0.609 g, 1.125 mmol), 2,5-diiodothiophene (0.084 g, 0.25 mmol), tetrakis(triphenylphosphine)palladium(0) (0.012 g, 0.01 mmol) and toluene (1.0 ml). The mixture is slowly heated to 50 °C for 1 h, then at 80 °C for 2 h and finally at 100–105 °C overnight, after which it is poured into water. The product is extracted with diethyl ether and the combined organic phases washed with sodium chloride solution, dried over magnesium sulfate and evaporated. The residue is purified by chromatography on silica gel using hexane as eluent, giving 0.069 g (47%) of the title compound as an orange solid.

3″,4″-Dinitro-α-quinqueterthienyl is prepared in 40% yield by palladium(II)-catalyzed coupling between 5-trimethylstannyl-2,2′-bithienyl and 2,5-dibromo-3,4-dinitrothiophene [87].

8.7.3.2 Coupling between two equivalents of 2-trialkylstannylthiophene and 5,5″-dihalo-α-terthienyls

Coupling between 2-*tert*-butyldimethylsilyl-5-tributylstannylthiophene and 5,5″-dibromo-α-terthienyl using tetrakis(triphenylarsine)palladium(0) in toluene as catalyst is used for the preparation of 5,5⁗-bis(dimethyl-*tert*-butylsilyl)-α-quinquethienyl [277].

5,5⁗-Bis(dimethyl-tert-butylsilyl)-2,2′:5′,2″:5″,2‴:5‴,2⁗-quinquethienyl [277]

To a solution of tetrakis(triphenylarsine)palladium(0) in toluene (20 ml) 5,5″-dibromo-2,5:5′,2″-terthienyl (0.406 g, 1.0 mmol) and 2-(dimethyl-*tert*-butylsilyl)-5-(tributylstannyl)thiopene (0.972, 2.0 mmol) are added. After heating at reflux for 12 h the reaction mixture is hydrolyzed with a saturated ammonium chloride solution and the product extracted with dichloromethane. The combined organic phases are dried over magnesium sulfate, concentrated, filtered through silica gel and evaporated. The residue is washed with diethyl ether giving 0.364 g (57%) of the title compound as an orange microcrystalline solid mp 221–212 °C.

Synthesis of an α-quinquethienyl with alternating 3,4-bis[(methoxycarbonyl)methyl] substituted rings is achieved by coupling of two equivalents of 2-tributylstannylthiophene and 5,5″-dibromo-3,4,3″,4″-tetra[(methoxycarbonyl)methyl]-α-terthienyl [224].

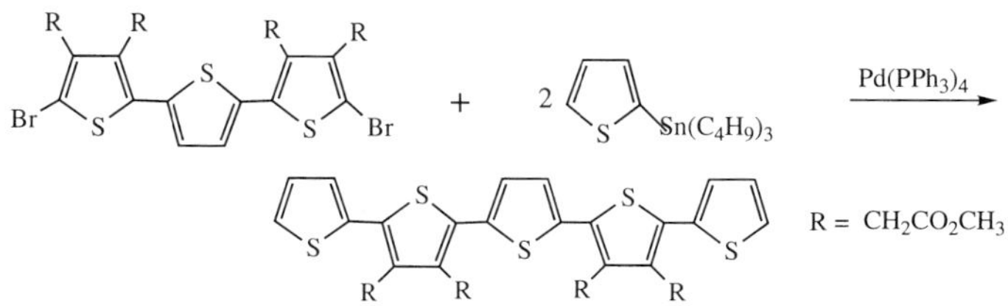

3′,4′,3‴,4‴-Tetra[(methoxycarbonyl)methyl]-2,2′:5′,2″:5″,2‴:5‴,2⁗-quinquethienyl [224]

A mixture of 3,4,3″,4″-tetra[(methoxycarbonyl)methyl]-2,2′:5′,2″-terthienyl (1.0 g, 1.44 mmol), 2-(tributylstannyl)thiophene (1.3 g, 3.47 mmol), tetrakis(triphenylphosphine)palladium(0) (0.058 g, 0.05 mmol) in anhydrous *N*,*N*-dimethylformamide (40 ml) is, under nitrogen and with stirring, heated at 105 °C for 1 h. The reaction is quenched with water and the product extracted with dichloromethane. The combined organic phases are dried over calcium chloride and evaporated. The residue is purified by chromatography using hexane/ethyl acetate (1:1) as eluent, giving 0.96 g (95%) of the title compound as an orange solid mp 170–171 °C after repeated crystallizations from methanol.

In connection with work on dendrimer-supported oligothiophenes 5-trimethylstannyl-2,2′-bithienyl was coupled with 2,5-dibromo-3-thiophene acetate having a dendrimer in the ester function [211].

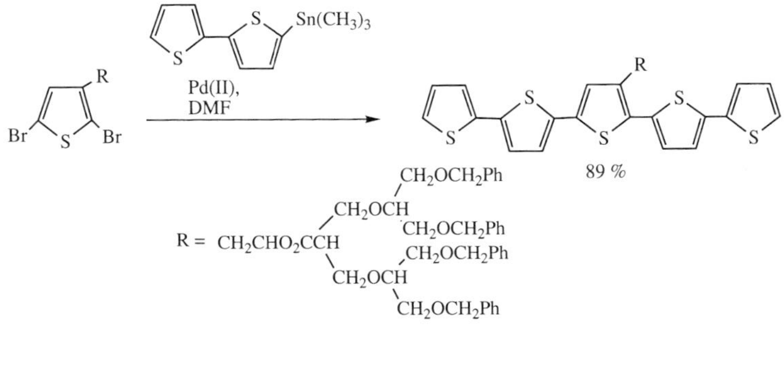

8.7.3.3 Coupling between 5,5″-trialkylstannyl-α-terthienyl and two equivalents of halothiophene

Coupling of 5,5″-bis(tributylstannyl)-α-terthienyl and 2-iodo-3-methoxythiophene gives 3,3⁗-dimethoxy-α-quinquethienyl. In a similar way 3,3″-dimethoxy-5,5″-tributylstannyl-α-terthienyl and 2-iodo-3-methoxythiophene give 3,4′,3‴,3⁗-tetramethoxy-α-quinquethienyl [28].

3,3⁗-Dimethoxy-2,2′:5′,2″:5″,2‴:5‴,2⁗-quinquethienyl [28]

An oven-dried tube is charged with 5,5″-bis(tributylstannyl)-2,2′:5′,2″-terthienyl (2.3 g, 2.8 mmol), 2-iodo-3-methoxythiophene (1.36 g, 5.7 mmol), bis(triphenylphosphine)palladium(II) chloride and toluene (10 ml).The reaction mixture is first bubbled with argon for 20 min and then the tube is sealed. The reaction mixture is stirred at 100–110 °C for 48 h before being cooled and poured into saturated ammonium chloride solution. The product is extracted with anhydrous ethyl acetate and the combined organic phases are washed with sodium chloride solution, dried over sodium sulfate and evaporated. The residue is purified by flash chromatography using ethyl acetate/hexane (2:8) as eluent giving 608 mg (46%) of the title compound as red crystals.

8.7.3.4 From two equivalent of 5-halo-2,2′-bithienyls and 2,5-trialkylstannyl-thiophenes

Palladium-catalyzed coupling of two equivalents of 5-bromo-5′-formyl-2,2′-bithienyl with 2,5-bis(tributylstannyl)thiophene gives diformylquinquethienyl shown below.

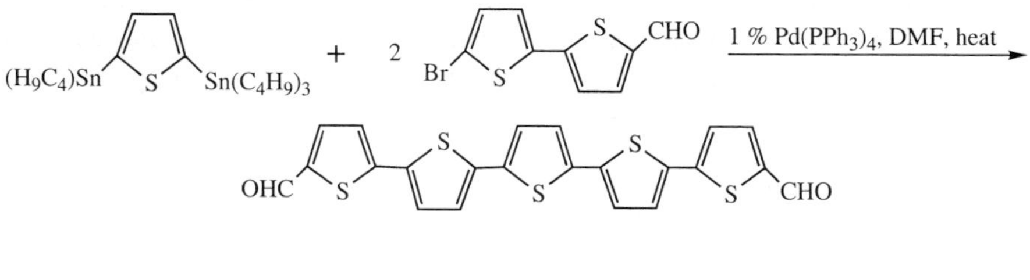

8.7.3.5 From 5-halo-α-terthienyls and trialkylstannyl-2,2′-bithienyl

The reaction of 5-bromo-4-octyl-5″-carbobenzyloxy-α-terthienyl with 5-trimethylstannyl-2,2′-bithienyl gives 5-carbobenzyloxy-4″ octyl-α-quinquethienyl in 90% yield [126,127].

8.7.3.6 From haloquaterthienyls and trialkystannylthiophenes

Reaction of polymer-bound 5-bromo-3′,4-dioctyl-α-quaterthienyl-5‴-carboxylate with 2-(trimethylstannyl)-4-octylthiophene is used for the preparation of the polymer-bound 4″,4‴,4⁗-trioctyl-α-quinquethienyl-5-carboxylate, which was transformed to the methyl ester [129].

8.7.4 Transition metal-catalyzed couplings according to Suzuki

8.7.4.1 Coupling between α-terthienyl-5-boronic acid derivatives and 5-bromo-2,2′-bithienyl derivatives

A 3,3″-dipentyl-α-terthienylboronic acid derivative was used for the preparation of the quinquethienyl shown below [84].

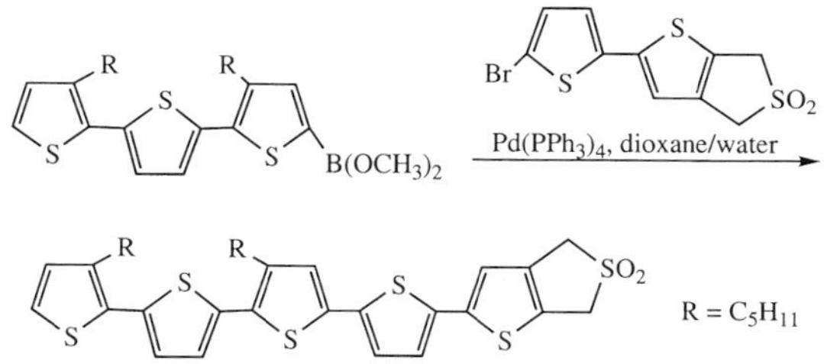

8.7.4.2 From thiopheneboronic acid derivatives and 5-halo-α-quaterthienyl

The following Suzuki coupling was used for the preparation of the quinquethienyl below [84].

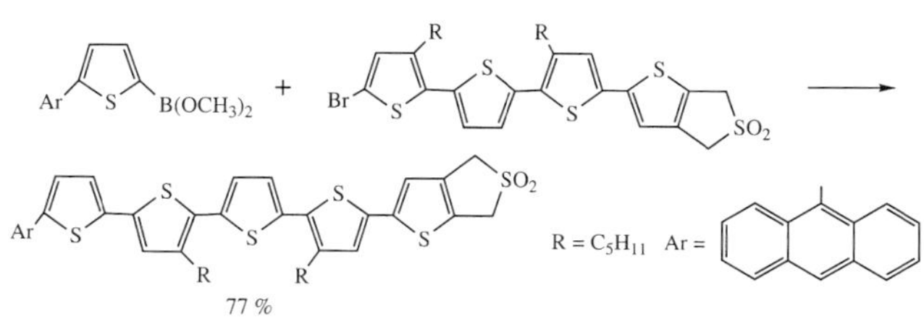

2-[5⁗-(9-Anthryl)-4′,3‴-dipentyl-5′,2″:5″,2‴:5‴,2⁗-quaterthienyl-2′-yl]-4,6-dihydrothieno[3,4b]thiophene-5,5-dioxide [84]

From butyllithium in hexane (0.25 ml, 0.4 mmol) and 2-anthryl-5-bromothiophene (125 mg, 0.38 mmol) in anhydrous tetrahydrofuran (5 ml), reaction time: 1 h; trimethyl borate (65 µl, 0.58 mmol) in anhydrous tetrahydrofuran (5 ml), reaction time 1–2 h; the bromo derivative (158 mg, 0.25 mmol) and tetrakis(triphenylphosphine)palladium(0) (33 mg) in dioxane/water (20:2 ml), reaction time 1 h; yield 156 mg (77%); orange solid mp 110–113 °C (dec) after recrystallization from ethyl acetate/hexane (1:3).

3,4′,4″,4‴,4⁗-Pentaoctyl-5⁗-chloro-α-quinquethienyl is prepared from 5-chloro-5‴-iodo-α-quaterthienyl and [1′,3′-(2′,2′-dimethylpropylene)]-3-octyl-2-thienylboronate [85,264].

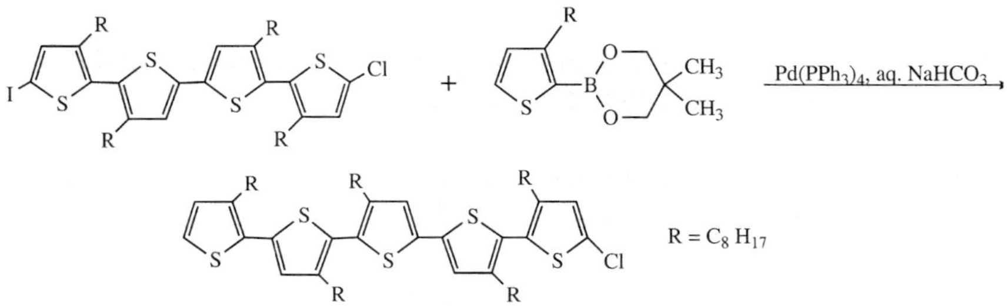

3,4′,4″,4‴,4⁗-Pentaoctyl-5⁗-chloro-2,2′;5′,2″:5″,2‴:5‴,2⁗-quinquethienyl [85]

A solution of 5-chloro-5‴-iodo-2,2′:5′,2″:5″,2‴-quatertienyl (1.42 g, 0.46 mmol) in 1,2-dimethoxyethane (3 ml) is carefully degassed and tetrakis(triphenylphosphine)palladium(II) (0.07 ml) is added. After stirring the mixture for 10 min at room temperature [1′,3′-(2′,2′-dimethylpropylene)]-3-octyl-2-thienylboronate (0.17 g, 0.55 mmol) and 1 *M* aqueous sodium bicarbonate (3 equiv.) are successively added. The reaction mixture is then refluxed under vigorous stirring for 4 h before being poured into water. The phases are separated and the aqueous phase extracted with diethyl ether. The combined organic phases are washed with water and sodium chloride solution, dried over sodium sulfate and evaporated. The residue is heated at 100 °C under vacuum overnight and the crude product purified by chromatography on silica gel using hexane as eluent or by semipreparative HPLC on reverse phase using chloroform/acetonitrile mixtures as eluents, giving 1.22 g (80%) of the title compound as an orange wax-like solid.

8.7.5 Preparation of quinquethienyls through ring-closure reactions

8.7.5.1 From 1,4-diketones

Cyclization of 1,4-diones prepared by Michael addition of thiophenealdehydes to Mannich bases by Lawesson's reagent can also be used for the preparation of the parent α-quinquethienyl [206].

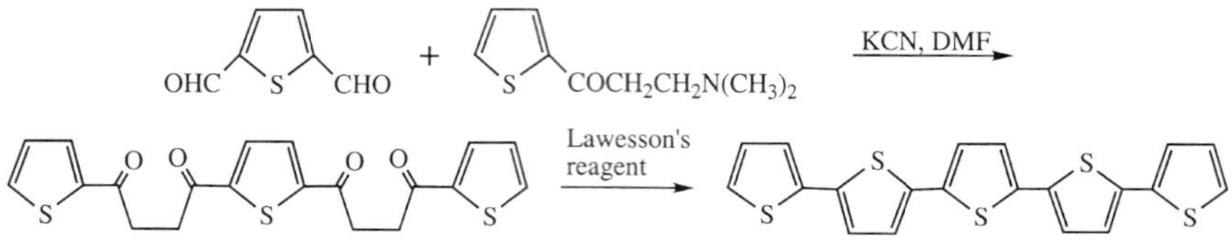

2,5-Bis[4-(2-thienyl)-1,4-butanedionyl]thiophene [206]

A solution of 2,5-diformylthiophene (200 mg, 1.43 mmol) in anhydrous *N,N*-dimethylformamide (6 ml) under nitrogen is added dropwise during 10 min to a suspension of potassium cyanide (91 mg, 1.5 mmol) in anhydrous *N,N*-dimethylformamide (6 ml). The mixture is stirred for 15 min and 3-dimethylamino-1-(2-thienyl)propanone (437 mg, 1.38 mmol) in anhydrous *N,N*-dimethylformamide (6 ml) is added during 30 min. The reaction mixture is stirred overnight, after which it is poured into water. The tan precipitate is filtered off, thoroughly washed with diethyl ether and recrystallized from dioxane giving 340 mg (68%) of the title compound as colorless crystals mp 186–187.5 °C.

In a similar way 3′,4′-dimethyl-α-quinquethienyl was prepared starting from 3′,4′-dimethyl-5-formyl-α-terthienyl and 3-dimethylamino-1-(2-thienyl)-propanone [206].

8.7.5.2 Various ring-closure reactions

α-Quinquethienyl is synthesized by the reaction of 5-chloroacetyl-2,2′-bithienyl with sodium sulfide in aqueous acetone to the diketone followed by reaction with Lawesson's reagent to the 1,4-dithiin, which upon reflux in *ortho*-dichlorobenzene gave the desired product in 56% yield together with small amounts of the isomeric quinquethienyl [315].

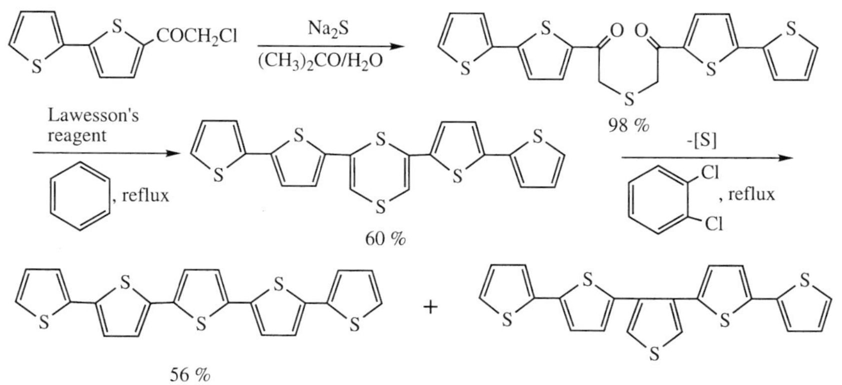

The isomer could be prepared in excellent yield by the reaction of the diketone shown below with a low valent titanium reagent in refluxing tetrahydrofuran followed by oxidation with 2,3-dichloro-5,6-dicyano-1,4-benzoquinone [315].

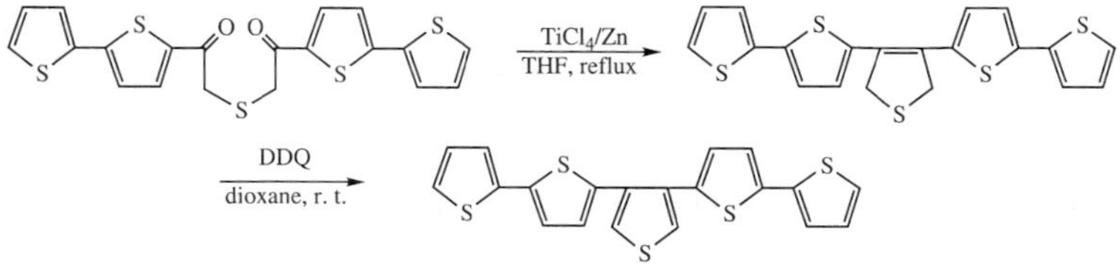

87 %

Heating of di(2-thienyl)acetylene in benzene in an autoclave gave 2,3,4,5-tetra(2-thienyl)thiophene [316,317].

2,3,4,5-Tetra(2-thienyl)thiophene [316]

A mixture of sulfur (48 mg, 1.5 mg-atoms) and di-(2-thienyl)acetylene (475 mg, 2.5 mmol) in benzene (100 ml) is heated at 205–210 °C for 14 h in an autoclave. After evaporation the residue is purified by chromatography on silica gel giving 24 mg (3%) of the starting material and 291 mg (57%) of the title compound as red crystals mp 189–190 °C after recrystallization from cyclohexane.

[2.2]Quinquethiophenophane is prepared according to the scheme below [70].

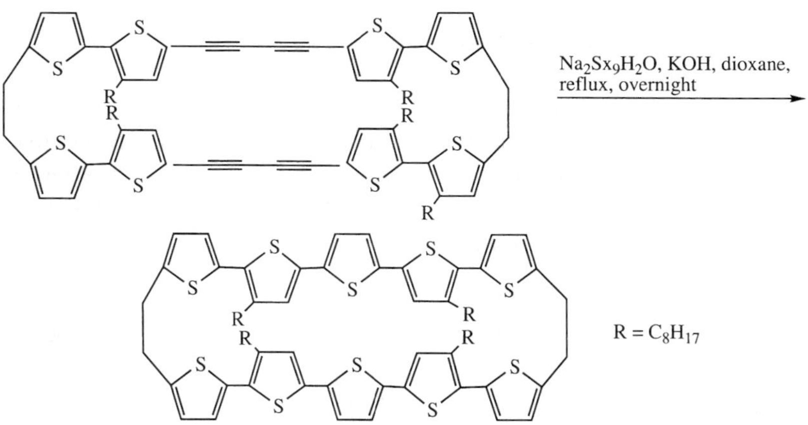

41 %

8.7.6 Substituted quinquethienyls *via* electrophilic substitutions

8.7.6.1 Formylation of α-quinquethienyls

Vilsmeyer formylation of 3,3''''-dipentyl- and 5-(9-anthryl)-3,3''''-dipentyl-α-quinquethienyl with phosphorus oxychloride and *N*,*N*-dimethylformamide in dichloromethane gave the 5-formyl derivatives in high yield [313,318].

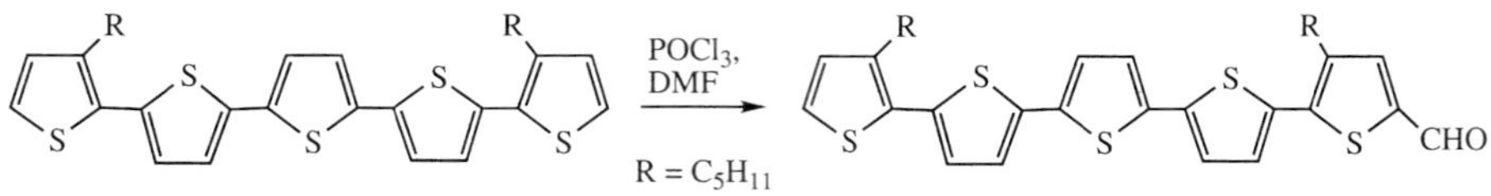

5-Formyl-3,3''''-dipentyl-2,2':5,2'':5'',2''':5''',2''''quinquethienyl [313]

In order to prepare the Vilsmeyer reagent a solution of phosphorus oxychloride (3.29 g, 21.46 mmol) and *N,N*-dimethylformamide (1.7 g, 23.26 mmol) in anhydrous dichloromethane (25 ml) is stirred for 2 h at room temperature. Of this solution 5 ml is added to a solution of 3,3''''-dipentyl-2,2',5',2'':5'',2''',5''',2''''-quinquethienyl (1.72 g, 3.11 mmol) in anhydrous dichloromethane (8 ml) and the reaction mixture is refluxed for 3 h. After hydrolyzing with 1 *M* sodium bicarbonate solution the phases are separated and the aqueous phase extracted with dichloromethane. The combined organic phases are washed with water, dried over sodium sulfate and evaporated. The residue is purified by chromatography on silica gel using dichloromethane as eluent giving 0.47 g (28%) of the starting material, 0.32 g (17%) of the diformyl derivative and 0.97 g (54%) of the title compound as an orange solid mp 79–80° C after recrystallization from methanol.

8.7.6.2 Bromination of α-quinquethienyls

Bromination of 5-benzyloxycarbonyl-4''-octylquinquethienyl with *N*-bromosuccinimide in *N,N*-dimethylformamide gives the 5''''-bromo derivative [126].

5-Benzyloxycarbonyl-4''-octyl-5''''-bromo-2,2':5',2''5'',2''':5''',2''''-quinquethienyl [126]

A mixture of 5-benzyloxycarbonyl-4''-octyl-2,2':5',2'':5'',2''':5''',2''''-quinquethienyl (1.266 g, 1.921 mmol) and *N,N*-dimethylformamide (80 ml) is warmed until the quinquethienyl is dissolved. In the absence of light *N*-bromosuccinimide (351 mg, 1.972 mmol) is added, after which stirring is continued for 2 h. During this period an orange precipitate is formed, which is filtered off and washed with cold *N,N*-dimethylformamide. After drying 1.374 g (97%) of the title compound is obtained as an orange solid.

8.7.6.3 Iodination of α-quinquethienyls

Iodination of 3'',4''-dihexyl-α-quinquethienyl-1'',1''-dioxide with *N*-iodosuccinimide in acetic acid/chloroform gives the 5,5''''-diiodo derivative [272].

2,5''''-Diiodo-3'',4''-dihexyl-2,2':5',2''5'',2''':5''',2''''-quinquethienyl-1'',1''-dioxide [272]

To a solution of 3'',4''-dihexyl-2,2':5',2'':5'',2''':5''',2''''-quinquethienyl-1'',1''-dioxide (0.340 g, 0.56 mmol) in acetic acid/chloroform (1:1) (50 ml) *N*-iodo-succinimide (0.252 g, 1.12 mmol) is added in portions at−20 °C. The reaction mixture is stirred overnight and worked up. The crude product is purified by chromatography on silica gel using hexane/ethyl acetate/dichloromethane (90:5:5) as eluent, giving 470 mg (98%) of the title compound as a red powder mp 148° C.

Iodination with iodine and mercuric oxide is used in the following preparation [248].

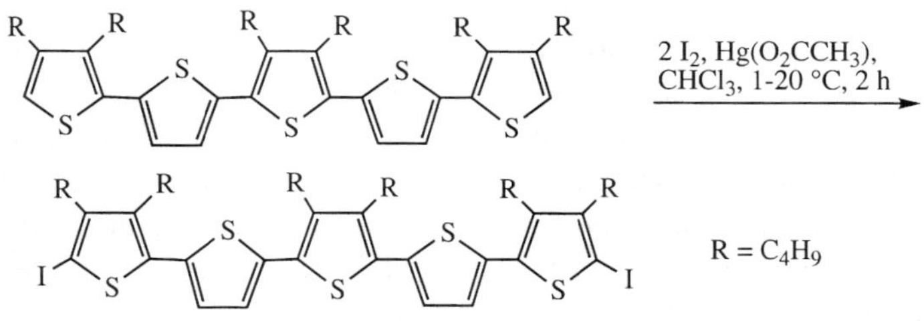

8.7.7 Substituted quinquethienyls *via* metalation reactions followed by electrophiles

Both 3,3''''- and 3,4',3''',3''''-tetramethoxy-α-quinquethienyl are dimetalated by butyllithium in the 5,5''''-positions and give upon reaction with dimethyl sulphate the 5,5''''-dimethyl derivatives in high yields [28].

3,3''''-Dimethoxy-5,5''''-dimethyl-2,2':5',2'':5'',2''':5''',2''''-quinquethienyl [28]

To a solution of 3,3''''-dimethoxy-2,2':5',2''5'',2''':5''',2''''-quinquethienyl (201 mg, 0.44 mmol) in anhydrous tetrahydrofuran (15 ml) at −78 °C butyl-lithium (0.36 ml, 0.90 mmol) is added dropwise. The mixture is, with stirring, warmed to 0 °C for 30 min and the recooled to −78 °C followed by addition of dimethyl sulfate (113 mg, 0.90 mmol). The reaction mixture is stirred at−78 °C for 3 h, warmed to room temperature for 1 h and poured into water. The product is extracted with diethyl ether, the combined organic phases are washed with sodium chloride solution, dried over sodium sulfate and evaporated. The residue is purified by flash chromatography using ethyl acetate/hexane (2:8) giving 190 mg (87%) of the title compound as dark-red crystals.

3,3''''-Dipentyl-α-quinquethienyl can be mainly monolithiated in the 5-position together with minor amounts of dilithiation with butyllithium and upon reaction with anthrone followed by treatment with hydrochloric acid in methanol/toluene 5-(9-anthryl)-3,3''''-dipentyl-α-quinquethienyl is obtained [318].

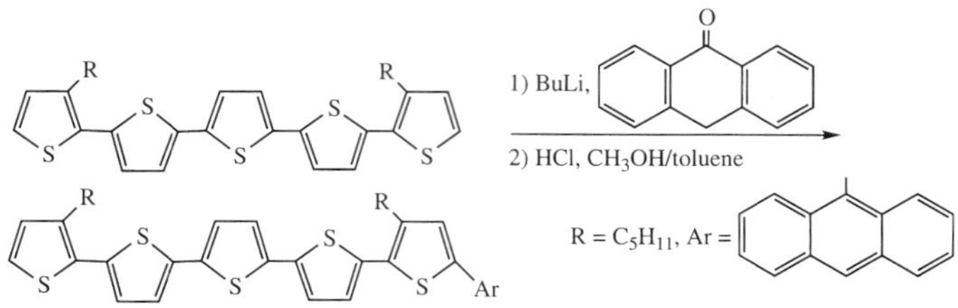

5-(9-Anthryl)-3,3''''-dipentyl-2,2':5',2''5'',2''':5''',2''''-quinquethienyl [318]

To a solution of 3,3''''-dipentyl-2,2':5',2'':5'',2''':5''',2''''-quinquethienyl (300 mg, 0.54 mmol) in anhydrous diethyl ether (10 ml) under argon, 1.6 *M* butyllithium in hexane (0.41 mmol) is added dropwise through a syringe over 10 min. The suspension is cooled to −78 °C and a solution of anthrone (84.84 mg, 0.44 mmol) in anhydrous diethyl ether (20 ml) cooled to −78 °C is added dropwise through a syringe over 20 min, after which anhydrous diethyl ether is added. The reaction mixture is stirred at −78 °C for 2.5 h, allowed to warm to room temperature for 16 h, and the poured into ice-cold water/hydrochloric acid (5:1, 50 ml). After stirring for 1 h the phases are separated and the aqueous phase extracted which dichloromethane (3 × 50 ml). The combined organic phases are concentrated and taken up in methanol/toluene (1:1, 40 ml). Concentrated hydrochloric acid (5 ml) is added and the reaction mixture is heated to 85° C for 30 min, dichloromethane (150 ml) is added followed by ice (150 g). The phases are separated and the aqueous phase extracted with dichloromethane (50 ml). The combined organic phases are washed with sodium bicarbonate solution and water, dried over sodium sulfate and evaporated. In order to remove unreacted anthrone the residue is chromatographed on silica gel using dichloromethane followed by hexane/dichloromethane (5:1) as eluents. The starting material, the title compound and the disubstituted derivative are separated by MPLC on silica gel using hexane/dichloromethane (15:1) as eluent (flow 30 ml/min, detection wavelength 260 and 430 nm) giving 95 mg (24%) of the title compound as an orange solid mp 110–112° C.

8.7.8 Modifications of side chains in quinquethienyls

8.7.8.1 From halo-α-quinquethienyls

Treatment of 3,4',4'',4''',4''''-pentaoctyl-5'''-chloro-α-quinquethienyl with tributyltin hydride and azo-bisisobutyronitrile gives 3,4',4'',4''',4''''-pentaoctyl-α-quinquethienyls in 70% yield [85].

8.7.8.2 Various side chain reactions

The reaction of 5-(9-anthryl)-5′′′′-formyl-3,3′′′′-dipentyl-α-quinquethienyl with pyrrole and boron trifluoride-etherate in diethyl ether/ethanol followed by oxidation with chloranil gave a light-harvesting system [318].

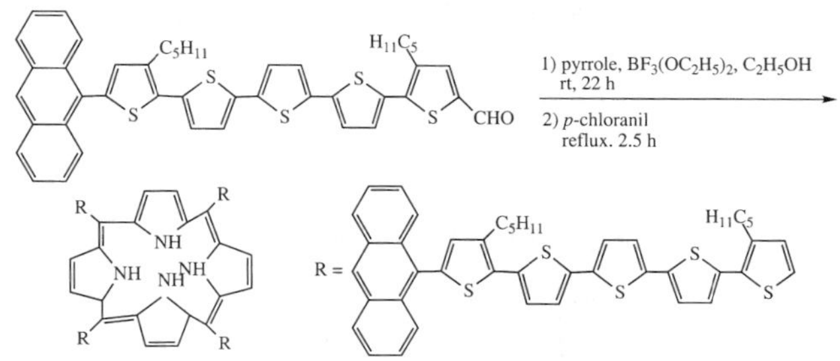

5,10,14,20-Tetrakis[5′′′′-(9-anthryl)-3,3′′′′-dipentyl-2,2′:5′,2′′:5′′,2′′′:5′′′,2′′′′-quinquethienyl-5yl]porphyrin [318]

A solution of 5′′′′-(9-anthryl)-5-formyl-3,3′′′′-dipentyl-2,2′:5′,2′′:5′′,2′′′:5′′′,2′′′′-quinquethienyl (205 mg, 0.27 mmol) in anhydrous dichloromethane (70 ml), pyrrole (18 mg, 0.27 mmol) and ethanol (0.5 ml) is stirred under argon at room temperature for 30 min in the absence of light. The condensation is initiated by addition of boron trifluoride etherate 2.5 *M* in dichloromethane (0.37 ml). The mixture is stirred for 8 h, after which boron trifluoride etherate (0.3 ml) and dichloromethane (200 ml) are added. The stirring is continued for another 14 h before *para*-chloranil (49 mg, 0.2 mmol) is added, the reaction mixture heated to reflux for 2.5 h, triethylamine is added and the stirring is continued for 45 min. After evaporation the residue is chromatographed on silica gel using dichloroethane as eluent. The porphyrin fractions purified by MPLC (nucleosil 1525 NO_2, flow 30 ml/min) with hexane/dichloromethane (66:34) as eluent. Crystallization is induced by addition of acetone to a saturated solution of the title compound in dichloromethane giving 7.6 mg (3.5%) of a blue-violet solid mp sintering > 89° C.

The diiodo derivative gave upon palladium-catalyzed coupling with trimethylsilylacetylene the trimethylsilyl-protected terminal acetylene, which was deprotected nearly quantatively under mild basic conditions to the quinquethienyldiyne [315].

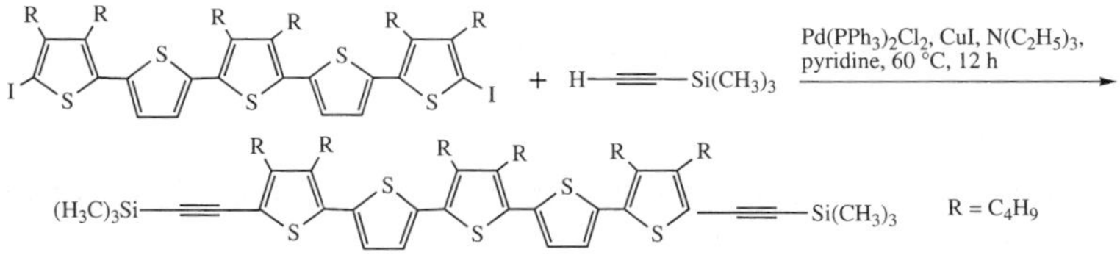

8.7.9 1,1-Dioxides of quinquethienyls

Defiened 1,1-dioxides are prepared by Stille coupling reactions. Thus reaction of 2,5-dibromothiophene-1,1-dioxide with two equivalents of 5-tributylstannyl-5′-butyldimethylsilyl-2,2′-bithienyl gives 5,5′′′′-bis(butyldimethylisilyl)-α-quinquethienyl-1′′,1′′-dioxide in 52% yield [227]. From two equivalents of 5-hexyl- and 5-butyldimethylsilyl-5′-tributylstannyl-2,2′-bithienyl-1,1-dioxide and 2,5-dibromothiophene-1,1-dioxide, 5,5′′′′-dihexyl- and 5-dibutyldimethylsilyl-α-quinquethienyl-1,3′′,3′′, 5′′′′,5′′′′-hexadioxide are obtained. Finally the coupling of 5-bromo-5′-butyldimethylsilyl-2,2-bithienyl with 5-butyldimethylsilyl-5′′-tributylstannyl-α-terthienyl is used for the preparation of 5,5′′′′-bis(butyldimethylsilyl)-α-quinquethienyl-1,1-dioxide [227].

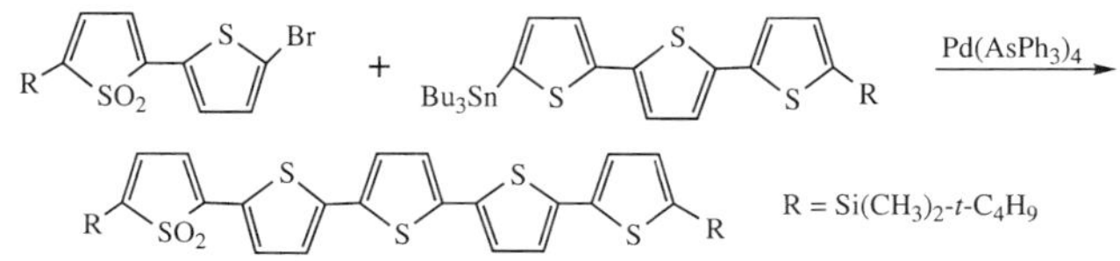

5,5′′′′-Bis(dimethyl-tert-butylsilyl)-2,2′:5′,2′′5′′,2′′′:5′′′,2′′′′-quinquethienyl-1,1-dioxide [227]

To a solution of tetrakis (triphenylarsine) palladium(0) (0.006 mmol) in toluene (10 ml) 5-(dimethyl-tert-butylsilyl)-5′′-(tributylstannyl)-2,2′:5′,2′′-terthienyl (0.138 g, 212 mmol) and 2-(dimethyl-tert-butylsilyl)-5′-bromo-2,2′-bithienyl (0.081, 0212 mmol) are added.

The reaction mixture is refluxed for 4 h, hydrolyzed with a saturated solution of ammonium chloride and worked up. The residue is washed with diethyl ether giving 0.069 g (47%) of the title compound as a red polycrystalline solid mp 216–217 °C.

3′′,4′′-Dihexyl-α-quinquethienyl-1′′,1′′-dioxide is also prepared by Stille coupling between 3,4-dihexyl-2,5-dibromothiophene-1,1-dioxide and 2-tributylstannyl-2,2-bithienyl in 76% yield [272].

8.8 HEXITHIENYLS

8.8.1 Symmetrical hexithienyls through the coupling of α-terthienyl derivatives with cupric chloride, ferric chloride or nickel complexes

8.8.1.1 Lithium derivatives with cupric chloride

Monolithiation of α-terthienyl with lithium diisopropylamide in tetrahydrofuran followed by cupric chloride gives α-hexithienyl in 73% yield [295].

3′,4′,3⁗,4⁗-Tetrabutyl-α-hexithienyl is obtained by treatment of 3′,4′-dibutyl-α-terthienyl with butyllithium followed by cupric chloride [314]. This method is preferred over ferric acetylacetonate for the dimerization of 5-hexyl-α-terthienyl [278]. Didodecyl-α-hexithienyls are also synthesized in this way [319].

8.8.1.2 Parent compounds or lithium derivatives with ferric chloride

α-Hexithienyl is prepared by coupling of 5-lithio-α-terthienyl with ferric acetylacetonate [275]. Similarly, 3,3″,4‴,3⁗-tetramethoxy-5,5′′′′′-dimethyl-α-hexithiophene is prepared by reaction of 3,3″-dimethoxy-5-methyl-α-terthienyl with butyllithium in tetrahydrofuran followed by ferric acetylacetonate [28].

3,3″,4‴3⁗Tetramethoxy-5,5′′′′′-dimethyl-2,2′:5′,2″:5″,2‴:5‴,2⁗:5⁗,2″-hexithienyl [28]

To a solution of 3,3″-dimethoxy-5-methyl-2,2′:5′,2″-terthienyl (110 mg, 0.34 mmol) in anhydrous tetrahydrofuran (10 ml), butyllithium (0.14 ml, 0.35 mmol) is added dropwise at 0° C. The mixture is stirred at 0° C for 1 h before it is transferred *via* cannula to a solution of ferric acetylacetonate (124 mg, 0.35 mmol) in anhydrous tetrahydrofuran (25 ml). The reaction mixture is refluxed for 24 h and then cooled to room temperature. The precipitate is filtered off and washed with dichloromethane. The combined organic phases are washed with saturated ammonium chloride solution, dried over sodium sulfate and evaporated. The residue is recrystallized from dichloromethane/methanol (1:9) giving 77 mg (70%) of the title compound as dark-red crystals.

The same methodology is also used to prepare 5,5′′′′′-bis (dimethyl-tert-butylsilyl)-α-hexithienyl from 5-(dimethyl-tert-butylsilyl)-α-terthienyl [277].

5,5′′′′′-Bis(dimethyl-tert-butylsilyl)-2,2′:5,2″:5″,2‴:5‴,2⁗:5⁗,2′′′′′-hexithienyl [277]

To a solution of 5-(dimethyl-*tert*-butylsilyl)-2,2′:5′,2″-terthienyl (0.40 g, 1.1 mmol), 2.5 M-butyllithium in hexane (0.44 ml) is added dropwise. After 5 min ferric acetylacetonate (0.39 g, 1.1 mmol) is added and the reaction mixture is stirred at room temperature for 3 h and then 3 *M* hydrochloric acid (30 ml) is added. The product is taken up in dichloromethane, the combined organic phases washed with sodium chloride solution, dried over magnesium sulfate and evaporated. The residue is washed with diethyl ether giving 0.23 g (58%) of the title compound as a dark-red microcrystalline compound mp 244–245 °C.

Oxidative coupling of heptadecyl-α-terthienyl with anhydrous ferric chloride gave 5,5'''''-diheptadecyl-α-hexithiophene [274] and from 3,3''-hexyl-α-terthienyl, 3,3'',4''',3'''''-tetrahexyl-α-hexithienyl is obtained [274].

α,ω-Diheptadecyl-α-hexithienyl [274]

To a solution of 3,3''-heptadecyl-2,2':5',2''-terthienyl (0.4 g, 0.82 mmol) in anhydrous benzene (25 ml), anhydrous ferric chloride (0.85 g, 5.2 mmol) is added. An immediate color change from pale-yellow to dark-blue is observed. After 10 min the benzene is evaporated, the mixture diluted with water/methanol (1:1) and the oxidized product is reduced with 40% aqueous hydrazine. After stirring in a sonicator for 1 h the product is filtered off, washed with water, methanol, dichloromethane, diethyl ether and dried. Purification is achieved by dissolving the red solid in a large amount of boiling benzene followed by hot filtration. At room temperature the title compound precipitates and after filtration, washing with benzene and drying a yield of 0.22 g (55%) of the compound is obtained mp 265 °C.

However, this method could not be used for the dimerization of the 5-heptadecyloxymethyl- and the 5-(heptadec-1-enyl) derivative. Anhydrous ferric chloride oxidation of 1,3-di-(2-thienyl)benzo[*c*] thiophene and of its 3-alkyl derivatives is a useful method for the preparation of the corresponding hexithiophenes [241].

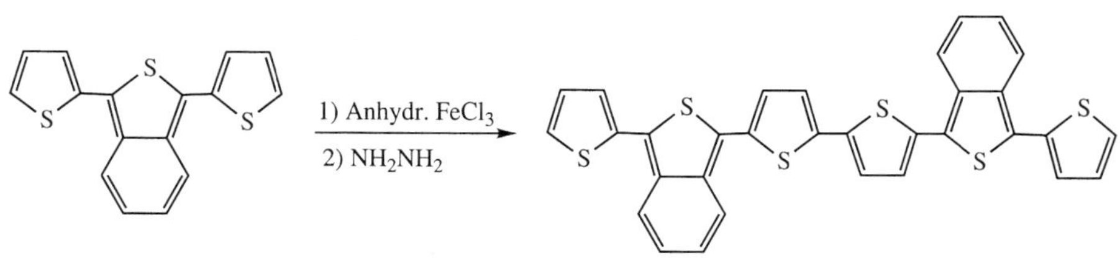

Dimer of 1,3-dithienylbenzo[c]thiophene [241]

A solution of 1,3-dithienylbenzo[*c*]thiophene (0.5 g, 1.68 mmol) in dichloromethane (40 ml) is treated with anhydrous ferric chloride (0.31 g, 1.8 mmol) under nitrogen. The reaction mixture is stirred at room temperature for 6 h, diluted with more dichloromethane (50 ml) and treated with a dilute solution of hydrazine hydrate. Evaporation affords a dark powder containing the title compound.

8.8.1.3 *Via* reductive coupling of bromo derivatives

Ullman coupling of 5-bromo-perfluoro-α-terthienyl with copper in *N,N*-dimethylformamide at 120 °C yields perfluoro-α-hexithienyl [91].

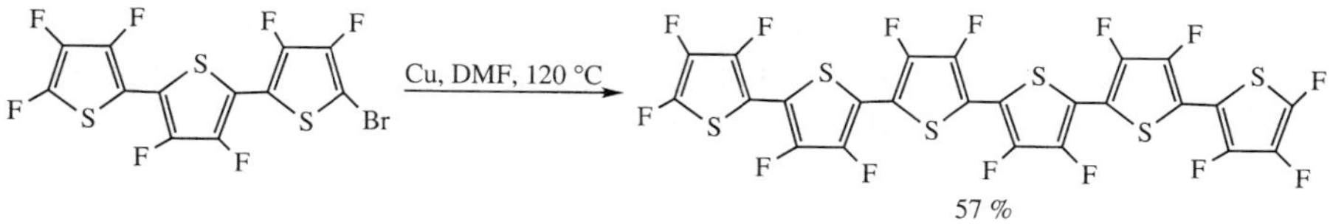

In some cases when the oxidative dimerization with ferric chloride failed, successful results were obtained by reductive dimerization of 5-bromo-α-terthienyl with zinc, triphenylphosphine, potassium iodide and dichlorobis (triphenylphosphine) nickel(II). Thus 5,5′′′′′-dihexadecyloxymethyl- and 5,5′′′′′-diheptadec-1-enyl-α-hexithineyls were obtained in over 80% yield from the 5-bromo-5″-hexadecyloxymethyl- and heptadec-1-enyl-α-terthienyl, respectively [274].

5,5′′′′′-Dihexadecyloxymethyl-2,2′:5′,2″:5″,2‴:5‴,2′′′′:5′′′′,2′′′′′-hexithienyl [274]

A flame-dried flask equipped with reflux condenser and flushed with nitrogen is charged with zinc (56.3 mg, 0.86 mmol), triphenylphosphine (451 mg, 1.7 mmol), potassium iodide (14 mg, 0.09 mmol), dichlorobis(triphenylphosphine) nickel(II) 563 mg, 0.86 mmol) and anhydrous oxygen-free *N*,*N*-dimethylformamide (6 ml). The mixture is heated at 60° C for 1 h with stirring under nitrogen, after which a solution of 5-bromo-5″-hexadecyloxymethyl-2,2′:5′,2″-terthienyl (0.5 g, 0.86 mmol) in anhydrous oxygen-free tetrahydrofuran is added *via* a syringe. The reaction mixture is stirred under nitrogen at 60° C for 24 h and after cooling poured into water. The precipitate is filtered off, washed with methanol and acetone and again with methanol and dried. This solid is extracted with carbon tetrachloride using a Soxhlet apparatus. The extract is cooled to room temperature, the precipitate is filtered off, washed with carbon tetrachloride and dried giving 0.35 g (81%) of the title compound as orange crystals mp 236° C.

This method was first used for the preparation of the parent α-hexithienyl in 58% yield by Nakayama and coworkers [244]. Another example of this useful approach is the preparation of 4′,3′′′′′,4‴3′′′′′-tetrahexyl-α-hexithienyl from 5-bromo-3′-dodecyl-α-terthienyl [62] and 5-bromo-3,3″-dihexyl-α-terthienyl, respectively [320].

4′,3′′′′-Didodecyl-2,2′:5′,2″:5″,2‴:5‴,2′′′′:5′′′′,2′′′′′-hexithienyl [62]

A mixture of anhydrous nickel dichloride (25.3 mg, 0.2 mmol), zinc (0.33 g, 5 mmol) and triphenylphosphine (0.33 g, 1.26 mmol) in *N*,*N*-dimethylformamide (5 ml) is heated at 70 °C for 1 h. The reddish-brown solution turns green upon addition of 5-bromo-3′-dodecyl-2,2′:5′,2″-terthienyl (0.8 g, 1.6 mmol).

The reaction mixture is heated at 70° C for 7 h and filtered while hot. Upon cooling the product precipitates, and the precipitate is filtered off and purified by chromatography on silica gel using hexane/dichloromethane (3:1) as eluent giving 0.23 g (33%) of the title compound as orange microcrystals mp 110–111° C.

Functionalized α-terthienyls such as 5-bromo-5-formyl-α-terthienyl are dimerized by this reductive coupling, giving 5,5′′′′′-diformyl-α-hexithienyl [102].

8.8.1.4 α-Hexithienyls *via* trimerization reactions

The reaction of 3-butylthio-2,2′-bithienyl with ferric chloride in chloroform/nitromethane (1:1) at room temperature gave 3,3′′′,3′′′′′-tris(butylthio)-α-hexithienyl [64,321].

3,3′′′,3′′′′′-Tris(butylthio)-2,2′:5′,2′′:5′′,2′′′:5′′′,2′′′′:5′′′′,2′′′′′-hexithienyl [64]

To a stirred solution of 3-butylthio-2,2′-bithienyl (4.0 g, 25.7 mmol) in distilled chloroform (240 ml) under nitrogen a solution of anhydrous ferric chloride (10.4 g, 64 mmol) in anhydrous nitromethane (240 ml) is added dropwise at room temperature during 2 h. The greenish-blue mixture is stirred at room temperature for 30 h and evaporated. The residue is stirred with a solution of hydrochloric acid/methanol (50 ml). The dark product is allowed to settle and the procedure repeated twice. Upon addition of methanol a brown solid is formed, which is filtered off, washed with methanol and dissolved in chloroform (200 ml). This chloroform solution is filtered to remove insoluble material, washed with 3% hydrazine solution (100 ml) and water and evaporated. The crude oil is treated with pentane (400 ml) and the pentane solution is evaporated. The residue is purified by chromatography on silica gel, neutralized with 2% triethylamine solution, using light petroleum/diethyl ether (9:1) as eluent giving 2.8 g (70%) of the title compound as a red oil.

8.8.2 Transition metal-catalyzed couplings according to Kumada

3,4′,3′′′′-3′′′′′-Tetraoctyl-α-hexithienyl is prepared by the [1,3-bis(diphenylphosphine)propane(II) chloride-catalyzed coupling of 5,5′′′-dibromo-3,3′′′-dioctyl-α-quaterthienyl with 3-octyl-2-thiophenemagnesium bromide [303].

The branched hexithienyl, 3,3′-bis(2,2′:5′,2′′-terthienyl), is prepared by [1,3-bis(diphenylphosphine)propane]nickel(II) chloride-catalyzed coupling of 3′-α-terthienylmagnesium bromide, obtained by halogen lithium exchange of 3′-bromo-α-terthienyl with butyllithium followed by magnesium bromide with 3-bromo-α-terthienyl [322].

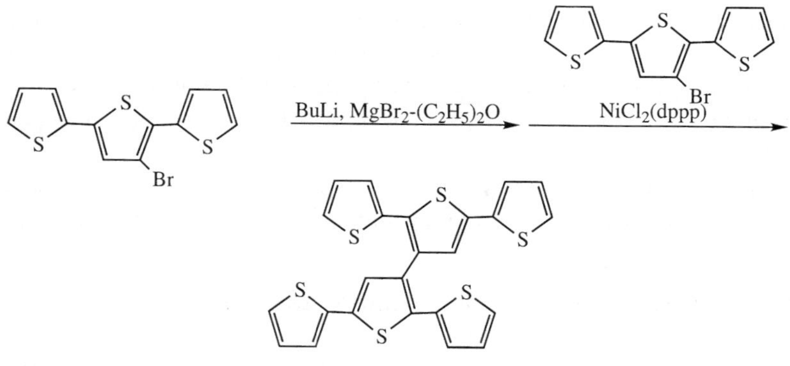

8.8.3 Transition metal-catalyzed couplings according to Stille

8.8.3.1 From 2-trialkylstannylthiophenes and 5-halo-α-quinquethienyls

The reaction of 5-bromo-4″-octyl-5⁗-carbobenzyloxy-α-quinquethienyl with 2-trimethyl-stannyl-4-octylthiophene gives 5-carbobenzyloxy-4″,4⁗-dioctyl-α-hexithienyl in 91% yield [126,127].

8.8.3.2 From two equivalents of 2-trialkylstannylthiophenes and 5,5‴-dibromo-α-quaterthienyls

A detailed investigation of the Stille coupling of 5,5‴-dibromo-3,4′,3″3‴-tetra(methylthio)-α-quaterthienyl with two equivalents of 2-tributylstannyl-thiophene and with 3-methylthio-2-trimethylstannylthiophene, using various palladium(0) catalysts and leading to 4,4″,3‴,3⁗-tetra(methylthio)- and 3,3′,4″,3‴,4⁗,3′⁗-hexa(methylthio)-α-hexithienyl in good yields is described [169,302].

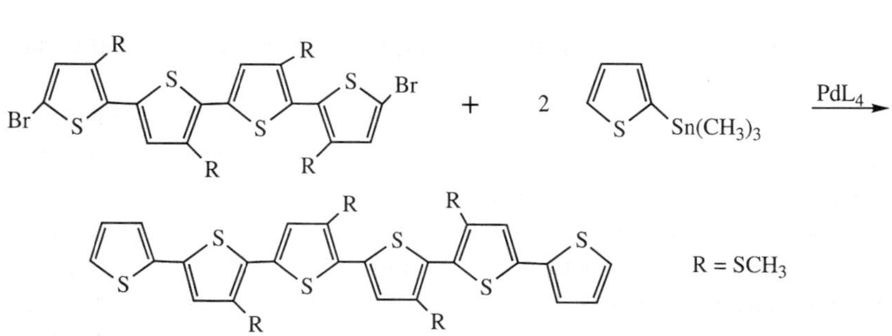

4,4″,3‴,3⁗-Tetra(methylthio)-2,2′:5′,2″:5″,2‴:5‴,2⁗:5⁗,2′⁗-hexithienyl [169]

To a solution of tris(benzylideneacetone)palladium(0) (15.5 mg, 0.05 mmol) in freshly distilled tetrahydrofuran (20 ml), triphenylarsine (18.4 mg, 0.20 mmol) is added. The mixture is stirred for 30 min, after which 5,5‴-dibromo-3,4′,3″2,3‴-tetra(methylthio)-2,2′,2″,2‴-quaterthienyl (400 mg, 0.60 mmol) and 2-tributylstannylthiophene (0.56 g, 1.50 mmol) are added. The reaction mixture is refluxed for 24 h and then additional tetrahydrofuran (20 ml) containing tris(benzylideneacetone)palladium(0) (155.5 mg) and triphenylarsine (18.4 mg) are added. The refluxing is continued for another 24 h, after which the reaction mixture is allowed to decrease to ambient temperature, treated with *M* hydrochloric acid (30 ml), neutralized with saturated sodium bicarbonate solution, washed twice with sodium chloride solution, dried over magnesium sulfate and evaporated. The residue is treated three times with methanol and three times with dichloromethane and then sublimed under high vacuum giving the title compound as a red-violet solid mp 163 °C.

8.8.3.3 From two equivalents of bromo-2.2′-bithienyls and 5,5′-distannyl-2,2′-bithienyls

3,3′,4″,3‴,4⁗,3′⁗-Hexakis[2-(tetrahydropyranyloxy)ethyl]-α-hexithienyl is prepared from 2,5-bis(tributylstannyl)-3,3′-[2-(tetrahydropyranyloxy)ethyl]-2,2′-bithienyl with two equivalents of 2-bromo-3,3′-bis[2-tetrahydropyranyloxy)ethyl]-2,2′-bithienyl [130].

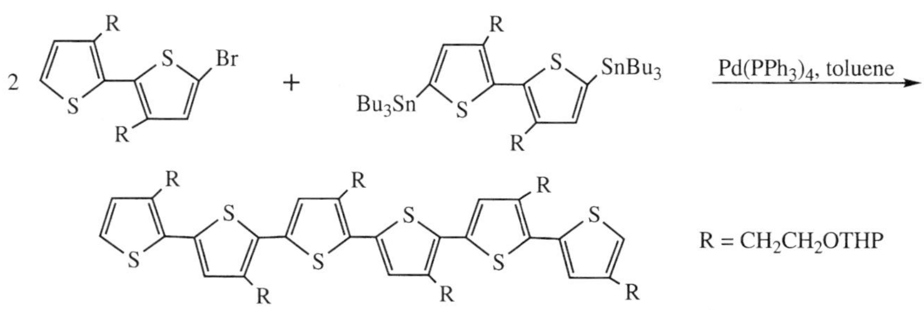

Similarly 5,5′⁗-dioctadecyl-α-hexithienyl is prepared from two equivalents of 5-octadecyl-5′-tributylstannyl-2,2′-bithienyl and 5,5′-dibromo-2,2′-bithienyl [106]. Also 5,5′⁗-bis(3-butoxypropyl)-α-hexithienyl and 5,5′⁗-bis(3-octyloxypropyl)-α-hexithienyl were prepared in similar ways [106].

The palladium(0)-catalyzed Stille coupling of 2,2′-dibromo-5,5′-dichloro-3,3′-bithienyl with two equivalents of 5,5′-dichloro-2-tributylstannyl-3,3′-bithienyl gives the hexachlorohexithiophene, which upon bromination with

N-bromosuccinimide followed by halogen–metal exchange and reaction with curpric chloride results in a rigid tubular hexithiophene [137].

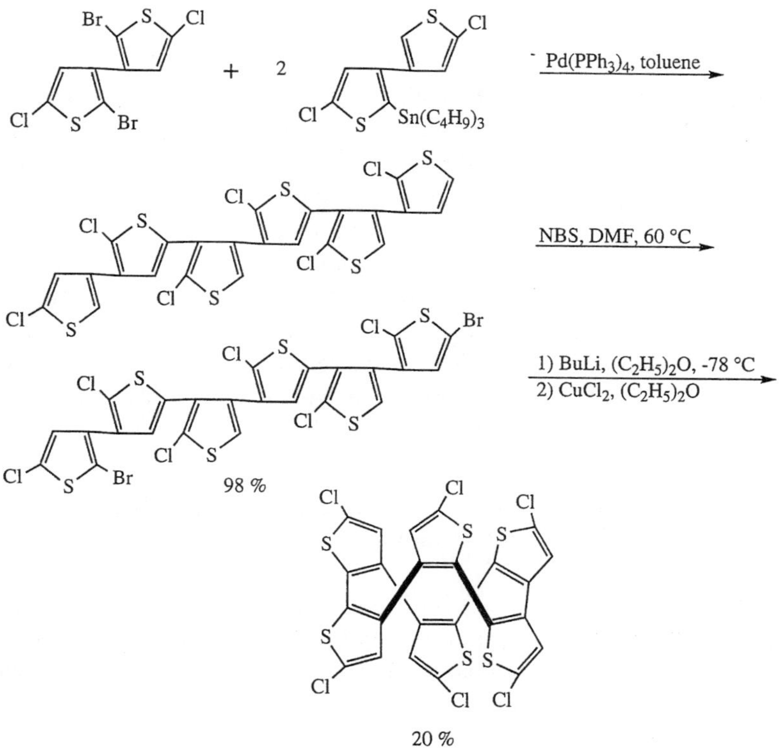

The following coupling using 5,5′-bis(trimethylstannyl)thiophene has been performed giving a chiral hexithienyl derivative [124].

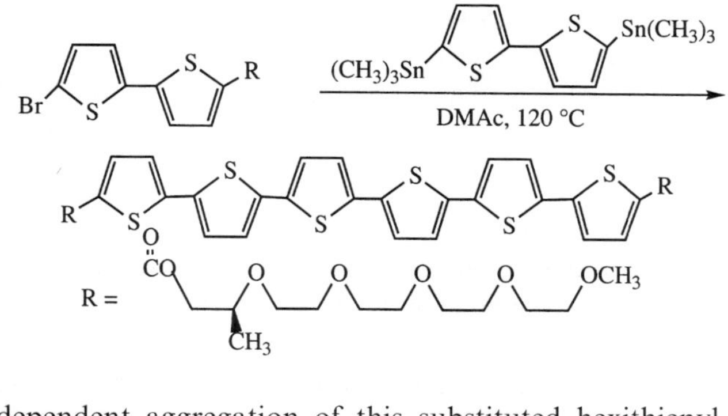

Solvent-dependent aggregation of this substituted hexithienyl was found, using nanospray FT-ICR mass spectroscopy [323].

8.8.3.4 From two equivalents of 2-trialkylstannyl-2,2′-bithienyl and 5,5′-dibromobithienyl

In connection with work on molecularly doped organic light-emitting devices the following Stille coupling has been performed [252].

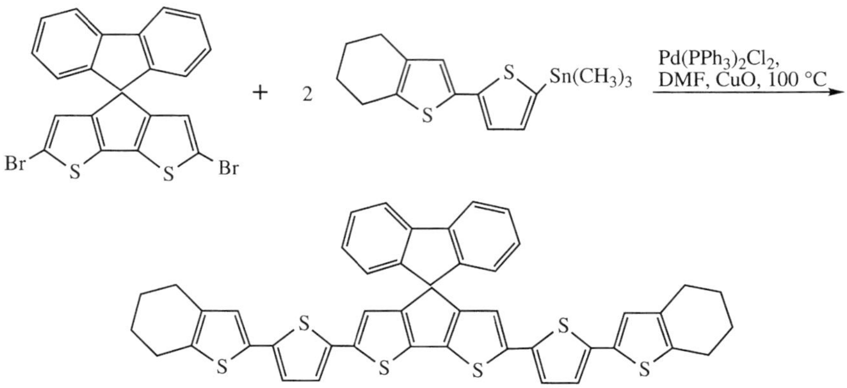

2,6-Bis[5-(4,5,6,7-tetrahydrobenzo[b]thiophen-2-yl]spiro[cyclopenta-[2,1-b:3,4-b′]dithiophene-4,9′-fluorene [252]

To a suspension of 2,6-dibromospiro[cyclopenta[2,1-*b*:3,4-*b*′]dithiophene-4,9′-fluorene] (500 mg, 1.03 mmol), bis(triphenylphosphino)palladium(II) dichloride (7.2 mg, 0.103 mmol) and copper(II) oxide (168 mg, 2.06 mmol) in anhydrous *N*,*N*-dimethylformamide (25 ml) at 100 °C, a solution of 5-(4,5,6,7-tetrahydro-benzo[*b*]thiophene-2-yl)-2-thienyl-2-trimethylstannylthiophene (854 mg, 2.26 mmol) in anhydrous *N*,*N*-dimethylformamide (10 ml) is added dropwise. The reaction mixture is refluxed for 20 h, allowed to cool to room temperature, hydrolyzed and extracted with dichloromethane. The combined organic phases are washed with saturated sodium bicarbonate solution, dried over magnesium sulfate and evaporated. The residue is purified by chromatography on silica gel using petroleum ether/dichloromethane (3:1) as eluent giving, after sublimation, 550 mg (70%) of the title compound as an amorphous bright-red solid mp 330–331 °C (decomp.)

8.8.3.5 From 5-trimethylstannyl-2,2′-bithienyl and 5-bromo-α-quaterthienyl

Reaction of the dendrimer-bound 5-bromo-5‴-α-quaterthienylcarboxylate with 5-trimethylstannyl-2,2′-bithienyl is used for the preparation of the following hexithienyl [211].

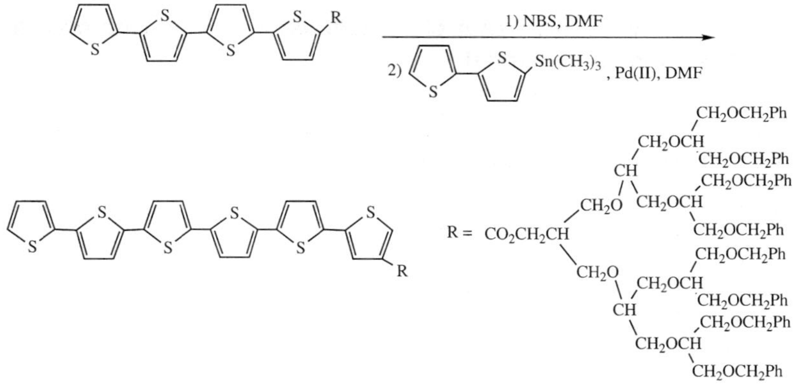

8.8.3.6 From 5-trialkylstannyl-α-terthienyls and 5-halo-α-terthienyls

5,5′′′′′-Bis(trimethylsilyl)-3′,3′′,4′′′,4′′′′-tetramethyl-α-hexithienyl is prepared by palladium(0)-catalyzed coupling of 3,4′dimethyl-5′′-(trimethylsilyl)-5-(trimethylstannyl)-α-terthienyl with 5-iodo-3,4′-dimethyl-5′′-(trimethylsilyl)-α-terthienyl [76].

5,5′′′′′-Bis(trimethylsilyl)-3′,3′′,4′′′,4′′′′-tetramethyl-2,2′:5′,2′′:5′′,2′′′:5′′′, 2′′′′:5′′′′,2′′′′′- hexithienyl [76]

A 25-ml round-bottomed flask is charged with 3,4′-dimethyl-5′′-(trimethylsilyl)-5-(trimethylstannyl)-α-terthienyl (0.213 g, 0.44 mmol), 5-iodo-3,4′-dimethyl-5′′-(trimethylsilyl)-α-terthienyl (0.103 g, 0.22 mmol), tetrakis(triphenylphosphine)-palladium(0) (0.005 g, 0.0044 mmol) and toluene (1.0 ml). The reaction mixture is heated at 100–105 °C overnight before being poured into water. The product is extracted with diethyl ether, the combined organic phases washed with sodium chloride solution, dried over magnesium sulfate and evaporated. The residue is purified by chromatography giving 0.079 g (52%) of the title compound as orange-red crystals.

Stille coupling between 5,5′-tributylstannyl-2,2′-bithienyl and two equivalents of 5-bromo-5′-perfluorohexyl-2,2′bithienyl gave in 95% yield the first n-type hexithienyl DFH-6T for thin film transistors [187].

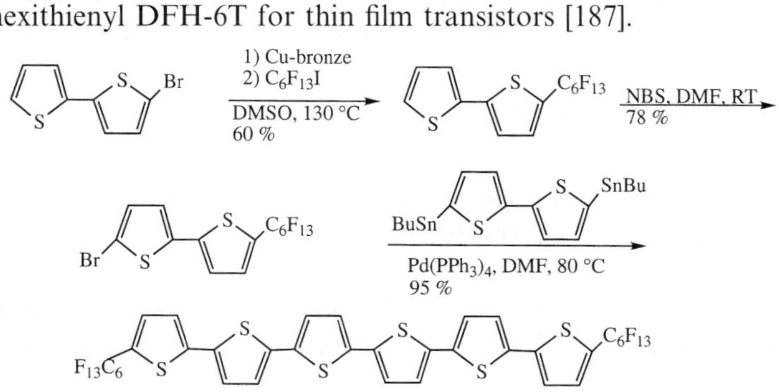

8.8.3.7 From 5,5″-trialkylstannyl-2,2′-bithienyl and two equivalents of 5-halo-2,2′-bithienyls

A good yield is obtained in the coupling of 5,5′-bis(tributylstannyl)-2,2′-bithienyl and two equivalents of 5-bromo-5′-formyl-2,2′-bithienyl to 5,5′′′′′-diformyl-α-hexithienyl [100].

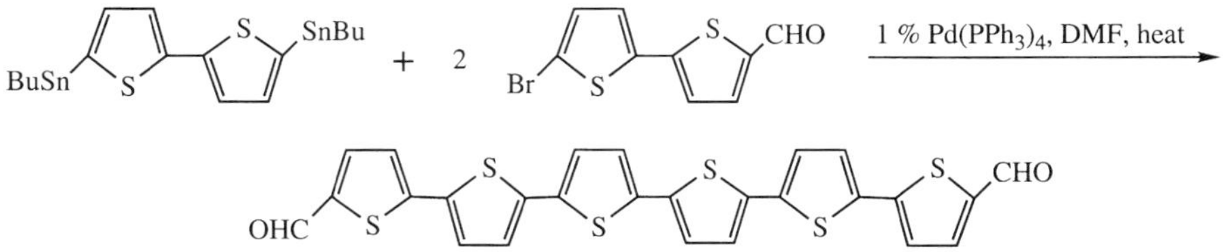

5,5′′′′′-Diformyl-2,2′:5′,2″:5″,2‴:5‴,2″″:5″″,2′′′′′-hexithienyl [100]

A solution of 5,5′-bis(tributylstannyl)-2,2′-bithienyl (730 mg, 0.98 mmol) and 5-bromo-5′-formyl-2,2′-bithienyl (535 mg, 1.96 mmol) in *N*,*N*-dimethylformamide (25 ml) is deaerated twice with argon. After addition of tetrakis(triphenylphosphine)palladium(0) (11 mg, 0.01 mmol) the reaction mixture is heated at 80 °C for 6 h under argon. The red suspension formed is cooled and filtered. The solid residue is washed with diethyl ether (200 ml), dichloromethane (200 ml) and tetrahydrofuran (200 ml) giving 430 mg (87%) of the title compound as a red solid.

8.8.3.8 From 5-trimethylstannyl-α-quaterthienyls and halo-2,2′-bithienyls

3′,4″,4″″-Tributyl-5-(trimethylsilyl)-α-hexithienyl is prepared by Stille coupling of 3′,3‴,4″″-tributyl-5-trimethylstannyl-5‴-trimethylsilyl-α-quaterthienyl with 3-butyl-5-iodo-2,2′-bithienyl in the presence of cuprous iodide [89].

3′,4″,4″″-Tributyl-5-(trimethylsilyl)- 2,2′:5′,2″:5″,2‴:5‴,2″″:5″″,2′′′′′-hexithienyl [89]

To a solution of diisopropylamine (0.410 g, 4.06 mmol) in anhydrous tetrahydrofuran (5.0 ml) 2.5 *M* butyllithium in hexane (1.35 ml, 3.38 mmol) is added dropwise at – 78 °C. The mixture is warmed to 0 °C for 5 min and recooled to – 78 °C. 3′,3‴,4″″-Tributyl-5-trimethylstannyl-5‴-trimethylsilyl-α-quaterthienyl (1.74 g, 3.38 mmol) is added dropwise *via* cannula. The mixture is warmed to 0 °C and recooled to – 78 °C, after which tributyltin chloride (1.10 g,

3.38 mmol) is added in one portion. The mixture is warmed to room temperature for 10 min and poured into water. The product is extracted with diethyl ether, the combined organic phases washed with sodium chloride solution, dried over magnesium sulfate and evaporated. The crude product is used directly in the next step. This crude product, assumed to be 3.38 mmol, 3-butyl-5-iodo-2,2′-bithienyl (1.30 g, 3.72 mmol), tertrakis(triphenylphosphine)-palladium(0) (0.059 g, 0.051 mmol), cuprous iodide (0.015 g, 0.077 mmol), is under nitrogen mixed in *N*,*N*-dimethylformamide (3.0 ml) sparged with nitrogen. The reaction mixture is stirred at room temperature and heated at 75–80 °C overnight, and poured into water. This solution is filtered through Celite, the phases are separated and the aqueous phase extracted with diethyl ether. The combined organic phases are washed with sodium chloride solution, dried over magnesium sulfate and evaporated. The residue is purified by chromatography on silica gel using hexane as eluent, giving 1.09 g (44%) of the title compound as a thick light-red liquid.

Very surprisingly the Stille coupling of 5,5′-dibromo-2,2′-bithienyl with 3-butylthio-2-trimethylstannylthiophene in the proportion of (1:1) in toluene using dichlorobis(triphenylphosphine)palladium(II) as catalyst gives a 45% yield of 3,3′′′′′-bis(butylthio)-α-hexithienyl [265].

3,3′′′′′-Bis(butylthio)-2,2′:5′,2′′:5′′,2′′′:5′′′,2′′′′:5′′′′,2′′′′′-hexithienyl [265]

A solution of 5,5′-dibromo-2,2′-bithienyl (0.24 g, 0.74 mmol), 3-butylthio-2-trimethylstannylthiophene (0.5 g, 1.5 mmol) dichlorobis(triphenylphosphine)-palladium(II) (0.052 g, 0.074 mmol) in anhydrous toluene (10 ml) is stirred and refluxed under a flow of anhydrous nitrogen for 20 h. The red reaction mixture is cooled, diluted with chloroform (40 ml), washed with water, dried over magnesium sulfate and evaporated. The residue is purified by chromatography on silica gel neutralized with 2% triamine solution using light petroleum/diethyl ether (2:1) as eluent. The last fraction, a brown solid, is stirred with pentane (20 ml), the precipitate is filtered off and recrystallized by reprecipitation from pentane into toluene giving 0.09 g (45%) of the title compound as red needles mp 112–114 °C.

8.8.4 Transition metal-catalyzed couplings according to Suzuki

8.8.4.1 From 5-halo-α-quinquethienyl and thiopheneboronic acid derivatives

3,4′,4′′,4′′′,4′′′′,4′′′′′-Hexaoctyl-5′′′′′-chloro-α-hexithienyl is prepared from 5-chloro-5′′′′-iodo-α-quinquethienyl and [1′,3′-(2′,2′-dimethylpropylene)]-3-octyl-2-thienylboronate [85,264].

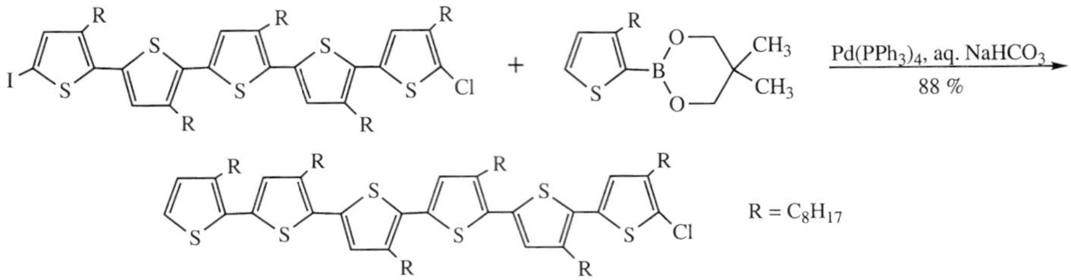

8.8.4.2 From 5-halo-quaterthienyls and 2,2′-bithienyls-5-boronic acid derivatives

5‴-iodo-4,3′,3″,3‴-tetradodecyl-α-quaterthienyl-5-carboxylic acid benzyl ester gives upon Suzuki coupling with 4,3′-didodecyl-2,2′-bithienyl-5-boronic acid propan-1,3-diol ester the 4,3′,3‴,3⁗,3′⁗-hexadodecyl-α-hexithienyl-5-carboxylix acid benzyl ester analogously to the quaterthienyl derivative [82].

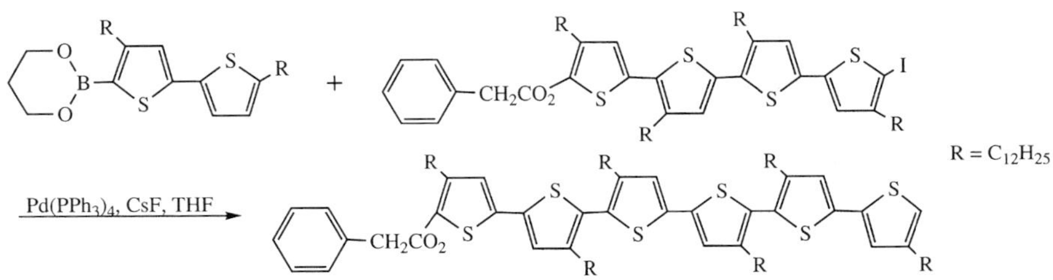

8.8.5 Preparation of hexithienyls *via* ring-closure reactions

8.8.5.1 From 1,4-diketones

α-Hexithienyl has been prepared in two ways either by the reaction of 1-(5-α-terthienyl)-4-(2,2′-bithienyl-5-yl)-1,4-butanedione or 5,5-bis[4-(2-thienyl)1,4-butanedionyl]-2,2′-bithienyl with Lawesson's reagent as shown for lower homologs [206].

2,2′:5′,2″:5″,2‴:5‴,2⁗:5⁗,2′⁗-hexithienyl [206]

Method A. A mixture of 1-(5-α-terthienyl)-4-(2,2′-bithienyl-5-yl)-1,4-butanedione (200 mg, 0.4 mmol) and Lawesson's reagent (97 mg, 0.24 mmol) in anhydrous toluene (200 ml) is refluxed under argon for 24 h. The red precipitate is filtered off and washed with ethanol giving 169 mg (86%) of the title compound mp 302–303 °C.

Method B. A mixture of 5,5-bis[4-(2-thienyl)1,4-butanedionyl]-2,2′-bithienyl (227 mg, 0.46 mmol) and Lawesson's reagent (221 mg, 0.56 mmol) in anhydrous toluene is refluxed under argon overnight. The precipitate is filtered off and thoroughly washed with ethanol giving 207 mg (91%) of the title compound mp 304–305 °C.

8.8.6 Substituted hexithienyls *via* electrophilic substitutions

8.8.6.1 Bromination of α-hexithienyls

Bromination of 5-benzycarbobenzoxy-4″,4″‴-dioctyl-α-hexithienyl with *N*-bromosuccinimide in *N*,*N*-dimethylformamide gives the 5⁗-bromo derivative in 98% yield [324].

8.8.7 Metalation of α-hexithienyls

Metalation of 5-chloro-3,4,4″,4‴,4⁗,4″‴-hexaoctyl-α-hexithienyl with lithium diisopropyl amide followed by reaction with trimethylstannyl chloride is used for the preparation of 5″‴-trimethylstannyl derivative [264].

8.8.8 Modifications of side chains in hexithienyls

8.8.8.1 From halo-α-hexithienyls

Treatment of 3,4′,4″,4‴,4⁗,4″‴-hexaoctyl-5-chloro-α-hexithienyl with tributyltin hydride and azo-bisisobutyronitrile gives 3,4,4″,4‴,4⁗,4″‴-hexaoctyl-α-hexithienyl in 70% yield [85]. The reaction of its 5-trimethylstannyl derivative with 1,3,5-tribromobenzene gives the (tristhienyl) substituted benzene derivative [264].

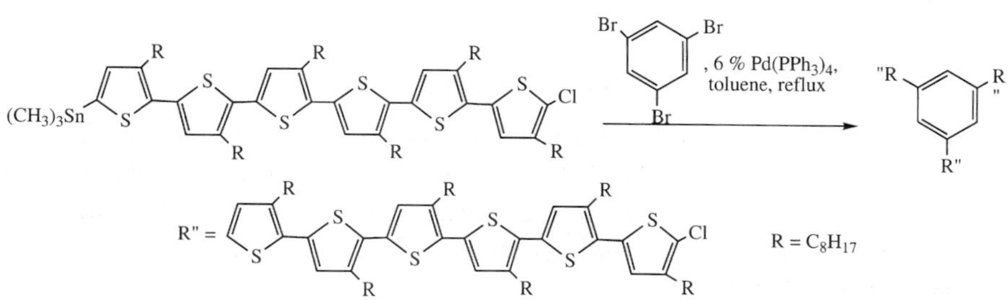

8.9 α-SEPTITHIENYLS AND LONGER DEFINED POLYTHIENYLS

8.9.1 α-Septithienyls

The parent α-septithienyls are prepared by reaction of α-terthienyl with chloroacetyl chloride under Friedel-Crafts condition giving 68% of 5-chloroacetyl-α-terthienyl, which upon reaction with sodium sulfide furnishes the diketosulfide. Refluxing with Lawesson's reagent in chorobenzene followed by heating gives the desired product [244].

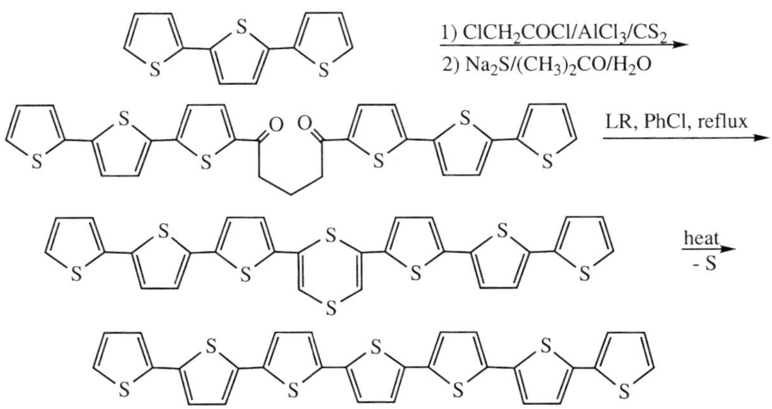

The isomeric septithiophene 3,4-bis-(α-terthienyl-5-yl)thiophene can be prepared from the diketosulfide by reacting with titanium tetrachloride/zinc followed by oxidation of the dihydrothiophene with 2,3-dichloro-5,6-dicyano-1,4-benzoquinone [315]. The same approach starting with 3,4-bis(2-thienyl)thiophene is used for the preparation of the α,β-type septithienyl shown below [244].

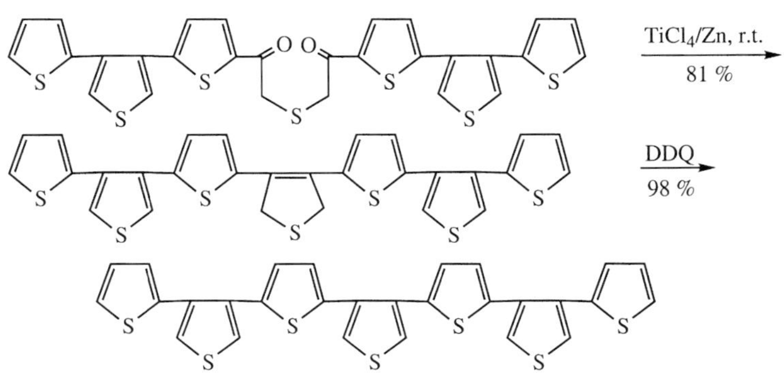

5,5′′′′′′-Bis(trimethylsilyl)-3′,3′′,4′′′′,4′′′′′-tetramethyl-α-septithienyl and 5,5′′′′′′-bis(trimethylsilyl)-3′,3′′,3′′′4′′′-4′′′′,4′′′′′-hexamethyl-α-septithienyl were prepared by Stille coupling of 3,4′-dimethyl-5-tributylstannyl-5′′-trimethylsilyl-α-terthienyl with 2,5-dibromothiophene and 2,5-diiodo-3,4-dimethylthiophene, respectively [76,325].

5,5′′′′′′-Bis(trimethylsilyl)-3′,3′′,4′′′′,4′′′′′-tetramethyl-α-septithienyl [76]

An oven-dried test tube is charged with 2,5-dibromothiophene (0.142 g, 0.59 mmol), 3,4′-dimethyl-5-(tributylstannyl)-5′′-(trimethylsilyl)-2,2′:5′,2′′-terthienyl (1.13 g, 1.75 mmol), tetrakis(triphenylphosphine)palladium(0) (0.068 g, 0.059 mmol) and toluene (4.00 ml). The reaction mixture is heated at 100 °C overnight before being poured into water. The product is extracted with diethyl ether and the combined organic phases washed with sodium chloride solution, dried over magnesium sulfate and evaporated. The residue is washed several times with hexane before being dried under reduced pressure giving 0.295 g (64%) of the title compound as dark-red crystals mp 181–183 °C.

Coupling of the core compound with the 5-tributylstannyl-3,4′ dimethyl-5′′-trimethylsilyl-α-terthienyl gives the orthogonal thiophene system [326].

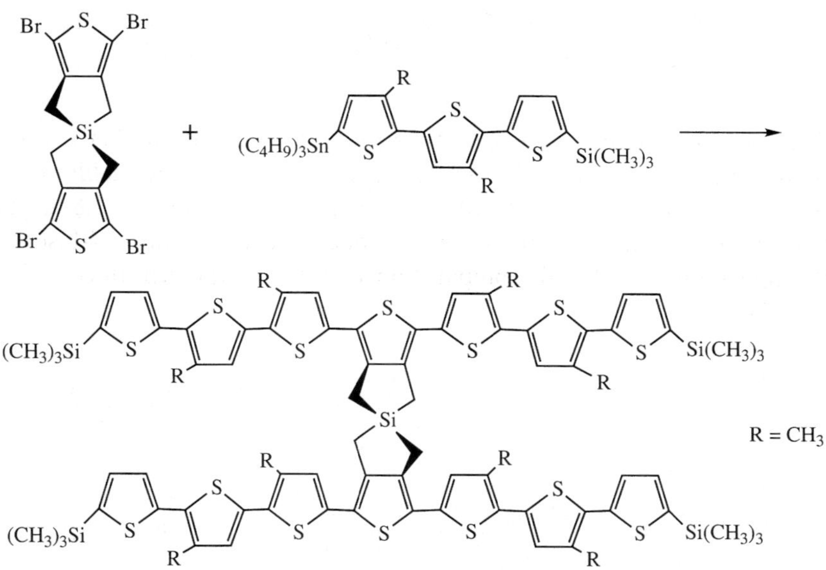

3′′′,4′′′-Dihexyl-α-septithienyl-1′′′,1′′′′-dioxide is prepared by Stille coupling of 5,5′′′′-diiodo-3′′,4′′-dihexyl-α-quinquethienyl-1′′,1′′-dioxide with two equivalents of 2-tributylstannylthiophene [272].

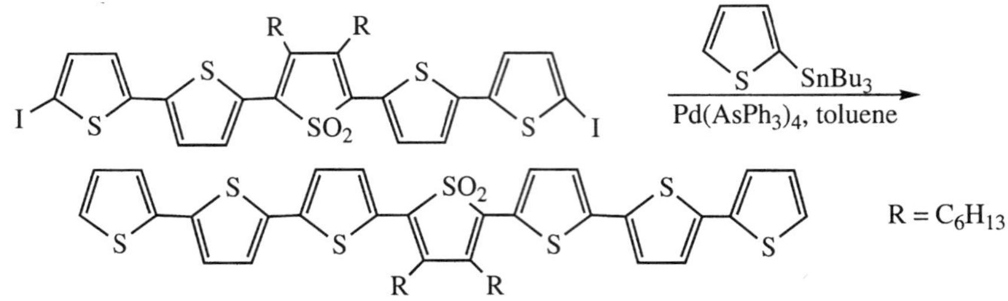

The molecular packing and photoluminescence efficiency in odd-membered oligothiophene *S,S*-dioxides, such as the above-mentioned heptamer, has been studied by single-crystal X-ray crystallography [327].

8.10 α-OCTITHIENYLS

8.10.1 Symmetrical octithienyls through the coupling of α-quaterthienyl derivatives with cupric chloride, ferric chloride or nickel complexes

Dimerization of α-quaterthienyls is a straightforward way to α-octithienyls. Thus the reductive coupling of 5-bromo-α-quaterthienyl with nickel chloride, zinc and triphenylphosphine yields α-octithiophene, albeit only in 18% yield [244]. Metalation of 5,5‴-dihexyl-α-quaterthienyl with one equivalent of lithium diisopropyl amide followed by cupric chloride gives a 37% yield of a mixture of tetrahexyloctithienyls [305].

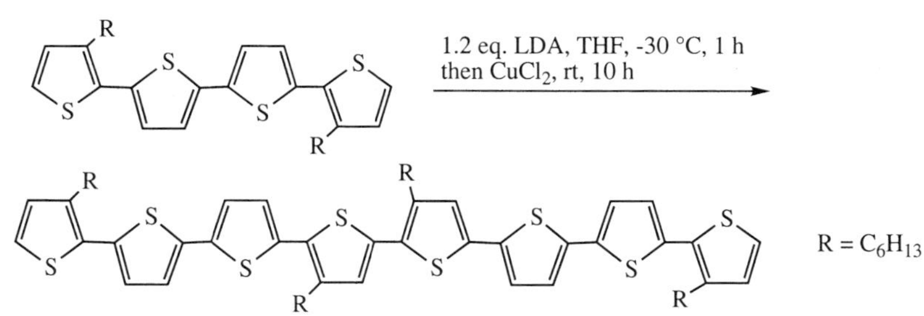

Through Vilsmeyer formylation this compound is transformed to the 5-formyl derivative and then treated with fullerene and *N*-methylglycine to give the fullereneoligothiophene-linked compound [305]. Metalation with butyllithium followed by cupric chloride was used for the dimerization of 3,3‴-didodecyl-α-quaterthienyl to the tetradodecyl-α-octithienyl [304].

Cerium(IV) oxidation of 4,4‴-bis(trimethylsilyl)-α-quaterthienyl gave the corresponding α-octithienyl in 66% yield accompanied only by small amounts of the α-dodecithienyl and α-hexadecithienyl [249]. The reaction of 3-octylthiophene with four equivalents of ferric chloride in chloroform surprisingly gives the following octamer in 41% yield [273].

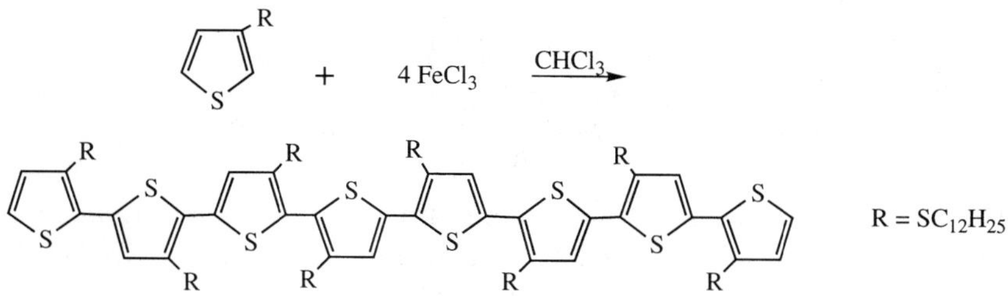

The reaction of 3,3‴-(butylthio)-2,2′:5′,2″:5″,2‴-quaterthienyl in chloroform/nitromethane (1:1) gave with anhydrous ferric chloride 3,3‴,4⁗,3‴‴′-tetra(butylthio)-α-octithienyl [64].

3,3‴,4⁗,3‴‴′-Tetra(butylthio)-2,2′:5′,2″:5″,3‴:5‴,2⁗:5⁗,2′′′′′:5′′′′′,2′′′′′′:5′′′′′′,2′′′′′′′-octithienyl [64]

To a stirred solution of 3,3‴-(butylthio)-2,2′:5′,2″:5″,2‴-quaterthienyl (0.6 g, 1.2 mmol) in distilled chloroform (30 ml) a solution of anhydrous ferric chloride (0.78 g, 4.8 mmol) in anhydrous nitromethane (30 m) is added under a flow of anhydrous nitrogen. The reaction mixture is stirred at room temperature for 30 h, after which it is evaporated and the residue treated with a solution of hydrochloric acid/methanol (50 ml). The dark product is allowed to settle and the procedure repeated twice. Addition of methanol led to formation of a brown solid which is filtered off, washed with methanol and dissolved in chloroform. This solution is filtered to remove insoluble material, washed with 3% hydrazine solution and water followed by evaporation. The crude product is treated with diethyl ether (200 ml). This solution is evaporated, the residue is stirred with pentane and the solid material filtered off giving 0.31 g (51%) of the title compound as a red solid mp 85–87 °C.

8.10.2 Transition metal-catalyzed coupling according to Stille

Stille coupling of 5⁗′-bromo-5-carbobenzyloxy-4″,4⁗-dioctyl-α-hexithienyl and 5-trimethylstannyl-2,2′-bithienyl gives the octamer shown below [126].

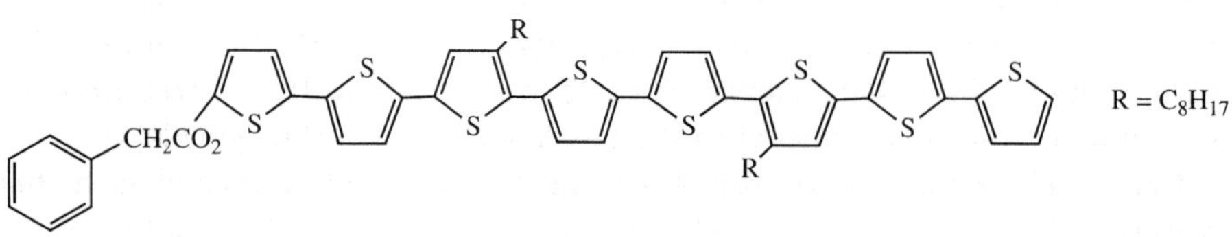

5,5‴‴′-Bis(trimethylsilyl)-3′,3″4‴‴,4‴‴′-tetramethyl-α-octithienyl is obtained by Stille coupling of 5,5′diiodo-2,2′-bithienyl and two equivalents of 3,4′-dimethyl-5-(tributylstannyl)-5′-trimethylsilyl-α-terthienyl [76].

5,5‴‴′-Bis(trimethylsilyl)-3′,3″4‴‴,4‴‴′-tetramethyl-2,2′:5′,2″:5″,2‴:5‴,2⁗:5⁗,2⁗′:5⁗′,2⁗″:5⁗″,2‴‴′-octithienyl [76]

An oven-dried test tube, washed with ammonium hydroxide, is charged with 3,4′-dimethyl-5-(tributylstannyl)-5″-trimethylsilyl-2,2′:5′,2″-terthienyl (0.73 g, 1.35 mmol), 5,5′-diiodo-2,2′-bithienyl (0.084 g, 0.2 mmol) and tetrakis(triphenylphosphine)palladium(0) (0.0093 g, 0.008 mmol) in toluene (1.0 ml). The reaction mixture is heated slowly to 50 °C for 1 h and 80 °C for 2 h and then at 100–105 °C overnight before being poured into water. The product is extracted with diethyl ether, the combined organic phases washed with sodium chloride solution, dried over sodium sulfate and evaporated. The residue is purified by chromatography on silica gel using hexane as eluent, giving 0.089 g (52%) of the title compound as a bright-red solid.

Stille coupling of two equivalents of 5-bromo-5‴-formyl-α-terthienyl with 5,5′-bis (tributylstannyl)-2,2′-bithienyl gives the α,ω-diformyl-α-octithienyl [100].

5,5‴‴′-Diformyl-2,2′:5′2″:5″,2‴:5‴,2⁗:5⁗,2⁗′:5⁗′,2⁗″:5⁗″,2‴‴′-octithienyl [100]

A solution of 5,5′-bis(tributylstannyl)-2,2′-bithienyl (0.50 g, 0.67 mmol) and 5-bromo-5‴-formyl-2,2′:5′2″-terthienyl (487 mg, 1.37 mmol) in *N*,*N*-dimethylformamide (25 ml) is deaerated twice with argon, after which tetrakis(triphenylphosphine)palladium(0) (11 mg, 0.01 mmol) is added. The reaction mixture is heated at 80–85 °C for 5.5 h under argon. After cooling the red suspension is filtered and the solid washed with diethyl ether (200 ml), dichloromethane (200 ml) and tetrahydrofuran giving 436 mg (91%) of the title compound as a dark red solid.

8.10.3 Transition metal-catalyzed coupling according to Suzuki

4,3′,3″,3‴,3⁗,3⁗′,3⁗″-Octadecyl-α-octithienyl-5-carboxylic acid benzyl ester is obtained by Suzuki coupling of 4,3′,3″,3‴,3⁗,3⁗′-5⁗″-iodo-α-sexithienyl-5-carboxylic acid benzyl ester with 4,3′-didodecyl-2,2′-bithienyl-5-boronic acid propan-1,3-diol ester as described for the hexathienyl above. This octamer could be iodinated using iodine and mercuric caproate in 90% yield [82].

Solid state synthesis on chloromethylated polystyrene employing a sequence of iodination and Suzuki coupling has been used for the preparation of the octamer, which upon treatment with tetrabuylammonium hydroxide and methyl iodide gives the methyl ester of the octamer [151].

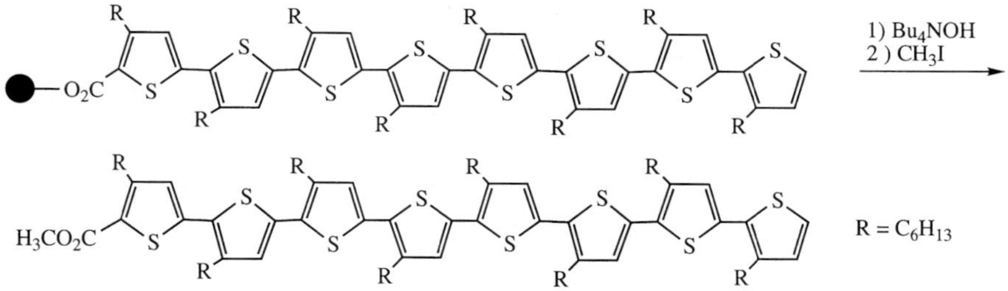

8.11 NOVITHIENYLS

Hexahexyl-α-novithienyls were obtained in low yields as a by-product in the reductive coupling of 5,5″-dibromo-3,3″-dihexyl-α-terthienyl [241]. The novithienyls below are formed as minor products in the ferric chloride oxidation of 1-(3-dodecylthienyl)-3-thienyl- and 1-(3-hexyl-2-thienyl)-3-thienylbenzo[*c*]thiophene [241].

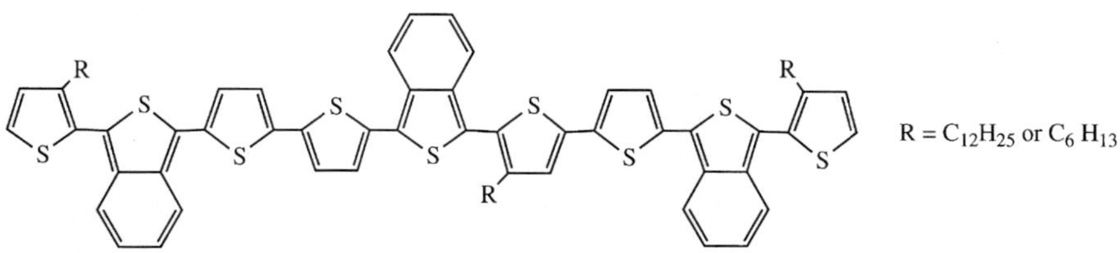

Stille coupling of the 5,5‴-dibromo derivative of the quinquethienyl with two equivalents of 5-trimethylstannyl-2,2′-bithienyl is used for the preparation of the dendrimer-supported compound below in 86% yield [211].

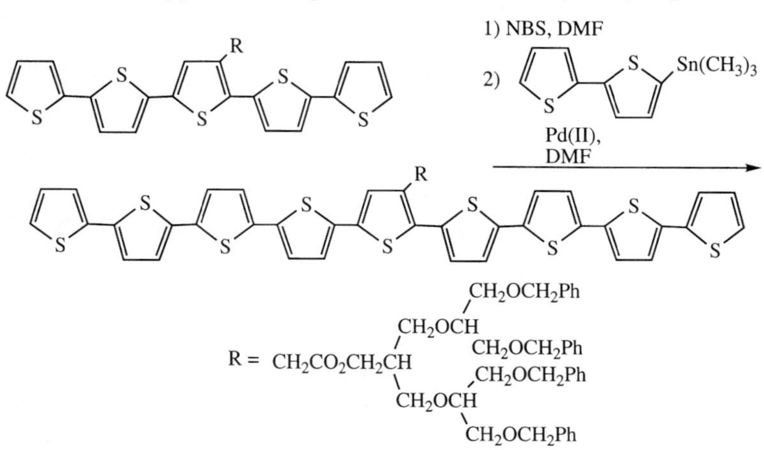

The branched novithienyl is prepared by heating bis(2,2′-bithienyl-5-yl)-acetylene with sulfur [328].

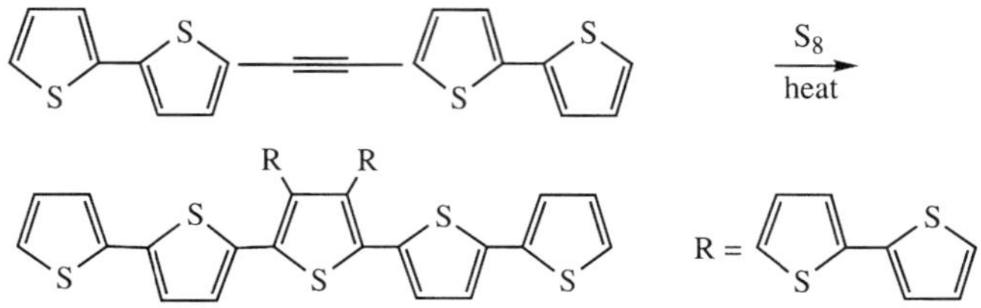

8.12 DECITHIENYLS

4,3′,3″,3‴,3′′′′,3′′′′′,3′′′′′′,3′′′′′′′,3′′′′′′′′,3′′′′′′′′′-Decadodecyl-α-decithienyl-5-carboxylic acid benzyl ester was prepared by Suzuki coupling of 5′′′′′′′-iodo-4,3′,3″,3‴,3′′′′,3′′′′′,3′′′′′′,3′′′′′′′-octadodecyl-α-octithienyl-5-carboxylic acid benzyl ester and 4,3′-didodecyl-2,2′-bithienyl-5-boronic acid propan-1,3-diol ester [82].

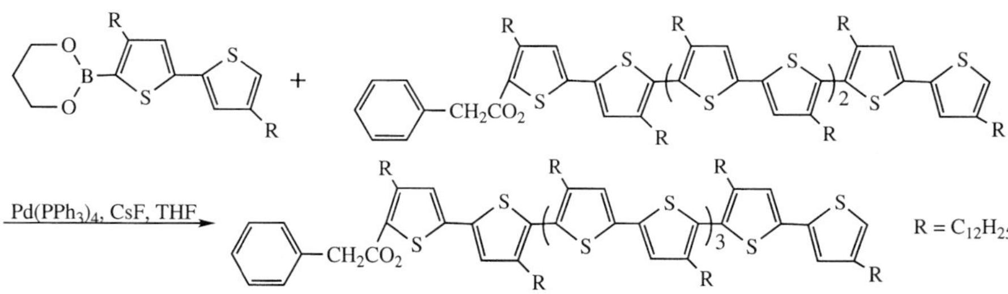

8.13 UNDECITHIENYLS

Butyl and dodecyl substituted undecithienyls were obtained *via* ring closure of the 1,4-diketone units to thiophene rings, by treatment with Lawesson's reagent [329].

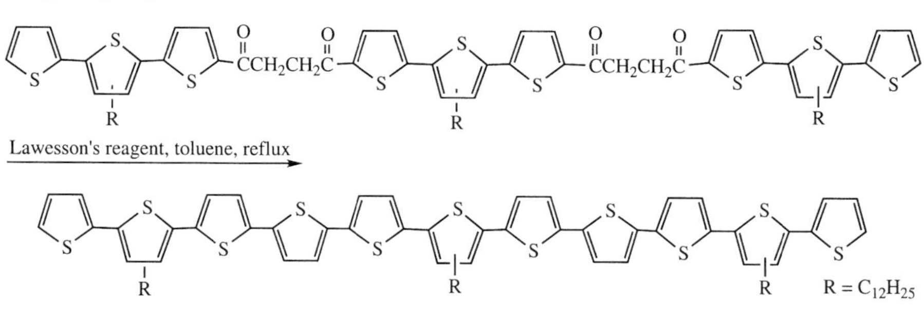

The polymer-bound carbonyl compound gave also upon reaction with Lawesson's reagent the undecithienyl in connection with the preparation of a polystyrene/oligothiophene/polystyrene triblock copolymer [330].

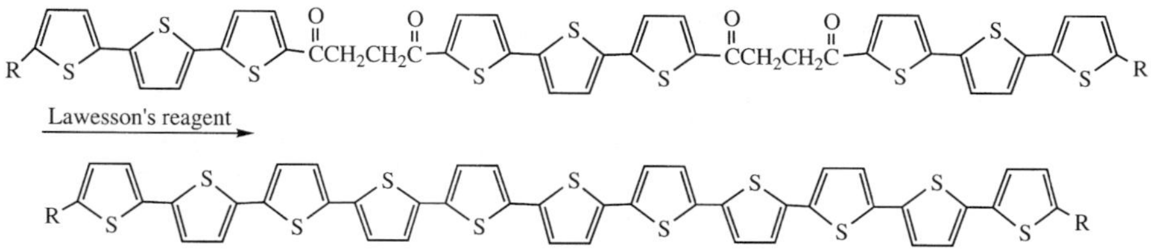

$R = (CH_2)_{10}$-O-PS

Stetter reaction to the tetraketone and its conversion [330]

A mixture of 5-(10-(polystyryloxy)decyl)-2,2′:5′,2″-terthiophene-5″-aldehyde (7.0 g, 1.9 mmol) and sodium cyanide (50 mg, 1.0 mmol) in *N*,*N*-dimethylformamide (20 ml) is stirred under argon for 30 min. Free Mannich base, 5,5″-bis[3-(dimethylamino)propionyl]-2,2′:5′2″-terthienyl (0.4 g, 0.89 mmol) in *N*,*N*-dimethylformamide (5 ml) is added over 1 h, after which the stirring is continued overnight. The reaction mixture is added dropwise to methanol (400 ml), the product isolated, dissolved in tetrahydrofuran and precipitated in methanol (400 ml). The crude tetraketone (4.0 g) is cyclized under argon with Lawesson's reagent (2 g, 10 equiv.) in toluene (300 ml) at 80 °C for 5 h. After evaporation the product is purified by chromatography on silica gel using toluene as eluent, then precipitated two times from tetrahydrofuran in methanol and dried under vacuum giving 2.15 g (30%) of the undecithienyl shown above.

The hydride dendrimer is prepared starting from the quinquethienyl, which is hydrolyzed and converted to the acid chloride, which was coupled to a third generation poly(benzylether) dendron, thus providing the monodendron functionalized pentamer. Finally Stille coupling between this compound and 2,5-bis(trimethylstannyl)thiophene gave the desired undecylthienyl [127].

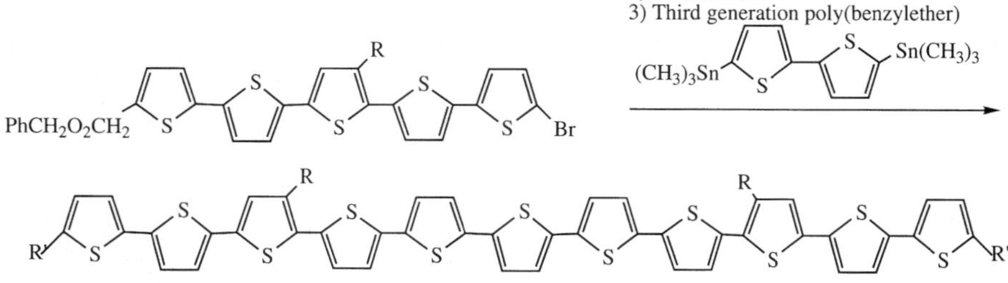

$R = C_8H_{17}$, $R' = CO_2CH_2C_6H_3$-poly(benzylether)

8.14 DUODECITHIENYLS

Hexyl-substituted duodecithienyls are obtained as byproduct in the nickel(0)-catalyzed coupling of 5,5″-dibromo-3,3″-dihexyl-α-terthienyls [320].

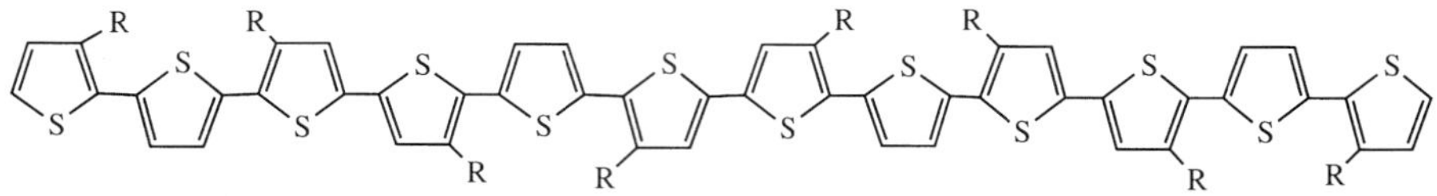

$R = C_6H_{13}$

Treatment with ferric chloride in dichloromethane generated a cation radical, the ESR-spectrum of which could be measured [331]. Metalation with butyllithium followed by cupric chloride was used for the dimerization of 3,3‴-didodecyl-α-quaterthienyl to the tetradodecyl-α-octithienyl in 30% and as a byproduct the hexadodecylduodecithienyl was isolated in 8% yield. Both compounds were obtained pure, due to their good solubility, by repeated chromatography [304].

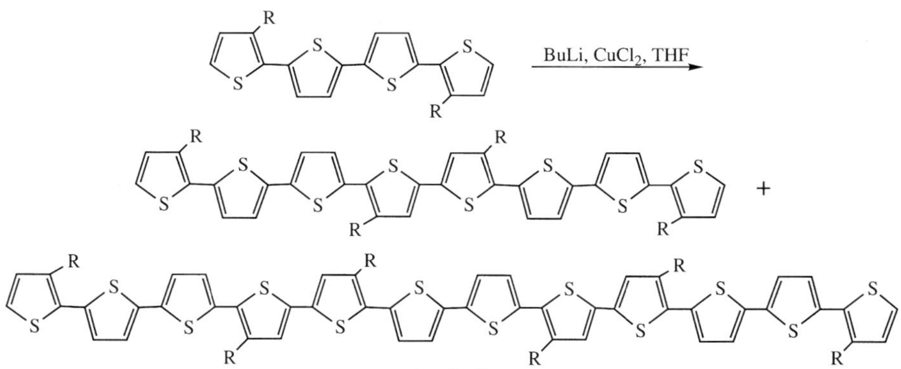

$R = C_{12}H_{25}$

Similarly, metalation of 3,3‴-dihexyl-α-quaterthienyl with lithium diisopropylamide followed by cupric chloride gave besides the octi-derivative, 11% of the duodecithienyl, which by Vilsmeyer fomylation is transformed to the 5-formyl derivative and then treated with fullerene and 1-methylglycine to give the fullerene duodecithienyl-linked compound [305].

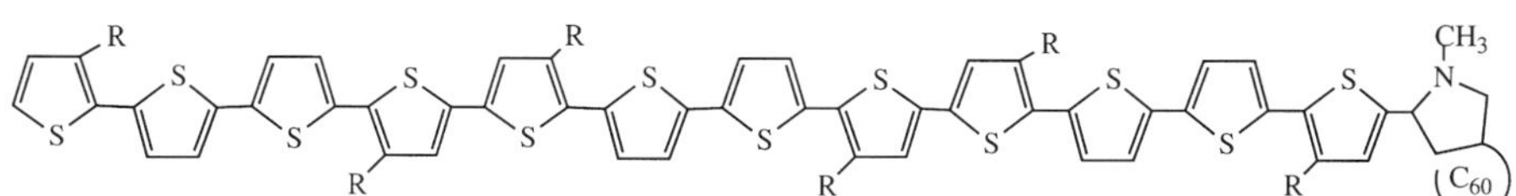

$R = C_6H_{13}$

Iodination of polymer-bound octithienyl with iodine and mercury hexanoate followed by Suzuki coupling with 3′,4-dihexyl-2,2′-bithienyl-5-boronic acid propanediyl ester two times yields the polymer-bound derivative, which upon treatment with tetrabutylammonium hydroxide and methyl iodide gave the duodecithienyl. Upon hydrolysis and decarboxylation with copper and quinoline at 180 °C the nonfunctionalized duodecithienyl could be prepared [151].

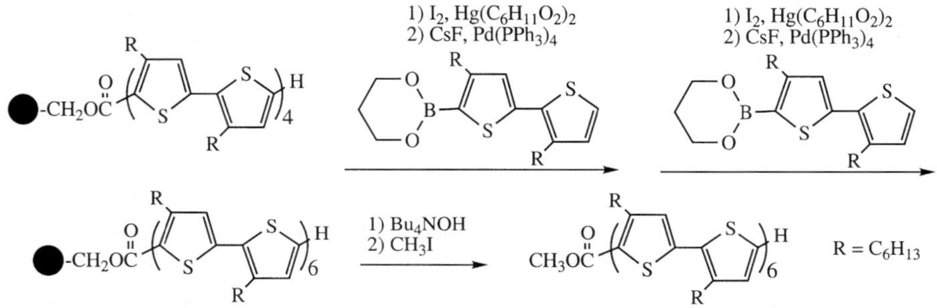

General cleavage of resin-bound oligothienyls [151]

The resin-bound oligothienyl (1 equiv.) is swollen in anhydrous tetrahydrofuran (30 ml per g of polymer) and after addition of tetrabutylammonium hydroxide (30 hydrate) (2 equiv.) the mixture is refluxed for 3 days and filtered to remove the polymer. The oligothienyl carboxylate is then converted *in situ* to the methyl ester with an excess of iodomethane (3 equiv.) at room temperature for 1 h. After evaporating of the solvent the residue is purified from homocoupled products by chromatography on silica gel using hexane/dichloromethane as eluent. The dodecamer ester is further purified by HPLC on a nitrophenyl column using hexane/dichloromethane (9:1) as eluent.

8.15 STILL LONGER OLIGOTHIENYLS

Hexyl-substituted quindecithienyls were obtained pure in low yield after chromatography of the product from the nickel(0)-catalyzed coupling reaction of 5,5″-dibromo-3,3″-dihexyl-α-terthienyl [332].

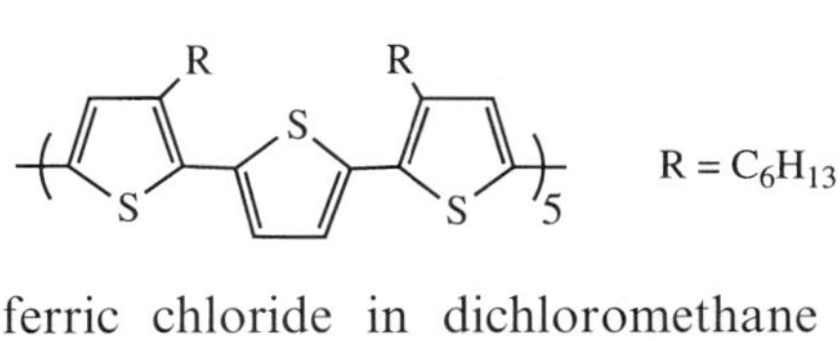

Treatment with ferric chloride in dichloromethane generated a cation radical, the ESR-spectrum of which could be measured [331]. The octadodecyl-substituted hexadecithienyl is prepared by monometalation of the octithienyl followed by cupric chloride analogous to lower homlogs [304].

The octithienyl oligomerized upon reaction with cerium(IV) sulfate in 1,2-dichloromethane at room temperature to give the hexadecithienyl ($n=2$) in 35% yields the tetracosithienyl ($n=3$) in 11% and a mixture of even higher oligomers (*ca* 20%) after column chromatography [249].

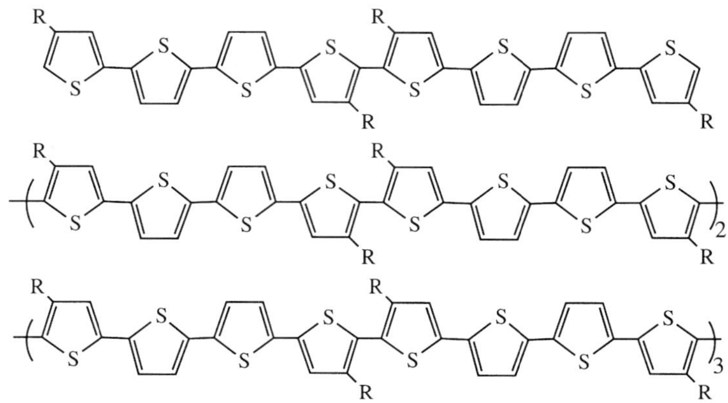

R = $Si(CH_3)_3$

The heptadecamer was prepared by Stille coupling of 2,5-bis(trimethylstannyl)thiophene with the bromo derivative of the octithienyl [127].

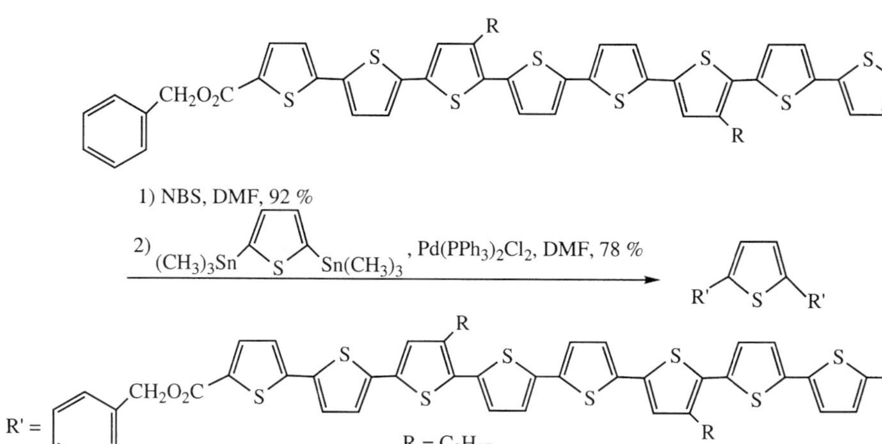

Among the longest oligothiophenes hitherto described are icosamers and heptacosamers containing a total number of 20 and 27 thiophene rings, respectively. The strategy for their preparation consists in the combination of two reactions, nickel(II)-catalyzed aryl Grignard cross-coupling for the preparation of the key intermediate, a tetraoctylsubstituted hexithiophene and Eglinton coupling followed by sodium sulfide induced 1,3-butadiyne cyclization. Bromination of the hexamer with one equivalent of *N*-bromosuccinimide led to the monobromo derivative, which upon palladium(0)-catalyzed coupling with tri(methylsilyl)acetylene gave the acetylene derivative, which was desilylated. Alternatively, the hexamer was dibrominated and converted to the diethynyl derivative. An Eglinton coupling reaction using a mixture of the two acetylene derivatives in a 3:1 molar ratio gave a mixture of polyacetylenes. These compounds were separated by gel permeation liquid chromatography and subsequent treatment of each component with sodium sulfide gave the desired oligothiophenes in good yields [303].

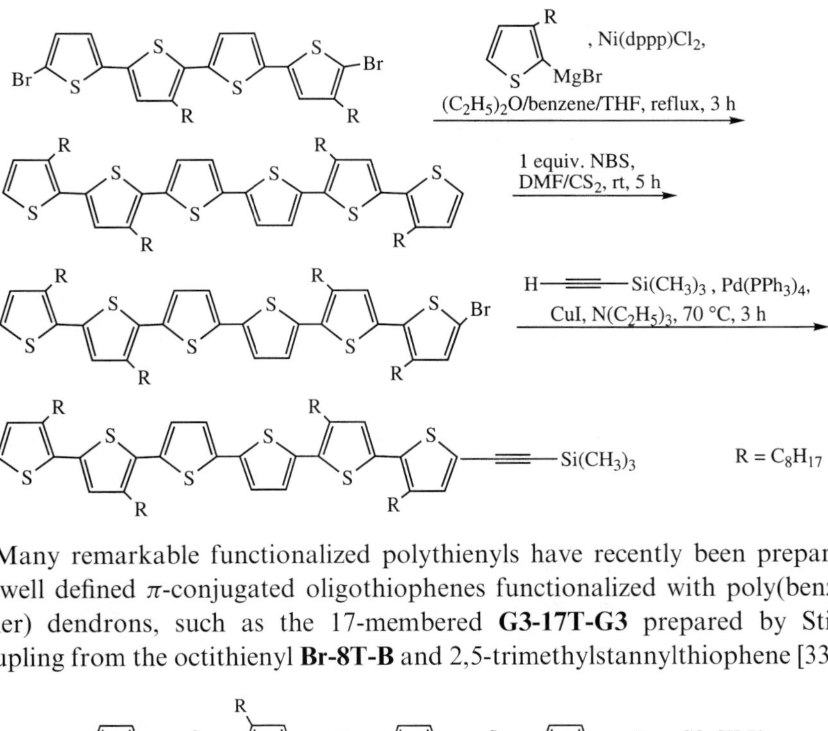

Many remarkable functionalized polythienyls have recently been prepared as well defined π-conjugated oligothiophenes functionalized with poly(benzyl ether) dendrons, such as the 17-membered **G3-17T-G3** prepared by Stille coupling from the octithienyl **Br-8T-B** and 2,5-trimethylstannylthiophene [333].

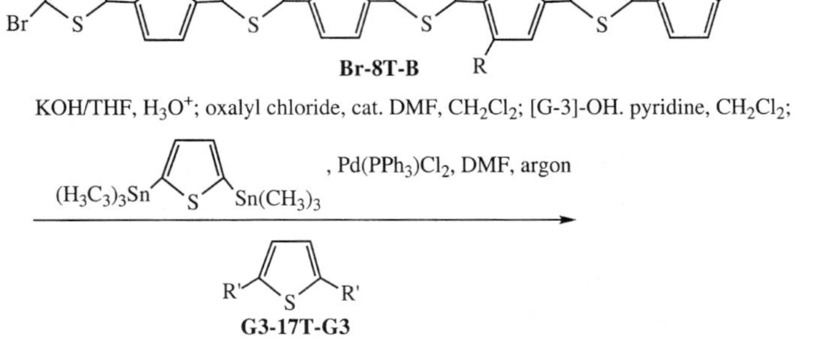

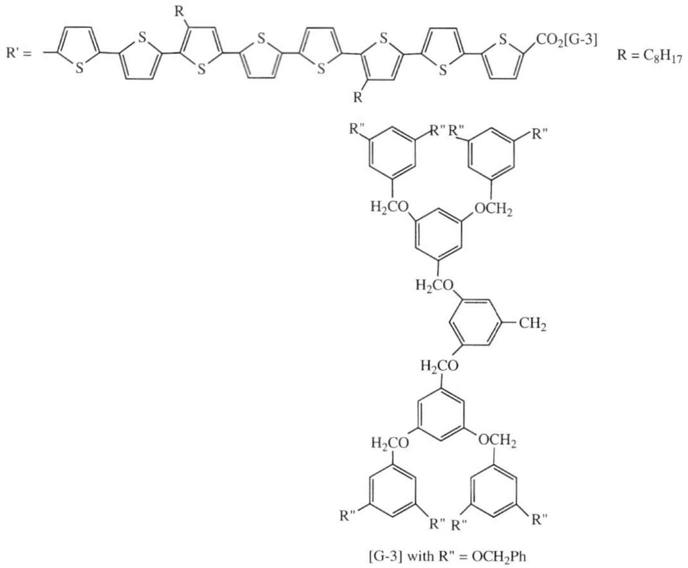

8.16 MACROCYCLIC OLIGOTHIOPHENES

The macrocyclic compounds of the type shown below were prepared by a modified Eglinton–Glaster coupling of diacetylenes under pseudo high dilution conditions. Refluxing the mixed cyclooligothiophene diacetylenes with sodium sulfide nonahydrate in 2-methoxyethanol/*para*-xylene for four hours gave the fully α-conjugated cyclo[*n*]thiophenes [248].

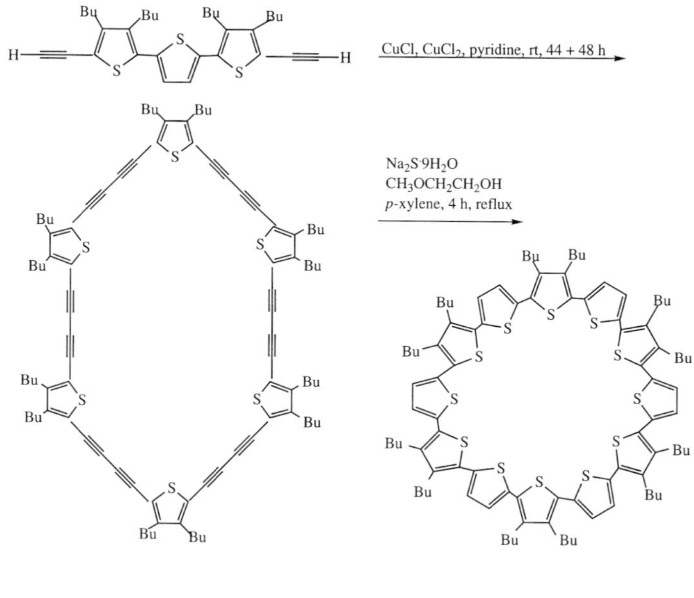

8.17 POLYTHIOPHENES BY POLYMERIZATION REACTIONS

8.17.1 Introduction

During recent years an explosion in the study of polythiophenes has occurred, due to the interest in conducting polymers and many other aspects of material sciences. They are used in specialized technical innovations such as optoelectronic devices, photoconduction, electroluminescence, photovoltaic and photolithography of semiconducting and electronic conducting polymer wires in the design of integrated circuitry and device fabrication. In this chapter only a brief review of the principal routes for the preparation of such polythiophenes will be given. The characterization and doping of such polymers will not be treated.

8.17.2 Electrochemical oxidation of thiophenes, bithienyls and higher oligothienyls

Polythienyl is formed from thiophene adsorbed on a roughened gold electrode [334]. Poly(3-methylthiophene) thin films with high conductivities have been prepared by electrochemical polymerization [334]. Electrochemically oxidative deposition of poly(3-hexylthiophene-2,5-diyl) gave a film in which p-doped molecules took an ordered structure [335]. α,ω-Bis(3-thienyl)hexane and -octane have a beneficial effect on the properties of poly(3-methylthiophene) [336]. Poly(3-arylthiophene)s were prepared by electropolymerization of 3-(4-fluorophenyl)-3-(4-cyanophenyl)-3-(4-methylsulfonyl) and 3-(3,4-difluorophenyl)thiophenes as active materials for electrochemical capacitor applications [337]. New conducting polymers have been synthesized by electropolymerization of (3-benzyloxyethyl)thiophenes [336]. Poly(3,4-ethylenedioxythiophene) has been electrosynthesized in an aqueous micellar medium (sodium dodecyl sulfate) at platinum electrodes, depositing conducting films at lower potential than in acetonitrile [339]. 3-Thiophenacetic acid is electrochemically oxidized and polymerized on indium tin oxide electrodes in a 0.05 mol/l tetrabutyl ammonium perchlorate acetonitrile solution [340].

A number of conducting poly(3,4-alkylidenedioxythiophenes) as fast electrochromics with high contrast ratios have been prepared electrochemically from alkylsubstituted and unsubstituted 3,4-dialkylenedioxythiophenes with different alkylidenedioxy rings [341].

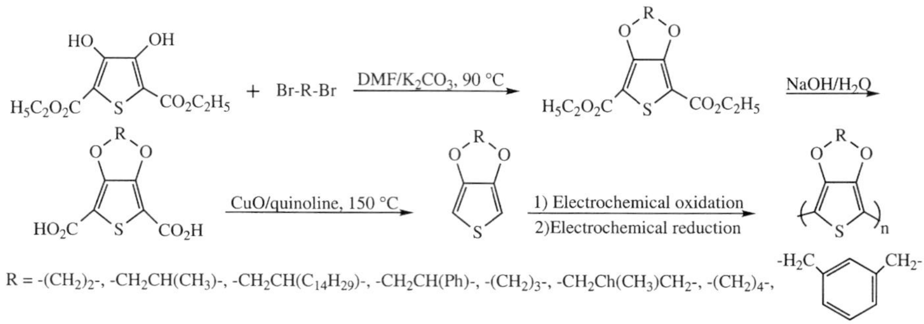

A series of novel bissalicyclidene metal complexes has been used for the preparation of polymers by electropolymerization [269].

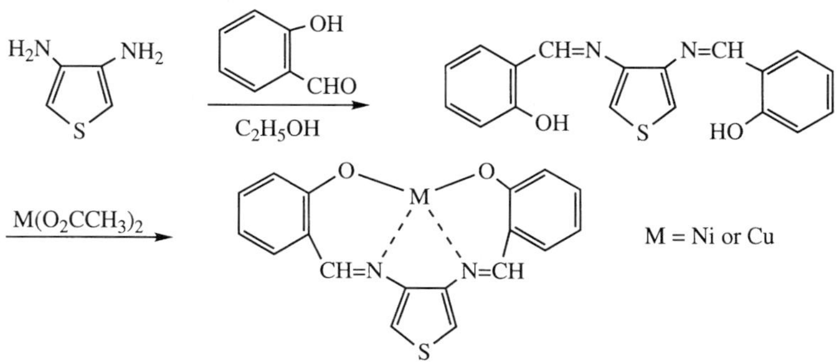

Enantioselective chiral polythiophenes have been prepared by electropolymerization [342].

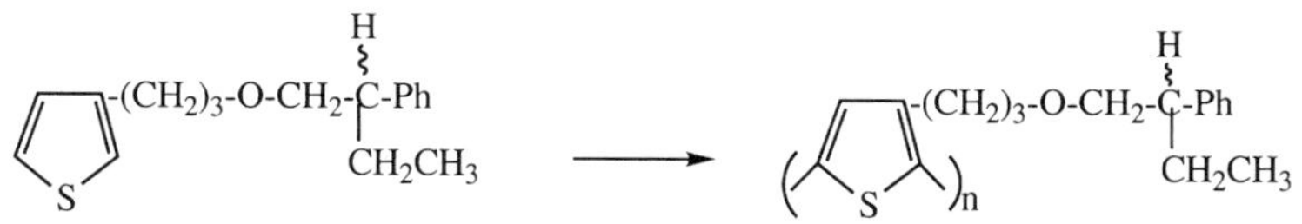

Electrochemical oxidation of symmetrical dimethyl-2,2′-bithienyls gives electroactive polymers with excellent cycling ability [7]. More special polythiophenes such as the following have been prepared by electrochemistry [343,344].

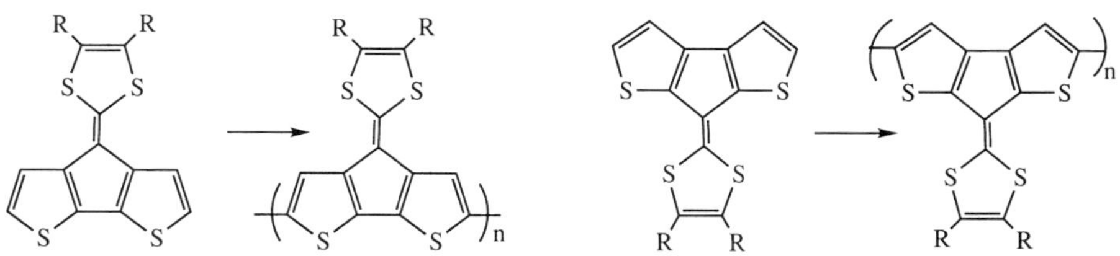

Electropolymerization of the compound below gives a polythiophene with extremely narrow bandgap [93].

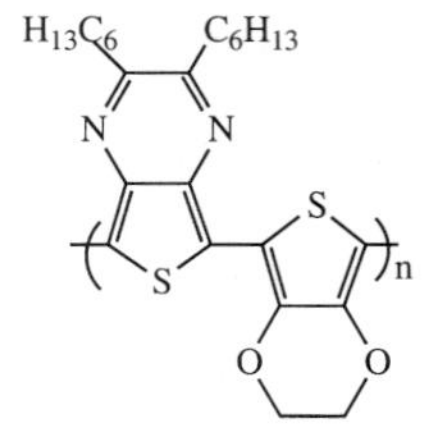

Electrochemical polymerization of 3′-(3-cyanopropyl)-α-terthienyl has recently been described [239].

3′-(3-Cyanopropyl)-2,2′:5′,2″-terthienyl [239]

A mixture of 1,4-di(2-thienyl)-2-(3-cynopropyl)butane-1,4-dione (0.350 g, 1.10 mmol), Lawesson's reagent (0.539 g, 1.32 mmol) and anhydrous toluene (15 ml) is slowly heated and maintained at reflux for 6 h. After cooling, the orange solution is poured into water (10 ml). The product is extracted with diethyl ether (3 × 15 ml), the combined organic phases are washed with water, dried over magnesium sulfate and evaporated. The residue, a brown oil, is purified by chromatography on silica gel using hexane/ethyl acetate (6:1) as eluent giving 0.282 g (82%) of the title compound as a yellow solid.

Electrochemical polymerization [239]

Electrochemical polymerization of 3′-(3-cyanopropyl)-2,2′:5′,2″-terthienyl in acetonitrile with 0.1 *M* tetrabutylammonium borontetrafluoride is performed both under galvanostatic control at constant current of *ca.* 0.1 mA/cm^2 and by potentiostatic control at *E ca.* 0.8 V *versus* the ferrocenium/ferrocene couple referenced internally. Gold foil (1.5 cm^2), platinum (0.03 cm^2) and glassy carbon (0.07 cm^2) disks are used as electrode substrates and the solution is throughly degassed with argon prior to polymerization.

Cyclopentabithiophenes functionalized with crown ether moieties, such as the compound below, were polymerized by anodic coupling in acetonitrile solution in the presence of 0.1 *M* tetraethylammonium perchlorate in order to obtain lithium- and sodium-sensing electrodes [345].

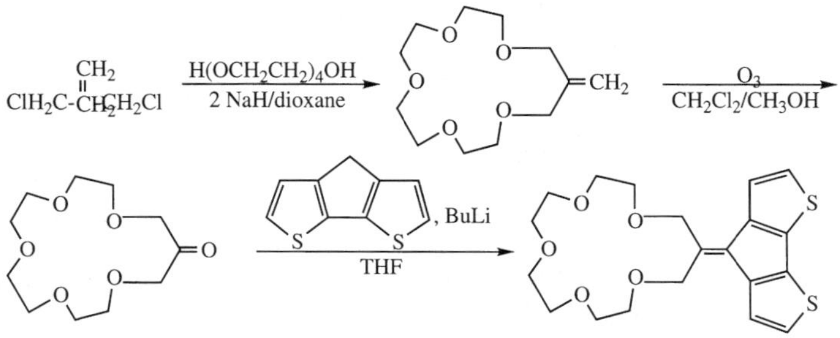

Polycyclopenta[2,2-*b*;3,4-*b'*]dithiophene-4-one gives upon electropolymerization an electroactive polymer with a lowered band gap [346]. The annelated TTF-thiophene derivatives can be polymerized electrochemically [347].

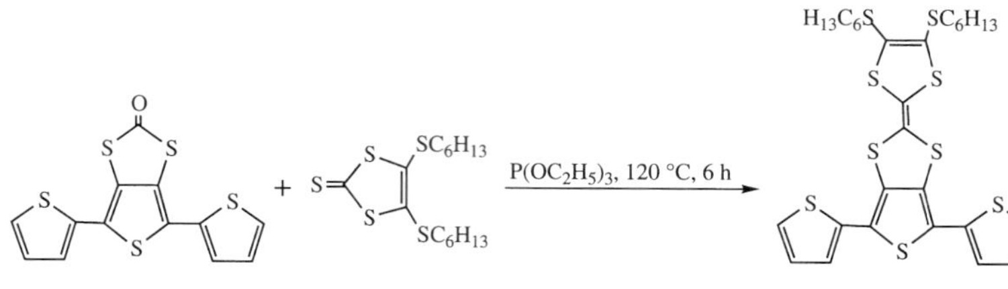

Anodic coupling of dipyrrole on thin films of poly(4-butanesulfonate)cyclopentadithiophene in acetonitrile produces polypyrrole within the polythiophene structure [348].

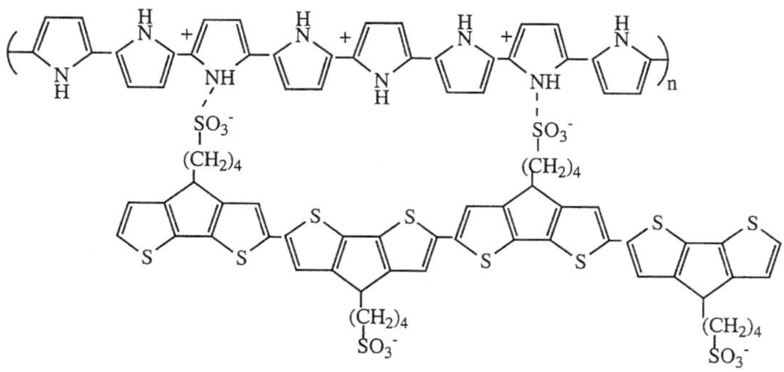

Copper(I) complexes of the following type gave a polymer entwined around copper centers and could then be polymerized [349].

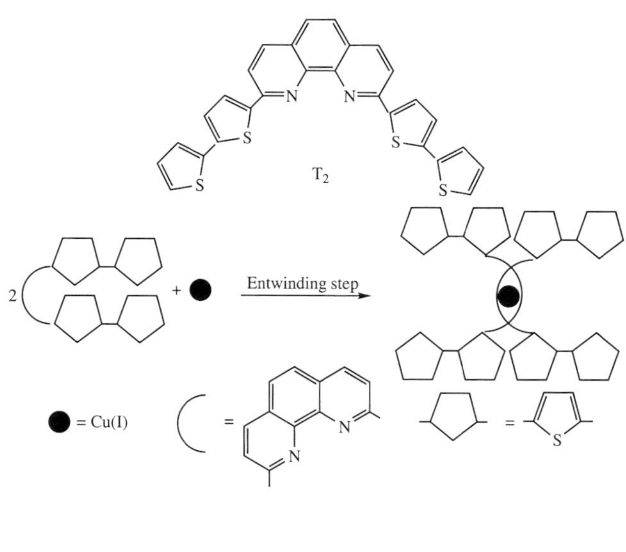

2-(3′,4′-dihexyl-2,2′:5′,2″-5-terthienyl)-1,10-phenanthroline [349]

A solution of diisopropylamine (425 µl, 2.9 mmol) in anhydrous tetrahydrofuran is cooled to − 78 °C, after which 1.77 *M* butyllithium (1.7 ml) is added dropwise. The mixture is stirred at − 78 °C for 15 min, allowed to warm to 5 °C for 20 min and them recooled to − 78 °C before it is added by means of the double-ended needle technique to a solution of 3′.4′-dihexyl-2,2′:5′,2″-terthienyl (1.06 g, 2.55 mmol) in anhydrous tetrahydrofuran (25 ml) at − 78 °C. This mixture is stirred at − 78 °C for 30 min and the temperature is increased to 0 °C before it is added to a colorless solution of 1,10-phenanroline (328 mg, 1.82 mmol) in anhydrous tetrahydrofuran (10 ml) at room temperature. A dark-red color appears immediately and the reaction mixture is stirred at room temperature. A saturated aqueous ammonium chloride solution is added slowly and after removal of the organic solvents dichloromethane (50 ml) is added. The phases are separated and the organic phase is, under vigorous stirring, treated with manganese dioxide (10 g) until there is complete rearomatization. Addition of magnesium sulfate (20 g) and additional stirring for 15 min is followed by filtration through Celite and evaporation of the filtrate. The residue is purified by chromatography on silica gel using a gradient of hexane/dichloromethane (9:1 to 1:99) as eluent, giving 442 mg (41%) of the title compound in the pure form.

2,2′-Bithienyl as an inclusion compound with hydroxypropyl-β-cyclodextrin has been electropolymerized in aqueous medium [350]. Poly(3,3″-didodecyl-α-terthienyl) is prepared by electrosynthesis [351].

Preparation of poly(3,3″-didodecyl-2,2′:5′,2″-terthienyl) [351]

The electrochemical synthesis of poly(3,3″-didodecyl-2,2′:5′,2″-terthienyl) is accomplished with a three electrode cell configuration. The depositon of poly (3,3″-didodecyl-2,2′:5′,2″-terthienyl) is carried out in acetonitrile/benzonitrile (4:1) containing 0.1 *M* tetrabutylammonium perchlorate and poly(3,3″-didodecyl-2,2′:5′,2″-terthienyl) (0.2–2 mmol). The solution is deaerated with nitrogen before the polymer synthesis. The polymer is electrodeposited in the potentiodynamic mode with the substrate potential cycled over a range $0 < E < 0.9$ versus silver/silver nitrate at 100 mV/s. After the synthesis the polymer film is rinsed with acetonitrile and dried under vacuum for 3 h at room temperature.

In order to build polaronic ferromagnetic polymers poly(*meta*-phenylene)-quaterthienylene) was prepared by electrochemical polymerization of 1,3-bis(2,2′-bithienyl-5-yl)benzene [352].

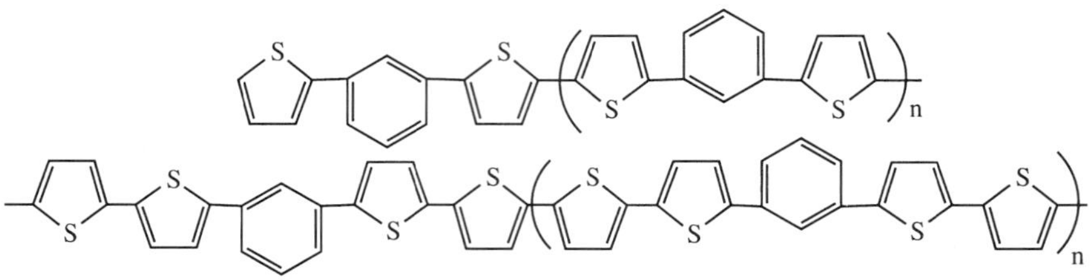

The polythiophene cobalt salen hybride polymer is anodically deposited from its monomer poly(1) on platinum or glassy carbon electrodes to give dark-yellow films [353].

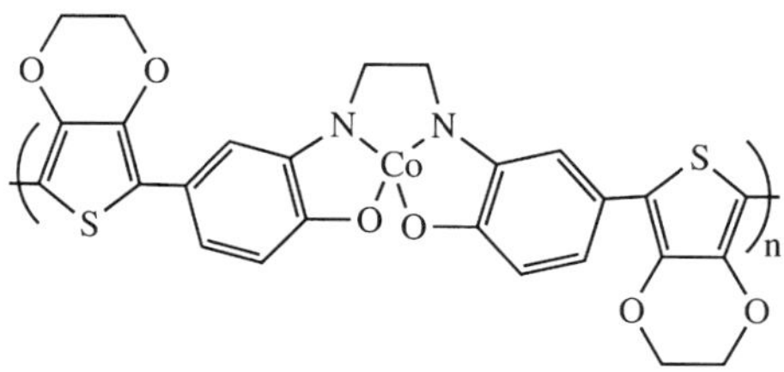

A 2-carboxyanthraquinone-bound 3-alkylthiophene has been electrochemically polymerized [354], and a fluorescent substituted thiophene is electrochemically polymerized to give the following polymer [355].

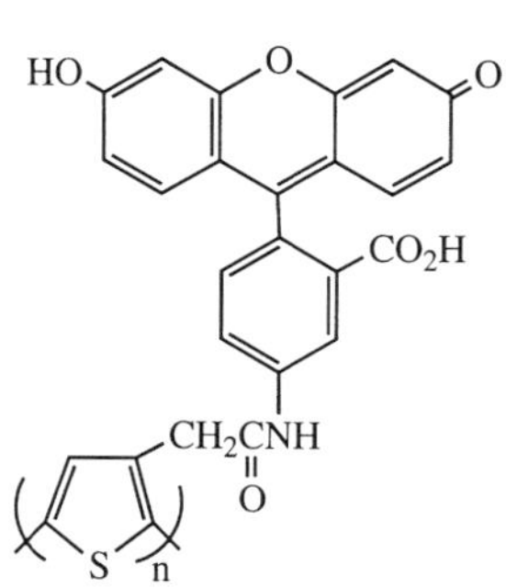

Electrochemical polymerization of thiophene oligomers ($n = 2$–4) bearing pentoxy groups in regiochemically defined positon (3 or 4 of the terminal thiophene rings) led to conducting polymers [356].

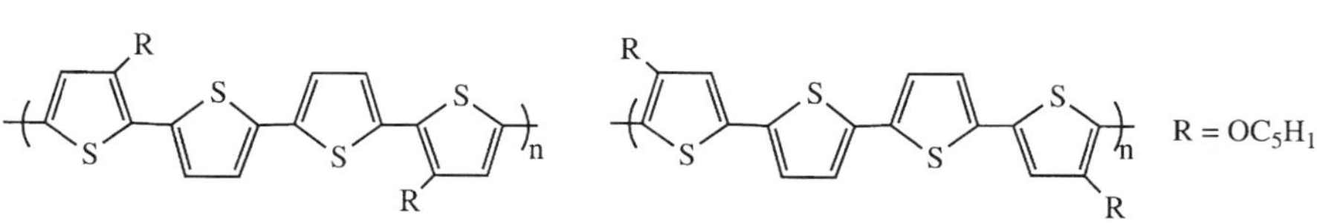

An electroactive polymer with a bandgap of 1.7 eV is obtained from 1,2-di(2-thienyl)benzo[*c*]thiophene [238].

Three poly(4,4″-dipentoxy-3′-alkyl)-α-terthienyls with dodecyl, octyl and hexyl as alkyl groups were prepared by cyclic voltammetry and the effect of the synthesis temperature and the length of the alkyl substituents on photoelectrical properties studied [357].

Polymers of the chiral compounds **1** and **2** were also prepared by electrochemical procedures [358].

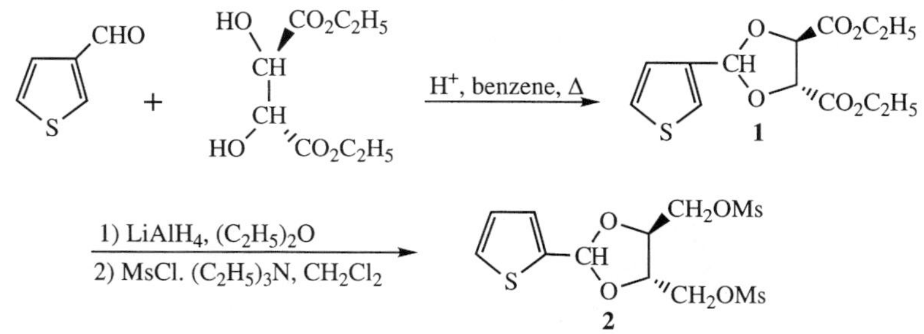

Several phenyl-substituted α-terthienyls such as 3,3″-diphenyl-α-terthienyl, 3,3′,3″triphenyl-α-tert-thienyl, 3′-phenyl-α-terthienyl and 3′,4′-diphenyl-α-terthienyl are conveniently electropolymerized at a platinum electrode in connection with a study of the effect of the number of thiophene rings and substitution pattern of the monomers on the spectroscopic and storage properties of the resulting polymers [251].

A low band gap conjugated metallopolymer has been prepared by electrochemical polymerization of bis[1,1-di(2-thienyl)-1,2-ethenedithiolene-nickel [359].

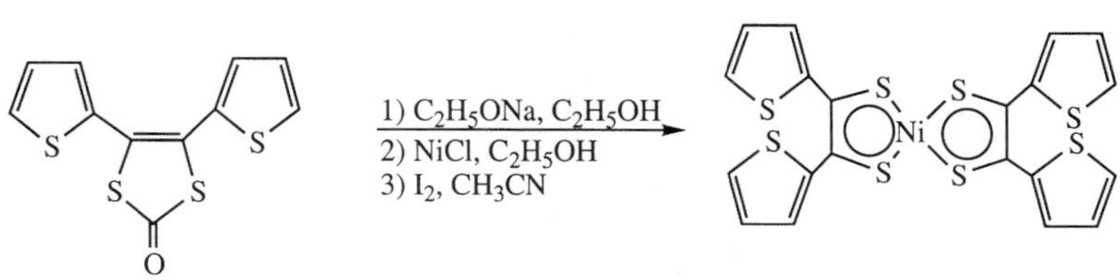

8.17.3 Chemical oxidation of thiophenes, bithienyls and higher oligothiophenes

Ferric chloride is most commonly used for the preparation of especially alkoxy-substituted poly(thiophenes). Thus soluble and conductive poly(3-methoxy-2,5-thiophenediyl) was obtained from 3-methoxythiophene [360]. 3,4-Dialkylthiophene gave nonplanar polymers exhibiting reduced conductivities. However, decreasing the steric hindrance by using alkoxy derivatives gave highly conducting poly(thiophenes) [361]. The oligo(ethylene oxide) substituted poly(thiophenes) below, when mixed with a salt, simultaneously act as a light-emitting layer and solid state electrolyte in light-emitting electrochemical cells and are prepared by treatment of the corresponding monomers with ferric chloride [192].

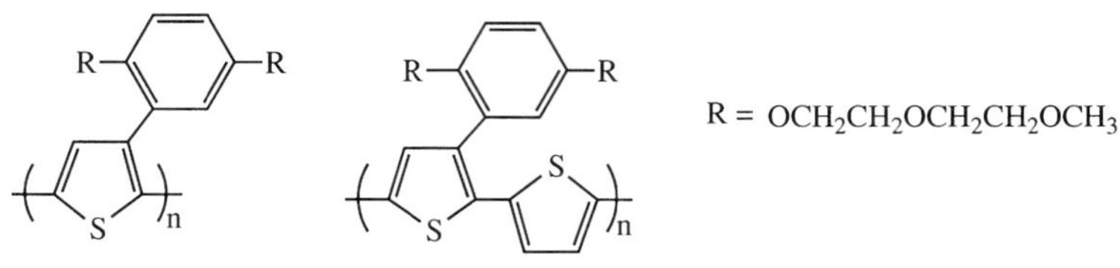

Poly(3-[2′,5′-bis(1″,4″,7″-trioxaoctyl)phenyl]thiophene) [192]

To a solution of 3-[2′,5′bis(1″,4″,7″-trioxaoctyl)phenyl]thiophene (0.57 g, 1.44 mmol) in chloroform (20 ml) is added a slurry of ferric chloride (2.9 g, 17.8 mmol) in chloroform (40 ml) over 2.5 h. The reaction mixture is stirred for another hour before being poured into methanol. The red precipitate is filtered off and redissolved in chloroform; this red solution is filtered through a sintered glass funnel and the filtrate is washed six times with concentrated ammonia, two times with 0.05 *M* ethylenediamine tetraacetic acid and two times with water. After removal of the solvent the residue is dissolved in a small volume of chloroform, which is added to methanol giving 250 mg of the title compound as a dark-red solid mp 135° C.

Polymerization of sodium 2-(4-methyl-3-thienyloxy) ethanesulphonate by iron trichloride in chloroform gives a self-doped highly conducting and nearly transparent water-soluble regular polythiophene [362, 363].

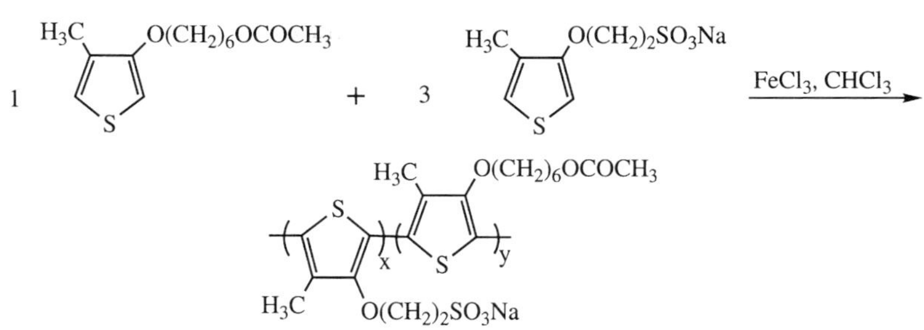

Chemical oxidation of 3,3′-dibutoxy-2,2′-bithienyl led to the synthesis of a nearly transparent conducting (2S/cm) polymer [364]. An optically actived polythiophene is prepared from 3,3′-di[(*S*)-(+)-2-methylbutyl]-α-terthienyl by using a specific oxidative polymerization method [356]. Poly[3,4-di((*S*)-2-methylbutoxy)thiophene] has also been synthesized and its circular dichromism and circular polarization of photoluminescence studied [365]. A copolymer of

1-(4-nitrophenyl)-2-(3-thienyl) ethane and 3-butylthiophene is prepared for the study of transient photoconductivity [366].

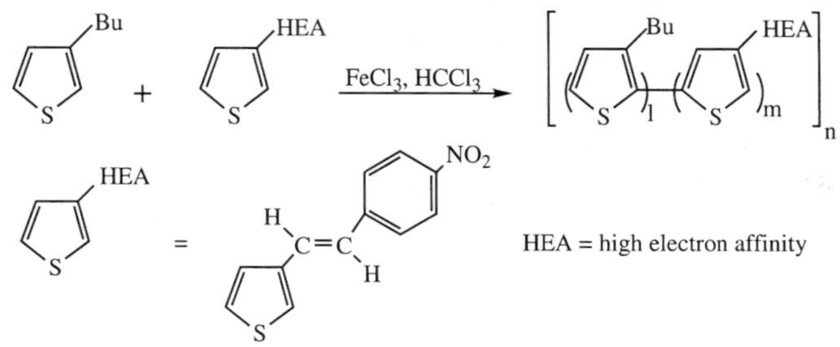

Water soluble sodium poly[2-(3-thienyloxy) ethanesulfonate] and sodium poly [2-(4-methyl-3-thienyloxy) ethanesulfonate [367], as well as sodium poly-3-(3′-thienyloxy) propanesulfonate and poly-3-(3′-thienyloxy) propyltriethylammonium were prepared by oxidation with ferric chloride [368].

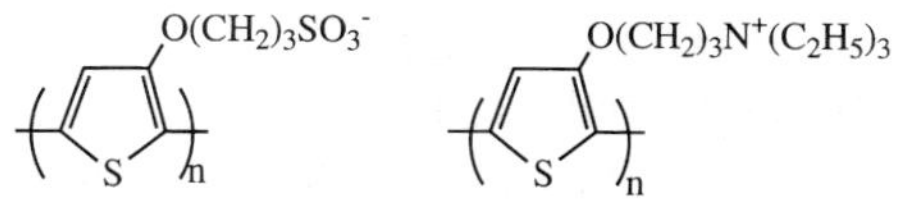

Polymerization of sodium 3-(3′-thienyloxy)propanesulfonate [368]

The polymerization is carried out with anhydrous ferric chloride (4:1) in anhydrous chloroform. The mixture is filtered after three days and the polymer dissolved in 2 *M* sodium hydroxide/10% hydrazine, after which iron hydroxide is removed by centrifugation. The neutralized crude polymer solution is dialyzed against water for two days, using a 3500 nominal molecular weight cutoff membrane. After concentration the solution was fractionated with a Sephadex G-50 column using water as eluent and the high weight fraction is collected. This fraction is concentrated and some sodium hydroxide solution and hydrazine are added in order to ensure conversion to the neural sodium salt. Ethanol is added to precipitate sodium poly-3-(3′-thienyloxy)propanesulfonate, which is collected and washed with ethanol.

A large number of dithienylcyanovinylene, carboxyvinylene and cyanobutadiene derivatives such as are shown below have recently been polymerized using ferric chloride [369].

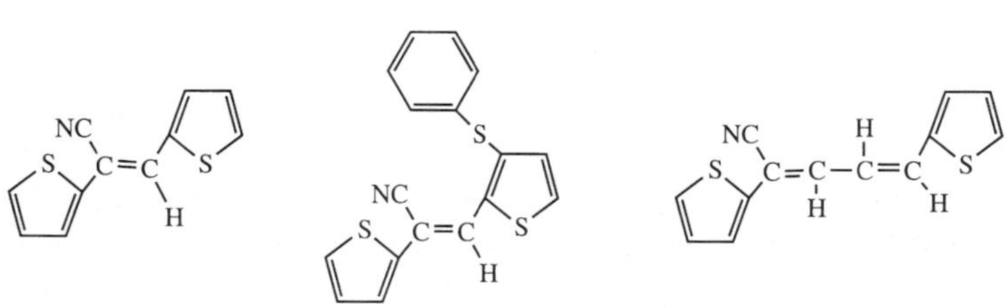

8.17.4 Organometallic polycondensation reactions

8.17.4.1 Introduction

During recent years organometallic polycondensations mediated by organo-transition metal complexes have been developed for the preparation of various π-conjugated polymers and a review article has recently been published [370–372].

Fully undoped oligo(3,4-ethylenedioxythiophene) has been synthesized by polycondensation of the corresponding dibromo monomer in the presence of catalytic Ni(0)-based complex in *N*,*N*-dimethylacetamide [373].

Oligo(3,4-ethylenedioxythiophene) [373]

Activated zinc (0.67 g, 10.3 mmol) and *N*,*N*-dimethylacetamide (6.6 ml) are mixed with nickel dibromide-2,2′-bipyridine (0.063 g, 0.166 mmol), triphenylphosphine (0.52 g, 2.0 mmol) and 2,5-dibromo-3,4-ethylenedioxythiophene (1.0 g, 3.3 mmol) in an argon atmosphere. The mixture is stirred at 95° C for 2.5. The stirring is continued for another hour, after which the mixture is poured into 5% sulfuric acid (200 ml). The black precipitate formed is filtered off, washed with water to neutral pH, diethyl ether (3 × 50 ml) and methanol (3 × 50 ml) and dried in vacuum giving the title compound (72%) as a black purple powder.

A 2,5-dibromothiophene having an ionic viologen mesoinic side chain in the 3-position is converted to a polythiophene by dehalogenative polycondensation using bis(1,5-cyclooctadiene)nickel(0) as catalyst [374].

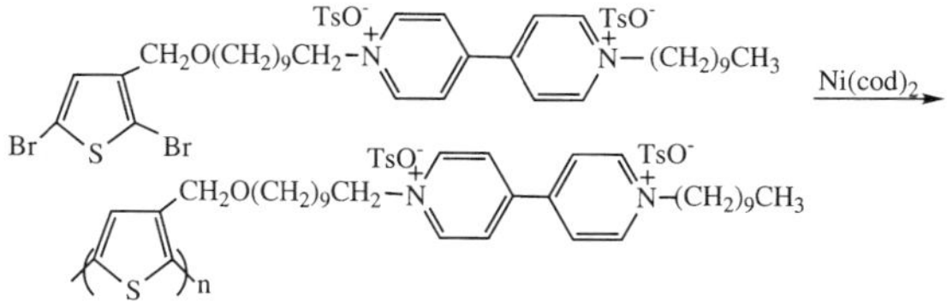

A novel soluble electroluminescent material with high absolute photouminescene quantum efficiency is obtained from the dibromo derivative below by dehalogenative polycondensation using a nickel catalyst system [375].

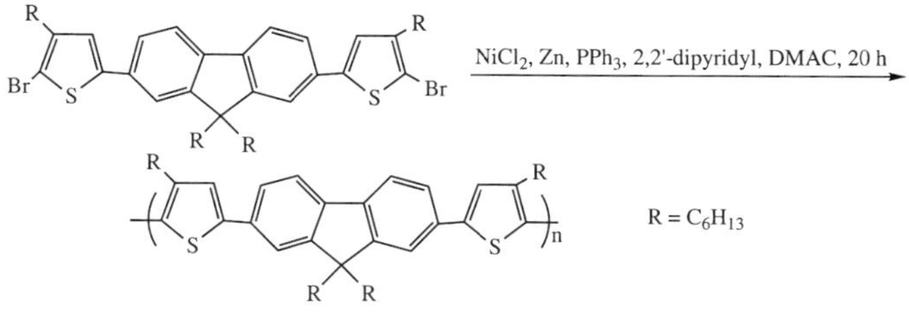

8.17.4.2 From dihalothiophenes, 5,5′-dihalo-2,2′bithienyl and higher α,ω-dihalothienyls and nickel(0) catalysts

Polymerization of 3-methoxy-, 3-butoxy- and 3-methoxyeloxy-2,5-dibromothiophene with nickel(II)1,5-cyclooctadienyl as catalyst is used for the preparation of poly(3-alkoxythiophene-2,5-diyl)s [376]. The following polymerization has been performed [271].

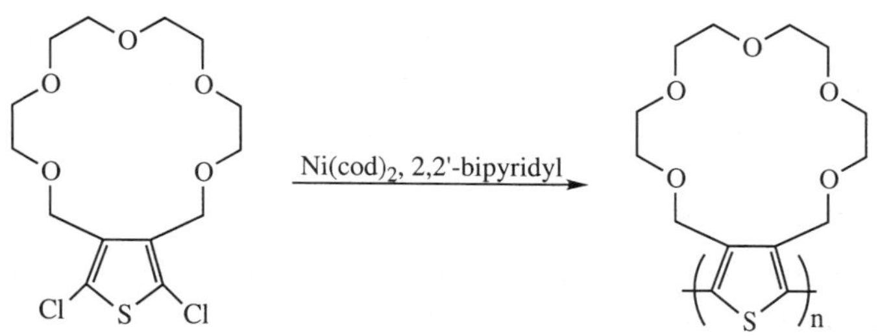

Preparation of the polymer [271]

A solution of the monomer (560 mg, 1.5 mmol) in *N*,*N*-dimethylformamide (5 ml) is added dropwise to a solution of bis (1,5-cylooctadiene)nickel(0) (550 mg, 2 mmol), 1,5-cyclooctadiene (μl) and 2.2′-bipyridyl (310 mg, 2.0 mmol). After stirring at 60° C for 48 h the reaction mixture is poured into methanol and the solvents are evaporated. The residue is dissolved in chloroform and the chloroform solution is washed twice with dilute hydrochloride acid, twice with an aqueous solution of disodium ethylenediaminetetraacetic acid, twice with an aqueous solution of disodium ethylenediaminetetraacetic acid under alkaline ammonia conditions, once with dilute aqueous ammonia, once with water, twice with dilute hydrochloric acid and once with water. Chloroform is removed and the residue dried under vacuum giving 86% of the polymer.

Using a mixture of the monomer and 2,5-dichlorothiophene, a mixture of randomized copolymers is obtained [271,377].

8.17.4.3 From 5-bromo-2-thiophenemagnesium halides and 5-halo-2,2′-bithienylmagnesium bromide *via* nickel(0) catalysis

The reaction of 2-bromo-3-dodecyl-5-thiophenemagnesium bromide obtained from the 5-lithium derivative and magnesium bromide with [1,3-bis(diphenylphosphine)propane]nickel(II) chloride as catalyst gives structurally homogeneous poly(3-dodecylthiophene) [378]. Regioregular head-to-tail poly[3-(6-bromohexyl)thiophene] is prepared in the same way by the reaction of 2-bromo-3-(6-bromohexyl)-5-thiophenemagnesium bormide. It reacted with carboxyanthraquinone to give a regioregular polythiophene containing a pendant functional group [379].

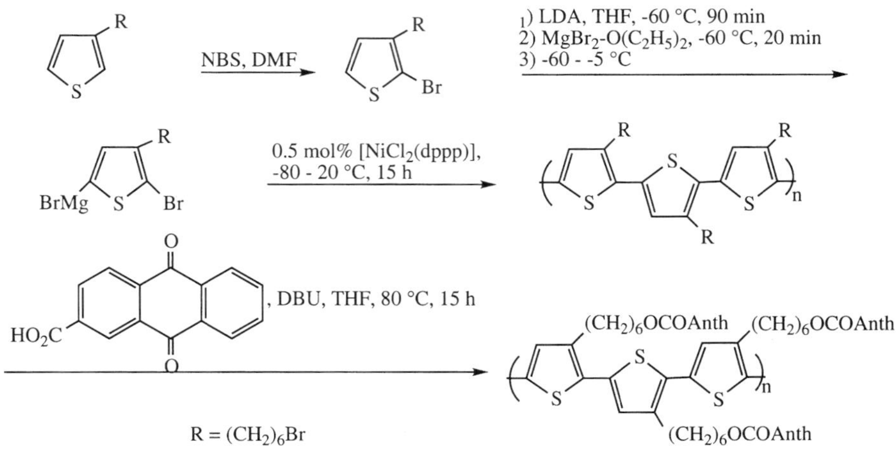

Preparation of the polymer [379]

To a solution of diisopropylamine (2.75 g, 27.11 mmol) in anhydrous tetrahydrofuran (85 ml) 1.6 *M* butyllithium in hexane (16.33 ml, 26.11 mmol) is added at room temperature. The mixture is stirred at room temperature for 15 min and then cooled to – 80 °C. A solution of 2-bromo-3-(6-bromohexyl)thiophene (8.52 g, 26.13 mmol) in anhydrous tetrahydrofuran (100 ml) is added at –80 °C, after which the temperature is allowed to rise slowly to – 40° C. The mixture is stirred at this temperature for 40 min and then cooled to – 60° C and magnesium bromide etherate (6.75 g, 26.14 mmol) is added. The mixture is stirred at – 60 °C for 20 min, at –40 °C for 20 min and then the temperature is allowed to slowly rise to – 5 °C. After recooling to – 80 °C [1,3-bis(diphenylphosphine)prophane]nickel(II) chloride (100 mg, 0.18 mmol) is added and the reaction mixture is allowed to warm to room temperature overnight, after which its volume is reduced to half by evaporation and poured into methanol (800 ml). The precipitate formed is filtered off, washed with methanol, water and methanol again and dried under high vacuum giving 4.0 g (62%) of the polymer as a deep-red powder.

It was found that the reaction of 3-butyl-, 3-hexyl, 3-octyl- and 3-decyl-2-bromo-5-thiophenemagenesium bromide, prepared by metalation of the 3-alkyl-2-bromothiophenes with lithium diisopropyl amide followed by reaction with magnesium bromide-etherate with catalytic amounts of [1,3-bis(diphenyl- phosphine)propane]nickel(II) chloride gave poly(3-alkyl) thiophene containing considerable amounts of bromine, which could be reacted with 5-trimethylsilyl-2-thiophenemagnesium bromide and catalyst in order to incorporate a 5-trimethylsilyl-2-thienyl end group. It is suggested that this procedure can be used to introduce end groups that are reactive to surfaces or to other functional groups, which opens the way to use these polymers in adhesion, block polymerization, electrooptical end groups and energy harvesting [380].

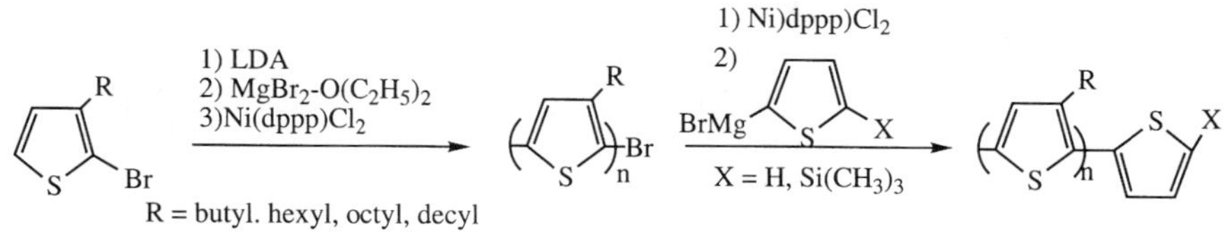

Regioregular polythiophene copolymers containing hexyl and 11-hydroxyundecyl side chains are synthesized by nickel-catalyzed cross-coupling of well-defined Grignard regents. The hydroxy groups were protected during the polymerization as tetrahydropyranyl ethers and subsequently transformed into azido groups [381].

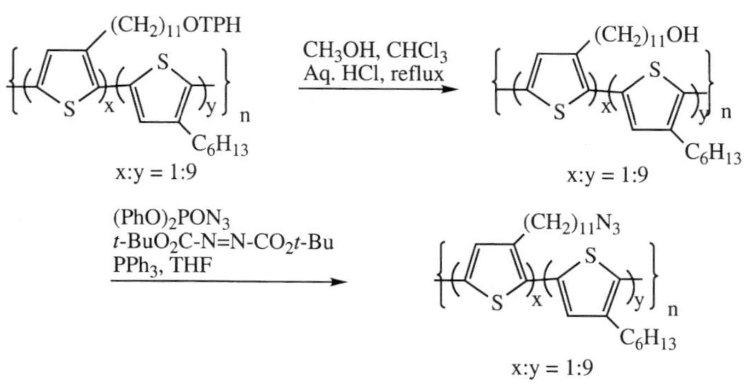

Regioregular 9:1 poly[3-hexylthiophene-co-3-(11-hydroxy)undecyl)thiophene] [381]

A solution of the tetrahydropyranyloxy protected polymer (107 mg, 0.06 mmol) in chloroform (2.5 ml) is added to methanol (12 ml) giving a deep purple suspension. 2 *M* Aqueous hydrochloric acid (15 ml) is added and the suspension stirred at reflux for 16 h and then allowed to cool. The precipitate is collected by centrifugation, washed twice with methanol, transferred to a flask with chloroform, and the chloroform is evaporated giving 95 mg of the title compound as a deep-purple solid.

Regioregular 9:1 poly[3-hexylthiophen-co-3-(11-azidoundecyl)thiophene] [381]

To a solution of regioregular 9:1 poly[3-hexylthiophene-co-3-(11-hydroxyundecyl)thiophene] (77 mg, 0.043 mmol) in anhydrous tetrahydrofuran (2 ml) triphenylphosphine (46 mg, 0.175 mmol), di-*tert*-butyl azodicarboxylate (43 mg, 0.187 mmol) and diphenylphosphoryl azide (39 ml, 0.181 mmol) are added. The red solution is stirred at room temperature in the dark for 5 days. The tetrahydrofuran in removed in a stream of nitrogen. The residue, a purple solid, is dissolved in chloroform (1.5 ml) and added to methanol (10 ml). The suspension formed is stirred for 30 min and the solid is isolated by centrifugation, washed with methanol, acetone and methanol, after which it is transferred to a flask with chloroform and the chloroform is evaporated giving 63 mg (81%) of the title compound as a deep-purple solid film.

The polymer shown below has been prepared [95].

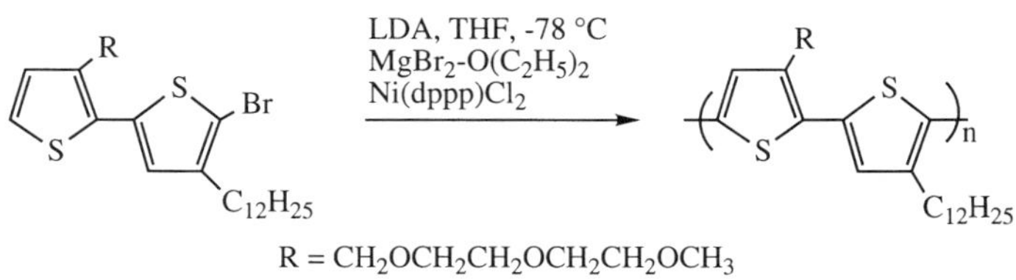

8.17.4.4 From 5-halo-2-thienylzinc halides and nickel catalysts

Reaction of 2,5-dibromo-3-alkylthiophene or 2-bromo-5-iodo-3-alkylthiophene with activated zinc gives the 2-bromo-5-bromozincio-3-alkylthiophene or 2-bromo-5-iodozincio-3-alkylthiophene, which are polymerized catalytically to a series of regiospecific poly(3-alkylthiophenes). The regioregularity of the polymer chain is solely controlled by the structure of the catalyst and an almost completely regioregular head-to-tail polymer is obtained by using [1,2-bis-(diphenylphosphino)ethane]nickel(II) chloride. A totally regiorandom polymer is afforded by using tetrakis(triphenylphosphino)palladium(0) as catalyst [382].

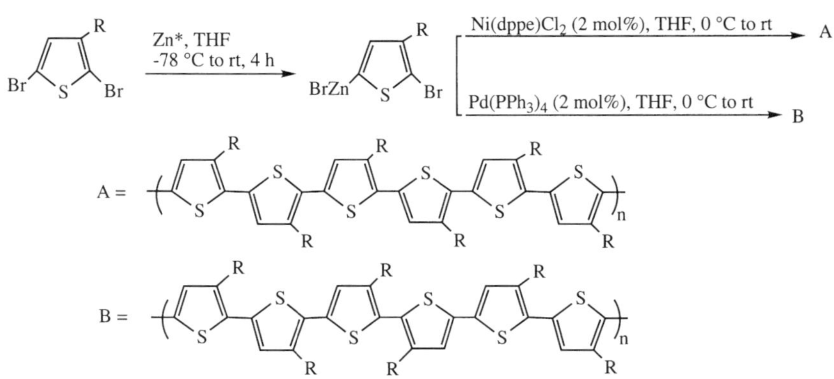

Preparation of regioregular HT poly(3-hexylthiophene) [382]

A solution of 2,5-dibromo-3-hexylthiophene (3.26 g, 10.0 mmol) in tetrahydrofuran (20 ml) is added *via* cannula to newly prepared Zn* (11.0 mmol) in tetrahydrofuran (40 ml) at − 78° C. The stirring is continued for 1 h, after which the mixture is allowed to warm to 0° C. At this temperature [1,2-bis (diphenylphosphino)ethane](II)chloride (11.0 mg, 0.02 mmol) in tetrahydrofuran (20 ml) is added *via* cannula. The reaction mixture is stirred at room temperature for 24 h and during this period a dark-purple precipitate is formed. After pouring the reaction mixture into a solution of methanol (100 ml) and 2 *M* hydrochloric acid (50 ml) the precipitate is filtered off, washed with methanol and 2 *M* hydrochloric acid and dried. Reprecipitation of the polymer by adding methanol to a chloroform solution gives 1.62 g (98%) of the title compound after drying under vacuum.

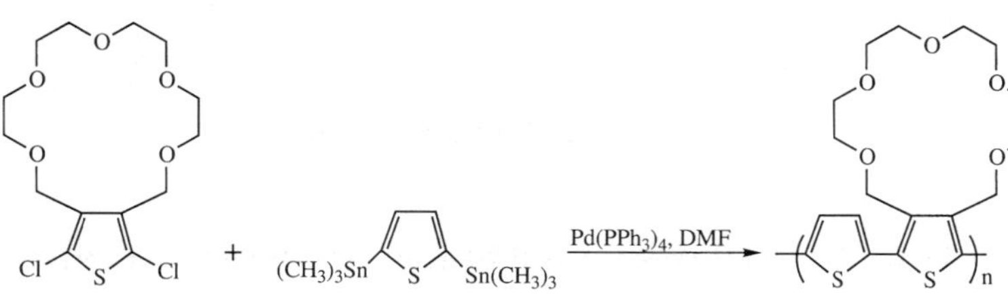

Coupling of the dichloro derivative with 2,5-(bistrimethylstannyl)thiophene or 5,5′-bis(trimethylstannyl)-2,2′-bithienyl gives the corresponding polymers [271]. Polythiophenes with alternating donor/acceptor repeat units are prepared according to the scheme below [383].

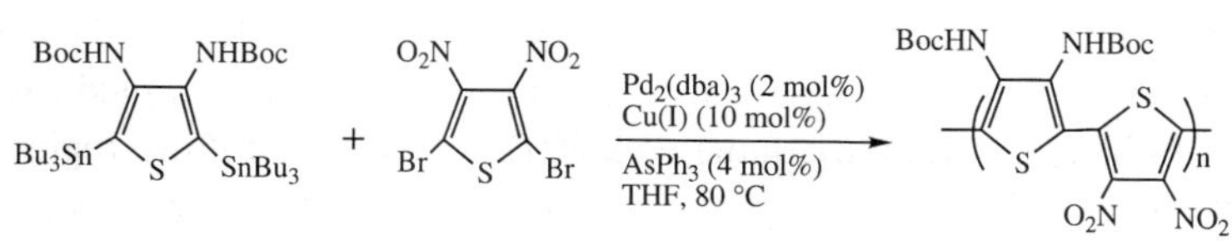

Stille coupling of 3,4-di-boc-amino-2,5-di(tributylstannyl)thiophene with 1,4-diacyl-2,5-dibromobenzene gives the phenyl-containing polymers and with 2,5-dibromo-3,4-diacylthiophene gives the polymers with bithienyl units [384].

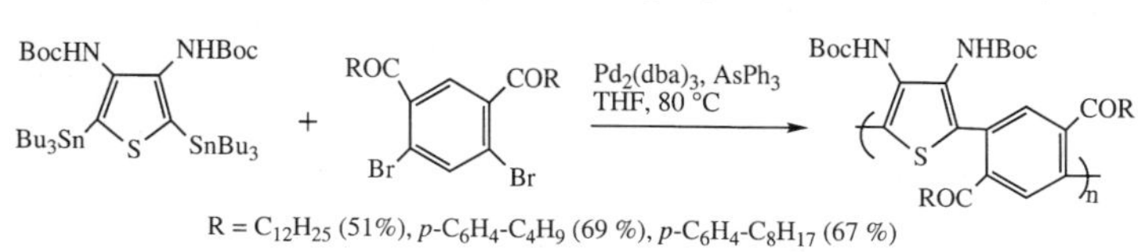

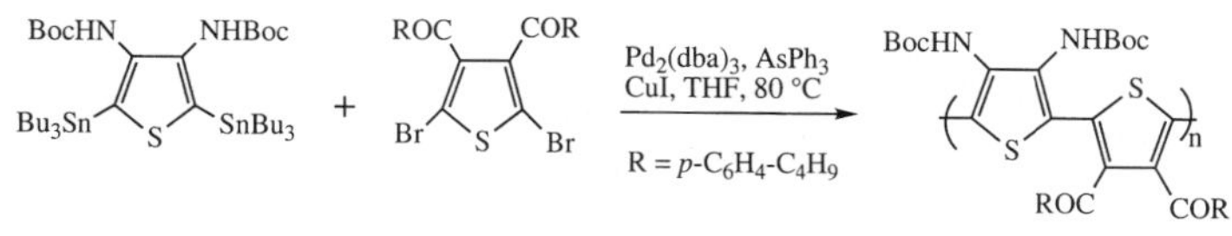

Polythiophenes with interesting non-linear optical properties are obtained by Stille coupling of dihalothieno[3,4-*b*]pyrazines and 5,5″-bis(trimethylstannyl)-terthienyl [385].

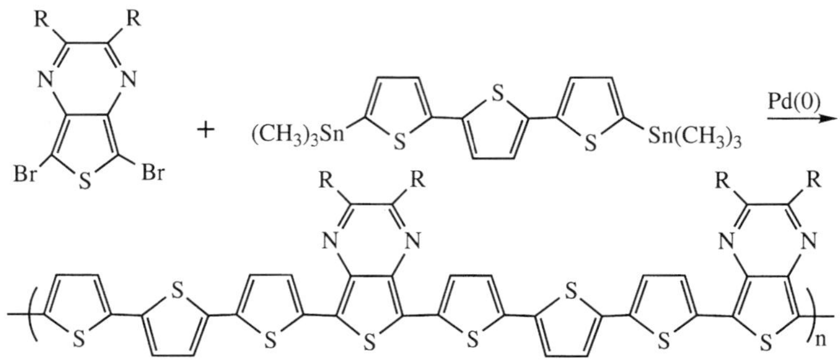

8.17.4.5 *Via* palladium (0)-catalyzed polymerization of 2-halo-5-tri(alkylstannyl)thiophenes and 5-halo-5′-trialkylstannyl-2,2′-bithienyls

Highly regioregular head-to-tail coupled 3-(amino)functionalized polythiophenes are prepared from amino-containing thiophenes by cupric oxide co-catalyzed Heck coupling using tris(dibenzylideneacetone)palladium(0) and triphenylphosphine in *N,N*-dimethylformamide at 100° C. Similarly a polythiophene with a long alkylamino group with improved solubility was also prepared [386].

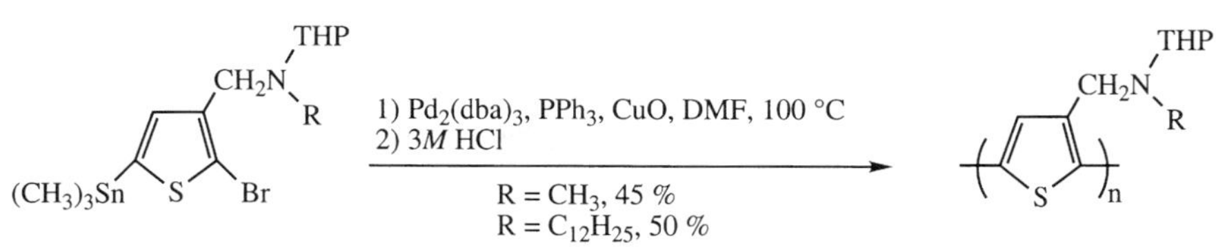

Regioregular poly(3-alkylthiophenes) with high molecular weight are prepared from 2-bromo-3-octyl-5-tributylstannylthiophene and from 5-bromo-4-octyl-5′-tributylstannyl-2,2′-bithienyl in tetrahydrofuran/*N,N*-dimethylformamide upon treatment with catalytic amounts tris(dibenzylideneacetone)palladium(0) and triphenylphosphine [72,387].

2-Bromo-3-octyl-5-tributylstannylthiophene [387]

A solution of 2-bromo-3-octylthiophene (8.5 g, 0.03 mol) in anhydrous tetrahydrofuran (30 ml) is cooled $-40°$ C, whereupon 2 *M* lithium diisopropylamide in tetrahydrofuran (15 ml, 0.03 mol) is added dropwise. The stirring is continued at $-40°$ C for 30 min and then tributylstannyl chloride (9.49 ml, 0.035 mol) is slowly added at the same temperature. The reaction mixture is allowed to warm to room temperature and stirred for 12 h, after which it is hydrolyzed. The phases are separated and the product extracted with diethyl ether. The combined ethereal extracts are dried over magnesium sulfate, evaporated, distilled and purified by chromatography on aluminia using chloroform as eluent, giving 13.4 g (79%) of the title compound bp 165° C/0.015 torr.

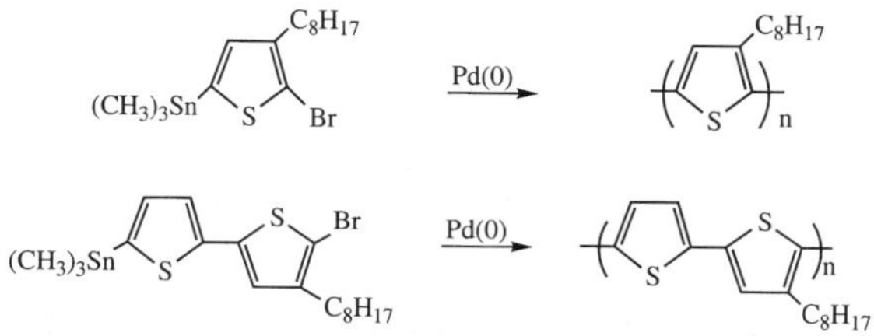

The similar reaction with tris(dibenzylideneactone)palladium(0) and triphenylphosphine in the presence of cupric oxide in *N*,*N*-dimethylformamide gives the polythiophenes below [388].

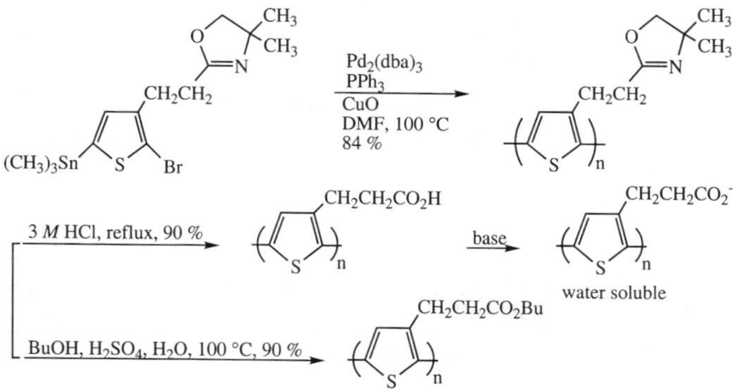

Polymerization in this way has also been applied to a 2,2′-bithienyl derivative [95].

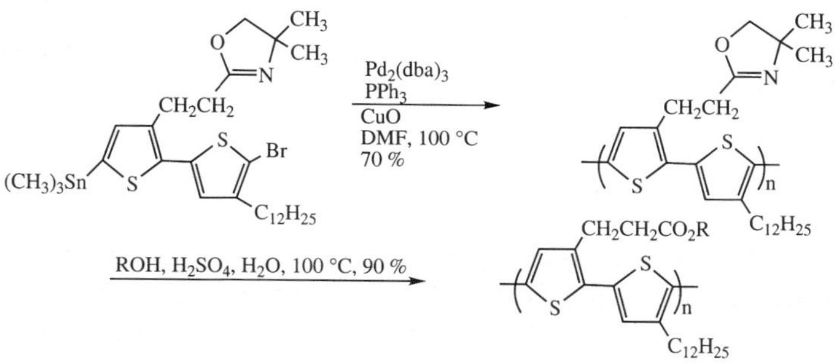

8.17.4.6 From Suzuki coupling between 2,5-dihalothiophene or 5,5′-dihalobithienyls and 2,7-fluorenediboronic acid derivatives

2,5-Dibromothiophene or 5,5′-dibromo-2,2′-bithienyl and 2,7-bis(4,4,5,5)-tetramethyl-1,3,2-dioxaborolane-2-yl)-9,9-dioctylfluorene and palladium (0) are polymerized in tetrahydrofuran and aqueous 2 *M* potassium carbonate [389].

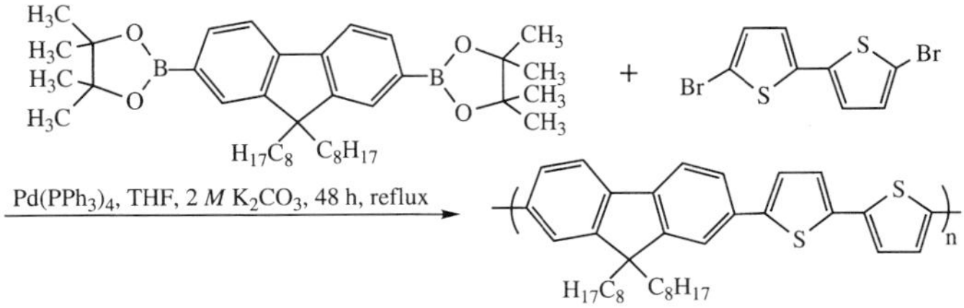

Polymerization [389]

Carefully purified dibromo derivatives (1.00 equiv.), 2,7-bis(4,4,5,5)tetramethyl-1,3,2-dioxaborolane-2-yl)-9,9-dioctylfluorene (1.00 equiv.) and tetrakis-(triphenylposphine)palladium(0) (0.5-1.5 mol%) are dissolved in a mixture of tetrahydrofuran/2 *M* aqueous potassium carbonate (1:1.5). After deaerating with argon the solution is refluxed under vigorous stirring for 48 h and poured into methanol (150 ml). The precipitate is collected on a Büchner funnel and washed with dilute hydrochloric acid. This solid material is washed for 24 h in a Soxhlet apparatus with acetone to remove oligomers and catalyst residues giving 85–95% of the polymers soluble in tetrahydrofuran and chloroform.

The polycondensation of 1,4-diynyl-2,5-dioctylocybenzene with the chiral bis-5-bromothienyl derivative of (1*R*, 2*R*)-1,2-diaminocyclohexane achieved in the presence palladium(0) and cuprous iodide gave the polymer shown below [390].

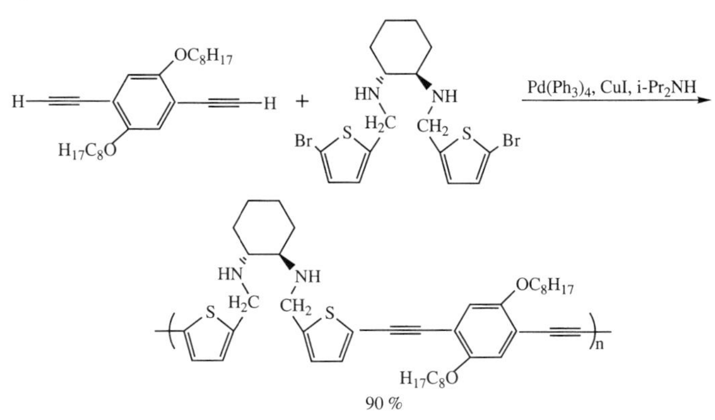

8.17.4.7 From Stille coupling of dihalothiophenes and di(trialkylstannyl)thiophenes

Coupling of 2,5-dibromothiophene and 3-octyl-2,5-dibromothiophene with (*E*,*E*,*E*)-1,6-bis(3′,4′-dioctyl-5′-tributylstanyl-2′-thienyl)hexa-1,3,5-triene is used for the preparation of copolymers [391].

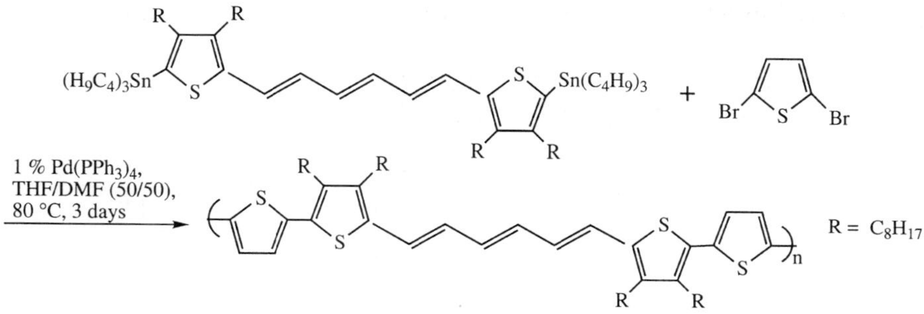

Preparation of the copolymer [391]

In a Schlenk tube equipped with a condenser, tetrakis(triphenylphosphine)-palladium(0) (11.5 mg, 0.01 mmol) is under nitrogen dissolved in tetrahydrofuran/*N*,*N*-dimethylformamide (50:50) at room temperature. To this solution (*E*,*E*,*E*)-1,6-bis(3′,4′-dioctyl-5′-tributylstannyl-2′-thienyl)hexa-1,3,5-triene(402 mg, 1.0 mmol) and 2,5-dibromothiophene (242 mg, 1.0 mmol) are introduced by a syringe. The stirring is continued at 80 °C for three days. The precipitate formed after cooling and addition of acetone is filtered off, washed with acetone and dried in *vacuo* giving 697 mg (90%) of the copolymer as a red powder.

The palladium-catalyzed reaction of 5,5′dibromo-3,3′-dioctyl-2,2′-bithienyl with hexabutyldistannane gives poly(3,3′-dioctyl-2,2′-bithienyl) [50].

REFERENCES

1. P. Bäuerle, In *The Synthesis of Oligothiophenes in Handbook of Oligo- and Polythiophenes* (ed. D. Fichou). Chapter 3, pp. 89–182, Wiley-VCH, Weinheim, 1999.
2. W. Steinkopf and J. Roch, *Liebigs Ann.* **482**, 251 (1930).
3. S. Gronowitz and H.-O. Karlsson, *Arkiv Kemi* **17**, 89 (1960).
4. A. Wiersema and S. Gronowitz, *Acta Chem. Scand.* **24**, 2593 (1970).
5. W. Schroth, E. Hintzsche, H. Jordan, T. Jende, R. Spitzner and I. Thondorf, *Tetrahedron* **53**, 7509 (1997).
6. S. Gronowitz and P. Pedaja, *Tetrahedron* **34**, 587 (1978).
7. B. Krische, J. Hellberg and C. Lilja, *J. Chem. Soc., Chem. Commun.* 1476 (1987).
8. T. Kauffmann and A. Woltermann, *Angew. Chem.* **84**, 824 (1972).
9. R. G. Hicks and M. B. Nodwell, *J. Am. Chem. Soc.* **122**, 6746 (2000).
10. J. Tormo, F. J. Moreno, J. Ruiz, L. Fajari and L. Julia, *J. Org. Chem.* **62**, 878 (1997).

11. M. Bouachrine, J.-P. Lère-Porte, J. J. E. Moreau and C. Terreilles, *J. Chim. Phys.* **95**, 1176 (1998).
12. T. Kauffmann, J. Jackisch, A. Woltermann and P. Röwemeyer, *Angew. Chem.* **84**, 826 (1972).
13. U. Dahlmann and R. Neidlein, *Helvetica Chim. Acta.* **79**, 755 (1996).
14. S. Gronowitz, *Acta Chem. Scand.* **15**, 1393 (1961).
15. G. Zotti, M. C. Gallazzi, G. Zerbi and S. V. Meille, *Synth. Metals* **73**, 217 (1995).
16. R. Håkansson and E. Wiklund, *Arkiv Kemi* **31**, 101 (1969).
17. A. Almqvist and R. Håkansson, *Chem. Scripta* **11**, 57 (1977).
18. B. E. Ayres, S. W. Longworth and J. F. W. McOmie, *Tetrahedron* **31**, 1755 (1975).
19. S. Gronowitz and J. E. Skramstad, *Arkiv Kemi* **28**, 115 (1967).
20. S. Gronowitz and S. Hagen, *Arkiv Kemi* **27**, 153 (1967).
21. S. Gronowitz and H. Frostling, *Tetrahedron Lett.* 604 (1961).
22. S. Gronowitz and H. Frostling, *Acta Chem. Scand.* **16**, 1127 (1962).
23. R. Håkansson, A. Ask and A. Almquist, *Chem. Scripta* **2**, 72 (1972).
24. E. Wiklund and R. Håkansson, *Chem. Scripta* **3**, 220 (1973).
25. S. Gronowitz and H. Beselin, *Arkiv Kemi* **21**, 349 (1963).
26. T. Sone, M. Kubo and T. Kanno, *Chem. Lett.* 1195 (1982).
27. T. Mitsumori, K. Inoue, N. Koga and H. Iwamura, *J. Am. Chem. Soc.* **117**, 2467 (1995).
28. L. Miller and Y. Yu, *J. Org. Chem.* **60**, 6813 (1995).
29. A. Bongini, F. Brioni and M. Panunzio, *J. Chem. Soc., Perkin Trans. 2* 927 (1997).
30. H. Higuchi, S. Yoshida, Y. Uraki and J. Ojima, *Bull. Chem. Soc. Japan* **71**, 2229 (1998).
31. H. Wynberg and A. Logothetis, *J. Am. Chem. Soc.* **78**, 1958 (1956).
32. R. Jente, G. A. Olatunji and F. Bosold, *Phytochemistry* **20**, 2169 (1981).
33. H. Wynberg and A. Bantjes, *J. Am. Chem. Soc.* **84**, 1421 (1959).
34. A. G. Mack, H. Suschitzky and B. J. Wakefield, *J. Chem. Soc., Perkin Trans. 1* 1682 (1980).
35. W. Steinkopf, R. Leitsmann, A. H. Müller and H. Wilhelm, *Leibigs Ann. Chem.* **541**, 271 (1939).
36. T. Sone, K. Sakai and K. Kuroda, *Bull. Chem. Soc. Japan* **43**, 1411 (1970).
37. W. Steinkopf, R. Leitsmann and K. H. Hoffmann, *Liebigs Ann. Chem.* **546**, 80 (1941).
38. W. Minnis, *J. Am. Chem. Soc.* **51**, 2143 (1939).
39. M. Sy, N. P. Buu-Hoi and N. D. Xuong, *J. Chem. Soc.* 1975 (1954).
40. S. Yoshida. M. Fujii, Y. Aso, T. Otsubo and F. Ogura. *J. Org. Chem.* **59**, 3077 (1994).
41. A. Lipkin, *J. Gen. Chem. USSR* **33**, 188 (1963).
42. C. Carpanelli and G. Leandri, *Ann. Chim. (Rome)* **51**, 181 (1961).
43. G. N. Jean and F. F. Nord, *J. Org. Chem.* **20**, 1363 (1955).
44. S. Gronowitz and K. Dahlgren, *Arkiv Kemi* **21**, 201 (1963).
45. R. Håkansson and A. Svensson, *Chem. Scripta* **7**, 186 (1975).
46. J. Cornforth, D. D. Ridley, A. F. Sierakowski, D. Uguen and T. W. Wallace, *J. Chem. Soc., Perkin Trans.* 2317 (1982).
47. M. K. Shepard, *J. Chem. Soc., Chem. Commun.* 880 (1985).
48. M. Nilsson and C. Ullenius, *Acta Chem. Scand.* **24**, 2379 (1970).
49. S. Zhang, D. Zhang and L. S. Liebeskind, *J. Org. Chem.* **62**, 2312 (1997).
50. J. Xu, S. C. Ng and H. S. O. Chan, *Tetrahedron Lett.* **42**, 5327 (2001).
51. T. Itahari, M. Hashimoto and H. Yumisashi, *Synthesis* 255 (1984).
52. J. Hassan, L. Lavenolt, C. Gozzi and M. Lemaire, *Tetrahedron Lett.* **40**, 857 (1999).
53. M. Kuroda, J. Nakayama, M. Hoshino, N. Furusho and S. Ohba, *Tetrahedron Lett.* **35**, 3957 (1994).
54. N. G. Andersen, S. P. Maddaford and B. A. Keay, *J. Org. Chem.* **61**, 9556 (1996).
55. I. Tabakovic, Y. Kunugi, A. Canavesi and L. Miller, *Acta Chem. Scand.* **52**, 131 (1998).
56. G. V. Tormos, K. A. Belmore and M. P. Cava, *J. Am. Chem. Soc.* **115**, 11512 (1993).
57. T. Sone, Y. Umetsu and K. Sato, *Bull. Chem. Soc. Japan* **64**, 864 (1991).
58. N. Jayasuriya and J. Kagan, *Heterocycles* **24**, 2261 (1986).

59. A. Amer, A. Burkhardt, A. Nkansah, R. Shabana, A. Galal, H. B. Mark, Jr and H. Zimmer, *Phosphorus, Sulfur and Silicon* **42**, 63 (1989).
60. H. Zimmer, A. Amer, R. Shabana, D. Ho, H. B. Mark, Jr, K. Sudsuansri and C. Striley, *Acta Chem. Scand.* **47**, 184 (1993).
61. G. Engelmann, G. Kossmehl, J. Heinze, P. Tschuncky, W. Jugelt and H.-P. Welzel, *J. Chem. Soc., Perkin Trans. 2* 169 (1998).
62. P. Bäuerle, F. Pfau, H. Schlupp, F. Würthner, K.-V. Gaudl, M. B. Caro and P. Fischer, *J. Chem Soc., Perkin Trans. 2* 489 (1993).
63. P. Bäuerle, F. Würthner, G. Götz and F. Effenberger, *Synthesis* 1099 (1993).
64. D. Iarossi, A. Mucci, L. Schenetti and V. Sodini, *J. Heterocycl. Chem.* **36**, 241 (1999).
65. R. Rossi and A. Carpita, *Gazz. Chim. Ital.* **115**, 575 (1985).
66. R. Rossi, A. Carpita, M. Ciofalo and J. L. Houben, *Gazz. Chim. Ital.* **120**, 793 (1990).
67. S. Hotta, S. A. Lee and T. Tamaki, *J. Heterocycl. Chem.* **37**, 25 (2000).
68. U. Folli, D. Iarossi, M. Montorsi, A. Mucci and L. Schenetti, *J. Chem. Soc., Perkin Trans. 1*, 537 (1995).
69. J. Nakayama, Y. Ting, Y. Sugihara and A. Ishii, *Heterocycles* **44**, 75 (1997).
70. T. Kaikawa, K. Takimiya, Y. Aso and T. Otsubo, *Org. Lett.* **2**, 4197 (2000).
71. J. Yu and S. Holdcroft, *Chem. Commun.* 1274 (2000).
72. J. P. Lère-Porte, J. J. E. Moteau and C. Torelles, *J. Chim. Phys.* **95**, 1250 (1998).
73. C. Branger, M. Lequan, M. Barzoukas and A. Fort, *J. Mater. Chem.* **6**, 555 (1996).
74. N. Ikemoto, I. Estevez, K. Nakanishi and N. Berova, *Heterocycles* **46**, 489 (1997).
75. F. Effenberger and F. Würthner, *Angew. Chem.* **32**, 719 (1993).
76. J. M. Tour and R. Wu, *Macromolecules* **25**, 1901 (1992).
77. A. F. M. Kilbinger and W. J. Feast, *J. Mater. Chem.* **10**, 1777 (2000).
78. A. Pelter, I. Jenkins and D. E. Jones, *Tetrahedron* **53**, 10357 (1997).
79. S. Gronowitz, V. Bobosik and K. Lawitz, *Chem. Scripta* **23**, 120 (1984).
80. S. Gronowitz and K. Lawitz, *Chem. Scripta* **22**, 265 (1983).
81. S. Gronowitz, A.-B. Hörnfeldt and Y. Yang, *Croatica Chem. Acta* **59**, 313 (1986).
82. T. Kirschbaum, R. Azumi, E. Mena-Osteritz and P. Bäuerle, *New J. Chem.* 241 (1999).
83. X.-S. Ye and H. N. C. Wong, *J. Org. Chem.* **62**, 1940 (1997).
84. F. Effenberger and G. Grube, *Synthesis* 1372 (1998).
85. G. Bidan, A. De Nicola, V. Enée and S. Guillerez, *Chem. Mater.* **10**, 1052 (1998).
86. G. T. Crisp, *Synth. Commun.* **19**, 307 (1989).
87. M. R. Kamal, S. A. Al-Taweel, M. M. El-Abadelah and K. M. Abu Ajaj, *Phosphorus, Sulfur and Silicon* **126**, 65 (1997).
88. F. Steybe, F. Effenberger, U. Gubler, C. Bosshard and P. Günter, *Tetrahedron* **54**, 8469 (1998).
89. R. Wu, J. S. Schumm, D. L. Pearson and J. M. Tour, *J. Org. Chem.* **61**, 6906 (1996).
90. L. Antolini, F. Goldoni, D. Iarossi, A. Mucci and L. Schenetti, *J. Chem. Soc., Perkin Trans. 1*, 1957 (1997).
91. Y. Sakomoto, S. Komatsu and T. Suzuki, *J. Am. Chem. Soc.* **123**, 4643 (2001).
92. G. Barbarella and M. Zambianchi, *Tetrahedron* **50**, 1249 (1994).
93. S. Akoudad and J. Roncali, *Chem. Commun.* 2081 (1998).
94. P. V. Bedworth, Y. Cai, A. Jen and S. R. Marder, *J. Org. Chem.* **61**, 2242 (1996).
95. T. Bjørnholm, D. R. Greve, N. Reitzel, T. Hassenkam, K. Kjaer, P. B. Howes, N. B. Larsen, J. Bøgelund, M. Jayraman, P. C. Ewbank and R. D. McCullough, *J. Am. Chem. Soc.* **120**, 7643 (1998).
96. R. Rossi, A. Carpita and T. Messeri, *Synth. Commun.* **21**, 1875 (1991).
97. R. Rossi, A. Carpita and P. Cossi, *Tetrahedron* **48**, 8801 (1992).
98. S.-K. Kang, H.-W. Lee, S.-B. Jang, T.-H. Kim and J.-S. Kim, *Synth. Commun.* **26**, 4311 (1996).
99. B. H. Lipshutz, F. Kayser and N. Maullin, *Tetrahedron Lett.* **35**, 815 (1994).

100. Y. Wei, Y. Yang and J.-M. Yeh, *Chem. Mater.* **8**, 2659 (1996).
101. E. Lescot, Ng. Ph. Buu-Hoi and Ng. Xuong, *J. Chem. Soc.* 3234 (1959).
102. Y. Wei, B. Wang, W. Wang and J. Tian, *Tetrahedron Lett.* **36**, 665 (1995).
103. M. D'Auria, *Synth. Commun.* **22**, 2393 (1992).
104. Y. Matsuki and Y. Miake, *Chem. Abstr.* **63**, 11475b (1965).
105. J. H. Uhlenbroek and J. D. Bijloo, *Rec. Trav. Chim.* **79**, 1181 (1960).
106. H. E. Katz, J. G. Laquindanum and A. J. Lovinger, *Chem. Mater.* **10**, 633 (1998).
107. S. Gronowitz and L. Karlsson, *Arkiv Kemi* **22**, 119 (1964).
108. W. Steinkopf and H. J. V. Petersdorff, *Liebigs Ann. Chem.* **543**, 119 (1940).
109. H. Wynberg and A. Bantjes, *J. Am. Chem. Soc.* **82**, 1447 (1960).
110. A. E. Lipkin, *J. Gen. Chem. USSR* **33**, 2216 (1963).
111. K. I. Vakreeva and A. E. Lipkin, *Chem. Heterocycl. Compd.* 857 (1974).
112. E. Khor, S. C. Ng, H. C. Li and S. Chai, *Heterocycles* **32**, 1805 (1991).
113. C. Dell'Erba, G. Garbarino and G. Guanti, *Gazz. Chim. Ital.* **100**, 916 (1970).
114. S. Gronowitz and P. Gustafson, *Arkiv Kemi* **20**, 289 (1963).
115. C. Dell'Erba, G. Guanti and G. Garbarino, *J. Heterocycl. Chem.* **11**, 1017 (1974).
116. T. Sone and Y. Abe, *Bull. Chem. Soc. Japan* **46**, 3603 (1973).
117. R. F. Curtis and G. T. Phillips, *J. Chem. Soc.* 5134 (1965).
118. Y. Geng, A. Fechtenkötter and K. Müllen, *J. Mater. Chem.* **11**, 1634 (2001).
119. R. M. Kellogg, A.-P. Schaap and H. Wynberg, *J. Org. Chem.* **34**, 343 (1969).
120. V. G. Nenajdenko, I. L. Baraznenok and E. S. Balenkova, *J. Org. Chem.* **63**, 6132 (1998).
121. M. Kuroda, J. Nakayama, M. Hoshino, N. Furosho, T. Kawata and S. Ohta, *Tetrahedron* **49**, 3735 (1993).
122. H. Higuchi, H. Koyama, H. Yokota and J. Ojima, *Tetrahedron Lett.* **37**, 1617 (1996).
123. H. Higuchi, Y. Uraki, H. Yokota, H. Koyama, J. Ojima, T. Wada and H. Sasabe, *Bull. Chem. Soc. Japan* **71**, 483 (1998).
124. A. F. M. Kilbinger, A. P. H. J. Schenning, F. Goldoni, W. J. Feast and E. W. Meijer, *J. Am. Chem. Soc.* **122**, 1820 (2000).
125. V. A. Smirnov and A. E. Lipkin, *Chem. Heterocycl. Compds.* 170 (1975).
126. P. R. Malenfant, L. Groenendaal and J. M. Fréchet, *Polymer Prep.* **39**, 135 (1998).
127. P. L. R. Malenfant, L. Groenendaal and J. M. J. Fréchet, *J. Am. Chem. Soc.* **120**, 10990 (1998).
128. P. R. L. Malenfant, L. Groenendaal and J. M. J. Fréchet, *Polymer. Prep.* **39**, 133 (1998).
129. P. R. L. Malenfant and J. M. J. Fréchet, *Chem. Commun.* 2657 (1998).
130. G. Barbarella, M. Zambianchi, A. Bongini and L. Antolini, *J. Org. Chem.* **61**, 4708 (1996).
131. J. Ohshita, M. Nodono, H. Kai, T. Watanabe, A. Kunai, K. Komaguchi, M. Shiotani, A. Adachi, K. Okita, Y. Harima, K. Yamashita and M. Ishikawa, *Organometallics*, **18**, 1453 (1999).
132. R. Håkansson, in *The Chemistry of Heterocyclic Compounds, Thiophene and its Derivatives*, (ed. S. Gronowitz). Vol. 44, Part 5, 755 (1992).
133. N. Gjøes and S. Gronowitz, *Acta Chem. Scand.* **21**, 2893 (1967).
134. H. Wynberg, G. J. Heeres, P. Jordens and H. J. M. Sinnige, *Rec. Trav. Chim.* **89**, 545 (1970).
135. R. Håkansson and E. Wiklund, *Acta Chem. Scand.* **24**, 2667 (1970).
136. R. Håkansson, *Acta Chem. Scand.* **25**, 1313 (1971).
137. M. J. Marsella, K. Yoon and F. S. Tham, *Org. Lett.* **3**, 2129 (2001).
138. R. Håkansson and E. Wiklund, *Acta Chem. Scand.* **25**, 2109 (1971).
139. R. Håkansson, *Chem. Scripta* **3**, 212 (1973).
140. E. Wiklund and R. Håkansson, *Chem. Scripta* **6**, 137 (1974).
141. F. de Jong and M. J. Janssen, *J. Org. Chem.* **36**, 1998 (1971).
142. D. Brown, J. Cymerman Craig and J. W. Westley, *J. Chem. Soc. (C)*, **89** (1966).
143. R. Rossi, A. Carpita and A. Lezzi, *Tetrahedron* **40**, 2773 (1984).
144. S. Gronowitz, J. E. Skramstad and B. Eriksson, *Arkiv Kemi* **28**, 99 (1967).
145. R. Håkansson and I. Jakobson, *Unpublished results.*

146. C. Alemán, E. Brillas, A. G. Davies, L. Fajarí, D. Giró, L. Juliá, J. L. Pérez and J. Rius, *J. Org. Chem.* **58**, 3091 (1993).
147. M. V. Lebedev, V. G. Nenajdenko and E. S. Balenkova, *Tetrahedron* **54**, 5599 (1998).
148. G. Barbarella, M. Zambianchi, A. Bongini and L. Antolini, *J. Org. Chem.* **61**, 4708 (1996).
149. S. C. Rasmussen, J. C. Pickens and J. E. Huchison, *J. Heterocycl. Chem.* **34**, 285 (1997).
150. M. S. Vollmer, F. Effenberger, T. Stümpfig, A. Hartschuh, H. Port and H. C. Wolf, *J. Org. Chem.* **63**, 5080 (1998).
151. T. Kirschbaum, C. A. Briehn and P. Bäuerle, *J. Chem. Soc. Perkin Trans. 1* 1211 (2000).
152. D. A. Forsyth and D. E. Vogel, *J. Org. Chem.* **44**, 3917 (1979).
153. J. A. E. H. van Haare, M. van Boxtel and R. A. J. Janssen, *Chem. Mater.* **10**, 1166 (1998).
154. J. Nakayama and J.-S. Lin, *Tetrahedron Lett.* **38**, 6043 (1997).
155. G. M. Tsivgoulis and J.-M. Lehn, *Chem. Eur. J.* 1399 (1996).
156. F. Cherioux, L. Guyard and P. Audebert, *Chem. Commun.* 2225 (1998).
157. C. Branger, M. Lequan, R. M. Lequan, M. Barzoukas and A. Fort, *J. Mater. Chem.* **6**, 555 (1996).
158. Y.-M. Kao and W.-S. Hwang, *J. Heterocycl. Chem.* **26**, 533 (1989).
159. C. Cai, I. Liakatas, M.-S. Wong, M. Bösch, C. Bosshard, P. Günter, S. Concilio, N. Tirelli and U. W. Suter, *Org. Lett.* **1**, 1847 (1999).
160. S. Kotani, K. Shiina and K. Sonogashira, *J. Organomet. Chem.* **429**, 403 (1992).
161. J. Nakayama and J.-S. Lin, *Tetrahedron Lett.* **38**, 6043 (1997).
162. R. Michalitsch, A. ElKassmi, A. Yassar and F. Garnier, *J. Heterocycl. Chem.* **38**, 649 (2001).
163. S. Inoue, S. Nishiguchi, S. Murakami, Y. Aso, T. Otsubo, V. Vill, A. Mori and S. Ujiie, *J. Chem. Research (S)* 596 (1999).
164. L. Groenendaal, M. J. Bruining, E. H. J. Hendrickx, A. Persoons, J. A. J. M. Vekemans, E. E. Havinga and E. W. Meijer, *Chem. Mater.* **10**, 226 (1998).
165. J. A. E. H. van Haare, M. van Boxtel and R. A. J. Janssen, *Chem. Mater.* **10**, 1166 (1998).
166. S. S. Zhu and T. M. Swager, *J. Am. Chem. Soc.* **119**, 12568 (1997).
167. H. E. Katz, J. G. Laquindanum and A. J. Lovinger, *Chem. Mater.* **10**, 633 (1998).
168. G. Barbarella, L. Favaretto, G. Sotgiu, M. Zambianchi, L. Antolini, O. Pudova and A. Bongini, *J. Org. Chem.* **63**, 5497 (1998).
169. G. Barbanella, M. Zambianchi, G. Sotgiu and A. Bongini, *Tetrahedron* **53**, 9401 (1997).
170. R. J. P. Corriu, J. J. E. Moreau, P. Thepot and M. W. C. Man, *Chem. Mater.* **4**, 1217 (1992).
171. C. Yoshina-Isshi, T. Asefa, N. Coombs, M. J. MacLachlan and G. A. Ozin, *Chem. Commun.* 2539 (1999).
172. J. Hock, A. M. W. Cargill Thompson, J. A. McCleverty and M. D. Ward, *J. Chem. Soc., Dalton Trans.* 4257 (1996).
173. L. Miller, Y. Yu, E. Gunic and R. Duan, *Adv. Mater.* **7**, 547 (1995).
174. A. Srinivasan, V. M. Reddy, S. J. Narayanan, B. Sridevi, S. K. Pushpan, M. Ravikumar and T. K. Chandrashekar, *Angew. Chem. Int. Ed. Engl.* **36**, 2598 (1997).
175. A. Neudeck, P. Audebert, L. Guyard, L. Dunsch, P. Guirec and P. Hapiot, *Acta Chem. Scand.* **53**, 867 (1999).
176. W. M. Albers, G. W. Canters and J. Reedijk, *Tetrahedron* **51**, 3895 (1995).
177. H. Kurata, H. Baba and M. Oda, *Chem. Letters* 571 (1997).
178. I. Tabakovic, T. Maki, L. L. Miller and Y. Yu, *Chem. Commun.* 1911 (1996).
179. K. Takahashi, S. Fujita, K. Akiyama, M. Miki and K. Yanagi, *Angew. Chem. Int. Ed.* **37**, 2484 (1998).
180. C. Wang and L. R. Dalton, *Tetrahedron Lett.* **41**, 617 (2000).
181. R. Håkansson, *Chem. Scripta* **2**, 109 (1972).
182. E. Wiklund and R. Håkansson, *Chem. Scripta* **6**, 76 (1984).
183. A. Almqvist and R. Håkansson, *Chem. Scripta* **10**, 117 (1976).
184. A. Almqvist and R. Håkansson, *Chem. Scripta* **11**, 180 (1977).

185. U. Dahlmann and R. Neidlein, *Synthesis* 1027 (1997).
186. T. Benincori, E. Cesarotti, O. Piccolo and F. Sannicolò, *J. Org. Chem.* **65**, 2043 (2000).
187. A. Facchetti, Y. Deng, A. Wang, Y. Koide, H. Sirringhaus, T. J. Marks and H. Friend, *Angew. Chem. Int. Ed.*, **39**, 4547 (2000).
188. U. Boas, A. Dhanabalan, D. R. Greve and E. W. Meijer, *Synlett* 634 (2001).
189. F. Goldoni, D. Iarossi, A. Mucci and L. Schenetti, *J. Heterocyl. Chem.* **34**, 1801 (1997).
190. M. Frigoli, C. Moustron, A. Samat and R. Guglielmetti, *Helv. Chim. Acta* **83**, 3043 (2000).
191. M. D'Auria, A. De Nico, F. D. Onofrio and G. Piancatelli, *J. Org. Chem.* **52**, 5243 (1987).
192. T. Johansson, W. Mammo, M. R. Andersson and O. Inganäs, *Chem. Mater.* **11**, 3133 (1999).
193. A. Rajca, H. Wang, V. Pawitranon, T. J. Brett and J. J. Stezowski, *Chem. Commun.* 1060 (2001).
194. S. Hotta, H. Kimura, S. A. Lee and T. Tamaki, *J. Heterocycl. Chem.* **37**, 281 (2000).
195. S. Yamaguchi, T. Endo, M. Uchida, T. Izumizawa, K. Furukawa and K. Tamao, *Eur. J. Chem.* **6**, 1683 (2000).
196. J. Kagan and S. K. Aurora, *J. Org. Chem.* **48**, 4317 (1983).
197. P. Pedaja and S. Gronowitz, *Chem. Scripta* **20**, 53 (1982).
198. M. Greenwald, D. Wessely, E. Katz, I. Willner and Y. Cohen, *J. Org. Chem.* **65**, 1050 (2000).
199. A. Almqvist and R. Håkansson, *Chem. Scripta* **10**, 120 (1976).
200. A. Srinivasan, S. K. Pushpan, M. R. Kumar, S. Mahajan, T. K. Chandreshekar, R. Roy and P. Ramamurthy, *J. Chem. Soc., Perkin 2* 961 (1999).
201. M. D'Auria and F. Volpe, *Gazz. Chim. Ital.* **123**, 527 (1993).
202. E. Wiklund and R. Håkansson, *Chem. Scripta* **6**, 174 (1974).
203. E. Wiklund and R. Håkansson, *Acta Chem. Scand.* **24**, 341 (1970).
204. M. D. Auria, A. de Mico, F. D. O'nofrio and G. Piancatelli, *Gazz Chim. Ital.* **116**, 747 (1986).
205. K. R. J. Thomas, J. T. Lin and K.-J. Lin, *Organometallics* **18**, 5825 (1999).
206. T. H. Luo and E. LeGoff, *J. Chinese Chem. Soc.* **39**, 325 (1992).
207. A. E. Lipkin and N. I. Putokhin, *Chem. Heterocycl. Compds.* **3**, 243 (1967).
208. O. C. Pfüller and J. Sauer, *Tetrahedron Lett.* **39**, 8821 (1998).
209. J. L. Sessler, M. Cyr and A. K. Burrell, *Tetrahedron* **44**, 9661 (1992).
210. U. Dahlmann, C. Krieger and R. Neidlein, *Eur. J. Org. Chem.* 525 (1998).
211. P. R. L. Malenfant, M. Jayaraman and J. M. J. Fréchet, *Chem. Mater.* **11**, 3420 (1999).
212. V. P. Zvolinskii, A. V. Yudashkin, V. P. Zakharov and L. A. Alekseeva, *J. Org. Chem. USSR* **26**, 334 (1990).
213. A. V. Yudashkin, A. E. Lipkin, V. P. Zvolinskii, V. F. Zhakarov and V. F. Ignatov, *Chem. Heterocycl. Compds.* 1057 (1979).
214. M. Turbiez, P. Frère, P. Blanchard and J. Roncali, *Tetrahedron Lett.* **41**, 5521 (2000).
215. K. Shin, C. Lim, C. Choi, Y. Kim and C. H. Lee, *Chem. Letters* 1331 (1999).
216. M. D'Auria, T. Ferri, G. Mauriello, A. Pesce and R. Racioppi, *Tetrahedron* **55**, 2013 (1999).
217. A. D'Agostini and M. D'Auria, *J. Chem. Soc., Perkin Trans. 1*, 1245 (1994).
218. M. D'Auria and T. Ferri, *J. Org. Chem.* **60**, 8360 (1995).
219. R. M. Kellogg, J. K. Dik, H. van Driel and H. Wynberg, *J. Org. Chem.* **35**, 2737 (1970).
220. M. D'Auria, C. Distefano, F. D. O'nofrio, G. Mauriello and R. Racioppi, *J. Chem. Soc., Perkin Trans. 1*, 3513 (2000).
221. T. Masquelin and D. Obrecht, *Tetrahedron Lett.* **35**, 9387 (1994).
222. E. Campaigne and R. L. White, Jr., *J. Heterocycl. Chem.* **25**, 367 (1988).
223. H. Wynberg, A. Logothetis and D. VerPloeg, *J. Am. Chem. Soc.* **79**, 1972 (1957).
224. A. Fazio, B. Gabriele, G. Salerno and S. Destri, *Tetrahedron* **55**, 485 (1999).
225. H. Hartmann, K. Eckert and A. Schröder, *Angew. Chem. Int. Ed.* **39**, 556 (2000).
226. G. Barbarella, O. Pudova, C. Arbizzani, M. Mastragostino and A. Bongini, *J. Org. Chem.* **63**, 1742 (1998).

227. G. Barbarella, L. Favaretto, G. Sotgiu, M. Ziambianchi, L. Antolini, O. Pudova and A. Bongini, *J. Org. Chem.* **63**, 5497 (1998).
228. K. E. Schulte, J. Reisch and L. Hoerner, *Chem. Ber.* **95**, 1943 (1962).
229. A. Carpita, R. Rossi and C. A. Veracini, *Tetrahedron* **41**, 1919 (1985).
230. J. P. Beny, S. N. Dhawan, J. Kagan and S. Sundlass, *J. Org. Chem.* **47**, 2201 (1982).
231. J.-P. Beny and J. Kagan, *J. Label. Comp. Radiopharm.* **XIX**, 313 (1982).
232. D. M. Perrine and J. Kagan, *Heterocycles*, **24**, 365 (1986).
233. J. Kagan, S. K. Aurora, I. Prakash and A. Ustunol, *Heterocycles* **20**, 1341 (1983).
234. H. J. Kooreman and H. Wynberg, *Rec. Trav. Chim. Pays-Bas* **86**, 37 (1967).
235. T. Asano, S. Ito, N. Saito and K. Hatahedo, *Heterocycles* **6**, 317 (1977).
236. H. Wynberg and J. Metselaar, *Synth. Commun.* **14**, 1 (1984).
237. R. M. Moriarty, O. Prakash and M. P. Duncan, *Synth. Commun.* **15**, 789 (1985)
238. S. Musmanni and J. P. Ferraris *J. Chem. Soc., Chem. Commun.* 172 (1993).
239. L. F. Schweiger, K. S. Ryder, D. G. Morris, A. Glidle and J. M. Cooper, *J. Master. Chem.* **10**, 107 (2000).
240. P. J. Skabara, I. M. Sebryakov, D. M. Roberts, I. F. Perepichka, S. J. Coles and M. B. Hursthouse, *J. Org. Chem.* **64**, 6418 (1999).
241. A. K. Mohanakrishnan, M. V. Lakshmikantham, C. Mc Dougal, M. P. Cava, J. W. Baldwin and R. M. Metzger, *J. Org. Chem.* **63**, 3105 (1998).
242. M. Kuroda, J. Nakayama, M. Hoshino, N. Furusho and S. Ohba, *Tetrahedron Lett.* **35**, 3957 (1994).
243. J. Nakayama, Y. Nakamura, T. Tajiri and M. Hoshino, *Heterocycles* **24**, 637 (1986).
244. J. Nakayama, T. Konishi, S. Murabayashi and M. Hoshino, *Heterocycles* **26**, 1793 (1987).
245. J. Nakayama, S. Murabayashi and M. Hoshino, *Heterocycles* **26**, 2599 (1987).
246. F. Boberg, U. Puttins, W. Schmidt and K.-F. Torges, *Phosphorus and Sulfur* **17**, 135 (1983).
247. L. DeWitt, G. J. Blanchard, E. Legoff, M. E. Benz, J. H. Liao and M. G. Kanatzidis, *J. Am. Chem. Soc.* **115**, 12158 (1993).
248. J. Krömer, I. Rios-Carreras, G. Fuhrmann, C. Musch, M. Wunderlin, T. Debaerdemaeker, E. Mena-Osteritz and P. Bäuerle, *Angew. Chem. Int. Ed.* **39**, 3481 (2000).
249. A. H. Mustafa and M. K. Sheperd, *Chem. Commun.* 2743 (1998).
250. K. Takahashi and T. Suzuki, *J. Am. Chem. Soc.* **111**, 5483 (1989).
251. E. Naudin, N, El Mehdi, C. Soucy, L. Breau and D. Belanger, *Chem. Mater.* **13**, 634 (2001).
252. U. Mitschke and P. Bäuerle, *J. Chem. Soc., Perkin Trans. 1* 740 (2001).
253. N. Jayasuriya and J. Kagan, *Heterocycles* **24**, 2261 (1986).
254. C. L. Jones and S. J. Higgins, *J. Mater. Chem.* **9**, 865 (1999).
255. J. Roncali, M. Giffard, P. Frère, M. Jubault and A. Gorgues, *J. Chem. Soc., Chem. Commun.* 689 (1993).
256. F. Andreani, L. Angiolini, D. Caretta and E. Salatelli, *J. Mater. Chem.* **8**, 1109 (1998).
257. G. Engelmann, R. Stösser, G. Kossmehl, W. Jugelt and H.-P. Welzel, *J. Chem. Soc., Perkin Trans. 2* 2015 (1996).
258. N. Jayasuriya and L. Kagan, *Heterocycles* **24**, 2901 (1986).
259. A. Mac Eachern, C. Soucy, L. C. Leitch, J. T. Arnason and P. Morand, *Tetrahedron* **44**, 22403 (1988).
260. M. Kuroda, J. Nakayama, M. Hoshino and N. Furosho, *Tetrahedron Lett.* **33**, 7553 (1992).
261. C. Soucy-Breau, A. Mac Eachern, L. C. Leitch, T. Arnason and P. Morand, *J. Heterocycl. Chem.* **28**, 411 (1991).
262. S. Gronowitz and D. Peters, *Heterocycles* **30**, 645 (1990).
263. S. Gronowitz and A. Svensson, *Isr. J. Chem.* **27**, 25 (1987).
264. J. Bras, S. Guillerez and B. Pépin-Donat, *J. Chim. Phys.* **95**, 1161 (1998).
265. L. Antolini, M. Borsari, F. Goldoni, D. Iarossi, A. Mucci, and L. Schenetti, *J. Chem. Soc., Perkin Trans. 1* 3207 (1999).

266. A. Yassar, C. Moustrou, H. K. Youssoufi, A. Samat, R. Guglielmetti and F. Garnier, *J. Chem. Soc., Chem. Commun.* 471 (1995).
267. A. Yassar, C. Monstrou, H. Korri Youssoufi, A. Samat, R. Guglielmetti and F. Garnier, *Macromolecules* **28**, 4548 (1995).
268. C. Kitamura, S. Tanaka and Y. Yamashita, *J. Chem. Soc., Chem. Commun.* 1585 (1994).
269. J. L. Reddinger and J. R. Reynolds, *Chem. Mater.* **10**, 1236 (1998).
270. Q. T. Zhang and J. M. Tour, *J. Am. Chem. Soc.* **120**, 5356 (1998).
271. T. Yamamoto, M. Omote, Y. Miyazaki, A. Kashiwazaki, B.-L. Lee, T. Kanbara, K. Osakada, T. Inoue and K. Kubota, *Macromolecules* **30**, 7158 (1997).
272. G. Barbarella, L. Favaretto, G. Sotgiu, M. Zambianchi, C. Arbizzani, A. Bongini and M. Mastragostino, *Chem. Mater.* **11**, 2533 (1999).
273. G. Barbarella, M. Zambianchi, R. di Toro, M. Colonna, Jr, D. Iarossi, F. Goldoni and A. Bongini, *J. Org. Chem.* **61**, 8285 (1996).
274. J. P. Parakka and M. P. Cava, *Tetrahedron* **51**, 2229 (1995).
275. H. E. Katz, *J. Mater. Chem.* **7**, 369 (1997).
276. K. Yui, Y. Aso, T. Otsubo and F. Ogura, *Bull. Chem. Soc. Japan* **62**, 1539 (1989).
277. G. Barbarella, P. Ostoja, P. Maccagnani, O. Pudova, L. Antolini, D. Casarini and A. Bongini, *Chem. Mater.* **10**, 3683 (1998).
278. H. E. Katz, A. Dodabalapur, L. Torsi and D. Elder, *Chem. Mater.* **7**, 228 (1995).
279. J. Nakayama and T. Fujimoro, *J. Chem. Soc., Chem. Commun.* 1614 (1991).
280. Y. Yu, E. Gunic, B. Zinger and L. L. Miller, *J. Am. Chem. Soc.* **118**, 1013 (1996).
281. T. Noda and Y. Shirota, *J. Am. Chem. Soc.* **120**, 9714 (1998).
282. F. Boberg, U. Puttins, C.-D. Czogalla and W. Schmidt, *Liebigs Ann.* 2029 (1983).
283. Y. Zhu, D. B. Millet, M. O. Wolf and S. J. Rettig, *Organometallics* **18**, 1930 (1999).
284. Y. Aso, T. Okai, Y. Kawaguchi and T. Otsubo, *Chem. Lett.* 420 (2001).
285. K. Yui, Y. Aso, T. Otsubo and F. Ogura, *J. Chem. Soc., Chem. Commun.* 1816 (1987).
286. F. Boberg and U. Puttins, *Phosphorus & sulfur* **20**, 121 (1984).
287. D. A. Weinberger, T. B. Higgins, C. A. Mirkin, L. M. Liable-Sands and A. L. Rheingold, *Angew. Chem. Int. Ed.* **38**, 2565 (1999).
288. D. A. Weinberger, T. B. Higgins, C. A. Mirkin, C. L. Stern, L. M. Liable-Sands and A. L. Rheingold, *J. Am. Chem. Soc.* **123**, 2503 (2001).
289. D. D. Graf, J. P. Campbell, L. L. Miller and K. R. Mann, *J. Am. Chem. Soc.* **118**, 5480 (1996).
290. D. D. Graf, R. G. Duan, J. P. Campbell, L. Miller and K. R. Mann, *J. Am. Chem. Soc.* **119**, 5888 (1997).
291. N. Jayasuriya, J. Kagan J. E. Owens, E. P. Kornak and D. M. Perrine. *J. Org. Chem.* **54**, 4203 (1989).
292. K. Imamura, D. Hirayama, H. Yoshimura, K. Takimiya, Y. Aso and T. Otsubo, *Tetrahedron Lett.* **40**, 2789 (1999).
293. K. Imamura, K. Takimiya, Y. Aso and T. Otsubo, *Chem. Commun.* 1859 (1999).
294. V. Rao, G. Anand, S. K. Pushpan, A. Srinivasan, S. J. Narayanan, B. Sridevi, T. K. Chandrasekar, R. Roy and B. S. Joshi, *Org. Lett.* **2**, 3829 (2000).
295. J. Kagan and S. K. Arora, *Heterocycles* **20**, 1937 (1983).
296. R. E. Atkinson, R. F. Curtis and G. T. Phillips, *J. Chem. Soc. (C)* 2001 (1967).
297. F. Garnier, R. Hajlaoui, A. El Kassmi, G. Horowitz, L. Laigre, W. Porzio, M. Armanini and F. Provasoli, *Chem. Mater.* **10**, 334 (1998).
298. A. K. Monakrishnan, A. Hucke, A. L. Michael, M. V. Lakshmikanthan and M. P. Cava, *Tetrahedron* **55**, 11745 (1999).
299. N. Jayasuriya, J. Kagan, D.-B. Huang and B. K. Teo, *Heterocycles* **27**, 1391 (1988).
300. M. Pasini, S. Destri, C. Botta and W. Porzio, *Tetrahedron* **55**, 14985 (1999).
301. H. Higuchi, T. Nakayama, H. Koyama, J. Ojima, T. Wada and H. Sasabe, *Bull. Chem. Soc. Japan* **68**, 2363 (1995).

302. G. Barbarella, M. Zambianchi, R. Di Toro, M. Colonna, L. Antolini and A. Bongini, *Adv. Mater.* **8**, 327 (1996).
303. H. Nakanishi, N. Sumi, Y. Aso and T. Otsubo, *J. Org. Chem.* **63**, 8632 (1998).
304. P. Bäuerle, T. Fisher, B. Bidlingmeier, A. Stabel and J. P. Rabe, *Angew. Chem, Int. Ed. Engl.* **34**, 303 (1995).
305. T. Yamashiro, Y. Aso, T. Otsubo, H. Tang, Y. Harima and K. Yamashita, *Chem. Letters* 443 (1999).
306. R. Azumi, G. Götz, T. Debaerdemaeker and P. Bäuerle, *Chem. Eur. J.* **6**, 735 (2000).
307. H. Mugurama, K. Kobiro and S. Hotta, *Chem. Mater.* **10**, 1459 (1998).
308. A. K. Mohanakrishnan, A. Hucke, M. A. Lyon, M. V. Lakshmikantham and M. P. Cava, *Tetrahedron*, **55**, 11745 (1999).
309. C. A. Briehn, T. Kirschbaum and P. Bäuerle, *J. Org. Chem.* **65**, 352 (2002).
310. J. Kagan and S. K. Arora, *Tetrahedron Lett.* **24**, 4043 (1988).
311. J. Kagan and S. K. Arora, *J. Org. Chem.* **48**, 4317 (1983).
312. D. Perrine, D. M. Bush, E. P. Kornak, M. Zhang, Y. H. Cho and J. Kagan, *J. Org. Chem.* **56**, 5095 (1991).
313. H. Müller, J. Petersen, R. Strohmaier, B. Gompf, W. Eisenmenger, M. S. Vollmer and F. Effenberger, *Adv. Mater.* **8**, 733 (1996).
314. J. C. Horne, G. J. Blanchard and E. LeGoff, *J. Am. Chem. Soc.* **117**, 9551 (1995).
315. J. Nakayama, Y. Nakamura, S. Murabayashi and M. Hoshino, *Heterocycles* **26**, 939 (1987).
316. J. Nakayama, R. Yomoda and M. Hoshino, *Heterocycles* **26**, 2215 (1987).
317. J. Nakayama, K. Sawada, A. Ishii and M. Hoshino, *Heterocycles* **34**, 1487 (1992).
318. M. S. Vollmer, F. Würthner, F. Effenberger, P. Emele, D. U. Meyer, T. Stümpfig, H. Port and H. C. Wolf, *Chem. Eur. J.* **4**, 260 (1998).
319. J. A. E. H. van Haare, E. E. Havinga, J. L. J. van Dongen, R. A. J. Janssen, J. Cornil and J.-L. Brédas, *Chem. Eur. J.* **4**, 1509 (1998).
320. M. Sato and M. Hiroi, *Chem. Letters* 745 (1994).
321. F. Goldoni, D. Iarossi, A. Mucci and L. Schenetti, *Chem. Commun.* 2175 (1997).
322. S. Tanaka and M. Kumei, *J. Chem. Soc. Chem. Commun.* 815 (1995).
323. A. F. M. Kilbinger, H. J. Cooper, L. A. McDonnell, W. J. Feast, P. J. Derrick, A. P. H. J. Schenning and E. W. Meijer, *Chem. Commun.* 383 (2000).
324. P. R. Malenfant, L. Groenendaal and J. M. Fréchet, *Polymer Prep.* **39**, 135 (1998).
325. J. Guay, A. Diaz, R. Wu and J. M. Tour, *J. Am. Chem. Soc.* **115**, 1869 (1993).
326. J. M. Tour, R. Wu and J. S. Schumm, *J. Am. Chem. Soc.* **113**, 7064 (1991).
327. L. Antolini, E. Tedesco, G. Barbarella, L. Favaretto, G. Sotgiu, M. Zambianchi, D. Casarini, G. Gigli and R. Cingolani, *J. Am. Chem. Soc.* **122**, 9006 (2002).
328. J. Nakayama, K. Sawada, A. Ishii and M. Hoshino, *Heterocycles* **34**, 1487 (1992).
329. W. ten Hoeve, H. Wynberg, E. E. Havinga and E. W. Meijer, *J. Am. Chem. Soc.* **113**, 5887 (1991).
330. M. A. Hempenius, B. M. W. Langeveld-Voss, J. A. E. van Haare, R. E. J. Janssen, S. S. Sheiko, J. P. Saptz, M. Möller and E. W. Meijer, *J. Am. Chem. Soc.* **120**, 2798 (1998).
331. M. Sato and M. Hiroi, *Chem. Letters* 1649 (1994).
332. M. Sato and M. Hiroi, *Chem. Letter* 985 (1994).
333. J. J. Apperloo, R. A. J. Janssen, P. R. L. Malenfant, L. Groenendaal and J. M. J. Frechet, *J. Am. Chem. Soc.* **122**, 7042 (2000).
334. W. Fujita, N. Teramae and H. Haraguchi, *Chem. Letters* 511 (1994).
335. T. Yamamoto and H. Kokubo, *Chem. Letter* 1295 (1999).
336. K. Kaeriyama, S. Satira and H. Masuda, *Makromol. Chem. Rapid Commun.* **11**, 37 (1990).
337. J. P. Ferraris, M. M. Eissa, I. D. Brosterston and D. C. Loveday, *Chem. Mater.* **10**, 3528 (1998).
338. J. Roncali, H. K. Youssoufi, R. Garreau, F. Garnier and M. Lemaire, *J. Chem. Soc., Chem. Commun.* 414 (1990).

339. N. Sakmeche, J. J. Aron, M. Fall, S. Aeiyach, M. Jouini, J. C. Lacroix and P. C. Lacaze, *Chem. Commun.* 2723 (1996).
340. K. Hara, K. Sayama and H. Arakawa, *Bull. Chem. Soc. Japan* **73**, 583 (2000).
341. A. Kumar, D. W. Welsh, M. C. Morvant, F. Piroux, K. A. Abboud and J. R. Reynolds, *Chem. Mater.* **10**, 896 (1998).
342. M. Lemaire, D. Delabouglise, R. Garreau, A. Guy and J. Roncali. *J. Chem. Soc., Chem. Commun.* 658 (1988).
343. H. Brisset, C. Thobie-Gautier, A. Gorgues, M. Jubeault and J. Roncali, *J. Chem. Soc., Chem. Commun.* 1305 (1994).
344. M. Kozaki, S. Tanaka and Y. Yamashita, *J. Org. Chem.* **59**, 442 (1994).
345. F. Sannicolò, E. Brenna, T. Benincori, G. Zotti, S. Zecchin, G. Schiavon and T. Pilati, *Chem. Mater.* **10**, 2067 (1998).
346. T. L. Lambert and J. P. Ferraris, *J. Chem. Soc., Chem. Commun.* 752 (1991).
347. P. J. Skabara, D. M. Roberts, I. M. Serebryakov and C. Pozo-Gonzalo, *Chem. Commun.* 1005 (2002).
348. G. Zotti, M. Musiani, S. Zecchin, G. Schiavon, A. Berlin and G. Pagani, *Chem. Mater.* **10**, 480 (1998).
349. P.-L. Vidal, B. Divisia-Blohorn, G. Bidan, J.-L. Hazemann, J.-M. Kern and J.-P. Sauvage, *Chem. Eur. J.* **6**, 1663 (2000).
350. C. Lagrost, K. I. Chane Ching, J.-C. Lacroix, S. Aeiyach, M. Jouini, P. C. Lacaze and J. Tanguy, *J. Mater. Chem.* **8**, 2351 (1999).
351. M. Tsionsky, A. J. Bard, D. Dini and F. Decker, *Chem. Mater.* **10**, 2120 (1998).
352. T. Sato, K. Hori and K. Tanaka, *J. Mater. Chem. Mater.* **8**, 589 (1998).
353. R. P. Kingsborough and T. M. Swager, *Chem. Mater.* **12**, 872 (2000).
354. J. A. Crayston, A. Iraqui, P. Mallon and J. C. Walton, *Synth. Met.* **55**, 867 (1993).
355. D. Millar, M. Uttamlal, R. Henderson and A. Keeper, *Chem. Commun.* 477 (1998).
356. G. Zotti, M. C. Gallazzi, G. Zerbi and S. V. Meille, *Synth. Metals* **73**, 217 (1995).
357. E. M. Girotto, G. Casalbore-Miceli, N. Camaioni, M. A. De Paoli, A. M. Fichera, L. Belobrzrckaja and M. C. Gallazzi, *J. Mater. Chem.* **11**, 1072 (2001).
358. P. Pellon, E. Deltel and J.-F. Pilard, *Tetrahedron Lett.* **42,** 867 (2001).
359. C. L. Kean and P. G. Pickup, *Chem. Commun.* 815 (2001).
360. S. Tanaka and K. Kaerimyama, *Bull. Chem. Soc. Japan* **62**, 1908 (1989).
361. M. Leclerc and G. Daoust, *J. Chem. Soc., Chem. Commun.* 273 (1990).
362. K. Faid and M. Leclerc, *Chem. Commun.* 2761 (1996).
363. K Faïd and M. Leclerc, *J. Am. Chem Soc.* **120**, 5274 (1998).
364. R. Cloutier and M. Leclerc, *J. Chem Soc., Chem. Commun.* 1194 (1991).
365. B. M. W. Langeveld-Voss, R. A. J. Janssen, M. P. T. Christiaans, S. C. J. Meskers, H. P. J. M. Dekkers and E. W. Meijer, *J. Am. Chem. Soc.* **118**, 4908 (1996).
366. Y. Greenwald, G. Cohen, J. Poplawski, E. Ehrenfreund, S. Speiser and D. Davidov, *J. Am. Chem. Soc.* **118**, 2980 (1996).
367. M. Chayer, K. Faïd and M. Leclerc, *Chem. Mater.* **9**, 2902 (1997).
368. J. Lukkari, M. Salomäki, A. Viinikanoja, T. Ääritalo, J. Paukkunen, N. Kochorova and J. Kankare, *J. Am. Chem. Soc.* **123**, 6083 (2001).
369. P. Soudan, P. Lucas, H. Ang Ho, D. Jobin, L. Breau and D. Bélanger, *J. Mater. Chem.* **11**, 773 (2001).
370. T. Yamamoto, *Bull. Chem. Soc. Japan* **72**, 621 (1999).
371. J. L. Segura and N. Martin, *J. Mater. Chem.* **10**, 2403 (2000).
372. D. Fichou, *J. Mater., Chem.* **10**, 571 (2000).
373. F. Tran-Van, S. Garreau, G. Louar, G. Froyer and C. Chevrot, *J. Mater. Chem.* **11**, 1378 (2001).
374. M. Kijima, K. Setoh and H. Shirakawa, *Chem. Letters* 936 (2000).
375. J. Pei, W.-L. Yu, W. Huang and A. J. Heeger, *Chem. Commun.* 1631 (2000).

376. Y. Miyazaki, T. Kanbara, K. Osakada and T. Yamamoto, *Chem. Letters* 415 (1993).
377. Y. Miyazaki and T. Yamamoto, *Chem. Letters* 41 (1994).
378. R. D. McCullough and R. D. Lowe, *J. Chem. Soc., Chem. Commun.* 70 (1992).
379. A. Iraqi, J. A. Crayston and J. C. Walton, *J. Mater. Chem.* **8**, 31 (1998).
380. B. M. V. Langeveld-Voss, R. A. J. Janssen, A. H. J. Spiring, J. L. J. van Dongen, E. C. Vonk and H. A. Claessens, *Chem. Commun.* 81 (2000).
381. K. A. Murray, A. B. Holmes, S. C. Moratti and G. Rumbles, *J. Mater. Chem.* **9**, 2109 (1999).
382. T.-A. Chen, X. Wu and R. D. Rieke, *J. Am. Chem. Soc.* **117**, 233 (1995).
383. Q. T. Zhang and J. M. Tour, *J. Am. Chem. Soc.* **120**, 5355 (1998).
384. Q. T. Zhang and J. M. Tour, *J. Am. Chem. Soc.* **119**, 9624 (1997).
385. W. Schrof, S. Rozouvan, T. Hartmann, H. Möhwald, V. Belov and E. Van Keuren, *J. Opt. Soc. Am. B* **15**, 889 (1998).
386. P. C. Ewbank, G. Nuding, H. Suenaga, R. D. McCullough and S. Shinkai, *Tetrahedron Lett.* **42**, 155 (2001).
387. J.-P. Lère-Porte, J. J. E. Moreau and C. Torreilles, *Eur. J. Org. Chem.* 1249 (2001).
388. R. D. McCullough, P. C. Ewbank and R. S. Loewe, *J. Am. Chem. Soc.* **119**, 633 (1997).
389. M. Ranger and M. Leclerc, *Can. J. Chem.* **76**, 1571 (1998).
390. J.-P. Lère-Porte, J. J. E. Moreau, F. Serein-Spirau and S. Wakim, *Tetrahedron Lett.* **42**, 3073 (2001).
391. F. Embert, J.-P. Lère-Porte, J. J. E. Moreau, F. Serin-Spiran, A. Righi and J.-L. Sauvajol, *J. Mater. Chem.* **11**, 718 (2001).

Substances Index